ORDERED POROUS SOLIDS

Recent Advances and Prospects

EDITED BY

VALENTIN VALTCHEV
SVETLANA MINTOVA
MICHAEL TSAPATSIS

Amsterdam • Boston • Heidelberg • London • New York • Oxford
Paris • San Diego • San Francisco • Singapore • Sydney • Tokyo

Elsevier
The Boulevard, Langford lane, Kidlington, Oxford, OX5 1GB, UK
Radarweg 29, PO Box 211, 1000 AE Amsterdam, The Netherlands

First edition **2009**

British Library Cataloguing in Publication Data
A catalogue record for this book is available from the British Library

Library of Congress Cataloging-in-Publication Data
A catalog record for this book is available from the Library of Congress

ISBN: 978-0-444-53189-6

For information on all Elsevier publications
visit our web site at books.elsevier.com

Printed and bound by CPI Group (UK) Ltd, Croydon, CR0 4YY

Transferred to Digital Print 2011

Contents

Contributors

Hristiyan A. Aleksandrov
Faculty of Chemistry, University of Sofia, 1126 Sofia, Bulgaria

Cindy Aquino
Laboratory of Advanced Catalysis for Sustainability, School of Chemistry—F11, The University of Sydney, NSW 2006, Australia

Florence Babonneau
Laboratoire de Chimie de la Matière Condensée de Paris, Université Pierre et Marie Curie, CNRS UMR 7574, 4 place Jussieu, 75252 Paris cedex 05, France

Dominique Bégin
Laboratoire des Matériaux, Surfaces et Procédés pour la Catalyse – LMSPC UMR 7515 du CNRS, Université Louis Pasteur, France

Jeroen A. van Bokhoven
ETH Zurich, Institute of Chemical and Bioengineering, HCI E115, 8093 Zurich, Switzerland

Josip Bronić
Ruđer Bošković Institute, Division of Materials Chemistry, Laboratory for the Synthesis of New Materials, P.O. Box 180, 10002 Zagreb, Croatia

Moises A. Carreon
Department of Chemical Engineering, University of Louisville, Louisville, Kentucky 40292, USA

Jiří Čejka
J. Heyrovský Institute of Physical Chemistry, AS CR, v.v.i., Dolejškova 3, 182 23 Prague 8, Czech Republic

Mart H.J.M. de Croon
Department of Chemical Engineering and Chemistry, Eindhoven University of Technology, 5600 MB Eindhoven, The Netherlands

Thomas Devic
Institut Lavoisier, CNRS-UMR 8180, Universite de Versailles St Quentin-en-Yvelines, 78035 Versailles, France

Junhang Dong
Department of Chemical and Materials Engineering, University of Cincinnati, Cincinnati, OH 45221, USA

Aidan M. Doyle
Division of Chemistry & Materials, Manchester Metropolitan University, Chester Street, Manchester M1 5GD, United Kingdom

Cristina Fernandez-Martin
Laboratoire de Chimie de la Matière Condensée de Paris, Université Pierre et Marie Curie, CNRS UMR 7574, 4 place Jussieu, 75252 Paris cedex 05, France

Fabrice Gaslain
Laboratoire de Matériaux à Porosité Contrôlée, Université de Haute Alsace, ENSCMu, CNRS, UMR 7016, 3 rue Alfred Werner, 68093 Mulhouse cedex, France

Vadim V. Guliants
Department of Chemical and Materials Engineering, University of Cincinnati, Cincinnati, Ohio 45221, USA

Konstantin I. Hadjiivanov
Institute of General and Inorganic Chemistry, Bulgarian Academy of Sciences, Sofia 1113, Bulgaria

Maguy Jaber
Laboratoire de Matériaux à Porosité Contrôlée, Université de Haute Alsace, ENSCMu, UMR CNRS 7016, 3 rue Alfred Werner, 68093 Mulhouse Cedex, France

Aldona Jankowska
Adam Mickiewicz University, Faculty of Chemistry Grunwaldzka 6, 60–780 Poznań Poland

Tatjana Antonić Jelić
Ruđer Bošković Institute, Division of Materials Chemistry, Laboratory for the Synthesis of New Materials, P.O. Box 180, 10002 Zagreb, Croatia

Venčeslav Kaučič
National Institute of Chemistry, Hajdrihova 19, 1000 Ljubljana, Slovenia

Stanisław Kowalak
Adam Mickiewicz University, Faculty of Chemistry Grunwaldzka 6, 60–780 Poznań Poland

Miron V. Landau
Blechner Center of Applied Catalysis and Process Development, Chemical Engineering Department, Ben-Gurion University of the Negev, Beer-Sheva 84105, Israel

Bénédicte Lebeau
Laboratoire de Matériaux à Porosité Contrôlée, Université de Haute Alsace, ENSCMu, CNRS, UMR 7016, 3 rue Alfred Werner, 68093 Mulhouse cedex, France

Marc-Jacques Ledoux
Laboratoire des Matériaux, Surfaces et Procédés pour la Catalyse – LMSPC UMR 7515 du CNRS, Université Louis Pasteur, France

Christopher M. Lew
Department of Chemical and Environmental Engineering, University of California – Riverside, Riverside, California 92521, USA

Yan Liu
Department of Chemical and Environmental Engineering, University of California – Riverside, Riverside, California 92521, USA

Raul F. Lobo
Department of Chemical Engineering, Center for Catalytic Science and Technology, University of Delaware, Newark, Delaware 19716, USA

Nataša Zabukovec Logar
National Institute of Chemistry, Hajdrihova 19, 1000 Ljubljana, Slovenia

Benoît Louis
Laboratoire des Matériaux, Surfaces et Procédés pour la Catalyse – LMSPC UMR 7515 du CNRS, Université Louis Pasteur, France

Duncan J. Macquarrie
Department of Chemistry, Centre of Excellence in Green Chemistry, University of York, Heslington, York, YO10 5DD, United Kingdom

Gregor Mali
National Institute of Chemistry, Hajdrihova 19, 1000 Ljubljana, Slovenia

Thomas Maschmeyer
Laboratory of Advanced Catalysis for Sustainability, School of Chemistry—F11, The University of Sydney, NSW 2006, Australia

Matjaž Mazaj
National Institute of Chemistry, Hajdrihova 19, 1000 Ljubljana, Slovenia

Jocelyne Miehé-Brendlé
Laboratoire de Matériaux à Porosité Contrôlée, Université de Haute Alsace, ENSCMu, UMR CNRS 7016, 3 rue Alfred Werner, 68093 Mulhouse Cedex, France

Martijn J.M. Mies
Department of Chemical Engineering and Chemistry, Eindhoven University of Technology, 5600 MB Eindhoven, The Netherlands

Boriana Mihailova
Department of Geosciences, University of Hamburg, Grindelallee 48, D-20146 Hamburg, Germany

Svetlana Mintova
Laboratoire de Matériaux à Porosité Contrôlée, Université de Haute Alsace, UMR-7016 CNRS, ENSCMu, 3, rue Alfred Werner, 68093 Mulhouse, France

Izabela Naydenova
Centre for Industrial and Engineering Optics Focas, DIT Kevin Street, Camden Row, Dublin 8, Ireland

Tina M. Nenoff
Surface and Interfacial Science, Sandia National Laboratories, MS 1415, Albuquerque, NM 87047, USA

Nataša Novak Tušar
National Institute of Chemistry, Hajdrihova 19, 1000 Ljubljana, Slovenia

Petko St. Petkov
Faculty of Chemistry, University of Sofia, 1126 Sofia, Bulgaria

Galina P. Petrova
Faculty of Chemistry, University of Sofia, 1126 Sofia, Bulgaria

Cuong Pham-Huu
Laboratoire des Matériaux, Surfaces et Procédés pour la Catalyse – LMSPC UMR 7515 du CNRS, Université Louis Pasteur, France

Pilar Pina
Chemical and Environmental Engineering Department, University of Zaragoza, Pedro Cerbuna 12, Zaragoza 50009, Spain

Gerhard D. Pirngruber
Institut Français du Pétrole, IFP-Lyon, Catalysis and Separation Division, Heterogeneous Catalysis Department, BP3, 69390 Vernaison, France

Evgeny V. Rebrov
Department of Chemical Engineering and Chemistry, Eindhoven University of Technology, 5600 MB Eindhoven, The Netherlands

Nan Ren
Department of Chemistry, Shanghai Key Laboratory of Molecular Catalysis and Innovative Materials, Fudan University, Shanghai 200433, P. R. China

Alenka Ristić
National Institute of Chemistry, Hajdrihova 19, 1000 Ljubljana, Slovenia

Jesús Santamaría
Chemical and Environmental Engineering Department, University of Zaragoza, Pedro Cerbuna 12, Zaragoza 50009, Spain

Wolfgang Schmidt
Max-Planck-Institut für Kohlenforschung, Kaiser-Wilhelm-Platz 1, D-45470 Mülheim an der Ruhr, Germany

Jaap C. Schouten
Department of Chemical Engineering and Chemistry, Eindhoven University of Technology, 5600 MB Eindhoven, The Netherlands

Christian Serre
Institut Lavoisier, CNRS-UMR 8180, Universite de Versailles St Quentin-en-Yvelines, 78035 Versailles, France

Boris Subotić
Ruđer Bošković Institute, Division of Materials Chemistry, Laboratory for the Synthesis of New Materials, P.O. Box 180, 10002 Zagreb, Croatia

Minwei Sun
Department of Chemical and Environmental Engineering, University of California – Riverside, Riverside, California 92521, USA

Yi Tang
Department of Chemistry, Shanghai Key Laboratory of Molecular Catalysis and Innovative Materials, Fudan University, Shanghai 200433, P. R. China

Vincent Toal
Centre for Industrial and Engineering Optics Focas, DIT Kevin Street, Camden Row, Dublin 8, Ireland

Lubomira Tosheva
Division of Chemistry & Materials, Manchester Metropolitan University, Chester Street, Manchester M1 5GD, United Kingdom

Miguel Urbiztondo
Chemical and Environmental Engineering Department, University of Zaragoza, Pedro Cerbuna 12, Zaragoza 50009, Spain

Valentin Valtchev
Laboratoire de Matériaux à Porosité Contrôlée, Université de Haute Alsace, UMR-7016 CNRS, ENSCMu, 3, rue Alfred Werner, 68093 Mulhouse, France

Georgi N. Vayssilov
Faculty of Chemistry, University of Sofia, 1126 Sofia, Bulgaria

Ajayan Vinu
National Institute of Materials Science, 1–1 Namiki, Tsukuba, Ibaraki, 305–0044, Japan

Leonid Vradman
Department of Chemistry, Nuclear Research Centre Negev, Beer-Sheva 84190, Israel

Alain Walcarius
Laboratoire de Chimie Physique et Microbiologie pour l'Environnement (LCPME), UMR 7564 CNRS, Universite Henri Poincare, Nancy I, 405, rue de Vandoeuvre, 54600 Villers-Les-Nancy, France

Junlan Wang
Department of Mechanical Engineering, University of California – Riverside, Riverside, California 92521, USA

Yushan Yan
Department of Chemical and Environmental Engineering, University of California – Riverside, Riverside, California 92521, USA

Yahong Zhang
Department of Chemistry, Shanghai Key Laboratory of Molecular Catalysis and Innovative Materials, Fudan University, Shanghai 200433, P. R. China

SECTION I

SYNTHESIS, MODIFICATION AND CHARACTERIZATION OF ORDERED POROUS MATERIALS

CHAPTER 1

A New Family of Mesoporous Oxides—Synthesis, Characterisation and Applications of TUD-1

Cindy Aquino *and* Thomas Maschmeyer

Contents

Abstract

There exists much work regarding the synthesis of mesoporous materials. TUD-1 is one recent development whereby a non-surfactant organic compound (triethanolamine) templates the formation of mesoporous oxides. Transition metals varying from atomically dispersed isolated centres to nanoparticulate oxides are incorporated easily and controllably in a one-pot synthesis mixture, and have shown considerable catalytic performance when evaluated in various reaction types. Furthermore, composite micro-/mesoporous systems have been synthesised and applied successfully in a range of acid-catalysed reactions. This chapter aims to provide an up to date review of TUD-1 and its derivatives, their synthesis, characterisation and application.

Keywords: Mesoporous oxides, Nanoparticles, Composite materials, Synthesis, Catalysis

Ordered Porous Solids
DOI: 10.1016/B978-0-444-53189-6.00001-9

Abbreviations

CTMA	Cetyltrimethylammonium
FTIR	Fourier Transform Infrared
MPV	Meerwein-Ponndorf-Verley
NMR	Nuclear Magnetic Resonance
SMPO	Styrene Monomer Propylene Oxide
TBHP	*tert*-Butylhydroperoxide
TEA	Triethanolamine
TEAOH	Tetraethylammonium hydroxide
TEOS	Tetraethoxysilane
TOF	Turnover frequency
(HR) TEM	(High Resolution) Transmission Electron Microscopy
UV-vis	Ultraviolet-Visible Spectroscopy
XRD	X-ray Diffraction

1. Introduction

Porous materials of varying chemical characteristics (basic, acidic, redox-active, inert, conducting, semi-conducting, etc.) are of fundamental importance in the areas of science and technology.[1,2] This is due to the presence of voids of controllable dimensions at the atomic, molecular and nanometre scale.[3] Their use in industry is extensive ranging from areas such as petroleum refining,[4] detergents,[5] medicinal applications[6] and separations.[7] The International Union of Pure and Applied Chemistry (IUPAC) divides porous materials into three classes based on their pore diameter (d): microporous $d < 2.0$ nm, mesoporous $2.0 \leq d \leq 50$ nm and macroporous $d > 50$nm.[8] Pore architectures (size, shape, connectivity) and the nature of the pore distribution, in combination with the chemical characteristics of the pore walls, determine the properties (and hence applications) of such materials.

Arguably, the most important group of porous materials so far is the microporous one, in particular zeolites. Although these crystalline aluminosilicates are used extensively in industry, they do suffer from significant drawbacks as a result of their small pore sizes. Zeolites can experience problems in mass transfer, affecting diffusivities of the reactants and products to and from the active sites. In addition, in some reactions, like catalytic cracking, this might lead to coke formation, degrading the catalytic activity of a zeolite. The inability of bulky substituents to make use of the extensive internal surface area restricts their use in important chemical processes. Much work has been carried out in the literature to address the limitations of zeolites, one of the methods being the synthesis of materials with larger sized pores. One of the most significant contributions has sprung from work involving mesoporous material synthesis, initiated by the groups led by Beck[9] and Kuroda.[10]

2. MCM-41 AND FSM-16

The best-known family of mesoporous materials is MS41, and from that group, MCM-41, first reported in 1992.[9] The material was synthesised in an effort to form an improved alternative to zeolites. MCM-41 material exhibits a regular, hexagonal arrangement of pores (*P6mm* symmetry) with one-dimensional parallel channels, formed as a result of liquid crystal templating. This mechanism is thought to be a result of either the silicate synthesis gel condensing around the pre-arranged micelles formed from the cetyltrimethylammonium ($CTMA^+$) template or by the silicate influencing the formation of the liquid crystal phase (Fig. 1.1),[11] depending on the concentrations used. Pore sizes obtained for this material can range from 1.5 to 10 nm.

FSM-16, a material with a similar pore arrangement was synthesised by Inagaki *et al.*[10] This method consists of interlayer cross-linking of a layered silicate (kanemite, $NaHSi_2O_5 \cdot 3H_2O$) that has undergone an ion exchange (usually alkyltrimethylammonium ions, such as $CTMA^+$). The silicate layers of the single-layered polysilicate are then able to condense to form silicate networks (Fig. 1.2). For the synthesis of FSM-16, a smaller amount of $CTMA^+$ is employed compared to MCM-41, since it is used as a swelling agent, rather than as a template, to direct the formation of the material.

Even though aluminosilicate versions of these materials with essentially amorphous walls were consequently no match for the activity associated with the high acidity of crystalline zeolites, it did open up a new door to mesoporous material synthesis. Early examples of these mesoporous materials exhibited low hydrothermal stability, thin walls, the use of an expensive template and the one-dimensional pore arrangement. Thus, much work has been targeted at either the improvement of these materials or the synthesis of novel mesoporous materials.

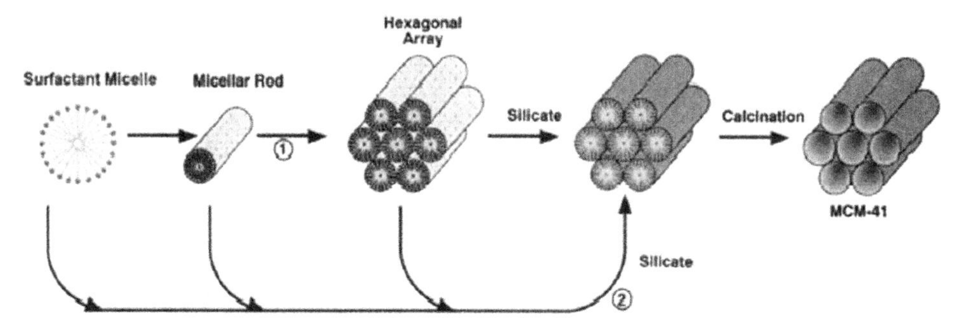

Figure 1.1 Possible mechanistic pathways for the formation of MCM-41: (1) liquid crystal phase initiated and (2) silicate anion initiated. Reprinted with permission from Beck *et al.*[11] Copyright (1992) The American Chemical Society.

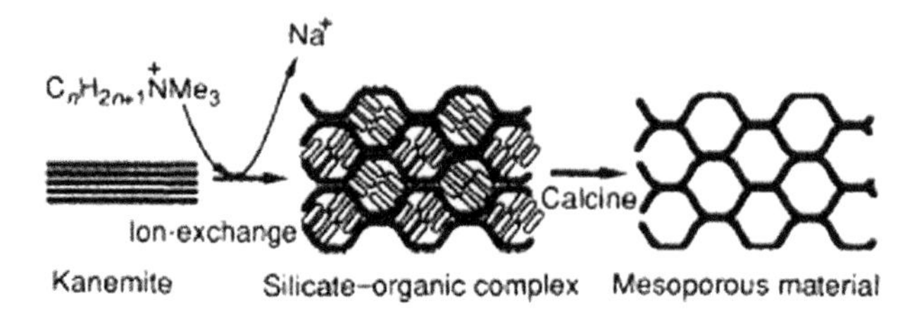

Figure 1.2 Schematic model for the formation of mesoporous material from kanemite.[10] Reproduced by permission of The Royal Society of Chemistry.

3. TUD-1

One such example of a novel, next-generation, mesoporous material set out to improve the short-comings of MCM-41 is TUD-1 (*T*echnische *U*niversiteit *D*elft), first reported in 2001 by one of us (T. Maschmeyer).[12] Along with its high surface area, thicker mesopore walls and hydrothermal stability, TUD-1 employs an inexpensive, non-surfactant structure directing agent, triethanolamine (TEA), to direct the formation of mesostructures during the polycondensation of inorganic species. The resulting material is one which exhibits high surface areas (up to 1000 $m^2\ g^{-1}$), tunable mesopore size distribution (25–250 Å) and a three-dimensional arrangement of interconnecting pores. The three-dimensional connectivity of TUD-1 was proven via carbon inverse replication, a method first reported by Ryoo *et al.*[13] to prove the interconnected pores of SBA-15.[13] The resulting carbonaceous material showed a similar foam-like mesopore structure to TUD-1 (Fig. 1.3).

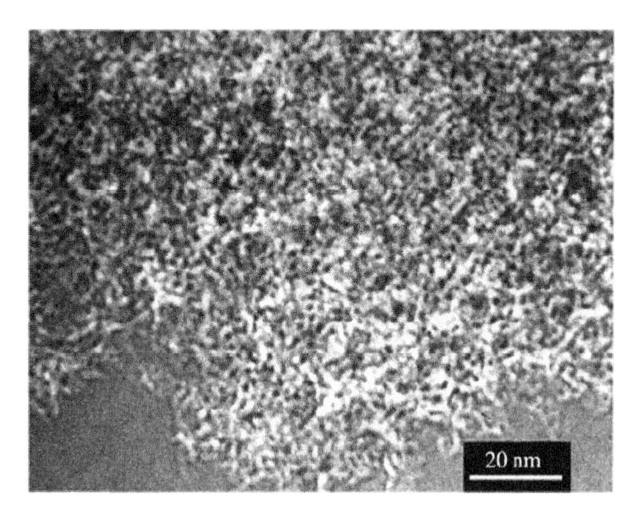

Figure 1.3 Mesoporous carbon templated by TUD-1 silica, yielding an inverse TUD-1 structure.[12] Reproduced by permission of The Royal Society of Chemistry.

For a typical synthesis of TUD-1, TEA and tetraethoxysilane (TEOS) are combined in the following ratios—1TEOS:0.25–2TEA:10–40H_2O. The gel is then aged, dried and treated hydrothermally for a certain amount of time. Since there exists a direct relationship between pore sizes and heating time, one can obtain a desired pore size simply by adjusting the heating time, as shown in Fig. 1.4. The template is removed via Soxhlet extraction with ethanol, followed by calcination in air, resulting in TUD-1.

A schematic mechanism of formation is outlined in Fig. 1.5. Firstly, the TEOS is hydrolysed to give $Si(OH)_4$ which condenses with TEA. Since C–O–Si bonds are more easily hydrolysed than Si–O–Si bonds, over time the TEA–silica linkages get repeatedly broken and the silica nucleus grows slowly, while being surrounded by TEA. Upon heating, the processes of TEA hydrolysis and silanol condensation are accelerated, leading to a controlled phase separation (i.e. microsyneresis). Subsequently, the TEA aggregates template the mesoporous silica framework as the silica species condense around them. Figure 1.6 is a transmission electron microscopy (TEM) image of the final calcined product where foam-like porosity is clearly evident.

In addition to directing pore formation, TEA is able to stabilise active metals in the synthesis gel, which then allows their incorporation into the siliceous framework. The synthesis of metal-incorporated mesoporous materials has been a field of active investigation, as these materials offer much better diffusion to and from the active sites than their microporous counterparts. The initial metal concentration for the synthesis of such materials must be taken into careful consideration: TEA can stabilise many active metals via complexation, thus it is possible to generate a heterometal-rich TEA phase that templates the mesopores. During calcination, the organic species are removed and the metal atoms are deposited into/onto the

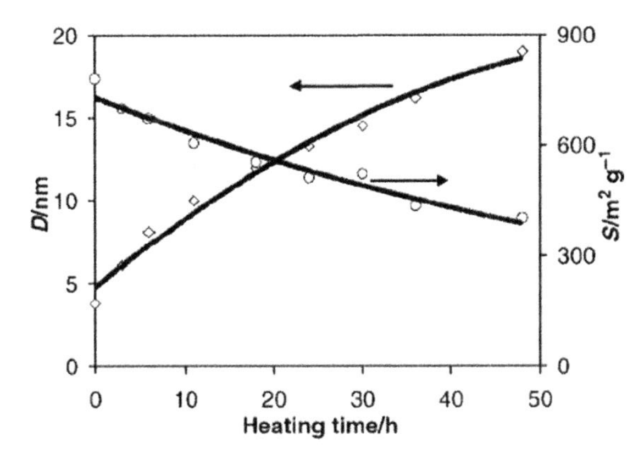

Figure 1.4 Typical graph obtained when tuning the mesoporosity by variation of the heating time during hydrothermal treatment (in this case a homogeneous synthesis mixture with a molar composition of TEOS:0.5TEA:0.1TEAPH:11H_2O was, after drying at 98 °C for 24 h, heated in an autoclave to 190 °C for different times). D is the mesopore diameter at maximum peak height calculated using the BJH model based on nitrogen desorption branch, S is the mesopore surface area calculated using the t-plot method.[12] Reproduced by permission of The Royal Society of Chemistry.

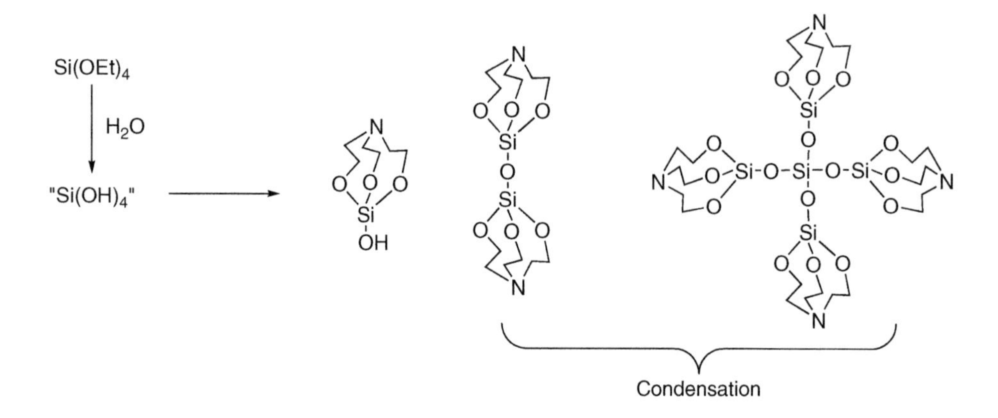

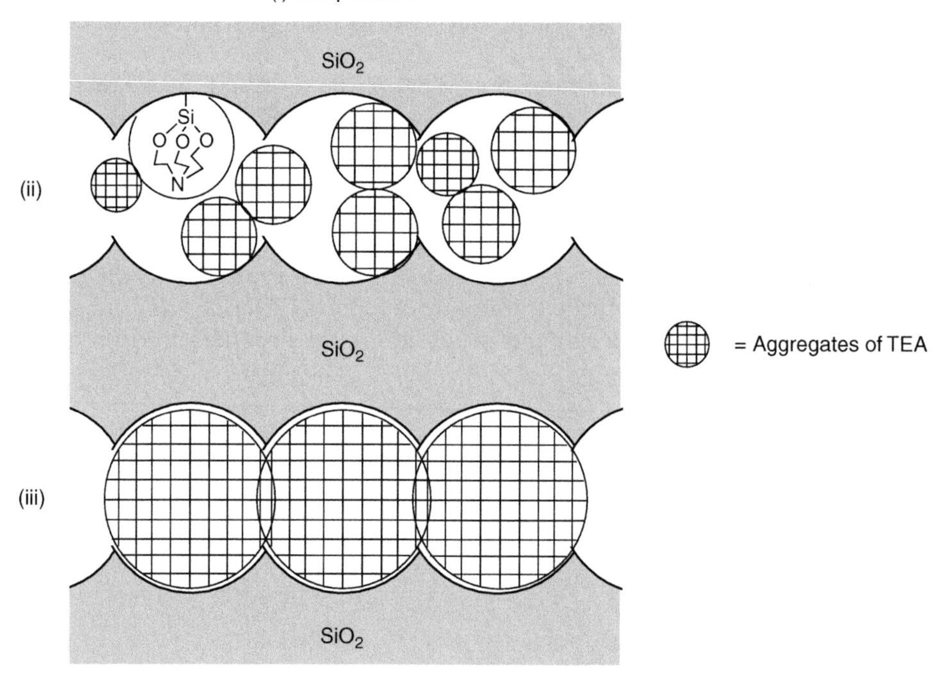

Figure 1.5 Mechanism of formation of TUD-1: (i) TEA hydrolysis and TEOS condensation from the initial silica nucleus, resulting in phase separation (ii) and pore formation (iii).

internal mesoporous surface (*in situ* grafting), depicted in Fig. 1.7a. At higher metal loadings, the possibility for hetero-metal-oxide formation increases and nanosized scale crystals can grow inside the pores, and/or bulky metal oxides are formed as extra framework crystals (Fig. 1.7b). Thus, with materials containing higher metal loadings, more than one type of catalytic site is likely to exist.

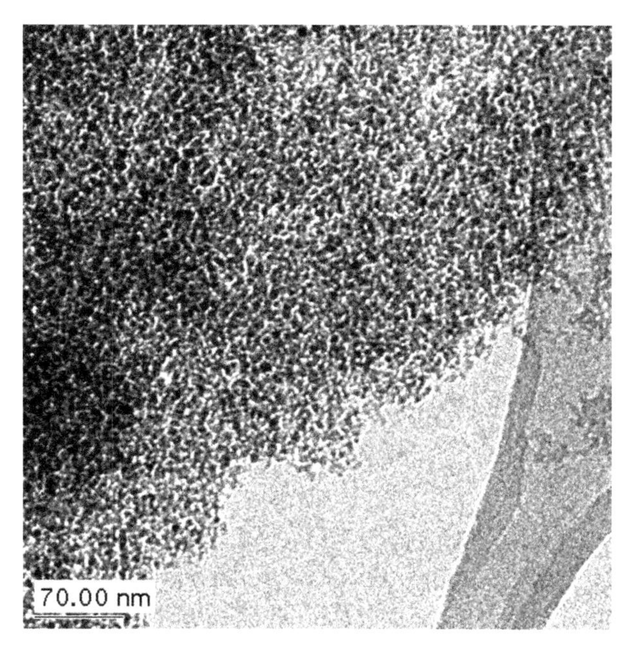

Figure 1.6 TEM image of TUD-1.

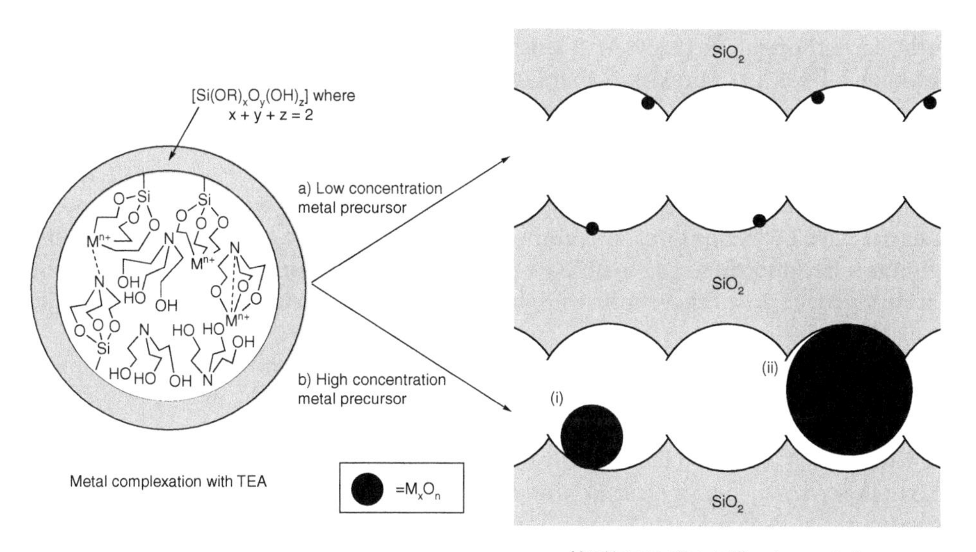

Figure 1.7 (a) Low concentration of the metal precursor allows the metal to be atomically dispersed and (b) higher concentration results in nanoparticles. The smaller sized particles (i) are controlled by concentration (lower end sizes) and the larger sized particles (ii) controlled by the material's pore sizes that can be adjusted via heating time.

Taking Ti as an example, TEA is known to form complexes more easily with Ti than Si (due to the interaction of the lone electron pair on nitrogen atom with empty d orbital of the Ti). Thus, Ti can be deposited in a controlled manner into the pores and onto the walls of TUD-1, as first reported by Shan *et al.*[14] The stabilisation of the active metal during the synthesis is extremely important as the starting materials (Ti(IV) alkoxides) have a tendency to form TiO_2 upon exposure to H_2O in a rapid and not easily controlled manner. Thus, not only would this lower the number of isolated and tetrahedrally co-ordinated metal sites, but also lower the integrity of the mesoporous structure as such.

4. Redox Metal Incorporation into Siliceous TUD-1 Framework

4.1. Ti-TUD-1

The motivation for the synthesis of a Ti-TUD-1 material has its origin in the desire to epoxidise bulky alkenes with peroxides and preferentially with hydrogen peroxide, analogous to titanosilicate-1 (TS-1). Initial interest was based on the SMPO (Styrene Monomer Propylene Oxide) process, which employs a titanium-on-silica (epoxidation) catalyst, first established in 1979,[15] and is a process still in use by Shell today. The report of epoxidation with hydrogen peroxide using TS-1 ignited the field.[16] Incorporation of Ti into the crystalline framework largely avoids the problem of leaching that typically occurs when metals are incorporated into zeolites with an ion-exchange procedure, but more importantly the bond angles due to the lattice are thought to prevent Ti-site deactivation by water and/or hydrogen peroxide. TS-1 has shown to exhibit remarkable activity due its structure and selective oxidations of alkanes,[17] alkenes[18] and alcohols.[18] Many other microporous titanosilicates and titanoaluminophosphates have also been synthesised, such as TS-2,[19] Ti-ZSM-48,[20] Ti-beta,[21] TAPO-5[22] and TAPO-11[22] all with tetrahedrally co-ordinated Ti centres. The small pore sizes, as with all microporous materials, are by definition the common problem, restricting their use with bulky substrates because these are unable to penetrate into the micropores and only the external surface area can be utilised—wasting everything but a small fraction of the available sites.

Thus, a number of mesoporous titanosilicates have been synthesised, for example using the M41S family, where Ti can be introduced into the mesoporous wall during synthesis or grafted onto the wall surface after synthesis of the material (e.g. Ti-MCM-41[23] and Ti-HMS[24]) to offer a solution to these problems. At loading above 0.5 wt.% Ti, a significant amount of Ti(IV) centre lies inaccessibly in the framework and dimmer/oligomer formation can occur, reducing the peroxide selectivity of the titanium centres. This problem was overcome by grafting an organometallic titanium catalytic-site precursor onto the inner walls after synthesis, leading to isolated and dispersed sites, at high loadings.[25]

Along with the drawbacks of MCM-41 mentioned previously, the one-dimensional pore structure can lead to problems associated with pore blockage

Table 1.1 Comparison of catalytic activities for cyclohexene epoxidation over Ti-TUD-1, Ti-MCM-41, Ti grafted TUD-1, and Ti grafted MCM-41

Catalysts	Ti^a (wt%)	$S_{BET}{}^b$ ($m^2 g^{-1}$)	D^c (nm)	TOF^d (h^{-1})	S^e (%)
Ti-TUD-1	1.50	917	4.5	20.2	94
Ti-MCM-41	1.82	921	3.1	3.6	–
Ti grafted TUD-1	1.87	561	10.1	27.7	81
Ti grafted MCM-41	1.79	1015	3.0	23.4	82

[a] Titanium loading.
[b] Surface area obtained from nitrogen adsorption.
[c] Mesopore diameter at maximum peak calculated by using the BJH model.
[d] Turnover frequency after 6 h reaction.
[e] Selectivity of TBHP after 6 h reaction.

and the stabilisation of the isolated, tetrahedral Ti-sites. However, Ti-TUD-1,[26] with its three-dimensional arrangement of pores, offers an alternative to such materials. It was tested as a potential catalyst and compared with Ti-MCM-41, Ti-grafted TUD-1 and Ti-grafted MCM-41 in the epoxidation of cyclohexene with *tert*-butylhydroperoxide (TBHP). The results are summarised in Table 1.1.

The turnover frequency (TOF) of Ti-TUD-1 was found to be about 5.6 times higher to that of framework-substituted Ti-MCM-41 and similar to that of Ti-grafted MCM-41. The selectivity of TBHP to cyclohexene epoxide was ~94% after 6 h of reaction. One advantage of Ti-TUD-1 over Ti-grafted MCM-41 is that the former titanosilicate can be synthesised in a one-pot synthesis. The higher activity of Ti-TUD-1 originates from both the high accessibility of the substrates to the catalytic site provided by the three-dimensional mesopore system and from the formation of isolated titanium centres on the surface of the mesopore wall, to give optimum usage of Ti.

4.2. Fe-TUD-1

In a similar vein to Ti-TUD-1, it has been possible to successfully immobilise Fe in siliceous TUD-1 to produce Fe-TUD-1. In addition, by adjusting the Si/Fe concentration one can also obtain nanoparticles of iron oxide, located preferentially inside the pores. The synthesis of Fe-TUD-1 was first reported by Hamdy *et al.*[27] and materials made with varying Si/Fe ratios were tested as a potential catalyst for the Friedel–Crafts benzylation of benzene (Table 1.2). Generally speaking, Friedel–Crafts alkylations are catalysed by Lewis acids in the liquid phase. As with other processes, there is a push to replace liquid acids by solid acid catalysts, as they are more easily separated from their reaction mixtures and conform more readily to the principles of 'Green Chemistry'. Fe-TUD-1 provides an effective alternative to the traditional $FeCl_3$ catalyst for the benzylation of benzene, as shown in Table 1.3. The catalytic test shows no reaction with pure siliceous TUD-1 or with pure iron oxide powder (Fe_2O_3), consistent with the need for dispersed nanoparticles in TUD-1 (for all active samples, the selectivity to diphenyl methane is 100%).

Table 1.2 Comparison between samples composition and activity (i.e. conversion of benzyl chloride)

Sample	Si/M ratio	Conversion (%)	Reaction time (min)	Temperature (°C)
TUD-1	∞	0	240	60
Fe_2O_3	0	0	240	60
Fe-TUD-1 (Fe-1)	113	86	240	60
Fe-TUD-1 (Fe-2)	54	57	240	40
Fe-TUD-1	54	100	180	60
Fe-TUD-1	54	98	30	80
Fe-TUD-1 (Fe-5)	21	100	10	60
Fe-TUD-1 (Fe-10)	10.1	100	<1.5	60
Ti-TUD-1	50	4.3	240	60
Fe-HMS[1]	50	73	240	60
Fe-MCM-41[1]	62	80	240	60
Fe-HMS[2]	23.8	100	120	75
Fe-HMS[2]	14	100	90	75
Fe-MFI[3]	16.5	90	25	80
Fe-AlMFI[3]	28.1	90	104	80

Table 1.3 Asymmetric hydrogenation of methyl 2-acetamidoacrylate using AlTUD-1 as catalyst[a]

Entry	Solvent	Time (min)	Conversion (%)	Enantiomeric excess (*ee*) (%)	Rh loss (mg/L) (%)[c]
Homogeneous	CH_2Cl_2	7	100	97	–
1	CH_2Cl_2	7	96	83	0.76 (2.9)
2	EtOAc	7	70	92	1.27 (5.5)
3	EtOAc	11	100	92	1.04 (4.5)
4[b]	EtOAc	1200	71	94	1.08 (35)
5	2-PrOH	7	39	91	2.30 (9.4)
6	2-PrOH	25	100	97	2.06 (8.4)
7	MTBE	7	11	94	0.32 (1.3)
8	MTBE	30	91	94	0.45 (1.8)
9	Water	35	75	95	0.21 (0.6)
10	Water	60	100	95	0.11 (0.3)

[a] 5 bar H_2, 50 mL solvent, [4a] = 0.05 M, 0.1 g catalyst with 1 wt% Rh.
[b] 10 bar H_2, 50 mL solvent, [4a] = 0.2 M, 0.01 g catalyst with 1 wt% Rh.
[c] Percentage of total amount of Rh determined by AAS of the filtrate.

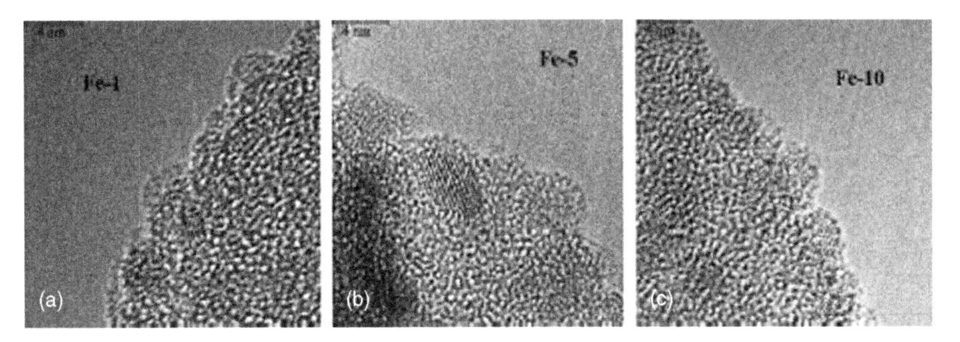

Figure 1.8 HR-TEM images of different Fe-TUD-1 samples. Reproduced from Hamdy *et al.*[27] by permission of Elsevier.

Fe-TUD-1 exhibits different types of catalytic sites; at lower Fe loadings isolated Fe(III) sites are consistent with diffuse reflectance UV–vis, TEM and XRD data, while at higher Fe loadings nanoparticles of iron oxide embedded into the mesopores of the TUD-1 silica matrix are observed by HR-TEM (Fig. 1.8). From the table above, it is clear that the sample with Si/Fe = 10 ratio was the most active.

By examining the time taken to reach for 90% as a function of Fe loading, we can observe that the increase in activity as a function of Fe loading is highly non-linear, pointing to a fundamental change in the nature of the active site (Fig. 1.9).

It has been suggested that one reason for the improved activity with higher loading is that the nanoparticles of iron oxide are less co-ordinatively saturated (due to lattice strain and defects) than isolated iron species or than the iron centres present in larger particles, thereby giving rise to more active centres. On comparing Fe-TUD-1 with other materials from the literature,[28–30] Fe-TUD-1 is more active by more than a factor of 15 when compared to the next best material (Fe-ZSM-5). This is likely due to the combination of improved local environment of active site

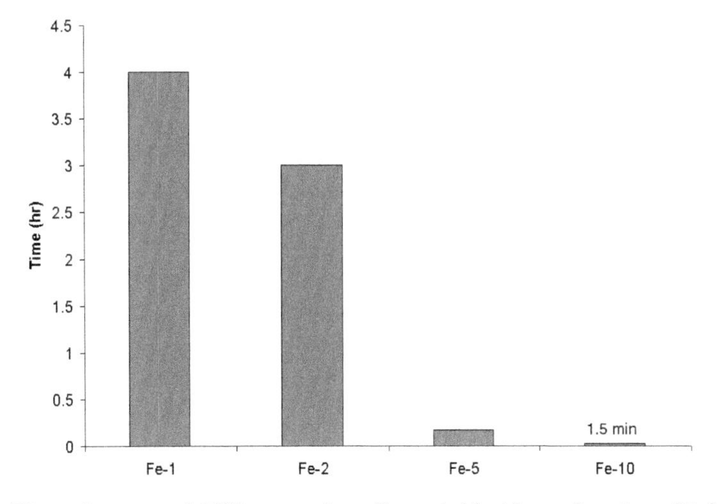

Figure 1.9 Time taken to reach 90% conversion of benzyl chloride as a function of Fe loading.

(strained/unsaturated nanoparticles) as well as the improved global structure of TUD-1, which is thought to be responsible for the higher accessibility of the reactants inside Fe-TUD-1.

4.3. Co- and Cr-TUD-1

The ease with which Fe and Ti could be selectively incorporated into the TUD-1 structure type (i.e. varying from atomically dispersed to stabilised nanoparticles inside the mesopores) paved the way for modifications with other transition metals. Co- and Cr-TUD-1 materials can also be synthesised in much the same way as their Fe- and Ti-TUD-1 analogues, as they also from relatively stable complexes with TEA. Thus, Cr- and Co-TUD-1 were investigated as suitable catalysts for the liquid phase oxidation of cyclohexane with TBHP and air as an oxidant.

Cyclohexane oxidation is an important industrial process, but as the product is more easily oxidised than the substrate, the cyclohexane conversions need to be kept low (~4%) such as to retain high selectivity for cyclohexanone and cyclohexanol (~70–85%). This has prompted much research into the development of new and more selective catalytic systems which work under mild conditions. One approach has been to replace the traditional homogeneous cobalt-based catalysts with heterogeneous catalysts, which prompted the synthesis of Co- and Cr-TUD-1 materials, containing a Si/M ratio of 100.[31] In the liquid-phase cyclohexane oxidation, at 70 °C, with TBHP as oxidant both the conversion of cyclohexane (Fig. 1.10a) and TBHP (Fig. 1.10b) were catalysed by all M-TUD-1 catalysts. By plotting the conversion of cyclohexane against selectivity for mono-oxygenated products[1] (Fig. 1.10c), both Co-TUD-1 and Cr-TUD-1 exhibited high activity and very high selectivity to these products (at conversion levels of 5–8%). Co-TUD-1 showed a very slow selectivity decline (even at higher conversion levels) of cyclohexane (compared to Cr-TUD-1). Cr-TUD-1 produced more cyclohexanone than cyclohexanol but Co-TUD-1 produced initially more of the alcohol (A), and then slowly oxidised it to the ketone (K) as shown in Fig. 1.10d. In summary, Cr and Co-TUD-1 showed >90% selectivity towards mono-oxygenated products and conversion levels as high as 8–10%. However, not unexpectedly, leaching was a problem, particularly in the chromium case.

In a further study, Co-TUD-1 was synthesised with different Si/Co ratios (100, 50, 20 and 10) and analysed under the same catalytic reaction.[32] Higher catalytic efficiency was observed in samples with a low Co loading (Co atoms in the framework). Figure 1.11 shows that the conversion of both cyclohexane and cyclohexanol decreases as the Co concentration in TUD-1 increases. Agglomeration of Co at higher metal loadings is thought to reduce the accessibility of individual Co atoms and therefore the number of active sites. Thus, it is most likely that the isolated Co(II) sites are responsible for the catalyst's activity highlighting the importance of metal site dispersion and the ability of TUD-1 to achieve this requirement.

[1] Mono-oxygenated products are cyclohexanol, cyclohexanone, cyclohexyl hydroperoxide and minor products such as cyclohexylformate and cyclohexyltertbutylper-ether.

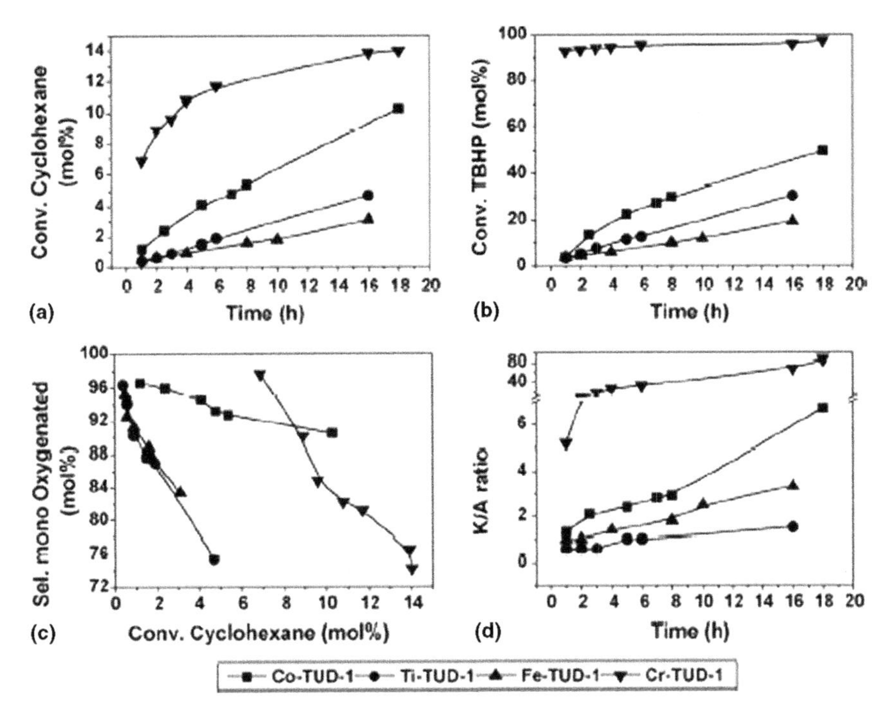

Figure 1.10 (a) Conversion of cyclohexane and (b) *tert*-butylhydroperoxide (TBPH) with TBPH as oxidant over M-TUD-1 at 70 °C. The conversion of cyclohexane versus selectivity for mono-oxygenated products (c) is also shown and (d) the formation of cyclohexanone (K)/cyclohexanol (A) against time. Reproduced from Anand *et al.*[31] by permission of Elsevier.

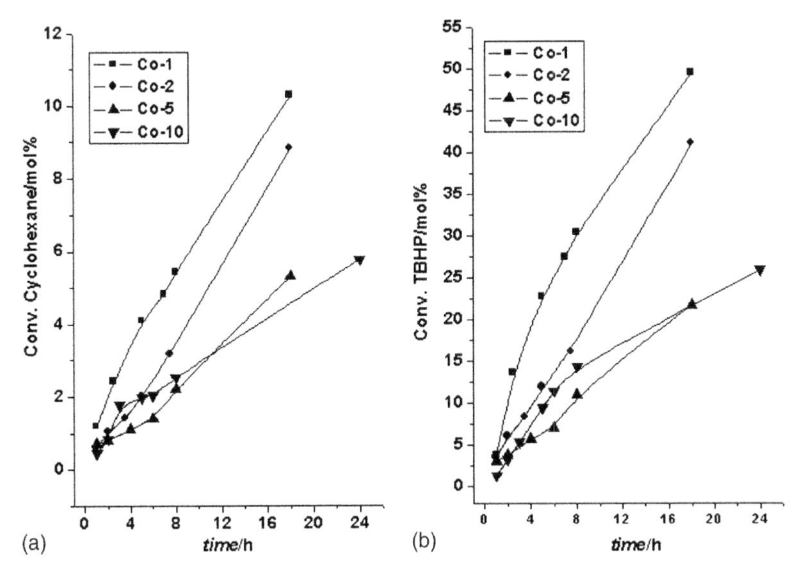

Figure 1.11 (a) Conversion of cyclohexane and (b) *tert*-butylhydroperoxide (TBHP) over Co-TUD-1 at various metal loadings. Reproduced from Hamdy *et al.*[32] by permission of Wiley-VCH.

4.3.1. Cr-TUD-1 as photocatalysts in the gas phase

There is an extensive body of work in which Cr has been incorporated into both mesoporous materials (such as HMS) and microporous materials (ZSM-5).[33–36] In particular, Cr^{6+} has shown a considerable amount of photocatalytic efficacy in the framework of both silica and silica–alumina supports. Therefore, overcoming the problem of leaching, while still making use of the catalytic potential of Cr-TUD-1, the selective photocatalytic oxidation of propane to acetone in the gas phase was investigated. Cr-TUD-1 (Si/Cr ratio of 130) was found to show high activity and high selectivity towards acetone under visible light irradiation.[37] Absorbed species arising from the propane oxidation were monitored using FTIR (Fig. 1.12a). The rate of total product formation is plotted as a function of exposure time (Fig. 1.12b).

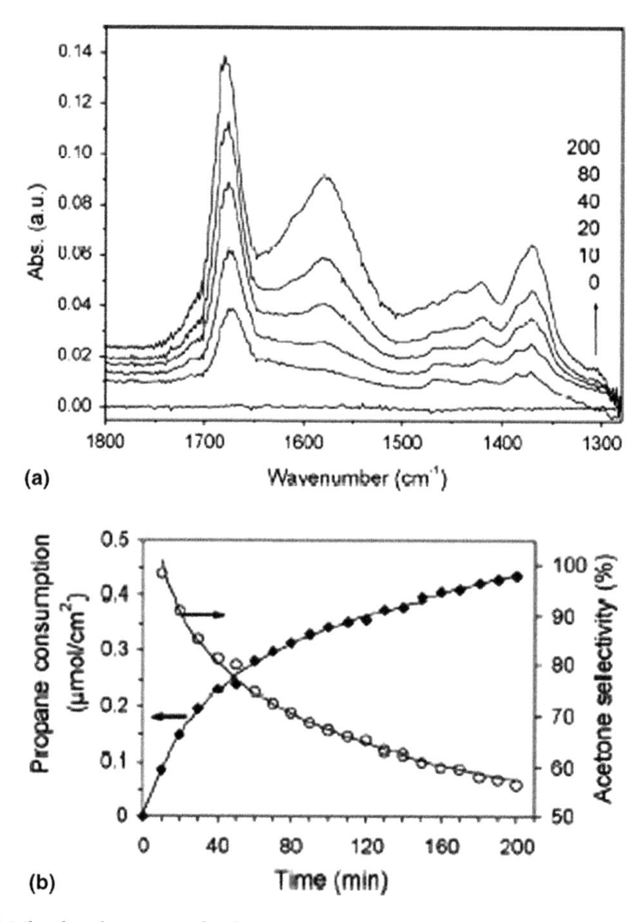

Figure 1.12 (a) The development of infrared absorbance spectroscopy during photocatalytic oxidation of propane over Cr-TUD-1 with 435 nm illumination. (b) Propane conversion and acetone selectivity plotted during 200 min of irradiation. Reproduced from Hamdy *et al.*[37] by permission of Elsevier.

Various Ti-TUD-1 materials (Si/Ti = 100, 20, 2.5, 1.6) were also investigated for the same reaction.[38] As a function of increasing metal loading, either isolated Ti atoms (when Ti loading ~2.5 wt.%) or combinations of isolated Ti atoms and anatase (TiO_2) nanoparticles are obtained. Propane oxidation occurred at an irradiation of $\lambda = 365$ nm (which selectively activates anatase nanoparticles). The rate of total product formation is plotted as a function of exposure time (Fig. 1.13). Clearly, Ti-TUD-1 is able to selectively photocatalyse the dehydrogenation of propane under blue light, especially at higher Si/Ti ratios. The selectivity is thought to arise from the 'de-tuned' catalytic activity that has been achieved by increasing the bandgap due to quantum confinement in the nanoparticle.

4.4. Ce-TUD-1

Ce-TUD-1 has shown to be an active catalyst for both the oxidation of *p-tert*-butyl toluene to *p-tert*-butyl benzaldehyde and peroxidative halogenation of Phenol Red. TUD-1 samples containing different Si:Ce ratios (20, 50, 100) were investigated. Complete incorporation of Ce into the framework of TUD-1 was achieved for Ce-TUD-1 (100) whereas in Ce-TUD-1 (20), a small portion of Ce as nanometre sized CeO_2 particles was found.[39] Regardless of the concentration of Ce incorporated in the TUD-1 synthesis gel, the conversion of *p-tert*-butyl toluene exhibits a conversion of 37%, with 57% selectivity towards the desired *p-tert*-butyl benzaldehyde product. These values are similar to those reported for Ce(III) acetate, suggesting the reaction mechanism over heterogeneous Ce-TUD-1 and homogeneous Ce(III) acetate catalysts is the same. With Ce-TUD-1 (20), leaching occurs (~7%); however, when the regenerated catalyst is reused, the substrate conversion increased to 90%, with an aldehyde selectivity of 55%. This surprising result is subject to further investigations.

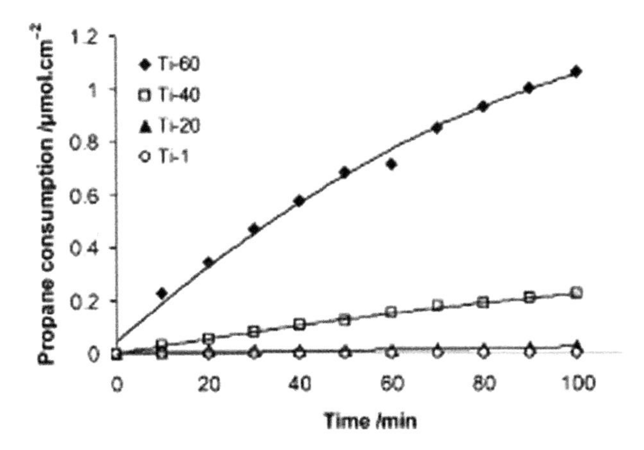

Figure 1.13 Propane consumption over different Ti-TUD-1 compositions. Reproduced from Hamdy *et al.*[38] by permission of Wiley-VCH.

4.5. Zr-TUD-1

Tetrahedrally co-ordinated zirconium has shown to be particularly active in the Meerwein–Ponndorf–Verley (MPV) reduction of aldehydes by Ramanathan *et al.*,[40] a mild redox reaction used extensively in steroid chemistry (Fig. 1.14).

The MPV reduction is typically catalysed by equimolar amounts of aluminium(III) isopropoxide. The proposed mechanism involves a complex in which both the reactants are coordinated to the Lewis acid metal centre and a hydride transfer from the alcohol to the carbonyl group occurs. Other Lewis acid metals such as Zr[41] and Ln[42] have shown their potential as homogeneous catalysts for the described reaction. More recently, the introduction of heterogeneous catalysts, in particular Al-free Sn-beta[43] and Al-free Zr-beta,[44–46] show significant improvements in activity. Al-free Zr-beta was found to be the superior catalyst of the two, as it exhibited higher regioselectivity and was less sensitive to moisture than Al-free Sn-beta. Work has also been conducted on grafting of Zr on MCM-41, MCM-48 and SBA-15, to extend the use to bulkier substituents, but the process is time consuming as grafting is performed post synthesis.[47] With Zr-TUD-1, the metal is able to be incorporated into the framework of TUD-1 using the one-pot synthesis approach described previously.[40] The catalyst displays high selectivity, in particular towards the sterically hindered steroids, compounds that are unable to be reduced with H-beta or Al-free Zr-beta. One particular example is the reduction of 5α-cholestan-3-one (Fig. 1.15). The keto-group was reduced almost quantitatively (> 95%) to yield 5α-cholestan-3-ol with α/β ratio of 1.6.

Thus, along with the Zr to act as an effective Lewis acid to catalyse the MVP reduction, the sponge-like mesoporous structure of TUD-1 allows for the conversion of large, bulky substituents.

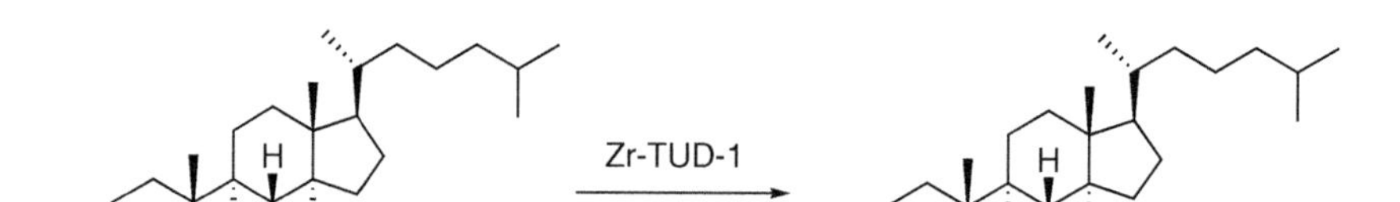

Figure 1.14 Proposed mechanism for the Meerwein–Ponndorf–Verley (MPV) reduction.

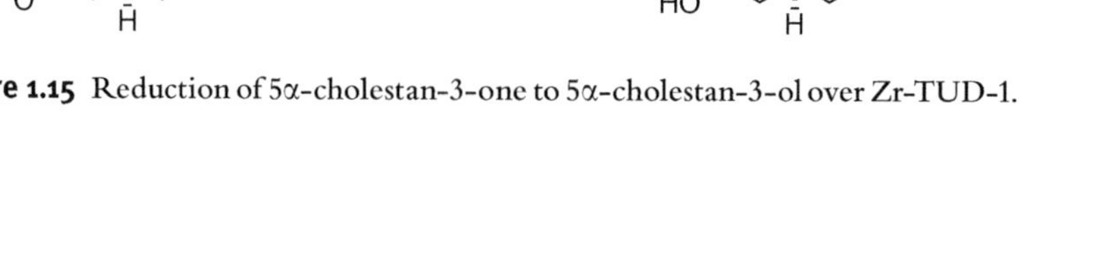

Figure 1.15 Reduction of 5α-cholestan-3-one to 5α-cholestan-3-ol over Zr-TUD-1.

5. Al_2O_3-TUD-1 AND Al-TUD-1

In 2003, Shan *et al.*[48] published results for the formation of a stable mesoporous alumina denoted Al_2O_3-TUD. In the same way that all-silica TUD-1 is templated by a small organic compound, Al_2O_3-TUD is synthesised with a similar compound, tetra(ethylene) glycol (TEG). In addition, due to the high reactivity of aluminium alkoxides with water, a controlled amount of water is used in the initial synthesis to avoid the formation of alumina particles. Surface areas of ~600 $m^2\ g^{-1}$ can be obtained with Al_2O_3-TUD.

Apart from forming a pure Al_2O_3-TUD, it is possible to synthesise a aluminosilicate TUD-1 (denoted Al-TUD-1), a method first described in Simons *et al.*[49] The charge imbalance, induced by the partial substitution of silicon with aluminium, allows this material to act as an effective anionic carrier onto which cationic metal complexes can be bound non-covalently. In addition, for high Brønsted acidity, a high Al/Si ratio (~0.25) is required and the aluminium should be preferentially tetrahedrally co-ordinated. It is important to note that in Al-TUD-1, we do not see the same dependence of hydrothermal treatment time and pore size distribution that we observe in TUD-1. It is thought that the Al-TUD-1 system is less dynamic during the early stages of heating because of the faster degree of Al–O–Si bond formation compared to that of Si–O–Si bonds. Thus, this results in a system that is less sensitive towards changes in pore size with temperature. In addition, TEG is not able to completely suppress the formation of hexa-coordinated Al. However, the resulting material does fulfil the requirements for an effective anionic carrier, with a high surface area and a good $Al_{tetrahedral}/Si$ ratio (= 0.11, total Al/Si = 0.25).

The ability of Al-TUD-1 to act as an efficient support was investigated by Simons *et al.*[49,50] whereby ($[Rh^{I}(cod)((R)\text{-MonoPhos})_2]BF_4$) (Fig. 1.16), a well-known asymmetric hydrogenation catalyst, was immobilised on Al-TUD-1 with a loading typically ~1% (Fig. 1.17) and tested with the asymmetric hydrogenation of methyl 2-acetamidoacrylate (Fig. 1.18).

Hydrogenising this traditional homogeneous catalyst allows for an easier and more complete separation of the products from the reaction mixture (Fig. 1.17). Importantly, a significant improvement on the reusability of this traditionally

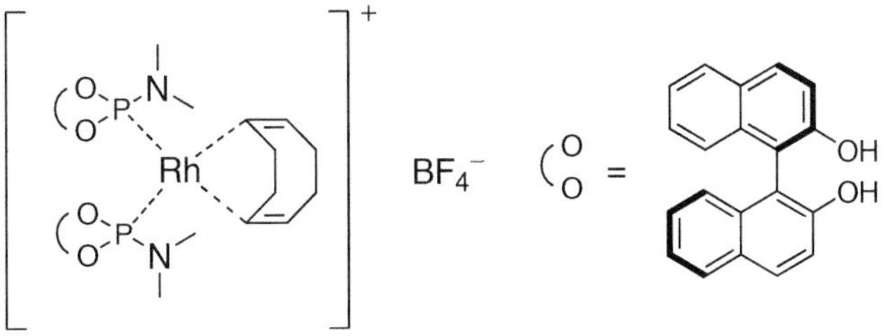

Figure 1.16 MonoPhos Rh catalyst.

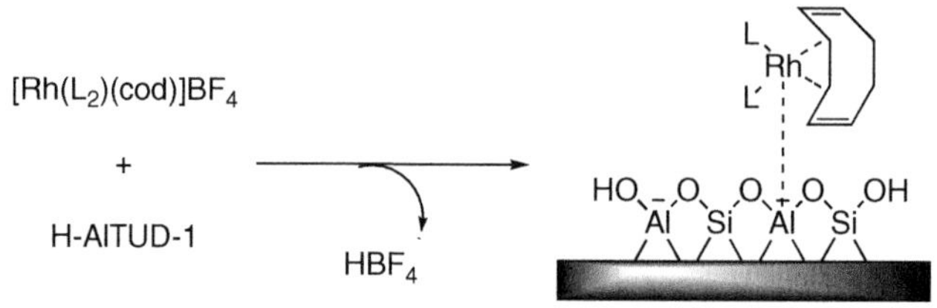

$[Rh(L_2)(cod)]BF_4$ + H-AlTUD-1 → (HBF_4)

Figure 1.17 Typical heterogenisation of an Rh catalyst on Al-TUD-1. Reproduced from Simons *et al.*[50] by permission of Springer.

Catalyst / Solvent, H_2

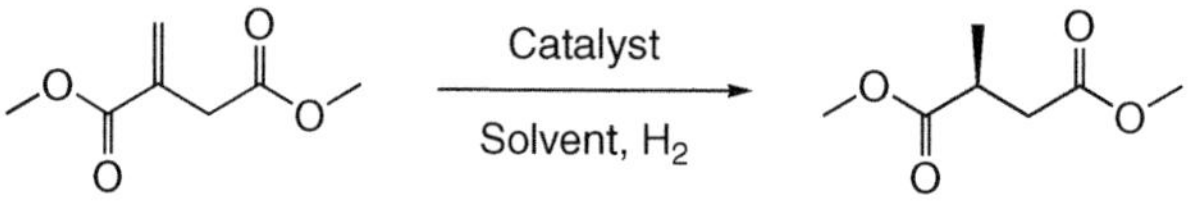

Figure 1.18 Asymmetric hydrogenation of methyl 2-acetamidoacrylate. Reproduced from Simons *et al.*[50] by permission of Springer.

expensive metal complex can be observed. Furthermore, as the complex is non-covalently bound to the aluminosilicate support, behaviour similar to that of the homogeneous catalyst could be observed, that is, the supported metal complex had not undergone significant changes in its catalytic performance.

Tests were carried out to examine the effect of the solvent used, in terms of conversion, selectivity and leaching.[50] The results are summarised in Table 1.3. It was found that once the catalyst was immobilised, the most problematic aspect was its susceptibility to leaching. With increasing solvent polarity, the solvent displays an increasing ability to stabilise charged species, therefore the amount of Rh lost in the system increases. Figure 1.19 shows the loss of Rh as a function of the polarity of the solvent.

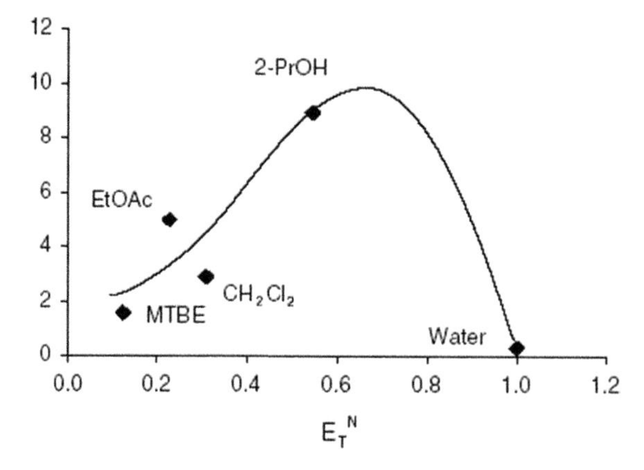

Figure 1.19 The correlation between the polarity of the solvent (E_T^N) and the loss of Rh for 5-Al-TUD-1. Reproduced from Simons *et al.*[50] by permission of Springer.

However, the problem of leaching could be overcome by using an extremely polar solvent, water, which is simply too polar and the support becomes again the preferential location of the cationic catalyst. By using water as the solvent, the catalyst exhibited excellent enantioselectivity, good activity and a very low loss of Rhodium. In addition, the catalyst can be reused without significant deterioration—proving that the phosphoramidite (MonoPhos) is very stable in aqueous reaction media.

In further work by Simons *et al.*,[51] a comparison of supports for the electrostatic immobilisation of Rhodium-MonoPhos was investigated. The metal complex was immobilised on four different anionic carrier materials: Al-TUD-1,[48] Nafion, a Nafion silica composite (SAC-13) and phosphotungstic acid on alumina (PWTUD). The efficiencies of these catalysts were examined, again, for the asymmetric reduction of methyl-2-acetamidoacrylate.

Although most of the catalysts are highly selective, the activity and loss of Rhodium is strongly dependent on the type of support. The bonding between Rh and the phosphotungstic acid (supposedly covalent) is thought to be the reason behind the catalyst's superior anchoring ability to PWTUD. This then translates to minimal leaching of the catalyst, regardless of which solvent system is employed. In addition, the enantioselectivity remained very high. The other supports did experience some leaching (as a result of a lower binding interaction between the support and catalyst) and in the case of SAC-13, a lower activity is also observed. When using Nafion, the catalysts were encapsulated, leading to transport limitations which resulted in low activity.

6. TUD-1 as Potential Drug Carriers

Mesoporous silica-based materials have been studied widely as potential drug delivery systems. Heikkilä *et al.*[52,53] studied the suitability of TUD-1 as a drug carrier, comparing the drug release kinetics of the drug loaded onto TUD-1, MCM-41 and of the solid drug by itself. When considering the pore diameters of TUD-1 and MCM-41, ibuprofen was selected as the model drug due to its molecular dimensions (1.0 nm×0.5 nm). Loading was accomplished via the 'soaking method' (ethanolic solution) where the carrier/drug ratio was 1:48 (w/w). Drug release experiments performed in acidic dissolution medium (mimicking conditions at the start of the small intestine) showed a rapid and almost complete (95%) liberation of ibuprofen from TUD-1 over 210 min. This is an acceptable time frame considering the drug transit time of the small intestine (often cited as 200 min). That the initial 60% of the drug was released quickly (15 min) is a positive result. The more restricted pore characteristic of MCM-41 is thought to be responsible for the slower initial release of ibuprofen (60% over 40 min). What is of interest to note is that both carriers released ibuprofen faster than is the case for the pure form of the drug (25% over 45 min) (Fig. 1.20). The improvement of dissolution is thought to be a result of the mesoporous material changing the solid state properties of the drug (crystal habit and/or amorphotisation, leading to faster dissolution rates).

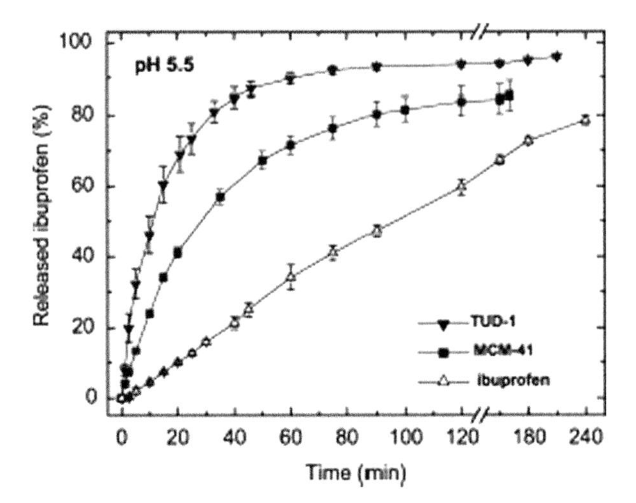

Figure 1.20 Ibuprofen release from TUD-1, MCM-41 and the pure crystalline form at HBSS pH 5.5 medium. Reproduced from Heikkilä *et al.*[52] by permission of Elsevier.

This effect is more pronounced in low pH environments as in neutral conditions the release times and amounts are not as significant since ibuprofen dissolves much more easily. Thus, the application of a siliceous mesoporous carrier diminished the pH dependency of the ibuprofen dissolution and demonstrated the potential for TUD-1 to act as a drug carrier.

7. Particle Incorporation

The formation of hierarchical materials by the combination of porous systems of different dimensions is of great interest, in particular to catalysis. The potential of TUD-1 to be a host for the stabilisation and dispersion of zeolite/zeotype nanoparticles has led to the design and synthesis of solid catalysts with very low diffusion barriers, enabling a deep level of control of transport phenomena.[54] The network of the mesoporous TUD-1 structure minimises diffusion barriers and allows for the incorporation and stabilisation of highly reactive, microporous zeolite and zeotype particles. These particles are very important structures, as they are the principle catalysts used in petrochemical conversions and represent by far the greatest tonnage of catalysts used today; however, they suffer from high intrinsic diffusion barriers due to their intricate and narrow pore systems. While reducing the size of the particles minimises this limitation, very small particles are unstable and tend to aggregate, negating their initial size advantage.

7.1. Zeolite Beta

The first reported instance of a composite zeolite Beta-TUD-1 material was reported in 2002:[55] Prior to the incorporation of zeolite Beta into the synthesis gel, the particles were homogeneously dispersed in aqueous ammonia to avoid initial aggregation of the particles. To this suspension were added TEOS, TEA, H_2O and finally TEAOH. The homogeneity of the dispersion was maintained (during and after gelation) due to the sudden and rapid increase in viscosity brought on by the base in the transition of liquid to thick gel. With a molar ratio of TEAOH/Si < 0.1, one observes gelation. However, with TEAOH/Si > 0.2, no gelation occurs and zeolite particles separate out.[54] X-ray Diffraction (XRD), NMR, nitrogen adsorption and HR-TEM can be used to confirm the incorporation of zeolite Beta into the mesoporous matrix and that the zeolites structure is maintained (Fig. 1.21).

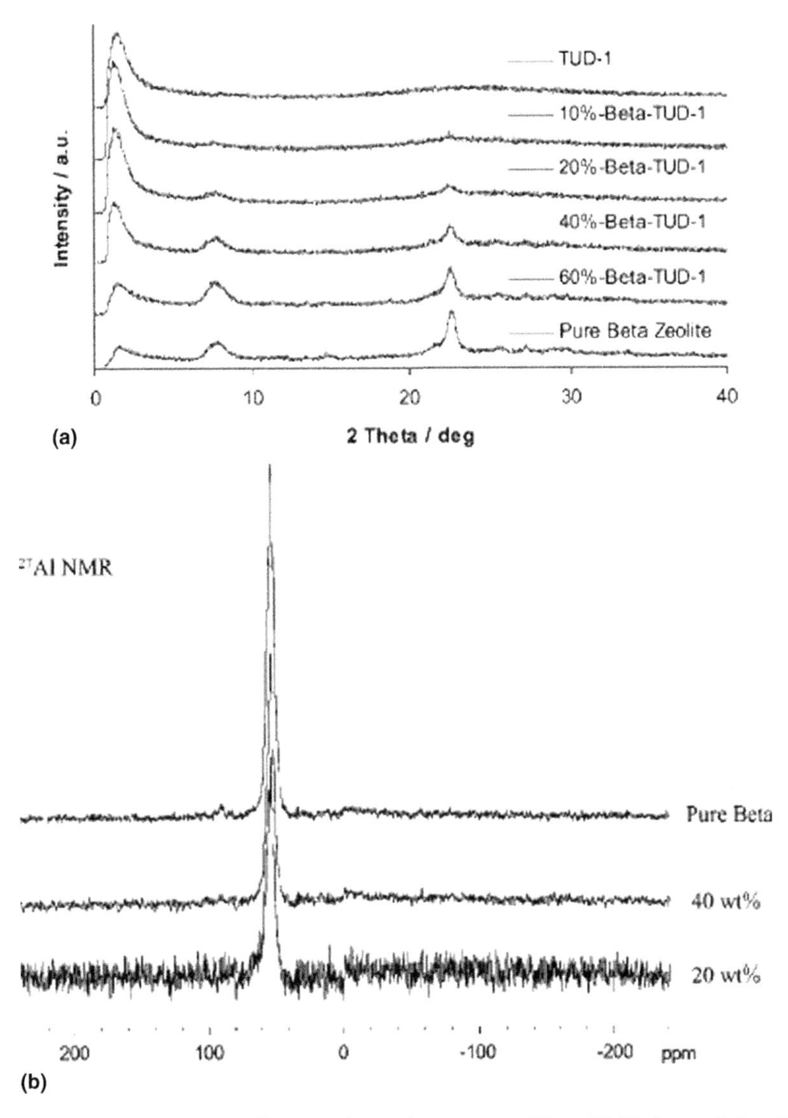

Figure 1.21 (a) X-ray powder diffraction (XRD) patterns of Beta-TUD-1 containing different amounts of zeolite Beta and pure zeolite Beta. (b) ^{27}Al NMR of pure zeolite Beta and Beta-TUD-1 containing 20 and 40 wt.% of zeolite Beta. Reproduced from Waller *et al.*[54] by permission of Wiley-VCH.

The XRD pattern (Fig. 1.21b) shows the characteristic peaks of zeolite Beta (7–8.5° and 22.4° in 2θ) are clearly evident, even at low zeolite loadings (20%), along with the characteristic mesoporous periodicity of TUD-1 (1.2° in 2θ).

^{27}Al NMR (Fig. 1.21b) indicates that all aluminium atoms in the composite material are four co-ordinate (as in pure zeolite Beta) thus one can assume that integrity and structure of the zeolite was maintained during the mesopore formation and subsequent calcination.

The catalytic activity of Beta-TUD-1 with different loadings was tested using *n*-hexane cracking, shown in Fig. 1.22. Zeolite Beta-TUD-1 with a zeolite loading of 40 wt.% showed the highest activity out of the composite materials, and was twice as active per gram of zeolite than either the pure zeolite Beta or the physical mixture (40 wt.%) of zeolite and TUD-1. This clearly illustrated the use of TUD-1 to prevent aggregation of the nanoparticles, but also points to an additional effect.

The enhanced activity of the Beta-TUD-1 composite (40 wt.% loading) is a result of the modification in the number and strength of acid sites resulting from the interactions between the amorphous TUD-1 and zeolite Beta. FTIR (employing CO and NH_3 adsorption) is able to differentiate between various BrØnsted acid sites with overlapping bands. Besides isolated and vicinal silanols, two other types of BrØnsted acid sites occur in the composites: (1) hydroxyl groups linked to partially extra-framework aluminium ions, sites of medium acidity and (2) bridged hydroxyl groups that form the strong BrØnsted acidity in zeolites. Figure 1.23 depicts NH_3 adsorption spectra whereby typical adsorption species and sites for zeolite Beta can be detected.

Additionally, NH_3 adsorption can also be used to identify strained surface siloxanes, due to its reaction with them leading to a surface group with a distinctive IR band at 1550cm^{-1}. In Fig. 1.24, NH_3 adsorption reveals the presence of distorted reactive siloxane bridges formed by the condensation of two vicinal silanols. It is of interest to note that the sample with the highest catalytic activity shows the most abundant presence of these distorted reactive siloxane bridges. It is thought that this synergistic effect of combining TUD-1 and zeolite Beta in such a way is responsible for the higher activity of the composite material.

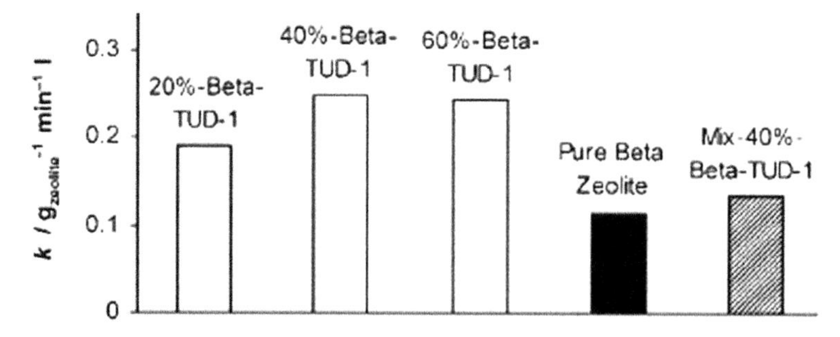

Figure 1.22 Pseudo-first-order reaction rate constants based on the mass of zeolite *n*-hexane cracking at 538 °C on zeolite Beta-TUD-1 catalysts, the pure zeolite Beta, and a physical mixture of 40 wt.% zeolite and TUD-1. Reproduced from Waller *et al.*[54] by permission from Wiley-VCH.

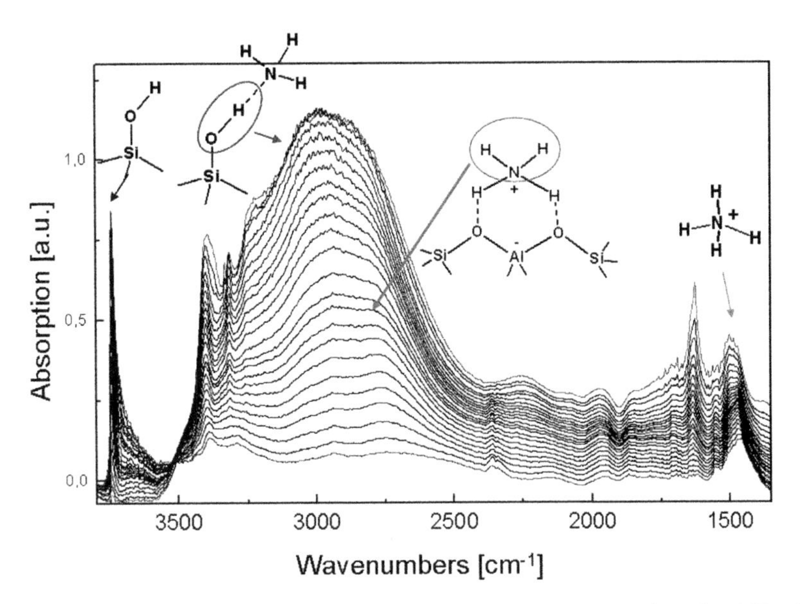

Figure 1.23 NH_3 adsorption spectra of zeolite Beta at room temperature (varying NH_3 dosage), showing typical adsorption sites/species. (See color insert.)

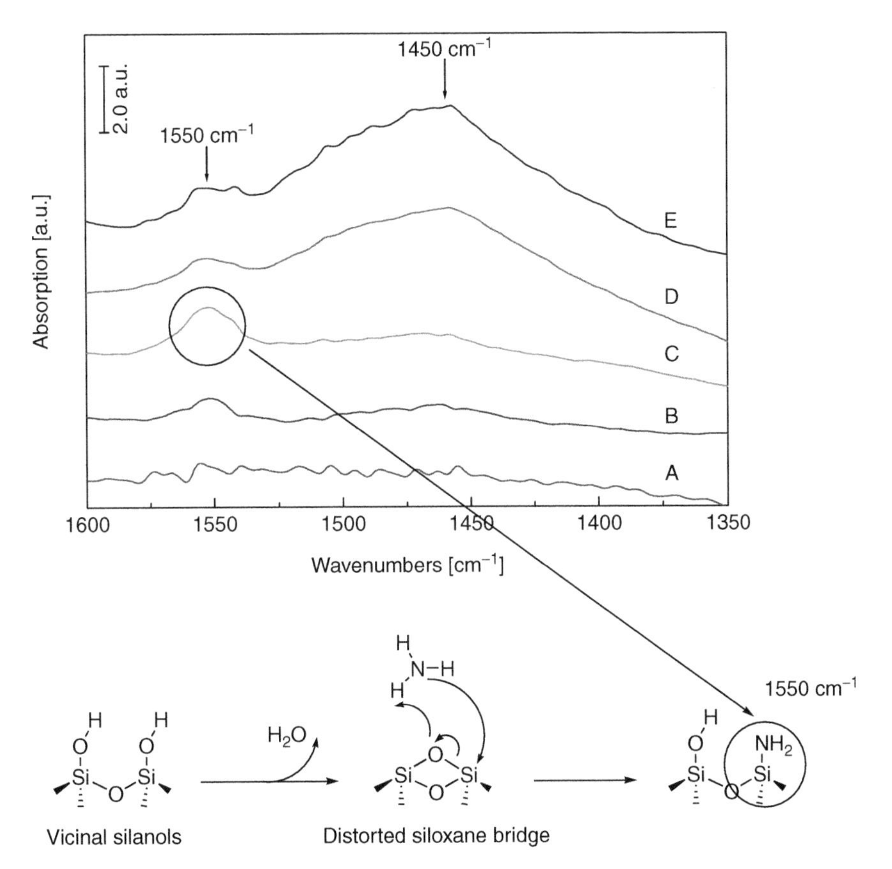

Figure 1.24 FTIR spectra in vacuum at room temperature after NH_3 adsorption: (a) TUD-1, (b) 20% Beta-TUD-1, (c) 40% Beta-TUD-1, (d) 60% Beta-TUD-1, and (e) zeolite Beta.

7.2. ITQ-2—Delaminated zeolite

Delamination is a common phenomenon in clays and layered materials. It occurs when the layers of a clay or layered material are swollen by the incorporation of a solvent or an organic compound, followed by the separation of the layers. Corma *et al.*[56] first utilised this property to separate layered sheets of a zeolitic precursor, MCM-22(P), by the cation exchange of one cation (HMI^{+}) for a larger, bulkier cation, $CTAB^{+}$, followed by ultra-sonication in a basic medium (pH 13.5), depicted in Fig. 1.25. Once the material is calcined, ITQ-2 is obtained, which exists as single sheets of the respective zeolite organised in a 'house-of-cards' arrangement. ITQ-2 exhibits well-structured accessible external surface area ($> 700\ m^2\ g^{-1}$), formed by 12-membered ring open cups with a microporous system made up of a sinusoidal 10-membered ring channel ($\sim$0.55 nm diameter) parallel to the main plain of the layer (Fig. 1.26).[57]

The ITQ-2 material proved to be a breakthrough due to the ease of accessibility of the material's active sites. The open cups (Fig. 1.26) provide easy entrance and exit points for the reactants and products. The delamination process, though severe, does not compromise the structural integrity of the active sites and thus the activity of the material.[56] One important reaction that demonstrates the benefits of the

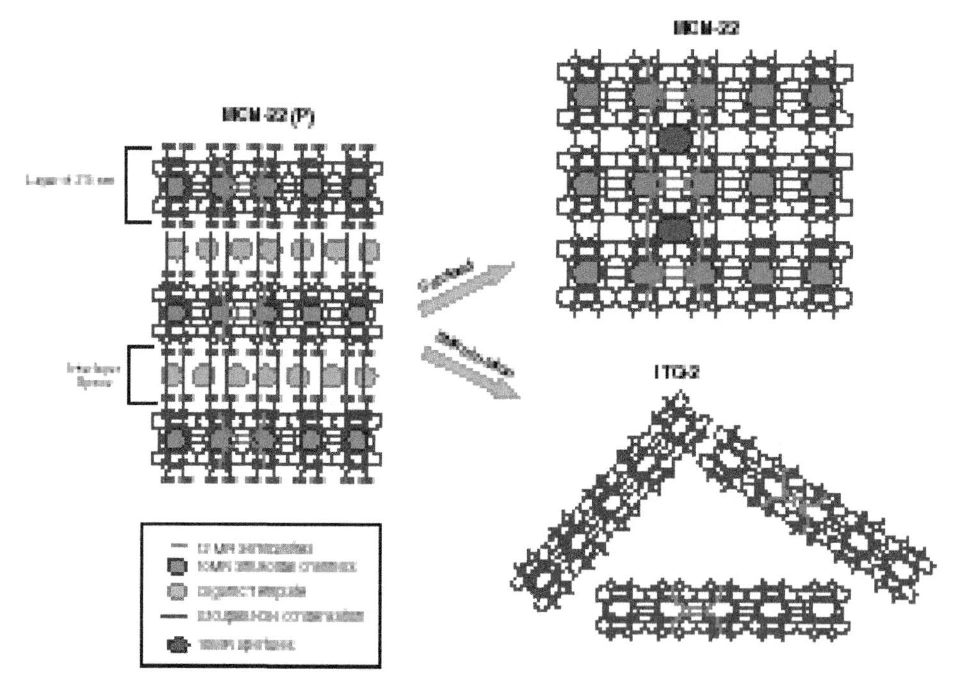

Figure 1.25 Structural schematic representation of MWW-type zeolites and ITQ-2 delaminated zeolite. Reproduced from Díaz *et al.*[58] by permission from Elsevier. (See color insert.)

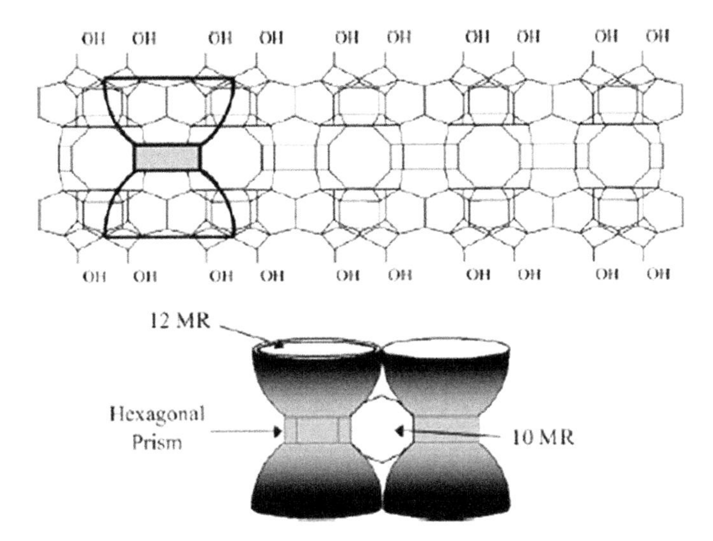

Figure 1.26 Crystal structure of ITQ-2 showing the 2.5 nm layered structure encompassing open cups and the 10-membered ring channels running between the major planes of the layer. Reproduced from Corma *et al.*[57] by permission from The Royal Society of Chemistry.

greater site accessibility and smaller diffusion path, among many, is the cracking of larger molecules such as di-isopropylbenzene and vacuum gasoil. Results show that along with ITQ-2 being much more active than the MCM-22, it is also more selective towards the desired gasoline and diesel products with a lower production of gas and coke.

Like the previous example of zeolite Beta incorporation into a TUD-1, a similar method can be used for ITQ-2. By introducing the material into the synthesis gel, one is able to disperse the sheets inside TUD-1, improving the handling of the material, much in the same way as with zeolite Beta.[59] However, an alternative method for ITQ-2 incorporation inside TUD-1 involves the *in situ* delamination of the layered MCM-22(P) in the synthesis gel during TUD-1 gel formation (see Fig. 1.27). This occurs with the addition of TEA to a suspension of MCM-22(P) in water. The MCM-22(P) is made following a static synthesis procedure.[60] It is suspected that TEA disrupts the layered material (assisted with sonication), which is held together by the electrostatic forces between the layers themselves and the $CTAB^+$ used to increase the interlayer spacing. Since this methodology can be applied during the synthesis of TUD-1 in a one-pot mixture, it avoids the use of the harsher conditions employed by Corma *et al.*[56] for the delamination, that is, a highly basic environment (pH ~13.5). TEM indicates partial, if not fully, delaminated species surrounded by TUD-1 (see Fig. 1.27). Though the XRD pattern of the composite is not sharp (due to the low loading of the initial MCM-22(P) and curved nature of the sheets), it is still possible to confirm the presence of ITQ-2 and TUD-1 in the same composite material (Fig. 1.28).

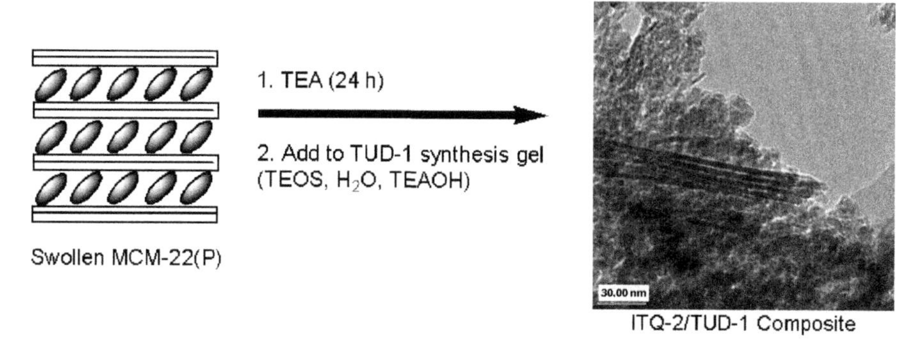

Figure 1.27 Schematic for the *in situ* delamination of MCM-22(P) in the synthesis gel of TUD-1 and the resulting TEM of the composite material.

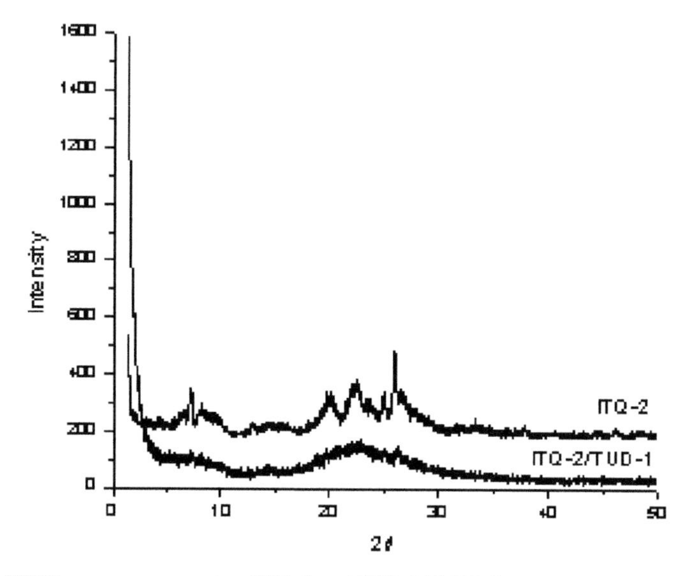

Figure 1.28 XRD pattern comparing ITQ-2 and ITQ-2/TUD-1.

8. Conclusion

In this short review, we have endeavoured to showcase the versatility of the TUD-1 mesoporous oxide family as a catalyst/drug carrier, anion-exchanger and donor material to micro-mesoporous composites.

In particular, TUD-1 offers an alternative to the regularly employed MCM-41. In addition to the advantage of higher hydrothermal stability and three-dimensional arrangement of pores, the structure directing agent, TEA, is inexpensive and the material allows for a tunable pore size.

Various metal-incorporated TUD-1 materials are easily synthesised in a one-pot synthesis gel, and the synthesis can be tuned to result in atomic dispersion of the desired metal in the framework, or the formation of stable metal oxide nanoparticles.

In addition, pre-formed particles, such as zeolites, can be readily included during the synthesis for adequate dispersion.

The potential for TUD-1 to act as an effective drug carrier has also been established, to give adequate release times for the model drug, ibuprofen.

TUD-1's ease of synthesis (even complex systems can be made in a 'one-pot approach') and the controllability of its compositional and structural details mean that it may serve as a very useful material for many current and future applications.

REFERENCES

[1] Davis, M. E., *Nature* **2002,** *417,* 813–821.
[2] Zhao, X. S., Su, F., Yan, Q., Guo, W., Bao, X. Y., Lv, L., Zhou, Z., *J. Mater. Chem.* **2006,** *16,* 637.
[3] Yang, P., Deng, T., Zhao, D., Feng, P., Pine, D., Chmelka, B. F., Whitesides, G. M., Stucky, G. D., *Science* **1998,** *282,* 2244–2246.
[4] Plank, C. J., Rosinski, E. J., Hawthorne, W. P., *Ind. Eng. Chem. Prod. Res. Dev.* **1964,** *3,* 165–169.
[5] Smolka, H. G., Schwuger, M. J., *Colloid Polym. Sci.* **1978,** *256,* 270–277.
[6] Vallet-Regí, M., Balas, F., Arcos, D., *Angew. Chem. Int. Ed.* **2007,** *46,* 7548–7558.
[7] Verweij, H., *J. Mater. Sci.* **2003,** *38,* 4677–4695.
[8] Rouquérol, J., Avnir, D., Fairbridge, C. W., Everett, D. H., Haynes, J. H., Pericone, N., Ramsay, J. D. F., Sing, K. S. W., Unger, K. K., *Pure Appl. Chem.* **1994,** *66,* 1739–1758.
[9] Kresge, C. T., Leonowicz, M. E., Roth, W. J., Vartuli, J. C., Beck, J. S., *Nature* **1992,** *359,* 710–712.
[10] Inagaki, S., Fukushima, Y., Kuroda, K., *J. Chem. Soc. Chem. Commun.* **1993,** 680–682.
[11] Beck, J. S., Vartuli, J. C., Roth, W. J., Leonowicz, M. E., Kresge, C. T., Schmitt, K. D., Chu, C. T.-W., Olson, D. H., Sheppard, E. W., McCullen, S. B., Higgins, J. B., Schlenker, J. L., *J. Am. Chem. Soc.* **1992,** *114,* 10834–10843.
[12] Jansen, J. C., Shan, Z., Marchese, L., Zhou, W., van der Puil, N., Maschmeyer, T., *Chem. Commun.* **2001,** 713–714.
[13] Ryoo, R., Joo, S. H., Jun, S., *J. Phys. Chem. B.* **1999,** *103,* 7743–7746.
[14] Shan, Z., Jansen, J. C., Marchese, L., Maschmeyer, T., *Microporous Mesoporous Mater.* **2001,** *48,* 181–187.
[15] Buijink, J. K. F., van Vlaanderen, J. J. M., Crocker, M., Niele, F. G. M., *Catal. Today* **2004,** *93–95,* 199–204.
[16] Taramasso, M., Perego, G., Notari, B. US Patent 4,410,501, 1983.
[17] Huybrechts, D. R. C., Debruycker, L., Jacobs, P. A., *Nature* **1990,** *345,* 240–242.
[18] Notari, B., *Stud. Surf. Sci. Catal.* **1988,** *37,* 413–425.
[19] Reddy, J. S., Kumar, R., *J. Catal.* **1991,** *130,* 440–446.
[20] Serrano, D. P., Li, H. X., Davis, M. E., *J. Chem. Soc. Chem. Commun.* **1992,** 745–747.
[21] Camblor, M. A., Corma, A., Martinez, A., Perez-Pariente, J., *J. Chem. Soc. Chem. Commun.* **1992,** 589–590.
[22] Ulagappan, N., Krishnasamy, V., *J. Chem. Soc., Chem. Commun.* **1995,** 373–374.
[23] Corma, A., Navarro, M. T., Perez-Pariente, J., *J. Chem. Soc. Chem. Commun.* **1994,** 147–148.
[24] Tanev, P. T., Chibwe, M., Pinnavaia, T. J., *Nature* **1994,** *368,* 321–323.
[25] Maschmeyer, T., Rey, F., Sankar, G., Thomas, J. M., *Nature* **1995,** *378,* 159–162.
[26] Shan, Z., Gianotti, E., Jansen, J. C., Peters, J. A., Marchese, L., Maschmeyer, T., *Chem. Eur. J.* **2001,** *7,* 1437–1443.
[27] Hamdy, M. S., Mul, G., Jansen, J. C., Ebaid, A., Shan, Z., Overweg, A. R., Maschmeyer, T., *Catal. Today* **2005,** *100,* 255–260.
[28] Bachari, K., Millet, J. M. M., Benaichouba, B., Cherifi, O., Figueras, F., *J. Catal.* **2004,** *221,* 55–61.
[29] Cao, J., He, N., Li, C., Dong, J., Xu, Q., *Stud. Surf. Sci. Catal.* **1998,** *117,* 461–467.

[30] Choudhary, V. R., Jana, S. K., Kiran, B. P., *Catal. Lett.* **1999,** *59,* 217–219.
[31] Anand, R., Hamdy, M. S., Gkourgkoulas, P., Maschmeyer, T., Jansen, J. C., Hanefeld, U., *Catal. Today* **2006,** *117,* 279–283.
[32] Hamdy, M. S., Ramanathan, A., Maschmeyer, T., Hanefeld, U., Jansen, J. C., *Chem. Eur. J.* **2006,** *12,* 1782–1789.
[33] Yamashita, H., Ariyuki, M., Higashimoto, S., Zhang, S. G., Chang, J. S., Park, S. E., Lee, J. M., Matsumura, Y., Anpo, M., *J. Synchrotron Radiat.* **1999,** *6,* 453–454.
[34] Yamashita, H., Ariyuki, M., Yoshizawa, K., Kida, K., Ohshiro, S., Anpo, M., *Res. Chem. Intermed.* **2004,** *30,* 235–245.
[35] Yamashita, H., Ohshiro, S., Kida, K., Yoshizawa, K., Anpo, M., *Res. Chem. Intermed.* **2003,** *29,* 881–890.
[36] Yamashita, H., Yoshizawa, K., Ariyuki, M., Higashimoto, S., Anpo, M., Che, M., *Chem. Commun.* **2001,** 435–436.
[37] Hamdy, M. S., Berg, O., Jansen, J. C., Maschmeyer, T., Arafat, A., Moulijn, J. A., Mul, G., *Catal. Today* **2006,** *117,* 337–342.
[38] Hamdy, M. S., Berg, O., Jansen, J. C., Maschmeyer, T., Moulijn, J. A., Mul, G., *Chem. Eur. J.* **2006,** *12,* 620–628.
[39] van de Water, L. G. A., Bulcock, S., Masters, A. F., Maschmeyer, T., *Ind. Eng. Chem. Res.* **2007,** *46,* 4221–4225.
[40] Ramanathan, A., Klomp, D., Peters, J. A., Hanefeld, U., *J. Mol. Catal. A: Chem.* **2006,** *260,* 62–69.
[41] Knauer, B., Krohn, K., *Liebigs Annalen* **1995,** 677–683.
[42] Namy, J. L., Souppe, J., Collin, J., Kagan, H. B., *J. Org. Chem.* **1984,** *49,* 2045–2049.
[43] Corma, A., Domine Marcelo, E., Nemeth, L., Valencia, S., *J. Am. Chem. Soc.* **2002,** *124,* 3194–3195.
[44] Zhu, Y., Chuah, G., Jaenicke, S., *Chem. Commun.* **2003,** 2734–2735.
[45] Zhu, Y., Chuah, G., Jaenicke, S., *J. Catal.* **2004,** *227,* 1–10.
[46] Zhu, Y., Chuah, G.-K., Jaenicke, S., *J. Catal.* **2006,** *241,* 25–33.
[47] Zhu, Y., Jaenicke, S., Chuah, G. K., *J. Catal.* **2003,** *218,* 396–404.
[48] Shan, Z., Jansen, J. C., Zhou, W., Maschmeyer, T., *Appl. Catal., A* **2003,** *254,* 339–343.
[49] Simons, C., Hanefeld, U., Arends, I. W. C. E., Sheldon, R. A., Maschmeyer, T., *Chem. Eur. J.* **2004,** *10,* 5829–5835.
[50] Simons, C., Hanefeld, U., Arends, I. W. C. E., Maschmeyer, T., Sheldon, R. A., *Top. Catal.* **2006,** *40,* 35–44.
[51] Simons, C., Hanefeld, U., Arends, I. W. C. E., Maschmeyer, T., Sheldon, R. A., *J. Catal.* **2006,** *239,* 212–219.
[52] Heikkilä, T., Salonen, J., Tuura, J., Hamdy, M. S., Mul, G., Kumar, N., Salmi, T., Murzin, D. Y., Laitinen, L., Kaukonen, A. M., Hirvonen, J., Lehto, V.-P., *Int. J. Pharm.* **2007,** *331,* 133–138.
[53] Heikkilä, T., Salonen, J., Tuura, J., Kumar, N., Salmi, T., Murzin, D. Y., Hamdy, M. S., Mul, G., Laitinen, L., Kaukonen, A. M., Hirvonen, J., Lehto, V.-P., *Drug Deliv.* **2007,** *14,* 337–347.
[54] Waller, P., Shan, Z., Marchese, L., Tartaglione, G., Zhou, W., Jansen, J. C., Maschmeyer, T., *Chem. Eur. J.* **2004,** *10,* 4970–4976.
[55] Shan, Z., Zhou, W., Jansen, J. C., Yeh, C. Y., Koegler, J. H., Maschmeyer, T., *Stud. Surf. Sci. Catal.* **2002,** *141,* 635–640.
[56] Corma, A., Fornes, V., Pergher, S. B., Maesen, T. L. M., Buglass, J. G., *Nature* **1998,** *396,* 353–356.
[57] Corma, A., Fornes, V., Sales Galletero, M., Garcia, H., Gomez-Garcia, C. J., *Phys. Chem. Chem. Phys.* **2001,** *3,* 1218–1222.
[58] Díaz, U., Fornés, V., Corma, A., *Microporous Mesoporous Mater.* **2006,** *90,* 73–78.
[59] Shan, Z., Gerhard, W. P. W., Maingay, B. G., Angevine, P. J., Jansen, J. C., Yeh, C. Y., Maschmeyer, T., Dautzenberg, F. M., Marchese, L., Pastore, H. D. O., Novel zeolite composite, method for making and catalytic application thereof. US Patent 2004138051 A1, July 15, 2004.
[60] Santos Marques, A. L., Fontes Monteiro, J. L. J. L., Pastore, H. O., *Microporous Mesoporous Mater.* **1999,** *32,* 131–145.

CHAPTER 2

Organoclays: Preparation, Properties and Applications

Maguy Jaber *and* Jocelyne Miehé-Brendlé

Contents

Abstract

The present chapter is devoted to recent developments in the area of layered inorganic–organic hybrid materials. The different approaches used for the preparation of organic nanoclays are summarized. These include intercalation, grafting and in situ incorporation of the organic moieties. Detail description of the methods is presented and their advantages/disadvantages are commented upon. Properties of the obtained hybrid materials are discussed. Examples demonstrating the potential of organoclays for various applications are given.

Keywords: Clays, Grafting, Intercalation, Ion exchange, One step synthesis, Organoclays

1. Introduction

A nature's remarkable feature is its ability to combine at the nanoscale level organic and inorganic components, thus allowing the construction of smart materials with a compromise between different properties or functions (mechanical, density, permeability, colour, hydrophobia, etc.). Current examples of natural organic–inorganic composites are crustacean carapaces or mollusc shells and bone or tooth tissues in vertebrates.

Ordered Porous Solids
DOI: 10.1016/B978-0-444-53189-6.00002-0

The possibility to combine properties of organic and inorganic components in material design and processing was investigated since the early ages of humanity (Egyptian inks, green bodies of china ceramics, prehistoric frescos, etc.). Maya blue is a good example of man made material that combines the colour of organic pigment (indigo) and the resistance of an inorganic host (clay minerals such as palygorskite) leading to a synergic solid with properties and performances beyond those of a simple mixtures.[1]

Lately, a lot of attention is paid to a new class of layered materials comprising an organic component. In this chapter, we will provide basic information about hybrid layered materials (organoclays). First structure particularities will be presented, followed by the synthesis methods and the potential applications.

Layered materials are built by the stacking of 'two-dimensional' units known as layers that are bond to each other through weak forces.[2,3] Layered structures can be classified in different categories depending on the type of the layer:

(i) **neutral** layers, e.g. brucite ($Mg(OH)_2$) and other hydroxides, phosphates and chalcogenides, and various metal oxides such as V_2O_5,
(ii) **negatively** charged layers with compensating cations in the interlayer space, e.g. widespread lamellar compounds in nature such as cationic clays (montmorillonite, hectorite, beidellite, etc.)
(iii) **positively** charged layers with compensating anions in the interlayer space,[4] the most common of which are the layered double hydroxides (LDH), also called anionic clays.

Layered materials from the second category are commonly found in nature (several minerals belong to the montmorillonite group—smectites), while LDH [hydrotalcite (HT)—anionic clays] are typically synthesized in laboratory conditions.[5]

Another interesting group of layered materials is the phyllosilicates whose members can be from natural or synthetic origin.[6,7] In general, the members from this group have a structure that consists of one layer stacking in which planes of oxygen atoms coordinate to cations such as Si^{4+}, $A1^{3+}$, Mg^{2+}, Fe^{3+} to form two-dimensional 'sheets'. The coordination of cations in adjacent sheets typically alternates between tetrahedral and octahedral. Tetrahedral sheets, which commonly contain Si^{4+}, consist of hexagonal or ditrigonal rings of oxygen tetrahedral linked by shared basal oxygens. The apical oxygens of these tetrahedra form the base of octahedral sheets having brucite-like or gibbsite-like structures that comprises Mg^{2+}, $A1^{3+}$, Li^{+}, Fe^{2+} or Fe^{3+} cations. A regular repeating assemblage of sheets (e.g., tetrahedral–octahedral or tetrahedral–octahedral–tetrahedral) is referred to as a layer.

Smectites, a subgroup of the phyllosilicates attract a lot of interest due to their swelling properties. Other subgroups include micas, kaolins, vermiculites, chlorites, talc, and pyrophyllite. These minerals are characterized by a 2:1 layer structure in which two tetrahedral sheets are attached on each side of an octahedral sheet via sharing of apical oxygens. As the apical oxygens from the tetrahedral sheet form ditrigonal or hexagonal rings, one oxygen from the octahedral sheet is located in the centre of each ring and is protonated, thus giving a structural hydroxyl (Fig. 2.1). In 2:1 phyllosilicates, isomorphous substitution of cations having different valences

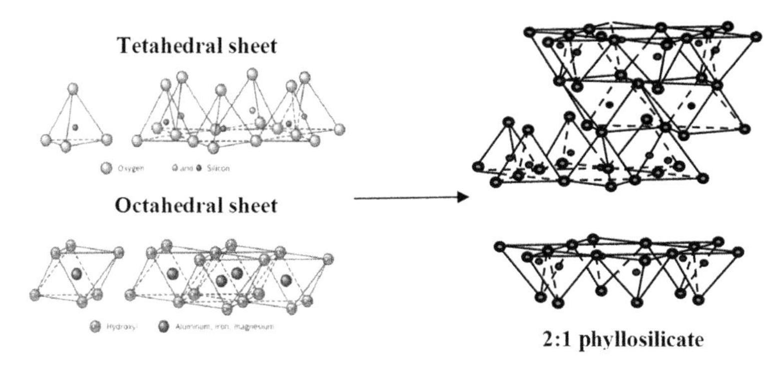

Figure 2.1 Schematic presentation of tetrahedral and octahedral sheets in layered materials (left) and a 2:1 phyllosilicate structure (right). (See color insert.)

can lead to charge imbalances within a sheet. These may be partly balanced by the opposite charge of the adjacent sheet (e.g. a positively charged octahedral sheet may offset some of the negative charge associated with a tetrahedral sheet). The net charge imbalance on a 2:1 layer, if it occurs, is negative. This charge is referred as *layer charge* of the mineral and is balanced by larger cations (e.g. Na^{+}, K^{+}, Ca^{2+} and Mg^{2+}) that coordinate to the basal surfaces of the tetrahedral sheets from adjacent layers. Because these charge-balancing cations are located between adjacent 2:1 layers they are referred to as 'interlayer cations', whose hydration is the reason for the high hydrophilicity of the phylosilicates.

Probably, the most widely exploited mineral from the phyllosilicate group is the montmorillonite due to its interesting physicochemical properties:

- particles of colloidal size;
- high specific surface area;
- presence of both Lewis and Brønsted acid sites;
- cation exchange capacity;
- presence of hydroxyl groups at the surface and at the edge of the layers.

The last two properties render clay minerals hydrophilic, the latter being highly desired for a number of applications. These are indeed necessary when the clay minerals are dispersed in less polar solvents, incorporated in polymers to form clay–polymer nanocomposites or to improve gas and liquid adsorption properties. Finding the way to perform this hydrophobization and, thus, to form organoclays is a challenge, which leads to a growing interest in the field of clay minerals.[8–10]

2. Preparation and Applications of Organoclays

Different ways of making organoclays have been reported, namely:

- intercalation of organic molecules or organic cations by adsorption or ion exchange;

- grafting of organic functionalities to the hydroxyl groups presence at the edge of the clay mineral layers;
- direct synthesis of hybrid inorganic–organic materials.

2.1. Organoclays prepared by intercalation of organic moieties in the interlayer space

The term 'intercalation' refers to a process whereby a guest molecule or ion is inserted into a host lattice. The reaction is usually reversible. The structure of the host or intercalated compound is only slightly changed by the guest species. In three-dimensional systems, the sizes of the guest species are constrained by the host lattice dimensions. In lower dimensional systems, no such restriction exists and the layers may adjust their separation freely to accommodate guest species of different sizes.[5] Since the number of appropriate organic guest species and layered materials is vast, the possibilities to combine organic matter with layered hosts are enormous.

Organoclays are usually prepared by modifying clays with cationic surfactants via ion exchange.[11,12] This is accomplished through the replacement of inorganic exchangeable cations, for instance Na^{+}, K^{+} and Ca^{2+}, within the clay mineral structure with organic cations, for example with quaternary ammonium cations, such as hexadecyltrimethylammonium bromide,[13] hexadecylpyridine bromide[14,15] and 2-mercaptobenzimidazole.[16] Cation exchange reactions are most commonly performed by mixing aqueous dispersions of clay minerals and solutions containing surfactants such as organoalkylammonium salts.[17,18] Reaction rates depend on many parameters such as the guest compound, temperature and concentration, type of smectite and particle size.[19–21] Products are separated either by centrifugation or filtration and washed repeatedly. Depending on the solubility of the guest species, water–alcohol solution mixtures can be used as solvents. Whatever the method used, the arrangement of the intercalated surfactant cations strongly depends on the layer charge and the alkyl chain length resulting in materials with different interlayer spacing.[21] Organoclays are used to remove toxic compounds from the environment and reduce the dispersion of pollutants in soil, water and air. However, as cationic surfactants are weakly bonded to the clay mineral layers through electrostactic forces, they can easily be released in aqueous medium[22] resulting in secondary pollution. This drawback can be circumvented by having a covalent linkage between organic moieties and the clay minerals (Fig. 2.2).

2.2. Organoclays prepared by grafting of organic moieties

Substantial efforts of research have been directed to functionalize clays in a durable way[23–25] to obtain organic–inorganic hybrids where the inorganic and organic components are linked via strong bonds (i.e. covalent or iono-covalent). This approach enables a durable immobilization of the reactive organic groups, preventing their leaching when they are used in liquid media. Such materials are expected to show superior properties in respect to unmodified clays when applied as adsorbents of heavy metals (low loading capacity, relatively weak binding strength and low selectivity).[26,27–34]

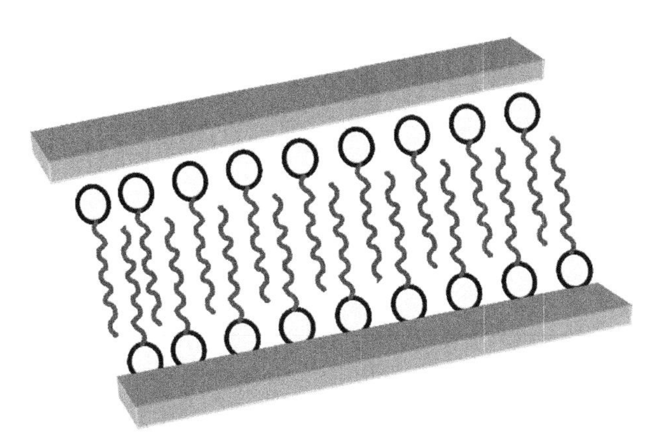

Figure 2.2 Schematic presentation of an alkylammonium compound intercalated between two clay layers. (See color insert.)

A covalent linkage can be achieved through the reaction between the layers reactive groups (hydroxyl groups) and an adequate molecule, which ensure higher chemical, structural and thermal stability for the compound. Alkylchlorosilane[35] or organoalcoxysilane[36] are usually used to provide functional organic moieties. Organoalcoxysilane have the following formula $RSi(OR')_3$ where R stands for organic function such as 3-aminopropyl, [3-(glycidyloxy)propyl] and R' for methoxy or ethoxy groups. Hydrolysis of the alkoxy groups yields in silanol groups that react with surface groups of the clay, thus, affording Si–O–Si or Al–O–Si ionocovalent bonds.[37,38] Reactions can be restricted to the crystal surface or on the external broken edges (the basal spacing remains unchanged) or layer surface (in this case an interlayer expansion occurs). Functionalization by grafting on the surface or clay minerals broken edges leads to the formation of an organic–inorganic hybrid compound combining, thus, the mechanical stability of the inorganic framework with the particular reactivity of the organofunctional group.

Silane coupling agents are preferred in the compounds in the preparation of layered hybrids since they may improve a set of important properties such as the adhesion, wetting and rheology. Despite these advantages, grafting has so far been scarcely applied to layered silicates' modifications compared to modification by ion exchange.[39–42] This can mainly be attributed to the low edge area to surface ratio, which results in a small number of reactive edge silanols. Consequently, the functionalization of layered clays remains rather inefficient, because of their relatively low density of accessible surface hydroxyl groups[28,33,43] and/or too great cohesive energy between the lamella, which induces restricted access to the interlayer region.[31,32]

The sepiolite is among the most suitable clay structures to be grafted with organics[33,44,45] due to its high content of silanol groups that are quite easily accessible to the grafting reactants.[24] Functionalization of swelling smectite-type aluminosilicate clays, such as montmorillonite, resulted generally in limited amounts of grafted organic groups that were localized at the outer surfaces and were bounded to the edges of the clay *via* condensation with edge –OH groups.[31,33,34] Grafting organic ligands on the interlamellar surface of this kind of clay remains limited and the resulting materials may suffer from hindered access to the binding sites, as pointed out by uncompleted uptake of heavy metal species with respect to the total number of ligands.[28,29]

The alkylsilane-surface bonding stability, bonding mode and the reaction mechanism of silylation have been thoroughly investigated.[46,47] The silylation of the interlayer surface of layered materials generated increased interest,[48] with a number of authors exploring direct bentonite grafting reactions.[49–50] However, according to Hofmann *et al.*[51] the Al–O octahedral sheet in montmorillonite is non reactive since there are only few OH groups at the edges of individual particles that can act as anchoring groups for modification. Because of the limited access of the alkylsilane to the octahedral sheet hydroxyls are inaccessible to alkylsilane, only limited amounts of organic groups can be covalently linked by direct silylation.[51] Nevertheless, interesting results were obtained by using solely specific organoalcoxysilanes or combination between ion-exchange and subsequent grafting. This double treatment method allowed the synthesis of multifunctional nanomaterials. It is worth mentioning that clays treated this way showed also a better dispersibility in non polar polymeric matrices.[52] On the contrary, good accessibility to the reactive groups was achieved by combining pillaring and subsequent grafting.[53]

In the following section, we will focus on work dealing with the covalent grafting of aminopropyl groups and the applications of these compounds.

2.1.1. Covalent grafting of clays with aminopropyl groups

Clay minerals can be grafted with aminopropyl groups by reaction between the surface silanol groups and 3-aminopropyltriethoxysilane.[54] Since alkylamine moities have acid-base properties, it is possible to create positive charges in the grafted solid just by protonation of the organic moieties. This opens a new field of application since such composite materials can bring anionic exchange properties. Grafting leads generally to a decrease of both specific surface area and pore volume, but the average pore size is nearly unaffected by the functionalization. This indicates that reaction is restricted to the edges of individual clay platelets and not extended to the interlayer region.

The chemical stability of the obtained grafted clay minerals is usually low in water and leads to leaching of the organic moieties and an increase of pH. This can be explained by the self-catalyzed breaking of Si–O–Si bonds. In order to prevent degradation of clay minerals during protonation, medium containing 0.005 M HCl are used. Indeed, for higher concentrations, clay minerals containing aluminium can undergo hydrolysis of their aluminium centers. Grafted clays demonstrated great potential as electrod modifiers, for instance Cameroon smectites and MX80 bentonite prove to be selective in respect to electroactive anions such as $Fe(CN)_6^{3-}$, while repelling cations.[55] Carbon paste electrodes modified by aminografted phyllosilicates were also applied to the voltametric analysis of heavy metal species such as mercury (II) by subsequent preconcentration.[56] This electrochemical sensor involving grafted clays minerals is a low cost alerting device for environmental monitoring. Applications can also be very interesting in the field of biosensing.

Indeed, amino-grafted clay minerals are also suitable as host materials for immobilization of enzymes onto electrode surface.[57] Entrapping of enzymes in clay minerals usually involves physical encapsulation within the three-dimensional structure originating from the special arrangement of layers upon drying and, thus, leads mainly to weak interactions. Having amounts of grafted aminopropyl groups as high as 1.5 mmol g^{-1} enables immobilization of glucose oxidase by covalent bonding. To avoid deterioration of the amine groups, a reaction with glutaraldehyde was first performed in order to transform them into imide groups. A similar approach was mentioned by Gopinak *et al.* in the case of grafted K10 montmorillonite.[58] Immobilization of glucoamylase and invertase led to a sharp decrease in pore volume as well as surface area. It was shown that protein backbones were situated at the periphery of the clay minerals, whereas the side chains penetrated between the layers. Moreover,[27] Al MAS NMR spectroscopy proved that tetrahedral Al atoms were involved in the adsorption process, whereas octahedral Al is influenced by grafting. Immobilized glucoamylase exhibited a very high activity and provided excellent results in continuous glucose production (Fig. 2.3).

Swelling ability and ion exchange characteristics are of paramount importance for the preparation of organoclays. As reported by He *et al.*,[59] grafting reaction between two clay minerals (synthetic fluorohectorite and natural montmorillonite) and 3-aminopropyltriethoxysilane can lead to different compounds. Using XRD, it was shown that 3-aminopropyltriethoxysilane was introduced into the interlayer space. The silane molecules adopted a parallel-bilayer arrangement model in natural montmorillonite and a parallel-monolayer arrangement in the synthetic fluorohectorite. As indicated by[29] Si CPMAS NMR spectroscopy, the parallel-bilayer arrangement resulted in the presence of T^2 ([Si(OSi)$_2$(OR')R] (R = $C_3H_6NH_2$, R' = H or C_2H_5) and T^1 molecular environments [Si(OSi)–(OR')$_2$R], whereas the parallel monolayer arrangements induced the formation of T^3 [Si(OSi)$_3$R] species. These results demonstrated that the silylation reaction and the interlayer structure of the grafted products strongly depend on the initial clay mineral. Indeed, in the case of a parallel monolayer, organic moieties were individually separated in order to lower the energy of the hybrid. Consequently, there was not a possibility for condensation between the silane that led to the presence of T^1 and T^2 units. In contrast, in the case of a parallel bilayer, condensation was easy to occur. This model differs from the one proposed by Herrera *et al.*[42] suggesting that the silane could only react with the broken edge of the clay rather than to be intercalated in the interlayer space. In order to verify the reliability of this model, grafting on commercial surfactant-modified fluorohectorite was performed. In this case, no grafting is

Figure 2.3 Schematic presentation of the edge grafting with an organoalkoxysilane.

occurred, as proved by TGA and ^{29} Si CPMAS NMR. It was, thus, shown that intercalation of silane molecules in the clay mineral interlayer occurred before the condensation between the silane molecule and clay surface. Condensation may then have happened between silane molecule and the interlayer surface rather than between silane molecule and the broken edge of the clay layer.

Dispersing medium has a significant influence on the grafting process. Many studies describe the behaviour of clay minerals in aqueous or organic medium, but there is only one that focused on the influence of the solvent used to graft aminopropyl moieties on clay minerals.[60] The influence of non polar and polar solvents was studied. Four solvents were chosen: toluene, which is commonly used to perform grafting, distilled water, tetrahydrofuran and ethylene glycol. It appeared that the surface energy of the solvent played a role in the degree of surface adsorption and in the intercalation of silane between the clay layers. In particular, it was shown that more silanes were adsorbed and/or intercalated with increasing solvent surface energy. It was shown that by using water, the interaction between the 3-aminopropyltriethoxysilane and the edges of clay platelets was less significant due to low-wetting phenomena and thus made the silane to diffuse in between the clay platelets. In this case, the grafting occurred as described by He *et al.*[59] X-ray diffractograms revealed that both the dispersive and polar components of surface energy influenced the average d-spacing. Additionally, deconvolution of DTG peaks in the temperature range from 200 to 750 °C reveals the appearance of four peaks, which were due to physically adsorbed, intercalated, surface-grafted and chemically grafted silanes between the clay platelets. Interaction between the trifunctional silane and clay took place by two different processes: adsorption of silane onto the broken edges of the clay platelets followed by a condensation reaction. It was concluded that the polar component of the surface energy led to hydrogen bonding with edge platelets in addition to the edge interaction of silane, and tactoids were thus formed. Depending on the solvent, silane could either be adsorbed onto the broken edges of the clay platelets or intercalated between the clay layers.

2.1.2. Covalent grafting of clays with miscellaneous functional organic moieties

Different clay minerals and various organosilanes have been used to produce organoclays. Table 2.1 gives some examples of organosilanes used for functionalization of clay minerals. Recently, attention was paid to the effects from the grafting methods and the synthesis media to elucidate the possible variations resulting from different preparation methods. Silane grafted montmorillonites were synthesized under silane vapour employing a trifunctional silane (γ-aminopropylchlorosilane) and a monofunctional silane (trimethylchlorosilane). The basal spacings of the silane grafted montmorillonites were significantly larger than that of Na-montmorillonite. This indicated that the silane has been intercalated into the montmorillonite interlayer space. However, the configuration of the silane had a prominent effect on the interlayer structure and consequential interlayer spacing. This study revealed that the grafting reaction from the silane vapour favours silane entering into the clay interlayer. This suggestion was evidenced by the increase in intensity of the

Table 2.1 Organosilanes employed in the preparation of layered hybrids

Clay mineral	Origin	Organosilane
Ca-Bentonite	Mongolia	$[(CH_3)_3SiCl]$ $[C_{18}H_{37}SiC_{13})$[61]
Na-Cloisite	Cameroon	*N*-trimethoxysilylpropyl-*N*,*N*,*N*-trimethylammonium chloride[62]
Na-Montmorillonite	China	3-aminopropyltriethoxysilane and trimethylchlorosilane[63]
Laponite RD	RockWood	(γ-MPTMS, $C_9H_{20}O_5Si$, 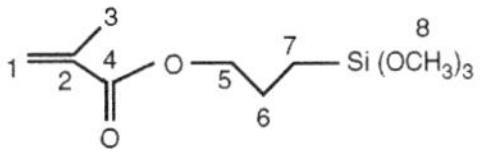 ς-MPDES, $C_{10}H_{22}O_3Si$[42] 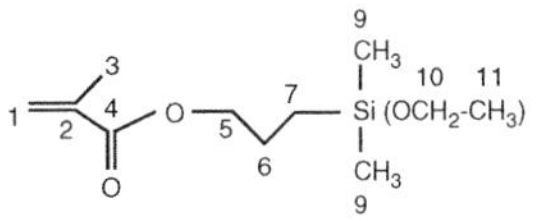
Synthetic fluorohectorite		$H_2N(CH_2)_3Si(OC_2H_5)_3$[64]
Natural Montmorillonite	Japan	
Laponite	Rockwood	Triethoxypropylamine silane $H_2N(CH_2)_3Si(OC_2H_5)_3$[65]
Montmorillonite K10		
Pyrophyllite	Turkey	3-(2-aminoethylamino)-propylmethyldimethoxysilane[66] (APMDS) R_ SiO(CH3) (OCH3)2
Palygorskite, Sepiolite		Dimethyldiethoxysilane[67]
Chrysotile		methylvinyldichlorosilane γ-methacryloxypropyltrimethoxysilane and delta-aminibutyldiethoxysilane
Laponite	Rockwood	3-aminopropyltriethoxysilane[68]
Beidellite		$(SH–C_3H_6–Si(OCH_3)_3$[53]
Hectorite		
Vermiculite		
Interstartified beidellite/ montmorillonite		

antisymmetric stretching mode of $–CH_3$ in TMCS, and antisymmetric and symmetric stretching vibrations of $–CH_2$ in γ-APS. DTG results provided further evidences for the silane grafting. The penetration of the methyl groups into the hexagonal rings of clay might be the reason for the observed decrease in the montmorillonite dehydroxylation, which is in good agreement with previous studies devoted to surfactant intercalated montmorillonites. Further work is currently underway employing molecular simulation and ^{13}C and ^{29}Si MAS NMR spectroscopies as reported by Shen *et al.*[63]

Boulinos *et al.*[40] grafted different clays by employing bis(trimethoxysilyl)hexane and the protonated salt of 3-aminopropyltriethoxysilane and performing the reaction in aqueous colloidal dispersion solution. Careful choice of the clay type and close control of the synthesis parameters allowed the preparation of clay platelets. Both, interlayer space and crystal edges were grafted by employing protonated aminosiloxane. It is worth mentioning that grafting of laponite with protonated aminopropyl group resulted in the formation optically transparent monoliths. Laponite was also used by Herrera *et al.*[69] to form γ-methacryloyloxypropyltriethoxysilane or γ-methacryloyloxypropyldimethylmethoxysilane grafted solids, which were then used as seeds for the synthesis of polymer–laponite nanocomposite particles through emulsion polymerization. The trifunctional silane coupling agent formed a polysiloxane coating that penetrated the interlayer space of laponite while the monofunctional silane formed monolayer coverage on the clay edge. In the first case, dispersion of the clay particles was not possible in water in the presence of anionic surfactant and a peptizing agent due to the irreversible coating created by the trifunctional silane, whereas monofunctional grafted laponite could easily be dispersed. This method enabled the preparation of nanocomposite latexes in which the clay platelets covered homogeneously the surface. However, the larger was the size of the clay aggregates, the poorer was the stability of the resulting latex suspension.

Dean *et al.*[70] prepared a series of acrylate and methacrylate clay composites. Combined XRD/TEM analysis revealed that the structures of clay minerals were namely intercalated with limited exfoliation.

Minerals with fibrous-like morphology, for example chrysotile, sepiolite and palygorskite were also grafted. The organosilicate derivatives of the latest are interesting materials, due to the higher number of hydroxyls at the crystal surface than that for 2:1 phyllosilicate. For instance, surface silanols of the sepiolite needle are spaced every 0.5 nm along the fibre length. These functional groups are easily available for coupling reactions and one may expect that other forces such as hydrogen bonding and van der Waals interactions are involved in the interface sorption phenomena. Functionalization of sepiolite was shown to be mainly limited to the external surface.[71] The use of heterofunctional silylating agents like chlorosilane, ethoxysilane or dimethylallylchlorosilane enables in this case the formation of stable covalent Si–O–Si–C bonds. This reaction occurs with or without pretreatment of the silicate with HCl. If during the reaction the monofunctional silanes are substituted by difunctional silanes, linear polymerization can take place.[67] Furthermore, fibrous materials with relatively high surface area are beneficial for the adhesion with polymeric matrices.

Removal of heavy metals from aqueous solutions is a priority in environmental remediation and cleans up. Thus, many strategies to improve the sorbing capacity of mineral solids have been developed.[72–77] Grafting of mercaptopropyltrimethoxysilane onto the surface of clay mineral is of particular interest due to the chelating effect of thiol function towards heavy metal cations such as Hg^{2+}, Pb^{2+}, Cu^{2+}.[22,54,78]

One interesting feature is also the capacity of this function to enable the fixation of electroactive cationic dye such as methylene blue, which can have various harmful effects on health.[79] Study performed with natural Cameroon clay mineral has shown the effectiveness of the grafting process. It appears that the pH has a dramatic influence on the accumulation of methylene blue on the thiol functionnalized clay. Indeed, being rather low in acidic medium (below 0.3 mmol $\cdot$ g^{-1} at pH value up to 5) the sorption capacity increases rapidly in the 6–8 pH range to reach a constant value close to 0.9 mmol g^{-1}. Comparison between these results and the one observed for pristine clay showed that in the latter case, the amount of adsorbed methylene blue did not vary significantly in the 3–12 pH range. Obviously different mechanisms control the accumulation process, which is most probably the reason for the observed variations. Indeed, when pH increases, thiol groups that accumulate *via* their unprotonated form interact with the positively charged clay through electrostatic interactions. It was also shown that a thiol functionalized clay can successfully be used as electrode modifier for methylene blue sensing. Removal of Pb^{2+} cations can also be performed using 3-(23-aminoethylamino) propyl-methyldimethoxysilane (APDMS) grafted pyrophyllite.[80] In this case, it was suggested that hydrogen bonds took place between the hydroxyl groups and/or oxygen atoms within the structure of pyrophyllite and hydroxyl groups of APDMS. It was observed that the grafted clay mineral lost its foliated structure and transformed into more isometric particles in respect to initial material. Adsorption of Pb^{2+} was found to be dependent on the solution pH. Maximum adsorption was achieved at pH 6.5 with an amount equal to three times the one of the pristine clay. It was also shown that the presence of cations such as Na^+, Fe^{2+}, Mn^{2+} and ligand type of lead complex added to the solution strongly affected the adsorption of Pb^{2+} from solution.

2.1.3. Multifunctional treatment

By taking advantage of the presence of exchangeable cations and reactive hydroxyl groups in clay minerals, it is possible to form multifunctional materials. Bursacchi *et al.*[52] subjected laponite to both surface modification by cation exchange and edge modification by grafting. Layers of 25 nm in diameter and a thickness of 0.92 nm make the laponite fairly suitable for grafting due to the relatively high edge area/surface ratio. Such laponite particles were grafted with (3-Trimethoxysilyl) propylmethacrylate (TSPM) and a double chain surfactant (dimethyldioctadecylammoniumchloride). Ion exchange tests were performed prior to grafting, and the following XRD analysis showed a decrease of the d_{001} value.[29] Si CPMAS NMR spectra showed the presence of signals with resonance assigned to T^1, T^2 and T^3 molecular environments. This confirmed the occurrence of a condensation reaction between ion-exchanged laponite and TSPM. The presence of T^1 units indicated that a substantial amount of non-anchored TSPM units was still remaining. The presence of an asymmetric signal with a resonance of −85.5 ppm was assigned to Q^2 silicon units which were present as $Si(OMg)(OSi)_2{-}OH$ silanols occurring at the clay particle edges. The evolution of this signal with the level of grafting showed that the treatment resulted in less ordered clay platelets with closer platelets. In conclusion, grafting occurred mainly at the clay edges but in an inhomogeneous way:

platelets were indeed kept together by alcoxysilane molecules or oligomers (one side brought closer two adjacent platelets but the other side twisted them with respect to the other).

It has also been shown, that reactive OH groups can be introduced by pillaring of large oligomeric polycations or hydrated Si species. The formation of reactive OH groups on the pillars provides the possibility of modification by silylation. By this way, it was possible to obtain adsorbents with high selectivity in respect to heavy metal cations. Diaz *et al.*[53] prepared and tested specific heavy metal sorbents by grafting thiol groups in the interlayer space of natural and synthetic silica pillared clays. The ultimate goal was to use them to reduce the concentration of heavy metals in polluted soils. Natural beidellite, hectorite and vermiculite were transformed into SH-grafted pillared clays but difficulties were encountered to keep a regular and significant interlayer distance until the last step.

2.1.4. Grafting of functional polymers

Polymer agents have also been employed in surface modification of layered materials.[81] The first approach, named one-step grafting method, consists of the condensation of polymers comprising reactive groups on a clay substrate. This method does not give highly dense polymer brushes because the chemisorptions of the first fraction of chains hinder the diffusion of the following chains to the surface for further attachments. Another approach, named two-step grafting method, that provides higher polymer densities have also been developed. In the latter case, a monolayer of polymerizable (macromonomer) or initiator (macroinitiator) molecules is covalently attached to the surface. After activation, the chains grow-up from the surface in outward direction and the only limit for propagation is the diffusion of monomers to the active species.[82]

The polycation-exchanged bentonites were prepared by the adsorption of polycations of the type $[CH_2\text{-}CHOHCH_2N(CH_3)_2]^{n+}$ onto a low-iron Texas bentonite named Westone-L (WL).[83] The influence of clay mineral, particle size and cation on the amount of adsorbed polycations was studied with variable temperature X-ray diffraction. The use of polycation exchanged clay minerals as scavengers of p-nitrophenol from water and the significant activity of acid-treated polycation exchanged clays for the catalytic isomersion of α-pinene to camphene and limonene were also reported. Lin *et al.*[84,85] reported the preparation of Na^+-montmorillonite with basal spacing as high as 58 or 92 Å by intercalating telechelic POP-diamines [poly(propylene glycol)-bis(2-aminopropyl ether)] with different molecular weights. The use of these species as intercalating agents allowed to tune the basal spacing of highly ordered montmorillonite. The expanded interlayer space encapsulating hydrophobic POP-amines rendered the silicates amphiphilic and self-assembling. Chu *et al.* used amine terminating Mannich oligomers (AMO) or polyamines for the exfoliation of montmorillonite.[86] Another interesting example is the adsorption of the cationic polymer, polyethyleneimine (PEI), on bentonite particles reported by Oztekin *et al.*[87] The adsorbed polyelectrolyte affected the rheological properties of the bentonite dispersions, improved the adsorption capacity and the adsorption rate. The influence of the polyelectrolyte on the rheological behaviour changed with the exchangeable ions, that is Na^+ and Ca^{2+}. The ammonium group-terminated i-polypropylene (PP-t-NH_3^+) polymers were prepared by the combination of rac-$Me_2Si[2\text{-}Me\text{-}4\text{-}Ph(Ind)]_2\text{-}ZrCl_2$/MAO catalyst and p-$NSi_2$-St/H2 (p-$NSi_2$-St: 4-{2-[*N*,*N*-bis(trimethylsilyl)amino]ethyl}styrene) chain transfer agent.[88] The results demonstrated the advantage of chain-end-functionalized PP (PP-t-NH_3^+) which resulted in an exfoliated montmorillonite structure.[89]

This overview will not be completed without the work of Lu *et al.* who prepared for the first time silane grafted polyethylene nanocomposites (vinyltrimethoxysilane VTMS-g-PE/hexadecyltrimethylammonium exchanged montmorillonite and dicumyl peroxide).[90] VTMS-grafted polymer chains were successfully intercalated into the interlayer space of montmorillonite.

2.1.5. Hydrophobization of clay mineral surfaces

Organoclay syntheses are usually performed in aqueous solution or in wet organic media with all the consequent problems related to environmental issues. To avoid wet chemistry approach a new method called dry plasma technique was recently developed.[91] The method allows to alter the hydrophilic character of the clay mineral, and the active species resulting from the glow discharge may be etched, sputtered or be deposited onto the substrates surface.[92,93] Two clay minerals with different aspect ratios and structural arrangements have been modified by acetylene cold plasma treatment. The following analysis revealed that the plasma treatment had two pronounced effects on the clay mineral surface: (i) modification of the SiO_4 environment; and (ii) grafting of hydrocarbon groups. The coating, constituted, namely, of CH_2 and CH_3 groups, is situated at the external surface of clay mineral particles. It was shown that the coating thickness can be controlled by tuning plasma parameters. However, plasma treatment has also undesired effects like dehydration and dehydroxylation of the clay mineral.

2.3. One step synthesis of organoclays

Lately, the direct synthesis of hybrid inorganic–organic materials having 2:1 phyllosilicate structure has attracted considerable attention. This synthesis route involves *in situ* incorporation of the organic component during the formation of layered material. The silica of the tetrahedral sheets is, thus, covalently bounded to the organic groups usually present in the interlayer space.

The formation of organotalc was probably most widely studied.[94–97] First experiments were conducted in water–methanol solution with Mg and Ni chloride hexahydrate and 3-methacryloxypropyltrimethoxysilane precipitated by sodium hydroxide.[94,98] Other organoalcoxydes have also been used. The syntheses was performed at room temperature or temperature range 100°–170 °C for 24 h. Thus, organotalcs with various organic functionalities useful for heavy metal retention[97,99–106] and clay–polymer nanocomposites preparation[103] were synthesized. These hybrid lamellar materials can also be used as environmental barriers and catalytic supports. Other potential applications include the improvement of optical and mechanical properties of polymer-containing composite materials[107] (Fig. 2.4).

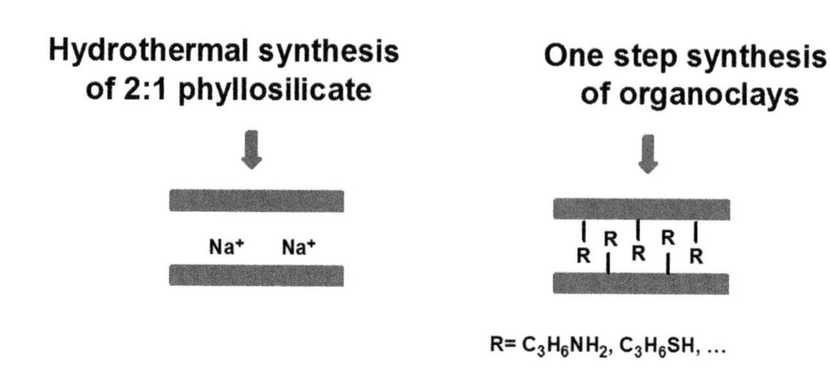

Figure 2.4 Schematic presentation of a clay (left) and an organoclay (right) obtained by one-step synthesis.

Ukrainczyck *et al.*[99] found similarities in the formation mechanisms controlling the formation of layered and ordered mesoporous hybrids. The mechanism is based on the micelles formation induced by the hydrophobic tails and hydrolysed silanols of organosilane. Further evidences of the layered hybrids formation were published by Guillot *et al.*,[108,109] who suggested the formation of a brucite-like structure and the grafting of $RSiO_3$ tetrahedra. These authors reported the synthesis of nickel and cobalt organotalc under autogenous pressure with or without employing fluoride as a mineralising agent. The X-ray absorption spectroscopy (XAS), providing information on the local environment, showed that all compounds exhibit a paramagnetic–ferromagnetic transition at low temperature.

For the materials containing essentially aluminum in the octahedral sheet, the synthesis lead to lamellar structures similar to those of MCM-50, whereas an octahedral sheet filled with magnesium gives a 2:1 organophyllosilicates. XRD patterns for these organoclays display the reflections characteristic of an organophyllosilicate structure: (001), (020, 110), (130, 220), (060). However, these reflections are broader than those observed for the parent phyllosilicates as previously mentioned in the literature.[110] This behaviour can be explained by the presence of organic chains in the interlamellar space.

The majority of this class of layered organosilicate compounds has magnesium as the octahedral cation, followed by nickel.[94,98–111] Other organically modified phyllosilicates have also been synthesized, for instance aluminium,[95,99,112] copper,[100] zinc,[102] and calcium.[113] Layered organosilane hybrids containing magnesium and nickel have a lamellar structure analogous to 2:1 trioctahedral phyllosilicates. When aluminium and copper metals are involved, structures similar to 2:1 dioctahedral phyllosilicate or 1:1 phyllosilicate are formed. A hectorite-like phyllosilicate derivative was synthesized under basic condition by combining magnesium and a silylating agent.[114] Employing a mixture of organosilane precursors resulted in the formation of multifunctional layered magnesium organosilicates.[115] Although the organic moiety attached to the inorganic backbone restricted the use of high temperature, these phyllosilicates can be applied as sorbents, environmental barriers, polymer fillers, catalytic supports or chemical sensors.[116] Ferromagnetic properties of a talc-like cobalt phyllosilicates were studied by Richard-Plouet *et al.*[117,118] A porphyrin derivative comprising phyllosilicate hybrid was used in alkane or alkene oxidation reactions by iodosylbenzene and hydrogen peroxide.[119] Finally, layered inorganic–organic aluminum (Sil-Al) and magnesium organosilicates (Sil-Mg) were successfully prepared by using a new alkoxysilane, (3-urea-dodecyl) propyltriethoxysilane, containing embedded urea groups. The study of chemical stability of such hybrids showed that Sil-Al is stable within 2–12 pH range, while Sil-Mg could be used in solutions with pH above 4. Thus, both compounds can be employed as complementary sorbent materials for solid phase extraction applications.[120]

The synthesis of covalently linked inorganic–organic lamellar composites with ion exchange properties and tailored hydrophobicity is of great interest.[110,112,121,122] Such syntheses were carried out using a single-step templating sol-gel procedure. The syntheses involved reaction of magnesium nitrate and aluminum acetylacetonate with octyltriethoxysilane. A series of materials having the following formula $Na_{4x}[(RSi)_{4(1-x)}Al_{4x}Mg_3O_{8+2x}(OH)_2]$, where x is ranging from 0 (talc) to $0.05 \leq x \leq 0.33$ (saponite), were prepared. It is important to note the variation of the d-value of the (060) peak in the XRD patterns, that decreased from 0.157 nm (pure trioctahedral character, i.e., three octahedral elements in the octahedral sheet) to 0.151 nm (pure dioctahedral character, i.e., two octahedral elements in the octahedral sheet and a vacancy), with the raise in aluminum content. In order to confirm the trioctahedral character of these materials, syntheses were carried out in fluorine in order to use ^{19}F as a local probe. ^{19}F MAS NMR spectroscopy confirmed the substitution of fluoride ions for hydroxyl groups. By using the empirical formula proposed by Huve *et al.*[123] for the determination of the octahedral environment ($\delta = 50e - 238$, where e is the average of the electronegativity of octahedral elements), it was shown that the peak at -164 ppm/$CFCl_3$ corresponds to a Mg-Al-Al fluorine environment and, respectively, a trioctahedral character. The positive charges of the octahedral sheet were partially balanced by the negatively charged tetrahedral sheet.

3. Outlook

In this chapter, the most important aspects of the synthesis and the application of organoclays were summarized. This new class of porous solids attracted the attention of researchers in recent years with their unique properties and potential areas of application, which are beyond those of conventional layered type materials. At the present, the research effort is aimed at developing of synthesis methods to better control the physicochemical characteristics of organoclays. The different synthetic strategies can be classified in three main groups:

- Ion-exchange intercalation of organic cations or charged organic species. Although interesting materials can be obtained employing this approach, it has a limited impact due to drawbacks related to the weak bonds between the clay support and the organic species.

- Grafting of organic molecules, namely silane coupling agents, via strong (covalent or iono-covalent) bonds. The organic groups are firmly fixed to the support using this method. A disadvantage of this approach is the relatively low concentration of organic moieties due to the limited number of available hydroxyls.
- *In situ* synthesis, where the organic part of the hybrid is grafted during the crystallization of the layered material. This approach allows both stable grafting and relatively high density of organic molecules.

Possible applications of organoclays were mentioned as well. Selected examples were given which revealed the enormous application potential of the layered hybrids. The applications vary from heavy metal remediation and catalysis to adsorption of organic pollutants and immobilization of biomolecules. The organoclays are also considered as promising materials as environmental barriers, polymer fillers, catalytic supports, chemical sensors and porous vehicles for chemotherapy drugs. Hence, one could predict that this class of porous solids will continue to be an active area of research for both fundamental and applied scientists.

REFERENCES

[1] Van Olphen, H. *Science* **1966,** *154,* 645–646.
[2] Schöllhorn, R. in: Müller-Warmuth, W. Schöllhorn, R. (Eds.), *Progress in Intercalation Research,* Kluver Academic Publishers, Dordrecht, 1994.
[3] Schöllhorn, R. in: Whittingham, M. S. Jacobson, A. J. (Eds.), *Intercalation Chemistry,* Academic Press, New York, 1982.
[4] Cool, P., Vansant, E. P., Poncelet, C., Schoonheydt, R., in: Schüth, F., Sing, K. S. W. (Eds.), *Handbook of Porous Solids,* Edition Wiley-VCH, 2002, pp. 1250–1255.
[5] Reichle, W. T., *Solid State Ionics* **1986,** *22(1),* 135.
[6] Grim, R. E. *Clay Mineralogy*, McGraw-Hill Book Company, Inc., New York, 1953.
[7] Brindley, G. W., Brown, G. in: *Crystal Structure of Clay Minerals and their X-Ray Identification,* Mineralogical Society, London, 1980.
[8] Boyd, S. A., Shaobai, S., Lee, J.-F., and Mortland, M. M., *Clays Clay Miner.* **1988,** *36(2),* 125–130.
[9] Hedley, C. B., Yuan, G., Theng, B. K. G., *Appl. Clay Sci.* **2007,** *35(3-4),* 180–188.
[10] Ruiz-Hitzky, E., Van Meerbeck, A., in: Bergaya, F., Theng, B. K. G., Lagaly, G., (Eds.), *Handbook of Clay Sciences,* Elsevier, Germany, 2006, pp. 583–622.
[11] Soule, N. M., Burns, S. E., *J. Geotech. Geoenviron. Eng.* **2001,** *127,* 363–369.
[12] Mortland, M. M., Shaobai, S., Boyd, S. A., *Clays Clay Miner.* **1986,** *34,* 581–585.
[13] He, H., Frost, R. L., Bostrom, T., Yuan, P., Duong, L., Yang, D., Xi, Y., Kloprogge, T. J., *J. Colloid Interface Sci.* **2005,** *11,* 1–7.
[14] Worren, O. S., Lee, S. C., *Stud. Surf. Sci. Catal.* **2001,** *133,* 429–433.
[15] Kozaka, M., Domkab, L., *J. Phys. Chem. Solids* **2004,** *65,* 441–445.
[16] Manohar, D. M., Krishnan, K. A., Anirudhan, T. S., *Water Res.* **2002,** *36,* 1609–1619.
[17] Janek, M., Lagaly, G., *Colloid Polym. Sci.* **2003,** *281,* 293–301.
[18] Kozak, M., Domka, L., *J. Phys. Chem. Solids* **2004,** *6,* 441–445.
[19] Othmani-Assamann, H., Benna Zayani, M., Geiger, S., Fraisse, B., Kbir-Ariguib, N., Trabelsi-Ayyadi, M., Ghermani, N. E., Grossiord, J. L., *J. Phys. Chem. C.* **2007,** *111(29),* 10869–10877.
[20] Xi, Y., Ding, Z., He, H., Frost, R. L., *Spectrochim. Acta Part A* **2005,** *61,* 515–525.
[21] Mermut, A. R., Lagaly, G., *Clays Clay Miner.* **2001,** *49,* 393–397.
[22] Lemic, T., Tomaševic-Canovic, M., Djuricic, M., Stani, T., *J. Colloid Interface Sci.* **2005,** *292,* 11–19.

[23] Prost, R., Yaron, B., *Soil Sci.* **2001,** *166,* 880–895.
[24] Alther, G., *Contamin. Soils* **2002,** *7,* 223–231.
[25] Carrado, K. A., *Appl. Clay Sci.* **2000,** *17,* 1–23.
[26] Domka, L., Krysztafkiewicz, A., Kozak, M., *Polym. Polym. Compos.* **2002,** *10(7),* 541–552.
[27] Baudu, M., Farkhani, B., Ayele, J., Mazet, J. M., *Environ. Technol.* **1993,** *14,* 247–256.
[28] Mercier, L., Detellier, C., *Environ. Sci. Technol.* **1995,** *29,* 1318–1323.
[29] Mercier, L., Pinnavaia, T. J., *Environ. Sci. Technol.* **1998,** *32,* 2749–2754.
[30] Mercier, L., Pinnavaia, T. J., *Microporous Mesoporous Mater.* **1998,** *20,* 101–106.
[31] Wasserman, S. R., Soderholm, L., Staub, U., *Chem. Mater.* **1998,** *10,* 556–559.
[32] Guimaraes, J. L., Peralta-Zamora, P., Wypych, F., *J. Colloid Interface Sci.* **1998,** *206,* 281–287.
[33] Celis, R., Hermosin, M. C., Cornejo, J., *Environ. Sci. Technol.* **2000,** *34,* 4593–4599.
[34] Song, K., Sandi, G., *Clays Clay Miner.* **2001,** *49,* 119–125.
[35] Letaief, S., Detellier, C., *Chem. Commun.* **2007,** *25,* 2613–2615.
[36] Guimaraes, A., Ciminelli, V. S. T., Vasconcelos, W. L., *Mater. Res.* **2007,** *10,* 37–41.
[37] Song, K., Sandi, G., *Clays Clay Miner.* **2001,** *49,* 119–125.
[38] Carrado, K. A., Xu, L., Csencsits, R., Muntean, J. V., *Chem. Mater.* **2001,** *13,* 3766–3773.
[39] Wheeler, P. A., Wong, J., Baker, J., Mathias, L. J., *Chem. Mater.* **2005,** *17,* 3012–3018.
[40] Bourlinos, A. B., Jiang, D. D., Giannelis, E. P., *Chem. Mater.* **2004,** *16,* 2404–2410.
[41] Wang, J., Wheeler, P. A., Jarett, W. L., Mathias, L. J., *Polym. Prepr.* **2005,** *46,* 546–565.
[42] Herrera, N. N., Letoffe, J.-M., Putaux, J.-L., David, L., Bourgeat-Lami, E., *Langmuir* **2004,** *20,* 1564–1570.
[43] Raussell-Colom, R., Serratosa, J. M., in: Newman, A. C. D. (Ed.), *Chemistry of Clays and Clay Minerals,* Longman, Essex, England, 1987, pp. 371–422.
[44] Ruiz-Hitzky, E., Fripiat, J. J., *Clays Clay Miner.* **1976,** *24,* 25–35.
[45] Hermosin, M. C., Cornejo, J., *Clays Clay Miner.* **1986,** *34,* 591–596.
[46] Waddell, A., Leyden, D. E., DeBello, M. T., *J. Am. Chem. Soc.* **1981,** *103,* 5303–5315.
[47] Nadiye-Tabbiruka, M. S., *Colloid Polym. Sci.* **2000,** *278,* 677–681.
[48] Ogawa, M., Okutomo, S., Kuroda, K., *J. Am. Chem. Soc.* **1998,** *120,* 7361–7362.
[49] Song, K., Sandí, G., *Clays Clay Miner.* **2001,** *49,* 119–125.
[50] Gieseking, J. E., *Adv. Agron.* **1949,** *1,* 59–60.
[51] Hofmann, A., Roussy, D., Filella, M., *Chem. Geol.* **2002,** *182,* 35–55.
[52] Bursacchi, S., Gemi, M., Ricci, L., Ruggeri, G., Veracini, C. A., *Langmuir* **2007,** *23,* 3953–3960.
[53] Diaz, M., Cambier, P., Brendlé, J., Prost, R., *Appl. Clay Sci.* **2007,** *37,* 12–22.
[54] Tonle, I. K., Nganemi, E., Njopwouo, D., Carteret, C., Walcarius, A., *Phys. Chem. Chem. Phys.* **2003,** *5,* 4951–4955.
[55] Tonle, I. K., Nganemi, E., Walcarius, A., *Electrochim. Acta* **2004,** *49,* 4951–4955.
[56] Tonle, I. K., Nganemi, E., Walcarius, A., *Sens. Actuators B.* **2005,** *11,* 195–203.
[57] Mbouguen, J. K., Nganemi, E., Walcarius, A., *Anal. Chim. Acta* **2006,** *578,* 145–155.
[58] Gopinak, S., Sugaran, S., *App. Clay Sci.* **2007,** *35,* 67–75.
[59] He, H. P., Duchet, J., Galy, J., Gerard, J. F., *J. Colloid Interface Sci.* **2005,** *288,* 171–176.
[60] Shanmugharaj, A. M., Rhee, K. Y., Ryu, S. H., *J. Colloid Interface Sci.* **2006,** *298,* 854–859.
[61] Zhu, A. L., Tian, S., Zhu, J., Shi, Y., *J. Colloid Interface Sci.* **2007,** *315,* 191–199.
[62] Mbouguen, J. K., Ngameni, E., Walcarius, A., *Biosens. Bioelectron.* **2007,** *23,* 269–275.
[63] Shen, W., He, H., Zhu, J., Yuan, Y., Frost, R. L., *J. Colloid Interface Sci.* **2007,** *313,* 268–273.
[64] He, H. P., Duchet, J., Galy, J., Gerard, J. F., *J. Colloid Interface Sci.* **2005,** *288,* 171–176.
[65] Frost Ray, L., Daniel Lisa, M., Zhu Huai Yong, *J. Colloid Interface Sci.* **2008,** *321*(2), 302–309.
[66] Erdemoglu, M., Erdemoglu, S., Sayilkan, F., Akarsud, M., Senerb, S., Sayilkan, H., *Appl. Clay Sci.* **2004,** *27,* 41–52.
[67] Frost, R. L., Mendelovici, E., *J. Colloid Interface Sci.* **2006,** *294,* 47–52.
[68] Guimaraes, A., Ciminelli, V. S. T., Vasconcelos, W. L., *Mater. Res.* **2007,** *10,* 37–41.
[69] Herrera, N. N., Putaux, J.-L., Bourgeat-Lami, E., *Prog. Solid State Chem.* **2006,** *34,* 121–137.
[70] Dean, K. M., Bateman, S. A., Simons, R., *Polymer* **2007,** *48,* 2231–2240.

[71] Tartaglione, G., Tabuani, D., Camino, G., *Microporous Mesoporous Mater.* **2007,** doi: 10.1016/j.compscitech.2007.06.023.
[72] Konishi, Y., Shimaoka, J., Asai, S., *React. Funct. Polym.* **1998,** *36,* 197–206.
[73] Reichert, J., Binner, J. G. P., *J. Mater. Sci.* **1996,** *31,* 1231–1241.
[74] Bonn, G., Reiffenstuhl, S., Jandik, P., *J. Chromatogr.* **1990,** *499,* 669–676.
[75] Bhattacharyya, D., Hestekin, J. A., Brushaber, P., Cullen, L., Bachas, L. G., Sikdar, S. K., *J. Membr. Sci.* **1998,** *141,* 121–135.
[76] Airoldi, C., Santos, M. R. M. C., *J. Mater. Sci.* **1994,** *4,* 1479–1485.
[77] Espinola, J. G. P., Oliveira, S. F., Lemus, W. E. S., Souza, A. G., Airoldi, C., Moreira, J. C. A., *Colloids Surf. A* **2000,** *166,* 45–50.
[78] Da Fonseca, M. G., Barone, J. S., Airoldi, C., *Clays Clay Miner.* **2000,** *48,* 638–645.
[79] Tonle, I. K., Nganemi, E., Tcheumi, H. L., Tchieda, V., Carteret, C., Walcarius, A., *Talanta* **2007,** *11,* doi:10.101016/jtalanta.2007.06.006.
[80] Erdemoglu, E., Erdemoglu, S., Sayiilkan, F., Akarsu, M., Sener, S., Sayiilkan, F. H., *Appl. Clay Sci.* **2004,** *27,* 41–52.
[81] Liu, P., *Appl. Clay Sci.* **2007,** doi:10.1016/j.clay.2007.01.004.
[82] Liu, P., Guo, J. S., *Colloids Surf. A Physicochem. Eng. Asp.* **2006,** *282,* 498–503.
[83] Breen, C., *Appl. Clay Sci.* **1999,** *15,* 187–219.
[84] Lin, J. J., Chen, Y. M., *Langmuir* **2004,** *20,* 4261–4264.
[85] Lin, J. J., Cheng, I. J., Wang, R., Lee, R. J., *Macromolecules* **2001,** *34,* 8832–8834.
[86] Chu, C. C., Chiang, L. M., Tsai, C. M., Lin, J. J., *Macromolecules* **2005,** *38,* 6240–6243.
[87] Oztekin, N., Alemdar, A., Gungor, A., Erim, N. F. B., *Mater. Lett.* **2002,** *55,* 73–76.
[88] Dong, J. Y., Wang, Z. M., Han, H., Chung, T. C., *Macromolecules* **2002,** *35,* 9352–9359.
[89] Wang, Z. M., Nakajima, H., Manias, E., Chung, T. C., *Macromolecules* **2003,** *36,* 8919–8922.
[90] Lu, H., Hu, Y., Li, M., Chen, Z., Fan, W., *Compos. Sci. Tech.* **2006,** *66,* 3035–3039.
[91] Celini, N., Bergaya, F., Poncin-Epaillard, F., *Polymer* **2007,** *48,* 658–679.
[92] d'Agostino, R., in: *Plasma Deposition, Treatment and Etching Polymers,* Academic Press, New York, 1990.
[93] Morosoff, N., *An Introduction to Plasma Polymerization.* Academic Press, New York, 1990.
[94] Fukushima, Y., Tani, M., *J. Chem. Soc. Chem. Commun.* **1995,** 241–242.
[95] Burket, S. L., Press, A., Mann, S., *Chem. Mater.* **1997,** *9,* 1071–1077.
[96] Whilton, N. T., Burkett, S. L., Mann, S., *J. Mater. Chem.* **1998,** *8,* 1927–1933.
[97] Fonseca, M., Da, G., Airoldi, C., *J. Chem. Soc. Dalton Trans.* **1999,** 3687–3682.
[98] Fukushima, Y., Tani, M., *Bull. Chem. Soc. Jpn.* **1996,** *69,* 3667–3671.
[99] Ukrainczyk, L., Bellman, R. A., Anderson, A. B., *J. Phys. Chem. B* **1997,** *101,* 531–537.
[100] Da Fonseca, M. G., Airoldi, C., *J. Mater. Chem.* **2000,** *10,* 1457–1462.
[101] Da Fonseca, M. G., Airoldi, C., *Thermochim. Acta* **2000,** *359,* 1–9.
[102] Da Fonseca, M., Da Silva, G., Filho, E. C., Machado Junior, R. S. A., Arakaki, L. N. H., Espinola, J. G. P., Airoldi, C., *J. Solid State Chem.* **2004,** *177(7),* 2316–2322.
[103] Da Fonseca, M. G., Barone, J. S., Airoldi, C., *Clays Clay Miner.* **2000,** *48,* 638–643.
[104] Cestari, A. R., Vieira, E. F. S., Bruns, R. E., Airoldi, C., *J. Colloid Interface Sci.* **2000,** *227,* 66–73.
[105] Da Fonseca, M. G., Silva, C. R., Barone, J. S., Airoldi, C., *J. Mater. Chem.* **2000,** *10,* 789.
[106] Whilton, N. T., Burkett, S. L., Mann, S., *J. Mater. Chem.* **1998,** *8,* 1927–1932.
[107] Sasai, R., Itoh, H., Shindachi, I., Shichi, T., Takagi, K., *Chem. Mater.* **2001,** *13,* 2012.
[108] Guillot, M., Richard-Plouet, M., Vilminot, S., *J. Mater. Chem.,* **2002,** *12,* 851.
[109] Richard-Plouet, M., Vilminot, S., Guillot, M., Kurmoo, M., *Chem. Mater,* **2002,** *14,* 3829–3836.
[110] Jaber, M., Miehe-Brendle, J., Le Dred, R., *Chem. Lett.* **2002,** *9,* 954.
[111] Burattin, P., Che, M., Louis, C., *J. Phys. Chem. B* **1999,** *103,* 6171–6178.
[112] Jaber, M., Miehe-Brendle, J., Delmotte, L., Le Dred, R., *Micropor. Mesopor. Mater.* **2003,** *65,* 155–163.
[113] Minet, J., Abramson, S., Bresson, B., Sanchez, C., Montouillout, V., Lequex, N., *Chem. Mater.* **2004,** *16,* 3955–3962.
[114] Carrado, K. A., Xu, L. Q., Csencsits, R., Muntean, J. V., *Chem. Mater.* **2001,** *13,* 3766–3773.

[115] Mizutani, T., Fukushima, Y., Okada, A., Kamigaito, O., *Bull. Chem. Soc. Jpn.* **1990,** *63,* 2094–2098.
[116] Sasai, R., Itoh, H., Shindachi, I., Shichi, T., Takagi, K., *Chem. Mater.* **2001,** *13,* 2012–2016.
[117] Richard-Plouet, M., Vilminot, S., *Solid State Sci.* **1999,** *1,* 381–393.
[118] Richard-Plouet, M., Vilminot, S., Guillot, M., *New J. Chem.* **2004,** *28,* 1073–1082.
[119] Faria, A. L., Airoldi, C., Doro, G. F., Fonseca, M. G., Assis, M. D., *Appl. Catal. A: Gen.* **2004,** *268,* 217–226.
[120] Moscofian Andrea, S. O., Silva César, R., Airoldi, C., *Micropor. Mesopor. Mate.* **2007,** doi:10.1016/j.micromeso.2007.05.047.
[121] Jaber, M., Miehé-Brendlé, J., Roux, M., Dentzer, J., Le Dred, R., Guth, J.-L., *New J. Chem.* **2002,** *16,* 1597–1600.
[122] Jaber, M., Miehé-Brendlé, J., Le Dred, R., *Mater. Res. Soc. Symp. Proc.* **2002,** *726,* 161–165.
[123] Huve, L., Le Dred, R., Saehr, D., Baron, J., *Clays Clay Miner.* **1992,** *40*(2), 186–191.

CHAPTER 3

Titanium-Based Nanoporous Materials

Wolfgang Schmidt

Contents

Abstract

Titanium as an integral compound is found in many different nanoporous materials. During the past decades, micro- and mesoporous titanosilicates became very prominent, especially with regard to their catalytic activity. The properties of these materials strongly depend on the coordination of the titanium. Fourfold coordination of the titanium by oxygen results in silicates with neutral frameworks but pronounced redox activity. Sixfold coordination results in two negative framework charges per titanium on framework sites. Such materials have similar properties as zeolites, that is, they may act as ion-exchangers, solid state acids and adsorbents, but they have no intrinsic redox activity. Another class of material is nanoporous titania obtained either by templating of by anodic oxidation of titanium. The latter process results in extended layers of mesoporous titania with layer thicknesses in the micrometre range. The regular pore arrangement and the high surface area of these materials make them very interesting for catalysis and sensor applications.

Keywords: Titania, Titanosilicates, Mesoporosity, Microporosity, Nanoporosity

Ordered Porous Solids
DOI: 10.1016/B978-0-444-53189-6.00003-2

1. Introduction

Nanoporous silicas are nowadays a very well-investigated class of materials. They include natural and synthetic zeolites as well as amorphous silica materials with defined porosity. Partial substitution of silicon by titanium in zeolites results in titanosilicates with titanium with the same coordination as the silicon in the silica, that is, tetrahedral coordination (denoted $Ti^{[4]}$). If the titanium is build in the titanosilicate in octahedral coordination (denoted $Ti^{[6]}$), structures unknown for pure silica are formed. In the following chapters, the properties of both types of materials will be addressed. The pore sizes of the different titanosilicates vary from micropore range (<2 nm) to mesopore range (2–50 nm). As a last class of materials, pure nanoporous titania will be addressed. A balanced discussion of all features of these very different materials would exceed the scope of the present chapter by far. Thus, only a brief overview can be given here. However, there exists comprehensive literature on the different topics addressed in the following chapters (e.g., Refs. 1–4).

2. Nanoporous Titanosilicates with $Ti^{[4]}$ Coordination

2.1. Titanium silicalite-1 (Ti-MFI, TS-1)

At present, the most prominent titanosilicate is known under the name TS-1. Because of its outstanding position, it will be introduced separately from other related materials in this first chapter. Titanium Silicalite-1 (TS-1) is isostructural to zeolite ZSM-5 with MFI structure.[5] The framework topology of the structure is illustrated in Fig. 3.1. The zeolite structure contains a three-dimensional (3D) pore system with straight channels along [010] intersecting with sinusoidal channels along [100]. The slightly elliptic pore openings have sizes of about 0.55 nm and are formed

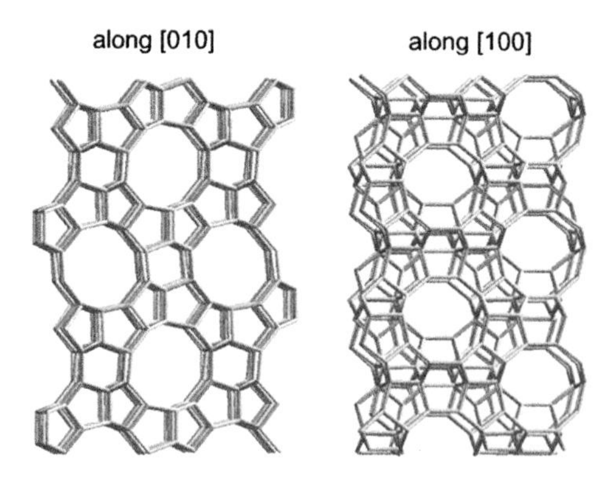

Figure 3.1 Framework topology of the MFI structure viewing into the straight and sinusoidal channels. Oxygen atoms are omitted for simplicity, that is, the lines directly connect silicon (titanium) sites.

by rings of 10 SiO_4 and/or TiO_4 tetrahedra (10-ring). In TS-1, up to 2% of the silicon can be substituted by titanium. Each Ti^{4+} cation is tetrahedrally coordinated by four O^{2-} anions and an integral part of the zeolite framework. Since Ti has the same oxidation state as the silicon in the framework, no framework charge is created by the substitution and thus, no extra-framework cations are present in the pores. The ionic radius of Ti^{4+} ($r_{Ti}{}^{4+} = 42$ pm) in tetrahedral coordination is significantly larger than the radius of Si^{4+}($r_{Si}{}^{4+} = 26$ pm).[6] Substitution of Si by Ti thus results in a linear expansion of the unit cell volume of the MFI structure.[7,8] Depending on the synthesis route, the unit cell volumes of TS-1 crystallites with identical Si/Ti ratios may differ significantly.[9,10] However, within a set of identically prepared samples differing only in the Si/Ti ratios of their frameworks, linearity of calibration curves is often observed, and calculation of Ti contents from unit cell volumes of TS-1 are thus possible. A typical IR band at 960 cm^{-1} is usually taken as an indication for titanium coordinated tetrahedrally by oxygen, that is, the titanium is integrated in the zeolite framework.[1,11]

TS-1 is usually obtained by a hydrothermal reaction in alkaline solution using a structure-directing agent (template), such as tetrapropylammonium (TPA) hydroxide. The first report on the synthesis of TS-1 was a patent filed by the Italian company Snamprogetti S.p.A., now a subsidiary of the Eni Group.[12] The discovery of this titanosilicate was a real breakthrough for oxidation catalysis. TS-1 was found to be an active catalyst for oxidation of a broad variety of organics using hydrogen peroxide as the oxidant. Up to this discovery, titanium-containing catalysts were known to work only in water-free environments. Thus, using anhydrous hydrogen peroxide in organic solvents was essential for such catalysts and generated water had to be removed immediately, a significant drawback for large scale applications.[13] With TS-1 as catalyst using aqueous hydrogen peroxide solutions became possible. The pore system of TS-1 is rather hydrophobic and strongly prevents water from interaction with the Ti sites inside the pores. This resistance of the TS-1 catalysts against water opened the door to large-scale application of the catalytic process.

Oxidation of propene to propene oxide was in the main focus of research but also other olefins have been oxidized with hydrogen peroxide using TS-1 as the catalyst. The solvent proved to play an important role for the efficiency of the catalyst. Polar solvents promote the oxidation reactions, and for alcohols as solvents, the epoxidation rate of propene decreases strongly in the order methanol > ethanol > *i*-propanol > *t*-butanol.[13–15] Methanol was found to be an excellent solvent with propene conversions exceeding 90% and selectivities for propene oxide in the range of 95% or more are reported.[14,15] However, also mixtures of methanol and water showed to be efficient solvents with comparable conversions and selectivities, provided the water content was not too high. At a water concentration of more than 50%, the rate of reaction was observed to decrease significantly.[14] The epoxidation reaction proceeds readily with 30–60 wt.% aqueous hydrogen peroxide solutions but also lower concentrations of hydrogen peroxide down to 1 wt.% of hydrogen peroxide were reported to result in sufficiently high conversions.

Also other olefins, such as butene, pentene, hexene and higher olefins have been successfully oxidized with hydrogen peroxide and TS-1 as the catalyst. The rate of oxidation for these olefins also depended strongly on the type of solvent used.[16,17]

There is evidence that the choice of solvent not only affects the distribution of the olefin in the solvent and the pore system of the titanosilicate but that polar solvents may as well play an important role in the catalytic reaction by interacting with the Ti site.[13,17]

Figure 3.2 shows one potential reaction mechanism for the epoxidation of propene involving Ti(η^1-OOH) species, in which one oxygen atom from the hydrogen peroxide is bound to the titanium. One Ti–O–Si bond is broken by reaction with H_2O_2, resulting in the formation of a silanol group. The reactive species is then titanium which is bound to three framework oxygen atoms and to one oxygen atom from the peroxide. The titanium is further coordinated by oxygen from alcohol (solvent) or water, forming a five-membered transition state.[13] The oxygen next to the titanium is then transferred to the double bond of the propene. One molecule of water and a Ti–OR bond are formed as an intermediate. After release of the water and the propene oxide, the active species (five-ring) is regenerated by reaction with fresh hydrogen peroxide.

A second type of reactive site was supposed to involve Ti(η^2-OOH) species, where the titanium interacts simultaneously with both oxygen atoms of a hydrogen peroxide molecule.[1,13,18,19] In the past, several other titanium peroxo species have been also proposed for TS-1. However, from computational and spectroscopic investigations it is now understood that two different Ti species are active in TS-1, that is, Ti(η^1-OOH) and Ti(η^2-OOH) as shown in Fig. 3.3.[20]

The stability of the complexes does not differ too much and both species have been verified by EXAFS investigations.[21] Water and/or alcohol molecules coordinating the Ti centre stabilize the transition state for both Ti-peroxo configurations. DFT calculations show that for all possible configurations, the preferred coordination number for Ti in TS-1 is six in the presence of hydrogen peroxide.[22] The reaction involving Ti(η^1-OOH) species is rather similar to that shown in Fig. 3.2.

The epoxidation of olefins is a very prominent reaction but there are also other oxidation reactions which are readily catalysed by TS-1. Figure 3.4 summarizes a selection of some of the most important reactions, including oxidation of ammonia to hydroxylamine, oxidation of secondary alcohols to ketones, hydroxylation of alkanes and phenols, ammoximation of cyclohexanone to cyclohexanone oxime, oxidation of sulphides to sulphoxides and oxidation of secondary amines to dialkylhydroxylamines.[23–25]

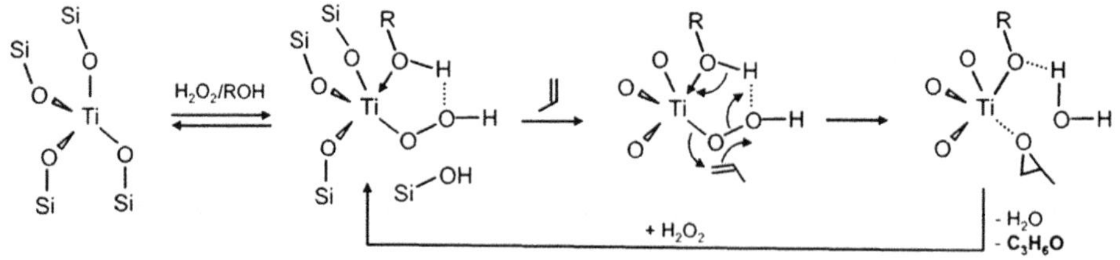

Figure 3.2 Propene epoxidation mechanism via Ti(η^1-OOH) species. (From Ref. 13 by permission of Urban-Verlag.)

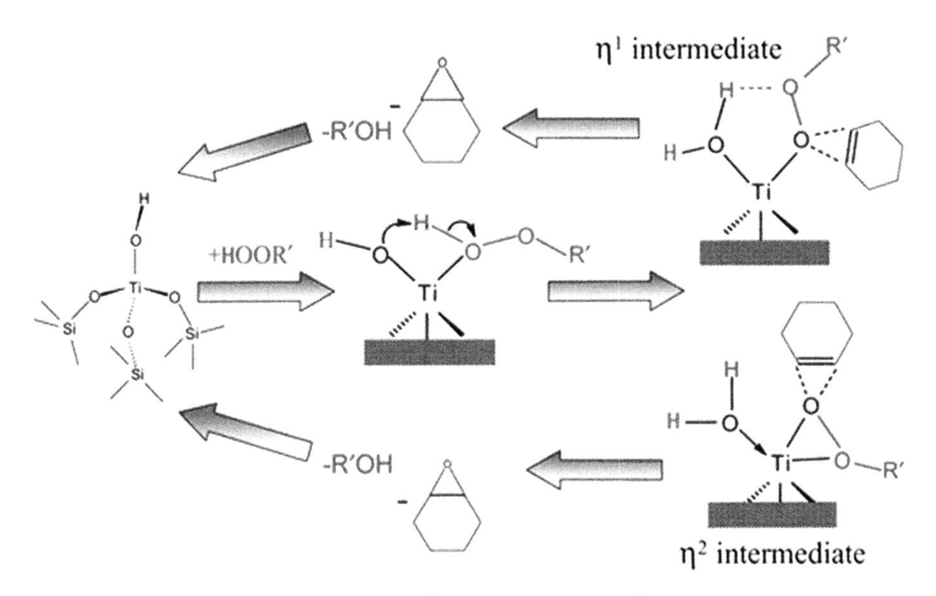

Figure 3.3 Epoxidation reactions via Ti(η^1-OOH) and/or Ti(η^2-OOH) species. (Reproduced from Ref. 20 by permission of The Royal Society of Chemistry.) (See color insert.)

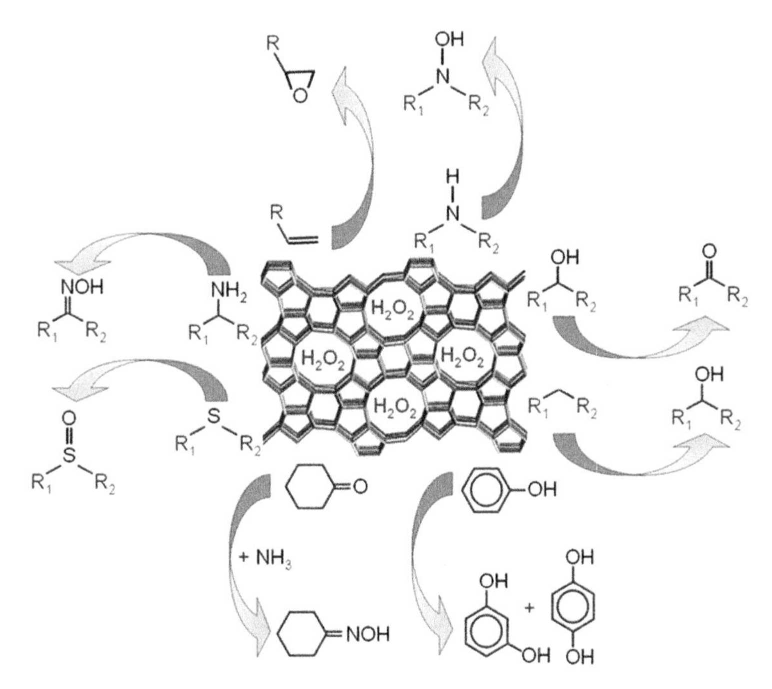

Figure 3.4 Oxidation reactions catalysed by TS-1 in the presence of 30% aqueous H_2O_2. (From Ref. 25 by permission of Springer-Verlag.) (See color insert.)

All these oxidation reactions can be performed with aqueous hydrogen peroxide solutions. An alternative is the *in situ* formation of the hydrogen peroxide,[13,26,27] for example, using a Pd/TS-1 catalyst and H_2 and O_2 along with propene as the feed according to:

$$H_2 + O_2 + C_3H_6 \xrightarrow{\text{Pd/TS-1}} C_3H_6O + H_2O$$

The high activity of TS-1 and the low concentration of H_2O_2 formed prevent the decomposition of the hydrogen peroxide on the Pd particles, but hydrogenation of the propene on the Pd catalyst is a clear drawback of this process. Also water is directly formed from the H_2 and O_2 on the noble metal catalyst. Thus, the yield of hydrogen peroxide is reduced. An alternative approach is applying a combination of the anthrahydroquinone (AO) process with TS-1 catalysis. The hydrogen peroxide which is consumed by the epoxidation reaction over TS-1 is produced by the AO process. Both reactions can be performed either in one single reactor or in separate ones.[13,28,29] In the latter case, the crude aqueous H_2O_2 solution is pumped into the epoxidation reactor without purification. The processes can be summarized by the following reactions:

$$O_2 + QH_2 \rightarrow Q + H_2O_2 \quad \text{(hydrogen peroxide formation)}$$

$$C_3H_6 + H_2O_2 \xrightarrow{\text{TS-1}} C_3H_6O + H_2O \quad \text{(epoxidation reaction)}$$

$$H_2 + Q \xrightarrow{\text{Pd}} QH_2 \quad \text{(regeneration of hydroquinone)}$$

In the last two decades, several processes were installed on an industrial scale using TS-1 as the catalyst.[25] EniChem started a plant for the hydroxylation of phenol with a capacity of 10,000 t/a and one for the ammoximation of cyclohexanone to cyclohexanone oxime with a capacity of 12,000 t/a in Porto Marghera, Italy.[30] A production of propene oxide on pilot plant scale with a capacity of about 2000 t/a was tested by EniChem. In 2003, Sumitomo commercialized the first process in which the ammoximation process and the Beckmann rearrangement reaction were combined to produce about 65,000 t/a of ϵ-caprolactam. The ammoximation of cyclohexanone with H_2O_2 and NH_3 over a TS-1 catalyst results in cyclohexanone oxime which is subsequently transformed to ϵ-caprolactam in the vapor phase Beckmann rearrangement over a high silica MFI zeolite as the catalyst.[31] Figure 3.5 shows the plant in Ehime, Japan, and the reactor scheme for the Sumitomo vapor phase Beckmann rearrangement reaction in which the cyclohexanone oxime from the EniChem process is fed.

In the same year, Sinopec announced the commercialization of the ammoximation process on a scale of 70,000 t/a.[32,33] The first propene oxide plant on the basis of the epoxidation with H_2O_2 using TS-1 as the catalyst is scheduled to start operation in Antwerpen in 2008. The process was developed jointly by BASF and Dow Chemical and is supposed to have an annual propene oxide capacity of 300,000 t.[13,34,35]

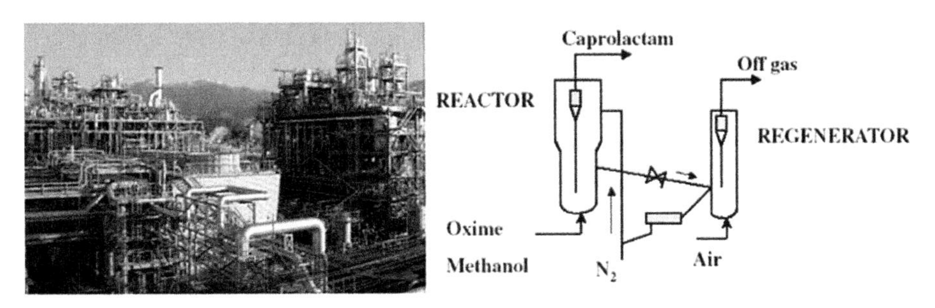

Figure 3.5 Production plant for ε-caprolactam and scheme of the fluidized bed reactor for the caprolactam production. (Reproduced from Ref. 31 by permission of The Chemical Society of Japan.) (See color insert.)

Another large-scale plant with a propene oxide capacity of 100,000 t/a will be installed by the Korean SKC in 2008 in Ulsan, South Korea.[36,37] The plant will run the HPPO (Hydrogen Peroxide Propene Oxide) process which also uses hydrogen peroxide and TS-1 as the catalyst. The process is licensed by Degussa/Uhde (now Evonik). One can resume that reactions on the basis of TS-1 catalysts are nowadays implemented in several production processes. However, the number of large-scale plants working with this new catalyst is not exceedingly high by now. On the other hand, new plants will be built and future will show how well TS-1 will be integrated in the list of commercially used standard catalysts.

Several innovative processes are evaluated at present and future may also show whether these processes will find their way to application. Two new developments shall serve as examples.

BASF developed a process for the production of propylene oxide starting from low cost propane instead of propene.[38] In a first step, propane is dehydrogenated into propene and hydrogen. The gas streams are separated and the hydrogen then reacts with oxygen to form hydrogen peroxide. In a next step, the hydrogen peroxide reacts with the propene obtained from the first step to form propene oxide over TS-1 and/or TS-2 (MEL structure type).

Since deactivation of the TS-1 catalyst during the olefin epoxidation is a severe problem, Degussa/Uhde developed a regeneration process for the catalyst in which the deactivated TS-1 is regenerated by washing with a methanol solvent at temperatures above 100 °C.[39]

2.2. Comparative evaluation of catalytic activity of $Ti^{[4]}$ titanosilicates

TS-1 is a highly selective catalyst for specific oxidation reactions but its application is limited to conversion of rather small molecules. Larger molecules cannot enter the micropore channels of TS-1 and are thus excluded from reaction. This shortcoming triggered investigations aiming on catalysts with comparable activity but larger pores. A considerable number of zeolite structures have been synthesized as titanosilicates by now, including MEL (ZSM-11), BEA (Zeolite Beta), MOR (Mordenite) and MWW (MCM-22) structure types, as illustrated in Fig. 3.6. Other structures are also reported for titanosilicates but for brevity will not be considered in this chapter.

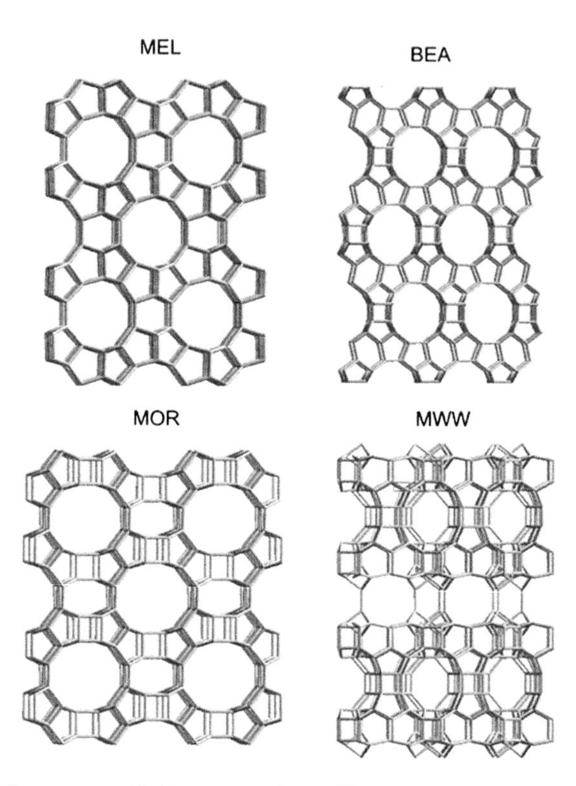

Figure 3.6 Selected structures of microporous titanosilicates.

The three-letter codes listed above refer to unique structure topologies of the different zeolite types.[5]

In the following, the different materials will be denoted as Ti-MEL, Ti-BEA, Ti-MOR and Ti-BEA for simplicity. The structure of Ti-MEL (= TS-2) is closely related to that of Ti-MFI (= TS-1). The major difference is that the MEL structure forms interconnected straight channels not only along [010] but also along [100]. Diffusion is therefore slightly enhanced in this pore system in comparison with the tortuous MFI pore system. The pore openings of Ti-MEL have sizes of about 0.54 nm and are formed by 10 SiO_4 and/or TiO_4 tetrahedra (10-ring). Somewhat larger pores are realized in the 3D pore system of Ti-BEA which is formed by 12-rings. The dimensions of the pores are about 0.77 × 0.66 nm and 0.56 nm. The structure can be considered as a stack of silicate layers which can be stacked in different sequences. Stacking disorder results in a high concentration of stacking faults and therefore XRD patterns of Ti-BEA do show only few defined reflections along with rather broad signals due to the disordered stacking. The structure of Ti-MOR on the other hand forms larger 12-ring pores with dimensions of 0.65 × 0.70 nm along [001] and smaller pores with dimensions of 0.26 × 0.57 nm along [010]. The pores of Ti-MWW are formed by 10-rings and have dimensions of about 0.40 × 0.55 nm and 0.41 × 0.51 nm. The last structure considered here is the MWW structure which is somewhat special. The structure of as-synthesized MWW comprises of layered silica sheets with only weak interaction between the individual sheets. The sheets are decorated with silanol groups which condensate during calcination. As the result, a rigid 3D framework structure with MWW topology is formed.

The titanosilicates can be synthesized either by direct hydrothermal synthesis with the addition of a titanium source to the reaction gel or by post-synthetic modification of a zeolite precursor. Aluminium or boron-containing zeolites are often used as starting materials because the aluminium or boron can be easily extracted from the zeolite frameworks, for example, by treatment with acidic solution. Silanol nests are formed which are a prerequisite for the subsequent incorporation of the titanium in the framework, for example, by reaction with $TiCl_4$.[40] For zeolites with somewhat smaller pores, such as MFI zeolites, the incorporation of Ti is probably restricted to the outer regions of the crystallites because the bulky $TiCl_4$ molecule cannot enter the pores of the zeolite.[2]

A common feature of all titanosilicates considered here is that the titanium species formed in contact with hydrogen peroxide solution are assumed to be more or less identical to those discussed above for Ti-MFI (TS-1). This requires that the titanium is in fact an integral part of the respective zeolite framework and tetrahedrally coordinated by oxygen atoms in the framework. The same holds for the class of amorphous mesoporous titanosilicates which have certain unique features. The zeolitic materials Ti-MEL,[41,42] Ti-BEA,[43] Ti-MOR[44] and Ti-MWW[45] crystallize from alkaline silica solutions with the aid of organic structure directing agents (template molecules). The resulting materials form crystalline 3D structures with a periodic arrangement of atoms and the pores are intrinsic parts of the crystal structures.

At the end of the 1980s, scientists from Mobil Oil synthesized a nanoporous silica with regular pore arrangement but amorphous pore walls.[46,47] In a certain range of concentration, surfactant molecules form rod-like micelles in aqueous solutions, as shown schematically in Fig. 3.7. These rod-like micelles aggregate in a 2D

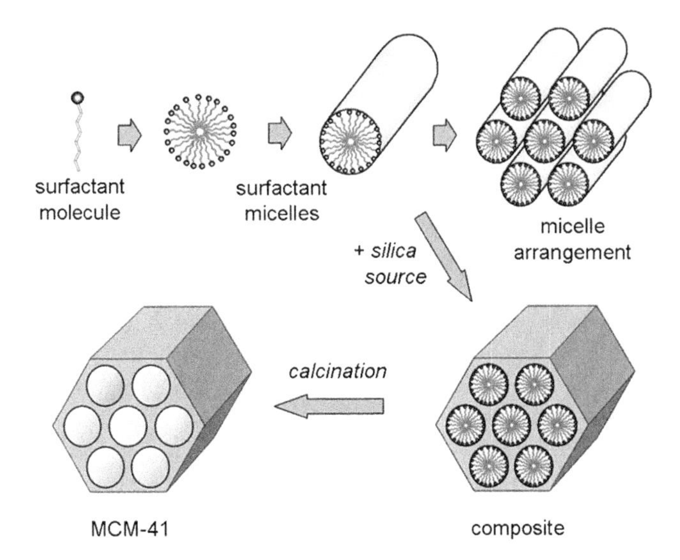

Figure 3.7 Formation of MCM-41 type materials. (See color insert.)

hexagonal arrangement. Addition of a silica source results in the formation of a silica coating of the micelles due to interaction of the negatively charged silica species and the positively charged head groups of the surfactant molecules (e.g., hexadecyltrimethylammonium). In alkaline solution, the silica condensates in the space between the micellar rods which serve as templates, forming amorphous silica walls. Calcination of the surfactant/silica composite results in combustion of the organics, and a silica material with regularly arranged mesopores, denoted as MCM-41, is formed. The pore sizes of this type of material are significantly larger than those of zeolites and may range from ~2 up to 5 nm. The sizes of the pores depend on the length of the lipophilic chain of the surfactant molecules and on the presence or absence of swelling agents.

Block copolymers of a certain composition with both hydrophilic and hydrophobic chains also form micellar rods and can be used as templates for the generation of another nanoporous silica.[48] After removal of the block-copolymer from the composite by calcination, an amorphous silica with regular pore arrangement is obtained, denoted as SBA-15. The pore sizes may range from ~6 up to 15 nm and are thus significantly larger than for MCM-41 materials. Even though the pore sizes might differ significantly, depending on the template used and synthesis conditions applied, the pore size distributions for both types of materials are rather sharp. The pore walls are usually thicker for SBA-15 (ca. 3–7 nm) than for MCM-41 (ca. 1 nm). Incorporation of titanium in tetrahedral coordination in such mesoporous silicates results in materials denoted as Ti-MCM-41 and Ti-SBA-15. Titanium is either built in the silica framework during the silica condensation process by adding alkoxytitanates to the reaction mixtures[49–51] or in a post-synthetic process by grafting Ti on the surface of the pore walls of the mesoporous silicas, for example, from titanocene dichloride.[52] There is certain evidence that the titanium is present in tetrahedral coordination also in these mesoporous titanosilicates and that it forms peroxo complexes similar to those in Ti-MFI (TS-1).[53] However, the mesoporous titanosilicates are hydrothermally less stable than the crystalline zeolite materials and leaching of heteroatoms, such as Ti, has to be faced. Treatment with water vapor results in a rather rapid decomposition of Ti-MCM-41[54] and reactions with hydrogen peroxide were observed to result in irreversible catalyst deactivation with time due to Ti leaching.[55] Using *tert*-butyl hydroperoxide as oxidant significantly reduced the leaching and the catalytic activity could be preserved.

The catalytic properties of titanosilicates will be briefly summarized in the following. However, the experimental database is extremely broad and only a few specific features can be highlighted here. Comprehensive reports on catalytic properties of titanosilicates in general are given in Refs. 1 and 2.

A general problem for catalysis in titanosilicates with titanium in tetrahedral coordination is TiO_2 impurities. As mentioned in the previous chapter, up to about 2% of Si can be substituted by Ti in the framework of MFI crystallites (TS-1). Higher concentrations of titanium easily result in the formation of extra-framework TiO_2 species. Similar observations were made for other $Ti^{[4]}$ titanosilicates. Such impurities have a significant effect on the performance of the catalysts for certain reactions.[2] For the oxidation of phenol, the selectivity is significant affected by TiO_2 whereas the oxidation of alkanes and alkenes is less susceptible to extra-framework TiO_2. Hydrogen peroxide is very efficiently decomposed in the presence of highly dispersed TiO_2 which of course has a strong effect on oxidation reactions using H_2O_2 as oxidant. Other reactions may also be catalysed by TiO_2 and the product composition of catalytic reactions thus can be significantly varied in the presence of such species. Extra-framework TiO_2 may as well be formed during catalytic reactions under more severe conditions and easily result in the degradation of the catalytic performance of a given titanosilicate catalyst.

Ti-MFI (TS-1) has comparably narrow pore openings and bulkier molecules have restricted access to its pore system or are even excluded. The pore size of the catalyst is thus a crucial factor for performance of a given catalyst. Table 3.1 illustrates the effect of structure type on the catalytic performance of three titanosilicate catalysts for the epoxidation of pentene. The reaction was performed at slightly higher temperature for Ti-BEA which might enhance the conversion due to enhanced diffusion within the pores at elevated temperature. However, diffusion constraints cannot explain the differences in epoxide selectivity. The selectivity for the large-pore Ti-BEA titanosilicate is significantly higher than for Ti-MFI.

Similar trends have been reported for other substrates. For bulkier alkenes, such as cyclohexene or cyclododecene, the epoxidation is enhanced using Ti-BEA. However, for linear alkenes up to long chain molecules such as dodecene, Ti-MFI is still the more active expoxidation catalyst.[56] Table 3.1 also demonstrates the intrinsically lower epoxidation selectivity of mesoporous Ti-MCM-41. At present, the non-crystalline mesoporous titanosilicates seem to be no real alternatives to the crystalline materials. Thus, attempts were made to create crystalline mesoporous titanosilicates. Embedding 18 nm sized carbon black pearls in Ti-MFI crystals admixing them to the synthesis gels resulted in mesoporous crystals after removal of the carbon by calcination.[57] An TEM image of such a mesoporous crystal is shown in Fig. 3.8 (left). The thus obtained material yielded a similar activity for the epoxidation of oct-1-ene as conventional Ti-MFI. However, for the bulky cyclohexene the mesoporous Ti-MFI performed significantly better due to less restricted diffusion as the consequence of larger transport pores (mesopores).

It should be mentioned that optimization processes can significantly improve the properties of Ti-containing catalysts. Optimization of Ti-MCM-41 and Ti-MCM-48 by application of high throughput technologies resulted in highly active olefin epoxidation catalysts.[58] The optimized materials showed much higher activities than their non-optimized analogues.

Table 3.1 Epoxidation of pentenol with H_2O_2 using different titanosilicates as catalysts

Catalyst	*T* (K)	Conversion (%)	Epoxide selectivity (%)
Ti-MCM-41	323	32	19
Ti-BEA	343	42	89
Ti-MFI	323	No data	76

Reproduced from Ref. 1 by permission of Elsevier.

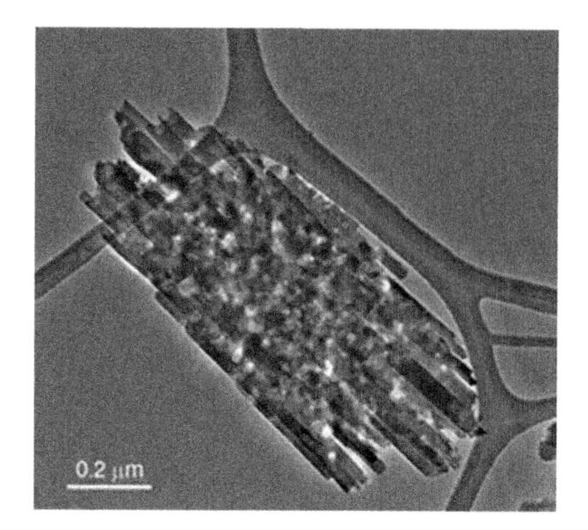

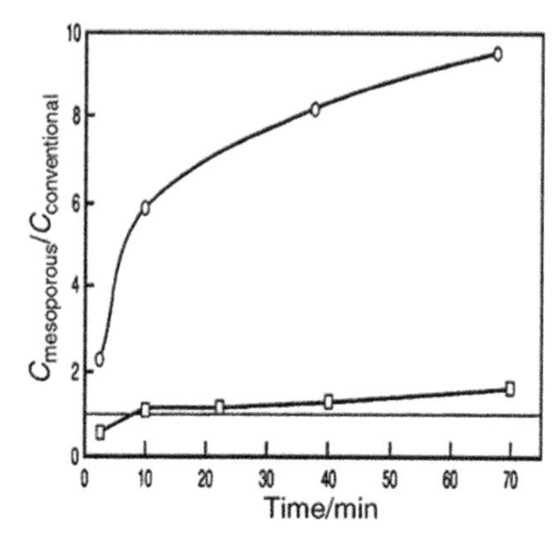

Figure 3.8 TEM image of mesoporous TS-1 (left) and a comparison of the ratios of product concentrations, that is, sum of epoxide and secondary products from oct-1-ene (□) and cyclohexene (○), obtained with conventional and mesoporous TS-1 as function of contact time. (Reproduced from Ref. 57 by permission of The Royal Society of Chemistry.)

Another interesting approach to convert a material with rather limited pore size into a catalyst with larger transport pores is delamination of crystalline zeolite precursors, such as for example, Ti-MWW. The structure of the precursor material demands specific properties, that is, the presence of layered silica sheets similar as in clay materials. One zeolite material which fulfils this requirement is non-calcined MWW zeolite. Upon calcination, Si–O–Si bonds are formed between neighbouring silica resulting in the rigid zeolite framework of MWW type, as found in MCM-22, ITQ-1 and SSZ-25.[5]

Delamination of non-calcined MCM-22 (MWW precursor) results in a material denoted as ITQ-2 in which the delaminated sheets form aggregates with larger transport pores and enhanced accessibility.[59] In the same way, a Ti-MWW precursor has been delaminated successfully and transformed in aggregates of Ti-MWW sheets, as shown schematically in Fig. 3.9.[60,61] These materials are denoted here as Del-Ti-MWW.

3D Ti-MWW has been shown to be a highly active catalyst for the epoxidation of alkenes with hydrogen peroxide.[61] Furthermore, a unique *trans*-selectivity for *cis*/*trans*-alkenes is reported for Ti-MWW catalysts.[62] Table 3.2 reflects these results. The 3D Ti-MWW catalyst shows higher conversions of 1-hexene than Ti-MFI, Ti-BEA and Ti-MCM-41 catalysts under the same conditions. The turn over numbers (TON) increased by a factor of almost 20 for Ti-MWW in comparison with Ti-MFI.

For the sample which had been delaminated under ultrasound, the 1-hexene conversion and TON were even slightly higher. 3D Ti-MWW and Del-Ti-MWW are also superior for the epoxidation of 2-hexene with H_2O_2, both conversion and TON again strongly enhanced for the delaminated samples.

The trend continues for the more bulky substrates cyclopentene and cyclododecene. The TON were again much higher for Del-Ti-MWW in comparison with the other titanosilicates. Especially for cyclododecene Del-Ti-MWW showed TON being about six times higher than 3D Ti-MWW.

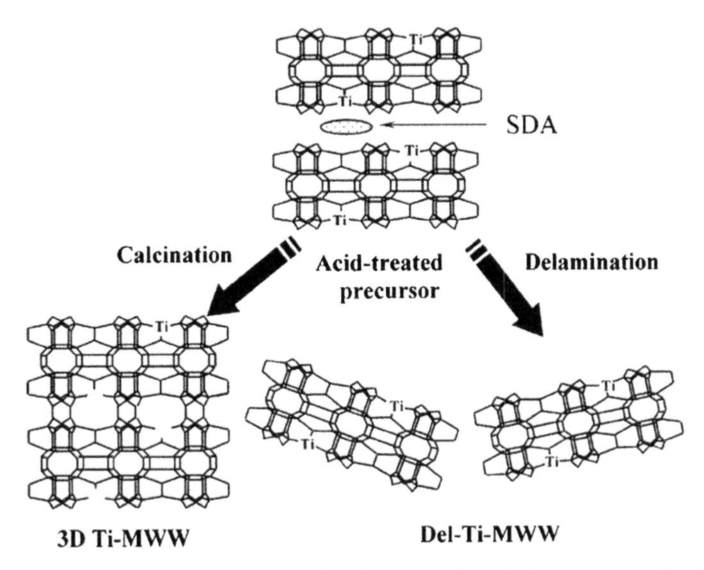

Figure 3.9 Structures of 3D Ti-MWW and delaminated Del-Ti-MWW obtained from acid-treated precursors. SDA = structure-directing agent. (From Ref. 61 with permission of American Chemical Society.)

Table 3.2 Alkene epoxidation with H_2O_2 over different titanosilicate catalysts

	1-Hexene		2-Hexene		Cyclopentene		Cyclododecene	
Catalyst	Conv.	TON	Conv.	TON	Conv.	TON	Conv.	TON
Del-Ti-MWW(a)	51.8	1390	89.5	2352	58.9	306	20.7	57
Del-Ti-MWW(b)	24.8	863	53.7	1305	34.7	163	16.4	40
3D Ti-MWW	29.3	934	40.5	1053	5.7	89	3.3	9
Ti-MFI	12.0	49	24.5	105	16.3	69	1.2	3
Ti-BEA	6.0	26	9.2	40	9.9	4	1.9	4
Ti-MCM-41	0.5	3	2.3	13	3.5	20	4.1	12

Conversions are given in mol% and TON in mol (mol Ti)$^{-1}$. Reproduced from Ref. 61 by permission of American Chemical Society.

[a] Delaminated with ultrasonic treatment and pH adjustment.

[b] Delaminated without ultrasonic treatment and pH adjustment.

It is noteworthy that Del-Ti-MWW not only showed a much higher activity than 3D Ti-MWW for the epoxidation of 2-hexene but a *trans*-selectivity which is comparable to that of 3D Ti-MWW.[61]

Other reactions in which Ti-MWW is reported to show remarkable catalytic activities even superior to that of Ti-MFI (TS-1) are the epoxidation of allyl alcohols with hydrogen peroxide and the ammoximation of cyclohexanone.[63,64] Thus, Ti-MWW is indeed a highly interesting catalyst with remarkable performance in the epoxidation of alkenes with H_2O_2. Future will show, whether this catalyst will find its way into technical applications.

3. Nanoporous Titanosilicates with $Ti^{[6]}$ Coordination

Titanosilicates with significantly different structural properties were synthesized in the laboratories of Engelhard Corp. in the late 1980s. Both materials could be crystallized from purely inorganic synthesis gels.[65,66] Addition of an organic additive, as indispensable for many zeolite syntheses, is not needed for the crystallization but addition of tetramethylammonium salts seems to enhance the crystallization rate for ETS-10.[67] In contrast to the titanosilicates considered in the previous chapter, these silicates contained titanium in octahedral coordination ($Ti^{[6]}$). Two materials became prominent because of their structural porosity. ETS-4 ($Na_9Si_{12}Ti_5O_{38}(OH){\cdot}12H_2O$) and ETS-10 ($M_2TiSi_5O_{13}{\cdot}H_2O$ with $M = Na, K$) both have microporous structures but significantly different pore sizes. ETS-10 has elliptical pores with 12-ring openings of $\sim$0.8 $\times$ 0.5 nm. ETS-4 has smaller pores with 8-ring openings of $\sim$0.35 nm. Chains of corner-sharing TiO_6 octahedra are a common feature of both structures. These titania chains are connected via SiO_4 tetrahedra. The structure of ETS-4 was solved by Rietveld refinement and showed to have topology related to that of the mineral zorite.[68,69]

Because of its small pore size, ETS-4 has not been investigated intensively as catalyst. However, it has a specific feature which makes it potentially attractive for size-selective adsorption of molecules.[70] The pore openings of ETS-4 can be contracted by dehydration. The contraction is caused by migration and reorientation of cations inside the pores. The pore size of this material thus can be tailored exactly to demands due to this gate effect. A drawback of ETS-4 is its rather low thermal stability. While ETS-10 can be heated to temperatures of about 600 °C without problems, ETS-4 decomposes basically completely at temperatures between 350 °C and 400 °C (Fig. 3.10).

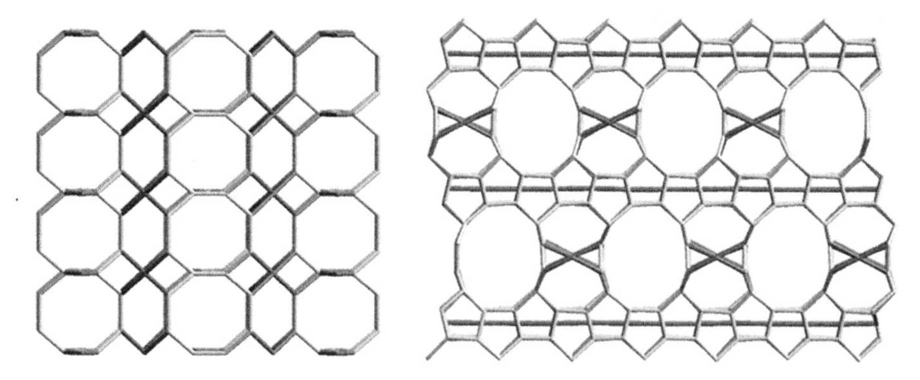

Figure 3.10 Structure frameworks of ETS-4 (left) and ETS-10 (right); oxygen atoms are omitted for clarity, that is, the lines directly connect silicon and/or titanium atoms. (See color insert.)

Solving the structure of ETS-10 from XRD data proved to be difficult because the structure is strongly disordered. The disorder is due to the formation of layers of $[TiSi_4O_{13}]$ chains. Stacking can be arbitrary from layer to layer; the structure nevertheless comprises basically of a fully connected titanosilicate framework. Stacking faults can also cause supermicropores in ETS-10, that is, missing $[TiSi_4O_{13}]$ chains form a larger pores. Because of the disorder of the layers, XRD patterns show broad signals along with very sharp reflections. Structural features of the ETS-10 structure have been solved by a combination of electron microscopy, X-ray diffraction and electron diffraction. As a working hypothesis, the structure was considered a mixture of two polymorphs, differing in the stacking of the individual titanosilicate layers.[71] The connectivity of the chains of titania octahedra and the silica tetrahedra have been correctly determined as has been shown later on by single crystal data.[72] In this work, the structure is described as a superposition of pores and rods of titania chains connected to silica tetrahedra. Selective omission of such rods in an alternating manner results in a realistic description of the ETS-10 structure.

While Si^{4+} is tetrahedrally coordinated by four O^{2-}, Ti^{4+} is octahedrally coordinated by six O^{2-} from the framework. Thus, two negative framework charges per titanium in the framework cannot be compensated by Ti^{4+} and Si^{4+}. These negative framework charges are compensated by extra-framework cations which are located in the pores of the titanosilicates. The extra-framework cations can be exchanged by other cations, similar as for zeolites. ETS-4 and ETS-10 are thus efficient ion exchangers.

Catalysis in ETS-10 and ETS-4 is significant different to that in titanosilicates with tetrahedral coordination of the titanium. The coordination is too high to achieve the catalytically active species for epoxidation with H_2O_2 which are discussed for TS-1 above. Partial leaching of titanium from the titanosilicate framework is assumed to reduce the coordination of the titanium to tetrahedral coordination to a certain extent. Ti(η^1-OOH) and/or Ti(η^2-OOH) may then form with hydrogen peroxide. As the result, slightly titanium-extracted ETS-10 shows a certain activity for the epoxidation of cyclohexene.[73] Protonation of ETS-10 results in significant structural damage and ETS-10 thus is not expected to be a very active catalyst with respect to high acidity.[74,75] Alkali cations inside the micropores cause a certain basicity of the materials and Cs-exchanged ETS-10 was used successfully as basic catalyst, for example, for the liquid-phase Knoevenagel reaction or the conversion of iso-propanol to acetone.[76,77] There are many other reports on base-catalysed reactions using alkali-modified ETS-10 as the catalyst which indicates the interest this material has created (e.g., Refs. 78,79).

Furthermore, ETS-10 has been shown to be an active photocatalyst. The O–Ti–O–Ti–O chains (titania chains) in ETS-10 are considered as semiconductor nanowires and are thus photoexcitable.[80–83] Irradiation with light excites one electron from the valence band to the conduction band where it becomes mobile and thus can be separated from the positive vacancy in the valence band. This charge separation and the successive transfer of the electron to the substrate are responsible for the photocatalytic activity of a semiconductor catalyst.[83] The intrinsic photocatalytic activity of ETS-10 seems to be somewhat lower than that of TiO_2.[84,85] However, ETS-10 is active in the photooxidative decomposition of organic compounds.[84,86] Exposed terminal ends of the isolated titania chains in ETS-10 are assumed to be the active sites for the photocatalytic reaction. In a non-defective material, such sites are basically only located at the external surface of the crystallites where the titania chains end at the crystal surface. Accordingly, it could be shown that the photodecomposition of larger 2,3-dihydroxynaphthalene molecules proceeds at the external surface of ETS-10 while smaller phenol molecules which can get adsorbed in the micropores of ETS-10 are protected from oxidation.[84] Photocatalytic conversion thus was restricted to the external surface of the catalyst and the photocatalytic activity thus indeed seems to correlate strongly with the concentration of terminal titania sites. In defective structures such sites can be also found at defects and/or stacking faults. The photocatalytic activity of ETS-10 should thus strongly benefit from an increase of defects along the titania chains.[85,87–89] This hypothesis was fostered by the observation that the photocatalytic activity of ETS-10 is indeed strongly correlated with the number of stacking faults and or otherwise induced defects in a given material. Basically defect-free ETS-10 catalyses the photopolymerization of ethene in the absence of oxygen.[81,85] In the presence of oxygen, partial oxidation to acetic acid and acetaldehyde is observed and the partially oxidized products remain strongly adsorbed in the pores of the catalyst. If the same reaction is performed on strongly defective ETS-10, the ethene is completely photooxidized to carbon dioxide. Defects thus strongly enhance the photocatalytic activity of ETS-10. To further increase this activity, structural defects can be induced by post-synthetic treatments. As for an example, exposure of ETS-10 to dilute hydrofluoric acid causes a partial dissolution of the silica around the titania chains and the resulting material shows a significantly increased activity for the photodegradation of organic molecules.[90]

Extraction of titania from the structure of ETS-10 can be also used for the generation of extended micropores and mesopores in the crystallites, important to overcome transport restrictions in narrow pores. Treatment with hydrogen peroxide solution results in extraction of [$TiSi_4O_{13}$] chains from the structure. By this process, larger micropores are created.[91] Applying microwave radiation during the hydrogen peroxide treatment results in a strongly enhanced extraction of [$TiSi_4O_{13}$] units.[92] This extraction results in the formation of mesopores and extended micropores (supermicropores). Furthermore, larger macropores are created as the result of dissolution of larger sections in the vicinity of the external crystal surfaces. It is interesting to note, that the extraction of the titania units obviously proceeds in such a manner that strongly aligned mesopores are formed, as shown in Fig. 3.11. Since these mesopores run parallel to the titania chains, that is, parallel to the micropores of the ETS-10 crystallites, one can assume that the extraction of the [$TiSi_4O_{13}$] units proceeds along the titania chains of the crystallites. The extraction of the titania chains goes along with a strongly increased number of silanol groups in the materials.

Protonated samples of this type of ETS-10 materials are active in the gas-phase Beckmann rearrangement reaction of cyclohexanone oxime to ϵ-caprolactam, a valuable educt for the production of nylon-6.[93] This reaction is catalysed by weakly acidic Bronsted sites and typically proceeds at or close to the external surface of the microporous catalyst. The creation of meso- and macropores provides both, transport pores and an increased external surface area of the ETS-10 crystallites. It could be shown that the productivity of ϵ-caprolactam strongly correlates with the size of the external surface area of the H-ETS-10 used.

Figure 3.11 Mesopores and supermicropores (inset) in ETS-10 treated with H_2O_2 under microwave irradiation. The parallel mesopores indicate extraction along [$TiSi_4O_{13}$] chains. (Reproduced from Ref. 92 by permission of The Royal Society for Chemistry.)

4. Nanoporous Titania

Most titania materials consisting of very small particles have a certain textural porosity caused by voids in between the particles. These pores are irregular but can have a very narrow pore size distribution if the particle size distribution is also narrow. However, no structural pores are present in this type of materials. Nevertheless, titania materials with structural porosity might be of high interest for applications in catalysis as well as for sensor, and solar cell technology. Nanocrystalline titanium oxide is a well-investigate semiconductor material and titanium oxide with regular porosity might be of similar interest to research.

Early attempts to synthesize mesoporous titania with surfactants, such as for example, cetylammonium bromide, as been successfully used for silica (MCM-41) failed. It was possible to synthesize surfactant/titania composites but the mesostructure collapsed during the template removal by calcination. The result was nonporous titania. However, phosphatation of the titania with phosphoric acid by impregnation of the surfactant/titania composite with phosphoric acid allowed calcination of the mesostructure with partial preservation of the regular mesoporosity.[94] The resulting solids were thus no pure titania but a mesoporous TiO_2/PO_4 material. Later on, it could be shown that the use of amphiphilic poly(alkylene oxide) block copolymers in non-aqueous solutions allowed the synthesis of a variety of mesoporous oxides including the titanium-containing materials TiO_2, $SiTiO_2$, $ZrTiO_2$ and Al_2TiO_2.[95,96] These oxides contained 2D-hexagonally ordered pores, as shown in Fig. 3.12, with pore sizes up to 14 nm and relatively thick pore walls comprising of crystalline nanoparticles. The pores are, however, not as well ordered as in SBA-type materials which is reflected in rather broad XRD reflections.

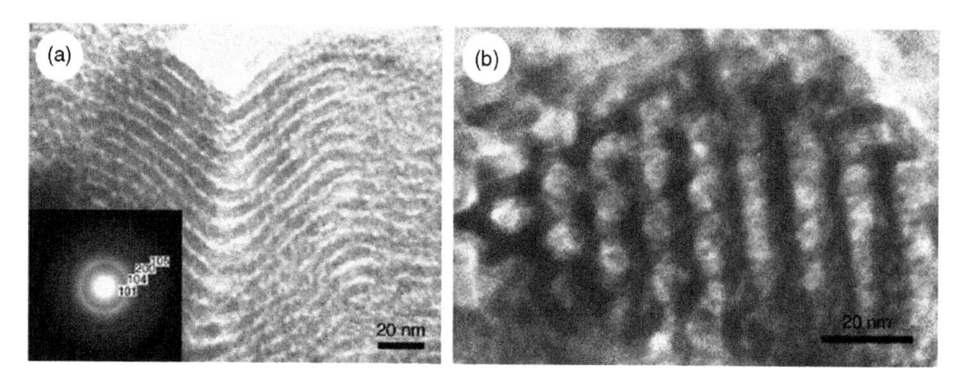

Figure 3.12 TEM image of mesoporous TiO_2 with hexagonally ordered pores. (Reprinted by permission from Macmillan Publishers Ltd. from Ref. 95, copyright 1998.)

Interestingly, the pore walls of mesoporous titania obtained by this method contained crystalline anatase nanoparticles with sizes of ~2–3 nm which is in strong contrast to SBA-15 where the pore walls are formed by entirely amorphous silica.

Condensation of titania around spherical polymer particles as solid templates can result in slightly larger anatase nanoparticles with sizes of ~5–20 nm as shown in Fig. 3.13. After removal of the polymer latex by calcination at 500 °C, inverse opals were obtained with macropores due to the regular arrangement of the latex particles and mesopores as the result of voids in between the titania nanoparticles. Thus, textural pores determine the mesopores of such materials. The mesopore size distribution of these materials was rather broad with pore sizes ranging from a few nanometres up to 160 nm in some cases.[97]

Another fascinating class of nanoporous materials is anodic oxidized titania.[98–100] This type of titanium dioxide only grows as thin film on substrates as shown in Fig. 3.14. The regularity of their pores and the degree of control with respect to tailoring pore sizes and shapes make these materials highly attractive. The oxide layer

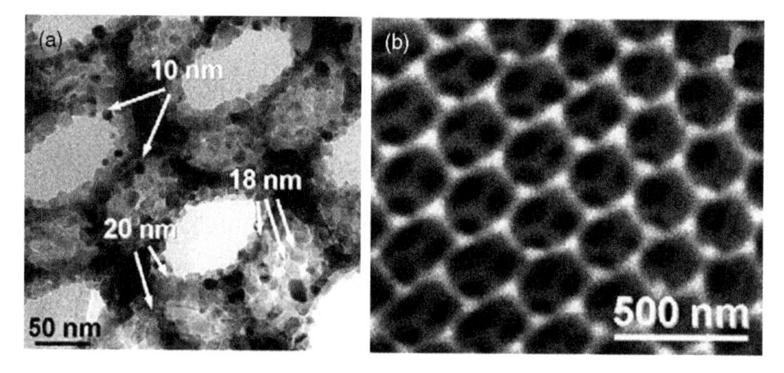

Figure 3.13 Titania particles forming inverse opals. (Reproduced from Ref. 97 by permission of Elsevier.)

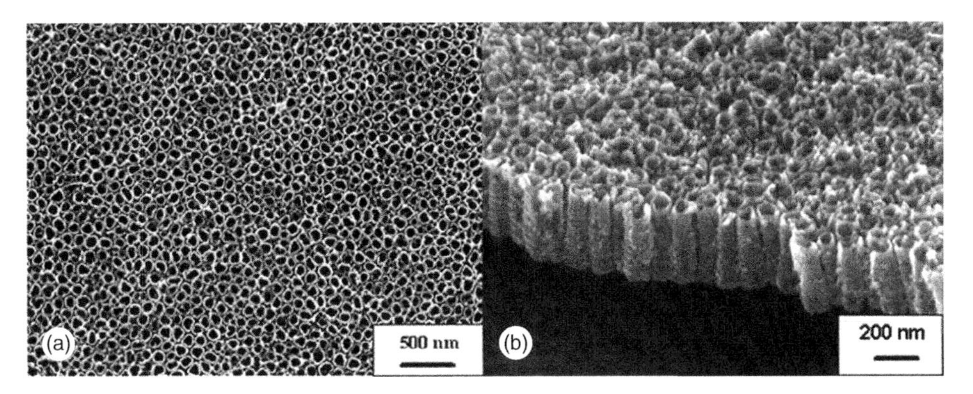

Figure 3.14 SEM image of nanoporous titanium oxide film as top view (left) and viewed from the side (right). (Reproduced from Ref. 99 by permission of Elsevier.)

grows on titanium foils in the presence of an electrolyte, typically containing HF and occasionally modifiers. The latter are molecules which allow a certain control of the structural properties during the anodization process.

Such mesoporous titania thin films gained a lot of interest in fields such as optics and sensor technology and as porous supports. Further fields in which they are investigated are solar cell technologies as well as photocatalysis and heterogeneous catalysis. It has been reported that vanadium oxide supported on mesoporous titania thin films is a highly active catalysts for the selective reduction of NO_x.[101]

5. Synthesis of Titanium-Based Nanoporous Materials

Most nanoporous materials described in the above chapters are by now well-investigated and a lot of progress has been made in the understanding of the properties of such materials. The synthesis of this type of materials is less well understood. Synthetic procedures are thus often based on empirical knowledge rather than on real understanding. A comprehensive discussion of all synthetic aspects which play a role in the formation of the different nanoporous materials would by far exceed the scope of the present chapter. Therefore, a few synthesis procedures will be described in the following to illustrate typical lab-scale preparation methods for a selection of materials.

5.1. Synthesis of Ti-MFI (TS-1)

In the following, the synthesis of TS-1 is described as reported in the original patent.[12] As for most syntheses of MFI-type crystallites, an organic structure directing agent (template) is added to the reaction mixture. The most common template for MFI materials is the TPA cation.

The chemical composition of the reaction gel can be varied but lies preferentially in the range of (0.015–0.03) TiO_2 : SiO_2 : (0.4–1) TPAOH : (60–100) H_2O. The preparation of the reaction gel is reported as follows. Fifteen grams of tetraethyltitanate are added under stirring to 455 g of tetraethylorthosilicate (TEOS) in a CO_2-free atmosphere. Then, gradually 800 g of 25 wt.% TPA hydroxide solution are added to this mixture and stirring is continued for another hour. The final mixture is heated to 80–90 °C to hydrolyse the alkoxides and to evaporate the released alcohol. After 5 h at this temperature, the volume of the mixture is increased to 1.5 litres by addition of distilled water and the homogeneous solution is reacted for 10 days at 175 °C under autogenous pressure in a stirred autoclave. The crystalline product is filtered and carefully washed with hot distilled water. After drying the product is calcined for 6 h at 550 °C. During the calcination under ambient air, template molecules which are incorporated in the pore system of the zeolite are removed by combustion.

5.2. Synthesis of Ti-BEA

Ti-containing zeolite beta can be synthesized either with of without aluminium present in the zeolite framework, and synthesis for both types of materials are described here. In zeolite syntheses, OH^- ions typically react with the silica source in a first step of the crystallization and allow the formation of soluble silica species. The pH of the respective synthesis gels is thus typically rather high. An alternative mineralizing agent are F^- ions which are also able to bring the silica source into a soluble form, typically at lower pH. In the following, two syntheses will be described for the crystallization of Ti-containing zeolite beta using either OH^- or F^- as mineralizers.[43]

(a) With F^- as mineralizer, Al-free Ti-BEA materials can be obtained from reaction gels of composition (0–2.5) TiO_2 : 25 SiO_2 : 14 TEAOH : 8.6 H_2O_2 : 189 H_2O : 14 HF. Optional crystal seeds of dealuminated zeolite beta may be added to the reaction mixture. Incorporation of aluminium in the zeolite is possible by reacting a gel of composition TiO_2 : 50 SiO_2 : 0.21 Al_2O : 28 TEAOH : 16.8 H_2O_2 : 361 H_2O : 28.4 HF.

(b) With OH^- as mineralizer, Al-free Ti-BEA materials can be obtained from reaction gels of composition (0.33–2) TiO_2 : 40 SiO_2 : 22 TEAOH : 13.5 H_2O_2 : 265 H_2O. Dealuminated zeolite beta crystals are added as seed crystals. For Al-containing material a gel of composition TiO_2 : 60 SiO_2 : 0.077 Al_2O_3 : 32.4 TEAOH : 613 H_2O can be used.

For reactions of type (a) and (b), TEOS is hydrolysed under stirring in an aqueous solution of tetraethylammonium hydroxide. Then hydrogen peroxide and tetraethylorthotitanate are added. Stirring of the mixture is continued to enable evaporation of the produced ethanol. If necessary, HF is added. Finally, the reaction gel is heated to 140 °C in a tumbling PTFE-lined autoclave. After the crystallization, the solid product can be separated by filtration or centrifugation, washed with distilled water. For removal of the organic template molecules, the material has to be calcined.

5.3. Synthesis of Del-Ti-MWW (Ti-ITQ-2)

Del-Ti-MWW can be prepared by synthesizing in a first step Del-MWW and then grafting Ti species on the surface of the thus obtained material.[60] For the synthesis of the layered silica precursor, a reaction gel of composition SiO_2 : 0.25 TMAdaOH : 0.31 HMI : 0.1 NaCl : 44 H_2O is prepared as follows: 6.164 g of hexamethyleneimine (HMI) and 1.88 g of NaCl are dissolved in 131.4 g of a 0.38 M solution of trimethyladamantammonium hydroxide (TMAdaOH). Then 37.56 g of H_2O and 12.02 g of SiO_2 (Aerosil 200) are added to the mixture. The resultant gel is stirred for 90 min and then reacted at 150 °C for 5 days. The lamellar precursor is filtered off and dried at 100 °C for 12 h.

Swelling of the lamellar material is achieved[102] by refluxing 27 g of a 20 wt.% slurry of the solid in a mixture of 105 g of an aqueous solution of 29 wt.% hexadecyltrimethylammonium bromide and 33 g of an aqueous solution of 40 wt.% TPA hydroxide for 16 h at 80 °C. One hour of ultrasonic treatment is applied to force apart the silica layers. After adjustment of the pH below 2 by addition of some drops of hydrochloric acid, the resulting solid is collected by centrifugation. Calcination of the solid at 540 °C removes the organics and the Del-MWW material is obtained.

In a consecutive step, Ti is now grafted on the surface of the thus obtained silica.[60] Ten grams of Del-MWW are dehydrated under vacuum at 300 °C over 2 h. An appropriate amount of titanocene dichloride in 90 g of $CHCl_3$ is then added to the solid and the mixture is stirred for 1 h under inert atmosphere. Triethylamine dissolved in 10 g of $CHCl_3$ (ratio $NEt3/TiCp_2Cl_2 = 1$) is then added. The solid is recovered by filtration from the yellow–orange suspension and washed with $CHCl_3$. Calcination is performed for 1 h at 540 °C in N_2 and then for 6 h in air for removal of all the organics.

5.4. Synthesis of Ti-MCM-41

Titanium can be also grafted on the surface of siliceous MCM-41, in a similar way as grafting on del-MWW is performed. However, direct synthesis of Ti-MCM-41 is also possible.[49,103] For this, 0.1 mol of TEOS (TEOS) is added to 0.001 mol of tetraisopropyl orthotitanate (TIPOT) in a mixture of 0.65 mol of ethanol and 0.1 mol of isopropyl alcohol. Under stirring, this solution is slowly added to a second solution containing 0.027 mol of dodecylamine, 3.6 mol of water and 0.002 mol of hydrochloric acid. Stirring is continued for 15 min and the solution is kept at ambient temperature. After the solid has formed, it is recovered by filtration, washed several times with distilled water and dried at room temperature. Organic molecules are removed by calcination in air at 650 °C for 6 h.

5.5. Synthesis of ETS-10

For the synthesis of ETS-10, no organic additive has to be added to the reaction mixture. It can be crystallized from an inorganic gel with a molar composition of TiO_2 : 5 SiO_2 : 3 NaOH : 0.75 KOH : 0.75 HF : 94 H_2O.[104,105] As titanium source a crystalline titanium oxide (e.g., P25, Degussa) can be used. The titanium oxide is dispersed in water, and sodium hydroxide and anhydrous potassium fluoride is added successively under stirring. After 5 min of stirring, the silica source, a colloidal silica (Ludox AS 40, Aldrich), is added and the mixture is then stirred for another 30 min. The resulting gel is transferred into Teflon-lined autoclaves and reacted at 200 °C under hydrothermal pressure for 7–10 days. The resulting solid can be separated by filtration or centrifugation and is then washed with de-ionized or distilled water and finally dried at 80–90 °C under ambient atmosphere.

5.6. Synthesis of TiO_2

In the following, a synthesis of nanoporous TiO_2 with SBA-15-type structure is described.[95] Block copolymers are used as structure-directing agents. They form micellar structures similar to those known from surfactants.

One gram of $HO(CH_2CH_2O)_{20}(CH_2CH(CH_3)O)_{70}(CH_2CH_2O)_{20}H$ block-copolymer ($EO_{20}PO_{70}EO_{20}$; Pluronic P-123, BASF) is dissolved in 10 g of ethanol. $TiCl_4$ (0.01 mol) is then added under vigorous stirring for 30 min. The resulting sol is then allowed to react at 40 °C in air for 1–7 days in an open Petri dish. After the solid is formed, the product is separated by filtration and washed with distilled water. Drying and calcination results in the final mesoporous TiO_2, as shown in Fig. 3.12.

6. Summary and Outlook

Titanium-based nanoporous materials have fascinating properties and are object of research in many different fields. The types of materials range from titanosilicates to pure titania and the types of pores found in these materials cover the whole range of pore types, from micropores (<2 nm) over mesoporous (2–50 nm) to macropores (>50 nm). These nanoporous materials are investigated with respect to many different characteristic properties, including catalytic activity, semiconductor properties and others, and they are discussed as potential candidates for various applications. However, applications of nanoporous Ti-based materials on larger scale are rather limited at present. The application of the titanosilicate TS-1 as epoxidation catalyst is presently probably the most prominent industrial process which makes use of such materials. However, as shown above, there are many different other titanium-containing porous materials with unique properties. The degree of control on the materials properties which is already achieved in the preparation of nanoporous titanium-containing solids is already impressive but it can be expected that knowledge and skills will further increase. Thus, the field is wide open and novel applications will emerge taking advantage of the high degree of control on the properties of titanium-based nanoporous materials. Research has been focussed mainly on classical fields of applications, especially of zeolitic materials, but there seems to exist no reason why these should be the only areas where such materials can be successfully applied. Ti-based materials have properties similar to zeolites (e.g., catalytic activity) but also quite different ones (e.g., semiconductor properties). Thus, novel applications may emerge in fields where zeolites are hardly of any use. Future will tell whether one or the other new material will make it into application. Meanwhile, the creation of novel Ti-based materials is a fascinating venture which surely will keep the scientific community rather busy during the years to come.

REFERENCES

[1] Ratnasamy, P., Srivinas, D., Knötzinger, H., *Adv. Catal.* **2004,** *48,* 1–169.
[2] Notari, B., *Adv. Catal.* **1996,** *41,* 253–334.
[3] Weitkamp, J., Puppe, L. (Eds.), *Catalysis and Zeolites, Fundamentals and Applications,* Springer, Berlin Heidelberg, 1999.
[4] Ertl, G., Knötzinger, H., Weitkamp, J. (Eds.), *Handbook of Heterogeneous Catalysis,* Wiley-VCH, Weinheim, 1997.
[5] Baerlocher, Ch., McCusker, L. B., Olson, D. H., *Atlas of Zeolite Framework Types*, 6th edition, Elsevier, Amsterdam, 2007.
[6] Shannon, R. D., *Acta Crystallogr.* **1976,** *A32,* 751–767.
[7] Thangaraj, A., Kumar, R., Mirajkar, S. P., Ratnasamy, P., *J. Catal.* **1991,** *130,* 1–8.
[8] Millini, R., Massara, E. P., Perego, G., Bellussi, G., *J. Catal.* **1992,** *137,* 497–503.
[9] Kleinsorge, M., *PhD thesis*, Ruhr-Universität Bochum, 2000.
[10] Höft, E., Kosslick, H., Fricke, R., Hamann, H. J., *J. Prakt. Chem.* **1996,** *338,* 1–15.
[11] Zecchina, A., Spoto, G., Bordiga, S., Ferrero, A., Petrini, G., Leofanti, G., Padovan, M., *Stud. Surf. Sci. Catal.* **1991,** *69,* 251–258.
[12] Taramasso, M., Perego, G., Notari, B., Patent US 4410501, 1983.
[13] Clerici, M. G., *Oil Gas—Eur. Mag.* **2006,** *2,* OG77–OG82.
[14] Clerici, M. G., Bellussi, G., Romano, U., *J. Catal.* **1991,** *129,* 159–167.
[15] Liu, X., Wang, X., Guo, X., Li, G., *Catal. Today* **2004,** *93–95,* 505–509.
[16] Clerici, M. C., Inagallina, P., *J. Catal.* **1993,** *140,* 71–83.
[17] Clerici, M. G., *Top. Catal.* **2001,** *15,* 257–263.
[18] Prestipino, C., Bonino, F., Usseglio, S., Damin, A., Tasso, A., Clerici, M., Bordiga, S., D'Acapito, F., Zecchina, A., Lamberti, C., *Chem. Phys. Chem.* **2004,** *5,* 1799–1804.
[19] Corà, F., Alfredsson, M., Barker, C. M., Bell, R. G., Foster, M. D., Saadoune, I., Simperler, A., Catlow, C. R. A., *J. Solid State Chem.* **2003,** *176,* 496–529.
[20] Thomas, J. M., Catlow, C. R., Sankar, G., *Chem. Commun.* **2002,** 2921–2925.
[21] Catlow, C. R. A., French, S. A., Sokol, A. A., Thomas, J. M., *Phil. Trans. R. Soc. A* **2005,** *363,* 913–936.
[22] Barker, C. M., Kaltsoyannis, N., Catlow, C. R. A., *Stud. Surf. Sci. Catal.* **2001,** *135,* 15–18.
[23] Sheldon, R. A., *Chemtech* **1991,** 566–576.
[24] Sheldon, A., *Top. Curr. Chem.* **1993,** *164,* 21–43.
[25] Bordiga, S., Damin, A., Bonino, F., Lamberti, C., *Top. Organomet. Chem.* **2005,** *16,* 37–68.
[26] Meiers, R., Dingerdissen, U., Hölderich, W. F., *J. Catal.* **1998,** *176,* 376–386.
[27] Jenzer, G., Mallat, T., Maciejewski, M., Eigenmann, F., Baiker, A., *Appl. Catal. A* **2001,** *208,* 125–133.
[28] Clerici, M. G., Ingallina, P., *Catal. Today* **1998,** *41,* 351–364.
[29] Wang, C., Wang, B., Meng, X., Mi, Z., *Catal. Today* **2002,** *74,* 15–21.
[30] Kane, L., Romanov, S., *Hydrocarb. Proc.,* Nov **2000,** 36.
[31] Izumi, Y., Ichihashi, H., Shimazu, Y., Kitamura, M., Sato, H., *Bull. Chem. Soc. Jpn.* **2007,** *80,* 1280–1287.
[32] *China Petroleum & Chemical Corporation*, Annual Report, 2003.
[33] Zhan, X., Wang, Y., Xin, F., *Appl. Catal. A* **2006,** *307,* 222–230.
[34] *Chem. Eng. News,* Sep 6 **2004,** *82,* 15.
[35] *Chem. Week,* Sep 8 **2004,** *166,* 14.
[36] *Press release Uhde GmbH,* May 10, 2006.

[37] *Elements, Degussa Science Newsletter* **2006,** *17,* 4–7.
[38] Bender, M., Zehner, P., Machhammer, O., Mueller, U., Harth, K., Schinder, G.-P., Junicke, H., Patent WO 04020423A1, **2004**.
[39] Haas, T., Brasse, C., Woll, W., Hofen, W., Jaeger, B., Stochniol, G., Ullrich, D., US 6878836, **2004**.
[40] Kraushaar, B., van Hooff, J. H. C., *Catal. Lett.* **1988,** *1,* 81–84.
[41] Bellussi, G., Carati, A., Clerici, M. G., Esposito, A., Millini, R., Buonomo, F., Belgium Patent. 1001038, **1989**.
[42] Reddy, J. S., Kumar, R., Ratnasamy, P., *Appl. Catal.* **1990,** *58,* L1–L4.
[43] Blasco, T., Camblor, M. A., Corma, A., Esteve, P., Guil, J. M., Martínez, A., Perdigón-Melón, J. A., Valencia, S., *J. Phys. Chem. B* **1998,** *102,* 75–88.
[44] Kim, G. J., Cho, B. R., Kim, J. H., *Catal. Lett.* **1993,** *22,* 259–270.
[45] Wu, P., Tatsumi, T., Komatsu, T., Yashima, T., *J. Phys. Chem. B* **2001,** *105,* 2897–2905.
[46] Beck, J. S., Chu, C. T. W., Johnson, I. D., Kresge, C. T., Leonowicz, M. E., Roth, W. J., Vartuli, J. C., Patent WO 9111390, **1991**.
[47] Kresge, C. T., Leonowicz, M. E., Roth, W. J., Vartuli, J. C., Beck, J. S., *Nature* **1992,** *359,* 710–712.
[48] Zhao, D., Huo, Q., Feng, J., Chmelka, B. F., Stucky, G. D., *J. Am. Chem. Soc.* **1998,** *120,* 6024–6036.
[49] Tanev, P. T., Cibwe, M., Pinnavaia, T. J., *Nature* **1994,** *368,* 321–323.
[50] Corma, A., Navarro, M. T., Pérez-Pariente, J., Sánchez, F., *Stud. Surf. Sci. Catal.* **1994,** *84,* 69–75.
[51] Corma, A., Navarro, M. T., Pérez Pariente, J., *Chem. Soc. Chem. Commun.* **1994,** 147–148.
[52] Maschmeyer, T., Rey, F., Sankar, G., Sankar, J. M., *Nature* **1995,** *378,* 159–162.
[53] Thomas, J. M., Catlow, C. R. A., Sankar, G., *Chem. Commun.* **2002,** 2921–2925.
[54] Koyano, K. A., Tatsumi, T., *Micropor. Mater.* **1997,** *10,* 259–271.
[55] Chen, L. Y., Chuah, G. K., Jaenicke, S., *Catal. Lett.* **1998,** *50,* 107–114.
[56] Corma, A., Camblor, M. A., Esteve, P., Martínez, A., Pérez Pariente, J., *J. Catal.* **1994,** *145,* 151–158.
[57] Schmidt, I., Krogh, A., Wienberg, K., Carlsson, A., Brorson, M., Jacobsen, C. J. H., *Chem. Commun.* **2000,** 2157–2158.
[58] Corma, A., Serra, J. M., Serra, P., Valero, S., Argente, E., Botti, V., *J. Catal.* **2005,** *229,* 513–524.
[59] Corma, A., Fornes, V., Pergher, S. B., Maesen, T. L. M., Buglass, J. G., *Nature* **1998,** *396,* 353–356.
[60] Corma, A., Díaz, U., Fornés, V., Jordá, J. L., Domine, M., Rey, F., *Chem. Commun.* **1999,** 779–780.
[61] Wu, P., Nuntasri, D., Ruan, J., Liu, Y., He, M., Fan, W., Terasaki, O., Tatsumi, T., *J. Phys. Chem. B* **2004,** *108,* 19126–19131.
[62] Wu, P., Tatsumi, T., *J. Phys. Chem. B* **2002,** *106,* 748–753.
[63] Wu, P., Tatsumi, T., *J. Catal.* **2003,** *214,* 317–326.
[64] Song, F., Liu, Y., Wu, H., He, M., Wu, P., Tatsumi, T., *J. Catal.* **2006,** *237,* 359–367.
[65] Kuznicki, S. M., Patent US 4853202, **1989**.
[66] Kuznicki, S. M., Patent US 4938939, **1990**.
[67] Valtchev, V., Mintova, S., *Zeolites* **1994,** *14,* 697–700.
[68] Cruciani, G., De Luca, P., Nastro, A., Pattison, P., *Micropor. Mesopor. Mater.* **1998,** *21,* 143–153.
[69] Philippou, A., Anderson, M. W., *Zeolites* **1996,** *16,* 98–107.
[70] Kuznicki, S. M., Bell, V. A., Nair, S., Hillhouse, H. W., Jacubinas, R. M., Braunbarth, C. M., Toby, B. H., Tsapatsis, M., *Nature* **2001,** *412,* 720–724.
[71] Anderson, M. W., Terasaki, O., Ohsuna, T., Malley, P. J. O., Philippou, A., MacKay, S. P., Ferreira, A., Rocha, J., Lidin, S., *Phil. Mag. B* **1995,** *71,* 813–841.
[72] Wang, X., Jacobson, A. J., *Chem. Commun.* **1999,** 973–974.
[73] Goa, Y., Yoshitake, H., Wu, P., Tatsumi, T., *Micropor. Mesopor. Mater.* **2004,** *70,* 93–101.
[74] Krisnandi, Y. K., Lachowski, E. E., Howe, R. F., *Chem. Mater.* **2006,** *18,* 928–933.

[75] Pavel, C. C., Zibrowius, B., Löffler, E., Schmidt, W., *Phys. Chem. Chem. Phys.* **2007,** *9,* 3440–3446.
[76] Goa, Y., Wu, P., Tatsumi, T., *J. Catal.* **2004,** *224,* 107–114.
[77] Philippou, A., Rocha, J., Anderson, M. W., *Catal. Lett.* **1999,** *57,* 151–153.
[78] Philippou, A., Anderson, M. W., *J. Catal.* **2000,** *189,* 395–400.
[79] Waghmode, S. B., Thakur, V. V., Sudalai, A., Sivasanker, S., *Tetrahedron Lett.* **2001,** *42,* 3145–3147.
[80] Lamberti, C., *Micropor. Mesopor. Mater.* **1999,** *30,* 155–163.
[81] Howe, R. F., Krisnandi, Y. K., *Chem. Commun.* **2001,** 1588–1589.
[82] Fox, M. A., Doan, K., Dulay, M. T., *Res. Chem. Intermed.* **1994,** *20,* 711–722.
[83] Corma, A., Garcia, H., *Chem. Commun.* **2004,** 1443–1459.
[84] Calza, P., Pazè, C., Pelizzeti, E., Zecchina, A., *Chem. Commun.* **2001,** 2130–2131.
[85] Krisnandi, Y. K., Southon, P. D., Adesina, A. A., Howe, R. F., *Int. J. Photoenerg.* **2003,** *5,* 131–140.
[86] Uma, S., Rodrigues, S., Martyanov, I. N., Klabunde, K. J., *Micropor. Mesopor. Mater.* **2004,** *67,* 181–187.
[87] Krisnandi, Y. K., Howe, R. F., *Appl. Catal. A* **2006,** *307,* 62–69.
[88] Uma, S., Rodrigues, S., Martyanov, I. N., Klabunde, K. J., *Micropor. Mesopor. Mater.* **2004,** *67,* 181–187.
[89] Krisnandi, Y. K., Southon, P. D., Adesina, A. A., Howe, R. F., *Int. J. Photoenerg.* **2003,** *5,* 131–140.
[90] Llabré i Xamena, F. X., Calza, P., Lamberti, C., Prestipino, C., Damin, A., Bordiga, S., Pelizzetti, E., Zecchina, A., *J. Am. Chem. Soc.* **2003,** *125,* 2264–2271.
[91] Pavel, C. C., Park, S.-H., Dreier, A., Tesche, B., Schmidt, W., *Chem. Mater.* **2006,** *18,* 3813–3820.
[92] Pavel, C. C., Schmidt, W., *Chem. Commun.* **2006,** 882–884.
[93] Pavel, C. C., Palkovits, R., Schüth, F., Schmidt, W., *J. Catal.* **2008,** *254,* 84–90.
[94] Blanchard, J., Schüth, F., Trens, P., Hudson, M., *Micropor. Mesopor. Mater.* **2000,** *39,* 163–170.
[95] Yang, P., Zhao, D., Margolese, D. I., Chmelka, B. F., Stucky, G. D., *Nature* **1998,** *396,* 152–155.
[96] Yang, P., Zhao, D., Margolese, D. I., Chmelka, B. F., Stucky, G. D., *Chem. Mater.* **1999,** *11,* 2813–2826.
[97] Carbjo, M. C., Ensio, E., Torralvo, M. J., *Colloid. Surf. A* **2007,** *293,* 72–79.
[98] de Tacconi, N. R., Chenthamarakshan, C. R., Yogeeswaran, G., Watcharenwong, A., de Zoysa, R. S., Basit, N. A., Rajeshwar, K., *J. Phys. Chem. B* **2006,** *110,* 25347–25355.
[99] Wei, X., *J. Cryst. Growth* **2006,** *286,* 371–375.
[100] Yu, X., Li, Y., Ge, W., Yang, Q., Zhu, N., Kalantar-zadeh, K., *Nanotechnology* **2006,** *17,* 808–814.
[101] Segura, Y., Chmielarz, L., Kustrowski, P., Cool, P., Dziembaj, R., Vansant, E. F., *J. Phys. Chem. B* **2006,** *110,* 948–955.
[102] Corma, A., Fornés, V., Pergher, S. B., Maesen, T. L. M., Buglass, J. G., *Nature* **1998,** *396,* 353–356.
[103] Gontier, S., Tuel, A., *Zeolites* **1995,** *15,* 601–610.
[104] Zibrowius, B., Weidenthaler, C., Schmidt, W., *Phys. Chem. Chem. Phys.* **2003,** *5,* 773–777.
[105] Liu, X., Thomas, J. K., *Chem. Commun.* **1996,** 1435–1436.

CHAPTER 4

Porous Metal Organic Frameworks: From Synthesis to Applications

Thomas Devic *and* Christian Serre

Contents

Ordered Porous Solids
DOI: 10.1016/B978-0-444-53189-6.00004-4

Abstract

This book chapter deals with a concise description of the porous Metal Organic Frameworks (MOFs), the youngest class of porous solids reported so far. After a brief analysis of their chemical composition, their key synthesis and structural aspects are given including some of their unusual flexible behaviours. Finally, a series of short reviews concerning most of their specific properties are proposed from gas storage, separation, catalysis, inclusion, physical properties and thin films to bioapplications.

Keywords: Adsorption, Catalysis, Drug Delivery, Inclusion, Magnetism, Metal Organic Frameworks, Optical Properties, Porous Solids, Separation, Synthesis, Thin Films

1. Introduction

Porous solids are important materials from an economic point of view as they are related to many industrial applications (catalysis, separation, fine chemistry, etc.). So far, research and applications have been focused on the zeolites, metalophosphates, activated carbons or mesoporous solids. It is only recently that attention has been driven to a new class of porous solids, that is, hybrid inorganic–organic compounds, denoted MOFs (Metal Organic Frameworks) or PCPs (Porous Coordination Polymers).[1–3] Numerous reviews dealing with their synthesis and properties have been published in the last decade.[4] We here focus only on few major examples of such solids, rather than report an exhaustive list of published MOFs. Moreover, the area of MOFs and computer simulation (ranging from structural elucidation, stability to gas uptake or molecule diffusion) will not be covered in this chapter. The reader interested in the field could refer to some recent references.[5,6]

The MOFs are built up of inorganic sub-units (clusters, chains, layers or 3D arrangement) connected to organic linkers possessing complexing groups (carboxylates, phosphonates, N-containing compounds) by strong ionocovalent or dative bonds (Fig. 4.1). This results in 3D hybrid networks where both inorganic and organic moieties are present. Almost all elements of the periodic table, from alkaline earths (Ca, Mg) to transition metals (Sc, Ti, . . ., Zn), *p* metals (Al, Ga, In) and lanthanides can be involved in the formation of one or several MOF structures. A huge number of organic linkers have been used so far including various organic spacers, aliphatics or aromatics, sometimes substituted by heteroelements (N, O, S, . . .), associated with one or several complexing functions (either anionic or neutral) such as carboxylates, phosphonates, sulphonates, imidazolates, amides, amines, pyridyls, nitriles groups or a combination of them. The number of MOFs reported in the literature is now increasing steadily due to the richness of organic chemistry associated with the diversity of the chemistry of metal elements of the periodic table. This results in almost one new MOF has been synthesized every week with various pore sizes, shapes and organic functionalities. Cavities or channels of MOFs are occupied by solvent molecules or free molecules of linker which are easily removed by heating and/or vacuum. Key advantages of MOFs are their

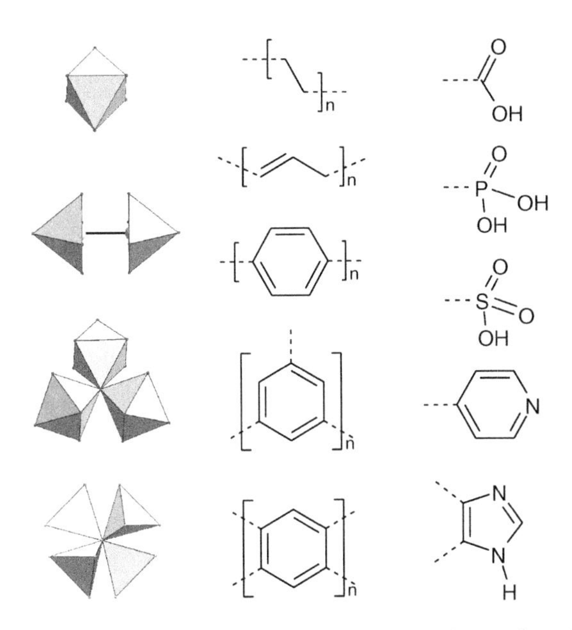

Figure 4.1 Typical inorganic sub-units, organic linkers and complexing functions encountered in MOFs.

low density, typically between 0.2 and 1 g cm^{-3}, their high BET surface areas, up to 4500 m^2 g^{-1} and large pore volume (up to 2 cm^3 g^{-1}).

MOFs are usually built up from dimers, trimers, tetramers or chains of polyhedra. Topical examples are the copper trimesate HKUST-1 and the zinc terephthalate MOF-5, resulting in large pore sizes and BET surface areas of 1800 and 3800 m^2 g^{-1}, respectively (Fig. 4.2).[7,8] Another interesting aspect of MOFs concerns the principle of isoreticular chemistry. This consists of increasing the size of the linker, whilst keeping the same structure type and, therefore, enlarging the pore size. This pioneering work was initiated by Clearfield *et al.* and then extended to metal carboxylates systems. Nice examples are the IRMOFs and the MIL-88 solids built up from either tetrameric zinc or trimeric iron sub-units and dicarboxylate linkers.[25,9]

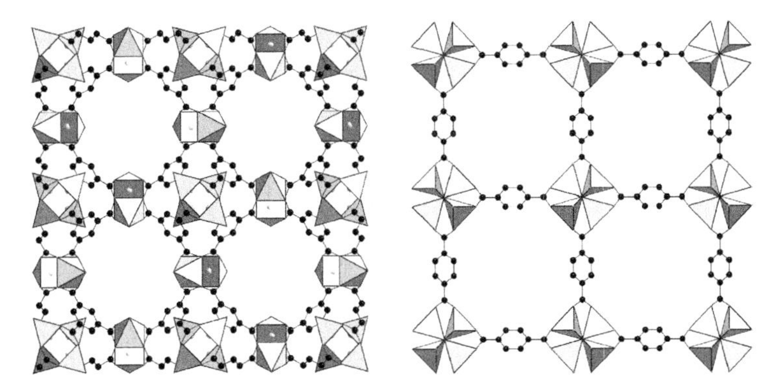

Figure 4.2 View of the structures of HKUST-1 and MOF-5.

Another important feature of MOFs is the possibility to modify the linker with various organic groups (apolar, polar, basic or acidic) whilst keeping the same structure type, in order to modify their properties. By taking advantage of the well-defined coordination geometries of metal centres as nodes, the structures of various minerals, such as quartz diamond,[10] perovskite,[11] rutile,[12] PtS[13] and feldspar,[14] with specific functionalities, have been artificially reproduced as MOF structures by replacing monoatomic anions (O^{2-}, S^{2-}) with polyatomic organic bridging ligands as linkers.

Other nice examples are the MOFs that possess a zeotype architecture (SOD, ANA, RHO, BCT, MTN, ...),[15–17] sometimes denoted Z-MOFs (Fig. 4.3).[17] These MOFs, which can be mesoporous, exhibit, unlike mesoporous silica, a crystalline framework.

Interesting examples are also those with the cubic MTN structure type[18] such as (Fig. 4.4) a cadmium amino solid,[19] chromium or iron carboxylates MIL-100 and MIL-101[15] and a terbium carboxylate,[16] which possess both one and 3D array of

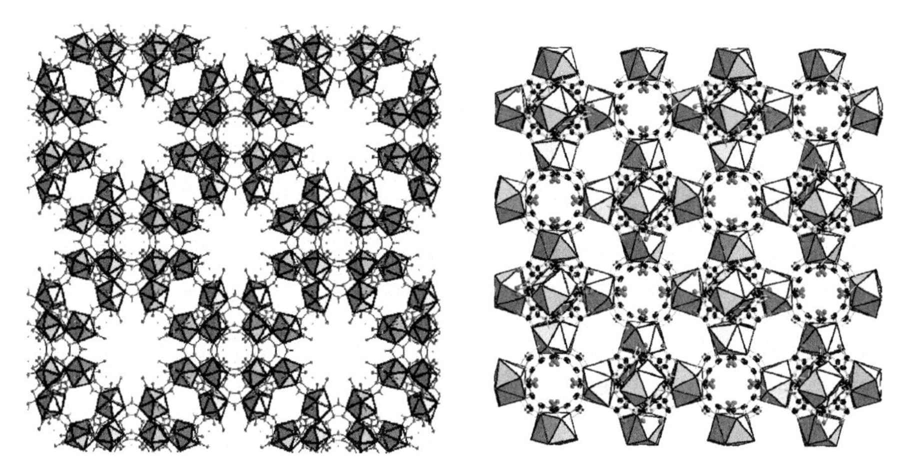

Figure 4.3 View of the structures of the ZMOFs with a sodalite and a zeolite-rho topology. (See color insert.)

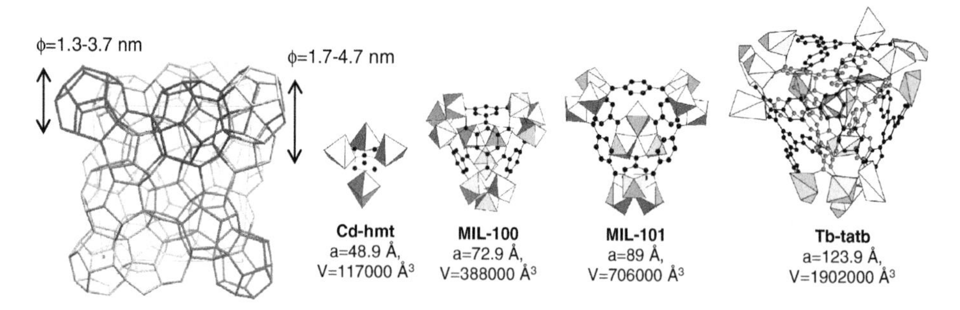

Figure 4.4 MOFs with a zeotype MTN architecture (left); hybrid supertetrahedra (right). (See color insert.)

cages of free diameter between 1.2 and 4.7 nm, accessible through pentagonal and hexagonal microporous windows. The small original SiO_4 tetrahedron from the initial ZSM-39 zeolite structure has been now replaced by larger hybrid supertetrahedra built up either from monomeric, trimeric or tetrameric metal polyhedra units and di- or tri-topic linkers. This results in very large BET surface areas (1000–4500 $m^2\ g^{-1}$) and high pore volumes (0.6–2.0 g cm^{-3}).

Generally, the thermal stability of MOFs is lower than that of the pure inorganic porous solids, typically between 200 and 400 °C under air. It is also commonly admitted that most MOFs are not very stable in water or in the presence of air moisture which may restrict their domains of applications.

2. Flexible MOFs

Crystalline porous solids are usually rigid, with small changes in volume reflecting the content of their pores. If most MOFs behave like their inorganic counterparts with rigid frameworks, an unusual feature of some MOFs is their flexibility or ability to swell or breathe under external stimuli (pressure, temperature, gas or solvent adsorption). Recently, Kitagawa *et al.* have interpretated the flexibility of MOFs by the role of their host–guest interactions.[20] This led to a classification of the known behaviours of breathing MOFs into six classes of different types of flexibility, according to the nature of interactions and the dimensionality of the inorganic subnetwork.[4d] This was later extended to other flexible MOFs that exhibit larger swelling amplitudes. These solids are built up from dicarboxylates linkers and either chains of corner-sharing octahedra, denoted MIL-53(Al, Cr),[21] or trimers of octahedra, denoted MIL-88.[9b,22] If MIL-53 exhibits a reversible breathing amplitude of 50% (Fig. 4.5) between the hydrated form and the dried one, the most spectacular example concerns the isoreticular class of flexible phases denoted MIL-88A-D (each letter corresponding to a different linear dicarboxylate), where each solid swell selectively in the presence of liquids with variations of volume between the open forms, filled with solvent and the dried forms up to 230%.

Such large and selective variations in volume might lead in a near future to new applications in terms of adsorption, storage and facile delivery of gases or biological molecules. The first use of flexible MOFs in gas or vapour sorption has shown that steps occur at various positions depending on the interaction between the adsorbed

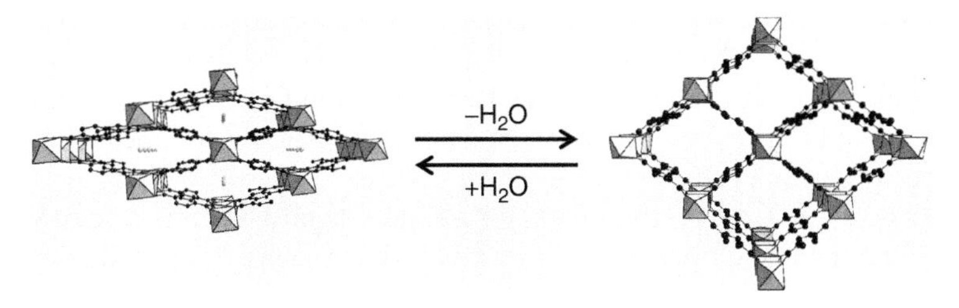

Figure 4.5 The breathing behaviour of MIL-53(Al, Cr).

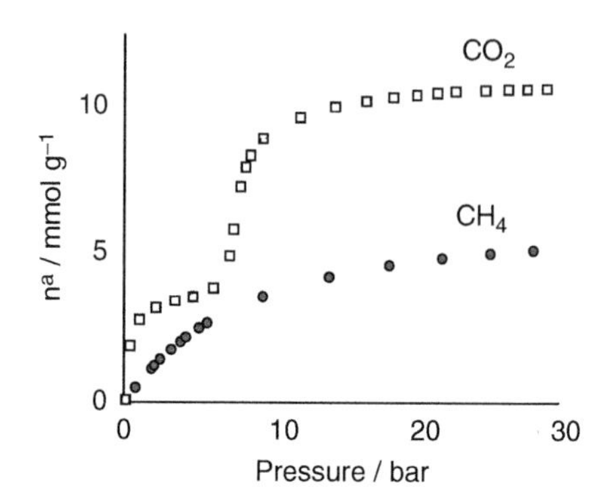

Figure 4.6 Methane and carbon dioxide adsorption isotherms in MIL-53(Cr) at 303 K.

species and the framework,[23] as shown in Fig. 4.6 during the adsorption of greenhouse gases in MIL-53.[24] If the non polar methane gas induces no change in the adsorption isotherm, the quadrupolar carbon dioxide molecules leads to a step in the isotherm, which is due to a breathing of the structure.

3. Synthesis

MOFs are usually produced by solution synthesis, starting from a metallic precursor and an organic polytopic ligand (or a mixture of linkers), between room temperature and 250 °C, either at room pressure in an open reactor or under autogeneous pressure in a closed vessel. Reaction times are comprised between a few minutes and a few days.

3.1. The reactants

The metal source is usually a salt (halide, nitrate, perchlorate or hydroxide) but can also be a metal which is further oxidized in the reaction vessel (if necessary upon addition of an oxidizing agent such as nitric or perchloric acid). In these cases, the inorganic sub-unit will form *in situ* during the reaction depending on the experimental conditions used. This is, for example, the method used to produce the archetypical IRMOF series from rigid and linear dicarboxylic linker and zinc nitrate source.[25] Alternatively, pre-built inorganic SBUs, such as monocarboxylate metallic clusters, can be used. The monocarboxylate linkers will be exchanged by polycarboxylate linkers, in order to produce the corresponding polymeric solid. This method was used to produce the MIL-88 series (see above) from linear dicarboxylic linkers and trimeric $Fe_3O(OAc)_6Cl$ units.[9]

In the case of a polycarboxylate linker, the ligand source is generally the corresponding polyacid, but other precursors have been proposed such as the ester, anhydride or nitrile form which can directly react with the metallic precursor to form the corresponding MOF. Compared with the parent carboxylic acid, a difference in solubility and/or rate of hydrolysis will directly affect the crystallinity of the resulting solid, and can even lead to new crystalline phases. Other organic reactions may also take place during the synthesis and lead to new linkers, as recently reviewed.[26,27] Physical properties of a MOF can be tuned through the use of functionalized linkers: the polarity and/or acidic–basic properties of the surface of the pores is modified through the introduction of chemical groups (halogen, alkyl, amine, etc.) on the linkers. Chiral linkers will lead to chiral networks;[28] whilst radical linkers will lead to magnetic materials.[29]

The solvent must solubilize, at least partially, both the organic and the inorganic precursors, and should be stable under the reaction conditions. Water is of course the solvent of choice, but amides (dimethyl- and diethyl-formamide) and alcohols (methanol, ethanol, propanol) or their mixtures are commonly used. Solvent molecules are generally occluded in the pores of the resulting MOFs (either free or coordinated), and may direct the synthesis; for example, a chiral dialcohol was thus used to prepare a homochiral MOF from achiral linkers.[30]

Like the case of zeolites or metallophosphates, the addition of hydrofluoric acid was shown to sometimes increase the crystallinity of the resulting materials, even when the fluoride anions are not finally introduced in the frameworks. Nevertheless, the effect of the addition of hydrofluoric acid in the synthesis of such hybrid solids was not systematically studied.

3.2. Temperature, pressure and pH

Synthesis of MOFs can be performed under room pressure in an open vessel or under autogeneous pressure in a closed vessel. The later technique is not only dedicated to hydro- or solvo-thermal conditions (i.e., above the boiling temperature of the solvent), but also give good results in sub-solvo-thermal conditions.

Few systematic studies of the influence of the temperature and pH[31,32] have been published, highlighting that higher pH and temperature lead to more condensed phases. In the case of the cobalt(II)/succinate system in water, the dimensionality of both the hybrid network and the inorganic part (M–O–M) increase with temperature (Fig. 4.7).[33]

More recently, high-throughput techniques were applied to the synthesis of MOFs which reduces the time-consuming process of phase discovery and put into light the high versatility of such systems. A semi automated system (reagents

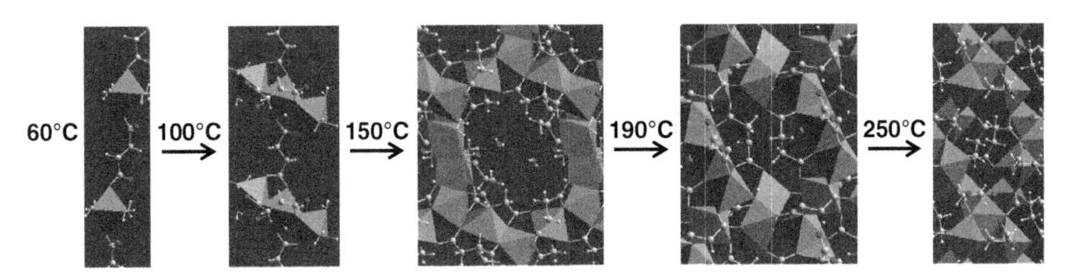

Figure 4.7 Progression of the five phases of cobalt succinate from low to high temperature. (See color insert.)

weighting, products recovering and characterization) was proposed by Stock *et al.* to perform 48 syntheses on a small scale (volume of the container = 0.25 ml), simultaneously, systematically varying the reaction parameters (stoichiometry, concentration, pH, . . .).[34] This system was used to screen the domain of existence of various complex systems such as the afore-mentioned cobalt(II)/succinate system (seven phases),[35] and a cadmium(II)/carboxyphosphonate system (five phases) using a temperature gradient reactor set-up (Fig. 4.8).[36]

The synthesis of MOFs in solution remains the most common method, some new methodologies have recently appeared and a brief overview of those methods is given in the following pages.

3.3. Heterogeneous reaction medium

Homogeneous reaction media, usually encountered in MOFs synthesis, can be changed to heterogeneous, mainly in order to tune the size of the crystallites, but also to induce new solids.

A biphasic mixture of solvent (water and cyclohexanol) was used by Cheetham *et al.* to produce a new copper adipate. Each non-miscible solvent solubilizes one of the reactants (metallic salt in water, carboxylic acid in alcohol) and the crystallization of the solid takes place at the interface of the two liquids.[37] This method was applied later to another system (terbium benzenetrisbenzoate) with success.[38]

The addition of a polymeric additive to the reaction mixture has been proposed, in order to tune the crystal size and morphology by controlling the rate of diffusion,[39] but also to afford new solids by changing the polarity of the polymer.[40]

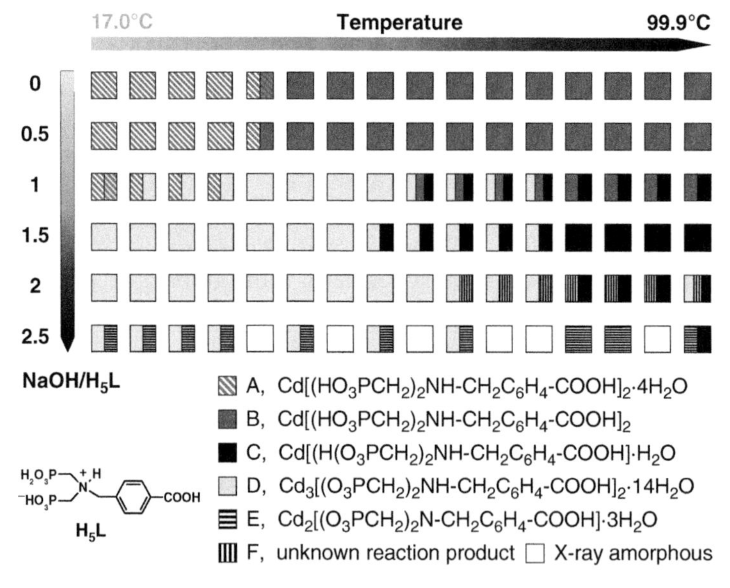

Figure 4.8 Phase diagram of a cadmium(II) chloride/carboxyphosphonate system/sodium hydroxide/water obtained by high-throughput investigation using a 24 h synthesis time.

3.4. Ionothermal synthesis

Ionic liquids (with advantages such as high thermal stability and negligible vapour pressure) are also profitable for the synthesis of porous solids. It was first applied to zeolites and phosphates, and recently to coordination polymers[41] and 3D networks.[42] Depending on the anion used (cation = imidazolium derivative), it was possible to control the structure of the resulting solid;[43] and recently a chiral anion was used to form a chiral MOF.[44]

3.5. Electrochemical synthesis

Mueller *et al.* have developed an electrochemical method to produce the copper trimesate HKUST-1,[45] which was formerly prepared by conventional solvothermal synthesis. The electrolysis of a copper plate in the presence of the trimesic acid leads rapidly to the pure corresponding solid. The interest of this method lies in its scalability and velocity as well as a higher purity due to the absence of counterions from the precursor during the synthesis.

3.6. Microwave synthesis

Microwave methodology finds its interest in many areas of chemistry, mainly as an invaluable technique to perform high speed synthesis. It also appeared as a good way to control the size and the shape of the resulting particles. It was thus applied to the synthesis of MOFs and allows the preparation, within minutes of various known porous phases such as some IRMOF materials,[46] a nickel glutarate[47] or the large pore MIL-101.[48] This technique was also recently coupled with ionothermal synthesis to produce a nickel trimesate.[49]

3.7. Solvent free synthesis

Solvent free synthesis is a topical area, mainly for the preparation of co-crystals and supramolecular solids. It was further extended to coordination polymers,[50] and to 1D[51] or 3D microporous solids,[52] as recently reviewed by Braga *et al.*[53] This technique simply consists of the grinding (mortal, ball milling) of reactants without any solvent. It is seen as a fast, easily scalable and environmentally friendly method for the preparation of MOFs, and could be attractive when large linkers of poor solubility are used. This synthetic route leads to polycrystalline products, which can be different from the one obtained from the solution or melt.

4. Thin Films

Thin films of microporous materials are required for many potential applications, in the area of membranes, chemical sensors and so on. In this prospect, thin films of zeolites have been developed over the last 10 years. Very recently, this type of work has been applied to MOFs with the pioneering work of Fischer *et al.*

in 2005. The preparation of polycrystalline films of MOF-5 on a Au(111) surface, grown on a SAM of alcanethiols terminated by carboxylic groups was reported.[54] This was further extended to other materials, such as HKUST-1 and $Zn_2(bdc)_2(dabco)$ and using other SAM modified surface such as silica or alumina.[55] A change of surface or terminal group on the alcanethiol leads to the oriented growth of HKUST-1. The same phenomenon was observed by Bein *et al.* using different terminal groups (hydroxyl or carboxylic acid), although the face selectivity was reported to be different, certainly due to different synthetic procedures.[56]

5. Adsorption

5.1. Green-house gases

Carbon dioxide (56%) and methane (18%) are the two of the principal greenhouse gases emitted today.[57] Porous MOFs or metal coordination polymers are currently being considered as alternatives to the zeolite and carbon-based adsorbents. A number of recent studies have been devoted to the adsorption of methane and carbon dioxide on such materials of which several have adsorption capacities that are equivalent or better than the current zeolite or activated carbon samples (Table 4.1).[58]

Table 4.1 Comparison of carbon dioxide adsorption in some reference porous solids and few MOFs (adapted from Ref. 63)

Adsorbent	Conditions	Maximum loading ($mmol\ g^{-1}$)	Maximum loading ($cm^3\ cm^{-3}$)	Reference
SBA-16	300 K/30 bars	Non-grafted 6	n.a.	59
		Grafted 3–4	n.a.	
$Cu(bpy)(BF_4)_2$	273 K/3.0 MPa	4	153	60
HKUST-1	298 K/4.2 MPa	10.7	210	61, 62
IRMOF-1	298 K/3.5 MPa	21.7	290	61
MOF-177	298 K/4.2 MPa	33.5	320	61
MIL53(Al, Cr)	302 K/2.5 MPa	10	225	24a
MIL-47(V)	302 K/2.0 MPa	11	250	24a
MIL-100(Cr)	304 K/5.0 MPa	18	280	63
MIL-101(Cr)	304 K/5.0 MPa	40	390	63
Silicalite	302 K/3.0 MPa	2.5	123	64
Zeolite NaX	302 K/3.0 MPa	7.8	147	65
Active carbon NORIT R1	298 K/3.0 MPa	10	96	66
Active carbon—Maxsorb	298 K/3.5 MPa	25	162	66

For instance, the metal carboxylates, MOF-177[61] and MIL-101[63], exhibit the highest carbon dioxide capacity with 33.5 mmol g^{-1}(298 K, 4.2 Mpa) and 40 mmol g^{-1} (303 K, 5 Mpa), respectively. This leads to enormous volumes of carbon dioxide per volume of adsorbent, with a value of 320 v/v for MOF-177 and 390 v/v for MIL-101. It should be noted that saturation occurs with large pore MOFs at much higher pressures than zeolites, which may be interesting for the recovery of CO_2 at high pressures from gas streams. Furthermore, regeneration of MOFs is generally possible under mild conditions which is a clear advantage with respect to zeolites or active carbons.

In the case of methane, several MOFs exhibit v/v capacities within the range or above the requirements of the DOE (180 $cm^3{}_{STP}.cm^{-3}$ at 289 K and 3.5 MPa).[24a,63,67] For instance, Seki *et al.* reported a series of isoreticular copper dicarboxylates-amine MOFs which equal or outreach the required capacities in v/v.[68] The authors claimed that these MOFs have ideal pore sizes and distributions for methane adsorption and have higher methane adsorption capacities than that of the theoretical maximum for activated carbon.

5.2. Hydrogen storage

The idea is commonly admitted that porous MOFs might be interesting materials for hydrogen storage, as a competitive alternative to other physisorption-based materials such as zeolites or activated carbons.[69] Since 2003, more than 50 MOFs have been tested for their ability to store hydrogen (Table 4.2).

Results have shown that the adsorption capacity of large pore MOFs,[79,81,88] can reach up 7.5 wt.% at 77 K and intermediate pressures (10–60 bars),[88a] and strongly depends on the surface area and pore volume. This opens up the possibility to reach the 2010 US DOE gravimetric adsorption goals (6 wt.%),[67] albeit at 77 K since most MOFs at room temperature do not exhibit capacities higher than 1 wt.% at high pressure (100 bars). At low pressure, the interactions between the surface and the hydrogen molecules dominate with pores remaining mostly unfilled. Small pores are more efficient than very large ones with an ideal pore size close to 5 Å, approximately twice the 2.8 Å kinetic diameter of H_2, which allows the H_2 molecule to interact with multiple portions of the framework, increasing the interaction energy between the framework and H_2. Several studies have highlighted that unsaturated metal sites in MOFs are a valuable way to increase the interactions with the dihydrogen molecule.[77,89] Although the metal sites are the preferential adsorption sites for hydrogen, the organic linker or the introduction of an electron-donating group (or groups) to the central portion of the ligand can also play an important secondary role in increasing the adsorption capacity.[88a,90] Recently, a method denoted 'hydrogen spillover', which consists of mixing a Pt/C catalyst with the MOF, allowing both the dissociative chemisorption of hydrogen onto the metal surface and the physisorption of H_2 in the pores of the MOF, has been reported by Li *et al.* and led to a strong enhancement of the hydrogen adsorption of MOF-5 and IRMOF-8, as much as eightfold.[91]

Table 4.2 Summary of hydrogen adsorption in MOFs at 77 K (from Ref. 81)

MOF	Surface area ($m^2 g^{-1}$)	H_2 uptake (wt.%)	Conditions	Reference
$Mn(HCO_2)_2$	297 (b)	0.9	1 atm	70
Ni(cyclam)(bpydc)	817	1.1	1 atm	71
$Zn_2(bdc)_2$(dabco)	1450 (a)	2.0	1 atm	72
$Ni_2(bpy)_3(NO_3)_4$ (M)	–	0.8	1 atm	73
$Ni_2(bpy)_3(NO_3)_4$ (E)	–	0.7	1 atm	73
$Ni_3(btc)_2$(3-pic)$_6$(pd)$_3$	–	2.1	14 bar	73
$Cu_2(pzdc)_2$(pyz),CPL-1	–	0.2	1 atm	74
Cu_2(bptc), MOF-505	1830	2.48	1 atm	75
$Sc_2(bdc)_3$	721 (a)	1.5	0.8 bar	76
$M_3[Co(CN)_6]_2$	720–870 (a)	1.4–1.8	1 bar	77
M = Mn, Fe, Co, Ni, Cu, Zn $Mg_3(ndc)_3$	190 (c)	0.48	1.17 bar	78
Al(OH)(bdc), MIL-53(Al)	1590	3.8	16 bar	79
Cr(OH)(bdc), MIL-53(Cr)	1500	3.2	16 bar	79
$Al_3O(OH)(btc)_3$, MIL-96(Al)	–	1.7	10 bar	80
$Cr_3OF(btc)_2$, MIL-100(Cr)	2800	3.3	25 bar	81
$Cr_3OF(bdc)_3$, MIL-101(Cr)	5500	6.1	60 bar	81
$Cr_3OF(ntc)_{1.5}$, MIL-102(Cr)	42	0.9	10 bar	82
$Cu_3(btc)_2$, HKUST-1	1958	3.6	50 bar	83,84
$Zn_4O(bdc)_3$, MOF-5 or IRMOF-1	2296 (a)	4.7	50 bar	83,84,85
$Zn_4O(cbbdc)_3$, IRMOF-6	3300	4.8	50 bar	86
$Zn_4O(ndc)_3$, IRMOF-8	1466	1.50	40 bar	85,86
$Zn_4O(hpdc)_3$, IRMOF-11	2340	3.5	35 bar	86
$Zn_4O(tmbdc)_3$, IRMOF-18	1501	0.89	40 bar	86
$Zn_4O(ttdc)_3$, IRMOF-20	4590	6.7	70 bar	86
$Zn_4O(BTB)_2$, MOF-177	5640	7.5	70 bar	86
Cu_2(bptc)	1670 (a)	4.02	20 bar	87
Cu_2(tptc)	2247 (a)	6.06	20 bar	87
Cu_2(qptc)	293 (a)	6.07	20 bar	87

Acronyms: bdc, benzene-1,4-dicarboxylate; ndc, naphthalene-2,6-dicarboxylate; hpdc, 4,5,9,10-tetrahydropyrene-2,7-dicarboxylate; tmbdc, 2,3,5,6-tetramethylbenzene-1,4-dicarboxylate; btb, benzene-1,3,5-tribenzoate, cyclam, 1,4,8,11-tetraazacyclotetradecane; bpydc, 2,2′-bipyridyl-5,5′-dicarboxylate; dabco, 1,4-diazabicyclo[2.2.2]octane, bpy, 4,4′-bipyridine; btc, benzene-1,3,5-tricarboxylate; 3-pic, 3-picoline; pd, 1,2-propanediol; ttdc, thieno[3,2-b] thiophene-2,5-dicarboxylate; bptc, 3,3′,5, 5′ biphenyl tetracarboxylate; tpdc, 3,3″,5, 5″ terphenyl tetracarboxylate; qpdc, quaterphenyl 3,3‴,5, 5‴ tetracarboxylate.

Apparent surface area calculated from N_2 adsorption data collected at 77 K using the Langmuir model except (a) (BET surface area from N_2 at 77 K), (b) (BET surface area from CO_2 at 195 K) and (c) (BET surface from O_2 adsorption at 77 K).

5.3. Other gases

A few studies concern the use of small pore MOFs for the size selective adsorption of small gas molecules (N_2, O_2, CO, . . .).[92] Flexible MOFs can also be used for gate opening separation involving a change in the size of the windows and/or the pores to allow or forbid the entrance of a gas molecule.[93] Kitagawa *et al.* reported the selective adsorption of acetylene over carbon dioxide.[94] In the latter case, the interactions between acidic protons of the acetylene molecules and non-coordinated oxygen atoms of the framework are at the origin of this difference.

6. MOFs and Separation

6.1. Separation of olefins

Separation is one of the most promising field of applications for MOFs,[95] considering that no high thermal or chemical stability is required, and that the wide range of possible chemical compositions will make easier the tuning of the separation properties. To date, most studies concern olefines.[96] For instance, Huang *et al.* recently reported selectivity during separation of *p*-xylene and *o*-xylene in a variant of MOF-5.[97a] Devos *et al.* used the flexible MIL-47 and MIL-53(Al) solids for the separation of liquid isomer of xylene; this resulted in a few selectivities higher than those of commercial zeolites.[97b] Another example concerns the separation of propane/propene which is of an utmost importance. Actual processes use energy-consuming cryogenic separation,[98] and a possible alternative could be the use of solid adsorbents such as MOFs with unsaturated metal sites. Ernst *et al.* have reported the use of the copper trimesate HKUST-1 for the dynamic separation of propane/propene[99] with a reasonable selectivity despite close isosteric heats of adsorption.

6.2. Separation of polar molecules

Several papers describe the separation of mixtures of polar solvents using MOFs. For instance, Eddaoudi *et al.* reported that MOF-4 could separate methanol from acetonitrile,[100] Chen *et al.* used a small pore zinc MOF for the size selective separation of methanol and diethylester,[101] Takamizawa *et al.* performed separation tests of alcohol molecules by evaporation at room temperature using a flexible MOF with a 1D chain skeleton.[102]

7. Inclusion and Reaction in MOFs

7.1. Metal nanoparticles

The synthesis of metallic nanoparticles in MOFs is appealing, as these solids, due the intrinsic regularity of their pores, appear as perfect moulds for the preparation of highly monodisperse particles often for catalytic purposes. Two methods of preparation have been proposed:

- Redox active frameworks (based on nickel(II) cyclam and di or tetracarboxylate) were dispersed in solution of silver(I)[103] or gold(III)[104] salt. Redox reactions involving the Ni^{II}/Ni^{III} and Ag^{0}/Ag^{I} or Au^{0}/Au^{III} couples afford Ag or Au nanoparticles under mild conditions, although their size was not directly correlated with the size of the pores.
- Various molecular metallic precursors (Pt, Pd, Au, Cu, Zn, Sn) were introduced in the porous MOF-5 by MOCVD (or alternatively by solution process in the case of Pd[105]), and then reacted under a reducing or oxidizing atmosphere to produce metallic (Pd, Au, Cu) [106] or oxide (ZnO) [107] nanoparticles.

7.2. Organic polymerization

The inclusion of organic monomers and further polymerization in porous MOFs was initiated by Kitagawa *et al.* recently. The main interest lied in the possibility offered by the monodispersity of the pores to strictly control the regio and stereoselectivity of the reaction, but also the 3D structure of the polymer. Radical polymerization of entrapped styrene[108] and divinylbenzene[109] in the presence of a radical initiator (AIBN) was reported, as well as the polymerization of monosubstituted acetylene catalysed by $C{\equiv}C{-}H{\cdots}O^{-}{-}C$ hydrogen bonds between the monomers and the network.[110] This work opens the way to the preparation of aligned polymers, or hybrid solids combining the properties of the host lattice and the polymer (case of π-conjugated systems).[111]

8. Catalytic Applications

MOFs offer many opportunities in catalysis. Their main advantages, compared to zeolites or modified mesoporous silica, lie in the direct incorporation of catalytically active metals and the easy modification of the neighbourhood of the catalytic sites through the functionalization of the linkers although their relatively low stability remains an important drawback.

Various examples of catalytic activity have been published up to now, ranging from NO decomposition or reduction (Cu^{I}-based MOF),[112] CO oxidation (Ni^{II}-based MOF),[113] hydrogenation of olefins (Rh^{II}-based MOF),[114] oxidation of alcohols ($Ru^{II/III}$-based MOF)[115] and thioethers (Zn^{II} [116] and Sc^{III}-based[117] solids), acetalization of aldehydes (Sc^{III} [118] and In^{III}-based[119] MOFs) or Friedel–Crafts benzylation of benzene (Fe^{III}-based solid).[120] In all these examples, either the redox behaviour or the Lewis acidity of the metallic cation is responsible for the activity. Base catalysis was also recently proposed by Kitagawa *et al.* The catalytic site lie on the linker (amide group) and the corresponding MOF exhibited activity for the Knoevenagel reaction.[121]

Heterogeneous enantioselective catalysis using chiral MOFs has been recently reviewed by Lin *et al.*[122,123] Kim *et al.* first showed that enantioselective transesterification (modest enantiomeric excess) using a porous MOF basic catalyst (pyridyl pendant group) was possible.[124] The acetalization of chiral aldehyde was also proposed (chiral Zn^{II}-based MOF). [125] Enantioselective catalysis using a metal not directly involved in the framework formation, but rather incorporated in the linker, either directly (Mn^{III} salen-based linker for epoxidation)[126] or by post-synthesis treatment (Ti^{IV} grafted on a dihydroxybinathyl-based linker for aldehyde reduction)[127,128] was also proposed.

9. Redox Activity

Because of their hybrid composition, MOFs exhibit various redox activities, based either on their inorganic (transition metal cation) or organic (redox active linker) parts. The use of oxidizable MOFs was previously discussed (Section 7.1). The reduction of porous MOFs through lithium inclusion (either chemically or electrochemically) has also been proposed with two main goals in mind:

- A theoretical work suggested that lithium reduction of porous MOFs may dramatically enhance their hydrogen sorption capacities, especially above 77 K.[129] A recent experimental report showed that partial reduction (3%) of the linkers of a Zn-based MOF already increased this sorption capacity, although the redox process was this case coupled with a displacement of the framework, which affect the sorption capacity.[130]
- A Fe^{III}-based porous MOF was electrochemically reversibly reduced up to 0.6 electron/Fe without destruction on the network. Although the capacity of this solid remains below the standard ones, this study illustrates the possible use of reducible MOFs as anode materials for lithium-based batteries.[131]

10. Magnetism

Many coordination polymers, or hybrid solids with 1D or 2D inorganic subnetworks present magnetic properties, related to the cations (mainly Ni, Co, Mn, Fe, Gd or their mixtures) or to the radical linkers (for example nitrosyl-nitroxide) involved in the frameworks. Among them, only few solids truly exhibit permanent porosity. We here only focus here on few examples, and on the interplay between porosity (pore filling) and magnetism.

Simple solids, such as $M_3(HCOO)_6$ (M = Mn, Fe, Co, Ni) exhibit both porosity and a variation of their magnetic properties upon pore evacuation or filling.[132] A few examples of porous Co^{II}-based solids showing guest induced modulation of the magnetic behaviour have been reported;[133] among which a squarate presenting a ferromagnetic/antiferromagnetic transition upon dehydration/hydration process,[134] and a Co^{II}-based solid exhibiting single chain magnetic behaviour which is transformed into a metamagnet upon dehydratation.[135] Porous Fe^{II}-based MOFs with spin crossover behaviour in their hydrated form may[136] or may not[137] maintain this property in the dehydrated state. A Ni^{II}-based hybrid solid with a 3D inorganic network was also proven to be both porous ($S = 300\ m^2\ g^{-1}$) and ferromagnetic at low temperature.[138]

An example of a porous MOF presenting both organic (extended radical linkers) and inorganic (Cu^{II} ions) spin carriers and variation of magnetic susceptibility upon evacuation and filling of the pores was reported by Veciana *et al.*[29]

11. Optical Properties

Permanent porosity and optical properties related to the metal or the linker can be combined in MOFs, for potential applications in chemical sensing. The use of coordination polymers as second order nonlinear optical materials (NLO) has been reviewed by Lin *et al.*; unsymmetrical linkers and metal centres with well-defined coordination (Zn, Cd) allowed the preparation of non-centrosymmetric networks, a prerequisite to Second Harmonic Generation (SHG).[139] Recent advances include cationic guest dependent NLO activity in a porous anionic hybrid network,[140] and combination of multiphoton up-conversion and red and blue NLO processes in a 2D terbium MOF.[141]

Numerous MOFs exhibiting ligand-based emission in the visible region (UV excitation) exist. The chemical modification of the linkers allows the preparation of polymeric networks presenting strong and tunable luminescence activity.[142] Nevertheless, only few studies on porous MOFs (mainly Cd[143] and Zn[144]) and guest dependence activity (solvent molecules) have been reported.

The second family of luminescent MOFs consists of the lanthanide (Eu, Tb, Dy) based materials. π-aromatic linkers act as antenna[145] to absorb the UV radiation; the energy is then transferred to the lanthanide which further emits in the visible or NIR region. Guests trapped in the pores may act the same,[146] and sensitization of a Eu^{III}-based MOF through a second lanthanide ion (Tb^{III}) has been proposed.[147] Only a restricted number of lanthanide-based MOFs are porous, and luminescence–porosity relationship studies are scarce. Depending on the nature of the MOF, dehydration/rehydration could lead to a non-luminescent/luminescent transformation (Tb^{III}- and Eu^{III}-based solids[148]) or may unaffect the luminescence properties.[149] The nature of the guests affects the decay time (Tb-MOF[150]) and the intensity (Eu^{III}-MOF)[151] of the emission for MOFs with open metal sites. Anion-sensing with a porous Tb^{III}-based MOF was recently described.[152] Eventually, the preparation of nanoparticles of Eu^{III}- or Tb^{III}-doped Gd^{III}-based MOFs and their use as biomarkers was proposed.[153]

12. Bioapplications

12.1. Drug delivery

There is a strong interest in the development of methods for the controlled drug release to deliver the entire dose needed over a prolonged time with only one administration.[154] The regular porosity and the presence of organic groups within the framework of MOFs, combined with the low toxicity of carboxylic acids and

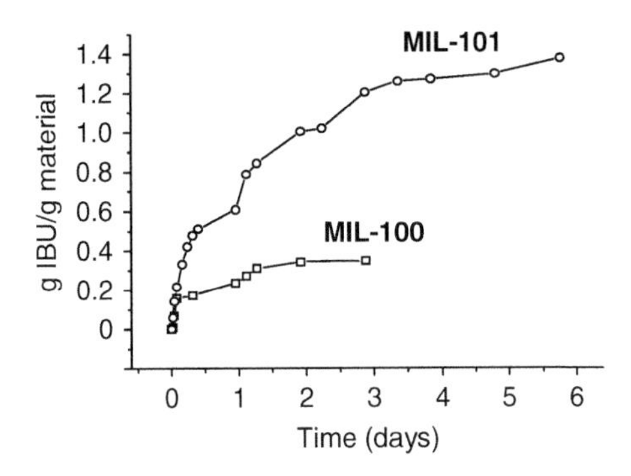

Figure 4.9 Ibuprofen delivery profile from MIL-100, MIL-101 in Simulated Body Fluid at 37 °C.

some transition metals (Fe, Zn, Ti, . . .) makes them attractive candidates for the controlled release of drugs. Horcajada *et al.* have reported the adsorption and delivery of Ibuprofen using model chromium solids,[155] despite their well-known toxicity, assuming that applications would concern less toxic MOFs. Both materials showed remarkable but drastically different Ibuprofen adsorption (0.35 and 1.4 g Ibu/g dehydrated material), due to their different pore size. The delivery of Ibuprofen was achieved from 3–6 days (Fig. 4.9). Taking into account it is a non-toxic material; MOFs can be altered for limitless applications in biomedicine in the future.

12.2. Controlled release of nitric oxide

There are huge possibilities for the use of exogenous NO in prophylactic and therapeutic processes for applications in anti-thrombogenic medical devices, improved dressings for wounds and ulcers and the treatment of infections.[156] Whilst the delivery of NO gas from a cylinder is not practical for most therapeutic applications, the lack of suitable materials is problematic, and thus it is vital to discover new materials able to produce, store and deliver NO. Morris *et al.* proved that NO could be stored and delivered from the copper trimesate HKUST-1 with unsaturated metal sites thus highlighting a prevention of platelet aggregation.[157,158]

13. Conclusions

MOFs or PCPs are a recent emerging class of porous solids. So far, more than a 1000 original structures have been reported and many are to come due to the almost endless possibilities of organic chemistry combined with the different chemistries of metals from all the periodic table. In addition, MOFs are a versatile class of solids with either new structures of zeotype architecture, frameworks that can be rigid or highly flexible, microporous or mesoporous, structures that exhibit either functional organic groups or accessible metal sites. Despite their lower thermal and chemical stability, MOFs can be considered as a new complementary class of porous solids relative to their inorganic analogues (zeolites, metal phosphates, ordered mesoporous oxides). If some applications generally dedicated to zeolites reveal lower performance (catalysis), MOFs sometimes exhibit better results (adsorption). It is too early to know if industrial applications will emerge from MOFs, there is still a multitude of new structures to be discovered, as well as exploring new possibilities of applications.

ACKNOWLEDGEMENTS

The authors wish to thank their past and present co-workers from the Institut Lavoisier, including Ph.D. students and post-docs, as well as their collaborators from other laboratories.

REFERENCES

[1] Clearfield, A., *Prog. Inorg. Chem.* **1998,** *47,* 371–510.
[2] See the special issues: *Acc. Chem. Res.* **2005**, *38*, 215–378; *J. Solid State Chem.* **2005**, *178*, 2409–2574.
[3] Eddaoudi, M., Moler, D. B., Li, H., Chen, B., Reineke, T. M., O'Keeffe, M., Yaghi, O. M., *Acc. Chem. Res.* **2001,** *34,* 319–330.
[4] Some major references in the field: (a) Yaghi, O. M., O'Keeffe, M., Ockwig, N. W., Chae, H. K., Eddaouddi, M., Kim, J., *Nature* **2003**, *423*, 705–714; (b) Rowsell, J. L. C., Yaghi, O. M., *Angew. Chem. Int. Ed. Engl.* **2005**, *44*, 4670–4679; (c) Rao, C. N. R., Natarajan, S., Vaidhyanathan, R., *Angew. Chem. Int. Ed. Engl.* **2004**, *43*, 1466–1496; (d) Kitagawa, S., Kitaura, R., Noro, S.-H., *Angew. Chem. Int. Ed. Engl.* **2004**, *43*, 2334–2375; (e) James, S. L., *Chem. Soc. Rev.* **2003**, *32*, 276–288; (f) Maspoch, D., Ruiz-Molina, D., Veciana, *J. Chem. Soc. Rev.* **2007**, *36*, 770–818; (g) Cheetham, A. K., Rao, C. N. R., *Chem. Commun.* **2006,** 4780–4795; (h) Kepert, C. J., *Chem. Commun.* **2006**, 695–700.
[5] Simulation and structure: (a) Mellot-Draznieks, C., Dutour, J., Férey, G., *Z. Anorg. Allg. Chemie* **2004**, *630*, 2599–2604; (b) Mellot-Draznieks, C., Dutour, J., Férey, G., *Angew. Chem. Int. Ed. Engl.* **2004**, *43*, 6290–6296; (c) Greathouse, J. A., Allendorf, M. D., *J. Am. Chem. Soc.* **2006**, *128*, 10678–10679; (d) Dubbeldam, D., Walton, K. S., Ellis, D. E., Snurr, R. Q., *Angew. Chem. Int. Ed. Engl.* **2007**, *46*, 4496–4499 and references therein.
[6] Simulation and adsorption: (a) Zhang, L., Wang, Q., Wu, T., Liu, Y.-C., *Chem. Eur. J.* **2007**, *13*, 6387–6396; (b) Mueller, T., Ceder, G., *J. Phys. Chem. B* **2005**, *109*, 17974–17983; (c) Klontzas, E., Mavrandonakis, A., Froudakis, G. E., Carissan, Y., Klopper, W., *J. Phys. Chem. C* **2007**, *111*, 13635–13640; (d) Yang, Q., Zhong, C., *J. Phys. Chem. B* **2006,** *110*, 17776–17783; (e) Nicholson, T. M., Bathia, S. K., *J. Phys. Chem. B* **2006**, *110*, 24834–24836; (f) Frost, H., Düren, T., Snurr, R. Q., *J. Phys. Chem. B* **2006**, *110*, 9565–9570; (g) Ramsahye, N. A., Maurin, G., Bourrelly, S., Llewellyn, P. L., Loiseau, T., Serre, C., Férey, G., *Chem. Commun.* **2007**, 3261–3263 and references therein.
[7] Chui, S. S. Y., Lo, S. M. F., Charmant, J. P. H., Orpen, A. G., Williams, I. D., *Science* **1999,** *283,* 1148–1150.
[8] Li, H., Eddaoudi, M., O'Keeffe, M., Yaghi, O. M., *Nature* **1999,** *402,* 276–279.
[9] (a) Serre, C., Millange, F., Surblé, S., Férey, G., *Angew. Chem. Int. Ed. Engl.* **2004,** *43,* 6285–6289; (b) Surblé, S., Serre, C., Mellot-Draznieks, C., Millange, F., Férey, G., *Chem. Commun.* **2006,** 284–286.
[10] Hoskins, B. F., Robson, R., *J. Am. Chem. Soc.* **1990,** *112,* 1546–1554.
[11] Carlucci, L., Ciani, G., Proserpio, D. M., Sirni, A., *Angew. Chem. Int. Ed. Engl.* **1995,** *107,* 2037–2040.
[12] Batten, S. R., Houskins, B. F., Robson, R., *J. Chem. Soc. Chem. Commun.* **1991,** 445–447.

[13] Gable, R. W., Houskins, B. F., Robson, R., *J. Chem. Soc. Chem. Commun.* **1990,** 762–763.
[14] Keller, S. W., *Angew. Chem. Int. Ed. Engl.* **1997,** *109,* 295–297.
[15] (a) Férey, G., Serre, C., Mellot-Draznieks, C., Millange, F., Surblé, S., Dutour, J., Margiolaki, I., *Angew. Chem. Int. Ed. Engl.* **2004,** *43,* 6296–6301; (b) Férey, G., Mellot-Draznieks, C., Serre, C., Millange, F., Dutour, J., Surblé, S., Margiolaki, I., *Science* **2005,** *309,* 2040–2042.
[16] Park, Y. K., Choi, S. B., Kim, H., Kim, K., Won, B.-H., Choi, K., Choi, J.-S., Ahn, W.-S., Won, N., Kim, S., Jung, D. H., Choi, S.-H., Kim, G.-H., Cha, S.-S., Jhon, Y. H., Yang, J. K., Kim, J., *Angew. Chem. Int. Ed. Engl.* **2007,** *46,* 8230–8233.
[17] (a) Liu, Y., Kravtsov, V. C., Larsena, R., Eddaoudi, M., *Chem. Commun.* **2006,** 1488–1490; (b) Su, C. Y., Lu, T. B., Jiang, L., Chen, J. M., *J. Am. Chem. Soc.* **2006,** *128,* 34–35; (c) Tian, Y.-Q., Zhao, Y.-M., Chen, Z.-X., Zhang, G.-N., Weng, L.-H., Zhao, D.-Y., *Chem. Eur. J.* **2007,** *13,* 4146–4154.
[18] http://www.iza-structure.org/databases/
[19] Huang, X.-C., Lin, Y.-Y., Zhang, J.-P., Chen, X.-M., *Angew. Chem. Int. Ed. Engl.* **2006,** *45,* 1557–1559.
[20] (a) Kitagawa, S., Uemura, K., *Chem. Soc. Rev.* **2005,** *34,* 109–119; (b) Uemura, K., Matsuda, R., Kitagawa, S., *J. Solid State Chem.* **2005,** *178,* 2420–2429.
[21] (a) Serre, C., Millange, F., Thouvenot, C., Noguès, M., Marsolier, G., Louër, D., Férey, G., *J. Am. Chem. Soc.* **2002,** *124,* 13519–13526; (b) Loiseau, T., Serre, C., Huguenard, C., Fink, G., Taulelle, F., Henry, M., Bataille, T., Férey, G., *Chem. Eur. J.* **2004,** *10,* 1373–1382.
[22] (a) Mellot-Draznieks, C., Serre, C., Surblé, S., Audebrand, N., Férey, G., *J. Am. Chem. Soc.* **2005,** *127,* 16273–16378; (b) Serre, C., Mellot-Draznieks, C., Surblé, S., Audebrand, N., Fillinchuk, Y., Férey, G., *Science* **2007,** *315,* 1828–1831.
[23] (a) Fletcher, A. J., Thomas, K. M., Rosseinsky, M. J., *J. Solid State Chem.* **2005,** *178,* 2491–2510; (b) Maji, T. K., Matsuda, R., Kitagawa, S., *Nat. Mater.* **2007,** *6,* 142–148.
[24] (a) Bourrelly, S., Llewellyn, P. L., Serre, C., Millange, F., Loiseau, T., Férey, G., *J. Am. Chem. Soc.* **2005,** *127,* 13519–13521; (b) Serre, C., Bourrelly, S., Vimont, A., Ramsahye, N., Maurin, G., Llewellyn, P. L., Daturi, M., Filinchuk, Y., Leynaud, O., Barnes, P., Férey, G., *Adv. Mater.* **2007,** *19,* 2246–2251.
[25] Eddaoudi, M., Kim, J., Rosi, N., Vodak, D., Wachter, J., O'Keeffe, M., Yaghi, O. M., *Science* **2002,** *295,* 469–472.
[26] Zhang, X.-M., *Coord. Chem. Rev.* **2005,** *249,* 1201–1219.
[27] Chen, X.-M., Tong, M.-L., *Acc. Chem. Res.* **2007,** *40,* 162–170.
[28] Dybstev, D. N., Yutkin, M. P., Peresypkina, E. V., Virovets, A. V., Serre, C., Férey, G., Fedin, V. P., *Inorg. Chem.* **2007,** *46,* 6843–6845.
[29] Maspoch, D., Ruiz-Molina, D., Wurst, K., Domingo, N., Cavallini, M., Biscarini, F., Tajada, J., Rovira, C., Veciana, *J. Nat. Mat.* **2003,** *2,* 190–195.
[30] Bradshaw, D., Prior, T. J., Cussen, E. J., Claridge, J. B., Rosseinsky, M. J., *J. Am. Chem. Soc.* **2004,** *126,* 6106–6114.
[31] Pan, L., Frydel, T., Sander, M. B., Huang, X., Li, J., *Inorg. Chem.* **2001,** *40,* 1271–1283.
[32] Go, Y. B., Wang, X., Anokhina, E. V., Jacobson, A. J., *Inorg. Chem.* **2005,** *44,* 8265–8271.
[33] Forster, P. M., Burbank, A. R., Livage, C., Férey, G., Cheetham, A. K., *Chem. Commun.* **2004,** 368–369.
[34] Stock, N., Bein, T., *J. Mater. Chem.* **2005,** *15,* 1384–1391.
[35] (a) Forster, P. M., Stock, N., Cheetham, A. K., *Angew. Chem. Int. Ed. Engl.* **2005,** *44,* 7608–7611; (b) Forster, P. M., Burbank, A. R., O'Sullivan, M. C., Guillou, N., Livage, C., Férey, G., Stock, N., Cheetham, A. K., *Solid State Sci.* **2005,** *7,* 1549–1555.
[36] Bauer, S., Stock, N., *Angew. Chem. Int. Ed. Engl.* **2007,** *46,* 6857–6860.
[37] Forster, P. M., Thomas, P. M., Cheetham, A. K., *Chem. Mater.* **2002,** *14,* 17–20.
[38] Devic, T., Serre, C., Audebrand, N., Marrot, J., Férey, G., *J. Am. Chem. Soc.* **2005,** *127,* 12788–12789.
[39] Uemura, T., Hoshino, Y., Kitagawa, S., Yoshida, K., Isoda, S., *Chem. Mater.* **2006,** *18,* 992–995.

[40] Grzesiak, A. L., Uribe, F. J., Ockwig, N. W., Yaghi, O. M., Matzger, A. J., *Angew. Chem. Int. Ed. Engl.* **2006,** *45,* 2553–2556.
[41] Liao, J.-H., Wu, P.-C., Huand, W.-C., *Cryst. Growth Des.* **2006,** *6,* 1062–1063.
[42] Dybtsev, D. N., Chun, H., Kim, K., *Chem. Commun.* **2004,** 1594–1595.
[43] Lin, Z., Wragg, D. S., Warren, J. E., Morris, R. E., *J. Am. Chem. Soc.* **2007,** *129,* 10334–10335.
[44] Lin, Z., Slawin, A. M. Z., Morris, R. E., *J. Am. Chem. Soc.* **2007,** *129,* 4880–4881.
[45] Mueller, U., Schubert, M., Teich, F., Puetter, H., Scierle-Arndt, K., Pastré, J., *J. Mater. Chem.* **2006,** *16,* 626–636.
[46] Ni, Z., Masel, R. I., *J. Am. Chem. Soc.* **2006,** *128,* 12394–12395.
[47] Jhung, S. H., Lee, J.-H., Forester, P. M., Férey, G., Cheetham, A. K., Chang, J.-S., *Chem. Eur. J.* **2006,** *12,* 7699–7705.
[48] Jhung, S. H., Lee, J.-H., Yoon, J. W., Serre, C., Férey, G., Chang, J.-S., *Adv. Mater.* **2007,** *19,* 121–124.
[49] Lin, Z., Wragg, D. S., Morris, R. E., *Chem. Commun.* **2006,** 2021–2023.
[50] Belcher, W. J., Longsrtaff, C. A., Neckenig, M. R., Steed, J. W., *Chem. Commun.* **2002,** 1602–1603.
[51] Braga, D., Curzi, M., Johansson, A., Polito, M., Rubini, K., Grepioni, F., *Angew. Chem. Int. Ed. Engl.* **2006,** *45,* 142–146.
[52] Pichon, A., Lazuen-Garay, A., James, S. L., *Cryst. Eng. Comm.* **2006,** *8,* 211–214.
[53] Braga, D., Giaffreda, S. L., Grepioni, F., Pettersen, A., Maini, L., Curzi, M., Polito, M., *Dalton Trans.* **2006,** 1249–1263.
[54] Hermes, S., Schröder, F., Chelmowski, R., Wöll, C., Fischer, R. A., *J. Am. Chem. Soc.* **2005,** *127,* 13744–13745.
[55] Zacher, D., Baunemann, A., Hermes, S., Fischer, R. A., *J. Mater. Chem.* **2007,** *17,* 2785–2792.
[56] Biemmi, E., Scherb, C., Bein, T., *J. Am. Chem. Soc.* **2007,** *129,* 8054–8055.
[57] (a) Kikkinides, E. S., Yang, R. T., Cho, S. H., *Ind. Eng. Chem. Res.* **1993,** *32,* 2714–2720; (b) Kikuchi, R., *Energy Environ.* **2003,** *14,* 383–395.
[58] Himeno, S., Komatsu, T., Fujita, S., *J. Chem. Eng. Data* **2005,** *50,* 369.
[59] Knöfel, C., Descarpentries, J., Benzaouia, A., Zelenák, V., Mornet, S., Llewellyn, P. L., Hornebecq, V., *Micropor. Mesopor. Mater.* **2007,** *99,* 79–85.
[60] Li, D., Kaneko, K., *Chem. Phys. Lett.* **2001,** *335,* 50–56.
[61] Millward, A. R., Yaghi, O. M., *J. Am. Chem. Soc.* **2005,** *127,* 17998–17999.
[62] Wang, Q. M., Shen, D., Bülow, M., Lau, M. L., Deng, S., Fitch, F. R., Lemcoff, N. O., Semanscin, J., *Micropor. Mesopor. Mater.* **2002,** *55,* 217–230.
[63] Llewellyn, P. L., Bourrelly, S., Serre, C., Vimont, A., Daturi, M., Hamon, L., De Weireld, G., Chang, J.-S., Hong, D. Y., Hwang, Y. K., Férey, G., *Langmuir* **2008**, in press, DOI: 10.1021/la800227x.
[64] Bourrelly, S., Maurin, G., Llewellyn, P. L., *Stud. Surf. Sci. Catal.* **2005,** *158,* 1121–1128.
[65] Maurin, G., Llewellyn, P. L., Bell, R. G., *J. Phys. Chem. B* **2005,** *109,* 16084–16091.
[66] Himeno, S., Komatsu, T., Fujita, S., *J. Chem. Eng. Data* **2005,** *50,* 369–376.
[67] http://www.eere.energy.gov/
[68] Seki, K., Mori, W., *J. Phys. Chem. B* **2002,** *106,* 1380–1385.
[69] (a) Lin, X., Jia, J., Hubberstey, P., Schröder, M., Champness, N. R., *Cryst. Eng. Comm.* **2007,** *9,* 438–448; (b) Thomas, K. M., *Catal. Today* **2007,** *120,* 389–398; (c) Collins, D. J., Zhou, H.-C., *J. Mater. Chem.* **2007,** *17,* 3154–3160. (d) Kim, D., Kim, J., Jung, D. H., Lee, T. B., Choi, S. B., Yoon, J. H., Kim, J., Choi, K., Choi, S.-H., *Catal. Today* **2007,** *120,* 317–323.
[70] Dybtsev, D. N., Chun, H., Yoon, S. H., Kim, D., *J. Am. Chem. Soc.* **2004,** *126,* 32–33.
[71] Lee, E. Y., Suh, M. P., *Angew. Chem. Int. Ed. Engl.* **2004,** *43,* 2798–2801.
[72] Dybtsev, D. N., Chun, H., Kim, D., *Angew. Chem. Int. Ed. Engl.* **2004,** *43,* 5033–5036.
[73] Zhao, X., Xiao, B., Fletcher, A. J., Thomas, K. M., Bradshaw, D., Rosseinky, M. J., *Science* **2004,** *306,* 1012–1015.
[74] Kubota, Y., Takata, M., Matsuda, R., Kitaura, R., Kitagawa, S., Kabo, K., Kobayashi, T. C., *Angew. Chem. Int. Ed. Engl.* **2005,** *44,* 920–923.

[75] Chen, B., Ockwig, N. W., Millard, A. R., Contreras, D. S., Yaghi, O. M., *Angew. Chem. Int. Ed. Engl.* **2005,** *44,* 4745–4749.
[76] Perles, J., Iglesias, M., Martin-Luengo, M.-A., Monse, M. A., Ruiz-Valero, C., Snejko, N., *Chem. Mater.* **2005,** *17,* 5837–5842.
[77] Kaye, S. S., Long, J. R., *J. Am. Chem. Soc.* **2006,** *127,* 6506–6507.
[78] Dinca, M., Long, J. R., *J. Am. Chem. Soc.* **2005,** *127,* 9376–9377.
[79] Férey, G., Latroche, M., Serre, C., Millange, F., Loiseau, T., Percheron-Guegan, A., *Chem. Commun.* **2003,** *24,* 2976–2977.
[80] Loiseau, T., Lecroq, L., Volkringer, C., Marrot, J., Férey, G., Haouas, M., Taulelle, F., Bourrelly, S., Llewellyn, P. L., Latroche, M., *J. Am. Chem. Soc.* **2006,** *128,* 10223–10230.
[81] Latroche, M., Surblé, S., Serre, C., Mellot-Draznieks, C., Llewellyn, P. L., Lee, J.-H., Chang, J.-S., Jhung, S. H., Férey, G., *Angew. Chem. Int. Ed. Engl.* **2006,** *45,* 8227–8231.
[82] Surblé, S., Millange, F., Serre, C., Düren, T., Latroche, M., Bourrelly, S., Llewellyn, P. L., Férey, G., *J. Am. Chem. Soc.* **2006,** *128,* 14889–14896.
[83] Panella, B., Hirscher, M., Pütter, H., Müller, U., *Adv. Funct. Mater.* **2006,** *16,* 520–524.
[84] Roswell, J. L. C., Yaghi, O. M., *J. Am. Chem. Soc.* **2006,** *128,* 1304–1315.
[85] Dailly, A., Vajo, J. J., Ahn, C. C., *J. Phys. Chem. B* **2006,** *110,* 1099–1101.
[86] Wong-Foy, A. G., Matzger, A. J., Yaghi, O. M., *J. Am. Chem. Soc.* **2006,** *128,* 3494–3495.
[87] Lin, X., Jia, J., Zhao, X., Mark Thomas, K., Blake, A. J., Walker, G. S., Champness, N. R., Hubberstey, P., Schröder, M., *Angew. Chem. Int. Ed. Engl.* **2006,** *45,* 7358–7364.
[88] (a) Rowsell, J. L. C., Millward, A. R., Park, K. S., Yaghi, O. M., *J. Am. Chem. Soc.* **2004,** *126,* 5666–5667. (b) Dinca, M., Dailly, A., Liu, Y., Brown, C. M., Neumann, D. A., Long, J. R., *J. Am. Chem. Soc.* **2006,** *128,* 16876–16883.
[89] Peterson, V. K., Liu, Y., Brown, C. M., Kepert, C. J., *J. Am. Chem. Soc.* **2006,** *128,* 15578–15579.
[90] (a) Rowsell, J. L. C., Yaghi, O. M., *J. Am. Chem. Soc.* **2006,** *128,* 1304–1315; (b) Chun, H., Dybtsev, D. N., Kim, H., Kim, K., *Chem. Eur. J.* **2005,** *11,* 3521–3529.
[91] (a) Li, Y., Yang, F. H., Yang, R. T., *J. Phys. Chem. C* **2007,** *111,* 3405; Li, Y., Yang, R. T., *J. Am. Chem. Soc.* **2006,** *128,* 726–727; (b) Li, Y., Yang, R. T., *J. Am. Chem. Soc.* **2006,** *128,* 8136–8137.
[92] (a) Wang, Q. M., Shen, D., Bülow, M., Lau, M. L., Deng, S., Fitch, F. R., Lemcoff, N. O., Semanscin, J., *Micropor. Mesopor. Mater.* **2002,** *55,* 217–230.
[93] (a) Kitaura, R., Seki, K., Akiyama, G., Kitagawa, S., *Angew. Chem. Int. Ed. Engl.* **2003,** *42,* 428–431. (b) Ma, S., Sun, D., Wang, X.-S., Zhou, H.-C., *Angew. Chem. Int. Ed. Engl.* **2007,** *46,* 2458–2462.
[94] Matsuda, R., Kitaura, R., Kitagawa, S., Kubota, Y., Belosludov, R. V., Tatsuo, C., Kobayashi, C., Sakamoto, H., Chiba, T., Takata, M., Kawazoe, Y., Mita, Y., *Nature* **2005,** *436,* 235–241.
[95] Snurr, R. Q., Hupp, J. T., N'Guyen, S. T., *AIChE J.* **2004,** *50,* 1090–1095.
[96] (a) Chen, B. L., Liang, C., Yang, J., Contreras, D. S., Clancy, Y. L., Lobkovsky, E. B., Yaghi, O. M., Daii, S., *Angew. Chem. Int. Ed. Engl.* **2006,** *45,* 1390–1393; (b) Pan, L., Olson, D. H., Ciemnolonski, L. R., Heddy, R., Li, J., *Angew. Chem. Int. Ed. Engl.* **2006,** *45,* 616–619; (c) Jiang, J. W., Sandler, S. I., *Langmuir* **2006,** *22,* 5702–5707; (d) Zhou, Y. Y., Yan, X.-P., Kim, K.-N., Wang, S.-W., Liu, M.-G., *J. Chromatogr. A* **2006,** *1116,* 172–178.
[97] (a) Huang, L., Wang, H., Chen, J., Wang, Z., Sun, J., Zhao, D., Yan, Y., *Micropor. Mesopor. Mater.* **2003,** *58,* 105–114. (b) Alaerts, L., Kirschhock, C. E. A., Maes, M., ven der Veen, M. A., Finsy, V., Depla, A., Martens, J. A., Baron, G. V., Jacobs, P. A., Denayer, J. F. M., De Vod, D. E., *Angew. Chem. Int. Ed. Engl.* **2007,** *46,* 4293–4297.
[98] Eldridge, R. B., *Ind. Eng. Chem. Res.* **1993,** *32,* 2208–2212.
[99] Wagener, A., Schindler, M., Rudolphi, F., Ernst, S., *Chem. Ingenieur Technik* **2007,** *79,* 851–855.
[100] Eddaoudi, M., Li, H., Yaghi, O. M., *J. Am. Chem. Soc.* **2000,** *122,* 1391–1397.

[101] Pan, L., Parker, B., Huang, X., Olson, D. H., Lee, J.-Y., Li, J., *J. Am. Chem. Soc.* **2006,** *128,* 4180–4181.
[102] Takamizawa, S., Kachi-Terajima, C., Kohbara, M., Akatsuka, T., Jin, T., *Chem. Asian J.* **2007,** *2,* 837–848.
[103] Moon, H. R., Kim, J. H., Suh, M. P., *Angew. Chem. Int. Ed. Engl.* **2005,** *44,* 1261–1265.
[104] Suh, M. P., Moon, H. R., Lee, E. Y., Jang, S. Y., *J. Am. Chem. Soc.* **2006,** *128,* 4710–4718.
[105] Sabo, M., Henschel, A., Fröde, H., Klemm, E., Kaskel, S., *J. Mater. Chem.* **2007,** *17,* 3827–3832.
[106] Hermes, S., Schröter, M.-K., Schmid, R., Khodeir, L., Muhler, M., Tissler, A., Fischer, R. W., Fischer, R. A., *Angew. Chem. Int. Ed. Engl.* **2005,** *44,* 6237–6241.
[107] Hermes, S., Schröder, F., Amirjalayer, S., Schmid, R., Fischer, R. A., *J. Mater. Chem.* **2006,** *16,* 2464–2472.
[108] Uemura, T., Kitagawa, K., Horike, S., Kawamura, T., Kitagawa, S., Mizuno, K., Endo, M., *Chem. Commun.* **2005,** 5968.
[109] Uemura, T., Hiramatu, D., Kubota, Y., Takata, M., Kitagawa, S., *Angew. Chem. Int. Ed. Engl.* **2005,** *46,* 4987–4990.
[110] Uemura, T., Kitaura, R., Otha, Y., Nagaoka, M., Kitagawa, S., *Angew. Chem. Int. Ed. Engl.* **2006,** *45,* 4112–4116.
[111] Uemura, T., Horike, S., Kitagawa, S., *Chem. Asian J.* **2006,** *1,* 36–44.
[112] Pârvulescu, A. N., Marin, G., Suwinska, K., Kravtsov, V. C., Andruh, M., Pârvulescu, V., Pârvulescu, V. I., *J. Mater. Chem.* **2005,** *15,* 4234–4240.
[113] Zou, R.-Q., Sakurai, H., Xu, Q., *Angew. Chem. Int. Ed. Engl.* **2006,** *45,* 2542–2546.
[114] Mori, W., Sato, T., Ohmura, T., Kato, C. N., Takei, T., *J. Solid State Chem.* **2005,** *178,* 2555–2573.
[115] Kato, C. N., Mori, W. C. R., *Chimie* **2007,** *10,* 284–294.
[116] Dybtsev, D., Nuzhdin, A. L., Chun, H., Bryliakov, K. P., Talsi, E. P., Fedin, V. P., Kim, K., *Angew. Chem. Int. Ed. Engl.* **2006,** *45,* 916–920.
[117] Perles, J., Iglesias, M., Martin-Luengo, M.-A., Monge, M. A., Ruiz-Valero, C., Snejko, N., *Chem. Mater.* **2005,** *17,* 5837–5842.
[118] Perles, J., Iglesias, M. I., Ruiz-Valero, C., Snejko, N., *Chem. Commun.* **2003,** 346–347.
[119] Gomez-Lor, B., Gutiérrez-Puebla, E., Iglesias, M., Monge, M. A., Ruiz-Valero, C., Snejko, N., *Chem. Mater.* **2005,** *17,* 2568–2573.
[120] Horcajada, P., Surblé, S., Serre, C., Hong, D.-Y., Seo, Y.-K., Chang, J.-S., Grenèche, J.-M., Margiolaki, I., Férey, G., *Chem. Commun.* **2007,** 2820–2822.
[121] Hasegawa, S., Horike, S., Matsuda, R., Furukawa, S., Mochizuki, K., Kinoshita, Y., Kitagawa, S., *J. Am. Chem. Soc.* **2007,** *129,* 2607–2614.
[122] Kesanli, B., Lin, W., *Coord. Chem. Rev.* **2003,** *246,* 305–326.
[123] Lin, W., *J. Solid State Chem.* **2005,** *178,* 2486–2490.
[124] Seo, J. S., Whang, D. H., Lee, S. I., Jun, J., Oh Jeon, Y. J., Kim, K., *Nature* **2000,** *404,* 982–986.
[125] Monge, A., Snekjo, N., Gutiérrez-Puebla, E., Medina, M., Cascales, C., Ruiz-Valero, C., Iglesias, M., Gomez-Lor, B., *Chem. Commun.* **2005,** 1291–1293.
[126] Cho, S.-H., Ma, B., Nguyen, S. T., Hupp, J. T., Albrecht-Schmitt, T. E., *Chem. Commun.* **2006,** 2563–2565.
[127] Wu, C.-D., Hu, A., Zhang, L., Lin, W., *J. Am. Chem. Soc.* **2005,** *127,* 8940–8941.
[128] Wu, C.-D., Lin, W., *Angew. Chem. Int. Ed. Engl.* **2007,** *46,* 1075–1078.
[129] Han, S. S., Goddard, W. A., III, *J. Am. Chem. Soc.* **2007,** *129,* 8422–8423.
[130] Mulfort, K. L., Hupp, J. T., *J. Am. Chem. Soc.* **2007,** *129,* 9604–9605.
[131] Férey, G., Millange, F., Morcrette, M., Serre, C., Doublet, M.-L., Grenèche, J.-M., Tarascon, J.-M., *Angew. Chem. Int. Ed. Engl.* **2007,** *46,* 3259–3263.
[132] Wang, Z., Zhang, B., Fujiwara, H., Kobayashi, H., Kurmoo, M., *Chem. Commun.* **2004,** *26,* 416–417. Wang, Z., Zhang, B., Zhang, Y., Kurmoo, M., Liu, T., Gao, S., Kobayashi, H., *Polyhedron* **2007,** *26,* 2207–2215.
[133] Cheng, X.-N., Zhang, W.-X., Lin, Y.-Y., Zheng, Y.-Z., Chen, X.-M., *Adv. Mater.* **2007,** *19,* 1494–1498.
[134] Kurmoo, M., Kumagai, H., Chapman, K. W., Kepert, C. J., *Chem. Commun.* **2005,** 3012–3014.

[135] Zhang, X.-M., Hao, Z.-M., Zhang, W. X., Chen, X.-M., *Angew. Chem. Int. Ed. Engl.* **2007,** *46,* 3456–3459.
[136] Neville, S. M., Moubaraki, B., Murray, K. S., Kepert, C. J., *Angew. Chem. Int. Ed. Engl.* **2007,** *46,* 2059–2062.
[137] Halder, G. J., Kepert, C. J., Moubaraki, B., Murray, K. S., Cashion, J. D., *Science* **2002,** *298,* 1762–1765.
[138] Guillou, N., Livage, C., Drillon, M., Férey, G., *Angew. Chem. Int. Ed. Engl.* **2003,** *42,* 5314–5317.
[139] Evans, O. R., Lin, W., *Acc. Chem. Res.* **2002,** *35,* 511–522.
[140] Liu, Y., Li, G., Li, X., Cui, Y., *Angew. Chem. Int. Ed. Engl.* **2007,** *46,* 6301–6304.
[141] Wong, K.-L., Law, G.-L., Kwok, W.-M., Wong, W.-T., Phillips, D. L., *Angew. Chem. Int. Ed. Engl.* **2005,** *44,* 3436–3439.
[142] Fu, R., Xiang, S., Hu, S., Wang, L., Li, Y., Huang, X., Wu, X., *Chem. Commun.* **2005,** 5292–5294.
[143] Huang, Y. Q., Ding, B., Song, H.-B., Zhao, B., Ren, P., Cheng, P., Wang, H.-G., Liao, D.-Z., Yan, S.-P., *Chem. Commun.* **2006,** 4906–4908.
[144] Bauer, C. A., Timofeeva, T. V., Settersten, T. B., Patterson, B. D., Liu, V. H., Simmons, B. A., Allendorf, M. D., *J. Am. Chem. Soc.* **2007,** *129,* 7136–7144. Lee, E. Y., Jang, S. Y., Suh, M. P., *J. Am. Chem. Soc.* **2005,** *127,* 6374–6381.
[145] Pellé, F., Surblé, S., Serre, C., Millange, F., Férey, G., *J. Lum.* **2007,** *122–123,* 492–495.
[146] De Lill, D. T., Gunning, N. S., Cahill, C. L., *Inorg. Chem.* **2005,** *44,* 258–266.
[147] De Lill, D. T., de Bettencourt-Dias, A., Cahill, C. L., *Inorg. Chem.* **2007,** *46,* 3960–3965.
[148] Zhu, W.-H., Wang, Z.-M., Gao, S., *Inorg. Chem.* **2007,** *46,* 1337–1342.
[149] Chandler, B. D., Yu, J. O., Cramb, D. T., Shimizu, G. K. H., *Chem. Mater.* **2007,** *19,* 4467–4473.
[150] Reineke, T. M., Eddaoudi, M., Fehr, M., Kelley, D., Yaghi, O. M., *J. Am. Chem. Soc.* **1999,** *121,* 1651–1657.
[151] Chen, B., Yang, Y., Zapata, F., Lin, G., Qian, G., Lobkovsky, E. B., *Adv. Mater.* **2007,** *19,* 1693–1696.
[152] Wong, K.-L., Law, G.-L., Yang, Y.-Y., Wong, W.-T., *Adv. Mater.* **2006,** *18,* 1051–1054.
[153] Rieter, W. J. L., Taylor, K. M., An, H., Lin, W., Lin, W., *J. Am. Chem. Soc.* **2006,** *128,* 9024–9025.
[154] (a) Langer, R., *Control J. Release* **1999,** *62,* 7–11; (b) Kokubo, T., Kushitani, H., Sakka, S., Kitsugi, T., Yamamuro, T. J., *Biomed. Mater. Res.* **1990,** *24,* 721–734.
[155] Horcajada, P., Serre, C., Vallet-Regí, M., Sebban, M., Taulelle, F., Férey, G., *Angew. Chem. Int. Ed. Engl.* **2006,** *45,* 5974–5978.
[156] Nitric Oxide - Therapeutics, Markets and Companies, Jain PharmaBiotech Report, March 2006.
[157] Wheatley, P. S., Butler, A. R., Crane, M. S., Rossi, A. G., Megson, I. L., Morris, R. E., *J. Am. Chem. Soc.* **2006,** *128,* 502–509.
[158] 'Pollution fighter turns clot buster', New Scientist, 2nd February 2005.

CHAPTER 5

Functionalisation and Structure Characterisation of Porous Silicates and Aluminophosphates

Nataša Zabukovec Logar, Nataša Novak Tušar, Alenka Ristić, Gregor Mali, Matjaž Mazaj, *and* Venčeslav Kaučič

Contents

Abstract

This chapter encompasses some recent achievements in preparation and structure characterisation of transition metal modified nanoporous silica- and phosphate-based framework materials for catalytic applications. Examples of successful preparation and functionalisation of new nanoporous solids using hydrothermal conventional and microwave procedures are Mn-, Fe- or Ti-containing microporous and mesoporous silicates and aluminophosphates, microporous/mesoporous silicate composites with nanosized zeolitic particles as well as mesoporous aluminophosphate thin films. Studies of structure-property relations using X-ray diffraction, spectroscopy (XAS, NMR), physical gas adsorption and electron microscopy characterisation techniques are briefly discussed.

Ordered Porous Solids
DOI: 10.1016/B978-0-444-53189-6.00005-6

Keywords: Aluminophosphates, Catalysts, EXAFS, Fe, Mesoporous, Microporous, Mn, Nanosized zeolites, NMR, SAXS, Silicates, Ti, XANES, XRD

1. Introduction

Nanoporous materials have ordered pore structures with high surface areas and large porosities. The definition of pore size in nanoporous compounds according to the International Union of Pure and Applied Chemistry (IUPAC) is that micropores are smaller than 2 nm in diameter, mesopores 2–50 nm and macropores larger than 50 nm.[1] Microporous materials are exemplified by crystalline framework solids such as aluminosilicate zeolites, whose crystal structure define channels and cages, that is, micropores, of strictly regular dimensions. Mesoporous materials, exemplified by the silicate MS41 materials family, are amorphous solids exhibiting highly ordered pore structures with narrow pore size distributions and large internal surface areas. The uniform arrangement of pores in microporous and mesoporous solids, their structural and compositional diversity, as well as the possibility to modify chemical properties, that is, functionalise the pore surfaces with specific organic ligands or metals and organometallic complexes, offers a wide variety of applications.[1,2] Both microporous and mesoporous materials have found great commercial applications as adsorbents, molecular sieves, and particularly as size- and shape-selective heterogeneous solid catalysts in the form of powders and thin films.[3]

Catalysis by microporous materials covers a broad range of economically important processes related to the upgrading of crude oil and natural gas as well as the profitable production of fine chemicals.[4] The reactions and conversions are based on the acid and redox properties and shape-selective behaviour of microporous solids. The most common reactions, where microporous acid catalysts are involved, are fluid catalytic cracking, isomerisation and transformation of aromatics. Redox microporous catalysts are increasingly used for a variety of selective oxidations on various substrates of synthetic hydrocarbons, alcohols and amines using H_2O_2 or molecular oxygen as oxidants.[5] The ordered mesoporous acid and redox catalysts have been prepared for catalytic reactions with larger molecules such as catalytic cracking of heavy oils and production of pharmaceuticals.[6,7]

In zeolites, the isolated atoms in silica matrix represent Lewis and/or Brønsted acid sites and generally do not exhibit any redox behaviour. Pure silica or aluminophosphate microporous or mesoporous frameworks are neutral and do not possess any significant catalytic activity at all. The incorporation of transition metals into silicate, aluminosilicate, aluminophosphate and similar inorganic microporous and mesoporous frameworks generates and/or moderates their redox and acid catalytic properties. The advantages of the so-prepared single-site catalysts with discrete active metal sites that mimic enzyme function are extremely high selectivities leading to the production of sharply defined molecular products.[8] Transition metal (Mn, Co, Fe, Ti, V, Cr, etc.) modified zeolites and microporous silicates catalyse a variety of selective oxidations and reductions under mild conditions with the advantage of facile recovering and recycling, if compared to homogeneous liquid-phase catalyst, like sulphuric acid.[9] The most important example is Ti-modified silicalite-1, which is an excellent catalyst with selective properties for epoxidation of olefins using H_2O_2.[10] Zeolite Beta and ZSM-5 modified with Fe, Mn or Co also show good catalytic activity for hydrocarbon oxidations and reductions.[11] Metal-modified $AlPO_4$-*n* microporous catalysts oxidise linear alkanes using molecular oxygen as reagent.[12] MeAPO-36 (Me = Mn, Co) is a bifunctional catalyst for converting cyclohexanone to ε-caprolactam, where mutually present Me^{2+} and Me^{3+} ions act as Brønsted and redox sites, respectively.[13] Direct oxidation of cyclohexene with aqueous H_2O_2 to adipic acid that is used in the production of nylon has been performed with different Ti- or Fe-substituted silicate and aluminophosphate microporous catalysts.[14] Transition metal modified mesoporous materials with aluminosilicate and aluminophosphate frameworks are reported to catalyse many acid- and redox-catalysed reactions. Among the most studied are Ti-, Fe- or Mn-modified silicate MCM-41 materials for oxidations of olefins, especially epoxidations.[7,15] More recently, Ti-, Fe-, Cu- and Cr-Al-modified mesoporous silicate SBA-15 materials with higher thermal stability compared to MCM-41 series were reported to catalyse oxidation and polymerisation reactions.[16] Potential in catalytic oxidations has been demonstrated also for mesoporous aluminophosphates containing Ti, Co, Mg, Fe and V.[17] So far, mesoporous catalysts did not exhibit catalytic properties comparable to those of microporous catalysts.[7,14] Low hydrothermal stability, hydrophilicity and leaching of metal species from the solid deactivate the mesoporous catalysts and disable their recovery. The preparation of microporous/mesoporous composite silicates and aluminophosphates, where we combine the high surface activity of microporous domains and better diffusion of reactants to the catalytically active sites in mesopores, is expected to overcome these problems.[18] Ti-modified microporous/mesoporous silicates show better hydrothermal and compositional stability than their mesoporous analogues, but the information on the catalytic properties of such materials is still limited.[19]

In this chapter, we will briefly describe synthetic procedures for the preparation of nanoporous silicates and aluminophosphates and some characterisation techniques for the determination of pore structure and nature of metal active sites. The emphasis is placed upon our recent results on Mn-, Fe- and Ti-modified microporous and mesoporous powders, nanosized zeolitic particles and microporous/mesoporous composites, as well as mesoporous thin films.

2. Hydrothermal Conventional and Microwave Synthesis of Nanoporous Silica- and Aluminophosphate-Based Materials

The synthesis of microporous materials is a complex process and has been discussed in many review papers and chapters in detail.[20] A brief summary is that microporous materials (zeolites and AlPOs) are usually prepared hydrothermally, sometimes solvothermally, from aqueous gels containing a source of the framework building elements (Si, Al, P), a mineraliser (OH^-, F^-) regulating the dissolution/condensation processes during the crystallisation, and a structure-directing agent or template, usually an organic amine or ammonium salt.[1,20] The most important synthesis parameters are gel composition, reaction components, sequence of addition, gel aging, seeding, pH, crystallisation temperature and time. Synthesis parameters direct the crystal assembly pathway and the final products formed and have to be well controlled. More recent literature describes a steam-assisted dry gel conversion techniques for the synthesis of TS-1,[21] the preparation of microporous products from layered silicates and aluminophosphates with kanemite-type layered structure as precursors,[22] the vapour-phase method for the preparation of large single crystals of the clathrate compound MTN (a zeolite-like material)[23] or the TS-1 synthesis using mechanochemical reaction by grinding titanium and silica powders with a planetary ball mill and succeeding hydrothermal treatment.[24] The majority of the syntheses of ordered mesoporous silica- and nonsilica-materials are based on the same principles, that is, the hydrothermal procedure using various structure-directing agents (surfactants), like cationic cetyltrimethylammonium hydroxide (synthesis of MCM-41) or nonionic block copolymers (synthesis of SBA-15).[1,25] Precise adjustments of synthetic parameters, like silica source, temperature and the time of crystallisation or pH, crucially affect the properties of final products, for example, pore size or hydrothermal stability. The diameter of the resulting mesopores is usually controlled by choosing an appropriate surfactant. In the recent report, the shrinkage in pore size of MCM-41 down to the submicroporous region was achieved by adding organic trialkoxysilanes (chloro propyl-, vinyl-, methyl-) in the usual synthesis mixture.[26] Transition metals can be incorporated into microporous or mesoporous materials by a post-synthetic ion-exchange treatment (impregnation) or by direct framework substitution by the addition of transition metal cations into the synthesis gel.[27] There are also some alternative routes reported, like the use of ultrasonic waves employed to incorporate ruthenium into the pore structure of SBA-15.[28] The strategies for the functionalisation of nanoporous matrix by organic groups and organometallic complexes, usually carried out by direct co-condensation method or post-synthetic grafting strategies,[2,26,27] will not be discussed here.

Synthesis of nanoporous materials is performed using conventional and/or microwave heating at elevated temperatures (80–200 °C) under autogeneous pressure in the time periods from several minutes to a couple of days. As-synthesized products are generally calcined in air or oxygen flow. By that step the template is removed from the pores, which makes them accessible. Usually, temperatures in the range of 450–600 °C are applied to remove the organic components, which is not critical for the stability of the majority of microporous zeolitic and also mesoporous compounds. The removal of the template by ion exchange is suitable only for small template molecules that are not hindered by the size of the pore openings.

Microwave technique is regarded as a novel synthesis tool for microporous and mesoporous materials because it offers several benefits, such as homogeneous nucleation, promotion of faster crystallisation, rapid synthesis, formation of uniform crystals and small crystallites, facile morphology control and avoidance of undesirable phases by shortening the synthesis time.[29] Recently, it was found that it provides an effective way to control the particle size distribution, crystal morphology, orientation and even the crystalline phase.[30]

Great attention has recently been focused on the synthesis of nanosized microporous particles (up to 200 nm). The major interest in nanosized zeolites is due to their use for the preparation of zeolite films and membranes as well as microporous/mesoporous composites. The reduction of particle size from micrometer to the nanometer scale leads to substantial changes in the properties of the materials; nanoparticles have large external surface areas and high surface activity. The external surface acidity is of importance when the zeolite is intended to be used as a catalyst in reactions involving bulky molecules. In addition, smaller zeolite crystals have reduced diffusion path lengths relative to conventional micrometer-sized zeolites. Many zeolitic materials have been prepared in the form of colloidal suspensions with narrow particle size distributions.[31] Silicalite-1 is the most frequently studied system. The synthesis of zeolite crystals with narrow particle size variation requires homogeneous distribution of viable nuclei in the system. The homogeneity of the starting clear solution together with the formation of precursor gel particles, and their transformation into crystalline zeolitic material, are very important. In general, very dilute systems containing high amounts of tetraalkylammonium hydroxides are used in the synthesis of zeolite nanocrystals to avoid the aggregation of the particles. All these factors together with proper choice of silica source allow the stabilisation of starting clear solutions, where only discrete gel particles are present. Discrete gel particles are often called colloidal zeolites.

For the preparation of microporous/mesoporous composites two synthetic procedures have been the most successful.[32,33] The first approach is based on the building of mesoporous materials from nanosized zeolitic species and the second on the creation of mesopores in the microporous crystals. For the design of mesoporous materials from nanosized zeolitic species, the best results were obtained when the synthesis was carried out in two steps. In the first step, zeolite seeds were prepared by the shortened hydrothermal treatment and then the second hydrothermal reaction was performed after adding surfactant, in order to direct the mesoporous phase.[34] The preparation of mesoporous zeolitic single crystals involve partial dissolution of microporous solid, dealumination procedures, such as steaming, acid and alkaline treatments[35] or a procedure, where mesoporous carbon matrix is impregnated with reaction mixture for zeolite ZSM-5 synthesis, and zeolite crystals are grown around the carbon particles.[36] The removal of encapsulated carbon matrix by calcination leads to isolated large zeolitic crystals with uniform mesoporous system. These and some other procedures, like a transformation of the preassembled walls of mesoporous materials such as MCM-41 and SBA-15 into zeolitic structures by post assembly treatment with microporous structure-directing template, coating of mesoporous materials with nanozeolite seeds using very diluted clear zeolite gels or formation of delaminated zeolites are also reported and reviewed in the literature.[37] A direct hydrothermal assembly process was recently developed to synthesize microporous/mesoporous aluminophosphates by the addition of organosilane surfactants into the conventional synthesis composition for crystalline microporous aluminophosphates.[38]

3. Determination of Structure Porosity and Acid Sites in Nanoporous Catalysts

Further progress in the field of nanoporous science and heterogeneous catalysis depends on the chemical and structural knowledge of the nature of surface active sites and the mechanisms of chemical reactions catalysed by these sites. Structural information that is essential for the understanding and designing of new syntheses is the size and connectivity of the channels and cavities and the coordination, location, oxidation state and strength of bonding of the ion that acts as the active site.[39] Besides, the structural data on working catalysts under *in situ* conditions is beneficial for the investigation of heterogeneous reaction systems.

3.1. X-ray diffraction

The conventional single-crystal X-ray diffraction gives the most complete answers about the structure properties of ordered crystalline materials. For many nanoporous materials, however, it cannot provide reliable structural information. This is the case, when the synthesized crystals or crystallites are too small with highly polycrystalline morphologies, when there is a low concentration and random distribution of metal active sites over the framework or extra-framework positions or simply when the material is not fully crystalline.

In the recent past, structure determination of microporous materials using X-ray diffraction techniques has experienced considerable developments in techniques and methodology. Rapid development of synchrotron radiation sources enabled the development of so-called microcrystallography, dealing with a few micron-sized single-crystals, and development of anomalous dispersion methods for the localization of active metal sites in the structures.[40] Synchrotron radiation has also enabled *in situ* diffraction studies of the structural changes during crystallisation and phase transitions as a function of temperature or pressure and *in situ* studies of reaction kinetics by following the structural changes during the catalysis and other processes that are taking place on the microporous surfaces.[41] We have recently written two review papers about the principles of X-ray anomalous dispersion and about the use of X-ray diffraction and anomalous dispersion method in the structure elucidation of microporous materials.[42,43] Maybe even more important is a constant development of powder X-ray diffraction methods that combine experimental data with crystal chemistry-based modelling. An interesting example is an algorithmic advance that facilitates combined analysis of powder diffraction and electron microscopy data to solve particularly insolvable zeolite structures. Using this method, the markedly complex 10-ring channel system forming the IM-5 zeolite, an active catalyst for hydrocarbon cracking and related reactions, was determined.[44]

X-ray powder diffraction is also frequently used for the characterisation of mesoporous materials. The cubic pore structure system of MCM-48 mesoporous silicate material has been characterised from X-ray diffraction data by applying recently developed methods of mesostructure analysis and full-profile refinement.[45] Techniques, like small angle X-ray scattering and grazing incidence reflectometry in combination with synchrotron radiation have been used for the examination of the kinetics of crystallisation in periodic templated mesoporous powders and thin films, and for the study of sorption and capillary condensation of an organic fluid in ordered mesoporous silica.[46]

Nevertheless, the powerful X-ray diffraction methods often fail for structural determination of complex mesoporous materials, nanoparticles or nanospecies encapsulated in mesoporous hosts. Generally, no broadly applicable and robust methods exist to replace crystallography for the characterisation in these cases. At the moment, complementary characterisation techniques, like high-resolution transmission electron microscopy (HRTEM) and physical gas adsorption, are used as a support in structural studies of amorphous mesoporous solids.[47] In the future, successful solutions to the nanostructure problems will probably have to involve interdisciplinary research that will combine modelling and experimental studies.[48]

3.2. High-resolution transmission electron microscopy and physical gas adsorption

The power of HRTEM is in gaining the real (direct)-space structure information now routinely at the atomic level, which can be combined with the electron diffraction information in reciprocal space.[49] For example, HRTEM and electron crystallography were successfully used for the determination of the structure of mesoporous MCM-48 material.[52] HRTEM was also used to confirm the presence of micropores and mesopores at the same area of investigation in (Ti,Al)-Beta/MCM-41 and (Ti,Al)-Beta/MCM-48 composites.[50,51] Additionally, good resolution images of nanosized Ti-Beta particles were collected using this method (Ch. 4.3). Physical gas adsorption is extensively used in the characterisation of micro- and mesoporous materials and is often considered as a straightforward-to-interpret technique.[53] The isotherm obtained from these adsorption measurements provides information on the surface area, pore volume and pore size distributions.[54] Probe gases like N_2, Ar and CO_2 are frequently used as adsorptives. N_2 adsorption at 77 K is a standard and widely used method, which provides valuable information about surface properties of porous adsorbents. If applied over a wide range of relative pressures, nitrogen adsorption isotherms provide information on size distributions in the micro-, meso- and macroporosity range. The presence of microporosity in microporous/mesoporous composites was proved with nitrogen adsorption measurements. The mass percentage (μ)[32] of the microporous material in the (Ti,Al)-Beta/MCM-48 [51] was ~6.5%, while ~45% mass percentages of microporous material were present in (Ti,Al)-Beta/MCM-41 composite.[50]

3.3. X-ray absorption spectroscopy

With the availability of synchrotron radiation sources, X-ray absorption spectroscopy (XAS) techniques have developed into a widely used tool for structural research of materials in any aggregate state. XAS analytical methods XANES (X-ray absorption near-edge structure) and EXAFS (extended X-ray absorption fine structure) provide microscopic structural information of a sample through the analysis of its X-ray absorption spectra of selected atoms. XANES identifies local symmetry and the average oxidation number of selected atom. EXAFS provides the description of a short-range order for selected atom in terms of the number of neighbours, distances, and thermal and static disorder within the range of those distances. Since XAS is selective towards a particular element and sensitive only towards short-range order, it is one of the most appropriate spectroscopic tools for characterisation in the field of catalysis.[55]

Metal ions, which generate catalytically active sites in metal-modified porous silicates and aluminophosphates, can isomorphously substitute framework elements (Al, P or Si) or can be attached to the aluminophosphate or silicate framework. Structure characterisation of such catalysts by XAS provides the information on the local environment of metal species as well as framework elements.[56] With the development of *in situ* methods, XAS also provides information on the formation process of the catalysts[57] and information on the behaviour of catalytically active sites during the reactions.[58] XAS techniques are decisive methods to follow of the synthesis pathways and also for the recognition of structural properties that are relevant to the overall optimal performance of a synthesis product as a potential catalyst.

Along these lines, we have studied local environment of metals using XAS techniques in several metal-modified porous aluminophosphates (MeAPO)[59–63] and metal-modified porous silicates.[50,51,64–66] For example, we have recently shown that the isomorphous substitution of framework aluminium by iron leads to redox sites in FeAPO-36[61] and FeHMA.[62]

The principles of the XAS analytical methods XANES and EXAFS are illustrated in Fig. 5.1 for the FeAPO-36 material.[61] The normalized Fe XANES spectra of the samples and reference compounds are shown on the left side of the Fig. 5.1. The zero energy is taken at the first inflection point in the corresponding metal spectrum (7112 eV), that is, at the 1s ionization threshold in the corresponding metal. The shape of the K-edge and the pre-edge resonances are characteristic of the local symmetry of the investigated atom and can be used as fingerprints in identification of its local structure. Tetrahedrally coordinated atoms, lacking an inversion centre, exhibit a single pre-edge peak, which can be assigned to 1s $\rightarrow$ 3d transition. Octahedral symmetry is evident from two weak resonances in the pre-edge region assigned to transitions of 1s electron into antibonding orbitals with octahedral symmetry. Characteristic tetrahedral resonance is present in both, the as-synthesized and template-free FeAPO-36, demonstrating that iron cations are incorporated into the tetrahedral sites. The pre-edge peak is weaker in the case of the as-synthesized sample, which indicates that the tetrahedral symmetry of the iron cations in this sample is slightly distorted. Changes in the valence state of metal cations in the samples during calcination can be deduced from the energy shift of the Fe absorption edge. A linear relation between the edge shift and the valence state was established for the atoms with the same type of ligands. From the spectra of the reference samples ($FeSO_4$ and $FePO_4$) with a known iron oxidation states, we found that the Fe K-edge shifts for 3.0 eV per valence state. The Fe XANES spectra of the FeAPO-36 samples clearly indicated oxidation of iron cations after the calcination: the Fe K-edge in the template-free sample is shifted for 1.5 eV to higher energies compared to the as-synthesized sample. From the energy shifts of the Fe K-edge, we obtained an average iron valence of 2.5 ± 0.1 in the as-synthesized and 3.0 ± 0.1 in the template-free sample. We thus concluded that the as-synthesized FeAPO-36 contained a mixture of Fe(II) and Fe(III) in the ratio 1:1, while during calcination all Fe(II) in the sample oxidised to Fe(III).

The Fe K-edge EXAFS spectra of FeAPO-36 material were quantitatively analysed for the coordination number, distance, and thermal and static disorder (Debye-Waller factor) of the nearest coordination shells of neighbour atoms. Fourier

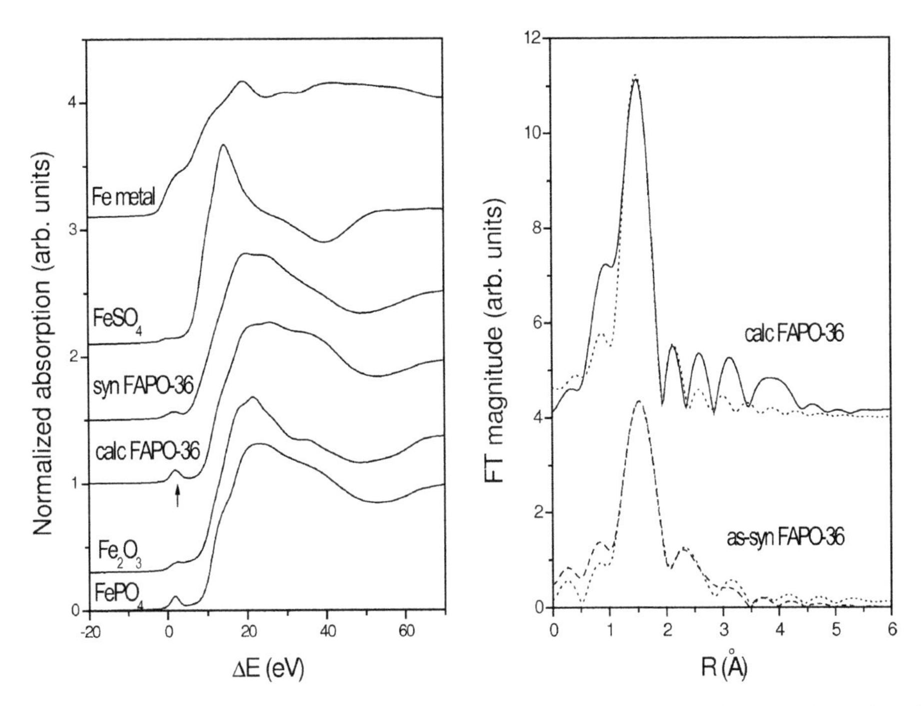

Figure 5.1 Normalised Fe K-edge XANES spectra of the as-synthesised (syn FAPO-36) and template-free FAPO-36 (calc FAPO-36) samples and references: $FePO_4$, $FeSO_4$, Fe_2O_3, and Fe metal (left). Fourier transforms of k^3-weighted Fe EXAFS spectra of the as-synthesised and template-free FAPO-36 (experiment – solid line, EXAFS model - dotted line).

transform magnitudes of Fe EXAFS spectra are shown on the right side in Fig. 5.1. An EXAFS model of the iron nearest neighbour environment composed of oxygen atoms was constructed. Best-fit parameters are listed in Table 5.1. The fit of the first coordination shell showed that in the as-synthesized sample iron was coordinated to four oxygens at 1.94(2) Å, which indicated the insertion of iron cations into the tetrahedral sites of the as-synthesized FeAPO-36 structure. In addition, two oxygen atoms at much longer distance of 2.49(4) Å with much larger Debye-Waller factor were found. This indicated the presence of coordinated water molecules in the

Table 5.1 Structural parameters of the first coordination shell around iron in the as-synthesized and template-free FeAPO-36: number of atoms (*N*); Fe–O distance (*R*); Debye-Waller factor (σ^2)

	As-synthesized			Template-free		
Sample	*N*	*R* (Å)	σ^2 (Å^2)	*N*	*R* (Å)	σ^2 (Å^2)
FAPO-36	4.2(9)	1.94(2)	0.006(2)	4.0(9)	1.86(1)	0.006(9)
	2.0(9)	2.49(4)	0.022(7)			

Uncertainties of the last digit are given in the parentheses.

pores. In the template-free sample, four oxygen neighbours at a shorter distances of 1.86(1) Å were found. There is, however, no evidence for the oxygen neighbours at longer Fe–O distances, which means that there are no water molecules in the pores coordinated to iron cations. From both, XANES and EXAFS results, we can conclude that in the template-free FeAPO-36 iron is incorporated into the tetrahedral framework sites in the form of Fe(III).

3.4. Nuclear magnetic resonance spectroscopy

Solid-state nuclear magnetic resonance (NMR) spectroscopy probes magnetic environment of nuclei in materials. The technique is very sensitive to changes in the coordination environment and can be used as an element-specific structure analysing tool. The environments of framework and extra-framework atoms of porous solids can be studied by spectroscopy of ^{29}Si, ^{27}Al, ^{31}P, ^{69}Ga or ^{71}Ga nuclei, or nuclei of charge-compensating ions like ^{1}H, ^{7}Li, ^{23}Na or ^{133}Cs. Such measurements can easily provide information about the number of inequivalent atomic sites and about the multiplicity of these sites. Recent development of techniques that enable one to probe internuclear distances increased the power of NMR even further. As an example, structure of two zeolites was recently solved using double-quantum dipolar recoupling NMR spectroscopy, which probed the distance-dependent dipolar interactions between ^{29}Si nuclei in the framework.[67] The potential of NMR spectroscopy for structural analysis was enhanced also because of recent progress in the field of *ab initio* computational methods, which can now readily predict NMR-observable parameters like chemical shift and quadrupolar coupling parameters.[68] First tests show that combination of NMR spectroscopy and *ab initio* calculations is very sensitive to structural variations and can complement X-ray powder diffraction information substantially.

The traditional power of NMR is certainly in its ability to study materials or motifs that lack long-range order.^{29} Si magic-angle-spinning (MAS) NMR spectroscopy in silicates and ^{27}Al and ^{31}P MAS NMR spectroscopy in aluminophosphates can establish the extent of molecular framework order and the degree of condensation. For example, linewidths in MAS NMR spectra of mesoporous materials can be an order of magnitude larger than linewidths of MAS NMR spectra of well-crystalline microporous materials. Still signals of nuclei in completely condensed $Si(OSi)_4$ or $P(OAl)_4$ units can be resolved from signals of $Si(OSi)_3(OH)$ or $P(OAl)_3(OH)$ units of disordered materials. Along these lines, we have studied the extent of framework order and the degree of condensation in mesoporous aluminophosphates and in aluminosilicate composites (Ti,Al)-Beta/MCM-48 and (Ti,Al)-Beta/MCM-41. In aluminophosphates, the framework consolidation process under the thermal treatment of thin films and powders was studied (Ch. 5) and in modified aluminosilicates framework order of composite materials was compared to the order observed in pure zeolitic (Ti,Al)-Beta and mesoporous MCM-48 and MCM-41 materials.[50,51]

For many years, NMR spectroscopy has also been extremely valuable for studying catalytic properties of porous materials. The fields of application of NMR spectroscopy ranged from the studies of guest–host interactions[69] and the *in situ* studies of reactions catalysed by zeolites,[70] to investigations of the structure of Brønsted acid sites.[71] In silicate materials, acid sites are usually generated by partial incorporation of aluminium into the silicate framework. In such materials, ^{29}Si chemical shifts depend sensitively on the number of silicon and aluminium atoms connected with a given SiO_4 tetrahedron. This allows one to quantify the framework Si/Al ratio.[72] The nature and the strength of acid sites can be further investigated by ^{1}H MAS NMR. In a similar way as acid sites in aluminosilicates, the catalytic centres of microporous and mesoporous alumino-phosphates can be obtained by partial substitution of framework aluminium with transition metal ions. ^{31}P NMR spectroscopy of aluminophosphates then plays a similar role as ^{29}Si NMR spectroscopy of aluminosilicates.[73] Recently, we have shown that broadline ^{31}P NMR can be employed for studying Ni(II), Co(II), Fe(II/III) and Mn(III) incorporation, when the extent of substitution, that is, Me/Al fraction is above 1%.[74] The principle is illustrated in Fig. 5.2, which presents high-resolution

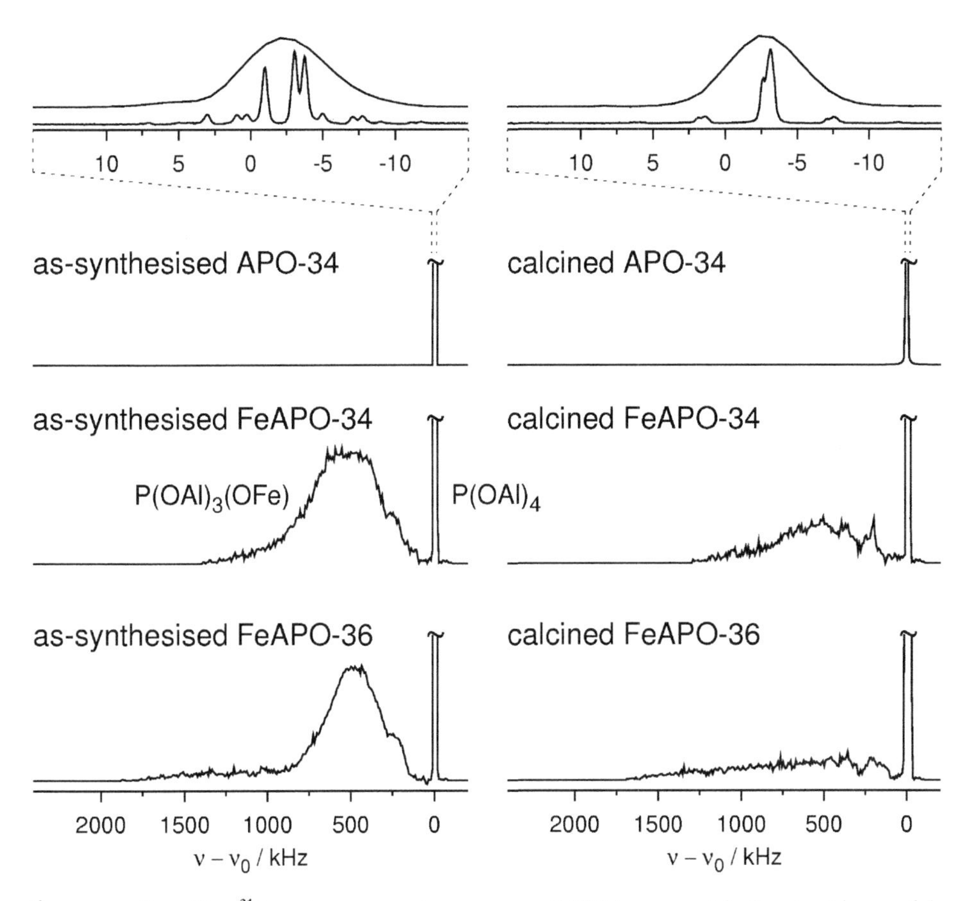

Figure 5.2 Broadline ^{31}P nuclear magnetic resonance (NMR) can provide direct evidence of the incorporation of iron into the aluminium framework sites of aluminophosphate molecular sieves. Quantitative analysis of NMR spectra can yield information about the distribution and the amount of the incorporated metals. In both FeAPO-34 and FeAPO-36 materials, substantial leaching of iron from the aluminophosphate framework was detected upon calcinations.[74]

^{31}P MAS NMR spectra of as-synthesized and calcined APO-34 aluminophosphates, and static broadline ^{31}P NMR spectra of as-synthesized and calcined APO-34, FeAPO-34, and FeAPO-36 aluminophosphates. While sharp NMR lines at ~0 kHz belong to phosphorus nuclei in familiar $P(OAl)_4$ environment, broad lines in the region between 200 kHz and 1500 kHz belong to phosphorus nuclei in $P(OAl)_3(OFe)$ environment and provide an evidence of the incorporation of Fe (II/III) ions into the aluminium framework sites. Intensities of these broad lines also carry information about the amount of the incorporated iron. Spectra in Fig. 5.2, for example, clearly demonstrate that calcination reduces the amount of iron in both FeAPO materials.

The potential of NMR spectroscopy is certainly not exhausted by the few listed topics and some additional fields of application are reviewed in reference.[75]

4. Mn-, Fe- and Ti-Functionalised Microporous and Mesoporous Silicates and Aluminophosphates: Case Studies

Functionalisation of silicate and/or aluminophosphate matrices with titanium, iron, manganese and/or cobalt has shown to be the most successful for the preparation of highly selective single-site catalysts.[9–19] Ti^{4+} most readily substitute for tetrahedral silicon in the microporous silicate frameworks, which results in the formation of stable redox catalytic sites. In mesoporous silicates, the isolated Ti^{4+} centres are mostly generated via grafting procedures.[8] The incorporation of Fe^{3+}, Mn^{3+} and Co^{3+} ions in the microporous and mesoporous silicate frameworks is more difficult. The stability of single acid sites is generally much lower than that of Ti^{4+}. For example, high temperature can cause migration of heteroatoms into extra-framework positions, where they can agglomerate.[8] The inclusion of Fe^{3+}, Mn^{3+}, Co^{3+}, as well as Fe^{2+}, Mn^{2+} and Co^{2+} species in the Al framework sites in the microporous aluminophosphate frameworks is usually successful by using direct-synthesis methods. In the so-prepared acid and/or redox catalysts, metal sites are quite stable and do not leach from the framework. Preparation of single-site catalysts based on mesoporous aluminophosphate frameworks is a challenge.

In the following text, we will show some examples of Mn-, Fe-, and Ti-functionalised microporous and mesoporous silicates and aluminophosphates with stable catalytic sites that we have prepared using different synthetic approaches. The emphasis was on the use of microwaves during the crystallisation period and on the preparation of nanosized metal-modified microporous particles that have also been included in the mesoporous matrix to form microporous/mesoporous composites. The use of microwaves showed particularly beneficial in the synthesis of metal-modified mesoporous aluminophosphates. The functionalisation of nanosized particles by transition metals turned out to be much more difficult then the incorporation of metals in the fully crystalline zeolitic particles and needed great attention during the synthesis including the use of alternative templates and framework metal sources.

4.1. Mn-modified silicates and aluminophosphates

Manganese-containing redox catalysts have received considerable attention for the selective oxidation of hydrocarbons using air/molecular oxygen as the oxidant.[76] Many Mn-containing microporous and mesoporous materials have been reported; most of them comprise manganese on extra-framework positions as manganese complexes, and very few with manganese in the framework positions. Two examples of single-site Mn-modified microporous compounds are MnAPO-5 and MnAPO-18 materials that were already found to be a good catalyst for the ethylation of benzene.[77] The importance of the location and nature of Mn in the mesoporous MCM-41 for the catalytic activity in oxidation reactions was shown for gas-phase grafted and impregnated Mn-MCM-41.[78] The uniqueness of the nanoporous manganese silicates is obvious from their superior activity and stability with respect to other well-known redox molecular sieve catalysts such as [Me]S-1 (Me = V, Ti), [Me]APO-5 (Me = Cr, Mn) or [Me]MCM-41 (Me = Cr, Co).[4,76]

4.1.1. Synthesis and characterisation of MnS-1, MnMCM-41 and MnHMA

We have prepared a series of Mn-modified microporous and mesoporous materials using direct synthetic procedures for the incorporation of Mn on framework positions. MnS-1 crystals with micrometer dimensions were synthesized hydrothermally in the presence of tetraethylammonium hydroxide (TEAOH) as template for the first time.[64] The use of different template and silica source, that is, tetrapropylammonium hydroxide (TPAOH) as a template and tetraethyl orthosilicate (TEOS) as silica source, lead to the formation of MnS-1 crystals with nanometer dimensions (Fig. 5.3). MnMCM-41 was prepared by direct hydrothermal method in the presence of cetyltrimethylammonium chloride (CTACl) as a template.[65] Thermally stable manganese-modified hexagonal mesoporous

Figure 5.3 MnS-1: Crystals with micrometer dimensions (~10 μm) (left) and crystals with nanometer dimensions (~100 nm) (right).

aluminophosphate (MnHMA) was synthesized hydrothermally in a microwave oven in the presence of CTACl as a template.[63]

The diffraction and nitrogen sorption studies revealed the high quality of the materials with respect to long-range order and specific surface area. The X-ray absorption spectroscopic methods (XANES, EXAFS), electron paramagnetic resonance (EPR) and electron spin-echo envelope modulation (ESEEM) techniques that were used for the study of local structure of all three samples revealed stable single metal framework sites. In silicate-based products (MnS-1, MnMCM-41), the isomorphous substitution of Si^{4+} by Mn^{3+} in the framework was determined.[64,65] The manganese cations in MnS-1 were coordinated to three oxygen atoms, two at a distance of 1.93(1) Å and one oxygen atom at a longer distance of 2.15(2) Å. Additionally, at a larger distance of 3.52(2) Å a presence of Si is indicated. Local environment of manganese in MnMCM-41 showed the coexistence of Mn^{2+} and Mn^{3+} cations. Mn^{3+} cations were incorporated into the MCM-41 framework and coordinated to three oxygens in the first coordination shell, two at a distance of 1.92(1) Å and one oxygen atom at a distance of 2.21(1) Å, similar to that in MnS-1. Manganese in aluminophosphate-based product (MnHMA) was present in the form of Mn^{2+} and Mn^{3+} ions in the ratio 40%/60% coordinated with four oxygen atoms.[63] EXAFS analyses revealed four oxygens in the first coordination sphere of manganese at the distances of 2.15(1) Å. Additionally, at a distance of 3.28(2) Å one Al atom was found. EPR and ESEEM measurements confirmed that Mn^{2+} was located into the mesoporous aluminophosphate framework.

4.1.2. Probing the catalytic activity of Mn-modified silicates

Catalytic tests that we have performed on the self prepared MnS-1, MnMCM-41 and MnMCM-48 showed that manganese(III)-containing microporous and mesoporous silicas selectively catalyse the oxidation of alkyl aromatics in benzylic position (ethylbenzene, 4-methylethylbenzene and diphenlymethane) to aromatic ketones in the absence of any initiator by using molecular oxygen as the terminal oxidant under mild, solvent-free, liquid-phase conditions. The presence of isolated Mn^{3+}-species in a hydrophobic environment on the catalyst surface favours high conversion rates. The results will be published elsewhere.

4.2. Fe-modified silicates and aluminophosphates

The substitution of iron is widely used to modify the original properties of porous catalyst hosts in microporous and mesoporous structures. Both the acidity and redox properties can be modified by inserting iron into the structure in the framework sites or as extra-framework iron complexes.[79] Iron-containing microporous heterogeneous catalysts are known for their remarkable activity in the reduction of nitrous oxides, oxidation of cyclohexane, oxidation of benzene to phenol and the selective oxidation of methane.[80] The most applicable iron-containing microporous catalysts are FeZSM-5 and FeS-1. Among microporous aluminophosphates, FeAPO-5, FeAPO-11 and FeVPI-5 were reported to be catalytically active in the oxidation of aromatic compounds, such as hydroxylation of phenol and benzene, and epoxidation of styrene as well as the oxidation of naphthol. The catalytic activity of FeAPO-5 and FeAPO-11 is in some reactions comparable with that of TS-1. The preparation method has a very great influence on the catalytic performances with respect to activity and stability since various iron species have been identified.[81] For example, iron silicates show high catalytic activity and stability for decomposition of N_2O to N_2 and O_2, that is regulated by the redox cycle Fe^{2+}/Fe^{3+}. Mesoporous FeMCM-41 was reported to catalyse oxidation of ketones and partial oxidation of methane. Mesoporous iron-functionalised catalysts have been reported to have high catalytic activity, for example, FeHMS (Fe-modified hexagonal mesoporous silica) and FeMCM-41 silicates for phenol hydroxylation or FeSBA-1 silicate for phenol *tert*-butylation.

4.2.1. Synthesis of FeAPO-34, FeAPO-36, FeVPI-5, FeHMA and FeTUD-1

We have prepared iron-substituted microporous aluminophosphates (FeAPO-34, FeAPO-36, FeVPI-5), mesoporous aluminophosphate FeHMA and mesoporous silicate FeTUD-1 using direct synthesis procedures for iron incorporation on framework positions. FeAPO-34, FeAPO-36 and FeVPI-5 were synthesized hydrothermally using piperidine, tripropylamine and dibutylamine as templates, respectively.[59,60,82] FeHMA was synthesized hydrothermally under microwave conditions in the presence of CTACl as a template.[62] A new templating method using small, inexpensive non-surfactant triethanolamine, reported for the preparation of mesoporous silicate with three-dimensional sponge-like pore structure denoted as TUD-1, was used also for the preparation of FeTUD-1 material.[66]

4.2.2. Characterisation of Fe-modified aluminophosphates FeAPO-34, FeAPO-36, FeVPI-5 and FeHMA

Table 5.2 shows that the framework iron within aluminophosphate products is mostly tetrahedrally coordinated. The octahedral coordination of iron in FeHMA thus indicates that iron occupies extra-framework positions. However, Mössbauer spectra demonstrate the tetrahedral coordination of iron after water desorption in vacuum at 427 °C and reversibility of $Fe^{3+} \leftrightarrow Fe^{2+}$ redox transitions.[62] Furthermore, iron environment within FeHMA is very similar to the iron environment within FeVPI-5,[60] for which acid and redox sites have been found. FeVPI-5 is an extra large-pore microporous aluminophosphate (on the border to mesoporous) with three inequivalent aluminium framework sites. One of them is octahedrally coordinated by four framework oxygen atoms and two water molecules. It is suggested that iron within FeVPI-5 occupies such octahedrally coordinated framework site. We suppose that similar situation occurs in FeHMA, since NMR spectra of template-free material also reveal the presence of six coordinated aluminium sites within the material.

4.2.3. Characterisation of Fe-modified silicate FeTUD-1

EXAFS analysis reveals that in microporous silicate FeS-1, the iron forms small particles of Fe_3O_4,[84] while in mesoporous silicate FeMCM-41 iron is present in an isolated form.[85,86] The literature data on iron local environment in FeTUD-1 describe different iron sites present, ranging from isolated iron atoms to iron oxide

Table 5.2 Structural parameters of the first coordination shell around Fe atom in the template-free microporous and mesoporous iron-modified aluminophosphate structures (FeAPO-*n*, *n* denotes structure type) compared to the reference samples ($FePO_4$, $FePO_4 \times 2H_2O$, $FeAl_2O_4$): average number of oxygen atoms (*N*), distance (*R*), and thermal/static disorder (Debye-Waller factor σ^2)

Structures	Reference	Pores (Å)	Fe valence state	*N* oxygen atoms	*R* (Å)	σ^2 (Å^2)
Microporous		**Pore opening**				
FeAPO-34	82	3.8 × 3.8	Fe^{3+}	4	1.86	0.004
FeAPO-18	83	3.8 × 3.8	Fe^{3+}	4	1.85	0.012
FeAPO-5	83	7.3 × 7.3	Fe^{3+}	4	1.85	0.011
FeAPO-36	61	7.4 × 6.5	Fe^{3+}	4	1.86	0.006
FeVPI-5	60	12	Fe^{3+}	6	1.98	0.011
Mesoporous		**Hex. unit cell parameter** a_0				
FeHMA	62	46.9	Fe^{3+}	6	1.97	0.008
Reference samples						
$FePO_4$	83	–	Fe^{3+}	4	1.86	–
$FePO_4 \times 2H_2O$	83	–	Fe^{3+}	6	1.95	–
$FeAl_2O_4$	83	–	Fe^{2+}	4	1.95	–

nanoparticles and/or bulk crystals of iron oxide.[87] Combined use of X-ray absorption and Mössbauer spectroscopies enable detailed characterisation of iron local environment in mesoporous silicate FeTUD-1.[66] The results of XANES and EXAFS analyses of the template-free FeTUD-1 show that the sample contains only isolated iron in Fe^{3+} form that is coordinated to six oxygens distributed at different distances in the R range from 1.86(1) Å to 2.00(2) Å. Mössbauer spectroscopy reveal that two distorted octahedral types of isolated Fe^{3+} cations exist in the ratio of 1:1. EXAFS and Mössbauer spectroscopy results also proved that there was no iron oxide present in prepared material.

4.3. Ti-modified silicates and aluminophosphates

The field of porous titanium silicates is one of the fastest developing areas of porous materials.[88] The materials possess remarkable catalytic activity in selective oxidation of organic compounds. Microporous titanium silicates, such as zeolites Ti-silicalite-1 and Ti-Beta, are extremely efficient catalysts for epoxidation of alkenes in the presence of aqueous H_2O_2 and *tert*-butyl hydroperoxide (TBHP) as oxidants, which is attributed to the unique architecture of titanium centres that are isolated in the silicate framework.[89] Titanium was also incorporated into mesoporous silica MCM-41, MCM-48 and SBA-15.[90] Ti-MCM-41 and Ti-MCM-48 have been found to catalyse the selective oxidation of bulky organic molecules. Mesoporous Ti-SBA-15 molecular sieve showed low catalytic activity in oxidation reactions with H_2O_2, but high catalytic activity for epoxidation of alkenes by TBHP. However, the problems of the titanium leaching and catalyst deactivation in the presence of aqueous H_2O_2 still considerably restricts the possibility of practical applications of these materials, that is, the metallic nanoparticles can aggregate over time leading to a loss of an active surface.

We have concentrated on the preparation of titanium-modified microporous/mesoporous catalysts to overcome the problem of unstable acid sites. The first approach that we have developed was the preparation of nanosized (Ti,Al)-Beta particles that were later organised in the mesoporous matrices of MCM-41 and MCM-48. The second approach was the post-synthesis deposition of different concentrations of nanosized Ti-Beta zeolite on the walls of mesoporous SBA-15. Both approaches resulted in new materials that did not suffer from Ti leaching during thermal treatments.

4.3.1. Synthesis and characterisation of (Ti,Al)-Beta/MCM-41, (Ti,Al)-Beta/MCM-48 and Ti-Beta/SBA-15

Thermally stable composites (Ti,Al)-Beta/MCM-48[51] and (Ti,Al)-Beta/MCM-41[50] were prepared by a hydrothermal two-step synthesis procedure. In the first and crucial synthetic step, colloidal zeolite (Ti,Al)-Beta particles were prepared using TEAOH as a template that were in the second step combined with surfactant(s) solution. (Ti,Al)-Beta/MCM-41 was synthesized in the presence of surfactant solution of cetyltrimethylammonium bromide (CTABr). (Ti,Al)-Beta/MCM-48 was synthesized in the presence of surfactants solution of CTABr and polyoxyethylene (8) isooctylhexyl ether (Triton® X-114). The presence of both micropores and mesopores in template-free (Ti,Al)-Beta/MCM-41 was detected by using X-ray diffraction, nitrogen adsorption/desorption analysis and a detailed HRTEM investigation (Fig. 5.4). XAS studies showed the presence of Ti^{4+} cations coordinated to four oxygens at a distance of 1.84(1) Å in the template-free sample. In the second coordination sphere one and two silicon atoms were found at the distances of 3.12(2) Å and 3.51(2) Å, respectively. These results indicated framework Ti and thus the presence of Ti oxidation centres within the composite material. No leaching of Ti from the framework was detected. NMR investigations showed the presence of Brønsted and Lewis acid sites related to framework aluminium, which were generated by the removal of potassium and sodium ions from the (Ti,Al)-Beta/MCM-41 pores. Elemental analysis and X-ray absorption studies of (Ti,Al)-Beta/MCM-48 showed the presence of tetrahedral Ti^{4+} species that also do not leach from the framework. Titanium was coordinated to four oxygen atoms in the first coordination shell, to one oxygen atom at 1.73(1) Å and three oxygen atoms at 1.86(1) Å. In the second coordination sphere, three silicon atoms were found at distances 3.18(2) Å and 3.48(2) Å, respectively. However, in (Ti,Al)-Beta/MCM-48 48 partial dealumination occurs during the thermal treatment, that is, when removing

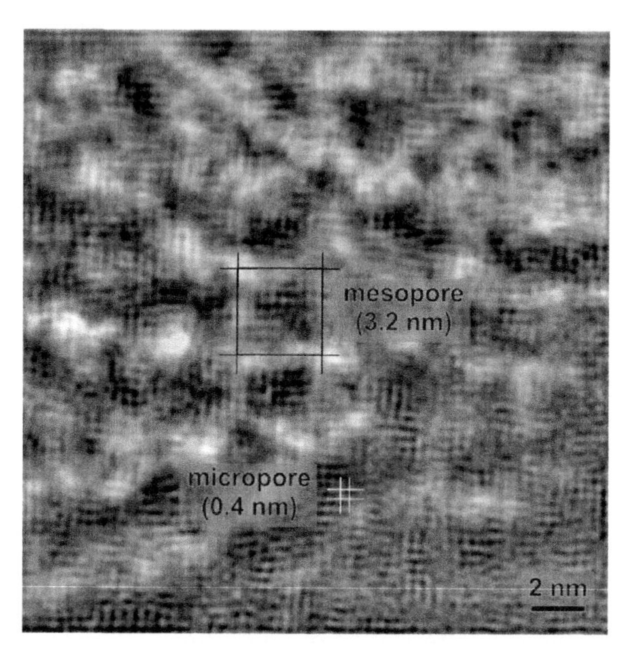

Figure 5.4 FT-filtered high-resolution transmission electron microscopy (HRTEM) image of (Ti,Al)-Beta/MCM-48 composite prepared by the organisation of nanosized (Ti,Al)-Beta particles into the mesoporous matrix. The image shows a delocalised micropore pattern having narrower lattice spacing than the hosting mesoporous matrix with a distorted cubic arrangement. Black square marks the mesopore. One micropore is enclosed with white lines.

template molecules from the pores by calcination. We thus have oxidation and not acid sites, what is an advantage, when one-step reactions are concerned.

The aluminium-free Ti-Beta/SBA-15 composite material was prepared by the post-synthesis dry-deposition of different amounts of Ti-Beta nanoparticle solution on SBA-15 matrix.[91] Hydrothermal synthesis of SBA-15 was performed using acidic aqueous solution of Pluronic® P123 triblock copolymer and TEOS. Ti-Beta nanoparticle solution was prepared hydrothermally using tetraethylorthotitanate as titanium source, tetraethylammonium hydroxide as template and fumed silica (Aerosil) as silica source. The presence of Ti-containing crystalline nanoparticles in the slab solution, used for impregnation on SBA-15 matrix, was confirmed by HRTEM measurements and energy dispersive X-ray (EDX) elemental analysis. Hexagonal mesopore arrangement of SBA-15 matrix was determined by XRD and TEM measurements. The presence and the deposition of Ti-incorporated zeolitic nanoparticles on the mesopore walls of SBA-15 matrix were proved by nitrogen sorption analysis (Fig. 5.5), infrared (IR) spectroscopy and thermogravimetric (TG) analysis. Local environment of titanium incorporated on zeolite Beta phase studied by XAS showed the presence of tetrahedrally coordinated framework Ti^{4+}. The product is thermally stable since it retains its porous and framework structure upon calcination process.

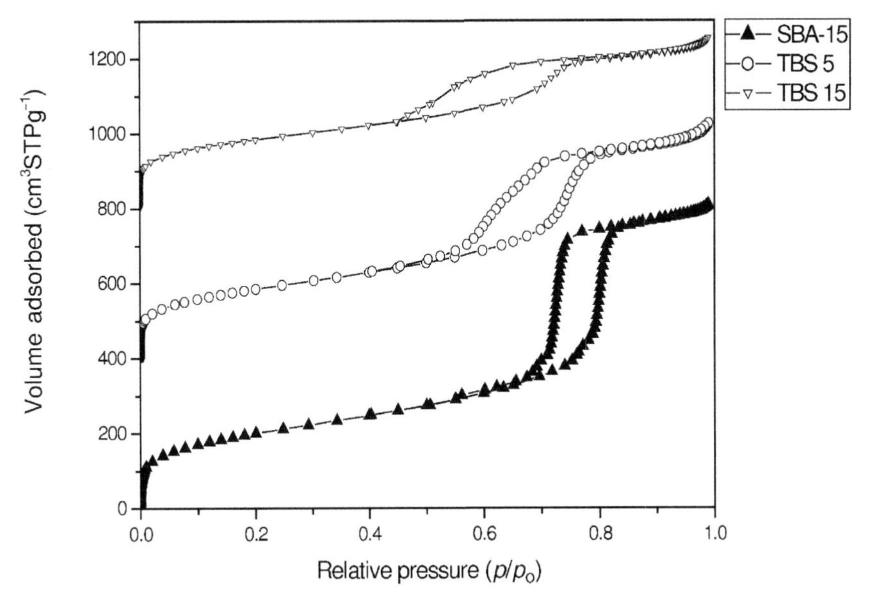

Figure 5.5 N_2-sorption isotherms of the Ti-Beta/SBA-15 products with 5 ml (TBS5), 15 ml (TBS15) of added Ti-Beta slab solution and pure SBA-15 material. TBS5 and TBS15 isotherms are offset vertically for 400, 800 cm^3/g, respectively.

5. Mesoporous Thin Films

The important feature of microporous and mesoporous solids based on various silica and metal oxides is also their ability to form thin films with nanometer-scale thickness, which can be deposited on flat or even complex shapes. They can act as potential hosts for pure metal/metal oxide nanoclusters or as a constitutive part of complex porous multilayer films.[92] The mesoporous thin films have potential applications as catalyst flow channel reactors and sensors.[93] Hydrothermal stability can be an issue in such thin films and thus continues to be a significant area of study.

5.1. Mesoporous aluminophosphate-based thin films with cubic pore arrangements

We have prepared novel thermally stable large-pore aluminophosphate-based mesoporous thin films with cubic (*Im3m*) pore arrangements by using nonionic block copolymer surfactants F127 and F108 as mesostructure-directing agents.[94] Aluminophosphate solution was deposited on a glass substrate under controlled conditions by dip-coating method. The mesostructure and thermal stability of thin films were investigated by Grazing Incidence Small Angle X-ray Scattering (GISAXS) and confirmed by X-ray diffraction and HRTEM. It has been shown that aluminophosphate-based thin films with thickness of 400 nm retained their ordered pore system up to 400 °C.

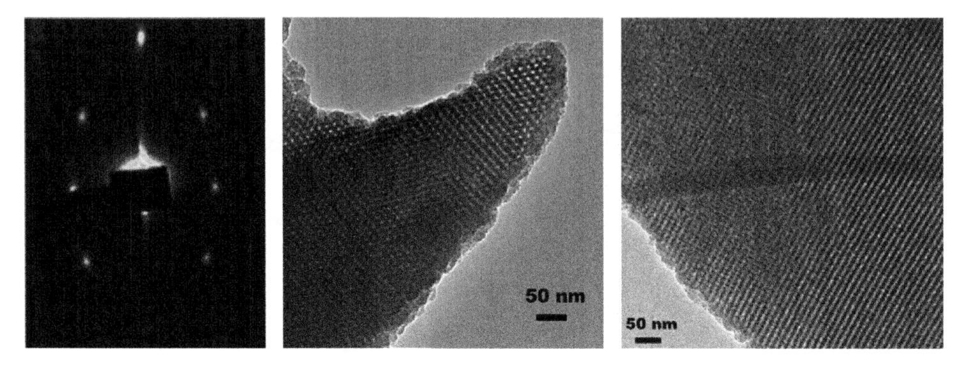

Figure 5.6 Grazing Incidence Small Angle X-ray Scattering (GISAXS) pattern of the F127-templated aluminophosphate thin film treated at 400 °C (left) and its transmission electron microscopy (TEM) images along [111] pore direction (middle) and [110] pore direction. (See color insert.)

Transition metal-modified aluminophosphate films interesting for catalytic applications could be produced under similar conditions (Fig. 5.6).

6. Conclusions

Transition metal functionalised nanoporous solids are efficient and environment friendly catalysts. The expansion of their use in chemistry and pharmacy is mostly limited by the size of the pores and sometimes by their deactivation at operating conditions. Regarding the pore size, the discovery of mesoporous solids 15 years ago was a real breakthrough in this field, but the enthusiasm was slowed down because of the low catalytic activity of these materials, which was attributed to the leaching of metal active sites from the amorphous framework and their agglomeration. The search for microporous zeolitic catalysts with larger pores resulted in a recent remarkable discovery of a new silicogermanate zeolite ITQ-33,[95] which exhibits straight large pore channels with a diameter of 1.22 nm, however, the surface properties of zeolitic materials significantly change when pores get bigger and become similar to layered aluminosilicates.[96] Better solution towards highly active large pore heterogeneous catalysts seems to be the preparation of metal-modified microporous/mesoporous composite catalysts.

Acknowledgements

This work was supported by the Slovenian Research Agency through the research programme P1-0021-0104 and research projects Z1-9744-0104 and J1-6350-0104.

REFERENCES

[1] Schüth, F., Sing, K. S. W., Weitkamp, J. (Eds.), *Handbook of Porous Solids*, Vol. I–IV, Wiley-VCH, Weinheim, Germany, 2002. Stein, A., *Adv. Mater.* **2003,** *15,* 763–775. Schuth, F., Schmidt, W., *Adv. Mater.* **2002,** *14,* 629–638.

[2] Bruhwiler, D., Calzaferri, G., *Microporous Mesoporous Mater.* **2004,** *72,* 1–23. Lu, A. H., Schuth, F., *Adv. Mater.* **2006,** *18,* 1793–1805. Fryxell, G. E., *Inorg. Chem. Commun.* **2006,** *9,* 1141–1150. Shylesh, S., Srilakshmi, Ch., Singh, A. P., Anderson, B. G., *Microporou Mesoporous Mater.* Published Online 24 Jul 2007. Shi, L. Y., Wang, Y. M., Ji, A., Gao, L., Wang, Y., *Mater. Chem.* **2005,** *15,* 1392–1396.

[3] Guinet, M., Gilson, J. P. (Eds.), *Zeolites for Cleaner Technologies,* Imperial College Press, London, UK, 2002. Schuth, F., *Annu. Rev. Mater. Res.* **2005,** *35,* 209–238. Mori, K., Imaoka, S., Nishio, S., Nishiyama, Y., Nishiyama, N., Yamashita, H., *Microporous Mesoporous Mater.* **2007,** *101,* 288–295. Logar, N. Z., Kaucic, V., *Slov. Chim. Acta.* **2006,** *53,* 117–135.

[4] Corma, A., García, H., *J. Chem. Soc., Dalton Trans.* **2000,** 1381–1394. Corma, A., García, H., *Chem. Rev.* **2002,** *102,* 3837–3892. Weitkamp, J., *Solid State Ionics* **2000,** *131,* 175–188.

[5] Frei, H., *Science* **2006,** *313,* 309–310. Wichterlova, B., Sobalík, Z., Dedecek, J., *Appl. Catal. B* **2003,** *41,* 97–114.

[6] Li, C., *Catal. Rev.* **2004,** *46,* 419–492. Zhao, X. S., Bao, X. Y., Guo, W. P., Lee, F. Y., *Mater. Today* **2006,** *9,* 32–39. Wang, Y. J., Caruso, F., *Chem. Mater.* **2005,** *17,* 953–961. Liu, C. J., Li, S. J., Pang, W. Q., Che, C. M., *Chem. Commun.* **1997,** 65–66.

[7] Taguchi, A., Schuth, F., *Microporous Mesoporous Mater.* **2005,** *77,* 1–45.

[8] Thomas, J. T., Raja, R., Lewis, D. W., *Angew. Chem. Int. Ed.* **2005,** *44,* 6456–6482. Hartman, M., Kevan, L., *Chem. Rev.* **1999,** *99,* 635. Bordiga, S., Damin, A., Berlier, G., Bonino, F., Richiardi, G., Zecchina, A., Lamberti, C., *Int. J. Mol. Sci.* **2001,** *2,* 167–182.

[9] Centi, G., Wichterlova, B., Bell, A. T. (Eds.), *Catalysis by Unique Metal Ion Structures in Solid Matrices: From Science to Application, NATO Science Series II,* Vol. 13, Kluwer Academic Press, New York, USA, 2001. Thomas, J. M., Raja, R., *Annu. Rev. Mater. Res.* **2005,** *35,* 315–350. Thomas, J. M., Raja, R., *Catal. Today* **2006,** *117,* 22–31. Ali, I. O., *Mater. Sci. Eng. A* **2007,** *459,* 294–302.

[10] Clerici, M. G. *Oil Gas Eur. Mag.* **2006,** *32,* 77–82. Lercher, J. A., Jentys, A. Application of microporous solids as catalysts, in: Schuth, F., Sing, K. S. W., Weitkamp, J. (Eds.), *Handbook of Porous Solids*, Vol. V, Wiley-VCH, Weinheim, Germany, 2002.

[11] Nowinska, K., Waclaw, A., Izbinska, A., *Appl. Catal. A* **2003,** *243,* 225–236. Dimitrova, R., Spassova, M., *Catal. Commun.* **2007,** *8,* 693–696. Yuranov, I., Bulushev, D. A., Renken, A., Minsker, L. K., *Appl. Catal. A.* **2007,** *319,* 128–136. Dzwigaj, S., Janas, J., Machej, T., Che, M., *Catal. Today* **2007,** *119,* 133–136. Akah, A., Cundy, C., Garforth, A., *Appl. Catal. B* **2005,** *59,* 221–226.

[12] Thomas, J. M., Raja, R., Sankar, G., Bell, R. G., *Nature* **1999,** *398,* 227–230. Thomas, J. M., Raja, R., *Catal. Today* **2006,** *117,* 22–31.

[13] Raja, R., Sankar, G., Thomas, J. M., *J. Am. Chem. Soc.* **2001,** *123,* 8153–8154.

[14] Schuchardt, U., Cardoso, D., Sercheli, R., Pereira, R., da Cruz, R. S., Guerreiro, M. C., Mandelli, D., Spinace, E. V., Pires, E. L., *Appl. Catal. A* **2001,** *211,* 1–17.

[15] De Vos, D. E., Dams, M., Sels, B. F., Jacobs, P. A., *Chem. Rev.* **2002,** *102,* 3615–3640. Ying, J. Y., Mehnert, C. P., Wong, M. S. *Angew. Chem. Int. Ed.* **1999,** *38,* 56–77. Xiao, F. S., *Topics Catal.* **2005,** *35,* 9–24. Jana, S., Dutta, B., Bera, R., Koner, S., *Langmuir* **2007,** *23,* 2492–2496. Chmielarz, L., Kustrowski, P., Dziembaj, R., Cool, P., Vansant, E. F., *Appl. Catal. B* **2006,** *62,* 369–380. Wang, Y., Yang, W., Yang, L. J., Wang, X. X., Zhang, Q. H., *Catal. Today* **2006,** *117,* 156–162. Selvaraj, M., Song, S. W., Kawi, S., *Microporous Mesoporous Mater.* Published online 5 July 2007.

[16] Wang, L. P., Kong, A. G., Chen, B., Ding, H. M., Shan, Y. K., He, M. Y., *J. Mol. Catal. A* **2005,** *230,* 143–150. Palkovits, R., Yang, C. M., Olejnik, S., Schuth, F., *J. Catal.* **2006,** *243,* 93–98. Aguado, J., Calleja, G., Carrero, A., Moreno, J., *Chem. Eng. J.* Available online 1 July 2007. Zhang, L. X., Hua, Z. L., Dong, X. P., Li, L., Chen, H. R., Shi, J. L., *J. Mol. Catal.*

A **2007,** *268,* 155–162. Wang, X. Q., Ge, H. L., Jin, H. X., Cui, Y. J., *Microporous Mesoporous Mater.* **2005,** *86,* 335–340. Ratnasamy, P., Srinivas, D., Knozinger, H., *Adv. Catal.* **2004,** *48,* 1–169. Byambajav, E., Ohtsuka, Y., *Appl. Catal. A* **2003,** *252,* 193–204.

[17] Pastore, H. O., Coluccia, S., Marchese, L., *Annu. Rev. Mater. Res.* **2005,** *35,* 351–395. Mohapatra, S. K., Hussain, F., Selvam, P., *Catal. Commun.* **2003,** *4,* 57–62. Mohapatra, S. K., Selvam, P., *Topic Catal.* **2003,** *22,* 17–22. Karthik, M., Vinu, A., Tripathi, A. K., Gupta, N. M., Palanichamy, M., Murugesan, V., *Micropor. Mesopor. Mater.* **2004,** *70,* 15–25. Selvam, P., Mohapatra, S. K., *J. Catal.* **2006,** *238,* 88–99. Subrahmanyam, C., Louis, B., Viswanathan, B., Renken, A., Varadarajan, T. K., *Appl. Catal. A* **2005,** *282,* 67–71.

[18] Vantomme, A., Leonard, A., Yuan, Z. Y., Su, B. A. L., *Colloid. Surface. A* **2007,** *300,* 70–78. Ivanova, I. I., Kuznetsov, A. S., Yuschenko, V. V., Knyazeva, E. E., *Pure Appl. Chem.* **2004,** *76,* 1647–1658. Srivastava, R., Choi, M., Ryoo, R., *Chem. Commun.* **2006,** 4489–4491. Meynen, V., Cool, P., Vansant, E. F., *Microporous Mesoporous Mater.* **2007,** *104,* 26–38. Mintova, S., Cejka, J. Micro/mesoporous Composites, in: Cejka, J., van Bekkum, H., Corma, A., Schuth, F. (Eds.), *Introduction to Zeolite Science and Practice,* 3rd revised edition, Elsevier, Amsterdam, 2007.

[19] Lin, K., Sun, Z., Lin, S., Jiang, D., Xiao, F. S., *Microporous Mesoporous Mater.* **2004,** *72,* 193–201. Meyen, V., Busuioc, A. M., Beyers, E., Cool, P., Vansant, E. F., Bilba, N., Mertens, M., Lebedev, O., van Tendeloo, G. Nanodesign of combined micro- and mesoporous materials for specific applications in adsorption and catalysis, in: Buckly, R. W. (Eds.) *Solid State Chemistry Research Trends,* Nova, New York, 2007. Solberg, S. M., Kumar, D., Landry, C. C., *J. Phys. Chem. B* **2005,** *109,* 24331–24337. Kholdeeva, O. A., Zalomaeva, O. V., Sorokin, A. B., Ivanchikova, I. D., Della Pina, C., Rossi, M., *Catal. Today* **2007,** *121,* 58–64. Kholdeeva, O. A., Zalomaeva, O. V., Shmakov, A. N., Melgunov, M. S., Sorokin, A. B., *J. Catal.* **2005,** *236,* 62–68.

[20] Cundy, C. S., Cox, P. A., *Microporous Mesoporous Mater.* **2005,** *82,* 1–78. van Santen, R. A., *Nature* **2006,** *444,* 46–47.

[21] Ke, X., Xu, L., Zeng, C., Zhang, L., Xu, N., *Microporous Mesoporous Mater.* Published online 25 Feb 2007.

[22] Pastore, H. O., Martins, G. A. V., Strauss, M., Pedroni, L. G., Superti, G. B., de Oliveira, E. C., Gatti, G., Marchese, L., *Microporous Mesoporous Mater.* Published online 24 Apr 2007.

[23] Dong, J., Tong, X., Yu, J., Xu, H., Liu, L., Li, J., *Mater. Lett.* Published online 20 Apr 2007.

[24] Yamamoto, K., García, S. E. B., Muramatsu, A., *Microporous Mesoporous Mater.* **2007,** *101,* 90–96.

[25] Oye, G., Sjoblom, J., Stocker, M., *Adv. Colloid. Interface* **2001,** *89–90,* 439–466. Ciesla, U., Schuth, F., *Microporous Mesoporous Mater.* **1999,** *27,* 131–149. Lang, N., Delichere, P., Tuel, A., *Microporous Mesoporous Mater.* **2002,** *56,* 203–217. Yu, C., Tian, B., Zhao, D., *Curr. Opin. Solid St. Mater.* **2003,** *7,* 191–197. Jana, S. K., Nishida, R., Shindo, K., Kugita, T., Namba, S., *Microporous Mesoporous Mater.* **2004,** *68,* 133–142. Davidson, A., *Curr. Opin. Colloid In.* **2002,** *7,* 92–106. Wang, S., Wu, D., Sun, Y., Zhong, B., *Mater. Res. Bull.* **2001,** *36,* 1717–1720. Zhao, X. S., Su, F. B., Yan, Q. F., Guo, W. P., Bao, X. Y., Lv, L., Zhou, Z. C., *J. Mater. Chem.* **2006,** *16,* 637–648.

[26] Shylesh, S., Srilakshmi, C. h., Singh, A. P., Anderson, B. G., *Appl. Catal. B* **2003,** *41,* 97–114. García, N., Benito, E., Guzman, J., Tiemblo, P., Morales, V., García, R. A., *Mater. Chem.* **2005,** *15,* 1392–1396.

[27] Kustrowski, P., Chmielarz, L., Dziembaj, R., Cool, P., Vansant, E. F., *J. Phys. Chem. B* **2005,** *109,* 11552–11558.

[28] Zhu, S. M., Zhou, H. S., Hibino, M., Honma, I., *J. Mater. Chem.* **2003,** *13,* 1115–1118.

[29] Tompsett, G. A., Conner, W. C., Yngvesson, K. S., *Chem. Phys. Chem.* **2006,** *7,* 296–319. Bandyopadhyay, M., Gies, H., *Compt. Rend. Chim.* **2005,** *8,* 621–626. Fantini, M. C. A., Matos, J. R., Cides da Silva, L. C., Mercuri, L. P., Chiereci, G. O., Celer, E. B., Jaroniec, M., *Mat. Sci. Eng. B* **2004,** *112,* 106–110. Newalkar, B. L., Katsuki, H., Komarneni, S., *Microporous Mesoporous Mater.* **2004,** *73,* 161–170.

[30] Park, S. E., Chang, J. S., Hwang, Y. K., Kim, D. S., Jhung, S. H., Hwang, J. S., *Catal. Surv. Asia* **2004,** *8,* 91–110.
[31] Kirschhock, C. E. A., Kremer, S. P. B., Vermant, J., Van Tendeloo, G., Jacobs, P. A., Martens, J. A., *Chem. Eur. J.* **2005,** *11,* 4306–4313. Tosheva, L., Valtchev, V. P., *Chem. Mater.* **2005,** *17,* 2494–2513. Larlus, O., Mintova, S., Bein, T., *Microporous Mesoporous Mater.* **2006,** *96,* 405–412.
[32] Prokesova-Fojtkova, P., Mintova, S., Cejka, J., Zilkova, N., Zukal, A., *Microporous Mesoporous Mater.* **2006,** *81,* 154–160.
[33] Mori, H., Uota, M., Fujikawa, D., Yoshimura, T., Kuwahara, T., Sakai, G., Kijima, T., *Microporous Mesoporous Mater.* **2006,** *91,* 172–180.
[34] Naik, S. P., Chiang, A. S. T., Thompson, R. W., Huang, F. C., Kao, H. M., *Microporous Mesoporous Mater.* **2003,** *60,* 213–224. Huang, L., Guo, W., Deng, P., Xue, Z., Li, Q., *J. Phys. Chem.* **2000,** *104,* 2817–2823. Liu, Y., Zhang, W., Pinnavaia, T. J., *Angew. Chem. Int. Ed.* **2001,** *40,* 1255–1258. Guo, W., Xiong, C., Huang, L., Li, Q., *J. Mater. Chem.* **2001,** *11,* 1886–1890. Liu, Y., Pinnavaia, T. J., *Chem. Mater.* **2002,** *14,* 3–5. Di, Y., Yu, Y., Sun, Y., Yang, X., Lin, S., Zhang, M., Li, S., Xiao, F. S., *Microporous Mesoporous Mater.* **2003,** *62,* 221–228.
[35] Dutartre, R., Menorval, L. C., Di Renzo, F., McQueen, D., Fajula, F., Schulz, P., *Microporous Mater.* **1996,** *6,* 311–320. McQueen, D., Chiche, B. H., Fajula, F., Auroux, A., Guimon, C., Fitoussi, F., Schulz, P., *J. Catal.* **1996,** *161,* 587–596.
[36] Jacobsen, C. H., Madsen, C., Houzvicka, J., Schmidt, I., Carlsson, A., *J. Am. Chem. Soc.* **2000,** *122,* 7116–7117.
[37] Goto, Y., Fukushima, Y., Ratu, P., Imada, Y., Kubota, Y., Sugi, Y., Ogura, M., Matsukata, M., *J. Porous Mater.* **2002,** *9,* 43–48. Kloestra, K. R., van Bekkum, H., Jansen, J. C., *Chem. Commun.* **1997,** 2281–2282. On, D. T., Kaliagine, S., *Angew. Chem. Int. Ed.* **2001,** *40,* 3248. Miyazawa, K., Inagaki, S., *Chem. Commun.* **2000,** 2121–2122. Van der Voort, P., Ravikovitch, P. I., De Jong, K. P., Neimark, A. V., Jansen, A. H., Benjelloun, M., Van Bavel, E., Cool, P., Weckhuysen, B. M., Vansant, E. F., *Chem. Commun.* **2002,** *39,* 1010–1011. Corma, A., Diaz, U., Domine, M. E., Fornes, V., *Angew. Chem. Int. Ed.* **2000,** *39,* 1499–1501.
[38] Choi, M., Srivastava, R., Ryoo, R., *Chem. Commun.* **2006,** 4380–4381.
[39] Cejka, J., Wichterlova, B., *Catal. Rev.* **2002,** *44,* 375–421.
[40] Helliwell, M., Helliwell, J. R., Kaucic, V., Logar, N. Z., Barba, L., Busetto, E., Lausi, A., *Acta Crystallogr. B* **1999,** *55,* 327–332.
[41] Helliwell, M., *J. Synchrotron Radiat.* **2000,** *7,* 139–147. Bazin, D., Guczi, L., Lynch, J., *Appl. Catal. A* **2002,** *226,* 87–113. Dalconi, M. C., Alberti, A., Cruciani, G., *J. Phys. Chem. B* **2003,** *107,* 12973–12980. Milanesio, M., Artioli, G., Gualtieri, A. F., Palin, L., Lamberti, C., *J. Am. Chem. Soc.* **2003,** *125,* 14549–14558. Norby, P., *Curr. Opin. Colloid. Interface Sci.* **2006,** *11,* 118–125.
[42] Cianci, M., Helliwell, J. R., Helliwell, M., Kaucic, V., Logar, N. Z., Mali, G., Tusar, N. N., *Crystallogr. Rev.* **2005,** *11,* 245–335.
[43] Helliwell, M., Jones, R. H., Kaucic, V., Logar, N. Z., *J. Synchrotron Radiat.* **2005,** *12,* 420–430.
[44] Baerlocher, C., Gramm, F., Massuger, L., McCusker, L. B., He, Z., Hovmoller, S., Zou, X., *Science* **2007,** *315,* 1113–1116.
[45] Solovyov, L. A., Belousov, O. V., Dinnebier, R. E., Shmakov, A. N., Kirik, S. D., *J. Phys. Chem. B* **2005,** *109,* 3233–3237.
[46] Kirsch, B. L., Richman, E. K., Riley, A. E., Tolbert, S. H., *J. Phys. Chem. B* **2004,** *108,* 12698–12706. Innocenzi, P., Malfatti, L., Kidchob, T., Costacurta, S., Falcaro, P., Piccinini, M., Marcelli, A., Morini, P., Sali, D., Amenitsch, H., *J. Phys. Chem. C* **2007,** *111,* 5345–5350. Zickler, G. A., Jahnert, S., Funari, S. S., Findenegg, G. H., Paris, O., *Appl. Cryst.* **2007,** *40,* s522–s526. Huang, Y. S., Jeng, U. S., Hsu, C. H., Torikai, N., Lee, H. Y., Shin, K., Hino, M., *Physica B* **2006,** *385–386,* 667–669.
[47] Solovyov, L. A., Belousov, O. V., Shmakov, A. N., Zaikovskii, V. I., Joo, S. H., Ryoo, R., Haddad, E., Gedeon, A., Kirik, S. D., *Stud. Surf. Sci. Catal.* **2003,** *146,* 299–302. de Jong, K. P., Koster, A. J., *Chem. Phys. Chem.* **2002,** *3,* 776–780.

[48] Billinge, S. J. L., Levin, I., *Science* **2007,** *316,* 561–565. Sauer, J., Marlow, F., Schuth, F., *Phys. Chem. Phys* **2001,** *3,* 5579–5584.
[49] Thomas, J. M., Gai, P. L. *Adv. Catal.* **2004,** *48,* 171–227. Gai, P. L., Calvino, J. J. *Annu. Rev. Mater. Res.* **2005,** *35,* 465–504. Tarasaki, O., Ohsuna, T., Liu, Z., Sakamoto, Y. Structural study of porous materials by electron microscopy, in: Cejka, J., van Bekkum, H., Corma, A., Schuth, F. (Eds.), *Introduction to Zeolite Science and Practice,* 3rd revised editions, Elsevier, Amsterdam, 2007.
[50] Mrak, M., Tusar, N. N., Logar, N. Z., Mali, G., Kljajic, A., Arcon, I., Launay, F., Gedeon, A., Kaucic, V., *Microporous Mesoporous Mater.* **2006,** *95,* 76–85.
[51] Mazaj, M., Logar, N. Z., Mali, G., Tusar, N. N., Arcon, I., Ristic, A., Recnik, A., Kaucic, V., *Microporous Mesoporous Mater.* **2007,** *99,* 3–13.
[52] Alfredson, V., Anderson, M. W., *Chem. Mater.* **1996,** *8,* 1141–1146.
[53] Groen, J. C., Peffer, J. A. A., Perez-Ramírez, J., *Microporous Mesoporous Mater.* **2003,** *60,* 1–17.
[54] Gregg, S. J., Sing, K. S. W. *Adsorption Surface Area and Porosity,* 2nd ed., Academic Press London, UK, 1982. Rouquerol, F., Rouquerol, J., Sing, K. S. W., *Adsorption by Powders and Porous Solids,* Academic Press, London, UK, 1999. Kruk, M., Jaroniee, M., *Chem. Mater.* **2001,** *13,* 3169–3183.
[55] Koningsberger, D. C., Mojet, B. L., van Dorssen, G. E., Ramaker, D. E., *Top. Catal.* **2000,** *10,* 143–155. van Bokhoven, J. A., Louis, C. T., Miller, J., Tromp, M., Safonova, O. V., Glatzel, P., *Angew. Chem. Int. Ed.* **2006,** *45,* 4651–4654.
[56] van Bokhoven, J. A., Sambe, H., Ramaker, D. E., Koningsberger, D. C., *J. Phys. Chem.* **1999,** *103,* 7557–7564. Penzien, J., Abraham, A., van Bokhoven, J. A., Jentys, A., Muller, T. E., Sievers, C., Lercher, J. A., *J. Phys. Chem. B* **2004,** *108,* 4116–4126.
[57] Grandjean, D., Beale, A. M., Petukhov, A. V., Weckhuysen, B. M., *J. Am. Chem. Soc.* **2005,** *127,* 14454–14465.
[58] van Bokhoven, J. A., van der Eerden, A. M. J., Prins, R., *J. Am. Chem. Soc.* **2004,** *126,* 4506–4507. Mathisen, K., Nicholson, D. G., Beale, A. M., Sanches-Sanches, M., Sankar, G., Brass, W., Nikitenko, S., *J. Phys. Chem. C* **2007,** *111,* 3130–3138.
[59] Tusar, N. N., Kaucic, V., Geremia, S., Vlaic, G., *Zeolites* **1995,** *15,* 708–713. Tuel, A., Arcon, I., Tusar, N. N., Meden, A., Kaucic, V., *Microporous Mater.* **1996,** *7,* 271–284. Tusar, N. N., Mali, G., Arcon, I., Kaucic, V., Ghanbari-Siahkali, A., Dweyer, J., *Microporous Mesoporous Mater.* **2002,** *55,* 203–216.
[60] Ristic, A., Tusar, N. N., Vlaic, G., Arcon, I., Thibault-Starzyk, F., Malicki, N., Kaucic, V., *Microporous Mesoporous Mater.* **2004,** *76,* 61–69.
[61] Ristic, A., Tusar, N. N., Arcon, I., Logar, N. Z., Thibault-Starzyk, F., Czyzniewska, J., Kaucic, V., *Chem. Mater.* **2003,** *15,* 3643–3649.
[62] Tusar, N. N., Logar, N. Z., Arcon, I., Mali, G., Mazaj, M., Ristic, A., Lazar, K., Kaucic, V., *Microporous Mesoporous Mater.* **2005,** *87,* 52–58.
[63] Logar, N. Z., Tusar, N. N., Mali, G., Mazaj, M., Arcon, I., Arcon, D., Recnik, A., Ristic, A., Kaucic, V., *Microporous Mesoporous Mater.* **2006,** *96,* 386–395.
[64] Tusar, N. N., Logar, N. Z., Arcon, I., Thibault-Starzyk, F., Ristic, A., Rajic, N., Kaucic, V., *Chem. Mater.* **2003,** *15,* 4745–4750.
[65] Tusar, N. N., Logar, N. Z., Vlaic, G., Arcon, I., Arcon, D., Daneu, N., Kaucic, V., *Microporous Mesoporous Mater.* **2005,** *82,* 129–136.
[66] Tusar, N. N., Ristic, A., Cecowski, S., Arcon, I., Lazar, K., Amenitsch, H., Kaucic, V., *Microporous Mesoporous Mater.* **2007,** *104,* 289–295.
[67] Brouwer, D. H., Darton, R. J., Morris, R. E., Levitt, M. H., *J. Am. Chem. Soc.* **2005,** *127,* 10365–10370.
[68] Pickard, C. J., Mauri, F., *Phys. Rev. B* **2001,** *63,* art. 245101.
[69] Shantz, D. F., Lobo, R. F., *Topic Catal.* **1999,** *9,* 1–11.
[70] Hunger, M., *Microporous Mesoporous Mater.* **2005,** *82,* 241–255.
[71] Hunger, M., *Catal. Rev. Sci. Eng.* **1997,** *39,* 345–393.
[72] Engelhardt, G., Michel, D., *High-Resolution Solid-State NMR of Silicates and Zeolites,* John Wiley & Sons, Chichester, UK, 1987.
[73] Canesson, L., Boudeville, Y., Tuel, A., *J. Am. Chem. Soc.* **1997,** *119,* 10754–10762.

[74] Mali, G., Ristic, A., Kaucic, V., *J. Phys. Chem. B* **2005,** *109,* 10711–10716.
[75] Epping, J. D., Chmelka, B. F., *Curr. Opin. Solid St. Mater.* **2006,** *11,* 81–117.
[76] Sheldon, R. A., Kochi, J. K., *Metal-Catalyzed Oxidations of Organic Compounds,* Academic Press, New York, 1981. Suresh, K. A., Sharma, M. M., Sridhar, T., *Ind. Eng. Chem. Res.* **2000,** *39,* 3958–3997. Moden, B., Zhan, B. Z., Dakka, J., Santiesteban, J., Iglesia, E., *J. Phys. Chem. C* **2007,** *111,* 1402–1411. Zhan, B. Z., Moden, B., Dakka, J., Santiesteban, J., Iglesia, E., *J. Catal.* **2007,** *245,* 316–325.
[77] Vijayaraghavan, V. R., Raj, K. J. A., *J. Mol. Catal. A* **2004,** *207,* 41–50.
[78] Zhang, Q., Wang, Y., Itsuki, S., Shishido, T., Takehira, K., *J. Mol. Catal. A* **2002,** *188,* 189–200. Yuan, Z. Y., Ma, H. T., Luo, Q., Zhou, W., *Mater. Chem. Phys.* **2002,** *77,* 299–303.
[79] Bachari, K., Miller, J. M. M., Bonville, P., Cherifi, O., Figueras, F., *J. Catal.* **2007,** *249,* 52–58.
[80] Kustov, A. L., Hansen, T. W., Kustova, M., Christenen, C. H., *Appl. Catal. B* **2007,** *76,* 311–319.
[81] Li, Y., Feng, Z., Lian, Y., Sun, K., Zhang, L., Jia, G., Yang, Q., Li, C., *Microporous Mesoporous Mater.* **2005,** *84,* 41–49. Kawabata, T., Ohishi, Y., Itsuki, S., Fujisaki, N., Shishido, T., Takaki, K., Zhang, Q., Wang, Y., Takehira, K., *J. Mol. Catal. A* **2005,** *236,* 99–106. Zhang, Q., Yang, W., Wang, X., Wang, Y., Shishido, T., Takehira, K., *Microporous Mesoporous Mater.* **2005,** *77,* 223–234. Holland, A. W., Li, G., Shahin, A. M., Long, G. J., Bell, A. T., Tilley, T. D., *J. Catal.* **2005,** *235,* 150–163. Choi, J. S., Yoon, S. S., Jang, S. H., Ahn, W. S., *Catal. Today* **2006,** *111,* 280–287. Vinu, A., Krithiga, T., Murugesan, V., Hartmann, M., *Adv. Mater.* **2004,** *16,* 1817–1821.
[82] Ristic, A., Tusar, N. N., Arcon, I., Thibault-Starzyk, F., Hanzel, D., Czyzniewska, J., Kaucic, V., *Microporous Mesoporous Mater* **2002,** *56,* 303–315. Arcon, I., Ristic, A., Tusar, N. N., Kodre, A., Kaucic, V., *Phys. Scr.* **2005,** *T115,* 753–755.
[83] Zenonos, C., Sankar, G., Cora, F., Lewis, Q. A., Pankhurst, D. W., Catlow, C. R. A., Thomas, J. M., *Phys. Chem. Chem. Phys.* **2002,** *4,* 5421–5429.
[84] Choi, S. H., Wood, B. R., Bell, A. T., Janicke, M. T., Ott, K. C., *J. Phys. Chem. B* **2004,** *108,* 8970–8975.
[85] Stockenhuber, M., Hudson, M. J., Joyner, R. W., *J. Phys. Chem. B* **2000,** *104,* 3370–3374.
[86] Stockenhuber, M., Joyner, R. W., Dixon, J. M., Hudson, M. J., Grubert, G., *Microporous Mesoporous Mater.* **2001,** *44,* 367–375.
[87] Hamdy, M. S., Mul, G., Wei, W., Anand, R., Hanefeld, U., Jansen, J. C., Moulijn, J. A., *Catal. Today* **2005,** *110,* 264–271. Anand, R., Hamdy, M. S., Gkourgkoulas, P., Maschmeyer, T. h., Jansen, J. C., Hanefeld, U., *Catal. Today* **2006,** *117,* 279–283. Hamdy, M. S., Mul, G., Jansen, J. C., Ebaid, A., Shan, Z., Overweg, A. R., Maschmeyer, T. h., *Catal. Today* **2005,** *100,* 255–260.
[88] Vayssilov, G. N., *Catal. Rev.* **1997,** *39,* 209–251. Perego, C., Carati, A., Ingallina, P., Mantegazza, M. A., Bellussi, G., *Appl. Catal. A* **2001,** *221,* 63–72. Notari, B., *Adv. Catal.* **1996,** *41,* 253–334.
[89] Lamberti, C., Bordiga, S., Arduino, D., Zecchina, A., Geobaldo, F., Spano, G., Carati, A., Villian, F., Vlaic, G., *J. Phys. Chem. B* **1998,** *102,* 6382–6390. Gleeson, D., Sankar, G., Catlow, C. R. A., Thomas, J. M., Spano, G., Bordiga, S., Zecchina, A., Lamberti, C., *Phys. Chem. Phys.* **2000,** *2,* 4812–4817.
[90] Arends, I. W. C. E., Sheldon, R. A., *Appl. Catal. A* **2001,** *212,* 175–187. Trukhan, N. N., Rumannikov, V. N., Shmakov, A. N., Vanina, M. P., Paukshtis, E. A., Bukhtiyarov, V. I., Kriventsov, V. V., Danilov, I. Y., Kholdeeva, O. A., *Microporous Mesoporous Mater.* **2003,** *59,* 73–84. Chiker, F., Nogier, J. P., Launay, F., Bonardet, J. L., *Appl. Catal. A* **2003,** *243,* 309–321. Tanev, P. T., Chibwe, M., Pinnavaia, T. J., *Nature* **1994,** *368,* 321–323.
[91] Mazaj, M., Stevens, W. J. J., Logar, N. Z., Ristic, A., Tusar, N. N., Arcon, I., Daneu, N., Cool, P., Vansant, E. F., Kaucic, V., sent for publication in *Microporous Mesoporous Mater.*
[92] Guliants, V. V., Carreon, M. A., Lin, Y. S., *J. Membr. Sci.* **2004,** *235,* 53–72. Pevzner, S., Regev, O., Yerushalmi-Rozen, R., *Curr. Opin. Colloid. Interface Sci.* **1999,** *4,* 420–427. Alberiu, P. C. A., Frindell, K. L., Hayward, R. C., Kramer, E. J., Stucky, G. D., Chmelka, B. F., *Chem. Mater.* **2002,** *12,* 3284–3294. Angelome, P. C., Fuertes, M. C., Soler-Illia, G. J. A. A., *Adv. Mater.* **2006,** *18,* 2397–2402.

[93] Muraza, O., Rebrov, E. V., Khimyak, T., Johnson, B. F. G., Kooyman, P. J., Lafont, U., de Croon, M. H. J. M., Schouten, J. C., *Chem. Eng. J.* Published online 10 July 2007. Doshi, D. A., Huesing, N. K., Lu, M. C., Fan, H. Y., Lu, Y. F., Simmons-Potter, K., Potter, B. J., Hurd, A. J., Brinker, C. J., *Science* **2000,** *290,* 107–111. Yang, G., Zhang, X., Liu, S., Yeung, K. L., Wang, J., *J. Phys. Chem. Solids* **2007,** *68,* 26–31.
[94] Mazaj, M., Costacurta, S., Logar, N. Z., Mali, G., Tusar, Innocenzi, P., Malfatti, L., Thibault-Starzyk, F., Amenitsch, H., Kaucic, V., Soler-Illia, G. J. A. A., sent for publication in *Langmuir.*
[95] Corma, A., Díaz-Cabanas, M. J., Jorda, J. L., Martínez, C., Moliner, M., *Nature* **2006,** *443,* 842–845.
[96] Lobo, R. F., *Nature* **2006,** *443,* 757–758.

CHAPTER 6

Theoretical and Practical Aspects of Zeolite Nucleation

Boris Subotić, Josip Bronić, *and* Tatjana Antonić Jelić

Contents

Abstract

Because of the great influence of zeolite crystal size on the mode and efficiency of their applications, the knowledge on the nucleation of zeolites has a crucial importance in the designing of the products having desired particulate properties needed for specific applications. For this reason, the existing concepts of zeolite nucleation are overviewed and critically evaluated. On the basis of the evaluation it is concluded that: (a) 'classical' homogeneous and heterogeneous nucleation in the liquid phase are neglected as the processes of the formation of zeolite nuclei, (b) secondary nucleation can occur only on the seed crystals, under specific synthesis conditions, (c) formation of nuclei in the gel matrix is relevant for heterogeneous (untemplated or templated hydrogels) and homogeneous (untemplated or templated, initially clear solutions) systems and (d) the general principles of zeolite nucleation do not depend on the system (homogeneous, heterogeneous, templated, untemplated) and crystallization conditions.

Keywords: Amorphous aluminosilicates, Autocatalytic nucleation, Clear solutions, Gels, Heterogeneous nucleation, Homogeneous nucleation, Hydrogels, Nuclei formation, Secondary nucleation, Zeolite

1. Introduction

Although most of applications of zeolites are closely related to their structural and chemical properties (i.e., type of zeolite, modification by ion exchange and/or isomorphous substitution etc.), size of zeolite crystals may play a crucial role in the

Ordered Porous Solids
DOI: 10.1016/B978-0-444-53189-6.00006-8

mode and efficiency of their application[1–3] as catalysts,[1,3–12] adsorbents[3,13–17] and cation exchangers.[3,18–25]

Most of zeolites used in the above mentioned 'classical' applications have, except in some rare cases, the crystal size in mainly micrometre to millimetre size range, that can be produced either by 'standard' synthesis procedures,[26,27] or by grinding of naturally occurred zeolites.[16,20,22–24] However, due to new and new application of zeolites, there is continuous requirement for finding out the synthesis procedures for production of both large (>20 μm) and small (<1 μm) zeolite crystals.

Large zeolite crystals with well defined morphology are used not only for fundamental studies (e.g., spatially resolved measurements of molecular diffusion in micropores, electron microscopic observations of crystal surface growing processes, single crystal structure determinations, determination of anisotropic magnetic and optical properties, etc.), but also they have a number of potential applications such as advanced functional materials in microelectronic and optics, materials for anisotropic ionic conductivity, matrices for creation of arraying micro clusters, catalysts, adsorbents, etc.[3,28] For these purposes, different techniques for the synthesis of large single crystals of zeolite A,[28,29] zeolite X,[28–30] zeolite Y,[31] ZSM-5,[29,32–34] ZSM-39,[32] analcime,[34] sodalite,[29,34,35] mordenite,[29,36] $AlPO_4$-5,[29] $AlPO_4$-34[29] and offretite[37] were developed.

On the other hand, although the existence of nanocrystalline zeolites has been well known since the early days of zeolite synthesis,[26] the use of colloidal science principles was consistently developed recently by Schoeman *et al.*[38] Since one would expect a significant change in the properties of these nanosized zeolites in comparison with the conventional molecular sieves, especially in the fields of sensing, membranes, microelectronics, catalysis and other applications,[3,39] there are many attempting to develop the methods and procedures for synthesis of zeolites having the crystal size in the nanometer range. These efforts resulted in the synthesis of many types of zeolites including A,[39–45] FAU,[38,39,44,46–50] L,[51,52] hydroxysodalite (HS),[53] beta,[44,54–57] $AlPO_4$-5,[58] CIT-6,[59] VPI-8[59] and especially MFI[44,60–72] in colloidal form with particle size in the nanometer range.

The above mentioned examples show that zeolite crystal size (distribution) has a crucial influence on the mode and efficiency of their general and specific applications. The following (i) population balance of a typical (well-mixed, isothermal, constant volume batch) zeolite crystallization:[73,74]

$$\partial N/\partial t + Q(\partial N/\partial L) = 0 \tag{1}$$

where $N = N(L,t)$ is the number density function representing crystal size distribution as a function of time, (ii) the corresponding moment equations:[73,74]

$$\mathrm{d}m_0/\mathrm{d}t = B \tag{2}$$

$$\mathrm{d}m_1/\mathrm{d}t = Qm_o \tag{3}$$

$$\mathrm{d}m_2/\mathrm{d}t = 2Qm_1 \tag{4}$$

$$\mathrm{d}m_3/\mathrm{d}t = 3Qm_2 \tag{5}$$

where

$$m_i = \int_0^\infty L^i[dN(L,t)dL] \quad (6)$$

and

$$B = dN/dt \quad (7)$$

is the rate of nucleation, where dN is the differential number of nuclei formed in the differential time dt, and

$$Q = dL/dt \quad (8)$$

is the rate of crystal growth, where L is crystal size at time t and (iii) the change in the differential mass, dm_z, of zeolite crystallized from a given reaction mixture, that can be expressed as:[73,75–80]

$$dm_z = G\rho L^3[dN(L,t)dL]dL \quad (9)$$

or [76,81–83]

$$dm_z = G\rho L^3(dN/dt)dt \quad (10)$$

where d$N(L,t)$/dL is the crystal size distribution at the crystallization time t, and G and ρ are geometrical shape factor and density of growing (zeolite) crystals, undoubtedly show that crystal size (distribution) at any crystallization time $t = t_c \leq t_c$(eq) depend on both the rates of nucleation and crystal growth, but that the final crystal size distribution [at $t_c = t_c$(eq)] in the batch crystallization strongly depends on the total number of nuclei formed during the crystallization and on the rate of their formation (rate of nucleation).[80–87] For this reason, the knowledge on the mechanism(s) of the particulate processes (nucleation, crystal growth)[73] has crucial importance in the control of particulate properties (crystal size, crystal size distribution, specific number of crystals, crystal shape), and thus on the designing of the product(s) having desired particulate properties needed for specific application(s).

2. Nucleation of Zeolites: An Overview and Critical Evaluation

While the mechanism of the crystal growth of zeolites is well defined by the reactions of active aluminate, silicate and/or aluminosilicate species from the liquid phase on the surface of growing zeolite crystals[3] as it is revealed by numerous AFM studies,[88–91] there is still much uncertainty regarding the relevant mechanisms of zeolite nucleation.[45,86,87,92–94] This is caused by several reasons: (a) Obtaining direct experimental data on nucleation processes is very difficult due to the extremely small fraction of the total mass involved in the problems of distinguishing this from surrounding reactants of very similar nature and composition.[86] For this reason, only the method developed by Zhdanov and Samulevich, enables the calculation of nucleation profiles from determinations of growth rate and crystal size distribution;[81,84] (b) there is a wide-spread meaning that mechanism(s) of nucleation of high-silica zeolites, obtained in the presence of organic templates, is(are) different than mechanism(s) of nucleation of low-silica zeolites crystallized from untemplated reaction mixtures; (c) there is also a wide-spread meaning that mechanism(s) of nucleation of zeolites in homogeneous systems (clear solutions) substantially differ(s) from the mechanism(s) of nucleation in heterogeneous systems (hydrogels) and (d) three different concepts of zeolite nucleation are present in literature: (i) nucleation in the liquid phase, (ii) nucleation in the gel phase and (iii) simultaneous nucleation in liquid and gel phase; this, depending on the type of zeolite and synthesis conditions generates numerous approaches to zeolite nucleation including homogeneous nucleation, heterogeneous nucleation, secondary nucleation, autocatalytic nucleation, etc. For the above mentioned reasons, in many papers, the nucleation of different types of zeolites is simply connected with so called 'induction time'.[92,95–101] This approach to the nucleation phenomena is based on the assumptions that[97] (i) nucleation and crystal growth are separated processes, (ii) the just 'induction time',t_i, that is, the time prior to the observation of X-ray crystallinity, is closely related to the nucleation of zeolite and (iii) the slope, S_n, of the Arrhenius plot ($\ln(1/t_i)$ vs. $1/T$, where T is absolute temperature), is proportional to the activation energy, E_n, of zeolite nucleation, that is, $S_n = E_n/R$, where R is gas constant. Analyzing the relationships between 'crystallization curve' and nucleation, den Ouden and Thompson concluded that zeolite crystals actually form and begin to grow prior to the onset of observable zeolite mass formation and that this observation undoubtedly stems from the detection limit of most analytical techniques which measure zeolite crystal mass or percent crystallinity in the solid phase.[102] In addition, an analysis of the relationships between $\ln(1/t_i)$ versus $1/T$ and $\ln(df_z/dt_c)$ versus $1/T$(f_z is the fraction of zeolite crystallized up to the time of crystallization t_c) has shown that the slope S_1 of the $\ln(1/t_i)$ versus $1/T$ straight line is almost the same as the slope S_2 of the $\ln(df_z/dt_c)$ versus $1/T$ straight line.[103] This means that the activation energy calculated from the slope S_1 corresponds to the activation energy of the entire crystallization process, rather than to the activation energy of nucleation. This finding supports the conclusion from Ref. 102, namely that 'from this result one must conclude that it is almost impossible to learn anything about the nucleation mechanism in zeolite crystallization from the analysis of the mass crystallization curve'. Hence, taking into considerations all the above mentioned reasons and facts, an overview and critical evaluation of zeolite nucleation will be done in accordance with the scheme shown in Fig. 6.1.

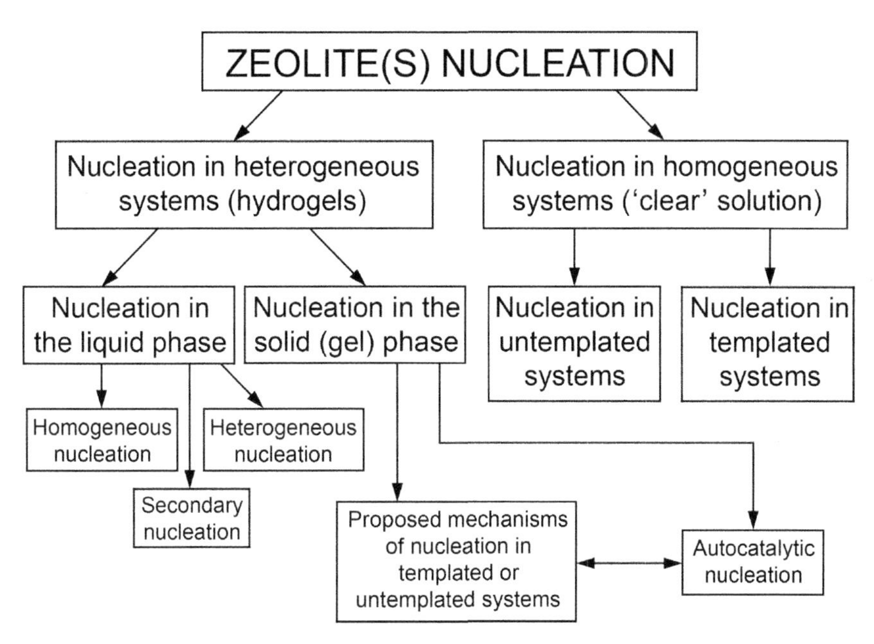

Figure 6.1 Schematic presentation of the overview and critical evaluation of zeolite nucleation.

2.1. Nucleation in heterogeneous systems (hydrogels)

2.1.1. Nucleation in the liquid phase

2.1.1.1. Homogeneous nucleation According to the classical theory of homogeneous nucleation from supersaturated solution,[104–106] the rate of homogeneous nucleation, $J_{\mathrm{hm}} = \mathrm{d}N\mathrm{d}t$, is usually written as:[105]

$$J_{\mathrm{hm}} = A\exp\left[-\beta\sigma^3 v^2/(kT)^3(\ln S)^2\right] \qquad (11)$$

where $A \approx 10^{30}$ cm^{-3} s^{-1} [105] is the pre-exponential factor which is a function of molecular diameter and diffusion rate of molecules or ions in the solution, β is the geometrical shape factor of critical nuclei, σ is the solid–liquid interface tension, v is the molecular volume of the solid phase, k is the Boltzman constant, T is the absolute temperature, and $S = C/C^{\star}$; C is the actual concentration of molecules or ions in the liquid phase and $C^{\star}$ is the concentration of molecules or ions which corresponds to the solubility of the solid phase at given conditions. Equation (11) shows that for constant values of β, σ, v and $C^{\star}$, characteristic for isothermal crystallization of a given compound, the rate of homogeneous nucleation is a function of the concentration C only, as can be expressed by Nielsen's empirical approximation for homogeneous nucleation,[105,107] that is

$$\mathrm{d}N/\mathrm{d}t = BC^n \qquad (12)$$

where B and n are empirical constants characteristic for a given compound and crystallization conditions, respectively. In accordance with Eq. (12), for approximately constant concentrations, C_{Si}, of silicon and, C_{Al}, of aluminium in the liquid phase during the main part of zeolite crystallization under different conditions, that is, $C = F(C_{Si}, C_{Al}) \approx$ constant,[3,76,83,108–113] the rate of nucleation is constant too during the main part of the crystallization process, that is

$$\mathrm{d}N/\mathrm{d}t_c = B = \text{constant} \tag{13}$$

and thus, number of nuclei increases linearly with the crystallization time t_c, that is

$$N = Bt_c \tag{14}$$

This probably was a reason that the first population balance of zeolite crystallization,[108] that is

$$W_c = K_1 S_a^{(n+1)} (t_c)^4 = K(\text{hm})(t_c)^4 \tag{15}$$

was derived on the basis of the Nielsen's approximation and a linear size-independent crystal growth of nuclei (crystals) at constant concentration difference $S_a - S_c$.[3] Here, W_c is the amount of crystalline phase (zeolite) crystallized up to the time t_c, $K_1 = 15D(S_a - S_c)/4\delta$, $K(\text{hm}) = K_1 S_a^{(n+1)}$, S_a and S_c are equilibrium concentration and concentration at crystal surface, respectively, of reactive aluminosilicate S-species, D is diffusion coefficient and δ is the diffusion film thickness on the surface of growing crystals. A difference between the measured kinetics of crystallization and the W_c versus t_c functions calculated by Eq. (15) (see Fig. 6 in Ref. 108) were explained by gradients around growing crystals, and thus by a reduced values of the power over t_c (=3.5–3.7) in measured kinetics[108] relative to the power (=4) calculated from the proposed model [see Eq. (15)].

Later on, homogeneous nucleation was identified as one of,[114–121] or only[102,122–127] mechanism of the formation of primary zeolite particles (nuclei) in many studies of zeolite crystallization, including ones in the presence of organic templates.[122,123,125] In all the mentioned studies, dissolution of amorphous aluminosilicate gel precursor in hot alkaline media and formation of low-molecular (monomer, dimmers) aluminate, silicate and/or aluminosilicate species in the liquid phase, was assumed as the first step of nucleation/crystallization of zeolites (see Fig. 7 in Ref. 121 as an example). Then, in accordance with the basic principles of homogeneous nucleation in solutions[104–106] and particularities of alkaline (alumino)silicate solutions,[26,109,126,128–138] nucleation (of low-silica zeolites, for example, having LTA and FAU structures) takes place by condensation of monomers and dimmers into more complex aluminosilicate structures (cyclic tetramers, cyclic hexamers; see Fig. 6.2; see also Fig. 7 in Ref. 121),[121,126] which in turn react to form, first the sodalite cage and then the zeolite structure.[126] The formation of sodalite cage is mediated by the presence of hydrated cations (see Fig. 25 in Ref. 139).[84,139]

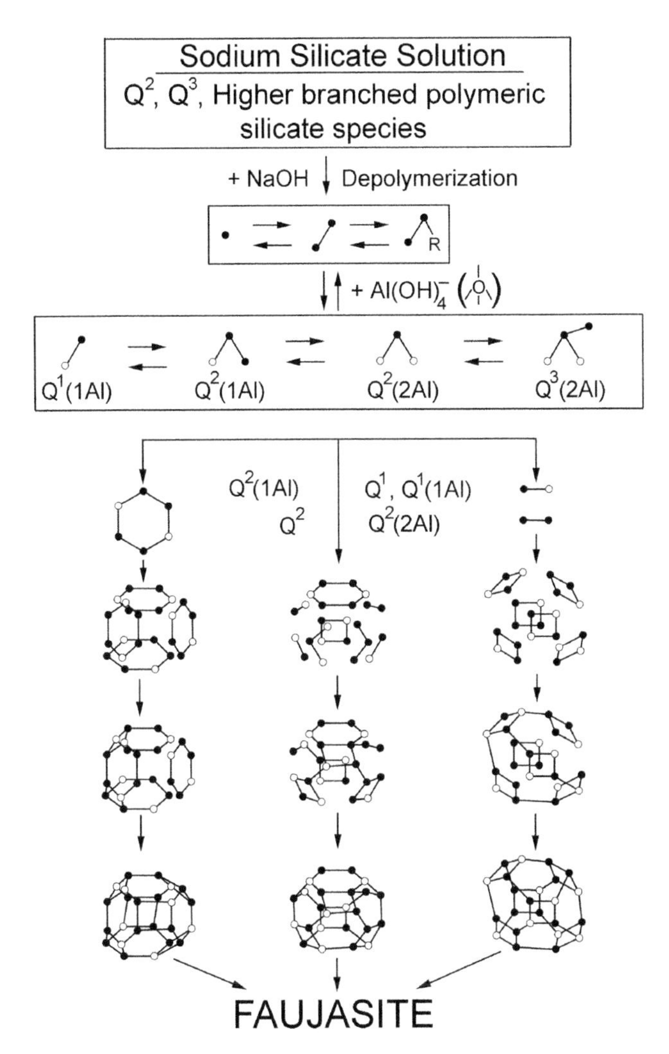

Figure 6.2 Probable scheme for the nucleation and crystallization of zeolite Y. (Reproduced from Ref. 126 with permission.)

A similar principle is valuable for homogeneous nucleation in templated syntheses;[122,123,125] when solutions of the aluminate and silicate or polysilicate anions are mixed to form the hydrogels, a strong base such as NaOH or TPAOH (tetrapropyl ammonium hydroxide) accelerate the dissolution of the gel materials. The dissolved aluminate and silicate ions can also undergo a polymerization processes to aluminosilicate or polysilicate ions. The soluble aluminosilicate or polysilicate species may regroup around the hydrated cations to form the nuclei of the ordered zeolite (see Fig. 6.3).[122,140]

Hence, the presence of 'structured' aluminosilicate blocks (secondary and tertiary building units; see Fig. 6.2; see also Fig. 7 in Ref. 121) are necessary for the

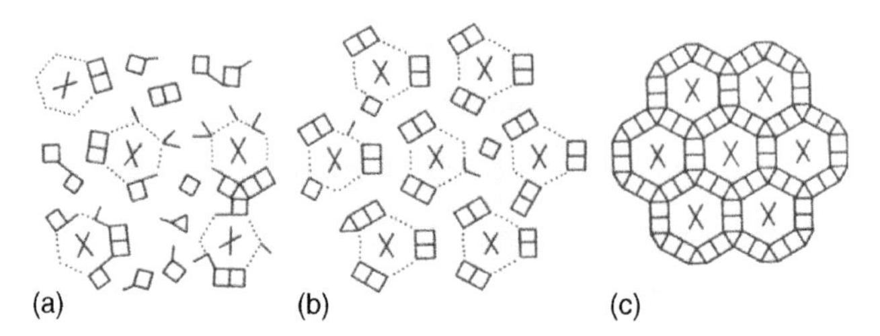

Figure 6.3 Schematic presentation of the first stages of growth of nuclei according to the C-model (can and cement). (A) An early stage of the reaction mixture of gel.'X cans'are water clusters with or without inorganic ions or charged template molecules. (B) Mutual repulsion among the X's causes an order. (C) A framework ('cement') is built between the X's and these are now the largest voids in a crystal. (Reproduced from Ref. 140 with permission.)

formation of nuclei in the liquid phase. However, these structures are not stable at increased temperatures and alkalinities usual for the synthesis of aluminium-rich zeolites, so that in neither case has there been evidence for a direct conversion of any of the proposed building units to the final zeolite structure.[131,141,142] This makes any possibility for occurrence of homogeneous nucleation in the liquid phase very dubious. Such conclusion can be supported by (i) disagreement between predicted (simulated) rate of homogeneous nucleation and the measured one (see Fig. 3a in Ref. 118 as an example), (ii) disagreement between the crystal size distribution(s) predicted (simulated) by the model of homogeneous nucleation and the measured crystal size distribution(s) (see Fig. 11 in Ref. 117 as an example) and (iii) the finding that the time required for first nuclei to appear through the process of homogeneous nucleation is infinite.[143] The last is probably due to the reason that supersaturation, typical for most of zeolite syntheses, was considerably lower than the critical supersaturation required for homogeneous nucleation of most inorganic compounds.[143]

2.1.1.2. Heterogeneous nucleation Heterogeneous nucleation generally refers to new particle formation resulting from the presence of foreign insoluble material.[73,104] The rate of heterogeneous nucleation, $J_{ht} = dN/dt$, can be expressed as:[73]

$$J_{ht} = N_a \exp\left[-\beta b \sigma^3 v / 3(kT)^2 (\ln S)^2\right] \quad (16)$$

where $N_a = N_a^o - N_{ht}$ is number of 'free' active centres, available for the formation of new zeolite nuclei[115] and b is a factor less than one and would account to the fact that σ would be less near solid–liquid interface. For a constant supersaturation, S, typical for most of isothermal zeolite syntheses,[3,76,83,108–113] $N_a \exp[-\beta b \sigma^3 v/3 (kT)^2 (\ln S)^2] = K_{ht} =$constant, and thus[115]

$$J_{ht} = dN_{ht}/d\tau = K_{ht} N_a = K_{ht}(N_a^o - N_{ht}) = K_{ht} N_a^o \exp(-K_{ht}\tau) \quad (17)$$

where K_{ht} is the rate constant of heterogeneous nucleation, N_a^o is the number of 'active centres' at $\tau = 0$ (τ is the time at which nucleation starts) and N_{ht} is the number of already formed zeolite nuclei. Now, applying Eq. (17), for a linear size-independent crystal growth of nuclei (crystals) at constant supersaturation, S^3 and for large values of K_{ht} (i.e., $K_{ht} \rightarrow \infty$; then $N_{ht} = N_a^o$ at $t_c \approx 0$) usual for assumed rapid heterogeneous nucleation of zeolites,[75,106,114–116,144–146] the mass fraction, $f_z(ht)$, of zeolite crystallized at time t_c can be expressed as:[76,82,114–116,135,145,146]

$$f_z(ht) = \lfloor G\rho N_a^o (K_g t_c)^3 \rfloor / m_z(eq) = K(ht)(t_c)^3 \tag{18}$$

where $K(ht) = G\rho N_a^o (K_g)^3 / m_z(eq) = \text{constant}$. When crystal growth is controlled by diffusion through an unstirred layer around the growing particles (crystals), the value δ in Eq. (15) is proportional to $r^{0.326}$,[107] and thus[75,145]

$$f_z(ht) = K(ht)'(t_c)^{2.263} \tag{19}$$

Eqs. (18) and/or (19) alone,[75,145] or in combination with the equations for homogeneous[114] and/or autocatalytic nucleation,[82,85,115,116,135,144,146,147] were frequently used for the analysis of nucleation processes during crystallization of zeolites A,[82,116] X,[116,147] P,[135,147] mordenite,[116] silicalite-1[144] and ZSM-5[146] as well as during transformation of zeolite A into HS.[75,114,145] An excellent agreement between the measured kinetics of transformation of zeolite A into HS and the kinetics calculated by Eqs. (18) and/or (19)[75,114,145] (see Fig. 1 in Ref. 145 as an example) indicates that a fast heterogeneous nucleation on the impurity particles always present in the liquid phase,[73,75,106,114–116,145] as schematically presented in Fig. 6.4, is probably the only process relevant for the formation of the primary particles (nuclei) of HS. On the other hand, the real role of the 'classical' heterogeneous nucleation in the synthesis of zeolites from hydrogels is still questioned.

Although in many cases measured kinetics of crystallization can be almost perfectly correlated with the kinetics of crystallization calculated by kinetic equations based on a combination of heterogeneous and autocatalytic nucleation,[82,85,115,116,135,146,147] the conclusions outlined on the basis of the mentioned kinetic analysis must be reconsidered. Namely, the used model of autocatalytic nucleation is based on the assumption that nuclei are homogeneously distributed through the gel matrix,[115,116] but in reality the distribution of nuclei in the gel matrix is heterogeneous,[77,119,120,148] and in many cases complex[76,77,79,80,83,149–152] and thus homogeneous distribution of nuclei in the gel matrix may appear only in rare specific cases (for details, see the Section 2.1.2.2). In addition, some recent studies of zeolite nucleation[3,76,77,79,80,83,149–154] and the principles of the gel 'memory effect'[80,150–154] show that most or even all zeolite nuclei are formed in the gel phase. Hence, although the formation of nuclei on the impurity particles always present in the liquid phase is possible,[149] it seems that number of nuclei eventually formed by this mechanism is very low as compared with the number of the nuclei formed by other mechanisms, so that the 'classical' heterogeneous nucleation in the liquid phase can be neglected as a mechanism relevant for the formation of zeolite nuclei.

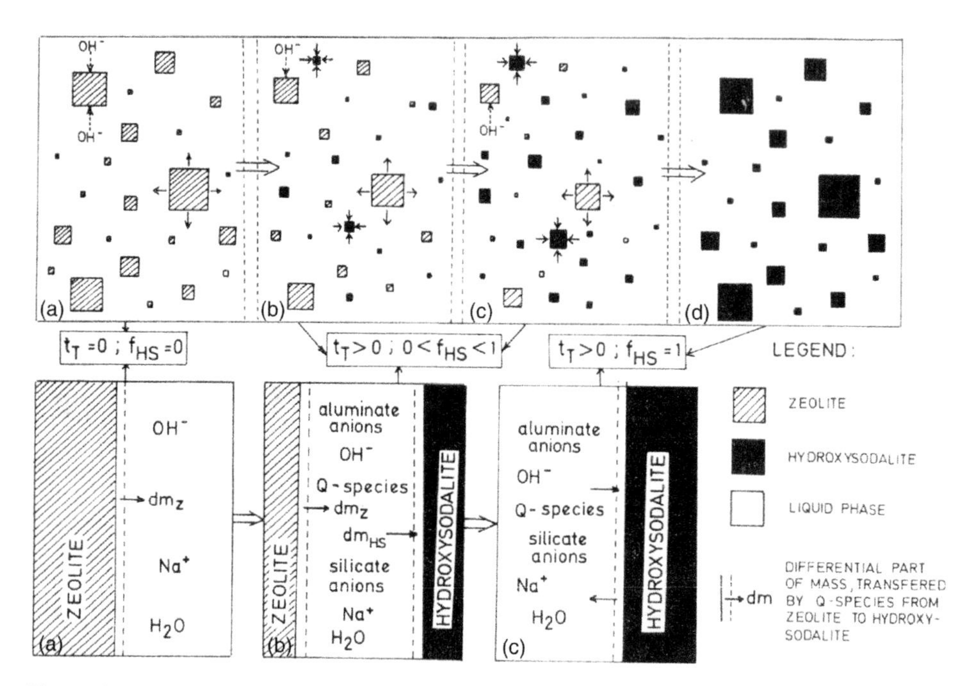

Figure 6.4 A schematic representation of the zeolite–HS transformation mechanism in the case when transformation takes place by dissolution of zeolite particles in caustic medium (NaOH), and by heterogeneous nucleation and crystal growth of HS particles from the solution supersaturated with Q-species. Upper part (a, b, c and d) shows the particulate processes occuring during the transformation. Lower part (A, B and C) shows the chemical composition and mass transport at different stages of the transformation process. (Reproduced from Ref. 114 with permission.)

Other approaches to 'heterogeneous' nucleation of zeolites assume formation of nuclei either on the particles of amorphous phase (gel)[155–157] or at the gel–liquid interface.[146,158–160] Since it is not easy to imagine how less soluble material (zeolite) can be formed (nucleated) at the more soluble substrate (gel), nucleation on the gel–liquid interface (see Fig. 6.5) seems more acceptable.

In contrast to a simple concept in which gel is presented in the form of dense particles (see Fig. 6.5),[158] in the another concept the precursor gel can be seen as a hierarchical structure involving micro-, meso- and macro-pores,[159,160] as schematically presented in Fig. 6.6. 'Heterogeneous' nucleation of zeolite takes place at the interface between the solution and the gel (see Fig. 6.6) where there is an abundance of precursors[146,159] which can rearrange before their release in solution.[159] Accordingly to this model, the rate of nucleation $dN/dt_c = B(t)$ can be expressed as:[159]

$$B(t) = k_3 \left\lfloor (G^{\star}/\varepsilon_p) - G_{eq}^{\star zeol} \right\rfloor S(q) \tag{20}$$

where k_3 is the nucleation rate constant, $G^{\star}$ is the nutrient concentration per unit volume of the crystallizing mixture, ε_p is volume fraction of the liquid phase in the gel microstructure, $G_{eq}^{\star zeol}$ is the zeolite equilibrium concentration and $S(q)$ is the

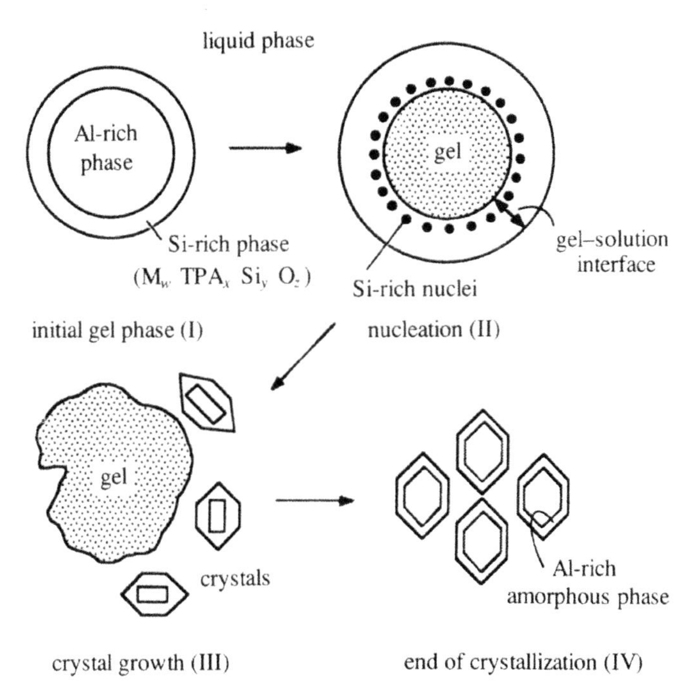

Figure 6.5 Schematic description of 'heterogeneous' nucleation and crystal growth of ZSM-5 crystals at the gel-solution interface. (Reproduced from Ref. 158 with permission.)

interfacial surface area. Equation (20) shows that the rate of nucleation is concentration (supersaturation) dependent, and thus, since the nucleation does not occur on the active centres of the impurity particles, the proposed mechanism of nucleation is rather homogeneous than heterogeneous. In this context, the criticism immanent for the homogeneous nucleation (see Section 2.1.1.1) can also be applied in this case. In addition, due to constant supersaturation during the main part of the crystallization (see Fig. 6.7b) and the starting constancy of the interfacial surface area (see Fig. 6.7c), the rate of nucleation calculated by Eq. (20) starts at the value which is about 50% of the maximum rate of nucleation (see Fig. 6.7a).

Usually, the rate of zeolite nucleation starts from zero value, that is, $dN/dt_c \approx 0$ at $t_c = 0$ (see Figs. 31, 32, 33 and 34b), so that the profile of zeolite nucleation shown in Fig. 6.7c is far from experimental[65,76,77,79,81,83,84,118,119,146,148–150,152,161,163] and theoretical[77–80,85,111,118,119,148,149,151,152,162] experiences. Taking into consideration the above mentioned arguments, it is quite clear that a possibility of occurrence of the 'heterogeneous' nucleation on the gel particles and at the gel–liquid interface, respectively, is dubious and that the conclusions made on the basis of such mechanisms need to be reconsidered.

2.1.1.3. *Secondary nucleation* While primary nucleation is driven only by the solution (see Section 2.1.1.1), or also driven by the solution and 'catalyzed' by active centres of insoluble particulates (see Section 2.1.1.2), secondary nucleation results

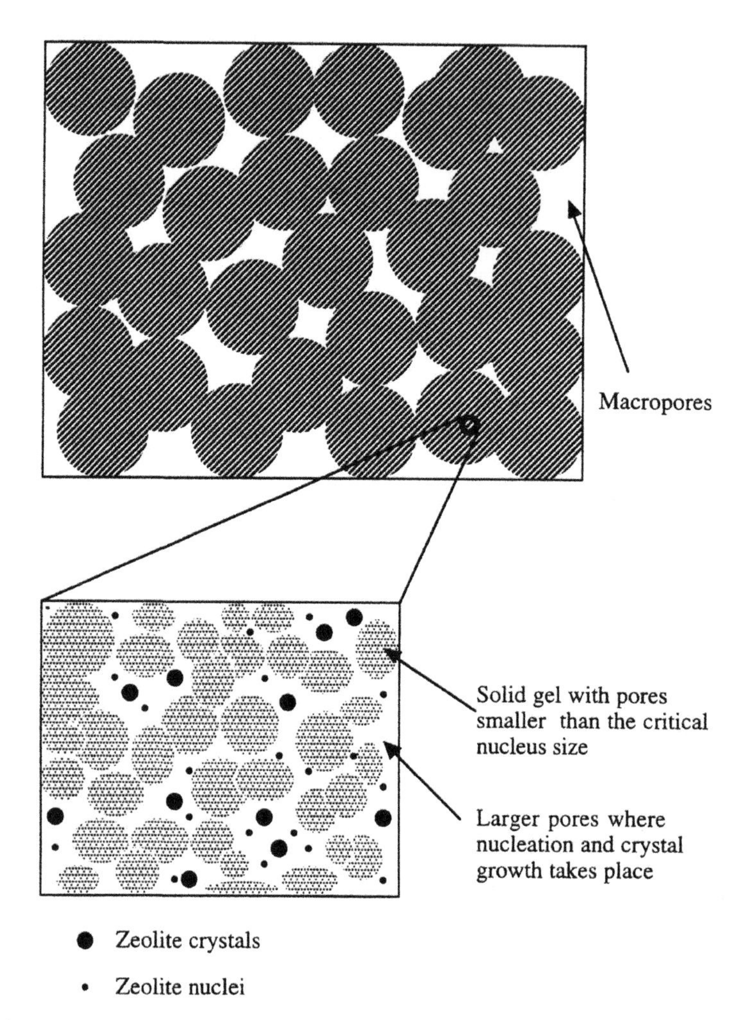

Figure 6.6 Schematic presentation of the gel microstructure crystals. (Reproduced from Ref. 159 with permission.)

from the presence of crystalline material (surface) in the medium and can be driven by different mechanisms such as surface (polycrystalline, initial) breeding, micro-attrition and fluid-shear activity.[73,118,164,165]

Polycrystalline breeding is formation of nuclei on the surface of crystals in a growth environment.[164] These 'surface' nuclei can then grow in the form of randomly shaped polycrystals. Such formed polycrystals can break up with agitation, and then grow independently in the solution.

Initial breeding results from microcrystalline dust being washed off seed surfaces into the crystallization solution.[164–166] The most possible way of the formation of the microcrystalline dust is schematically represented in Fig. 6.8.

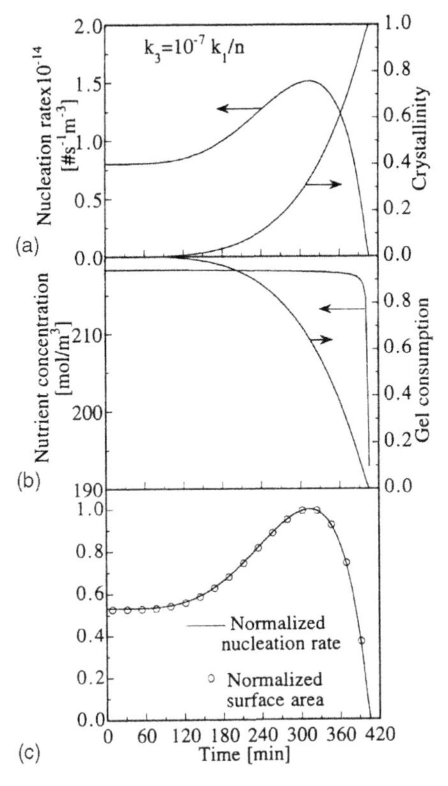

Figure 6.7 Simulation results for (a) nucleation rate and crystallinity, (b) nutrient concentration and gel consumption and (c) normalized nucleation rate and surface area. (Reproduced from Ref. 159 with permission.)

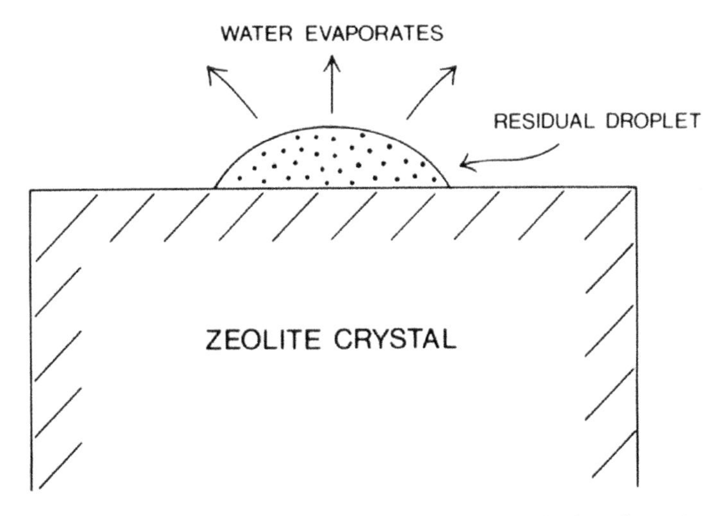

Figure 6.8 Schematic presentation of how initial bred nuclei might form by moisture evaporation. (Reproduced from Ref. 166 with permission.)

During drying of (seed) crystals, any residual droplet on the crystal surface might have dissolved reagents in it. These ingredients become more concentrated as the solvent evaporates and can eventually precipitate as crystallites. These microcrystalline fragments, initially held on seed surfaces by weak electrostatic forces, can become viable growing entities in fresh solution, and thus can be appeared as a new population of crystals once they have grown to observable sizes.[166] Similar microcrystallites may also be formed in the dried state if the product powder is handled in such a way that microcrystalline dust is formed by breakage.[164]

Intense agitation of the reaction mixture can sometimes promote nuclei formation by micro-attrition, that is, by causing microcrystalline fragments to be broken off of existing growing crystals in the medium.[118] These fragments arise from crystal contacts with the stirrer, other crystals, or the walls of the reactor, and may become growing entities in a supersaturated solution. It has been speculated that nuclei can also be created by fluid passing by the surface of growing crystals with sufficient velocity to sweep away quasi-crystalline entities (clusters) incorporated on the crystalline surface.[118] If these clusters are swept away into a sufficiently supersaturated environment, they will have the thermodynamic tendency to grow, and become viable crystals.

One of the first experimental evidence of the secondary nucleation of zeolite was observed by Culfaz and Sand.[132] They showed that nucleation of mordenite occurred on the surface of mordenite seed crystals. Later on, Thompson and co-workers studied the effects of seeding on crystallization of silicalite-1,[164,166–168] HS[164] and zeolite A.[164,169,170] It has been shown that the surfaces of silicalite-1 seed crystals catalyzed the nucleation of new silicalite-1 crystals (see Fig. 6.9). Monomodal crystal size distribution obtained in the presence of the as-synthesized wet silicalite-1 seed crystals and bimodal crystal size distributions obtained in the

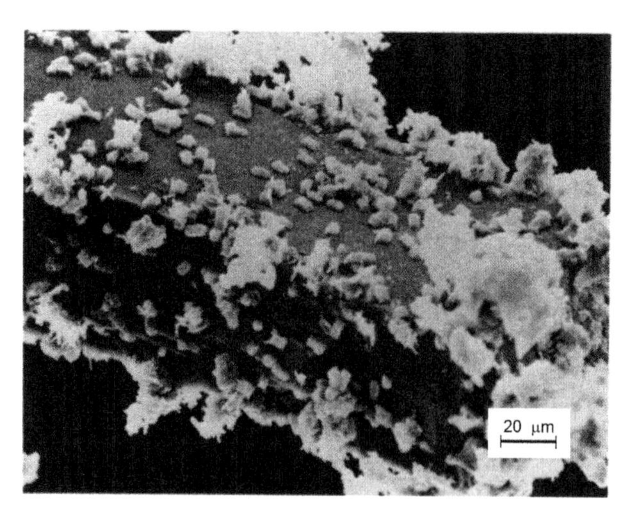

Figure 6.9 SEM (scanning electron microscopy) photograph of overall view of a seed reacted for 6 days in a solution of TPA_2O:89.7$(NH_4)_2O$:1123H_2O with calcined Al-free Na^+-ZSM-5. (Reproduced from Ref. 167 with permission.)

presence of dried and/or calcined seed crystals (see Fig. 7 in Ref. 166) clearly indicate that a new population of silicalite-1 crystals was formed by initial breeding (by the mechanism schematically presented in Fig. 6.8) on the surfaces of dried and/or calcined silicalite-1 seed crystals and not on the surface of the as-synthesized wet seed crystals.[166,167]

Investigation of the influence of the size of seed crystals of zeolite A on the formation of secondary nuclei in the clear aluminosilicate solution showed that small seed crystals (1–3 μm) do not promote nucleation of a new population of zeolite crystals, but that much larger zeolite A crystals (about 40 μm; see Fig. 6.10) promote zeolite crystal nucleation by an initial breeding mechanism (see Fig. 6.11).[170]

On the other hand, when HS seed crystals (Fig. 6.12a) were added to a clear aluminosilicate solution at 90 °C,[164] the product was predominantly HS, but the well defined habit (Fig. 6.12a) was lost and a polycrystalline mass was formed over the seed crystals (see Fig. 6.12b). In this situation, polycrystalline breeding occurred on the HS crystals in the growth solution.[164]

Finally, under given (specific) conditions, nuclei of one type of zeolite can be formed on the surfaces of the crystals of another type of zeolite; typical examples are surface nucleation of zeolite Losod on the surfaces of the seed crystals of zeolite A added in the clear aluminosilicate solution at 86 °C (see Fig. 6.13)[164] as well as surface nucleation of zeolite omega on the surfaces of octahedral crystals of faujasite

Figure 6.10 SEM photographs of 40 μm zeolite A crystals used as seeds. Initial bred nuclei are clearly shown on the surface of crystal. (Reproduced from Ref. 170 with permission.)

Figure 6.11 SEM photographs of product obtained by using 40 μm zeolite A seed crystals. (Reproduced from Ref. 170 with permission.)

(see Fig. 6.14) crystallized at 100 °C from a TMAOH (tetramethylammonium hydroxide)-containing aluminosilicate hydrogel.[171]

The elaborated examples show that secondary nucleation of zeolites is not relevant for most of the 'standard' zeolite syntheses, and that it takes place only under specific synthesis conditions, that is, when the dried, calcined and/or mechanically treated seed crystals were added into clear (alumino)silicate solutions. Since, (i) zeolite crystals are rarely grown larger than about 50–100 μm[165] and (ii) many of zeolite synthesis are carried out with no agitation, or very mild agitation,[118]

Figure 6.12 SEM photographs of: (a) hydroxysodalite (HS) seeds grown in the presence of triethanolamine and (b) growth on HS seeds from (a) when placed in clear aluminosilicate solution. (Reproduced from Ref. 164 with permission.)

micro-attrition breeding may be viewed as not universally important in zeolite crystallizations.[118,165] Also, it is not expected that fluid-shear-induced nucleation will be relevant to zeolite synthesis, due to the viscosity of the solutions, and because it is not believed to be important except at quite high agitation rates, or quite high fluid velocities relative to crystals in the medium.[118] Hence, the mechanisms by which the secondary nucleation occur are limited to the initial breeding,[164,166–168,170] polycrystalline breeding[97,164] and/or surface nucleation of one type of zeolite on the surfaces of the (seed) crystals of another type of zeolite.[164,171]

Figure 6.13 SEM photographs of: (a) Zeolite NaA seeds used in clear aluminosilicate growth experiment and (b) zeolite NaA seeds from (a) after 9 h in clear aluminosilicate solution. (Reproduced from Ref. 164 with permission.)

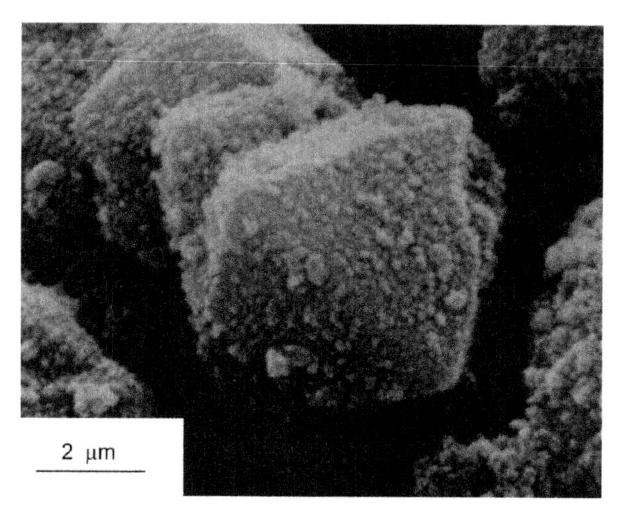

Figure 6.14 SEM photograph of crystals of zeolite omega on the surface of octahedral crystals of faujasite after 5 days of reaction at 100 °C. (Reproduced from Ref. 171 with permission.)

2.1.2. Nucleation in the gel phase

Although the earliest hypotheses on zeolite nucleation assumed formation of zeolite nuclei in the gel phase,[81,84,172–174] the first experimental evidence on the presence of structurally ordered phase in the matrix of X-ray amorphous aluminosilicate gel appeared in the year 1971,[175] and was the only till the year 1981, when the first direct observation of the structurally ordered phase ('quasi-crystalline' phase) appeared in the literature.[176] Since then, the presence of medium- and long-range ordered (partially crystallized, 'quasi-crystalline') phase, comparable with zeolite structure (potential nuclei) in amorphous aluminosilicates (gels) was assumed and directly and/or indirectly detected by using different experimental techniques, as it is elaborated below.

2.1.2.1. *Experimental evidences on the presence of structurally ordered phase in the gel matrix and/or the proposed mechanisms of nucleation* The already mentioned first experimental evidence on the presence of the structurally ordered phase in the X-ray amorphous phase[175] was made by application of electron diffraction (ED) technique. Knowing that ED of zeolite skeleton is intense for agglomerates of even few unit cells,[177] later on, the identification of structurally ordered phase in the gel matrix by ED was published by Tsuruta *et al.*,[177] Gora *et al.*,[40,41] Subotić *et al.*[178] and Kosanović *et al.*;[179] the appearance of the few faint diffraction circles of the ED pattern of gel (see Fig. 6.15) undoubtedly indicates that the X-ray amorphous solid contains particles of partially or 'quasi-crystalline' phase having a size below the X-ray diffraction (XRD) detection limit, but above the ED detection limit.[177,178]

Nine years after the first experimental evidence of structurally ordered phase in the gel matrix,[175] Thomas and Bursill published[176] the first direct evidence of the presence of partially crystalline phase ('quasi-crystalline' zeolite A) in the the X-ray amorphous aluminosilicate precursor (gel) for zeolite A synthesis. HRTEM (high-resolution transmission electron-microscopy) image in Fig. 6.16 undoubtedly shows the 'islands' of 'quasi-crystalline' zeolite A having the size about 5–10 nm, surrounded by amorphous material.

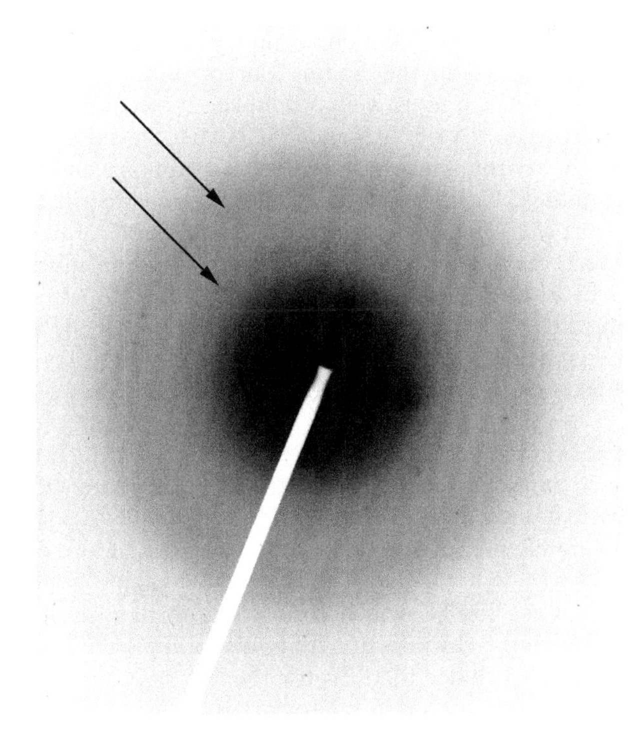

Figure 6.15 Electron-diffraction pattern of the X-ray amorphous aluminosilicate precursor (Na_2O:Al_2O_3:2.576SiO_2:2.28H_2O) separated from freshly prepared aluminosilicate hydrogel (3.51Na_2O:Al_2O_3:2.1SiO_2:85.2H_2O). (Reproduced from Ref. 179 with permission.)

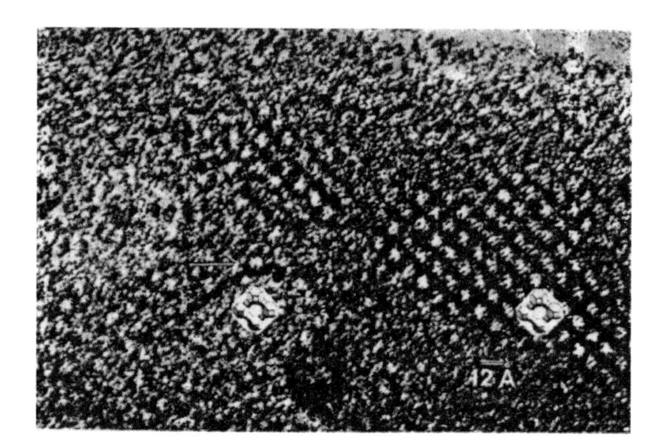

Figure 6.16 'Quasi-crystalline' zeolite A, showing 'rafts' of ordered structure surrounded by amorphous material. (Reproduced from Ref. 176 with permission.)

Unfortunately, at least by our knowledge, this is unique direct observation of 'quasi-crystalline' zeolite in the gel matrix isolated from heterogeneous system (hydrogel).

A wide application of the solid-state nuclear magnetic resonance (NMR) techniques in the investigation of the solid phases appeared at different stages of zeolite crystallization[180,181] reflected in the finding out of different stages of structural arrangements of Si and Al atoms as well as template molecules in the investigated solids. Using the solid-state ^{23}Na-, ^{27}Al- and ^{29}Si-NMR, Dewaelle *et al.* have shown that Na-, Al- and Si-environments in the initial (amorphous) and final (zeolite Y) solid phases are similar.[109] Using ^{13}C-NMR and differential thermal analysis (DTA) techniques, Gabelica *et al.*[182] have found that a large number of nuclei are formed within the hydrogel, due to the high concentration of the relative aluminate and silicate species which are in intimate interaction with TPA^+structure directing cations. Analyzing the ^{31}P-NMR and ^{27}Al-NMR spectra of the reactive $AlPO_4$ gels aged at room temperature for different times, Prasad *et al.*[183] concluded that the framework precursor species are generated in the gel, but only after addition of template.

Thermal analysis is found as a very sensitive method for detection of very small structurally ordered units having a zeolite structure distributed through the gel matrix. In the thermal study (differential thermal analysis—DTA) of the solid phase separated from the freshly prepared system: $(TPA)_2O$–$(NH_4)_2O$–Al_2O_3–SiO_2–H_2O, Nastro *et al.*[123] showed that the DTA curve of the sample is characterized by exothermic peak at about 450 °C which corresponds to thermal decomposition of TPA species occluded in the channels of MFI type zeolites (see Fig. 6.17).

Hence, the authors concluded that 'Even if the lack of X-ray diffraction pattern does not strictly exclude that ZSM-5 microcrystals are present in the amorphous material, it appears very likely from the DTA data that in systems without alkalies TPA may be present in the precursor gel'. As already mentioned, similar conclusion based on the DTA and ^{13}C-NMR analyses of gels was done by Gabelica *et al.*[182]

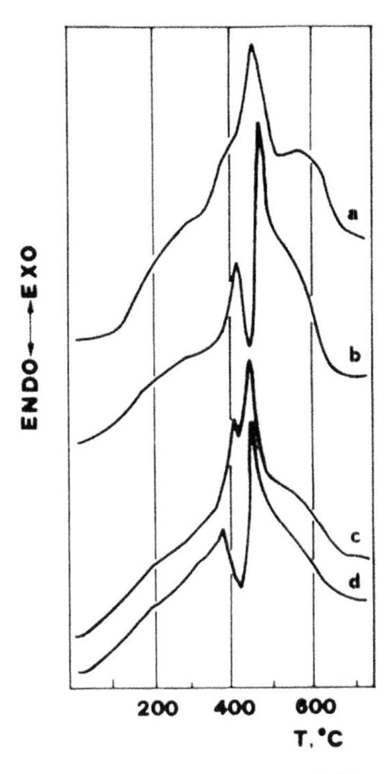

Figure 6.17 DTG curves of: (a) No alkali added precursor gel; (b) no alkali added, 85% crystallinity; (c) 1.5 moles Na_2O, 94% crystallinity; and (d) 1.5 moles K_2O, 100% crystallinity. (Reproduced from Ref. 123 with permission.)

Studying the influence of the type of alkali cations on thermal properties of aluminosilicate gels, Aiello *et al.*[184] have found that the DTG (differential thermal gravimetry) curves of Na-gel (G1) and (Li,Na)-gel (G6) have two endothermic peaks; the first 'low-temperature' peak at about 75 °C which corresponds to the removal of loosely held moisture from within the solid microstructure and the second 'higher-temperature' one at about 180 °C (see Fig. 6.17). Increase of the molar ratio $R_{K,Na}=[K^+/(K^++Na^+)]$ in the (K,Na)-gels causes increase of the intensity of the 'low-temperature' peak and simultaneous decrease of the intensity of the 'higher-temperature' one, so that the DTG curve of the pure K-gel (G5) have the 'low-temperature' peak only (see Fig. 6.18).

Since the position of the 'higher-temperature' peak in the DTG curve of the gels G1 and G6 is similar to the position of the minimum in the DTG curves of the crystalline end products (HTPG1, HTPG5—zeolite A) obtained by heating of the corresponding amorphous solids in 2 M NaOH at 80 °C (see Fig. 6.19), it was concluded that the presence of 'structure-forming' ions (Li^+, Na^+) in the batch induces the formation of structural subunits or even more complex structures, resembling those in the crystalline products, inside the gel matrix during its formation.

Later on, the similar conclusions based on the TG/DTG analyses of amorphous aluminosilicate gel precursors were outlined by Subotić *et al.*,[178] Krznarić *et al.*[185] and Kosanović *et al.*[179]

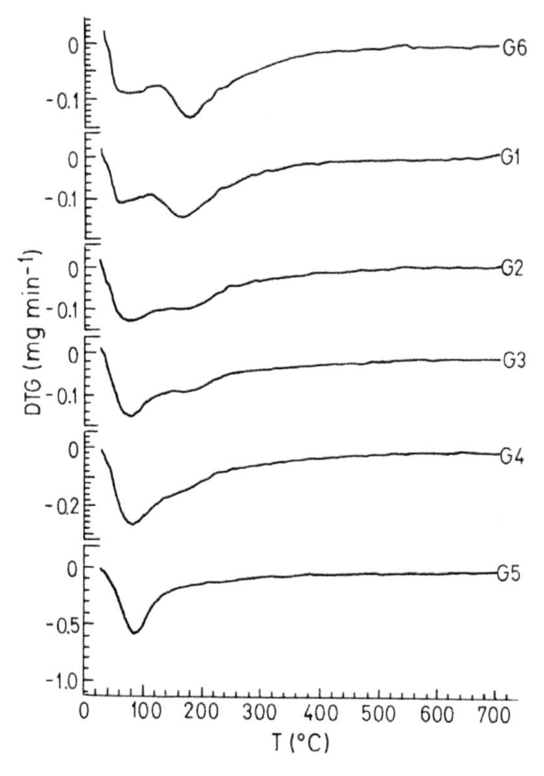

Figure 6.18 DTG curves of the amorphous solids G1 ($R_{K,Na} = 0$), G2 ($R_{K,Na} = 0.365$), G3 ($R_{K,Na} = 0.580$), G4 ($R_{K,Na} = 0.768$), G5 ($R_{K,Na} = 1$) and G6 ($R_{Li,Na} = 0191$). $R_{K,Na} = [K^+/(K^+ + Na^+)]$ in the (K, Na)-gels and $R_{Li,Na} = [Li^+/(Li^+ + Na^+)]$ in the (Li, Na)-gel. (Reproduced from Ref. 184 with permission.)

Knowing that infrared (IR) vibrations of zeolite skeleton are intense for agglomerates of even few unit cells,[186] vibrational techniques (IR, laser Raman) were frequently used in the structural studies of the solid phases appeared during zeolite crystallization. From the appearance of weak absorbances at ~560, ~680 and ~760 cm^{-1} (which are similar to those of faujasite type zeolite) in the IR spectra of the nucleation gel aged at ambient temperature for ≥115 h (see Fig. 6.20), Evmiridis and Yang concluded that nuclei or precursor of faujasite zeolite are gradually developed in the nucleation gel during ageing.[187]

The weak IR band at 575 cm^{-1}, close to IR band at 556 cm^{-1} which is assigned to external vibrations related to D4 rings in zeolite A (see Fig. 6.21) was recently found in the IR spectrum of X-ray amorphous aluminosilicate solid separated from freshly prepared hydrogel.[179] It is interesting that the same weak absorbance at about 570 cm^{-1} appeared in the X-ray amorphous solid obtained by mechanochemical amorphization of zeolite A (see Fig. 8 in Ref. 188). In both the cases, the weak absorbance at about 570 cm^{-1} is ascribed to 'quasi-crystalline' zeolite phase formed in the gel matrix during its precipitation,[179] or by the

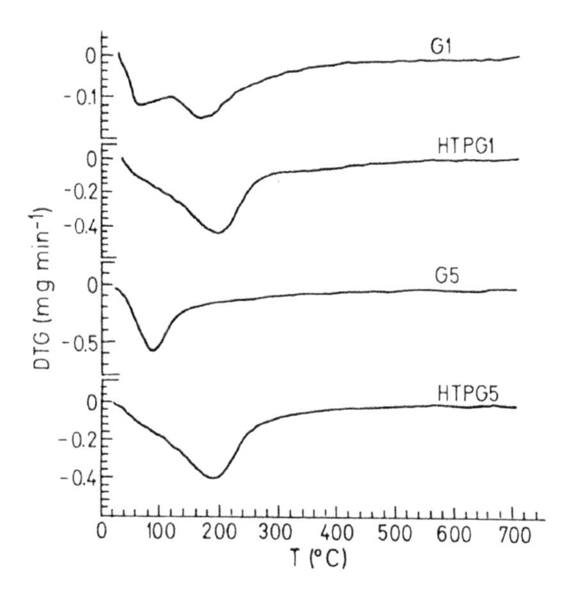

Figure 6.19 DTG curves of amorphous solids G1 and G5 and of the crystalline products HTPG1 (obtained by hydrothermal treatment of the amorphous solid G1) and HTPG5 (obtained by hydrothermal treatment of the amorphous solid G5). (Reproduced from Ref. 184 with permission.)

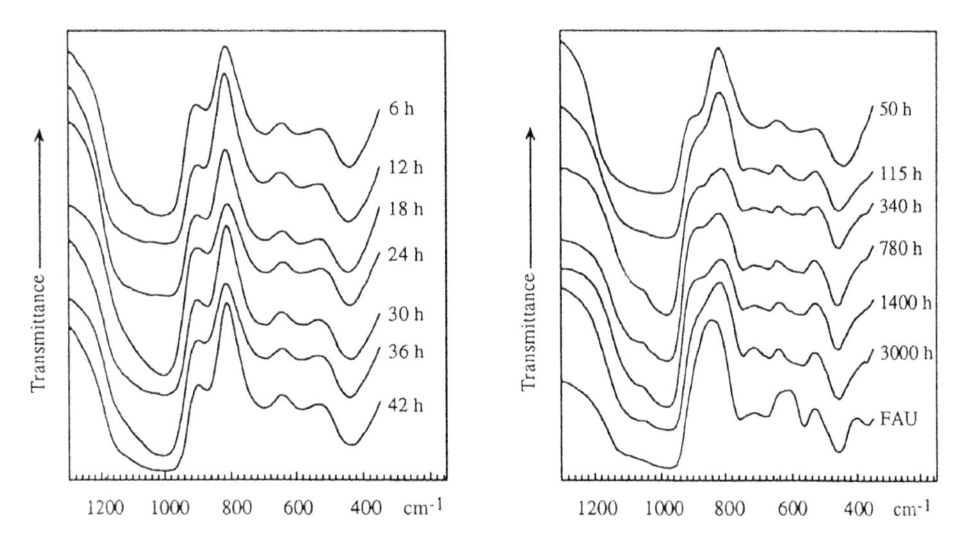

Figure 6.20 IR spectra of the solid samples obtained by filtration from nucleation gels with various periods of ageing (hour), washed with distilled water and dried at ambient temperature for 1 week. IR spectrum of a FAU type zeolite is also shown in the figure. (Reproduced from Ref. 187 with permission.)

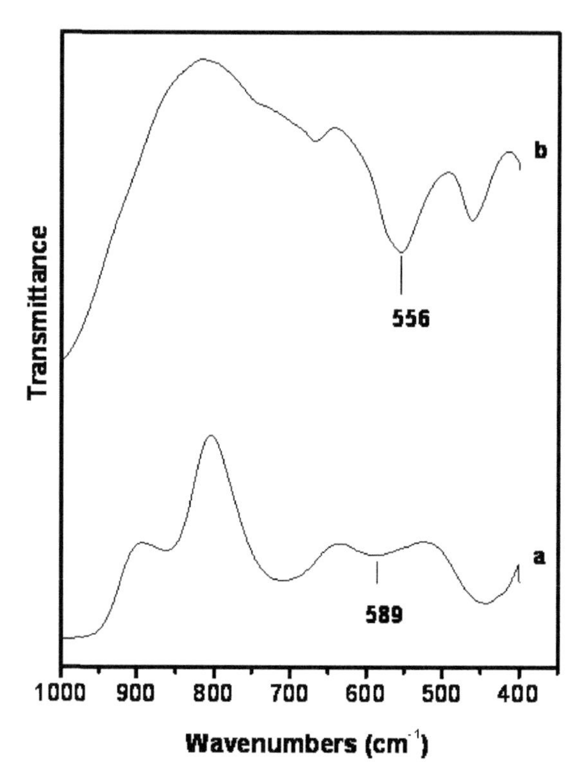

Figure 6.21 IR spectrum of the X-ray amorphous aluminosilicate (Na_2O:Al_2O_3:2.576-SiO_2:2.28H_2O) separated from freshly prepared aluminosilicate hydrogel (3.51Na_2O:Al_2O_3:2.1-SiO_2:85.2H_2O). (Reproduced from Ref. 179 with permission.)

presence of a small fraction of (quasi) crystalline phase (zeolite A) which has not been transformed into amorphous phase and which amount and/or particle size is below the X-ray detection limit.[188]

Recently,[179] three different entities were observed in the AFM (atomic force microscopy) images of the freshly prepared gel; aggregate of disc-shaped particles (having the mean diameter of about 80 nm and mean height of about 15 nm, see Fig. 6.22A), 'transition', probably partially crystalline, features (particles of 'quasi-crystalline' phase, see Fig. 6.22B) and aggregates of 'pyramidal-shape' features which look like fully crystalline material (see Fig. 6.22C).

Observation of isolated 'pyramidal-shaped' particles in the AFM images of the X-ray amorphous solid phase extracted from hydrogel after its heating at 80 °C for 30 min (see Fig. 6.23A), is in accordance with the principle of the autocatalytic nucleation, namely that the less soluble, more structurally ordered entities 'released' from the gel matrix dissolved during heating of hydrogel. Hence, it can be concluded that just the 'pyramidal-shaped' particles are zeolite nuclei which grow during the process of crystallization at 80 °C and form the near cubic crystals at $t_c = 90$ min (see Fig. 6.23B; t_c is the time of crystallization) and typical zeolite A crystals at $t_c = 240$ min (Fig. 6.23C).

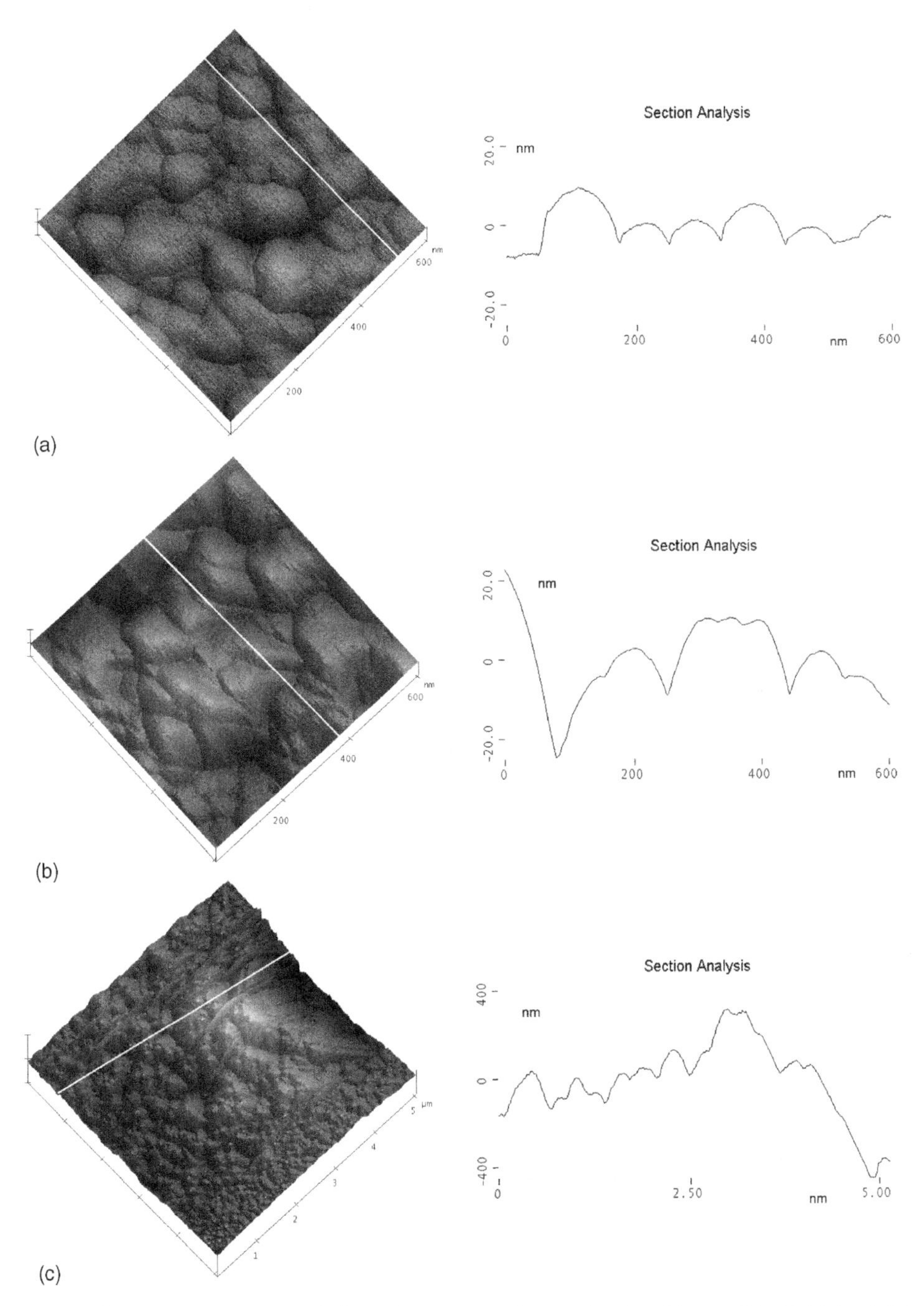

Figure 6.22 AFM images of the X-ray amorphous aluminosilicate (Na_2O:Al_2O_3:2.576-SiO_2:2.28H_2O) separated from freshly prepared aluminosilicate hydrogel (3.51Na_2O:Al_2O_3:2.1-SiO_2:85.2H_2O) showing (A) aggregate of disc-shaped particles (having the mean diameter of about 80 nm and mean height of about 15 nm, (B) 'transition', probably partially crystalline, features (particles of 'quasi-crystalline' phase and (C) aggregates of 'pyramidal-shaped' features which look like fully crystalline material (see Fig. 37). (Reproduced from Ref. 179 with permission.) (See color insert.)

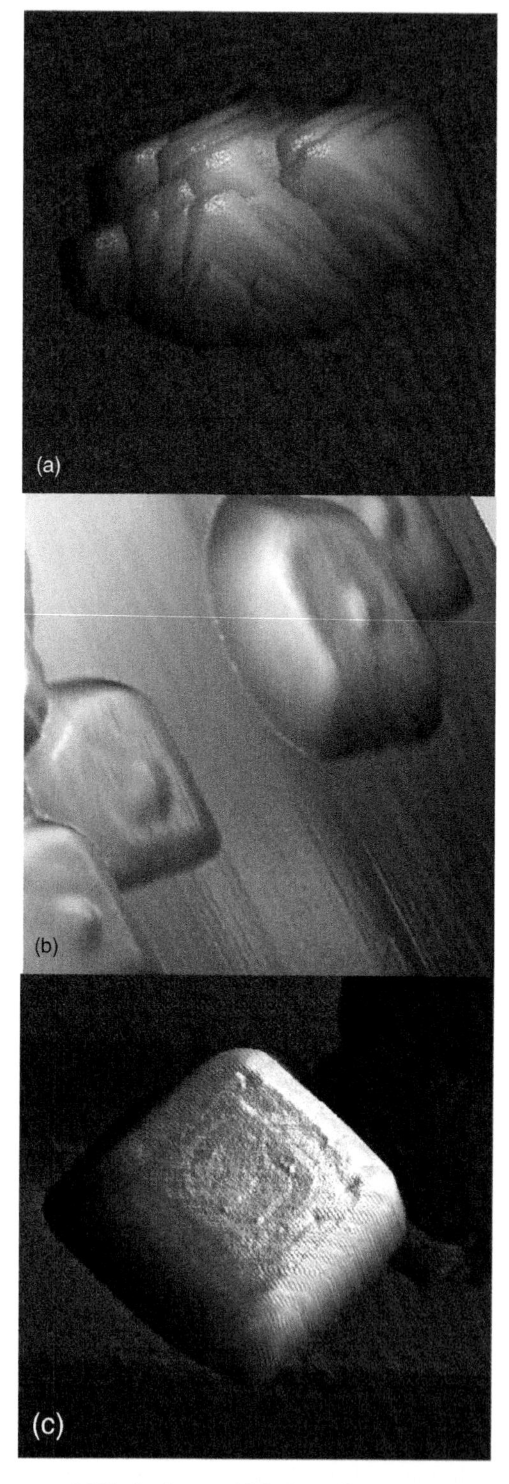

Figure 6.23 AFM images of (A) the 'pyramidal-shaped' particles obtained in the solid phase drawn off the reaction mixture (hydrogel) after its hydrothermal treatment at 80 °C for 30 min, (B) near-cubic-shaped crystals contained in the solid phase drawn off the reaction mixture after its hydrothermal treatment at 80 °C for 90 min and (C) typical cubic crystals of zeolite A contained in the solid phase drawn off the reaction mixture after its hydrothermal treatment at 80 °C for 240 min. (Reproduced from Ref. 179 with permission.) (See color insert.)

Based on the above described findings[40,41,109,123,147,175–179,182–187] as well as on the other kinds of experimental evidences, many authors concluded that a great part or even all nuclei needed for growth of different types of zeolites are formed in the gel phase; [45,49,50,190–203] consequently, a number of models of nucleation (and/or crystal growth) of zeolites in the gel phase appeared in literature.

From the analysis of the adsorption isotherms and diffuse reflectance spectra (DRS) of methylene blue on differently aged nucleation gel and well-crystallized zeolite X seed crystals, Richards and Pope[189] concluded that both of these materials have rather similar DRS spectral characteristics which are intermediate between those of amorphous gel and well-crystallized zeolite X, and hence that nucleation gel may be composed of essentially crystalline domains which are too small to produce a characteristic diffraction pattern.

Based on the results of study of crystallization of FAU-type zeolite at room temperature by different experimental techniques such as HRTEM, *in situ* synchrotron XRD, IR, dynamic light scattering (DLS), N_2 adsorption measurements and chemical analyses, Valthev and Bozhilov[49] concluded that the process of zeolite formation can be divided into four general stages as it is schematically represented in Fig. 6.24: (i) 0–1.5 h, formation of an amorphous gel with highly variable composition, local formation of stable zeolite nuclei and nanometre sized metastable hydrated silica phase; (ii) 1.5 h to 10 days, chemical evolution of the gel composition that is coupled with structural rearrangements of the amorphous gel and further development of zeolite nuclei; (iii) 10–14 days, spontaneous mass transformation of the amorphous gel into spherical aggregates of small (10–20 nm) crystals around individual crystallization centres of viable nuclei and collapse of the metastable phase and (iv) 17–38 days, crystal growth involving agglomeration, dissolution and regrowth of individual nanoparticles around spherical aggregates.

Analyzing the results obtained by different experimental methods (high-resolution solid-state ^{29}Si and ^{27}Al MAS NMR spectroscopy, SAXS (small-angle X-ray scattering), XRD and SEM (scanning electron microscopy)) during crystallization

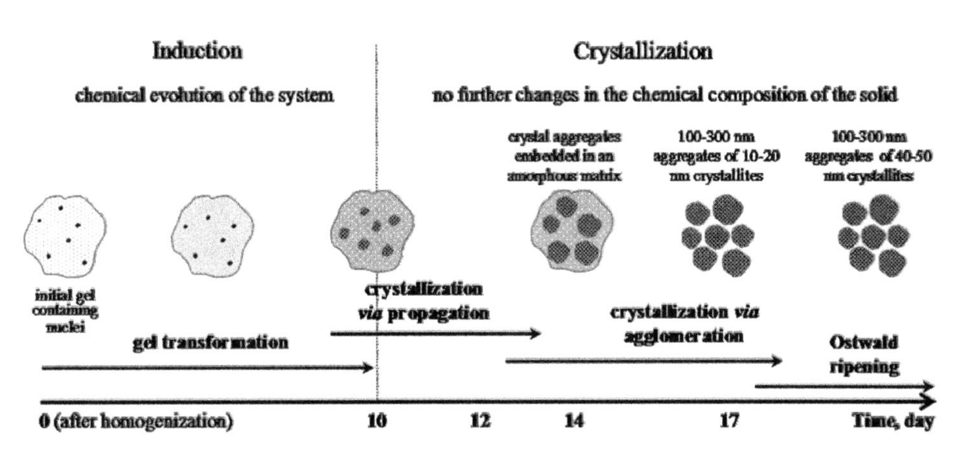

Figure 6.24 Schematic illustration of the crystallization mechanism of FAU-type zeolite under ambient conditions. (Reproduced from Ref. 49 with permission.)

of zeolite A, Smaihi *et al.*[45] concluded that the first crystallization stage proceeds by reorganization of the amorphous aluminosilicate units formed during the mixing of precursors. The reorganization involves an ordering of Si–O–Al bond angles and operates on aggregates of constant volume. The crystallization process takes place in the volume of aggregates by propagation through the gel phase. After this stage, the growth in the system is dominated by the solution-mediated transport, which leads to formation of well-shaped zeolite A crystals.

On the basis of the study of crystallization of zeolite TS-1 from amorphous wetness impregnated SiO_2–TiO_2 hydrogels using different methods (XRD, IR, diffuse reflectance UV–vis spectroscopy, *n*-hexane adsorption, SEM), Serrano *et al.*[190] concluded that the crystallization proceeds by a non-conventional mechanism as it is schematically represented in Fig. 6.25. In accordance with this mechanism, TS-1 is crystallized mainly by solid–solid transformation which allows the raw amorphous material to be reorganized yielding TS-1. In contrast with the two-step process (nucleation, crystal growth), here it is not true crystal growth, but the crystals are formed by zeolitization of amorphous particles as whole. The main roles of the solution phase are to provide the additional TPAOH necessary for crystallization and to favour migration and isolation of the secondary particles.

Derouane *et al.*[191] and Gabelica *et al.*[192] have found that an addition of acidic Al-sulphate solution containing TPABr into aqueous Na-silicate solution favour a rapid nucleation of ZSM-5 in the gel matrix (see Fig. 6.26). Structure directing TPA^+cations still present all thought gel, can interact intimately with the numerous reactive aluminosilicate anions and a direct recrystallization process involving the solid hydrogel phase transformation is expected (see Fig. 6.26). During heating, a rapid growth yielding a large number of small crystallites which present a homogeneous Al radial distribution was confirmed experimentally.[191]

Similar model of nucleation of ZSM-5 was developed by Chang and Bell.[193] Based on the ion-exchange and ^{29}Si-MASNMR, they surmised that in TPA^+gel

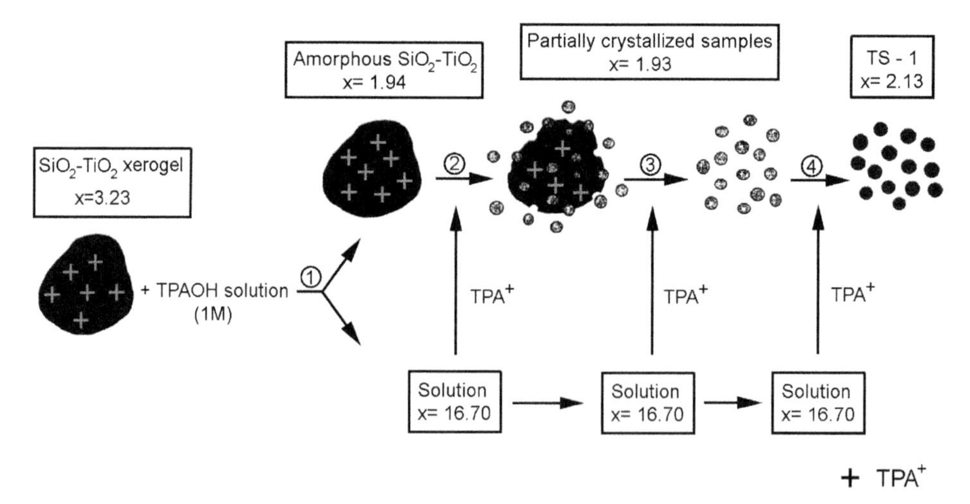

Figure 6.25 Proposed mechanism for TS-1 crystallization [$x = (100 \times Ti)/(Si + Ti)$]. (Reproduced from Ref. 190 with permission.)

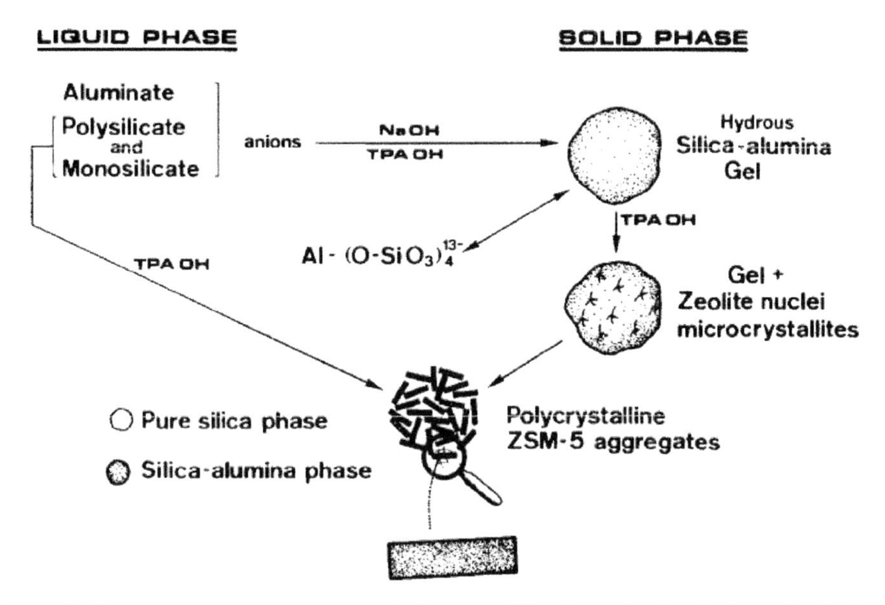

Figure 6.26 Schematic representation of the crystallization of ZSM-5 from the hydrogel prepared by addition of Al-sulfate solution containing TPABr into aqueous Na silicate solution. (Reproduced from Ref. 192 with permission.)

system, embryonic structures are formed rapidly upon heating by following steps: (1) The formation of clathrate-like water structure around the template, (2) isomorphous substitution of silicate for water in these cages, which resemble ZSM-5 channel intersections and (3) progressive ordering of these entities into the final crystal structure (see Fig. 6.27).

The models of nucleation/crystallization of ZSM-5 presented in Figs. 6.26 and 6.27 and described in the Refs. 191–193 includes a rapid reaction of TPA^+ ions with numerous reactive (alumino)silicate anions and formation of amorphous gel,[191,192] or formation of clathrate-like water structure around the template and isomorphous substitution of silicate for water.[193] Later on, based on cryo-TEM and SAXS analyses of the liquid phase of silicalite-1 hydrogel, Regev *et al.* identified so-called globular structural units (GSU) in the freshly prepared synthesis hydrogels.[194] The authors assumed that each GSU, having a diameter of about 5 nm, is composed of several 'tetrapods' constructed of an (alumino)silicate skeleton wrapped around TPA^+ cation, so that the 'tetrapods' have a similar structure to those connecting the straight and sinusoidal channels in the final crystalline MFI-type zeolites, as it is schematically presented in Fig. 6.28. For this reason, the authors assumed that the GSU may be either amorphous or crystalline. One year later, based on the ^{1}H-^{29}Si CP MAS NMR analyses of the solid phases in the systems $(TPA)_2O$:SiO_2:D_2O, it was found that a close contact between the protons of TPA and silicon atoms of the inorganic phase is established by the van der Waals interaction, prior to the formation of the long-range ordering of crystalline zeolite structure.[195,196]

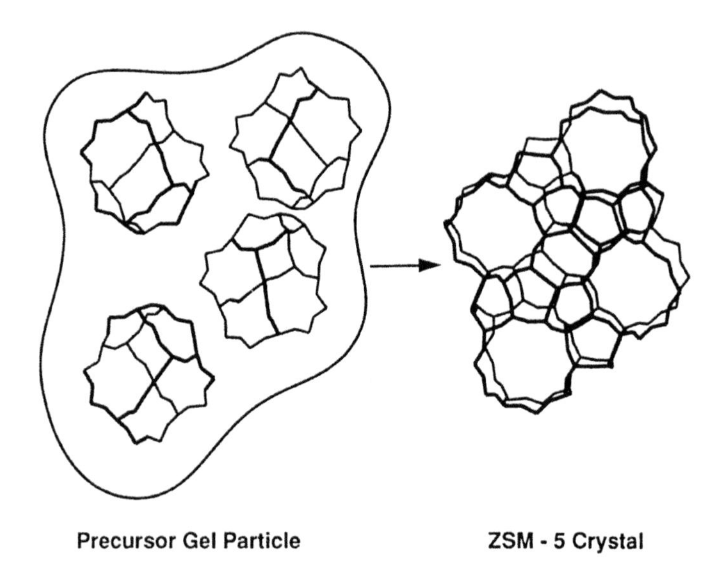

Figure 6.27 Representation of ZSM-5 gel nucleation mechanism. (Reproduced from Ref. 193 with permission.)

It was proposed that silicate is closely associated with the TPA molecules, thus forming inorganic–organic composite species (IOCS) that are the key species for the self-assembly of MFI structure (see Fig. 6.29).[195,196] Hence, there is no doubt that the 'tetrapods' shown in Fig. 6.28 represent an 'idealized picture' of the IOCS.[194] Both groups of authors[194–196] suggested that the key step in the synthesis of MFI-type zeolites is formation of 'tetrapods' [194] and/or IOCS[195,196] that then provide the fundamental units for nucleation and crystal growth.

Based on these findings and the recent studies of crystallization of mainly zeolites ZSM-5 and silicalite-1 applying DLS, SANS (small-angle neutron scattering), SAXS and WAXS (wide-angle X-ray scattering), formation of nuclei of MFI structural type in the gel phase may be imaged as follows: The hydrophobic hydration sphere that is formed around TPA in aqueous solutions are partially or completely replaced by silica as previously proposed by Chang and Bell[193] and later on by Twomey *et al.*[60], thus forming IOCS ('tetrapods') (see Figs. 6.28 and 6.29). Since the replacement of water by silica can be realized only when a sufficient amount of soluble silicate species are available, the IOCS can be formed even at room temperature when TEOS or soluble Na-silicates are used as silicon source, but the IOCS can be formed only upon heating of the reaction mixture when condensed silica (e.g., Cab–O–Sil) was used as silica source.[195,196] The next critical step in the crystallization of MFI-type zeolites in heterogeneous systems is formation of primary units, having the size of about 3 nm,[42,197–200] most probably by aggregation of several IOCS [200] (see Fig. 6.29) at the start of the crystallization process even at room temperature.[194,197–199] The presence of the primary units is independent on the structure directing agent, alkalinity and presence of gel phase.[200,202] Therefore, several authors[51,62,63,194,197–203] concluded that just the primary subcolloidal

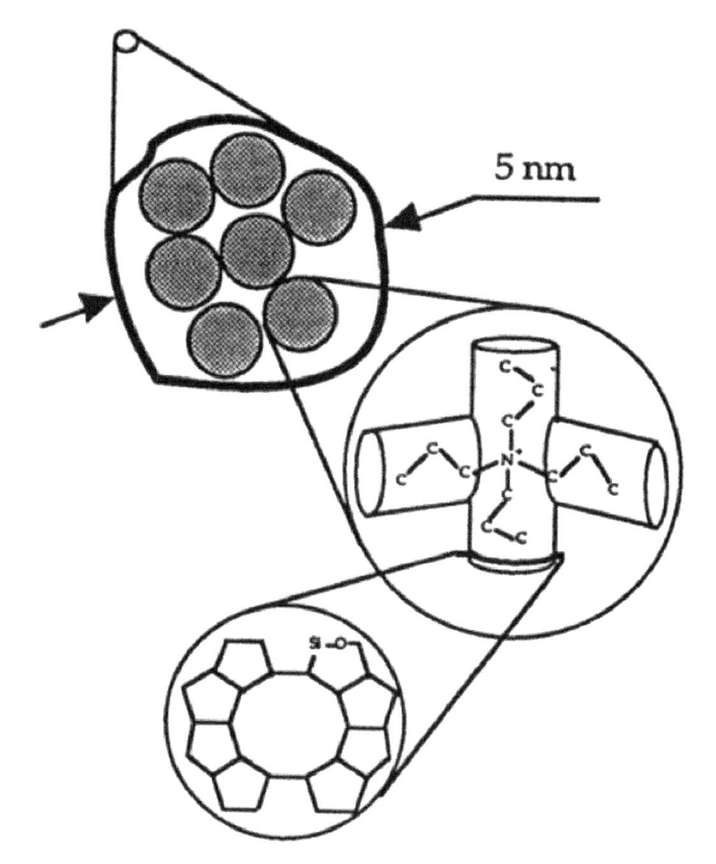

Figure 6.28 Schematic representation of the structure of the structural globular unit. (Reproduced from Ref. 194 with permission.)

particles are precursors for nucleation and crystal growth of silicalite and other siliceous zeolites. In addition, secondary units having the size 5–10 nm are formed by a stepwise aggregation of primary (≈3 nm) units (see Fig. 6.29).[194,198,200] The secondary units probably correspond to the GSU[194] as well as to the fractal aggregates (6.4–7.2 nm) identified by SANS, SAXS and WAXS analyses of the reaction mixtures at the early stages of crystallization of silicalite-1 from both heterogeneous (gel)[198] and homogeneous (clear solution) systems.[197,198] In more 'concentrated' heterogeneous systems, the secondary units may rapidly be agglomerated into large(r) amorphous aggregates having the size 40–600 nm.[200] After the amorphous aggregates reach a 'critical' size (e.g., ≥10 nm), part of gel nutrient transforms into crystalline phase (viable nuclei)[194,198,200] (see Fig. 6.29) by reorganization and condensation of the amorphous aggregates.[192–194,197,198,200–203] Recently,[57] the crystallization of zeolite beta from the system xTEAOH:50 SiO_2:Al_2O_3:750H_2O ($x = 10.5$, 20 and 30) was explained by the same mechanism.

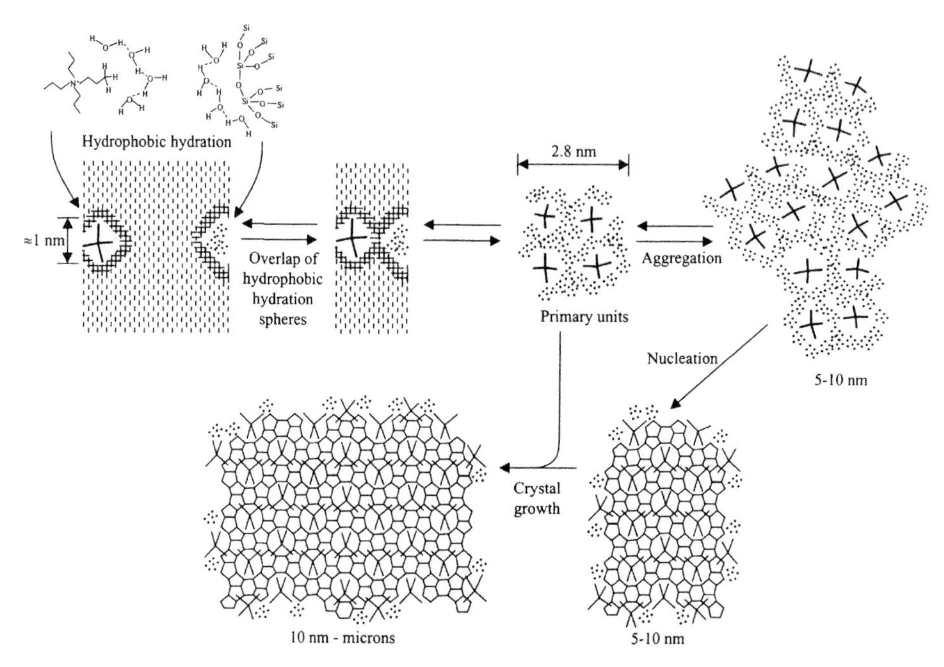

Figure 6.29 Scheme for the crystallization mechanism of Si-TPA-MFI. (Reproduced from Ref. 200 with permission.)

Although the formation of nuclei in the gel matrix is not inquisitive,[40,41,49,50,109,123,147,175–179,182–187,190–203] this process is not quite clear in the models described in the cited papers and schematically presented in Fig. 6.29. Evidences of structurally ordered entities (nuclei) in freshly prepared (unheated) gel precursors [49,109,128,147,176–179,182–185,190–192,195,196,200] indicate that the above described processes, including formation of viable nuclei in the gel matrix (see Figs. 6.24, 6.25, 6.26 and 6.29), can take place at ambient temperature when a sufficient amount of soluble silicate species is available (e.g., when TEOS or soluble Na-silicates are used as silicon sources).[191,192,195,196] Otherwise, for example, when amorphous silica (fumed or precipitated) is used as silicon source, the mentioned processes can take place only under hydrothermal conditions,[191,192,195,196] that is, after a part of amorphous silica has been dissolved, thus producing a sufficient amount of soluble silicate species needed for the formation of an aluminosilicate gel precursor and/or IOCS needed for the formation of (alumino)silicate gel precursors in templated systems.[195,196]

Most of the models of zeolite nucleation presented in this section assume that both nucleation and crystal growth of zeolite(s) take place in the gel phase without the participation of solution,[109,190–200] or that crystallization process is solution-mediated only partially, usually at the end of the process.[50,109] Although, due to high supersaturation of gel with the constituents (Na, Al, Si, template)[40–42,182,204] nuclei may be formed very rapidly (e.g., during gel precipitation), the increase of number of nuclei during room-temperature ageing of hydrogels[49,79,80,83,147,186,187,205] indicates that reorganization and condensation reactions that form the viable nuclei inside the gel matrix are time-dependent processes. Hence, while the formation of (new) nuclei in the gel matrix can take place either during the room-temperature ageing or hydrothermal treatment of hydrogels, it seems that the possibility of their growth inside the gel matrix considerably depend on the size of 'primary' gel particles and the 'density' of their packaging in larger gel particles. For instance, when zeolites A and X are crystallized from initially clear solutions, amorphous phase was formed as discrete particles, having 40–80 nm in size, during room-temperature ageing.[42,46] During hydrothermal treatment, small, nano-sized crystals grow inside the gel particles (see more in Section 2.2.1). Hence, it is quite possible that density of these amorphous particles is low enough for 'free' transport of material through the gel-solution interface as well as inside the gel matrix.[179] On the other hand, in a milky-white low-viscosity gel, the grain-like 'structure' composed of particles with sizes ranging between 10 and 50 nm, is formed.[49] In contrast to formation of only one nucleus in each of 'low-density primary gel particle',[42,46] several nuclei can be formed in each of the 10–50 nm 'primary' gel particle, even at room temperature (see Fig. 6.24),[49] hence it is quite reasonable to conclude that the number of nuclei formed inside each 'primary' gel particle increases with its aluminosilicate concentration. As already shown, (see Fig. 6.24), the room-temperature ageing of this milky-white, low-viscosity hydrogel results in the growth of nuclei inside the gel matrix and formation of crystal aggregates (100–300 nm) of 10–20 nm zeolite crystallites. During the initial stage of hydrothermal treatment of the same reaction mixture (aged at room temperature for 24 h before heating at 90 °C),[50] a growth via propagation through the amorphous network co-exists with the typical solution-mediated hydrothermal crystallization, while 'the later stages of development of the systems followed the classical for zeolite-yielding systems crystallization that could be described by autocatalytic nucleation' (see more in the Section 2.1.2.2). This indicates that the rate of crystal growth inside the gel matrix is low even at 90 °C, so that the relatively large FAU crystals (1–4 μm) could be formed in relatively short time (600 min at 90 °C) only by the solution-mediated growth of the nuclei and nano-crystals released from the gel dissolved during the crystallization process. Finally, it is really to assume that due to high aluminosilicate concentration, the primary gel particles formed from usual heterogeneous systems[108,109,117,123,128,147,164,176–179,182–187,206] have higher density than those formed in milky-white low-viscosity gels[73,75] and/or in the gels formed in initially clear solutions;[42,46] these primary particles have a strong tendency to form dense packed agglomerates having the size in the micrometre range (see also Figs. 6.16 and 6.22).[108,117,164,206] Under such conditions, the transport of material through the gel-solution interface as well as inside the gel particles is limited only to thin surface/subsurface layers of gel particles,[179] and thus the growth of nuclei inside the gel matrix is considerably retarded.[207] Hence, the nuclei formed inside the gel particles formed in heterogeneous systems (hydrogels) can grow only after their release from the gel dissolved during the crystallization, that is, when they are in full contact with the liquid phase (autocatalytic nucleation; see below). Such occurrence of crystallization process (solution-mediated growth of nuclei and crystals in a direct contact with the liquid phase) can be clearly argued by the existence of distinctly separated growing zeolite crystals and dissolving gel particles during the entire process of crystallization,[45,108,117,122,164,190,206,208–210] in spite of maintaining the gel-zeolite particle size in an assumed growth inside gel matrix. [42,46,109,190–200]

2.1.2.2. Autocatalytic nucleation Analyses of assumed homogeneous (see Section 2.1.1.1) and heterogeneous nucleation (see Section 2.1.1.2) of zeolites showed that the fraction, f_z, of crystallized zeolite can be generally expressed as:

$$f_z = K(t_c)^q \tag{21}$$

with $q \leq 4$ for homogeneous nucleation (see Eq. (15)), $q \leq 3$ for heterogeneous nucleation (see Eqs. (18) and (19)) and consequently, $2.263 \leq q \leq 4$ when nuclei are simultaneously formed by homogeneous and heterogeneous nucleation in the liquid phase. However, analysis of many zeolite crystallizing systems showed that the power q in Eq. (21) is higher than 4.[76,77,81–85,115,116,135,150,211] This indicates that the rate of nucleation increases with the time of crystallization, at least during the induction period of the crystallization process.[81,84] Such an increase in the nucleation rate was firstly explained by Zhdanov,[84] who postulated '... that not only the aluminosilicate blocks formed in the liquid phase but also the similar blocks with ordered structure in the gel skeleton can be the nuclei of crystals. The number of such blocks passing into solution and coming out at the surface of gel particles for a unit of time must increase with increasing dissolution rate of the gel skeleton during the autocatalytic stage of crystallization'. Hence, the formation of particles of structurally ordered phase (potential nuclei, when they are 'hidden' in the gel matrix) in the gel during its precipitation and/or ageing, and their growth after 'release' from the gel dissolved during the crystallization, that is, when they are being in full contact with the liquid phase (see Fig. 6.30), are the necessary conditions for occurrence of autocatalytic nucleation.[76,79,81,83,84,111,115,116,147,150,211]

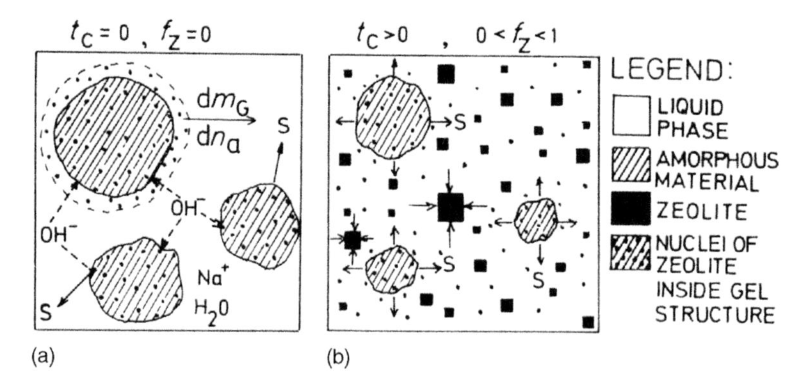

Figure 6.30 Schematic representation of crystallization of zeolites from aluminosilicate gels. (a) Dissolution of gel particles in hot alkaline medium and 'releasing' of zeolite nuclei from the dissolved part(s) of gel. (b) Simultaneous dissolution of gel particles, 'releasing' of zeolite nuclei from dissolving gel particles and growth of zeolite nuclei (crystals). (Reproduced from Ref. 116 with permission.)

From this condition is evident that rate of autocatalytic nucleation is proportional to the rate of gel dissolution,[76,77,82–85,115,116,118–120,135,147,148,162,165,178,212] that is

$$B = \mathrm{d}N/\mathrm{d}t_c = F(\mathrm{N})'(\mathrm{d}m_G^\star/\mathrm{d}t_c) \tag{22}$$

where $F(\mathrm{N})' = N_s(\mathrm{d}f_N^\star/\mathrm{d}f_G^\star)$ is the density function of distribution of nuclei in the gel matrix, $f_N = N/N_0$ is the fraction of nuclei released from the mass fraction $f_G^\star = m_G^\star/m_G^\star(0)$ of dissolved gel, N_0 is total number of nuclei in the gel matrix, $m_G^\star(0)$ is the entire amount of the dissolved gel and N_s is the specific number of nuclei, that is, number of nuclei 'released' from a unit mass of dissolved gel. Since the mass, m_z, of crystallized zeolite is proportional to the mass, $m_G^\star$, of dissolved gel,[79,80,149–152], the rate of autocatalytic nucleation can also be expressed as:

$$B = \mathrm{d}N/\mathrm{d}t_c = F(\mathrm{N})(\mathrm{d}m_z/\mathrm{d}t_c) \tag{23}$$

where $F(\mathrm{N}) = N_s(\mathrm{d}f_N/\mathrm{d}f_z)$ is the modified density function of distribution of nuclei in the gel matrix showing the number, N (or fraction f_N), of nuclei released from the mass, $m_G^\star$ (or fraction $f_G^\star$), of gel needed for crystallization of the mass, m_z (or fraction $f_z = m_z/m_z(\mathrm{eq})$, where $m_z(\mathrm{eq})$ is the mass of zeolite at the end of crystallization process, that is, $t_c = t_c(\mathrm{eq})$) of crystallized zeolite. The first, 'primitive' models of autocatalytic nucleation assumed homogeneous distribution of nuclei in the gel matrix (see Fig. 6.30),[115,116,212] that is

$$\mathrm{d}f_N/\mathrm{d}f_G^\star \propto \mathrm{d}f_N/\mathrm{d}f_z = K \tag{24}$$

where K is constant close to 1, and thus in accordance with Eq. (22) and (23), respectively

$$B = \mathrm{d}N/\mathrm{d}t_c = N_s(\mathrm{d}m_G^\star/\mathrm{d}t_c) = N_s(\mathrm{d}m_z/\mathrm{d}t_c) \tag{25}$$

This implies that the maximum rate of autocatalytic nucleation corresponds to the maximum rate of crystallization, that is, that maximum rate of nucleation is reached at the crystallization time at which about 50% of gel is transformed into zeolite.[85] This is, in the most cases, in disagreement with the measured kinetics of nucleation in which the maximum appear during 'induction period', or at early stage of conversion[65,77,81,84,146,148,150,161,163] even in the case when the kinetics of crystallization can be almost perfectly correlated with the values calculated by the kinetic equations derived from Eq. (25) (see Fig. 6.31).[146] From this reason, in literature appeared serious critiques of one of fundamental assumptions concerning the early model of autocatalytic nucleation, namely that distribution of nuclei in the gel matrix is homogeneous,[118–120,148,165,212]; even Gonthier *et al.*[119,148] suggested a modification of the originally posed concept of autocatalytic nucleation, as expressed by Eq. (25).[115,116]

The modification is based on the assumption that nuclei are not distributed throughout the gel phase uniformly, but that they are rather located preferentially

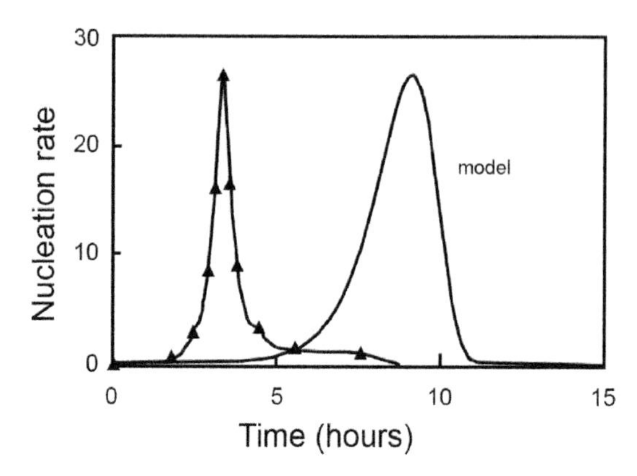

Figure 6.31 Comparison of the nucleation rate for experiment B (data taken from Ref. 146) with predictions using the autocatalytic nucleation model as originally posed. (Reproduced from Ref. 119 with permission.)

near the outer 'surface' of the gel particles. In this case, nuclei would be activated much earlier in the synthesis process than previously predicted, consistent with observations.[65,77,81,84,146,148,150,161,163] To quantify this modification, an empirical function for the distribution of nuclei in the gel matrix, which describe a distribution of nuclei near the outer rim of the gel particles was suggested in the form

$$N = N_s \exp(-K m_G^{\star}) \tag{26}$$

and thus the rate of autocatalytic nucleation is

$$\mathrm{d}N/\mathrm{d}t_c = K N_s \left[\exp(-K m_G^{\star})\right] (\mathrm{d}m_G^{\star}/\mathrm{d}t_c) \tag{27}$$

Simulation results from this model[119,120] were at least consistent, in trend with the measured nucleation rate values (see Fig. 6.32), whereas the crystallization curve and the increase in the largest crystal sizes with time were fitted reasonably well.[119]

Later on, applying the modified model of autocatalytic nucleation (see Eqs. (26) and (27)) for the analysis of crystallization of ZSM-5 at different temperatures by population balance, Falamaki *et al.* have found almost perfect correlation between measured and calculated kinetics of nucleation (see Fig. 6.33).[77]

Accepting the arguments that distribution of nuclei in the gel matrix is not homogeneous, authors of the original (the 'primitive' one) model of autocatalytic nucleation developed the method for determination of distribution of nuclei in the gel matrix:[76,79,80,83,149,150,152] Kinetics of nucleation (symbols in Fig. 6.34b) is calculated from the corresponding kinetic of crystal growth (see Fig. 6.34c) and crystal size distribution (see Fig. 6.35) using the well known method described by Zhdanov and Samulevich.[81] Integration of the nucleation curve results in the change of total number, N, of nuclei and/or its fraction, $f_N = N/N_s$, during crystallization (dashed curves in Fig. 6.36). Now, if the number, N, or fraction,

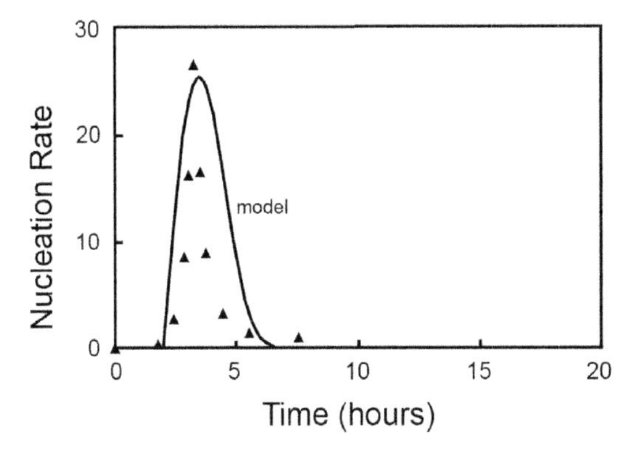

Figure 6.32 Comparison of the nucleation rate for experiment B (data points taken from Ref. 146) and the predicted values from the model assuming that nuclei are preferentially located near the outer surface of amorphous gel particles [Eqs. (26) and (27)]. (Reproduced from Ref. 119 with permission.)

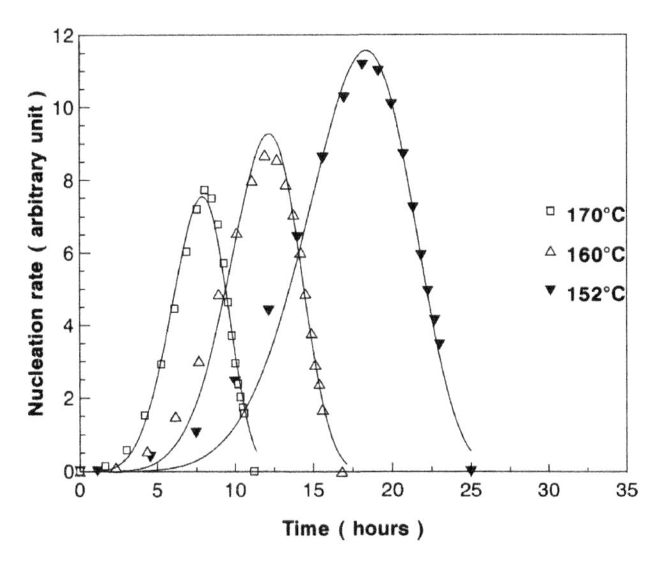

Figure 6.33 Measured kinetic of nucleation (points) and the kinetics of nucleation calculated by Eq. (27) (solid curves) which correspond to crystallization of zeolite ZSM-5 at different temperatures. (Reproduced from Ref. 77 with permission.)

f_N, of nuclei released from the part of gel dissolved up to various crystallization times, t_c, are plotted versus the corresponding fractions, $f_c(=f_z)$ of zeolite crystallized up to the same crystallization times, then the corresponding N versus f_c (see Fig. 6.37a) or f_N versus f_c (see Fig. 6.37b) are obtained.

Analyses of f_N versus $f_G^\star \propto f_N$ versus $f_c = f_z$ functions in the gels prepared under different conditions, by the above described method, showed that in many cases the distribution is more complex[76,77,79,80,83,149–152] (see Fig. 37 as well as Fig. 12 in Ref.

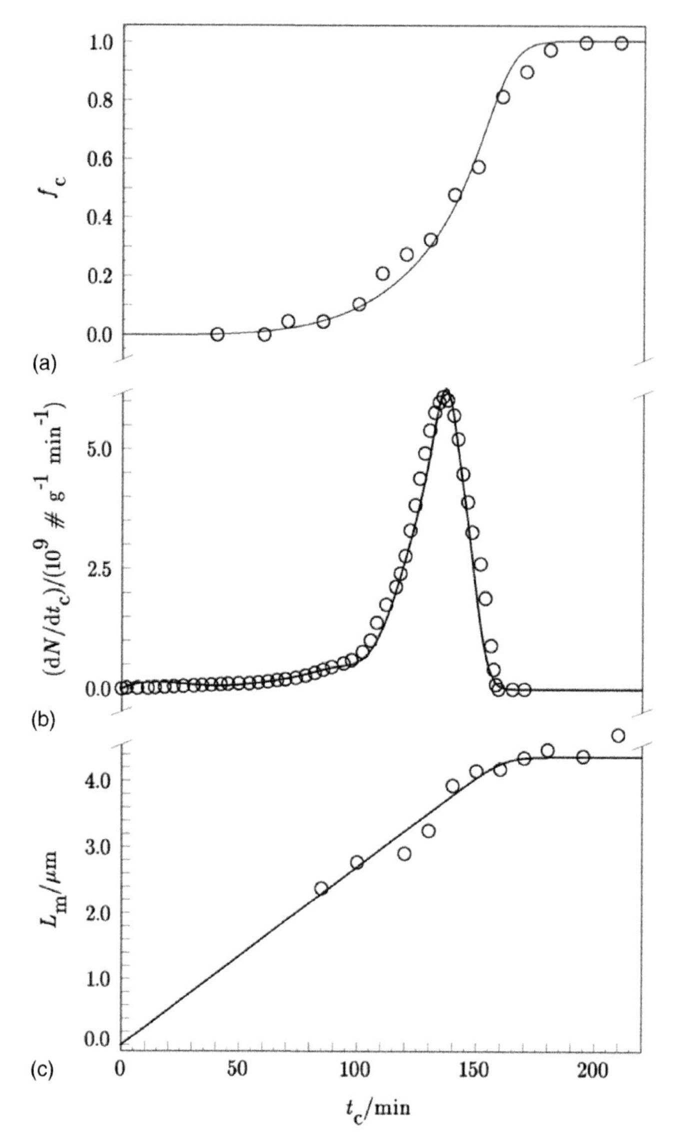

Figure 6.34 Simulated (curves) and measured (symbols) changes in (a) fractions, f_c, of crystalline phase (zeolite A), (b) absolute rate, dN/dt_c, of the nucleation and (c) dimension, L_m, of the largest crystals during the crystallization of zeolite A from hydrogel: $4.72Na_2O$:Al_2O_3:1.93 SiO_2:254.86H_2O at 80 °C. (Reproduced from Ref. 152 with permission.)

76, Figs. 4 and 8 in Ref. 79, Fig. 6 in Ref. 80, Fig. 8 in Ref. 83, Fig. 1 in Ref. 149 and Figs. 9–11 in Ref. 150) than this defined by Eq. (26). In the other words, the homogeneous distribution ($f_N/f_G^\star$ = constant) as well as the distribution defined by Eq. (26) represent only special cases. Taking into consideration that all nuclei are formed in the X-ray amorphous aluminosilicate precursor (gel) during its

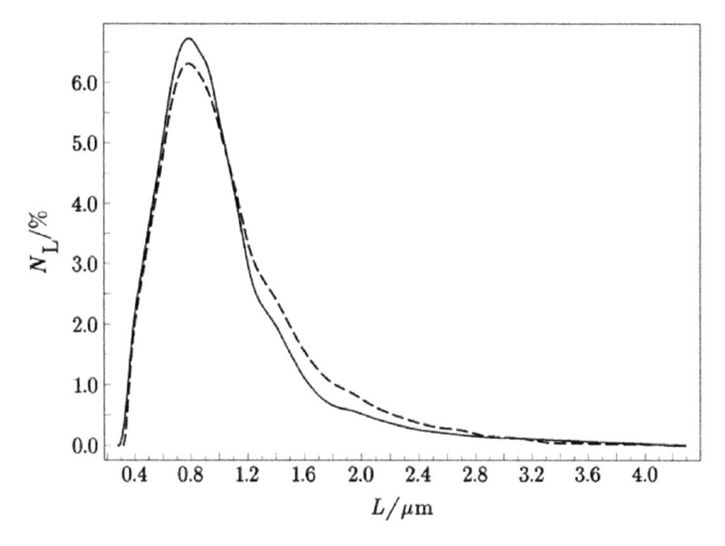

Figure 6.35 Crystal size distribution (solid curve) of the crystalline end product obtained during the crystallization of zeolite A from hydrogel: $4.72Na_2O$:Al_2O_3:$1.93SiO_2$:$254.86H_2O$ at 80 °C. (Reproduced from Ref. 152 with permission.)

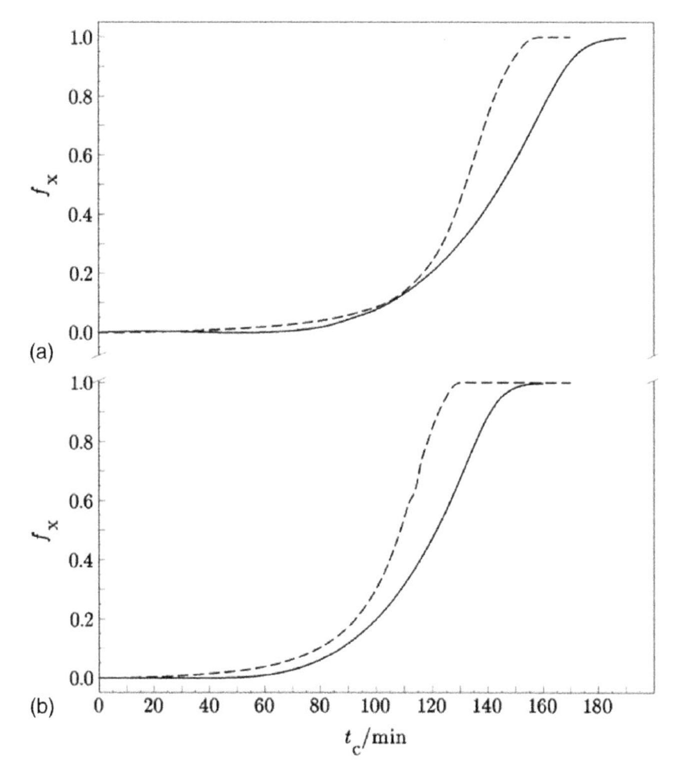

Figure 6.36 Changes in fractions, $f_x = f_c$ of zeolite A (solid curve) and $f_x = f_N$ (dashed curve) of the number of nuclei (crystals) during crystallization of zeolite A from hydrogel: $4.72Na_2O$: Al_2O_3:1.93 SiO_2:$254.86H_2O$ at 80 °C (lower part). (Reproduced from Ref. 152 with permission.)

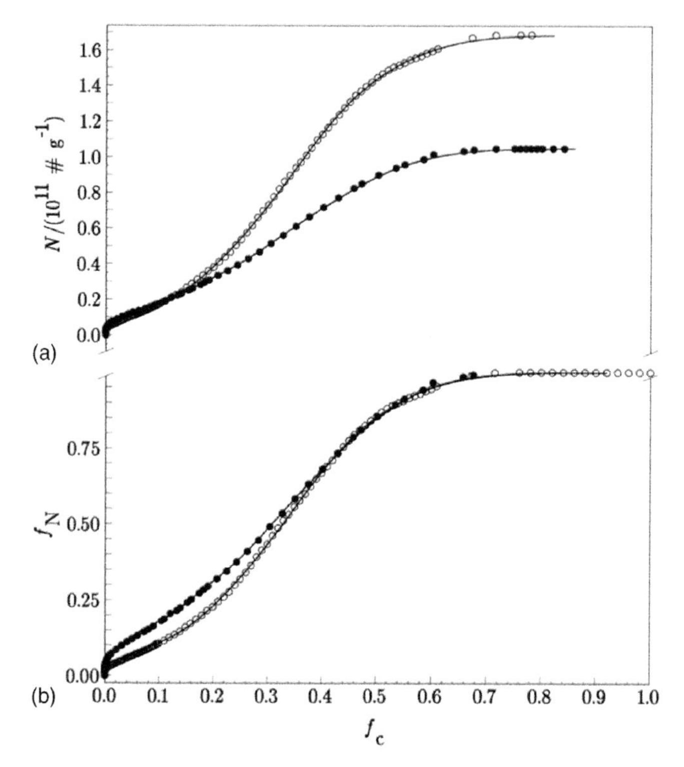

Figure 6.37 Measured (symbols, ○) and calculated (by Eq. (29); curves) N vs. $f_c = f_z$ plots (a) and f_N vs. $f_c = f_z$ plots (b) which correspond to crystallization of zeolite A from hydrogel: $4.72Na_2O{:}Al_2O_3{:}1.93\ SiO_2{:}254.86H_2O$ at 80 °C (lower part). (Reproduced from Ref. 152 with permission.)

precipitation[49,83–85,109,113,115,116,123,135,146,147,149,150,176–179,182,185,189,192,206] and/or room-temperature ageing,[79,80,83,147,187,205] the corresponding N versus $f_c = f_z \propto N$ versus $f_G^\star$ and f_N versus $f_c = f_z \propto f_N$ versus $f_G^\star$ plots represent a measure of the distribution of nuclei in the gel matrix.[76,79,80,83,149–152] Recent analyses of the relationship between f_N and $f_c \propto f_G^\star$ showed that it may generally be expressed by four-parameter equation,[111,214,218] that is

$$f_N = f_0 + \sum_{i=1}^{p} f_i\{1 - \exp[-k_i(f_z)^{n_i}]\} \tag{28}$$

and thus

$$F(N) = N_s(df_N/df_z) = N_s \sum_{i=1}^{p} f_i k_i n_i (f_z)^{(n_i-1)} \exp[-k_i(f_z)^{n_i}] \tag{29}$$

where $f_o + \sum_{i=1}^{p} f_i = 1$ and $f_o \langle\langle \sum_{i=1}^{p} f_i$. Thus, in accordance with Eq. (23) the rate of autocatalytic nucleation is

$$B = \mathrm{d}N/\mathrm{d}t_c = N_s \left\{ \sum_{i=1}^{p} f_i k_i (f_z)^{(n_i-1)} \exp[-k_i (f_z)^{n_i}] \right\} (\mathrm{d}m_z/\mathrm{d}t_c) \qquad (30)$$

Figure 6.34 shows that the measured kinetics of crystallization (symbols in Fig. a), nucleation (symbols in Fig. b) and crystal growth (symbols in Fig. c) are almost perfectly correlated with the values calculated (simulated) by population balance [see Eqs. (2)–(5)], using the nucleation rate equation [Eq. (30)] based on the modified model of autocatalytic nucleation; similar perfect correlations between measured and calculated (simulated) data can also be found in Refs. 77,79 and 149. Besides the mentioned evidences of autocatalytic nucleation revealed by partial,[76,81,82,84,85,119,120,135,146,147,165,178,179,184,205,206,211,213] or complete agreement between measured and calculated (simulated) data,[77,79,80,83,149,152] as well as the experimental evidences of the gel 'memory' effect,[79,80,150,152–154,213] which is relevant only for autocatalytic nucleation,[78,151] many authors explain the obtained results by the principles of autocatalytic nucleation,[41,50,57,121,154] for example, 'Rather it is likely that initial-bred nuclei were associated with the gel particles, either inside the particles or on their surfaces and that these initial-bred nuclei grew in the sodium aluminosilicate solution after the amorphous part of the gel particles dissolved';[41] 'Both observations highlight how the dissolution of starting material can have a large influence on the crystal growth of zeolites and possibly provides new evidence for the autocatalytic mechanism, widely postulated in literature';[121] 'The viable nuclei are surrounded by the amorphous "shell" and lie dormant until they are released into solution by dissolution and thus become active growing crystals';[57] 'During the initial stage of reaction a growth via propagation through the amorphous network co-existed with the typical solution-mediated hydrothermal crystallization. The later stages of development of the system followed the classical for zeolite-yielding systems crystallization that could be described by autocatalytic nucleation'.[75] On the basis of these results, it is really to assume that autocatalytic nucleation is a dominant mechanism of the appearance of primary zeolite particles (nucleus) at least in heterogeneous systems.

2.2. Nucleation in homogeneous systems ('clear' solutions)

Obtaining of near monodisperse crystalline products (zeolites) in many syntheses of zeolites from clear aluminosilicate solutions[38,40,41,53,55,60,199,214–221] implied that most of the clear solution syntheses produce a single burst of nuclei in a short time,[165,219,220] for example, by a rapid heterogeneous nucleation in the liquid phase, and in the most cases, promoted by impurities in silica source.[60,220–222] Hence, for a long time, a general meaning was that crystallization of zeolites from clear (alumino)silicate solutions takes place in a direct way, that is, by nucleation and growth of zeolite crystals in/from the liquid phase,[118,220,222] and thus without the formation of an intermediate amorphous (alumino)silicate. On the other hand, many recent studies of crystallization of different types of zeolites, such as A,[40,42,43,157,223] faujasite(s),[46,48,187] L,[51] CIT-6,[59] VPI-8,[59] ZSM-5,[64,194,224,225] ZSM-12,[69,156] silicalite-1[65,66,71,72,92,198,200–204,226,227] and TS-2,[228] from clear (alumino)silicate solutions demonstrated the presence of (nano-scale) amorphous (alumino)silicate agglomerates formed during the room-temperature ageing[40,42,46,66,71,72,187,225] or at the early stages of hydrothermal treatment of the synthesis solutions.[43,48,51,56,59,64,69,92,156,157,194,198,201–204,223,224,226–228] Although, this phenomenon is observed in both untemplated and templated systems, due to some specifies caused by the specific interaction between template molecules and (alumino)silicate species, nucleation processes in these two groups of systems will be considered separately.

2.2.1. Nucleation in untemplated homogeneous systems

As already mentioned, the earliest study of crystallization of zeolites from clear aluminosilicate solutions showed that the first solid formed from the initially clear solution appeared in the form of X-ray amorphous lamellae.[175] ED of single particles indicated that crystal nucleation occurs within the amorphous lamellae. Almost 20 years later, Evrimidis and Yang[187] have found that the 'nucleation gel' ($15Na_2O$: Al_2O_3:15 SiO_2:$320H_2O$), appeared to be a transparent liquid at early stages of ageing turned into a more and more turbid mixture after about 5 h of ageing at 35 °C. Based on the analysis of transmittance IR spectra of the 'nucleation gel' (see Fig. 6.20) the authors concluded '... that nuclei or precursors of faujasite zeolite are gradually developed in the nucleation gel during ageing'. Tsapatsis *et al.* have found that the formation of zeolite L nanoclusters from a clear aluminosilicate solution starts with the formation of an X-ray amorphous gel phase.[51] HRTEM studies of the solid phases at different stages of the crystallization process did not justify the assumption that growth takes place by aggregation of individual zeolite L nanocrystals but rather by propagation of more isolated nucleation events through the gel network. *In situ* monitoring, by quasi-elastic light scattering spectroscopy (QELSS), of particulates formed during room temperature (25 °C) of a clear aluminosilicate solution ($50Na_2O$:Al_2O_3:$5SiO_2$:$1000H_2O$) showed presence of small species about 1 nm in size,[40] previously observed by SANS measurements in the clear solution syntheses of zeolite A and ZSM-5.[229] Larger particles were first detected at 5 h and had a diameter of about 170 nm; according to QELSS, the size of the product did not change during next 30 h of ageing (see Fig. 6.38). Thereafter, the average particle size increases approximately linearly with time of ageing, so that the product obtained after 73.5 h combined a wide spectrum of particle size; the smallest particles were about 20 nm in diameter, and all other particles, which had sizes up to 1.1 μm, looked like agglomerates of the smaller ones.

Appearance of a very broad and weak circle in the electron-diffraction pattern of the sample aged for 44.5 h at 25 °C indicates the presence of some crystalline material. Figure 6.39 shows the effect of ageing of the same initially clear solution on subsequent crystallization at 60 °C.

It is evident that induction time decreased with increasing time of ageing at 25 °C; while the induction time (about 45 h) for the freshly prepared (non-aged) solution correspond to the appearance of the circle in the electron-diffraction pattern, the induction time for the synthesis from a solution aged for 6 days was

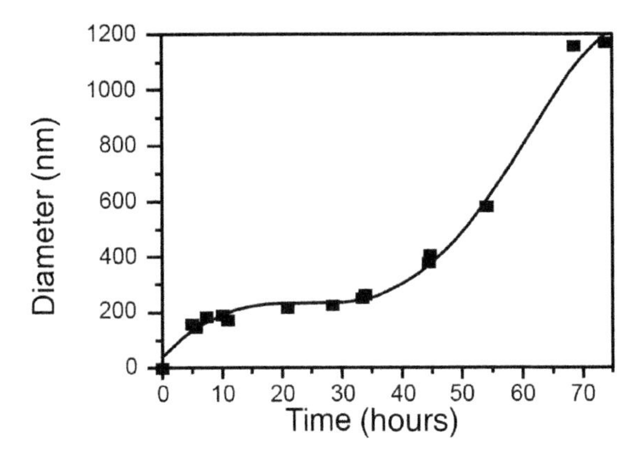

Figure 6.38 Increase in diameter of product from the standard solution ($50Na_2O$:Al_2O_3:5-SiO_2:$1000H_2O$) measured by quasi-elastic light scattering spectroscopy (QELSS) at 25 °C. (Reproduced from Ref. 40 with permission.)

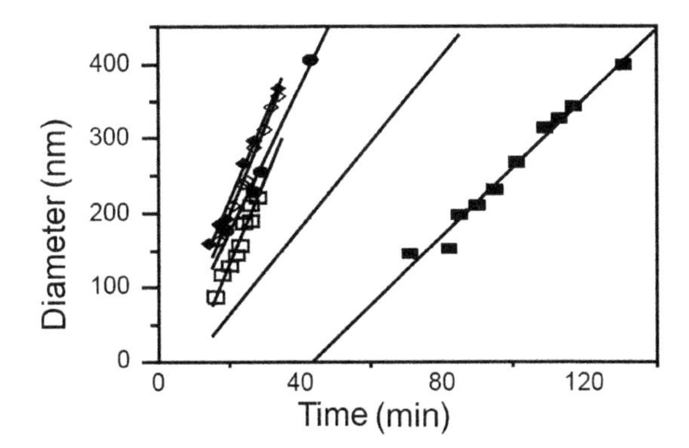

Figure 6.39 Effect of ageing the standard solution ($50Na_2O$:Al_2O_3:$5SiO_2$:$1000H_2O$) at 25 °C on subsequent crystallization kinetics at 60 °C. (■), Nonaged; (○), aged for 7 h; (□), aged for 10 h; (●), aged for 1 day; (◇), aged for 2 days; and (◆), aged for 6 days. (Reproduced from Ref. 40 with permission.)

close to 0 min. Based on this strong influence of the room-temperature ageing on the nucleation event at 60 °C, authors assumed that the particles present in an aged solution are aluminosilicate flocks that form with heating of the solution at the reaction temperature,[40] or a source of semi-crystalline particles covered by aluminosilicate gel. Mintova *et al.* found that a freshly prepared aluminosilicate solution ($0.167Na_2O$:Al_2O_3:$6.25SiO_2$:$389H_2O$) contained spherical objects of about equal size with a diameter of 5–10 nm.[42] After adding tetramethyl ammonium hydroxide ($(TMA)_2O/Al_2O_3 = 7.44$) and mixing for 5 min, aggregates in the size range of 40–80 nm were formed. The XRD pattern and IR spectrum of this sample showed the presence of an amorphous phase. The crystalline zeolite structure emerged from the gel during the first 3 days of ageing at room temperature; extremely small crystallites of zeolite A, which were only 10–30 nm in diameter were embedded in amorphous

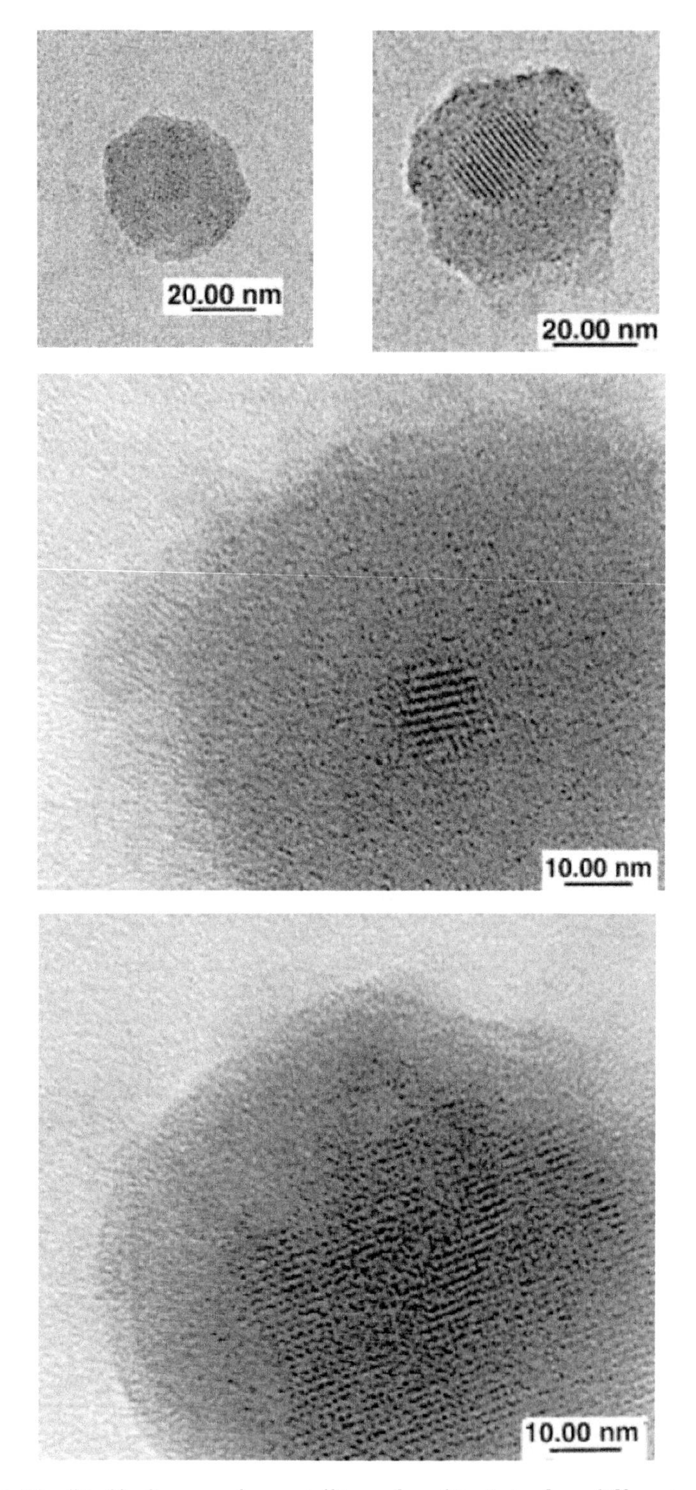

Figure 6.40 The 'birth' of nanoscale crystallites of zeolite A in four different gel particles, obtained after 3 days of crystallization at room temperature. (Reproduced from Ref. 42 with permission.)

gel aggregates of ≈ 30–60 nm (see Fig. 6.40). The amorphous gel-zeolite particles maintained their average size over the course of complete conversion into (denser) zeolite, suggesting that mass transfer from solution supplies some of precursor material (see Fig. 6.41). On heating this suspension, at 80 °C, solution-mediated transformation resulted in additional substantial crystal growth (see Figs. 6.40 and 6.41). Crystallization of nano-sized zeolite Y takes place by similar mechanism from a clear aluminosilicate solution ($0.0652Na_2O$:$2.39(TMA)_2O$:Al_2O_3:$4.35SiO_2$:$247.8H_2O$) at 100 °C.[46] These results represent the first direct evidence on the formation of nuclei in the amorphous gel formed in initially clear solutions during its ageing at room temperature,[42] or during hydrothermal treatment at 100 °C.[46]

Here it is interesting that each amorphous aggregate nucleates only one zeolite crystals (see Figs. 6.40 and 6.41) either near the centre of the gel particle (see Figs. 6.40 and 6.41)[42] or at the periphery of the gel phase (see Figs. 2 and 6 in Ref. 46).

The formation of (nano-scale) amorphous aluminosilicate agglomerates in the starting clear (alumino)silicate synthesis solutions during their room-temperature ageing[40,42,46,187] and/or hydrothermal treatment of the initially clear solutions[43,48,51,157,175,223] as well as experimental evidences on the formation of structurally ordered phase (nuclei) in the gel matrix[40,42,46,51,175,187] show that the critical process of zeolite crystallization is formation of the amorphous agglomerates. Hence, it is very certain that the formation of nuclei in the amorphous phase formed during room-temperature ageing and/or hydrothermal treatment of initially clear

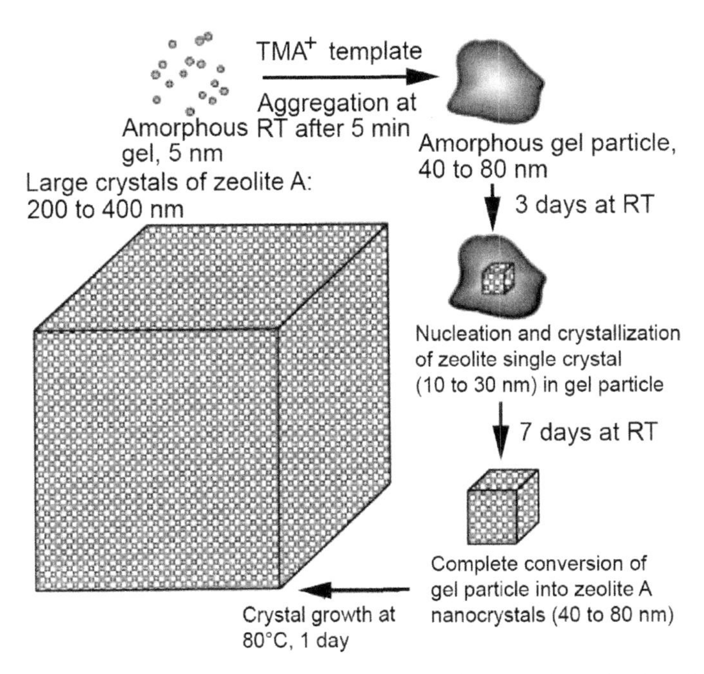

Figure 6.41 A schematic representation of the proposed zeolite growth mechanism. (Reproduced from Ref. 42 with permission.)

aluminosilicate solutions takes place in the same way and by the same mechanism(s) as the formation of nuclei in heterogeneous systems, namely by reorganization and condensation reactions that form the viable nuclei inside the gel matrix (see Fig. 6.2). The temperature at which the process is going to start and the rate of nuclei formation depend on the variety of physico-chemical conditions under which the systems are prepared and treated—however, the general mechanism of the formation of nuclei in the gel matrix does not depend on the conditions.

2.2.2. Nucleation in templated homogeneous systems

The formation of an amorphous phase, preceded to the crystallization (of silicalite-1) from a templated, initially clear Na_2O–$(TPA)_2O$–SiO_2–H_2O solution was firstly observed by Cundy *et al.*[92] Based on the results of optical and electron microscopy, the authors proposed that silicalite nucleation occurred on, or 'in' amorphous gel 'rafts'. Almost simultaneously, Di Renzo *et al.* observed that 'In the case of the more silicic stoichiometry the systems initially consist of clear solution.[156] Gelling, and recovery of any solid fraction from sampling, begins only after 1 day at the synthesis temperature'. Four years later, Dokter *et al.* investigated the formation of nano-sized silicalite-1 crystals from a clear $(TPA)_2O$:SiO_2:D_2O solution by *in situ*, time-resolving WAXS and SAXS methods.[198] They observed presence of subcolloidal primary units with the average size of 2–4 nm at early stage of synthesis (after mixing of reactants at room temperature and/or immediately after heating of the reaction solution) (see Fig. 6.42).

The existence of the subcolloidal primary units was firstly observed by Regev *et al.*[229] Later on, the presence of the subcolloidal primary units was detected in many templated clear solutions.[63,66,71,72,199–203] Very recently, Fedeyko *et al.*,[67,230,231] Rimer *et al.*[232–234] and Cheng and Shantz,[68–70] have found that the spontaneous formation of silica nanoparticles is a general phenomenon in basic solutions of small tetra alkyl ammonium (TAA) cations. The particles have a core-shell structure with silica at the core and the TAA cations at the shell. The particle core size is nearly independent of the size of the TAA cation but decreases with pH, suggesting that electrostatic forces are a key element controlling their size and stability. The nanoparticles formation is a reversible process at low temperatures, in several ways similar to surfactant aggregation into micelles. Upon heating of the reaction solution, the subcolloidal primary units aggregate into amorphous primary fractal aggregates having the size 6.4 nm (see Fig. 6.42). At prolonged reaction time the fractal aggregates become denser with the size of 7.2 nm. At these stages of reaction, only short-range intermolecular interactions exist between the protons of TPA and silicon atoms of zeolite precursor. After the densification, the 7.2 nm particles aggregate into a new, secondary fractal aggregates. Formation of the secondary fractal aggregates is followed by the appearance of the crystalline structures inside amorphous aggregates (see Fig. 6.42). Accordingly to Docter *et al.*,[198] crystal growth occurs by combination of densified primary aggregates into kinetically determined secondary aggregates, indicative for growth by combination of already growing nuclei, which subsequently densify into more favourable, smooth particles. Here it is interesting that the authors stated that the same mechanism is relevant for heterogeneous systems.[197,198] Similar conclusions, based on investigation of crystallization of silicalite-1 from clear solutions were obtained by Watson *et al.*[61] SAXS

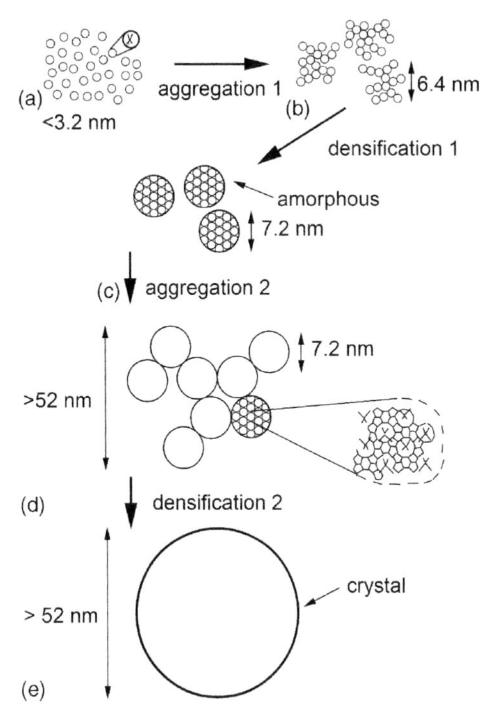

Figure 6.42 Mechanism of microstructural random packing, subsequent ordering and crystallization. (a) Silicate/TPA clusters in solution; (b) primary fractal aggregates formed from the silicate/TPA clusters (6.4 nm, Fig. 1a); (c) densification of these primary fractal aggregates (Fig. 1b); (d) combination of the densified aggregates into secondary fractal structure and crystallization (Fig. 1c); and (e) densification of the secondary aggregates and crystal growth. (Reproduced from Ref. 198 with permission.)

pattern of freshly prepared, unheated clear solution indicated the presence of particles having the size les than 1 nm. After heating for 20 min, depending on temperature (90–115 °C), particles with a radius of gyration of the order 3.7–6.1 nm were observed. The size of these particles corresponds to the size of particles (6.4 nm) observed by Dokter *et al.* under similar conditions.[198] The radius of gyration slightly increased during prolonged heating; the solution remained clear until after approximately 14 h when it became faintly cloudy. At this point, radius of gyration reached about 9–10 nm, and was constant at least during 34 h of heating. Continued heating of the solution resulted in the more concentrated sol with sediment being first observed after 24 h. A heating for 24 h produced particles that were substantially sedimentated. The DLS data are in agreement with the SAXS data. Here it is interesting that population of $\approx$ 10 nm particles increased up to 14 h, and then, their 'concentration' decreased during prolonged heating. The SANS contrast variation data showed that 9.4 nm particles have a density which is nearly that of the fully densified silicalite particle containing the TPA. This finding was revealed by the presence of the IR band at 560 cm^{-1} (which is characteristic for MFI type zeolite) appearing in the IR spectra of the solids formed for $t_c \geq 11$ h. Since at this time the crystallizing system contained only X-ray amorphous 9.4 nm particles, it was concluded that nucleation of silicalite-1 appeared in just these 9.4 nm particles. It can be observed that the described mechanisms[61,198] are very similar to this, schematically represented in Fig. 6.29; the subcolloidal primary units (<3.2 nm) in Fig. 6.42 correspond to the primary units (2.8 nm) in Fig. 6.29; the primary fractal aggregates (6.4 nm) in Fig. 6.42 correspond to the secondary units (5–10 nm) in Fig. 6.29 as well as to the 3.7–6.1 nm particles found by Watson *et al.*;[61] densification of the primary fractal aggregates (Fig. 6.42) and secondary units (Fig. 6.29), with or without a small change in the size (Fig. 6.42) as well as 9.4 nm particles described in Ref. 61 is immanent for all three models. Also it can be observed that the ≈10 nm particles appeared in all three models (Fig. 29 in Ref. 200; Fig. 42 in Ref. 198; Ref. 61) correspond to the GSU (see Fig. 6.28) found by Regev *et al.*[194,229] in the homogeneous systems[229] by SAXS as well as in the mother liquor of a heterogeneous silicalite-1 synthesis[194] by Cryo-TEM and SAXS. Later on, de Moor *et al.*[200–202] showed that the mechanism, previously described in the Section 2.1.2.1, and schematically represented in Fig. 6.29, is also relevant for the nucleation in the templated homogeneous systems. The process starts with the formation of primary units with an average diameter of 2.8 nm (see Figs. 6.29 and 6.42) by aggregation of several IOCS (see Fig. 6.29).[195,196] In the next step of the process, the primary 2.8 nm units aggregate into 10 nm amorphous particles (aggregates) (see Figs. 6.29 and 6.42; see also Ref. 61). In contrast to very rapid formation of the 2.8 nm primary units, their aggregation into 10 nm aggregates is substantially slower process as measured by SAXS (see Fig. 6.43).[201] It was assumed that the 2.8 nm primary units used for the formation of 10 nm aggregates are compensated by the unreacted

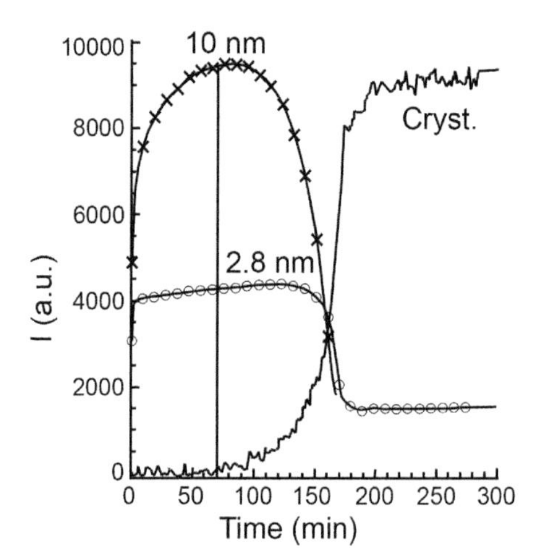

Figure 6.43 Time-dependent scattering intensity at fixed angles, corresponding with *d* spacing of 2.8 nm (primary units) and 10 nm (secondary units—aggregates), together with the area of the Bragg reflections of the product Si-TPA-MFI crystals for Si-TPA-MFI synthesis mixture with Si/OH ratio 2.72. The scattered intensity of the aggregates (×) was plotted because their presence could be demonstrated clearly from the scattering curve and was divided by 2 for clarity. (Reproduced from Ref. 201 with permission.)

IOCS. In this way, the 'concentration' of the primary units keeps constant or even increases slightly during the main part of crystallization process. After the amorphous aggregates reach the 'critical' size (e.g., 10 nm), part of gel nutrient transforms into crystalline phase. The strong decrease of the scattered intensity from the aggregates shows that only a small fraction transforms into the crystalline phase and that the vast majority dissolves to 2.8 nm primary units.[200]

In this way, nuclei are surrounded by amorphous 'shell', as it is clearly shown in Figs. 6.40 and 6.41 (see also Refs. 42 and 46). Although the initial growth of the nuclei most probably takes place by propagation through amorphous matrix, prolonged heating of the reactant solution causes dissolution of the amorphous 'shell' around nuclei. This dissolution is evidenced by the decrease in the scattering intensity of the 10 nm aggregates at the time when only a small fraction of amorphous phase has been transformed into crystalline one (see Fig. 43 as well as Fig. 5 in Ref. 61). When the amorphous 'shell' is completely dissolved, the nuclei (crystals), grown by propagation through the gel matrix, are in full contact with the liquid phase (autocatalytic nucleation?); the nuclei (crystals) having the size lower than the critical size dissolve together with the amorphous phase, whereas the nuclei (crystals) having the size larger than the critical size start to grow as it is indicated by the increase of crystallinity that takes place simultaneously with the decrease in the 'concentration' of 10 nm aggregates (see Fig. 6.43). Recent results of crystallization of zeolites from templated homogeneous systems published by Van Grieken *et al.*,[58] Mintova *et al.*,[66] Serrano *et al.*,[59,228] Cheng and Shantz,[70] Hsu *et al.*[71] and Davis *et al.*[72] confirm the general features of the nucleation/crystallization processes described above,[61,194–198,200–202,229] and schematically represented in Figs. 6.29 and 6.42. Hence, formation of (nano-scale) amorphous (alumino)silicate agglomerates in the starting clear (alumino)silicate solutions,[59,64–66,69,71,72,92,156,194,198,200–204,225–228] abundant experimental evidence on the formation of structurally ordered phase in the amorphous phases precipitated from the clear solutions[59,61,64,66,70–72,92,194,198,200–202,228] and the correlation between presence of (nano-scale) amorphous (alumino)silicate agglomerates and nucleation processes undoubtedly show that nucleation in the templated homogeneous systems is similar or even the same as in the templated heterogeneous systems (see Section 2.1.2.1).

3. Summary and Outlook

Particulate properties (crystal size distribution, average crystal size, particle shape) of zeolites have considerable influence on the mode and efficiency of their applications. Since the final size of zeolite crystals obtained in a typical batch crystallization strongly depends on the total number of nuclei formed during the crystallization and on the rate of their formation, it is quite clear that the knowledge on the nucleation events has a crucial importance in the control of particulate properties of zeolites, and thus on the designing of the products having desired particulate properties needed for specific applications. For this reason, in this

Chapter, numerous concepts of zeolite nucleation are shown and/or critically evaluated in accordance with the scheme shown in Fig. 6.1. However, before this, it was clearly shown that it is almost impossible to learn anything about the nucleation mechanism in zeolite crystallization from the analysis of the mass crystallization curve ('induction time').

In Section 2.1.1.1, the general principles of homogeneous nucleation in the liquid phase as well as the specific conditions for homogeneous nucleation of zeolites are represented, and the papers in which homogeneous nucleation in the liquid phase is assumed as one of, or only mechanism of the formation of zeolite nuclei are reviewed. Also, it is shown that due to the supersaturation which is considerably lower than the critical supersaturation needed for the formation of zeolite nuclei in the liquid phase as well as the lack of any experimental evidence of the existence of 'structured' aluminosilicate blocks (e.g., secondary and tertiary building units) necessary for the formation of nuclei in the liquid phase, the possibility for occurrence of homogeneous nucleation in the liquid phase is very dubious.

In Section 2.1.1.2, are explained the basic principles of heterogeneous nucleation, and kinetics of the heterogeneous nucleation is mathematically expressed by appropriate kinetic equations. Although many investigators recognized 'heterogeneous' nucleation as dominant mechanism of the formation of primary zeolite particles, it is shown that 'classical' heterogeneous nucleation (fast formation of nuclei on the impurity particles always present in the liquid phase) is probably relevant only for the solution-mediated transformation of one (less stable) type of zeolite into another (more stable) one, for example, transformation of zeolite A into HS. On the other hand, during the synthesis of zeolites from hydrogels, most or even all nuclei are formed in the gel phase (see also the Section 2.1.2), so that the classical heterogeneous nucleation in the liquid phase of heterogeneous systems can be neglected as a mechanism relevant for the formation of zeolite nuclei. The concept of formation of nuclei on the surface of gel particles, or at the gel–liquid interface is in many papers identified as 'heterogeneous' nucleation. While the concept of formation of nuclei on the surface of gel particles can be easily ruled out, because nobody can explain how a less soluble material (zeolite) can be formed (nucleated) at the more soluble substrate (gel), the concept of the formation of nuclei at the gel–liquid interface is closer to homogeneous than heterogeneous nucleation. In this context, a possibility of occurrence of the 'heterogeneous' nucleation on the gel particles and on the gel–liquid interface is dubious, or in a wider context, heterogeneous nucleation (regardless to the concept) has a minor importance in the formation of primary zeolite particles during the crystallization of zeolites from hydrogels.

In Section 2.1.1.3 was shown that secondary nucleation is not relevant for most of the 'standard' zeolite syntheses; among different possible mechanisms of secondary nucleation (surface polycrystalline breeding, surface initial breeding, micro-attrition and fluid-shear activity) only surface nucleation (polycrystalline breeding, initial breeding and nucleation of one type of zeolite on the surface of the (seed) crystals of another type of zeolite) can take place under specific synthesis conditions, that is, when the dried, calcined and/or mechanically treated seed crystals were added into clear (alumino)silicate solution.

The Section 2.1.2.1 (experimental evidences on the presence of structurally ordered phase in the gel matrix and/or proposed mechanisms of nucleation) shows that the X-ray amorphous (alumino)silicates precipitated in hydrogels during mixing together the reaction solutions are not only the source of aluminate, silicate and/or aluminosilicate species needed for subsequent growth of zeolite crystals, but that also have a crucial role in the formation of nuclei. The presence of medium- and long-range ordered phase comparable with zeolite structure (potential or real nuclei) was evidenced by various experimental methods such as electron diffraction, HRTEM, ^{13}C-, ^{23}Na-, ^{27}Al-, ^{29}Si- and ^{31}P-NMR (nuclear magnetic resonance), DTA, IR spectroscopy, AFM, adsorption of gases and vapours, diffuse reflectance, UV–vis spectroscopy and SAXS in the amorphous solid phases of both templated and untemplated heterogeneous systems (hydrogels). Although, based on these findings, the formation of nuclei in the gel matrix is not inquisitive, mechanisms of the formation of structurally ordered phase and its developing into viable nuclei or even nanocrystals inside the gel matrix, is not quite clear at present. However, it is reliable to assume, that due to high supersaturation of gel with the constituents (inorganic cations, Al, Si, template), the polycondensation reactions similar to these schematically represented in Fig. 6.2 can be realized easier and faster in the gel phase than in the liquid phase. Although the polycondensation reactions represented in Fig. 6.2 are primarily valuable for untemplated heterogeneous systems, it is reliable to assume that formation of nuclei in the gel phase of templated heterogeneous systems is based on the principles similar to these relevant to the formation of nuclei in the gel phase of untemplated heterogeneous systems, of course, taking into consideration specific interactions between active (alumino)silicate anions and template cations (see Figs. 6.3 and 6.26–6.29). The conditions (temperature and time) at which the structurally ordered phase would be formed and developed to viable nuclei considerably depend on the availability of soluble (alumino)silicate species: When TEOS or soluble silicates are used as silicon source, viable nuclei in the gel matrix can be formed at ambient temperature. Otherwise, for example, when amorphous silica is used as silicon source, the mentioned processes can take place only under hydrothermal conditions, that is, after a part of amorphous silica has been dissolved, thus producing a sufficient amount of soluble silicate species needed for the formation of aluminosilicate gel precursor and/or IOCS needed for the formation of an (alumino) silicate gel precursor in templated heterogeneous systems.

The Section 2.1.2.2 shows a 'historical' development of autocatalytic nucleation of zeolites in heterogeneous systems (hydrogels), from Zhdanov's postulation that '. . . not only the aluminosilicate blocks formed in the liquid phase, but also a similar blocks with ordered structure in the gel skeleton can be nuclei of crystals through the "primitive" models of autocatalytic nucleation, based on homogeneous distribution of nuclei in the gel matrix to the actual model of autocatalytic nucleation based on heterogeneous distribution of nuclei in the gel matrix'. While the 'primitive' models of autocatalytic nucleation assume the possibility of nucleation events in both the liquid phase (homogeneous, heterogeneous) and in the gel phase, the actual model of autocatalytic nucleation is based on the formation of nuclei in the gel phase only. However, regardless to the distribution of nuclei in the gel matrix, all the presented models of autocatalytic nucleation are based on the assumption that the nuclei formed in the gel phase cannot grow in the gel matrix, so that they are potential nuclei when they are 'hidden' in the gel matrix and can start to grow after their 'release' from the gel dissolved during the crystallization, that is, when the nuclei are in full contact with the liquid phase. Besides the evidences of autocatalytic nucleation revealed by partial agreement between measured and calculated data and explanation of experimental results by the principles of autocatalytic nucleation as well as experimental evidences of the gel 'memory effect', which is relevant only for autocatalytic nucleation, the actual model of autocatalytic nucleation was revealed by complete aggrement between the measured and the calculated (simulated) data.

In Section 2.2 was shown that many recent studies of crystallization of different types of zeolites from both untemplated and templated clear (alumino)silicate solutions demonstrate the presence of (nano-scale) amorphous (alumino)silicate agglomerates formed during the room-temperature ageing or at the early stages o hydrothermal treatment of the synthesis solutions. Formation and evolution of structurally ordered phase (nuclei) in the amorphous solids precipitated from both untemplated and templated homogeneous systems was evidenced by various experimental methods such as electron diffraction, HR-TEM, Cryo-TEM, ^{27}Al-NMR, ^{29}Si-NMR, DTA, IR, SAXS, WAXS, SANS, etc. Based on these findings, it can be concluded that the formation of nuclei in the amorphous phase formed during room-temperature ageing and/or hydrothermal treatment of initially clear (alumino)silicate solutions takes place in the similar or even same way as in heterogeneous systems, that is, by the polycondensation reactions, schematically represented in Fig. 6.2 (untemplated systems), and by the processes schematically presented in Figs. 6.29 and 6.42 (templated systems).

Hence, (i) the formation of (nano-scale) amorphous (alumino)silicate agglomerates in the starting clear (alumino)silicate synthesis solutions, (ii) abundant experimental evidence on the formation of structurally ordered phase in the gel matrix of both heterogeneous (hydrogels) and homogeneous (clear solutions) reaction mixtures and (iii) the correlation between presence of (nano-scale) amorphous (alumino)silicate agglomerates and nucleation processes, show that the critical process of zeolite nucleation/crystallization is formation of the amorphous agglomerates, and thus it is reliable to assume that general principles of zeolite nucleation do not depend on the system (homogeneous, heterogeneous, templated, untemplated) and crystallization conditions.

REFERENCES

[1] Petrik, L. F., O'Connor, C. T., Schwartz, S., in: Beyer, H. K., Karge, H. G., Kiricsi, I., Nagy, J. B. (Eds.), *Catalysis by Microporous Materials, Proceedings of ZEOCAT '95*, Szombathely, Hungary, July 09–13, 1995; Studies in Surface Science and Catalysis 94; Elsevier, Amsterdam, 1995, pp. 517–524.

[2] Di Renzo, F., *Catal. Today* **1998,** *41,* 37–40.

[3] Subotić, B., Bronić, J.Theoretical and practical aspects of zeolite crystal growth, in: Auerbach, S. M., Carrado, K. A., Dutta, P. K. (Eds.), *Handbook of Zeolite Science and Technology,* Marcel Dekkker Inc., New York—Basel, 2003, pp. 129–203.

[4] Vanderpol, A. J. H. P., Vanhooff, J. H. C., *Appl. Cattal. A* **1993,** *106,* 97–113.

[5] Ishida, H., Akagishi, K., *Nippon Kagaku Kaishi* **1996,** 290–297.

[6] Weitkamp, J., *Solid State Ionics* **2000,** *131,* 175–188.

[7] Climent, M. J., Corma, M. J., Garcia, A., Ibarra, M., Prime, S., *J. Appl. Catal. A* **1995,** *130,* 5–12.
[8] Loendres, R., Jacobs, P. A., Martens, J. A., J. Catal. **1998,** *176,* 545–551.
[9] Wang, W., Liu, S., Meng, X. Y., Li, B. S., Jiang, D. Z., Xiao, F. S., Qin, S. L., *Chem. J. Chin. Univ.* **2003,** *24,* 205–207.
[10] Namba, S., Yashima, T., *J. Japan Petrol Inst.* **1984,** *27,* 567–569.
[11] Rajagopalan, K., Peters, A. W., Edwards, G. C., *Appl. Catal.* **1986,** *23,* 69–80.
[12] Bonetto, L., Camblor, M. A., Corma, A., Perez-Pariente, J., *J. Appl. Catal.* **1992,** *82,* 37–50.
[13] Yucel, H., Ruthven, D. M., *J. Chem. Soc. Faraday I* **1980,** *76,* 60–70.
[14] You, Y. S., Kim, J.-H., Seo, G., *Polym. Degrad. Stab.* **2000,** *70,* 365–371.
[15] Upadek, H., Kattwitz, B., Schreck, B., *Tenside Surfactants Deterg.* **1996,** *33,* 385–392.
[16] Chang, C. Y., Tsai, W. T., Ing, C. H., Chang, C. F., *J. Colloid Interface Sci.* **2003,** *260,* 273–279.
[17] Wang, Y. Q., Liu, S., Xu, J. Z., Han, T. W., Chuan, S. T., Zhu, T., *J. Hazard. Mater.* **2006,** *136,* 735–740.
[18] Nagy, J. B., Bodart, P., Hannus, I., Kiricsi, I., *Synthesis, Characterization and Use of Microporous Materials,* DecaGen Ltd., Szeged, Hungary, 1998, pp. 93–118.
[19] Dragčević, Z., Subotić, B., Bronić, J., *Tekstil* **1993,** *42,* 267–274.
[20] Singh, B., Alloway, B. J., Bocherean, F. J. M., *Commun. Soil Sci. Plant Anal.* **2000,** *31,* 2775–2786.
[21] Singh, G., Prasad, B., *Water Environ. Res.* **1997,** *69,* 157–161.
[22] Wen, D. H., Ho, Y. S., Tang, X. Y., *J. Hazard. Mater.* **2006,** *133,* 252–256.
[23] Ali, A. A. H., Elbishtawi, R., *J. Chem. Technol. Biotechnol.* **1997,** *69,* 27–34.
[24] El-Bishtawi, R. F., Ali, A. A. H., *J. Environ. Sci. Health Part A Tox./Hazard. Subst. Environ. Eng.* **2001,** *36,* 1055–1072.
[25] Shawabkeh, R., Al-Harahsheh, A., Al-Otoom, A., *Oil Shale* **2004,** *21,* 1613–1622.
[26] Barrer, R. M., *Hydrothermal Chemistry of Zeolites,* Academic Press, London, 1982, pp. 133–247.
[27] Guth, J.-L., Caullet, P., *J. Chim. Phys.* **1986,** *83,* 155–175.
[28] Yang, X., Albrecht, D., Caro, J., *Micropor. Mesopor. Mater.* **2006,** *90,* 53–61.
[29] Qiu, S., Yu, J., Zhu, G., Terasaki, O., Nozue, Y., Pang, W., Xu, R., *Micropor. Mesopor. Mater..* **1998,** *21,* 245–251.
[30] Kornatowski, J., Finger, G., Schmitz, W., *Cryst. Res. Technol.* **1990,** *25,* 17–23.
[31] Ferchiche, S., Warzywoda, J., Sacco, A., Jr., *Int. J. Inorg. Mater.* **2001,** *3,* 773–780.
[32] Daqing, Z., Shilun, Q., Wenqin, P., *J. Chem. Soc. Chem. Commun.* **1990,** *19,* 1313–1314.
[33] Chen, S. Z., Huddersman, K., Keir, D., Rees, L. V. C., *Zeolites* **1998,** *8,* 106–109.
[34] Shimizu, S., Hamada, H., *Angew. Chem. Int. Ed. Engl.* **1999,** *38,* 2725–2727.
[35] Shiraki, T., Wakihara, T., Sadakata, M., Yoshimura, M., Okubo, T., *Micropor. Mesopor. Mater.* **2001,** *42,* 229–234.
[36] Warzywoda, J., Dixon, A. G., Thompson, R. W., Sacco, A. Jr., *J. Mater. Chem.* **1995,** *5,* 1019–1025.
[37] Gao, F., Li, X., Zhu, G., Qui, S., Wei, B., Shao, C., Terasaki, O., *Mater. Lett.* **2001,** *48,* 1–7.
[38] Schoeman, B. J., Sterte, J., Otterstedt, J.-E., *Zeolites* **1994,** *14,* 110–116.
[39] Larlus, O., Mintova, S., Bein, T., *Micropor. Mesopor. Mater.* **2006,** *96,* 405–412.
[40] Gora, L., Sterletzky, K., Thompson, R. W., Phillies, G. D. J., *Zeolites* **1997,** *18,* 119–131.
[41] Gora, L., Streletzky, K., Thompson, R. W., Philies, D. J., *Zeolites* **1997,** *19,* 98–106.
[42] Mintova, S., Olson, N. H., Valtchev, V., Bein, T., *Science* **1999,** *283,* 958–960.
[43] Caputo, D., De Gennaro, B., Liguori, B., Testa, F., Carotenuto, L., Piccolo, C., *Mater. Chem. Phys.* **2000,** *66,* 120–125.
[44] Valtchev, V., Mintova, S., *Micropor. Mesopor. Mater.* **2001,** *43,* 41–49.
[45] Smaishi, M., Barida, O., Valtchev, V., *Eur. J. Inorg. Chem.* **2003,** 4370–4377.
[46] Mintova, S., Olson, N. H., Bein, T., *Angew. Chem. Int. Ed. Engl.* **1999,** *38,* 3201–3204.
[47] Lassinantti, M., Hedlund, J., Sterte, J., *Micropor. Mesopor. Mater.* **2000,** *38,* 25–34.
[48] Holmberg, B. A., Wang, H., Norbeck, J. M., Yan, Y., *Micropor. Mesopor. Mater.* **2003,** *59,* 13–28.
[49] Valtchev, V. P., Bozhilov, K. N., *J. Phys. Chem. B* **2004,** *108,* 15587–15598.
[50] Valtchev, V., Rigolet, S., Bozhilov, K. N., *Micropor. Mesopor. Mater.* **2007,** *101,* 73–82.

[51] Tsapatsis, M., Lovallo, M., Davis, M. E., *Micropor. Mesopor. Mater.* **1996,** *5,* 381–388.
[52] Nikolakis, V., Vlachos, D. G., Tsapatsis, M., *J. Chem. Phys.* **1999,** *111,* 2143–2150.
[53] Schoeman, B. J., Sterte, J., Otterstedt, J.-E., *Zeolites* **1994,** *14,* 208–216.
[54] Landau, M. V., Tavor, D., Regev, O., Kaliya, M. L., Herskowitz, M., Valtchev, V., Mintova, S., *Chem. Mater.* **1999,** *11,* 2030–2037.
[55] Schoeman, B. J., Babouchkina, E., Mintova, S., Valtchev, V. P., Sterte, J., *J. Porous Mater.* **2001,** *8,* 13–22.
[56] Mintova, S., Valtchev, V., Onfroy, T., Marichal, C., Knoezinger, H., Bein, T., *Micropor. Mesopor. Mater.* **2006,** *90,* 237–245.
[57] Ding, L., Zhang, Z., Ring, Z., Chen, J., *Micropor. Mesopor. Mater.* **2006,** *94,* 1–8.
[58] Mintova, S., Mo, S., Bein, T., *Chem. Mater.* **1998,** *10,* 4030–4036.
[59] Serrano, D. P., van Grieken, R., Davis, M. E., Melero, J. A., Garcia, A., Morales, G., *Chem. Eur. J.* **2002,** *8,* 5153–5160.
[60] Twomey, T. A. M., Mackay, M., Kuipers, H. P. C. E., Thompson, R. W., *Zeolites* **1994,** *14,* 162–168.
[61] Watson, J. N., Iton, L. E., Keir, R. I., Thomas, J. C., Dowling, T. L., White, J. W., *J. Phys. Chem. B* **1997,** *101,* 10094–10104.
[62] Tsay, C. S., Chiang, A. S. T., *Micropor. Mesopor. Mater.* **1998,** *26,* 89–99.
[63] Nikolakis, V., Kokkoli, E., Tirrell, M., Tsapatsis, M., Vlachos, D. G., *Chem. Mater.* **2000,** *12,* 845–853.
[64] Van Grieken, R., Sotelo, J. L., Menendez, J. M., Melero, J. A., *Micropor. Mesopor. Mater.* **2000,** *39,* 135–147.
[65] Li, Q., Mihailova, B., Creaser, D., Sterte, J., *Micropor. Mesopor. Mater.* **2000,** *40,* 53–62.
[66] Mintova, S., Olson, N. H., Senker, J., Bein, T., *Angew. Chem. Int. Ed. Engl.* **2002,** *41,* 2558–2561.
[67] Fedeyko, J. M., Rimer, J. D., Lobo, R. F., Vlachos, D. G., *J. Phys. Chem. B* **2004,** *108,* 12271–12275.
[68] Cheng, C.-H., Shantz, D., *Curr. Opin. Colloid Interface Sci.* **2005,** *10,* 188–194.
[69] Cheng, C. H., Shantz, D. F., *J. Phys. Chem. B* **2005,** *109,* 19116–19125.
[70] Cheng, C.-H., Shantz, D., *J. Phys. Chem. B* **2005,** *109,* 13912–13920.
[71] Hsu, C.-Y., Chiang, A. S. T., Selvin, R., Thompson, R. W., *J. Phys. Chem. B* **2005,** *109,* 18804–18814.
[72] Davis, T. M., Drews, T. O., Ramanan, H., He, C., Dong, J., Schnablegger, H., Katsoulakis, M. A., Kokkoli, E., McCormick, A. V., Penn, R. L., Tsapatsis, M., *Nat. Mater.* **2006,** *5,* 400–408.
[73] Randolph, A. D., Larson, M. A., *Theory of Particulate Processes,* Academic Press, NewYork, NY, 1971, pp. 12–63.
[74] Thompson, R. W., Dyer, A., *Zeolites* **1985,** *5,* 202–210.
[75] Subotić, B., Mašić, N., Šmit, I. in: Držaj, B., Hočevar, S., Pejovnik, S. (Eds.), *Zeolites; Synthesis, Structure, Technology and Application, Proceedings of an International Symposium,* Portorož, Slovenia, Sept. 03–08, 1984; Studies in Surface Science and Catalysis 24; Elsevier, Amsterdam, 1985, pp. 207–214.
[76] Subotić, B., Antonić, T., Šmit, I., Aiello, R., Crea, F., Nastro, A., Testa, F., in: Occelly, M. L., Kessler, H. (Eds.), *Synthesis of Porous Materials: Zeolites, Clays and Nanostructures,* Chemical Industries: A series of Reference Books and Textbooks 69; Marcel Dekker Inc., New York, N.Y., 1996, pp. 35–58.
[77] Falamaki, C., Edrissi, M., Sohrabi, M., *Zeolites* **1997,** *19,* 2–5.
[78] Subotić, B., Antonić, T., *Croat. Chem. Acta* **1998,** *71,* 929–948.
[79] Subotić, B., Antonić, T., Bosnar, S., Bronić, J., Škreblin, M., in: Kiricsi, I., Pal-Borbely, G., Nagy, J. B., Karge, H. G. (Eds.), *Porous Materials in Environmentally Friendly Processes, Proceedings of the First International FEZA Conference,* Eger, Hungary, Sept. 01–04, 1999; Studies in Surface Science and Catalysis 125; Elsevier, Amsterdam, 1999, pp. 157–164.
[80] Bosnar, S., Antonić, T., Bronić, J., Krznarić, I., Subotić, B., *J. Cryst. Growth* **2004,** *267,* 270–282.

[81] Zhdanov, S. P., Samulevich, N. N., in: Rees, L. V. C. (Ed.), *Proceedings of the Fifth International Conference on Zeolites*, Naples, Italy, June 02–06, 1980; Heyden, London-Philadelphia-Rheine, 1980, pp. 75–84.
[82] Bronić, J., Subotić, B., Šmit, I., Despotović, L. J. A., in: Grobet, P. J., Mortier, W. J., Vansant, F. F., Schulz-Ekloff, G. (Eds.), *Innovation in Zeolite Materials Science, Proceedings of an International Symposium*, Nieuwpoort, Belgium, September 13–17, 1998; Studies in Surface Science and Catalysis 37; Elsevier, Amsterdam, 1988, pp. 107–114.
[83] Antonić, T., Subotić, B., Stubičar, N., *Zeolites* **1997,** *18,* 291–300.
[84] Zhdanov, S. P., in: Sand, L. B., Flanigen, E. M. (Eds.), *Molecular Sieve Zeolites—I,* Advances in Chemistry Series 101; American Chemical Society, Washington, DC, 1971, pp. 20–43.
[85] Subotić, B., Bronić, J., in: Von Ballmoos, R., Higgins, J. B., Treacy, M. M. J. (Eds.), *Proceedings of Ninth International Zeolite Conference*, Montreal, Canada, May 05–10, 1992; Butterworth-Heinemann, Boston, MA, 1992, pp. 321–328.
[86] Cundy, C. S., Cox, P. A., *Micropor. Mesopor. Mater.* **2005,** *82,* 1–78.
[87] Sameen, S., Zaidi, A., Rohani, S., *Rev. Chem. Eng.* **2005,** *21,* 265–306.
[88] Anderson, M. W., Agger, J. R., Thornton, J. T., Forsyth, N., *Angew. Chem. Int. Ed. Engl.* **1996,** *35,* 1210–1213.
[89] Agger, J. R., Pervaiz, N., Cheetham, A. K., Anderson, M. W., *J. Am. Chem. Soc.* **1998,** *120,* 10754–10759.
[90] Dumrul, S., Bazzana, S., Warzywoda, J., Biederman, R. R., Sacco, A., Jr., *Micropor. Mesopor. Mater.* **2002,** *54,* 79–88.
[91] Wakihara, T., Okubo, T., *J. Chem. Eng. Jpn.* **2004,** *37,* 669–674.
[92] Cundy, C. S., Lowe, M., Sinclair, M., *J. Cryst. Growth* **1990,** *100,* 189–202.
[93] Cundy, C. S., Cox, P. A., *Chem. Rev.* **2003,** *103,* 663–701.
[94] Cheng, C.-H., Shantz, D. F., *J. Phys. Chem. B* **2005,** *109,* 7266–7274.
[95] Marui, Y., Irie, R., Takiyama, H., Uchida, H., Matsuoka, M., *J. Cryst. Growth* **2002,** *237,* 2448–2452.
[96] Zhang, Y., Li, Y., Zhang, Y., *J. Cryst. Growth* **2003,** *254,* 156–163.
[97] Culfaz, A., Sand, L. B., in: Meier, W. M., Uytterhoeven, J. B. (Eds.), *Molecular Sieves, Proceedings of The Third International Zeolite Conference*, Zurich, Switzerland, September 03–07, 1973; Advances in Chemistry Series 121; American Chemical Society, Washington, DC, 1973, pp. 140–151.
[98] Wang, L. Q., Wang, X. S., Guo, X. W., Li, G., Xiu, J. H., Liu, S., *Chin. J. Catal.* **2003,** *24,* 132–136.
[99] Uguina, M. A., Serrano, D. P., Ovejero, G., Van Grieken, R., Camacho, M., *Appl. Catal.* **1995,** *124,* 391–408.
[100] Uzcategui, D., Gonzalez, G., *Catal. Today* **2005,** *107–108,* 901–905.
[101] Cundy, C. S., Lowe, B. M., Sinclair, D. M., *Faraday Discuss.* **1993,** *95,* 235–252.
[102] den Ouden, C. J. J., Thompson, R. W., *Ind. Eng. Chem. Res.* **1992,** *31,* 369–373.
[103] Subotić, B., Bronić, J., Antonić, T., in: Treacy, M. M. J., Marcus, B. K., Bisher, M. E., Higgins, J. B. (Eds.), *Proceedings of 12th International Zeolite Conference*, Baltimore, MA, July 05–10, 1998; Materials Research Society, Warrendale, PA, 1998, pp. 2057–2064.
[104] Brečević, L. J., Kralj, D., in: Kallay, N. (Ed.), *Interfacial Dynamics,* Marcel Deker, New York, 1999, pp. 435–474.
[105] Nielsen, A. E., *Kinetics of Precipitation*, Pergamon Press, Oxford, 1964.
[106] Walton, A. G., *The Formation and Properties of Precipitates*, Interscience Publishers, New York, London, Sydney, 1967.
[107] Nielsen, A. E., *Croat. Chem. Acta* **1980,** *53,* 255–279.
[108] Ciric, J., *J. Colloid Interface Sci.* **1968,** *28,* 315–324.
[109] Dewaele, N., Bodart, P., Nagy, J. B., *Acta Chim. Hungarica* **1985,** *119,* 233–244.
[110] Cundy, C. S., Henty, M. S., Plaisted, R. J., *Zeolites* **1995,** *15,* 342–352.
[111] Antonić, T., Subotić, B., in: Treacy, M. M. J., Marcus, B. K., Bisher, M. E., Higgins, J. B. (Eds.), *Proceedings of 12th International Zeolite Conference*, Baltimore, MA, July 05–10, 1998; Materials Research Society, Warrendale, PA, 1998, pp. 2049–2056.
[112] Bosnar, S., Subotić, B., *Micropor. Mesopor. Mater.* **1999,** *28,* 483–493.

[113] Bosnar, S., Bronić, J., Subotić, B., in: Kiricsi, I., Pal-Borbely, G., Nagy, J. B., Karge, H. G. (Eds.), *Porous Materials in Environmentally Friendly Processes, Proceedings of the First International FEZA Conference*, Eger, Hungary, Sept. 01–04, 1999; Studies in Surface Science and Catalysis 125; Elsevier, Amsterdam, 1999, pp. 69–76.
[114] Subotić, B., Škrtić, D., Šmit, I., Sekovanić, L., *J. Cryst. Growth* **1980,** *50,* 498–508.
[115] Subotić, B., Graovac, A., Sekovanić, L., in: Sersale, R., Colella, C., Aiello, R. (Eds.), *Recent Progress Reports and Discussions, Proceedings of the Fifth International Conference of Zeolites*, Naples, Italy, June 02–06, 1980; Giannini, Naples, 1981, pp. 54–57.
[116] Subotić, B., Graovac, A., in: Držaj, B., Hočevar, S., Pejovnik, S. (Eds.), *Zeolites; Synthesis, Structure, Technology and Application, Proceedings of an International Symposium*, Portorož, Slovenia, Sept. 03–08, 1984; Studies in Surface Science and Catalysis 24; Elsevier, Amsterdam, 1985, pp. 199–206.
[117] Hu, H. C., Chen, W. H., Lee, T. Y., *J. Cryst. Growth* **1991,** *108,* 561–571.
[118] Thompson, R. W., *Molecular Sieves Vol. 1.* Springer-Verlag, Berlin Heidelberg, 1998, pp. 1–33.
[119] Gonthier, S., Gora, L., Güray, I., Thompson, R. W., *Zeolites* **1993,** *13,* 414–418.
[120] Sheikh, A. Y., Jones, A. G., Graham, P., *Zeolites* **1996,** *16,* 164–172.
[121] Walton, R. I., O'Hare, D., *J. Phys. Chem. B* **2001,** *105,* 91–96.
[122] Padovan, M., Leofanti, G., Solari, M., Moretti, E., *Zeolites* **1984,** *4,* 295–299.
[123] Nastro, A., Aiello, R., Colella, C., *Ann. Chim.* **1984,** *74,* 579–587.
[124] Dixon, A. G., Thompson, R. W., *Zeolites* **1986,** *6,* 154–160.
[125] Moudafi, L., Massiani, P., Fajula, F., Figueras, F., *Zeolites* **1987,** *7,* 63–66.
[126] Thangaray, A., Kumar, R., *Zeolites* **1990,** *10,* 117–120.
[127] Song, H., Ilegbusi, O. J., Sacco, A., Jr., *Matter Lett.* **2005,** *59,* 2668–2672.
[128] Engelhardt, G., Fahlke, B., Mägi, M., Lippmaa, E., *Zeolites* **1983,** *3,* 292–294.
[129] Bell, A. T., McCormic, A. V., Hendricks, W. M., Radke, C. J., *Chem. Express* **1986,** *1,* 687–690.
[130] Fahlke, B., Müller, D., Wieker, W., *Z. Anorg. Allg.Chem.* **1988,** *526,* 141–144.
[131] Szostak, R., *Molecular Sieves: Principles of Synthesis and Identification,* Van Nostrand-Reinhold, New York, N.Y., 1989, pp. 51–202.
[132] Harvey, G., Dent Glasser, L. D., in: Occelli, M. L., Robson, H. E. (Eds.), *Zeolite Synthesis,* ACS Symposium Series No. 398; American Chemical Society, Washington, D.C., 1989, pp. 49–65.
[133] Bell, A. T., in: Occelli, M. L., Robson, H. E. (Eds.), *Zeolite Synthesis,* ACS Symposium Series No. 398; American Chemical Society, Washington, D.C., 1989, pp. 66–82.
[134] McCormick, A. V., Bell, A. T., Radke, C. J., *J. Phys. Chem.* **1989,** *93,* 1741–1744.
[135] Katović, A., Subotić, B., Šmit, I., Despotović, L. J. A., *Zeolites* **1990,** *10,* 634–641.
[136] Martlock, R. F., Bell, A. T., Radke, C. J., *J. Phys. Chem.* **1991,** *95,* 7847–7851.
[137] Šefčik, J., McCormck, A. V., in: Treacy, M. M. J., Marcus, B. K., Bisher, M. E., Higgins, J. B. (Eds.), *Proceedings of 12th International Zeolite Conference*, Baltimore, MA, July 05–10, 1998; Materials Research Society, Warrendale, PA, 1998, pp. 1595–1602.
[138] Šefčik, J., McCormck, A. V., *Chem. Eng. Sci.* **1999,** *54,* 3513–3519.
[139] Breck, D. W., *J. Chem. Educ.* **1964,** *41,* 678–689.
[140] Brunner, G. O., *Zeolites* **1992,** *12,* 428–430.
[141] Anderson, M. W., Agger, J. R., Pervaiz, N., Weigel, S. J., Chetam, A. K., in: Treacy, M. M. J., Marcus, B. K., Bisher, M. E., Higgins, J. B. (Eds.), *Proceedings of 12th International Zeolite Conference,* Baltimore, MA, July 05–10, 1998; Materials Research Society, Warrendale, PA, 1998, pp. 1487–1494.
[142] Shüth, F., *Curr. Opin. Solid State Mater. Sci.* **2001,** *5,* 389–395.
[143] Bronić, J., Subotić, B., *Micropor. Mater.* **1995,** *4,* 239–242.
[144] Crea, F., Nastro, A., Nagy, J. B., Aiello, R., *Zeolites* **1998,** *8,* 262–267.
[145] Subotić, B., Sekovanić, L., *J. Cryst. Growth* **1986,** *75,* 561–572.
[146] Golemme, G., Nastro, A., Nagy, J. B., Subotić, B., Crea, F., Aiello, R., *Zeolites* **1991,** *11,* 776–783.
[147] Katović, A., Subotić, B., Šmit, I., Despotović, L. J. A., Ćurić, M., in: Occelli, M. L., Robson, H. E. (Eds.), *Zeolite Synthesis,* ACS Symposium Series No. 398; American Chemical Society, Washington, D.C., 1989, pp. 124–139.

[148] Gonthier, S., Thompson, R. W., in: Jansen, J. C., Stöcker, M., Karge, H. G., Weitkamp, J. (Eds.), *Advanced Zeolite Science and Applications,* Studies in Surface Science and Catalysis 85; Elsevier, Amsterdam, 1994, pp. 43–50.
[149] Subotić, B., Antonić, T., Bronić, J., in: Galameau, A., di Renzo, F., Fajula, F., Vedrine, J. (Eds.), *Zeolites and Mesoporous Materialsat the Down of the 21th Century. Proceedings of the 13th International Zeolite Conference,* Montpellier, France, July, 08–13, 2001; Studies in Surface Science and Catalysis 135; Elsevier, Amsterdam, 2001, poster 02-P-24.
[150] Antonić Jelić, T., Bosnar, S., Bronić, J., Subotić, B., Škreblin, M., *Micropor. Mesopor. Mater.* **2003,** *64,* 21–32.
[151] Subotić, B., Antonić Jelić, T., Bronić, J., in: Xu, R., Gao, Z., Chen, J., Yan, W. (Eds.), *From Zeolites to Porous MOF Materials – The 40th Anniversary of International Zeolite Conference. Proceedings of the 15th International Zeolite Conference,* Beijing, P.R. China, August, 12–17, 2007; Studies in Surface Science and Catalysis 170A; Elsevier, Amsterdam, 2007, pp. 233–241.
[152] Antonić Jelić, T., Bronić, J., Hadžija, M., Subotić, B., Marić, I., *Micropor. Mesopor. Mater.* **2007,** *105,* 65–74.
[153] Bronić, J., Subotić, B., Škreblin, M., *Micropor. Mesopor. Mater.* **1999,** *28,* 73–82.
[154] Brar, T., France, P., Smirniotis, P. G., *Ind. Eng. Chem. Res.* **2001,** *40,* 1133–1139.
[155] Di Renzo, F., Remoue, F., Massiani, P., Fajula, F., Figueras, F., Des Courieres, T., *Zeolites* **1991,** *11,* 539–548.
[156] Di Renzo, F., Albizane, A., Nicolle, M.-A., Fajula, F., Figueras, F., Des Courieres, T., in: Öhlmann, G., Pfeifer, H., Fricke, R. (Eds.), *Catalysis and Adsorption by Zeolites, Proceedings of ZEOCAT 90,* Leipzig, Germany, August 20–23, 1990; Studies in Surface Science and Catalysis 65; Elsevier, Amsterdam, 1991, pp. 603–612.
[157] Grizetti, R., Artioli, G., *Micropor. Mesopor. Mater.* **2002,** *54,* 105–112.
[158] Nagy, J. B., Bodart, P., Collette, H., Fernandez, C., Gabelica, Z., Nastro, A., Aiello, R., *J. Chem. Soc. Faraday Trans 1* **1989,** *85,* 2749–2769.
[159] Nikolakis, V., Vlacho, D. G., Tsapatsis, M., *Micropor. Mesopor. Mater.* **1998,** *21,* 337–346.
[160] Valtchev, V. P., Bozhilov, K. N., *J. Am. Chem. Soc.* **2005,** *127,* 16171–16177.
[161] Budd, P. M., Myatt, G. J., Price, C., *Zeolites* **1994,** *14,* 188–202.
[162] Myatt, G. J., Budd, P. M., Price, C., Hollway, F., Carr, S. W., *Zeolites* **1994,** *14,* 190–197.
[163] Li, Q. H., Mihailova, B., Creaser, D., Sterte, J., *J. Micropor. Mesopor. Mater.* **2001,** *43,* 51–59.
[164] Edelman, R., Kudalkar, D. V., Ong, T., Warzywoda, J., Thompson, R. W., *Zeolites* **1989,** *9,* 496–502.
[165] Thompson, R. W., in: Catlow, C. R. A. (Ed.), *Modelling of Structure and Reactivity in Zeolites,* Academic Press, London, 1992, pp. 231–255.
[166] Hou, L.-Y., Thomson, R. W., *Zeolites* **1989,** *9,* 526–530.
[167] Warzywoda, J., Edelman, R. D., Thomson, R. W., *Zeolites* **1991,** *11,* 318–324.
[168] Tsokanis, E. A., Thompson, R. W., *Zeolites* **1992,** *12,* 369–373.
[169] Warzywoda, J., Thomson, R. W., *Zeolites* **1991,** *11,* 577–582.
[170] Gora, L., Thomson, R. W., *Zeolites* **1995,** *15,* 526–534.
[171] Fajula, F., Vera-Pacheco, M., Figueras, F., *Zeolites* **1987,** *7,* 203–208.
[172] Flanigen, E. M., Breck, D. W., in: *137th Meeting of the ACS, Division of Inrganic Chemistry,* Cleveland OH, April 1960, Paper No. 82.
[173] McNicol, B. D., Pott, G. T., Loos, K. R., *J. Phys. Chem.* **1972,** *76,* 3388–3390.
[174] Breck, D. W., *Zeolite Molecular Sieves,* Wiley, New York, N.Y., 1974, pp. 342–343.
[175] Aiello, R., Barrer, R. M., Kerr, I. S., in: Sand, L. B., Flanigen, E. M. (Eds.), *Molecular Sieve Zeolites—I.* Advances in Chemistry Series 101; American Chemical Society, Washington, DC, 1971, pp. 44–50.
[176] Thomas, J. M., Bursill, L. A., *Angew. Chem. Int. Ed. Engl.* **1980,** *19,* 755–756.
[177] Tsuruta, Y., Satoh, T., Yoshida, T., Okumura, O., Ueda, S., in: Murakami, Y., Iijima, A., Ward, J. W. (Eds.), *New Development in Zeolite Science, Proceedings of the 7th Zeolite Conference,* Tokyo, Japan, August 17–22, 1986; Studies in Surface Science and catalysis 28; Elsevier, Amsterdam, 1986, pp. 1001–1007.
[178] Subotić, B., Tonejc, A. M., Bagović, D., Čižmek, A., Antonić, T., in: Weitkamp, J., Karge, H. G., Pfefer, H., Hoelderich, W. (Eds.), *Zeolites and Related Microporous Materials:*

State of Art 1994, Proceedings of 10th International Zeolite Conference, Gramisch-Partenkirchen, Germany, July 17–22, 1994; Studies in Surface Science and Catalysis 84A; Elsevier, Amsterdam, 1994, pp. 259–266.

[179] Kosanović, C., Bosnar, S., Subotić, B., Svetličić, V., Mišić, T., Dražić, G., Havancsák, K., *Micropor. Mesopor. Mater.* **2008,** *110,* 177–185.

[180] Epping, J. D., Cmelka, B. F., *Curr. Opin. Colloid Interface Sci.* **2006,** *11,* 81–117.

[181] Gates, B. C., Knozinger, H., *Adv. Catal.* **2006,** *50,* 149–225.

[182] Gabelica, Z., Nagy, J. B., Debras, G., Derouane, E. G., *Acta Chim. Hungarica* **1985,** *119,* 275–284.

[183] Prasad, S., Chen, W.-H., Liu, S.-B., *J. Chin. Chem. Soc.* **1995,** *42,* 537–542.

[184] Aiello, R., Crea, F., Nastro, A., Subotić, B., Testa, F., *Zeolites* **1991,** *11,* 767–775.

[185] Krznarić, I., Antonić, T., Subotić, B., Babić-Ivančić, V., *Thermochim. Acta* **1998,** *317,* 73–84.

[186] Jacobs, P. A., Derouane, E. G., Weitkamp, J., *J. Chem. Soc. Chem. Commun.* **1981,** 591–593.

[187] Evmiridis, N. P., Yang, S., in: Beyer, H. K., Karge, H. G., Kiricsi, I., Nagy, J. B. (Eds.), *Catalysis by Microporous Materials, Proceedings of ZEOCAT '95,* Szombathely, Hungary, July 09–13, 1995; Studies in Surface Science and Catalysis 94; Elsevier, Amsterdam, 1995, pp. 341–348.

[188] Kosanović, C., Bronić, J., Subotić, B., Šmit, I., Stubičar, M., Tonejc, A., Yamamoto, T., *Zeolites* **1993,** *13,* 261–268.

[189] Richards, M. D., Pope, C. G., *J. Chem. Soc. Faraday Trans.* **1996,** *92,* 317–323.

[190] Serrano, D. P., Uguina, M. A., Ovejero, G., Van Grieken, R., Camacho, M., *Micropor. Mesopor. Mater.* **1996,** *7,* 309–321.

[191] Derouane, E. G., Detremmerie, S., Gabelica, Z., Blom, N., *Appl. Catal.* **1981,** *1,* 201–224.

[192] Gabelica, Z., Derouane, E. G., Blom, N., in: Whyte, T. E., Jr., Dalla Betta, R. A., Derouane, E. G., Baker, R. T. K. (Eds.), *Catalytic Materials: Relationship Between Structure and Reactivity,* Advances in Chemistry Series 101; American Chemical Society, Washington, DC, 1984, pp. 219–251.

[193] Chang, C. D., Bell, A. T., *Catal. Lett.* **1991,** *8,* 305–316.

[194] Regev, O., Cohen, Y., Kehat, E., Talmon, Y., *Zeolites* **1994,** *14,* 314–319.

[195] Burkett, S. L., Davis, M. E., *Chem. Mater.* **1995,** *7,* 920–928.

[196] Burkett, S. L., Davis, M. E., *Chem. Mater.* **1995,** *7,* 1453–1463.

[197] Dokter, W. H., Beleen, T. P. M., van Garderen, H. F., Rummens, C. P. J., van Santen, R. A., Ramsay, J. D. F., *Colloids Surf. A* **1994,** *85,* 89–95.

[198] Dokter, W. H., van Garderen, H. F., Beleen, T. P. M., van Santen, R. A., Bras, V., *Angew. Chem. Int. Ed. Engl.* **1995,** *34,* 73–75.

[199] Schoeman, B. J., in: Chon, H., Ihm, S.-K., Uh, Y. S. (Eds.), *Progress in Zeolite and Microporous Materials, Proceedings of the 11th International Zeolite Conference,* Seoul, South Korea, August 12–17, 1996; Studies in Surface Science and Catalysis 105; Elsevier, Amsterdam, 1997, pp. 647–654.

[200] de Moor, P.-P. E. A., *The Mechanism of Organic-Mediated Zeolite Crystallization,* Ph.D. Thesis; Technical University of Eindhoven, Eindhoven, The Netherlands, 1998.

[201] de Moor, P.-P. E. A., Beelen, T. P. M., van Santen, R. A., *J. Phys. Chem. B* **1999,** *103,* 1639–1650.

[202] de Moor, P.-P. E. A., Beleen, T. P. M., Komanschek, B. U., Beck, L. W., Wagner, P., Davis, M. E., van Santen, R. A., Beck, L. W., *Chem. Eur.* **1999,** *5,* 2083–2088.

[203] de Moor, P.-P. E. A., Beleen, T. P. M., van Santen, R. A., Beck, L. W., Davis, M. E., *J. Chem. Phys. B* **2000,** *104,* 7600–7611.

[204] Yan, Y., Chaudhuri, S. R., Sarkar, A., *Chem. Mater.* **1996,** *8,* 473–479.

[205] Čižmek, A., Subotić, B., Kralj, D., Babić-Ivančić, V., Tonejc, A., *Microporous Mater* **1997,** *12,* 267–280.

[206] Katović, A., Subotić, B., Šmit, I., Despotović, L. J. A., *Zeolites* **1989,** *9,* 45–53.

[207] Kacirek, H., Lechert, H., *J. Phys. Chem.* **1975,** *79,* 1589–1593.

[208] Zhang, H., Kamotani, Y., Ostrach, S., *J. Cryst. Growth* **1993,** *128,* 1288–1292.

[209] Fulcher, M., Warzywoda, J., Sacco, A., Jr., Thompson, R. W., Dixon, A. G., *Micropor. Mesopor. Mater.* **1997,** *10,* 199–209.

[210] Marui, Y., Matsuoka, M., Uchida, H., Takiyama, H., *J. Chem. Eng. Jpn.* **2003,** *36,* 616–622.

[211] Subotić, B., in: Occelli, M. L., Robson, H. E. (Eds.), *Zeolite Synthesis,* ACS Symposium Series No. 398; American Chemical Society, Washington, D.C., 1989, pp. 110–123.
[212] Warzywoda, J., Thomson, R. W., *Zeolites* **1989,** *9,* 341–345.
[213] Bosnar, S., Bronić, J., Krznarić, I, Subotić, B., *Croat. Chem. Acta* **2005,** *78,* 1–8.
[214] Schoeman, B. J., Sterte, J., Otterstedt, J.-E., *J. Chem. Soc. Chem. Commun.* **1993,** 994–995.
[215] Persson, A. E., Schoeman, J. B., Sterte, J., Otterstedt, J.-E., *Zeolites* **1994,** *14,* 557–567.
[216] Schoeman, J. B., Sterte, J., Otterstedt, J.-E, *Zeolites* **1994,** *14,* 568–575.
[217] Brock, A. A., Link, G. N., Potras, P. S., Thompson, R.W, *J. Mater. Chem.* **1993,** *3,* 907–908.
[218] Di Renzo, F., Fajula, F., Espiau, P., Nicolle, M.-A, Dutartre, R, *Zeolites* **1994,** *14,* 256–261.
[219] Schoeman, J. B., Sterte, J., Otterstedt, J.-E., *J. Colloid Interface Sci.* **1995,** *170,* 449–456.
[220] Wiersma, G. S., Thompwson, R. W., *J. Mater. Chem.* **1996,** *6,* 1693–1699.
[221] Gora, L., Thompson, R. W., *Zeolites* **1997,** *18,* 132–141.
[222] Subotić, B., Aiello, R., Bronić, J., Testa, F., in: Aiello, R., Giordano, G., Testa, F. (Eds.), *Impact of Zeolites and Other Porous Materials on the New Technologies at the Beginning of the New Millenium, Proceedings of the 2nd International FEZA Conference,* Taormina, Italy, September 01–05, 2002; Studies in Surface Science and Ctalysis 142A; Elsevier, Amsterdam, 2002, pp. 423–430.
[223] Andac, Ö., Tatier, M., Sikrecloglu, A., Ece, I., Erdem-Senatalar, A., *Micropor. Mesopor. Mater.* **2005,** *79,* 225–233.
[224] Kim, S. D., Noh, S. H., Park, J. W., Kim, W. J., *Micropor. Mesopor. Mater.* **2006,** *92,* 181–188.
[225] Dougherty, J., Iton, L. E., White, J. W., *Zeolites* **1995,** *15,* 640–649.
[226] de Moor, P.-P. E. A., Beleen, T. P. M., Van Santen, R. A., *Micropor. Mesopor. Mater.* **1997,** *19,* 117–130.
[227] Ban, T., Mitaku, H., Suzuki, C., Kume, T., Ohya, Y., Takahashi, Y., *J. Porous Mater.* **2005,** *12,* 255–263.
[228] Serrano, D. P., Uguina, D. P., Sanz, R., Castillo, E., Rodriguez, A., Sanchez, P., *Micropor. Mesopor. Mater.* **2004,** *69,* 197–208.
[229] Regev, O., Kohen, Y., Kehat, E., Talmon, Y., *J. Phys. IV* **1993,** *3,* 397–400.
[230] Fedeyko, J. M., Vlachos, D. G., Lobo, R. F., *Langmuir* **2005,** *21,* 5197–52106.
[231] Fedeyko, J. M., Egolf-Fox, H., Fickel, D. W., Vlachos, D. G., Lobo, R. F., *Langmuir* **2007,** *23,* 4532–4540.
[232] Rimer, J. D., Vlachos, D. G., Lobo, R. F., *J. Phys. Chem. B* **2005,** *109,* 12762–12771.
[233] Rimer, J. D., Lobo, R. F., Vlachos, D. G., *Langmuir* **2005,** *21,* 8960–8971.
[234] Rimer, J. D., Fedeyko, J., MVlachos, D. G., Lobo, R. F., *Eur. Chem. J.* **2006,** *12,* 2926–2934.

CHAPTER 7

Modern Spectroscopic Methods Applied to Nanoscale Porous Materials

Boriana Mihailova

Contents

Abstract

Recent achievements in studying the structure of nanoscale zeolite-type materials by applying Raman scattering and Fourier Transform infrared (FTIR) spectroscopy are reviewed. Studies on microporous materials of different framework topology and composites with a hierarchical porous system demonstrate the ability of these methods for unambiguous identification of small-sized crystalline porous particles. The detection of formation of thin zeolite films, analysis of precursor clusters and crystal-growth processes are also discussed. The type and degree of structural defects, guest–host interactions and surface modified particles and investigations of sub-nanometric clusters embedded into zeolite matrices studied by these techniques are presented as well.

Keywords: ATR-IR and RAIR Spectroscopy, Microporous Materials, Raman Scattering

Ordered Porous Solids
DOI: 10.1016/B978-0-444-53189-6.00007-X

1. Introduction

The majority of advanced functional materials are nanoscale materials, which can be divided into two groups: 'monolitic' samples with a complex nanoscale domain structure and (ii) nanosized materials like nanoparticles, thin films and nanowires. The physicochemical properties of the latter group are strongly influenced by the fact that at least one dimension is of order of 100 nm or less. Examples of the former groups are relaxor ferroics, whose structure is composed of ferroic nanoclusters embedded inside a para-matrix, materials consisting of nanosized spatial regions of one phase incorporated into a host matrix of another phase, inorganic–organic nanocomposites, etc. The advantageous properties for these 'monolitic' material stems from their intrinsic structural inhomogeneity.

Structural characterization of nanoscale materials is not an easy task. Conventional diffraction methods often fail to reveal the structural peculiarities of nanoscale materials because they probe time-and-space average structure of materials with long-range order. Hence, diffraction methods are sensitive to static spatial regions possessing translational symmetry with a linear size at least 10 times larger than the period of translation. Modern functional materials usually comprise specific atomic clustering within spatial regions sized a few unit cells or even within one unit cell, as in the case of zeolite-type materials. Besides, instantaneous structural fluctuations and dynamical bonding may exist in nanoscale materials. In this regard, spectroscopy provides excellent tools for structural characterization of nanoscale functional materials. Spectroscopic methods have much higher length- and time-scale sensitivity than diffraction methods do because spectroscopy is based on the change in energy of the probe radiation, rather than on the change in direction as in the case of diffraction. Hence, periodicity is not the necessary condition to form a signal and crystalline and amorphous solids can be equally well analysed. By spectroscopy one can distinguish static as well as dynamic atomic clustering on nano- and even subnanometric scale. Spectroscopic methods are especially beneficial in studying nanoscale porous materials because they provide statistically relevant (in contrast to transmission electron microscopy) information on the framework structure as well as on the guess–host interactions, that is, the manner of incorporation and the strength of bonding between molecules and ions encapsulated into pores.

Among the great variety of spectroscopic methods, those using electromagnetic waves as incident radiation is most frequently used due to the relatively easily accessible instrumentation and handling of measurements.

Nuclear magnetic resonance (NMR) spectroscopy is based on the response of the nuclear spins to radio-waves under an external magnetic field and is very applicable for studying the first and the second coordination sphere of a certain type of atom. NMR is observed only for those atomic nuclei having non-zero spin; however, almost for each chemical element there is an isotope that meets this requirement. The magic-angle-spinning technique allows the successful application of NMR spectroscopy not only to liquids but also to solids. Hence, NMR is widely used for studying the structure of intermediate species in solidification processes of nanozeolites,[1–3] external and internal surface active sites of porous nanomaterials,[4,5] atomic coordination and guest–host interactions in aluminosilicate, aluminophosphate, and mixed tetrahedral–octahedral porous materials.[6–13]

Electron spin resonance (ESR) spectroscopy is based on the response of electron spins to microwaves under an external magnetic field. Its application is restricted to materials containing transition elements of low concentration. However, the latter condition for performing ESR experiments is in fact rather beneficial to study structural states and active centres in transition-metal-modified micro- and mesoporous nano-materials.[14–17]

Fourier Transform infrared (FTIR) and Raman spectroscopy are the two principle varieties of vibrational spectroscopy with electromagnetic radiation. FTIR spectroscopy is based on absorption of infrared radiation from atomic vibrations, whereas Raman spectroscopy consists of inelastic visible/near-visible light from atomic vibrations. Both techniques provide information on short- and intermediate-range ordering, that is, the geometry of polyhedra representing the atomic nearest neighbourhood and the manner of linkage of these polyhedra into larger clusters. Thus FTIR and Raman spectroscopic methods are very efficient for studying kinetic processes and precursor clusters as well as to identify small-sized crystalline particles in the final product.[10,13,18–33] In addition, FTIR and Raman spectroscopy are very sensitive to atomic bonding and thus gives the opportunity to study active sides and interactions between embedded functional molecules and framework atoms in porous materials.[5,12,33–42] Raman spectroscopy compares favourably to FTIR spectroscopy because of the better signal-to-noise ratio in the low-energy spectral range and the better spatial resolution when a microscope is used. Besides, Raman spectroscopy does not require any special sample preparation, which makes this technique a truly non-invasive and non-destructive method for structural analysis. On the other hand, surface-sensitive FTIR methods give better opportunity to study thin porous films.[43–49] Another advanced application of Raman spectroscopy is the characterization of nanocomposites of carbon and metal-oxide wires and clusters encapsulated into the porous system of a zeolite-type host.[50–57]

Visible/near-visible (optical absorption and laser-induced luminescence) spectroscopy, which probes valence-electron states, is an excellent tool to distinguish between cations in oxidation state and sub-nanometric metal-oxide clusters embedded into the framework of nanoscale porous materials.[10,17,33,50,51] The utilization of these spectroscopic methods for structural analysis is often hindered from the broad width and the overlapping of spectral signals; however, the combination of visible/near-visible spectroscopy with vibrational and/or NMR spectroscopy is beneficial for characterization of photoactive materials. Besides, the recently developed time-resolved luminescence spectroscopy gives unique opportunity to characterize the distribution of sub-nanometric guest species into micrometer-sized zeolite crystals used solar-cell and photonic devices.[58,59]

X-ray absorption spectroscopy with its three varieties: pre-edge, X-ray absorption near the edge structure (XANES) and extended X-ray absorption fine structure (EXAFS), probes the core-electron states. The main advantage of this method is that it is strictly element-sensitive and thus provides information on short-, intermediate-

and long-range ordering related to a particular type of cation occupying a certain crystallographic site. The method is especially advantageous in studying active centres in functionalized by appropriate doping porous materials; however, its proper utilization requires a synchrotron radiation.[15,39,52,60,61]

Mößbauer spectroscopy is based on recoilless resonance absorption/emission of gamma-rays in solids. Its utilization for structural analysis stems from the fact that the principle energy levels of atomic nuclei are strongly influenced by the positions of the surrounding atoms. The methods are characterized by extremely high spectral resolution and hence very weak changes in the short-range order are detectible by the method. Because of the available sources of incident radiation, the application of Mößbauer spectroscopy in materials science is mainly restricted to Fe-containing systems. However, Mößbauer spectroscopy is an excellent tool to investigate Fe ions and Fe-containing molecular groups or clusters incorporated into porous materials and, in addition, doping with Fe allows us to study the local structure of nanoscale porous materials.[17,62]

The brief overview of the spectroscopic methods shows that in materials science Raman scattering, FTIR spectroscopy and NMR are most extensively used spectroscopic methods for structural analysis due to their applicability to a wide range of materials, their high sensitivity to short- and intermediate-range ordering and easily-recognized relationships between the observed spectral features and nanoscale atomic arrangements. In particular, vibrational spectroscopy is easy handling, requires only a small amount of sample and the performance of *in situ* experiments is relatively simple. The recent development of high-sensitive, less expensive Raman spectrometers and improved accessories for reflectance infrared spectroscopy further renewed the interest of the materials-science community to vibrational spectroscopic methods. Hence, this chapter aims to provide a link between vibrational spectroscopy and nanoscale porous materials science and to review the recent achievements in utilizing Raman scattering and FTIR spectroscopy for studying the structure and guest–host interactions in nanoscale porous materials.

2. Vibrational Spectroscopy: Fundamentals and Practical Hints

The basic of solid-state vibrational spectroscopy is the interaction of light propagating through a solid with phonons—quasi-particles representing collectivized atomic vibrations that occur in crystals. Each phonon or phonon mode is characterized by wavenumber, wavevector and symmetry, which is determined by the corresponding set of atomic vibrations. Only optical phonon modes, that is, having a dipole moment, interact with light and thus contribute to vibrational spectra. Depending on the symmetry, a certain phonon mode generates a peak in the IR and/or Raman spectrum. The peak positions are determined from the phonon wavenumbers, while in general the intensity of the peaks is determined from the atomic vector displacements.

The main interaction processes between light and atomic vibrations are schematically represented in Fig. 7.1.

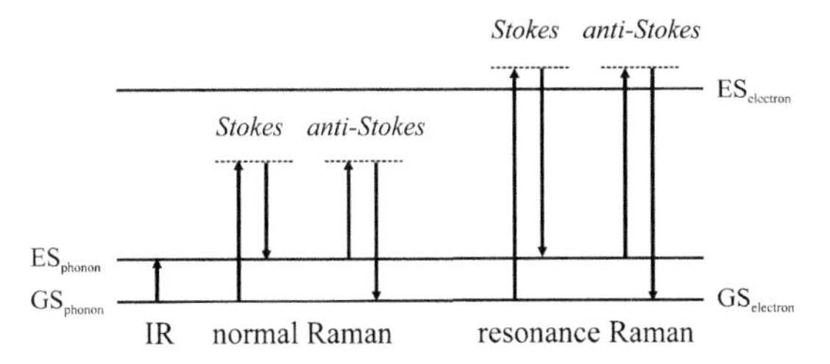

Figure 7.1 Schematic presentation of interaction processes between light and solid; GS and ES stand for 'ground state' and 'excited state', respectively.

IR spectroscopy is a pure absorption phenomenon: the solid absorbs the incoming light photon with wavelength of the order of tens microns (the range of infrared electromagnetic waves) and, as a result, a transition from the ground state (GS) to an excited state (ES) of phonon occurs. The energy conservation law is

$$\hbar\omega_{\text{photon}} = E_{\text{ES}}^{(\text{phonon})} - E_{\text{GS}}^{(\text{phonon})} \tag{1}$$

and hence

$$\omega_{\text{photon}} = \Omega \tag{2}$$

where Ω is the phonon wavenumber.

Raman spectroscopy is an inelastic light scattering process, consisting in a change in energy and wavevector of the propagating photon due to its interaction with the atomic vibrations. As a result, the scattered light can have either a higher or a lower frequency than that of the incident light, depending on whether phonon absorption of phonon emission is involved, that is, if the phonon transition is from the ground to an ES (Stokes process) or from an excited to the GS (anti-Stokes). From the conservation laws of energy and wavevector, it follows that for Stokes processes

$$\omega_s = \omega_i - \Omega \text{ and } \mathbf{k}_s = \mathbf{k}_i - \mathbf{K} \tag{3}$$

while for anti-Stokes processes

$$\omega_s = \omega_i + \Omega \text{ and } \mathbf{k}_s = \mathbf{k}_i + \mathbf{K} \tag{4}$$

In both Eqs. (3) and (4), ω_i and $\mathbf{k}_i$ are the frequency and the wavevector of the incident photon, ω_s and $\mathbf{k}_s$ are the frequency and the wavevector of the scattering photon and Ω and $\mathbf{K}$ are the wavenumber and the wavevector of the phonon involved in the corresponding scattering process, respectively. The probability of Stokes scattering processes is much higher than the probability of anti-Stokes processes because usually the occupation of the GS is much higher than the occupation of the excited state. Hence, usually Stokes Raman scattering is measured and used for material characterization. However, in some cases it is useful to record also the anti-Stokes scattering. For example, when the experiment requires a precise control of the temperature T in the sample bulk, one can use the intensity ratio between the Stokes (I_S) and anti-Stokes (I_{AS}) Raman scattering

$$I_S/I_{AS} = \exp(\hbar\Omega/k_B T) \tag{5}$$

Another example is when photoluminescence signals contribute to the Raman spectral range of interest and one need to distinguish between luminescence and Raman peaks; a Raman signal positioned at Ω always has a corresponding peak at $-\Omega$, which is not the case for photoluminescence.

Normal Raman scattering occurs when the energy of the incident photon is lower than the energy difference between the first excited electron level and the ground electron level. If the exciting radiation energy coincides with an electron energy transition, $\hbar\omega_{\text{incident}}^{(\text{photon})} \approx \Delta E^{\text{electron}}$, resonance Raman scattering is realized.

Note that Eqs. (2)–(4) describe one-phonon processes, that is, only one phonon is involved in the interaction between the propagating light and the solid. Because of the high probability to occur, one-phonon processes determine the principle IR absorption and Raman scattering. Multi-phonon processes involve simultaneous energy transitions of two or more phonons; they are of low probability and generate peaks of very low intensity, positioned at wavenumbers, which are proportional or a linear combination of the position of the peaks arising from one-phonon processes.

The IR intensity is determined from the dipole moment directly induced by the change in the atomic positions. Expanding the dipole moment $\boldsymbol{\mu}(Q)$ in a Taylor series of normal coordinates Q_k representing the atomic displacements, we obtain

$$\boldsymbol{\mu}(Q) = \boldsymbol{\mu}_0 + \sum \frac{\partial \boldsymbol{\mu}}{\partial Q_k} Q_k + \ldots \tag{6}$$

where the first term is the static dipole moment, while the second term is the vibration-induced dipole moment and, hence, the infrared absorption intensity is

$$I_{IR} \propto \left| \frac{\partial \boldsymbol{\mu}}{\partial Q_k} \right|^2 \tag{7}$$

The Raman intensities are related to the dependence of the field-induced dipoles on the normal coordinates. Field-induced dipoles arise from the changes of electron shells under electric field. The light scattering intensity is proportional to the square of the so-called polarizability tensor α_{ij}, i, j,=1,2,3, which is the proportional coefficient between the induced polarization $\mathbf{P}$ and the electric filed $\mathbf{E}$, $\mathbf{P} = \boldsymbol{\alpha}.\mathbf{E}$. The expansion of the polarizability tensor $\boldsymbol{\alpha}$ in Taylor series of normal coordinates gives

$$\boldsymbol{\alpha}(Q) = \boldsymbol{\alpha}_0 + \sum \frac{\partial \boldsymbol{\alpha}}{\partial Q_k} Q_k + \ldots \tag{8}$$

where the first term accounts for the Rayleigh (elastic) scattering, while the second for the Raman (inelastic) scattering.

The symmetry of the phonon mode determines the selection rules, that is, whether the first derivatives of $\boldsymbol{\mu}$ and $\boldsymbol{\alpha}$ are non-zero and, hence, if the phonon mode is active in IR and/or Raman spectra or it is inactive. It is worth noting that the Raman cross-section depends on the deformability of electron shells. Thus transition metal–oxygen bonds, which involve d- or f-electrons, are highly polarizable and generate much stronger Raman peaks than Si–O bonds do. Consequently, Raman spectroscopy is very efficient in studying transition metal–oxygen clusters or sub-nanoparticles embedded in a porous silicate matrix. Also, bonds of organic molecules have much larger Raman cross-section than cation–oxygen bonds in inorganic substance. Hence, solidification and crystal-growth processes when an organic template is used can be followed by the Raman spectral fingerprints of the corresponding organic molecule. Besides, the strong intensity of the Raman signals generated by organic molecules encapsulated in inorganic porous materials facilitates the study of guest–host and guest–guest interactions.

One should underline that the selection rules for resonance Raman scattering differ from those for normal Raman scattering because resonance Raman processes also involve transitions between electron states. Thus the Raman spectra of a certain material excited with incident light of different wavelength may be very different. Normal Raman scattering occurs if the energy of the incident light is well below the energy of the principle optical absorption edge. For inorganic silicate and phosphate porous materials, this condition is fulfilled for near-IR and visible lasers. Resonance Raman scattering may occur for transition metal-containing porous materials when UV radiation is used, which is also very helpful in analyzing transition metal–oxygen bonding. The utilization of UV lasers is also efficient to overcome the strong luminescence background, which is often observed for porous materials when visible lasers are used to incite the Raman scattering. Luminescence-free Raman spectra may be also collected using near-IR lasers, but UV laser radiation is preferred because the Raman intensity I is much stronger for shorter laser wavelength λ, $I \sim 1/\lambda^4$. However, when UV laser radiation is used, preliminary optical absorption experiments have to be conducted to verify if non-resonance or resonance scattering conditions are realized for the corresponding material and laser source.

The wavenumbers and the atomic vector displacements of normal phonon modes correspond to the eigenvalues and the eigenvectors of the so-called dynamical matrix, which is determined from the atomic masses and the second derivatives of the crystal potential. Thus, the dynamical matrix and, consequently, the phonon wavenumbers and atomic vector displacements depend on the atomic coordinates, interatomic force constants and the atomic masses. The dependence on the atomic coordinates means that IR and Raman spectra directly examine the structure. The dependence on atomic masses allows for studying compositional disorder and for distinction between two isovalent ions occupying the same crystallographic site; in general, peaks generated from phonon modes involving heavier atoms are positioned at lower wavenumbers and vice versa. The dependence on force constants allows for probing the atomic bonding, for example, formation of H-bonds, compositional disorder involving ions of different valency, etc.

The number of optical phonon modes in a crystal is $3N-3$, where N is the number of atoms in the primitive unit cell. The number of IR- and Raman-active phonon modes can be determined by performing group-theory analysis. Zeolite-type materials usually have a huge number of atoms, which results in a huge number of peaks that should exit in the IR and Raman spectra. However, only several peaks are experimentally observed because zeolites have large unit cells and the translation-symmetry restrictions on lattice dynamics are somehow suppressed. Crystal normal phonons are grouped in bands generating a relatively small number of peaks assigned to vibrations of characteristic structural species building the framework of the crystal. In particular, the Raman spectra of framework silicates (crystalline and amorphous) can be interpreted in terms of modes of rings of SiO_4 tetrahedra and the range between 250 and 650 cm^{-1} is dominated by ring modes involving in-phase vibrations of the bridging oxygen atoms (see Fig. 7.2).

The number of tetrahedra composing the ring strongly influences the corresponding Raman peak position. The Raman scattering of various framework silicates reveal that a Raman signal near 650–600 cm^{-1} is related to three-membered rings, Raman signals between ~560 and 490 cm^{-1} to four-membered rings, between ~490 and 400 cm^{-1} to six-membered rings and between ~400 and 300 cm^{-1} to five-membered rings. The peak positions also depend on ring geometry, which is specific for different frameworks. Thus the Raman spectral range 250–650 cm^{-1} is very useful for kinetic studies.

The finite size of nanoparticles influences the vibrational spectra in two aspects: (i) the number of surface atoms is comparable with the number of bulk atoms and thus the existence of surface phonons or surface species with their own vibrations has to be taken into account and (ii) the bulk phonons are affected because of the confined volume. The latter effect is valid for isolated particles as well as for particles embedded into a host matrix if the phonon dispersion curves of the encapsulated particle do not overlap with that of the host. Thus optical phonon confinement effects are substantial for carbon or metal-oxide particles embedded into zeolite-

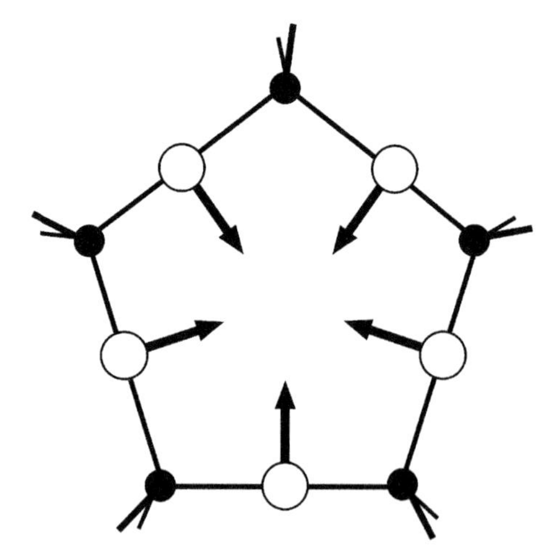

Figure 7.2 SiO_4 ring mode.

type matrices and the mean size of the incorporated particles can be derived from Raman spectra. The average size of zeolitic nanoparticles synthesized so far approaches 30 nm. Thus optical phonon confinement phenomena are not so significant for the available porous nanoparticles themselves and have not yet been explored. However, the large external as well as internal surface is a precondition for a large number of surface species, the contribution of which may be observed in IR and Raman spectra.

The Raman signal may be amplified by several orders of magnitude for molecules adsorbed on roughened surface or colloidal particles of transition metals (mostly silver). This technique is known as Surface-Enhanced Raman Spectroscopy (SERS) and has the potential to study thin films on metal surfaces. However, the method requires much sample preparation to optimize the experimental conditions and additional measurements to interpret correctly the data collected. Thus SERS is seldom applied to solid films and is mainly used in biology. Surface-sensitive FTIR spectroscopic techniques turned to be more appropriate for studying zeolite films and membranes. Two methods should be considered in this regard: reflectance–absorbance infrared (RAIR) spectroscopy and attenuated total reflectance infrared (ATR-IR) spectroscopy.

RAIR spectroscopy is applied to films deposited on strongly reflected surface, for example, Au, Ag, Pt, Cu and sometimes steal. The presence of metallic surface changes the boundary conditions in the Maxwell equations for electromagnetic wave propagation through a solid and thus the characteristic depth of penetration of the method is from a monomolecular layer to $\sim$1 μm. Hence, the method is suitable to probe the structure of thin zeolite-type films of thickness between a few to 1000 nm. The metallic surface also imposes strict selection rules on the induced dipole moments and only dipole moments with a non-zero component normal to the surface give rise to peaks. This makes the RAIR method advantageous for determining the orientation of the specimens in the layer. The incident light should be polarized in the plane of incidence (p-polarization) and the absorption factor is largest at grazing incidence angles. For silicate and phosphate films on a gold support, the optimum conditions for RAIR spectroscopy are achieved at an incident angle of $\sim$85°. If the polarization of the incident light is perpendicular to the plane of incidence (s-polarization), no absorption from phonons should be observed. The positions of the peaks observed in RAIR spectra are typically slightly higher than the IR peak positions measured in transmittance regime.

ATR-IR spectroscopy is a reflection technique in which the sample surface is in contact with a crystal (ATR crystal) of high index of refraction. Germanium, silicon, diamond, ZnSe, etc. are used as ATR crystal. The precondition for ATR-IR experiments is the refractive index of the sample to be lower than that of the used ATR crystal. For porous silicates and aluminophosphates (APOs), this precondition is fulfilled for all commercially available ATR crystals. The incident beam is directed through the ATR crystal and is totally reflected at the ATR crystal-sample contact surface. However, the sample absorbs some of the radiation because of the appearance of evanescent wave penetrating through the sample. Thus, the total reflectance is attenuated and the reflected beam is modulated by the IR absorption from the sample. The depth of penetration d_p depends on the refractive index of the sample n_S, of the ATR crystal n_{ATR}, and the angle of incidence θ. At a given wavelength λ, the penetration depth is

$$d_p = \frac{\lambda}{2\pi n_{ATR}\sqrt{\sin^2\theta - n_S{}^2/n_{ATR}{}^2}} \tag{9}$$

For silicates, the penetration depth at 1000 cm^{-1} when a Ge ATR crystal and $\theta = 45°$ are used is 650 nm. By changing the ATR crystals and the incident angle, one can vary the penetration depth and thus to do depth profile on zeolite-type membranes. For quantitative analyses, the spectra should be normalized to a constant penetration depth, for example, at 1000 cm^{-1}, using the relation

$$I_{ATR} = I_{measured} \cdot X/1000 \tag{10}$$

where X stands for the corresponding wavenumber. The peak positions in ATR-IR spectra are slightly shifted to lower wavenumbers as compared to those of the IR peaks measured in transmittance.

3. Application of Raman and IR Spectroscopy to Zeolite-Type Nano-Materials

3.1. Solidification and crystal growth processes

The growth mechanism of microporous materials is still not well understood and it is a great challenge to identify the structural species that assemble during the transformation from precursor sols or gels into crystalline nanoparticles. Clear organic-template containing aluminosilicate solutions can serve as a model system for studying the entire crystallization processes of microporous materials. The application of Raman spectroscopy to as-synthesized as well as calcined products of hydrothermal treatment (HT) for different periods of time is rather efficient to follow precursor atomic clustering. The Raman spectra of as-synthesized samples reveal how organic-template molecules alter during the process of pore formation, whereas Raman spectra of calcined samples directly show how the solidified amorphous substance is transformed into a crystalline zeolite structure during the HT. As an example, the formation of MFI structure in clear solutions containing tetrapropylammonium (TPA) ions is monitored by Raman spectra of non-calcined and calcined samples (Figs. 7.3 and 7.4, respectively).

According to X-ray diffraction (XRD) analysis conducted on the same series, crystalline domains are formed after 14 h of HT and the crystallization process is completed within 18 h, while after 2 h of HT crystalline regions sized ~10 nm, sporadically distributed in an amorphous matrix are seen by for porous silicates and APOs.[23] The Raman scattering generated from C–H bond stretching modes track the strength of bonding between TPA^+ and aluminosilicate framework. The shift of the peaks to lower wavenumbers reveals that after 2 h of HT treatment, the TPA

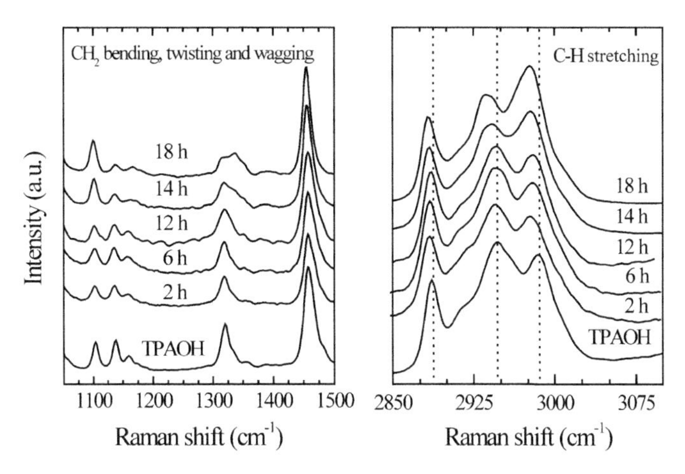

Figure 7.3 Raman spectra of as-synthesized MFI samples obtained for different times of hydrothermal treatment and of TPAOH.

ions are incorporated into solidified inorganic matrix. The intensity ratios of the peaks generated by TPA in the entire spectral range are the same as those for free TPAOH molecules, indicating that the majority of encapsulated TPA ions are non-twisted, that is, the inorganic substance is abundant of pores and cavities providing enough room for non-deformed TPA molecules, which is a sign for a non-periodic T–O network (T = Si, Al). The shape of the spectrum profile for C–H bond stretching and H–C–H bond bending modes is changed after 14 h of HT treatment, that is, when X-ray detectible crystallites are formed, because the conformation of TPA^+ is altered to adapt the pore structure typical of MFI. The change is most pronounced for the band near 1320 cm^{-1}. Using these spectral fingerprints of the organic template, MFI nanocrystallites can be distinguished from TPA-containing amorphous particles/substance. The effect of aging on the nucleation kinetics for the synthesis of nanosized TPA-silicalite-1 with various silica sources was also investigated on the basis of organic molecule spectral features.[19]

Figure 7.4 compares the Raman scattering spectra of calcined, template-free samples with that of amorphous silica. The main spectral features of amorphous silica are a broad band near 450 cm^{-1} and signals at 495 and 606 cm^{-1}, generated by six-, four- and three-membered rings, respectively. The Raman scattering of the 2-h intermediate product shows that the structure contains a large amount of four- and three-membered rings. Further, the specific broad band near 290 cm^{-1} is indicative for 10-membered rings of TO_4 tetrahedra formed around the TPA ions embedded in the amorphous T–O network. These structural entities play the role of nuclei for the formation of the MFI pore structure upon further HT. After 6 h of HT, a subtle increase in the Raman scattering near 380 cm^{-1} is observed, suggesting an enlargement of the number of MFI-type five-membered rings in the amorphous matrix. The spectrum of 12-h HT treated sample, which is still amorphous according to XRD, reveals abundance of five-membered rings. Besides, the simultaneous appearance of the features at 603 and 977 cm^{-1} shows the occurrence of quasi ***1D*** atomic

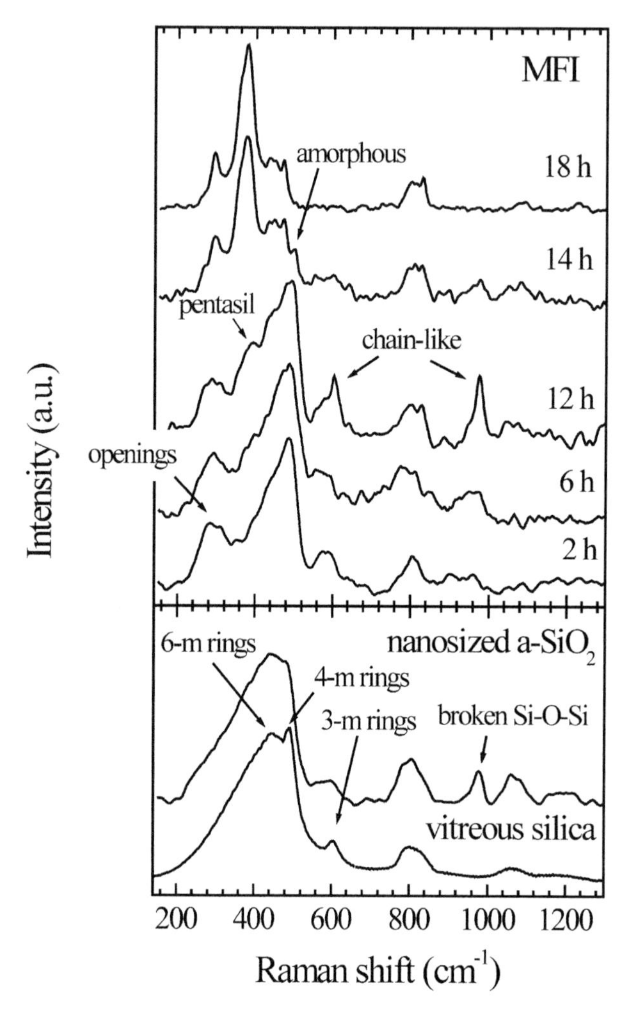

Figure 7.4 Raman spectra of calcined MFI samples obtained for different times of hydrothermal treatment and of amorphous silica.

arrangements. These atomic formations are related to a development of the MFI-type pentasil chains parallel to the *c*-axis. Therefore, after 12 h of crystallization the amorphous T–O network adapts the intermediate-range order of MFI structure, thus facilitating the growth and aggregation of the initially formed MFI nuclei. After 14 h of crystallization, only a small amount of amorphous phase still exists and after 18 h of HT, the solidified substance completely converts crystalline microporous material. Thus, Raman spectroscopic analysis shows that the crystallization of MFI-type nanoparticles from a clear TPAOH–Al_2O_3–SiO_2–H_2O solution passes through the following stages: (i) the formation of a porous disordered T–O network with a skeleton rich of four- and three-membered rings of TO_4 tetrahedra; (ii) formation of chain-like entities of five-membered rings typical of MFI medium-range order; (iii) the enlargement of the correlation length between the pentasil structural species in the plane perpendicular to the pentasil chains, thus facilitating the amorphous-to-crystalline phase transition, and an appearance of XRD-detectible MFI domains and (iv) a complete transformation of the amorphous microporous T–O network into MFI crystalline structure.

A similar impact of the porous crystalline framework on the vibrations on the corresponding organic template is also observed for other compounds. Figure 7.5 depicts the Raman spectra of as-synthesized nanoscale zeolites of type LTA, FAU and GIS in the spectral range dominated by C–N bond stretching modes of tetramethylammonium (TMA) ions used as organic additives to the synthesis solution of those zeolites.[21]

The framework of LTA and FAU zeolites is composed of sodalite-type structural units. Thus, for both zeolite types TMA ions are occluded in the sodalite cages as well as in the corresponding pore system. Since the TMA^+ size approximates the size of the sodalite-unit void, the embedded organic species experience strong elastic stress along the N–C bonds and, as a consequence the peaks arising from the antisymmetrical and symmetrical N–S bond stretching modes shift to higher wavenumber. The effect is better pronounced for the symmetrical N–C bond stretching mode and leads to a well-resolved splitting of the Raman peak near 760 cm^{-1}; the lower-wavenumber component corresponds to TMA^+ in the pore system providing enough room for TMA^+, while the higher-wavenumber component to TMA^+ confined in the sodalite cages. Thus, the appearance of an additional peak near 770 cm^{-1} is indicative for occurrence of sodalite-type zeolite crystals. The GIS framework topology is quite dissimilar from that of sodalite-type zeolites and,

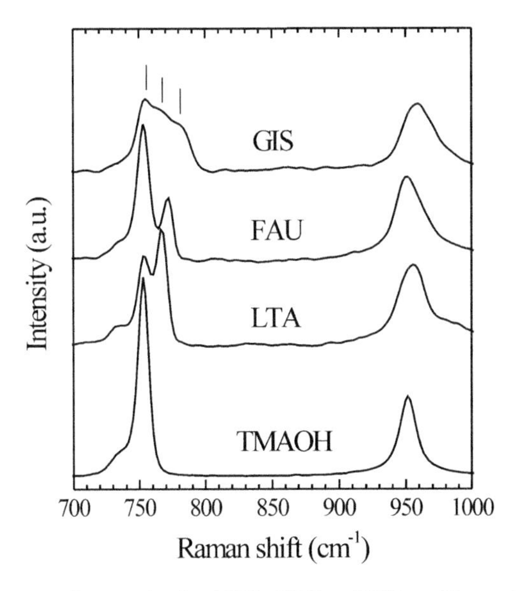

Figure 7.5 Raman spectra of as-synthesized LTA, FAU and GIS zeolites and of TMAOH.

hence, the organic template has rather different response to the surrounding inorganic matrix. In TMA/GIS type structure, the nitrogen and carbon atoms occupy certain crystallographic positions, that is, TMA species form a periodic sub-lattice inside the channel regular array of GIS structure. Thus, the establishment of long-range ordering transforms the internal modes of the TMA species into normal modes of TMA/GIS crystalline structure, which results in a specific three-component shape of the band related to the N–C symmetrical stretching and a shift of the C–H bond stretching modes to higher wavenumbers. The former effect is due to symmetry constrains, while the latter to coupling of the C–H stretchings in the periodic system of TMA^+.

BEA-type zeolites are commonly synthesized using tetraethylammonium (TEA) hydroxide as an organic template and similar template-framework interaction effects are observed for the TEA/BEA system.[24] The size of free TEA^+ is near the same as that of the BEA framework openings and, hence, the change in the shape of the bands near 670, 780–830 and 900 cm^{-1} (see Fig. 7.6), arising from stretching modes in the N–C–C chains, is due to the specific conformation and hardening of the vibrational modes of TEA^+ incorporated into BEA micropores. Besides, due to the occurrence of H-bonding in the guess/host system a characteristic splitting of the methylene C–H stretching at 2950 cm^{-1} is observed when a pore system typical of BEA framework topology is formed.

The use of an organic template is also critical for the formation of microporosity in most AlPOs. Crystalline MeAlPO-34 as well as AlPO-5 is usually synthesized in the presence of TEAOH. Time-resolved *in situ* Raman spectroscopy highlighted the organic/inorganic interactions leading to structure-directed microporous APO crystallization.[27] Based on the intensity ratio of the Raman peaks near 655 and 665 cm^{-1}, which corresponds to two different conformations of TEA^+ existing in

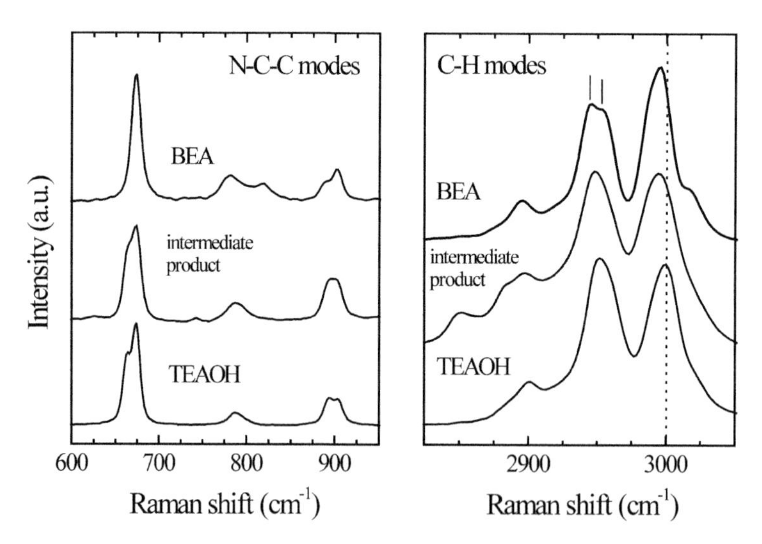

Figure 7.6 Raman spectra of as-synthesized BEA zeolite, X-ray amorphous intermediate product of synthesis and TEAOH.

solutions, it was revealed that both the location and the conformation of the template molecule are critical for the formation of MeAlPO-34, while the formation of AlPO-5 is controlled by the TEA^+ conformation alone.

Complementary application of Raman and NMR spectroscopy helped to construct a plausible scheme of the formation of BEA structures via steam-assisted crystallization.[31] The presence of Raman peak at 503 cm^{-1} evidenced the appearance of stable double four-membered rings in the dry gel. Thus the combined spectroscopic data revealed that double three-membered silicate rings stimulated by TEA^+ ions exist in the mother TeA-silicate solution to prepare a dry gel from which BEA-type zeolites are prepared using the steam-assisted crystallization method. In the course of preparing the dry gel, the double three-membered units are transformed into double four-membered rings, which play a significant role in the BEA zeolite formation. During the steam-assisted crystallization, these double four-membered units rearrange and induce a long-range order typical of BEA framework topology.

3.2. Structure control of colloidal zeolites stabilized in water

The changes in the geometry and the atomic surroundings of small quaternary ammonium ions encapsulated in the specific pores and cages of the zeolite framework provide an excellent opportunity to use Raman scattering and liquid-state ^{13}C NMR for structural control of nanosized zeolites stabilized in aqueous suspensions.[25] The results obtained for LTA, FAU, GIS, BEA and MFI framework topologies show that the crystallinity of nanosized particles with monomodal particle size distribution stabilized in water can be examined using the Raman and magnetic resonance spectral features of the organic-template molecules occluded in the zeolite framework. Raman scattering compares favourably to liquid-state ^{13}C NMR due to the shorter time of measurement, easier handling of the experimental set-up and the potential for a more precise distinction between precursor amorphous and crystalline zeolite nanoparticles. On the other hand, liquid-state ^{13}C NMR spectroscopy can be applied to characterize colloidal zeolite particles stabilized in various solvents. Therefore, depending on the corresponding needs one can preferably utilize the Raman or the liquid-state NMR analysis for non-destructive probing of the structure of the zeolite nanoparticles stabilized in solvents. Thus, by applying a structure control via spectroscopic methods the zeolitic nanoparticles can be further used for preparation of two- and three-dimensional constructs without the necessity of drying the samples prior to their characterizations.

3.3. Structural characterization of assemblies of microporous nanocrystallites

The spectral features of both the encapsulated organic-template molecules and the inorganic framework can be used to characterize the structural features of nanozeolite-based self-assemblies with a hierarchical porous system.

Raman spectroscopy was applied to probe the degree of structural disorder in micron-sized self-assemblies of closely packed zeolitic nanocrystallites (BEA and

MFI), prepared via transformation of porous amorphous silica under thermal treatment.[22,26] Insights into structural defects are gained from the spectra collected from calcined samples. The Raman scattering near 960 cm^{-1} is indicative for point defects in the silicon–oxygen framework (violated Si–O–Si linkages), while the percentage of non-crystalline domains in the samples prepared by different methods can be quantified by fitting the shape of the multi-component band in the range 250–650 cm^{-1}, generated by SiO_4-ring modes.

Raman spectroscopy was also applied to analyze the interlayer stacking disorder in BEA zeolite nanoparticles as well as in zeolite Beta nanocrystals embedded into a host mesoporous matrix.[24] It is shown that the intensity ratio ρ between the Raman signals at 314 and 343 cm^{-1} is most sensitive to the degree of periodicity faults along the [001]-direction. A larger value of ρ indicates a larger size of polymorph stacking sequence, that is, improvement of the stacking faultlessness. The Al content is significant for the concentration of point defects, but is less important for the interlayer stacking sequences in colloidal zeolite Beta. The occurrence of BEA-type interlayer connections in BEC-type materials can be detected via the appearance of Raman scattering near 343 cm^{-1}. Raman spectroscopic data on a composite material built of confined zeolite Beta nanocrystals well-dispersed into a host mesoporous matrix reveal a very high degree of periodicity faultlessness along the crystallographic *c*-axis, which is most probably due to the small average size (~15 nm) of the crystalline domains.

Raman spectroscopy also shed light on the synthesis mechanism of microporous self-bonded mm-size macrospheres prepared using resin beads as shape-directing macrotemplates.[18,20] Silicalite-1 macrostructures of spherical shape and of high crystallinity can be produced by a two-step synthesis procedure. A primary HT at lower temperature is necessary to form an efficient number of structural species typical of the medium-range order in silicalite-1, while a secondary HT at higher temperature is required to develop these incipient species into defect-poor crystal lattice. Prolonged HT at lower temperature induces a competitive formation of undesired metastable phases and leads to insufficient amount of MFI structure, whereas a direct HT at higher temperature results in high degree of extended defects in the MFI structure and thus hampers the formation of crystallites of good quality. Zeolite Beta spheres can be synthesized without preliminary HT, because the atomic arrangement in the amorphous silica used as a silica source in the synthesis solution resembles the medium-range order in zeolite Beta.

IR spectroscopy is also widely applied to differentiate between the atomic-level ordering in crystalline zeolites and amorphous substance. A strong, prominent IR band near 550 cm^{-1} is typical of zeolite frameworks containing five-membered TO_4-rings. The structure of novel class of Ti-bearing microporous/mesoporous composites prepared by subsequent addition to the same synthesis mixture of the corresponding organic templates for the mesoporous and microporous phase was characterized by IR spectroscopy.[10] The appearance of a strong IR peak neat 550 cm^{-1} as well as two weak peaks near 590 and 630 cm^{-1} unambiguously revealed the formation of nanocrystalline TS-1 particles dispersed into mesoporous Ti-MCM-48 matrix. The existence of an IR band at 550 cm^{-1} also evidenced the formation of zeolitic secondary building units containing five-membered rings in a mesoporous aluminosilicate framework and, in such a manner, confirmed the occurrence of a hierarchical porous system in a highly stable molecular sieve (MMS-H) having a structure analogous to MCM-48, but containing zeolite building units.[13]

3.4. Active centres

While structural features of the framework (precursor atomic arrangements, topological defects, crystallinity) are better revealed by Raman scattering, IR spectroscopy is most useful to study the hydration level and therefore acidic centres in porous materials. Especially beneficial for obtaining detailed information about internal and external surface sites in nanocrystalline zeolites is to combine FTIR and solid-state NMR spectroscopy because the two spectroscopic techniques independently probe the adsorbed hydrous species and $Si(OSi)_n(OH)_{4-n}$ or $Si(OSi)_n(OAl)_{4-n}$ units in the framework.[5] However, FTIR spectroscopy alone can unambiguously identify various types of surface hydroxyl groups and thus to quantify the number of the corresponding active sites.[34] In addition, using probe molecules as H_2, N_2, CO, C_2H_2, CH_3CN, CH_3OH, etc., one can estimate the acid strength of the active acidic centres.[34] Commonly, the O–H bond stretching mode of the surface OH groups is used to analyze the acid sites. Two sharp IR peaks are observed in the range 3550–3800 cm^{-1}: near 3745 cm^{-1} which arises from terminal silanol group and near 3615 cm^{-1} generated by Si(OH)Al Brønsted acid sites. A broad band centred at ca. 3250 cm^{-1} can appear in the IR spectra; the band originates from a second group of Brønsted sites for which the hydrogen atom interact with one neighbouring oxygen atom, thus forming H-bond. Probe basic molecules also form hydrogen bonding with the Brønsted-site hydrogen atom, which results in a lower-wavenumber of the peak at 3615 cm^{-1} coming from unperturbed Brønsted sites. The peak shift depends on the proton affinity of the base molecules. For example, the negative peak shift increases on passing from H_2 to N_2 and to CO adducts. The full-width-at-half-maximum also increases on passing from unperturbed Brønsted groups to H_2, N_2 and to CO. The lower wavenumber and the larger width are common results from the decrease of the force constant induced by the polarization of the O–H bond and by the mode coupling. The silanol peak is also disturbed by the presence of base molecules, but the effect is much wearer as compared to that for the Brønsted-site peak. The strength of Lewis sites also can be estimated by the perturbation of the corresponding IR peaks from the interaction with probe molecules.[34] IR spectroscopy is very helpful to study photocatalytic active centres in mixed octahedron–tetrahedron titanosilicates.[33] The IR peaks of silanol and titanol groups not involved in hydrogen bonding appear near 3740 and 3705 cm^{-1}, respectively, whereas a weak feature near 3490 cm^{-1} is attributed to titanol groups located at internal defect sites where the –O–Ti–O–Ti– changes are interrupted. It is worth noting that hydrous surface species can also be detected by Raman spectroscopy; however, the Raman signals are much weaker than the corresponding IR signals and partial dehydration due to laser-induced sample overheating may occur during the experiment.

3.5. Guest–host interactions

Raman scattering is frequently used to probe the guest–host interactions in photoactive porous nanomaterials and/or systems with catalytic applications. A combined Diffuse Reflection UV-Visible Absorption and Raman scattering study on naphthalene–ZSM-5 systems revealed that naphthalene is sorbed as intact molecule in non-Brønsted acidic M_nZSM-5 ($M = Li^+$, Na^+, K^+, Rb^+, Cs^+) and close proximity between M^+ and the occluded naphthalene molecules.[36] Strongest guest–host interactions are observed for $Na_{0.66}$ZSM-5 due to the combined effect of confinement and electrostatic field.

The manner of incorporation of 2-(2′-hydroxyphenyl)benzothiazole (HBT) in nanosized FAU-type particles was analyzed by Raman, IR and ^{13}C solid-state NMR spectroscopies.[12] The occlusion of HBT molecules into the nanosized zeolite leads to changes in the local structure of the FAU skeleton, partially destroying the sodalite-type cages and thus affecting the atomic surroundings of the TMA ions. At high concentration of HBT a large fraction of the sodalite cages are destroyed and the HBT and TMA molecules are located in the subsequently formed cavities. For high degree of dye-molecule loading the mechanism of inclusion appears to involve complex interactions between TMA and HBT molecules that occur during the templating synthesis of FAU nanocrystals.

The adsorption of *n*-pentane in several representative zeolites was also explored by Raman scattering in combination with thermogravitric analysis.[37] The framework topology determines the conformation of the sorbed *n*-pentane molecules. In small-pore zeolites conformers that fit best into the channel are preferential, whereas in large-pore zeolites various conformers exist, indicated that the framework imposes only slightly on the conformational equilibrium. The dynamics of the guest molecule is affected remarkable by the existence of extra-framework charge-balancing cations and the cation-guest interactions contribute to the static disorder of the sorbed *n*-pentane molecules.

Resonance Raman spectroscopy is also sufficient to detect and analyze low-concentration guest molecules occluded in the structure of porous nanomaterials. Resonance enhancement of the Raman intensity when UV excitations is used made possible to detect carbenium ions (1,3-dimethylcyclopentenyl) of low concentration adsorbed in zeolite H-MFI.[38] The chemical nature of impurity iron species and photoactive sites in commercially available HZSM-5 zeolites of different Si/Al ratios was studied by Raman spectroscopy using two different excitation wavelengths in the UV range.[39] Resonance Raman conditions are achieved at 244 nm due to $O^{2-}-Fe^{3+} \rightarrow O^{2}-Fe^{2+}$ ligand to metal charge-transfer transition. As a result, the band near 1000–1280 cm^{-1} attributed to symmetrical stretching of $[Fe(O–Si)_4]$ units is enhanced, similarly to other zeolites of MFI framework topology with inserted transition metal elements.[40] The structural information obtained by resonance Raman spectroscopy together with the results from other analytical methods revealed that photocatalytically active sites in MFI zeolites are the isolated tetrahedrally coordinated heteroatoms rather than metal-oxide and hydroxide clusters dispersed on the pore surface. Several excitation wavelengths were used to analyze the energy state of *trans*-Stilbene (*t*-St) incorporated as an intact molecule the channels of non-acidic aluminium rich $Na_{6.6}ZSM-5$ zeolite.[41] Raman scattering monitored unusually slow decay of photoinduced *t*-St^{+} and pointed out unique hole transfer from the internal surface of the zeolite channel.

3.6. Thin films

Surface-sensitive FTIR spectroscopy is very helpful to detect the formation of thin zeolite-type films, and to study kinetic processes, structural state of the framework and molecules sobbed in the film pores. MFI-type films grown on gold surfaces seeded with colloidal crystals were investigated by RAIR to follow up the structural development of the seeded nanocrystals into films defects and to determine the optimal synthesis conditions for preparation of defect-poor silicalite-1 films.[43,44] The spectral features observed in the range 1000–1300 cm^{-1} revealed that the initial enlargement of the seeds accompanied by a gradual change in the topology of the five-membered rings, leading to formation of dislocations in the silicalite-1 structure upon further HT and, hence, inducing void spaces and incipient cracks Using 60 nm-sized colloidal crystals the seeding method is capable to produce defect-poor silicalite-1 films with thickness ranging from 100 to 300 nm, which are predominantly oriented with the *a*-axis perpendicular to the surface.

Zeolite films grown directly on the ATR crystal, used as the highly refractive unit in ATR-IR experiments, are very appropriate model systems to investigate *in situ* adsorption processes and catalytic reactions that take place within the molecular sieve structure.[45] The efficiency of this novel method was demonstrated by investigating silicalite-1 films grown on Si ATR crystal and detecting a low amount of organic molecules in a gas flaw.[45] Such sensors of zeolite-coated ATR crystals have a great potential for industrial application.

ATR-IR spectroscopy evidenced the formation of homogeneous, ultra-thin silicalite 1 and zeolite Beta films prepared on Si wafers by the Langmuir–Blodgett method.[47–49] The ATR-IR studies on the initial zeolite nanoparticles stabilized in different alcohol suspensions revealed that methanol modifies the surface of the particles, forming a coating layer of $-OCH_3$ groups. Thus the methanol modification increases the hydrophobicity of the particles and facilitates the synthesis of highly ordered zeolite films.

In some cases, when the support has poor absorption in the spectral range of interest, thin zeolite films can be studied via FTIR spectroscopy in transmittance mode. Using this technique, mesoporous pure-silica-zeolite films prepared by spin-coating the precursor on a Si wafer were characterized.[63] The FTIR transmittance method is not so informative for framework structural features of zeolite films and less reliable for phase identification as compared to RAIR and ATR-IR spectroscopies, but is very helpful in analyzing surface chemical bonding and adsorbed hydrous species.[63]

FTIR spectroscopy with a microscope operating in a reflectance mode can also gain information about hydrous and organic molecular species incorporated into the zeolite framework of thin films deposited on various supports. Besides, the utilization of a microscope allows for probing different spatial regions in films with micron-scale in inhomogeneities. FTIR micro-spectroscopy was applied to MFI thin films on a Si wafer in which a spatially-controlled removal of the organic structure-directing template was achieved via UV/ozone illumination with a physical mask.[64] Infrared spectra collected from UV-exposed and non-exposed spatial regions unambiguously confirmed the ability of the method for a designed removal of the organic template.

Utilization of a microscope is also possible for recording RAIR and ATR-IR spectra. RAIR micro-spectroscopy was successfully applied to characterize layer-by-layer zeolite/protein assembly films grown on pyrolytic graphite electrodes.[65] The shape and the position of the two infrared amide bands, at 1700–1600 cm^{-1} caused by C=O stretching and at 1600–1500 cm^{-1} resulting from a combination of H–N–H in-plane banding and C–N stretching of the peptide group, provides detailed information about protein conformation and thus the two IR bands were used to study the structural feature of zeolite/protein assembly films.

3.7. Zeolite-based nanocomposites

Recently, a novel concept for zeolite-based photovoltaic cell was developed. Zeolite particles serve as a host matrix to encapsulate transition metal-oxide sub-nanoparticles.[50–52] Zeolites provide well defined and well-ordered porous system to confine the metal–oxygen clusters. Thus the size and configuration of the encapsulated clusters can be tailored using different microporous materials. In addition to their potential application in optoelectronics, such nanocomposites may be used as catalysts. Due to the high Raman response of transition metal–oxygen bonds, the Raman technique is an excellent tool for probing the structure of the confined sub-nanoparticles and their interaction with the host matrix. In such a manner, this technique can be used to find the optimum synthesis conditions for obtaining nanocomposites with desired properties. The utilization of UV lasers to excite the Raman scattering in these systems is beneficial since it is not only enhances the Raman intensity but also successfully avoids the interference of fluorescence. Zeolite matrices with encapsulated ZnS and Cd_xZn_yS nanoclusters also have potential applications as window materials in solar cells and in photocatalytic hydrogen production using visible light. The structure of such nanocomposites can be successfully characterized using IR and Raman spectroscopy.[53] Raman spectroscopy turned to be very sufficient also for structural characterization of iodine and single-wall carbon nanotubes formed in the channels of AlPO-5 zeolite single-crystal.[54–56]

4. Perspectives

4.1. Synchrotron sources

The benefits of Raman and IR spectroscopy for studying nanoscale porous materials can further be explored by applying synchrotron radiation. The high brilliance of synchrotron IR radiation is essential for applying RAIR and ATR spectroscopies to characterize ultra-thin zeolite films, since weak IR peaks indicative for the formation of a zeolite crystalline phase are observable. In addition, the use of synchrotron-based IR radiation enhances the spatial resolution when a microscope is used and thus mapping of structural inhomogeneities in porous films and membranes could be performed. The recently initiated projects and achievements in developing VUV free-electron laser sources are encouraging for performing time-resolved spectroscopy on zeolite-based functional materials with applications in photosensitive devices.

4.2. High-pressure studies

In situ phase-transition monitoring by IR and Raman spectroscopy has a great potential for studying the structure of nanoscale porous materials. In particular high-pressure experiments can reveal the energetically preferable structural species in the zeolite framework as well as the dependence of compressibility on framework and extra-framework cations. Also, high-pressure experiments allow us to study the effect of different specimens encapsulated into the pore system on the stability and mechanical properties of zeolite-based nanocomposites. Although the great advantage of Raman spectroscopy to study phase transitions and amorphization processes, pressure-induced structural transformations in zeolites have been mainly investigated by XRD so far and high-pressure spectroscopic studies are only sporadic.[66]

REFERENCES

[1] Fyfe, C. A., Dalton, R. J., Schneider, C., Scheffler, F., *J. Phys. Chem. C* **2008,** *112,* 80–88.
[2] Cheng, C.-H., Shantzm, D. F., *J. Phys. Chem. B* **2006,** *110,* 80–88.
[3] Kirschhock, C. E. A., Liang, D., Aerts, A., Aerts, C. A., Kremer, S. P. B., Jacobs, P. A., Tendeloo, G. V., Martens, J. A., *Angew. Chem. Int. Ed.* **2004,** *43,* 4562–4564.
[4] Fernandez, A. B., Lezcano-Gonzalez, I., Boronat, N., Blasco, T., Corma, A., *J. Catal.* **2007,** *249,* 116–119.
[5] Larsen, S. C., *J. Phys. Chem. C* **2007,** *111,* 18464–18474.
[6] Ferreira, P., Nunes, C. D., Pires, J., Carvalho, A. P., Brandao, P., Rocha, J., *J. Mater. Sci. Forum.* **2006,** *514–516,* 470–474.
[7] Egger, C. C., Anderson, M. W., Tiddy, G. J. T., Casci, J. L., *Phys. Chem. Chem. Phys.* **2005,** *7,* 1845–1855.
[8] Labouriau, A., Higley, T. J., Earl, W. L., *J. Phys. Chem. B* **1998,** *102,* 2897–2904.
[9] Solberg, S. M., Kumar, D., Dandry, C. C., *J. Phys. Chem. B* **2005,** *109,* 24331–24337.
[10] Balmer, M. L., Bunker, B. C., Wang, L. Q., Peden, C. H. F., Su, Y., *J. Phys. Chem. B* **1997,** *101,* 9170–9179.
[11] Antonijevic, S., Ashbrook, S. A., Biedasek, S., Wlaton, R. I., Wimperis, S., Yang, H., *J. Am. Chem. Soc.* **2006,** *128,* 8054–8062.
[12] Mintova, S., De Waele, V., Hölzl, M., Mihailova, B., Schmidhammer, U., Riedle, E., Bein, T., *J. Phys. Chem. A* **2004,** *108,* 10640–10648.
[13] Sakthivel, A., Huang, S.-J., Chen, W.-H., Lan, Z.-H., Chen, K. H., Lin, H.-P., Mou, C.-Y., Lui, S.-B., *Adv. Funct. Mater.* **2005,** *15,* 253–258.
[14] Murdoch, M., Yeates, R., Howe, R., *Micropor. Mesopor. Mater.* **2007,** *101,* 184–190.
[15] Beale, A. M., Sankar, G., Catlow, C. R. A., Anderson, P. A., Green, T. L., *Phys. Chem. Chem. Phys.* **2005,** *7,* 1856–1860.
[16] Decyk, P., *Catal. Today* **2006,** *114,* 142–153.
[17] Dzwigaj, S., Stievano, L., Wagner, F. E., Che, M., *J. Phys. Chem. Solids* **2007,** *68,* 1885–1891.

[18] Tosheva, L., Mihailova, B., Valtchev, V., Sterte, J., *Micropor. Mesopor. Mater.* **2000,** *39,* 91–101.
[19] Li, Q., Mihailova, B., Creaser, D., Sterte, J., *Micropor. Mesopor. Mater.* **2001,** *43,* 51–59.
[20] Tosheva, L., Mihailova, B., Valtchev, V., Sterte, J., *Micropor. Mesopor. Mater.* **2001,** *48,* 31–37.
[21] Kecht, J., Mihailova, B., Karaghiosoff, K., Mintova, S., Bein, T., *Langmuir* **2004,** *20,* 5271–5276.
[22] Mintova, S., Hölzl, M., Valtchev, V., Mihailova, B., Bouizi, Y., Bein, T., *Mater. Chem.* **2004,** *16,* 5452–5459.
[23] Mihailova, B., Wagner, M., Mintova, S., Bein, T., *Stud. Surf. Sci. Catal.* **2004,** *154,* 163–170.
[24] Mihailova, B., Valtchev, V., Mintova, S., Petkov, N., Faust, A.-C., Bein, T., *Phys. Chem. Chem. Phys.* **2005,** *7,* 2756–2763.
[25] Mihailova, B., Mintova, S., Karaghiosoff, K., Metzger, T., Bein, T., *J. Phys. Chem. B* **2005,** *109,* 17060–17065.
[26] Majano, G., Mintova, S., Ovsitser, O., Mihailova, B., Bein, T., *Micropor. Mesopor. Mater.* **2005,** *80,* 227–235.
[27] O'Brien, M. G., Beale, A. M., Catlow, R. A., Weckhuysen, B. M., *J. Am. Chem. Soc.* **2006,** *128,* 11744–11745.
[28] Llabres, i., Xamena, F. X., Damin, A., Bordiga, S., Zecchina, A., *Chem. Commun.* **2003,** 1514–1515.
[29] Groen, J. C., Hamminga, G. M., Moulijn, J. A., Perez-Ramirez, J., *Phys. Chem. Chem. Phys.* **2007,** *9,* 4822–4830.
[30] Tompsett, G. A., Panzarella, B., Conner, W. C., Yngvesson, K. S., Lu, F., Suib, S. L., Jones, K. W., Bennet, S., *Rev. Sci. Instrum.* **2006,** *77,* 124101/1–124101/10.
[31] Inagaki, S., Nakatsuyama, K., Saka, Y., Kikuchi, E., Kohara, S., Matsukata, M., *J. Phys. Chem. B* **2007,** *111,* 10285–10293.
[32] Halasz, I., Agarwal, M., Li, R., Miller, N., *Catal. Lett.* **2007,** *117,* 34–42.
[33] Usseglio, S., Calza, P., Damin, A., Minero, C., Bordiga, S., Lamberti, C., Pelizzetti, E., Zecchina, A., *Chem. Mater.* **2006,** *18,* 3412–3424.
[34] Zecchina, A., Spoto, G., Bordiga, S., *Phys. Chem. Chem. Phys.* **2005,** *7,* 1627–1642.
[35] Crupi, V., Longo, F., Majolino, D., Venuti, V., *J. Phys. Condens. Matter* **2006,** *18,* 3563–3580.
[36] Moissette, A., Marquis, S., Cornu, D., Vezin, H., Bremard, C., *J. Am. Chem. Soc.* **2005,** *127,* 15417–15428.
[37] Wang, H., Turner, E. A., Huang, Y., *J. Phys. Chem. B* **2006,** *110,* 8240–8249.
[38] Chua, Y. T., Stair, P. C., Nicholas, J. B., Song, W., Haw, J. F., *J. Am. Chem. Soc.* **2003,** *125,* 866–867.
[39] Yan, G., Long, J., Wang, X., Li, Z., Wang, X., Xu, Y., Fu, X., *J. Phys. Chem. C* **2007,** *111,* 5192–5202.
[40] Ricchiardi, G., Damin, A., Bordiga, S., Lamberti, C., Spanò, G., Rivetti, F., Zecchina, A., *J. Am. Chem. Soc* **2001,** *123,* 11409–11419.
[41] Moissette, A., Bremard, C., Hureau, M., Vezin, H., *J. Phys. Chem. C* **2007,** *111,* 2310–2317.
[42] Lede, B., Demortier, A., Gobeltz-Hautecoeur, N., Lelieur, J.-P., Picquenard, E., Duhayon, C., *J. Raman Spectrosc.* **2007,** *38,* 1461–1468.
[43] Mihailova, B., Engström, V., Hedlund, J., Holmgren, A., Sterte, J., *Micropor. Mesopor. Mater.* **1999,** *32,* 297–304.
[44] Engström, V., Mihailova, B., Hedlund, J., Holmgren, A., Sterte, J., *Micropor. Mesopor. Mater.* **2000,** *38,* 51–60.
[45] Wang, Z., Larsson, M. L., Grahn, M., Holmgren, A., Hedlund, J., *Chem. Commun.* **2004,** 2888–2889.
[46] Li, G., Zhou, H., Zhu, G., Liu, J., Yang, W., *J. Membr. Sci.* **2007,** *297,* 10–15.
[47] Tosheva, L., Valtchev, V. P., Mihailova, B., Doyle, A. M., *J. Phys. Chem. C* **2007,** *111,* 12052–12057.
[48] Tosheva, L., Wee, L. H., Wang, Z., Mihailova, B., Vasilev, C., Doyle, A. M., *Stud. Surf. Sci. Catal.* **2008,** in press.
[49] Wang, Z., Wee, L. H., Mihailova, B., Edler, K. J., Doyle, A. M., *Chem. Mater.* **2007,** *9,* 5806–5808.
[50] Chen, J., Feng, Z., Ying, P., Li, C., *J. Phys. Chem. B* **2004,** *108,* 12669–12676.

[51] Alvaro, M., Carbonell, E., Atienzar, P., Garcia, H., *Chem. Phys. Chem.* **2006,** *7,* 1996–2002.
[52] Lacheen, H. S., Cordeiro, P. J., Iglesia, E., *Chem. Eur. J.* **2007,** *13,* 3048–3057.
[53] Raymond, O., Villavicencio, H., Flores, E., Petranovski, V., Siqueiros, J. M., *J. Phys. Chem. C* **2007,** *111,* 10260–10266.
[54] Ye, J. T., Tang, Z. K., Sui, G. G., *Appl. Phys. Lett.* **2006,** *88,* 073114/1–073114/3.
[55] Ye, J. T., Tang, Z. K., *Phys. Rev. B* **2005,** *72,* 045414/1–045414/4.
[56] Ye, J. T., Zhai, J. P., Tang, Z. K., *J. Phys. Condens. Matter* **2007,** *19,* 44503/1–44503/18.
[57] Paredes, J. I., Martinez-Alonso, A., Yamazaki, T., Matsuoka, K., Tascon, J. M. D., Kyotani, T., *Langmuir* **2005,** *21,* 8817–8823.
[58] Hashimoto, S., Uehara, K., Sogawa, K., Takada, M., Fukumura, H., *Chem. Phys. Chem.* **2006,** *8,* 1451–1458.
[59] Shi, J., Chen, J., Feng, Z., Chen, T., Wang, X., Ying, P., Li, C., *J. Phys. Chem. B* **2006,** *110,* 25612–25618.
[60] Bhargava, S. K., Akolekar, D. B., Foran, G., *Radiat. Phys. Chem.* **2006,** *75,* 1909–1912.
[61] Beale, A. M., van der Eerden, A. D. M. J., Grandjean, D., Petukhov, A. V., Smith, A. D., Weckhuysen, B. M., *Chem. Commun.* **2006,** 4410–4412.
[62] Oliverira, L. C. A., Pekowicz, D. I., Smaniotto, A., Pergher, S. B. C., *Water Res.* **2004,** *38,* 3699–3704.
[63] Seo, T., Yoshinoro, T., Cho, Y., Hata, N., Kikkawa, T., *Jpn. J. Appl. Phys.* **2007,** *46,* 5742–5746.
[64] Li, Q., Amweg, M. L., Yee, C. K., Navrotsky, A., Parich, A. N., *Micropor. Mesopor. Mater.* **2005,** *87,* 45–51.
[65] Xie, Y., Liu, H., Hu, N., *Biochemistry* **2007,** *70,* 311–319.
[66] Ovsyuk, N., Goryainov, S., *Appl. Phys. Lett.* **2006,** *89,* 134103/1–134103/3.

CHAPTER 8

Computational Modelling of Nanoporous Materials

Georgi N. Vayssilov, Hristiyan A. Aleksandrov,
Galina P. Petrova, *and* Petko St. Petkov

Contents

Abstract

This chapter reviews the contemporary computational approaches based on quantum chemical or hybrid methods, which are used for modelling of nanoporous materials such as zeolites and other molecular sieves. It also includes various examples for the applications of these methods that show their ability to help in understanding of the structures and spectral features, as well as of the mechanisms and origins of the chemical and physical processes at molecular and atomic levels.

Keywords: Active sites, Acidity, Adsorption, Basicity, Carbon monoxide, Computational chemistry, Density functional theory, Deprotonation energy, Hybrid methods, Iridium, Iron, Metal cations, Metal clusters, Methanol, Proton affinity, QM/MM methods, Rhodium, Sodium, Transition metals, Zeolites, Zinc

Ordered Porous Solids
DOI: 10.1016/B978-0-444-53189-6.00008-1

Glossary

Hybrid QM/MM methods	Computational method combining higher-level quantum chemical method for the active site and molecular mechanical method for surrounding
T atoms	The atoms located in tetrahedral framework positions as Si or Al
DFT	Density functional theory
ONIOM	hybrid QM/MM method named by Morokuma and co-workers as "our own N-layered integrated molecular orbital + molecular mechanics" method
SCREEP	"Surface Representation of the Electrostatic Embedding Potential" method developed by E.V. Stefanovich and T.N. Truong
XPS	X-ray photoelectron spectroscopy

1. Introduction

The substantial extension of the application areas of the molecular sieves and the development of new types of nanoporous materials promoted extensive investigation of the structure and properties of different micro- and mesoporous materials from various aspects. Notable contribution in these research efforts has been provided by modern methods for computational modelling, in particular in combination with related experimental information. The reliable application of the computational approaches to real problems of practical interest became feasible in the last two decades due to the rapid progress in computers and the fast development of the efficient and reliable computational techniques and procedures. Nowadays, such computational modelling provides not only insights towards the understanding of the nature, structure, and behavior of existing systems, but also allows simulation of the new imaginary species, structures, and processes that play a key role in the design and tailoring of new materials with predefined properties. In this chapter, we summarize the computational approaches, based on quantum chemical or hybrid methods, which are used for modelling of nanoporous materials and presented some examples for the applications of these methods showing their ability to help in understanding of the structures and spectral features, as well as of the mechanisms and origins of the chemical and physical processes at molecular and atomic levels. Because of the limited space, we do not provide complete computational details for the described studies, but mention only those characteristics of the calculations that are important to evaluate the reported results and their reliability with respect to comparison with experimental results.

2. Quantum Chemical and Hybrid Methods for Modelling of Nanoporous Materials

2.1. Computational methods

The computational methods for modelling of micro- and mesoporous materials and various processes on them are divided in two groups according to the level at which the interactions in the system are described – molecular mechanical and quantum chemical methods. The combination between them, the so-called hybrid QM/MM methods, is also applied in various cases, but it is usually considered together with the higher level method – quantum chemical. The selection of the method for description of a certain problem is based on (i) the ability of the method to describe well the dominant types of the interactions and process and (ii) the size of the system. Typically, molecular mechanical methods are applied for problems that do not include formation or breaking of a chemical bond (including also coordination and hydrogen bonds) such as the structure of the material, its solid state properties and diffusion of guest species. These methods allow modelling of large systems and longer simulation times in case of molecular dynamics. On the contrary, the quantum chemical methods are used for description of problems and processes connected with chemical interactions and reactions, redistribution of electron density, simulation of spectral features connected with the electronic structure of the material, its active centres or guest species. Since the quantum chemical approaches are orders of magnitude more demanding computationally, they are applied for smaller systems or systems with smaller unit cells compared to molecular mechanical methods. The hybrid QM/MM methods include modelling of the important (with respect to the investigated chemical or spectral features) part of the system with quantum chemical method and accounting for the influence of the rest of the system (representing large part of the structure of the nanoporous material) by molecular mechanical approach.

The quantum chemical methods themselves can be performed at different levels– semi-empirical, *ab initio* Hartree-Fock (HF), various post-HF methods, or density functional theory (DFT)-based methods. Again, the selection of the method depends on the investigated problem and the size of the modelled system, and as a rule, for certain computational resources and time, the accuracy of the method decreases when the size of the modelled system (or of the unit cell in periodic approaches) increases. The balance between accuracy and performance of these groups of methods, their advantages and disadvantages, and applicability to certain systems and problems are extensively studied in the specialized literature[1,2] and will not be considered here.

While in 1980s and 1990s the semi-empirical methods helped to shed light on various questions related to local structure of the zeolites and their properties, in the last years these methods are used only as a low-level method included in hybrid calculations instead of molecular mechanical methods. The *ab initio* HF methods are also rarely applied alone, they are often used for structural optimization or preoptimization followed by refinement of the electronic structure of the system by a method that takes into account the electron correlation as post-HF methods. Most of the model studies of micro- and mesoporous materials nowadays are performed by DFT methods since they intrinsically account for the electron correlation and are computationally efficient. In principle DFT methods, however, do not describe well the dispersion interactions, which are often important for zeolite systems, in particular for molecular sieves with narrow pores or cavities.

2.2. Models

The selection of the model depends on the structure of the system and the properties to be investigated, as well as on the computational method. For crystalline microporous materials, three types of models are used – isolated cluster models, hybrid embedded cluster models, and periodic models (the features of the models are described below). Both latter types of models consider whole zeolite pores or cavities, while with isolated clusters one can include either one side of the pore or model the whole pore. For simulation of mesoporous materials the periodic models are not relevant since the material, building the pore walls is amorphous. Therefore, only isolated or embedded clusters can be applied as models. In addition, the inclusion of the whole channel or cavity of the mesoporous material in the model is not feasible due to the rather large size of these channels. By this reason, the modelling of mesoporous materials is essentially the same as the modelling of extended open surface composed by the material forming the pore walls as silica, aluminosilicates, etc.

2.2.1. Isolated cluster models

In these models only an isolated fragment from the zeolite lattice is considered (Fig. 8.1). Since the fragment is cut from the real zeolite framework by (formally) cleaving Si–O or Al–O bonds, 'dangling' covalent bonds appear at the border atoms of the cluster. Usually, these bonds are saturated by artificial H atoms, which are situated in direction towards the neighbouring atom in the real zeolite lattice. In different studies, the additional H atoms replace real atoms, either Si or O (Fig. 8.1). The first case is more realistic since the electronegativity values of Si and H are closer compared to the electronegativities of O and H. Since the isolated cluster models are intended to simulate a real zeolite framework, their full unrestricted geometry optimization would lead to unrealistic structures. In order to keep the models close to the real zeolite structure, some geometry restrictions are imposed to the clusters as constraining the positions of terminating H atoms and all or part of the T atoms (T atom denotes the atoms located in tetrahedral framework positions

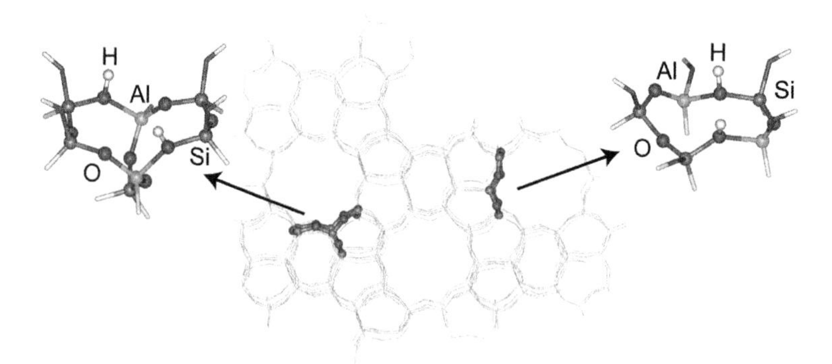

Figure 8.1 Example for selection of isolated cluster model as a fragment from the zeolite framework. (See color insert.)

as Si or Al) during the geometry optimization. Another drawback of the isolated cluster approach is that it neglects the influence of the surrounding crystal lattice on the reaction centre, namely the steric hindrance from the pore walls and the electrostatic effect of the zeolite lattice.

However, despite of the above-mentioned problems, when the investigated interactions or properties are localized at a certain part of the framework, then in various cases with isolated cluster models one can obtain reliable information for the geometry, electronic structure, energetic and spectral characteristics of the corresponding species. Important points in the success of such calculations are the selection of relevant zeolite fragments as cluster models (their size and location) and application of appropriate constraints.[3–8] For example, the interaction of charge-compensating protons with zeolite oxygen centre and forming bridging OH groups can be described well with relatively small clusters, since H^+ is bound via covalent bond to one O centre from $[AlO_4]^-$ tetrahedron. However, the correct modelling of extra-framework metal cations in zeolites requires larger cluster models (typically one zeolite ring characteristic for the concrete zeolite structure) since metal cations interact with more O atoms from one or more $[AlO_4]^-$ or $[SiO_4]$ tetrahedra. When adsorption of large molecules, metal complexes, or metal clusters in molecular sieves is modelled, the selected zeolite fragment should originate from the part of the framework that is exposed to the channels where it is accessible for the guest species.

Some of the studies based on large isolated cluster models employ mixed basis sets to reduce the computational cost. As a simplified version of the hybrid approaches described below, in this scheme the active site is described with larger basis sets, whereas the rest atoms of the cluster are described with smaller ones.

2.2.2. Hybrid QM/MM models

The idea for the application of hybrid QM/MM methods for modelling of zeolites[9–21] is to overcome the main disadvantages of isolated cluster models without substantial increase of the computational efforts. In this approach, the active centre (including catalytic site, guest particles, defect of the framework, or other species of interest) represented again as a cluster model is embedded in the rest of the zeolite framework described at lower level of theory. With this approach one aims at proper accounting for the (i) steric hindrance due to limited space in the pores or cavities, (ii) structural/mechanical constraints and (iii) electrostatic field of the framework. In QM/MM methods, the whole system (S) is divided on two parts: active centre denoted as inner part (I) or QM cluster and described at higher level method, and the rest of zeolite lattice denoted as outer part (O) or MM part, which is treated with molecular mechanical method (Fig. 8.2). Some implementations of the embedded cluster approach allow utilizing of different computational methods such as higher and lower level methods, for example, outer part (O) can be described at semi-empirical level or DFT. Such examples are the hybrid MP2/DFT calculations of Tuma and Sauer[22] and the three-layer ONIOM approach of Sillar and Burk[23] using the following combination of methods: B3LYP/6–311+G**:B3LYP/3–21G*:MNDO (the highest level is on the left hand side).

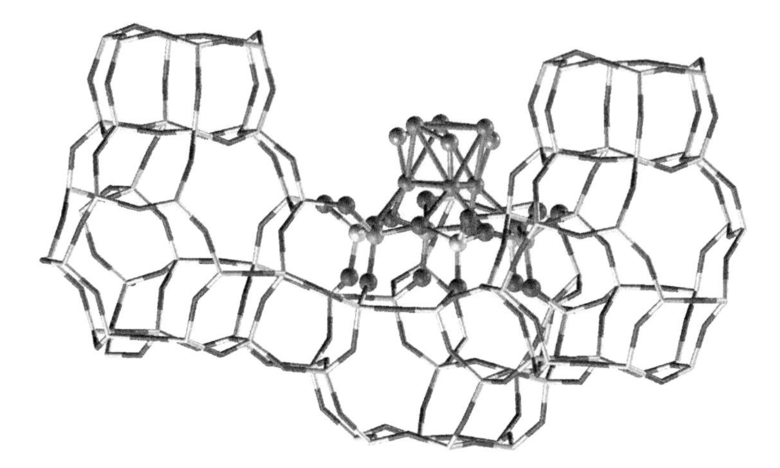

Figure 8.2 Schematic representation of QM/MM hybrid model: atoms included in the inner part (I) are shown as balls, while the outer part (O) is shown only as a framework. (See color insert.)

Different hybrid schemes have several specific features: (i) calculation of the total energy of the system; (ii) termination of the dandling bonds of the QM cluster; (iii) treatment of the electrostatic interaction between the subsystems; (iv) self-consistence of the computational approach. In the next section, we shortly comment on some of these features for schemes relevant to modelling of zeolites or mesoporous materials.

2.2.2.1. Calculation of the total energy of the system According to the calculation of total energy of the whole system the hybrid methods are divided into two groups: subtraction and summation schemes, defined according to the following expressions:

$$E^S = E^S_{\mathrm{MM}} + E^I_{\mathrm{QM}} - E^I_{\mathrm{MM}} \quad \text{subtraction scheme}$$

$$E^S = E^I_{\mathrm{QM}} + E^O_{\mathrm{MM}} - E^{I-O}_{\mathrm{int}} \quad \text{summation scheme}$$

In these expressions, the upper index to the energy represents the part of the system for which the energy is calculated, while the lower one shows the method used for this calculation. In the subtraction scheme, the energy of the whole system is obtained by replacing the energy of the inner part of the system calculated at lower computational level with the energy of this part calculated at higher QM level. This construction of the hybrid method allows separate calculation for the three different energies (systems) and combining the results from these separate calculations to obtain the final QM/MM result. By this reason, the subtraction scheme is very flexible with respect to the choice of low- and high-level methods including also hybrid methods of QM/QM type. It can be shown easily that in this scheme the interaction between the two subsystems is accounted for at the lower-level method.

This feature causes a substantial disadvantage for description of the electrostatic interactions in this type of schemes since the polarization of the inner part is accounted for only at MM level, that is, the electronic wave function of the QM cluster is not influenced by the electrostatic field of the zeolite framework. On the contrary, this feature is an advantage when one models system in which the dispersion interactions of the adsorbed species with zeolite wall are important since these interactions are described well with properly parameterized force field, while HF and the standard DFT methods fail to give correct results. The construction of these schemes is proposed by Morokuma and co-workers[24–26] in their classical ONIOM method, which is applied also for zeolite-related problems.[23,27–31] Similar scheme, denoted as QM-pot, was adapted especially for zeolite systems in 1990s by Sauer and co-workers.[9–11]

In the summation scheme, one can select which types of interactions between the two subsystems to be accounted for in the calculations. In particular, with such scheme the polarization of the electron density of the QM cluster can be included. However, implementation of these schemes is more complicated since they depend on the computational methods applied for inner and outer subsystems and on the type of termination of the QM cluster. An advantage of this scheme is that it does not require calculation of the inner part (QM cluster) by the MM method; this is particularly important in the case of simulation of transition states and processes including bond breaking/formation. Three different hybrid QM/MM methods based on summation scheme have been developed in the last years for zeolites and similar systems: method included in ChemShell software,[32,33] the scheme developed by Sulimov *et al.*[18], and covEPE method.[19–21]

A hybrid MP2/DFT method was proposed by Tuma and Sauer[22] for theoretical simulation of problems that involve both bond rearrangements and van-der-Waals interactions. This method combines second-order Møller–Plesset perturbation theory (MP2) as high-level method for the reaction site with DFT as low-level method for the large part of the zeolite system under periodic boundary conditions. This MP2/DFT hybrid approach was applied for studying the adsorption of isobutene in zeolite H-ferrierite (FER) and formation of different adsorption complexes – π-complex, isobutoxide, tert-butoxide and tertbutyl carbenium ion. The obtained results and the comparison with data from simulations at pure DFT (PBE functional) level show that the MP2 corrections to the interaction energies are substantial and differ substantially for the different hydrocarbon species in the zeolite. In fact, when dispersion is included at the MP2 level, the heat of adsorption changes from -12 kJ/mol at PBE level to the more realistic values, from -74 ± 10 kJ/mol.

2.2.2.2. Termination of the QM cluster One of the main problems in all QM/MM schemes, used for covalent oxides and zeolites, is the construction of the border region between the two subsystems. Similarly to the isolated cluster models, in most of the proposed schemes these dangling bonds are saturated by H atoms, denoted as link atoms. In order to avoid artifacts (due to the presence of these artificial atoms) during geometry optimization, special restrictions are imposed for the length and orientation of the saturated bonds.[9–13] In order to ensure precise subtraction of the energy contribution of the link atoms, Sauer and co-workers parameterized suitable interatomic potentials used in MM calculations, to reproduce exactly the structure and the energetic changes due to the deformation of the border of the cluster, calculated at QM level.[9–11] Their QM-pot method is developed in versions based on HF and DFT.

Usually, hydrogen as a link atom is used in the subtraction schemes, it is also applied in some of the summation schemes as that included in ChemShell software. In this method, the electrostatic potential of MM region participates in electronic Hamiltonian of the cluster and the polarization of the lattice due to the changes in the electron density of QM cluster is also taken into account. However, the presence of artificial link atoms too close to some of the atoms from the lattice (Fig. 8.3a) requires special rearrangement of the charges of the atoms from MM part, located close to link atoms.[12,13] As a result, the electrostatic potential close to the border between QM and MM parts is strongly perturbed, which disturbs the electronic structure of the QM cluster and its interaction with the sorbates.

Another approach for termination of the QM border that overcomes the problems, connected with presence of artificial H atoms, is utilization of special type of border centres representing real atoms from the lattice described as reparameterized pseudopotential. In order to be used for saturation of dangling bonds of the QM cluster these centres are modified to be monovalent. In the scheme proposed by Sulimov *et al.*[18], the border centres between QM and MM parts imitate Si atoms and have a special sp orbital bound to the O atoms from the inside of the QM cluster. In covEPE scheme[19–21] the border centres (O*, Fig. 8.3b) represent O atoms. In order to avoid strong polarization of the border O centres (when they carry flexible orbital basis set) in the covEPE scheme the border region was extended by addition in it of the neighbouring Si centres from the MM part as pseudopotentials without explicit representation of valence electrons.[21]

2.2.2.3. Treatment of the electrostatic interaction Some of the QM/MM methods consider only the electrostatic effect of the lattice on the QM cluster and mechanistic interactions are not taken into account.[14–16] The electrostatic potential of the environment is included in the Hamiltonian of the QM cluster by two different strategies. The first one is addition of point charges, located in the

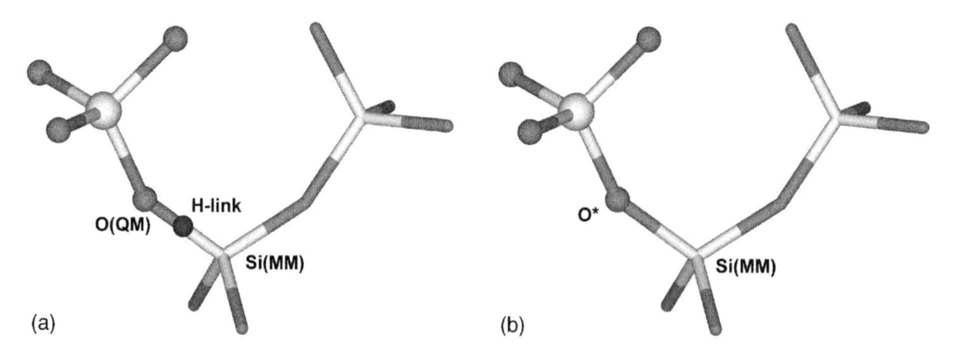

Figure 8.3 Schematic representation of QM/MM border: termination by H as link atoms (a); termination by special border centres as used in covEPE scheme (b). (See color insert.)

corresponding crystallographic positions; and the second one is addition of correction potential, which is obtained from periodic calculations on unperturbed lattice. However, both strategies require rearrangement of charges around border region, due to the presence of artificial link H atoms at the border of QM cluster.[14–16] Other disadvantage is that the structural relaxation of the cluster is not taken into account correctly, as mechanistic connection is neglected.

A computationally efficient technical approach for accounting the long-range electrostatic interaction was suggested by Stefanovich and Truong.[34] In their method, denoted as SCREEP, they suggested initial projection of the charges of ions from an infinite lattice on a sphere and to use the charge density distribution on the sphere in further calculations with embedded QM cluster.

2.2.3. Periodic models

Since zeolites and other types of molecular sieves are crystalline materials, periodic models can be used for their simulation. These models are based on the translation symmetry of the framework, that is, the unit cell translates periodically in the three crystallographic directions. The main advantage of these models is that all atoms in the zeolite systems are treated at one computational level, since the model includes the whole (ideal) zeolite crystal lattice. In this way, both the electrostatic effect and the proper flexibility of the zeolite lattice are taken into account. The periodic quantum chemical calculations can be performed with two types of representation of the electronic wave function: atomic-centred basis sets[35] or plane waves.[36] In the first case, the wave function of the system represents a linear combination of functions in the atomic basis set, so the basis set used in the periodic model could be the same as those used in molecular calculations. Plane waves are natural choice for description of systems with periodicity. Their simple mathematical expression makes the calculations of electron integrals in Schrödinger equation quite fast. On the contrary, large number of wave functions should be used in linear expansion of one-electron functions, the number of the plane waves used in the calculations can be controlled by cutoff value for the kinetic energy. The smaller this value, the smaller basis set is necessary for the calculation. Since the description of the core level electrons requires plane waves with very high kinetic energy, in order to save computational time core electrons are described by pseudopotentials.[37]

Computational time in periodic models depends not only on the number of electrons in the model but also on the volume of the unit cell, since the number of plane waves to be included in the calculations increases with the increasing volume of the unit cell for a given energy cutoff. Hence, quantum chemical calculations with basis sets of plane waves are applied routinely for zeolites with relatively small unit cells, while the computational efforts for zeolites with large unit cells increase considerably. Additional increase of the size of the unit cell and the computational efforts used in the periodic calculations arises from the irregular distribution of Al in zeolite lattice, which breaks the symmetry of the zeolite lattice. The same problem leading to reduction of the symmetry of the lattice appears also from the random position of metal ions and protons, which compensate the charge of the zeolite lattice.

Quantum chemical calculations with plane wave basis sets can be performed only with DFT with local or nonlocal functionals, while for atomic-centred basis sets software, which allows application of HF and MP2 methods, is available.

3. Selected Areas and Research Problems

3.1. Structure and isomorphic substitution of T atoms

Since optimization of the *unperturbed* zeolite model cluster or periodic structure is the initial step in the computational modelling, essentially all studies reported below contain structural information. For pure silicalite, modelled with embedded cluster method using DFT-BP86 method, the calculated Si–O bond distances vary between 163.0 and 163.8 pm, which is by about 2 pm longer than experimental Si–O distances in MFI zeolites.[38] Si–O bond lengths of zeolites, calculated with different quantum chemical methods, deviate up to 5 pm from crystallographic values, depending on the computational method and the quality of the modelling.[10,39,40]

The local structure of the zeolite framework and substitution of the Si atoms by Al, Ti, or other elements is also a subject of computational modelling. In order to be reliable, these calculations should be performed with periodic quantum chemical method or with hybrid embedded cluster approach. For example protonic forms of faujasite, chabasite (CHA) and ZSM-5 zeolite were studied by covEPE method. The obtained structural parameters, Al–O distances and O–Al–O angles,[20] differ by at most 2% from the experimental values reported by van Bokhoven *et al.*[41] and Joyner *et al.*[42] from extended X-ray absorption fine structure (EXAFS) studies of the surrounding of Al centres in zeolites. Both experiment and calculations predict that the distance Al–O_b, between an Al centre and the O atom of the OH group, is by about 20 pm longer than the other Al–O distances.

Modelling of the isomorphic substitution of Si by Ti in MFI structure[43] was also performed with the same method, which is related to the structure of TS-1 molecular sieves. The real material TS-1 is notable for its catalytic activity and selectivity in various oxidation reactions.[44] Despite extensive experimental characterization of this material, it is not clear whether titanium is randomly distributed in the framework or it preferentially occupies some of the crystallographic T atom positions. By this reason, the energy for substitution of Si by Ti in the 12 distinct T sites of MFI was estimated using embedded clusters containing 5 T atoms.[43] The obtained substitution energies suggest that incorporation of isolated Ti ions into MFI framework is energetically most favourable at the T12 position. A tentative evaluation of the Ti population of different crystallographic sites for a temperature of 448 K, typical for zeolite synthesis, yielded 71% at the T12 site, that is, if the Ti distribution in the framework was governed by thermodynamic stability of the final material, the experimental studies should observe Ti mainly at this position. This is, however, not the case. In addition, different experimental investigations obtained different results for the Ti distribution. Therefore, it was concluded that the actual distribution of Ti centres in TS-1 is not governed by the stability of the pure TS-1 framework, but by other thermodynamic or kinetic factors.[43]

Ti atoms, incorporated at tetrahedral sites of a siliceous MFI lattice, cause substantial changes of the crystal geometry, mainly because the T–O bond lengths increase from 163 pm to 181 pm. This local geometry of Ti sites agrees very well with experimental EXAFS results for Ti–O = 181 ± 1 pm.[44] Similar results have been obtained in DFT calculations of isolated clusters with different steric restriction, 178–184 pm.[45–48] The extension of the Ti–O bonds induces an elongation of the interatomic Ti–Si distances compared to Si–Si distances of the pure silica form of MFI on average by 10 pm.

3.2. Bridging hydroxyl groups

One of the specific active sites in zeolites are the bridging OH groups formed in vicinity of each Al (or other three-valence metal) centre in the protonic forms of zeolites. These centres act as Brønsted acid sites and they are one of the origins of the high catalytic activity of zeolites in various transformations of hydrocarbons. Several factors influence the acidity of the zeolite OH groups: (i) aluminum content of the zeolite framework; (ii) type of three-valence T atom connected to the bridging OH group (e.g. Al, Ga and Fe); (iii) structure of the framework; (iv) crystallographic position of the OH group; (v) presence of other charge-compensating cations in the vicinity of the OH group. The clarification of the effect of these factors on the acidity of Brønsted acid sites in zeolites in many cases is based on relevant computational model studies, some of which are described shortly below. As a measure of the acidity of the OH group, the calculations provide the value of its deprotonation energy (DE) to corresponding negatively charged form of the zeolite fragment/framework (higher DE corresponds to lower acidity).

The effect of the aluminum content of the zeolite is considered in two ways–influence of additional Al atoms close to the bridging OH group and influence of distant Al centres by variation of the total Si/Al ratio in the framework. According to calculations with QM-pot method,[49] when additional Al atoms are located in neighbouring T atom position to the Si of the OH group, its acidity was found to decrease with the number of Al atoms in such position increases. However, the influence of the total aluminum content in the zeolite framework on the acidity is essentially negligible, as calculated by the covEPE embedding scheme, the DE increases by only 6 kJ/mol (from 1265 to 1271 kJ/mol) for increase of the Si/Al ratio from 11 to infinity.[20]

The influence of the type of the three-valence centre substituting Si in the zeolite framework was studied[50] using isolated cluster models. The following order of decreasing acidity was found: Al > Ga > Fe.

The effect due to a different local structure of the zeolite framework was found larger than that due to changes of the Al content. The model studies with covEPE method suggested that the DE energies of the acid sites decrease by about 20–25 kJ/mol along the series faujasite > CHA > HZSM-5.[19,20] This observation is in agreement with experimental trend in the acidity of these zeolites tentatively estimated from measured shifts of the OH frequency upon adsorption of a probe CO molecule.[51]

The presence of additional metal cation in vicinity to the bridging OH groups has substantial influence on the Brønsted acidity. This effect was studied for some alkaline and alkaline-earth cations (Li^+, Na^+, K^+, Mg^{2+} and Ca^{2+}).[52] The presence of an alkali or alkaline-earth cation leads to a decrease of the DE of the bridging OH group; thus, mixed fragments containing both protons and metal cations for charge compensation are more acidic than the pure protonic forms. The lowest DE value was found for the bridging OH group of the cluster with Ca^{2+} cation; it is by 154 kJ/mol lower than the DE value of the corresponding completely protonic form of the model cluster. This suggests that zeolites containing Ca^{2+} and H^+ close to each other are considerably more acidic than the corresponding protonic forms. The sodium cations also increase the Brønsted acidity of the zeolite with respect to the corresponding protonic form, DE decreases by about −50 kJ/mol. The established effect is explained by the stabilization of the final deprotonated (anionic) form of the cluster when a metal cation is present. Because of its high mobility, after deprotonation of a bridging OH group the cation can move and partially compensate the excess of the negative charge at the 'deprotonated' O centre.[52]

3.3. Extra-framework metal ions and their complexes

Because of presence of three-valence Al ions in aluminosilicates instead of four-valence Si, crystal lattice of zeolites is negatively charged. The negative charge is distributed over four O atoms, connected with Al,[53,54] and by this reason different charge-compensating metal ions in zeolites are located in vicinity of the Al centres. Most of the interesting properties of the zeolites are due to presence of such counterions. However, due to the complexity of the interactions, the explanation of experimental observations often needs support from relevant computational modelling. Using quantum chemical calculation one can estimate the relative stability of metal cations at different positions in the zeolite lattice with respect to corresponding protonic form of the system or to a virtual negatively charged framework. In addition, simulations allow to obtain valuable information for the structural and spectral characteristics for a cation in certain position or state and to compare this information with available experimental data. By this reason various theoretical investigations are performed using all considered above models: isolated or embedded clusters, and periodic models.

Rice *et al.*[55] modelled coordination of some two-valence metal cations (Co^{2+}, Cu^{2+}, Fe^{2+}, Ni^{2+}, Pd^{2+}, Pt^{2+}, Rh^{2+}, Ru^{2+}, and Zn^{2+}) in ZSM-5 zeolite. They used isolated cluster models, containing one and two Al centres and representing different fragments of MFI framework (denoted according to the size of the cluster and its location in the framework – Z4, I5, S5, Z5, S6, and Z6). The stability of different coordination positions was estimated by the reaction energy for reduction of the metal-exchanged form of the zeolite to its protonic form after interaction with gaseous H_2 using the following reaction sequence:

$$O_z^-/M^{2+}/O_z^- + H_2(g) \rightarrow 2O_z^-/H^+ + M(g)$$

$$M(g) \rightarrow M(s) \quad \text{(experimental value)}$$

$$O_z^-/M^{2+}/O_z^- + H_2(g) \rightarrow 2O_z^-/H^+ + M(s) \quad \text{(overall)}$$

According to the obtained results, the stability of the cations strongly depends on their location. For instance, Cu^{2+}, Co^{2+}, Fe^{2+}, Ni^{2+} cations prefer coordination to five-membered rings, containing two Al centres and located in the main channels of MFI structure, while Pd^{2+}, Pt^{2+}, Ru^{2+} and Zn^{2+} prefer coordination to six-membered rings again in the main channels. The authors also concluded that this preference depends on the optimal length of the M–O bonds. When the average M–O bond distance is less than 202 pm (Co^{2+}, Cu^{2+}, Fe^{2+}, Ni^{2+}), coordination to five-membered ring is favourable, while for metals with average M–O bond distance more than 202 pm, coordination to six-membered rings is preferred.

The ions $M^{2+}(OH)^-$ and $[M\text{-}O\text{-}M]^{2+}$ were also modelled. The stability for reduction of $M^{2+}(OH)^-$ ions is estimated with respect to reduction by H_2 to metal and water. This reaction was found exothermic process in almost all studied cases and stability with respect to reduction decreases in the order: Zn > Co > Fe > Ni > Cu > Rh > Pd > Ru > Pt. The formation of dimmer $[M\text{-}O\text{-}M]^{2+}$ cation from two $M^{2+}(OH)^-$ ions was calculated endothermic for all modelled ions, hence $[M\text{-}O\text{-}M]^{2+}$ should be unstable with respect to hydrolysis.

McMillan *et al.*[56], using isolated clusters, considered several divalent cations: Mg^{2+}, Mn^{2+}, Fe^{2+}, Co^{2+}, Ni^{2+}, Cu^{2+}, Zn^{2+} coordinated at six-membered ring and two coupled five-membered rings from ferrierite structure. They found that calculated natural bond orbital (NBO) charge is highest for Zn^{2+}, but for the rest of the series decreases with the increase of the atomic number from Mn^{2+} to Cu^{2+}. They also calculated the skeleton vibrational modes for T-O-T bridges perturbed due to the coordination of the metal cation. These vibrations are in the range of 920–980 cm^{-1} and depend on the coordinated metal ion, which is close to available experimental values.[57,58]

The structure and relative stability of different Zn-containing cationic species in ZSM-5 zeolite: Zn^{2+}, $ZnOH^+$, $Zn(H_2O)^{2+}$ and $ZnOZn^{2+}$, were studied both with isolated and embedded (covEPE) clusters. The model zeolite fragments represented one and two five-membered rings, M5 and M7, respectively.[59,60] The relative stability of different forms was evaluated by interconversion reactions between the studied Zn species. The main conclusion from this evaluation is that on the zeolite fragments containing two Al centres the most stable Zn species are $Zn(H_2O)^{2+}$, while on fragment with isolated Al centres $ZnOH^+$ species exist. The adsorption energy of water on Zn^{2+} species is estimated between -130 and -184 kJ/mol, depending on the zeolite fragment and the method. In the presence of a bridging OH group in vicinity of $ZnOH^+$ species spontaneous formation of $Zn(H_2O)^{2+}$ is predicted due to the high exothermicity of the reaction, $-34 \div -117$ kJ/mol. It should be noted that the results for the stability of the different Zn species are qualitatively similar with the two isolated and embedded cluster models. The obtained results allow also comparison with the available experimental data from EXAFS.[61] These EXAFS data show that Zn ions in Zn, H–ZSM5 after treatment of sample at 773 K have 4.1 O contacts at average distance about 195 pm. Close to the experimental values are the computational results obtained for $Zn(H_2O)^{2+}$ species at M5 fragment at covEPE level, which has four Zn–O contacts with average distance 199 pm. Other covEPE structures with four Zn–O contacts are $ZnOH^+$ at M5 cluster and $Zn(H_2O)^{2+}$ and Zn^{2+} at the larger

M7 cluster. All modelled isolated cluster structures have maximum three Zn–O contacts, due to the imposed geometry restrictions on the zeolite clusters.

Barbosa *et al.*[62] investigated the stability of Zn^{2+} ions on different four-, six- and eight-membered rings in CHA structure, using periodic DFT calculations. They found that the most stable location of Zn^{2+} is at six-membered rings. They also modelled interaction of Zn^{2+} with H_2O. Its adsorption energy to Zn^{2+} is between −68 and −96 kJ/mol, when Zn^{2+} is located on six-membered rings, and −127 and −192 kJ/mol, when Zn^{2+} is located on four- and eight-membered rings, respectively. In another periodic DFT study, Kachurovskaya *et al.*[63] modelled Zn^{2+} in ferierrite with two Al centres in the unit cell and traced how the stability of the system changes when the distance between the two Al centres increases. The calculations showed that when Al centres are at distance 900 pm, the structure is destabilized by about 200 kJ/mol with respect to the structure with both Al centres in one zeolite ring and close to the Zn^{2+} ion.

Benco *et al.*[64], using DFT periodic calculations, modelled Zn^{2+} ions in mordenite. The cations were coordinated at five-, six- and eight-membered rings containing one or two Al centres. The most stable structure for Zn^{2+} ions was found at five-membered ring containing two Al centres. The coordination of Zn^{2+} at six-membered rings containing again two Al centres is by 15–36 kJ/mol less stable compared to coordination to the five-membered ring. The structures, where Zn^{2+} are coordinated close to only one isolated Al centre, are significantly less stable by 67–179 kJ/mol compared to those with Zn^{2+} interacting with two Al centres.

Kucera *et al.*[65] modelled alkali metal cations in different coordination positions in ZSM-5 using hybrid QM-pot method. For K^+, they found that K^+–O distances in the most stable structures are in the range 280–290 pm and above 300 pm for O atoms from AlO_4 and SiO_4 tetrahedra, respectively. Because of its large size, K^+ cation preferentially coordinates in the more open space on the channel intersections. In the most stable structures Na^+–O distances are in the range 230–240 pm and above 250 pm for O atoms from AlO_4 and SiO_4 tetrahedra, respectively. Here almost equal stabilities are found for various positions. The distances between Li^+ and framework oxygen atoms are in the range 190–200 pm and above 220 pm for O atoms from AlO_4 and SiO_4 tetrahedra, respectively. These results fit well to the distances obtained earlier with isolated cluster models representing six-membered rings (with different Al content) from faujasite structure.[52,66]

Nachtigallova *et al.*[67], also using QM-pot approach, modelled Cu^{2+} and Cu^+ in zeolite fragments (with two Al centres) from ZSM-5. The reduction of Cu^{2+} to Cu^+ is endothermic process with 16–110 kJ/mol, depending on the zeolite fragment and type of Cu^{2+} coordination. Comparing the results obtained for Cu^+ and Ag^+,[68,69] the authors found several differences: (i) Ag^+ ions occupy almost exclusive sites on the channel intersection while both site types (on the channel intersection and on the channels wall) are occupied by Cu^+ ions; (ii) the average coordination number of Ag^+ is lower than that of Cu^+ in ZSM-5; (iii) the average M^+–O bond lengths are about 30 pm longer for Ag^+ than for Cu^+; (iv) the interaction energy of the cation with the zeolite framework is 84 kJ/mol lower for Ag^+ compared to Cu^+ ions.

In addition to the clarification of the positions and the local structure of the metal cations, different computational approaches are applied to investigate the interaction of probe molecules with the active sites of zeolites and the spectral features of the adsorbed species. Carbon monoxide is among the most studied probes since it is routinely used for determination of the type, oxidation and coordination state of various metal cations in different materials.[70] For example, computational studies with isolated cluster models[66] helped in the assignment of the infrared (IR) peaks of CO on sodium exchanged faujasite. In the experimental IR spectra of CO on NaX zeolite (with high Al content and high concentration of extra-framework sodium cations), two main bands of almost equal intensity are observed at 2175 and 2165 cm^{-1} with shifts of 32 and 22 cm^{-1} with respect to the gas phase CO.[71] According to the computational modelling, these bands are assigned to CO adsorbed on sodium cations in SIII and in SII sites, respectively. The assignment of the CO bands for NaY zeolite was more complicated since the computational modelling suggested that the frequency shift of CO adsorbed on Na^+ cations in zeolites depends not only on the position of the cation but also on the number of Al atoms in the neighbouring zeolite ring.

Nachtigall and co-workers examined the concept proposed by Busca and co-workers[72,73] for coordination of CO probe molecule to multiple cation sites in alkali-exchanged zeolites. The computational studies for Al-rich Na-A, Na-FER and K-FER zeolites were performed by isolated clusters and periodic models.[74–76] Because of the large amount of extra-framework cations in these materials, the authors expected that CO adsorption on the cations should occur in dual or multiple fashions. The specific geometry of adsorption complexes depends on the topology of the particular zeolite framework, but their formation is expected to be predominant in all Al-rich zeolites having their extra-framework cations close. Optimum distance between cation centres forming multiple sites depends on cation size. For dual sites on alkali-exchanged zeolites, the optimal distances are estimated to 750, 650 and 550 pm for K^+, Na^+ and Li^+, respectively. Formation of dual-site coordinated CO in lithium exchanged zeolites is unlikely because of the smaller intercation distance needed and because of the location of the Li^+ ions in the plane of surrounding framework oxygen centres[52] that renders them less adapted for CO bridging. Large concentration of dual and multiple cation sites leaves only small fraction of cations available for formation of isolated single-site monocarbonyls. The concept of assigning each IR absorption band appearing in the spectra to a specific type of single cation site, that proved to be fruitful for high-silica zeolites,[70] was questioned for the CO spectra in Al-rich zeolites due to substantial contribution of bands originating from CO adsorption on more complex cation sites.

The EXAFS and IR measurements of uniform $Rh^+(CO)_2$ complex in DAY zeolite[5] gave an opportunity to compare the interatomic distances and vibrational frequencies derived from these experimental techniques with the structural data and frequencies obtained in computational modelling. The computational modelling, based on isolated zeolite fragment, allowed to identify the real structure of the supported complex among three possible structural models suggested from the analysis of the EXAFS. The close correlation between the experimental (IR and EXAFS) parameters and the optimized structure of the $Rh^+(CO)_2$ complex on zeolite four-ring shows that the obtained structure corresponds to the real one in the system $Rh^+(CO)_2$/DAY. The complex $Rh(CO)_2{}^+$ is planar with a rhodium cation bound to two zeolite oxygen centres and both CO ligands oriented along the continuation of Rh–O_{zeo} bonds. For this structure, the distances differ at most by 4 pm for the nearest neighbours and by 7 pm for the next-nearest neighbours. The difference in the IR CO frequency shifts of the adsorbed species with respect to gas phase CO molecule is also small, 4–11 cm^{-1}. With the clarification of the location of $Rh(CO)_2{}^+$ species in dealuminated Y zeolite, it became possible to determine also the structure of less stable rhodium carbonyl species in zeolite and to assign their IR frequencies.[6] Using the same computational approach and structural models, the assignment of the experimentally observed weak IR band at 2093 cm^{-1} [77,78] was examined. This band was assumed to correspond to monocarbonyl $Rh(CO)^+$ species. The computational modelling, however, suggested that this band should be assigned to rhodium carbonyl species which contain additional hydrogen ligands, for example, $Rh(CO)(H_2)^+$ or $Rh(CO)(H)_2{}^+$, whereas the zeolite supported complex $Rh(CO)^+$ is characterized by the IR band observed at 2014 cm^{-1},[77] with a frequency shift relative to dicarbonyl similar to the experimental IR shift of the corresponding molecular Rh^+ carbonyl complexes.[79] These corrected assignments were later confirmed experimentally.[80]

Ivanova *et al.*[81] reported combined experimental IR and computational study of CO, NO, and mixed CO/NO complexes of Rh^+ and Rh^{2+} located in ZSM-5 zeolite. The binding energy (BE) of CO to Rh^+ (193 kJ/mol per molecule) is more than twice larger than to Rh^{2+} (91 kJ/mol). These results are consistent with experimental observations that $Rh^{2+}(CO)_2$ species are thermally less stable than $Rh^+(CO)_2$.[82] The BE of NO molecule in $Rh^+(NO)_2$ was calculated at 228 kJ/mol, that is, NO binds 35 kJ/mol per molecule stronger than CO. Taking into account these values, one can easily rationalize why $Rh^+(CO)_2$ can be transformed into $Rh^+(NO)_2$ species, as observed experimentally.[81,82] Following the experimental result for formation of mixed complexes of Rh^{2+} with CO and NO ligands, such complexes were modelled with one or two NO ligands: $Rh^{2+}(NO)(CO)_2$ and $Rh^{2+}(NO)_2(CO)_2$. However, only the former structure was found stable; in it the $Rh^{2+}(NO)(CO)_2$ moiety is located similarly to the zeolite fragment as the $Rh^{2+}(CO)_2$ with the NO ligand coordinated perpendicularly to the plane of the $Rh^{2+}(CO)_2$ complex.

While the CO frequencies in the dicarbonyl complexes of Rh^+ and Rh^{2+} on zeolite cluster models are reasonably close to the experimental values, the calculated IR frequencies of NO and CO in the mixed complex $Rh(NO)(CO)_2{}^{2+}$ on the zeolite fragment deviate considerably from the experimental values: the CO and NO frequencies are by 57–64 and 128 cm^{-1}, respectively, lower than in the experiment. Since these differences are beyond the accuracy of the applied theoretical and experimental approaches, it was assumed that the $Rh^{2+}(NO)(CO)_2$ complex is bound to some smaller fragments as $(OH)_3AlOSi(OH)_3{}^-$ and $(OH)_3SiO^-$, as well as HO^-. Indeed, the calculated vibrational CO and NO frequencies for the species $Rh^{2+}(OH)(NO)(CO)_2$ are found very close to the experimental values.

Benco *et al.*[83] performed an extensive study of NO adsorption over the Fe-exchanged ferrierite. Adsorption energy of NO on Fe^{2+} cation is in the range

180–240 kJ/mol. In the most stable β-site, the Fe^{2+} cation can adsorb two and in all other locations three NO molecules. In the case of adsorption of several NO ligands, two NO molecules are adsorbed directly on the Fe^{2+} cation, a third molecule adsorbs to an Al site, and the fourth binds again to the cation. With four NO molecules, the Fe site is completely saturated. The calculated NO stretching frequencies agree with the experimentally observed peaks in the IR spectra and support the established assignment of spectral bands. Adsorption of a single NO molecule produces a single band centred at 1876 cm^{-1}. Two molecules adsorbed in a transconfiguration and completing an octahedral coordination of the Fe^{2+} cation exhibit extremely downshifted frequencies. Besides the band of a single NO molecule, the most intense bands are those originating from trinitrosyl species with a tetrahedral coordination of the Fe^{2+} cation at ~1925, 1824 and 1893 cm^{-1}. The adsorption of NO at increased pressure leads to the complete saturation of the coordination sphere of the Fe^{2+} cation. The $Fe(NO)_3$ tetrahedron binds to one O atom of the framework and contains three adsorbed NO molecules. Interestingly, good agreement of calculated stretching frequencies is observed only for di- and trinitrosyl species with an additional NO molecule adsorbed on the Al site. This indicates that at increased NO pressure both the Fe^{2+} cations and Al sites are saturated. The adsorption power of the Al site develops when two molecules adsorb on the Fe^{2+} cation producing an $Fe(NO)_2{}^+$ complex. The decrease of the charge of the extra-framework particle leads to concerted oxidation of the third NO molecule adsorbed on the Al site producing a NO^+ cation long lived on the inner surface of the zeolite.

The proper interpretation of the results in many computational studies, aimed at interpretation of experimental problem, is connected with evaluation of the acceptable difference between calculated and experimentally derived parameters. On the one hand, this difference depends on the accuracy of the experimental measurement, which is known from the experiment. On the other hand, one has to consider several sources of potential errors that may arise from the calculations. On the example of the IR frequencies, these potential errors are (i) the anharmonicity of the vibration; (ii) an intrinsic (systematic) error of the computational method and (iii) the limited size of the zeolite model and specific restrictions applied to it. The former two effects are to a large extent accounted for by a correction of the calculated frequency with a value determined from the difference between experimental and calculated vibrational frequencies of a free ligand molecule in the gas phase. For example, if the experimental and calculated values for CO are 2144 and 2112 cm^{-1}, respectively, then all calculated vibrational frequencies of CO adsorbed in different positions (using the same computational approach) should be corrected by +32 cm^{-1}.

The error in the calculated frequency due to the limited size of the zeolite model could be caused by (i) the lack of the long-range electrostatic field of the crystal environment, (ii) steric restrictions due to the zeolite framework, which are not represented in the isolated models and (iii) the restricted flexibility of the model. These errors can be estimated by comparison with results of extended models of the zeolite framework as embedded cluster or periodic supercell models. Results from calculations based on the covEPE embedded cluster scheme [19–21] suggest that these short- and long-range effects of the zeolite environment affect the vibrational frequency of a CO probe, adsorbed at an acidic OH group in HY and HZSM-5 zeolites at most by 5 cm^{-1}.[21] A slightly larger shift, 13 cm^{-1}, was calculated as environmental effect on the frequency of CO on a Cu^{+} cation in ferrierite.[84,85] Direct comparison of calculated vibrational frequencies with experimental IR results for species, whose structure in the zeolite pores is known, provides another way to estimate the influence of the zeolite framework that remains unaccounted with isolated cluster models. An example is the $Rh^{+}(CO)_2$ geminal dicarbonyl complex in dealuminated Y zeolite, which has been characterized by IR, EXAFS and DFT methods. For that complex, experimental values and computational results from isolated cluster differed by 10–15 cm^{-1}, after the corrective shift of 32 cm^{-1} (described above) had been applied. Therefore, after the correction for anharmonicity and systematic errors of the method, experimental and calculated frequencies are expected to differ at most about 15–20 cm^{-1}. This value could be higher for the zeolites with narrow pores.

3.4. Basic oxygen centres

In addition to the Brønsted and Lewis acidic centres, substantial role in sorption and catalytic properties of zeolites play the oxygen centres bound to Al atoms of the framework. Because of the excess of negative charge around Al, these oxygen centres act as basic sites. The basicity of the oxygen centres can be experimentally evaluated indirectly by the adsorption of probe molecules or by some observable spectral features as O1s core level shift measured by XPS. A direct measure for the basicity of the oxygen centres is their proton affinity (PA), namely the energy gained after attachment of a proton to the corresponding oxygen centre. Since the PA of zeolite oxygen centres cannot be measured directly, it is useful to clarify its connection with experimentally measurable quantities mentioned above.

The PA values of the individual oxygen atoms in a series of model six rings with different Al content (and different basicity) were found to correlate very well with the calculated core level shifts ΔE_b(O1s) of the atom.[53] From the slope of the correlation line one can conclude that a negative shift of the O1s binding energy by 1 eV (towards a less stable O1s level) implies an increase of the PA by 82 kJ mol^{-1}. This correlation can be used for estimation of the basicity of the oxygen centres in zeolites. However, since the signal measured by XPS is averaged over all oxygen centres of different basicity in the sample, one cannot determine easily the PA of the most basic oxygen centres.

Computational studies with isolated cluster models demonstrated that the shifts of the stretching and the deformation O–H vibrational frequencies of methanol adsorbed on alkali-exchanged zeolites and forming a hydrogen bond to zeolite oxygen centres correlate with the calculated PA of those centres.[4] When the calculated PA values are calibrated with respect to relevant experimental gas phase PA values,[54] the results are directly applicable to experimental estimation of PA without further calculations. In this way, the experimental PA values of zeolite oxygen centres can be derived from measured IR frequency shifts[86] of the O–H stretching ($\Delta\nu$) and deformation ($\Delta\delta$) bands of methanol adsorbed on alkali-exchanged zeolites according to the following expressions:[54]

$$PA(\nu) = 776 - 0.282\Delta\nu(OH) \tag{1}$$

$$PA(\delta) = 776 + 1.52\Delta\delta(OH) \tag{2}$$

In the above equations PA is in kJ/mol and the frequency shifts are in cm^{-1}. The frequency shifts vanish at $PA = 772 \pm 5$ kJ mol^{-1}. As specific advantage, this approach allows a direct 'experimental' determination of the PA value of a particular type of oxygen centres using two different criteria, Eqs. (1) and (2), which reduce the error of the measurement.

3.5. Zeolite-supported transition metal clusters

Size-selected transition metal clusters supported in zeolite cavities are convenient model system for precise investigation (both experimental and computational) of catalytic processes occurring in such bifunctional systems. In particular, supported clusters of late transition metals were found active catalysts for various reactions including hydrocarbon transformations, hydrogen activation, oxidation/reduction, etc., in which the state of the metal species and their interaction with the support are crucial for the catalytic activity.[87] In a series of papers, Gates and co-workers[87–89] reported preparation of bare transition metal clusters with defined nuclearity, as Rh_4, Rh_6, Ir_4, Ir_6, in zeolites or on oxide surfaces. However, by means of density functional calculations at isolated cluster level it was found that the experimentally observed metal–metal distances are by more than 10–20 pm longer than the optimized ones.[90] For example, Rh–Rh and Rh–O distances were found to fit very well not to adsorbed bare Rh_6 but to a zeolite-supported Rh_6H_3 cluster. Thus, Rh_6 clusters anchored at a hydroxylated zeolite cavity carry hydrogen heteroatoms which were transferred from the zeolite support due to reverse hydrogen spillover process.[8] Moreover, the hydrogenated cluster, assigned as Rh_6H_3/zeo, was calculated to be notably more stable than the corresponding supported bare cluster Rh_6 by 123 kJ/mol per transferred proton. Expanding the theoretical investigation further to 12 metals M of the groups 8–11 supported as M_6 in a hydroxylated faujasite cage, M_6/zeo(3H), it was shown that the hydrogenated state, M_6H_3/zeo, is energetically preferred over the bare form M_6/zeo (3H).[91] The energies, E_{RS}, of reverse hydrogen spillover of zeolite supported M_6 clusters are presented on Fig. 8.4. Along each period of the periodic system, the energy E_{RS} decreases, with strongest decrease for 5d metals. All 4d metals studied, except for Ag, feature lower values of the reverse spillover energy than the 3d and 5d metals of the corresponding triad.

The preference for formation of hydrogen-covered form of the metal clusters was also studied by the more accurate embedded cluster approach covEPE for three metals, M = Rh, Ir and Au.[92] For rhodium and iridium the QM/MM approach predicted reverse hydrogen spillover to be energetically favourable, as the previous model with isolated zeolite fragment, but the reverse spillover energy decreases by

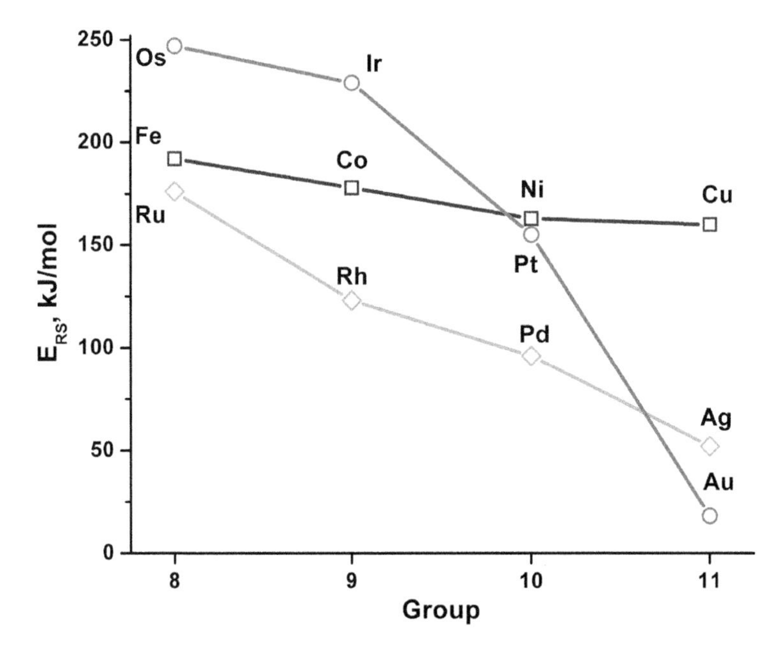

Figure 8.4 Energy E_{RS} (per OH group) gained due to reverse H spillover on M_6/zeo(3H) according to the values reported in Vayssilov and Rösch[91]: $E_{RS} = \{E[Me_6/zeo(3H)]-E[Me_6H_3/zeo]\}/3$. (See color insert.)

about 30%. For the zeolite supported gold species, however, the bare form of the cluster, Au_6/zeo(3H), is energetically preferred over the hydrogenated form Au_6H_3/zeo, that is, the reverse hydrogen spillover from less acidic OH groups onto Au_6 was calculated to be endothermic, on average by 29 kJ/mol per transferred proton.[92]

The reverse spillover process was also studied for the smaller Ir_4 cluster supported on zeo(3H) fragment.[93] The stepwise hydrogen transfer from bridging OH groups of the support to Ir_4 clusters was calculated exothermic, in total -421 kJ/mol for the three steps studied as spillover energy decreases with the number of transferred protons (Fig. 8.5).

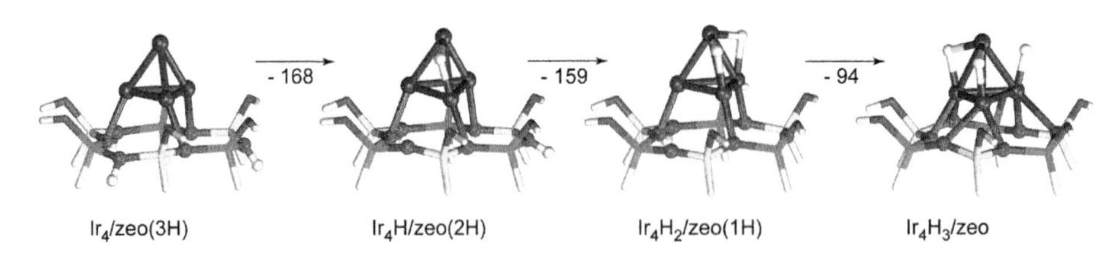

Figure 8.5 Stepwise proton transfer from the zeolite fragment, zeo(3H), to the supported Ir_4 cluster. Reaction energies in kJ/mol, according to the values reported in Petrova et al.[93] (See color insert.)

Mikhailov *et al.*[94] reported a DFT study at isolated cluster model level of metal–support interaction in the case of Pt_6 cluster adsorbed on supports with different properties: γ-Al_2O_3 and zeolites faujasite and H–ZSM-5. While in the case of Pt_6/γ-Al_2O_3, the hydrogen spillover is energetically favourable with respect to the reverse process, for the interaction of the platinum cluster with both types of zeolite fragments the reverse hydrogen spillover is preferred. For Pt_6/faujasite system (with the intact OH groups of the support), the step-wise transfer of the three protons from the Brønsted acid sites of the zeolite fragment to the platinum cluster was studied. As in the case of the smaller Ir_4 cluster, each step of the proton transfer was found exothermic and the transfer energy decreases accordingly: 96, 50 and 38 kJ/mol for the first, second and third proton, respectively. For the Pt_6/H–ZSM-5 system, the processes of proton transfer to the platinum cluster was also found exothermic with the transfer energy equal to 196 kJ/mol, which exceeds the energy gained by the transfer of three protons in faujasite, 184 kJ/mol. It should be noted that no activation barrier for the proton transfer from the bridging OH group to the metal clusters was observed.

Focusing on a concrete experimental problem for the structural features of Ir_4 cluster in faujasite,[95] the successive adsorption of H_2 from the gas phase on supported Ir_4 clusters was simulated. Two states of the support were considered, hydroxylated and dehydroxylated zeolite fragment. The results[96] suggested that the adsorption of H_2 on the small cluster Ir_4, either in the gas phase or supported, is a favourable process up to at least 12 hydrogen atoms at the metal cluster, corresponding to a ratio $H/Ir = 3$. The structural characteristics of the metal moiety in the model complexes indicate that the average Ir–Ir distances are almost independent on the presence of a support or its type (hydroxylated or dehydroxylated). Rather, the average Ir–Ir distance of the cluster models increases with the number of hydrogen atoms adsorbed on the metal moiety, by about 1.9 pm per adsorbed hydrogen atom. On the basis of available structural information from EXAFS, one can conclude that the tetrairidium species in Y zeolite, produced experimentally and investigated by EXAFS, correspond to supported hydrogenated moieties Ir_4H_x containing 9–12 hydride centres.

3.6. Adsorption and chemical transformations of guest molecules in zeolites

3.6.1. Adsorption and dissociation of H_2

In a periodic DFT study, Benco *et al.*[97] investigated molecular and dissociative adsorption of H_2 on various cations: Na^+, Cu^+, Ag^+, Zn^{2+}, Cu^{2+}, Ga^{3+} and Al^{3+} located in zeolite with mordenite structure. From the obtained results they differentiate three groups of active sites:

(1) Weak sites, which cannot dissociate H_2. This group comprises the Brønsted acid site, weak Lewis sites (Na^+), three coordinated Al and extra-framework cations: Al^{3+}, coordinated in zeolite fragment with three Al centres and Cu^{2+}, coordinated to zeolite fragment with two Al centres.

(2) Lewis sites with intermediate activation power, which can dissociate H_2 to H^+ and H^-, as H^+ is located on basic O centre from the zeolite framework, while H^- is adsorbed on the metal cation. This group includes polyvalent cations Zn^{2+}, Al^{3+}, located close to one and two Al centres and Ga^{3+}, located close to one Al centre.
(3) Monovalence cationic Lewis sites Cu^+ and Ag^+, which activated strongly H_2, but dissociation of H_2 does not occur. These cations have large H_2 adsorption energies, 60–90 kJ/mol. The dissociation of H_2, however, is endothermic due to the production of neutral species CuH and AgH.

Barbosa *et al.*[98], using also periodic DFT calculations, modelled interaction of H_2 with Zn^{2+} located in three different zeolite frameworks – CHA, mordenite and ferrierite, in order to investigate the influence of the zeolite structure on the reactivity of Zn^{2+} located at eight-membered rings. In CHA and MOR either molecular or dissociative adsorption of H_2 is preferred depending on the relative location of the two Al centres in eight-membered ring. In FER only molecular adsorption was found favourable. As expected, the molecular adsorption of hydrogen on the Zn^{2+} species results in red shifts of the H–H frequency. This negative frequency shift on different modelled zeolite frameworks decreases in the following order: FER (Al-far-Al) > CHA (AlSiAl) > MOR (AlSiAl) ≈ FER (AlSiSiSiAl) ≈ FER (AlSiAl). While in FER (Al-far-Al) model, the strong frequency shift is due to the interaction only with Zn^{2+}, the large $\Delta\nu$ found for CHA (AlSiAl) is probably due to combined interaction of H_2 with Zn^{2+} and basic O centres from the zeolite structure.

The dissociative adsorption of H_2 on different types of Zn species, Zn^{2+}, $ZnOH^+$, $Zn(H_2O)^{2+}$ and $ZnOZn^{2+}$, was also investigated using isolated cluster models representing five- and six-membered rings of ZSM-5 zeolite.[99] Dissociation of H_2 is exothermic process on Zn^{2+} and $ZnOZn^{2+}$ ions, while on the cationic species with additional ligands, $ZnOH^+$ and $Zn(H_2O)^{2+}$, the process is endothermic. The calculated Zn–H vibrational frequency, when H_2 is heterolytically dissociated on Zn^{2+} ions, 1935 and 1943 cm^{-1} (depending on the zeolite cluster), fits best to the available experimental data, 1934–1936 cm^{-1}.[100,101]

3.6.2. Adsorption and dissociation of hydrocarbons

Pidko *et al.* modelled the mechanism of ethane dehydrogenation on zinc and gallium exchanged zeolites. As active sites they considered Zn^{2+} ions, located close to one and two Al centres, $ZnOZn^{2+}$ species,[102] and gallyl ions (GaO^+).[103] The calculations were performed with isolated zeolite fragments representing two coupled five-membered rings. Among the considered Zn species, Zn^{2+} ions located close to only one Al centre feature lowest activation barrier for ethane dehydrogenation; the calculated apparent activation energy is 153 kJ/mol. The most favourable mechanism includes heterolytic dissociation of the ethane molecule to $C_2H_5^-$ (coordinated to Zn^{2+} ion) and H^+ (bound to basic zeolite O centre). Afterwards ethene and H_2 are obtained in one step.

Two different reaction paths were found for the initial C–H activation on GaO^+ active sites.[103] The first one is the heterolytic dissociation of ethane on the gallyl ion (via transition state with low activation barrier), which results in formation of a very stable intermediate $[Ga(C_2H_5)(OH)]^+$ that can be decomposed via ethene desorption to form $[Ga(H)(OH)]^+$ species. Further regeneration of the active site via H_2 desorption is strongly disfavoured both thermodynamically and kinetically. The second modelled reaction path for ethane dehydrogenation is initiated by partial oxidation of an ethane molecule by the extralattice oxygen atom. The resulting ethanol molecule then interacts with the adsorption site (Ga^+), leading to formation again of the very stable intermediate $[Ga(OC_2H_5)(H)]^+$ which is a product of 'carbenium' mechanism. Since both considered reaction paths lead to formation of rather stable intermediates, gallyl ions cannot be considered as catalytically active sites for dehydrogenation of light alkanes.

3.6.3. Catalytic decomposition of N_2O

Heyden *et al.*[104] used a combination of DFT and transition state theory to determine the rate coefficients for an ensemble of over 100 elementary reactions that might be involved in NO-promoted decomposition of N_2O on Fe–ZSM-5 at temperatures below 700 K. Under these conditions, highly active $Z^-[FeO]^+$ sites react with small amounts of water vapor to form $Z^-[Fe(OH)_2]^+$ sites, which are inactive for N_2O decomposition. NO can react with such species to form HNO_2 and $Z^-[FeOH]^+$ sites. These newly formed sites are catalytically active for N_2O decomposition to N_2 and O_2. The reaction pathway involves also formation and decomposition of NO_2. Since the activation barrier for the formation of $Z^-[FeOH]^+$ sites is low, 56 kJ/mol, the number of active iron sites for the N_2O decomposition increases significantly in the presence of small amounts of NO. In this way the N_2O decomposition rate increases. The authors also considered N_2O decomposition on another catalytic species, namely, binuclear oxygen bridged extra-framework iron sites.[105] At low temperatures the hydroxylated iron site, $Z^-[HOFeOFeOH]^{2+}Z^-$, was found to dominate the catalytic surface. This site shows low activity for N_2O decomposition. At higher temperatures, water can desorb that allows N_2O decomposition on the formed active site $Z^-[FeOFe]^{2+}Z^-$ to occur. The mechanisms of N_2O decomposition on $Z^-[FeOFe]^{2+}Z^-$ and $Z^-[FeOH]^+$ are similar and both types of sites are poisoned by small amounts of water. As a result, the nuclearity (mono- or binuclear iron site) of the catalytically active site for N_2O decomposition appears to be less important than catalyst poisoning by traces of water in the gas streams. Antiferromagnetic coupling was shown to have no effect on the reaction mechanism.

3.6.4. Adsorption of chloroform

Interaction of chloroform and other chlorocarbons with active sites in alkali exchanged zeolites was modelled by molecular simulation techniques – Monte Carlo simulations with Metropolis scheme in the canonical ensemble (at a constant volume V and temperature T).[106] Note that these simulations were not based on quantum chemical method but on classical approach. Three zeolite hosts were considered: (i) siliceous faujasite, (ii) NaY with 16 Na^+ cations in site SI and 32

Na^+ cations in SII and (iii) NaX, placing 32 Na^+ cations in SI, 32 in SII, and the rest 24 cations in site SIII. In each system, the number of guest molecules, N, was varied from a single molecule per unit cell up to 30–40 molecules per unit cell to reproduce the variation of loading in a fashion similar to the calorimetric measurements.

The Monte Carlo simulations in the three hosts capture remarkably well the experimental findings from the calorimetric heats of adsorption of chloroform at room temperature. Two dominant features of chloroform adsorption are revealed: (i) an increase in adsorption energies over the whole loading range in the sequence siliceous faujasite < NaY < NaX; and (ii) an increase in adsorption energies of more than 10 kJ/mol upon loading in siliceous faujasite and NaY, in contrast to the chloroform/NaX system which shows a relatively flat profile. To elaborate these two features of chloroform adsorption, the total host–guest interaction energies were decomposed into their short-range and long-range components. Since the short-range interactions are very similar in magnitude in all three hosts, the higher affinity of chloroform to the more polar zeolites comes from the stronger electrostatic interactions occurring in these cation containing systems. At zero loading, the electrostatic interaction contributes 47% in the total adsorption heat in NaX, 24% in NaY, and 11% in siliceous faujasite.

In NaY, calculations show clear evidence for the three types of interatomic guest–host interactions[107]: (i) short-range Cl–O_{zeo} repulsive interaction; (ii) electrostatic Cl–Na(SII) interaction and (iii) H–O_{zeo} hydrogen bonding. In the case of NaX, the results point to the possibility that the Na^+ cations in site SIII are involved in additional interactions with the sorbate, in comparison with NaY. Indeed, the location of chloroform in the NaX structure investigated by docking and energy minimization calculations (at 0 K) showed new features in comparison with siliceous faujasite and NaY: some of the chlorine atoms are attracted towards the Na^+ cations in site SIII that controls the orientation of the molecule in the 12-ring window.

3.6.5. Adsorption of methanol

Methanol is among the IR probe molecules suggested for estimation of the basicity of oxygen centres. The experimental studies[86] had shown that the position and intensity of the IR O–H band in methanol adsorbed on alkali exchanged zeolites depends on (i) the type of the alkali cation, (ii) the Al content (which is connected to the basicity) of the zeolite and (iii) the methanol loading. Isolated clusters studies were employed to clarify the adsorption modes of methanol on alkali zeolites with three model clusters: Na^+ located in six-membered rings, containing one and two Al centres, and K^+ located in six-membered ring, containing one Al centre.[4] Comparison of CH_3OH adsorption on the first two models allows one to evaluate the effect of the Al content and of the zeolite basicity on the adsorption mode. The third model was used to analyze the influence of the type of the alkali cation that also affects the basicity of the zeolite O atoms.[53]

Methanol molecule is adsorbed on the model clusters in two ways. The first type of adsorption on the zeolite cluster models represents O-bound methanol forming only a coordination bond of its O atom to the alkali cation. The second adsorption mode comprises O,H-bound species forming in addition an H-bond of the methanol OH group to a zeolite O centre. For O-bound and O,H-bound adsorbates, the coordination bond between the methanol O atom and the metal cation furnishes the main contribution to the adsorption interaction with Na- or K-exchanged zeolites. Species bound only by a coordination bond exhibit adsorption energies of 41–56 kJ/mol, whereas an H-bond contributes in addition 12–18 kJ/mol. The energy increases with the Lewis acidity of the cation for both types of adsorption. On the contrary, the strength of the hydrogen bond increases with the PA[53] of the basic O centre of the zeolite model. Following these observations, one would expect that methanol will adsorb at available Na^+ or K^+ cations in the zeolites and will form a hydrogen bond with a nearby basic zeolite O atom.[4] The dominant role of the coordination bond to the cation influences also the chemical reactivity of adsorbed methanol species on Na-exchanged zeolites. This bond activates CH_3OH by weakening the C–O bond, but it stabilizes the C–H bonds. Thus, Na-exchanged zeolites act similar to weak acidic catalysts rather than basic ones. The decrease of the Lewis acidity of the cation and the simultaneous increase of the basicity of the zeolite O centres when going from Li- to Cs-exchanged zeolites alters the relative contribution of the coordination and the H-bonds to the adsorption energy. In this way, the catalytic activity of zeolites exchanged with alkali cations varies with the atomic number of the cation.

4. Outlook

With selected examples, we have shown that computational methods are reliable approach for investigation of nanoporous and sub-nanoporous systems of practical interest. In particular, the quantum chemical methods are used for description of problems and processes connected with chemical interactions and reactions, redistribution of electron density, simulation of spectral features connected with the electronic structure of the material, its active centres or guest species. In many cases, these methods show very good agreement of the computed properties with available experimental data and provide complementary information that cannot be derived only by experimental means. The most suitable techniques for modelling of zeolite systems are periodic models; however, they are applicable so far only for zeolites with relatively small unit cell. However, most of the zeolites used in different industrial applications have large unit cell and in these cases the hybrid QM/MM approaches are more suitable. They allow proper accounting for the flexibility and the electrostatic field of the whole zeolite framework using relevant force field which decreases computational efforts. The simplest way for modelling of processes in zeolites and mesoporous materials are isolated cluster models, which have several drawbacks, but in many cases the obtained results are reasonable. Application of one of these techniques or combination of them allows a deeper understanding of the structure and properties of the studied molecular sieves and plays a key role in the design and tailoring of new materials with predefined properties.

ACKNOWLEDGEMENTS

The authors thank the Ministry of Education and Science (Bulgaria) and the Bulgarian-French bilateral programme Rila for the financial support.

REFERENCES

[1] Koch, W., Holthausen, M. C., *A Chemist's Guide to Density Functional Theory*, Wiley-VCH, Weinheim, 2001.
[2] Szabo, A., Ostlund, N. S., *Modern Quantum Chemistry: Introduction to Advanced Electronic Structure Theory*, Dover, New York, 1996.
[3] Vayssilov, G. N., Hu, A., Birkenheuer, U., Rösch, N., *J. Mol. Catal. A* **2000,** *162,* 135–145.
[4] Vayssilov, G. N., Lercher, J. A., Rösch, N., *J. Phys. Chem. B* **2000,** *104,* 8614–8623.
[5] Goellner, J. F., Gates, B. C., Vayssilov, G. N., Rösch, N., *J. Am. Chem. Soc.* **2000,** *122,* 8056–8066.
[6] Vayssilov, G. N., Rösch, N., *J. Am. Chem. Soc.* **2002,** *124,* 3783–3786.
[7] Neyman, K. M., Vayssilov, G. N., Rösch, N., *J. Organomet. Chem.* **2004,** *689,* 4384–4394.
[8] Vayssilov, G. N., Gates, B. C., Rösch, N., *Angew. Chem. Int. Ed.* **2003,** *42,* 1391–1394.
[9] Eichler, U., Kölmel, C. M., Sauer, J., *J. Comput. Chem.* **1998,** *18,* 463–477.
[10] Sierka, M., Sauer, J., *Faraday Discuss.* **1997,** *106,* 41–62.
[11] Sauer, J., Sierka, M., *J. Comput. Chem.* **2000,** *21,* 1470–1493.
[12] Sherwood, P., de Vries, A., Collins, S. J., Greatbanks, S. P., Burton, N. A., Vincent, M. A., Hiller, I. H., *Faraday Discuss.* **1997,** *106,* 79–92.
[13] de Vries, A., Sherwood, P., Collins, S. J., Rigby, A. M., Rigutto, M., Kramer, G. J., *J. Phys. Chem. B* **1999,** *103,* 6133–6141.
[14] Teunissen, E. H., Jansen, A. P. J., van Santen, R. A., *J. Phys. Chem.* **1995,** *99,* 1873–1879.
[15] Kyrlidis, A., Cook, S. J., Chakraborty, A. K., Bell, A. T., Theodorou, D. N., *J. Phys. Chem.* **1995,** *99,* 1505–1515.
[16] Vollmer, J. M., Stefanovich, E. V., Truong, T. N., *J. Phys. Chem. B* **1999,** *103,* 9415–9422.
[17] Lopez, N., Pacchioni, G., Maseras, F., Illas, F., *Chem. Phys. Lett.* **1998,** *294,* 611–618.
[18] Sulimov, V. B., Sushko, P. V., Edwards, A. H., Shluger, A. L., Stoneham, A. M., *Phys. Rev. B* **2002,** *66,* 024108-1–024108-14.
[19] Nasluzov, V. A., Ivanova, E. A., Shor, A. M., Vayssilov, G. N., Birkenheuer, U., Rösch, N., *J. Phys. Chem. B* **2003,** *107,* 2228–2241.
[20] Ivanova Shor, E. A., Shor, A. M., Nasluzov, V. A., Vayssilov, G. N., Rösch, N., *J. Chem. Theory Comp.* **2005,** *1,* 459–471.
[21] Shor, A. M., Ivanova Shor, E. A., Nasluzov, V. A., Vayssilov, G. N., Rösch, N., *J. Chem. Theory Comp.* **2007,** *3,* 2290–2300.
[22] Tuma, C., Sauer, J., *Phys. Chem. Chem. Phys.* **2006,** *8,* 3955–3965.
[23] Sillar, K., Burk, P., *Phys. Chem. Chem. Phys.* **2007,** *9,* 824–827.
[24] Maseras, F., Morokuma, K., *J. Comput. Chem.* **1995,** *16,* 1170–1179.
[25] Matsubara, T., Sieber, S., Morokuma, K., *Int. J. Quantum Chem.* **1996,** *60,* 1101–1109.
[26] Svensson, M., Humbel, S., Morokuma, K., *J. Chem. Phys.* **1996,** *105,* 3654–3661.
[27] Bobuatong, K., Limtrakul, J., *Appl. Catal. A* **2003,** *253,* 49–64.
[28] Yuan, S. P., Wang, J. U., Duan, Y. B., Li, Y. W., Jiao, H. J., *J. Mol. Catal. A* **2006,** *256,* 130–137.
[29] Joshi, A. M., Delgass, W. N., Thomson, K. T., *J. Phys. Chem. B* **2006,** *110,* 16439–16451.
[30] Jansang, B., Nanok, T., Limtrakul, J., *J. Mol. Catal. A* **2007,** *264,* 33–39.
[31] Barone, G., Casella, G., Giuffrida, S., Duca, D., *J. Phys. Chem. C* **2007,** *111,* 13033–13043.
[32] Sokol, A. A., Bromley, S. T., French, S. A., Catlow, C. R. A., Sherwood, P., *Int. J. Quantum Chem.* **2004,** *99,* 695–712.

[33] Van Dam, H. J. J., Guest, M. F., Sherwood, P., Thomas, J. M. H., van Lenthe, J. H., van Lingen, J. N. J., Bailey, C. L., Bush, I. J., *J. Mol. Str. (Theochem)* **2006,** *771,* 33–41.
[34] Stefanovich, E. V., Truong, T. N., *J. Phys. Chem. B* **1998,** *102,* 3018–3022.
[35] Dovesi, R., Orlando, R., Civalleri, B., Roetti, C., Saunders, V. R., Zicovich-Wilson, C. M., *Z. Kristallographie* **2005,** *220,* 571–573.
[36] Kresse, G., Hafner, J., *Phys. Rev. B* **1993,** *47,* 558–561; ibid. **1994,** *49,*14251–14269.
[37] Payne, M. C., Teter, M. P., Allan, D. C., Arias, T. A., Joannopoulos, J. D., *Rev. Mod. Phys.* **1992,** *64,* 1045–1097.
[38] Marra, G. L., Artioli, G., Fitch, A. N., Milanesio, M., Lamberti, C., *Microporous Mesoporous Mater.* **2000,** *40,* 85–94.
[39] Fois, E., Gamba, A., Spano, E., *J. Phys. Chem. B* **2004,** *108,* 154–159.
[40] Zicovich-Wilson, C. M., Dovesi, R., *J. Mol. Catal. A* **1997,** *119,* 449–458.
[41] Van Bokhoven, J. A., van der Eerden, A. M. J., Prins, R., *J. Am. Chem. Soc.* **2004,** *126,* 4506–4507.
[42] Joyner, R. W., Smith, A. D., Stockenhuber, M., van den Bergy, M. W. E., *Phys. Chem. Chem. Phys.* **2004,** *6,* 5435–5439.
[43] Deka, R. C., Ivanova Shor, E. A., Shor, A. M., Nasluzov, V. A., Vayssilov, G. N., Rösch, N., *J. Phys. Chem. B* **2005,** *109,* 24304–24310.
[44] Vayssilov, G. N., *Catal. Rev. Sci. Eng.* **1997,** *39,* 209–251.
[45] de Man, A. J. M., Sauer, J., *J. Phys. Chem.* **1996,** *100,* 5025–5034.
[46] Sinclair, P. E., Sankar, G., Catlow, C. R. A., Thomas, J. M., Maschmeyer, T., *J. Phys. Chem. B* **1997,** *101,* 4232–4237.
[47] Vayssilov, G. N., van Santen, R. A., *J. Catal.* **1998,** *175,* 170–174.
[48] Damin, A., Bordiga, S., Zecchina, A., Lamberti, C., *J. Chem. Phys.* **2002,** *117,* 226–237.
[49] Sierka, M., Eichler, U., Datka, J., Sauer, J., *J. Phys. Chem. B* **1998,** *102,* 6397–6404.
[50] Strodel, P., Neyman, K. M., Knözinger, H., Rösch, N., *Chem. Phys. Lett.* **1995,** *240,* 547–552.
[51] Coliccia, S., Marchese, L., Martra, G., *Microporous Mesoporous Mater.* **1999,** *30,* 43–56.
[52] Vayssilov, G. N., Rösch, N., *J. Phys. Chem. B* **2001,** *105,* 4277–4284.
[53] Vayssilov, G. N., Rösch, N., *J. Catal.* **1999,** *186,* 423–432.
[54] Vayssilov, G. N., Rösch, N., *Phys. Chem. Chem. Phys.* **2002,** *4,* 146–148.
[55] Rice, M. J., Chakraborty, A. K., Bell, A. T., *J. Phys. Chem. B* **2000,** *104,* 9987–9992.
[56] McMillan, S. A., Snurr, R. Q., Broadbelt, L. J., *Microporous Mesoporous Mater.* **2004,** *68,* 45–53.
[57] Sobalik, Z., Tvaruzkova, Z., Wichterlova, B., *J. Phys. Chem. B* **1998,** *102,* 1077–1085.
[58] Sobalik, Z., Sponer, J. E., Tvaruzkova, Z., Vondrova, A., Kuriyavar, S., Wichterlova, B., *Stud. Surf. Sci. Catal.* **2001,** *135,* 136–140.
[59] Aleksandrov, H. A., Vayssilov, G. N., Rösch, N., *Stud. Surf. Sci. Catal.* **2005,** *158,* 593–600.
[60] Aleksandrov, H. A., Vayssilov, G. N., Rösch, N., **2008,** *to be published.*
[61] Biscardi, J. A., Meitzner, G. D., Iglesia, E., *J. Catal.* **1998,** *179,* 192–202.
[62] Barbosa, L. A. M. M., van Santen, R. A., Hafner, J., *J. Am. Chem. Soc.* **2001,** *123,* 4530–4540.
[63] Kachurovskaya, N. A., Zhidomirov, G. M., van Santen, R. A., *Res. Chem. Intermed.* **2004,** *30,* 99–103.
[64] Benco, L., Bucko, T., Hafner, J., Toulhoat, H., *J. Phys. Chem. B* **2005,** *109,* 20361–20369.
[65] Kucera, J., Nachtigall, P., *Phys. Chem. Chem. Phys.* **2003,** *5,* 3311–3317.
[66] Vayssilov, G. N., Staufer, M., Belling, T., Neyman, K. M., Knozinger, H., Rösch, N., *J. Phys. Chem. B* **1999,** *103,* 7920–7928.
[67] Nachtigallova, D., Nachtigall, P., Sauer, J., *Phys. Chem. Chem. Phys.* **2001,** *3,* 1552–1559.
[68] Nachtigallova, D., Nachtigall, P., Sierka, M., Sauer, J., *Phys. Chem. Chem. Phys.* **1999,** *1,* 2019–2026.
[69] Silhan, M., Nachtigallova, D., Nachtigall, P., *Phys. Chem. Chem. Phys.* **2001,** *3,* 4791–4795.
[70] Hadjiivanov, K. I., Vayssilov, G. N., *Adv. Catal.* **2002,** *47,* 307–511.
[71] Hadjiivanov, K., Knözinger, H., *Chem. Phys. Lett.* **1999,** *303,* 513–520.
[72] Salla, I., Montanari, T., Salagre, P., Cesteros, Y., Busca, G., *Phys. Chem. Chem. Phys.* **2005,** *7,* 2526–2532.
[73] Salla, I., Montanari, T., Salagre, P., Cesteros, Y., Busca, G., *J. Phys. Chem. B* **2005,** *109,* 915–922.

[74] Nachtigallova, D., Bludsky, O., Otero Arean, C., Bulanek, R., Nachtigall, P., *Phys. Chem. Chem. Phys.* **2006,** *8,* 4849–4852.
[75] Garrone, E., Bulanek, R., Frolich, K., Otero Arean, C., Rodriguez Delgado, M., Turnes, G., Palomino, G., Nachtigallova, D., Nachtigall, P., *J. Phys. Chem. B* **2006,** *110,* 22542–22550.
[76] Otero Arean, C., Rodriguez Delgado, M., Lopez Bauca, C., Vrbka, L., Nachtigall, P., *Phys. Chem. Chem. Phys.* **2007,** *9,* 4657–4661.
[77] Miessner, H., *J. Am. Chem. Soc.* **1994,** *116,* 11522–11530.
[78] Hadjiivanov, K., Ivanova, E., Dimitrov, L., Knözinger, H., *J. Mol. Struc.* **2003,** *661,* 459–463.
[79] Dougherty, T. P., Grubbs, W. T., Heilweil, E. J., *J. Phys. Chem.* **1994,** *98,* 9396–9399.
[80] Wang, X. L., Wovchko, E. A., *J. Phys. Chem. B* **2005,** *109,* 16363–16371.
[81] Ivanova, E., Mihaylov, M., Aleksandrov, H. A., Daturi, M., Thibault-Starzyk, F., Vayssilov, G. N., Rösch, N., Hadjiivanov, K. I., *J. Phys. Chem. C* **2007,** *111,* 10412–10418.
[82] Ivanova, E., Mihaylov, M., Thibault-Starzyk, F., Daturi, M., Hadjiivanov, K., *J. Catal.* **2005,** *236,* 168–171.
[83] Benco, L., Bucko, T., Grybos, R., Hafner, J., Sobalik, Z., Dedecek, J., Sklenak, S., Hrusak, J., *J. Phys. Chem. C* **2007,** *111,* 9393–9402.
[84] Neyman, K. M., Strodel, P., Ruzankin, S. P., Schlensog, N., Knözinger, H., Rösch, N., *Catal. Lett.* **1995,** *31,* 273–285.
[85] Bludsky, O., Silhan, M., Nachtigall, P., Bucko, T., Benco, L., Hafner, J., *J. Phys. Chem. B* **2005,** *109,* 9631–9638.
[86] Rep, M., Palomares, A. E., Eder-Mirth, G., van Ommen, J. G., Rösch, N., Lercher, J. A., *J. Phys. Chem. B* **2000,** *104,* 8624–8630.
[87] Argo, A. M., Odzak, J. F., Lai, F. S., Gates, B. C., *Nature* **2002,** *415,* 623–626.
[88] Alexeev, O. S., Li, F., Amiridis, M. D., Gates, B. C., *J. Phys. Chem. B* **2005,** *109,* 2338–2349.
[89] Argo, A. M., Odzak, J. F., Goellner, J. F., Lai, F. S., Xiao, F.-S., Gates, B. C., *J. Phys. Chem. B* **2006,** *110,* 1775–1786.
[90] Ferrari, A. M., Neyman, K. M., Mayer, M., Staufer, M., Gates, B. C., Rösch, N., *J. Phys. Chem. B* **1999,** *103,* 5311–5319.
[91] Vayssilov, G. N., Rösch, N., *Phys. Chem. Chem. Phys.* **2005,** *7,* 4019–4026.
[92] Ivanova Shor, E. A., Nasluzov, V. A., Shor, A. M., Vayssilov, G. N., Rösch, N., *J. Phys. Chem. C* **2007,** *111,* 12340–12351.
[93] Petrova, G. P., Vayssilov, G. N., Rösch, N., *Chem. Phys. Lett.* **2007,** *444,* 215–219.
[94] Mikhailov, M. N., Kustov, L. M., Mordkovich, V. Z., *Russ. Chem. Bull. Int. Ed.* **2007,** *56,* 397–406.
[95] Kawi, S., Gates, B. C., *J. Phys. Chem.* **1995,** *99,* 8824–8830.
[96] Petrova, G. P., Vayssilov, G. N., Rösch, N., *J. Phys. Chem. C* **2007,** *111,* 14484–14492.
[97] Benco, L., Bucko, T., Hafner, J., Toulhoat, H., *J. Phys. Chem. B* **2005,** *109,* 22491–22501.
[98] Barbosa, L. A. M. M., van Santen, R. A., *J. Phys. Chem. C* **2007,** *111,* 8337–834.
[99] Aleksandrov, H. A., Vayssilov, G. N., Rösch, N., *J. Mol. Catal. A* **2006,** *256,* 149–155.
[100] Kazansky, V. B., Serykh, A. I., *Phys. Chem. Chem. Phys.* **2004,** *6,* 3760–3764.
[101] Kazansky, V. B., Serykh, A. I., Anderson, B. G., van Santen, R. A., *Catal. Lett.* **2003,** *88,* 211–217.
[102] Pidko, E. A., van Santen, R. A., *J. Phys. Chem. C* **2007,** *111,* 2643–2655.
[103] Pidko, E. A., Hensen, E. J. M., van Santen, R. A., *J. Phys. Chem. C* **2007,** *111,* 13068–13075.
[104] Heyden, A., Hansen, N., Bell, A. T., Keil, F. J., *J. Phys. Chem. B* **2006,** *110,* 17096–17114.
[105] Hansen, N., Heyden, A., Bell, A. T., Keil, F. J., *J. Phys. Chem. C* **2007,** *111,* 2092–2101.
[106] Mellot, C. F., Cheetham, A. K., Shani Harms, S., Scott Savitz, S., Gorte, R. J., Myers, A. L., *J. Am. Chem. Soc.* **1998,** *120,* 5788–5792.
[107] Mellot, C. F., Davidson, A., Eckert, J., Cheetham, A. K., *J. Phys. Chem.* **1998,** *102,* 2530–2535.

CHAPTER 9

Intermolecular Forces in Zeolite Adsorption and Catalysis

Raul F. Lobo

Contents

Abstract

The concepts of classical intermolecular forces are used to understand isosteric heats of adsorption of small molecules adsorbed within zeolite pores. This is used as an organizing principle to comprehend the complex and multifaceted set of chemical properties of zeolites in general. We correlate molecular properties to the magnitude of molecule–zeolite forces and these to measurable quantities such as heats of adsorption. Several key thermodynamic concepts are reviewed and then the most relevant types of intermolecular forces as applied to microporous materials are described. The interaction of small molecules with zeolite Brønsted acid sites is then described as a chemical interaction, rather than in terms of physical forces. We conclude with a discussion of hydrophobicity as applied to siliceous zeolites.

Keywords: Adsorption, Intermolecular forces, Calorimetry, Catalysis, Zeolites

1. Introduction

As molecules travel down the pore structure of a zeolite, they encounter cations and other chemical groups that expose the molecules to important classes of forces. These forces, and more frequently, the difference between forces for different molecules, ultimately lead to the useful properties of zeolites as adsorbents

Ordered Porous Solids
DOI: 10.1016/B978-0-444-53189-6.00009-3

and catalysts, and should be understood as a foundation to the analysis of material properties in molecular terms. In this chapter, we provide the basis for this understanding with emphasis on the origin and magnitude of the various forces relevant to zeolites. We intend this approach to be an organizing principle to comprehend the complex and multifaceted set of chemical properties of zeolites in general. We explicitly attempt to correlate molecular properties to the magnitude of the forces and these to measurable quantities such as heats of adsorption. To keep the length of this chapter reasonable, it is assumed throughout that we are working at the dilute limit, and that molecule–molecule forces are insignificant. We start with some basic thermodynamic concepts, and then proceed to describe the most relevant types of intermolecular forces as applied to microporous materials and finish with a discussion of hydrophobicity as applied to siliceous zeolites.

2. Basic Thermodynamic Concepts[a]

The *pair potential* $w(r)$—in vacuum—or *potential of mean force*—in a medium—describes the interaction potential between two molecules.[1] The force between the molecules is given by $F = -\mathrm{d}w(r)/\mathrm{d}r$ and shows that the work $w(r)$ that can be done by the force F is also a measure of the available energy or *free energy* of this system of two molecules. The *medium* in the case of microporous materials is the micropore and is different from more classical media such as a liquid solvent. First, in a liquid, the solute molecules can perturb the local structure of the solvent; in a zeolite the framework atoms are essentially fixed in space and barely move in response to the presence of an adsorbate in the pores. Second, when a molecule is introduced into a liquid, work must be done to create a cavity to accommodate the guest molecule; in zeolites, the cavity (the pores) is already there and no energy is needed to form it. A zeolite is, however, also like a liquid in the sense that for two solute molecules their pair potential includes molecule–molecule interactions, and also solute–solvent interactions (or equivalently molecule–molecule and molecule–zeolite interactions). A zeolite, just like a solvent, can also change the properties of the adsorbed molecule (dipole moment and polarizability, geometry) with respect to the molecular values in vacuum.

A molecule in medium has a *cohesive energy* μ^i in a medium i given by the sum of its interactions with the surroundings. Inside a siliceous zeolite pore, these interactions are mainly the interaction between the molecule and Si–O–Si groups that surround the pore. It is possible to relate the cohesive energy to the pair potential $w(r)$. For instance, consider the pair potential given by a power law such as

$$\begin{aligned} w(r) &= -C/r^n \quad \text{for } r > \sigma \\ w(r) &= \infty \quad\quad\;\; \text{for } r \le \sigma \end{aligned} \tag{1}$$

[a] The description of intermolecular forces used here follows closely the treatment used by Israelachvili [1].

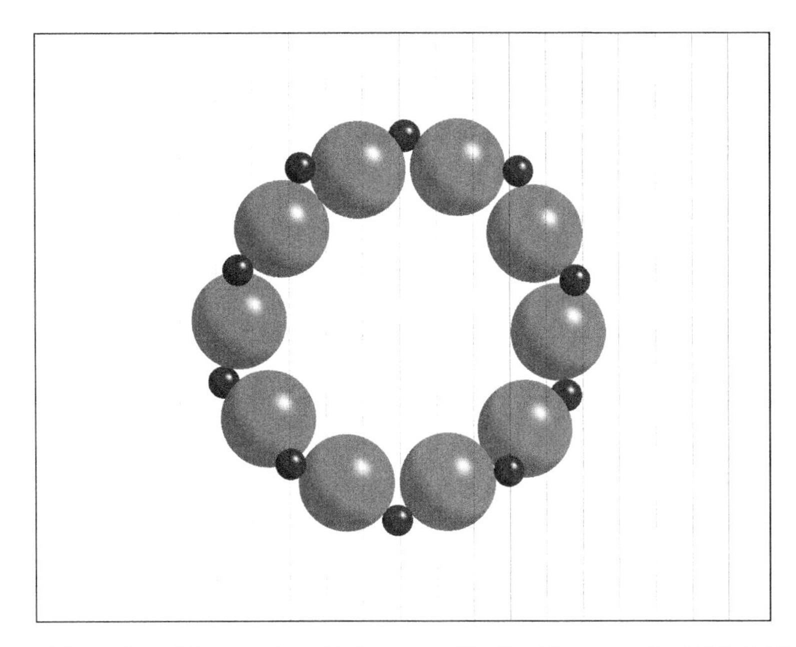

Figure 9.1 Illustration of the pore size of a 10-ring zeolite (in this case zeolite ZSM-5). The effective pore opening is 5.5Å and the oxygen-to-oxygen distance is 8.1 Å. (See color insert.)

where σ is the hard sphere diameter of the molecule and r is the distance between the centre of the molecule and another atom (or molecule). We can calculate μ^i by summing all the pair potentials over all space, and then make use of the fact that zeolites are crystalline and atoms are located at specific positions in space. Alternatively, we can estimate this value by considering only the first row of zeolite oxygen atoms that surround the molecule and placing the molecule in the centre of the pore. For a 10-ring zeolite (see Fig. 9.1), the pore opening is ~5.5 Å (see the effective pore dimensions of the MFI[b] framework type in the Atlas of zeolite structure-types[2] or the web[3]) and the distance between oxygen atoms at opposite ends of the pore is $2 \times r_{\text{pore}} \sim 5.5 + 2 \times 1.3 = 8.1$Å (the effective pore diameter reported in the Atlas plus twice the ionic oxygen radius of 1.3 Å). The net energy change by taking a molecule from the gas phase to the centre of a zeolite 10-ring pore is given then by

$$\mu^i_{\text{zeo}} \approx 10\ w(r_{\text{pore}}) \quad r_{\text{pore}} = 4.05\text{Å} \tag{2}$$

where $w(r)$ describes, in particular, the pair potential between zeolite oxygen atoms and the adsorbed molecule. This value is a lower bound of the cohesive energy because the off-axis position may be more favourable than the centre, and because contributions from other zeolite oxygen and silicon atoms further away

[b] We will use the three-letter-code described in the Atlas to designate unambiguously the topology of the framework of the various zeolite materials that will be discussed in this chapter.

have not been included. This function already tells us, for instance, that the cohesive energy of a molecule in a zeolite decreases as the effective pore size increases, since $w(r)$ is a monotonically decreasing function of r at the relevant molecular distances. Consequently, the molar cohesive energy U of an adsorbed molecule can be calculated simply by multiplying by Avogadro's number $U = -N_A\mu^i_{zeo}$.

The cohesive energy of a molecule can have different values in different parts of a system (say, in the zeolite pores and in the gas phase). Then, at equilibrium, the concentrations of the molecules in the two regions is given by the *Boltzmann distribution*

$$X_1 = X_2 \exp\left[-\left(\mu^i_1 - \mu^i_2\right)/k_B T\right] \quad (3)$$

where X_1 and X_2 are dimensionless concentrations (usually mole fractions or volume fractions) and k_B is the Boltzmann constant. One can expect differences in the concentration of species i if the differences in chemical potential are substantially larger than the thermal energy (at 300 K $k_B T$ is ~2.5 kJ/mol and at 600 K is ~5 kJ/mol). This equation can be restated as

$$\mu^i_1 + k_B T \ln X_1 = \mu^i_2 + k_B T \ln X_2 \quad (4)$$

Here, we assume the phases mix ideally and the activity coefficients can then be assumed to be one. Also, this equation is valid only if the condition of equilibrium between the two states is fulfilled. This expression can be extended to a system with many states by

$$\mu^i_n + k_B T \ln X_n = \mu = \text{constant for all states } n = 1, 2, 3 \ldots$$

In such a system, molecules will flow between different states until these n equations are satisfied. The term μ is also known as the *chemical potential*, the total free energy per molecule including its interactions with its medium and its thermal energy. The term $k_B \ln X_n$ is the entropic contribution to the total chemical potential (again assuming ideal mixing).

3. Dispersion (London) Forces

Dispersion forces are one of the three components to the total van der Waals force (in addition to induced dipole and orientation polarization forces).[1] They are very important because they are always present, independently of the molecular and geometric properties of the molecules. Dispersion forces are, in general, complex to calculate precisely because their magnitude can be affected by the presence of other molecules, that is, it is a non-additive interaction. The origin of this force can be thought of to be a quantum-mechanical induced polarizability force, and it is described for two different atoms 1 and 2 by London's expression

$$w(r) = -\frac{3}{2}\frac{\alpha_{0,1}\alpha_{0,2}}{(4\pi\varepsilon_0)^2 r^6}\frac{I_1 I_2}{(I_1 + I_2)} \tag{5}$$

where $\alpha_{0,i}$ is the electronic polarizability of molecule i, ε_0 is the permittivity of free space (8.854×10^{-12} $C^2\,J^{-1}\,m^{-1}$) and I is the ionization potential (in J). Table 9.1 summarizes the values of the electronic polarizability for small molecules and a few molecular groups, and Table 9.2 lists the ionization potentials of a selected group of molecules. Polarizability values span roughly two orders of magnitude for this group of molecules, and thus this force can vary greatly, depending on the identity of the molecules in question. Note that water molecules have a rather low polarizability similar to the polarizability of argon.

The energetics of adsorption of non-polar molecules in high-silica zeolites is dominated by dispersion forces, and thus these forces dictate the value of cohesive energy gained by a molecule as it enters a zeolite from the gas phase. The isosteric heat of adsorption (H_{st}) in the limit of zero coverage, a quantity that can be obtained experimentally, is a good estimate of this change in cohesive energy. Fortunately, Savitz *et al.*[4] have conducted a comprehensive study of the adsorption of hydrocarbons and other simple molecules on siliceous zeolites. These data will be used to further elucidate the role of dispersion forces in zeolite–adsorbate interactions. Table 9.3 shows the zeolites used in this study and their relevant characteristics. Three unidimensional zeolites of different pore sizes ZSM-22 ([SiO_2]-TON, 10-ring), ZSM-12 ([SiO_2]-MTW, 12-ring) and UTD-1 ([SiO_2]-DON, 14-ring) were used in addition to multidimensional 10-ring ([SiO_2]-MFI) and 12-ring

Table 9.1 Electronic polarizabilities[a] α_o of atoms, molecules and molecular groups[b]

Atoms and molecules					
He	0.208	CO	1.953	C_2H_4	4.188
H_2	0.787	NH_3	2.103	CH_3OH	3.081
H_2O	1.501	HCl	2.515	Cl_2	4.4
O_2	1.562	CO_2	2.507	$CHCl_3$	8.2
N_2	1.710	CH_4	2.448	C_6H_6	10.453
Ar	1.664	C_2H_6	4.226	CCl_4	10.002
		C_3H_8	5.921		
Molecular groups					
	C–O–H	1.28	CH_2	1.84	
	C–O–C	1.13	Si–O–Si	1.39[c]	
	C–NH_2	2.03	Si–OH	1.60	

[a] Polarizabilities given in units of $(4\pi\varepsilon_0)\text{Å}^3=(4\pi\varepsilon_0)\,10^{-30}\,m^3=1.11\times10^{-40}\,C^2\,m^2\,J^{-1}$.
[b] Data from NIST Chemistry Webook[8] if available (experimental values). Otherwise from compilation in Israelachvili.
[c] This value is probably an underestimation. See discussion below.

Table 9.2 Ionization potential[a] of selected atoms and molecules (in eV)[b]

Atoms and molecules					
He	24.59	CO	14.01	C_2H_4	10.51
H_2	15.42	NH_3	10.07	CH_3OH	10.84
H_2O	12.62	HCl	12.74	Cl_2	11.48
O_2	12.07	CO_2	13.78	$CHCl_3$	11.37
N_2	15.58	CH_4	12.61	C_6H_6	9.24
Ar	15.56	C_2H_6	11.52	CCl_4	11.47

[a] Data from NIST Chemistry Webook.[8]
[b] $1\ J = 6.242 \times 10^{18}$ eV.

Table 9.3 Properties of high-silica zeolites used for adsorption measurements of hydrocarbons in Savitz *et al.*[4a]

Zeolite framework type	V_{pore} (cm^3/g)[b]	Si/Al ratio	Pore type	Pore size (*n*-rings)	Pore/Cavity dimensions (Å)
TON	0.069	52	1-D	10	5.5×4.4
MTW	0.097	140	1-D	12	6.2×5.5
DON	0.106	∞	1-D	14	10×7.5
FER	0.105	22	2-D	8, 10	4.8×3.5, 5.4×4.2
MFI	0.173	300	2-D	10, 10	5.3×5.6, 5.5×5.1
FAU	0.245	∞	3-D spherical		12.6

[a] Reproduced with permission from Savitz *et al.*[4]
[b] V_{pore} = Pore volume.

([SiO_2]-FAU) materials. Ferrierite (FER) has a two-dimensional pore system with 10- and 8-ring pores running perpendicular to each other.

The isosteric heats of adsorption for methane, ethane and propane in these zeolites (Table 9.4) follow the trends expected from the pore size of the zeolites and the polarizability of the molecules. For all molecules, the heat of adsorption decreases as the size of the pore or cavity increases. For a given structure, the heat of adsorption increases as the size of the molecule increases. In fact, the increments in heat of adsorption follow quite closely the changes in electronic polarizability (α_0) of the adsorbate. For example, for silicalite-1 ([SiO_2]-MFI) the heats of adsorption increase in the ratios 1:1.5:2, while the polarizabilities increase in the ratios 1:1.7:2.4, a consequence of London's formula. Also note that the cohesive energy of the solid hydrocarbons (sum of the latent heats of melting and vapourization, $\Delta H_m + \Delta H_v$) is substantially lower than the isosteric heat of adsorption for *all* zeolites, thus highlighting the strength of the van der Waals forces between the zeolite and the hydrocarbon.

Table 9.4 Zero-coverage isosteric heats of adsorption of small hydrocarbons in high-silica zeolites at 25 °C[a]

Zeolite framework type	H_{st} (kJ/mol)			CH_4 H_{st} (kJ/mol) from theory
	CH_4	C_2H_6	C_3H_8	
FER	27.7	41.7	53.3	
TON	27.2	39.0	48.8	22.4
MFI	20.9	31.1	41.4	
MTW	20.9	29.5	37.6	19.7
UTD	14.2	22.2	28.1	14.2
FAU	11.0	19.7	26.2	
$\Delta H_m+\Delta H_v$	9.1	17.5	18.6	–

[a] Reproduced with permission from Savitz *et al.*[4]

The cohesive energy is closely related to the change in internal energy per mole upon adsorption, a quantity that is called *differential heat of adsorption* (U_d). In turn, the differential and *isosteric heats* of adsorption are related by

$$H_{st} = U_d + ZRT \tag{6}$$

where $Z = PV/RT$ is the compressibility factor in the bulk gas phase (for an ideal gas $Z = 1$). On the basis of this information, for instance, we can attempt to estimate the isosteric heat of adsorption of methane in zeolite ZSM-22 (TON) for methane. $CH_4 = 1$, Si–O–Si = 2, $\alpha_{0,1} / (4\pi\varepsilon_0) = 2.448 \times 10^{-30}$ m, $I_1 = 12.61$ eV $= 2.02\times10^{-18}$J, $\alpha_{0,2} / (4\pi\varepsilon_0) = 1.39\ \ 10^{-30}$ m, $I_2 \approx 2 \times 10^{-18}$J (we use a typical value), $r_{pore} = 3.825 \times 10^{-10}$ m. Using London's equation we obtain $w(r) = -1.64 \times 10^{-21}$ J. Consequently, $U_{st} = N_A\ 10\ w(r) = -9.86$ kJ/mol and $H_{st} = -7.37$ kJ/mol.

Compared to experiment, this estimate is quite low, reflecting first that these estimates are a minimum value since they exclude interaction with many nearby atoms, but probably also reflect that the value of $\alpha_{0,2}$ for Si–O–Si groups is underestimated (on the basis of the London formula it is not possible to reach the observed isosteric heats otherwise).

Alternatively, one can consider an *average* potential energy over a layer of oxygen atoms distributed uniformly on the wall of a cylindrical pore as reported by Savitz *et al.*[4] based on the model of Everett and Powl,[5] using the Lennard-Jones 12–6 potential, with parameters ε_{1s} and σ_{12}. The potential energy of this model is given by

$$w(r) = \frac{5}{2}\pi\varepsilon_{1s}\left\{\frac{21}{32}\left(\frac{\sigma_{12}}{R}\right)^{10}\left[\sum_k^{\infty}\alpha_k\left(\frac{r}{R}\right)^{2k}\right] - \left(\frac{\sigma_{12}}{R}\right)^{4}\left[\sum_k^{\infty}\beta_k\left(\frac{r}{R}\right)^{2k}\right]\right\} \tag{7}$$

where α_k and β_k are defined by

$$\alpha_k^{1/2} = \frac{\Gamma(-4.5)}{\Gamma(-4.5-k)\Gamma(k+1)}$$
$$\beta_k^{1/2} = \frac{\Gamma(-1.5)}{\Gamma(-1.5-k)\Gamma(k+1)} \tag{8}$$

The gas–solid energy parameter ε_{1s} is given by $\varepsilon_{1s} = 6/5\pi\rho_{os}\varepsilon_{12}\sigma_{12}$ where ρ_{Os} is the number of oxygen atoms per unit of area on the pore surface, and the collision diameter (sum of molecular radii) is σ_{12}=3.195 Å using values of 2.7 Å for the diameter of an oxygen atom and 3.691 Å for a methane molecule. The results of the calculation (Fig. 9.2) show that the shape of the potential energy as a function of the distance from the centre of the pore is different for the three zeolites. For ZSM-22 (TON), the minimum is at the centre of the pore while the minima for ZSM-12 (MTW) and UTD-1 (DON) are closer to the wall. In fact, the shape of the curve for DON is quite similar to the potential for a flat wall indicating that the transition from a microporous environment to a mesoporous molecular environment (i.e. one in which interactions resemble a flat surface) starts to be observed even for 14-ring zeolites.

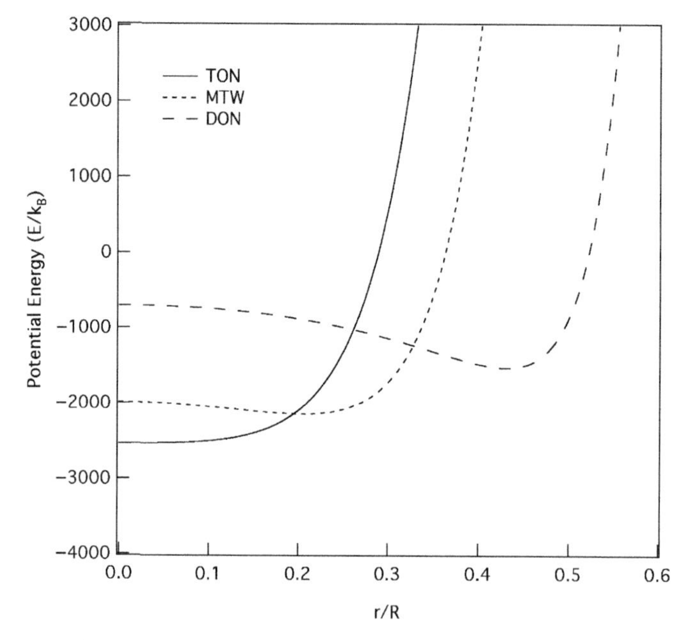

Figure 9.2 Potential energy of a methane molecule in 1-D zeolites ZSM-22, ZSM-12 and UTD-1 ([SiO_2]-TON, MTW and DON) as a function of distance from the pore centre. The value of the well depth was obtained from an independent calculation ϵ_{12}/k_B=133.33. The average radii for the three zeolites (framework-types TON, MTW and DON) are 3.825, 4.275 and 5.725 Å, respectively, and the average densities of oxygen atoms on the pore surfaces are 0.166, 0.179 and 0.184 O atoms/Å^2.

The desired results are predictions of the isosteric heats of adsorption, and these can be obtained from the Boltzmann weighted average energy

$$\overline{w} = \frac{\int_0^R w(r)e^{-w(r)/k_B T} r\, dr}{\int_0^R \left(e^{-w(r)/k_B T} - 1\right) r\, dr} \tag{9}$$

and by the following equation

$$H_{\text{st}} = -N_A\, \overline{w} + RT \tag{10}$$

The comparison of this theory to experiment is very reasonable (Table 9.4). The average difference is 8%, larger than experimental error but small nevertheless, with the highest deviation for ZSM-22. The best agreement is found for UTD-1, a zeolite with pores that are quite uniform—that is, framework oxygen atoms are at the same distance from the pore centre as one moves along the pore direction. The internal surface of the pores of ZSM-22 and ZSM-12, on the contrary, is corrugated and the difference between theory and experiment may reflect this deviation from the model assumptions. The important conclusion is that simple models based on van der Waals interactions can capture semiquantitatively the isosteric heats of adsorption of simple non-polar molecules on zeolite microporores. For large molecules like dodecane and above, van der Waals forces often dominate the interaction molecule–zeolite over other forces, and can be used as a first estimate of the selectivity of adsorption respect to the gas phase.

Finally, in Fig. 9.3 the isosteric heat of adsorption at zero coverage for several non-polar molecules is plotted versus the electronic polarizability for zeolite $[SiO_2]$-MFI. These data show that there is indeed a very good correlation between the heats of adsorption and the electronic polarizability. For larger molecules, this correlation is expected to break down as London's model assumes the molecules can be represented by one point, and clearly this is already incorrect even for molecules as small as propane. Nevertheless, the correlation is excellent.

4. Induced Dipole and Quadrupole–Charge Interactions

We next consider the interaction of molecules with cations coordinated to the zeolite pore walls. An electric field induces a dipole, u_{ind}, on all molecules (polar and non-polar) that is proportional to the polarizability and the strength of the electric field ($u_{\text{ind}} = \alpha E$). The size of this induced dipole can be substantial (of the order of 1 Debye), and thus this interaction can be quite important in determining the overall energetics of an adsorbed molecule inside a zeolite pore. For non-polar molecules, the polarizability is composed only of the *electronic* polarizability term such that $\alpha = \alpha_0$, but for polar molecules the polarizability contains another term denoted *orientational polarizability*

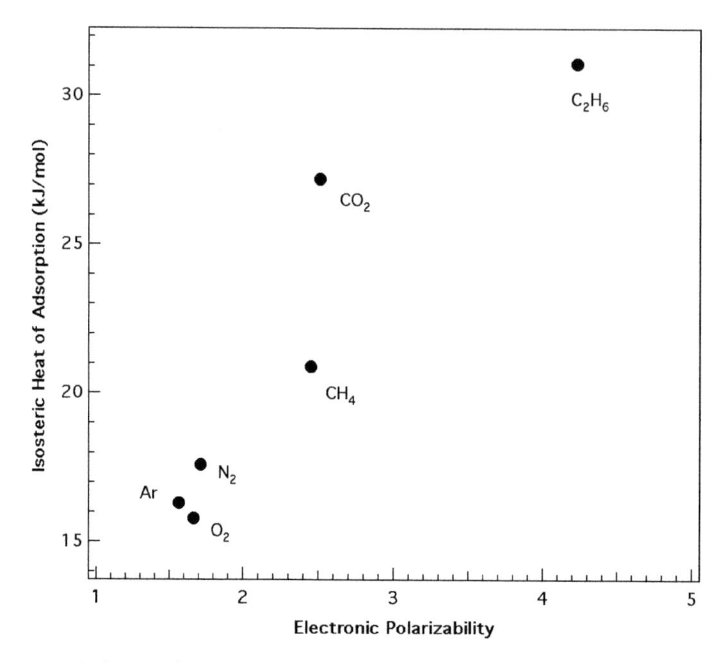

Figure 9.3 Isosteric heats of adsorption on [SiO_2]-MFI zeolites versus electronic polarizability. Note the close correlation between molecular properties such as electronic polarizability and observable macroscopic properties such as isosteric heats of adsorption. (Data from Dunne *et al.*).[22,23]

that arises from the effect of an external field on the average orientation of the molecule in space. The net ion-induced dipole interaction free energy is given by

$$w(r) = -\frac{(ze)^2\alpha}{2(4\pi\varepsilon_0\varepsilon)^2 r^4} = -\frac{(ze)^2}{2(4\pi\varepsilon_0\varepsilon)^2 r^4}\left(\alpha_0 + \frac{u^2}{3k_B T}\right) \tag{11}$$

where e is the charge of the electron, z is the valence of the cation, u is the permanent dipole of the molecule and ε is the dielectric constant of the medium. First, note that this interaction depends on the negative fourth power of r, and thus it is a longer-range interaction than the dispersion forces. Second, inside the zeolite pores it is not obvious what one should use as the value of the dielectric constant ε. The dielectric constant of quartz is 4.8, while the one of amorphous silica glass is 3.8, and one may think that a density-scaled dielectric constant could be used as a good approximation.[c] This is adequate within mesoscopic and larger length scales; however, in the limit of zero coverage, the length scale of interest are distances comparable to the pore diameter or smaller, so a more appropriate value of ε *inside* the zeolite pore is $\varepsilon \sim 1$ (i.e. comparable to air or vacuum). At the same time, the weaker dependence on distance makes it doubly important that we understand the role of

[c] To the best of our knowledge the dielectric constant of siliceous zeolites has not been measured.

other adsorbed molecules can have on this force. If the pore is filled with water molecules ($\varepsilon_{H_2O} = 88$ in the bulk), for instance, the magnitude of this interaction will decrease by a factor of ~80–90.[d]

A non-uniform distribution of charge within the molecule gives rise to a second type of interaction with an electric field even when a molecule has no permanent dipole. This distribution of charge is called a quadrupole Θ, and the interaction between a quadrupole and the electric field generated by a cation can be very important in zeolites as we will see shortly. In general, the quadrupole of a molecule is a tensor quantity, and its interaction with a charge (a cation) is angle dependent.[6] Frequently though, we will be interested in molecules (such as N_2 and O_2) that have a linear distribution of charge; and in such cases, the quadrupole can be thought of as a two equidirectional but opposed dipoles (−++−). The energy of interaction $w(r)$ between a linear molecule with quadrupole moment and a point charge is given by

$$w(r) = \frac{ze}{(4\pi\varepsilon_0\varepsilon)} \frac{\Theta(3\cos^2\theta - 1)}{2r^3} \tag{12}$$

This interaction can be attractive or repulsive depending on molecular orientation and the sign of the charge and quadrupole. For N_2 $\Theta = -1.39$ D Å and the interaction is a maximum when the molecule axis is aligned along the cation-molecule direction ($\theta = 0, (3\cos^2\theta - 1) = 2$), and minimized with a different sign when the molecule is perpendicular to the cation-molecule direction ($\theta = \pi/2, (3\cos^2\theta - 1) = -1$). Since the angle average of $\langle\cos^2\theta\rangle = 1/3$, the average quadrupole interaction can be neglected at high temperatures when molecules tumble isotropically in the zeolite pores (i.e. $\langle 3\cos^2\theta - 1\rangle = 0$).

Cations in a zeolite can be highly exposed to molecules in the pores, can be partially blocked in such a way that only small molecules have direct contact with the cation, or can be coordinated to the zeolite framework such that there is no—or only very weak—interaction with adsorbed molecules. Because of this distribution of cation coordination environments, adsorbed molecules will find a complex landscape of adsorption sites. Moreover, different zeolites will show different zero-coverage heats of adsorption precisely because the distribution of coordination environments is different in each case. Because of the negatively charged zeolite oxygen atoms and due to charge transfer processes to the cation from the framework oxygen atoms, the effective charge of the cation is substantially lower than the charge of an isolated cation (i.e. charge $< e$). To clarify this point, Table 9.5 shows the isosteric heat of adsorption of cation-containing zeolites compared to their all-silica analogues. We can indeed see that the presence of sodium cations in the pores substantially increases the heat of adsorption of the molecules. Figure 9.4 compares the sodium cation affinity of several of the molecules in the table to the

[d] The local dielectric constant of a medium is dependent on the structure of the solvent and inside a micropore such structure is bound to be quite different from the bulk structure of the solvent. The effective dielectric constant will be different from 88, although it is not possible to evaluate the local dielectric constant without recourse to atomistic models of the solvent in the pores. It should be dependent on pore size and zeolite composition.

Table 9.5 Isosteric heats of adsorption of small molecules on MFI-type and FAU-type zeolites[a]

	H_{st} (kJ/mo)				
Adsorbate	$[SiO_2]$-MFI	$\|H^+\|$ $[SiAlO_2]$-MFI	$\|Na^+\|$ $[SiAlO_2]$-MFI	$[SiO_2]$-FAU	$\|Na^+\|$ $[SiAlO_2]$-FAU (NaX)
Ar	15.8		18.0		12.7
CH_4	20.9		26.5	11.0	19.2
SF_6	34.4	35.2	42.0		28.2
O_2	16.3		~17[b]		15.0
N_2	17.6	20.7	24.1		19.9
C_2H_6	31,1	33.3	38.0	19.7	27.0
CO_2	27.2	38.0	50.0		49.1

[a] Data from Dunne and Savitz *et al.*[7,22–24]
[b] This datum estimated from figure in ref. 7.

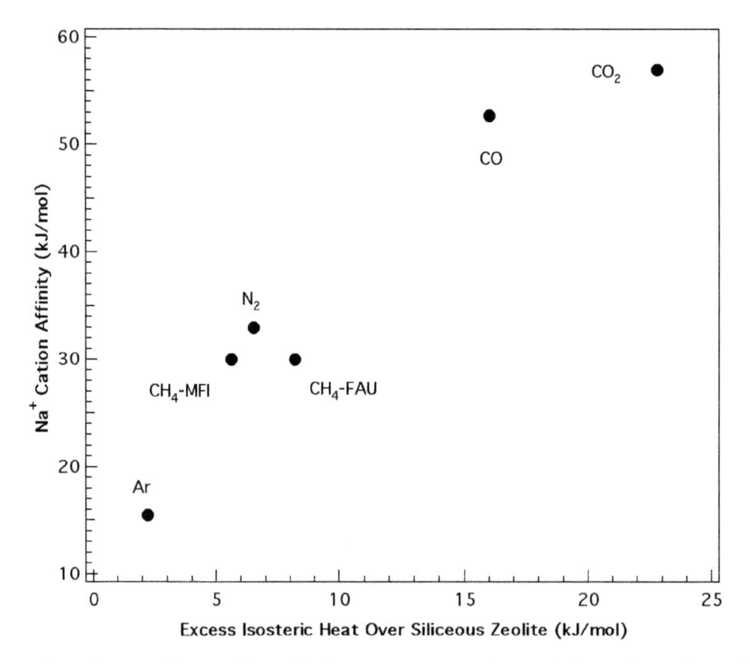

Figure 9.4 Gas-phase sodium cation affinity versus excess isosteric heat above the value of the siliceous zeolite. The data show excellent correlation between the two sets of numbers indicating that the interaction of small molecules with cations inside zeolite pores resemble energetically (and probably structurally) the molecule-cation adducts formed in the gas phase. Sodium cation affinity data from NIST Chemistry Webook and from Table 9.5.

excess isosteric heat of adsorption due to the presence of cations on the zeolite (the difference between the sodium-cation containing zeolite and its all-silica analogue). We can clearly observe a very good trend between the two quantities. Within some scatter, the trend is linear, (although notice that the excess heat for methane in Na–ZSM-5 and in Na-X (FAU) differ by more than 3 kJ/mol). This linear trend is an indication that the geometry of the adsorbed molecule on the cation, and type of interaction in the gas phase and in the zeolite pores are quite similar. However, the magnitude of the excess heat of adsorption is substantially lower than the ΔH_r of the interaction of the molecule with the cation in the gas phase. This large difference is the consequence of the coordination of the negatively charged framework to the cation, that is, the effective weaker charge of the cation.

To illustrate this further, let's consider a specific example of a cation coordinated to an $AlO_{4/2}^-$ tetrahedron on the surface of a high-silica zeolite. For simplicity, we will consider an idealized model of an exposed cation as depicted in Fig. 9.5. This

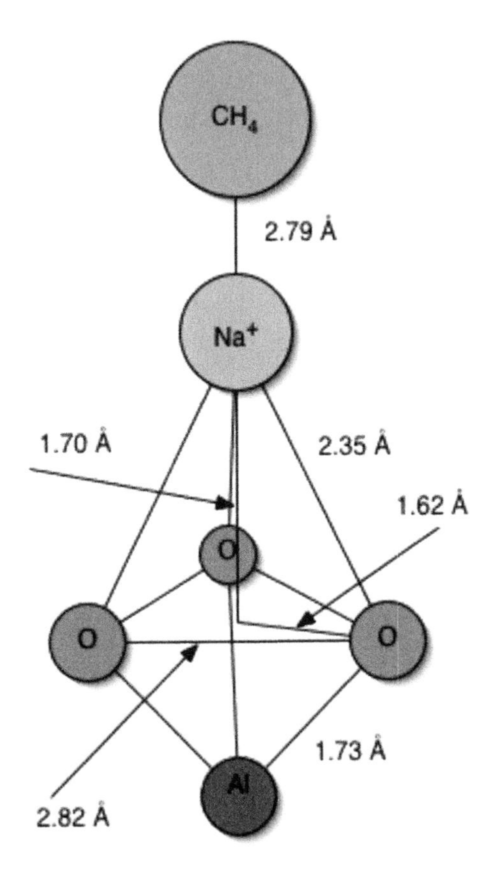

Figure 9.5 Schematic of the coordination of sodium cations to a model $AlO_{4/2}^-$ tetrahedron. Distances are not to scale. Since this site has a threefold axis of symmetry, the Al–O bond distance is 1.73 Å and the angle between the O–Al–O atoms is 109.47° (the tetrahedral angle), all distances can be calculated using elementary geometry. (See color insert.)

figure shows the sodium cation coordinated to three of the oxygen atoms of the alumina tetrahedron. The interaction energy at contact between a methane molecule and the sodium cation is given by the formulae above. For a Na^+-CH_4 distance of 2.79 Å and an effective charge of +1 *e*, $w(r)$=27.8 kJ/mol, much larger than the excess isosteric heat of 8.2 kJ/mol. This is of course because the effective charge on the cation is smaller due to the effect of the negatively charged oxygen atoms. Using 8.2 kJ/mol as the benchmark, the effective charge can be back calculated to be ~0.54 *e*. Using this effective charge, we can then calculate the total interaction for N_2, a molecule with a substantial quadrupole moment. Using a distance of 3.17 Å between Na^+ and the centre of mass of N_2, and assuming N_2 is aligned along the molecule-cation axis, the induced dipole and quadrupole interactions are 3.4 and 6.6 kJ/mol, respectively. The total 10 kJ/mol is very much in agreement with the excess isosteric heat between silicalite ([SiO_2]-MFI) and NaZSM-5 (|Na^+|[$SiAlO_2$]-MFI) of 9 kJ/mol (Table 9.5).

5. Dipole–Charge Interactions

The interaction of polar molecules—molecules with permanent dipole moments—with cations on the surface of a zeolite pore is the last type of electrostatic interaction that will be reviewed here. To start, let's recall that the dipole moment of a polar molecule u is defined as $u=q\,l$ where l is the distance between two charges $+q$ and $-q$. Small molecules have dipole moment of the order of 1 Debye $= 3.336\times10^{-30}$ C m. The energy of interaction between an ion and a dipole is given by

$$w(r) = -\frac{(ze)u\cos\theta}{4\pi\varepsilon_0\varepsilon r^2} \tag{13}$$

where the symbols are as defined before, and θ is the angle between the direction of the dipole moment u and the line joining the centre of mass of the molecule and the ion. This interaction can be attractive or repulsive depending on the angle θ. It is maximum when the angle is 0°. In practice, the interactions between an ion and a polar molecule can be divided into strong and weak depending on whether the interaction at $\theta = 0°$ is large compared to k_BT. In such a case, the molecule will tend to be oriented towards the ion and would be coordinated to it for long periods of time. This is expected for small and multivalent cations interacting with small molecules containing a sizable dipole moment. For this case the energy is given by

$$w(r) = -\frac{(ze)u}{4\pi\varepsilon_0\varepsilon r^2} \tag{14}$$

Table 9.6 Comparison of calculated electrostatic interactions of CO and water molecules with sodium cations in the gas phase to isosteric heat of adsorption in Na–ZSM-5

Interaction	CO kJ/mol	H_2O kJ/mol
Dipole–charge	–3.65	–96.7
Quadrupole–charge	–32.0	–2.9
Induced dipole	–17.9	–34.1
Total	–53.5	–134
ΔH_{st} Silicalite-1	17	30
ΔH_{st} Na–ZSM-5	33[7]	95[25]
$\Delta\Delta H_{st}$	16	65

For weak interactions, one can assume that the molecule tumbles with isotropic motion in the vicinity of the ion and one requires the angle-averaged charge–dipole interaction given by[1]

$$w(r) \approx -\frac{q^2u^2}{6(4\pi\varepsilon_0\varepsilon)^2 k_B T r^4} \quad \text{for } k_B T > \frac{qu}{4\pi\varepsilon_0\varepsilon r^2} \tag{15}$$

For instance, the heat of adsorption of CO on NaZSM-5 (33 $\pm$ 1 kJ/mol at 195 K) and Silicalite-1 (17 $\pm$ 1 kJ/mol)[7] can be compared to the enthalpy of reaction of CO and Na^+(to form $CO{\cdot}Na^+$) in the gas phase ($\Delta H_r = 52.7$ kJ/mol).[5,e] The difference between the two zeolites reflects the additional contribution of the sodium cations to the heat of adsorption, that is, 16 kJ/mol or about 30% of the gas-phase value. This is comparable to what we observed when we estimated the effective charge of sodium cations using the nitrogen molecule. This heat of adsorption on sodium cations has two contributions: one from the dipole–charge interaction, and another from the induced dipole moment on CO by the electric field of the cation. Using the following data ($r_{CO} = 0.2$ nm, $r_{Na} = 0.095$ nm, $q = ze = 1.6 \times 10^{-19}$ C, $\alpha_0 = 1.95 \times 1.11 \times 10{-}40$ C^2 m^2 J^{-1}, $T = 300$ K, $k_B = 1.38 \times 10^{-23}$ J K^{-1}, dipole moment of CO $= 0.11 \times 3.336 \times 10^{-30}$ C m, $\Theta = -9.47 \times 10{-}40$ C m^2), the magnitude of the various interactions can be determined (Table 9.6).

Water is an example of a small molecule with a large dipole moment. Water, in its simplest form, can be treated as a spherical molecule of 0.14 nm in radius. It has a dipole moment $u = 1.85$ D and it can adsorb into a sodium cation (0.095 nm in radius). Calculation of the maximum interaction energy for this complex in the gas phase gives a total of $w(r) = 134$ kJ/mol, including dipole, polarizability and quadrupole interactions. The dipole-charge contribution is by far the most important one. (Note that this value is higher than the value of 95 $\pm$ 7.9 kJ/mol that has been measured experimentally for the reaction $Na^+ + H_2O = Na{\cdot}H_2O^+$ in the gas phase.)[8] The heat of adsorption of water in Na–ZSM-5 and NaA zeolites has

[e] Values of 32 $\pm$ 7.9 kJ/mol have also been reported.

been measured by Dubinin and co-workers[9] giving values in the range of 95–105 kJ/mol. This energy of adsorption is dominated by the dipole–charge interaction (see Table 9.6). For the zeolites, the contribution of the sodium cation corresponds to ~65 kJ/mol, somewhat higher than the expected value on the basis of the partial charge of the sodium cation on the zeolite pores. This is the case because the water molecule can coordinate by hydrogen bonding to the oxygen atoms of the zeolite, adding further to the heat of adsorption.

6. Interactions of Adsorbed Molecules with Acid Sites

The interaction of molecules with Brønsted acid sites in zeolites deserves a separate discussion because the binding of protons to the zeolite framework is very different from the binding of cations like sodium and potassium. Likewise, the coordination of small molecules to acid sites can be quite different from the coordination to alkali-metal cations. Figure 9.6 illustrates the structure and geometry of the acid sites in zeolites as has been determined for high-silica zeolites.[10] The key element of the site is the presence of an Al–OH–Si group. The bond between the hydrogen atom and the zeolite oxygen atom is predominantly covalent in nature, and there is only a small effective charge on the hydrogen atom. This site can coordinate or donate the proton to adsorbed molecules forming 1:1, 2:1 and

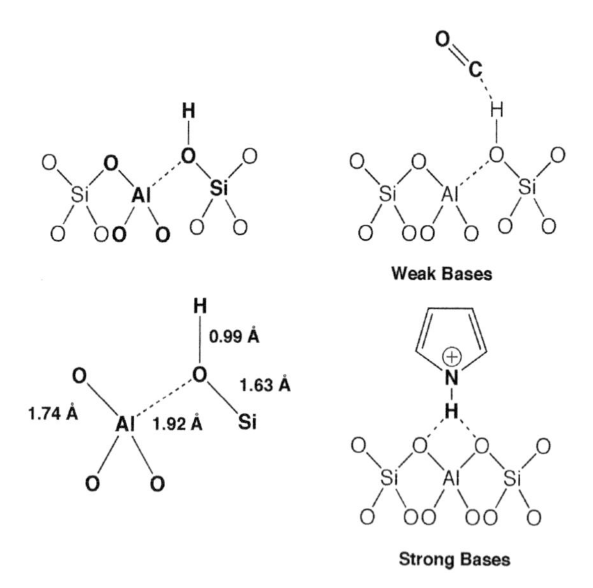

Figure 9.6 Schematic of the structure and geometry of the classical Brønsted acid site in high-silica zeolites. The structures on the right illustrate the coordination of weak and strong bases to the acid site. For strong bases, there is proton transfer and charge separation between the framework and the adsorbed molecule.

sometimes higher adducts. The oxygen atom in the acid site is tri-coordinated, and since it should keep a bond order of 2, there is a weakening of the three bonds, but particularly the O–H bond; this allows for facile transfer of the proton to adsorbed molecules. The top section of Fig. 9.6 also highlights the presence of three additional oxygen atoms coordinated to the aluminum that have Lewis basic character and that can coordinate through hydrogen bonding to molecules that exhibit both hydrogen-acceptor and hydrogen-donating groups, such as water or ammonia. Usually, two out of the three oxygen atoms—but sometimes only one—are physically accessible to the adsorbed molecules due to the specific configuration of the framework atoms with respect to the zeolite pore (see Fig. 9.5). These geometric differences appear to have no major thermochemical consequence[11] and the heat of adsorption of *strong* bases on high-silica zeolites is the same—within experimental error—regardless of the specific geometry of the site.

Figure 9.6 also shows the various bond lengths of the acid site. Note that the Al–OH bond is substantially longer than the other three bonds (1.92 Å vs 1.73 Å); and as a consequence the coordination environment of the aluminum atom is not tetrahedral (as is the case for framework aluminum coordinated to alkali-metal cations), but an intermediate structure between tetrahedral and planar trigonal. The long aluminum–oxygen bond reflects, of course, the fact that this bond is weaker than the other three. Because of the elongation and weakening of the O–H bond, there is a change in the infrared (IR) absorption frequency from 3749 cm^{-1} (for isolated Si–OH groups) to 3614 cm^{-1} (Al–OH–Si).[12] This change in vibrational frequency is also an indication that the ability to donate the proton of a zeolite Brønsted acid site is much larger than the ability of an isolated silanol group.

There are two extremes of interaction of a zeolite acid site *ZOH* with a base *B*. For *weak* bases the adduct forms a hydrogen bond to the hydroxyl group ($ZOH\cdots B$), while for *strong* bases there is complete proton transfer and charge separation ($ZO^{-}\cdots HB^{+}$). And there are, of course, intermediate cases that fall between these two cases. A series of calorimetric, computational and spectroscopic investigations have shown that the molecular property that correlates more closely to the observed thermochemical and spectroscopic measurements is the proton affinity (PA) of the base. If the PA is below 858 kJ/mol for the 1:1 adduct, there is no proton transfer and the complex is described more closely as ZOH•••B. If the PA is above ~858 kJ/mol, there is proton transfer and the complex is better described as $ZO^{-}\cdots HB^{+}$. Ammonia is right at this limit with a PA = 858 kJ/mol, and definitely forms the ammonium ion upon adsorption on acid zeolites.[13] Figure 9.7 shows the IR spectra of 1:1 adducts on zeolite beta at ~1:1 coverage for bases of different PA. The perturbations of the IR spectra increase as the PA increases (see Table 9.7) up to pyridine, where there is proton transfer as confirmed by the observation of the pyridinium ion signatures in the IR spectrum.

As illustrated in Fig. 9.6, the adsorption of weak bases slightly perturbs, but not substantially changes the geometry of the acid site. This weak hydrogen bond interaction is mainly of electrostatic character. It is weak because the effective charge of the hydrogen atom in the ZOH group is small. This is evident from Table 9.5 where we can observe that the excess isosteric heat for HZSM-5 versus NaZSM-5 is small, regardless of the small effective radius of the hydrogen atom.

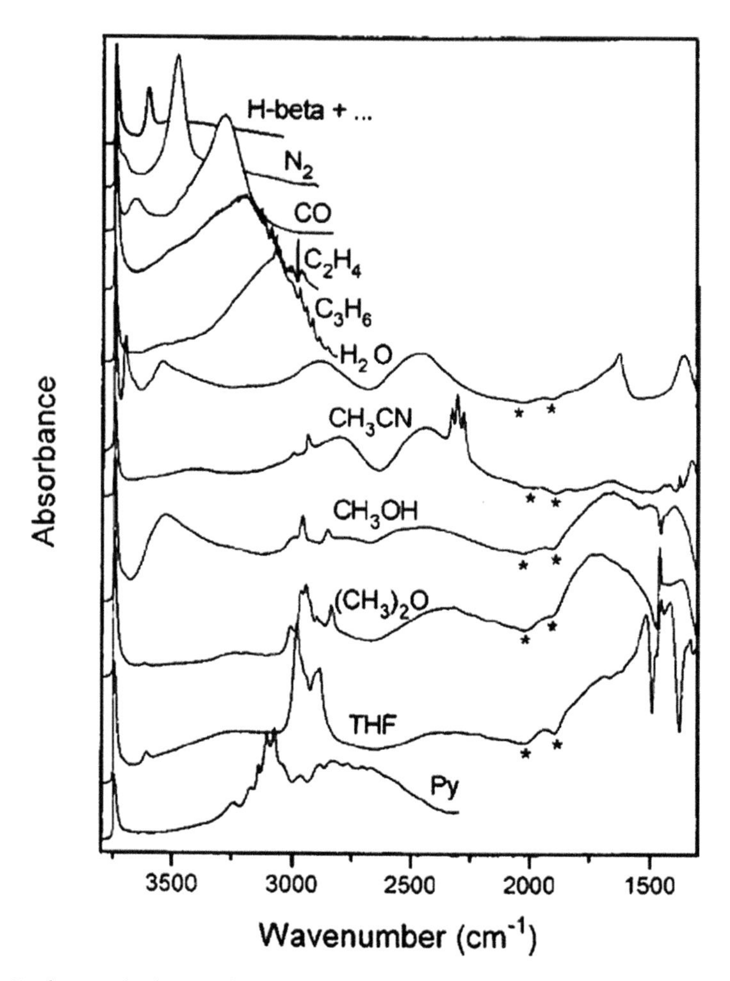

Figure 9.7 Background subtracted infrared (IR) spectra of bases of different proton affinity (PA). Nitrogen and CO conducted at ~100 K, C_3H_6 at 180 K and the rest at room temperature. The perturbation of the IR spectra of clean H-beta increases with PA. The spectra become increasingly more complex due to mixing and doubling of bending and stretching modes. Reproduced with permission from Paze *et al.*[12]

The abstraction of the proton by a strong base, in contrast, changes the geometry of the aluminum from a distorted tetrahedron to a symmetric tetrahedron where all Al–O bonds are similar and about 1.75 Å. For strong bases, both the differential heat of adsorption and the IR spectra of the adducts are independent of the zeolite structure, indicating that the complexes formed by a molecule on all zeolites are very similar (provided there is space to form the complex).

Molecules with hydrogen accepting and donating groups are somewhat different because they can form 1:1, 2:1, and higher order complexes with acid sites in the zeolite pores (see Fig. 9.8). At low coverage (based on the Brønsted acid site ratio), IR spectra indicate that the 1:1 complex dominates for water and methanol, but even before the B:ZOH site ratios approach 1:1, evidence for clusters of molecules

Table 9.7 Comparison of adsorption enthalpies in H–ZSM-5 and silicalite with gas-phase proton affinities

Molecule	Proton affinity (kJ/mol)	H–ZSM-5 ΔH_{ads} (kJ/mol)	Silicalite ΔH_{ads} (kJ/mol)
N_2	494	20.7	17.6
CO	594	26	17
C_2H_4	680		
H_2O	724	90	30
CH_3CN	798	110	75
CH_3OH	774	115	65
$(C_2H_5)_2O$	838	135	70
NH_3	858	145	
$N(CH_3)_3$	938	205	
Pyridine	922	200	

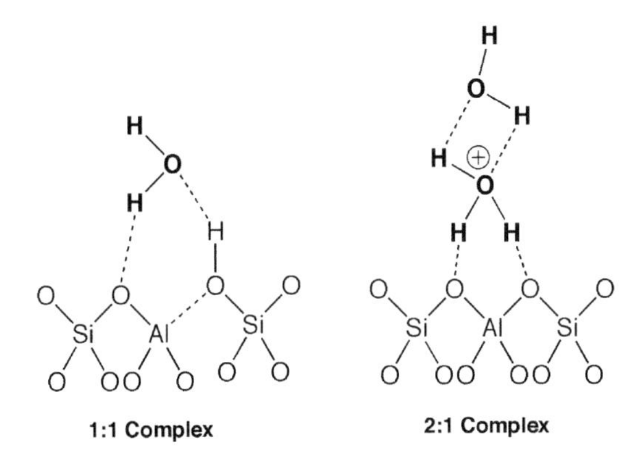

Figure 9.8 Coordination of 1:1 and 2:1 water–zeolite complexes on acid sites.

are observed. For water and methanol above low coverages, 2:1 and higher order adducts are clearly observed, and here proton transfer is occurring, as may have been guessed from PA data. For instance, the gas phase PA of H_2O is 723 kJ/mol (*weak*), but the PA of $(H_2O)_2$ is ~850 kJ/mol, similar to ammonia (*strong*). Calorimetric investigations indicate that there is dynamic equilibrium between the 1:1 and 2:1 adducts (again, for water and methanol) at room temperature, even before the total coverage reaches 1:1 B:ZOH site ratios.[14,15]

The interaction of alkenes with acid sides is briefly discussed next because of its particular importance in petrochemical catalytic processes.[16] The precise nature of this interaction remains a topic of ongoing research,[17] yet important insights have been gained that start to clarify how organic molecules bind to zeolites. Alkenes are different from the two cases discussed above because there can be proton transfer to the alkene, yet a neutral species could be the result of this chemisorption process. The main question is then whether alkenes form silil-alkoxide species

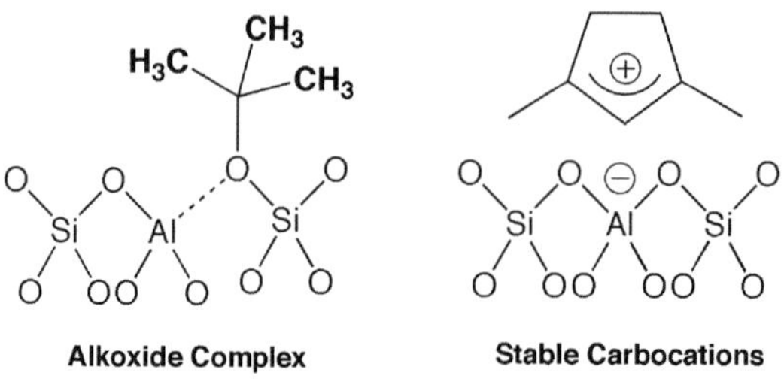

Alkoxide Complex

Stable Carbocations

Figure 9.9 Comparison of the structure alkoxide complexes and stable carbocations formed on zeolite acid sites. The left side shows the structure formed by alkenes reacting with zeolite Brønsted acid sites. The right side shows an example of a stable carbocation formed by protonation of dimethyl-cyclopentadiene.

upon reaction with an acid site, or is there charge separation to form carbenium ions (Fig. 9.9). The evidence indicates that it depends on the structure of the organic molecule.[18] Ethene is a weak base (PA = 680 kJ) and at low temperature it forms a hydrogen-bonded complex with acid sites in zeolites, but upon reaction it is predicted to form the neutral ethoxide group, covalently bonded to the zeolite framework. The complex formed by propene has been difficult to establish because it rapidly forms oligomeric species.[19] The tert-butyl alkoxide (from iso-butene, PA = 802 kJ/mol) has been isolated in some zeolites, but there are indications that steric hindrances can shift the equilibrium towards the tert-butyl cation. In summary, the bulk of the evidence indicates that primary and secondary alkenes will form covalent (neutral) complexes with zeolite acid sites. Tertiary alkenes can form covalent or charged complexes, depending on the specific structure of the alkene and the zeolite.

In the last decade, there has been accumulating evidence demonstrating that organic molecules with very high PAs (such as 1,3-dimethylcyclopentadiene, PA = 878 kJ/mol) do indeed form stable ions inside the zeolite pores.[10,17,20,21] These stable carbocations can (Fig. 9.9), in fact, be very important in the conversion of hydrocarbons such as the methanol to gasoline and methanol to olefins processes. In addition to 1,3-dimethylcyclopentadiene, various stable carbocations have been identified inside the zeolite pores using nuclear magnetic resonance (NMR) spectroscopy, UV/vis spectroscopy and other techniques.

Summarizing, the interaction of small molecules with zeolite acid sites is fundamentally a *chemical* interaction that can be broadly described as Lewis-base assisted, Brønsted acid–base interaction. Because it is a chemical interaction, often chemical bonds are made and broken and the end result of this process depends on the specific structure of the adsorbate. All acid sites in high-silica zeolites appear to be structurally very similar and the complexes they form with small molecules are, in thermochemical and spectroscopic terms, similar to each other (within a zeolite structure and between different zeolite structures as well). The gas-phase PA of the adsorbed molecule is the molecular property that gives the most useful information towards predicting the nature of the adduct formed upon adsorption.

7. Hydrophobic Interactions in High-Silica Zeolites

Water is unique because of its ability to form three-dimensional networks of hydrogen bonds, and this specific interaction leads to a remarkably high enthalpy of vapourization ($\Delta H_{vap} = 43.0$ kJ/mol at 298 K) as compared to molecules of similar size like methane ($\Delta H_{vap} = 8.5$ J/mol at 99 K). This is despite the fact that water molecules have a rather low electronic polarizability, and thus have relatively weak dispersion forces with other molecules. For instance, water has a polarizability of 1.5 $4\pi\varepsilon_0$ Å^3 as compared to methane that has a polarizability of 2.5 of $4\pi\varepsilon_0$ Å^3 (see Table 9.1).

Inside a purely siliceous zeolite, at low coverage, the only force that governs the interaction of water molecules with the zeolite framework is the dispersion or London force. Because of its low polarizability, this force is low in magnitude, and it is especially low compared to its high enthalpy of vapourization. For comparison, methane's heat of adsorption in siliceous MFI (silicalite) is 20.9 kJ/mol, much higher than its heat of vapourization, yet the heat of adsorption of water is smaller than the heat of adsorption of methane. For this reason, high-silica zeolites can be used to selectively adsorb hydrocarbons even in the presence of high vapour pressures of water, a property that it is nearly unique among readily available industrial sorbents. This is very different from high-surface-area porous silicas that contain a surface covered with silanol groups and are hydrophilic (in the absence of surface treatments).

In fact, purely siliceous zeolites (without defects such as internal silanol groups or tetrahedral vacancies) belong to a class of materials called hydrophobic solids. In these materials, water condensation occurs at a pressure above the saturation vapour pressure. For instance, the condensation of water in defect-free silicalite-1 is about 100 MPa at room temperature. For zeolites with larger pores, the condensation pressure is still high: for siliceous zeolite beta (with a three-dimensional-12-ring pore system) it is above 50 MPa at room temperature. This additional pressure can be thought to be the work that is needed to accommodate the massive disruption of the hydrogen bond network that is unavoidable upon the incorporation of water molecules into a microporous, hydrogen-bond-free environment such as the one presented in the pores of siliceous zeolites.

8. Final Remarks

This chapter has presented an overview of the important classes of zeolite–molecule forces that determine the adsorption and some catalytic properties of zeolite materials. Our intention is to relate molecular properties of the adsorbates to measurable quantities such as heats of adsorption. The data reveal that qualitatively, this approach helps to understand differences in the adsorption of small molecules for multiple classes of zeolite structures and compositions. At the same time, often differences are observed between the predictions of the simple models used here and the actual measurements of heat of adsorption. These reveal the limitations of this simplistic approach, yet also point to aspects of zeolite–molecule interactions that are not well understood. In particular, further calorimetric measurements of hydrogen-bond accepting and donating groups are needed to clarify the separate contribution of OH bonds and electrostatic forces.

REFERENCES

[1] Israelachvili, J. N., *Intermolecular and Surface Forces*. 2nd ed. Academic Press: London, 1991; p 450.
[2] Baerlocher, C., Meier, W. M., Olson, D. H., *Atlas of Zeolite Framework Types*. 5th ed.; Elsevier: Amsterdam, 2001.
[3] Baerlocher, C., McCusker, L. B. Database of Zeolite Structures. http://www.iza-structure.org/databases/
[4] Savitz, S., Siperstein, F., Gorte, R. J., Myers, A. L. "Calorimetric study of adsorption of alkanes in high-silica zeolites", *J. Phys. Chem. B* **1998,** *102*, (35), 6865–6872.
[5] Everett, D. H., Powl, J. C., "Adsorption in slit-like and cylindrical micropores in Henry's Law region: Medol for microporosity of carbons", *J. Chem. Soc. Faraday Trans. I* **1976,** *72,* 619–636.
[6] Papai, I., Goursot, A., Fajula, F., Plee, D., Weber, J., "Modeling of N-2, and O-2 Adsorption in Zeolites", *J. Phys. Chem.* **1995,** *99*, (34), 12925–12932.
[7] Savitz, S., Myers, A. L., Gorte, R. J., "A calorimetric investigation of CO, N-2, and O-2 in alkali-exchanged MFT", *Micropor. Mesopor. Mater.* **2000,** *37*, (1-2), 33–40.
[8] NIST Chemistry Web book. http://webbook.nist.gov/chemistry/
[9] Dubinin, M. M., Rakhmatkariev, G. U., Isirikyan, A. A., "Differetial heats of adsorption of water vapor on NaA zeolite at 300 and 450 K", *Izest. Ak. Nauk SSSR, S. Khim.* **1989,** *12,* 2877–2778.
[10] Beck, L. W., Xu, T., Nicholas, J. B., Haw, J. F., "Kinetic Nmr And Density-Functional Study Of Benzene H/D Exchange In Zeolites, The Most Simple Aromatic-Substitution", *J. Am. Chem. Soc.* **1995,** *117,* 11594–11595.
[11] Savitz, S., Myers, A. L., Gorte, R. J., White, D., "Does the cal-ad method distinguish differences in the acid sites of H-MFI?", *J. Phys. Chem. B* **1998,** *120*, (23), 5701–5703.
[12] Paze, C., Bordiga, S., Lamberti, C., Salvalaggio, M., Zecchina, A., Bellussi, G., "Acidic properties of H-beta zeolite as probed by bases with proton affinity in the 118-204 kcal mol (-1) range: A FTIR investigation", *J. Phys. Chem. B* **1997,** *101*, (24), 4740–4751.
[13] Zecchina, A., Marchese, L., Bordiga, S., Paze, C., Gianotti, E., "Vibrational spectroscopy of NH4+ ions in zeolitic materials: An IR study", *J. Phys. Chem. B* **1997,** *101*, (48), 10128–10135.
[14] Haase, F., Sauer, J., "Interaction Of Methanol With Bronsted Acid Sites Of Zeolite Catalysts - An Ab-Initio Study", *J. Am. Chem. Soc.* **1995,** *117*, (13), 3780–3789.
[15] Haase, F., Sauer, J., "Ab initio molecular dynamics simulation of methanol interacting with acidic zeolites of different framework structure", *Micropor. Mesopor. Mater.* **2000,** *35-6,* 379–385.
[16] Haw, J. F., Song, W. G., Marcus, D. M., Nicholas, J. B., "The mechanism of methanol to hydrocarbon catalysis", *Acc. Chem. Res.* **2003,** *36*, (5), 317–326.
[17] Clark, L. A., Sierka, M., Sauer, J., "Stable mechanistically-relevant aromatic-based carbenium ions in zeolite catalysts", *J. Am. Chem. Soc.* **2003,** *125*, (8), 2136–2141.
[18] Nicholas, J. B., Haw, J. F., "The prediction of persistent carbenium ions in zeolites", *J. Am. Chem. Soc.* **1998,** *120*, (45), 11804–11805.
[19] Gorte, R. J., White, D., "Interactions of chemical species with acid sites in zeolites"., *Topics Catal.* **1997,** *4*, (1-2), 57–69.

[20] Boronat, M., Viruela, P. M., Corma, A., "Reaction intermediates in acid catalysis by zeolites: Prediction of the relative tendency to form alkoxides or carbocations as a function of hydrocarbon nature and active site structure", *J. Am. Chem. Soc.* **2004,** *126,* (10), 3300–3309.
[21] Song, W. G., Nicholas, J. B., Haw, J. F.,"Acid-base chemistry of a carbenium ion in a zeolite under equilibrium conditions: Verification of a theoretical explanation of carbenium ion stability", *J. Am. Chem. Soc.* **2001,** *123,* (1), 121–129.
[22] Dunne, J. A., Mariwals, R., Rao, M., Sircar, S., Gorte, R. J., Myers, A. L., "Calorimetric heats of adsorption and adsorption isotherms .1. O-2, N-2, Ar, CO2, CH4, C2H6 and SF6 on silicalite", *Langmuir* **1996,** *12,* (24), 5888–5895.
[23] Dunne, J. A., Rao, M., Sircar, S., Gorte, R. J., Myers, A. L., "Calorimetric heats of adsorption and adsorption isotherms .2. O-2, N-2, Ar, CO2, CH4, C2H6, and SF6 on NaX, H-ZSM-5, and Na-ZSM-5 zeolites", *Langmuir* **1996,** *12,* (24), 5896–5904.
[24] Savitz, S., Myers, A. L., Gorte, R. J., "Calorimetric investigation of CO and N-2 for characterization of acidity in zeolite H-MFI", *J. Phys. Chem. B* **1999,** *103,* (18), 3687–3690.
[25] Dubinin, M. M., Rakhmatkariev, G. U., Isirikyan, A. A., "Energy of adsorption of water vapor adsorption on high and pure-silica zeolites", *Izest. Ak. Nauk USSR, S. Khim.* **1989,** *12,* 2862–2864.

CHAPTER 10

Application of Isotopically Labelled IR Probe Molecules for Characterization of Porous Materials

Konstantin I. Hadjiivanov

Contents

Abstract

In this chapter, the application of isotopically labelled infrared (IR) probe molecules for characterization of porous materials is considered. Initially, a brief background on the spectra of isotopically labelled molecules is provided. Then the application of isotopes for determination of the chemical composition of adsorbed species is considered, in particular: Does a surface species contain a definite atom or not? What is the number of these atoms? To which kinds of other atoms is it bound? Examples on spectral identification of NO^+, $[N_2O_2]^+$, nitrogen-oxo anion, nitriles, and isocyanates are provided. In a separate section, the use of labelled compounds for determination of polyligand species (dicarbonyls, dinitrosyls, tricarbonyls, etc.) is considered. Finally, some other particular applications are given, for example, the use of d_3-acetonitrile (CH_3CN) for measuring surface acidity.

Keywords: Adsorption, ^{13}C, Deuterium, FTIR spectroscopy, Isotopes, ^{15}N, ^{18}O, Porous materials, Probe molecules, Zeolites

Ordered Porous Solids
DOI: 10.1016/B978-0-444-53189-6.00010-X

Glossary

f—interaction constant between two ligands, N m^{-1}
F—force constant, N m^{-1}
i—isotopic shift factor
I—intensity of an IR band, a.u.
μ—reduced mass, g
ν—stretching frequency, s^{-1}
$\tilde{\nu}$—wave number, cm^{-1}

1. Introduction

One of the most informative and widely used techniques for characterization of powdered materials is infrared (IR) spectroscopy.[1–8] This technique is not expensive and is widely available, highly sensitive and selective. Vibrational spectra give information on the chemical bonds and are thus very useful for understanding the molecular structure of species. Usually, the first step of IR characterization is to obtain the sample spectrum after dilution, for example, using the so-called KBr technique. In this way information on the skeletal vibrations and hence, the structure of the material is obtained. However, very important is the so-called surface characterization based on the IR spectrum of activated pellets made of pure material and the IR spectra of adsorbed probe molecules.[1–8] Here, we shall concentrate on this application.

All samples pre-exposed to air contain adsorbed water, organic contaminants, and so on. To obtain clean surfaces the contaminants have to be removed. This is achieved by the so-called activation. The first step of this procedure is usually heating in oxygen in order to oxidize eventual organic contaminants. The heating temperature is typically 673–773 K but can vary depending on the particular system. Then the samples are evacuated, usually at the same temperature, in order to remove water, carbonates and other adspecies.

On the basis of IR spectrum of the sample pellet only, one can obtain poor information. This concerns mainly the hydroxyl coverage and the existence of stable surface species such as sulfates and carbonates. However, even in these cases information is restricted. Thus, no unambiguous conclusions about the acidity of the OH groups can be derived from the background spectra only.[7] That is why, the so-called probe molecules are widely applied.[1–8] These are molecules that specifically interact with definite kinds of sites, thus providing information about their concentration, surroundings, acidic properties and so on. There are several requirements for a molecule to be a good probe.[2,4,8] For instance: (i) some spectral parameters of the molecule (usually frequency) must be sensitive to the strength of its interaction with the surface; (ii) the extinction coefficients of the informative bands must be high enough; (iii) the probe molecule should not cause any chemical modification of the surface; (iv) the functional group or the atom through which the molecule is bonded to the surface should be well known; (v) the probe molecules should occupy the same kind of sites during adsorption on different samples and form complexes with similar structures and (vi) the informative absorption bands of the surface species should be in regions where the sample is transparent.

The technique described is applied to a wide range of powdered materials, both porous and nonporous. The term *surface characterization* is not always appropriate for studies of porous materials. However, in these cases there are also sites inaccessible to adsorption. That is why this term has gained popularity and shall be used hereafter.

Porous materials often possess some peculiar properties, for example:

(i) H-forms of zeolites and other porous materials are characterized by the so-called bridged hydroxyls which are much more acidic than the usual OH groups observed with nonporous systems, for example, metal oxides.[4,7]
(ii) Cations exchanged in zeolites or in other porous materials are characterized by an unusual low coordination and are thus able to accommodate several guest molecules simultaneously.[4,9]

Sometimes, when using IR spectroscopy of surface species, one can be hindered in the interpretation of some spectra and the identification of the respective species. In many of these cases, the problems could be resolved by the application of isotopically labelled molecules. Briefly, the applications of isotopically labelled molecules can be used to answer the following questions: (i) Does a surface species contain a definite atom (A) or not? (ii) What is the number of these atoms? (iii) To which kinds of other atoms A is bound? (iv) What is the number of known ligands coordinated to one cation?

There are also cases where the utilization of labelled molecules is motivated simply by the fact that they are registered in a more convenient spectral region. For instance, the use of deuterated compounds allows a significant shift of the working region to lower wave numbers where the noise level is lower.

The aim of this work is to summarize the principles of the use of isotopically labelled probe molecules and to help the reader with application.

2. Brief Background of the Vibrational Spectra of Isotopically Labelled Molecules

The stretching frequency, ν, of a diatomic molecule AB depends on the force constant (k) and the reduced mass (μ) of the two atoms:[10,11]

$$\nu = \frac{1}{2\pi}\sqrt{\frac{k}{\mu}} \tag{1}$$

The reduced mass can be calculated according to the equation:

$$\mu = \frac{M_A M_B}{M_A + M_B} \tag{2}$$

where M_A and M_B are the masses of the two atoms, respectively.

Equations (1) and (2) show that the stretching frequency depends on the reduced mass of the system, that is, it will be changed after replacement of one of the atoms in the molecule by an isotope (for instance A by A′). Defining the isotopic factor, i, as the ratio of stretching frequencies, it can be calculated as follows:

$$i = \frac{\nu_{AB}}{\nu_{A'B}} = \sqrt{\frac{\mu_{A'B}}{\mu_{AB}}}. \tag{3}$$

In Eq. (3), the stretching frequency can be replaced by the wave number (which is used in the practice) because they are linked by the equation $\nu = c\tilde{\nu}$ (where c is the light speed).

Although the above analysis involves two atoms only, very often the stretching vibrations are isolated and the isotopic shifts can be calculated according to Eq (3). In any case, if one vibration involves a definite atom, it should be affected after replacement of this atom by its isotope.

The calculated isotopic shifts with the most common chemical bonds used in IR spectroscopy of probe molecules are presented in Table 10.1. It is seen that H–D exchange affects most strongly the wave number because addition of a neutron to the light hydrogen atom changes the reduced mass dramatically. For instance, OH stretching modes at 3740 cm^{-1} should be shifted to 2721 cm^{-1} after H-D exchange, whereas ^{14}N-^{16}O modes at 1900 cm^{-1} should be shifted to 1868 cm^{-1} only after replacement of ^{14}N with ^{15}N.

Equation 1 also shows that knowing the wave number (in cm^{-1}), it is easy to calculate the force constant, F. To have dimensions in N m^{-2}, the following equation should be used:[11]

$$F = 5.8915 \times 10^{-5}\tilde{\nu}^2\mu \tag{4}$$

Table 10.1 Calculated isotopic shifts of some stretching modes of practical importance

Mode 1	Mode 2	Isotopic shift factor	Mode 1	Mode 2	Isotopic shift factor
$^{16}O-H$	$^{18}O-H$	0.99673	$^{12}C-^{16}O$	$^{13}C-^{16}O$	0.97777
	$^{16}O-D$	0.72761		$^{12}C-^{18}O$	0.97590
	$^{18}O-D$	0.72310		$^{13}C-^{18}O$	0.95311
$^{14}N-H$	$^{15}N-H$	0.99778	$^{12}C-^{14}N$	$^{13}C-^{14}N$	0.97907
	$^{14}N-D$	0.73030		$^{12}C-^{15}N$	0.98450
	$^{15}N-D$	0.72725		$^{13}C-^{15}N$	0.96323
$^{12}C-H$	$^{13}C-H$	0.96076	$^{14}N-^{14}N$	$^{14}N-^{15}N$	0.98319
	$^{12}C-D$	0.73396		$^{15}N-^{15}N$	0.96609
	$^{13}C-D$	0.72982	$^{14}N-^{16}O$	$^{15}N-^{16}O$	0.98206
$^{12}C-^{12}C$	$^{12}C-^{13}C$	0.98058		$^{14}N-^{18}O$	0.97373
	$^{13}C-^{13}C$	0.96065		$^{15}N-^{18}O$	0.95530

3. Chemical Composition and Structure of Surface Species

The determination of the chemical composition and structure of surface species is the most commonly used application of isotopically labelled IR probe molecules. There are many examples that an IR band is registered in a region where more than one bond could be vibrating. For instance, the region between 2200 and 2100 cm^{-1} is typical of adsorbed CO, but C–N, N–O, and other stretching vibrations could also be observed there. The use of isotopically labelled molecules can prove the existence of a definite atom in a surface species, but can also answer the question about the nature of atoms to which this atom is bonded and about the number of these bonds.

Here below we shall provide an example of using isotopically labelled molecules for the assignment of a band at 2133 cm^{-1}. We shall focus on this band because it is now well established that it is due to NO^+ species occupying cationic positions in zeolites, and is formed with the participation of the zeolite bridging hydroxyls.[12]

There was, after 1995, a boom in the investigations on the nature of surface nitrogen-oxo compounds[5] which was mainly caused by the discovery of the selective catalytic reduction of NO_x by hydrocarbons.[13] Many researches reported that a band at 2133 cm^{-1} appeared after NO_x adsorption on H–ZSM-5, Me–ZSM-5 and other zeolites. The band was assigned to NO_2,[14] N_2O[15] and so on. Unambiguous assignment of this band required utilization of different isotopic probes.

Hoost *et al.*[16] detected the 2133 cm^{-1} band after NO adsorption on a Cu–H–ZSM-5 sample. After adsorption of ^{15}NO, they observed that the band was shifted to 2095 cm^{-1}. The appearance of the isotopic shift itself was an unambiguous proof that the 2133 cm^{-1} species contained at least one nitrogen atom. Moreover, the calculation of the isotopic shift of N–O vibrations (on the basis of a factor of 0.98206, see Table 10.1) showed that the expected wave number was 2094.7 cm^{-1}, that is, it coincided with the experimentally observed one. On this basis, the authors concluded that the species contained N–O bond(s).

The above observations are illustrated in Fig. 10.1. Coadsorption of ^{14}NO and O_2 on H–ZSM-5 results in a band at 2134 cm^{-1} (Fig. 10.1, spectrum a), while the use of ^{15}NO leads to a band at 2096 cm^{-1} (Fig. 10.1, spectrum b). Here again, the observed isotopic shift factor of 0.98218 corresponds, within accuracy, to the theoretically calculated one.

The authors assigned the band at 2133 cm^{-1} to NO_2^+ species.[16] However, the results left open the question about the number of oxygen atoms in these species and did not exclude a NO^+ structure. Hence, application of other isotopically labelled probes was necessary to establish the structure of the 2133 cm^{-1} species.

If the species were NO_2^+, they should have displayed two N–O modes: symmetric and antisymmetric. However, as often happens in surface studies, one of the modes could have been masked by the zeolite's own strong adsorbance. Thus, the spectra do not exclude the NO_2^+ assignment. To establish the structure of the 2133 cm^{-1} species, we have performed further investigations involving isotopically labelled oxygen.[12] NO was coadsorbed with $^{18}O_2$ on H–ZSM-5. Initially, the band

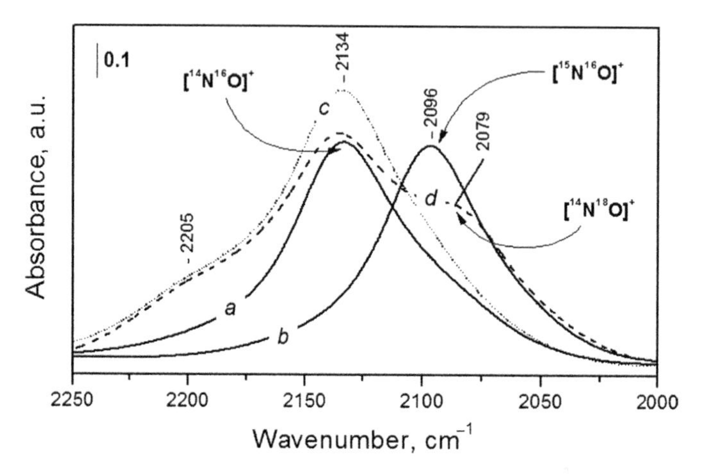

Figure 10.1 Fourier transform infrared (FTIR) spectra of NO and O_2 with different isotopic compositions coadsorbed on H–ZSM-5. 512 Pa $^{14}N^{16}O$ + 133 Pa $^{16}O_2$ (a); 512 Pa $^{15}N^{16}O$ + 133 Pa $^{16}O_2$ (b); 512 Pa $^{14}N^{16}O$ + 512 Pa $^{16}O_2$ (c); and 512 Pa $^{14}N^{16}O$ + 512 Pa $^{18}O_2$ followed by a short evacuation (d).

at 2133 cm^{-1} appeared and then a band at 2077 cm^{-1} started to develop. The exchange of ^{16}O with ^{18}O in NO should lead to the same shift, namely to 2077 cm^{-1}. These results confirm the conclusion of Hoost *et al.*[16] about the existence of an N–O bond in the species. However, if the NO_2^+ structure is correct, one should register two bands: one for $[N^{16}O^{18}O]^+$ and one for $[N^{18}O_2]^+$ species. The fact that only one band has been registered, unambiguously demonstrates that the species contain one oxygen atom only. The use of a deuteroxylated surface proved the absence of H atoms in the structure and finally the band at 2133 cm^{-1} was assigned to NO^+ species.[12]

The above considerations are demonstrated on Fig. 10.1. It is clearly seen that, after $^{14}N^{16}O$ and $^{18}O_2$ coadsorption, there is no band between the bands at 2133 cm^{-1} (here observed at 2135 cm^{-1}) and that at 2079 cm^{-1} (Fig. 10.1, spectrum d). To be precise, one should discuss a weak band at 2205 cm^{-1}. It is due to another kind of NO^+ species formed in excess oxygen.[17] The same band was also registered in studies without isotopically labelled molecules (Fig. 10.1, spectrum c).

More recently[18] it was reported that NO^+ species could be formed with sodium forms of zeolites (NaY, NaMOR) replacing Na^+. The assignments were confirmed by isotopically labelled molecules (^{14}NO and $^{18}O_2$).

It was also reported[17] that the NO^+ species were converted, in presence of NO and at low temperature, into another kind of species displaying two bands, at 2000 and 1687 cm^{-1}. The structure of the latter was established using ^{15}N and ^{18}O labelled molecules. At first, the use of ^{15}NO resulted in bands at 1965 and 1653 cm^{-1} instead of the bands at 2000 and 1687 cm^{-1}. The experimental isotopic shifts (factors of 0.9825 and 0.9798, respectively) were close to the shift expected for NO (0.98206). However, the small deviations indicated some coupling of the respective vibrations. Two additional bands, at 1987 and 1665 cm^{-1}, appeared when ^{14}NO and ^{15}NO were coadsorbed. These were assigned to the ^{14}NO and ^{15}NO modes of the same species. This result proved the existence of two nitrogen atoms in the structure. On this basis the species under consideration were assigned to $[ONNO]^+$ species.

Another example of the use of ^{15}NO is to attribute unambiguously bands in the 1650–1200 cm^{-1} region to nitrogen-oxo anions (nitro, nitrite, nitrato, etc. species).[19]

To understand the mechanism of selective catalytic reduction of nitrogen oxides with hydrocarbons, many researchers have tried to identify species observed under reaction conditions and absorbing in the 2300–2000 cm^{-1} region.[20,21] The problem is complex and sometimes the results are not definite even after application of isotopically labelled reactants. This is due to the diversity of the possible species involving N, C, O and H atoms, as well as the high number of isomers. A good background is provided by Yeom *et al.*[20] Briefly, –CN and –NC species show similar $^{14}N \rightarrow {}^{15}N$ and $^{12}C \rightarrow {}^{13}C$ isotopic shifts. On this basis bands at 2225 and 2173 cm^{-1} have been assigned to nitriles.[21] The asymmetric stretching modes of the N-C-O group shift by ca. 10 cm^{-1} as a result of a $^{14}N \rightarrow {}^{15}N$ substitution and by ca. 60 cm^{-1} when ^{12}C is substituted for ^{13}C. On this basis bands at 2270–2258 cm^{-1} have been assigned to isocyanates.[20,21]

4. Identification of Polyligand Species

One of the principal applications of isotopically labelled IR probe molecules is the spectral identification of polyligand species as well as the number of these ligands. This is particularly important for metal-exchanged zeolites since, as was already mentioned, cations in zeolites are characterized by a low coordination number. Usually when polyligand species are formed, the ligands interact vibrationaly and their stretching modes are split.[11] We shall regard dicarbonyls in detail, then dinitrosyls and other dilgand species briefly, and finally, species with more than two ligands.

4.1. Dicarbonyls

One of the most used IR probe molecules is CO.[1–4,6] The CO stretching modes are highly sensitive to the bond that CO forms with metal cations. There are three kinds of interactions that are essential for the M-CO bond: electrostatic interaction, σ-bond and back π-bond. The two former bonds are relatively weak and lead to an increase of the CO stretching frequency. On the contrary, the back π-bond results in a decrease of the CO modes and also reflects in a strong enhancement of the CO extinction coefficient. This bond is important with metals and with univalent cations rich in d-elecrons.[1,4] Analysis of the literature data shows that with two-valent cations the π-bond is rarely important. Exceptions are some ions of heavy elements, such as Rh^{2+} and Pt^{2+}, where a restricted π-back donation is realized.[4]

Table 10.2 Spectral parameters of various dicarbonyl species

Sample	Species	$\tilde{\nu}_s$, cm^{-1}	$\tilde{\nu}_{as}$, cm^{-1}	$\Delta\tilde{\nu}$, cm^{-1}	f, N m^{-2}	References
Ir–ZSM-5	$Ir^+(CO)_2$	2105	2033	72	60.2	22
Co–ZSM-5	$Co^+(CO)_2$	2113	2042	71	59.6	23
Rh–ZSM-5	$Rh^+(CO)_2$	2114	2048	66	55.9	24
Rh/DAY[a]	$Rh^+(CO)_2$	2115	2050	65	54.7	25
Ru/DAY	$Ru^{2+}(CO)_2$	2082	2016	66	54.6	26
Pt–ZSM-5	$Pt^+(CO)_2$	2122	2092	40	46.9	27
Ir–ZSM-5	$Ir^{2+}(CO)_2$	2173	2129	44	38.2	22
Ni–ZSM-5	$Ni^+(CO)_2$	2136	2092	44	37.6	28
Ni-BEA	$Ni^+(CO)_2$	2138	2095	43	36.8	29
Pt–MOR	$Pt^{3+}(CO)_2$	2205	2167	38	33.6	30
Pt–ZSM-5	$Pt^{3+}(CO)_2$	2212	2176	36	31.9	27
Rh–ZSM-5	$Rh^{2+}(CO)_2$	2176	2142	34	29.7	24
Pt–MOR	$Pt^+(CO)_2$	2135	2101	34	29.1	30
Cu–ZSM-5	$Cu^+(CO)_2$	2178	2150.5	27.5	24.0	31,32
Pd–ZSM-5	$Pd^{3+}(CO)_2$	2221	2200	21	18.8	33
Ag–ZSM-5	$Ag^+(CO)_2$	2192	2186	6	5.3	34
Mn-ZSM-5	$Mn^{2+}(CO)_2$	2202	2202	0	0	35
CaY	$Ca^{2+}(CO)_2$	2191	2191	0	0	36
NaY	$Na^+(CO)_2$	2175	2175	0	0	37

[a] DAY means dealuminated Y zeolite
Note: The examples are ordered according to decreasing interaction constant, f.

When two CO molecules are adsorbed at the same cationic site and when the cation–CO interaction is rather important (not electrostatic) the two CO molecules interact vibrationally and the CO stretching modes are split into symmetric and antisymmetric (see Table 10.2). The higher the CO interaction constant, the higher the split value.

If coupling between metal–carbon and carbon–oxygen vibrations is assumed to only negligibly affect the frequencies, the force (F) and interaction constants (f) can be calculated. The theoretical background is rather complex and large and is described in the literature.[11] For the sake of brevity, here we shall only provide practical hints helping with the calculations.

Initially, one calculates the parameter λ for each of the two bands:

$$\lambda_i = 5.8915 \times 10^{-5} \tilde{\nu}_i^2 \qquad (5)$$

then F and f are obtained according to the formulas:

$$F = \frac{\lambda_1 + \lambda_2}{\mu_1 + \mu_2} \qquad (6)$$

$$f = \sqrt{F^2 - \frac{\lambda_1 \lambda_2}{\mu_1 \mu_2}} \tag{7}$$

Let us suppose that CO adsorption leads to formation of well-defined dicarbonyls. This is illustrated in Fig. 10.2 showing the spectra registered after CO adsorption on CO-reduced Ni–ZSM-5. In the presence of some CO equilibrium pressure, two bands, at 2136 and 2092 cm^{-1}, are dominant (Fig. 10.2, spectrum a). These two bands decrease in intensity in concert and a new band at 2151 cm^{-1} develops instead when the coverage decreases due to evacuation (Fig. 10.2, spectra b–h). The bands at 2136 and 2092 cm^{-1} are due to the symmetric and antisymmetric C–O modes of $Ni^{+}(CO)_2$ dicarbonyl species.[28] These lose one CO ligand during evacuation, being thus converted into linear carbonyls (2109 cm^{-1}).

The spectra presented in Fig. 10.2 are a good proof of the existence of dicarbonylic species. The bands at 2136 and 2092 cm^{-1} change in concert. The existence of two isosbestic points indicates direct conversion of these species into monocarbonyls, as indicated by a band at 2109 cm^{-1}. However, there are other principal assignments. For instance, the bands at 2136 and 2092 cm^{-1} could correspond to tricarbonyl species with a C_{3v} symmetry (thus they should manifest two bands only) that lose simultaneously two CO ligands, forming monocarbonyls. In many cases, the bands could overlap with other carbonyl bands which hinders the interpretation of the spectra. In other words, to assign unambiguously a set of two bands to dicarbonylic species, one should involve isotopically labelled molecules in the investigations. In what follows we shall briefly describe the theoretical background of these studies.

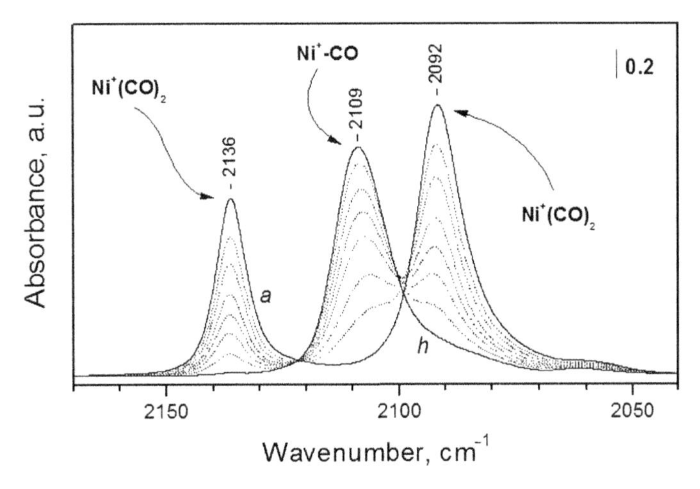

Figure 10.2 Fourier transform infrared (FTIR) spectra of CO adsorbed on a CO-reduced Ni–ZSM-5 sample. Initial CO equilibrium pressure of 266 Pa (a) and development of the spectra in the conditions of dynamic vacuum (b–h).

Now consider what happens if a 1:1 molar mixture of ^{12}CO and ^{13}CO is adsorbed on the sample under conditions when dicarbonyls are exclusively formed (high coverage). The following species are expected: $Ni^+(^{12}CO)_2$, $Ni^+(^{12}CO)(^{13}CO)$ and $Ni^+(^{13}CO)_2$. Simple calculations show that the molar ratio between these species should be 1:2:1. In other words, the mixed ligand species will be dominant.

What the spectral behaviour of the new species formed will be? Using the ^{12}CO–^{13}CO isotopic shift factor (see Table 10.1) it is easy to calculate that the $Ni^+(^{13}CO)_2$ dicarbonyls will manifest symmetric modes at 2088.5 cm^{-1} and anti-symmetric ones at 2045.5 cm^{-1}. The respective bands should be slightly less intense than the bands of $Ni^+(^{12}CO)_2$ species since the extinction coefficient of ^{13}CO is a little lower than that of ^{12}CO.

Indeed, adsorption of a ^{12}CO–^{12}CO isotopic mixture produces bands of $Ni^+(^{12}CO)_2$ at 2136 and 2092 cm^{-1} and $Ni^+(^{13}CO)_2$ at 2089 and 2046 cm^{-1} (Fig. 10.3, spectrum b). The exact band positions are taken from the second derivatives and the computer deconvolution of the spectrum. It is also seen that the two most intense bands in the spectrum are at 2123 and 2059 cm^{-1}. On the basis of above consideration, one can assign these bands to $Ni^+(^{12}CO)(^{13}CO)$ species.

The wave numbers of the dicarbonyls containing one ^{12}CO and one ^{13}CO ligand can be calculated on the basis of the force (F) and interaction (f) constants found from the Eqs (6) and (7). Here again, the model will be only briefly described. On the basis of values of F and f, one can calculate λ_3 and λ_4 for the two vibrations.

$$\lambda_3 = \frac{(\mu_1 + \mu_2)F}{2} + \frac{1}{2}\sqrt{(\mu_1 + \mu_2)^2 F^2 - 4\mu_1\mu_2(F^2 - f^2)} \tag{8}$$

$$\lambda_4 = \frac{(\mu_1 + \mu_2)F}{2} - \frac{1}{2}\sqrt{(\mu_1 + \mu_2)^2 F^2 - 4\mu_1\mu_2(F^2 - f^2)} \tag{9}$$

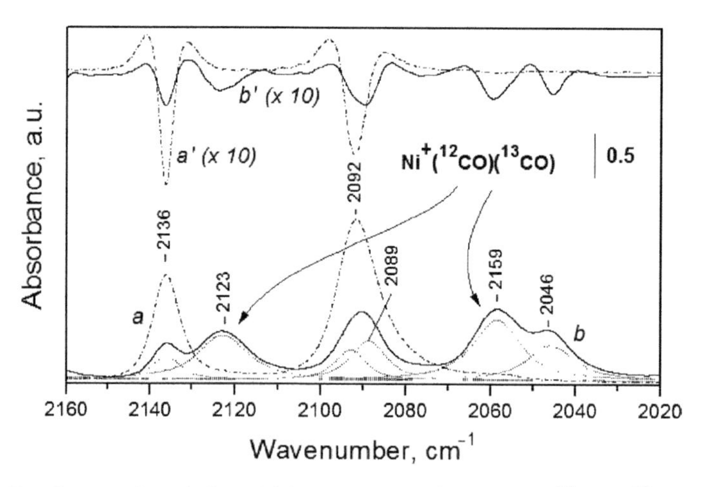

Figure 10.3 Fourier transform infrared (FTIR) spectra of CO (a) and $^{12}CO + ^{13}CO$ (b) adsorbed on a CO-reduced Ni–ZSM-5 sample. The respective second derivatives are denoted as a′ and b′. Computer deconvolution of spectrum b is presented by dotted lines.

Finally,

$$\tilde{\nu}_i = 0.76576\sqrt{\lambda_i} \tag{10}$$

For dicarbonyls with ^{13}CO, one could use the following reduced mass: $\mu(^{12}CO) = 0.1459$ and $\mu(^{13}CO) = 0.1394$.

On the basis of above formulas, the $Ni^+(^{12}CO)(^{13}CO)$ species should manifest two modes, at 2122.5 and 2058.5 cm^{-1}. These values agree very well with the experimentally observed ones (Fig. 10.3, spectrum b). Table 10.3 shows calculated and experimentally observed wave numbers of the mixed ligand dicarbonyls for a series of $M^{n+}(^{12}CO)(^{13}CO)$ species.

Analysis of the data in Table 10.2 shows that the split strongly depends on the cation and, to a smaller extent, on the matrix. In the case of $Ag^+(CO)_2$ species the split is very low (6 cm^{-1} only).[34] The last line of Table 10.2 shows that no split occurs when CO forms dicarbonyls with Na^+.[37] In fact, the Na^+—CO interaction is essentially electrostatic and the two CO molecules cannot interact vibrationally or the interaction is too weak to be detected experimentally. There are many similar examples when dicarbonyls are formed but characterized by only one CO stretching mode (as, e.g. Ca^{2+}, $Sr^{2+}$and Ba^{2+}cations in Y zeolite,[36] Na^+, K^+, Rb^+, Cs^+ in EMT,[38] Mn^{2+} in ZSM-5,[35]). In all these cases the interaction is essentially electrostatic. Unfortunately, here the application of isotopically labelled molecules is useless.

There is, however, a case when dicarbonyls can display only one mode even with strong interaction. The intensities of the symmetric and antisymmetric modes are not related to the strength of interaction between the CO molecules. The ratio between these intensities depends on the angle between the CO molecules (2Θ) according to the formula:[11]

$$\frac{I_s}{I_{as}} = \tan^2\Theta \tag{11}$$

where I_s and I_{as} are the intensities of the bands due to the symmetric and antisymmetric CO modes, respectively.

It is seen that when the angle between the CO ligands is 90°, the intensity of the symmetric modes is zero. In this case, however, the dicarbonyl structure can again be revealed by isotopically labelled molecules. With some big cations, for example, Pt^+, the angle can decrease below 90° and the symmetric modes become more intense.[27]

There are two general types of dicarbonyl adspecies: site specifed and complex specified.[9] The complex-specified dicarbonyls are formed with particular cations: Rh^{2+}, Ir^{2+}, Rh^+, Ir^+, Pt^{3+}, Pd^{3+}, and Ru^{n+}. A peculiarity of these species is that, when decomposed, they do no produce a measurable fraction of monocarbonyls, that is, they are more stable than the monocarbonyls. When complex-specified dicarbonyls are formed and cannot be converted into tricatbonyls, the isotopic exchange with ^{13}CO is hindered. However, when the dicarbonyls can be converted into tricarbonyls, a fast isotopic exchange occurs.

Table 10.3 Stretching frequencies of some dicarbonyls presented in Table 10.2 and of the respective $M(^{12}CO)(^{13}CO)$ mixed ligand species. The calculated values are presented in brackets

Sample	Species	$M(^{12}CO)_2$		$M(^{12}CO)(^{13}CO)$		References
		$\tilde{\nu}_s$, cm^{-1}	$\tilde{\nu}_{as}$, cm^{-1}	$\tilde{\nu}$ (^{12}CO), cm^{-1}	$\tilde{\nu}$ (^{13}CO), cm^{-1}	
Cu–ZSM-5	$Cu^+(CO)_2$	2177	2151	2169 (2167.7)	2112 (2112.1)	31
Ag–ZSM-5	$Ag^+(CO)_2$	2192	2186	2189 (2189.2)	2140 (2140.0)	34
Co–ZSM-5	$Co^+(CO)_2$	2113	2042	2097 (2096.4)	2013 (2012.3)	23
Ni–ZSM-5	$Ni^+(CO)_2$	2136	2092	2123 (2122.5)	2058 (2058.4)	28
Ni-BEA	$Ni^+(CO)_2$	2138	2095	2125 (2124.7)	2062 (2061.2)	24
Rh–ZSM-5	$Rh^+(CO)_2$	2114	2048	2098 (2097.8)	2018 (2017.8)	24
Rh–ZSM-5	$Rh^{2+}(CO)_2$	2176	2142	2164 (2164.3)	2106 (2105.6)	24
Rh/DAY	$Rh^+(CO)_2$	2118	2053	2102 (2101.9)	2022 (2022.6)	26
Pd–ZSM-5	$Pd^{3+}(CO)_2$	2221	2200	2210 (2210.9)	2158 (2157.9)	33
Ir–ZSM-5	$Ir^{2+}(CO)_2$	2173	2129	2157 (2158.4)	2090 (2093.7)	22
Ir–ZSM-5	$Ir^+(CO)_2$	2105	2033	2088 (2087.4)	2002 (2003.5)	22
Pt–ZSM-5	$Pt^{3+}(CO)_2$	2212	2176	2199 (2199.8)	2138 (2139.3)	27
Pt–ZSM-5	$Pt^+(CO)_2$	2122	2092	2112 (2111.3)	2052 (2055.8)	27
Pt–MOR	$Pt^{3+}(CO)_2$	2205	2167	2193 (2192.4)	2129 (2130.9)	30
Pt–MOR	$Pt^+(CO)_2$	2135	2101	2123 (2123.4)	2069 (2065.4)	30

Note: The calculated wave numbers are presented in brackets.

Site-specified dicarbonyls are produced as a result of low coordination of the metal cation and are decomposed passing through monocarbonyls. From the examples presented in Table 10.1, site-specified dicarbonyls are produced with the cations of Cu^{+}, Ag^{+}, Ni^{+} and all geminal species where the interaction constant is zero are also site specified. In these cases fast isotopic exchange occurs.

In conclusion, the use of isotopically labelled molecules can unambiguously prove dicarbonyl structures providing the C–O modes are split into symmetric and antisymmetric (strong interaction between the ligands).

4.2. Dinitrosyls

Nitrogen monoxide can easily dimerize and, consequently, complex-specified dinitrosyls are often observed. For instance, such dinitrosyls are always formed when NO interacts with Co^{2+}, V^{3+}, V^{4+}, W^{4+} and Mo^{4+} ions.[5] There are few described cases of site-specified dinitrosyls (also called true dinitrosyls): with Cu^{+}[39,40] and Cr^{2+}[41] cations. The number of known site-specified dinitrosyls is much more restricted than that of similar dicarbonyls since no electrostatic interaction is typical of NO.

As in the case of dicarbonyls, the dinitrosyl structure can be established by the use of isotopically labelled molecules, for example, by adsorption of a ^{14}NO–^{15}NO isotopic mixture. This is illustrated in Fig. 10.4. Adsorption of NO on a reduced Cr–ZSM-5 sample leads to the appearance of two main bands in the nitrosyl region at 1905 and 1782 cm^{-1} (Fig. 10.4, spectrum a). Supposing the bands are characterizing dinitrosyls, we have calculated, using the approximate force field model,[11] the wave numbers of the mixed $Cr^{n+}(^{14}NO)(^{15}NO)$ complexes. According to the calculation, the $\nu(^{14}N$–$O)$ and $\nu(^{15}N$–$O)$ modes should be located at 1890 and 1761 cm^{-1}, respectively. Indeed, the spectra recorded after coadsorption of a ^{14}NO–^{15}NO mixture (Fig. 10.4, spectrum b) have shown two intense bands at 1891 and 1759 cm^{-1}. Thus, the results have proved the dinitrosyl structure Table 10.4.

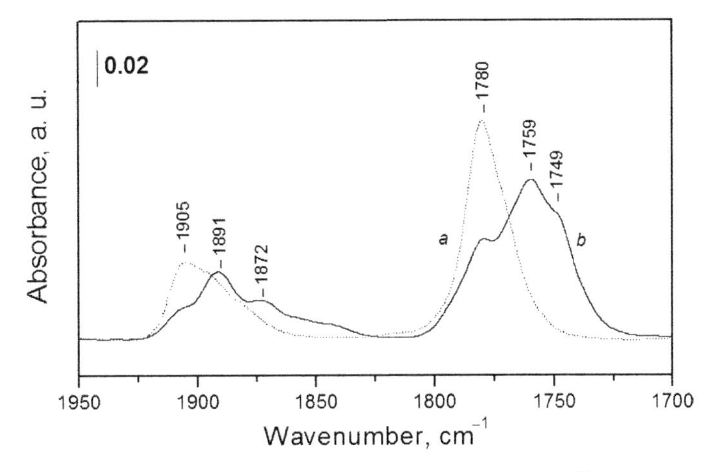

Figure 10.4 Fourier transform infrared (FTIR) spectra of adsorbed ^{14}NO (a) and coadsorbed ^{14}NO and ^{15}NO (b) on a Cr–ZSM-5 sample reduced with hydrogen at 773 K. Equilibrium ^{14}NO (^{14}NO and ^{15}NO) pressure of 170 Pa, followed by evacuation.

Table 10.4 Stretching frequencies of some dinitrosyls and of the respective $M(^{14}NO)(^{15}NO)$ mixed ligand species. The calculated values are presented in brackets

Sample	Species	$M(^{14}NO)_2$		$M(^{14}NO)(^{15}(NO)$		References
		$\tilde{\nu}_s$, cm^{-1}	$\tilde{\nu}_{as}$, cm^{-1}	$\tilde{\nu}(^{14}NO)$, cm^{-1}	$\tilde{\nu}(^{15}NO)$, cm^{-1}	
Co–ZSM-5	$Co^{2+}(NO)_2$	1900	1812	1882 (1886)	1790 (1792)	19
Cr–ZSM-5	$Cr^{3+}(NO)_2$	1905	1780	1891 (1890)	1759 (1761)	41
Rh–ZSM-5	$Rh^{+}(NO)_2$	1862	1785	1848 (1849)	1765 (1766)	42
Fe–silicalite	$Fe^{2+}(NO)_2$	1839	1765	1827 (1826)	1744 (1746)	43

Note: The calculated wave numbers are presented in brackets.

4.3. Other diligand species

Among the other diligand species, maybe the most studied ones are the complexes with two N_2 ligands. Similar species have been observed with Rh/DAY[25] and Ru/DAY[26] samples. The $Rh^{+}(N_2)_2$ species have been found at 2243 and 2217 cm^{-1} and the $Ru^{2+}(N_2)_2$ species have been reported at 2207 and 2173 cm^{-1}. In both cases, the structures have been confirmed by coadsorption with $^{15}N_2$.

In many cases, the complexes with dinitrogen are weak and there is no split of the N–N modes into symmetric and antisymmetric.[45] As with the carbonyls, the application of isotopic mixtures in these cases is useless. For some experiments, $^{15}N_2$ has been used in order to shift the N–N modes out of the region where gaseous CO_2 absorbs.[45]

4.4. Tri- and tetracarbonyls

There are cases when one cationic site can accommodate up to three or four CO molecules. The tetracarbonyls are not often observed, The only known tetracarbonyls proven after CO adsorption on cationic sites are the $Rh^{+}(CO)_4$ [25,44] and $Co^{+}(CO)_4$[23] species. Tricarbonyls are more often observed. When having C_{3v} symmetry, they are expected to display two IR active modes. However, due to steric hindrance, the symmetry of tricarbonyls is usually distorted and they manifest three bands.

Assuming that the tricarbonyls have C_{3v} symmetry (two IR active bands), coadsorption of ^{12}CO and ^{13}CO should result in the formation of the following species: (i) $M^{n+}(^{12}CO)_3$; (ii) $M^{n+}(^{12}CO)_2(^{13}CO)$; (iii) $M^{n+}(^{12}CO)(^{13}CO)_2$; and (iv) $M^{n+}(^{13}CO)_3$. Each of the mixed ligand species should display three IR bands, so the total number of bands will be 10. This leads to overlapping of many bands, which hinders detailed analysis. In the practice, the situation is even much more complex because of following several reasons:

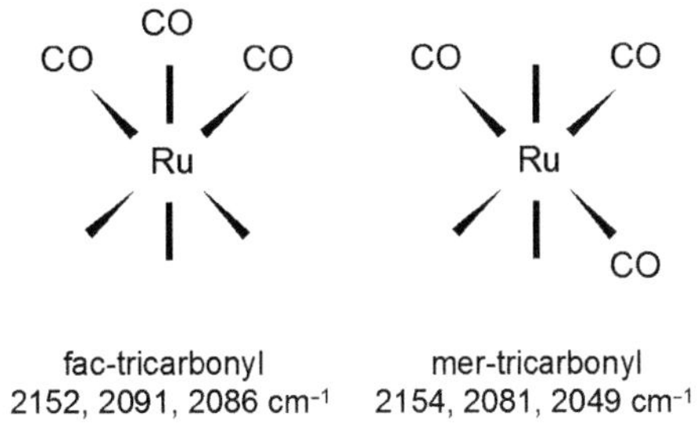

fac-tricarbonyl
2152, 2091, 2086 cm^{-1}

mer-tricarbonyl
2154, 2081, 2049 cm^{-1}

Scheme 10.1 Structure and spectral behaviour of fac- and mer-tricarbonyls of Rh^{2+} formed in Rh/DAY. Adopted from Miessner and Richter.[26]

- The symmetry is usually distorted, that is, the CO ligands are not equivalent. This increases the number of mixed ligand species and the number of observed bands, respectively.
- Overlapping with bands due to other carbonyl species usually occurs.
- The commercially available ^{13}CO usually contains $^{13}C^{18}O$ as a contaminant, which additionally complicates the spectra.

However, the use of isotopic mixtures can unambiguously reject the dicarbonyl structure and provide evidence of the existence of more than two CO ligands. The use of different ^{12}CO:^{13}CO ratios can supply information on the spectral behaviour of mixed ligand carbonyls.[23,26,28]

In some cases with well-defined systems, isotopic substitution can provide information on the structure of the tricarbonyl species. In fact, using isotopes and calculations, Miessner and Richter[26] have distinguished between fac- and mer-tricarbonyls of Ru^{2+} in Ru/DAY, as shown on Scheme 10.1.

A new horizon in the studies with isotopically labelled carbonyls is the fact that $^{13}C^{18}O$ is already commercially available. A look at Table 10.1 shows that the isotopic shift factor is much higher. For instance, a carbonyl band at 2200 cm^{-1} should be shifted to 2151 cm^{-1} using ^{13}CO, but the shift with $^{13}C^{18}O$ should reach 2097 cm^{-1}. Therefore, the use of $^{12}C^{16}O$–$^{13}C^{18}O$ isotopic mixtures should result in a much better separation of the various carbonyl bands and will probably help with answering some still open questions.

5. Use of D_3-Acetonitrile for Determination of Surface Acidity

There are many probe molecules for determination of surface acidity. Among the most used are NH_3, pyridine and CO.[1–4] However, depending on the aims, one should choose an appropriate probe molecule. For instance, the molecule of pyridine is relatively big and not always appropriate for porous materials. Acetonitrile (CH_3CN) is a small molecule and is a soft base. With protonic acid sites, CH_3CN either forms strong hydrogen bonds or can become protonated, depending on the strength of the acid site.[46–50] However, the use of CH_3CN as an IR probe is accompanied by some difficulties. Interpretation of the spectra can be complicated for coordinated or H-bonded species because of Fermi resonance between the ν_2 stretching mode and the $(\nu_3+\nu_4)$ combination mode.[48] To overcome this, the use of deuterated CH_3CN is recommended.[48] CD_3CN displays a single ν(C-N) band at 2263 cm^{-1} in the liquid phase.[46–50]

The fundamental $\nu_s(CD_3)$ and $\nu_{as}(CD_3)$ of CD_3CN in liquid phase are at 2114 and 2250 cm^{-1}, respectively. These modes are hardly affected by adsorption on acid sites. However, the ν(C-N) stretching mode is shifted to higher wave numbers when CH_3CN is bound to acid sites. Moreover, the value of this shift provides information on the strength of the acid site. Thus, characteristic ν(C-N) bands of d d_3-CH_3CN adsorbed on different sites are as follows:

- Strong Al^{3+} Lewis sites, 2332–2320 cm^{-1};[46,48,50]
- Weak Lewis sites, for example, Co^{2+}, $Mg^{2+}$and Zn^{2+}, located in framework sites of AlPO-5 or zeolite cationic sites, about 2310 cm^{-1};[49]
- Bridging OH groups in high silica zeolites, 2302–2296 cm^{-1};[46,49,50]
- Terminal Si–OH groups, 2280–2274 cm^{-1}.[46,49,50]
- Physisorbed CH_3CN, 2265 cm^{-1}.[49]

6. Use of Isotopic Exchange for Changing the Spectral Region

The light scattering is important at higher wave numbers and particularly with samples characterized by big particles. As a result, in some cases the registering of the spectra in the O–H stretching region is practically impossible. In addition, the

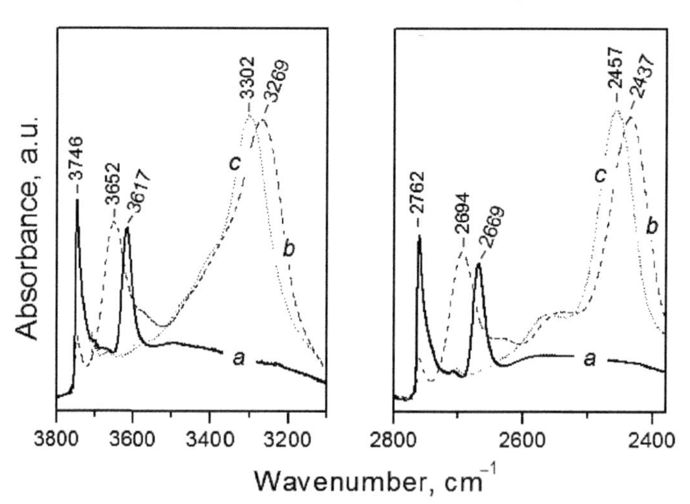

Figure 10.5 Fourier transform infrared (FTIR) spectra of H-D-ZSM-5 (a), after adsorption of CO at 85 K and 1 kPa equilibrium pressure (b) and after decrease of the CO pressure to ca. 0.1 Pa at 85 K (c).

spectrum of water in air can often lead to undesirable noise in the OH region. These inconveniences can be overcome by working in the OD region. As already mentioned, the D → H exchange results in a significant decrease of the wave numbers of the vibrations involving H/D atoms.

A widespread technique for measuring the surface acidity of OH groups is their interaction with soft bases. CO adsorption at low temperature is often used for this purpose.[2,4] As a results of CO interaction with OH groups, the O–H bands shift to lower wave numbers. The higher the acidity, the larger the shift. Figure 10.5 illustrates that information on the surface acidity is essentially the same considering both, O–H and O–D regions.

7. Conclusions

The use of isotopically labelled molecules is very useful in the IR spectroscopy of adsorbed species. It can answer the following questions:

- Does a surface species contain a definite atom or not?
- What is the number of these atoms?
- To which kinds of other atoms is it (are they) bound?
- What is the number of known ligands coordinated to one cation?

In some cases, however, additional approaches should be applied in order to assign some IR bands unambiguously.

ACKNOWLEDGEMENTS

The financial support by the European Commission (project INCO 016414) is greatly acknowledged.

REFERENCES

[1] Davydov, A., *Molecular Spectroscopy of Oxide Catalyst Surfaces*, Wiley, Chichester, 2003.
[2] Knözinger, H., in: Ertl, G., Knözinger, H., Weitkamp, J. (Eds.), *Handbook of Heterogeneous Catalysis*, Wiley-VCH, Weinheim, 1997, vol. 4, p. 707.
[3] Riczkowski, J., *Catal. Today* **2001,** *68,* 263–381.
[4] Hadjiivanov, K., Vayssilov, G., *Adv. Catal.* **2002,** *47,* 307–511.
[5] Hadjiivanov, K., *Catal. Rev.—Sci. Eng.* **2000,** *42,* 71–144.
[6] Zecchina, A., Scarano, D., Bordiga, S., Spoto, G., Lamberti, C., *Adv. Catal.* **2001,** *46,* 265–397.
[7] Busca, G., *Catal. Today* **1998,** *41,* 191–206.
[8] Kustov, L., *Top. Catal.* **1997,** *4,* 131–144.
[9] Hadjiivanov, K., Ivanova, E., Klissurski, D., *Catal. Today* **2001,** *70,* 59–63.
[10] Nakamoto, K., *Infrared Spectra of Inorganic and Coordination Compounds,* 2nd edition. Wiley, New York, **1970**.
[11] Bratermann, P. S., *Metal Carbonyl Spectra*, Academic Press, London, 1975.
[12] Hadjiivanov, K., Saussey, J., Freysz, J.-L., Lavalley, J.-C., *Catal. Lett.* **1998,** *52,* 103–108.
[13] Iwamoto, M., Hamada, H., *Catal. Today* **1991,** *10,* 57–71.

[14] Valyon, J., Keith Hall, W., *Stud. Surf. Sci. Catal.* **1993,** *74,* 1339–1344.
[15] Cheung, T., Bhargava, S., Hobday, M., Foger, K., *J. Catal.* **1996,** *158,* 301–310.
[16] Hoost, T. E., Laframboise, K. A., Otto, K., *Catal. Lett.* **1995,** *33,* 105–116.
[17] (a) Hadjiivanov, K., Penkova, A., Daturi, M., Saussey, J., Lavalley, J.-C., *Chem. Phys. Lett.* **2003,** *377,* 642–646; (b) Penkova, A., Hadjiivanov, K., Mihaylov, M., Daturi, M., Saussey, J., Lavalley, J.-C., *Langmuir* **2004,** *20,* 5425–5431.
[18] Henriques, C., Marie, O., Thibault-Starzyk, F., Lavalley, J.-C., *Microporous Mesoporous Mater.* **2000,** *50,* 167–171; (b) Szanyi, J., Kwak, J. H., Moline, R. A., Peden, C. H. F., *Phys. Chem. Chem. Phys.* **2003,** *5,* 4045–4051.
[19] Beutel, T., Adelman, B. J., Sachtler, W. M. H., *Appl. Catal. B* **1996,** *9,* L1–L10.
[20] Yeom, Y. H., Wen, B., Sachtler, W. M. H., Weitz, E., *J. Phys. Chem. B* **2004,** *108,* 5386–5404.
[21] (a) Lobree, L. J., Aylor, A. W., Reimer, J. A., Bell, A. T., *J. Catal.* **1997,** *169,* 188–193; (b) Poignant, F., Saussey, J., Lavalley, J. C., Mabilon, G., *Catal. Today* **1996,** *29,* 93–97.
[22] Mihaylov, M., Ivanova, E., Thibault-Starzyk, F., Daturi, M., Dimitrov, L., Hadjiivanov, K., *J. Phys. Chem. B* **2006,** *110,* 10383–10389.
[23] (a) Hadjiivanov, K., Tsyntsarski, B., Venkov, Tz., Daturi, M., Saussey, J., Lavalley, J.-C., *Phys. Chem. Chem. Phys.* **2003,** *5,* 243–245; (b) Hadjiivanov, K., Tsyntsarski, B., Venkov, Tz., Klissurski, D., Daturi, M., Saussey, J., Lavalley, J.-C., *Phys. Chem. Chem. Phys.* **2003,** *5,* 1695–1702.
[24] Ivanova, E., Mihaylov, M., Thibault-Starzyk, F., Daturi, M., Hadjiivanov, K., *J. Catal.* **2005,** *236,* 168–171.
[25] Miessner, H., *J. Am. Chem. Soc.* **1994,** *116,* 11522–11530.
[26] Miessner, H., Richter, K., *J. Mol. Catal. A* **1999,** *146,* 107–115.
[27] Chakarova, K., Mihaylov, M., Hadjiivanov, K., *Catal. Commun.* **2005,** *6,* 466–471.
[28] Hadjiivanov, K., Knözinger, H., Mihaylov, M., *J. Phys. Chem. B* **2002,** *106,* 2618–2624.
[29] Penkova, A., Dzwigaj, S., Kefirov, R., Hadjiivanov, K., Che, M., *J. Phys. Chem. C* **2007,** *111,* 8623–8631.
[30] Mihaylov, M., Chakarova, K., Hadjiivanov, K., Marie, O., Daturi, M., *Langmuir* **2005,** *21,* 11821–11828.
[31] (a) Hadjiivanov, K., Kantcheva, M., Klissurski, D., *J. Chem. Soc. Faraday Trans.* **1996,** *92,* 4595–4600; (b) Hadjiivanov, K., Knözinger, H., *J. Catal.* **2000,** *191,* 480–485.
[32] (a) Pieplu, T., Poignant, F., Vallet, A., Saussey, J., Lavalley, J.-C., *Stud. Surf. Sci. Catal.* **1995,** *96,* 619–626; (b) Spoto, G., Zecchina, A., Bordiga, S., Ricchiardi, G., Martra, C., Leofanti, G., Petrini, G., *Appl. Catal. B* **1994,** *3,* 151–172.
[33] Chakarova, K., Ivanova, E., Hadjiivanov, K., Klissurski, D., Knözinger, H., *Phys. Chem. Chem. Phys.* **2004,** *6,* 3701–3709.
[34] (a) Hadjiivanov, K., Knözinger, H., *J. Phys. Chem. B* **1998,** *102,* 10936–10940; (b) Hadjiivanov, K., *Microporous Mesoporous Mater.* **1998,** *24,* 41–49.
[35] Hadjiivanov, K., Ivanova, E., Kantcheva, M., Cifitikli, E., Klissurski, D., Dimitrov, L., Knözinger, H., *Catal. Commun.* **2002,** *3,* 313–319.
[36] Hadjiivanov, K., Ivanova, E., Knözinger, H., *Microporous Mesoporous Mater.* **2003,** *58,* 225–236.
[37] Hadjiivanov, K., Knözinger, H., *Chem. Phys. Lett.* **1999,** *303,* 513–520.
[38] Hadjiivanov, K., Massiani, P., Knözinger, H., *Phys. Chem. Chem. Phys.* **1999,** *1,* 3831–3838.
[39] Spoto, G., Bordiga, S., Scarano, D., Zecchina, A., *Catal. Lett.* **1992,** *13,* 39–44.
[40] Davydov, A., Lokhov, Yu., *React. Kinet. Catal. Lett.* **1978,** *8,* 47–52.
[41] Mihaylov, M., Penkova, A., Hadjiivanov, K., Daturi, M. J., *Mol. Catal. A* **2006,** *249,* 40–46.
[42] Ivanova, E., Mihaylov, M., Aleksandrov, H., Daturi, M., Thibault-Starzyk, F., Vayssilov, G., Rösch, N., Hadjiivanov, K., *J. Phys. Chem. C* **2007,** *111,* 10412–10418.
[43] Spoto, G., Zecchina, A., Berlier, G., Bordiga, S., Clerici, M. G., Basini, L., *J. Mol. Catal. A* **2000,** *158,* 107–114.
[44] Ivanova, E., Hadjiivanov, K., *Phys. Chem. Chem. Phys.* **2003,** *5,* 655–661.
[45] Hadjiivanov, K., Knözinger, H., *Catal. Lett.* **1999,** *58,* 21–26.
[46] Daniell, W., Topsøe, N.-Y., Knözinger, H., *Langmuir* **2001,** *17,* 6233–6239.
[47] Lercher, J. A., Grundlich, C., Eder-Mirth, G., *Catal. Today* **1996,** *27,* 353–376.

[48] Sobalik, Z., Tvarůžková, Z., Wichterlová, B., Filab, V., Špatenka, Š., *Appl. Catal. A* **2003,** *253,* 271–282.
[49] Sobalik, Z., Belhekar, A., Tvarůžková, Z., Wichterlová, B., *Appl. Catal. A* **1999,** *188,* 175–186.
[50] Paze, C., Zecchina, A., Spera, S., Spano, G., Rivetti, F., *Phys. Chem. Chem. Phys.* **2000,** *2,* 5756–5760.

CHAPTER 11

Organically Modified Ordered Mesoporous Siliceous Solids

Bénédicte Lebeau, Fabrice Gaslain,
Cristina Fernandez-Martin, *and* Florence Babonneau

Contents

Abstract

This chapter aims to give an overview of the main advances in the development of ordered mesoporous siliceous solids functionalized by organic groups including periodic organo-silica mesoporous materials. In the first part, the main strategic routes developed to prepare these hybrid mesoporous solids will be presented: the post-synthesis grafting methods and the direct synthesis routes by co-condensation and homo-condensation of bridged silsesquioxanes. The second part deals with the structural and the physico-chemical characteristics of these solids: porous characteristics, organic group content,

Ordered Porous Solids
DOI: 10.1016/B978-0-444-53189-6.00011-1

localization of organic groups and morphology. The shape modelling of these materials is presented in a third part. Finally, the main potential applications of these hybrid solids are reviewed in a fourth part.

1. Introduction

The pore surface functionalization by organic groups was already mentioned in the first reports related to the discovery of silica-based ordered mesoporous solids by the Mobil company[1] and Yanagisawa *et al.*[2] Nevertheless, their aptitude to be functionalized by anchoring organic groups on their pore surface started to be seriously considered with the early works of Brunel *et al.*[3] and Maschmeyer *et al.*[4] The growing interest in such organic modifications is related to the characteristics of ordered mesoporous silica (OMS) materials such as spatially organized and regular porosity, pore size ranges, high specific surface area and large pore volumes. Moreover, the amorphous character of their frameworks offers a large number of available silanol groups that can easily react with various functional groups and allows the strong and stable coupling of organic moieties to the silica network. A second direct method based on the co-condensation of silica precursors with organoalkoxysilanes in a structuring medium was quickly reported in the same time by Burkett *et al.*[5] and Macquarrie *et al.*[6] Both functionalization procedures were extended to numerous organic functions and compared in terms of loading rates, organic group localization, structural and textural characteristics. Related methods have been also the subject of numerous researches with the aim to anchor particular organic groups or improve the organic function loadings.

A third important step in this research area has been the incorporation of organic groups in the silica wall of the ordered mesoporous solid by using bridged silsesquioxane precursors leading to the so-called periodic mesoporous organosilicas (PMOs).[7–9] The main progresses in the development of these materials will not be too much detailed here since they have been recently presented in a review.[16]

All these hybrid inorganic–organic materials have been widely characterized in terms of spatial pore arrangement, porosity (pore size and pore volume), specific surface area, organic function loading, integrity, localization and accessibility, pore surface properties, bulk material properties (mechanical, thermal and hydrothermal stability, hyphobic/hydrophilic character, acidity/basicity, etc.), morphology and particle size. The main current characterization techniques employed have been X-ray diffraction (XRD), N_2-sorption, thermal analyses, elemental analyses, infrared (IR) spectroscopy, solid-state nuclear magnetic resonance (NMR), transmission electronic microscopy (TEM) and scanning electronic microscopy (SEM).

An important attention has been also given to the shape modelling of these hybrid materials with a particular effort on the elaboration of thin films and monoliths.

The design of organically modified ordered mesoporous solids has been mainly motivated for application fields such as catalysis, adsorption, sensing and more recently drug delivery systems with controlled release. However, other more specific application fields have been also targeted such as optics (energy transfer, luminescence), electrochemistry (matrices for solid electrodes), confined supports for nanoparticles growth, ceramic precursors...

This chapter gives an overview of main progresses in the research field of organically modified mesoporous siliceous solids which has been already the subject for several reviews.[10–16] It will mainly concern the synthesis methods, the structural and textural characterization, the shape modelling and the targeted applications.

2. Functionalization Methods

Three routes are commonly followed to organically modify OMS materials. Organic pendant groups can be anchored to the silica backbone by post-synthesis reactions with residual Si–OH groups (post-syntheis grafting) or by a direct synthesis process based on co-condensation of silica precursors with organoalkoxysilanes. The third route consists in introducing organic moieties as main chain groups by using bridged silsesquioxane precursors for the hybrid organic–inorganic framework. The principle of these three methods is summarized in Fig. 11.1.

2.1. Post-synthesis grafting method

Organic functionalization by using a post-synthesis grafting procedure was firstly reported by Kresge and co-authors[1] and Kuroda and co-authors [2] who synthesized M41S and FSM-16 OMS-based materials, respectively. This organic functionalization route started to be developed with Brunel *et al.*[3] and Maschmeyer *et al.*[4] works. The main strategy consists in first preparing the OMS solid by conventional surfactant templating method. In a second step, after surfactant removal, the silanol groups present on the pore surface are reacted with organosilane precursors (F–SiX_3 with X = Cl, OR, NH_2, etc.) to introduce functional F groups. The efficiency of this grafting process very much depends on the number, reactivity and accessibility of silanol groups. It also depends on the nature of the grafting agent and the

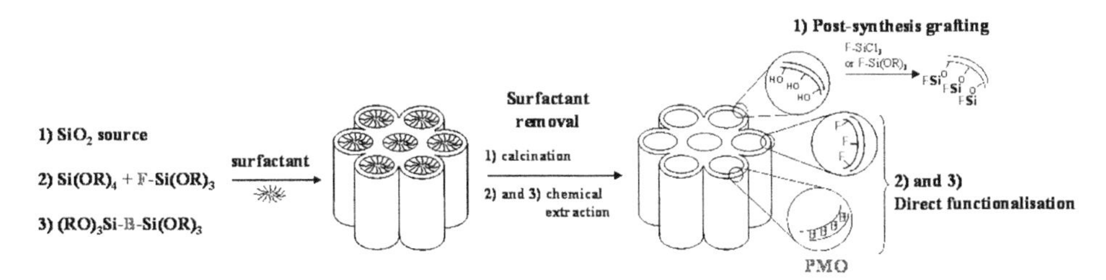

Figure 11.1 Schematic representation of the three main routes to prepare organically modified ordered mesoporous silica (OMS).

experimental conditions. Related post-synthesis methods have been developed such as the attachment of functional molecules to mesoporous support similarly to the preparation of self-assembled monolayers (SAMMs).[17] This last method referred to as coating, lead to high organic function loadings. A surfactant-silyl exchange process that has been proposed by Antochshuk *et al.*[18] consists in using organosilanes to modify the pore surface and replace the surfactant molecules at the same time. The organosilanes can be used as solvent or in a mixture with ethanol.[19] This method, which found to be favoured when applied to mesoporous silica prepared under acidic conditions, was extended to numerous organosilanes (chloro or alkoxy).[20]

2.2. Direct synthesis route

The direct functionalization process is based on the co-condensation of the silica precursor with an organotrialkoxysilane precursor (F–Si(OR)$_3$) under basic,[5,21] acidic[22,23] or neutral[6,24] conditions. In this process, the organosilane precursor has a double role to participate to the silica-based framework formation and to anchor an organic group to the backbone. The presence of organic functions in the as-synthesized solids necessitates surfactant removal procedure under soft conditions. Although some organic functions allow using a moderate calcination temperature sufficient to remove the surfactant molecules,[25] a chemical extraction procedure is preferably used. Depending on the nature of the surfactant molecule and synthetic process this chemical extraction is usually performed by refluxing in EtOH/HCl[5] or EtOH,[22] using Soxhlet method[24a] or under strongly H_2SO_4 acidic conditions.[26] It is noteworthy that the EtOH/HCl chemical extraction method leads to an ethoxylation of the residual silanol groups.[27,28]

Swelling agents, such as polypropylene glycol, trimethylbenzene and tetradecane, have sometimes been used to obtain direct organo-functionalized OMS with larger pores.[29,30]

A stepped templated sol-gel method has been proposed as a variant of the direct functionalization of OMS.[31] The basic aqueous solution containing the surfactant and the tetraalkoxysilane precursor is first hydrothermally treated for a time long enough to allow the formation of the MCM-41-type mesostructure. The organotrialkoxy silane is then added and the hydrothermal is prolonged. This originally stepped templated sol-gel technology was found to gather superior properties of the two conventional post-synthesis and direct synthesis methods. It increases the condensation degree of the framework with abundant content of functional groups.

Another direct synthesis methods using condensable surfactant have been reported.[32] Following this approach, the use of a 'Lizard' template, whose tail can be cleaved off after anchoring is particularly original.[33] Indeed, the functional groups are brought by the condensable templating agent and thus anchored on the pore surface and the porosity is opened by removing the surfactant hydrophobic tail.

2.3. Periodic mesoporous organosilicas

In 1999, three research groups, independently synthesized a new class of mesoporous materials designated as PMOs.[7–9] PMOs are prepared with bridged organotrialkoxysilane precursors (RO)$_3$Si–B–Si(OR)$_3$ (Fig. 11.1, B for bridging group).

Unlike organically functionalized mesoporous hybrids, synthesized *via* post-synthetic grafting or co-condensation direct synthesis, organic groups in PMOs are a part of the silica matrix walls so that high organic loadings can be reached, with a greater avoidance of pore blocking compared to organosilicas with terminal groups. Catalysis, sensing, separation as well as microelectronics are the main fields of targeted applications.

The first PMOs solids were synthesized from alkyl chain bridges in the presence of cationic surfactants as structure-directing agents in basic media. Further studies were focussed on the enlargement of the pore diameter by using non-ionic tri-block copolymers as structure-directing agents. The most common ones are the triblock Pluronic P123 ($EO_{20}PO_{70}EO_{20}$) and Pluronic F127 ($EO_{106}PO_{70}EO_{106}$). Bifunctional PMOs have been also synthesized by either co-condensation of bridged bis(trialkoxysilyl)organosilanes and terminal (trialkoxysilyl)organosilanes[34] or co-condensation of two different bissilylated precursors[35] in the presence of a structure-directing agent.

2.4. Role and nature of organic functional groups

For post-synthesis grafting process organo-alkoxysilane, -chlorosilane and -silazane[36] are the most common grafting agents used. They can be mono-, bi- or trifunctional.[37] An anhydrous medium is preferably required to avoid self-coupling. However, for some grafting method such as coating[17] (see Section 2.1) a controlled amount of water is preferred.

Some other post-synthesis grafting processes have also been developed using non-silylated reactants such as Grignard reagents and organolithium coumpounds.[38–41] Angloher *et al.*[42] have reported a low-cost metal-organic modification procedure that leads to the bonding of organic functional groups with an aromatic core as spacer. An interesting approach has been also reported by Anwander *et al.*[43] who have anchored an organometallic lanthanide silylamide for the surface modification of OMS.

Concerning the direct synthesis process, organotrialkoxysilanes are usually employed. However, organodialkoxysilane was also used[44] and recently the co-condensation process has been extended to bridged silsesquioxane precursors.[45,46] Most of the syntheses have been realized starting with tetraalkoxysilane as silica precursor in order to favour co-condensation, but some studies have been devoted to the use of low-cost silica precursor such as sodium silicate.[47,48]

For both post-synthesis and direct synthesis methods, the choice of functional groups is large: chloro,[48] iodo,[49] amine derivatives,[6,37,50] urea,[51] alkyl groups,[27,28] vinyl groups,[52] cyclame rings,[45] cyclodextrins,[53] calixarenes,[54,55] fullerene C60,[56] metal complexes,[57,58] ferrocen derivatives,[59] dyes,[60,61] dipeptides,[62] aromatic fluorinated compounds,[63] hydrosilane groups,[64] polymer precursors,[65] phosphonates,[66] phosphonic acid groups[67].

Three main strategies can be identified: (i) post-synthesis or direct anchoring of organic groups, (ii) grafting onto a previously functionalized surface and (iii) post-modification of the anchored organic groups. The approach (iii) has been widely followed to produce OMS functionalized with acidic sulfonic[23] and carboxylic[68]

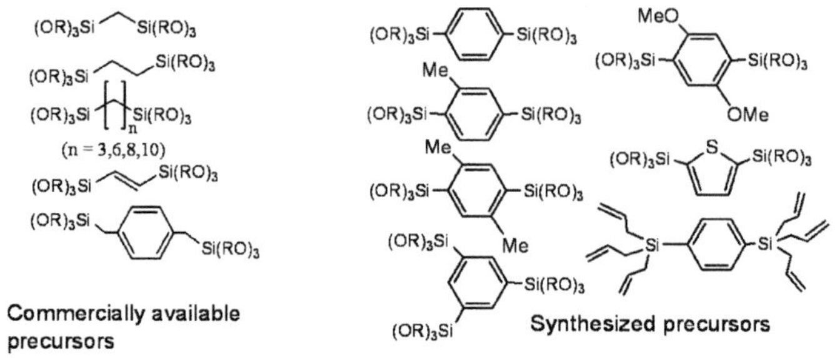

Commercially available precursors

Synthesized precursors

Figure 11.2 Overview of precursors used to prepare periodic mesoporous organosilicas(PMOs). (R $=CH_3$, C_2H_5).

groups from the oxidation of thiol and cyano groups, respectively. Following these strategies, the limits of complexity of the functional groups was greatly enlarged.

Both post-synthesis and direct synthesis methods also present the advantage to allow the multifunctionalization, which is realized by a multiple step approach[69] or by using a mixture of organosilanes.[70–73] The combination of these two functionalization methods was also achieved to fabricate bifunctional ordered mesoporous materials.[74]

Multifunctionalization has also been realized by selective post-synthesis grafting of different organic functions. One strategy consists in first grafting the external particle surface and then the internal pore surface.[75,76] A second strategy was developed for the selective grafting of micropores and mesopores of silica SBA-15.[77]

The first PMO materials were synthesized using 1,2-bis(trimethoxysilyl)ethane (BTME)[7] and 1,2-bis(triethoxysilyl)ethylene[8,9] precursors and cationic surfactants as structure-directing agents in basic media. Then the family of the PMOs extended to materials with aromatic bridges such as 1,4-bis(triethoxysilyl)benzene (BTEB) (Fig. 11.2). Like in the previous examples, the bridges present certain rigidity which seems to be a prerequisite for allowing the formation of highly ordered periodic mesoporous materials.[16,78]

Recently, the group of Ozin has also reported two new classes of PMOs formed from dendrimers building blocks precursors to yield periodic mesoporous dendrisilicas (PMDs) and from siloxane-disilsesquioxane or siloxy-trisilsesquioxane units to yield Bifunctional PMOs denoted as MT^3-PMO and DT^2-PMO.[79,80]

3. Characterization

3.1. Structural and textural characteristics

Organically modified OMS prepared by post-synthesis or by direct synthesis have been compared in terms of structural and textural characteristics, organic group loadings, localization and accessibility.[52,81–83] The spatial pore arrangement, which is mainly characterized by XRD in the low-angle range was found to be similar for organically modified OMS prepared by the post-grafting process and their parent solids. This result shows that this functionalization method allows maintaining the initial porous network structure. However, it was observed that the post-synthesis grafting led to materials with lower specific surface area, pore size and volume, these textural characteristics are related to the pore surface coverage with organic functions, spacers and silicon anchoring site. Moreover, the organic functions are preferably located at the pore entrance.[82] As this functionalization method depends highly on the number, nature and accessibility of hydroxyl groups present on the pore surface, solid state ^{29}Si,^{1}H NMR, Fourier transform infrared (FTIR) and thermo-gravimetric/differential thermal analyzer (TG/DTA) techniques have been used to analyze these reactive groups.[84–85] There are three types of surface silanols, which are single and geminal hydroxyl groups, as well as hydrogen-bonded groups. Only free-Si(OH) and geminal-$Si(OH)_2$ groups take parts in silylation reactions, since hydrogen-bonded silanol groups are less accessible to modification because they form hydrophilic networks.

Differences in porous network structure, specific surface area and porosity is usually observed if one compares organically modified OMS prepared by direct synthesis method to the corresponding pure OMS. In this case, pure silica reference materials cannot be considered as parent materials but as pure silica analogues. It is often reported that the nature of the organic function and/or the organic content greatly influence the pore network structure.[86,87] As the organosilane precursor participates to the silica-based framework formation, its presence modifies interactions between silicate species and surfactant molecules. Moreover, in most cases, the organic groups are hydrophobic and present steric constraints. As a consequence, they present disruptive effects on the mesostructure depending on their nature.[82,88] A phase transition of the porous network structure is sometime observed with increasing organic loading but in any case, high loading leads to a disordered porous network and total loss of porosity. However, control of experimental conditions (nature of the surfactant, solvent, pH, T, etc.) has allowed reaching high organic function loadings (up to 50% of the silicon sites).[89] Fine-tuning of the organic functionalization degree of mesoporous materials has also been achieved by electrostatically matching various organoalkoxysilanes with the cationic cetyltrimethylammonium bromide surfactants in a based catalyzed condensation reaction with tetraethoxysilane.[90] In this latter approach, disulfide-containing organotrimethoxysilanes have been used for the formation of organically functionalized mesoporous silica materials. After surfactant removal, the formation of thiol-functionalized mesoporous silica materials is achieved *via* disulfide reduction.

For both post-grafting and direct synthesis functionalized OMS, integrity and local environment of organic groups and modification of the pore surface have been mainly studied by solid-state ^{29}Si and ^{13}C NMR.[91,92]

The pore arrangement symmetry of PMOs synthesized from alkyl chain bridges is mainly two-dimensional hexagonal, even if three-dimensional hexagonal packing was also reported.[7] Interestingly, the groups of Inagaki[93] and Sayari[94] prepared PMO materials with cubic Pm3n symmetry starting from BTME in the presence of cetyltrimethylammonium chloride (CTAC) as structure-directing agent under basic conditions. Particles with well-defined octadecahedron shapes were observed (Fig. 11.3). The specific surface area was $\sim$850 m^2/g with pore diameter around

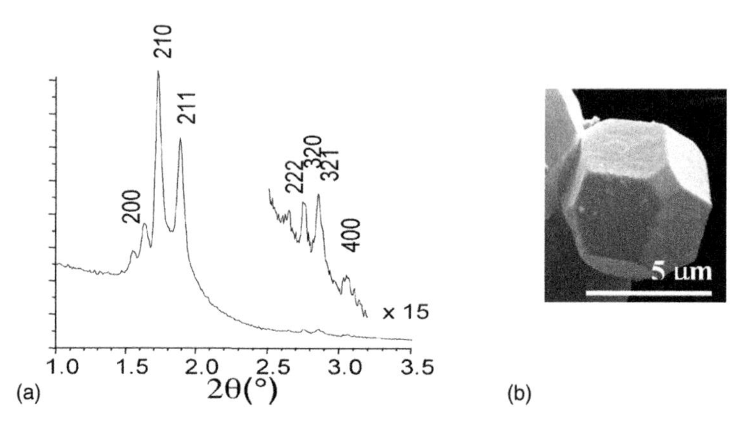

Figure 11.3 X-ray diffraction (XRD) pattern (a) and SEM image (b) of periodic mesoporous organosilica (PMO) sample with a cubic Pm3n structure prepared from 1,2-bis(trimethoxysilyl) ethane (BTME) in the presence of cetyltrimethylammonium chloride (CTAC) as structure directing agent.[98]

3–4 nm. PMOs with aromatic bridges have also been prepared. In the case of benzene-bridged PMOs, crystal-like organization of the aromatic-bridge groups within the pore walls have been revealed by the presence of diffraction peaks in the wide-angle region of XRD patterns characteristic of molecular scale periodicity in addition to the mesoscale periodicity.[95] For the first time, the presence of crystallinity within the walls was demonstrated, and other PMOs with similar structural characteristic were then reported, mostly with aromatic bridges.[96,97] It seems that Si–C bonds within the aromatic bridges present a certain rotational freedom arranging and orienting themselves within the pore walls showing structural periodicity.

The group of Fröba[99] reported in 2001 the first syntheses of large-pore Pluronic P123-templated PMOs using BTME as precursor under acidic conditions. Pore size of 6.4 nm was reported with a two-dimensional hexagonal arrangement. Using non-ionic polyoxyethylene alkyl ethers as Brij 56 ($C_{16}H_{33}(EO)_{10}OH$) and Brij 76 ($C_{18}H_{37}(EO)_{10}OH$) smaller pore sizes were observed since they favour the formation of thicker organosilica pore walls. For instance, Burleigh *et al.*[100] reported highly ordered two-dimensional hexagonal PMOs with a pore size of 4.5 nm using BTME precursor in the presence of Brij76 as structure-directing agent. However, using the shorter Brij type (Brij56), a smaller pore diameter (Ø = 3.9 nm) was found with poor ordering. Examples of PMOs prepared with the most common precursors and with various structure-directing agents are presented in Table 11.1. It is, however, worth mentioning that large improvement in the degree of ordering of non-ionic surfactant-templated PMOs could be obtained through the addition of ionic salts that enhance interaction between the template and the inorganic framework.[101–104]

Besides experimental parameters, such as temperature, pH value, concentrations, type of templating agents, which affect formation of PMOs, the most important factor affecting the mesostructure is the geometry of the organic bridges. PMOs reveal different mesostructures and crystalline pore walls, depending on the rigidity

Table 11.1 Examples of periodic mesoporous organosilica (PMO) syntheses reported in the literature

Precursor	Template	pH	Ø (nm)	Mesophase	References
$(MeO)_3Si-CH_2CH_2-Si(OMe)_3$	CTAC	basic	2.9	Cubic Pm 3n	93
	Brij56	acidic	4.5	2D hex.	105
	P123	acidic	6.0	2D hex.	99
	F127	acidic	5.6	Cubic Im3m	104
$(RO)_3Si-C_6H_4-Si(OR)_3$	OTAC	basic	3.9	2D hex.	95
	Brij56	acidic	3.5	2D hex.	106
	P123	acidic	7.4	2D hex.	107

CTAC, cetyltrimethylammonium chloride.

of organic-bridged units during hydrolysis and condensation. For instance, PMOs synthesized from alkyl chain bridges with more than two carbons do not show ordered periodic porous structures since they are too flexible.[108]

3.2. Stability

The organic functionalization of ordered mesoporous siliceous materials has been often considered for improving their chemical and mechanical stability.[109–110] The incorporation of organic groups directly attached to the backbone was found to enhance the structure stability towards water and mechanical pressure as a result of an increase of hydrophobicity.[110] Indeed, adsorption of water onto the silanols groups leads to the collapse of the mesostructure.

Concerning the PMOs, more complex systems synthesized from organic bridge units bound to more than two silicon atoms of the framework have been introduced by Ozin and co-authors[111] in order to improve rigidity, mechanical and thermal stability.

3.3. Morphology

The morphology and size of the primary particles can be of great importance for the bulk material properties for sorbent or catalyst use, for example. Indeed, these characteristics are suspected to play on the diffusion of species and thus matter transport within the porous network. They can be modulated (spheres, fibres, etc.) by experimental conditions such as the nature and the concentration of organic function, solvent, pH, T.[112–113]

Some efforts have been devoted in the production of spherical particles of organically modified mesoporous silica with controlled size from few nanometres to several micrometres. Colloidal suspensions of organically modified mesoporous silica nanoparticles have been prepared by controlling the extent of condensation reactions during the network silica–siloxane formation.[114] A Stöber *et al.*[115] derived method has been developed to prepare thiol- and amine-functionalized mesoporous silica spheres with 300–500 nm mean diameter.[116] The direct synthesis of thiol-functionalized mesoporous silica spheres with 100 m mean diameter was achieved under acidic conditions.[117] A vesicular morphology was also reported for thiol-functionalized mesostructured silica.[118]

Concerning PMOs, they are usually precipitated as powder consisted of particles with non-uniform sizes and irregular shapes. However, particles with well-defined octadecahedron shapes were observed for PMO materials with cubic Pm3n symmetry.[93,94] The first investigations of Park *et al.*[119] on the influence of various experimental parameters on the morphology of ethane-bridged PMOs, have led to series of different complex morphologies such as rope, spiral, gyroid and spheroid The generation of these morphologies was interpreted in terms of the degree of curvature and the accretion type induced by various topological defects.[120–122] Since then, PMOs with spherical,[123,124] rod-like[125] and fibres[126] morphologies have been reported depending on the pH, surfactant concentration and the presence of salt along with its concentration. The type and the concentration of co-precursors were also found to play a role in the morphology of direct-synthesized organically modified PMO particles prepared under basic conditions that adopt various morphologies, including rod-shaped and worm-like particles that have different degrees of bending.[127]

When using alkylammonium surfactant, variations in the surfactant chain length (C_n) allowed control of both the pore size and particle morphology of ethylene-containing mesoporous organosilica materials without compromising molecular-scale or structural ordering.[128] The particle morphology of the obtained PMOs gradually changed from monodisperse spheres (C_{12}) to rod- or cake-like particles (C_{14}) and elongated rope-like particles for longer chain surfactants (C_{16} and C_{18}).

4. Shape Modelling

The sol-gel process with its sol state offers large opportunities for the shape-modelling of the final material. Although most of the OMS (organically modified or not) are precipitated as powder with micrometer-size particles, the shape-modelling of these materials at the macroscopic scale with a continuous silica-based network is required for some applications such as sensors. The evaporation-induced self-assembly (EISA) process is well adapted for this purpose.[129] In particular, this method has been adapted to prepare mesoporous hybrid silica-based materials processed as spray-dried spheres,[130,131] monoliths,[132] dip-coated[133–135] or spin-coated[136] thin films, patterned functional nanostructures[137] and also to produce hierarchically ordered systems with macroporous architecture by dual templating.[138]

5. Targeted Applications

5.1. Catalysis

Catalysis has been the most studied application for the organically modified mesoporous siliceous solids. Historically, the discovery of these materials is the result of researches devoted to the synthesis of zeolites or related solids with larger pores for catalysis application. During the genesis of ordered mesoporous materials, their organic modification has mainly contributed to enhance their potential for such application. Initial aims of such modification were to improve stability to moisture and high-temperature treatments, acid or basic surface properties, to use them as high specific surface solid catalyst supports. Multifunctionality has also been considered and contributes to promote these materials as good candidates for heterogeneous catalysis. As a consequence of the permanent interest in their use for catalysis applications, they have been subjects of numerous reviews.[11,139–142]

Organically modified OMSs have been designed for acid, basic, oxidative and chiral heterogeneous catalysis. Solid-state acid catalysts have been mainly prepared by sulfonic functionalization of the pore surface. OMSs were first functionalized by thiol groups and a post-oxidative treatment allows to form sulfonic groups.[143–145] The direct sulfonic functionalization was also achieved.[146] Amino-OMS derivatives have been studied as base catalysts.[147,148] Although few examples of organically functionalized OMS have been reported for oxidative catalysis, these materials are suitable for chiral catalysis.[149] For this purpose, OMS solids have been functionalized with metal salen complexes, alkaloids and alkaloid derivates, amino alcohols as well as proline and benzylpenicillin derivatives.[150–153]

Although the potential advantages of PMOs in catalysis are claimed in almost each published work, few of them have reported on their catalytical properties. For this type of hybrid OMS, the acid catalytic properties of sulfonic functionalized PMOs have been mainly described.[154–156]

The OMS pore size range, in particular large pores such as in SBA-15 type OMS, is well adapted to the immobilization of small enzymes or proteins for biocatalytical applications. Although the chemisorption of these biological systems, which is generally realized with functionalized OMS, may result in partial or total lost of activity, it is preferable to the physisorption in order to avoid leaching during recycling.[157,158]

5.2. Adsorption

Porous materials are commonly employed for the adsorption of molecules or elements and can be found in everyday applications. For example, zeolites are used in double glazing windows to avoid the formation of steam patches; activated carbons are use in filters to remove undesirable odors, etc. From the beginning of their discovery, organic–inorganic OMSs were considered to have great potentials towards adsorption applications. In addition, these materials take advantages of both the inorganic framework (non-swelling and stable under acidic conditions, which can be prepared with controlled porosity and high surface area) and the complexation capacity of the organic functional groups anchored onto the silica surfaces.

One of the major adsorption applications if not the most important one for this type of materials is their use as pollutant adsorbents especially for heavy metals. Indeed, environmental pollution resulting from technological developments is a serious concern for ecology and represents a significant threat to the ecosystem and especially to people. Before the discovery of OMSs and their organically modified forms in the late 1990s, grafted silicas were conceived for such applications and extensively studied. A recent review by Mishra and co-authors[159] describes the variety of functionalizing entities that have been employed. On the basis of those screening, scientists like Feng, Fryxell *et al.*,[17a,160] Mercier and Pinnavaia[161] started to develop effective mesoporous sorbents based on mesoporous materials as supports. The essence of their methodology was to coat surfaces of mesoporous silica with organic functional groups to enhance their affinities for the targeted metal ions. Surface modification strategies by post-synthesis grafting and co-condensation (direct synthesis) methods with organoalkoxysilanes have been largely investigated. Some reviews have reported advances in this field.[13,14,17b,162]

5.3. Adsorption of metals

Among all the possible applications for functionalized OMS as sorbents, heavy metal remediation has been the historical and major topic of interest. Thiol groups are studied for their high affinities towards Hg(II) species, while amino-based ligands are preferentially specific towards Cr(III), Ni(II), Cu(II), Zn(II) and Cd(II) species.

Methacryloxypropyltrimethoxysilane (MPTMS) and methacryloxypropyltriethoxysilane (MPTES) precursors have been reported as suitable organic sources in the synthesis of thiol-functionalized mesoporous silica. Such materials have been used for selective mercury adsorption.[160,161,163–167] Feng *et al.*[160] and Mercier and Pinnavaia[161] developed in 1997 mesostructured siliceous materials with a high-selective capacity of removing mercury by grafting the silica surface of MCM-41- and hexagonal mesoporous silica (HMS)-type materials, respectively, with mercaptopropyl groups. Since then, the mercury adsorption capacity on mercaptopropyl-functionalized HMS,[163] MCM-41,[164,165] and SBA-15[166] obtained by grafting of the preformed mesoporous silica has been reported. These new adsorbents present a significant adsorption capacity because of high amounts of functional groups that are accessible to the mercuric species due to the uniform open-framework mesoporosity. Unfortunately, it is necessary to employ a large excess of thio-organic precursor in the grafting procedure. Moreover, to achieve the highest sulfur incorporation, a pretreatment of the calcined mesoporous silica is required. So, the material presenting the highest mercury capacity (FMMS) was obtained by repeated hydrolysis and silylation processes.[160]

More recently, adsorption of mercury on mesostructured silica MCM-41,[167] HMS,[168,169] and MSU[169–171] obtained by the co-condensation method has been studied. MPTMS was used in all cases. This method has gained importance because of the simplicity of the procedure, smaller consumption of organic precursor and shorter synthesis time. The resulting materials exhibit a high content and good dispersion of functional groups, although for some of them a constriction of pores into the micropore range has been observed. This phenomenon can prevent the total access of mercury species to the active sorption sites.[168] For instance, thiol-MCM-41 obtained by the co-condensation procedure contains a high sulfur content, but mercury adsorption capacity (2.1 mmol.g^{-1} of Hg) is only about half of the total sulfur content of the material. A similar behaviour has been observed for the one-step synthesized thiol-HMS material. Conversely, Brown *et al.*[170] developed a thiol-functionalized MSU material with a high sulfur content that preserved the mesostructured channel network, so allowing full access of mercury to every binding site.

Later, Aguado *et al.*[172] studied thiol-functionalized SBA-15-type silica materials synthesized by the co-condensation procedure. SBA-15 materials with the largest pore size found for siliceous mesostructured materials and thicker silica walls were expected to impart significant thermal and mechanical stabilities. The adsorbents made were extremely efficient in removing Hg(II) (max. loading of ca. 2.9 mmol g^{-1}) and the efficiency was found to remain almost constant from neutral pH down to 0.75.

Jaroniec and co-authors[173] also studied the recyclability of such materials with benzoylthiourea-modified mesoporous silica for mercury (II) adsorption from aqueous solutions. Immobilization of benzothiourea ligand on mesoporous silica support was carried out through a two-step reaction: first step consisted in grafting 3-aminopropyltriethoxysilane (APTES) onto M41S materials followed by a second step where these materials were reacted with benzoyl isothiocyanate. Since interactions between the Hg(II) and 1-benzoyl-3-propylthiourea are relatively weak, regeneration of the mesoporous adsorbent can be done with hydrochloric acid under mild conditions. Sierra and co-authors.[174] have recently screened over a rich variety of thiol-derived functionalities for OMS materials, to improve the heavy metal adsorption capacity and the materials recyclability. The adsorption capacity of Hg(II) species were found to be dependant on the type of the thiol-derived functionality and to be higher in OMS functionalized by the direct synthesis method since higher loadings of thiol-derived functionalities are achieved.

As for the adsorption of other metals, 3-aminopropyltrimethoxysilane (APTMS) and APTES-functionalizing agents are the most commonly used due to their relatively good chelating performances, wide availability and cheapness. In a series of articles, Walcarius *et al.* looked at the adsorption of several amino-functionalized materials using various matrices such as MCM-41, MCM-48, SBA-1 prepared by grafting or co-condensation methods.

To go one step further, some research groups started to evaluate the main differences between the functionalization methods used, post-synthesis versus direct synthesis. In this frame, Tatsumi and co-authors[37] compared mono-, di-, and triamino-functionalized MCM-41 silicas [using APTMS, N-(2-aminoethyl)-3-aminopropyltrimethoxysilane (AEAPTMS) and (3-trimethoxysilylpropyl)-diethylenetriamine (TMSPDTA) grafting agents] synthesized directly by co-condensation or *via* a post-synthesis grafting. It was revealed that all amino moieties incorporated into direct synthesized OMS were not present at the surface, but some of them were in the wall of the hexagonal channels. The authors found that the adsorption capacity of Co(II) and Fe(III) increased on amino-functionalized samples prepared via the co-condensation method with an increase in the surface density of amino groups regardless of the amino-organoalkoxysilanes. When mono- and di-amino silanes were grafted onto the surface of MCM-41, the adsorption capacity increased with an increase in the surface density of amino groups. However, when tri-amino silane was anchored to the surface, the behaviour was reversed. It has to be mentioned that when using amino-based functionalities, the materials obtained tend to degrade noticeably in contact with water. In fact, the amino groups that locally induce basicity, dissolve the silica framework and slowly free themselves, accelerating the dissolution of the inorganic framework.[175]

Different strategies to improve the affinity and selectivity of these materials have been applied by increasing the density of functional groups while retaining the open porosity.[81,82,89,176–178] One strategy is to select the most adequate framework system with high-performance ligands such as 1,4-bis(triethoxysily)propane tetrasulfide (TESPTS).[179,180] Sierra and co-author also looked at novel functionalities associated with the right mesostructure. 5-mercapto-1methyl-1-H-tetrazol (MTTZ) was used to chemically modify HMS supports for the adsorption of Pb(II)[181] and MCM-41 for the adsorption of Zn(II).182 This group also studied the adsorption by 2-mercaptopyrimidine (MP)-modified OMS of Cd(II) contained from aqueous media.[183]

A major drawback that came up to light recently is the difficulty to comparatively extract the results found in literature since everyone reports about similar but, however, slightly different materials (differences in terms of size and shapes of the particles) or experimental conditions (different concentrations of pollutants, pH, etc.) and no standard exists. As we have already seen in the previous paragraph, the efficiency of a functionalized material is usually assessed by its capacity to retain pollutants over a certain period; however, only few studies looked into the accessibility of pollutant species to the active sites[184] and mass-transfer reaction behaviours for these systems;[17a,81,161,164–166,168,185,186] and again their results were impossible to compare. Such data are of importance since none of all the active sites located inside the mesoporous structure are accessible[184] and the speed at which solution-phase species are allowed to move to these centres is the rate-determining step. The latest point has been highlighted by Walcarius *et al.* using electrochemical analysis at chemically modified electrodes[185] or pollutant removal from diluted solutions.[17a,81,161,164–166,168,186] However, even if some reports have tried to present a qualitative or quantitative approach to answer this question of paramount importance, most lacked from direct comparativeness of their results mainly due to the inhomogeneity of the materials. Walcarius and co-authors [187] have started to give answers by following and comparing Hg(II) uptake with various materials of similar particle size and shape under well-controlled conditions.

PMOs have also been functionalized with different chelating or complexing agents for adsorption applications. The use of a bridged tetrasulfide precursor co-condensed with TEOS allowed to produce PMO with very high affinity for Hg(II).[188] Anion-exchange resin based on PMOs has also been developed for adsorption of environmental pollutants such as perrhenates, perchlorates and pretechnetates.[189]

5.4. Adsorption of other pollutants and novel applications

Adsorption of metal ions is not the only application that has been found for these functionalized OMS supports. Yoshitake *et al.*[190] prepared mono-, di-, and triamino- functionalized mesoporous silicas (using APTMS-, AEAPTMS- and TMSPDTA-grafting agents) with MCM-41 and SBA-1 structure types for the sequestration of chromates and arsenates. They exhibited relatively good stability and good performances for adsorption of oxyanions, and thus appeared to be candidates as inexpensive adsorbents for reducing toxic anion pollution. Later, for the same purpose, they developed a series of ethylenediamine (EDA)-, diethylenetriamine (DETA)-, triethylenetetramine (TETA)-, tetraethylenepentamine (TEPA)- and pentaethylenehexamine (PEHA)-functionalized mesoporous silica (MCM-41 and SBA-15) by the reaction of pre-grafted 3-chloropropyltrimethoxysilane (CPTMS) with polyamine.[191] They highlighted a clear correlation between the number of amines in the polyamine, and the adsorption capacity. A comparison with adsorbents prepared by the same procedure but with nonporous silica showed that the porosity is also a key parameter for reaching a high adsorption capacity. In a recent work, Crudden and co-authors[192] successfully employed aminopropyltriethoxy silane-derivatized silicates for the removal of residual ruthenium contaminants after olefin metathesis reactions and asymmetric hydrogenations.

Organic molecules can also be removed from water using organically grafted OMS materials. Stein and Lim[82] studied vinyl-functionalized MCM-41 samples prepared by either post-synthesis grafting or direct co-condensation and looked at their performances to adsorb non-polar solvents from aqueous solutions or emulsions. However, those materials could only adsorb relatively small organic molecules. Strategies to expand the mean pore size diameter were developed to accommodate relatively bulky organic molecules. Mercier and co-author[53] used cyclodextrins, large cyclic oligosaccharides, to functionalize HMS silica in order to selectively remove water-soluble aromatic molecules. Sayari *et al.*[193] proposed to use a co-surfactant to enlarge the pores and improve the adsorption of organic pollutants from wastewater. Inumaru *et al.*[194] reported an octylsilane-grafted MCM-41 mesoporous silica to remove low-concentrated nonylphenol, an endocrine disrupter (one of the major estrogenic contaminants to induce feminising effects in fish), in water with high efficiency comparable to that of activated carbon. This team carried on this topic and prepared other alkyl-grafted MCM-41 to adsorb 4-*n*-heptylaniline and 4-nonylphenol.[195b]

Yeung and co-authors[196] prepared OMS adsorbents by grafting amino- and carboxylic-containing functional groups onto MCM-41 for the selective removal of acid blue 25 and methylene blue dyes from wastewater. Better selectivity of the OMS-based adsorbents means longer operating life and less maintenance. In a similar research frame, Li and co-authors[197] used OMS with different pore sizes and a high density of carboxylic acid groups as adsorbents, for the removal of three basic dyes (methylene blue, phenosafranine and night blue) from wastewater.

In the quest of environmentally friendly, efficient and low-cost industrial processes for the recovery of carbon dioxide, SBA-16 silicas grafted with N'-[3-(trimethoxysilyl)propyl]-ethylenediamine (TMSPEDA) were developed by Hornebecq and co-authors.[198] Earlier, Yogo and co-authors[199] performed CO_2 adsorption in the presence of water vapour using modified SBA-15 with APTES-, N-(2-aminoethyl)-3-aminopropyltriethoxysilane (AEAPTES)-or TMSPDTA-grafting agents.

In another environmentally friendly field, FSM-type mesoporous silica were modified with 1,4-butanediol by Kuroda and co-authors[200] and the esterified material was used as the adsorbent to accommodate chlorophyll molecules. Pheophytinization was effectively suppressed by the presence of the surface organic groups grafted onto the mesoporous silica, whereas pheophytinization occurred in unmodified mesoporous silica. These results contribute to the development of photochemical and photophysical studies using chlorophylls in porous heterogeneous media and may lead to the construction of *in vitro* biomimetic solar energy conversion and storage systems.

Matsumoto *et al.*[201] reported a very interesting comparative study of various tailored APTES grafted MCM-48 silica spheres using a silane-coupling agent, *n*-octyldimethylchlorosilane. Changes in adsorption features of water and acetaldehyde were elucidated by adsorption calorimetry. While, the MCM-48 surface exhibited a hydrophobic feature after the chemical modification, which prevented hydrolysis and degradation of the silica matrix in water, the APTES treatment drastically increased adsorptivity of acetaldehyde vapour. This example of a double grafting treatment showed that grafting can not only improve the adsorbing capacity of an ordered mesoporous material, but also the material life time. Such strategy is of prime importance for making efficient adsorbers with a long lifetime (column packing materials for chromatographic separation).

Soon after the initial publications mentioning the use of mesoporous silica for enzyme immobilization, Ackerman *et al.*[195a] prepared functionalized mesoporous silica (FMS) grafted with COOH or NH_2 groups. COOH-functionalized FMS showed excellent capability to immobilize organophosphorus hydrolase and enhanced the overall material stability. Chong *et al.* applied this concept to the immobilization of Penicillin G acylase (PGA)[202] and prepared supports that can be potential enhanced biocatalysts. Another possible bio-application is the protein size selective adsorption with thiol-functionalized mesoporous molecular sieves.[203]

5.5. Sensors

Organically modified OMS with their regular porosity and large pore diameters are good candidates for sensing applications. Moreover, the easy shape modelling for preparing macroscopic bulk materials with high transparency made them even more attractive for the development of optical sensing devices. Sensing systems required several criteria such as selectivity, sensitivity, detectability, rapid probe response, and stability (in time and in the conditions of use). Response of the probe to any changes in its environment is usually detected by optical, electrochemical and mechanical signals.

The design of OMS for sensing applications is mainly based on the anchoring of probes since it was quickly recognized that the covalent coupling of the probes to the silica backbone is preferred to avoid leaching and improve their chemical and thermal stability.

As already mentioned, OMS can be functionalized by organic groups, such as thiol groups, able to bond heavy metal species.[185c–f,186b,c] The resulting solid was successfully used as electrode modifier for pre-concentration of target analytes such as Hg(II) before electrochemical detection. Organically modified OMS-based sensors have been developed for optical detection of pH changes,[204] biogenic molecules such as dopamine and glucosamine,[205] anions,[206] metal species,[54,207,208] phenol derivatives[209] and gas[210]. A particular attention has been given to the development of thin films for electrochemical,[211,212] optical[207] and mechanical[213] sensors.

5.6. Drug delivery systems

Because of their structural and textural characteristics, organically modified OMS have been identified as suitable solid carriers for controlled drug delivery systems.[214] The drug vectorisation is an application field in full emergence where OMS and, in particular, organically modified OMS have been lately identified as good potential candidates. Han *et al.*[215] were the first to report the ability of functionalized OMS to sequester and release proteins. Organic functionalization was found to be preferable for better adsorption properties and also for controlled release of the active molecules.[216] The drug molecules can also be covalently bonded to the silica walls of the OMS to control its release.[217] Since these early reports, works in this area have exponentionally increased.[218–220] One important contribution in control release system is the photo-switched storage and release of guest molecules in the porosity of coumarine-modified MCM-41.[221] A photosensitive coumarine derivative is grafted on the pore outlet of the OMS and act as a gatekeeper. Indeed, under UV-irradiation the coumarine photodimerizes and closes the pore outlet with cyclobutadiene dimer.

5.7. Ceramic precursors

Most applications that are largely cited for PMOs exploit the presence of functional groups in the porous framework. However, the presence of Si–O and Si–C bonds in the framework makes them very suitable as precursors for porous silicon oxycarbide (SiCO) glasses. Most of the studies have been so far focussed on the preparation of dense SiCO glasses, in which tetravalent C atoms substitute divalent O atoms in a silica network. As demonstrated by previous studies, such substitutions bring unique high-temperature properties compared to vitreous silica in terms of mechanical strength[222] and chemical stability[223]. Additionally, if these materials can exhibit a periodic porous network, then they can find applications as filters, catalysts or membranes under severe operating conditions. The challenge here is to convert PMOs in SiCO glasses without a total collapse of the porous network during pyrolysis. PMOs with ethane bridges and cubic Pm3n symmetry were successfully converted into porous SiCO glasses after pyrolysis at 1000 °C under Ar with a specific surface area above 700 m^2/g and pore size around 2.5 nm.[224,225]

5.8. Miscellaneous

Besides these previously discussed major applications, some more specific ones have been considered. Sulfonic functionalized mesoporous materials have been studied not only for their surface acidity in catalysis,[145,146] but also for their high proton conduction properties that make them promising solid electrolytes.[226] Coupled with their ability for water retention, they are also considered a good candidate for the elaboration of fuel cell composite membranes.[227]

The functionalization of pore surface of OMS has been also performed to favour the growth of nanoparticles such as platinum,[228] gold,[229,230] ZnS,[231] CdS[232] and

Co_2P[233] within the pores. Indeed, the pore surface modification results in an increase of affinity towards the nanoparticle precursors and favour the preferential nanoparticle formation inside the porous network.

Functionalization of OMS with monomers has allowed preparing mesoporous silica-reinforced polymer nanocomposites.[234,235]

Grafting of rare-earth complexes has been studied to produce materials with luminescence properties.[236–239] In this context, the formation of some Europium (III) complex was observed in ordered SBA-15 mesoporous silica containing chelating groups while the complex formation was not possible in solution.[240] Energy transfers have also been observed between chromophores in mesopores when donor–acceptor pairs were grafted.[241,242]

Organically modified OMS were also found suitable as stationary phases for chromatography.[243,244]

One immediate application for organically modified OMS films is for low dielectric constant (k) layers in microelectronic systems. The large increasing density of electronic components in modern miniaturized integrated circuits requires insulators with sufficiently small k values. A new generation of ultra-low-k materials (<2.0) are required for device feature sizes in the range of 100 nm.[245] Silica thin films with large pore volume ($k_{air} \approx 1.0$) are very attractive low-k materials if water adsorption can be prevented. This is the reason why the introduction of hydrophobic organic functions in such porous silica thin films has been explored. One can cite the work of the group from Philips research laboratories[246] on methyl-modified silica thin films who reached values as low as 1.7. PMO-based thin films have also been tested and Lu *et al.*[247] reported k value around 2.0. Hatton *et al.* works have also contributed to show the potential of PMOs in this application field.[248]

6. Outlook and Prospects

In more than 10 years, the research area of hybrid inorganic–organic ordered mesoporous siliceous materials has exploded resulting in a considerable number of literature reports, including reviews. Three main strategies for the preparation of these materials have been developed and have quickly showed their main advantages and limits depending on the targeted applications. Early studies in this field have been mainly devoted to the characterization of the structure, texture and physico-chemical properties of the resulted organically modified OMS solids. Based on the results of these investigations, recent studies have evolved towards the development of new organic-modification methods, the functionalization with more complex organic groups, the selective grafting to elaborate multifunctional materials, the preparation of sophisticated functional materials (photo-switched storage and release of guest molecules,[221] gatekeeping layer effect,[249] controlled pore size photoresponsive nanocomposites[250]).

The large potential of these materials in terms of structural and textural characteristics, surface and bulk properties, processing, functionalities and so on make them suitable materials for a large panel of applications. However, current laboratory scale for the preparation, environmental and ecological problems related to the elaboration process and production costs are still barriers for their industrial development. These drawbacks have to be highly considered for future developments.

REFERENCES

[1] Beck, J. S., Vartuli, J. C., Roth, W. J., Leonowicz, M. E., Kresge, C. T., Schmidt, K. D., Chu, C. T. W., Olson, D. H., Sheppard, E. W., McCullen, S. B., Higgins, J. B., Schlenker, J. L., *J. Am. Chem. Soc.* **1992,** *114,* 10834–10843.
[2] Yanagisawa, T., Shimizu, K., Kuroda, K., Kato, C., *Bull. Chem. Soc. Jpn.* **1990,** *63,* 988–992.
[3] Brunel, D., Cauvel, A., Fajula, F., DiRenzo, F., *Stud. Surf. Sci. Catal.* **1995,** *97,* 173–180.
[4] Maschmeyer, T., Rey, F., Sankar, G., Thomas, J. M., *Nature* **1995,** *378,* 159–162.
[5] Burkett, S. L., Sims, S. D., Mann, S., *Chem. Commun.* **1996,** 1367–1368.
[6] Macquarrie, D., *Chem. Commun.* **1996,** 1961–1962.
[7] Asefa, T., MacLachan, M. J., Coombs, N., Ozin, G. A., *Nature* **1999,** *402,* 867–871.
[8] Inagaki, S., Guan, S., Fukushima, Y., Ohsuna, T., Terasaki, O., *J. Am. Chem. Soc.* **1999,** *121,* 9611–9614.
[9] Melde, B. J., Holland, B. T., Blanford, C. F., Stein, A., *Chem. Mater.* **1999,** *11,* 3302–3308.
[10] Moller, K., Bein, T., *Chem. Mater.* **1998,** *10,* 2950–2963.
[11] Maschmeyer, T., *Curr. Opin. Solid. State. Mater.* **1998,** *3,* 71–78.
[12] Impens, N. R. E. N., van der Voort, P., Vansant, E. F., *Microporous Mesoporous Mater.* **1999,** *28,* 217–232.
[13] Stein, A., Melde, B. J., Schroden, R. C., *Adv. Mater.* **2000,** *12,* 1403–1419.
[14] Sayari, A., Hamoudi, S., *Chem. Mater.* **2001,** *13,* 3151–3168.
[15] Vinu, A., Hossaim, K. Z., Ariga, K., *J. Nanosci. Nanotech.* **2005,** *5,* 347–371.
[16] Hoffman, F., Cornelius, M., Morell, J., Fröba, M., *Angew. Chem. Int. Ed.* **2006,** *45,* 3216–3251.
[17] (a) Liu, J., Feng, X., Fryxell, G. E., Wang, L.-Q., Kim, A. Y., Gong, M., *Adv. Mater.* **1998,** *10,* 161–165. (b) Liu, J., Shin, Y., Nie, Z., Chang, J. H., Wang, L. Q., Fryxell, G. E., Samules, W. D., Exarhos, G. J., *J. Phys. Chem. A* **2000,** *1004,* 8328–8339. (c) Fryxell, G. E., Mattigod, S. V., Lin, Y., Wu, H., Fiskum, S., Parker, K., Zheng, F., Yantasee, W., Zemanian, T. S., Addleman, R. S., Liu, J., Kemmer, K., Kelly, S., Feng, X., *J. Mater. Chem.* **2007,** *17,* 2863–2874.
[18] Antochshuk, V., Jaroniec, M., *Chem. Commun.* **1999,** 2373–2374.
[19] Lin, H.-P., Yang, L. Y.-L., Mou, C. Y., Liu, S. B., Lee, H. K., *New J. Chem.* **2000,** *24,* 253–255.
[20] Liu, Y.-H., Lin, H.-P., Mou, C.-H., *Langmuir* **2004,** *20,* 3231–3239.
[21] Fowler, C. E., Burkett, S. L., Mann, S., *Chem. Commun.* **1997,** 1769–1770.
[22] Babonneau, F., Leite, L., Fontlupt, S., *J. Mater. Chem.* **1999,** *9,* 175–178.
[23] Margolese, D., Melero, J. A., Christiansen, S. C., Chmelka, B., Stucky, G. D., *Chem. Mater.* **2000,** *12,* 2448–2459.
[24] (a) Richer, R., Mercier, L., *Chem. Commun.* **1998,** 1775–1776. (b) Mercier, L., Pinnavaia, T., *Chem. Mater.* **2000,** *12,* 188–196.
[25] Kumar, R., Chen, H.-T., Escoto, J. L. V., Lin, V. S.-Y., Pruski, M., *Chem. Mater.* **2006,** *18,* 4319–4327.
[26] Yang, C. M., Zibrowius, B., Schüth, F., *Chem. Commun.* **2003,** 1772–1773.
[27] Goletto, V., Imperor, M., Babonneau, F., *Stud. Surf. Sci. Catal.* **2001,** *135,* 1129–1136.
[28] Lesaint, C., Lebeau, B., Marichal, C., Patarin, J., *Microporous Mesoporous Mater.* **2005,** *83,* 76–84.
[29] Bambrough, C. M., Slade, R. C. T., Williams, R. T., *J. Mater. Chem.* **1998,** *8,* 569–571.
[30] Park, B.-G., Guo, W., Cui, X., Park, J., Ha, C.-S., *Microporous Mesoporous Mater.* **2003,** *66,* 229–238.
[31] Zhang, C., Zhou, W., Liu, S., *J. Phys. Chem. B* **2005,** *109,* 24319–24325.
[32] Shimojima, A., Kuroda, K., *Angew. Chem. Int. Ed.* **2003,** *42,* 4057–4060.

[33] Zhang, Q., Ariga, K., Okabe, A., Takuzo, A., *J. Am. Chem. Soc.* **2004,** *126,* 988–989.
[34] Burleigh, M. C., Markowitz, M. A., Spector, M. S., Garber, B. P., *J. Phys. Chem. B* **2001,** *105,* 9935–9942.
[35] Zhu, H., Jones, D. J., Zajac, J., Dutartre, R., Romari, M., Rozière, J., *Chem. Mater.* **2002,** *14,* 4886–4894.
[36] Anwander, R., Nagi, I., Widenmeyer, M., Engelhardt, G., Groeger, O., Palm, C., Röser, T., *J. Phys. Chem. B* **2000,** *104,* 3532–3544.
[37] Yokoi, T., Yoshitake, H., Tatsumi, T., *J. Mater. Chem.* **2004,** *14,* 951–957.
[38] Yamamoto, K., Tatsumi, T., *Microporous Mesoporous Mater.* **2001,** *44–45,* 459–464.
[39] Lim, J. E., Shim, C. B., Kim, J. M., Lee, B. Y., Yie, J. E., *Angew. Chem. Int. Ed.* **2004,** *43,* 3839–3842.
[40] Ramirez, A., Sierra, L., Lebeau, B., Guth, J.-L., *Microporous Mesoporous Mater.* **2007,** *98,* 115–122.
[41] Angloher, S., Kecht, J., Bein, T., *Chem. Mater.* **2007,** *19,* 3568–3574.
[42] Angloher, S., Bein, T., *J. Mater. Chem.* **2006,** *16,* 3629–3634.
[43] Anwander, R., Roesky, R., *J. Chem. Soc., Dalton Trans.* **1997,** 137–138.
[44] Joo, J., Hyeon, T., Hyeon-Lee, J., *Chem. Commun.* **2000,** 1487–1488.
[45] (a) Corriu, R. J. P., Mehdi, A., Reyé, C., Thieuleux, C., *Chem. Commun.* **2002,** 1382–1383. (b) Corriu, R. J. P., Medhi, A., Reyé, C., Thieuleux, C., *Chem. Commun.* **2003,** 1564–1565.
[46] Corriu, R. J. P., Mehdi, A., Reyé, C., Thieuleux, C., *New J. Chem.* **2003,** *27,* 905–908.
[47] Yu, N., Gong, Y., Wu, D., Sun, Y., Luo, Q., Liu, W., Deng, F., *Microporous Mesoporous Mater.* **2004,** *72,* 25–32.
[48] Corriu, R. J. P., Mehdi, A., Reyé, C., Thieuleux, C., *Chem. Commun.* **2004,** 1440–1441.
[49] Alauzun, J., Mehdi, A., Reyé, C., Thieuleux, C., *New J. Chem.* **2007,** *31,* 911–915.
[50] Hicks, J. C., Jones, C. W., *Langmuir* **2006,** *22,* 2676–2681.
[51] Gong, Y. J., Li, Z. H., Wu, D., Sun, Y. H., Deng, F., Luo, Q., Yue, Y., *Microporous Mesoporous Mater.* **2001,** *49,* 95–102.
[52] Lim, M., Blanford, C. A., Stein, C. A., *J. Am. Chem. Soc.* **1997,** *119,* 4090–4091.
[53] Huq, R., Mercier, L., *Chem. Mater.* **2001,** *13,* 4512–4519.
[54] Métivier, R., Lebeau, B., Leray, I., Valeur, B., *J. Mater. Chem.* **2005,** *15,* 2965–2973.
[55] Liu, C., Naismith, N., Fu, L., Economy, J., *Chem. Commun.* **2003,** 2472–2473.
[56] Lee, C.-H., Lin, T.-S., Lin, H.-P., Zhao, Q., Liu, S.-B., Mou, C. -Y., *Microporous Mesoporous Mater.* **2003,** *57,* 199–209.
[57] Diaz, J. F., Balkus, K. J., Jr., Bedioui, F., Kurshev, V., Kevan, L., *Chem. Mater.* **1997,** *9,* 61–67.
[58] Jurnes, C. D., Pillinger, M., Valente, A. A., Gonçalves, I. S., Rocha, J., Ferreira, P., Kühn, F. E., *Eur. J. Inorg. Chem.* **2002,** 1100–1107.
[59] MacLchlan, M. J., Aroca, P., Coombs, N., Manners, I., Ozin, G. A., *Adv. Mater.* **1998,** *10,* 144–149.
[60] Fowler, C. E., Lebeau, B., Mann, S., *Chem. Commun.* **1998,** 1825–1826.
[61] Seçkin, T., Gültek, A., *J. Appl. Polym. Sci.* **2003,** *90,* 3905–3911.
[62] Walcarius, A., Sayen, S., Gérardin, C., Hamdoune, F., Rodehüser, L., *Coll. Surf. A: Physicochem. Ang. Aspects* **2004,** *243,* 145–151.
[63] Lebeau, B., Marichal, C., Mirjol, A., Soler-Illia, G. J., de, A. A., Buestrich, R., Popall, M., Mazerolles, L., Sanchez, C., *New J. Chem.* **2003,** *27,* 166–171.
[64] Mehdi, A., Mutin, H., *J. Mater. Chem.* **2006,** *16,* 1606–1607.
[65] Dig, J., Hudalle, C. J., Cook, J. T., Walsh, D. P., Boissel, C. E., Iraneta, P. C., O'Gara, J. E., *Chem. Mater.* **2004,** *16,* 670–681.
[66] Elbhiri, Z., Chevalier, Y., Chovelon, J.-M., Jaffrezic-Renault, N., *Talanta* **2000,** *52,* 495–507.
[67] Corriu, R. J. P., Datas, L., Guari, Y., Mehdi, A., Reyé, C., Thieuleux, C., *Chem. Commun.* **2001,** 763–764.
[68] (a) Yang, C.-M., Zibrowius, B. F., Schüth, F., *Chem. Commun.* **2003,** 1772–1773. (b) Yang, C.-M., Wang, Y., Zibrowius, B., Schüth, F., *Phys. Chem. Chem. Phys.* **2004,** *6,* 2461–2467.
[69] Park, M., Komarneni, S., *Microporous Mesoporous Mater.* **1998,** *25,* 75–80.
[70] Hall, S. R., Fowler, C. E., Lebeau, B., Mann, S., *Chem. Commun.* **1999,** 201–202.

[71] Díaz, I., Márquez-Alvarez, C., Mohino, F., Pérez-Pariente, J., Sastre, E., *J. Catal.* **2000,** *193,* 283–294.
[72] Yang, H., Zhang, G., Hong, X., Zhu, Y., *Microporous Mesoporous Mater.* **2004,** *68,* 119–125.
[73] Mouawia, R., Mehdi, A., Reyé, C., Corriu, R., *New J. Chem.* **2006,** *30,* 1077–1082.
[74] Zhang, W.-H., Lu, X.-B., Xiu, J.-H., Hua, Z.-L., Zhang, L.-X., Robertson, M., Shi, J.-L., Yan, D.-S., Holmes, J. D., *Adv. Funct. Mater.* **2004,** *14,* 544–552.
[75] Shepard, D. S., Zhou, W., Maschmeyer, T., Matters, J. M., Roper, C. L., Parsons, S., Johnson, B. F. G., Duer, M. J., *Angew. Chem. Int. Ed.* **1998,** *37,* 2719–2723.
[76] de Juan, F., Ruiz-Hitzky, E., *Adv. Mater.* **2000,** *12,* 430–432.
[77] Yang, C.-M., Lin, H.-A., Zibrowius, B., Spliethoff, B., Schüth, F., Liou, S.-C., Chu, M.-W., Chen, C.-H., *Chem. Mater.* **2007,** *19,* 3205–3211.
[78] Yoshina-Ishii, C., Asefa, T., Coombs, N., MacLachlan, M. J., Ozin, G. A., *Chem. Commun.* **1999,** 2539–2540.
[79] Landskron, K., Ozin, G. A., *Science* **2004,** *306,* 1529–1532.
[80] Hunks, W. J., Ozin, G. A., *Adv. Funct. Mater.* **2005,** *15,* 259–266.
[81] Zhao, X. S., Lu, G. Q., *J. Phys. Chem.* **1998,** *102,* 1556–1561.
[82] Lim, M. H., Stein, A., *Chem. Mater.* **1999,** *11,* 3285–3295.
[83] Antochschuk, V., Jaroniec, M., *J. Phys. Chem. B* **1999,** *103,* 6252–6261.
[84] Landmesser, H., Kosslick, H., Stiorek, W., Kricke, R., *Solid State Ionics* **1997,** *101–103,* 271–277.
[85] Wouters, B. H., Chen, T., Dewilde, M., Grobet, P. J., *Microporous Mesoporous Mater.* **2001,** *44–45,* 453–457.
[86] Goletto, V., Dagry, V., Babonneau, F., *Mat. Res. Symp. Proc.* **1999,** *576,* 229–234.
[87] Wang, Y. Q., Yang, C. M., Zibrowius, B., Spliethoff, B., Linden, M., Schüth, F., *Chem. Mater.* **2003,** *15,* 5029–5035.
[88] Chong, A. S. M., Zhao, X. S., Kustedjo, A. T., Qiao, S. Z., *Microporous Mesoporous Mater.* **2004,** *72,* 33–42.
[89] Mori, Y., Pinnavaia, T., *Chem. Mater.* **2001,** *13,* 2173–2178.
[90] Radu, D. R., Lai, C.-Y., Huang, J., Shu, X., Lin, V. S.-Y., *Chem. Commun.* **2005,** 1264–1266.
[91] Sutra, P., Fajula, F., Brunel, D., Lentz, P., Daelen, G., Nagy, J. B., *Colloids. Surf. A: Physicochem. Eng. Aspects* **1999,** *158,* 21–27.
[92] Kao, H.-M., Chang, P.-C., Wu, J. D., Chiang, A. S. T., Lee, C.-H., *Microporous Mesoporous Mater.* **2006,** *97,* 9–20.
[93] Guan, S., Inagaki, S., Ohsuna, T., Terasaki, O., *J. Am. Chem. Soc.* **2000,** *122,* 5660–5661.
[94] Sayari, A., Hamoudi, S., Yang, Y., Moudrakovski, I. L., Ripmeester, J. R., *Chem. Mater.* **2000,** *12,* 3857–3863.
[95] Inagaki, S., Guan, S., Ohsuna, T., Terasaki, O., *Nature* **2002,** *416,* 304–307.
[96] Kapoor, M. P., Yang, Q. H., Inagaki, S., *J. Am. Chem. Soc.* **2002,** *124,* 15176–15177.
[97] Bion, N., Ferreira, P., Valente, A., Goncalves, I. S., Rocha, J., *J. Mater. Chem.* **2003,** *13,* 1910–1913.
[98] Toury, B., Blum, R., Goletto, V., Babonneau, F., *J. Sol–Gel. Sci. Tech.* **2005,** *33,* 99–102.
[99] Muth, O., Schellbach, C., Froba, M., *Chem. Commun.* **2001,** 2032–2033.
[100] Burleigh, M. C., Markowitz, M. A., Spector, M. S., Gaber, B. P., *J. Phys. Chem. B* **2002,** *106,* 9712–9716.
[101] Guo, W. P., Park, J. Y., Oh, M. O., Jeong, H. W., Cho, W. J., Kim, I., Ha, C. S., *Chem. Mater.* **2003,** *15,* 2295–2298.
[102] Guo, W. P., Kim, I., Ha, C. S., *Chem. Commun.* **2003,** 2692–2693.
[103] Wang, W. H., Xie, S. H., Zhou, W. Z., Sayari, A., *Chem. Mater.* **2004,** *16,* 1756–1762.
[104] Zhao, L., Zhu, G. S., Zhang, D. L., Di, Y., Chen, Y., Terasaki, O., Qiu, S. L., *J. Phys. Chem. B* **2005,** *109,* 764–768.
[105] Hamoudi, S., Kaliaguine, S., *Chem. Commun.* **2002,** 2118–2119.
[106] Wang, W. H., Zhou, W. Z., Sayari, A., *Chem. Mater.* **2003,** *15,* 4886–4889.
[107] Goto, Y., Inagaki, S., *Chem. Mater.* **2002,** 2410–2411.
[108] Hoffmann, F., Cornelius, M., Morell, J., Froba, M., *J. Nanosci. Nanotechnol.* **2006,** *6,* 265–288.
[109] Koyano, K. A., Tatsumi, T., Tanaka, Y., Nakata, S., *J. Phys. Chem. B* **1997,** *101,* 9436–9440.

[110] Igarashi, N., Hashimoto, K., Tatsumi, T., *J. Mater. Chem.* **2002,** *12,* 3631–3636.
[111] Kuroki, M., Asefa, T., Whitnal, W., Kruk, M., Yoshina-Ishii, C., Jaroniec, M., Ozin, G. A., *J. Am. Chem. Soc.* **2002,** *124,* 13886–13895.
[112] Huh, S., Wiench, J. W., Yoo, J.-C., Pruski, M., Lin, V. S.-Y., *Chem. Mater.* **2003,** *15,* 4247–4256.
[113] Hodgkins, R. P., Garcia-Bennett, A. E., Wright, P. A., *Microporous Mesoporous Mater.* **2005,** *79,* 241–252.
[114] Fowler, C. E., Kushalani, D., Lebeau, B., Mann, S., *Adv. Mater.* **2001,** *13,* 649–652.
[115] Stöber, W., Fink, A., Bohn, E., *J. Colloid Interface Sci.* **1968,** *26,* 62–69.
[116] Etienne, M., Lebeau, B., Walcarius, A., *New J. Chem.* **2002,** *26,* 384–386.
[117] Kosuge, K., Murakami, T., Kikukawa, N., Takemori, M., *Chem. Mater.* **2003,** *15,* 3184–3189.
[118] Jainisha, S., Pinnavaia, T. J., *Chem. Commun.* **2005,** *12,* 1598–1600.
[119] (a) Lee, C. H., Park, S. S., Choe, S. J., Park, D. H., *Microporous Mesoporous Mater.* **2001,** *46,* 257–264. (b) Park, S. S., Lee, C. H., Cheon, J. H., Park, D. H., *J. Mater. Chem.* **2001,** *11,* 3397–3403.
[120] Kapoor, M. P., Inagaki, S., *Chem. Lett.* **2004,** *33,* 88–89.
[121] Rebbin, V., Jakubowski, M., Pötz, S., Fröba, M., *Microporous Mesoporous Mater.* **2004,** *72,* 99–104.
[122] Kim, D.-J., Chung, J.-S., Ahn, W.-S., Kang, G.-W., Cheong, W.-J., *Chem. Lett.* **2004,** *33,* 422.
[123] Xia, Y. D., Yang, Z. X., Mokaya, R., *Chem. Mater.* **2006,** *18,* 1141–1148.
[124] Cho, E. B., Kim, D., Jaroniec, M., *Langmuir* **2007,** *23,* 11844–11849.
[125] (a) Qiao, S. Z., Yu, C. Z., Xing, W., Hu, Q. H., Djojoputro, H., Lu, G. Q., *Chem. Mater.* **2005,** *17,* 6172–6176. (b) Qiao, S. Z., Yu, C. Z., Hu, Q. H., Jin, Y. G., Zhou, X. F., Zhao, X. S., Lu, G. Q., *Microporous Mesoporous Mater.* **2006,** *91,* 59–69.
[126] Wahab, M. A., Imae, I., Kawakami, Y., Ha, C. S., *Microporous Mesoporous Mater.* **2006,** *92,* 201–211.
[127] Wahab, M. A., Imae, I., Kawakami, Y., Ha, C. S., *Chem. Mater.* **2005,** *17,* 2165–2174.
[128] Xia, Y. D., Mokaya, R., *J. Phys. Chem. B* **2006,** *110,* 3889–3894.
[129] (a) Lu, Y., Ganguli, R., Drewien, C. A., Anderson, M. T., Brinker, C. J., Gong, W., Guo, Y., Soyez, H., Dunn, B., Huang, M. H., Zink, J. I., *Nature* **1997,** *389,* 364–368. (b) Brinker, C. J., Lu, Y., Sellinger, A., Fan, H., *Adv. Mater.* **1999,** *11,* 579–585.
[130] Alonso, B., Clinard, C., Durand, D., Véron, E., Massiot, D., *Chem. Commun.* **2005,** 1746–1748.
[131] Ji, X., Hu, Q., Hampsey, J. E., Qiu, X., Gao, L., He, J., Lu, Y., *Chem. Mater.* **2006,** *18,* 2265–2274.
[132] Lebeau, B., Fowler, C. E., Hall, S. R., Mann, S., *J. Mater. Chem.* **1999,** *9,* 2279–2281.
[133] Liu, N., Assink, R. A., Smarsly, B., Brinker, C. J., *Chem. Commun.* **2003,** 1146–1147.
[134] (a) Cagnol, F., Grosso, D., Sanchez, C., *Chem. Commun.* **2004,** 1742–1743. (b) Nicole, L., Boissière, C., Grosso, D., Quash, A., Sanchez, C., *J. Mater. Chem.* **2005,** *15,* 3598–3627.
[135] Ogawa, M., *Chem. Comm.* **1996,** 1149–1150.
[136] Wong, E. M., Markowitz, M. A., Qadri, S. B., Golledge, S. L., Castner, D. G., Gaber, B. P., *Langmuir* **2002,** *18,* 972–974.
[137] Fan, H., Lu, Y., Stump, A., Reed, S. T., Baer, T., Schunk, R., Perez-Luna, V., López, G. P., Brinker, C. J., *Nature* **2000,** *405,* 56–60.
[138] Lebeau, B., Fowler, C. E., Mann, S., Farcet, C., Charleux, B., Sanchez, C., *J. Mater. Chem.* **2000,** *10,* 2105–2108.
[139] (a) Brunel, D., Bellocq, N., Sutra, P., Cauvel, A., Laspéras, M., Moreau, P., Di Renzo, F., Galarneau, A., Fajula, F., *Coord. Chem. Rev.* **1998,** *178–180,* 1085–1108. (b) Brunel, D., Blanc, A. C., Galarneau, A., Fajula, F., *Catal. Today* **2002,** *73,* 139–152.
[140] Wight, A. P., Davis, M. E., *Chem. Rev.* **2002,** *102,* 3589–3614.
[141] De Vos, D. E., Dams, M., Sels, B. F., Jacobs, P. A., *Chem. Rev.* **2002,** *102,* 3615–3640.
[142] Taguchi, A., Schüth, F., *Microporous Mesoporous Mater.* **2005,** *77,* 1–45.
[143] (a) Das, D., Lee, J.-F., Cheng, S., *Chem. Commun.* **2001,** 2178–2179. (b) Das, D., Lee, J.-F., Cheng, S., *J. Catal.* **2004,** *223,* 152–160.

[144] Shimizu, K., Hayashi, E., Hatamachi, T., Kodama, T., Higuchi, T., Satsuma, A., Kitayama, Y., *J. Catal.* **2005,** *231,* 131–138.
[145] Sow, B., Hamoudi, S., Zahedi-Niaki, M. H., Kaliaguine, S., *Microporous Mesoporous Mater.* **2005,** *79,* 129–136.
[146] Dufaud, V., Davis, M. E., *J. Am. Chem. Soc.* **2003,** *125,* 9403–9413.
[147] Corma, A., Iborra, S., Rodriguez, I., Sanchez, F., *J. Catal.* **2002,** *211,* 208–215.
[148] Macquarrie, D. J., Maggi, R., Mazzacani, A., Sartori, G., Sartorio, R., *Appl. Catal. A* **2003,** *246,* 183–188.
[149] Brunel, D., Fajula, F., Nagy, J. B., Deroide, F., Verhoef, M. F. J., Veum, L., Peters, J. A., van Bekkum, H., *Appl. Catal. A* **2001,** *213,* 73–82.
[150] Brunel, D., *Microporous Mesoporous Mater.* **1999,** *27,* 329–344.
[151] Motorina, I. C. M., Crudden, *Org. Lett.* **2001,** *3,* 2325–2328.
[152] Whang, M. S., Kwon, Y. K., Kim, G.-J., *J. Ind. Eng. Chem.* **2002,** *8,* 262–267.
[153] Dhar, D., Beadham, I., Chandasekaran, S., *Proc. Indian Acad. Sci. Chem. Sci.* **2003,** *115,* 365–372.
[154] Yuan, X., Lee, H. I., Kim, J. W., Yie, J. E., Kim, J. M., *Chem. Lett.* **2003,** *32,* 650–651.
[155] (a) Yang, Q., Liu, J., Yang, J., Kapor, M. P., Inagaki, S., *J. Catal.* **2004,** *228,* 265–272. (b) Yang, Q., Kapor, M. P., Inagaki, S., Shirokura, N., Kondo, J. N., Domen, K., *J. Mol. Catal. A* **2005,** *230,* 85–89.
[156] Hamoudi, S., Royer, S., Kaliaguine, S., *Microporous Mesoporous Mater.* **2004,** *71,* 17–25.
[157] Hartmann, M., *Chem. Mater.* **2005,** *15,* 4577–4593.
[158] Yiu, H. H. P., Wright, P. A., *J. Mater. Chem.* **2005,** *15,* 3690–3700.
[159] Jal, P. K., Patel, S., Mishra, B. K., *Talanta* **2004,** *62,* 1005–1028.
[160] Feng, X., Fryxell, G. E., Wang, L. Q., Kim, A. Y., Liu, J., Kemner, K. M., *Science* **1997,** *276,* 923–926.
[161] Mercier, L., Pinnavaia, T. J., *Adv. Mater.* **1997,** *9,* 500–503.
[162] Zhao, X. S., Lu, G. Q., Whittaker, A. J., Millar, G. J., Zhu, H. Y., *J. Phys. Chem. B* **1997,** *101,* 6525–6531.
[163] Mercier, L., Pinnavaia, T. J., *Environ. Sci. Technol.* **1998,** *32,* 2749–2754.
[164] Chen, X., Feng, X., Liu, J., Fryxell, G. E., Gong, M., *Sep. Sci. Technol.* **1999,** *34,* 1121–1132.
[165] Mattigod, S. V., Feng, X., Fryxell, G. E., Liu, J., Gong, M., *Sep. Sci. Technol.* **1999,** *34,* 2329–2345.
[166] Liu, A. M., Hidajat, K., Kawi, S., Zhao, D. Y., *Chem. Commun.* **2000,** 1145–1146.
[167] Lim, M. H., Blanford, C. F., Stein, A., *Chem. Mater.* **1998,** *10,* 467–470.
[168] Brown, J., Mercier, L., Pinnavaia, T. J., *Chem. Commun.* **1999,** 69–70.
[169] Billinge, S. J. L., McKimmy, E. J., Shatnawi, M., Kim, H.-J., Petkov, V., Wermeille, D., Pinnavaia, T. J., *J. Am. Chem. Soc.* **2005,** *127,* 8492–8498.
[170] Brown, J., Richer, R., Mercier, L., *Microporous Mesoporous Mater.* **2000,** *37,* 41–48.
[171] Bibby, A., Mercier, L., *Chem. Mater.* **2002,** *14,* 1591–1597.
[172] Aguado, J., Arsuaga, J. M., Arencibia, A., *Ind. Eng. Chem. Res.* **2005,** *44,* 3665–3671.
[173] (a) Antochshuk, V., Jaroniec, M., *Chem. Commun.* **2002,** 258–259. (b) Antochshuk, V., Olkhovyk, O., Jaroniec, M., Park, I.-S., Ryoo, R., *Langmuir* **2003,** *19,* 3031–3034. (c) Olkhovyk, O., Antochshuck, V., Jaroniec, M., *Colloids Surf., A* **2004,** *236,* 69–72.
[174] (a) Pérez-Quintanilla, D., del Hierro, I., Fajardo, M., Sierra, I., *J. Environ. Monit.* **2006,** *8,* 214–222. (b) Pérez-Quintanilla, D., del Hierro, I., Fajardo, M., Sierra, I., *J. Hazard. Mater.* **2006,** *134,* 245–256. (c) Pérez-Quintanilla, D., del Hierro, I., Fajardo, M., Sierra, I., *Microporous Mesoporous Mater.* **2006,** *89,* 58–68.
[175] Etienne, M., Walcarius, A., *Talanta* **2003,** *59,* 1173–1188.
[176] Kruk, M., Asefa, T., Jaroniec, M., Ozin, G. A., *J. Am. Chem. Soc.* **2002,** *124,* 6383–6392.
[177] Beaudet, L., Hossain, K.-Z., Mercier, L., *Chem. Mater.* **2003,** *15,* 327–334.
[178] Wei, Q., Nie, Z., Hao, Y., Chen, Z., Zou, J., Wang, W., *Mater. Lett.* **2005,** *59,* 3611–3615.
[179] Zhang, L., Zhang, W., Shi, J., Hua, Z., Li, Y., Yan, J., *Chem. Commun.* **2003,** 210–211.
[180] Liu, J., Yang, J., Yang, Q., Wang, G., Li, Y., *Adv. Funct. Mater.* **2005,** *15,* 1297–1302.
[181] Pérez-Quintanilla, D., Sanchez, A., del Hierro, I., Fajardo, M., Sierra, I., *J. Sep. Sci.* **2007,** *30,* 1556–1567.

[182] Pérez-Quintanilla, D., Sanchez, A., del Hierro, I., Fajardo, M., Sierra, I., *J. Colloid Interface Sci.* **2007,** *313,* 551–562.
[183] Pérez-Quintanilla, D., del Hierro, I., Fajardo, M., Sierra, I., *J. Mater. Chem.* **2006,** *16,* 1757–1764.
[184] Hodgkins, R. P., Garcia-Bennett, A. E., Wright, P. A., *Microporous Mesoporous Mater.* **2005,** *79,* 241–252.
[185] (a) Walcarius, A., *Electroanalysis* **1998,** *10,* 1217–1235. (b) Walcarius, A., *Chem. Mater.* **2001,** *13,* 3351–3372. (c) Sayen, S., Etienne, M., Bessière, J., Walcarius, A., *Electroanalysis* **2002,** *14,* 1521–1525. (d) Walcarius, A., Etienne, M., Sayen, S., Lebeau, B., *Electroanalysis* **2003,** *15,* 414–421. (e) Walcarius, A., Delacôte, C., Sayen, S., *Electrochim. Acta* **2004,** *49,* 3775–3783. (f) Walcarius, A., Delacôte, C., *Anal. Chim. Acta* **2005,** *547,* 3. (g) Lesaint, C., Frebault, F., Delacôte, C., Lebeau, B., Marichal, C., Walcarius, A., Patarin, J., *Stud. Surf. Sci. Catal.* **2005,** *156,* 925–932.
[186] (a) Walcarius, A., Etienne, M., Bessière, J., *Chem. Mater.* **2002,** *14,* 2757–2766. (b) Walcarius, A., Etienne, M., Lebeau, B., *Chem. Mater.* **2003,** *15,* 2161–2173. (c) Walcarius, A., Delacôte, C., *Chem. Mater.* **2003,** *15,* 4181–4192.
[187] Gaslain, F., Delacôte, C., Lebeau, B., Marichal, C., Patarin, J., Walcarius, A., *Stud. Surf. Sci. Catal.* **2007,** *165,* 417–420.
[188] Zhang, L., Zhang, W., Shi, J., Hua, Z., Li, Y., *Chem. Commun.* **2003,** 210–211.
[189] Lee, B., Bao, L.-L., Him, H.-J., Dai, S., Hagaman, E. W., Lin, J. S., *Langmuir* **2003,** *19,* 4246–4252.
[190] Yoshitake, H., Yokoi, T., Tatsumi, T., *Chem. Mater.* **2002,** *14,* 4603–4610.
[191] Yoshitake, H., Koiso, E., Horie, H., Yoshimura, H., *Microporous Mesoporous Mater.* **2005,** *85,* 183–194.
[192] McEleney, K., Allen, D. P., Holliday, A. E., Crudden, C. M., *Org. Lett.* **2006,** *8,* 2663–2666.
[193] Sayari, A., Hamoudi, S., Yang, Y., *Chem. Mater.* **2005,** *17,* 212–216.
[194] Inumaru, K., Kiyoto, J., Yamanaka, S., *Chem. Commun.* **2000,** 903–904.
[195] (a) Lei, C. S., Shin, Y. S., Ackerman, E. J., *J. Am. Chem. Soc.* **2002,** *124,* 11242–11243. (b) Maria Chong, A. S., Zhao, X. S., *Catal. Today* **2004,** *93–95,* 293–299.
[196] Ho, K. Y., McKay, G., Yeung, K. L., *Langmuir* **2003,** *19,* 3019–3024.
[197] Yan, Z., Tao, S., Yin, J., Li, G., *J. Mater. Chem.* **2006,** *16,* 2347–2353.
[198] Knöfel, C., Descarpentries, J., Benzaouia, A., Zeleňák, V., Mornet, S., Llewellyn, P. L., Hornebecq, V., *Microporous Mesoporous Mater.* **2007,** *99,* 79–85.
[199] Hiyoshi, N., Yogo, K., Yashima, T., *Chem. Lett.* **2004,** *33,* 510–511.
[200] Murta, S., Hata, H., Kimura, T., Sugahara, Y., Kuroda, K., *Langmuir* **2000,** *16,* 7106–7108.
[201] Matsumoto, A., Tsutsumi, K., Schumacher, K., Unger, K., *Langmuir* **2002,** *18,* 4014–4019.
[202] Maria Chong, A. S., Zhao, X. S., *Catal. Today* **2004,** *93–95,* 293–299.
[203] Yiu, H. H. P., Botting, C. H., Botting, N. P., Wright, P. A., *Phys.Chem. Chem. Phys.* **2001,** *3,* 2983–2985.
[204] Wirnsberger, G., Scott, B. J., Stucky, G. D., *Chem. Commun.* **2001,** 119–120.
[205] Lin, V. S.-Y., Lai, C.-Y., Huang, J., Song, S.-A., Xu, S., *J. Am. Chem. Soc.* **2001,** *123,* 11510–11511.
[206] Descalzo, A. B., Jimenez, D., Marcos, M. D., Manez, R.-M., Soto, J., El Haskouri, J., Guillém, C., Beltran, D., Amoros, P., Borrachero, V., *Adv. Mater.* **2002,** *14,* 966–969.
[207] Nicole, L., Boissière, C., Grosso, D., Hasemann, P., Moreau, J., Sanchez, C., *Chem. Commun.* **2004,** 2312–2313.
[208] Fryxell, G. E., *Inorg. Chem. Commun.* **2006,** *9,* 1141–1150.
[209] (a) Wenxiang, X., Dan, X., Hongyan, Y., *Sens. Lett.* **2007,** *5,* 445–449. (b) Wenxiang, X., Dan, X., *Talanta* **2007,** *72,* 1288–1292.
[210] Fiorilli, S., Onida, B., Barolo, C., Viscardi, G., Brunel, D., Garrone, E., *Langmuir* **2007,** *23,* 2261–2268.
[211] Yantasee, W., Lin, Y., Li, X., Fryxell, G. E., Zemanian, T. S., Viswanathan, V. V., *Analyst* **2003,** *128,* 899–904.
[212] Etienne, M., Walcarius, A., *Electrochem. Commun.* **2005,** *7,* 1449–1456.

[213] Palaniappan, A., Li, X., Tay, F. E. H., Li, J., Su, X., *Sens. Actuators B: Chem. B* **2006,** *119,* 220–226.
[214] Munoz, B., Ramila, A., Perez-Pariente, J., Diaz, I., Vallet-Regi, M., *Chem. Mater.* **2003,** *15,* 500–503.
[215] Han, Y.-J., Stucky, G. D., Butler, A., *J. Am. Chem. Soc.* **1999,** *121,* 9897–9898.
[216] Song, S.-W., Hidajat, K., Kawi, S., *Langmuir* **2005,** *21,* 9568–9575.
[217] Tourné-Péteilh, C., Brunel, D., Bégu, S., Chiche, B., Fajula, F., Lerner, D. A., Devoisselle, J.-M., *New J. Chem.* **2003,** *27,* 1415–1418.
[218] Slowing, I. I., Trewyn, B. G., Giri, S., Lin, V. S.-Y., *Adv. Funct. Mater.* **2007,** *17,* 1225–1236.
[219] Giri, S., Trewyn, B. G., Lin, V. S.-Y., *Nanomed.* **2007,** *2,* 99–111.
[220] (a) Vallet-Regi, M., Balas, F., Arcos, D. L., *Angew. Chem. Int. Ed.* **2007,** *46,* 7548–7558. (b) Vallet-Regi, M., Balas, F., Colilla, M., Manzano, M., *Solid State Sci.* **2007,** *9,* 768–776.
[221] Mal, N. K., Fujiwara, M., Tanaka, Y., Taguchi, T., Matsukata, M., *Chem. Mater.* **2003,** *15,* 3385–3394.
[222] Rouxel, T., Massouras, G., Sorarù, G. D., *J. Sol-Gel Sci. Technol.* **1999,** *14,* 87–92.
[223] Soraru, G. D., Modena, S., Guadagnino, E., Colombo, P., Egan, J., Pantano, C., *J. Am. Ceram. Soc.* **2002,** *85,* 1529–1535.
[224] Toury, B., Babonneau, F., *J. European Ceram. Soc.* **2005,** *25,* 129–135.
[225] Masse, S., Laurent, G., Babonneau, F., *J. Non-Cryst. Solids* **2007,** *353,* 1109–1119.
[226] Mikhailenko, S., Desplatier-Giscard, D., Danumah, D. C., Kaliaguine, S., *Microporous Mesoporous Mater.* **2002,** *52,* 29–37.
[227] Marshall, R., Bannat, I., Caro, J., Wark, M., *Microporous Mesoporous Mater.* **2007,** *99,* 190–196.
[228] Yang, C.-M., Liu, P.-H., Ho, Y.-F., Chiu, C.-Y., Chao, K.-J., *Chem. Mater.* **2003,** *15,* 275–280.
[229] Guari, Y., Thieuleux, C., Mehdi, A., Reyé, C., Corriu, R. J. P., Gomez-Gallardo, S., Philippot, K., Chaudret, B., *Chem. Mater.* **2003,** *15,* 2017–2024.
[230] Ghosh, A., Patra, C. R., Mukherjee, P., Sastry, M., Kular, R., *Microporous Mesoporous Mater.* **2003,** *58,* 201–211.
[231] Zhang, W.-H., Shi, J. L., Chen, H.-R., Hua, Z.-L., Yan, D.-S., *Chem. Mater.* **2001,** *13,* 648–654.
[232] Wellmann, H., Rathousky, J., Wark, M., Zukal, A., Schulz-Ekloff, G., *Microporous Mesoporous Mater.* **2001,** *44–45,* 419–425.
[233] Schweyer-Tihay, F., Braunstein, P., Estournès, C., Guille, J. L., Lebeau, B., Paillaud, J.-L., Richard-Plouet, M., Rosé, M., *J. Chem. Mater.* **2003,** *15,* 57–62.
[234] Moller, K., Bein, T., Fisher, R. X., *Chem. Mater.* **1999,** *11,* 665–673.
[235] Ji, X., Hampsey, J. E., Hu, Q., He, J., Yang, Z., Lu, Y., *Chem. Mater.* **2003,** *15,* 3656–3662.
[236] Li, H. R., Lin, J., Fu, L. S., Guo, J. F., Meng, Q. G., Liu, F. Y., Zhang, H. J., *Microporous Mesoporous Mater.* **2002,** *55,* 103–107.
[237] Xu, Q., Dong, W., Li, H., Li, L., Feng, S., Xu, R., *Solid State Sci.* **2003,** *5,* 777–782.
[238] Fernandes, A., Dexper-Ghys, J., Gleizes, A., Galarneau, A., Brunel, D., *Microporous Mesoporous Mater.* **2005,** *83,* 35–46.
[239] Peng, C., Zhang, H., Meng, Q., Li, H., Yu, J., Guo, J., Sun, L., *Inorg. Chem. Commun.* **2005,** *8,* 440–443.
[240] Corriu, R. J. P., Mehdi, A., Reyé, C., Thieuleux, C., Frenkel, A., Gibaud, A., *New J. Chem.* **2004,** *28,* 256–260.
[241] Furukawa, H., Watanabe, T., Kudoda, K., *Chem. Commun.* **2001,** 2002–2003.
[242] Quash, A., Escax, V., Nicole, L., Goldner, P., Guillot-Noël, O., Aschehoug, P., Hesemann, P., Moreau, J., Gourier, D., Sanchez, C., *J. Mater. Chem.* **2007,** *17,* 2552–2560.
[243] Salesch, T., Bachmann, S., Brugger, S., Rabelo-Schaefer, R., Albert, K., Steinbrecher, S., Plies, E., Mehdi, A., Reyé, C., Corriu, R. J. P., Lindner, E., *Adv. Funct. Mater.* **2002,** *12,* 134–142.
[244] Zhao, J., Gao, F., Fu, Y., Jin, W., Yan, P., Zhao, D., *Chem. Commun.* **2002,** 752–753.
[245] Hatton, B., Landskron, K., Whitnall, W., Perovic, D., Ozin, G. A., *Acc. Chem. Res.* **2005,** *38,* 305–312.
[246] Balkenende, A. R., de Theije, F. K., Kriege, J. C. K., *Adv. Mater.* **2003,** *15,* 139–143.

[247] Lu, Y., Fan, H., Doke, N., Loy, D. A., Assink, R. A., LaVan, D. A., Brinker, C. J., *J. Am. Chem. Soc.* **2000,** *122,* 5258–5261.
[248] Hatton, B. D., Landskron, K., Whitnall, W., Perovic, D. D., Ozin, G. A., *Adv. Funct. Mater.* **2005,** *15,* 823–829.
[249] Radu, D. R., Lai, C.-Y., Wiench, J. W., Pruski, M., Lin, V. S.-Y., *J. Am. Chem. Soc.* **2004,** *126,* 1640–1641.
[250] Liu, N., Chen, Z., Dunphy, D. R., Jiang, B., Assink, A., Brinber, C. J., *Angew. Chem. Int. Ed.* **2003,** *42,* 1731–1734.

SECTION II

BI- AND THREE-DIMENSIONAL ORGANISED POROUS CONSTRUCTS

CHAPTER 12

Hydrothermal Synthesis of Zeolitic Coatings for Applications in Micro-structured Reactors

Evgeny V. Rebrov, Martijn J. M. Mies, Mart H. J. M. de Croon, *and* Jaap C. Schouten

Contents

Abstract

An elegant way to produce catalytic micro-reactors is by applying a coating of zeolite crystals onto a micro-structured substrate. The synthesis methods, including substrate modifications, are reviewed and illustrated with a number of particular examples. There exists a narrow range of synthesis conditions (alkalinity, ionic strength, temperature and heating rate) leading to zeolite coating formation rather than to homogeneous crystallization. The progress made in understanding the fundamental mechanisms that control the growth of the zeolitic coatings is presented. Potential application fields are discussed.

Keywords: Hydrothermal synthesis, Micro-structured reactor, Surface roughness, Zeolitic coatings

Ordered Porous Solids
DOI: 10.1016/B978-0-444-53189-6.00012-3

Glossary

Dry	reactive ion etching is a technique for micro-fabrication of channel structures by plasma under vacuum.
Knoevenagel	condensation is a modification of the Aldol condensation in which a nucleophilic addition of an organic compound to a carbonyl group is followed by a dehydration reaction.
Lay	is a constant waviness profile on a surface obtained in a roll casting process.
Pervaporation	is a method for the separation of mixtures by partial vaporization of one compound through a (micro-)porous membrane.

1. Introduction

1.1. Micro-structured reactors

A micro-structured reactor is a chemical reactor with a typical channel diameter of 50–500 μm and a channel length of 5–50 mm. The small dimensions result in an inherently large surface area-to-volume ratio, which leads to remarkable differences in reaction conditions as compared to large scale reactors. The improved heat and mass transfer properties typical of micro-fluidic systems enable the use of more intensive reaction conditions that result in higher yields than those obtained with conventional size reactors. Residence time and heat management in micro-reactors can be exactly tuned to avoid secondary reactions. This leads to a higher selectivity of desired products, especially for a series of sequential reactions. Finally, micro-reactors offer the capability to carry out flexible on-site production of fine chemicals at the point of demand. This is important because most of them are not stable products. Therefore, a new approach to process synthesis and scale-up becomes apparent, in which diverse and interacting small-scale processing units are integrated into a complex micro-structured device.

Micro-reaction technology provides smart and attractive methods for environmentally benign processing in the twenty-first century. The major breakthrough is the proof of chemical production with these devices, announced about 20 times publicly during 2005.[1–3] Usually, the geometric surface of the micro-channels is not sufficient for performing catalytic reactions. The efficient use of micro-structured catalytic reactors requires a shaping of the catalyst usually by deposition of thin catalytic coatings at the walls of the reactor channels. Zeolite crystals can be organized within a micro-channel framework with high spatial precision on the micrometer scale. As a result, a more intensified contacting of the reactants with zeolitic coatings has been obtained in micro-structured reactors, showing promising results in a series of different applications.[4–7] For many industrial applications, crystalline structures containing pores in the size range of above 0.6 nm and exhibiting thermal and hydrothermal stability are highly desirable, especially if the pore system is more than one dimensional. Often, the target is to obtain a single layer of zeolite crystals to avoid mass transfer limitations in the macro-pores.

1.2. Zeolitic coatings

One of the main problems in using micro-structured reactors for heterogeneously catalyzed reactions is the synthesis of the catalytic active phase. An elegant way to produce catalytic micro-reactors is by applying a coating of zeolite crystals onto a micro-structured substrate. Advantages of coatings compared to micro pellet fixed beds or extrudates are an improved accessibility of the crystallites as well as the catalyst sites, an effective heat removal from the reaction zone via a metal substrate, and the absence of macro-pores in a single layer of crystals. The past decade has seen significant advances in the ability to synthesize different types of coatings with ordered structures from a wide range of different precursors. This has resulted in materials with unusual properties and broadened their application range toward sensors,[8] host materials to integrate molecular electronics, and other molecular devices within their frameworks.[9] Several zeolitic coatings (Linde A, Linde Y, mordenite, ZSM-5) were synthesized on cordierite supports either by a direct hydrothermal synthesis[10] or a secondary growth method after seeding the support.[10–12] Zeolite A coatings were grown on a copper substrate by hydrothermal synthesis after seeding,[13] and on a rutile TiO_2 substrate (Trumem^{TM}) after UV irradiation.[14] Zeolite Beta coatings were synthesized on a TiO_2-modified stainless steel substrate.[15] ZSM-5 films with a thickness of 0.15–2.3 m were prepared on alumina, quartz and soda glass substrates using a seeded method[16,17] and on micro-structured AISI 316 stainless steel substrates by direct hydrothermal synthesis.[4,18] Ferrierite type (FER) zeolitic coatings were grown on a pre-oxidized FeCrAl ferritic steel substrate.[19]

Zeolitic coatings can be deposited on a substrate by means of slurry coating, wash coating, and seed or direct hydrothermal synthesis. The advantage of this latter method is that a homogeneous, thin coating with a strong adhesion to the support can be achieved. Prior to the hydrothermal synthesis, the substrate surface should be modified to improve hydrophilicity, to increase the number of surface irregularities and stability. An improved hydrophilicity and the number of defect sites increase the number of positions where the crystals can nucleate and anchor.

1.3. Influence of synthesis parameters on the properties of zeolitic coatings

The relationship between the Si/Al ratio in the synthesis gel and the characteristics of the resulting zeolitic coating have been reported for ZSM-5 coatings.[4,20] The addition of aluminum to the ZSM-5 synthesis solution accelerates the formation of a precursor gel layer, which serves as the primary source of nuclei,[21] on the substrate. However, the rates of nucleation and crystallization processes in synthesis gels depend strongly on gel composition. These processes are relatively slow in Al-rich gels used for ZSM-5 synthesis as compared to Al-depleted ones.[21,22] On the contrary, nucleation and crystallization occur much faster in Al-rich gels used for zeolite Beta synthesis.[23] Therefore, to obtain highly crystalline zeolitic coatings with high coverages, the synthesis conditions have to be optimized depending on the Si/Al ratio of the synthesis gel.

Zeolite growth on a substrate involves a complex interaction between the precursor mixture and the support, which depends strongly on the gel composition, the synthesis conditions, and the roughness and hydrophilicity of the substrate surface.[5,14,24] Crystal size is a function of the ratio between the rate of nucleation and the rate of growth. Both rates increase with super-saturation, but the exponential law of the nucleation rate rises more sharply than the low-order power law of the growth rate.[25–27] As a consequence, smaller crystals and faster syntheses are observed at a higher super-saturation. At lower super-saturation growth is favoured over nucleation. The choice of the silica or alumina starting material can influence the size and morphology of zeolite crystals. More reactive silicon sources, for example silica sol, will provide a supersaturated solution very rapidly and thus the simultaneous formation of many nucleation sites, increasing crystallization rate.[28]

To increase the coating thickness, the synthesis conditions should be chosen such that nucleation and crystallization are favoured at the substrate surface and the rate of exhaustion of the nutrients in the bulk phase is minimized. Several parameters (alkalinity, ionic strength and temperature) influence the silicate and aluminate solubilities, and thus the concentration levels in the synthesis system. To suppress bulk crystallization, coating syntheses of zeolite 4A[29], HS, P[30] and ZSM-5[31] on stainless steel substrates were carried out via the substrate heating method. In these experiments, the reaction mixture was kept below 100 °C, while the metal plates were heated to above 100 °C. They resulted in the formation of continuous coatings of different textures with different crystal morphologies and void fractions. A closed layer of zeolite 4A with a thickness of about 1.5 μm was obtained at 60 °C after 10 h. In the presence of ultrasound at 35 kHz, the synthesis time and the thickness of the coating were decreased to 3 h and 0.6 μm, respectively.[32] Stirring of the synthesis mixture effectively modifies the crystallization kinetics.[24] The presence of an amorphous gel affects diffusion kinetics, which is especially important for the transport of nutrients to the substrate surface. Nucleation can be affected by other procedures, like aging of the synthesis medium, seeding, and selective crystallization poisoning; potassium is well known to act as a crystallization poison for zeolite Beta.[33]

Obtaining a uniform thickness of the coating requires a careful control of the synthesis mixture characteristics, and of the procedures of coating and substrate treatments. In this chapter, the role of different factors in the hydrothermal synthesis of the zeolitic coatings will be discussed and illustrated with a number of particular examples. It will be shown that usually there exists a narrow range of synthesis conditions leading to preferable zeolite coating formation rather than to homogeneous crystallization. In conclusion, the performance of catalytic micro-structured reactors will be compared with that of packed bed reactors with pelletted zeolite catalysts.

2. MATERIAL CHOICE

The choice of substrate materials for the micro-reactor is to a great extent based on: (i) the surface hydrophilicity of substrates, (ii) material restrictions on the manufacturing side, (iii) the need for fast development in the testing phase, (iv) the corrosion activity of the substrate in the expected chemical environment and (v) the possible interaction between substrate components and catalyst leading to quicker catalyst deactivation.[34] It is thus clear that only a limited number of materials can be used for application in micro-structured reactors at elevated temperatures, because of either a relatively low melting point, for example, in case of aluminum and copper, or a low mechanical stability, for example, quartz and nickel. For highly exothermic reactions, the choice is further limited to materials that have high thermal conductivity, for example, silicon and some of the refractory metals, which allows near-isothermal reactor operation by the effective dissipation of reaction heat.[35,36] Silicon, however, has several drawbacks such as a low wettability of the surface and a low fracture toughness that complicates the deposition of catalytic coatings. Several refractory metals satisfy most of the design criteria and can be applied in high-temperature processes involving large heat effects. For example, molybdenum was used for deposition of zeolitic coatings; however, molybdenum substrates oxidize above 350 °C in the presence of oxygen.[37] Furthermore, molybdenum partially dissolves in the highly alkaline synthesis mixture during the hydrothermal growth of zeolitic coatings.[24] One approach to overcome these difficulties, while still utilizing the otherwise desirable intrinsic properties of molybdenum, is to apply a thin protective film on the metal surface.[37–40] For reaction with only moderate heat effects, AISI 316 stainless steel substrates can perform as well as molybdenum.[4]

Metal substrates are not, however, compatible with many fine chemical synthesis applications. Pyrex glass seems to be an ideal substrate for zeolitic coatings in low-temperature liquid phase chemical syntheses. It contains a relatively high concentration of surface hydroxyl groups, has high corrosion resistance and is chemically inert toward many organic molecules. Furthermore, micro-channels with a precise control of both the cross section and surface roughness can be easily etched in glass by dry reactive ion etching. However, most zeolite syntheses take place at 140–150 °C at pH above 13.9 over several days, presenting conditions that would result in severe dissolution of glass. Therefore, the glass substrate has to be protected before the hydrothermal synthesis by, for example, the deposition of a 700 nm non-porous zirconia film.[41]

3. EFFECT OF SURFACE ROUGHNESS ON COVERAGE

Surfaces and interfaces are very important for the physical properties of any physical object. For thin zeolitic coatings, mainly those used in micro-structured reactors, this influence of surface/interface properties becomes crucial as the material volume reduces dramatically in comparison to surface area. It has been shown that formation of silicalite-1 and zeolite Y can be improved on a plastically deformed or mechanically destroyed copper substrate because of the higher number of surface imperfections, which are supposed to provide nucleation centres for zeolite crystals.[42–43]

After the synthesis, it is important to determine the (periodic) micro- and nanoscale dimensions as well as the macroscopic character of the films. In the last 10 years, atomic force microscopy (AFM) and laser scanning confocal microscopy (LSCM) have become some of the most widely used methods for characterizing the surface structure of solid materials at the nanometer scale. An atomic force microscope usually collects data in the form of a square grid, typically having dimensions in the range of 10×10 nm^2 with a resolution of 0.2 nm. As AFM produces three-dimensional data related with the surface geometry, it allows determination of surface parameters using statistical methods. However, AFM is limited by the size of measurement region and cannot provide the whole planar structural character over a large region (several hundreds of millimeters) required to characterize the overall uniformity of the films. As opposed to AFM, LSCM possesses many advantages such as high resolution over a larger substrate area (300×300 μm^2), no need of vacuum environment, simple specimen preparation and non-destructive measurements.[5]

Usually, a metal surface has a constant waviness profile (lay) as a result of the roll casting process. Striations or peaks and valleys are usually observed on a metal foil in the direction that the tool was drawn across the surface. The difference in altitude between the highest and lowest points of a metal substrate is usually 1–1.5 μm. Surface roughness measurements are usually carried out in the direction parallel to the direction of the lay of the surface. The digitized two-dimensional (2-D) surface profile (Fig. 12.1a) is then corrected for the intrinsic waviness of the substrate (Fig. 12.1b). The average surface roughness (R_a) values are determined by Eq. (1)

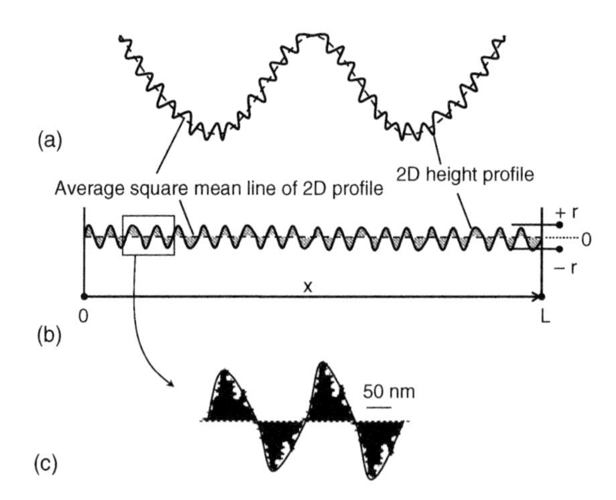

Figure 12.1 (a) A 2-D surface profile of a substrate showing intrinsic waviness. (b) The average roughness value, R_a, is calculated from the 2-D profile after correction for the waviness. The R_a value is calculated from the integral of the absolute value of the roughness profile (shaded area) divided by the length L [Eq. (1)]. (c) The R_a value does not provide information about the nanoscale roughness.

from three 2-D height profiles taken at three different locations on the substrate parallel to the direction of the lay of the surface,

$$R_a = \frac{1}{L}\int_0^L |r(x)|dx \tag{1}$$

where $r(x)$ is the difference between the absolute height and the least squares mean of the 2-D surface profile at position x, and L is the total length of the analysed profile. However, R_a does not include information about the nanometric roughness of the surface (Fig. 12.1c). Therefore, a second parameter is introduced to characterize the nanometric surface roughness by taking the weight loss per area (g m^{-2}) after pre-treatment steps, which are usually applied to clean the surface.[5,44]

Generally, larger differences between the coefficients of thermal expansion of a substrate and a coating requires a rougher substrate surface. The surface roughness can be increased by a micropowder jet treatment or by application of an etching procedure.[5,44] The results of our experimental investigations of the effect of surface roughness on the zeolitic coating coverage are presented below in this section. The LSCM scans of the metal substrates with different nanometric surface roughnesses corresponding to a weight loss of 0 (S_0), 4 (S_4) and 55 (S_{55}) g m^{-2} are shown in Fig. 12.2 a, c and e, respectively. The corresponding Scanning Electron Microscopy (SEM) images of the zeolite Beta coatings are shown in Fig. 12.2 b, d and f. The results show uncovered areas on the non-treated substrate S_0. A closed layer of zeolite crystals was obtained on S_4, for which the nanometric surface roughness was increased. The increased coverage (35%) is attributed to the increased number of surface irregularities, and clearly shows the effect of surface roughness on the coverage of zeolitic coatings. Further increase in the surface roughness leads to a case where the maximum difference between the top and bottom parts of the substrate surface becomes larger than the average diameter of the zeolite crystals. In this situation, a second crystal layer is deposited on top of the crystals located in the substrate pockets. Thus, to obtain a single layer of zeolite crystals, the surface morphology has to contain a large number of imperfections, while the size of the available pockets should not exceed the crystal size.

4. Effect of Surface Hydrophobicity on Coverage

The presence of lay improves the hydrophilic properties of a surface. However, the substrate wettability can further be improved by surface treatments[14] or by deposition of a hydrophilic film.[24] Thin hydrophilic coatings on the silicon, glass, or metal substrates can be achieved through the use of atomic layer deposition (ALD), which is an advanced variant of chemical vapour deposition (CVD), where precursor vapours are delivered in stepwise pulses that are separated by inert gas purges. Unlike conventional chemical vapour deposition, ALD operates via a surface-limited mechanism whereby each repetition of the pulse sequence

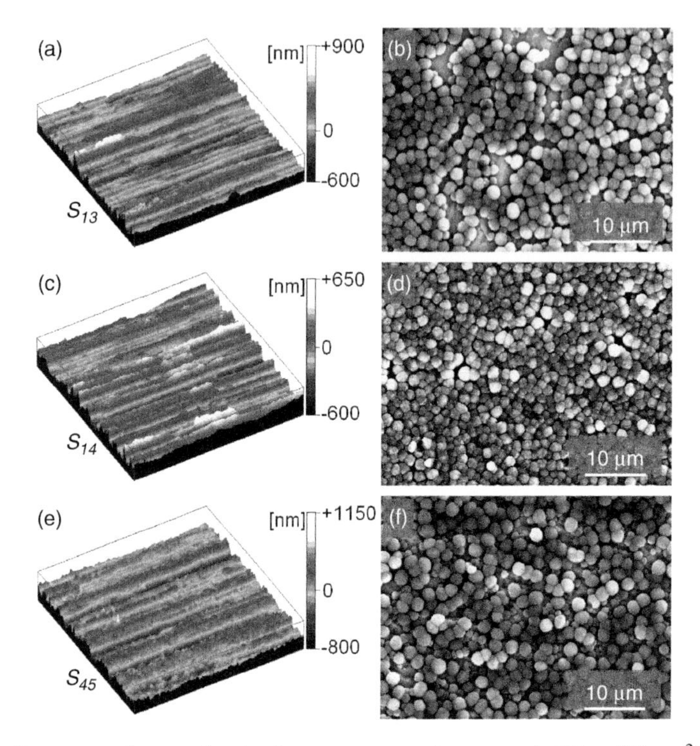

Figure 12.2 Laser scanning confocal microscopy scans (scan area is 320 × 320 μm^2) of the substrates: (a) S_0, (c) S_4, (e) S_{55} and the corresponding scanning electron micrographs of the as synthesized zeolite Beta coatings (b, d and f, respectively). The subscripts denote nanometric surface roughness corresponding to a weight loss of 0, 4, and 55 g m^{-2} after the etching treatment. The black arrows in the laser scanning confocal microscopy scans indicate the direction of the lay of the substrate parallel to which the 2-D height profiles were digitized for the determination of the R_a value. (See color insert.)

affords up to a monolayer of the desired material. ALD has received considerable attention due to its ability to provide films with excellent thickness control, uniformity and film conformality on micro-structured substrate features.[45,46] The difference between the thermal expansion coefficients of film and substrate has to be as small as possible to avoid crack formation. For the same reason, thinner films are preferred.[47] Deposition experiments of TiO_2 thin films by ALD can be performed using a $TiCl_4/H_2O$ process.[48] The growth of TiO_2 ALD films can be performed in the temperature range between 300 and 500 °C. The highest temperature in the zeolite post treatments is usually 500 °C. Therefore, the ALD films should preferably be deposited at this temperature. A 50-nm anatase layer can be grown after 750 deposition cycles with repeating pulses of $TiCl_4$ and H_2O.[5] Prior to the zeolitic synthesis, the anatase film can be made super-hydrophilic (water contact angle $<5°$ corresponding to >15 OH nm^{-2}) by UV irradiation.[49,50] The increased hydrophilicity of the substrate results in a better wettability with the precursor gel.

The effect of surface wettability on the zeolite coverage was also investigated. A substrate without UV irradiation has coverage of ca. 50% of a single crystal layer, while the UV irradiated substrate was completely covered with a single layer of

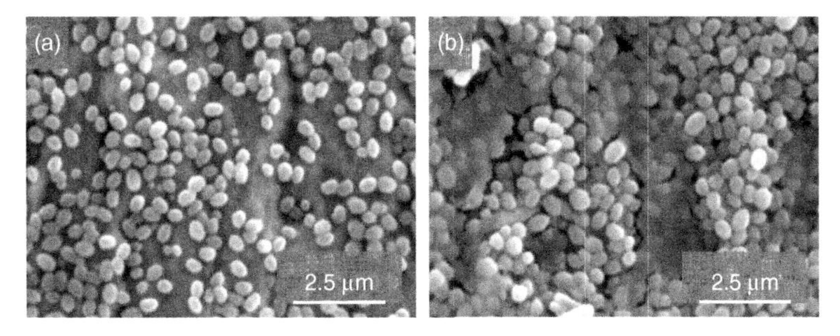

Figure 12.3 Scanning electron micrographs of zeolite Beta coatings (a) without and (b) after UV irradiation of the S_{55} substrate. The subscript denotes the surface roughness of the substrate calculated by Eq. (1). Synthesis conditions: $T = 150\ ^{\circ}C$, $t = 16$ h. Adapted from Ref. [44] with permission from Elsevier.

zeolite crystals (Fig. 12.3). The zeolite crystals in the coating evolved from an amorphous gel layer,[44] which was initially attached to the surface of the substrate, as will be discussed further below. This clearly demonstrates that the wettability of the surface of the substrates influenced the zeolitic coating properties. Similar findings were reported in a study on the improvement of zeolite NaA nucleation sites on (001) rutile by means of UV radiation.[14]

Figure 12.4 summarizes the proposed method for the hydrothermal synthesis of zeolitic coatings on a metal substrate for application in micro-structured reactors.

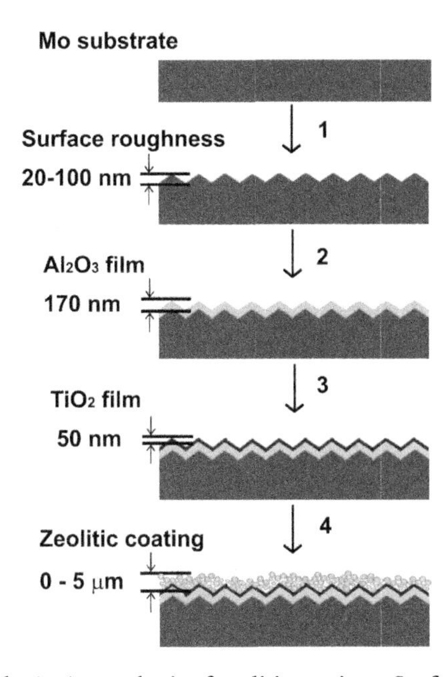

Figure 12.4 Method for the *in situ* synthesis of zeolitic coatings. Surface modification of a metal substrate: (1) cleaning and etching of the substrate, (2) protective (alumina, zirconia) film deposition, (3) hydrophilic (titania) film deposition, and (4) *in situ* crystallization of zeolitic coatings (after a UV treatment of the surface).

In the first step, the number of surface irregularities on the substrate is increased (Step 1) to enhance both the number of nucleation sites and the positions of zeolite crystals for anchoring. Subsequently, two thin films of Al_2O_3 (Step 2) and TiO_2 (Step 3) are successively deposited by ALD. The internal Al_2O_3 film protects the metal substrate from oxidation in an oxidative environment. High wettability of the external TiO_2 film after UV irradiation increases zeolite nucleation at its surface. Finally, a perfectly ordered single layer of zeolite crystals can be obtained (Step 4).

5. Effect of Synthesis Conditions on Coverage

To optimize the synthesis conditions for a specific zeolitic coating, and generalize this method to other types of zeolitic coatings, a fundamental understanding of the coating formation mechanism is of importance. Heterogeneous nucleation at the substrate can occur either from a thin gel layer of primary subcolloidal particles,[51] or directly from the synthesis mixture,[52] the latter being more likely in a mixture with a low degree of super-saturation. Aggregates are more essential for nucleation, while primary units are involved in the crystal growth.[53,54] A high ratio between the primary units and aggregates is achieved at a high degree of super-saturation of the nutrients. In this case, condensation occurs rather quickly, slowing down or even preventing the gel rearrangement.[55] Such conditions can be realized during fast heating of precursor mixtures with low aluminum content.

Attention should be devoted to analysis of the aluminum content in the coatings. It should be noted that the Si/Al ratios, which may be obtained from X-ray photoelectron spectroscopy (XPS), do not provide direct evidence of the chemical composition of coatings. XPS provides information at the nanometer level, but not at the micrometer scale, as is required to determine the long range average Si/Al ratio. In many cases, the application of energy dispersive X-ray (EDX) analysis could not give reliable information because it also detected Al from the protective alumina layer and from the metal substrate (which is often doped with aluminum). As a result, the Si/Al ratios can substantially be underestimated. Inductively coupled plasma atomic emission spectrometry (ICP-AES) seems to be the best method for reliable determination of the chemical composition of zeolitic coatings. In this method, the coatings are wetted with a droplet of demineralized water and hand scrubbed using a scouring pad.

5.1. Effect of heating rate, Si/Al ratio and synthesis temperature

The effect of synthesis mixture heating rate on the coating properties was investigated by the following experiment.[44] A molybdenum substrate with protective alumina and hydrophilic titania layers is positioned vertically in a 50-ml PEEK^{TM} (Polyetheretherketone) insert filled with 35 ml of the synthesis mixture with Si/Al ratio of either 17 or 23. The insert is sealed in a stainless steel outer shell—these compose the autoclave (m=1.50 kg). The stainless steel shell is kept either at room temperature or preheated to 180 °C to determine an initial heating rate of the precursor gels. The unheated shell results in a heating rate of 1 °C min^{-1}, whereas the preheated shell results in a heating rate of 10 °C min^{-1}. These will respectively be referred to as slow and fast heating. After the insert is positioned in the shell, the autoclave was immediately sealed and placed in a convection oven maintained at either 140 or 150 °C. The gel must be heated in the convection oven for a period of 4 or 0.5 h for slow and fast heating, respectively, to reach the synthesis temperature to within 10 °C of the set. In the latter case, a rather constant heating rate is maintained by heat transfer from the preheated outer shell of the autoclave toward the synthesis gel providing high reproducibility of these experiments. Within this period, the heating rate does not depend on the oven temperature set value, commonly referred to as the synthesis temperature, because the heat flux between the outer shell and the oven is negligible comparing to that between the outer shell and the synthesis gel (via the wall of the insert). This is due to two reasons: first, the heat transfer coefficient from the outer shell toward the oven is considerably smaller than that toward the synthesis gel and second, the initial temperature gradient toward the oven (30–40 °C) is much smaller than that toward the synthesis gel (160 °C). As a result, the initial heating rate and overall synthesis time can precisely be controlled by the temperature of the outer shell.

It has been shown that fast heating almost doubles the duration of the induction period compared to slow heating when the nutrients from the precursor mixture have sufficient time to rearrange into the amorphous gel aggregates (Fig. 12.5 and Table 12.1). Furthermore, the rate of heterogeneous nucleation on the substrate increased during fast heating as can be seen from the larger number of small nanocrystals present on the surface in the initial stage of the crystallization process

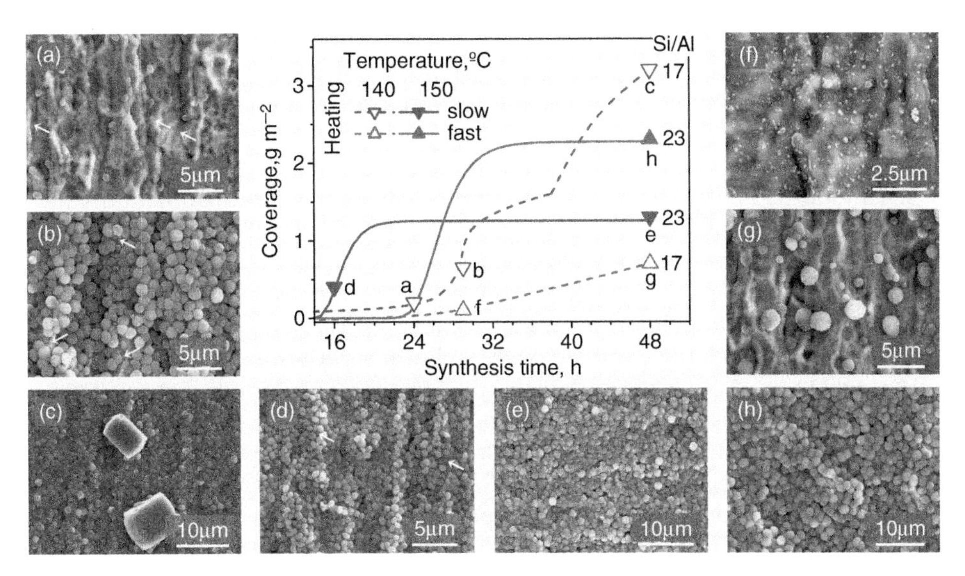

Figure 12.5 Crystallization curves and corresponding SEM images of the coatings grown from the synthesis gels with Si/Al ratios of 17 and 23 at 140 and 150 °C after different synthesis times. (See color insert.)

Table 12.1 Properties of the zeolitic coatings obtained from the synthesis mixture with 46 SiO_2 : *x* Al_2O_3 : 11.4 $(TEA)_2O$: (0.88+0.68*x*) Na_2O : 0.93 K_2O : 1.8 HCl : 710 H_2O, where *x* = 1.35 or 1

	Temperature and Al content in terms of Si/Al ratios			
	140 °C, Si/Al = 17		150 °C, Si/Al = 23	
	After induction period	After 48 h	After induction period	After 48 h
Slow heating	Bimodal CSD: 0.15–0.25 μm and 0.5–1.0 μm after 24 h (Fig. 12.5a) Single layer after 29h (Fig. 12.5b)	MER crystals 8–15 μm on top of the zeolite Beta single layer (Fig. 12.5c)	Single layer with an average size of 0.7 μm after 16 h (Fig. 12.5d)	Single layer with an average particle size of 1.2 μm (Fig. 12.5e)
Fast heating	Non-closed layer CSD: 0.05–0.15 μm after 29 h (Fig. 12.5f)	Non-closed layer, CSD: 1–4 μm. (Fig. 12.5g)	Non-closed layer, CSD: 0.05–0.25 μm after 24 h (not shown)	Single layer with an average particle size of 1.6 μm (Fig. 12.5h)

CSD, crystal size distribution

(Fig. 12.5f) as compared to the appearance of only a few nanocrystals attached to the substrate at the end of the induction period under slow heating conditions (Fig. 12.5a).

A stable gel layer cannot be formed on the substrate under fast heating conditions at Si/Al ratio of 17 (and below) due to the low degree of super-saturation, which results in a low rate of formation of gel particles on the substrate surface and a rather high rate of nucleation and crystallization (to zeolite) in the liquid phase.[23] Consequently, the formation of coatings proceeds by heterogeneous nucleation and growth on the substrate resulting in a discontinuous coating with a large crystal size distribution (CSD), for which the coverage increases linearly with time (Fig. 12.5g).

A single layer of zeolite Beta crystals on a substrate corresponds to a coverage of 1.0–1.3 g m^{-2}, depending on the average crystal size, and can obtained under fast heating conditions after 48 or 26 h hydrothermal treatment at 140 and 150 °C, respectively (Fig. 12.5). A mixture of BEA and MER type structures was formed under slow heating at 140 °C. As the temperature increases from 140 to 150 °C, the induction period decreases while the crystallization rate increases, yielding a larger crystal size with a narrower unimodal CSD (Fig. 12.5b and e). The crystallization process has to be quenched when the precursor mixture is depleted of aluminum, to prevent the transformation of the zeolite Beta coatings into more stable phases such as MER (Fig. 12.5c). The outer layer of the zeolitic coatings possesses substantially lower Si/Al ratios than the crystals in the synthesis mixtures[44]. The coatings obtained at 140 °C showed lower Si/Al ratios than those obtained at 150 °C, indicating a higher efficiency of Al incorporation at lower temperatures. As the Si/Al ratio increases from 15 to 23, the coating coverage decreases from ca. 2 to 1 monolayers at 140 °C, while it increases from ca. 0.5 to 2 monolayers at 150 °C.[44] These results suggest that there is an optimum temperature at which the crystallization process on the substrate dominates and that this temperature depends strongly on the Si/Al ratio in the synthesis mixture.

5.2. Growth model of continuous zeolitic coatings

This section describes a general growth model of continuous zeolitic coatings. Various steps in a growth mechanism are schematically shown in Fig. 12.6. The individual steps in the growth of zeolite Beta coatings from a gel layer are illustrated by the evolution of a coating with a Si/Al of 23 at 150 °C in Fig. 12.6 a. Figure 12.6b shows the heterogeneous nucleation and growth of zeolite Beta crystals with a Si/Al ratio of 17 at 140 °C.

The narrow CSD of the coatings (Fig. 12.5 b, e and h) indicates that these are grown from a gel layer initially deposited at the surface of the substrate. Initially, all crystals grew quickly due to high concentration of nutrients both in the gel layer and in the liquid phase. The crystal growth rate reduced as the concentration of the nutrients was depleted in the gel layer. Therefore, zeolite crystals with similar size could be formed. In the initial stage of the synthesis, the dissolution and rearrangement of the precursor mixture into amorphous gel particles dominate (Fig. 12.6, Steps I and II).[53] The thermodynamic potential and liquid motions cause transport of these gel particles from the liquid phase to the substrate,[56] forming a discontinuous layer of primary particles on its super-hydrophilic surface.[5] At low super-saturation, these particles agglomerate into a thin homogeneous gel layer.[56,57] Heterogeneous nucleation at the substrate/gel interface yields the zeolite nanocrystals (Step II). At the same time, primary subcolloidal gel particles of ca. 3 nm in diameter can assemble in the liquid phase and aggregate to give amorphous clusters up to 10 nm in size.[53,54,58] In synthesis mixtures with high super-saturation, these clusters further aggregate into amorphous aluminosilicate particles up to 50 nm in size, which are strongly enriched in aluminum, sodium and potassium.[53,54,59] These particles sediment slowly due to gravity and can only attach to the substrate if formed in the vicinity of the substrate surface (Step a-III). In the fourth Step (a-IV), zeolite nuclei are formed via a reaction between dissolved aluminum and silicon with TEA-silicate species. This reaction has the highest rate at the gel/liquid interface because the gel layer is enriched in aluminum, while the bulk liquid is enriched in silicon and tetraethyl ammonium cation (TEA) species.[60] The aluminum concentration in the liquid is regulated by the total content of alkali cations.[33,60] In the last step, the nucleation is followed by fast crystallization via condensation reactions of the precursors into the zeolite framework at both the substrate/gel and gel/liquid interfaces (Step a-V), resulting in a continuous coating of zeolite crystals (Fig. 12.5h). The bimodal CSD is related to a disparity of crystallization rates at the gel/liquid and substrate/gel interfaces

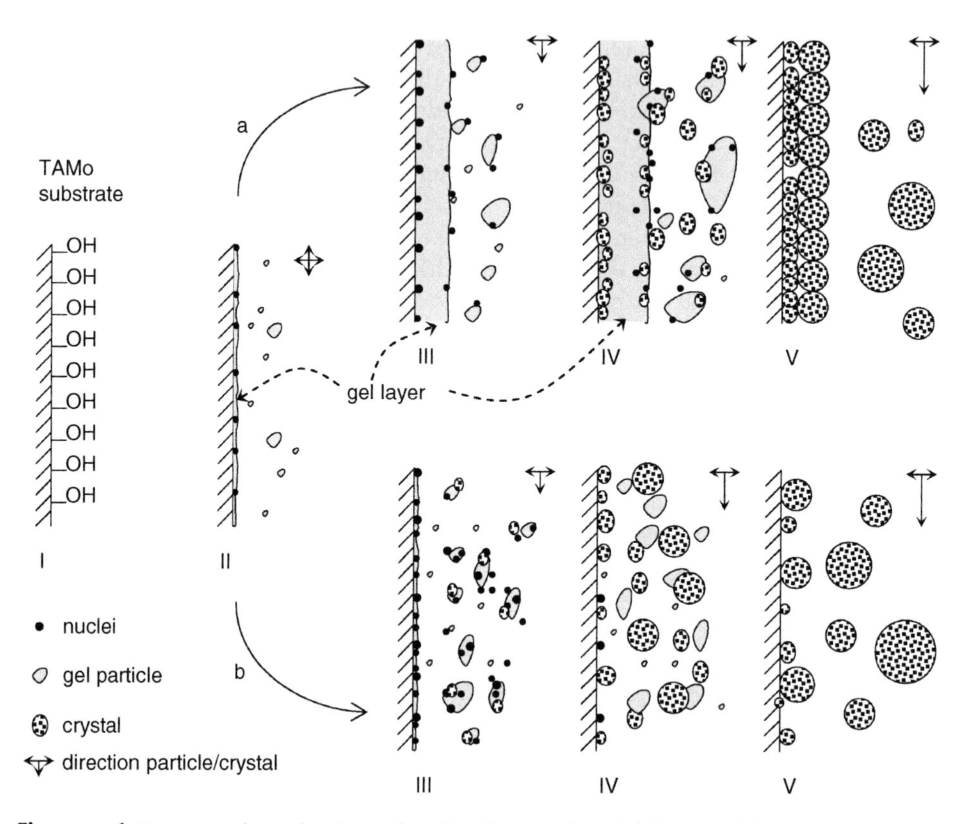

Figure 12.6 Two growth mechanisms of zeolite Beta coatings: (a) from a gel layer and (b) from a diluted precursor mixture. A good wetting of the super-hydrophilic surface of a metal substrate with the precursor mixture is reached after the UV treatment (Step I). A thin homogeneous amorphous aluminosilicate gel layer is deposited on the surface of the substrate in the initial stage of the synthesis process (Step II). Then, heterogeneous nucleation occurs at the substrate/gel interface. Depending on the synthesis conditions, either (a) the gel particles agglomerate in the bulk liquid into large lumps which in turn are responsible for the growth of the gel layer at the surface (a-III and a-IV) or (b) the nucleation and crystallization are enhanced in the bulk liquid (b-III, b-IV), which prevents the formation of a stable gel layer at the substrate and results in a discontinuous coating with a large crystal size distribution (b-V). In the first case, however, nuclei are formed not only in the bulk liquid but also at the both sides of the gel layer (a-III and a-IV). This is followed by fast crystallization at the gel/liquid interface, and a relatively slow crystallization at the substrate/gel interface (a-V). As a result, the crystals show a narrow crystal size distribution. (See color insert.)

(higher at the gel/liquid interface due to enhanced mass transfer of silicon and TEA species). Thus, the crystals in the top layer are considerably larger than those crystallized at the substrate/gel interface. Nevertheless, the crystals at the substrate/gel interface could slowly grow further in size in the course of the synthesis process due to the porous nature of the gel layer.[61]

A stable gel layer cannot be formed under fast heating conditions in Al-rich synthesis mixtures due to the low degree of super-saturation. This results in a low rate of formation of gel particles (Step b-III), while their consumption rate toward nucleation and crystallization of zeolite crystals in the liquid phase is rather high due to higher concentrations of both aluminum and sodium (Steps b-IV and b-V). Therefore, the formation of coatings proceeds through heterogeneous nucleation and growth on the substrate, resulting in a discontinuous coating with a large CSD whose coverage increased linearly in time (Fig. 12.5f and g).

Both a high level of super-saturation within the synthesis mixture and an effective Al incorporation should be realized to obtain a continuous amorphous gel layer, which is important for the subsequent synthesis of a continuous zeolitic coating, at least for BEA type zeolites. The super-saturation decreases at slow heating, while the efficiency of Al incorporation is higher at lower temperatures. Therefore, as the Si/Al ratio increases, synthesis conditions have to be changed from slow heating and a synthesis temperature of 140 °C to fast heating and a higher synthesis temperature of 150 °C.

6. Microwave-Assisted Synthesis of Zeolitic Coatings

High-speed synthesis with microwaves has attracted much attention in recent years due to considerable enhancement of reaction rates, especially in the area of organic synthesis.[62] Recently, this technology was successfully applied for the synthesis of thin films: Kim *et al.* have demonstrated that microwave-assisted hydrothermal synthesis (MAHyS) with NH_4F as mineralizing agent results in a highly crystalline zeolite Beta (91% crystallinity) after 8 h at 150 °C.[63] Titanosilicate (ETS-4) thin films were successfully synthesized by microwave heating at 235 °C within 50 min.[64] Microwave irradiation shortened the synthesis time considerably as compared to 36–48 h required for the traditional ETS-4 synthesis. The coupling of microwave heating with a hydrothermal synthesis requires application of special non-polar substrate materials such as quartz, pure aluminum oxide (corundum), special glass types and plastics. While these are not exotic materials for making micro-reactors and indeed have been applied for laboratory prototypes,[2] there is presently no clear methodology for performing hydrothermal synthesis of zeolitic coatings on such substrate materials. An approach for the development of glass based micro-structured devices is to apply a thin protective film consisting of a more alkali-resistant material such as zirconia on the glass surface. The wettability can be improved by surface treatments or by deposition of a titania hydrophilic film by ALD, as described in Section 4. On glass substrates, the zeolite Beta synthesis time can be decreased by several factors by using the additive effect between fluoride ions and the zeolite seed solution in microwave-assisted hydrothermal synthesis (MaHyS).[41] After pre-seeding the substrate with zeolite Beta nanocrystals, complete crystallization of the Beta zeolite is achieved after 10 h with the combination of fluoride ions and seeding under microwave irradiation. Additional cross-linking with the surface hydroxyl groups of the substrate is achieved during the course of crystallization. However, complete coverage of the substrate can not be achieved without pre-seeding the substrate with zeolite Beta nanocrystals. It is apparent from those studies that microwave heating accelerates the crystallization while avoiding substrate dissolution and the formation of undesired zeolite phases.

7. Large-Scale Synthesis of Zeolitic Coatings

Under laboratory conditions, it is typical to prepare zeolitic coatings on small substrates of 1×1 cm^2 or on longer metal foil plates with a length of 4 cm (total geometric area of 8 cm^2). The implementation of zeolitic coatings in micro-reactors requires the development of a scale-up procedure to simultaneously synthesize identical coatings on a large number of reactor plates or channels. With increasing autoclave size, additional measures against the occurrence of density and temperature gradients during the coating crystallization must be taken. To eliminate the gradients during an initial period, the solution can be continuously stirred. A doubling of coating selectivity was reported for ZSM-5 hydrothermal synthesis on stainless steel monoliths under stirring conditions.[65] Stirring increases the chance that nuclei or crystals formed in the homogeneous liquid phase far away from the substrate surface collide with, or are carried onto the substrate or the zeolitic coating that has already formed on the surface. After an initial period, the synthesis can be continued under static conditions as in the small autoclaves. We developed a dedicated plate holder for 72-substrate plates, in which the distances between the plates were optimized within a total volume of 3 liter (Fig. 12.7). The holder consists of two identical layers. Each layer consists of three symmetrical sets of 12 identical positions. The open structure of the holder provides equal accessibility of the nutrients to the substrate surfaces during the crystallization process. The feed-to-support area should be chosen to provide a Si conversion below 50%, ensuring that the synthesis mixture is not locally depleted so as to generate concentration gradients.[24]

The average coverage and the total deviation per plate for six symmetrical positions in the holder are depicted in Fig. 12.8. The average zeolite coverage is 14.8 g/m^2 and the standard deviation over 72 plates is ±0.4 g/m^2. These small and non-systematic differences between the separate sets proved that no concentration gradients were present in the synthesis volume in the 3 liter autoclave.

8. Applications in Micro-structured Reactors

8.1. Bulk chemicals processing

In one of the first applications of zeolitic coatings in micro-reactors, 500-μm wide channels made in an AISI 316 stainless steel substrate were coated with a ZSM-5 single layer and, after ion-exchange with ceria, studied the resulting micro-reactor in the selective catalytic reduction of NO with ammonia. In this way, mass transfer limitations were effectively avoided and higher reaction rates toward

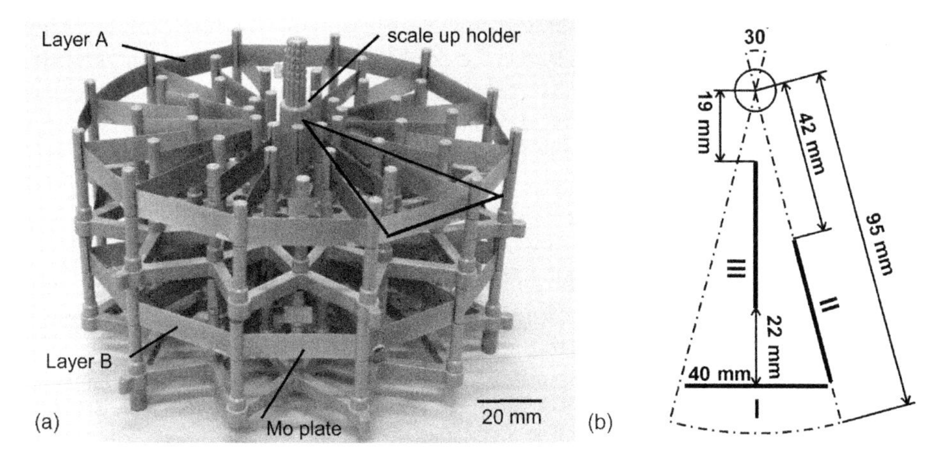

Figure 12.7 Holder for large scale synthesis of zeolitic coatings on 72 plates of 40×9.8 mm^2: (a) image of the holder consisting of two identical layers (layers A and B), which are positioned on top of each other; (b) schematic top view of the holder geometry in which three different symmetry positions of the plates (I, II and II) are shown. This is an enlargement of the triangle indicated in (a), $^1/_{12}$ of the complete holder is depicted.

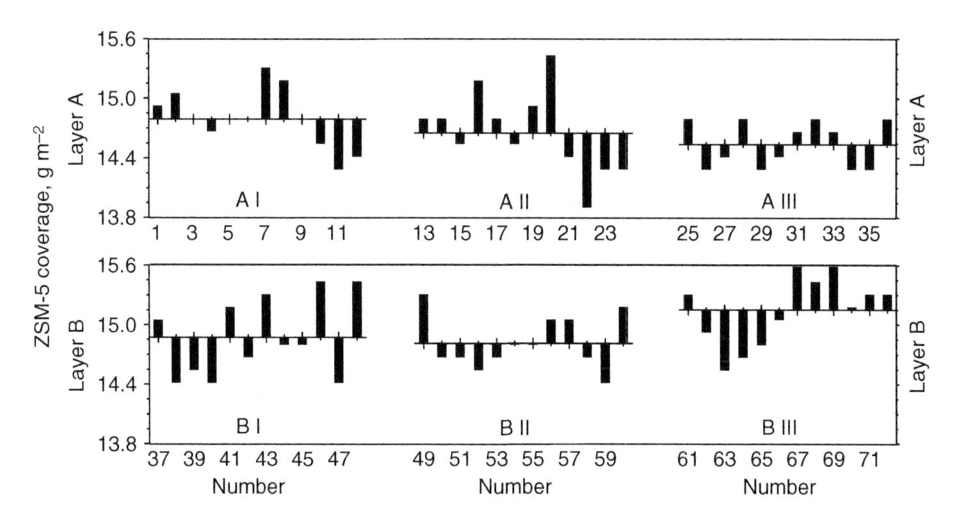

Figure 12.8 ZSM-5 coverage on the 72 metal substrates after the scale-up procedure. The average coverage per symmetry set is given for both layers A and B as well as the deviation per individual plate. Conditions: 150 °C, 48 h, Si/Al = 25, TPA/Al = 2.0, H_2O/Si = 50. The synthesis mixture was stirred during the initial 2 h of the heating up period.

formation of nitrogen were observed in comparison with a conventional pelletted Ce-ZSM-5 catalyst[4]. The bonding of the ZSM-5 crystals with the surface was strong, as the coating did not show any signs of de-lamination or deterioration after 24 h on-stream at 250 °C in a corrosive mixture of 683 ppm NO, 800 ppm NH_3, 5.0 vol.% O_2, balanced by helium. The sum of the intensities of the strongest

XRD peaks appearing at 22–25° 2θ was only 2% less than in the fresh samples. More recently, the activity and selectivity of Co-ZSM-5 and Co-BEA coatings with a thickness of 1 μm deposited on a Mo substrate were determined in the ammoxidation of ethylene in a micro-structured reactor. A maximum reaction rate, expressed in terms of turnover frequency (TOF), of 0.8 s^{-1} was obtained at 500 °C, with selectivity over acetonitrile of 58%. Again, the zeolitic coatings showed higher reaction rates than a pelletted Co-ZSM-5 catalyst.[5,66] Zeolite Beta coatings were more selective toward acetonitrile formation than the ZSM-5 coatings. The rate of acetonitrile formation is increased by a factor of 6 when the Si/Al ratio was changed from 23 to 15.[66]

Besides metallic substrates with micro-channels, metallic wires with a diameter of 100 μm with deposited zeolitic coatings can be considered as micro-structured catalytic reactors.[71] High thermal conductivity of the sintered metal fibre (SMF) matrix provides an efficient heat transfer, allowing isothermal operation of an exothermic reaction. The metal framework serves as a static micro-mixer. These materials are advantageous for their use as building units in micro-structured reactors. Yuranov *et al.* grew thin zeolite coatings with controlled thickness on the SMF.[67] The coating consisted of highly inter-grown crystals about 1 μm in size with the MFI morphology. The activated Fe-ZSM-5 coatings were tested for catalytic activity in N_2O decomposition in a mixture of 2 vol.% of N_2O in helium. Full conversion was achieved at 520 °C. The coating was stable without any deactivation for 20 h on stream. The performance of the Fe-ZSM-5/SMF coatings was evaluated in hydroxylation of benzene in a mixture of 1 vol.% of C_6H_6, 5 vol.% of N_2O and 94 vol.% of He. This industrially important reaction was chosen as a model reaction to test the newly developed composite materials under highly exothermic conditions. The overall reaction enthalpy of the hydroxylation of benzene to phenol at 400 °C is 259 kJ mol^{-1}. Selectivity toward phenol formation above 98% at 270 °C was reported.[67] At higher temperatures, catalyst deactivation was observed. The coatings were reactivated by calcination at 500 °C in an air flow. Similar zeolitic coatings of MFI-type (ZSM-5, Silicalite-1) deposited on SMF-micro-structured substrates were applied as micro-adsorbers for purification of low-content volatile organic compound (VOC) gas streams[68]. When VOC breakthrough occurs, the coatings can be regenerated by desorbing the VOC in a carrier gas suitable for their catalytic oxidation. The adsorption rate was shown to be governed by gas diffusion via zeolitic coatings rather than by internal diffusion in zeolite crystals.

Yang *et al.* obtained a separation factor of 10,000 and a flux of 0.04 kg $m^{-2}h^{-1}$ for water pervaporation from a 3 wt.% water–benzaldehyde solution at a pressure difference of 165 mbar at room temperature in a stainless steel micro-reactor with zeolitic coatings.[69] In this study, a monolayer of the NaA zeolite seeds was covalently bonded to the surface of the micro-channels via alkoxysilane linkers (3-chloropropyltrimethoxysilane and 3-aminopropyltrimethoxysilane) grafted on the wall of the micro-channels. The NaA zeolite coating was grown on the micro-channels by hydrothermal synthesis. This method can be used to seed and grow other zeolite types including MFI, FAU and MOR.[69]

8.2. Fine chemicals synthesis

Fine chemicals and pharmaceuticals are high value added products that are produced in relatively small quantities. Recent advances in the design and fabrication of micro-mixers, micro-separators and micro-reactors bring closer the realization of desktop micro-factories. Application of micro-porous catalysts in fine chemical production has a large potential in the future. It allows avoiding the complex separation steps needed to recover the homogeneous catalysts from the reaction mixture, resulting in cost and energy savings. A few reports concerning zeolite-coated micro-reactor concepts with a continuous-flow mode are discussed below. They generally deal with the benefits associated with rate enhancements, higher yields and greater product selectivity.

Wan *et al.* synthesized Ti-silicalite (TS-1) coatings on silicon substrates for epoxidation of 1-pentene to 1,2-epoxypentane with hydrogen peroxide in a micro-reactor.[70,71] The reaction rate increased with decreasing crystal size in the zeolite films indicating the presence of mass transfer resistances. Unfortunately, irreversible deactivation, attributed to leaching of framework titanium by water and hydrogen peroxide, was observed during the course of the experiments.[71] The selective oxidation of aniline to azoxybenzene in a multi-channel micro-reactor, with or without water removal, employing TS-1 nanozeolite coatings on a ZSM-5 zeolite membrane, was reported.[72] Aniline oxidation, being an exothermic reaction, benefits from the good heat transfer properties of a micro-reactor. The risk of dealing with very toxic chemicals such as aniline and strong oxidizing agents like hydrogen peroxide was minimized by confining the hazardous chemicals in small volumes. A 25-μm thick ZSM-5 coating was grown on the back of a porous multi-channel plate to serve as a pervaporation membrane with a cross section of 25×25 mm^2. A rather thick zeolite layer was needed to bridge the large pores in the porous stainless steel substrate and to obtain a leak-free behaviour. Membrane operation was carried out at a pressure difference of 165 mbar. A reactant mixture containing 2.4 M aniline and 0.6 M hydrogen peroxide dissolved in chloroform was fed to the micro-reactor maintained at 70 °C. The zeolite membrane supported on the porous stainless steel substrate displayed excellent adhesion and could withstand harsh operating conditions. Azoxybenzene and azobenzene were the only products in the micro-reactor, while nitrosobenzene was reported to be the main by-product in a conventional batch reactor.[73] The azoxybenzene yield was improved by 35% as compared with the case without pervaporation. The catalyst deactivation rate was also slowed by ca. 20% when water was removed. A thinner membrane grown on the walls of the micro-channel is expected to increase the rate of water removal from the reaction and further retard catalyst deactivation by water. In general, zeolite-coated micro-reactor applications led to improved reaction conversion, higher product purity and reduced rate of catalyst deactivation.

Knoevenagel condensation between carbonylic compounds and methylene malonic esters on a CsNaX zeolitic coating in a micro-reactor demonstrated an order of magnitude higher productivity as compared with a traditional packed bed reactor while the selectivity remained the same in both reactors.[74,75] A nearly fourfold increase in reactant conversion was obtained for the micro-reactor when NH_2 modified CsNaX zeolitic coatings were applied.[76] The conversion was further improved when zeolitic coatings were grown onto a stainless steel membrane (0.2 μm pores) inserted in the micro-reactor.[77] The water produced by the condensation reaction was selectively pervaporated through the zeolitic membrane. The placing of the CsNaX catalyst onto the membrane increased the product yield considerably. This is because the Knoevenagel condensation reaction is constrained by thermodynamic equilibrium. The continuous and selective removal of water from the reaction led to an increase in conversion in the micro-reactor. The micro-reactor also benefits from simpler process optimization, rapid design implementation, better safety, and easier scale-up by the replication of repeated units.

9. Summary and Outlook

Substantial progress has been made recently in the preparation of single layer zeolitic coatings on different substrate materials for application in micro-structured reactors.[78] Such reactors result in higher reaction rates and, in many cases, to higher selectivities toward the desired products compared to conventional methods. The modification of the substrate prior to hydrothermal synthesis is an important issue in the preparation of thin zeolitic coatings. Protective coatings may be required to prevent corrosion and/or to avoid diffusion of elements from the substrate material to the coatings during high-temperature treatments. Control of both surface roughness in the nanometer range and hydrophilicity of the surface is important to reduce the defect density in zeolitic coatings. Scale-up to the 0.1 m^2 regime seems feasible even in the laboratory. The reproducibility of the synthesis methods is acceptable for the micro-reactor application, as the thickness deviation from the average value does not exceed 3% over a large set of identical (micro-structured) substrates. The synthesis time required to obtain stable defect-free zeolitic coatings remains rather long (36–48 h). However, the synthesis time can be substantially reduced by using a fast heating mode by preheating the autoclave or by applying microwave heating. Current trends focus on the development of stable zeolitic coatings by microwave-assisted hydrothermal synthesis (MaHyS), and on improvement of high-temperature hydrothermal stability of the coatings by isomorphous substitution. Furthermore, in the future, micro-reactors must be integrated with sensors for process control purposes. Zeolitic thin films grown on a conductive support can be applied to reach this goal.

ACKNOWLEDGEMENTS

The authors thank Mr. Oki Muraza from Eindhoven University of Technology for MaHyS of zeolitic coatings, Dr. Milja Mäkelä from Nanoscale Oy (Finland) for her cooperation in the optimization of the ALD procedures, and Prof. Koos

(J.C.) Jansen from Delft University of Technology for the many fruitful discussions. The authors thank the Dutch Technology Foundation (STW), the Netherlands Organization for Scientific Research (NWO), Shell International Chemicals B.V., Akzo Nobel, and Avantium Technologies B.V., for financial support.

REFERENCES

[1] Thayer, A. M., Harmessing microreactions, *C&EN Coverstory,* 83, May 30, 2005, pp. 43–52.
[2] Hessel, V., Löwe, H., "Organic synthesis with microstructured reactors", *Chem. Eng. Technol.* **2005,** *28,* 267–284.
[3] Hessel, V., Löwe, H., "From microreactor design to microreactor process design", *Chem. Eng. Technol.* **2005,** *28,* 243.
[4] Rebrov, E. V., Seijger, G. B. F., Calis, H. P. A., de Croon, M. H. J. M., van den Bleek, C. M., Schouten, J. C., "The preparation of highly ordered single layer ZSM-5 coating on prefabricated stainless steel microchannels 4", *Appl. Catal. A* **2001,** *206,* 125–143.
[5] Mies, M. J. M., Rebrov, E. V., Jansen, J. C., de Croon, M. H. J. M., Schouten, J. C., "Method for the in situ preparation of a single layer of zeolite Beta crystals on a molybdenum substrate for microreactor applications", *J. Catal.* **2007,** *247,* 328–338.
[6] Coronas, J., Santamaria, J., "The use of zeolite films in small-scale and micro-scale applications 6", *Chem. Eng. Sci.* **2004,** *59,* 4879–4885.
[7] Urbiztondo, M. A., Valera, E., Trifonov, T., Alcubilla, R., Irusta, S., Pina, M. P., Rodríguez, A., Santamaría, J., "Development of microstructured zeolite films as highly accessible catalytic coatings for microreactors", *J. Catal.* **2007,** *250,* 190–194.
[8] Scandella, L., Binder, G., Mezzacasa, T., Gobrecht, J., Berger, T., Lang, H. P., Gerber, C. H., Gimzewski, J. K., Koegler, J. H., Jansen, J. C., "Combination of single crystal zeolites and microfabrication: Two applications towards zeolite nanodevices", *Microporous Mesoporous Mater.* **1998,** *21,* 403.
[9] Davis, M. E., Ordered porous materials for emerging applications, *Nature* **2002,** *417,* 813–821.
[10] Li, L., Xue, B., Chen, J., Guan, N., Zhang, F., Liu, D., Feng, H., "Direct synthesis of zeolite coatings on cordierite supports by in situ hydrothermal method", *Appl. Catal. A* **2005,** *292,* 312–321.
[11] Ulla, M. A., Mallada, R., Coronas, J., Gutierrez, L., Miró, E., Santamaria, J., "Synthesis and characterization of ZSM-5 coatings onto cordierite honeycomb supports", *Appl. Catal. A* **2003,** *253,* 257–269.
[12] Öhrman, O., Hedlund, J., Sterte, J., "Synthesis and evaluation of ZSM-5 films on cordierite monoliths", *Appl. Catal. A.* **2004,** *270,* 193–199.
[13] Bonaccorsi, L., Freni, A., Proverbio, E., Restuccia, G., Russo, F., "Zeolite coated copper foams for heat pumping applications", *Microporous Mesoporous Mater.* **2006,** *91,* 7–14.
[14] Van den Berg, A. W. C., Gora, L., Jansen, J. C., Maschmeyer, T., "Improvement of zeolite NaA nucleation sites on (0 0 1) rutile by means of UV-radiation", *Microporous Mesoporous Mater.* **2003,** *66,* 303–309.
[15] Maloncy, M. L., van den Berg, A. W. C., Gora, L., Jansen, J. C., "Preparation of zeolite beta membranes and their pervaporation performance in separating di- from mono-branched alkanes", *Microporous Mesoporous Mater.* **2005,** *85,* 96–103.
[16] Hedlund, J., Öhrman, O., Msimang, V., van Steen, E., Böhringer, W., Sibya, S., Möller, K., "The synthesis and testing of thin film ZSM-5 catalysts", *Chem. Eng. Sci.* **2004,** *59,* 2647–2657.
[17] Öhrman, O., Hedlund, J., Msimang, V., Möller, K., "Thin ZSM-5 film catalysts on quartz and alumina supports", *Microporous Mesoporous Mater.* **2005,** *78,* 199–208.
[18] Louis, B., Reuse, P., Kiwi-Minsker, L., Renken, A., "Synthesis of ZSM-5 coatings on stainless steel grids and their catalytic performance for partial oxidation of benzene by N_2O", *Appl. Catal. A.* **2001,** *210,* 103–109.

[19] Włoch, E., Łukaszczyk, A., Żurek, Z., Sulikowski, B., "Synthesis of ferrierite coatings on the FeCrAl substrate", *Catal. Today* **2006,** *114,* 231–236.
[20] Ulla, M. A., Mallada, R., Coronas, J., Gutierrez, L., Miró, E., Santamaria, J., "Synthesis and characterization of ZSM-5 coatings onto cordierite honeycomb supports", *Appl. Catal. A* **2003,** *253,* 257–269.
[21] Lai, R., Yan, Y., Gavalas, G. R., "Growth of ZSM-5 films on alumina and other surfaces", *Microporous Mesoporous Mater.* **2000,** *37,* 9–19.
[22] Persson, A. E., Shoeman, B. J., Sterte, J., Otterstedt, J.-E., "Synthesis of stable suspensions of discrete colloidal zeolite (Na, TPA)ZSM-5 crystals", *Zeolites* **1995,** *15,* 611–619.
[23] Mintova, S., Valtchev, V., Onfroy, T., Marichal, C., Knözinger, H., Bein, T., "Variation of the Si/Al ratio in nanosized zeolite Beta crystals", *Micorporous Mesoporous Mater.* **2006,** *90,* 237–245.
[24] Mies, M. J. M., van den Bosch, J. L. P., Rebrov, E. V., Jansen, J. C., de Croon, M. H. J. M., Schouten, J. C., "Hydrothermal synthesis and characterization of ZSM-5 coatings on a molybdenum support and scale-up for application in micro reactors", *Catal. Today* **2005,** *110,* 38–46.
[25] Di Renzo, F., "Zeolites as tailor-made catalysts: Control of the crystal size", *Catal. Today* **1998,** *41,* 37–40.
[26] Feoktistova, N. N., Zhdanov, S. P., Lutz, W., Büllow, M., "On the kinetics of crystallization of silicalite I" *Zeolites* **1989,** *9,* 136–139.
[27] Cundy, C. S., Lowe, B. M., Sinclair, D. M. J., "Crystallisation of zeolitic molecular sieves: direct measurements of the growth behaviour of single crystals as a function of synthesis conditions" *Faraday Discuss.* **1993,** *95,* 235–252.
[28] Schmidt, W., Toktarev, A., Schüth, F., Ione, K. G., Unger, K. K., in: Galarneau, A., Di Renzo, F., Fajula, F. and Vedrine, J (Eds.), "The influence of different silica sources on the crystallization kinetics of zeolite beta in *Stud. Surf. Sci. Catal".* Vol. 135, Elsevier, Amsterdam, 2001 pp. 311–318 or CD version*: Stud. Surf. Sci. Catal.* 135, 190 (2001), full text CD A02p23.
[29] Erdem-Senatalar, A., Tatlier, M., Ürgen, M., "Crystallisation of zeolitic molecular sieves: direct measurements of the growth behaviour of single crystals as a function of synthesis conditions" *Microporous Mesoporous Mater.* **1999,** *32,* 331–343.
[30] Erdem-Şenatalar, A., Öner, K., Tatlier, M., in: van Steen, E., Callanan, L. H. and Claeys, M. (Eds.), "Preparation of zeolite coatings by direct heating of the substrates", *Stud. Surf. Sci. Catal.*, Vol. 154, Elsevier, Amsterdam, 2004, p. 667.
[31] Tatlier, M., Demir, M., Tokay, B., Erdem-Şenatalar, A., Kiwi-Minsker, L., "Substrate heating method for coating metal surfaces with high-silica zeolites: ZSM-5 coatings on stainless steel plates" *Microporous Mesoporous Mater.* **2007,** *101,* 374–380.
[32] Andaç, Ö., Telli, Ş.M., Tatlier, M., Erdem-Şenatalar, A., "Effects of ultrasound on the preparation of zeolite A coatings" *Microporous Mesoporous Mater.* **2006,** *88,* 72–76.
[33] Camblor, M. A., Pérez-Pariente, J., "Crystallization of zeolite beta: Effect of Na and K ions" *Zeolites* **1991,** *11,* 202–210.
[34] Mies, M. J. M., Rebrov, E. V., de Croon, M. H. J. M., Schouten, J. C., "Design of a molybdenum high throughput microreactor for high temperature screening of catalytic coatings" *Chem. Eng. J.* **2004,** *101,* 225–235.
[35] Rebrov, E. V., de Croon, M. H. J. M., Schouten, J. C., "Design of a microstructured reactor with integrated heat-exchanger for optimum performance of a highly exothermic reaction" *Catal. Today* **2001,** *69,* 183–192.
[36] Groppi, G., Ibashi, W., Tronconi, E., Forzatti, P., "Structured reactors for kinetic measurements in catalytic combustion" *Chem. Eng. J.* **2001,** *82,* 57–71.
[37] Kuznetsov, S. A., Kuznetsova, S. V., Rebrov, E. V., Mies, M. J. M., de Croon, M. J. H. M., Schouten, J. C., "Synthesis of molybdenum borides and molybdenum silicides in molten salts and their oxidation behavior in an air-water mixture" *Surf. Coat. Technol.* **2005,** *195,* 182–188.
[38] Kuznetsov, S. A., Rebrov, E. V., Mies, M. J. M., de Croon, M. H. J. M., Schouten, J. C., "Synthesis of protective Mo-Si-B coatings in molten salts and their oxidation behavior in an air–water mixture" *Surf. Coat. Technol.* **2006,** *201,* 971–978.
[39] Groner, M. D., Elam, J. W., Fabreguette, F. H., George, S. M., "Electrical characterization of thin Al_2O_3 films grown by atomic layer deposition on silicon and various metal substrates" *Thin Solid Films* **2002,** *413,* 186–197.

[40] Hoivik, N. D., Elam, J. W., Linderman, R. J., Bright, V. M., George, S. M., Lee, Y. C., "Atomic layer deposited protective coatings for micro-electromechanical systems" *Sens. Actuators A* **2003,** *103,* 100–108.

[41] Muraza, O., Rebrov, E. V., Chen, J., Putkonen, M., Niinistö, L., de Croon, M. H. J. M., Schouten, J. C., "Microwave-assisted hydrothermal synthesis of zeolite beta coatings on ALD-modified borosilicate glass for application in microstructured reactors" *Chem. Eng. J.* **2008,** *135,* 117–120.

[42] Valtchev, V., Mintova, S., Konstantinov, L., "Influence of metal substrate properties on the kinetics of zeolite film formation" *Zeolites* **1995,** *15,* 679–683.

[43] Valtchev, V, Mintova, S, "The effect of the metal substrate composition on the crystallization of zeolite coatings" *Zeolites* **1995,** *15,* 171–175.

[44] Mies, M. J. M., Rebrov, E. V., Jansen, J. C., de Croon, M. H. J. M., Schouten, J. C., "Hydrothermal synthesis of a continuous zeolite Beta layer by optimization of time, temperature and heating rate of the precursor mixture" *Microporous Mesoporous Mater.* **2007,** *106,* 95–106.

[45] Kim, H., "Atomic layer deposition of metal and nitride thin films: Current research efforts and applications for semiconductor device processing" *J. Vac. Sci. Tech. B* **2003,** *21,* 2231–2261.

[46] Niinistö, L., Päiväsaari, J., Niinistö, J., Putkonen, M., Nieminen, M., "Feature Article: Advanced electronic and optoelectronic materials by Atomic Layer Deposition: An overview with special emphasis on recent progress in processing of high-*k* dielectrics and other oxide materials" *Phys. Stat. Sol. A* **2004,** *201,* 1443–1452.

[47] Krautheim, G., Hecht, T., Jakschik, S., Schröder, U., Zahn, W., "Mechanical stress in ALD-Al_2O_3 films" *Appl. Surf. Sci.* **2005,** *252,* 200–204.

[48] Ritala, M., Leskelä, M., Nykänen, E., Soininen, P., Niinistö, L., "Growth of titanium dioxide thin films by atomic layer epitaxy" *Thin Solid Films* **1993,** *225,* 288–295.

[49] Sirghi, L., Hatanaka, Y., "Hydrophilicity of amorphous TiO_2 ultra-thin films" *Surf. Sci.* **2003,** *530,* 323–327.

[50] Wang, X., Yu, Y., Hu, X., Gao, L., "Hydrophilicity of TiO_2 films prepared by liquid phase deposition" *Thin Solid Films* **2000,** *371,* 148–152.

[51] Clet, G., Jansen, J. C., van Bekkum, H., "Synthesis of a zeolite Y coating on stainless steel support" *Chem. Mater.* **1999,** *11,* 1696–1702.

[52] Oudshoorn, O. L., "Zeolitic coatings applied in structured catalyst packings", Ph.D. Thesis, University, Delft University of Technology, 1998.

[53] Cundy, C. S., Cox, P. A., "The hydrothermal synthesis of zeolites: Precursors, intermediates and reaction mechanism" *Microporous Mesoporous Mater.* **2005,** *82,* 1–78.

[54] de Moor, P. P. E. A., Beelen, T. P. M., van Santen, R. A., Tsuji, K., Davis, M. E., "SAXS and USAXS investigation on nanometer-scaled precursors in organic-mediated zeolite crystallization from gelating systems" *Chem. Mater.* **1999,** *11,* 36–43.

[55] Slangen, P. M., Jansen, J. C., van Bekkum, H., "The effect of ageing on the microwave synthesis of zeolite NaA" *Microporous Mater.* **1997,** *9,* 259–265.

[56] Koegler, J. H., van Bekkum, H., Jansen, J. C., "Growth model of oriented crystals of zeolite Si-ZSM-5" *Zeolites* **1997,** *19,* 262–269.

[57] den Exter, M. J., van Bekkum, H., van Rijn, C. J. M., Kapteijn, F., Moulijn, J. A., Schellevis, H., Beenakker, C. I. N., "Stability of oriented silicalite-1 films in view of zeolite membrane preparation" *Zeolites* **1997,** *19,* 13–20.

[58] de Moor, P. P. E. A., "The mechanism of organic-mediated zeolite crystallization", Ph.D. Thesis, Eindhoven University of Technology, 1998.

[59] de Moor, P. P. E. A., Beelen, T. P. M., Komanschek, B. U., Beck, L. W., Wagner, P., Davis, M. E., van Santen, R. A., "Imaging the assembly process of the organic-mediated synthesis of a zeolite" *Chem. Eur. J.* **1999,** *5,* 2083–2088.

[60] Pérez-Pariente, J., Martens, J. A., Jacobs, P. A., "Factors affecting the synthesis efficiency of zeolite BETA from aluminosilicate gels containing alkali and tetraethylammonium ions" *Zeolites* **1988,** *8,* 46–53.

[61] Yan, Y., Chaudhuri, S. R., Sarkar, A., "Synthesis of oriented zeolite molecular sieve films with controlled morphologies" *Chem. Mater.* **1996,** *8,* 473–479.

[62] Kappe, C. O., "Controlled microwave heating in modern organic synthesis" *Angew. Chem. Int. Ed.* **2004,** *43,* 6250–6284.
[63] Kim, D.-S., Chang, J.-S., Hwang, J.-S., Park, S.-E., Kim, J. M., "Synthesis of zeolite beta in fluoride media under microwave irradiation" *Microporous Mesoporous Mater.* **2004,** *68,* 77–82.
[64] Coutinho, D., Losilla, J. A., Balkus, K. J., Jr., "Microwave synthesis of ETS-4 and ETS-4 thin films" *Microporous Mesoporous Mater.* **2006,** *90,* 229–236.
[65] Shan, Z., van Kooten, W. E. J., Oudshoorn, O. L., Jansen, J. C., van Bekkum, H., van den Bleek, C. M., Calis, H. P. A., "Optimization of the preparation of binderless ZSM-5 coatings on stainless steel monoliths by in situ hydrothermal synthesis" *Microporous Mesoporous Mater.* **2000,** *34,* 81–91.
[66] Mies, M. J. M., Rebrov, E. V., Schiepers, C. J. B. U., de Croon, M. H. J. M., Schouten, J. C., "High-throughput screening of Co-BEA and Co-ZSM-5 coatings in the ammoxidation of ethylene to acetonitrile in a microstructured reactor" *Chem. Eng. Sci.* **2007,** *62,* 5097–5101.
[67] Yuranov, I., Renken, A., Kiwi-Minsker, L., "Zeolite/sintered metal fibers composites as effective structured catalysts" *Appl. Catal. A* **2005,** *281,* 55–60.
[68] Nikolajsen, K., Kiwi-Minsker, L., Renken, A., "Structured fixed-bed adsorber based on zeolite/sintered metal fibre for low concentration VOC removal" *Chem. Eng. Res. Des. Trans. IchemE* **2006,** *84*(7 A), 562–568.
[69] Yang, G., Zhang, X., Liu, S., Yeung, K. L., Wang, J., "A novel method for the assembly of nano-zeolite crystals on porous stainless steel microchannel and then zeolite film growth" *J. Phys. Chem. Solids* **2007,** *68,* 26–31.
[70] Wan, Y. S. S., Gavriilidis, A., Yeung, K. L., "1-Pentene epoxidation in titanium silicalite-1 microchannel reactor: experiments and modelling" *Chem. Eng. Res. Des.* **2003,** *81,* 753–759.
[71] Wan, Y. S. S., Chau, J. L. H., Yeung, K. L., Gavriilidis, A., "1-Pentene epoxidation in catalytic microfabricated reactors" *J. Catal.* **2004,** *223,* 241–249.
[72] Wan, Y. S. S., Yeung, K. L., Gavriilidis, A., "TS-1 oxidation of aniline to azoxybenzene in a microstructured reactor" *Appl. Catal. A.* **2005,** *281,* 285–293.
[73] Gontier, S., Tuel, A., "Oxidation of aniline over TS-1, the titanium substituted silicalite-1" *Appl. Catal. A* **1994,** *118,* 173–186.
[74] Lai, S. M., Ng, C., Po Martin-Aranda, R., Yeung, K. L., "Knoevenagel condensation reaction in zeolite membrane microreactor" *Microporous Mesoporous Mater.* **2003,** *66,* 239–252.
[75] Lau, W. N., Zhang, X. F., Yeung, K. L., Martin-Aranda, R., "Zeolite membrane microreactor for fine chemical synthesis" *Stud. Surf. Sci. Catal.* **2005,** *158,* 1335–1342.
[76] Zhang, X., Lai, E. S. -M., Martin-Aranda, R., Yeung, K. -L., "An investigation of Knoevenagel condensation reaction in microreactors using a new zeolite catalyst" *Appl. Catal. A* **2004,** *261,* 109–118.
[77] Yeung, K. L., Zhang, X. F., Lau, W. N., Martin-Aranda, R., "Experiments and modeling of membrane microreactors" *Catal. Today,* **2005,** *110,* 26–37.
[78] Mies, M. J. M., "Application of zeolitic coatings in microstructured reactors", Ph.D. Thesis, Eindhoven University of Technology, **2006**.

CHAPTER 13

Pure-Silica-Zeolite Low-Dielectric Constant Materials

Christopher M. Lew, Minwei Sun, Yan Liu, Junlan Wang, *and* Yushan Yan

Contents

Abstract

The intrinsic porous and crystalline nature of pure-silica-zeolites allows them to have many of the desired features necessary for a low-dielectric constant (low-*k*) material for next-generation microprocessors. Several different zeolite films have been prepared,

Ordered Porous Solids
DOI: 10.1016/B978-0-444-53189-6.00013-5

including *in situ*, seeded-growth, spin-on and zeolite/polymer hybrid materials, and they have been characterized for k value, mechanical properties, hydrophobicity and pore structure. This chapter focuses on the latest developments in the synthesis and characterization of pure-silica-zeolites as low-k materials.

Keywords: Hydrophobic, Low-dielectric constant, Mechanical properties, Zeolite

Glossary Entries

Dielectric constant (k)	the relative permittivity of a dielectric material; $k = \varepsilon\varepsilon_o^{-1}$, where ε_o is the vacuum permittivity and is defined as 8.8542×10^{-12} F m^{-1} and ε is the permittivity of the sample
Framework density	the number of T atoms (e.g. Si, Al, B, Ge) per nm^3
Hydrophobic	Repels water
Mesopore	a pore with a diameter between 2 and 50 nm; for spin-on zeolite films, a mesopore is the intercrystalline space
Micropore	a pore with a diameter of less than 2 nm; for zeolites, a micropore is the intracrystalline space
Zeolite	a microporous crystalline oxide

1. Introduction

Gordon Moore first proposed Moore's Law in 1965, and in his seminal paper, he predicted that the number of transistors on a chip would double every two years.[1] While the time frame has since been modified to every 18 months, the semiconductor industry's development of microprocessors has been remarkably consistent with these predictions. The Intel 4004 was released in 1971 and had 2250 transistors, while the Dual-Core Intel Itanium 2 Processor from 2006 contained over 1 billion transistors. Figure 13.1 shows how the production of microprocessors has kept pace with Moore's Law.

Unfortunately, as the industry currently reaches into the sub-32 nm range, the limits of the materials are being tested. Smaller feature sizes are accompanied by RC (resistance–capacitance) delay, which slows down processing speed, and new materials must be developed in order to continue progress along Moore's Law. To reduce the resistance component of the RC delay, the semiconductor industry has developed a Damascene process to switch from aluminum ($R = 3$ μΩ cm) to copper ($R = 1.7$ μΩ cm). The capacitance comes into play in the dielectric material, which has traditionally been dense silica with a dielectric constant (k) of about 4. A suitable alternative low-k material, however, has not yet been found. Target k values for the year 2012 are between 1.8 and 2.1 for a dynamic random access memory ½ pitch of 36 nm, and manufacturable materials are known only for values 2.1–2.4 or higher.[2]

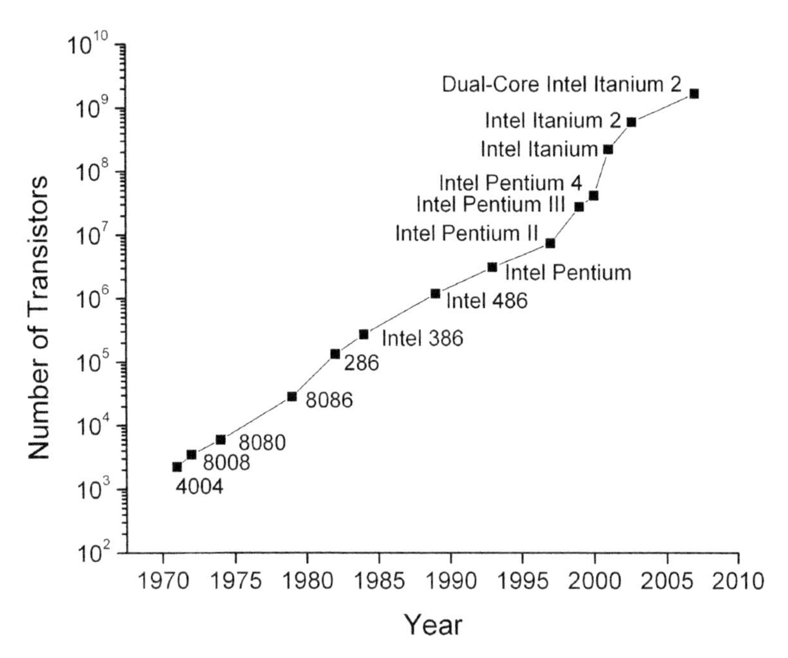

Figure 13.1 A plot of Moore's law showing how the number of transistors on a chip has increased over time.

1.1. Alternative low-*k* materials

The semiconductor industry has been evaluating a number of new low-*k* materials with a variety of chemical compositions. Two important factors in their exploration are the polarizability and number density of the materials' molecules; ideally, both quantities should be low.[3,4] Three major classes of materials have been studied: organic polymers, carbon-doped oxides and silica-based materials.

Organic polymers have generated a large amount of interest because of their ability to achieve low *k* values. Studies have shown that saturated C–C bonds and C–F bonds have a low electronic polarizability.[5] For example, SiLK^{TM} is an organic low-*k* material developed by Dow Chemical Company and has a *k* value of 2.65.[6] Since the high electronegativity of fluorine decreases the polarizability of a material, polymers such as polytetrafluoroethylene have also been studied as potential low-*k* alternates.[7] SiLK^{TM}, however, like many other organic polymers, has low mechanical strength and its coefficient of thermal expansion does not allow for integration with copper wires. Recent efforts by Lin *et al.* have shown that integrating silica nanoparticles into the SiLK^{TM} materials helps to improve the mechanical reliability.[8] Strong mechanical properties are necessary for the material to survive the chemical–mechanical polishing (CMP) processes (see Section 5).

Because C–Si and H–Si bonds have a lower polarizability than O–Si, carbon-doped oxides have been studied extensively. Furthermore, C–Si–O materials are chemically more similar to silica than organic polymers and can be relatively easily deposited through plasma-enhanced chemical vapour deposition.[9–14] Several trademarked low-k materials, such as Black Diamond, Aurora and Coral were developed by Applied Materials, ASM International N.V. and Novellus Systems, respectively, and they are in use in commercial microprocessors. Incorporating porosity into the carbon-doped oxides has been shown to be effective in lowering the k values by taking advantage of the low dielectric constant of air ($k \approx 1$). Second-generation carbon-doped oxides with even lower k values, such as Black Diamond 2, Aurora 2.7 and Aurora ULK, have been produced. Su *et al.* prepared and characterized polysilsesquioxanes that were deposited by spin-coating.[15] The k value was measured to be between 2.1 and 2.7.

Since dense silica was the traditional dielectric material, other silica-based materials are a natural alternative to study. Porous silicas, including sol-gel silica[16,17] and surfactant-templated mesoporous silica,[18–21] have been considered as potential candidates for low-k materials. However, the undesirable trade-off between k value and the mechanical strength of these silicas raises concerns.[22] Furthermore, amorphous silicas also tend to have low heat conductivity and hydrophilic properties, both of which are not desired in low-k applications.[4]

1.2. Zeolite low-*k* materials

Zeolites are microporous crystalline oxides that have traditionally been used as catalysts in the oil refining industry, and they contain a number of desirable attributes that make them a suitable replacement low-k material. Recently, new and novel applications for zeolite thin films have been developed, including but not limited to membranes,[23–28] membrane reactors,[29–32] catalysis,[33–36] sensors,[37,38] heat pumps,[39] thermoelectrics,[40] hydrophilic anti-microbial coatings,[41,42] hydrophilic heat transfer coatings,[43] corrosion-resistant coatings,[44–46] proton conductors[47] and low-k films.[48–51] The micropores are uniform in size and are generally small (<2 nm), while the crystallinity of the material naturally gives the material high mechanical strength, heat conductivity and hydrophobic properties. The pure-silica form of the zeolites is also chemically compatible with the existing semiconductor manufacturing infrastructure that relied on silica, and there is no concern for ion exchange capacity. Furthermore, by synthesizing the zeolites in a nanoparticle suspension with nanocrystals of about 50–70 nm in diameter, the zeolites easily fit into an industry-friendly spin-on process.

2. Pure-Silica-Zeolite Film Synthesis and Characterization

Zeolite films can be prepared through several methods including *in situ*, seeded growth and spin-on techniques. The films are characterized for k value, pore structure, hydrophobicity (Section 4) and mechanical properties (Section 5).

2.1. *In situ* growth *b*-oriented pure-silica-zeolite MFI films

Pure-silica-zeolite (PSZ) low-k films were first investigated in 2001 by Wang *et al.* using a PSZ MFI film grown through an *in situ* crystallization method.[50,51] In this process, a silicon substrate was vertically submerged in a synthesis solution of molar composition 0.32 TPAOH:1 TEOS:165 H_2O contained in a Teflon®-lined autoclave, which was then placed into a pre-heated oven at 165 °C for 2 h (TPAOH is tetrapropylammonium hydroxide and TEOS is tetraethylorthosilicate). The resulting film was about 430 nm thick and was oriented in the *b*-direction (the *b*-axis of the MFI crystals were oriented perpendicular to the substrate). The crystals were well intergrown. Transmission electron microscopy studies on self-supporting films revealed that the crystals formed through homogeneous nucleation and growth in the bulk solution and then self-assembled onto the substrate in a close-packed structure with a *b*-axis orientation.[52] The resulting *k* value of the calcined films was 2.7. The films could be further smoothed by a brief polishing process, and the film did not crack or delaminate, which indicated that the film was well-adhered to the substrate.

2.2. Seeded-growth PSZ MFI films

PSZ MFI films prepared by the method of Sterte *et al.*[53] were studied by Gaynor.[54] Gold was deposited onto a TiN-covered silicon substrate and a layer of PSZ MFI nanocrystalline seeds[55] was adsorbed onto the surface. The substrates were then placed into a precursor solution of molar composition 3 TPAOH:25 SiO_2:1500 H_2O:100 EtOH and polycrystalline films were synthesized at 100 °C for 20 min (EtOH is ethanol). The *k* value was measured to be 2.84.

Li *et al.* also studied seeded PSZ MFI films.[56] First, nanocrystalline PSZ MFI seeds were synthesized following the work of Li *et al.*[57] and dip-coated onto a silicon substrate. The silicon substrate was then placed into a precursor solution of molar composition 9 TPAOH:40 SiO_2:9500 H_2O:160 EtOH in Teflon®-lined autoclaves and placed in a preheated oven at 140 °C for 48 h. The samples were polished and calcined, and the resulting *k* value was measured to be 3.1. These films were oriented along their *c*-axis.

2.3. Spin-on PSZ MFI films

Since the *in situ* crystallization process is not conducive to microprocessor manufacturing, PSZ MFI films prepared through a spin-on method were also studied by Wang *et al.*[49,51] In this process, a zeolite nanoparticle suspension with crystal sizes less than 100 nm was hydrothermally synthesized and spun onto silicon substrates. The precursor solution had a molar composition of 1 TPAOH:2.8 SiO_2:22.4 EtOH:40 H_2O and was synthesized in an oil bath at 80 °C with stirring for 1–5 days. Large particles were removed by centrifugation at 5000 rpm, and the resulting nanoparticle suspension was spun onto low-resistivity silicon substrates at 3300 rpm. The resulting films were calcined at 450 °C to remove the organic structure-directing agent and underwent a brief polishing step to improve surface smoothness.

Although PSZ MFI is intrinsically hydrophobic,[58] the nanocrystalline films contained defects, such as silanol groups, that render the film hydrophilic.[50,59] Furthermore, the spin-on films are actually a composite of crystalline zeolite and an amorphous silica component. To make the films hydrophobic, vapour-phase silylation was performed using trimethylchlorosilane (TMCS). The *k* value of the unsilylated films increased from 2.3 to 3.9 after 1 h of exposure to ambient conditions (50–60% relative humidity) due to moisture adsorption, while the silylated films had a relatively stable *k* value of 2.1 that increased less than 10% after exposure to air. The added methyl functionalization after silylation showed that the zeolites can be modified to increase the hydrophobicity of the resulting films.

Gaynor also studied nanocrystalline PSZ MFI combined with methylsilsesquioxane, hydridosilsesquioxane, TEOS, or tetramethylorthosilicate as a binding material.[54] The zeolite was synthesized following a number of different methods.[53,55,60–62] The binder was either mixed into the PSZ MFI suspension or deposited in a separate step, and it helped to increase the mechanical properties and reduce the hydrophilicity of the zeolite by binding with terminal hydroxyl groups. The zeolite crystals constituted at least 90% of the total mass of the material in the film. Although an experimental *k* value was not given, Gaynor expected the *k* to be in the range of 2.2–2.6.

The decreased *k* value of the spin-on films can be attributed to the added intercrystalline mesoporosity. Nitrogen adsorption/desorption measurements reveal the existence of a bimodal pore size distribution consisting of the micropores within the zeolite and the mesoporous void spaces between the crystals.[49]

2.4. Measurement of the dielectric constant

There are several well-developed techniques to measure the dielectric constant. Parallel plate is the most general characterization method for thin, flat sheets. This technique has high accuracy and is best at low frequencies. *k* values can be determined from capacitance measurements of metal–insulator–semiconductor (MIS) or metal–insulator–metal (MIM) structures at low frequencies (e.g. 1 MHz).[63] A typical measurement system using the parallel plate method consists of an LCR meter or impedance analyser and a fixture. For example, an insulating layer with a thickness of several hundred nanometers to a few micrometers is grown or spun onto a low-resistance silicon wafer. Aluminum dots are then deposited onto the low-*k* layer and on the backside of the wafer to form an MIM structure. This technique can also be explored to measure plate-like single crystals. The *k* value can also be estimated by the square of the index of reflection.[64] Other methods, such as coaxial probe,[65] transmission line,[66,67] free space[68] and resonant cavity,[69,70] are under investigation for *k* measurements of other geometries.

2.5. Pore characterization

The properties of the pore structure within the films can be approximated by nitrogen adsorption/desorption measurements on zeolite powders that have been thermally calcined in a similar way to the thin films.[49] PSZ MFI nanoparticle suspensions were dried and characterized by Wang *et al.* using the Horvath-Kawazoe model[71] modified by Foley[72] for the micropore diameter and the Barrett–Joyner–Halenda (BJH) plot[73] for the mesopores.[49] The micropore value was 0.55 nm, which matched well with the expected value from crystallography (0.55 nm),[74] and mesopore diameters were 2.63–2.81 nm.

Several non-destructive on-wafer techniques are also available to characterize the pore structure of low-*k* films. These include positronium annihilation lifetime spectroscopy (PALS),[75–82] ellipsometric porosimetry (EP)[83] and small-angle neutron scattering spectroscopy combined with specular X-ray reflectivity (SANS–SXR).[84] In PALS, positrons are implanted into the film and positronium is formed. By measuring the vacuum lifetime of the annihilated orthopositronium, information can be gained on the pore sizes. Li *et al.* used PALS to determine that PSZ MFI *in situ* films have a pore diameter of 0.52–0.58 nm, while PSZ MFI spin-on films have a micropore diameter of 0.55 nm and a mesopore diameter of 2.3–2.6 nm.[85] The micropore values are consistent with crystallographic data,[74] and the spin-on values match well with PSZ MFI nanoparticle powder values measured by nitrogen adsorption/desorption.[49] Tanaka *et al.* studied the capping layer that is necessary for the spin-on films.[86] EP relies on changes in the optical properties of the films during adsorption and desorption of an adsorbate. The pore size distribution of PSZ MFI films were characterized by EP toluene adsorption isotherms by Eslava *et al.*[87] The mesopore size distribution ranged from 2 to 8 nm in UV-assisted cured films to 2–50 nm for thermally calcined films, and total porosities were 35 and 53 vol%, respectively.

3. Strategies to Reduce the Dielectric Constant

Using the PSZ MFI spin-on films as a starting point, three strategic aspects of the films were evaluated and specifically engineered to lower the *k* value: the mesoporosity, partial crystallinity and microporosity. Each of these features increases the total porosity of the films, and thus lowers the *k* value. However, unlike several amorphous porous silicas, increasing the porosity of the zeolite films does not lead to a significant decrease in the mechanical properties because of the natural crystalline state of the zeolites.

3.1. Mesoporosity

Li *et al.* increased the mesopore space between the crystals by adding a porogen into the PSZ MFI nanoparticle suspension before the spin-on process.[88] They chose γ-cyclodextrin because it easily bonds to the terminal hydroxyl groups on the external surfaces of the PSZ MFI nanocrystals, thus allowing the porogen to be highly soluble in the zeolite suspension. By varying the porogen loading amount, they were able to tune the mesoporosity, and therefore the *k* value, to a desired value. As shown in Fig. 13.2, the total porosity increased from 45% to 61% as the γ-cyclodextrin amount increased from 0% to 15%.

Nitrogen adsorption/desorption measurements showed that the increase in total porosity was due to higher mesopore volumes, rather than micropore volumes. At the highest loading amount, the *k* value reached 1.8.

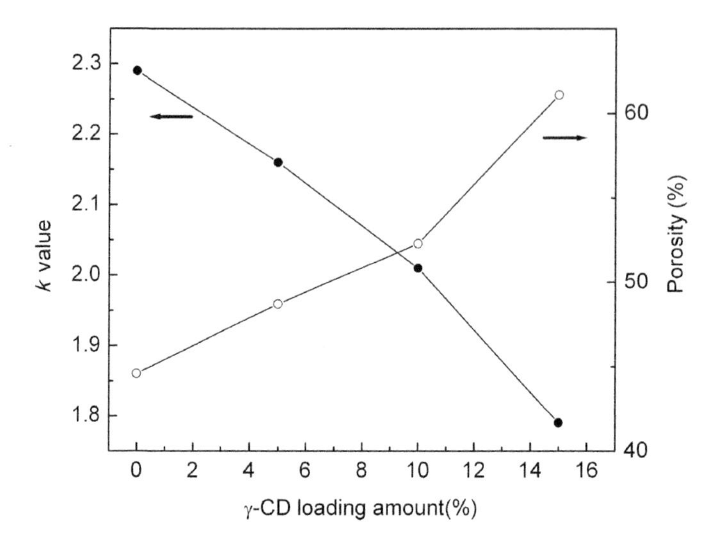

Figure 13.2 Effect of porogen loading on *k* value and porosity of the pure-silica-zeolite (PSZ) MFI spin-on film.[88] (S. Li, Z.J. Li, and Y.S. Yan: "Ultra-Low-*k* Pure-Silica Zeolite MFI Films Using Cyclodextrin as Porogen". *Advanced Materials*. 2003. Volume 15. Pages 1528–1531. Copyright Wiley-VCH Verlag GmbH & Co. KGaA. Reproduced with permission.)

3.2. Crystallinity

Increasing the relative crystallinity of the zeolite, usually reflected by an increase in the micropore volume, is shown to lead to a lower *k* value.[57] Moreover, higher crystallinity will result in higher mechanical strength, higher thermal conductivity and a higher degree of hydrophobicity, all of which are essential for low-*k* materials. For the nanocrystalline PSZ MFI system, one common way to increase the crystallinity is to increase the synthesis time. However, longer synthesis times inevitably lead to larger crystal sizes, which subsequently result in poor film quality, such as higher surface roughness. Thus, the ideal suspension would have a high yield of small nanocrystals with high crystallinity. While other studies have investigated the reuse of the mother liquor[89] or the use of additives to the synthesis solution,[90] Li *et al.* employed a two-stage synthesis[57] following the work of Li *et al.*[91] In this process, a low-temperature first stage focused on nucleation, while a higher-temperature second stage looked at crystal growth. By isolating the development of nuclei without significant crystal formation, Li *et al.* were able to synthesize a PSZ MFI nanoparticle suspension with a high yield (77%) and small crystal diameter (<80 nm). The precursor solution with a molar composition of 1 TPAOH:3 SiO_2:52.4 H_2O:25.1 EtOH was aged at room temperature for 3 days. The first stage occurred in an oil bath at 60 °C for 2 days with stirring. For the second stage, the solution was quickly transferred to Teflon®-lined autoclaves and placed into a preheated oven at 100 °C for varying amounts of time. The resulting suspension was then centrifuged and spun-on in the same manner as the previous PSZ MFI suspension of Wang *et al.*,[49] and the films were also calcined and silylated with TMCS. By increasing the second-stage synthesis time, Li *et al.* were able to increase the crystallinity of the films and keep

Table 13.1 Surface area, pore volume and relative crystallinity of pure-silica-zeolite (PSZ) MFI nanoparticles[a]

Synthesis time (h)	BET surface area (m^2/g)	Micropore area (m^2/g)	Total pore volume (cm^3/g)	Micropore volume[b] (cm^3/g)	Relative crystallinity[c] (%)
8	71	35	0.066	0.016	9.41
10	361	152	0.241	0.067	39.41
20	489	240	0.372	0.105	61.76
26	465	266	0.372	0.116	68.24
69	488	291	0.382	0.126	74.12

[a] Z.J. Li, S. Li, H.M. Luo, and Y.S. Yan: "Effects of Crystallinity in Spin-On Pure-Silica-Zeolite MFI Low-Dielectric-Constant Films". *Advanced Functional Materials*. 2004. Volume 14. Pages 1019–1024. Copyright Wiley-VCH Verlag GmbH & Co. KGaA. Reproduced with permission.
[b] The micropore volume is calculated by the *t*-plot method.
[c] Crystallinty=micropore volume/0.17; 0.17 cm^3/g is the micropore volume of fully crystalline PSZ MFI microcrystals.

the particle size small. As shown in Table 13.1 and Fig. 13.3A, the increasing crystallinity led to higher micropore volumes and lower k values. Figure 13.3B plots the particle size versus yield for one- and two-stage syntheses and shows that smaller particles with high yields can be produced with the two-stage synthesis method. The lowest k value reported through this synthesis technique was 1.6.

3.3. Microporosity

While increasing the mesoporosity decreases the k value, it also leads to weaker mechanical properties.[3,4] In order to take advantage of the strong crystalline nature of the zeolites, increasing the microporosity may be the best way to lower the k value while retaining high elastic modulus. Figure 13.4 shows 175 zeolite framework types (all currently known frameworks except RWY[92]).

The zeolites in the lower-left-hand corner are the best candidates for low-k materials because of their low framework density (number of T atoms per nm^3) and small pore window size. Filled black squares represent known zeolite types, while underlined, filled blue squares are pure-silica analogues. For low-k applications, PSZs with a known nanoparticle recipe (i.e., MFI, MEL and BEA) are useful because they can be easily spun-on into thin films (filled red squares). Future work in this area should continue for pure-silica-zeolites in the direction of lower framework densities. For example, Corma *et al.* published a recipe for PSZ LTA,[93] and if the particle size can be decreased below 100 nm, nanocrystalline PSZ LTA films should have the lowest known zeolite k value.

Mitra *et al.* synthesized PSZ BEA films using an *in situ* crystallization method.[94] Stainless steel substrates were placed in Teflon®-lined autoclaves and were covered with a synthesis gel with a final molar composition of 0.6 TEAOH:0.6 HF:1 SiO_2:9.8 H_2O (TEAOH is tetraethylammonium hydroxide and HF is hydrofluoric acid). The autoclaves were placed into a preheated oven at 130 °C for 14 days. The as-synthesized

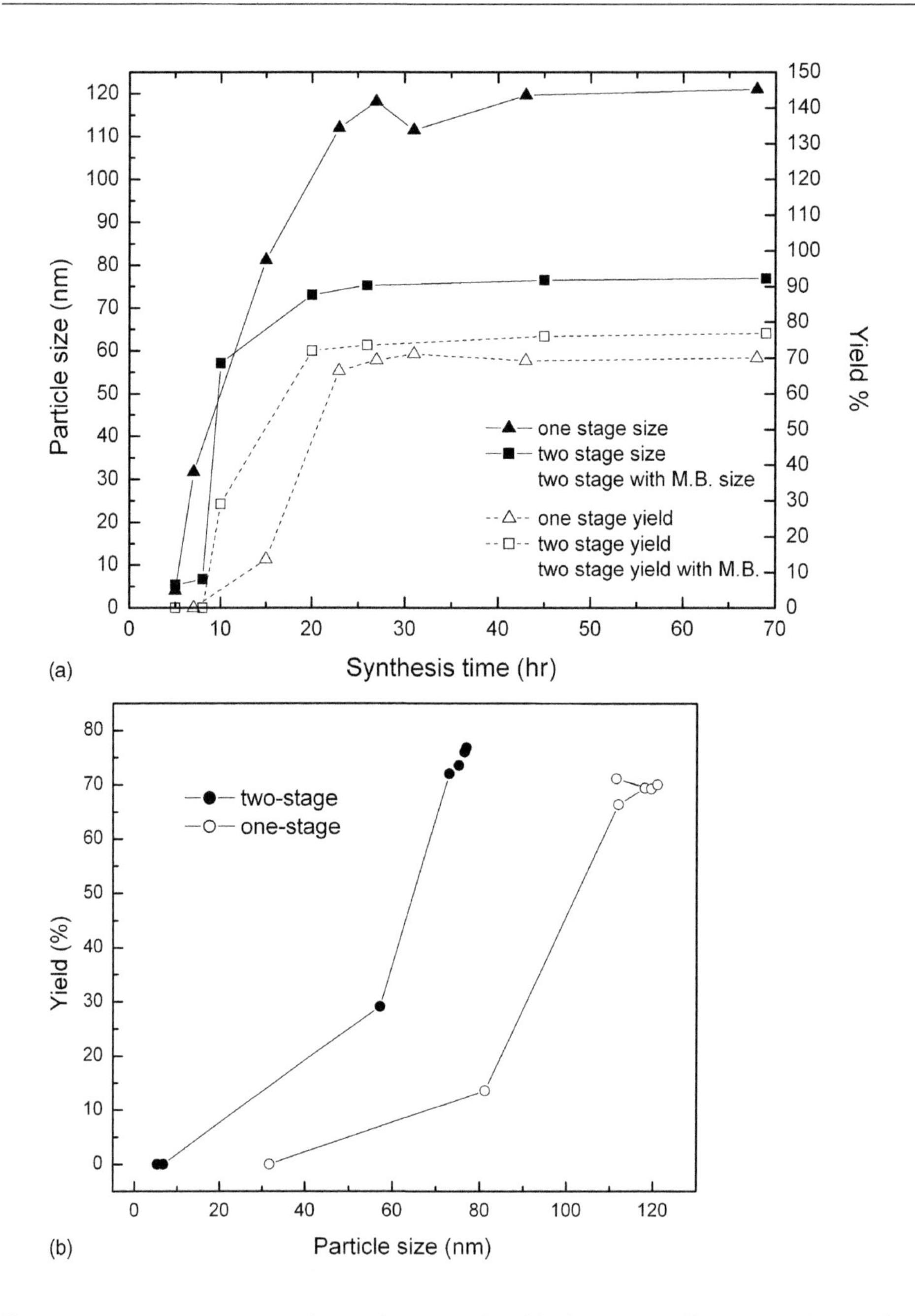

Figure 13.3 (A) Comparison of particle size and yield of nanocrystalline pure-silica-zeolite (PSZ) MFI from one-stage and two-stage hydrothermal syntheses. Synthesis time for the two-stage synthesis refers to the synthesis time of the second stage.[57] (Z.J. Li, S. Li, H.M. Luo, and Y.S. Yan: "Effects of Crystallinity in Spin-On Pure-Silica-Zeolite MFI Low-Dielectric-Constant Films". *Advanced Functional Materials*. 2004. Volume 14. Pages 1019–1024. Copyright Wiley-VCH Verlag GmbH & Co. KGaA. Reproduced with permission.) (B) Particle size vs. yield for one- and two-stage syntheses.

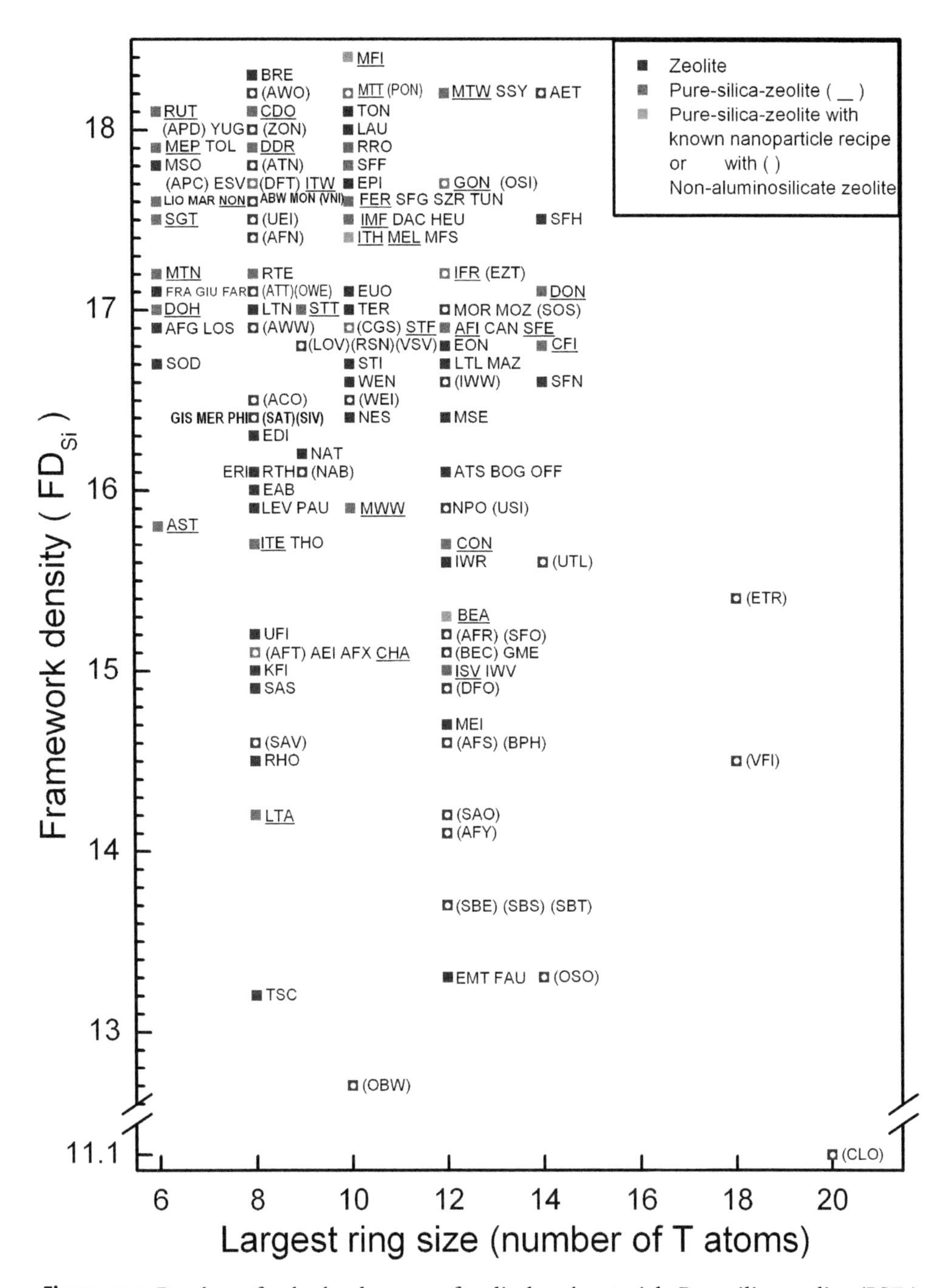

Figure 13.4 Road map for the development of zeolite low-*k* materials. Pure-silica-zeolites (PSZs) with a known nanoparticle recipe are marked with red boxes, other PSZs are marked with blue boxes with—, non-aluminosilicate zeolites are marked with hollow boxes with ( ), and all others are marked as dark boxes.[48] (Adapted with permission from *The Journal of Physical Chemistry B*, 2005, 109(18):8653. Copyright 2005 American Chemical Society.) (See color insert.)

films were washed and polished, and although crack-free calcined films were not able to be prepared, comparisons between MFI and BEA films quickly reveal the importance of the microporosity in lowering the *k* value. The as-synthesized PSZ BEA films had a *k* value of 2.3, while the as-synthesized PSZ MFI films had a *k* value of 3.4,[50] and it is reasonable to assume that the *k* value of calcined PSZ BEA films should be less than 2.3. The only significant difference between the films is the framework density.

The influence of microporosity of PSZ spin-on films on *k* value was investigated by Li *et al.* by moving to the MEL framework.[48] Similar to the PSZ MFI synthesis, they utilized a two-stage method to hydrothermally synthesize PSZ MEL with a high yield and small particle size. The precursor solution with a molar composition of 0.3 TBAOH:1 SiO_2:4 EtOH:10 H_2O was aged for 1 day at room temperature with stirring (TBAOH is tetrabutylammonium hydroxide) and was then placed into an oil bath at 80 °C with stirring for 2 days (first stage). The solution was then quickly transferred to Teflon®-lined autoclaves and placed into a preheated oven at 114 °C for the second stage. The resulting nanoparticle suspension was spun onto silicon substrates at 3300 rpm, calcined at 400 °C and subjected to a vapour-phase silylation with TMCS. As with the PSZ MFI films, longer second-stage synthesis times resulted in films with higher micropore volume and relative crystallinity, and lower *k* values. Suspensions with crystal sizes of 50 nm had yields as high as 57%. For a second-stage synthesis time of 69 h, the *k* value was 1.5.[48] A recipe for nanocrystalline PSZ BEA, with a framework density of 15.3 T nm^{-3} versus 17.4 T nm^{-3} for MEL and 18.4 T nm^{-3} for MFI, was reported by Mintova *et al.* in 2003.[95]

4. Hydrophobicity

Since water has a *k* value of about 80, moisture adsorption from air is a serious concern for porous low-*k* materials. The original work on PSZ MFI by Flanigen *et al.* showed that the dense crystalline frameworks of PSZs are hydrophobic.[58] Terminal hydroxyl groups on the surface of the nanocrystals, however, render the PSZ low-*k* films hydrophilic.[50,59] Several methods have been used to functionalize the zeolite films and increase the hydrophobicity of the material.

Hydrophobicity can be characterized through a variety of techniques. Static water contact angle measurement is an easy method and consists of placing a drop of water onto the surface of the film.[87,96] The angle between the liquid/solid and liquid/vapour interface is then measured. An angle greater than 150° is considered super-hydrophobic, while an angle less than 5° is super-hydrophilic.[97] For low-*k* applications, a higher water contact angle indicates less moisture adsorption and a lower *k* value. Despite their ease, contact angle measurements only determine the hydrophobicity of the surface of the films. Water adsorption isotherms using ellipsometric porosimetry can shed light onto the hydrophobic nature of the internal pore surfaces of the films.[87,98] At high pressures, a low amount of water uptake shows that the internal volume is hydrophobic. Finally, simply measuring the *k* value against exposure time to ambient conditions is a reliable method to determine whether or not moisture uptake has been minimized.[48,57,87,96] A film that is hydrophobic will possess a *k* value that is stable over time.

One common way to reduce moisture adsorption is to replace the hydroxyl groups with hydrophobic molecules, such as methyl groups. Wang *et al.* used a vapour-phase silylation with trimethylchlorosilane after calcination to add methyl functionalization to PSZ MFI films.[49] Gaynor also noted the use of hexamethyldisilazane, hexaphenyldisilazane and dichlorodimethylsilane after calcination to render PSZ MFI films hydrophobic.[54] Li *et al.* grafted methyl groups onto PSZ MFI by adding methyltrimethoxysilane directly into the synthesis precursor solution.[96] For the purposes of microprocessor manufacturing, each processing step can be very costly and time-consuming, and thus this direct synthesis method easily adds hydrophobic functionalization without the extra silylation step. Both of these techniques have been shown to reduce the moisture uptake and k. Figure 13.5 compares the k values of as-calcined, vapour-phase silylated and pre-synthesis functionalized PSZ MFI spin-on films over exposure time to ambient conditions.[96] Both the silylated and the functionalized films exhibit lower final k values than the as-calcined film.

Ultraviolet-assisted curing has also been shown by Eslava *et al.* to be an easy method to make PSZ MFI films hydrophobic.[87] In their work, they used a UV-curing step instead of a thermal calcination to remove the organic structure-directing agent. As seen in Fig. 13.6, IR shows that methyl groups from the tetrapropylammonium hydroxide replaced the terminal hydroxyl groups, and water contact angle measurements of 137° revealed a high degree of hydrophobicity. Water adsorption isotherms by ellipsometric porosimetry also showed that the internal surfaces of the film resisted moisture uptake. Furthermore, UV-curing reduced the chances of crack formation and increased the elastic modulus over thermally calcined films.

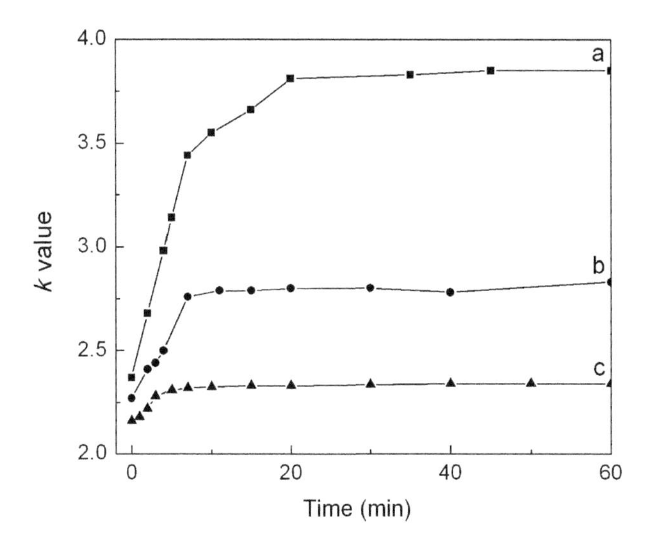

Figure 13.5 k vs. exposure time to air with 50–60% relative humidity: (a) non-silylated pure-silica-zeolite (PSZ) MFI spin-on film after calcination in air; (b) PSZ MFI spin-on film pre-synthesisfunctionalized with methyltrimethoxysilane calcined in nitrogen at 355°C; (c) PSZ MFI spin-on film vapour-phase silylated with TMCS.[96] (Reprinted with permission from *Chemistry of Materials*, 2005, 17(7):1854. Copyright 2005 American Chemical Society.)

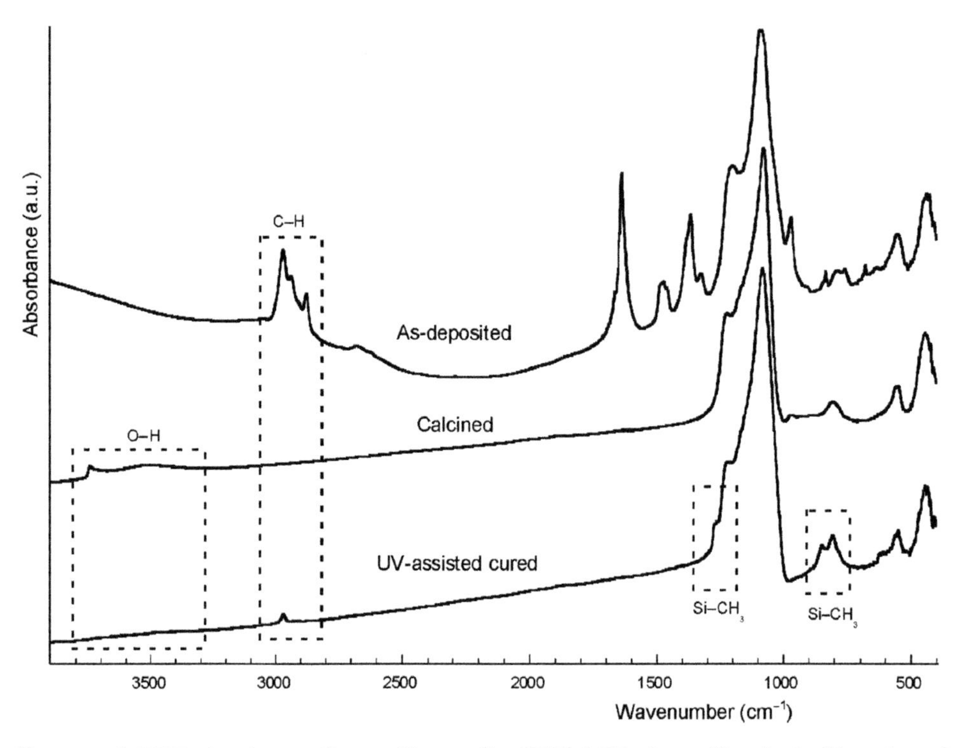

Figure 13.6 FTIR absorbance of pure-silica-zeolite (PSZ) MFI spin-on films in the IR region of 3900–400 cm^{-1}.[87] (Reprinted with permission from *Journal of the American Chemical Society*, 2007, 129(30):9288. Copyright 2007 American Chemical Society.)

5. Mechanical Properties

Low-*k* materials have to sustain significant stresses during the packaging[99] and CMP processes. Therefore, characterization of the mechanical properties of low-*k* thin films is extremely important to ensure that the materials will function properly throughout the lifetime of an integrated circuit. The most common metrics used by the semiconductor industry are the elastic modulus, which measures mechanical stiffness, and hardness. The minimum threshold elastic modulus value for low-*k* materials is generally around 6 GPa.[50]

Adding porosity to a material usually decreases the mechanical strength of a material, and thus porous low-*k* materials often suffer weak mechanical properties. Xu *et al.* showed that there is a strong trade-off between porosity, which decreases the *k* value, and elastic modulus for amorphous porous silicas. As seen by the blue curve in Fig. 13.7, at the ultra-low-*k* values below 2.0, the elastic modulus is significantly lower than 6 GPa, thus putting the reliability of many amorphous porous silica-based materials into doubt.

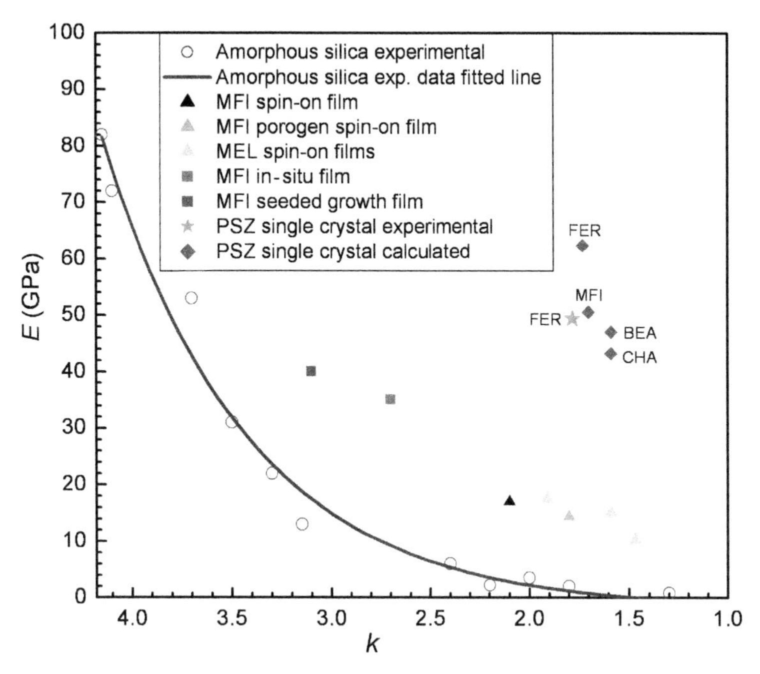

Figure 13.7 E vs. k for amorphous silicas and pure-silica-zeolites (PSZs). The open circles and solid data fitted line correspond to experimental data taken from Xu et al.[22] Squares correspond to experimental data for polycrystalline PSZ MFI films; triangles correspond to experimental data for spin-on PSZ MFI and MEL films; diamonds correspond to calculations of PSZ single crystals; and the star corresponds to experimental data for FER single crystals.[56] (Z.J. Li, M.C. Johnson, M.W. Sun, E.T. Ryan, D.J. Earl, W. Maichen, J.I. Martin, S. Li, C.M. Lew, J. Wang, M.W. Deem, M.E. Davis and Y.S. Yan: "Mechanical and Dielectric Properties of Pure Silica Zeolite Low-*k* Materials". *Angewandte Chemie International Edition*. 2006. Volume 45. Pages 6329–6332. Copyright Wiley-VCH Verlag GmbH & Co. KGaA. Reproduced with permission.) (See color insert.)

Li *et al.* hypothesized that the microporous yet crystalline nature of the zeolites allows the PSZ low-*k* films to have higher mechanical properties than amorphous porous silicas at any given *k* value.[56] This advantage is one of the strongest assets that zeolites have over other alternative low-*k* materials because the mechanical reliability is of utmost importance to the semiconductor industry.

Until recently, the mechanical properties of zeolites have not been extensively studied, and the literature still lacks comprehensive zeolite mechanical characterization. Nonetheless, several mechanical studies have been performed on zeolite films and single crystals, and the results are discussed in this section. Zeolite elastic modulus and hardness are often measured using nanoindentation, although other techniques for thin film mechanical characterization exist, such as micro-tensile testing of free-standing thin films,[100] micro-beam deflection,[99,101] wafer curvature tests[102] and laser-generated surface acoustic wave techniques.[103]

5.1. Nanoindentation

The elastic modulus and hardness of the zeolite films are typically characterized by nanoindentation in the following way.[104,105] Multiple indentations using a pyramidal indenter tip are performed at different locations on the sample with varying loading force and indent depths. For each indentation, a trapezoidal loading/force profile (linear loading, holding and linear unloading) is employed, as shown in Fig. 13.8. A typical resulting load/displacement curve is shown in Fig. 13.9.

The hardness and reduced elastic modulus are determined from the unloading portion of the load/displacement curve. For each indentation, hardness, H, is calculated by

$$H = \frac{P_{\max}}{A(h)} \tag{1}$$

where $P_{\max}$ is the maximum loading force and A is the contact area as a function of contact depth, h. The area function demonstrates the correlation between contact area and contact depth and is calibrated on a fused quartz standard before measurement. A schematic of the indentation process is shown in Fig. 13.10.

The reduced elastic modulus, E_r, is calculated by

$$E_r = \frac{\sqrt{\pi}}{2\sqrt{A(h)}} S \tag{2}$$

where S is the measured stiffness of the unloading curve starting from a specific point (specified by the operator according to the curve feature). The measured reduced elastic modulus results from elastic deformations of both the film and the indenter probe. The relationship between the measured reduced elastic modulus and the elastic modulus of the sample is demonstrated by

$$\frac{1}{E_r} = \frac{1 - \nu_i^2}{E_i} + \frac{1 - \nu_s^2}{E_s} \tag{3}$$

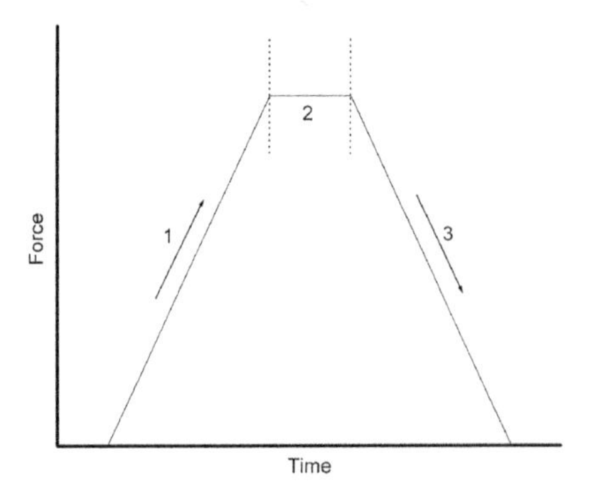

Figure 13.8 Loading force profile.

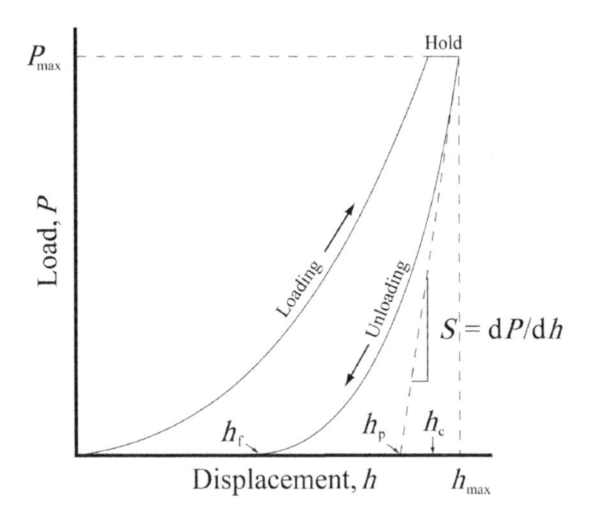

Figure 13.9 Resulting load/displacement curve. See Fig. 13.10 for the definitions of h_f, h_p and h_c.

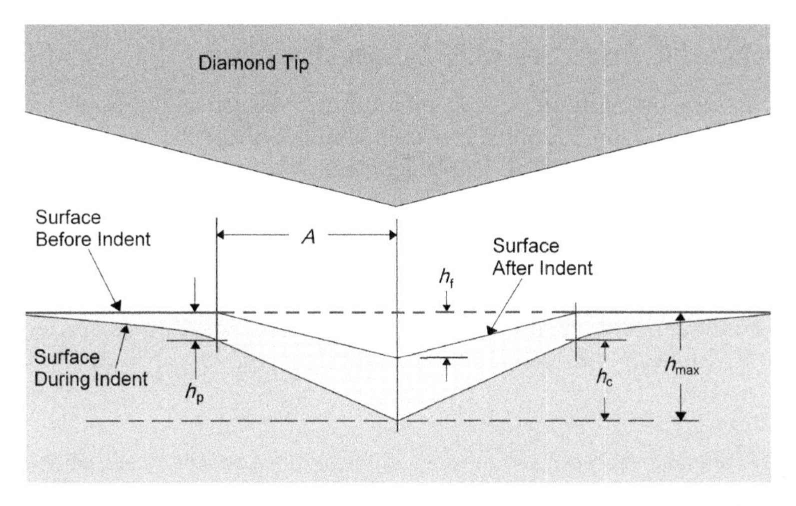

Figure 13.10 Schematic of the film surface before, during and after an indentation.[104] (See color insert.)

where ν_i and E_i are the Poisson's ratio and Young's modulus of the probe and ν_s and E_s are the values for the sample. For a diamond probe, the Poisson's ratio and Young's modulus are 0.07 and 1140 GPa, respectively. For glassy materials, ν_s is 0.25, which is a reasonable approximation for PSZ.[56] Given ν_s, E_s can be calculated from E_r. Ideally, the calculated elastic modulus and the hardness should be the same for each indentation, regardless of the indentation loading force and contact depth. The final value is determined by the average of all the indentations.

5.2. Mechanical properties of PSZ spin-on films

Wang *et al.* performed nanoindentation on PSZ MFI spin-on low-*k* films and obtained moduli between 16 and 18 GPa,[49] while the films of Li *et al.* employing γ-cyclodextrin as a porogen had an elastic modulus of 14.3 GPa.[88] Wang *et al.*'s films were 0.42-μm thick and encountered the substrate effect at contact depths greater than 10% penetration.[106] Because crystalline silicon has an elastic modulus of about 130 GPa, the substrate tends to influence the mechanical measurements on films less than 1 μm.[56] The indenter tip, thickness of the film and material properties determine when the substrate effect will occur, and a thicker film will generally eliminate these inaccuracies.[56,107] For example, Johnson *et al.* found that using a sharper cube-corner tip versus a flatter Berkovich tip eliminated serious concerns about the substrate effect when they measured PSZ MEL thin films.[107] Furthermore, they found that the mechanical, chemical, microstructural and electrical properties were all correlated to the second-stage synthesis time. The results are summarized in Table 13.2, and Fig. 13.11 shows that with the exception of the 8-h sample, the PSZ MEL films have significantly higher elastic moduli than organo-silicate glass. For the lowest *k* value of 1.47, the elastic modulus of the PSZ MEL film is 10.1 GPa.

5.3. Mechanical properties of PSZ crystalline films

Because the spin-on films are a crystalline/amorphous silica composite, measurements on purely crystalline zeolite films are essential to show that the intrinsic nature of the zeolites gives them the necessary mechanical stiffness for low-*k* applications. PSZ MFI *in situ* films were tested with nanoindentation by Wang *et al.* and had elastic moduli between 30 and 40 GPa,[50] while the elastic modulus of PSZ MFI films prepared by secondary growth, or seeded, methods had an E of 41.8 GPa.[56] The seeded PSZ MFI films were also greater than 6 μm, so the results were more reliable than those that were influenced by the substrate. Similar tests on seeded PSZ MFI films by Johnson *et al.* verified these results by achieving an E of 43.4 GPa.[108]

Table 13.2 Correlation between second-stage synthesis time, porosity, yield, particle size, modulus/hardness and surface roughness[a]

Second-stage synthesis time (h)	Total porosity (%)	Particle yield (%)	Particle size (nm)	Reduced modulus (GPa)	Hardness (GPa)	Average surface roughness, R_a (nm)	k
8	20.3	0	0	17.6 ± 1.1	1.22 ± 0.03	0.88	3.70
20	64.8	57	50	17.5 ± 1.4	0.73 ± 0.05	6.2	1.91
40	63.0	77	85	15.0 ± 1.0	0.75 ± 0.04	5.6	1.59
69	61.5	77	83	10.1 ± 1.9	0.36 ± 0.04	11.8	1.47

[a] Reprinted from *Thin Solid Films*, Vol. 515, M.C. Johnson, Z.J. Li, J. Wang, and Y.S. Yan, "Mechanical Characterization of Zeolite Low Dielectric Constant Thin Films by Nanoindentation", pg. 3164–3170, 2007, with permission from Elsevier.

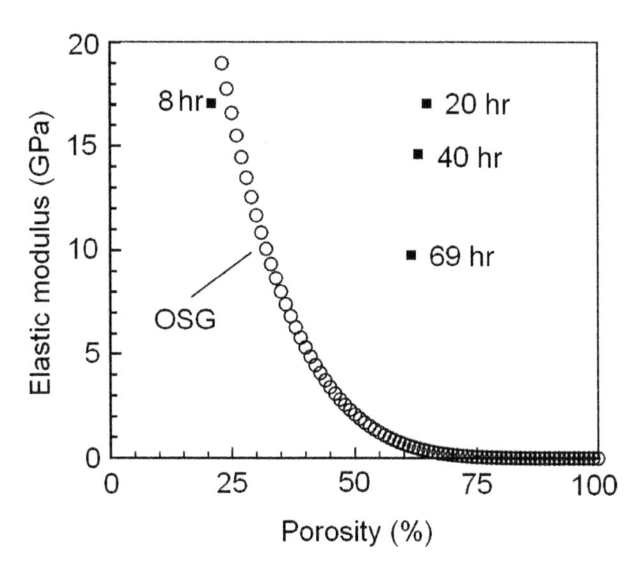

Figure 13.11 Elastic modulus as a function of porosity for pure-silica-zeolite (PSZ) MEL spin-on films.[107]

These experiments showed that the crystalline nature of zeolites does indeed help improve the mechanical properties over amorphous porous silicas.

The effects of calcination and grain boundaries were studied by Johnson *et al.* on PSZ MFI seeded growth films.[108] They found that the elastic modulus and hardness of the calcined films were 3.6% and 21.8% lower than those of the as-synthesized films, respectively. The negligible difference in the elastic modulus between the calcined and as-synthesized samples was consistent with results obtained by Wang *et al.*[109] and Brabec *et al.*[110] in which they performed experiments on PSZ MFI single crystals. Johnson *et al.* hypothesized that the organic structure-directing agent in the as-synthesized films added some resistance to plastic deformation and contributed to the higher hardness values over the calcined films. Furthermore, both the E and H of their polycrystalline films were lower than values obtained on PSZ MFI single crystals, thus showing that the existence of grain boundaries and interzeolitic defects contribute to lower mechanical properties.

5.4. Mechanical properties of zeolite single crystals

The effects of grain boundaries can be eliminated by studying single zeolite crystals. Li *et al.* studied zeolite single crystals FER, MFI and CHA, and obtained E values of 49.4, 53.9 and 48.9 GPa, respectively.[56] Johnson *et al.* measured uncalcined PSZ MFI single crystals and obtained E and H values of 57.4 and 7.2 GPa, respectively.[108] E values between 36.3 and 40.7 GPa obtained by Brabec *et al.* on PSZ MFI single crystals[110] were lower but still showed that zeolites possess considerable stiffness over amorphous porous silicas. Micromechanical compression studies on PSZ MFI single crystals performed by Wang *et al.* yielded an average Young's modulus of about 4 GPa.[109]

5.5. Pore dimensionality and symmetry

Following the work of Fan *et al.*,[111] we hypothesize that the pore dimensionality and symmetry will also be important in determining the ideal framework for low-*k* applications. Fan *et al.* found that for any density, the Young's modulus, *E*, for disordered (D), hexagonal (H), and cubic (C) nanostructured porous silicas are in the following order: $E_C > E_H > E_D$. In other words, for any given density, the mechanical stiffness will be higher for the structure with a cubic phase. Furthermore, a larger concentration of smaller ring sizes increased the Young's modulus because the smaller rings were stiffer than the weaker larger rings. Given these results, we plot the pore dimensionality versus framework density of zeolites with the largest ring sizes of 6 and 8 in Figs. 13.12 and 13.13, respectively. In these figures, the cubic

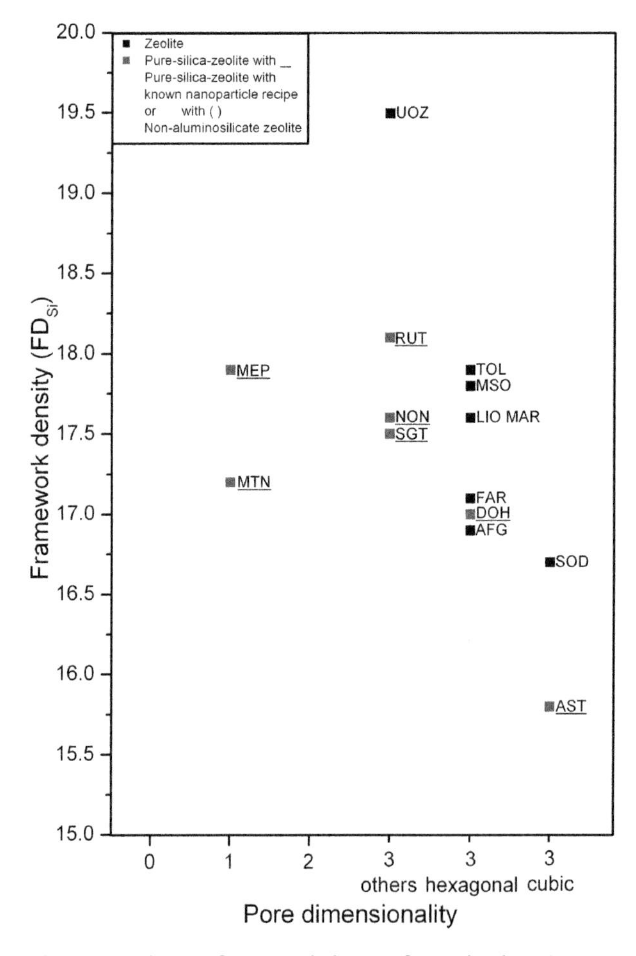

Figure 13.12 Pore dimensionality vs. framework density for zeolite low-*k* materials with the largest ring size of 6. Pure-silica-zeolites (PSZs) with a known nanoparticle recipe are marked with red boxes, other PSZs are marked with blue boxes with _, non-aluminosilicate zeolites are marked with hollow boxes with ( ) and all others are marked as dark boxes. (See color insert.)

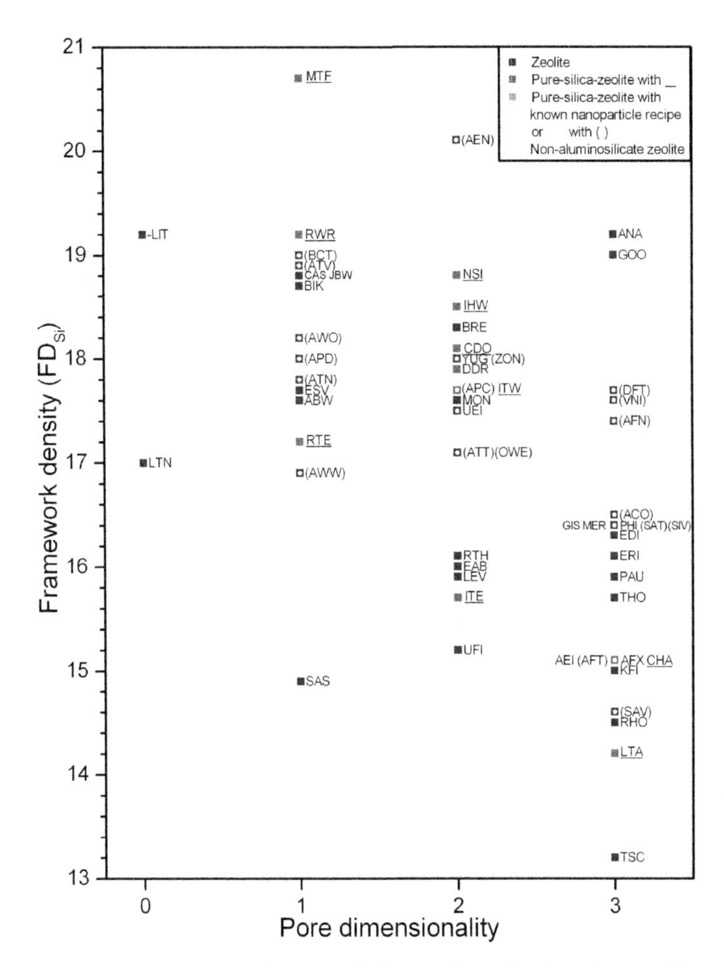

Figure 13.13 Pore dimensionality vs. framework density for zeolite low-*k* materials with the largest ring size of 8. Pure-silica-zeolites (PSZs) with a known nanoparticle recipe are marked with red boxes, other PSZs are marked with blue boxes with _, non-aluminosilicate zeolites are marked with hollow boxes with ( ) and all others are marked as dark boxes. (See color insert.)

phases should have the strongest mechanical properties, while the frameworks with the lowest framework densities should have the lowest *k* values. For example, in Fig. 13.12, AST should have the lowest *k* value, while maintaining good mechanical properties because of its cubic phase; a similar conclusion can be made for TSC in Fig. 13.13.

5.6. Interfacial adhesion studies

Interfacial adhesion of zeolite low-*k* thin films to relevant substrates is critical for the final integration of the material. The adhesion of the spin-on, *in situ* and seeded-growth PSZ MFI films to silicon substrates was investigated using a laser spallation

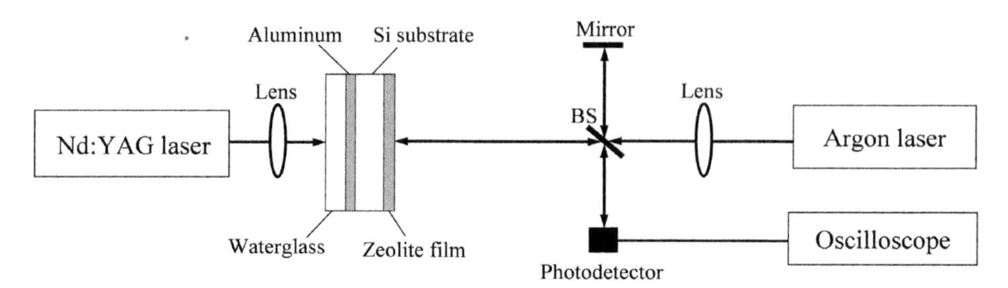

Figure 13.14 Schematic of the laser spallation setup.[112]

technique by Hu *et al.*[112] In this method, a Nd:YAG laser provides a high-energy short-duration laser pulse with a wavelength of 1064 nm that is absorbed by an energy absorbing layer confined between a transparent confining layer and the back surface of the substrate. Upon absorbing the laser energy, the sudden expansion and exfoliation of the energy absorbing layer induces a compressive stress wave that travels through the substrate and the film to the free-film surface and is reflected back to the film–substrate interface. The interface is loaded in tension and the out-of-plane displacement of the film is detected by a Michelson interferometer. The interfacial stress is derived from the transient displacement of the film using wave mechanics and the threshold stress at delamination initiation is regarded as the adhesion strength of the film on the substrate. The setup is shown schematically in Fig. 13.14.

The adhesion strength of the spin-on, *in situ* and seeded-growth films were measured to be 111 ± 29 MPa, 324 ± 17 MPa and 801 ± 68 MPa, respectively. Hu *et al.* hypothesized that the *in situ* and seeded-growth films have higher adhesion strengths because they are fully crystalline and adhere to the substrate during the hydrothermal synthesis, whereas the spin-on films are only partially crystalline and are bonded to the substrate post-synthesis during the 400 °C annealing step. The initial seed layer deposited during the synthesis of the seeded-growth films may help to enhance the adhesion strength over the *in situ* films.

Gaynor studied the adhesion of a dense polycrystalline PSZ MFI film using stud pull measurements.[54] In these two tests, the silicon substrate shattered at forces of 35 and 51 MPa but the film was still attached to the wafer, thus allowing Gaynor to conclude that the adhesive/cohesive strength was at least 51 MPa.

5.7. Tribological properties

The friction and wear properties of hydrophilic and hydrophobic PSZ MEL spin-on films were studied by Johnson *et al.* using a nanoscratch technique with a conical diamond indenter tip (radius ~5 μm).[113] The as-synthesized films were hydrophilic, and some of the films were made hydrophobic through the TMCS vapour-phase silylation process described in Section 4. The coefficient of friction in ramped load tests for both films is the same up to a load of 500 μN, at which point it increases linearly for the hydrophilic film and remains constant at 0.17 for the hydrophobic film. The rate of plastic deformation increases dramatically at a critical load of 450 μN for the hydrophilic film, while the hydrophobic film did not have a significant increase in plastic deformation up to normal loads of 2000 μN. Furthermore, the hardness of the hydrophobic films was 61% higher than the hydrophilic films between indentation depths of 15 and 60 nm. These results imply that the hydrophobic films have greater wear resistance than the hydrophilic films, and the authors hypothesize that the methyl groups have a lubricating effect on the surface that reduces friction.

6. Hybrid Materials

Zeolite-polymer composites have garnered research interest because blending the two combines the unique properties of each material into one. For example, such composites have been studied as proton-exchange membranes in fuel cells[114–116] and as gas separation membranes.[117,118] Larlus *et al.* investigated a two-component film consisting of PSZ MFI and acryl latex for low-*k* applications.[64] First, one of three thin layers of PSZ MFI nanocrystals was deposited: discrete crystals of either 30 nm (S1) or 60 nm (S2) separated by centrifugation, or a film of nanoparticle suspension with no purification with particle sizes of about 100 nm (S3). After high temperature treatment at 450 °C, a 3 wt% solution of acrylate latex particles was spin-coated onto the PSZ MFI film. Combinations of either the centrifuged 30 nm crystal film or the unpurified 100 nm film with the latex particles were compared with the individual zeolite and polymer films, as well as a PSZ MFI film grown by a seeded method. The polymer material binds the zeolite together via the hydroxyl groups and decreases the surface roughness. The final *k* values of the PSZ MFI and polymer composite films are both 2.00, as measured by the refractive index, and they are lower than the individual zeolite, latex and seeded films. The authors hypothesized that the lower *k* values are due to the low preferred orientation of the zeolite crystals and the higher hydrophobicity that results from the binding of the hydroxyl groups.

7. Future Considerations and Conclusions

PSZ thin films have been shown to be promising candidates for new low-*k* materials. Their spin-on deposition process and silica chemical composition allow PSZs to be easily integrated into the semiconductor manufacturing process. A *k* value as low as 1.5 has been obtained, which meets the necessary requirements for a low-*k* material through 2020, as described by the International Technology Roadmap for Semiconductors 2006 Update—Interconnect.[2] Moreover, the mechanical properties of these particular films exceed the minimum threshold value of 6 GPa; most other porous silicas do not have the necessary mechanical reliability. Finally, several research groups have shown that PSZs can be further engineered to reduce the *k* value and provide hydrophobic functionality. Table 13.3 provides a summary of the *k*, *E*, and *H* values of the zeolite thin films and single crystals discussed in this chapter.

Table 13.3 *k*, *E* and *H* of zeolite films and crystals discussed in this chapter

Sample	Calcined[a] (Y/N)	*k*	Modulus (GPa)	Hardness (GPa)	Reference
PSZ MFI *in situ* (*b*-oriented)	Y	2.7	30–40	–	50
PSZ MFI *in situ* (*b*-oriented)	N	3.4	–	–	50
PSZ MFI seeded (*c*-oriented)	Y	3.1	43.4 ± 1.6–53.9 ± 0.5	4.3 ± 0.4	56, 108
PSZ MFI seeded (*c*-oriented)	N	–	45.0 ± 2.0	5.5 ± 0.3	108
PSZ MFI seeded	–	2.84	–	–	54
PSZ MFI spin-on (one stage)	Y	2.1	6.2 ± 1.5–18	0.5 ± 0.2	49, 87
PSZ MFI spin-on (one stage)	Y (UV-assisted curing)	2.16 ± 0.02	10.7 ± 3.2	1.1 ± 0.4	87
PSZ MFI spin-on (one-stage, with binder)	Y	~2.2–2.6	–	–	54
PSZ MFI spin-on (one-stage, with porogen)	Y	1.8	14.3	–	88
PSZ MFI spin-on (two-stage)	Y	1.6	–	–	57
PSZ MEL spin-on (two-stage)	Y	1.5	10.1 ± 1.9	0.36 ± 0.04	48, 107
PSZ BEA *in situ*	N	2.3	–	–	94
PSZ MFI/polymer composite	Y	2.00	–	–	64
PSZ MFI single crystal	Y	–	37.7 ± 2.3–39.7 ± 1.3	3.7 ± 0.2–4.1 ± 0.4	110
PSZ MFI single crystal	N	–	36.3 ± 8.3–57.4 ± 0.6	2.1 ± 0.6–7.2 ± 0.2	108, 110
PSZ FER single crystal	N	2.18	49.4 ± 1.1	–	56
CHA single crystal	Y	–	48.9 ± 1.2	–	56

[a] Calcination is by thermal treatment unless otherwise noted.
PSZ, pure-silica-zeolite.

Before they can be completely accepted as a suitable low-*k* alternative, other properties of PSZs must be fully characterized and optimized. These include, but are not limited to, thermal properties (e.g. coefficient of thermal expansion and thermal conductivity), electrical properties (e.g. breakdown voltage), and chemical compatibility (e.g. copper drift).

There are also other integration and characterization issues that researchers must address for PSZs. For example, low-*k* materials must withstand corrosion from several chemicals during the manufacturing process. The semiconductor industry often tests silica-based materials with HF, since this is an extremely corrosive substance to silicon dioxide, and therefore PSZs must be proven to survive HF attack for several minutes. One possibility for imparting HF resistance to PSZs is to increase the hydrophobicity of the material. Since HF (and many other corrosive chemicals) is diluted in water, protecting the zeolite from water may also render the material impervious to HF.

Although the micropores are uniform, the non-uniformity of mesopores of spin-on films may present a problem. If the mesopore size distribution varies from 2 to 50 nm,[48,57,87] the range is too large, and the synthesis solution or the deposition process must be modified to narrow the range of the mesopore size. A large pore distribution may result in non-uniform electrical, thermal and mechanical properties; pore interconnectivity; and etching difficulties, among other problems.

The large crystal sizes of PSZ spin-on films remain a concern. As the feature size decreases into the sub-65 nm range, etching patterns into a film that contains 70 nm crystals is not realistic. Ideally, the largest crystals should be 20% of the size of the critical dimension of the smallest feature.[54] Thus, for the 32 nm process, the crystal sizes would have to be about 6 nm. Although several groups have studied methods to reduce the crystal size,[89,90] none approach the level necessary for process integration. While pressurized, *in situ*, hydrothermal synthesis is not practical for low-*k* applications, a similar approach in which synthesis and crystal growth occurs around the chip features may be a possible solution.

Despite these challenges, PSZs are still a strong low-*k* alternative. The issues discussed above are *engineering* challenges, rather than *inherent* faults of zeolites. In other words, there has yet to be found an intrinsic 'killer' weakness of zeolites that would prevent them from process integration, and we predict that the issues above can be solved with clever engineering modifications. Indeed, PSZs offer some of the strongest mechanical properties of most low-*k* alternatives at ultra-low-*k* values below 2.0. Given the importance of mechanical reliability, PSZs are natural low-*k* materials that warrant exceptional future research and consideration.

ACKNOWLEDGEMENTS

We would like to acknowledge financial support over the past 9 years from NSF (Grant CTS-0404376), Advanced Micro Devices Inc., Intel Corporation, Asahi Kasei, Engelhard and UC Discovery Grant. We thank the former members of the Yan Group, Professor Zhengbao Wang, Professor Huanting Wang, Dr. Anupam Mitra, Dr. Limin Huang, Dr. Shuang Li, Dr. Zijian Li, Mr. Tiegang Cao, Ms. Dora

Medina and former members of the Wang Group, Mr. Mark Johnson and Dr. Lili Hu, for their contributions to the zeolite low-*k* research at UCR.

REFERENCES

[1] Moore, G. E., 'Cramming more components onto integrated circuits', *Electronics* **1965,** *38,* 114–117.
[2] International Technology Roadmap for Semiconductors - Interconnect, 2006 Update, **2006**.
[3] Maex, K., Baklanov, M. R., Shamiryan, D., Iacopi, F., Brongersma, S. H., Yanovitskaya, Z. S., 'Low dielectric constant materials for microelectronics', *J. Appl. Phys.* **2003,** *93,* 8793–8841.
[4] Morgen, M., Ryan, E. T., Zhao, J. H., Hu, C., Cho, T. H., Ho, P. S., 'Low dielectric constant materials for ULSI interconnects', *Annu. Rev. Mater. Sci.* **2000,** *30,* 645–680.
[5] Hougham, G., Tesoro, G., Shaw, J., 'Synthesis and properties of highly fluorinated polyimides', *Macromolecules* **1994,** *27,* 3642–3649.
[6] Martin, S. J., Godschalx, J. P., Mills, M. E., Shaffer, E. O., Townsend, P. H., 'Development of a low-dielectric-constant polymer for the fabrication of integrated circuit interconnect', *Adv. Mater.* **2000,** *12,* 1769–1778.
[7] Morgen, M., Ryan, E. T., Zhao, J. H., Hu, C. A., Cho, T. H., Ho, P. S., 'Low dielectric constant materials for advanced interconnects', *JOM-US* **1999,** *51,* 37–40.
[8] Lin, Q. H., Cohen, S. A., Gignac, L., Herbst, B., Klaus, D., Simonyi, E., Hedrick, J., Warlaumont, J., Lee, H. J., Wu, W. L., 'Low dielectric constant nanocomposite thin films based on silica nanoparticle and organic thermosets', *J. Polym. Sci. Pol. Phys.* **2007,** *45,* 1482–1493.
[9] Grill, A., Patel, V., 'Low dielectric constant films prepared by plasma-enhanced chemical vapor deposition from tetramethylsilane', *J. Appl. Phys.* **1999,** *85,* 3314–3318.
[10] Han, L. C. M., Pan, J. S., Chen, S. M., Balasubramanian, N., Shi, J. N., Wong, L. S., Foo, P. D., 'Characterization of carbon-doped SiO2 low k thin films - Preparation by plasma-enhanced chemical vapor deposition from tetramethylsilane', *J. Electrochem. Soc.* **2001,** *148,* F148–F153.
[11] Loboda, M. J., Seifferly, J. A., Dall, F. C., 'Plasma-enhanced chemical-vapor-deposition of A-SiC-H films from organosilicon precursors', *J. Vac. Sci. Technol. A* **1994,** *12,* 90–96.
[12] Naik, M., Parikh, S., Li, P., Educato, J., Cheung, D., Hashim, I., Hey, P., Jeng, S., Pan, T., Redeker, F., Rana, V., Tang, B., Yost, D., in: *Proceedings of the IEEE International Interconnect Technology Conference*, San Francisco, CA, May 24–26, **1999**.
[13] Xu, P., Huang, K., Patel, A., Rathi, S., Tang, B., Ferguson, J., Huang, J., Ngai, C., Loboda, M., in: *Proceedings of the IEEE International Interconnect Technology Conference,* San Francisco, CA, May 24–26, **1999**.
[14] Yamada, N., Takahashi, T., 'Methylsiloxane spin-on-glass films for low dielectric constant interlayer dielectrics', *J. Electrochem. Soc.* **2000,** *147,* 1477–1480.
[15] Su, R. Q., Muller, T. E., Prochazka, J., Lercher, J. A., 'A new type of low-kappa dielectric films based on polysilsesquioxanes', *Adv. Mater.* **2002,** *14,* 1369–1373.
[16] Hrubesh, L. W., Keene, L. E., Latorre, V. R., 'Dielectric-properties of aerogels', *J. Mater. Res.* **1993,** *8,* 1736–1741.
[17] Seraji, S., Wu, Y., Forbess, M., Limmer, S. J., Chou, T., Cao, G. Z., 'Sol-gel-derived mesoporous silica films with low dielectric constants', *Adv. Mater.* **2000,** *12,* 1695–1698.
[18] Doshi, D. A., Huesing, N. K., Lu, M. C., Fan, H. Y., Lu, Y. F., Simmons-Potter, K., Potter, B. G., Hurd, A. J., Brinker, C. J., 'Optically, defined multifunctional patterning of photosensitive thin-film silica mesophases', *Science* **2000,** *290,* 107–111.
[19] Landskron, K., Hatton, B. D., Perovic, D. D., Ozin, G. A., 'Periodic mesoporous organosilicas containing interconnected [Si(CH2)](3) rings', *Science* **2003,** *302,* 266–269.
[20] Pai, R. A., Humayun, R., Schulberg, M. T., Sengupta, A., Sun, J. N., Watkins, J. J., 'Mesoporous silicates prepared using preorganized templates in supercritical fluids', *Science* **2004,** *303,* 507–510.
[21] Zhao, D., Yang, P., Melosh, N., Feng, J., Chmelka, B. F., Stucky, G. D., 'Continuous mesoporous silica films with highly ordered large pore structures', *Adv. Mater.* **1998,** *10,* 1380–1385.
[22] Xu, G., He, J., Andideh, E., Bielefeld, J., Scherban, T., in: *Proceedings of the IEEE International Interconnect Technology Conference,* San Francisco, CA, June 3–5, **2002**.

[23] Kapteijn, F., Bakker, W. J. W., Vandegraaf, J., Zheng, G., Poppe, J., Moulijn, J. A., 'Permeation and separation behavior of a silicalite-1 membrane', *Catal. Today* **1995,** *25,* 213–218.
[24] Nishiyama, N., Ueyama, K., Matsukata, M., 'A defect-free mordenite membrane synthesized by vapor-phase transport method', *J. Chem. Soc., Chem. Commun.* **1995,** *19,* 1967–1968.
[25] Sano, T., Yanagishita, H., Kiyozumi, Y., Mizukami, F., Haraya, K., 'Separation of ethanol-water mixture by silicalite membrane on pervaporation', *J. Membrane Sci.* **1994,** *95,* 221–228.
[26] Tuan, V. A., Li, S. G., Falconer, J. L., Noble, R. D., 'Separating organics from water by pervaporation with isomorphously-substituted MFI zeolite membranes', *J. Membrane Sci.* **2002,** *196,* 111–123.
[27] Vroon, Z., Keizer, K., Gilde, M. J., Verweij, H., Burggraaf, A. J., 'Transport properties of alkanes through ceramic thin zeolite MFI membranes', *J. Membrane Sci.* **1996,** *113,* 293–300.
[28] Xomeritakis, G., Tsapatsis, M., 'Permeation of aromatic isomer vapors through oriented MFI-type membranes made by secondary growth', *Chem. Mater.* **1999,** *11,* 875–878.
[29] Salomon, M. A., Coronas, J., Menendez, M., Santamaria, J., 'Synthesis of MTBE in zeolite membrane reactors', *Appl. Catal. A-Gen.* **2000,** *200,* 201–210.
[30] Tanaka, K., Yoshikawa, R., Ying, C., Kita, H., Okamoto, K., 'Application of zeolite T membrane to vapor-permeation-aided esterification of lactic acid with ethanol', *Chem. Eng. Sci.* **2002,** *57,* 1577–1584.
[31] Wan, Y. S. S., Chau, J. L. H., Gavriilidis, A., Yeung, K. L., 'Design and fabrication of zeolite-based microreactors and membrane microseparators', *Micropor. Mesopor. Mat.* **2001,** *42,* 157–175.
[32] Yawalkar, A. A., Pangarkar, V. G., Baron, G. V., 'Alkene epoxidation with peroxide in a catalytic membrane reactor: a theoretical study', *J. Membrane Sci.* **2001,** *182,* 129–137.
[33] Jansen, J. C., Koegler, J. H., van Bekkum, H., Calis, H. P. A., van den Bleek, C. M., Kapteijn, F., Moulijn, J. A., Geus, E. R., van der Puil, N., 'Zeolitic coatings and their potential use in catalysis', *Micropor. Mesopor. Mat.* **1998,** *21,* 213–226.
[34] Li, J. Q., Dong, J. X., Liu, G. H., Stud. Surf. Sci. Catal. **1994,** *90,* 327–331.
[35] Li, J. Q., Dong, J. X., Liu, G. H., Shi, Y. J., Cao, J. H., Xu, W. Y., Wu, F., 'Preparation and properties of zeolite coating on metal-surface', *React. Kinet. Catal. L.* **1992,** *47,* 287–291.
[36] Louis, B., Kiwi-Minsker, L., Reuse, P., Renken, A., 'ZSM-5 coatings on stainless steel grids in one-step benzene hydroxylation to phenol by N2O: Reaction kinetics study', *Ind. Eng. Chem. Res.* **2001,** *40,* 1454–1459.
[37] Yan, Y. G., Bein, T., 'Molecular recognition on acoustic-wave devices - sorption in chemically anchored zeolite monolayers', *J. Phys. Chem.* **1992,** *96,* 9387–9393.
[38] Yan, Y. G., Bein, T., 'Molecular-sieve sensors for selective ethanol detection', *Chem. Mater.* **1992,** *4,* 975–977.
[39] Tatlier, M., Erdem-Senatalar, A., Stud. Surf. Sci. Catal. **1999,** *125,* 101–108.
[40] Hillhouse, H. W., Tuominen, M. T., 'Modeling the thermoelectric transport properties of nanowires embedded in oriented microporous and mesoporous films', *Micropor. Mesopor. Mat.* **2001,** *47,* 39–50.
[41] McDonnell, A. M. P., Beving, D., Wang, A. J., Chen, W., Yan, Y. S., 'Hydrophilic and antimicrobial zeolite coatings for gravity-independent water separation', *Adv. Funct. Mater.* **2005,** *15,* 336–340.
[42] O'Neill, C., Beving, D. E., Chen, W., Yan, Y. S., 'Durability of hydrophilic and antimicrobial zeolite coatings under water immersion', *AIChE J.* **2006,** *52,* 1157–1161.
[43] Munoz, R. A., Beving, D., Yan, Y. S., 'Hydrophilic zeolite coatings for improved heat transfer', *Ind. Eng. Chem. Res.* **2005,** *44,* 4310–4315.
[44] Beving, D. E., McDonnell, A. M. P., Yang, W. S., Yan, Y. S., 'Corrosion resistant high-silica-zeolite MFI coating - One general solution formulation for aluminum alloy AA-2024-T3, AA-5052-H32, AA-6061-T4, and AA-7075-T6', *J. Electrochem. Soc.* **2006,** *153,* B325–B329.
[45] Cheng, X. L., Wang, Z. B., Yan, Y. S., 'Corrosion-resistant zeolite coatings by in situ crystallization', *Electrochem. Solid St. Lett.* **2001,** *4,* B23–B26.
[46] Mitra, A., Wang, Z. B., Cao, T. G., Wang, H. T., Huang, L. M., Yan, Y. S., 'Synthesis and corrosion resistance of high-silica zeolite MTW, BEA, and MFI coatings on steel and aluminum', *J. Electrochem. Soc.* **2002,** *149,* B472–B478.

[47] Holmberg, B. A., Hwang, S. J., Davis, M. E., Yan, Y. S., 'Synthesis and proton conductivity of sulfonic acid functionalized zeolite BEA nanocrystals', *Micropor. Mesopor. Mat.* **2005,** *80,* 347–356.

[48] Li, Z. J., Lew, C. M., Li, S., Medina, D. I., Yan, Y. S., 'Pure-silica-zeolite MEL low-k films from nanoparticle suspensions', *J. Phys. Chem. B* **2005,** *109,* 8652–8658.

[49] Wang, Z. B., Mitra, A. P., Wang, H. T., Huang, L. M., Yan, Y. S., 'Pure silica zeolite films as low-k dielectrics by spin-on of nanoparticle suspensions', *Adv. Mater.* **2001,** *13,* 1463–1466.

[50] Wang, Z. B., Wang, H. T., Mitra, A., Huang, L. M., Yan, Y. S., 'Pure-silica zeolite low-k dielectric thin films', *Adv. Mater.* **2001,** *13,* 746–749.

[51] Yan, Y. S., Wang, Z. B., International Patent WO 02/07191 A3, **2002**.

[52] Li, S., Li, Z. J., Bozhilov, K. N., Chen, Z. W., Yan, Y. S., 'TEM investigation of formation mechanism of monocrystal-thick b-oriented pure silica zeolite MFI film', *J. Am. Chem. Soc.* **2004,** *126,* 10732–10737.

[53] Sterte, J., Mintova, S., Zhang, G., Schoeman, B. J., 'Thin molecular sieve films on noble metal substrates', *Zeolites* **1997,** *18,* 387–390.

[54] Gaynor, J. F. U.S. Patent 6,329,062, **2001**.

[55] Persson, A. E., Schoeman, B. J., Sterte, J., Otterstedt, J. E., 'The Synthesis of Discrete Colloidal Particles of TPA-Silicalite-1', *Zeolites* **1994,** *14,* 557–567.

[56] Li, Z. J., Johnson, M. C., Sun, M. W., Ryan, E. T., Earl, D. J., Maichen, W., Martin, J. I., Li, S., Lew, C. M., Wang, J., Deem, M. W., Davis, M. E., Yan, Y. S., 'Mechanical and dielectric properties of pure-silica-zeolite low-k materials', *Angew. Chem. Int. Ed.* **2006,** *45,* 6329–6332.

[57] Li, Z. J., Li, S., Luo, H. M., Yan, Y. S., 'Effects of crystallinity in spin-on pure-silica-zeolite MFI low-dielectric-constant films', *Adv. Funct. Mater.* **2004,** *14,* 1019–1024.

[58] Flanigen, E. M., Bennett, J. M., Grose, R. W., Cohen, J. P., Patton, R. L., Kirchner, R. M., Smith, J. V., 'Silicalite, a New Hydrophobic Crystalline Silica Molecular-Sieve', *Nature* **1978,** *271,* 512–516.

[59] Li, S., Wang, X., Beving, D., Chen, Z. W., Yan, Y. S., 'Molecular sieving in a nanoporous b-oriented pure-silica-zeolite MFI monocrystal film', *J. Am. Chem. Soc.* **2004,** *126,* 4122–4123.

[60] Grose, R. W., Flanigen, E. M. U.S. Patent 4,061,724, **1975**.

[61] Mintova, S., Olson, N. H., Valtchev, V., Bein, T., 'Mechanism of zeolite a nanocrystal growth from colloids at room temperature', *Science* **1999,** *283,* 958–960.

[62] Otterstedt, J. E., Sterte, J., Schoeman, B. J., International Patent WO 94/05597, **1994**.

[63] Ho, P. S., Lee, W. W., Leu, J. J. *Overview on Low Dielectric Constant Materials for IC Applications,* "Low dielectric constant materials for IC applications", Vol. 9. Springer, New York, **2003**.

[64] Larlus, O., Mintova, S., Valtchev, V., Jean, B., Metzger, T. H., Bein, T., 'Silicalite-1/polymer films with low-k dielectric constants', *Appl. Surf. Sci.* **2004,** *226,* 155–160.

[65] Dinkel, J. A., 'Universal Capacitance Probe Liquid Level Measuring System', *Rev. Sci. Instrum.* **1966,** *37,* 1549–1554.

[66] Fellner-Feldegg, H., 'Measurement of Dielectrics in Time Domain', *J. Phys. Chem.* **1969,** *73,* 616–623.

[67] Topp, G. C., Davis, J. L., Annan, A. P., 'Electromagnetic Determination of Soil-Water Content - Measurements in Coaxial Transmission-Lines', *Water. Resour. Res.* **1980,** *16,* 574–582.

[68] Keller, W. C., 'Permittivity Measurements Using Spherical Sample Microwave Scattering Technique', *Rev. Sci. Instrum.* **1966,** *37,* 1211–1213.

[69] Bleaney, B., Loubser, J. H. N., Penrose, R. P., 'Cavity Resonators for Measurements with Centimetre Electromagnetic Waves', *Proc. Phys. Soc. (London)* **1947,** *59,* 185–199.

[70] Jen, C. K., 'A Method for Measuring the Complex Dielectric Constant of Gases at Microwave Frequencies by Using a Resonant Cavity', *J. Appl. Phys.* **1948,** *19,* 649–653.

[71] Horvath, G., Kawazoe, K., 'Method for the calculation of effective pore-size distribution in molecular-sieve carbon', *J. Chem. Eng. Jpn.* **1983,** *16,* 470–475.

[72] Saito, A., Foley, H. C., 'Argon porosimetry of selected molecular-sieves - experiments and examination of the adapted Horvath-Kawazoe model', *Micropor. Mater.* **1995,** *3,* 531–542.

[73] Barrett, E. P., Joyner, L. G., Halenda, P. P., 'The determination of pore volume and area distributions in porous substances. 1. Computations from nitrogen isotherms', *J. Am. Chem. Soc.* **1951,** *73,* 373–380.

[74] Olson, D. H., Kokotailo, G. T., Lawton, S. L., Meier, W. M., 'Crystal-structure and structure-related properties of ZSM-5', *J. Phys. Chem.* **1981,** *85,* 2238–2243.
[75] Dull, T. L., Frieze, W. E., Gidley, D. W., Sun, J. N., Yee, A. F., 'Determination of pore size in mesoporous thin films from the annihilation lifetime of positronium', *J. Phys. Chem. B* **2001,** *105,* 4657–4662.
[76] Eldrup, M., Lightbody, D., Sherwood, J. N., 'The temperature-dependence of positron lifetimes in solid pivalic acid', *Chem. Phys.* **1981,** *63,* 51–58.
[77] Gidley, D. W., Frieze, W. E., Dull, T. L., Sun, J., Yee, A. F., Nguyen, C. V., Yoon, D. Y., 'Determination of pore-size distribution in low-dielectric thin films', *Appl. Phys. Lett.* **2000,** *76,* 1282–1284.
[78] Gidley, D. W., Frieze, W. E., Dull, T. L., Yee, A. F., Ryan, E. T., Ho, H. M., 'Positronium annihilation in mesoporous thin films', *Phys. Rev. B* **1999,** *60,* R5157–R5160.
[79] Petkov, M. P., Wang, C. L., Weber, M. H., Lynn, K. G., Rodbell, K. P., 'Positron annihilation techniques suited for porosity characterization of thin films', *J. Phys. Chem. B* **2003,** *107,* 2725–2734.
[80] Sun, J. N., Gidley, D. W., Hu, Y., Frieze, W. E., Ryan, E. T., 'Depth-profiling plasma-induced densification of porous low-k thin films using positronium annihilation lifetime spectroscopy', *Appl. Phys. Lett.* **2002,** *81,* 1447–1449.
[81] Tao, S. J., 'Positronium annihilation in molecular substances', *J. Chem. Phys.* **1972,** *56,* 5499–5510.
[82] Wang, Y. Y., Nakanishi, H., Jean, Y. C., Sandreczki, T. C., 'Positron-annihilation in amine-cured epoxy polymers - pressure-dependence', *J. Polym. Sci. Pol. Phys.* **1990,** *28,* 1431–1441.
[83] Baklanov, M. R., Mogilnikov, K. P., Polovinkin, V. G., Dultsev, F. N., 'Determination of pore size distribution in thin films by ellipsometric porosimetry', *J. Vac. Sci. Technol. B* **2000,** *18,* 1385–1391.
[84] Wu, W. L., Wallace, W. E., Lin, E. K., Lynn, G. W., Glinka, C. J., Ryan, E. T., Ho, H. M., 'Properties of nanoporous silica thin films determined by high-resolution x-ray reflectivity and small-angle neutron scattering', *J. Appl. Phys.* **2000,** *87,* 1193–1200.
[85] Li, S., Sun, J. N., Li, Z. J., Peng, H. G., Gidley, D., Ryan, E. T., Yan, Y. S., 'Evaluation of pore structure in pure silica zeolite MFI low-k thin films using positronium annihilation lifetime spectroscopy', *J. Phys. Chem. B* **2004,** *108,* 11689–11692.
[86] Tanaka, H. K. M., Kurihara, T., Mills, A. P., 'Evaluation of the diffusion barrier continuity on porous low-k films using positronium time of flight spectroscopy', *Phys. Rev. B* **2005,** *72,* 195006.
[87] Eslava, S., Iacopi, F., Baklanov, M. R., Kirschhock, C. E. A., Maex, K., Martens, J. A., 'Ultraviolet-Assisted Curing of Polycrystalline Pure-Silica Zeolites: Hydrophobization, Functionalization, and Cross-Linking of Grains', *J. Am. Chem. Soc.* **2007,** *129,* 9288–9289.
[88] Li, S., Li, Z. J., Yan, Y. S., 'Ultra-low-k pure-silica zeolite MFI films using cyclodextrin as porogen', *Adv. Mater.* **2003,** *15,* 1528–1531.
[89] Song, W., Grassian, V. H., Larsen, S. C., 'High yield method for nanocrystalline zeolite synthesis', *Chem. Commun.* **2005,** *23,* 2951–2953.
[90] Lew, C. M., Li, Z. J., Zones, S. I., Sun, M. W., Yan, Y. S., 'Control of size and yield of pure-silica-zeolite MFI nanocrystals by addition of methylene blue to the synthesis solution', *Micropor. Mesopor. Mat.* **2007,** *105,* 10–14.
[91] Li, Q., Creaser, D., Sterte, J., 'The nucleation period for TPA-silicalite-1 crystallization determined by a two-stage varying-temperature synthesis', *Micropor. Mesopor. Mat.* **1999,** *31,* 141–150.
[92] Zheng, N. F., Bu, X. G., Wang, B., Feng, P. Y., 'Microporous and photoluminescent chalcogenide zeolite analogs', *Science* **2002,** *298,* 2366–2369.
[93] Corma, A., Rey, F., Rius, J., Sabater, M. J., Valencia, S., 'Supramolecular self-assembled molecules as organic directing agent for synthesis of zeolites', *Nature* **2004,** *431,* 287–290.
[94] Mitra, A., Cao, T. G., Wang, H. T., Wang, Z. B., Huang, L. M., Li, S., Li, Z. J., Yan, Y. S., 'Synthesis and evaluation of pure-silica-zeolite BEA as low dielectric constant material for microprocessors', *Ind. Eng. Chem. Res.* **2004,** *43,* 2946–2949.
[95] Mintova, S., Reinelt, M., Metzger, T. H., Senker, J., Bein, T., 'Pure silica BETA colloidal zeolite assembled in thin films', *Chem. Commun.* **2003,** *3,* 326–327.

[96] Li, S., Li, Z. J., Medina, D., Lew, C., Yan, Y. S., 'Organic-functionalized pure-silica-zeolite MFI low-k films', *Chem. Mater.* **2005,** *17,* 1851–1854.
[97] Xu, L. B., Chen, W., Mulchandani, A., Yan, Y. S., 'Reversible conversion of conducting polymer films from superhydrophobic to superhydrophilic', *Angew. Chem. Int. Ed.* **2005,** *44,* 6009–6012.
[98] Baklanov, M. R., Mogilnikov, K. P., Le, Q. T., 'Quantification of processing damage in porous low dielectric constant films', *Microelectron. Eng.* **2006,** *83,* 2287–2291.
[99] Nix, W. D., 'Mechanical-Properties of Thin-Films', *Metall. Mater. Trans. A* **1989,** *20,* 2217–2245.
[100] Read, D. T., Dally, J. W., 'A New Method for Measuring the Strength and Ductility of Thin-Films', *J. Mater. Res.* **1993,** *8,* 1542–1549.
[101] Weihs, T. P., Hong, S., Bravman, J. C., Nix, W. D., 'Mechanical Deflection of Cantilever Microbeams - a New Technique for Testing the Mechanical-Properties of Thin-Films', *J. Mater. Res.* **1988,** *3,* 931–942.
[102] Elbrecht, L., Storm, U., Catanescu, R., Binder, J., 'Comparison of stress measurement techniques in surface micromachining', *J. Micromech. Microeng.* **1997,** *7,* 151–154.
[103] Xiao, X., Hata, N., Yamada, K., Kikkawa, T., 'Mechanical property determination of thin porous low-k films by twin-transducer laser generated surface acoustic waves', *Jpn. J. Appl. Phys. 1* **2004,** *43,* 508–513.
[104] Oliver, W. C., Pharr, G. M., 'An Improved Technique for Determining Hardness and Elastic-Modulus Using Load and Displacement Sensing Indentation Experiments', *J. Mater. Res.* **1992,** *7,* 1564–1583.
[105] Shen, L., Zeng, K. Y., 'Comparison of mechanical properties and non-porous low-k dielectric of porous films', *Microelectron. Eng.* **2004,** *71,* 221–228.
[106] Bushan, B. Introduction-Measurement Techniques and Applications, *Handbookof Micro/Nanotribology,* CRC Press: Boca Raton, FL, 1996.
[107] Johnson, M., Li, Z. J., Wang, J. L., Yan, Y. S., 'Mechanical characterization of zeolite low dielectric constant thin films by nanoindentation', *Thin Solid Films* **2007,** *515,* 3164–3170.
[108] Johnson, M. C., Wang, J. L., Li, Z. J., Lew, C. M., Yan, Y. S., 'Effect of Calcination and Polycrystallinity on Mechanical Properties of Nanoporous MFI Zeolites', *Mater. Sci. Eng. A* **2007,** *456,* 58–63.
[109] Wang, Z. M., Lobo, R. F., Lambros, J., 'The mechanical properties of siliceous ZSM-5 (MFI) crystals', *Micropor. Mesopor. Mat.* **2003,** *57,* 1–7.
[110] Brabec, L., Bohac, P., Stranyanek, M., Ctvrtlik, R., Kocirik, M., 'Hardness and elastic modulus of silicalite-1 crystal twins', *Micropor. Mesopor. Mat.* **2006,** *94,* 226–233.
[111] Fan, H. Y., Hartshorn, C., Buchheit, T., Tallant, D., Assink, R., Simpson, R., Kisse, D. J., Lacks, D. J., Torquato, S., Brinker, C. J., 'Modulus-density scaling behaviour and framework architecture of nanoporous self-assembled silicas', *Nat. Mater.* **2007,** *6,* 418–423.
[112] Hu, L. L., Wang, J. L., Li, Z. J., Li, S., Yan, Y. S., 'Interfacial adhesion of nanoporous zeolite thin films', *J. Mater. Res.* **2006,** *21,* 505–511.
[113] Johnson, M. C., Lew, C. M., Yan, Y. S., Wang, J. L., 'Hydrophobicity-dependent friction and wear of spin-on zeolite thin films', *Scripta Mater.* **2008,** *58,* 41–44.
[114] Chen, Z. W., Holmberg, B., Li, W. Z., Wang, X., Deng, W. Q., Munoz, R., Yan, Y. S., 'Nafion/zeolite nanocomposite membrane by in situ crystallization for a direct methanol fuel cell', *Chem. Mater.* **2006,** *18,* 5669–5675.
[115] Libby, B., Smyrl, W. H., Cussler, E. L., 'Polymer-zeolite composite membranes for direct methanol fuel cells', *AIChE J.* **2003,** *49,* 991–1001.
[116] Sancho, T., Soler, J., Pina, M. P., 'Conductivity in zeolite-polymer composite membranes for PEMFCs', *J. Power Sources* **2007,** *169,* 92–97.
[117] Husain, S., Koros, W. J., 'Mixed matrix hollow fiber membranes made with modified HSSZ-13 zeolite in polyetherimide polymer matrix for gas separation', *J. Membrane Sci.* **2007,** *288,* 195–207.
[118] Jiang, L. Y., Chung, T. S., Kulprathipanja, S., 'Fabrication of mixed matrix hollow fibers with intimate polymer-zeolite interface for gas separation', *AIChE J.* **2006,** *52,* 2898–2908.

CHAPTER 14

Highly Selective Zeolite Membranes

Tina M. Nenoff *and* Junhang Dong

Contents

Abstract

Much effort has recently been devoted to the synthesis and potential application of inorganic membranes in the domains of gas separation, pervaporation, reverse osmosis, or in the development of chemical sensors and catalytic membranes. Zeolite membranes, in particular, combine pore size and shape selectivity with the inherent mechanical, thermal and chemical stability necessary for continuous long-term separation processes. A variety of methods have been employed to produce high-quality zeolite membranes for a variety of separations applications. The integrity of the membrane, the choice of framework type and the ability to modify the surface of the zeolites allows for fine tuning of the selectivity of membranes. Both inorganic and organic surface modifications of the zeolite membranes are described in full. This, combined with optimization of process parameters, results in improved selectivities. Results presented show that excellent separations can be achieved for a variety of light gases, hydrocarbon mixtures and reverse osmosis processes.

Keywords: zeolite membranes, p-Xylene purification, CO_2 sequestration, catalytic zeolite membrane reactors, RO membrane

Ordered Porous Solids
DOI: 10.1016/B978-0-444-53189-6.00014-7

1. Introduction to Zeolite Membranes

Much effort has recently been devoted to the synthesis and potential application of inorganic membranes in the domains of gas separation, pervaporation, reverse osmosis, or in the development of chemical sensors and catalytic membranes.[1] Inorganic membranes, which have good thermal stability and chemical inertness, have advantages over polymer membranes for many industrial applications. Improved membrane integrity and manufacturing costs are constant factors, which are the focus of many research efforts. Zeolite membranes, in particular, combine pore size and shape selectivity with the inherent mechanical, thermal and chemical stability necessary for continuous long-term separation processes. The effective pore size distribution of the zeolite membrane, and hence its separation performance, is intrinsically governed by the choice of the zeolitic phase(s). This applies when molecular size exclusion sieving is dominant and no other diffusion pathways bypass the network of well-defined zeolitic pores/channels; otherwise, viscous flow through grain boundaries prevails. The optimum thickness of the zeolite film is always a compromise between separation performance and overall trans-membrane flux and is often tailored to the specific needs of the envisioned application.

Zeolites are crystalline inorganic framework structures that have uniform, molecular-sized pores. They have been used extensively as bulk catalysts and adsorbents. The zeolite structure is made up of TO_2 units, with T = a tetrahedral framework atom (Si, Al, B, Ge, etc.). In all cases other than neutral silica zeolite frameworks, the net overall charge of the framework is negative and is charge balanced by cations (either inorganic or organic). The cations reside in the pores of the framework; the size of the pore is categorized by the number of T atoms in that ring. Small-pore zeolites include those structures made up of eight-member oxygen rings, medium-pore zeolites have 10-member rings and large-pore zeolites have 12-member rings.[2] More recently, membranes of continuous polycrystalline zeolite layers have been deposited on porous supports. As described in the review by Bowen *et al.*, the first zeolite membranes were reported in 1987[3,6]. Since then, significant progress has been made to expand the types of zeolites utilized in membranes, improve membrane quality and widen their range of applications. Today, a large number of zeolite framework structures have been used as H_2 selective separation membranes including MFI,[4–12] LTA,[13–15] MOR[16–18] and FAU.[19–22] The commonly used zeolite is the MFI structure that is typically used in zeolite membranes because of its pore size and ease of preparation, and this structure includes silicalite-1 (pure silica) and ZSM-5 (aluminosilicate).[6]

Significant progress has been made in developing new membranes, their synthetic preparation and understanding transport and separation fundamentals over the last decade. Several reviews of zeolite membranes[2,23–37] have focused mainly on membrane synthesis and gas separation application. This progress suggests that many applications of zeolite membranes in commercially valuable enterprises, like separations, are promising. Gas and liquid separation on zeolite membranes is primarily

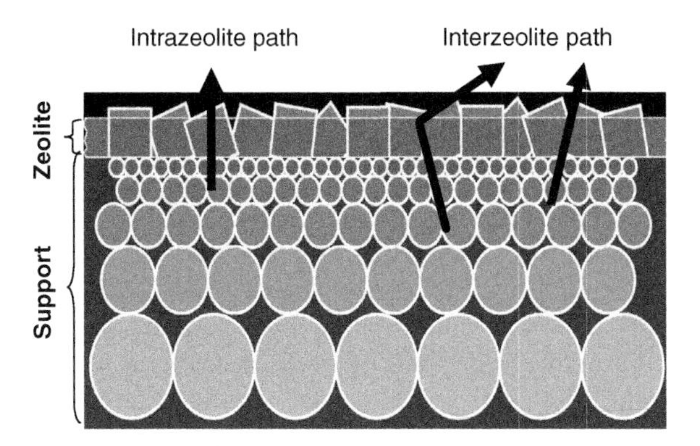

Figure 14.1 Not-to-Scale representation of a zeolite membrane on a non-selective oxide support (i.e., Al_2O_3), showing possible permeation pathways, either inter-zeolite crystals or intra-zeolite crystals. Reprinted with permission from Ref. [1]. Copyright (2007), American Chemical Society.

governed by competitive adsorption and diffusion mechanisms. When the zeolite pore size distribution falls between the molecular sizes of the feed components, a size exclusion mechanism can dominate the separation process.[38,39] However, one of the main challenges in zeolite membrane development is the minimization of inter-crystal pores formed inherently in polycrystalline zeolite films (see Fig. 14.1). The existence of inter-crystal pores with sizes larger than the zeolitic pores is the major cause for decline in molecular separation efficiency.[38] The elimination of inter-crystalline pores is essential for having high separation factors viable for industrial applications. Currently, research is on-going to resolve the inter-crystalline diffusion path issue by using mixed matrix membranes. Readers are directed to references 6, 40 and 41 for further information on this area.

2. Zeolite Membrane Growth Methods

Syntheses of MFI membranes, described recently, can be broadly classified into two categories: *in situ* and secondary (or seeded) growth.[24,42,43] In the *in situ* technique, the support surface is directly contacted with an alkaline solution containing the zeolite precursors and subjected to hydrothermal conditions. Under appropriate conditions, zeolite crystals nucleate on the support and grow to form a continuous zeolite layer. At the same time, reactions occurring in the solution lead to deposition of nuclei and crystals on the surface followed by their incorporation into the membrane, thus minimizing inter-crystal pore contributions. MFI films grown *in situ* may exhibit a preferred orientation that depends on the synthetic protocol and associated interplay of nucleation and growth phenomena.[42] However, because of the insufficient understanding of nucleation and growth processes in hydrothermal systems, the success of *in situ* methods in yielding uniformly oriented MFI films is limited.

In the secondary (or seeded) growth technique, zeolite nucleation is largely decoupled from zeolite growth by depositing a layer of zeolite seed crystals on the support surface prior to membrane growth. The layer of seed crystals can be deposited with precise control over which crystallographic axis is oriented perpendicular to the support (see Fig. 14.2).[38] The seeded surface is then exposed to the membrane growth solution and hydrothermal conditions, whereupon the seed crystals grow into a continuous film. Although this method offers greater flexibility in controlling the orientation of the zeolite crystals and the micro-structure of the zeolite membrane, it is done so at the expense of additional processing steps. In principle, the orientation and morphology of the membrane can be manipulated by changing the morphology and orientation of the deposited seed layer and then performing secondary growth under appropriate conditions.

3. Permeation and Gas Transport

The following five-step model can be used to describe the gas-molecule transport through a zeolite membrane:[44,45] (1) adsorption from bulk phase to zeolite external surface, (2) diffusion from surface to inside of zeolite channels, (3) diffusion inside the zeolite channels, (4) diffusion from the zeolite channel to the external surface and (5) desorption from the external surface to the gas phase. The actual mechanism of gas permeation through an MFI-type zeolite membrane depends on the gas adsorption properties on the zeolite. For non-adsorbing gases, molecules may directly enter the zeolite pores from the gas phase. The separation factor of a non-adsorbing gas mixture is determined by the mobility of the molecules inside the zeolite pores and the probability of the molecules entering the zeolitic pores.[8] Gas molecules with small size and high mobility tend to permeate through the zeolite membrane, while those with larger size and lower mobility tend not to permeate. For strongly adsorbing gases, permeation through an MFI membrane is controlled by either adsorption or activated diffusion, or both, depending on the operation conditions (temperature and pressure) on both sides of the membrane. The maximum value of flux with respect to temperature can be observed for strongly adsorbing gases, because the apparent activation energy is the sum of adsorption heat (negative) and diffusion activation energy (positive).[9,45] The temperature of maximum flux increases with the adsorption strength of the substance.[4] For permeation of binary gas mixtures, when both components are non-adsorbing, the separation factor is α

$$\alpha = \frac{(\gamma_1/\gamma_2)_p}{(\gamma_1/\gamma_2)_f} \qquad (1)$$

The separation factor, α, is defined as the enrichment factor of one component in the permeate as compared to the feed composition ratio, where γ_1 and γ_2 are the mole fractions of components 1 and 2, respectively, and subscripts p and f refer to the permeate side and feed side, respectively [see Eq. (1)].[45] When one or two strongly

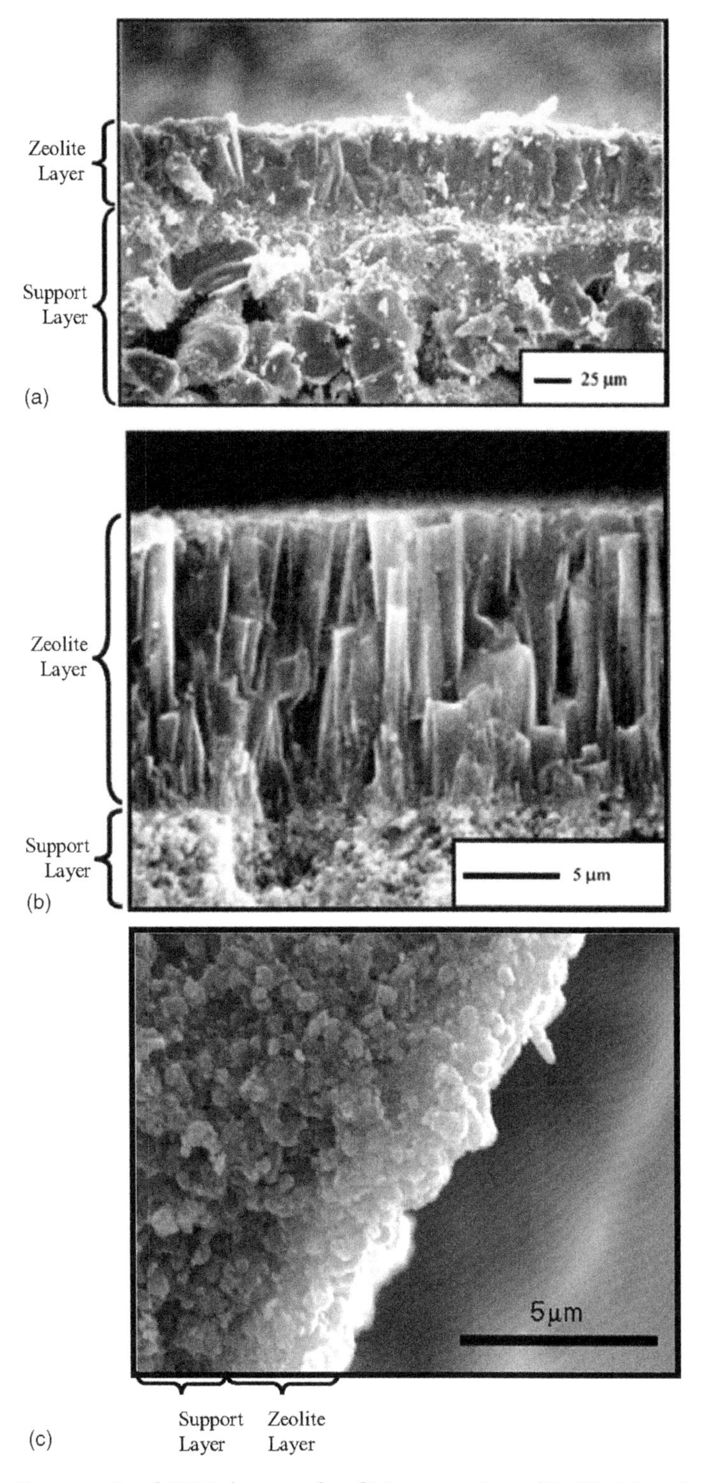

Figure 14.2 Cross-sectional SEM photographs of (a) a non-oriented B-ZSM-5 zeolite membrane on an α-Al_2O_3 coated SiC porous support (b) an oriented silicalite-1 membrane on silica coated α-Al_2O_3 porous support; (c) a non-oriented silicalite-1 MFI on tubular α-Al_2O_3 substrates (Pall Corp., New York, USA). Reprinted with permission from Ref. [1]. Copyright (2007), American Chemical Society.

adsorbing components are involved, there is no correlation between the permselectivity and the separation factor. For gas mixtures containing strongly adsorbing components, the separation factor strongly depends on the operation conditions, that is, temperature and pressure.[45] Molecular simulations and Maxwell–Stefan (M–S) modelling of multi-component diffusion through zeolite pores indicate that in some mixtures, slower larger molecules inhibit the diffusion of faster smaller molecules.[46,47] In addition, detailed studies by several groups have been reported over the last 5 years[48–52] and reviewed by Sholl last year.[53]

4. Defect-Sites Diffusion/Non-Zeolitic Pores

Polycrystalline zeolite membranes contain transport pathways within the intercrystalline regions, or non-zeolite pores. As described by Bowen *et al.*, the synthesis procedure, type of zeolite, and the calcination conditions affect the number and size of non-zeolite pores.[6] There is a different adsorption between molecules that interact with zeolite pores and those that interact with the non-zeolite pores. However, that difference is difficult to quantify due to variations in size and shape. Usually, only non-zeolite pores that are larger than the zeolite pores are considered, but non-zeolite pores have a size distribution, and pores smaller than the zeolite pores may also affect flux and selectivity.[6] Transport through non-zeolite pores that are larger than zeolite pores has contributions from both surface diffusion and Knudsen diffusion, and might also have viscous flow contributions. Knudsen diffusion requires that the pores are smaller than the mean free path of the diffusing molecules,[54] whereas viscous flow requires a pressure gradient across the membrane and sufficient interactions between diffusing molecules that their motions are driven by the pressure gradient.[55]

5. Thin Films

Recent advances in preparing thin zeolite membranes have dramatically increased gas permeation fluxes while maintaining good selectivities. Recently, ultra thin silicalite-1 membranes with a thickness of 0.5 μm were made and had light gas fluxes that are one to two orders of magnitude higher than other silicalite-1 membranes reported in the literature.[5] In another report, Lai *et al.* [38] prepared 1-μm thick oriented silicalite-1 membranes that performed significantly better for xylene isomer gas-phase separations than previously reported membranes. They obtained *p*-/*o*-xylene separation factors as high as 500 with a permeance of 2×10^{-7} mol/m^2 s Pa at 200 °C.

6. Zeolite Membrane Modification

In an effort to further improve zeolite membranes surface modification techniques have been developed by a number of research groups.[56] The majority of the techniques are post-treatment methods that include: inorganic silylation to decrease pore size[57] and to increase hydrophobicity[58,59] and defect treatments to fill non-zeolite pores by chemical vapour deposition (CVD),[39] atomic layer deposition (ALD)[60] or coking. [61,62] Recently, we reported on a new method of online membrane modification by carbonization of 1,3,5-triisopropylbenzene in the feed stream was found to be effective for reducing the MFI inter-crystalline pores and improving the PX separation (see Fig. 14.2c and PX separations section below).[63]

The silylation method of modifying the effective pore opening of a zeolite was first reported by Masuda *et al.*[57] In this method, methyldiethoxysilane (MDES) compounds are pre-adsorbed on active sites within the MFI zeolite, and then catalytically cracked leaving coke that contains Si atoms on the active sites. After calcination, mono SiO_2 units are formed on active sites, thereby reducing the size of the pores. After the silation modification of the membrane, a mixture of varying gas ratios of H_2/N_2 was tested (fraction of H_2 in retentate gas: $H_2/(H_2+N_2) = 0.2–0.8$; 110 °C, 101.9 kPa steady state pressure). The separation factor of H_2 was calculated at 90–140 for the treated membrane. This is about 50 times larger than that of the fresh membrane (1.5–4.5). Similar results were obtained for mixture gases of H_2 and O_2 (separation factor = 110–120).[57]

This method was borrowed and applied[64] to modification of B-ZSM-5 and SAPO-34 whose pores are ~0.4 nm and, thus, too small for the silyation compound to penetrate. The MDES reacted in the B-ZSM-5 pores and reduced their effective pore diameter, and their H_2 selectivity greatly increased. The H_2/CO_2 separation factor at 200 °C increased from 1.4 to 37, whereas the H_2/CH_4 separation factor increased from 1.6 to 33. Although silylation decreased the H_2 permeances in the B-ZSM-5 membranes, at temperature (400 °C), the H_2 permeance increases and the H_2/CO_2 separation factor was 47. In contrast, MDES does not fit into SAPO-34 pores. However, the silylation decreased the pore size of the non-zeolite pores in the SAPO-34 membranes. The only drastic changes in permeation occurred with the CH_4 because it primarily permeated through non-SAPO-34 pores.[64] The H_2 permeances and H_2/CO_2 and H_2/N_2 separation selectivities were almost unchanged in the SAPO-34 membranes because H_2, CO_2 and N_2 permeated mainly through SAPO-34 pores. This was confirmed by mixed gas separations studies.

The synthesis of small-pore zeolite membranes has also been pursued for the separation of small light gas molecules. Zeolite A membranes have shown H_2 permeances ranging from 10^{-10} to $<10^{-11}$ mol/m^{2}s kPa, with a maximum of H_2/N_2 separation factor of 4.8 between 35 and 125 °C.[14] Changes in the charge balancing cation result in changes in the H_2 permeance, and followed the order of K < Na < Ca, which is consistent with the order of the pore size of the A zeolite.[65] The highest H_2/N_2 separation factor of 9.9 was obtained for a KA membrane. For $AlPO_4$-5 membranes (pore size of 0.73 nm), the H_2 permeance was 2×10^{-10} mol/m^{2}s kPa at 35 °C[66]. The H_2/CO_2 ideal selectivity α_{H_2,CO_2} was 24, and the separation factor for an equimolar H_2/CO_2 mixture was 9.7 at 35 °C.

More recently, we explored using zeolite membranes for the separation of hydrogen from multi-component reforming streams.[67–69] Using methods developed by Dong *et al.*,[70] we synthesized silicalite-1 membranes and tested their H_2 separation abilities with varying temperatures (70–300 °C) and feed compositions.[71] The feed composition of the dry stream was H_2, CO_2, CO, CH_4, H_2S in the ratio

70.8:8.7:5.79:14.69:0.03; the wet stream was H_2, CO_2, CO, CH_4, H_2S, H_2O in the ratio 50:10:6:4:0.02:30. At lower temperatures in both experiments, H_2 had low permeation due to pore blockage by adsorbing components such as H_2O, CO_2, CH_4 and CO. H_2 permeance increased with temperature throughout 70–300 °C with a separation factor varying from 0.13 to 0.4. However, the H_2 separation value for the 5-component stream increases to 2 when water is not included, with permeances around 3×10^{-11} mol/m^{2}s kPa.

7. Applications

Zeolite membranes are being applied to a wide variety of applications. The following subsections outline just a few of those processes and are not meant to be a complete listing.

7.1. CO_2 sequestration

Many H_2 membrane separation technologies are based on the most widely used method of hydrogen production: the steam reforming of light hydrocarbons in which H_2 purification ultimately equates with CO_2 removal. The common process is steam–methane reforming (SMR), consists of two basic steps. In the initial reforming step, methane (CH_4) and excess steam (H_2O) react to form carbon monoxide (CO) and hydrogen (H_2) at about 820 °C (reaction 1). Additional H_2 is obtained by the subsequent reaction of CO with H_2O in the water-gas shift (WGS) reaction (reaction 2). For each mole of CH_4 consumed the overall SMR process (reaction 3) theoretically yields 4 moles H_2 and 1 mole of CO_2, although in practice this is seldom achieved. The H_2 product composition prior to purification depends on the exact nature of the shift process used. Typically, in a high temperature shift reactor operating at 350 °C, a product stream composition is 73.9% H_2, 17.7% CO_2, 6.9% CH_4 and 1.0% CO. However, a second shift process involving a lower temperature (190–210 °C) shift reaction is often used with a resulting product composition of 74.1% H_2, 18.5% CO_2, 6.9% CH_4 and 0.1% CO.[1]

Initial reforming: $CH_4 + H_2O \rightarrow CO + 3\ H_2$ (reaction 1)
Water–gas shift (WGS): $CO + H_2O \rightarrow CO_2 + H_2$ (reaction 2)
Steam–methane reforming (SMR): $CH_4 + 2\ H_2O \rightarrow CO_2 + 4\ H_2$ (reaction 3)

CO_2 separation is one of the most studied applications for FAU-type zeolite membranes.[72,73] Dong *et al.* investigated FAU membranes for the purification of CO_2 from 50/50 mixtures of CO_2/N_2 under dry and moist conditions in a temperature range of 23–200 °C at atmospheric pressure.[74] At room temperature, the CO_2 selectivity was about 31.2 for the CO_2/N_2 dry gas mixture with a CO_2 permeance of 2.1×10^{-11} mol/m^{2}s kPa. The addition of water to the stream significantly enhanced the CO_2 selectivity at 110–200 °C but drastically lowered the CO_2 selectivity below 80 °C. At 200 °C, with increasing water partial pressure, the CO_2 selectivity increased then decreased after reaching a maximum of 4.6 at a water partial pressure of 12.3 kPa.

In another study, Noble and Falconer showed that their silica/aluminophosphate (SAPO-4) zeolite membranes were used to separate CO_2 from CH_4 under a variety of pressures and temperatures.[75] The highest CO_2/CH_4 separation factor (270) was measured at −20 °C (pressure drop of 0.14 MPa), and the CO_2 permeance was 2.0×10^{-7} mol/m^2 s Pa. There is an inverse relationship between the selectivity of the membranes and the temperature. The separation factor decreases as temperature increases because separation is partially due to the preferential adsorption of CO_2 (which inhibits CH_4 adsorption). Furthermore, it was shown that CH_4 slows the rate of CO_2 diffusion. The overall effect is that separation selectivities with these membranes are higher than the ideal selectivities. Most of the observed pressure dependence for CO_2 flux in mixtures was predicted and confirmed by Maxwell–Stefan diffusion with an extended Langmuir isotherm.[75]

Recent advances by the group have further utilized Maxwell–Stefan (M–S) diffusion formulations to predict quantitatively binary light gas mixture permeation fluxes across a SAPO-34 membrane using only data on pure-component adsorption isotherms and diffusivities.[76] Good agreement between model and experiments was generated with the application of specific mathematical assumptions.

7.2. Hydrocarbon separations

In an effort to eliminate, or at least, decrease the amount of energy needed for hydrocarbon separations via distillation, much research is being focused on the use of membranes for hydrocarbon feedstock separations. Examples include the (1) separation of branched C5 hydrocarbons from linear C5 hydrocarbons and (2) the separation of isomers such as *p*-xylene (PX) from the *o*-xylene (OX) and *m*-xylene (MX). It appears that a combination of size selectivity and tuned surface adsorption in modified zeolite membranes allows for enhanced separations under approximate industrial process conditions. In particular, the method of modifying the zeolite membrane has direct effects on separations abilities.

7.2.1. C5-hydrocarbon (HC) separations

Isoprene, 2-methyl-1,3-butadiene, is used as a monomer in the Ziegler–Natta polymerization to produce artificial rubber (cis-1,4-polyisoprene). Industrially, isoprene must be purified from a mixture that contains similar C5 hydrocarbon compounds including *n*-pentane, 1,3-pentadiene, cyclopentadiene and dicyclopentadiene. These components have similar relative volatilities, and isoprene forms an azeotrope with *n*-pentane at 307 K (101 kPa, isoprene mole fraction of 0.72).[77,78] Extractive or azeotropic cryogenic distillation, which are high-energy processes, has been used for these separations.[79]

Falconer *et al.* recently reported on the metal-doped aluminosilicate MFI membranes grown on stainless steel tubular supports that exhibited high selectivity for *n*-pentane/isoprene mixtures.[77] By doping with boron (B-ZSM-5), the membrane achieved a selectivity of 74 at 60 °C. Separation is due to faster diffusion of the linear alkane. The single-component permeance of *n*-pentane was higher than its permeance in the mixture due to interference with the branched (and slower diffusing) isoprene. No effects due to isoprene polymerization were observed, and the acid sites in B-ZSM-5 did not affect the separations.[77]

Concurrently, our studies on surface modified tubular silicalite membranes synthesized by *in situ* crystallization method were used for testing isoprene separation from *n*-pentane.[80] The MFI membrane was selective toward *n*-pentane. The temperature and feed pressure dependencies of the *n*-pentane flux and selectivity suggest that the permeation is controlled by an adsorption–diffusion mechanism. For the equimolar *n*-pentane/isoprene mixture, at 50 °C and feed pressure of 32 kPa, the *n*-pentane flux and its separation factor over isoprene were 4.0×10^{-4} mol/m^{2}s and ~25, respectively.

The regenerated zeolite membranes were carbonized by exposure to an inert carrier gas stream containing controlled amounts of the HC mixture (either 50/50 or 80/20 isoprene/*n*-pentane) at elevated temperatures. The modification procedure for the membranes was preformed online in the modification unit and was derived from the procedure for the bulk zeolite modification.[62] The procedure for an online modification of the membrane is as follows: the membrane was calcined in air at 550 °C for 4 h with a slow ramp and cooling rate and then placed in the stainless steel tube at the centre of the modification reactor, and a 50/50 mixture of isoprene/*n*-pentane is flowed through it at 450 °C for 1.5 h. The resulting modified MFI membrane was selective toward *n*-pentane. The highest enrichment value of this sample was 4.1%.

Another modification method resulted in highly selective zeolite membranes for isoprene purification from a multi-component stream. In this study, our zeolite membrane was modified by an online carbonization of 1,3,5-triisopropylbenzene (TIPB)[63] and found to moderately improve the separation performance. The limited improvement was primarily due to the blockage of large inter-crystal pore by carbon deposition, but not the modification of zeolitic pores. The modified membrane was used to test separation of a quaternary mixture (composition provided by Goodyear Chemicals) containing isoprene (IP), *n*-pentane (NP), 1,3-pentadiene (PD), 2-methyl-2-butene (MB) mixture with IP, NP, PD, MB molar ratio of 40:40:2.15:17.85.[80] The total feed partial pressure was 14.7 kPa. The membrane exhibited high selectivity for *n*-pentane and 1,3-pentadiene. The highest *n*-pentane/isoprene selectivity was ~14, which was observed at 50 °C. The 1,3-pentadiene/isoprene selectivity increased continuously with temperature and reached a very high value of ~50 at 180 °C. However, there was almost no separation achieved on the MFI membrane for 2-methyl-2-butene/isoprene.

7.2.2. Xylene separations

p-Xylene is an important chemical stock for synthesis of terephthalic acid, a starting material in synthetic plastics. Industrial production of pure *p*-xylene is costly to a large extent because of the difficulties associated with the separation of xylene isomers. The major aromatic compounds in the streams from commonly used catalytic reforming or isomerization processes include *p*-xylene (PX), *m*-xylene (MX), *o*-xylene (OX) and ethylbenzene (EB). The xylene isomers have similar molecular structures and close boiling points; hence, they are difficult to separate by distillation. Currently, xylene isomers are separated mainly by fractional crystallization and zeolitic adsorption processes, both of which are energy intensive and require batch operations.

High separation of xylene isomers on MFI zeolite membranes can be obtained by shape selectivity at the entrance of zeolite pores but not through selective adsorption or preferential diffusion mechanisms.[38] Since the MFI pore geometry can be distorted to readily accommodate the MX and OX molecules under high sorption levels of PX,[81] it is critical for PX separation that proper operating conditions are employed to keep the xylene sorption at a sufficiently low level. Therefore, operating under low feed pressures and/or at relatively high temperatures would be appropriate for enhancing the selectivity of PX over MX and OX by permeation through the MFI membranes. The specific feed pressure for effective xylene separation is dependent on the operating temperature because xylene sorption is a function of both temperature and pressure.

The xylene isomers are strongly adsorbing to MFI zeolites with comparable heats of adsorption and large loading numbers at high vapour pressures because of the flexibility of the MFI framework when in contact with aromatic molecules.[81a,82,83] The diffusion behaviours of xylenes in zeolitic pores are strongly affected by the presence of other isomers due to strong sorbate–framework and sorbate–sorbate interactions. The diffusion rate of OX can be dramatically enhanced by PX, while the diffusion rate of PX can be reduced by OX molecules in highly siliceous MFI zeolites.[81a,84] The critically sized C8 aromatic molecules are difficult, if not impossible, to pass by each other in the MFI channels. Therefore, the MFI membranes are unlikely to achieve high selectivities for PX over MX and OX in separation of their mixtures based on competitive diffusion when the operating conditions favour a high xylene sorption level.

Keizer *et al.*[85] obtained a PX/OX selectivity (αPX/OX) of >200 at 125 °C, but αPX/OX = 1 (no selectivity) at 25 °C on MFI zeolite membranes when the feed partial pressures of PX and OX were 0.31 and 0.26 kPa, respectively. Because the critically sized C8 aromatic molecules diffuse predominately through the *b*-direction straight channels of silicalite,[86] the orientation of the zeolite membrane has a significant effect on xylene separation. Tsapatsis has[42,43,87,88] demonstrated that high permeance and high separation factor for PX could be obtained at low feed pressures in defect-free *b*-oriented (vs. a- and c-oriented) MFI films. Gump *et al.*[89] compared silicalite-1, ZSM-5 and boron-substituted ZSM-5 membranes for separation of PX/OX mixtures. They found that at 150 °C and a xylene partial pressure of 2.1 kPa, the boron-substituted ZSM-5 membrane exhibited a higher separation factor than did the other two membranes. The substitution of boron for silica in the zeolite framework decreased the unit cell size and made the framework more rigid, possibly reducing the zeolite structure distortion. To date, it has been verified by different research groups that high separation factors for xylene isomers are attainable through vapour permeation in MFI zeolite membranes at very low feed pressures and elevated temperatures.[63,85] However, the extremely low xylene partial pressures in the feed resulted in impractically low *p*-xylene fluxes.

The TIPB method we recently developed was also utilized for MFI zeolite membranes, synthesized on tubular α-alumina substrates, to investigate the separation of *p*-xylene (PX) from *m*-xylene (MX) and *o*-xylene (OX) in binary, ternary and simulated multi-component mixtures in wide ranges of feed pressure and operating temperature.[63] The online membrane modification by carbonization of 1,3,5-triisopropylbenzene in the feed stream was found to be effective for reducing the inter-crystalline pores and improving the PX separation in a multi-component stream (see Fig. 14.3). The results demonstrated that separation of PX from MX and OX through the MFI membranes relies primarily on shape selectivity when the xylene sorption level in the zeolite is sufficiently low.

The temperature dependences of the PX flux and its selectivities over MX and OX were similar to those found with the binary and ternary systems. However, the maximum PX permeance and PX selectivity occurred at higher temperatures for the multi-component system of higher xylene partial pressures in the feed. In particular, an eight-component mixture containing hydrogen, methane, benzene, toluene, ethylbenzene (EB), PX, MX and OX had a PX/(MX+OX) selectivity of 7.71 with a PX flux of 6.8×10^{-6} mol/m^{2}s and was obtained at 250 °C and atmospheric feed pressure (87 kPa) (see Fig. 14.4). By comparison, at 275 °C, the highest PX/(MX+OX) selectivity of 6.4 and PX flux of 7.6×10^{-6} mol/m^2 s was obtained. However, at 275 °C, the PX flux and selectivity were found to decrease gradually with time and were unable to stabilize in more than 6 h. Thus, further experiments on the multi-component mixtures were conducted at 250 °C at which the permeation behaviour tended to stabilize normally within 6 h.

Although the kinetic diameters of benzene, toluene, and EB molecules are similar to that of PX, almost no separation effects were observed on the MFI membrane for benzene, toluene and EB. The high selectivity of PX over benzene,

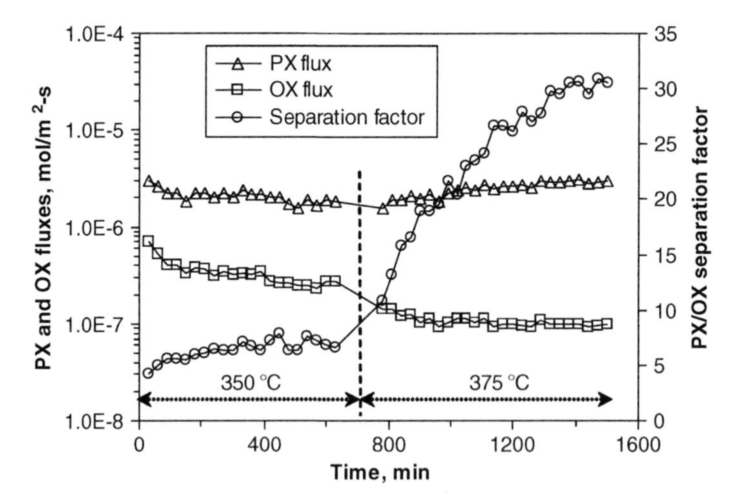

Figure 14.3 Result of online membrane modification by TIPB carbonization/deposition. Reprinted with permission from Ref. [63]. Copyright (2006), Elsevier.

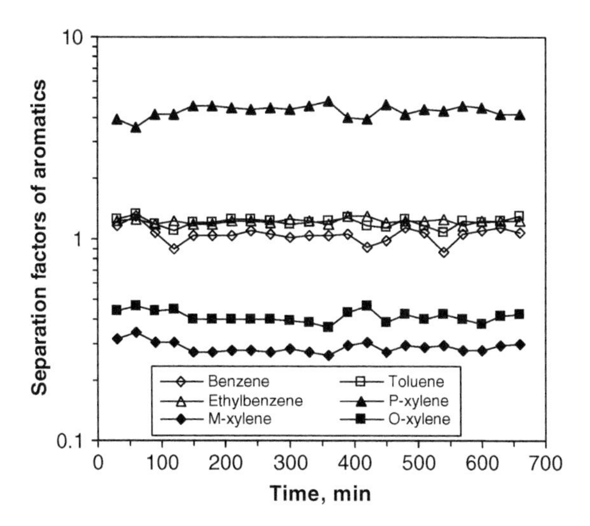

Figure 14.4 Fluxes and separation factors for the eight-component mixture at 250 °C and atmospheric feed pressure (0.87 bar). Reprinted with permission from Ref. [63]; Copyright (2006), Elsevier.

toluene and ethylbenzene may be attributed to the higher heat of adsorption and also faster diffusion of PX in MFI zeolites compared to these aromatics [90]. At 300 °C, the PX flux and PX/(MX+OX) selectivity declined drastically from 4.2×10^{-6} to 1.7×10^{-6} mol/m^2 s and 3.5 to 1.4 mol/m^2 s, respectively, in 5 h of operation. Further increase of the temperature to 325 and 350 °C caused only slight changes in PX flux and selectivity. Large decrease in fluxes was also observed for benzene, toluene and ethylbenzene but not for MX and OX. In fact, changes in fluxes of MX and OX were rather small as compared to the PX in the entire temperature range of 200–350 °C. The unexpected decreases of PX flux and separation factor at 300 °C might be caused by the adsorption of large impurity molecules contained in the chemicals used, which blocked the zeolite channels. The nearly constant low fluxes of MX and OX in the temperature range suggested that these two large molecules penetrated the membrane mainly through the nanoscale inter-crystalline pores as their chances to enter the zeolite pores were limited by the small molecules, including the smaller aromatic components, hydrogen and methane. When the zeolite channels are blocked, the PX molecules also transported mainly through the non-selective inter-crystal pores. Thus, the PX flux reduced to about half of the MX flux but slightly higher than the OX flux at above 300 °C because the flux rates of the isomers through the non-selective pores are proportional to their respective partial pressures in the feed.

After operating at 350 °C, the membrane was retested for separation at 250 °C. The results showed that the loss of separation capability for the MFI membrane was not reversible. The membrane was then treated at 400 °C for 24 h under air flowing over both sides. The separation of the eight component mixture was performed again on the regenerated membrane at 250 °C. Both the PX flux and PX/(MX+OX) selectivity were fully recovered as compared to the values obtained before operating at 300 °C.

Note, the addition of a small quantity of nonane (C9) to the multi-component mixture caused drastic decreases in the fluxes of aromatic components and the PX separation factor because of the preferential adsorption of nonane in the zeolite channels. The nanoscale inter-crystalline pores also caused serious decline in the PX separation factor.

7.3. Reverse osmosis by zeolite membranes

Zeolite membranes, due to their excellent chemical and thermal stabilities, may be particularly useful for treating various kinds of wastewater that cannot be handled effectively by polymeric RO membranes, such as solutions containing organic solvents and radioactive ions and/or those requiring operation at elevated temperatures. Recently, molecular dynamic simulation has shown that zeolite membranes are theoretically suitable for ion removal from aqueous solutions by reverse osmosis (RO) processes.[91] The simulation revealed that 100% Na^+ rejection could be achieved on a perfect (single crystal), all silica, ZK-4 membrane through RO. The separation mechanism of the perfect ZK-4 zeolite membranes is the size exclusion of hydrated ions, which have kinetic sizes (0.8–1.0 nm for $[Na(H_2O)_x]^+$)[92] significantly larger than the aperture of the ZK-4 zeolite (diameter 0.42 nm). Kumakiri *et al.*[93] reported using an A-type zeolite membrane in RO separation of water/ethanol mixtures. The hydrophilic A-type zeolite (pore size ≈ .4 nm) membranes showed 44% rejection of ethanol and a water flux of 0.058 kg/m^{2}h under an applied feed pressure of 1.5 MPa. However, very little experimental data of RO desalination on zeolite membranes has been reported so far.

Existing zeolite membranes possess an imperfect polycrystalline structure[94] and may be of different types with various pore sizes and Si/Al ratios.[30] Therefore, experimental investigation of the effectiveness of RO desalination using real zeolite membranes is necessary. Our recent work demonstrated RO separation using an α-alumina-supported MFI-type zeolite membrane for solutions containing a single cation and multiple cations. The selection of MFI zeolite types in this early stage of research was made based on three considerations, namely (1) ease to synthesize defect-free thin membranes, (2) medium pore size to achieve high ion rejection with reasonable permeability and (3) possibility to obtain pure-silica zeolite structure for simplification in mechanism study.

RO experiments were conducted at room temperature for a 0.1 M NaCl solution and a multi-component solution of various chloride salts, respectively. Water flux and ion rejection were measured as functions of operation time,[95] with an experimental study on the RO desalination of aqueous solutions using α-alumina-supported MFI-type zeolite membranes. A Na^+ rejection of 76.7% with a water flux of about 0.112 kg/ m^{2}h was obtained for a 0.1 M NaCl feed solution under an applied pressure of 2.07 MPa. For a complex feed solution containing 0.1 M NaCl+0.1 M KCl+0.1 M NH_4Cl+0.1 M $CaCl_2$+0.1 M $MgCl_2$, rejections of Na^+, K^+, NH_4^+, Ca^{2+} and Mg^{2+} reached 58.1%, 62.6%, 79.9%, 80.7% and 88.4%, respectively, with a water flux of 0.058 kg/m^{2}h, after 145 h of operation at an applied pressure of 2.4 MPa.

Results of RO separation for the multiple-salt solution showed that the rejection rates of the bivalent Ca^{2+} and Mg^{2+} and the polyatomic NH_4^+ ion were higher than those of the univalent Na^+ and K^+. This was verified by additional experiments for the same solution with extended time of operation until Na^+ rejection tended to stabilize. The bivalent Ca^{2+} and Mg^{2+} have higher charge density, hence greater ability to polarize the neighbouring water molecules and form larger and more rigid hydrated complexes compared to K^+ and Na^+.[96,97] The polyatomic ion NH_4^+ has the same hydration number as K^+ and Na^+, but the former forms a larger hydration shell because of its rigid tetrahedral four-point charge model.[98]

7.4. Catalytic zeolite membranes

Zeolite membranes have a promising future for applications in catalytic reactors[28,99,100] due to their selectivity and permeability. As an example, a two-layered H-ZSM-5-mordenite membrane was developed by Coronas' group for the esterification of acetic acid with ethanol. It was a successful way to merge the catalytic activity of H-ZSM-5 and the selectivity for water separation of mordenite membranes, therefore giving rise to bi-functional membranes.[101] Additionally, catalytic zeolite membranes incorporate both the high conversions and yields of nanoparticle catalysts plus the high throughput of thin film membranes. The catalytic membrane has been described by Bernardo as a 'nanoengineered' catalyst because catalytically active nanoparticles are entrapped in a thin zeolite layer (a few microns thick).[102] For that reason, they are viewed as a potential substitute for traditional reactors.[103]

In the fabrication of catalytic membranes, large-pored zeolites are useful for the ability to load nanoparticles of catalysts directly into the zeolite pores. Faujasite (FAU, X or Y) zeolites are large pored and ideally suited for catalytic membrane reactor applications. The pore opening and super-pore size allows for light gases and small hydrocarbon molecules to easily enter and react. A growing number of applications have been reported for zeolite catalytic membranes, in both the areas of hydrocarbon and light gas conversions. We present two examples and their performance here.

The first example deals with the effectively conversion of methane to higher hydrocarbons (C_{2+}) and hydrogen.[104] This is of major interest because of the tremendous industrial interest associated with methane as it is renewable through bioprocesses. The transition metal-catalysed non-oxidative CH_4 conversion to C_{2+} and H_2 may offer an alternative to oxidative methods because of its outstanding advantages of low reaction temperature, high selectivity and zero CO_2 emission. However, practical consideration of the non-oxidative system has been discouraged by the limitation of its two-step reaction. In the first step, CH_4 is decomposed on the metal surface into methyl radicals ($CH_{x(0\leq x\leq 3)}$) and hydrogen; in the second step, the chemisorbed CH_x species are re-hydrogenated and oligomerized into C_{2+}. A method to overcome the two-step limitation is to utilize catalytic zeolite membranes. Recently, we developed Pt–Co/Na–Y zeolite membranes for the non-oxidative conversion of CH_4.[104]

The Pt–Co bimetallic clusters were loaded inside the zeolite channels by successive ion exchange in a 0.065 M $Pt(NH_3)_4(NO_3)_2$ solution and a 0.06 M $Co(NO_3)_2$ solution under refluxing at 80 °C, respectively. The ion exchange time was varied to control the Pt–Co loading level, with a determined optimized time of 15 min. The membrane was tested for continuous isothermal CH_4 conversion at 300 °C and 0.86 bar, and compared to a packed-bed reaction of a Pt–Co/Na–Y particulate catalyst with the same metal loading level. The conversion of CH_4 to C_{2+} products was optimized at 300 °C and increased H_2 in the feed stream. Comparison with the packed-bed reactor indicates that 1910 mol of CH_4 was converted to C_{2+} on the membrane, which corresponds to 893 two-step cycles needed for equivalent production with the packed bed.

The principle of CH_4 conversion through the catalytic membrane is proposed to have three basic steps as illustrated in Fig. 14.5: (1) decomposition of CH_4 on the CH_4 feed side surface, (2) surface diffusion of chemisorbed CH_x through the Pt–Co-loaded zeolite channels to the H_2 sweep side driven by the gradient of surface coverage and (3) re-hydrogenation of the CH_x species to form C_{2+} under H_2 sweep.

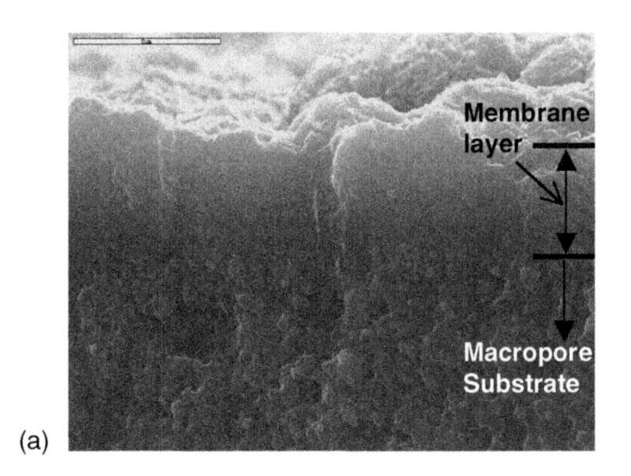

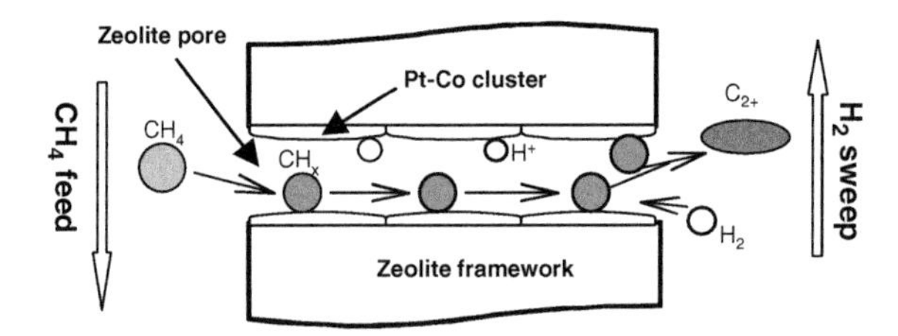

(b) $$\left(\sum_{n=2}^{N} n\upsilon_n\right) CH_4 \xleftrightarrow{\text{Pt-Co cat,~300°C}} \sum_{n=2}^{N}(\upsilon_n C_n H_{2n+2}) + \left[\sum_{n=2}^{N} \upsilon_n (n-1)\right] H_2 + \Delta H_r$$

Figure 14.5 (a) SEM images of The Pt–Co/Na–Y membrane cross-section, with catalyst loaded by 15 min ion exchange and (b) schematic showing the principle of the proposed membrane process.[104] With kind permission from Springer Science and Business Media.

A limited back diffusion of H_2 can maintain a low level of H_2 on the CH_4 feed side surface and inside the catalyst-loaded zeolite pores that suppresses deep dissociation of CH_x into inactive carbonaceous species and benefits the C_{2+} selectivity.[105]

The second example of a process that benefits from the use of zeolite catalytic membranes is the selective oxidation of CO (Selox). This process is valuable for its ability to remove CO from H_2-rich gas streams found in (1) reforming processes and (2) multiuse ethylene plants.[102] A selective catalyst is necessary to avoid H_2 consumption, because two competitive reactions take place, CO and H_2 oxidation [see Eqs. (2) and (3)]:

$$2CO + O_2 \rightarrow 2CO_2 \quad (2)$$

$$2H_2 + O_2 \rightarrow 2H_2O \quad (3)$$

A number of groups have studied the Selox process with zeolite-based catalytic membranes.[106–108] In general, these Y (FAU) zeolite-based membranes were impregnated with catalysts metals Pt, Ru, Rb, Co, Ni, Cu and Ag; the Pt-Y membrane showed the highest CO oxidation rate.[108] By comparison, Bernardo *et al.* used an ion exchange process to load the catalyst into the zeolite, resulting in Pt-Y membranes. These membranes were run under similar reaction conditions to the previous ones.[102]

Using a flow through configuration for the membrane, and a variety of reaction conditions (temperature, pressure, composition), they showed that the zeolite membranes were able to improve on both conversion of CO and selectivity for CO_2 over traditional and micro-reactors. Under hydrogen-free conditions (CO:O_2:N_2 = 10.5:8.7:80.8), the membranes reached 100% CO conversion at 200 °C. In the presence of hydrogen, (H_2:CO:O_2 = 80.8:10.5:8.7), the membrane had high CO conversion (98%) and CO selectivity (62%). When comparing Goerke's results[109] using Au-loaded metal oxide catalysts (α-Fe_2O_3 and CeO_2) in micro-reactors to Bernardo's Pt-Y membrane reactor (using similar reaction conditions), the membrane reactor was more efficient. Further improvements to the CO conversion and CO_2 yield were obtained with slight increases in temperature and pressure.

8. Fabrication and Manufacturing

Zeolite membrane manufacturing is still an industry in the making. Most membranes are still fabricated in lab-scale sizes and quantities. Furthermore, the technology needs to be able to commercialize large-scale continuous films without inter-crystalline pores for successful high separation factor. However, recently, there has been a big leap forward in the commercialization of this technology. Currently, only Mitsui Engineering and Shipbuilding Company in Japan have commercialized a process using zeolite membranes.[37] It is a pervaporation process using NaA zeolite membranes for organic dehydration (see Fig. 14.6). The membranes are 20–30 m thick on porous, tubular ceramic supports. The plant processes alcohols up to

Figure 14.6 Zeolite membrane pervaporation module layout for a Mitsui Engineering & Shipbuilding Co. large-scale solvent dehydration plant. Reprinted from Ref. [37]. Copyright (2001), with permission from Elsevier.

530 liter/hr with separation factors up to 10,000. Manufacturers of zeolite membranes listed by Bowen *et al.*[37] include Smart Chemical Co., Ltd and Christison Scientific, both in the UK and Artisan Industries, Inc., USA.

Current estimated costs per zeolite membrane gas separation module have been approximated around $400/ft^2$. Although this is an estimate, it compares favourably with metal membranes and modules ($1500/ft^2$).[110] It is safe to assume that once in mass use and production, those costs will decline significantly to about $100/ft^2$, allowing zeolite membranes to compete both in economics and on performance.

Significant progress has been made in the synthesis or various types of crystalline zeolite membranes. Good quality zeolite membranes can be prepared by several methods, including *in situ* synthesis, secondary growth and vapour phase transportation. To be considered useful for gas separations applications, these membranes will have to be synthesized without macro-pore-sized defects or pinholes. Furthermore, the ability to surface modifies zeolite membranes (both internal pore surfaces and external surfaces) through silation or carbonization, allows for the fine tuning in selectivity. Gas separations[24] through the membranes are governed by mechanisms of preferential adsorption, selectively configurationally diffusion or molecular sieving. Gas permeation through these micro-porous inorganic membranes is an activated process that can be predicted through gas diffusion theories (i.e., Maxwell–Stephans equations that govern gas permeation and separation).

Zeolite membranes have chemical, mechanical and thermal stability not observed in many types of membranes. The trends in zeolite membrane research show clearly the improvements in selectivity, fabrication methodology and energy-production applications. In the near future, the ability to inexpensively fabricate these membranes for tuned selectivity will put them at the forefront of separations technology. For the time being, their stability at high temperatures and their ability to be regenerated without loss to performance make them interesting candidates for a variety of streamlined light gas, hydrocarbon and cation separation processes.

9. Conclusion

Zeolite membranes are a very important emerging area of separations media due to their inherent chemical, mechanical and thermal stability. The trends in zeolite membrane research show clearly the improvements in selectivity, fabrication methodology and energy-production applications. In the near future, the ability to inexpensively fabricate these membranes for tuned selectivity will put them at the forefront of separations technology. Currently, they are interesting candidates for streamlined light gas and hydrocarbon feed stocks production via multi-component streams because of their stability at high temperatures and their ability to be regenerated without loss to performance. Furthermore, in specific conditions, zeolite membranes can be applied as RO membranes for desalination processes. As a means to streamline industrial processes that combine catalytic reactions and separations, zeolite catalytic membranes are of great interest. They are chemically, mechanically and thermally durable. Plus, they can utilize their pore sizes to incorporate both nanoparticle catalysts and be selective in the product yield. Concerns associated with inorganic membranes centre on their fabrication reproducibility. Compared with organic membranes, inorganic membranes are currently expensive to manufacture. However, introduction into large-scale production facilities should result in more competitive production costs.

ACKNOWLEDGEMENT

Sandia is a multi-program laboratory operated by Sandia Corporation, a Lockheed Martin Company, for the US DOE's NNSA, contract DE-AC04-94-Al85000. We thank Drs. N. W. Ockwig and X. Gu for continued contributions to this work.

REFERENCES

[1] Ockwig, N. W., Nenoff, T. M., *Chem. Rev.* **2007,** *107*(10), 4078–4110 (and references within).
[2] den Exter, M. J., Jansen, J. C., van de Graaf, J. M., Kapteijn, F., Moulijn, J. A., van Bekkum, H., *Recent Adv. New Horizons Zeolite Sci. Technol.* **1996,** *102,* 413–454.
[3] Suzuki, H., Composite membrane having a surface layer of an ultrathin film of cage-shaped zeolite and processes for production thereof, U.S. Patent 4,699,892; **1987**.
[4] Bakker, W. J. W., van den Broeke, L. J. P., Kapteijn, F., Moulijn, J. A., *AIChE J.* **1997,** *43,* 2203–2214.
[5] Hedlund, J., Sterte, J., Anthonis, M., Bons, A.-J., Carstensen, B., Corcoran, N., Cox, D., Deckman, H., Gijnst, W. D., de Moor, P.-P., Lai, F., McHenry, J., Mortier, W., Reinoso, J., Peeters, J., *Microporous Mesoporous Mater.* **2002,** *52,* 179–189.
[6] Bowen, T. C., Kalipcilar, H., Falconer, J. L., Noble, R. D., *J. Membr. Sci.* **2003,** *215,* 235–247.
[7] Sano, T., Yanagishita, H., Kiyozumi, Y., Mizukami, F., Haraya, K., *J. Membr. Sci.* **1994,** *95,* 221–247.
[8] Lovallo, M. C., Gouzinis, A., Tsapatsis, M., *AIChE J.* **1998,** *44,* 1903–1911.
[9] Burggraaf, A. J., Vroon, Z. A. E. P., Keizer, K., Verweij, H., *J. Membr. Sci.* **1998,** *44,* 77–86.
[10] Noack, M., Kolsch, P., Caro, J., Schneider, M., Toussaint, P., Sieber, I., *Microporous Mesoporous Mater.* **2000,** *35,* 253–265.
[11] Wegner, K., Dong, J. H., Lin, Y. S., *J. Membr. Sci.* **1999,** *158,* 17–27.
[12] Lin, X., Kita, H., Okamoto, K., *Chem. Commun.* **2000,** *19,* 1889–1890.
[13] Gardner, T. Q., Flores, A. I., Noble, R. D., Falconer, J. L., *AIChE J.* **2002,** *48,* 1155–1167.
[14] Aoki, K., Kusakabe, K., Morooka, S., *J. Membr. Sci.* **1998,** *141,* 197–205.
[15] Kita, H., Horii, K., Ohtoshi, Y., Tanaka, K., Okamoto, K. I., *J. Mater. Sci. Lett.* **1995,** *14,* 206–208.
[16] Hedlund, J., Schoeman, B., Sterte, J., *Chem. Commun.* **1997,** 1193–1194.
[17] Navajas, A., Mallada, R., Téllez, C., Coronas, J., Menéndez, M., Santamaría, J., *Desalination* **2002,** *148,* 25–29.
[18] Nishiyama, N, Ueyama, K, Masahiko, M, *Chem. Commun.* **1995,** 1967–1968.
[19] Yamazaki, S., Tsutsumi, K., *Adsorption* **1997,** *3,* 165–171.
[20] Kita, H., Asamura, H., Tanaka, K., Okamoto, K.-I., Kondo, M., *Abstr. Papers Am. Chem. Soc.* **1997,** *214,* 269.
[21] Kusakabe, K., Kuroda, T., Uchino, K., Hasegawa, Y., Morooka, S., *AIChE J.* **1999,** *45,* 1220–1226.
[22] Li, S., Tuan, V. A., Falconer, J. L., Noble, R. D., *Microporous Mesoporous Mater.* **2002,** *53,* 59–70.
[23] Iler, R. K., *The Chemistry of Silicas,* Wiley, New York, 1979.
[24] Lin, Y. S., Kumakiri, I., Nair, B. N., Alsyouri, H., *Sep. Purif. Methods* **2002,** *31,* 229–379.
[25] Kapteijn, F., Rodriguez, F., Mirasol, J., Moulijn, J. A., *Appl. Catal. B Environ.* **1996,** *9*(1–4), 25–64.
[26] Geus, E. R., Den Exter, M. J., Van Bekkum, H., *J. Chem. Soc., Faraday Trans.* **1992,** *88,* 3101–3109.
[27] van Bekkum, H., Geus, E. R., Kouwenhoven, H. W., *Stud. Surf. Sci. Catal.* **1994,** *85,* 509–542.
[28] Coronas, J., Santamaria, J., *Sep. Purif. Methods* **1999,** *28,* 127–177.
[29] Bein, T., *Chem. Mater.* **1996,** *8,* 1636–1653.
[30] Caro, J., Noack, M., Kolsch, P., *Adsorption* **2005,** *11*(3–4), 215–227.
[31] Chiang, A. S. T., Chao, K.-J., *J. Phys. Chem. Solids* **2001,** *62,* 1899–1910.
[32] Matsukata, M., Kikuchi, E., *Bull. Chem. Soc. Jpn.* **1997,** *70,* 2341–2356.
[33] Mizukami, F., *Stud. Surf. Sci. Catal.* **1999,** *125,* 1–12.
[34] Tavolaro, A., Drioli, E., *Adv. Mater.* **1999,** *11,* 975–976.
[35] Davis, M. E., *Nature* **2002,** *417,* 813–821.
[36] Nair, S., Tsapatsis, M., in: Auerbach, S. M, Carrado, K. A, Dutta, P. K (Eds.), *Handbook of Zeolite Science and Technology,* Marcel Dekker, New York, 2003.
[37] Bowen, T. C., Noble, R. D., Falconer, J. L., *J. Membr. Sci.* **2004,** *245,* 1–33 (and references therein).

[38] Lai, Z. P., Bonilla, G., Diaz, I., Nery, J. G., Sujaoti, K., Amat, A. M., Kokkoli, E., Terasaki, O., Thompson, R. W., Tsapatsis, M., Vlachos, D. G., *Science* **2003,** *300,* 456–460.
[39] Cui, Y., Kita, H., Okamoto, K., *J. Mater. Chem.* **2004,** *14,* 924–932.
[40] Husain, S., Koros, W. J., *J. Membr. Sci.* **2007,** *288,* 195–207.
[41] Pechar, T. W., Kim, S., Vaughan, B., Marand, E., Tsapatsis, M., Jeong, H. K., Cornelius, C. J., *J. Membr. Sci.* **2006,** *277,* 195–202.
[42] Choi, J., Ghosh, S., Lai, Z., Tsapatsis, M., *Angew. Chem. Int. Ed.* **2006,** *45,* 1154–1158.
[43] Lai, Z., Tsapatsis, M., Nicolich, J. P., *Adv. Funct. Mater.* **2004,** *14,* 716–729.
[44] Barrer, R. M., *J. Chem. Soc., Faraday Trans.* **1990,** *86,* 1123–1130.
[45] Bakker, W. J. W., Kapteijn, F., Poppe, J., Moulijn, J. A., *J. Membr. Sci.* **1996,** *117,* 57–78.
[46] Jost, S., Bar, N. K., Fritzsche, S., Haberlandt, R., Karger, J., *J. Phys. Chem. B* **1998,** *102*(33), 6375–6381.
[47] Krishna, R., Paschek, D., *Phys. Chem. Chem. Phys.* **2002,** *4,* 1891–1898.
[48] Sanborn, M. J., Snurr, R. Q., *Sep. Purif. Technol.* **2000,** *20,* 1–13.
[49] Sanborn, M. J., Snurr, R. Q., *AIChE J.* **2001,** *47,* 2032–2041.
[50] Skoulidas, A. I., Sholl, D. S., Krishna, R., *Langmuir* **2003,** *19,* 7977–7989.
[51] Skoulidas, A. I., Bowen, T. C., Doelling, T. C., Falconer, J. L., Noble, R. D., Sholl, D. S., *J. Membr. Sci.* **2003,** *227,* 123–136.
[52] Skoulidas, A. I., Sholl, D. S., *AIChE J.* **2005,** *51,* 867–877.
[53] Sholl, D. S., *Acc. Chem. Res.* **2006,** *39,* 403–411.
[54] Cussler, E. L., *Diffusion Mass Transfer in Fluid Systems,* 2nd edition, Cambridge University Press, Cambridge, UK, 1997.
[55] Burggraaf, A. J., *Transport and separation properties of membranes with gases and vapors*, Elsevier, Amsterdam, 1996.
[56] Makhlouf, M. M., Sisson, R. D., *Metall. Mater. Trans. A* **1991,** *22,* 1001–1006.
[57] Masuda, T., Fukumoto, N., Kitamura, M., Mukai, S. R., Hahimoto, K., Tanaka, T., Funabiki, T., *Microporous Mesoporous Mater.* **2001,** *48,* 239–245.
[58] Park, D. H., Nishiyama, N., Egashira, Y., Ueyama, K., *Ind. Eng. Chem. Res.* **2001,** *40,* 6105–6110.
[59] Sano, T., Hasegawa, M., Ejiri, S., Kawakami, Y., Yanagishita, H., *Microporous Microporous Mater.* **1995,** *5,* 179–184.
[60] Falconer, J. L., George, S. M., Ott, A. W., Klaus, J. W., Noble, R. D., Funke, H. H., Modification of zeolite or molecular sieve membranes using atomic layer controlled chemical vapor deposition, U.S. Patent 6,043,177, **2000**.
[61] Yan, Y. S., Davis, M. E., Gavalas, G. R., *J. Membr. Sci.* **1997,** *123,* 95–103.
[62] Nenoff, T. M., Kartin, M., Thoma, S. G., Enhanced Selectivity of Zeolites by Controlled Carbon Deposition, U.S. Patent 7,041,616, **2006**.
[63] Gu, X., Dong, J. H., Nenoff, T. M., Ozokwelu, D. E., *J. Membr. Sci.* **2006,** *280,* 624–633.
[64] Hong, M., Falconer, J. L., Noble, R. D., *Ind. Eng. Chem. Res.* **2005,** *44,* 4035–4041.
[65] Guan, G., Kusakabe, K., Morooka, S., *Sep. Sci. Technol.* **2001,** *36,* 2233–2245.
[66] Guan, G. Q., Tanaka, T., Kusakabe, K., Sotowa, K. I., Morooka, S., *J. Membr. Sci.* **2003,** *214,* 191–198.
[67] Mitchell, M. C., Autry, J. D., Nenoff, T. M., *Mol. Phys.* **2001,** *99,* 1831–1837.
[68] Mitchell, M., Gallo, M., Nenoff, T. M., *J. Chem. Phys.* **2004,** *121,* 1910–1916.
[69] Gallo, M., Nenoff, T. M., Mitchell, M. C., *Fluid Phase Equilib.* **2006,** *247,* 135–142.
[70] Dong, J., Liu, W., Lin, Y. S., *AIChE J.* **2000,** *46,* 1957–1966.
[71] Adams, K. L., Li, L., Gu, X., Dong, J. H., Mitchell, M. C., Nenoff, T. M., *J. Membr. Sci.* **2008,** in preparation.
[72] Seike, T., Matsuda, M., Miyake, M., *J. Mater. Chem.* **2002,** *12,* 366–368.
[73] Bernal, M. P., Coronas, J., Menéndez, M., Santamaría, J., *AIChE J.* **2004,** *50,* 127–135.
[74] Gu, X., Dong, J. H., Nenoff, T. M., *Ind. Eng. Chem.* **2005,** *44,* 937–944.
[75] Li, S., Martinek, J. G., Falconer, J. L., Noble, R. D., Gardner, T. Q., *Ind. Eng. Chem. Res.* **2005,** *44,* 3220–3228.
[76] Li, S., Falconer, J. L., Noble, R. D., Krishna, R., *Ind. Eng. Chem. Res.* **2007,** *46*(12), 3904–93911.

[77] Arruebo, M., Falconer, J. L., Noble, R. D., *J. Membr. Sci.* **2006,** *269,* 171–176.
[78] Lee, L., Huang, J., Yang, W., Chiang, A. S. T., *Fluid Phase Equilib.* **1994,** *102,* 257–273.
[79] Lindner, A., Wagner, U., Volkamer, K., Rebafka, W., Recovery of isoprene from a C5-hydrocarbon mixture, US Patent 4,647,344, **1987**.
[80] (a) Nenoff, T. M., Ulutagay-Kartin, M., Bennett, R., Johnson, K., Gray, G., Anderson, T., Arruebo, M., Noble, R., Falconer, J., Gu, X., Dong, J., *Novel Modified Zeolites for Energy-Efficient Hydrocarbon Separations,* Sandia National Laboratories, SAND2006–6892, November 2006; (b) Gu, X., Dong, J., Nenoff, T. M., *J. Membr. Sci.* **2008,** in preparation.
[81] (a) Karsh, H., Culfaz, A., Yucel, H., *Zeolites* **1992,** *12,* 728–732; (b) Mohanty, S. H., Davis, T., McCormick, A. V., *Chem. Eng. Sci.* **2000,** *55,* 2779–2792.
[82] Chempath, S., Snurr, R. Q., Low, J. J., *AIChE J.* **2004,** *50,* 463–469.
[83] Snurr, R. Q., Bell, A. T., Theodorou, D. N., *J. Phys. Chem.* **1993,** *97,* 13742–13752.
[84] Baertsch, C. D., Funke, H. H., Falconer, J. L., Noble, R. D., *J. Phys. Chem.* **1996,** *100,* 7676–7679.
[85] Keizer, K., Burggraaf, A. J., Vroon, Z. A. E. P., Verweij, H., *J. Membr. Sci.* **1998,** *147,* 159–172.
[86] Ruthven, D. M., Eic, M., Richard, E., *Zeolites* **1991,** *11,* 647–653.
[87] Mabande, G. T. P., Ghosh, S., Lai, Z., Schwieger, W., Tsapatsis, M., *Ind. Eng. Chem. Res.* **2005,** *44,* 9086–9095.
[88] Choi, J., Ghosh, S., King, L., Tsapatsis, M., *Adsorption* **2006,** *12*(5–6), 339–360.
[89] Gump, C. J., Tuan, V. A., Noble, R. D., Falconer, J. L., *Ind. Eng. Chem. Res.* **2001,** *40,* 565–577.
[90] Song, L., Sun, Z., Duan, L., Gui, J., McDougall, G. S., *Microporous Mesoporous Mater.* **2007,** *104,* 115–128.
[91] Lin, J., Murad, S., *Mol. Phys.* **2001,** *99,* 1175–1181.
[92] Murad, S., Oder, K., Lin, J., *Mol. Phys.* **1998,** *95,* 401–408.
[93] Kumakiri, I., Yamaguchi, T., Nakao, S., *J. Chem. Eng. Jpn.* **2000,** *33,* 333–336.
[94] Dong, D., Lin, Y. S., Hu, M. Z. C., Peascoe, R. A., Payzant, E. A., *Microporous Mesoporous Mater.* **2000,** *34,* 241–253.
[95] Li, L., Dong, J. H., Nenoff, T. M., *Sep. Purif. Technol.* **2007,** *53,* 42–48.
[96] Kiriukhin, M., Collins, K. D., *Biophys. Chem.* **2002,** *99,* 155–168.
[97] Pavlov, M., Siegbahn, P. E., Sandstrom, M., *J. Phys. Chem. A* **1998,** *102,* 219–228.
[98] Palinkas, G., Radnai, T., Szasz, G. I., Heinzinger, K., *J. Chem. Phys.* **1981,** *74,* 3522–3526.
[99] Miachon, S., Dalmon, J.-A., *Top. Catal.* **2004,** *29*(1–2), 59–65.
[100] Caro, J., Noack, M., Kölsch, P., Schäfer, R., *Microporous Mesoporous Mater.* **2000,** *38*(1), 3–24.
[101] de la Iglesia, O., Irusta, S., Mallada, R., Menendez, M., Coronas, J., Santamarıa, J., *Microporous Mesoporous Mater.* **2006,** *93,* 318–324.
[102] Bernardo, P., Algieri, C., Barbieri, G., Drioli, E., *Catal. Today* **2006,** *118,* 90–97.
[103] Barbieri, G., Drioli, E., Golemme, G., *Chem. Eng. J.* **2002,** *85,* 53–59.
[104] Gu, X., Dong, J., Nenoff, T. M., Li, L., Lee, R., *J. Catal.* **2005,** *102*(1–2), 9–13 (and references within).
[105] Guczi, L., Borko, L., *Catal. Today* **2001,** *64,* 91–96.
[106] Hasegawa, Y., Kusakabe, K., Morooka, S., *J. Membr. Sci.* **2001,** *190,* 1–8.
[107] Sotowa, K.-I., Hasegawa, Y., Kusakabe, K., Morooka, S., *Int. J. Hydrogen Energy* **2002,** *27,* 339–346.
[108] Hasegawa, Y., Sotowa, K.-I., Kusakabe, K., Morooka, S., *Microporous Mesoporous Mater.* **2002,** *53,* 37–43.
[109] Goerke, O., Pfeifer, P., Schubert, K., *Appl. Catal. A* **2004,** *263,* 11–18.
[110] DOE/H_2 Multi-Year Research, Development and Demonstration Plan, available electronically at: http://www.1.eere.energy.govhydrogenandfuelcells/

CHAPTER 15

Gas Sensing with Silicon-Based Nanoporous Solids

Miguel Urbiztondo, Pilar Pina, *and* Jesús Santamaría

Contents

Abstract

In this chapter, we discuss how Si-containing nanoporous solids can be used to improve the performance of gas sensors. Therefore, materials such as porous silicon, zeolites, mesoporous silica and ordered mesoporous materials are dealt with, and the sensing applications are classified according to the type of transducer used. Far from attempting a comprehensive review, we have tried to give a broad description on the uses and prospects of Si-based nanoporous solids in gas sensing, using selected examples to illustrate specific developments or applications.

Keywords: Advanced Materials, Gas Sensing, Nanoporous Solids, Si-Based

Abbreviations

CMOS Complementary Metal–Oxide–Semiconductor
IDC Interdigital Capacitor

Ordered Porous Solids
DOI: 10.1016/B978-0-444-53189-6.00015-9

MEMS	Microelectromechanical Systems
NEMS	Nanoelectromechanical Systems
NR	Nile Red
PEG	Polyethylene Glycol
PL	Photoluminiscence
Pph3	Triphenylphosphine
QCM	Quartz Crystal Microbalance
SAW	Surface Acoustic Wave
SPV	Surface Photo-Voltage
TNT	Trinitrotoluene
β-CD	Alkenyl-β-Cyclodextrin

1. Introduction

The two most important qualities of a gas sensor are its sensitivity (defined as the capability of sensing ever smaller amounts of a given compound) and its selectivity (the ability to respond only, or at least predominantly, to certain components in a mixture, and not to others). It is clear that the progress achieved during the last decade in terms of sensitivity has been outstanding, with ultra-sensitive nanoelectromecanical devices (NEMS) now reaching attogram (10^{-18} g) level resolution.[1] However, comparable advances have not been achieved regarding sensor selectivity, and therefore the sensing of a specific component in a gas mixture where other components may interfere still remains a challenging task. This is especially important given the increasing demand for sensors capable of discriminate specific components in complex mixtures, and has prompted a strong research effort in this direction. Very recent examples of investigations addressed to the development of sensor systems capable of analysing mixtures in the food,[2,3] safety/security,[4,5] health,[6,7] process control[8,9] and environmental monitoring[10,11] fields can easily be found in the literature.

In the last two decades, the traditional classification of materials (metals, semiconductors, polymers, ceramics and composites) has become outdated because many of the most promising sensors include hybrids, nanomaterials and biomolecular materials.[12,13] Very often, gas sensing is a surface-controlled process, and from this point of view, hybrid devices are of especial interest, with a suitable bulk material acting as support of a functionalized surface where most of the sensing tasks are located. Also, as we progress towards single-molecule analysis, it becomes increasingly clear that the importance of nanostructured materials in sensors will increase. On the one hand, in these materials, the transducer has a physical structure in which one or more dimensions are in the size range of the molecules to be analysed, and this often can be used to develop surface–molecule interactions that increase the sensor selectivity. On the other, the relative number of sensible sites exposed on the surface increases as the dimensions of the system are reduced,[14] leading to an increase of sensor sensitivity. Finally, in general, the response time also decreases as we move to the micro- and then to the nanoscales, due to the acceleration of mass and heat transfer as the surface to volume ratio increases, together with lower thermal inertia (allowing faster temperature cycling) and a more rapid renewal of the fluid phase in contact with the sensor surface. Thus, it can generally be said that scaling down from micro to nano should lead to new, highly efficient sensor platforms, able to measure faster and more accurately than traditional devices. However, scalability, reliability and connection to the macroworld are still challenging issues, as noted by French *et al.*,[15] who pointed out that in many cases scaling down is still application-specific with system integration being one of the main practical obstacles.

By definition, nanostructured materials have at least one dimension smaller than 100 nm, but often the characteristic sizes are well below this limit. Nanosized structures are typically understood as items in the form of particles, wires/tubes or thin films, whose *external* dimensions fulfil the above size requirement. However, there are also intrinsically nanostructured materials that *contain* nanosized entities. Thus, it is widely accepted that nanostructures proliferate in biological systems. Also, and directly concerned with this chapter, there are natural and synthetic materials whose structure contains internal cavities or pores. When the dimensions of these pores are between a few angstroms and a few nanometres (therefore, in the micropore or low mesopore region), guest-host force fields develop,[16] and the resulting interactions can be used to implement molecular recognition functions. Nanoporous solids are abundant, and of a varied nature, including carbon, silicon, silicates, polymers, organosilicas, ceramics, metallic minerals and metal-organic frameworks.[17–20] As could be expected, the distribution of sizes, shapes and volumes of the void spaces in porous materials, coupled with their chemical structure and composition, have a direct influence on their performance as receptors in sensing applications. In particular, materials with tailor-made porosity and chemical specificity hold considerable promise for the development of chemical sensors, as will be discussed below.

1.1. Increasing sensor selectivity with nanoporous solids

In view of the molecular recognition properties that nanoporous solids can afford, the strategy followed by many research groups to increase sensor selectivity involves the use of nanoporous solids as added elements on already existing platforms of sufficient sensitivity [e.g., quartz crystal microbalances, (QCMs), surface acoustic wave (SAW) devices or cantilevers]. To this end, the nanoporous material(s) are deployed on the sensor with the aim of serving either as a target for the desired species or as a barrier for interfering components. The added nanoporous materials may be in the form of isolated entities or be a fully grown film completely coating the sensor surface. For the purposes of this chapter, the term *coating* will mainly be used for the first case, that is, when the sensor is covered by a layer made of largely independent crystals or nanoparticles, *film* will mainly be employed for well-covered surfaces, in which there may be some degree of intergrowth among the individual particles, and finally, the term *membrane* will be used to describe well-intergrown layers forming continuous films, normally in barrier-type applications.

The layers on the sensors can also be classified according to the dimensions of the individuals that constitute them. Thus, the so-called zero-dimensional nanoporous materials used in sensors include colloidal particles and atomic or molecular clusters, which are nanodimensional in the three spatial directions. These will often be combined into patterned single layers, self-assembled multilayers and hierarchical structures using a variety of preparation methods that allow direct assembly of zero-dimensional nanobuilding blocks into complex architectures with advanced functionalities. By progressively removing the nanodimension limitations, we move from 0-D to 1-D structures (fibres, wires, springs) where the nanorestriction applies to two spatial dimensions and 2-D structures (films and membranes), which are nanorestricted in only one dimension (thickness) (see Fig. 15.1). It should be noted that many of the nanoporous solids used today (such as catalyst particles) can be viewed as 3-D structures constituted by zero-dimensional building blocks (or alternatively, as 3-D structures containing a network of 1-D void entities).

Hundred per cent selective gas sensing, in which a sensor responds solely to one gas phase molecule and is not affected by the presence of other chemical species, is rarely accomplished and usually takes place only in especial situations, where the number and nature of the gas phase components is limited. A more realistic situation involves multicomponent mixtures and sensors that respond, at least partly, to more than one species. In this case, the strategy of choice often follows a biomimetic approach with the so-called electronic or artificial "nose" consisting of a chemical

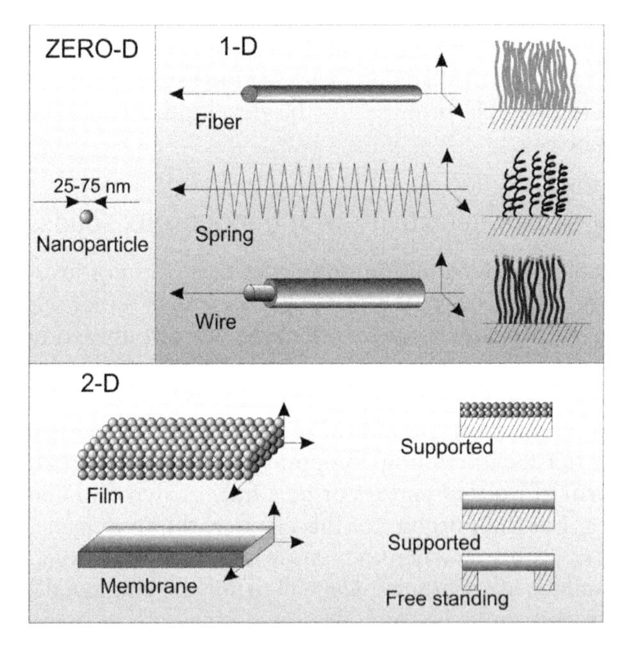

Figure 15.1 Nanoporous structures usually prepared by *bottom-up* procedures: from zero-dimensional to 2-D assemblies. (See color insert.)

sensor array and a pattern recognition system (e.g., an artificial neural network). In this way, sensor arrays are analogous to olfaction systems containing multiple receptors whose responses are interpreted by brain-controlled odour recognition processes. Each vapour presented to the sensing system produces a signature or 'fingerprint', and thus by using the pattern recognition system and the fingerprints database, the identification of each chemical becomes possible.

It is clear that a workable electronic nose requires miniaturisation of its components, for economic as well as for practical (system integration) reasons. The manufacture of miniature devices (down to the *micro* and to the *nano* levels, as defined above) is still primarily based on *top-down* approaches,[21] starting from large building blocks (e.g., a whole Si wafer), and reducing them to smaller elements through a well defined series of etching and micromachining steps (see Fig. 15.2). Standard silicon processing technology allows the integration of microelectronic and micromechanical components on a single wafer with patterns whose dimensions are close to the micro–nano frontier. Further scaling down can be achieved by laboratory techniques such as electron or ion beam lithography. However, there is a clear need in many sensing applications to go beyond silicon processing, to materials that are more amenable for sensing.[22] On the other hand, the *bottom-up* manufacturing methods involve the direct use of atoms, ions, molecules or clusters for the construction of larger functional structures (belonging to the zero, 1-D or 2-D categories already discussed). These can then be used in a hybrid approach (see Fig. 15.3), which has often been used in the publications reviewed in this work: the sensor is created using a top-down approach (usually on silicon), and then it is functionalized by added elements that enhance its selectivity. These elements are commonly prepared by bottom-up manufacturing and then coated or patterned onto the sensor surface.

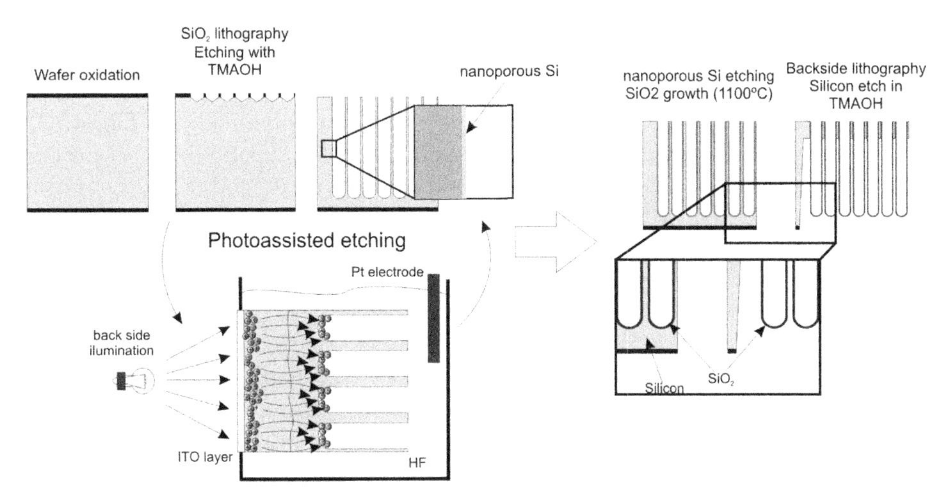

Figure 15.2 An example of a *top-down* approach for the fabrication of nanoporous structures: from an Si wafer to an array of silicon nanoporous pillars(adapted from Rodriguez *et al.*[23]). (See color insert.)

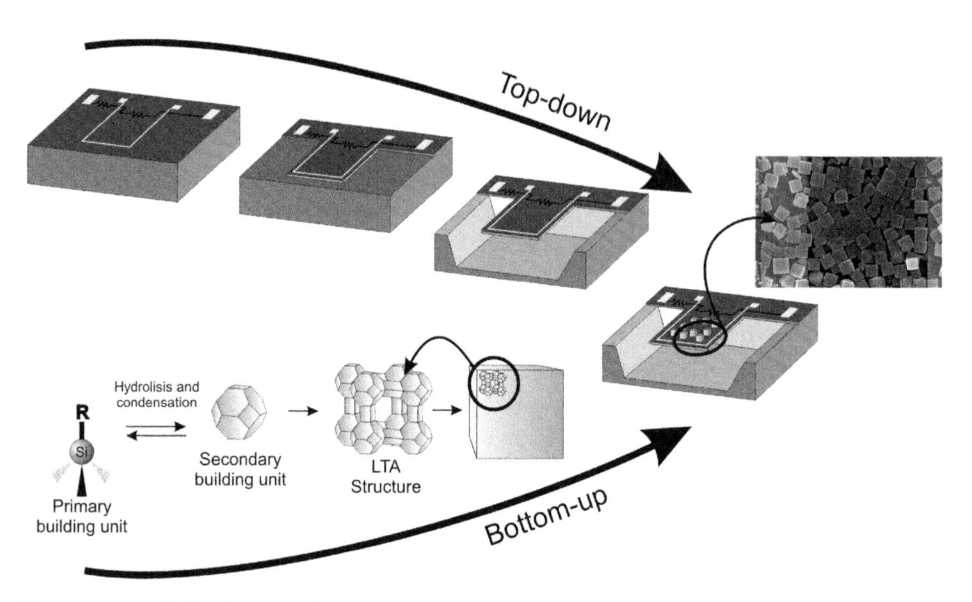

Figure 15.3 A hybrid approach to prepare humidity sensors: Si cantilevers are fabricated with standard *top-down* procedures. Then a selective coating is deposited on the cantilever using NaA zeolite crystals prepared by a *bottom-up* synthesis. (See color insert.)

2. Porous Silicon

2.1. Fabrication and structural properties of porous silicon

Porous silicon[24,25] has attracted a strong interest of the scientific community since the discovery of room temperature photoluminescence (denoted as PL) in 1990. Anodisation of crystalline silicon in solutions of hydrofluoric acid is now a well-established preparation method where the type of substrate doping and density determine the general morphology of the porous system obtained (see Fig. 15.4), but the detailed characteristics of the product obtained also depend on other process parameters such as temperature, current density, illumination and etchant concentration. Thus, as-generated amorphous porous layers can appear as macro-, meso- or microporous structures.

Anodic etching allows the fabrication of structures with characteristic dimensions covering three orders of magnitude in a process that does not need masking layers or nanolithograpy. Moreover, Si-based devices are amenable to the creation of microfabricated arrays with integrated complementary metal-oxide semi-conductor circuits, opening up an easy route to a cost-effective mass production of microsystems. For industrial application, the thickness, porosity, pore size distribution and specific surface area are the key parameters controlling the chemical, mechanical and thermal behaviour of these materials. Spectroscopic ellipsometry has become the non-destructive characterisation method most often chosen to study microstructural (pore size distribution and surface area measurement) and optical properties of

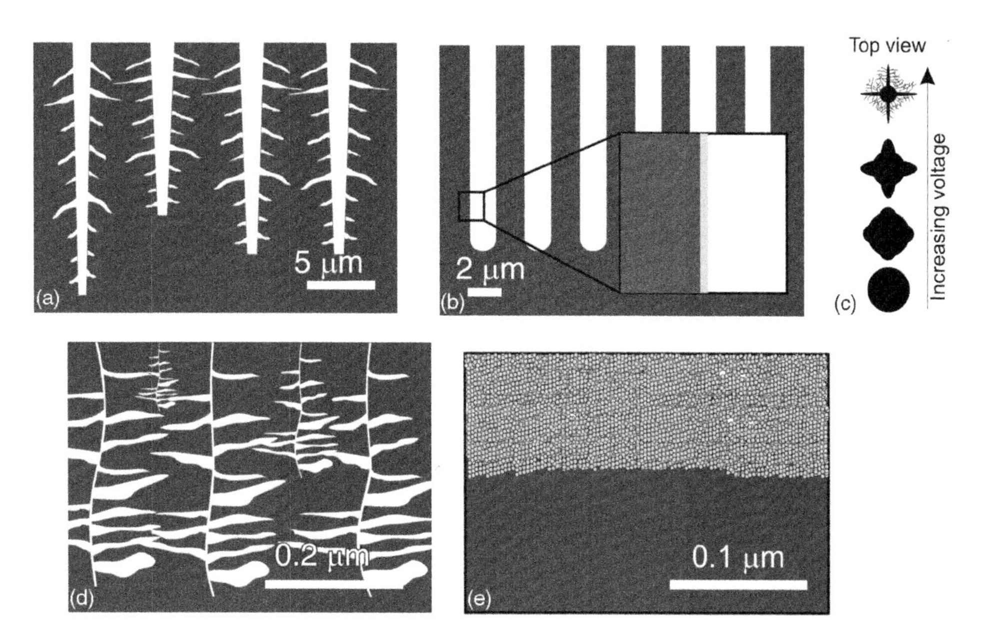

Figure 15.4 Morphologies of porous silicon: (a) formation of branched macroporosity onto n-type [100] oriented Si in the dark; (b) straight and regular macropore deep trenches formed on pre-etched n-type [100] oriented Si; (c) schematic evolution of the macropores with increasing current density; (d) branched mesoporosity developed on n^+-type [100] oriented Si; (e) microporosity formation on top of a p-doped silicon (adapted from Thomas *et al.*[24]).

porous silicon.[26] In this technique, water is normally used as adsorbate to determine pore size distributions due to its efficient penetration in adsorbents with small pores or "ink-bottle" shaped pores. On the other hand, the optical changes in porous silicon layers due to adsorption from different vapours can be used for sensing purposes as will be described when dealing with ellipsometry-based optical sensors.

The large internal surface area of anodized microporous silicon layers ca. 200 m^2/cm^3 and good chemical reactivity are excellent starting points for gas sensing applications.[27] However, the interest in the use of porous silicon for chemical sensors lies mainly in the change that its optical (i.e., PL) and electrical (resistivity, capacitance, conductance and so on) properties experiment when the charge state of the surface is modified by the presence of gas molecules that are chemically or, more often, physically adsorbed or undergo capillary condensation.

The high sensitivity provided by porous silicon sensors in optical and electrical transducers will be briefly discussed in Sections, 2.2 and 2.3, respectively. However, a sponge-like structure is also generally formed that retards the gas-vapour transport, leading to relatively long response times. To alleviate this problem, some authors have attempted a hybrid macroporous (1–20 μm in diameter, 20–100 μm in depth)–microporous layer (around 200 nm in thickness) on silicon wafers previously etched by standard lithographic techniques to generate an ordered array of etch pits,[23] thus facilitating a rapid transduction of changes in the analyte gas concentration[28]. On a similar concept, Li and co-workers, prepared hierarchical structures combining a regular array of micron-sized silicon pillars on which a nanometre scale layer were deployed to be used as thin film sensors.[29–31]

2.2. Optical transducers

PL quenching on porous silicon forms the basis of simple and inexpensive chemical sensor devices. Both chemisorbed and physically adsorbed chemicals are capable of PL quenching.[32]

Chemical sensors taking advantage of this phenomenon at room temperature have been developed for organic solvents,[33] nitroderivates,[34] NO_x,[35,36] SO_2 and halogens,[37,38] CO[39–40] and CH_4.[40] Zhang *et al.*[32] have surveyed detection limits for this type of sensors ranging from ppt (parts per thousand) to ppb (parts per billion) levels. It is worthwhile mentioning the approach used by Sailor *et al.*[34] to increase the sensor selectivity when discriminating between nitroderivates and aromatic compounds. In this case, selectivity was achieved by oxidation of nitroderivates to NO_2 in an upstream reactor with noble metal catalysts.

While it is true that different photophysical processes result in PL quenching, the mechanism is still the subject of debate, with more than one likely contributing phenomenon (energy-transfer, surface chemisorption, charge-transfer and so on).[41] In addition, the metastability of the initial surface upon exposure to ambient air has to be taken into account in view of the changes that it may induce in the structural, optical and electrical properties of the sensor. Thus, it has been demonstrated that extended storage in air at room temperature converts the hydrophobic, hydrogen-terminated surface of freshly anodized layers into a contaminated native oxide. Because of this, different modifications to the porous silicon surface by oxidation[37,38] or silylation[33] have been proposed not only to obtain higher stability but also to increase selectivity because the surface interactions can be tuned by specific chemical transformations. Sailor *et al.*[42] have integrated stabilized porous silicon chips in artificial noses using PL quenching as one of the operating transduction modes. They were able to attain a much faster sensor response and baseline recovery compared with commercial devices based on metal oxide sensors, while maintaining a similar discrimination capability. Improved performance was expected for upcoming sensors with a variety of doping types, pores sizes and chemical surface modifications.

As previously pointed out, the ellipsometry spectroscopy data can be used for gas sensing.[26,43,44] The operating principle is based on the changes in the polarisation of light reflected by the sample when exposed to vapours (water, acetone, alcohols and so on). Optical analysis directly detects the presence of adsorbed or condensed vapour (depending of the relative pressure) in the micropores. Normally, four-layer optical models with the porosity changing slightly from one sublayer to the next[26] are sufficient to represent the porous silicon structure used in these devices. A structure with a thickness typically ranging from 300 to 700 nm where the vapour to be detected can penetrate by diffusion, replacing the air filling the pores and introducing additional dipole moments, results in substantial optical changes that can be measured, with a considerably improved sensitivity. This, however, comes at the expense of higher baseline recovery times, depending on transport resistances in the pore network. Similarly to PL sensors, in recent years, a considerable effort has been devoted to post-processing of porous silicon to fine-tune the surface properties (hidrophilicity/organophilicity). Also, polymers[43] or metals[44] have been introduced, avoiding pore blockage, to increase the sensor selectivity and sensitivity.

2.3. Electrical transducers

Electrical sensors based on porous silicon generally use the variation of conductance or capacitance induced by changes in the composition of the gas atmosphere in contact with the sensor. The observed responses can be explained by the effect of replacing the air in the pores by the analyte, with significantly different physical properties (dielectric constant, dipole moment and electronic polarisability).[45–47] Application of this concept is straightforward in scenarios where the concentration of the analyte is relatively important (e.g., ethanol concentration in breath analysers).[46]

Nevertheless, conductance variations at room temperature are appreciable even at lower gas concentrations because charge transfer mechanisms are also involved. In addition, the variations in the electrical conductivity of porous silicon with gas atmosphere as a transduction mechanism is probably one of the easiest and simplest ways to realize a gas sensor. However, implementation of these systems is hindered by practical problems, such as the difficulties in establishing a reliable, low-resistance electrical contact due to the fragility of the porous silicon surface; also, deposition procedures involving several steps are often required to produce devices for specific applications. Finally, special attention must be paid to 'wet' photolithographic steps because of the high rate of oxidation of Si microstructures that may result.[28]

In the usual configuration, the sensor is biased between two electrodes at a constant dc voltage while the current is measured. Very promising results for NO_2 sensing[48–52] have been reported in the literature using microporous silicon with detection limits in the ppb range. In this case, the electrical conductivity of porous silicon is strongly enhanced by the presence of adsorbed NO and NO_2 species, due to the electrostatic interaction between the polar molecules and interfacial defects on the porous surface. This allows overcoming the effect of interfering species with lower dipole moment, such as organic vapours (alcohols), hydrocarbons and environmental pollutants (benzene, ozone, CO). On the other hand, the presence of humidity leads to a substantial decrease of conductivity[49] explained by the donor-like character of water molecules adsorbed at surface defects.

The formation of hybrid macroporous (3 μm diameter conically shaped)/microporous (around 2 nm in diameter) layers on silicon wafers by specific etching procedures (see Fig. 15.4b) provides a high surface area platform that has been successfully applied for rapid detection of reducing gases NH_3, HCl, NO, CO and SO_2.[28,53] Stabilisation of the surface by HCl cleaning and deposition of metals (Au, Sn) by electroless metallisation have allowed to reach detection limits in the low ppm (below 5 ppm of CO and NO_x) and ppb (500 ppb of NH_3) levels. The sensitivity and response times are comparable to those of SnO_2 solid state sensors, with the added advantage of operating at room temperature.

On a similar approach, a rapid capacitive sensor for humidity (5–15 s) and ethanol (15–30s) has been developed by Li *et al.*[29,30] based on a regular array of silicon nanoporous pillars built on single crystal silicon wafers. The inclusion of $Fe(NO_3)_3$ during hydrothermal etching improves the stability at high temperatures and under harsh environments. In a further refinement, the same authors deposited a thin film of nanocrystalline magnetite[31] to enhance the output signal intensity at the expense of slightly higher response time.

3. Si-Based Microporous Solids: Zeolites and Related Materials

The well-defined porous structure of zeolitic materials combined with their tunable adsorption properties and ion-exchange capacity make them ideal candidates as added-on selectivity enhancers in gas sensors, a role also assisted by their high thermal stability and good chemical resistance. As such, zeolite coatings, films and membranes have been used in a variety of applications in which the discriminating role of zeolites is based on their use as molecular sieving agents (i.e., when zeolites operate in the size exclusion regime), as selective adsorbents or both. In addition, at the operating temperatures of some types of sensors, the catalytic activity of zeolites is already evident, and this can also be used to remove interfering components. Different uses of zeolites for gas sensors were reviewed by Xu *et al.*[54]

In addition to zeolites, in this section, we have also included other ordered microporous structures, different from classical zeolites, containing tetrahedrally co-ordinated phosphorus, such as $AlPO_4$[55,56], SAPO-34[57–59], and also Ti and V in their structure. Both ETS-4 with a pore size of 0.3–0.4 nm and ETS-10 with a larger pore size (0.49 × 0.76 nm) have been prepared as films and membranes on porous supports,[60–63] thus opening their possible deployment in sensor applications. Mixed octahedral–tetrahedral oxides present novel possibilities of isomorphous framework substitution in principle allowing fine-tuning of the catalytic and adsorption properties, while preserving the microporous structure. Finally, a microporous titanosilicate ($K_2TiSi_3O_9 \cdot H_2O$), with the structure of umbite and a pore size of 0.3 nm, has been prepared as a continuous ca. 5- μm-thick membrane on porous TiO_2 tubular asymmetric supports.[64] The small pore size of the membrane allowed H_2/N_2 separation factors close to 50, which may be useful in H_2 sensing applications.

Finally, new larger pore zeolite materials that have not yet been used in sensors present interesting properties that are likely to be tested in future applications for sensing intermediate size molecules. As examples in this group we find wide pore zeolite such as ITQ-21 that is accessible through six circular 0.74 nm openings,[65] and extra-large-pore materials, like the phosphate-based VPI-5 or the more stable silicas UTD-1[66] and CIT-5[67] with 0.8–1.2 nm openings.

3.1. Electrochemical/electrical transducers

One of the earliest sensor-related uses of zeolites took place with chemical electrodes, in which the addition of zeolites often allowed the improvement of the electrode surface by increasing their electroanalytical performance and stability. These are still among the major challenges remaining before these devices can make the transition out of the laboratory into commercial use.[2]

Zeolites have also been used as off-sensor agents, often using zeolite beds as catalyst or barriers to reduce the influence of non-targeted elements either by chemical reaction or by adsorption.[68–70] In this case, relatively thick filters (typically several hundred milligrams) were typically used, leading to a slow sensor response. A more sophisticated approach was used by Fukui and Nishida,[71] who used commercial crystals of FAU and FER zeolites with colloidal silica as a binder to cover the sensing layer of La_2O_3-Au/SnO_2 sensors, and could remove ethanol interference during CO sensing. Different zeolite masks, also consisting of commercial crystals, were used to minimize the interference of O_2 in yttria-stabilizsed zirconia sensors used to analyse nitric oxide.[72]

In previous works of our laboratory,[73,74] MFI and LTA films (in fact most of the MFI films and some of the LTA films prepared could be considered as membranes because they presented a high degree of intergrowth) were directly synthesized on a Pd-doped SnO_2 layer (see Fig. 15.5). These molecular filters decreased interference of undesired molecules mainly via a selective adsorption mechanism that hindered the diffusion of less strongly adsorbed species. Depending on the operating conditions used, the sensitivity of the zeolite-modified sensors towards certain analytes could be reduced, and in some cases suppressed altogether, thus strongly increasing the sensing selectivity. More recently, the system has been scaled down by a factor of 50, from the mm^2 (the SnO_2 surface

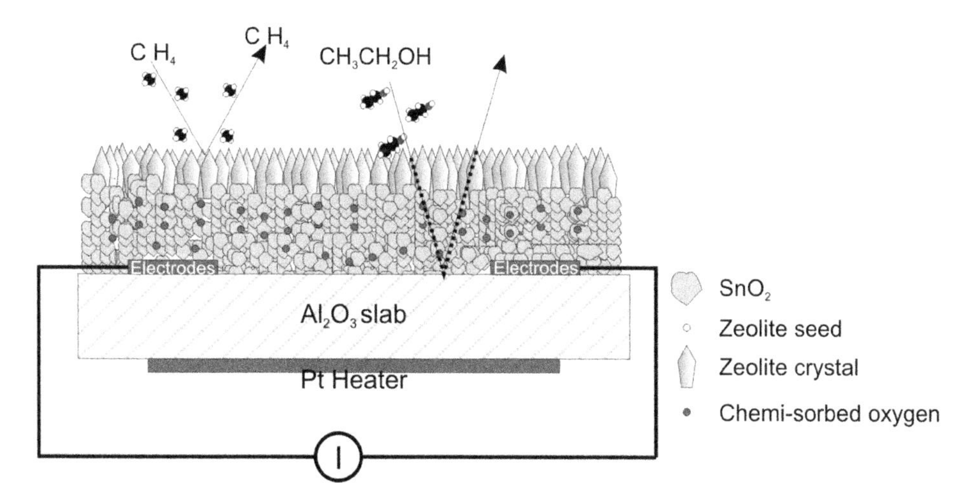

Figure 15.5 Schematic cross-section of a zeolite-modified Pd/SnO_2 sensor in operation. The zeolite layer eliminates the sensor sensitivity to methane, even in high concentrations.[73] (See color insert.)

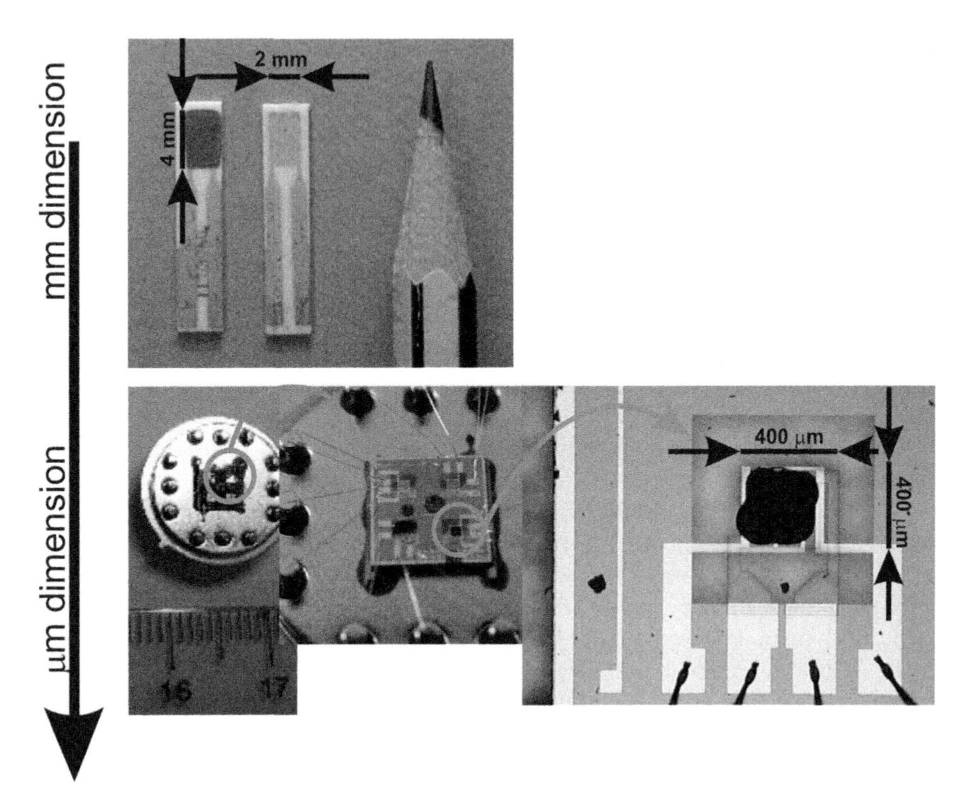

Figure 15.6 Zeolite-modified Pd/SnO_2 sensors of different dimensions. Top: a 2 mm × 4 mm sensor screen-printed on alumina. On top of the Pd/SnO_2 surface, a zeolite layer was grown hydrothermally.[73] Bottom: Detail of a micromachined sensor containing four microsensors, each 400 microns of lateral size. In this case, zeolite crystals were deposited by a microdropping technique.[74] (See color insert.)

in the works cited was screen-printed on a 2.5 × 4 mm surface) to the μm^2 (400 × 400 μm) range (see Fig. 15.6) without loss of sensor performance.[74]

A different approach is based on the fact that the adsorption of a gas produces a change of the dielectric constant of zeolites whose effects can be measured. In this way, chemical sensors based on interdigital capacitors (IDCs) using zeolite layers as sensitive coatings find many different applications depending on the type of zeolite, its chemical treatments (e.g., ion exchange) and the working temperature of the zeolite-coated IDC sensor. Zeolite-coated IDCs have been tested for sensing *n*-butane[75] and also, NH_3, NO and CO[76,77] on Na-Y and NaPtY zeolite-based sensors at temperatures high enough (above 200 °C) to take advantage of the catalytic properties of zeolites. The response time was of the order of seconds and one of the advantages of working at high temperatures was that water condensation was avoided, thereby reducing the cross-sensitivity to water. Under certain conditions, the selectivity of these reactive chemical sensors is remarkable. Thus, the detection of 10 ppm of *n*-butane with an NaPtY interdigitated capacitor with no response to CO and H_2 has been reported.[76] Similarly, Moos *et al.*[78] described a

ZSM-5-based capacitor sensor with on-chip heating for temperatures up to 450 °C capable of detecting NH_3 with no cross-sensitivity to COx, hydrocarbons and O_2. In our laboratory, we have used different zeolites to coat IDC sensors working at room temperature. Again in this case, the objective is to use the zeolite crystals to provide sensor selectivity by acting as the discriminating agent among gas phase molecules. Both simple coating with preformed zeolite seeds, with or without chemical bonding to the silicon surface and *in situ* growth of zeolite films by hydrothermal synthesis were used. Figure 15.7 shows an example of the sensitivity imparted by modifying an IDC-type capacitor (10 μm in thickness) with zeolites.

3.2. Mass/piezoelectric transducers

Piezoelectric sensor devices (QCMs) have also been modified with zeolites to take advantage of the molecular sieving effects and selective surface interactions provided by different zeolites in gas sensing applications (see examples in the patent literature[79,80]).

Zeolite-modified QCMs were used to sense ethanol (using MFI zeolite[81]), humidity (using LTA and BEA zeolites[82]), and NO, SO_2 and water (using zeolite A, silicalite-1 and sodalite[83,84]). In a recent work, Vilaseca *et al.*[85] showed an example of the discriminating capacity of zeolite-coated QCM sensors. Either hydrophilic $AlPO_4$-18 or hydrophobic silicalite-1 layers were deposited on the QCMs that were then exposed to propane–water vapour mixtures. Each of the sensors was able to substantially eliminate the interference of the opposing species (i.e., the silicalite sensor showed a strong sensitivity for propane and almost null for water, when sensed in the presence of propane, while the opposite was true for the $AlPO_4$-18-loaded sensor).

SAW devices, consisting of a single-crystal quartz substrate with interdigital transducers, can also operate as highly sensitive piezoelectric balances that respond to small amounts of gas adsorbed via frequency changes of an oscillator circuit. The application of zeolite coatings (H-ZSM-5, zeolite Y, chabazite and zeolite A) on SAW devices was pioneered by Bein *et al.*[86,87] for humidity and vapour sensing

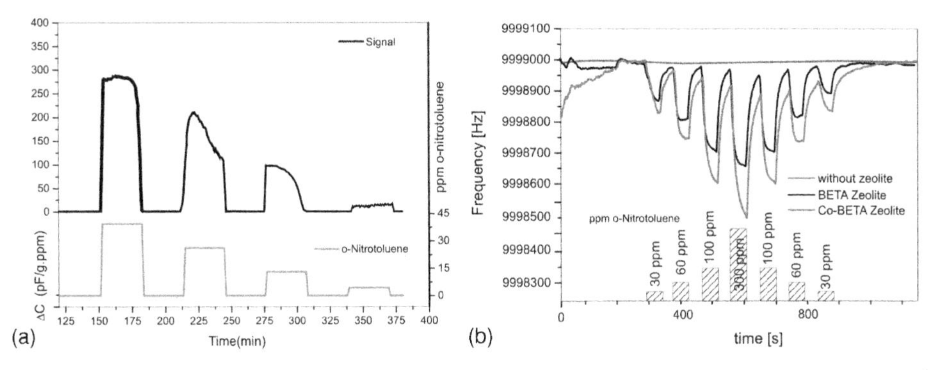

Figure 15.7 Room temperature capacitance of a BEA zeolite-modified IDC-type capacitor as a function of the concentration of *o*-nitrotoluene vapours. (See color insert.)

(methanol, ethanol, propanol, isooctane, pyridine and perfluorotributylamine). These authors[87] found out that a porous silica overlayer enhanced the mechanical stability of the system providing a reactive surface in which organosilane molecules could be grafted via siloxane linkages forming a molecular gate controlling the permeation of larger molecules.

Similarly to QCMs, cantilever-based sensors can also be configured to work as tiny microbalance in which the analysis of changes in the resonance frequency or in the deflection of the cantilever can be used to determine mass loading, as shown schematically in Fig. 15.8A and B, respectively.

Standard cantilevers are theoretically capable of detecting a minimum mass loading of 50 fg with a short response time (milliseconds), although more advanced cantilevers can detect even smaller amounts, as shown above. As with previous sensors, the main challenge with cantilevers is not attaining a high sensibility (which in most cases is sufficient for the desired applications), but the ability to respond solely to the species to be sensed. This becomes even more critical as the sensor sensitivity increases, and ever smaller amounts of other species are able to interfere with the measurement. Scandella *et al.*[89] showed how the sensitivity towards certain compounds could be enhanced by simply depositing crystals of a suitable zeolite. Based on this concept, ZSM-5 crystals (around 500 ng) have been chemically anchored to detect humidity[88,89] and freon-12,[90] achieving a satisfactory performance in determining mass loadings at the nanogram scale. Moreover, arrays of such devices have been fabricated[89] to conduct complex analysis of vapours by using various microporous materials (ZSM-5, VZSM-5) attached to the sensors. Figure 15.9 shows a SIL-1-loaded spade-shaped cantilever used in our laboratory for hydrocarbon sensing at ppm levels.[91] The fabrication procedure has already been schematized in Fig. 15.3.

The application of etching techniques (see Fig. 15.10a) borrowed from the microelectronics processing allows the release of freestanding zeolite structures[92] that could give rise to zeolite-only cantilevers (see Fig. 15.10b).

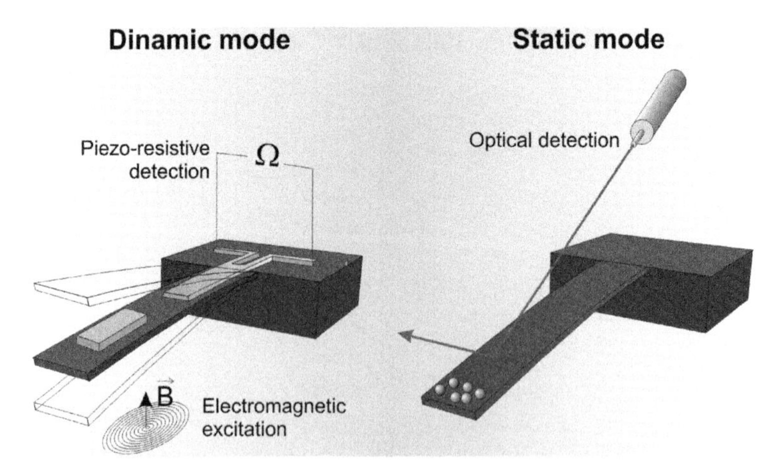

Figure 15.8 Scheme showing cantilever mass sensors operating in the resonant-dynamic (left) or deflection (right) modes. (See color insert.)

Figure 15.9 Left: Optical micrograph of a spade-shaped (the spade form increases the sensible area) cantilever sensor. Right: SEM micrographs detailing the zeolite-coated surface of the cantilever. (See color insert.)

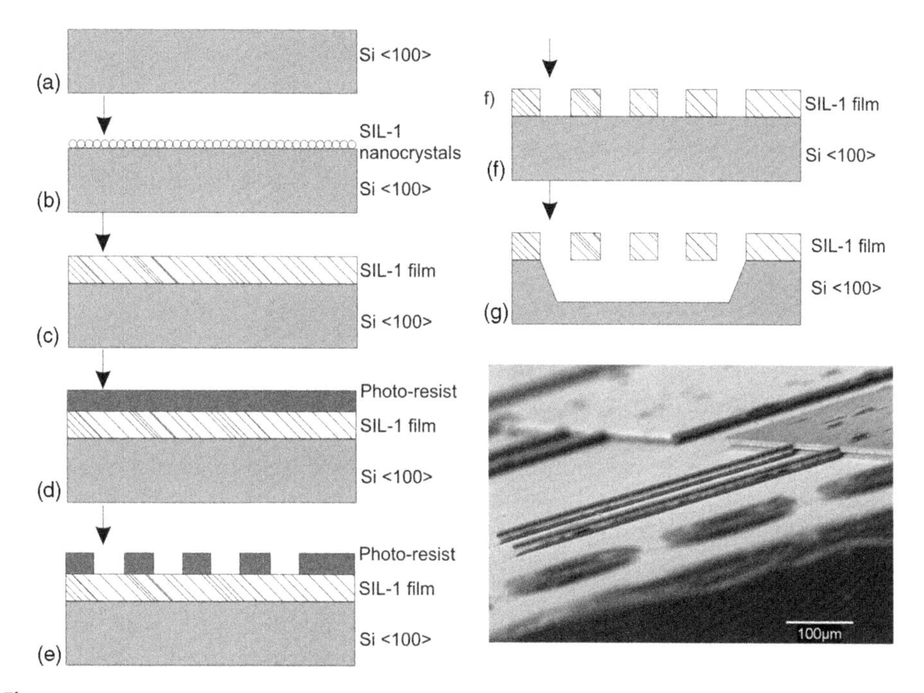

Figure 15.10 (a) Starting Si wafer; (b) spin coating of SIL-1 nanocrystals (4 wt.% in ethanol); (c) hydrothermal synthesis of a zeolite; (d) deposition of a TI-35 ES reversal photoresist; (e) UV photolithography, reversal bake process and resist development; (f) BHF etching of the SIL layer; (g) TMAOH etching of Si and zeolite cantilever release, as shown in the SEM micrograph.

3.3. Optical transducers

Optical sensors are among the most important chemical sensors produced for real-time monitoring of diverse analytes.[93] Over the last few years, more attention is being paid to sensors based on the optical properties of nanomaterials because of several advantages over conventional sensors: resistivity to electromagnetic noise, larger potential for miniaturisation and the capability to remote control and information transfer through an optical fibre network almost instantly.

A typical optical chemical sensor employs an analyte-sensitive species, whose optical properties vary in the presence of the molecule to be sensed. The main problem in this type of optical sensors is often related to the photo-unstability, meaning decomposition or evaporation on extended exposure to light.[94] Avoiding this problem is difficult due to the fact that the sensor response depends on exciting the sample. Moreover, as many of the sensors are designed for use in fibre optic systems, the sensor mass is small and a relatively low number of molecules must handle the intense light exposures necessary for adequate signal-to-noise ratio.

The high thermal and mechanical stability of zeolites as well as their optical transparency in the visible region and fast response towards ultraviolet radiation are some of the main features of zeolites invoked for their use in the encapsulation of organic chromophores to provide a confined environment that increases their stability. Indeed, molecular-sieve encapsulated dye molecules are attracting considerable attention with respect to new photonic devices and optical sensor applications[95,96] due to an increased stability against chemical attack, durability under light, migration stability, photobleaching and thermal decomposition (up to two orders of magnitude) with respect to their unconfined counterparts.

Moreover, the structural features of zeolite materials enable the incorporation of optically active guest molecules in crystallographically defined positions or highly organised arrangements resulting in peculiar host–guest interactions. Controlling the morphology and the size of the zeolite host, the optimisation of dye-zeolite composites for a specific application can be attained.

A variety of synthetic methods for the inclusion of dyes in molecular sieves, including dye synthesis within the nanopores and cages of the zeolite, or encapsulation during the hydrothermal synthesis, have been reported in the literature.[95] For instance, the encapsulation of the solvatochromic dye Nile red (NR) inside the pores of de-aluminated zeolite Y by a 'ship in a bottle' synthesis procedure and subsequent dye adsorption has been investigated by Meinershagen and Bein[97] for optical sensing of acetone and ethanol using spectral absorbance in the UV-Vis range. The sensing principle is based on the chromophore spectral change as a function of the polarity/dielectric constant of a solvent.[98] NR, a hydrophobic fluorescent and solvatochromic dye of molecular formula $C_{20}H_{18}$-N_2O_2, with a length of 10 Å and an equivalent diameter of ~6 Å, is one of the most interesting molecules, being highly sensitive to the polarity of the microenvironment.[99] It has been described as displaying positive solvatochromism leading to a bathochromic shift of 128 nm on measuring the UV-Vis spectra in hexane (a non-polar solvent) and water (a polar solvent).[100] This substantial shift is due to changes in the dipole moment of the first excited singlet state.[101] The structure of FAU-type zeolite, with a unit cell of that consists of 8 supercages 11 Å in diameter and 16 windows 7.4 Å in diameter, allows diffusion of NR molecules through the free diameter of the channels. When dye molecules are hosted in the FAU-type zeolite supercages, solvent loading in combination with the molecular sieving properties as well as hydrophilic-hydrophobic interactions control the spectral signature of the dye.

In our recent work,[102] the same concept has been used for continuous water and hexane detection with a robust and low-cost configuration based on standard plastic optical fibre. In this case, the reflection of the light on the surface of the zeolite layer deposited onto a glass substrate placed at the end of an optical fibre has been continuously measured as a function of the atmosphere composition. As the analyte molecules are adsorbed on the dye-zeolite composite, the reflectance spectrum of the embedded NR changes. The magnitude of these variations can be related to the vapour concentration, providing the basis for the design of an optochemical sensor. An analogous approach was used by Shorabnezhad *et al.*[103] who incorporated methylene blue in H-mordenite for humidity-sensing purposes. In this case, sensor operation was based on the change of colour on protonation/deprotonation of dye molecules associated with the dehydration/hydration of the zeolite.

Other research efforts are focused on the development of integrated optical sensors, in particular towards optical fibre sensors operating at standard telecommunication wavelengths.[104] One of the approaches followed creates coatings of NaY, LiY and BaY zeolite crystals on quartz fibres using TEOS as a binding agent[105,106] for asymmetric photoreactions and also for phosphorescence sensing of aromatic molecules.

4. Mesoporous Silica

For chemical sensing purposes, materials with tailor-made porosity are desirable. In this respect, mesoporous silica has drawn particular attention because of its chemical stability and well-established preparation chemistry, mainly via the sol–gel route. This involves simultaneous hydrolysis and condensation processes starting with a suitable alkoxide precursor $Si(OR)_4$, where R is typically an alkyl group, in a liquid medium (water and alcoholic or low molecular weight organic solutions). During the sol–gel transformation, the viscosity of the solution gradually increases as the sol (colloidal suspension of small particles <100 nm in size) becomes the gel, a semi-rigid network. After drying the system to remove water and alcohol from the pores, a tetrahedral SiO_2 network (dry gel) is formed whose properties (i.e., average pore size, pores size distribution, pore shape, surface area, etc.) are strongly influenced by the sol–gel process parameters.[107] Polymers and surfactants are commonly used as pore-generating agents, and both conventional (thermal) and plasma calcination are used for gelation. Compared to conventional thermal calcinations,[108] the use of argon[109–112] or oxygen[113] plasma is more attractive due to the advantages of low processing temperature and shorter processing times that allow the preparation of mesoporous films while preserving the properties of the transducer element. In addition, oxygen treatment creates silanol groups on the coatings, which often result useful for subsequent funtionalisation purposes.

In an interesting variation of standard preparation procedures, sol–gel materials can also be molecularly imprinted. This approach basically involves assembling a polymeric network around a suitable template. Upon removal of the template, cavities with a specific size, shape and/or chemical functionality remain in the cross-linked host and in this way a mesostructure can be created.[114] The sol–gel route to prepare molecularly imprinted polymers (vs organic polymerisation) is favourable because the control of thickness, porosity and surface area is easier. Imprinted functionalised silica has been deposited onto suitable transducers for fluorescence, piezoelectric, radioactive and electrochemical-based detection in the liquid phase.[20] When used for sensing applications, the usual objective consists in using the target molecule to leave an imprint in the structure prepared in the sol–gel process, in such a way that it can be uniquely recognised by the same molecule during sensing. As an example, Marx *et al.*[115] used parathion imprinted in a sol–gel matrix that was applied on QCM resonators for parathion sensing. While the cavity formed exhibited high specificity in the liquid phase, the same authors pointed out the difficulties in applying the same method to gas phase sensing.

Finally, Vycor glass, which is fabricated by a completely different procedure that involves chemical etching, has also been used in optical sensor applications. As an example, organic-impregnated mesoporous silica glass (Vycor 7930) with a nominal 4 nm pore size diameter and 200 m^2/g of surface area has been used by the group of Hayashi[116–118] as a low-cost optical sensor for NO_2 detection at ppb level. The low detection limits achieved, even in presence of humidity are due to colouration reactions between nitrogen dioxides and diazo-coupling reagents in the nanopores.

4.1. Mass/piezoelectric transducers

Polyethylene glycol-doped mesoporous silica has been used as sensible coating in QCMs for gas sensing applications (aromatics and alcohols) by Su *et al.*[110–112] Argon plasma treatment was been used instead of thermocalcination to avoid alterations in the piezoelectric properties of the quartz disk as well as possible detachment of the film from the surface due to the difference in the thermal expansion coefficients. The higher surface area of the three-dimensional mesoporous silica films was used to immobilise alkenyl-α-cyclodextrin (α-CD) modified through alkenylation by covalent linkage to the silica film previously thiolated.[110–111] The inner diameter of the α-CD toroid cavity (6.4–15.4 Å in diameter), its pronounced hydrophobic properties and the weak van der Waals forces with guest molecules were found to be suitable for benzene and ethanol detection (from 0.6 ppm V to 60 ppm V). In a step further, the same group used photolithography techniques to develop a four-channel QCM array in which it was possible to discriminate benzene and ethanol via the selective interactions with β-CD and triphenylphosphine hosted in the silica network.[112]

5. Ordered Mesoporous Materials: M41S and PMOs Families

The first synthesis of an ordered mesoporous material was described in the patent literature in 1969. However, since 1992, when researchers of the Mobil Oil Corporation discovered the remarkable features of the M41S family[119,120], a strong interest has developed based on the possible applications of these materials in membranes, catalysts, sensors, optical devices, insulating layers for microelectronics and drug delivery vectors, among other possibilities. As a matter of fact, many reviews[121–128] have been published covering various aspects of ordered mesoporous materials, such as their synthesis, surface modifications, application as host materials and catalysis.

These new class of materials contain amorphous silica walls spatially arranged with hexagonal or cubic symmetry displaying regular mesopores in the range of 2–30 nm (for SBA-15) after template removal. In general, highly porous ordered solids with extremely high surface area (from 700 up to 1400 m^2/g), mesopore volume (from 0.7 up to 2.5 g/cm^3) and narrow pore size distribution are obtained, which are obviously attractive for sensing applications.

In their interesting review paper, Hayward *et al.*[129] addressed examples in which mesostructured materials with "guest" regions possessing chemical, electrical, optical or magnetic properties within a "host matrix" with large surface area, mechanical stability, electrical conductivity, dialectical properties and optical transparency could provide new opportunities and compete with state of the art technologies in selected applications. In particular for sensors, they pointed out that these materials permit the confinement of optically or electrochemically active molecules within a well-defined matrix that can be constructed with a variety of electrical or optical properties. In addition, the versatile surface chemistry of silanol groups from MCM materials[126] was recognised in a patent for sensing application shortly after their invention.[130] Indeed the reactivity of silica allows these solids to be derivatised with a wide range of organic groups to give mesoporous organic–inorganic hybrids.[13,20] This can be achieved by post-synthesis grafting (reaction of organosilanes on pre-fabricated mesoporous materials) or, alternatively, in a single step by co-condensation of a tetra-alkoxysilane and one or more organoalkoxysilane(s) during synthesis. Either method can be optimised to obtain a better control of the amount of incorporated species and their short-range organisation, as well as various morphologies (particles, spheres, films and monoliths) could be achieved (see, for instance, the works by Brinker *et al.*[131] and Kaliaguine *et al.*[132]). In any case, the sheer weight of the large surface area of mesostructured materials is often the main driver in their use as performance enhancers of sensing devices.[133]

A related group of materials are the periodic mesoporous organosilicas, obtained by Inagaki *et al.*[134] In 2002, these authors prepared an ordered mesoporous benzene–silica composite containing a hexagonal array of mesopores with a lattice constant of 5.25 nm and crystalline pore walls, featuring structural periodicity with a spacing of 0.76 nm along the channel direction. Following this work, several alkyl groups as well as heteroatom-containing organic functional moieties have been successfully incorporated into these composites. The periodical arrangement of hydrophilic silicate layers and hydrophobic organic layers is capable to promote structural orientation of guest molecules or clusters, opening new possibilities such as novel catalytic processes and opto-electrical applications.

In summary, the field of ordered mesoporous silica-related materials present extremely attractive properties from the point of view of sensor development. Most of these, however, have not yet materialised as specific devices for gas sensing applications.

5.1. Electrochemical/electrical transducers

Different methods have been used to deposit mesoporous silica-related materials on electrode surfaces,[20] from a mere mechanically mixing of as-synthesised particles to sol–gel processing in the presence of a template. With the simplest approach,[135] improvements in H_2 sensitivities compared with the pure SnO_2 counterparts have been attained, due to the faster H_2 diffusion into the porous structure of sensor particles compared with larger molecules (i.e., CH_4, CO). The same group[136] developed a surfactant-templating synthetic method to increase up to 100 m^2/g the surface area of SnO_2 sensors [3 m^2/g for the commercial SnO_2]. These authors explained the linear dependence of H_2 and CO sensitivity with surface area by the total amount of surface active sites for catalytic oxidation of H_2 and CO over the surface of the as prepared SnO_2 material.[137]

Surface photovoltage (SPV) gas sensors based on SBA16[138] and MCM41[139] on metal-insulator semi-conductor structures have been proposed for NO_x detection. The changes in capacitance of the insulator layer (mesoporous film + Si_3N_4 + SiO_2) due to gas adsorption constitute the operating principle of this type of sensors. Thin films of SBA-16 deposited by spin coating over SPV systems exhibited high sensitivity (in comparison with bare SPV systems) towards NO detection at room temperature. The incorporation of transition metals such as Sn, V and W on MCM-41 materials has led to a sensitive detector of NO_2 gas at concentrations down to several hundred parts per billion.[139]

The electrical response to changes of relative humidity of 1-dH mesophase[140] and cubic-like[141] mesoporous silica films deposited by dip coating on alumina and silica on Si substrates with comb-type gold electrodes have also been evaluated for sensing applications. The large porosity of the receptor materials combined with a highly active surface coated with silanol groups adsorbs water molecules increasing the current intensity related to protonic conduction.

5.2. Optical transducers

The luminescence properties of rhodamine dyes covalently anchored at the walls of mesoporous Si-MCM-41 have been used by Wark *et al.*[142,143] to develop an SO_2 optical sensor on a glass carrier. As the pore diameter of the host is large enough to anchor almost any dye molecule, selective targeting of gas phase molecules becomes possible. The same authors have dispersed Sn-oxide clusters in the mesoporous matrix for detection of reducing gases CO, H_2, NH_3 and for O_2 detection by *in situ* diffuse reflectance UV-Vis spectroscopy, with good results regarding response times, sensitivity and stability of the sensors.

Mesoporous cubic silica functionalised with α-diketones via a one-pot synthesis has been proposed for BF_3 detection by taking advantage of the optical properties (fluorescence) of the difluoride adducts formed[144] in the selective reaction between the hybrid organo-silica and the halogenated gas.

Hierarchical porous films[145] containing both ordered macroporous structures surrounded with hexagonally mesostructured walls of porphyrin-functionalised silica (prepared by spin-coating infiltration) have been recently tested for trinitrotoluene detection at ppb level. After excitation at 420 nm, the nanocomposite films show a high fluorescence quenching sensitivity towards trinitrotoluene besides other nitroderivates. These results indicate that an appropriate combination of macro- and mesopores to simultaneously achieve a high permeability and a high density of interaction sites may be a suitable arrangement for rapid detection of explosives.

6. Conclusions and Outlook

Materials containing micro- and mesopores, such as silicon-based nanoporous solids, are attracting substantial interest in the gas sensing field. This could be expected, given properties such as a high surface area (an excellent base to increase sensor sensitivity) and pores in the molecular size range which, together with a tunable surface composition, often enable substantial selectivity enhancements. Thus, considerable improvements of sensor selectivity have been achieved through the deployment of (often functionalised) Si-based films as targets for the desired gas phase analytes, or as barriers to avoid the effect of interfering molecules. Adding to this are the considerable refinements occurred during the last decade on the synthesis of porous silicon-based films (such as porous silicon, zeolites and mesoporous silica). Coatings of these materials can now be grown/deposited on most surfaces of interest and the fine tuning of their thickness and porous structure has become possible.

However, the most significant advances will likely continue to take place in the microelectromechanical system/nanoelectromechanical system area, which will benefit from the expertise gathered on the development of micro- and nanostructures. Nanoporous Si-based materials will find many opportunities to bestow exquisite sensing selectivity upon these structures. Finally, it is worth noting that the full potential of hybrid materials comprising organic (or bioorganic) functional groups and nanoporous inorganic structures has not yet been realised in the sensor field. In the case of Si-based materials such as silica and zeolites, the porous structure of the solid combined with the rich chemistry related to the presence of hydroxyl surface groups will allow the hosting/grafting of specific molecules, opening up a wealth of sensing possibilities.

References

[1] Li, M., Tang, H. X., Roukes, M. L., *Nat. Nanotechnol.* **2006,** *2,* 114–120.
[2] Pathange, L. P., Mallikarjunan, P., Marini, R. P., O'Keefe, S., Vaughan, D., *J. Food Eng.* **2006,** *77,* 1018–1023.
[3] Rudnitskaya, A., Delgadillo, I., Legin, A., Rocha, S. M., Costa, A. M., Simoes, T., *Chemom. Intell. Lab. Syst.* **2007,** *88,* 125–131.
[4] Singh, S., *J. Hazard. Mater.* **2007,** *144,* 15–28.
[5] Joo, B. S., Huh, J. S., Lee, D. D., *Sens. Actuators B Chem.* **2007,** *121,* 47–53.
[6] Balasubramanian, S., Panigrahi, S., Logue, C. M., Doetkott, C., Marchello, M., Sherwood, J. S., *Food Control* **2004,** *19,* 236–246.
[7] Chen, X., Xu, F., Wang, Y., Pan, Y., Lu, D., Wang, P., Ying, K., Chen, E., Zhang, W., *Cancer* **2007,** *110,* 835–844.

[8] Nicolas, J., Romain, A. C., Ledent, C., *Sens. Actuators B Chem.* **2006,** *116,* 95–99.
[9] Bhattacharyya, N., Seth, S., Tudu, B., Tamuly, P., Jana, A., Ghosh, D., Bandyopadhyay, R., Bhuyan, M., Sabhapandit, S., *Sens. Actuators B Chem.* **2007,** *122,* 627–634.
[10] Fernandez, M. J., Fontecha, J. L., Sayago, I., Aleixandre, M., Lozano, J., Gutierrez, J., Gracia, I., Cané, C., Horrillo, M. D. C., *Sens. Actuators B Chem.* **2007,** *127,* 277–283.
[11] De Vito, S., Massera, E., Quercia, L., Di Francia, G., *Sens. Actuators B Chem.* **2007,** *127,* 36–41.
[12] He, L., Toh, C.-S., *Anal. Chim. Acta* **2006,** *556,* 1–15.
[13] Sánchez, C., Julián, B., Belleville, P., Popall, M., *J. Mater. Chem.* **2005,** *15,* 3559–3592.
[14] Kuchibhatla, S. V. N. T., Karakoti, A. S., Bera, D., Seal, S., *Prog. Mater. Sci.* **2007,** *52,* 699–913.
[15] French, P. J., Yang, C. K., *Sens. Transducers J.* **2007,** *October Special Issue,* 1–9.
[16] Smit, B., *Comput. Model. Microporous Mater.* **2004,** (2), 25–47.
[17] Davis, M. E., *Nature* **2002,** *417,* 813–821.
[18] Rocha, J., Anderson, M. W., *Eur. J. Inorg. Chem.* **2000,** *200*(5), 801–818.
[19] Fajula, F., Galerneau, A., Di Renzo, F., *Microporous Mesoporous Mater.* **2005,** *82,* 227–239.
[20] Walcarius, A., Mandler, D., Cox, J. A., Collinson, M., Lev, O., *J. Mater. Chem.* **2005,** *13,* 3663–3689.
[21] Madou, M. J., *Fundamentals of Microfabrication,* 2nd edition, CRC Press, 2002.
[22] Wilson, S. A., Jourdain, R. P. J., Zhang, Q., Dorey, R. A., Bowen, C. R., Willander, M., Wahab, Q. U., Willander, M., Al-hilli, S. M., Nur, O., Quandt, E., Johansson, C., Pagounis, E., Kohl, M., Matovic, J., Samel, B., van der Wijngaart, W., Jager, E. W. H., Carlsson, D., Djinovic, Z., Wegener, M., Moldovan, C., Iosub, R., Abad, E., Wendlandt, M., Rusu, C., Persson, K., *Mater. Sci. Eng.* **2007,** *R56,* 1–129.
[23] Rodriguez, A., Molinero, D., Valera, E., Trifonov, T., Marsal, L. F., Pallarès, J., Alcubilla, R., *Sens. Actuators B* **2005,** *109,* 135–140.
[24] Thomas, D. F., Porous silicon in handbook of nanostructured materials and nanotechnology, in: H. S. Nalwa (Ed.), *Optical Properties,* Vol. 4, Academic Press, **2000**.
[25] Föll, H., Christophersen, M., Carstensen, J., *Mater. Sci. Eng.* **2002,** *R39,* 93–141.
[26] Wongmanerod, C., Zangooie, S., Arwin, H., *Appl. Surf. Chem.* **2001,** *172,* 117–125.
[27] Zangooie, S., Bjorklund, R., Arwin, H., *Sens. Actuators B* **1997,** *43,* 168–174.
[28] Lewis, S. E., DeBoer, J. R., Gole, J. L., Hesketh, P. J., *Sens. Actuators B* **2005,** *110,* 5–65.
[29] Xu, Y. Y., Li, X. J., He, J. T., Hu, X., Wang, H. Y., *Sens. Actuators B* **2005,** *105,* 219–222.
[30] Li, X. J., Chen, S. J., Feng, C. Y., *Sens. Actuators B* **2007,** *123,* 461–465.
[31] Wang, H. Y., Li, X. J., *Sens. Actuators B* **2005,** *105,* 260–263.
[32] Shi, J., Zhu, Y., Zhang, X., Baeyens, W. R. G., García-Campaña, A. M., *Trends Analyt. Chem.* **2004,** *23,* 351–360.
[33] Holec, T., Chvojka, T., Jelinek, I., Jindrich, J., Nemec, I., Pelant, I., Valenta, J., Dian, J., *Mater. Sci. Eng. C Biomimetic Supramol. Syst.* **2002,** *19,* 251–254.
[34] Content, S., Trogler, W. C., Sailor, M. J., *Chem. A Eur. J.* **2000,** *6,* 2205–2213.
[35] Harper, J., Sailor, M. J., *Anal. Chem.* **1996,** *68,* 3713–3717.
[36] Quercia, L., Cerullo, F., La Ferrara, V., Di Francia, G., Baratto, C., Faglia, G., *Phys. Stat. Sol. A Appl. Res.* **2000,** *182,* 473–477.
[37] Kelly, M. T., Bocarsly, A. B., *Coord. Chem. Rev.* **1998,** *171,* 251–259.
[38] Kelly, M. T., Bocarsly, A. B., *Chem. Mater.* **1997,** *9,* 1659–1664.
[39] Baratto, C., Comini, E., Faglia, G., Sberveglieri, G., Di Francia, G., De Filippo, F., La Ferrara, V., Quercia, L., Lancelloti, L., *Sens. Actuators B* **2000,** *65,* 257–259.
[40] Di Francia, G., La Ferrara, V., Quercia, L., *J. Porous Mater.* **2000,** *7,* 287–290.
[41] Kelly, M. T., Bocarly, A. B., *Coord. Chem. Rev.* **1998,** *171,* 251–259.
[42] Letant, S. E., Content, S., Tan, T. T., Zenhausern, F., Sailor, M. J., *Sens. Actuators B* **2000,** *69,* 193–198.
[43] Bakker, J. W. P., Arwin, H., Wang, G., Jarrendahl, K., *Phys. Stat. Sol. A Appl. Res.* **2003,** *197,* 378–381.
[44] Wang, G., Arwin, H., *Sens. Actuators B* **2002,** *85,* 95–103.
[45] Archer, M., Christophersen, M., Fauchet, P. M., *Sens. Actuators B Chem.* **2005,** *106,* 347–357.
[46] Kim, S. J., Jeon, B. H., Choi, K. S., Min, N. K., *J. Solid State Electrochem.* **2000,** *4,* 363–366.

[47] Galeazzo, E., Peres, H. E. M., Santos, G., Peixoto, N., Ramirez-Fernandez, F. J., *Sens. Actuators B Chem.* **2003,** *93,* 384–390.
[48] Boarino, L., Baratto, C., Geobaldo, F., Amato, G., Comini, E., Rossi, A. M., Faglia, G., Lerondel, G., Sberveglieri, G., *Mater. Sci. Eng. B Solid State Mater. Adv. Technol.* **2000,** *69,* 210–214.
[49] Baratto, C., Faglia, G., Comini, E., Sberveglieri, G., Taroni, A., La Ferrara, V., Quercia, L., Di Francia, G., *Sens. Actuators B Chem.* **2001,** *77,* 62–66.
[50] Pancheri, L., Oton, C. J., Gaburro, Z., Soncini, G., Pavesi, L., *Sens. Actuators B Chem.* **2003,** *89,* 237–239.
[51] Massera, E., Nasti, I., Quercia, L., Rea, I., Di Francia, G., *Sens. Actuators B Chem.* **2004,** *102,* 195–197.
[52] Di Francia, G., Castaldo, A., Massera, E., Nasti, I., Quercia, L., Rea, I., *Sens. Actuators B Chem.* **2005,** *111–112,* 135–139.
[53] Seals, L., Gole, J. L., Tse, L. A., Hesketh, P. J., *J. Appl. Phys.* **2002,** *91,* 2519–2523.
[54] Xu, X., Wang, J., Long, Y., *Sensors* **2006,** *6,* 1751–1764.
[55] Guan, G., Tanaka, T., Kusakabe, K., Sotowa, K., Morooka, S., *J. Membr. Sci.* **2003,** *214,* 191–198.
[56] Vilaseca, M., Mintova, S., Valtchev, V., Metzger, T. H., Bein, T. J., *Mater. Chem.* **2003,** *13,* 1526–1528.
[57] Zhang, L. X., Jia, M. D., Min, E. Z., *Stud. Surf. Sci. Catal.* **1997,** *105,* 2211–2216.
[58] Poshusta, J. C., Noble, R. D., Falconer, J. L., *J. Membr. Sci.* **2001,** *186,* 25–40.
[59] Li, S. G., Falconer, J. L., Noble, R. D., *J. Membr. Sci.* **2004,** *241,* 121–135.
[60] Braunbarth, C. M., Boudreau, L. C., Tsapatsis, M., *J. Membr. Sci.* **2000,** *174,* 31–42.
[61] Guan, G. Q., Kusakabe, K., Morooka, S., *Microporous Mesoporous Mater.* **2001,** *50,* 109–120.
[62] Guan, G., Kusakabe, K., Morooka, S., *Sep. Sci. Tech.* **2002,** *37,* 1031–1039.
[63] Lin, Z., Rocha, J., Navajas, A., Tellez, C., Coronas, J., Santamaria, J., *Microporpus Mesoporous Mater.* **2004,** *67,* 79–86.
[64] Sebastián, V., Lin, Z., Rocha, J., Tellez, C., Santamaría, J., Coronas, J., *Chem. Commun.* **2005,** (24), 3036–3037.
[65] Corma, A., Díaz-Cabañas, M. J., Martínez-Triguero, J., Rey, F., Rius, J., *Nature* **2002,** *418,* 514–517.
[66] Lobo, R. F., Tsapatsis, M., Freyhardt, C. C., Khodabandeh, S., Wagner, P., Chen, C. Y., Balkus, K. J. Jr., Zones, S. I., Davis, M. E., *J. Am. Chem. Soc.* **1997,** *102,* 7139–7147.
[67] Yoshikawa, M., Wagner, P., Lovallo, M., Tsuji, K., Takewaki, T., Chen, C. Y., Beck, L. W., Jones, C., Tsapatsis, M., Zones, S. I., Davis, M. E., *J. Phys. Chem. B* **1998,** *102,* 7139–7147.
[68] Hugon, O., Sauvan, M., Benech, P., Pijolat, C., Lefebvre, F., *Sens. Actuators B* **2000,** *67,* 235–243.
[69] Kaneyasu, K., Otsuka, K., Setoguchi, Y., Sonoda, S., Nakahara, T., Aso, I., Nakagaichi, N., *Sens. Actuators B* **2000,** *66,* 56–58.
[70] Szabo, N. F., Dutta, P. K., *Sens. Actuators B* **2003,** *88,* 168–177.
[71] Fukui, K., Nishida, S., *Sens. Actuators B* **1997,** *45,* 101–106.
[72] Szabo, N. F., Du, H., Akbar, S. A., Soliman, A., Dutta, P. K., *Sens. Actuators B* **2002,** *82,* 142–149.
[73] Vilaseca, M., Coronas, J., Cirera, A., Cornet, A., Morante, J. R., Santamaría, J., *Catal. Today* **2003,** *82,* 179–185.
[74] Vilaseca, M., Coronas, J., Cirera, A., Morante J. R., Santamaria, J., *Sens. Actuators B* submitted for publication.
[75] Alberti, K., Fetting, F., *Sens. Actuators B* **1994,** *21,* 39–50.
[76] Plog, C., Kurzweil, P., Maunz, W., *Sens. Actuators B* **1995,** *25,* 403–406.
[77] Plog, C., Maunz, W., Kurzweil, P., Obermeier, E., Scheibe, C., *Sens. Actuators B* **1995,** *25,* 653–656.
[78] Moos, R., Müller, R., Plog, C., Knezevic, A., Leye, H., Irion, E., Braun, T., Marquardt, K. J., Binder, K., *Sens. Actuactors B* **2002,** *83,* 181–189.
[79] Bein, T., Brown, K. D., Frye, G. C., Brinker, C. J., *Molecular sieve sensors for selective detection at the nanogram level.* US Patent 5,151,110, September 21, **1992.**

[80] Grimes, C. A., Stoyanov, P. G., Remote magneto-elastic analyte, viscosity and temperature sensing apparatus and associated methods of sensing. US Patent 6,397,661, June 4, **2002**.
[81] Yan, Y., Bein, T., *Chem. Mater.* **1992,** *4,* 975–977.
[82] Mintova, S., Bein, T., *Microporous Mesoporous Mater.* **2001,** *50,* 159–166.
[83] Osada, M., Sasaki, I., Nishioka, M., Sadakata, M., Okubo, T., *Microporous Mesoporous Mater.* **1998,** *23,* 287–294.
[84] Sasaki, I., Tsuchiya, H., Nishioka, M., Sadakata, M., Okubo, T., *Sens. Actuators B* **2002,** *86,* 26–33.
[85] Vilaseca, M., Yagüe, C., Coronas, J., Santamaría, J., *Sens. Actuators B* **2006,** *117,* 143.
[86] Bein, T., Brown, K., Frye, G. C., Brinker, C. J., *J. Am. Chem. Soc.* **1989,** *111,* 7640–7641.
[87] Yan, Y., Bein, T., *Microporous Mater.* **1993,** *1,* 413–422.
[88] Berger, R., Gerber, C. H., Lang, H. P., Gimzewski, J. K., *Microelectron. Eng.* **1997,** *35,* 373–379.
[89] Scandella, L., Binder, G., Mezzacasa, T., Gobrecht, J., Berber, R., Lang, H. P., Gerber, C. H., Gimsewski, J. K., Koegler, J. H., Hansen, J. C., *Microporous Mesoporous Mater.* **1998,** *21,* 403–409.
[90] Zhou, J., Li, P., Zhang, S., Long, Y., Zhou, F., Huang, Y., Yang, P., Bao, M., *Sens. Actuators B* **2003,** *94*(3), 337–342.
[91] Pellejero, I., Urbiztondo, M., Villarroya, M., Sesé, J., Irusta, S., Dufour, I., Rébiere, D., Pina, M. P., Proceedings of the 3èmes Jornadas Hispano Francesas Cmc2 –Ibernam, San Sebastian, España, 9–10 Nov, 2006.
[92] Pellejero, I., Urbiztondo, M., Villarroya, M., Sesé, J., Pina, M. P., Santamaria, J., *Microporous Mesoporous Mater.* **2007,** in press.
[93] Shi, J., Zhu, Y., Zhang, X., Baeyens, W. R. G., García-Campaña, A. M., *Trends Analyt. Chem.* **2004,** *23,* 351–360.
[94] Demas, J. N., DeGraff, B. A., *Coord. Chem. Rev.* **2001,** *211,* 317.
[95] Schulz-Ekloff, G., Wöhrle, D., Duffel, B., Schoonheydt, R. A., *Microporous Mesoporous Mater.* **2002,** *51,* 91.
[96] Calzaferri, G., Huber, S., Maas, H., Minkowski, C., *Angew. Chem.* **2003,** *42,* 3732.
[97] Meinershagen, J. L., Bein, T., *J. Am. Chem. Soc.* **1999,** *121,* 448.
[98] Li, D., Mills, C., Cooper, J., *Sens. Actuators B* **2003,** *92,* 73.
[99] Levistsky, I., Krivoshlykov, S., Grate, J., *Anal. Chem.* **2001,** *73,* 3441.
[100] Reichardt, C., *Chem. Rev.* **1994,** *94,* 2319.
[101] Ghoneim, N., *Spectrochim. Acta A* **2000,** *56,* 1003.
[102] Pellejero, I., Urbiztondo, M., Izquierdo, D., Irusta, S., Salinas, I., Pina, M. P., *Ind. Eng. Chem. Res.* **2007,** *46,* 2335–2341.
[103] Sohrabnezhaz, S., Pourahman, A., Sadjadi, M. A., *Mater. Lett.* **2007,** *61,* 2311–2314.
[104] Bariain, C., Matias, I. R., Fernández-Valdivieso, C., Arregui, F. J., Rodríguez-Méndez, M. L., de Saja, J. A., *Sens. Actuators B* **2003,** *93,* 153.
[105] Pradhan, A. R., Macnaughtan, M. A., Raftery, D., *J. Am. Chem. Soc.* **2000,** *122,* 404–405.
[106] Pradhan, A. R., Uppili, S., Shailaja, J., Sivaguru, J., Ramamurthy, V., *Chem. Commun.* **2002,** (6), 596–597.
[107] Brinker, J., Scherer, G., *Sol–gel Science*, Academic Press, New York, 1989.
[108] Higuchi, T., Kurumada, K., Nagamine, S., Lothogkum, A. W., Tanigaki, M., *J. Mater. Sci.* **2000,** *35,* 3237–3243.
[109] Zhang, J., Palaniappan, A., Su, X., Tay, F. E. H., *Appl. Surf. Sci.* **2005,** *245,* 304–309.
[110] Palaniappan, A., Li, X., Tay, F. E. H., Li, J., Su, X., *Sens. Actuators B* **2006,** *119,* 220–226.
[111] Palaniappan, A., Su, X., Tay, F. E. H., *J. Electroceram.* **2006,** *16,* 503–505.
[112] Palaniappan, A., Su, X., Tay, F. E. H., *IEEE Sens. J.* **2006,** *6,* 1676–1682.
[113] Gomez-Vega, J. M., Teshima, K., Hozumi, A., Sugimura, H., Takai, O., *Surf. Coat. Technol.* **2003,** *169–170,* 504–507.
[114] Díez García, M. E., Badía Laiño, R., *Microchim. Acta* **2005,** *149,* 19–36.
[115] Marx, S., Zaltsman, A., Turyan, I., Mandler, D., *Anal. Chem.* **2004,** *76,* 120–126.
[116] Maruo, Y. Y., Tanaka, T., Ohyama, T., Hayashi, T., *Sens. Actuators B Chem.* **1999,** *57,* 135–141.

[117] Tanaka, T., Guilleux, A., Ohyama, T., Maruo, Y. Y., Hayashi, T., *Sens. Actuators B* **1999,** *56,* 247–253.
[118] Ohyama, T., Maruo, Y. Y., Tanaka, T., Hayashi, T., *Sens. Actuators B* **2000,** *64,* 142–146.
[119] Beck, J. S., Vartuli, J. C., Roth, W. T., Leonowicz, M. E., Kresge, C. T., Schmitt, K. D., Chu, C. T. W., Olson, D. H., Sheppard, E. W., Higging, J. B., Schelenker, J. L., *J. Am. Chem. Soc.* **1992,** *114,* 10834–10843.
[120] Kresge, C. T., Leonowicz, M. E., Roth, W. J., Vartulli, J. C., Beck, J. S., *Nature* **1992,** *359,* 710–712.
[121] Lindén, M., Schacht, S., Schüth, F., Steel, A., Unger, K.K., *J. Porous Mater.* **1998,** *5,* 177–193.
[122] Selvam, P., Bhatia, S. K., Sonwane, C. G., *Ind. Eng. Chem. Res.* **2001,** *40,* 3237–3261.
[123] Liu, Y., Pinnavaia, T. J., *J. Mater. Chem.* **2002,** *12,* 3179–3190.
[124] Wight, A. P., Davis, M. E., *Chem. Rev.* **2002,** *102,* 3589–3614.
[125] Taguchi, A., Schüth, F., *Microporous Mesoporous Mater.* **2005,** *77,* 1–45.
[126] Moller, K., Bein, T., *Chem. Mater.* **1998,** *10,* 2950–2963.
[127] Trong On, D., Desplantier-Giscard, D., Danumah, C., Kaliaguine, S., *Appl. Catal. A Gen.* **2001,** *222,* 299–357.
[128] Liu, Y., Zang, W., Pinnavaia, T. J., *Angew. Chem. Int. Ed.* **2001,** *40,* 1255.
[129] Hayward, R. C., Alberius-Henning, P., Chmelka, B. F., Stucky, G. D., *Microporous Mesoporous Mater.* **2001,** *44–45,* 619–624.
[130] Olson, D. H., Stucky, G. D., Vartuli, J. C., US Patent 5.364.797. Nov 15, 1994.
[131] Lu, Y., Fan, H., Stump, A., Ward, T. L., Rieker, T., Brinker, C. J., *Nature* **1999,** *398,* 223–226.
[132] Danumah, C. H., Vaudreuil, S., Boneviot, L., Bousmina, M., Giasson, S., Kaliaguine, S., *Microporous Mesoporous Mater.* **2001,** *44–45,* 241–247.
[133] Gollas, B., Elliott, J. M., Bartlett, P. N., *Electrochim. Acta* **2000,** *45,* 3711–3724.
[134] Inagaki, S., Guan, S., Ohsuna, O., Terasaki, O., *Nature* **2002,** *416,* 304.
[135] Li, G.-J, Kawi, S., *Sens. Actuators B Chem.* **1999,** *59*(1), 1–8.
[136] Li, G.-J., Kawi, S., *Mater. Lett.* **1998,** *1–2,* 99–102.
[137] Li, G.-J., Zhang, X.-H., and Kawi, S., *Sens. Actuators B Chem.* **1999,** *60*(1), 64–70.
[138] Yamada, T., Zhou, H. S., Uchida, H., Tomita, M., Ueno, Y., Honma, I., Asai, K., Katsube, T., *Microporous Mesoporous Mater.* **2002,** *54,* 269–276.
[139] Yuliarto, B., Honma, I., Katsumura, Y., Zhou, H., *Sens. Actuators B* **2006,** *114,* 109–119.
[140] Innocenzi, P., Martucci, A., Guglielmi, M., Bearzotti, A., Traversa, E., *Sens. Actuators B* **2001,** *76,* 299–303.
[141] Bearzotti, A., Bertolo, J. M., Innocenzi, P., Falcaro, P., Traversa, E., *J. Eur. Ceram. Soc.* **2004,** *24,* 1969–1972.
[142] Ganschow, M., Wark, M., Wöhrle, D., Schulz-Ekloff, G., *Angew. Chem. Int. Ed.* **2002,** *39,* 161–166.
[143] Wark, M., Rohlting, Y., Altindag, Y., Wellmann, H., *Phys. Chem. Chem. Phys.* **2003,** *5,* 5188–5194.
[144] Banet, P., Legagneux, L., Hesemann, P., Moreau, J. J. E., Nicole, L., Quach, A., Sanchez, C., Tran-Thi, T.-H., *Sens. Actuators B* **2007,** doi:10.1016/j.snb.2007.07.103.
[145] Tao, S., Shi, Z., Li, G., Li, P., *Chem. Phys. Chem.* **2006,** *7,* 1902–1905.

CHAPTER 16

Mesostructuring of Metal Oxides Through EISA: Fundamentals and Applications

Moises A. Carreon *and* Vadim V. Guliants

Contents

Abstract

This chapter discusses the fundamentals of mesostructuring metal oxides by the coassembly between inorganic entities and liquid crystalline phases on evaporation. We discuss relevant structural, textural, and compositional properties of silicate and non-silicate systems synthesized by evaporation induced self-assembly. Finally, emerging applications of these phases are presented.

Keywords: Mesoporous phases, Evaporation-induced self-assembly, Nanofunctional applications

Glossary

Brij-56	Non-ionic surfactant. Polyethylene glycol hexadecyl ether ($C_{16}H_{33}(OCH_2CH_2)_nOH$, $n \sim 10$)
Brij-58	Non-ionic surfactant. Polyethylene glycol hexadecyl ether ($HO(CH_2CH_2O)_{20}C_{16}H_{33}$)
Brij-78	Non-ionic surfactant. Polyethylene glycol hexadecyl ether ($C_{58}H_{118}O_{21}$)
Critical micelle concentration	Concentration of surfactants above which micelles are spontaneously formed

Ordered Porous Solids
DOI: 10.1016/B978-0-444-53189-6.00016-0

CTAB	Cationic surfactant. Cethyltrimethyl ammonium bromide $((C_{16}H_{33})N(CH_3)_3Br)$
CTACl	Cationic surfactant. Cethyltrimethyl ammonium chloride $((C_{16}H_{33})N(CH_3)_3Cl)$
F127	Triblock copolymer $EO_{106}PO_{70}EO_{106}$ (where EO = poly(ethylene oxide), PO = poly(propylene oxide))
Isoelectric point	pH at which a particular molecule or surface carries no net electrical charge
KLE	Non-ionic surfactant composed of (poly(ethylene–co–butylene)-b-poly(ethylene oxide) blocks
Liquid crystals	Phases that exhibit properties between those of a conventional liquid and those of a solid crystal. In a liquid crystal molecules are arranged and/or oriented in a crystal-like way
Mesostructure	Ordered porous structure with unimodal pore size and high surface area in the mesoporous regime (2–50 nm) synthesized employing structure directing agents
Micelle	Aggregate of surfactant molecules dispersed in a liquid
Modulable steady state	EISA stage at which the inorganic framework is partially condensed, allowing control of the final mesostructure by modifying relative humidity conditions
P123	Triblock copolymer $EO_{20}PO_{70}EO_{20}$ (where EO = poly(ethylene oxide), PO = poly(propylene oxide))
Self-assembly	Spontaneous organization of materials mainly through hydrogen bonding, Vander Waals forces, and electrostatic forces without external influence
Surfactant	Surface acting phase employed as a structure directing agent for the synthesis of mesostructures
Tamman temperature	Measurement of the mobility of metal ions or atoms in a crystalline metal oxide is defined as 0.5–0.52 T_m (where T_m = metal oxide melting point)
Thin film	A material in a specific planar geometry, which does not allow selective transport of mass species
Triton X-100	Non-ionic surfactant with a hydrophilic polyethylene oxide group (~9.5 ethylene oxide units), and a hydrocarbon lipophilic or hydrophobic group $(C_{14}H_{22}O(C_2H_4O)_n)$
Wormhole	Disordered mesoporous structure, lacking of periodic ordered arrays of pores

1. Self-Assembly of Mesoporous Materials

Before discussing fundamentals and applications of the evaporation induced self-assembly (EISA) process, it is important to briefly review the structural characteristics of mesostructured oxides and the main mechanisms that have been proposed to explain the formation of these mesoporous phases. The original

M41S family of mesoporous molecular sieves is typically synthesized by reacting a silica source [e.g., tetraethylorthosilicate (TEOS), Ludox, fumed silica, sodium silicate], an alkyltrimethylammonium halide surfactant [e.g., cetyltrimethylammonium bromide (CTAB)], a base [e.g., sodium hydroxide or tetramethylammonium hydroxide (TMAOH)], and water at ≥100 °C for 24–144 h. The as-synthesized product contains occluded organic surfactant, which is removed by calcination at 500 °C in air to yield an ordered mesoporous material. The 2D hexagonal phase with *P6mm* symmetry, templated by hexagonal close-packed cylindrical arrays of surfactants (liquid crystalline phases), is the most common ordered mesostructure. Lamellar phases are unstable on surfactant removal, which causes the collapse of the expanded layered structures. Various cubic phases have also been reported. In the case of mesoporous silica, the bicontinuous cubic gyroid phase with *Ia3d* symmetry is found in alkali-catalyzed syntheses. This phase, with its system of interconnected pores, is much more attractive than the 2D hexagonal phase for applications requiring diffusion of species into and out of the pore network. The *Pm3n* phase templated by spherical surfactant micelles in a cubic close-packed arrangement is found in both acid- and alkaline-catalyzed syntheses of mesoporous silica.[1]

Two main mechanistic routes have been proposed for the formation of these materials. Figure 16.1 shows the 'liquid crystal templating' (LCT) formation

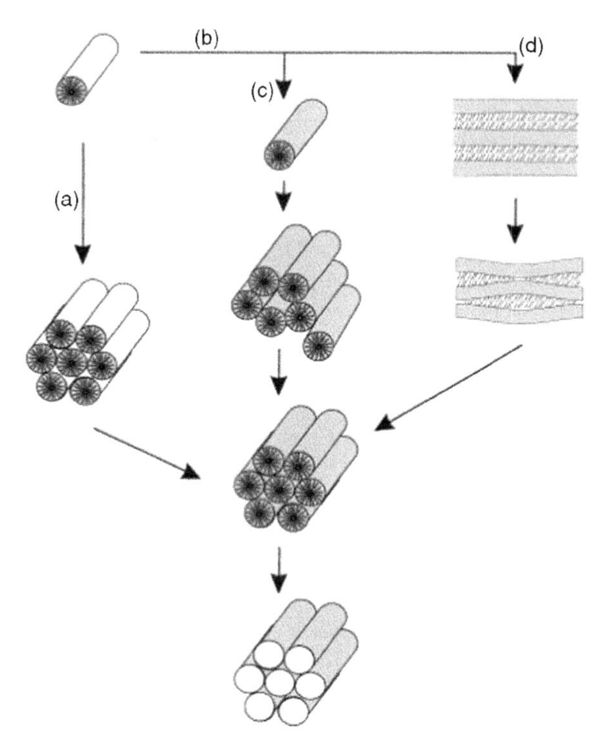

Figure 16.1 Formation of mesoporous phases by liquid crystal templating (LCT) mechanism. See text for detailed description.

mechanism proposed by the Mobil researchers, based on the similarity between liquid crystalline surfactant assemblies (i.e., lyotropic phases) and M41S.[1] The formation of these mesophases takes place by the following pathways: (a) via preexisting liquid crystal phase, (b) via the formation of silica-coated surfactant species, (c) either form micelles which agglomerate to form ordered and disordered arrays or, (d) a lamellar phase which undergoes a phase transition into the final hexagonal phase. An ordered mesoporous phase exhibiting a hexagonal array of ordered pores is obtained after template removal.

Two mechanistic pathways were proposed by the Mobil researchers to explain formation of MCM-41 containing 2D hexagonal arrays of cylindrical mesopores: (1) via condensation and cross-linking of the inorganic species at the interface with a preexisting hexagonal lyotropic liquid crystal (LC) phase, and (2) via ordering of the surfactant molecules into the 2D hexagonal mesophase mediated by the inorganic species and followed by their condensation and cross-linking.

In both pathways, the inorganic species, which are negatively charged at the high synthesis pH, preferentially interact with the positively charged alkyl ammonium head-groups of the surfactants and condense into a solid, continuous framework. The resulting organic–inorganic mesostructure could be viewed as a hexagonal array of surfactant micellar rods embedded in a silica matrix. The surfactant removal produces the open ordered mesoporous MCM-41 framework. These mesophases with pore sizes > 2.5 nm generally display type IV nitrogen adsorption–desorption isotherms at 77 K characteristic of mesoporous materials with ordered unimodal pore size distribution. However, most likely pathway 1 did not take place in the Mobil syntheses because the surfactant concentrations used were far below the critical micelle concentration (CMC) required for the hexagonal LC formation.[2] The second mechanistic pathway of LCT was proposed as a cooperative self-assembly of the alkyl ammonium surfactant and the silicate precursor species below the CMC. It is well established that no preformed LC phase is necessary for the MCM-41 formation. Recently, we have reviewed and discussed in detail the various proposed formation mechanisms of periodic ordered mesostructured phases.[3]

The high thermal stability of mesoporous phases is perhaps the most critical requirement for their use in several functional applications. In general, the thermal stability of mesostructured metal oxide phases will depend on (a) the degree of charge-matching at the organic–inorganic interface, (b) the interaction strength between inorganic species and surfactant head-groups, (c) the Tamman temperature of the metal oxide, (d) the flexibility of the M–O–M bond angles in the constituent metal oxides, and (e) the occurrence of redox reactions in the metal oxide wall.

The charge matching at the organic–inorganic interface allows control over the wall composition, and facilitates cross-linking of the inorganic species into a robust mesostructured framework. The knowledge of the electrokinetic behaviour of the inorganic species in solution (i.e., the isoelectric points) is required for fine-tuning electrostatic and other interactions at the inorganic–organic interface in order to obtain thermally stable mesoporous phases. The presence of strong covalent bonds between metal oxide species and surfactant head-groups, for example metal–N bonds, means that harsh conditions, such as combustion, are required for surfactant removal, leading typically to collapse of the mesostructure.

On the contrary, metal oxide species should possess low lattice mobility at elevated temperatures in order to prevent transformation of the mesostructured metal oxides into more thermodynamically stable dense phases. The mobility of metal ions or atoms in a crystalline metal oxide increases considerably in the vicinity of its Tamman temperature (defined as 0.5–0.52 T_m, where T_m is the metal oxide melting point. Therefore, the low Tamman temperature of several transition metal oxides (i.e., V_2O_5 472 K, MoO_3 534 K, WO_3 873 K, Nb_2O_5 892 K, Fe_2O_3 919 K, MnO 962 K, TiO_2 1064 K, NiO 1129 K, ZrO_2 1492 K, MgO 1563 K)[4] translates into a limited thermal stability of the corresponding mesostructures. Non-flexible M–O–M bond angles that are unable to accommodate the curvature of the inorganic–organic interface may result in the formation of only lamellar or dense metal oxide phases. Finally, the structural collapse of mesophases may be caused by redox reactions occurring in the metal oxide wall during surfactant removal or catalytic reaction.

2. Evaporation-Induced Self-Assembly Origins

Seminal work on the synthesis of mesoporous silica thin films by EISA was first reported by Ozin's[5] and Brinker's groups.[6] In these first examples, mesoporous silica films were grown on cleaved mica substrates and at the air–water interface.[5] When mica was used as substrate, first, elongated mesoporous silica film islands formed on the surface of the substrate, implying that the crystal growth was regulated by charge and structure matching at the interface. Then, the crystalline islands expanded in size, and continuous assembly of the silica-micellar species led to the growth of the mesoporous film. Mesoporous films were grown also at the air–liquid interface; in this case, the authors proposed that the film formation involved polymerization of silicates in the surfactant head-group regions of the hexagonal mesophase, which concentrated with time at the liquid–air surface interface. Preexisting incipient silica mesostructures present at solid–liquid and liquid–vapor interfaces at $C_o < CMC$ (where C_o is the initial surfactant concentration and CMC is the critical micelle concentration) helped to nucleate and orient the mesophase development. Therefore, this preorganized surfactant layer was gradually formed with time and facilitated the organization of the ordered mesoporous film. Later, the EISA concept was formally introduced.[6] In this report, it was demonstrated that a homogeneous solution of polymeric silica sol with an initial surfactant concentration below the critical micelle concentration, led to ordered mesoporous films. By changing the initial surfactant concentration, the film symmetry (i.e., hexagonal or cubic) was selectively tailored. The proposed formation mechanism involved cooperative assembly of silica-surfactant micellar species with supramolecular cylindrical or hemicylindrical micelles that self-assembled at the substrate–liquid interface at the initial stage of the deposition method (dip-coating). In this case, the surfactant enrichment by gradual solvent evaporation exceeded the critical micelle concentration leading to a cooperative self-assembly between the inorganic precursor and the surfactant micelles forming a well-defined mesostructure. The initial surfactant concentration for these reports was far below the critical micelle concentration.

3. Fundamentals

EISA is a well-suited approach to prepare ordered mesoporous thin films. When a homogeneous solution of soluble silica and surfactant is prepared in the presence of ethanol or other volatile solvents and water with $C_o << CMC$, preferential evaporation of ethanol concentrates the film in surfactant and silica species. Then, a progressive increase in surfactant concentration drives the self-assembly of silica-surfactant micelles leading to a final organization into liquid crystalline mesophases. Through variation of the initial alcohol/water/surfactant mole ratio it is possible to obtain different mesostructures.[7] When preparing mesostructured films, via EISA, we may consider several parameters that influence the self-assembly process. Sanchez's group[8] divided these parameters in two categories: (a) the chemical parameters, related to sol–gel hydrolysis–condensation reactions and relative quantities of surfactant and inorganic precursor; and (b) the processing parameters related to the diffusion of alcohol, water, catalysts (HCl, HNO_3, H_2SO_4) to or from the film.

Let us analyze first the chemical parameters. The knowledge of the electrokinetic behaviour (i.e., the isoelectric points) of the inorganic species in solution, the nature, and the density of chemical groups of inorganic species is required for fine-tuning electrostatic and other interactions at the inorganic–organic interface. For instance, the formation of the mesostructure is favoured when neutral templates such as block copolymers are combined with inorganic precursor solutions in the vicinity of the isoelectric point. Above the isoelectric point, the metallic species carry a net negative charge and therefore cationic templates favour the self-assembly process, below the isoelectric point, metallic species carry net positive charge, and anionic templates favour the formation of well-defined mesostructures. It is of great importance that the dimensions of the inorganic species present in the initial solution do not exceed the mesostructure wall thickness (which are typically in the 1–10 nm range).

The sol–gel chemistry reactions (hydrolysis and condensation) need to be controlled by adjusting the relative quantities of its components, such as water, the amount of catalyst, pH, etc. For example, the rapid organization of thin films mesophases requires suppression of inorganic polymerization during the coating process.[7b] For the most common system (i.e., silicates), this is achieved under acidic conditions (near the isoelectric point of silica, pH $\sim$ 2). Therefore, by turning off siloxane species condensation, the cooperative silica-surfactant self-assembly can take place.[7b] Obviously, the molar ratios of inorganic precursors and surfactant need to be fixed to allow the organization of the mesostructure. Typical surfactant volume ratios are in the 20–80% vol. range based on surfactant–water–alcohol phase diagrams.[9] For example, it has been shown that the CTAB/TEOS molar ratios between 0.1 and 0.35 lead to the formation of ordered periodic silica mesostructures with *Pm3n* cubic, *p6m* 2D hexagonal, *P6₃/mmc* 3D hexagonal, or lamellar symmetries.[8] Other chemical parameters that need to be considered are the volatility and wettability of the alcohol with hydrophobic substrates. Typically, highly volatile alcohols, such as ethanol and butanol, with high wettability to substrates are desired for the preparation of homogeneous continuous films. Finally, the solution aging time is another important parameter to be controlled, since the extent of condensation of the inorganic species influences the formation of the final mesostructure.

The processing parameters play a critical role once the initial solution is deposited on the substrate via dip-coating or spin-coating. Evaporation of volatile components takes place at the air–film interface as soon as the inorganic precursor-template solution is deposited. At this point, preferential evaporation of volatile species occurs at the air–film interface. Gradual evaporation of volatile species leads to an increase in the surfactant and silica concentration in the film. At some point, the surfactant concentration reaches the critical micelle concentration leading to the formation of micelles with the subsequent formation of the organized mesostructure. It has been demonstrated that the critical processing parameter for the formation of ordered mesostructures is the relative humidity (RH).[10] In other words, the quantity of water in the film changes with RH and determines the final mesostructure. Obviously, the content of water and alcohol in the film can be adjusted by changing the relative humidity conditions in the media in which the film is being aged. It has been demonstrated that an inorganic precursor-surfactant solution with low viscosity evaporates faster than more viscous solutions.[8] As a result, there is a short time needed for volatile species to diffuse through the film, leaving less time for intermediate phases to form. Therefore, by employing diluted solutions we can avoid the formation of intermediate multiphase formation. The formation of diverse mesophases by EISA strongly depends on the kinetics of competitive processes of condensation versus organization. Both processes are influenced by the diffusion of volatile species from or to the film surface.

The resultant as-synthesized mesostructure is treated thermally to eliminate the organic structure directing agents. This step is used to stabilize the inorganic framework and to obtain the open porous mesostructure. Calcination temperatures will depend on the decomposition temperature of the structure directing agent. Typical values are in the 300–500 °C range. Alternatively, the structure directing agents can be eliminated by solvent extraction[11] or ultraviolet (UV) photodegradation.[12]

Figure 16.2a summarizes the steps involved in the organization of ordered periodic mesostructures. The initial solution contains the inorganic species, surfactant, and volatile components. As the evaporation of volatiles proceeds, surfactant concentration increases, and it aggregates to form micellar structures above the critical micelle concentration (EISA). When the mesostructure is initially formed, the inorganic framework is not fully condensed, and the control of the final mesostructure can be adjusted by modifying the RH conditions (i.e., water molecules can diffuse in and out of the film). At this point, the flexible mesostructure experiences a transition state known as modulable steady state (MSS).[10] Finally, the inorganic framework fully condenses, forming the final mesostructure.

One notable characteristic of EISA is the fact that in the mesostructured films made by dip-coating, the boundaries (vapor–solid) of the liquid impose alignment of the surfactant structures, and the presence of concentration gradients in the film and

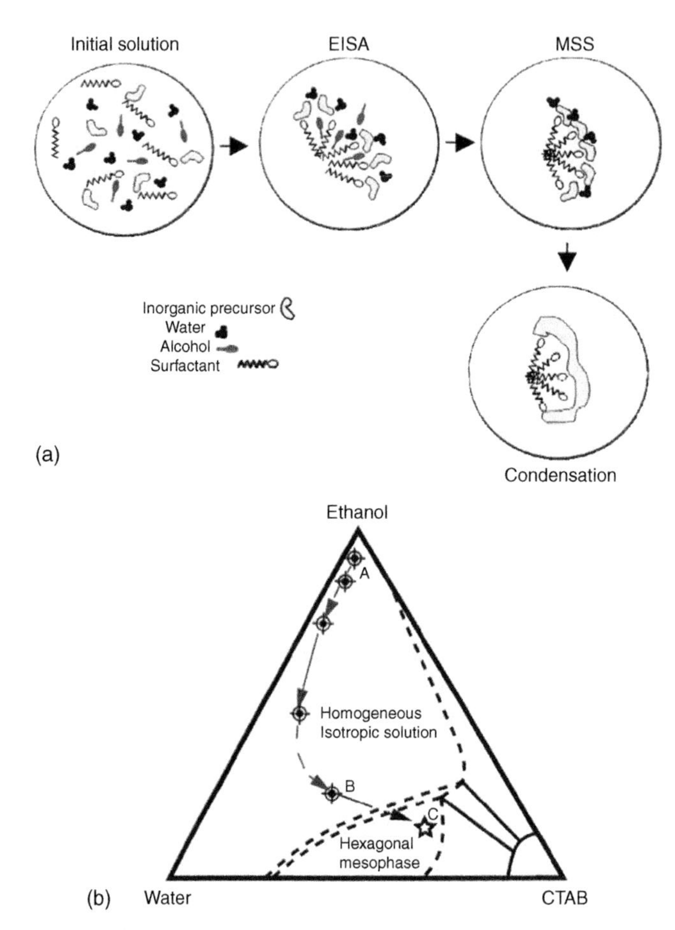

Figure 16.2 (a) Schematic representation of evaporation-induced-self-assembly for the preparation of long-range ordered mesostructures. (b) Typical EISA trajectory for the formation of mesostructured films by dip-coating.[7b] (See color insert.)

flow velocity may help in aligning the final mesostructure along certain preferred direction.[13] Therefore, it is possible to envision the preparation of oriented mesoporous thin films, which may represent ideal candidates for several important nanotechnological applications in bioseparations. A typical pathway for the formation of a final hexagonal mesostructure via EISA is shown in Fig. 16.2b. Point A is the initial solution composition, point B is the region in the vicinity of the drying line, and point C corresponds to the final dried film.[7b] By changing the chemical parameters, such as initial solution concentration, and processing parameters, such as water content in the film, it is possible to follow different trajectories in this compositional space and, therefore, synthesize mesoporous materials with different space groups and pore symmetries. The physics of EISA, based on a quantitative description of the local environment of a fluid of element during the deposition step (i.e., dip-coating) has been successfully explained from hydrodynamics and capillary processses.[13]

4. Preparation Methods of Mesoporous Films

Most of the successful syntheses of mesoporous phases by EISA have been done on films. Therefore, before discussing relevant examples of mesoporous phases prepared via EISA, first, we briefly discuss the most common methods to prepare mesoporous films. We have reviewed these methods in detail elsewhere.[14]

4.1. Solvent evaporation techniques

The solvent evaporation techniques involve formation of a liquid film containing the solvent, surfactant, and silica precursor followed by evaporation of the solvent. Several methods can be used to form liquid films. These include dip-coating, spin-coating, and film casting. Solvent evaporation has been suggested as the driving force for the organization of surfactant species into mesoscale aggregates (i.e., lamellar, hexagonal, cubic) around which condensation of silicate species takes place.

In *dip-coating*, the substrates are withdrawn from a homogeneous precursor solution and the dip-coated solution is allowed to drain to a particular thickness.[15] The thickness of the film is mainly determined by the rate of evaporation of the solvent and the viscosity of the solution. One of the advantages of this method is the facile formation of films on non-planar surfaces. *Casting* is another solvent evaporation method that has been used for the preparation of mesostructured films. In this method, the solution is dropped on to the substrate and allowed to solidify, resulting in much thicker films.

Spin-coating has been widely used for the preparation of mesostructured films by solvent evaporation. Four stages of spin-coating can be distinguished as deposition of the surfactant/inorganic solution, spin-up, spin-off, and evaporation.[15] Initially, an excess of liquid is deposited on the surface of the substrate during the first stage. In the spin-up stage, the liquid flows radially outward by centrifugal force. In the spin-off stage, the excess of liquid flows to the perimeter and leaves in the form of droplets. In the final stage, evaporation takes place leading to the formation of uniform thin films. One of the most important advantages of this method is that the film tends to become very uniform in thickness. The main disadvantage of this method is that it can only be used with flat substrates.

4.2. Growth from solution

The basic principle for the synthesis of ordered mesoporous films by growth from solution is to bring the synthesis solution (including a solvent, surfactant, and inorganic precursor) in contact with a second phase, for example, solid (ceramic), gas (air), or another liquid (oil). The two-phase system is kept under specific conditions, and the ordered film is formed at the interface. When the second phase is solid, it is the support on which the ordered film or membrane is grown. When the second phase is air or oil, the solid films are self-standing. Ordering of inorganic-surfactant species can occur spontaneously at the solid support interface submerged in the synthesis solution.

4.3. Other deposition techniques

Alternative deposition techniques reported to date include electrodeposition and pulsed laser deposition (PLD). Electrodeposition of mesoporous thin films has been employed to create continuous thin films of mesostructured materials on a variety of substrates. The electrodeposition method has an advantage that it can be used to deposit thin films onto non-planar substrates. However, it is limited to electrodeposition of metals under mild conditions (temperature, pH, reactant concentration). PLD is another alternative deposition technique for creating continuous mesoporous thin films on non-planar surfaces. The laser ablation process involves irradiating a bulk mesoporous target with an intense laser beam, which results in the ejection and deposition of nanoscale fragments of the mesoporous structure on a substrate surface. Although the PLD method can produce good quality-ordered mesoporous films, sophisticated equipment is required for film production. However, it is unlikely that this method will be able to compete with solvent evaporation techniques in speed and degree of control over resulting film structures and pore orientations.

5. Mesostructured Silicate Systems

In the following paragraphs, we describe relevant work done in the preparation of silicate-based films and particles via EISA. As we have discussed previously, the EISA concept was formally introduced in 1997. In this report, the authors showed a simple and continuous process for the formation of hexagonal and cubic silica mesostructures in thin film form. Spatially resolved fluorescence–depolarization experiments were employed to gain improved mechanistic understanding of the *in situ* micelle formation during the film deposition. For the initial diluted solution ($C_o << CMC$), it was found that the degree of polarization or molecular anisotropy was zero; then, as evaporation proceeded, the polarization increased abruptly to 0.18 (the point which was identified as the CMC), which indicated that the silica-surfactant micellar structures self-assembled by induced evaporation of the volatiles. This polarization change was associated with the nearly isotropic nature of the micelles.

Later, it was demonstrated that not only films but also spherical particles displaying long-range order hexagonal and cubic symmetries could be prepared via EISA employing an aerosol process.[7] It was proposed that evaporation during aerosol processing created in each droplet a radial gradient in surfactant concentration, with a maximum concentration at the surface of the droplet. As the evaporation proceeds, the CMC is exceeded from the surface throughout the droplet. This surfactant enrichment caused the organization of the final mesostructure. As shown in Fig. 16.3, depending on the surfactant nature, mesoporous particles with different symmetries were prepared. For example, CTAB led to a hexagonal mesophase, Brij-58 favoured the cubic mesophase, and P-123 resulted in the formation of a vesicular mesophase. Alternatively, mesoporous films can be prepared by the aerosol

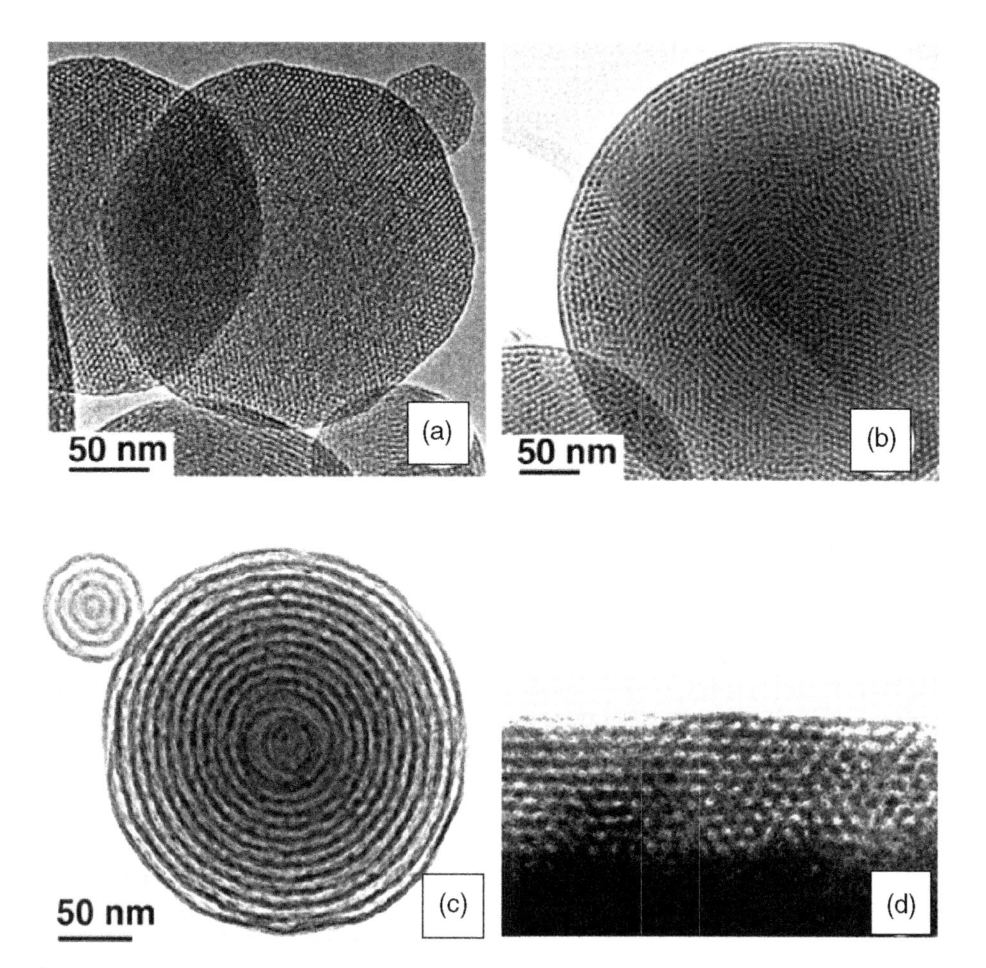

Figure 16.3 TEM images of mesoporous silica (a-c) aerosol particles and (d) aerosol deposited film prepared via EISA. Calcined particles displaying (a) hexagonal, (b) cubic, (c) vesicular mesophases, and (d) Cross sectional view of a mesoporous film deposited on Si substrate.[7b]

method via coalescence of aerosol droplets (Fig. 16.3d). This method was successfully used with a variety of inorganic precursors (TEOS, silsesquioxane monomers) and surfactants (ionic: CTAB, SDS; copolymer: P-123; and non-ionic: Brij-56, Brij-58).[6,7,16] For example, Brinker's group reported the preparation of polybridged silsesquioxane films and particle-like mesostructures with organic moieties incorporated in the mesostructured inorganic framework via EISA.[16] Spin-coating and dip-coating were used to prepare homogeneous films, and an aerosol dispersion was used to prepare particles. In both approaches, the preferential evaporation of ethanol concentrated the sol in water, surfactant, and organic polysilsesquioxane species. The progressively increase in surfactant concentration led to the self-assembly of polysilsesquioxane surfactant micelles and their further organization into liquid crystalline phases. Typically, the formation of these mesostructured films or particles oriented with respect to the solid–liquid or liquid–air interfaces takes place in several seconds. Furthermore, modification of the surfactant shape, charge, and initial concentration led to different film or particle mesostructures.

Sanchez's group[8] demonstrated that an alternative way to prepare mesoporous silicate materials with different symmetries (3D hexagonal, cubic, 2D hexagonal, and lamellar) is to vary the initial surfactant/Si molar ratio (chemical parameter), and the relative amounts of water entering and departing the film boundaries (processing parameters), as shown in Fig. 16.4. From this texture diagram, one can see the effect of surfactant volume fraction and the effect of water in the formation of different mesostructures. In general, it can be seen that when the surfactant concentration increases and the water content decreases, the surfactant packing parameter g ($g = V/a_o l$[17], where the term V is the volume occupied by the surfactant chains and cosolvent organic molecules located between the chains, a_o is the surfactant head-group area at the micelle surface, and l is the surfactant tail length) increases and, therefore, there is a decrease in the micellar curvature. In other words, the presence of water at the inorganic–organic interface promotes intercalation of water molecules between the surfactant polar head-groups, pushing them apart and thus increasing a_o which consequently lead to a decrease in g. On the contrary, when the surfactant volume fraction is decreased by increasing the water content, highly curved mesophases are favoured due to a decrease in g.

Through EISA, it has been possible to molecularly design silicates with diverse chemical, physical, morphological, and structural properties. For example, a modified aerosol process (vibrating orifice aerosol generator) based on evaporation-driven self-assembly for the synthesis of spherical silicate porous particles was proposed.[18] In this method, all the chemical species in solution are initially confined to microdroplets which experience evaporation-driven concentration changes. These concentration gradients led to self-assembled liquid crystalline surfactant structures that serve as templates for the organization of the inorganic precursors. Finally, the inorganic silica-based framework solidified by condensation, left a

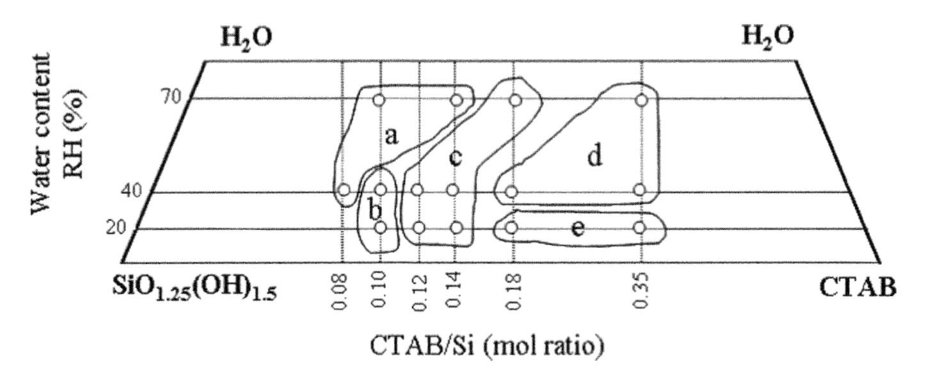

Figure 16.4 EISA adapted texture diagram for the CTAB/Si/EtOH/H_2O system showing the different synthesized[0] mesophases as a function of water content and CTAB/Si ratio: (a) disordered, (b) 3D hexagonal (*P6₃/mmc*), (c) cubic (*Pm3n*), (d) 2D hexagonal (*p6m*), and (e) lamellar mesophases. (Adapted from Ref. 8)

mesoporous silica mesostructure after template removal. Interestingly, this method offered control over particle size (1–50 μm) and monodispersity, as well as control over internal mesostructure and pore size (2.2–2.8 nm). CTAB and Brij-58 were used as structure directing agents, TEOS as inorganic precursor and ethanol and water as solvents.

The same research group reported the preparation of smaller mesoporous spherical silica particles (0.2–1 μm) with pore sizes of 2.7–3.1 nm by EISA of aerosols.[19] These spherical submicron particles displayed 2D hexagonally ordered mesostructures with 3D interconnected network of pores and BET surface areas up to 1300 m^2/g. The mesostructures were successfully obtained from acidic aqueous solutions (pH = 2) of TEOS and CTAB. This pH near the isoelectric point of silica slows the rate of silica condensation, allowing a more efficient self-assembly during EISA. The CTAB/Si molar ratio was a critical chemical parameter to form hexagonally ordered mesostructures. These mesostructures were formed only in the $0.09 < \text{CTAB/Si} < 0.28$ range. The spheroidal shape, broad particle size regime, high accessible surface area, monodisperse pore size, and mesoporous nature make these materials highly attractive for its use as catalytic supports, sorbents, and chromatographic media.

Mesostructured silicate films with spherical mesopores possessing average pore sizes of 5–7 nm and displaying 3D connectivity have been prepared using non-conventional silica and surfactant sources, such as methyl triethoxy silane and PS (35)-b-PEO(109) diblock copolymer, respectively.[20] TEM measurements showed that the films consisted of isolated spherical voids with almost no interconnecting microporosity. Interestingly, these materials showed closed porosity (due to retention of methyl moieties after calcination) and hydrophobicity. In principle, CH_3 moieties would reduce the framework dielectric constant, while the cubic arrangement of isolated pores would minimize interdiffusion of metallic species, making these materials highly attractive to be used as low-*k* dielectrics.

Smaller pore size (1.6–1.8 nm) mesoporous silica with 2D hexagonal (*p6mm*), 3D cubic (*Pm-3n*), and 3D hexagonal (*p63/mmc*) pore structures have been prepared via EISA employing a polyoxyethylene alkyl amine surfactant (PN-430) and tetrahydrofuran (THF) to direct the synthesis.[21] The resultant mesophases displayed BET surface areas up to 730 m^2/g and pore volume of 0.36 cm^3/g. Interestingly, solvothermal post-treatment with *n*-hexane at 70 °C led to phase transformation from 3D cubic to 3D hexagonal mesophase. The structural order of these mesophases was improved by incorporating non-ionic co-surfactants (Brij-78 or Triton X-100).

EISA has been used as an alternative route to confine mesoporous silica in the channels of anodic alumina membranes.[22] The sol mixtures containing the silica precursor and the structure directing agent were introduced in the pores of anodic alumina membranes (pore diameter 120–200 nm) by soaking the membranes at room temperature. As a result of gradual evaporation of the solvent, the self-assembly between inorganic species and liquid crystalline phases takes place and well-shaped mesoporous silica nanofilaments are formed in the vertical hexagonal pores of the anodic alumina membranes. These molecularly engineered nanocomposites are highly attractive for bioseparations. Cationic CTAB and neutral Brij-56 and P-123 were used as structure directing agents, TEOS as inorganic precursor and ethanol–water mixture as solvent. Interplanar spacings of these mesostructures were effectively controlled depending on the nature of the structure directing agent. For example, $d_{spacing} = 4.3–4.5$ nm (CTAB), $d_{spacing} = 6.0–6.4$ nm (Brij-56), and $d_{spacing} = 11.0–11.5$ nm (P-123).

Periodic hexagonal mesoporous organosilica (PMOs) thin films employing silsesquioxanes as inorganic precursors were also obtained via EISA.[23] CTACl was used as structure directing agent, ethanol as solvent, and silsesquioxanes of the type $(C_2H_5O)_3Si$-R-$Si(OC_2H_5)_3$ or R_1-$[Si(OC_2H_5)_3]_3$, where R = methane, ethylene, ethane, or 1,4-phenylene, and R_1 = 1,3,5-phenylene. Interestingly, the dielectric constants in these films can be fine-tuned depending on the organic group nature and content. Because of the low dielectric constants, good mechanical and thermal stability and hydrophobic nature, these films could find potential applications as low *k* layers in microelectronics. Table 16.1 summarizes some of the relevant work done on the preparation of mesoporous silica-based films and particles by EISA.

6. Mesostructured Non-Silicate Systems

EISA method has been successfully extended to transition metal oxides. Mesostructuring transition metal oxides is still challenging due to their higher reactivities towards condensation, lower Tamman temperatures of the constituent metal oxides, less flexible metal–oxygen–metal bond angles as compared to Si and the occurrence of redox reactions in the metal oxide wall. Mesostructured transition metal oxides are of great interest because of their unique catalytic, electrical, optical, and magnetic properties.[24] To date, most of the research efforts on mesostructured transition metal oxides by EISA have been done on titania.[8,25–30] Titania possesses very interesting catalytic, photochemical, and electrical properties, which make it an attractive system for several applications. For instance, TiO_2 is commonly used as a photocatalyst.[31] Crystalline titania modified with sulfate ions is an active catalyst for low-temperature esterification, isomerization, alkylation, and cracking of hydrocarbons.[32] Other applications of titania are: as an electrode material in electrochemistry, a capacitor in electronics, in humidity and gas sensors, and in photovoltaic solar cells.[33]

2D hexagonal mesoporous titania displaying high surface areas (up to 370 m^2/g) and 2.0–2.5 nm pore size was prepared by EISA.[25a] The authors proposed a formation pathway for these mesophases, in which hydrolysis of Ti precursor leads to the formation of Ti-hydroxochloroalkano species which condensed into colloidal Ti-oxo nanobuilding blocks (NBB). Evaporation of the solvent allowed the formation of a liquid crystalline mesophase by the assembly of the Ti-oxo nanobuilding blocks around the tubular micelles. Bromide and chloride anions helped to counterbalance the positive charges of the surfactant CTAB and NNB. The formation of these mesostructures depended on the CTAB/Ti ratio. At low surfactant-to-metal ratios, wormlike mesostructures were formed. The authors suggested that the template molecules could not act cooperatively enough to impose curvature to the Ti-oxo

Table 16.1 Mesostructured silica films and particles prepared by EISA

Structure	Precursors	Substrate	Deposition method	Reference
3D hexagonal and cubic films	TEOS, CTAB, EtOH, H_2O, HCl	(100) Si	Dip-coating	6
Hexagonal and cubic particles	TEOS, CTAB, Brij-56, Brij-58, P-123, EtOH, H_2O, HCl	None	Aerosol dispersion	7a
1D hexagonal, and cubic (fcc). film particles	Silsesquioxane TEOS, CTAB, SDS, P-123 EtOH, H_2O, HCl	(100) Si	Spin-coating Dip-coating Aerosol dispersion	16
Hexagonal particles	TEOS, CTAB, Brij-58, EtOH, H_2O, HCl	None	Vibrating orifice aerosol generator	18
2D hexagonal particles	TEOS, CTAB, H_2O, HCl	None	Aerosol generator	19
Hexagonal and cubic (fcc) films	Methyl triethoxy silane, PS35-b-PEO copolymer, H_2O, HCl, THF	Si wafers	Dropwise	20
2D hexagonal (*p6m*), 3D hexagonal (P63/mmc) cubic (Pm3n) and lamellar films	TEOS, CTAB, EtOH, H_2O, HCl	Si wafers	Dip-coating	8
2D hexagonal (*p6m*), 3D cubic (Pm3n), 3D hexagonal (P63/mmc) particles	TEOS, polyoxyethylene alkylamine (PN-430), Brij-78, Triton X-100, HCl, H_2O, THF	None	Casting	21
Hexagonal filaments	TEOS, CTAB, Brij-56, P-123, EtOH, H_2O, HCl	Anodic alumina membrane	Soaking	22
Hexagonal films	Silsesquioxanes, TEOS, CTACl, HCl, EtOH, H_2O	Si wafers	Spin-coating	23

NBB. For intermediate CTAB/Ti ratios, the mesostructures are well organized in hexagonal symmetries. Finally, high surfactant-to-metal ratios produced inverse micelles with Ti-oxo cores. The amorphous nature of the resultant mesophase limited the use of this material for practical applications in which a well-developed crystalline structure is required.

Later, the same research group reported the synthesis of mesoporous nanocrystalline anatase films displaying wormlike, 2D hexagonal and 3D cubic symettries.[25b] In their approach, poly(ethylene oxide)-based surfactants (F-127, P-123, Brij-56, Brij-58) were used to direct the synthesis. For Pluronic-templated titania films, the crystallization of the inorganic framework started at ~ 350 °C with the formation of anatase. At 400 °C, few anatase crystallites of ~ 5 nm appeared in the amorphous matrix. Further heating (600 °C) led to the growth of these nanocrystallites to ~ 12 nm, with anatase as the only present phase. The authors suggested that at this temperature fully crystallized inorganic mesoporous framework was formed. A careful control over synthesis conditions and post-synthesis treatments led to the formation of anatase nanocrystals in the mesoporous walls. The authors proposed the water content entering and departing the films as the key variable that determined the final mesophase.

Alternatively, phosphorus could be introduced in the titania framework to form self-standing mesoporous titania–phosphorus oxide films containing ~ 2.4 nm anatase nanocrystals via EISA.[26] The resultant hexagonal mesostructures displayed pore diameter of ~ 4.3 nm and BET surface area of 330 m^2/g. In this case, the presence of phosphate prevented excessive crystal growth and improved the thermal stability of the final mesostructure. As in the previous example, a triblock copolymer surfactant (P123) was employed to direct the synthesis, ethanol–water as solvent, titanium tetraisopropoxide, and phosphorus pentachloride as inorganic sources. The coexistence of amorphous and rutile phases in mesoporous titania films synthesized by EISA have been reported.[27] In this report, a solution of $TiCl_4$, Brij-58, water, and ethanol was prepared and deposited in Si wafers and glass slides by dip-coating. The mesostructure formation was driven by EISA. Unfortunately, these materials were composed of lamellar mesophases which limits their practical applications. Various synthesis modifications have led to improved mesoporous nanocrystalline titania phases. For example, it has been shown recently that post-synthesis modification of mesoporous titania with NH_4OH or NaOH helps in stabilizing anatase nanocrystals in the mesostructure.[28] The resultant mesostructures had high surface areas (up to 573 m^2/g), large pore volumes (0.71 cm^3/g), and high sorption capacities for rhodamine 6G.

Ozin's group reported the synthesis of highly crystalline and well-ordered periodic mesoporous nanocrystalline anatase films displaying 2D hexagonal and 3D cubic pore structures.[29] The high-quality synthesized mesostructures were obtained by controlling the temperature and external relative humidity, as well as by using 1-butanol as a solvent which enhanced the microphase separation between the surfactant and the inorganic precursor, and consequently created a highly condensed inorganic framework. 1-butanol also acted as a swelling agent for the template increasing the resultant pore diameter of the mesostructure. The resultant 3D hexagonal (*R-3m*) mesostructure was composed of large domains of periodic arrays of interconnected elliptically shaped 20 nm × 6 nm cages with titania framework composed of 5.8 nm anatase nanocrystals.[29a] 2D hexagonal mesostructures (*p6m*) with anatase crystal size in the 7–9.9 nm range and cubic mesostructure (*fm3m*) with crystal size in the 7.8–9.8 nm range have been reported.[29b] In these examples, P-123 and 1-butanol were used as structure directing agent and solvent, respectively.

Figure 16.5 shows representative HRSEM and STEM images of these ordered periodic nanocrystalline hexagonal meso-TiO_2 and cubic meso-TiO_2 films. The top surface of the hexagonal TiO_2 films displays a swirling pattern originating from topological defects in the precursor mesophase often observed for the channel director of the 2D hexagonal mesostructure (Fig. 16.5a). The channels are well-ordered and are aligned parallel to the substrate (Fig. 16.5b). For the cubic TiO_2

Figure 16.5 HRSEM and STEM images of hexagonal meso-nc-TiO_2 (a and b) and cubic meso-nc-TiO_2 (c and d).[29b] See text for detailed description.

films, the surface of the film displays hexagonally close-packed open pores (Fig. 16.5c). STEM images shown in Fig. 16.5D confirm the surface pores comprising the termini of an internal network of 3D interconnected cage-type pores. The insets (Fig. 16.5a and c) show the graphical illustrations of pore architectures and lattice parameters of hexagonal and cubic mesostructures, respectively.[29b]

Recently, EISA has been used to prepare ultrathin mesostructured nanocrystalline anatase films.[30] In this approach, diblock copolymer KLE ($H(CH_2CH_2CH_2(CH)$-$CH_2CH_3)_{89}$-$(OCH_2CH_2)_{79}OH$) and Brij-58 led to spherical and cylindrical micelles, respectively. The synthesis approach employed by the authors led to thin (17–18 nm) titania films displaying a monolayer or bilayer of highly ordered mesopores arranged on a 2D cubic lattice in which the inorganic matrix was composed of $\sim$ 6–10 nm anatase crystals. Other successful examples of mesoporous nanocrystalline titania prepared by EISA and displaying anatase crystals with pore sizes in the 5–23 nm range and crystal size in the 4.5–23 nm range have been reported.[34–37] Cubic mesoporous films deposited on Si wafers and displaying 7.5 nm anatase crystals and pore size of $\sim$ 5.5 nm were prepared by EISA employing F-127 as template.[34] The films were used for the photocatalytic degradation of methylene blue and lauric acid. Mixed inorganic titania precursors (titanium chloride and titanium isopropoxide) combined with P-123 were used to prepare wormlike anatase mesoporous films.[35] These mesophases displaying BET surface areas up to 190 m^2/g, pore size of $\sim$ 7.4 nm, and anatase nanocrystal size of 4.5 nm were employed as dye-sensitized solar cells. Mesoporous titania films synthesized with F-127 as template has been successfully employed as sensitized solar cells.[36] Recently, mesoporous nanocrystalline nitrided-titania films have been reported.[37] These N-doped mesoporous films were prepared by annealing mesoporous nanocrystalline anatase under ammonia flow. The mesoporous structure composed of 10–14 nm anatase nanocrystals was maintained at temperatures as high as 700 °C. Nitrogen doping is particularly useful in decreasing the effective bandgap of anatase. Table 16.2 summarizes some of the relevant work done on the preparation of mesoporous titania films and particles by EISA.

Other transition metal oxides that have been successfully mesostructured by EISA are summarized in Table 16.3.[30,38–45] Brezesinski *et. al.* reported the synthesis of large pore (10–25 nm) highly ordered mesosotructured WO_3, CeO_2, and MoO_3 ultrathin films deposited on Si wafers and galss slides and employing KLE and Brij-58 as structure-directing agents.[30] Interestingly, these three mesostructures displayed well-developed nanocrystalline monoclinic, cerianite and *fcc* phases for WO_3, CeO_2, and MoO_3, respectively. The bulk phases of these oxides have found several interesting applications, namely, WO_3 as a semiconductor material,[24b] CeO_2 as oxygen sensing material,[46] and Mo-based oxides as catalytic phases in the selective oxidation and reduction of hydrocarbons.[47] Recently, crystalline mesostructured Y_2O_3 films doped with rare earth metals and displaying nanocrystals in the 10–17 nm range and average pore sizes between 8.5 and 14 nm have been reported.[38] These mesofilms were prepared employing a polyethylene oxide–based copolymer structuring agent (KLE-22 (poly(ethylene–co–butylene)-b-poly(ethylene oxide). Because of its high dielectric constant, low absorption, and high thermal conductivity, bulk Y_2O_3 has been used in luminescent displays.[48]

Table 16.2 Mesostructured titania films and particles prepared by EISA

Structure	Precursors	Substrate	Deposition method	Reference
2D hexagonal and wormlike particles	$Ti(EtO)_4$, CTAB, HCl, H_2O	None	Casting	25a
2D hexagonal, 3D cubic films	$TiCl_4$, Ti $(EtO)_4$, F-127, P-123, Brij-56, Brij-58, EtOH, H_2O	Si wafers, fused silica or glass	Dip-coating	25b
Hexagonal films	$Ti(EtO)_4$, PCl_5, P-123, EtOH, HCl, H_2O	None	Casting	26
Lamellar films	$TiCl_4$, Brij-58, EtOH, H_2O	Si wafers, (100) Si, glass	Dip-coating	27
Mesoporous not periodic particles	Titanium tetra-isopropoxide, HDA, CTAB, EtOH, NH_4OH, H_2O	None	Casting	28
3D hexagonal, 2D hexagonal, 3D cubic films	$Ti(OEt)_4$, P-123, BuOH, HCl, H_2O	Glass	Spin-coating	29
1D hexagonal, 2D cubic films	$TiCl_4$, KLE, EtOH, THF, H_2O	Si, Glass	Dip-coating	30
Cubic films	$TiCl_4$, F-127, EtOH, H_2O	Si wafers	Dip-coating	34
Wormlike films	Titanium isopropoxide, $TiCl_4$, P-123, EtOH	F-doped SnO_2 coated glass	Doctor blading technique	35
Mesoporous films	$TiCl_4$, F-127, EtOH, H_2O	F-doped SnO_2 coated glass	Spin-coating	36
Mesoporous films[a]	$TiCl_4$, F-127, EtOH, H_2O	Si wafers, quartz	Dip-coating	37

[a] Nitrided titania films.

Table 16.3 Mesostructured metal oxide films and particles prepared by EISA

Structure	Precursors	Substrate	Deposition method	Reference
WO_3, MoO_3, CeO_2 1D hexagonal, 2D cubic films	WCl_2, $MoCl_5$, $CeCl_3$ $7H_2O$, KLE, EtOH, THF, H_2O	Si wafers, glass	Dip-coating	30
Y_2O_3 doped with Eu, Sm, Er mesoporous not periodic films	YCl_3 $6H_2O$, $EuCl_3$ $6H_2O$, KLE-22, EtOH, NH_4OH, H_2O	Si wafers	Dip-coating	38
CeO_2, ZrO_2, CeO_2-ZrO_2 cubic (bcc) films	$CeCl_3$ $7H_2O$, $ZrCl_4$, KLE block copolymer, EtOH, THF, H_2O	Si wafers	Dip-coating	39
Al_2O_3 cubic (fcc) films	$AlCl_3$ $6H_2O$, KLE22, KLE23, EtOH, H_2O, NH_4	Si wafers	Dip-coating	40
Fe_2O_3, FeOOH mesoporous films	$FeCl_3$, PIB-PEO, KLE, EtOH, THF, H_2O	Si wafers, glass	Dip-coating	41
$SrTiO_3$, $MgTa_2O_6$, $Co_xTi_{1-x}O_{2-x}$ 3D and 2D pore films	$SrCl_2$ $6H_2O$, $TiCl_4$, $CoCl_2$, $Mg(OH)_2$, $Ta(OC_2H_5)_5$, THF, EtOH, H_2O	Si wafers	Dip-coating	42
Nb–M oxides (M=V, Mo, Sb) particles	$NbCl_5$, V_2O_5, $VOSO_4$, MoO_3, Sb_2O_3, P-123, EtOH	None	Casting	43
Co-Ni oxide particles	$(Ni_{0.33}Co_{67})$ (OMe)(acac)(MeOH) Brij-35, BuOH	None	Casting	44
WO_3–TiO_2 cubic films	$Ti(O\textit{i}PC)_4$,W pentaethoxide, F-127, EtOH, HCl, H_2O	Pirex glass	Spin-coating	45
SiO_2–TiO_2 nanowires	TEOS, $Ti(O\textit{i}PC)_4$, PS-*b*-PEO, EtOH, HCl	Porous alumina	Immersion of sol in porous alumina	52

Nanocrystalline mesostructured CeO_2, ZrO_2 and CeO_2-ZrO_2 thin films with ordered arrays of ~ 9–10 nm mesopores and displaying crystals sizes in the 2–10 nm range have been synthesized.[39] The successful preparation of these mesophases was mainly attributed to the KLE surfactant employed in synthesis. It was proposed that KLE accelerates the mesostructure formation, avoiding the insufficient build-up or loss of mesostructure typically observed when other surfactants are used.[39] Also, KLE polymer is more thermally stable, and, therefore, it may maintain the mesostructure even beyond the onset of the metal oxide crystallization. Bulk ZrO_2 is a particularly interesting catalytic system, which found many uses in chemical and petrochemical processes because it possesses acidic, basic, and redox properties. For example, it is an effective catalyst for the selective formation of isobutane and isobutene from the synthesis gas (Fischer-Tropsch process).[49]

Mesostructured nanocrystalline γ-Al_2O_3 with ellipsoidal disk-shaped pores (~9–29 nm) and interconnecting windows of 2 or 5 nm in diameter have been synthesized employing KLE-22 and KLE-23 surfactants.[40] The stabilization of the γ-Al_2O_3 nanoparticles into a well-ordered mesostructure was attributed to the stability of the KLE-22 template above the dehydration temperature of the inorganic phase. Furthermore, when the thermal treatment reached the decomposition temperature of the template (~350 °C) the amorphous inorganic framework was rigid enough to prevent the mesostructure collapse. Alumina has a broad range of applications as an adsorbent, catalyst, support, and ultrafiltration media.[47,50] Mesoporous films composed of 7–10 nm crystals of α-Fe_2O_3 with average pore diameter of ~ 10 nm were synthesized employing a novel PIB-PEO (poly(isobutylene)-block-poly-(ethylene oxide)) copolymer and KLE.[41] The retention of the structural order in these films was attributed to the high thermal stability of these surfactants. The films retained the well-ordered mesostructure up to 450 °C. Interestingly, when F-127 was employed as template, the resulting α-Fe_2O_3 mesostructure collapsed at 350 °C. Some of the most relevant applications that iron oxides have found are as magnetic data storage devices, catalysts, magnetic-optical devices, electrodes, and sensors.[51]

Recently, SiO_2–TiO_2 hybrid nanowires consisting of a silica core containing linear arrays of mesocages and titania shell have been prepared by EISA.[52] The silica inorganic precursor-template (TEOS- PS-*b*-PEO copolymer) solution was infiltrated into porous alumina.The gradual evaporation of ethanol by EISA and subsequent gelation and calcination produce mesostructured silica ~ 50–60 nm nanowires containing spherical mesocages of ~ 15–40 nm. When PS(9500)-*b*-PEO(9500) template was used small mesocages of 15 nm were obtained; when PS(9500)-*b*-PEO(18000) template was used bigger mesocages of 40 nm were obtained. The nanowires were released from the alumina by wet chemical etching, and finally coated with TiO_2 by atomic layer deposition. This method allowed internal mesoscopic fine structure (by EISA) and compositional control over outer nanowire surface (atomic layer deposition).

Mesoporous cubic WO_3–TiO_2 films have been prepared via EISA employing titanium tetraisopropoxide and tungsten pentaethoxide as inorganic sources and triblock copolymer F-127 as template.[45] It was found that the incorporation of WO_3 hindered the crystallization of TiO_2 during calcination at 400 °C and improved the ordering of the resultant mesostructure. The pore size of the resultant composite mesostructure was ~ 10 nm with BET surface area of 130 m^2/g.

Examples of mesostructured multimetallic mixed oxides prepared by EISA have been reported by Sanchez's group.[42] In this report, ordered nanocrystalline mesoporous films composed of $SrTiO_3$, $MgTa_2O_6$, and $Co_xTi_{1-x}O_{2-x}$ were prepared employing a designed non-ionic block copolymer surfactant composed of hydrogenated poly (butadiene)-block-poly(ethylene oxide) designated as KLE3739. Optimal conditions of film deposition deduced from *in situ* time resolved SAXS, optimal conditions for annealing deduced from *in situ* time resolved simultaneous WAXS and SAXS, and the use of KLE3739 allowed the formation of these highly crystalline mesostructures. Bulk phases of $SrTiO_3$ find applications as piezoelectrics, actuators, and ferromagnetic data storage devices.[53] $MgTa_2O_6$ and $Co_xTi_{1-x}O_{2-x}$ are used as photocatalysts for solar energy conversion and as ferromagnetic semiconductors, respectively.

Mixed mesoporous Nb-M (M = V, Mo, Sb) metal oxides with surface areas up to 200 m^2/g, pore sizes in the 5–14 nm range and high pore volumes (0.46 cm^3/g) have been prepared by EISA.[43] In this report, oxides (Sb_2O_3, V_2O_5, MoO_3) and salts ($NbCl_5$, $VOSO_4$) were used as inorganic sources, and P-123 as the structure directing agent. Bulk phases of these mixed metal oxides are active and selective catalysts for alkane oxidation.[47] Uniform nanocrystalline mesoporous cobalt–nickel oxide spinels have been prepared by employing a heterometallic alkoxide precursor ($Ni_{0.33}Co_{0.67}$)(OMe)(acac)(MeOH) and oligomeric alkyl-ethylene oxide surfactant (Brij-35) as a structure-directing agent via EISA at 50 °C and 60% RH.[44] In this synthesis approach, the desired metal oxide stoichiometries were introduced on a molecular level by reacting the heterometallic alkoxide precursor in the presence of supramolecular liquid crystalline phases. The resultant mesoporous phases displayed unimodal pores in the 7–12 nm range and relatively high specific surface areas up to 83 m^2/g. The mesoporous phases were composed mainly of ~ 8–11 nm Co–Ni–O spinel nanocrystals.

7. Emerging Applications

In the next paragraphs we briefly describe emerging functional applications of EISA-based mesostructured metal oxides in photocatalysis, dye sensitized solar cells (DSSCs), and heterogeneous catalysis.

The photocatalytic activity of mesoporous nanocrystalline titania films has been reported by independent research groups.[28,29b,34,37,45] In these reports, superior photocatalytic efficiency has been observed for the mesoporous nanocrystalline films prepared via EISA as compared to conventional and non-periodic non-porous nanocrystalline titania. For example, it was found that NH_4OH modified mesoporous nanocrystalline titania degraded rhodamine 6G faster than commercially available Degussa P25 titania.[28] Periodic ordered hexagonal and cubic nanocrystalline anatase films photodegraded methylene blue ~ 2× faster than conventional nanocrystalline titania displaying randomly organized pore network.[29b] In particular, the 3D cubic mesophase exhibited the best photocatalytic performance. The observed superior photocatalytic activity of this mesoporous phase originated from the higher photoactivity of anatase nanocrystallites comprising the more open cubic framework as well as geometrical advantages, such as a larger surface area and less obstructed 3D diffusion paths of guest molecules. Mesoporous titania comprised of ~ 7.5 nm crystals displayed improved photocatalytic performance for methylene blue degradation as compared to a TiO_2 control sample.[34] Superior photocatalytic activity in all these systems may be related to the periodic open porous architectures with organized framework of nanocrystals, which provide facile diffusion pathways for guest molecules. Recently, nanocrystalline mesoporous nitrided-titania films were used as effective photocatalysts for the photodegradation of lauric acid and methylene blue.[37] The incorporation of optimal concentration of nitrogen in the mesostructure led to superior photocatalytic performance as compared to non-doped mesostructures. It was claimed by the authors that nitrogen substituted oxygen in the inorganic anatase framework, creating oxygen vacancies (which does not play an important role in promoting the recombination of photogenerated electrons and holes), and, therefore, improving the photocatalytic efficiencies.

WO_3–TiO_2 was employed as photocatalyst for the gas phase decomposition of 2-propanol.[45] Its photocatalytic activity was ~ 2.2× higher than that of pure mesoporous TiO_2 and ~ 6.1× higher than that of conventional non-porous TiO_2. The enhanced photocatalytic performance of WO_3–TiO_2 was attributed not only to the periodic ordered architecture, but also to the increase in surface acidity due to the incorporation of W.

DSSCs have been fabricated from mesoporous nanocrystalline titania prepared by EISA.[35,36] For example, a DSSC composed of mesoporous crystalline anatase and Ru-based dye exhibited high light-to-electricity conversion efficiency up to 5.31%.[35] In another example, a solid state DSSC composed of mesoporous titania as *n*-type semiconductor, Ru complex dye for visible light absorption, and poly(3-octylthiophene) as hole conductor displayed solar conversion efficiency of ~ 0.52%.[36]

Thermally stable mesoporous mixed metal Nb–M (M = V, Mo, Sb) oxide phases prepared via EISA were employed as catalysts in the oxidative dehydrogenation (ODH) of propene to propylene.[43] It was found that the mesoporous Nb–Mo–O catalyst containing molecularly dispersed MoO_x species displayed higher propane ODH activity than supported MoO_x/Nb_2O_5, in which molybdenum oxide was present as MoO_x species and MoO_3 crystals. Mesoporous nanocrystalline cobalt-nickel oxides were studied as catalysts in the combustion of propane.[44] These mesoporous phases composed of 8–11 nm Co–Ni–O spinel nanocrystals displayed a higher rate of propane combustion to CO and CO_2 as compared to a conventional dense spinel. This improved catalytic behaviour may be related to their morphological and structural features, such as large surface area and mesoporous structure, and to the relatively small size of spinel crystallites.

8. Concluding Remarks

During the last decade, EISA has emerged as a powerful synthetic approach to design technologically relevant and functional oxides in the fiber particle and film form at the nanoscale. The method relies in using very dilute surfactant initial concentration from which a liquid crystalline mesophase is gradually developed upon solvent evaporation. The slow co-assembly between the inorganic network and the liquid crystalline phase leads to the formation of long-range order, well-defined mesostructures. One of the great advantages of EISA is that it is a highly efficient and flexible method that allows one to rationally target a particular mesostructure by adjusting appropriately chemical and processing parameters. A wide variety of metal oxides displaying highly ordered pore nanoarchitectures, flexible compositions, and nanocrystallinity have been prepared by this method.

For many structure-sensitive applications, it is highly desirable to have a well-defined crystalline phase. In this respect, EISA is an ideal approach for the molecular design of nanocrystals possessing remarkable order in the mesoscale regime. Although the chemical and physical phenomena that take place during EISA are well-understood, it is still challenging to fully describe the EISA process in quantitative terms (i.e., quantification of mass transfer at the liquid–air interface, concentration gradients, and kinetics of condensation of inorganic species and organization of liquid crystalline phases) because of the relatively short period of time in which these processes occur. To date, most of the research efforts have been centered on synthesizing oxide mesostructures with different compositions and pore nanostructures, however, only few examples on the application of these materials have been reported. The use of these novel materials in specific functional applications such as catalysis, optics, magnetism, sensing, and separations would lead to an improved fundamental understanding of their structure-properties relationships. EISA is a highly promising approach for the preparation of other important functional materials displaying polymeric, semiconducting and fully crystalline zeolite frameworks.

REFERENCES

[1] (a) Kresge, C. T., Leonowicz, M. E., Roth, W. J., Vartuli, J. C., Beck, J. S.,"Ordered mesoporous molecular sieves synthesized by a liquid-crystal template mechanism" *Nature* **1992,** *359,* 710–712; (b) Beck, J. S., Vartuli, J. C., Roth, W. J., Leonowicz, M. E., Kresge, C. T., Schmitt, K. D., Chu, C. T.-W., Olson, D. H., Sheppard, E. W., McCullen, S. B., Higgins, J. B., Schlenker, J. L.,"A new family of mesoporous molecular sieves prepared with liquid crystal templates" *J. Am. Chem. Soc.* **1992,** *114,* 10834–10843; (c) Kresge, C. T., Leonowicz, M. E., Roth, W. J., Vartuli, J. C.,"Synthetic mesoporous crystalline materials having high adsorption capacity for benzene" US Patent 5,098,684, **(1992)**, to Mobil Oil Corp.

[2] Vartuli, J. C., Kresge, C. T., Leonowicz, M. E., Chu, A. S., McCullen, S. B., Johnson, I. D., Sheppard, E. W.,"Synthesis of mesoporous materials: liquid-crystal templating versus intercalation of layered silicates" *Chem. Mater.* **1994,** *6,* 2070–2077.

[3] Carreon, M. A., Guliants, V. V.,"Ordered meso and macroporous binary and mixed metal oxides" *Eur. J. Inorg. Chem.* **2005,** *1,* 27–43.

[4] Weast, R. C., Astle, M. J., Beyer, W. H., CRC Handbook of Chemistry and Physics: A Ready-Reference Book of Chemical and Physical Data, Section 4: Properties of the elements and inorganic compounds, 67th ed., CRC Press, Boca Raton, FL, 1986.

[5] (a) Yang, H., Kuperman, A., Coombs, N., Mamiche-Afara, S., Ozin, G. A.,"Synthesis of oriented films of mesoporous silica on mica" *Nature* **1996,** *379,* 703–705; (b) Yang, H., Coombs, N., Sokolov, I., Ozin, G. A., "Free-standing and oriented mesoporous silica films grown at the air–water interface" *Nature* **1996,** *381,* 589–592.

[6] Lu, Y., Ganguli, R., Drewien, C. A., Anderson, M. T., Brinker, C. J., Gong, W., Guo, Y., Soyez, H., Dunn, B., Huang, M. H., Zink, J. I.,"Continuous formation of supported cubic and hexagonal mesoporous films by sol-gel dip-coating" *Nature* **1997,** *389,* 364–368.
[7] (a) Lu, Y., Fan, H., Stump, A., Ward, T. L., Rieker, T., Brinker, C. J.,"Aerosol-assisted self-assembly of mesostructured spherical nanoparticles" *Nature* **1999,** *398,* 223–226; (b) Brinker, C. J., Lu, Y., Sellinger, A., Fan, H.,"Evaporation-induced self-assembly: nanostructures made easy" *Adv. Mater.* **1999,** *11,* 579–585.
[8] Grosso, D., Cagnol, F., Soler-Illia, G. J. de A. A., Crepaldi, E. L., Amenitsch, H., Brunet-Bruneau, A., Burgeois, A., Sanchez, C.,"Fundamentals of mesostructuring through evaporation-induced self-assembly" *Adv. Funct. Mater.* **2004,** *14,* 309–322.
[9] (a) Fontell, K., Khan, A., Lindstrom, B., Maciejweska, D., Puang-Ngern, S. P.,"Phase equilibria and structures in ternary systems of a cationic surfactant (C_{16}TABr or (C_{16} TA)$_2$SO$_4$), alcohol, and water" *Colloid Polym. Sci.* **1991,** *269,* 727–742; (b) Soler-Illia, G. J. de, A. A., Crepaldi, E. L., Grosso, D., Sanchez, C.,"Block copolymer-templated mesoporous oxides" *Curr. Opin. Colloid Interface Sci.* **2003,** *8,* 109–126.
[10] Cagnol, F., Grosso, D., Soler-Illia, G. J. de A. A., Crepaldi, E. L., Babonneau, F., Amenitsch, H., Sanchez, C.,"Humidity-controlled mesostructuration in CTAB-templated silica thin film processing. The existence of a modulable steady state" *J. Mater. Chem.* **2003,** *13,* 61–66.
[11] Grosso, D., Balkenende, A. R., Albouy, P. A., Lavergne, M., Mazerolles, L., Babonneau, F., "Highly oriented 3D-hexagonal silica thin films produced with cetyltrimethylammonium bromide" *J. Mater. Chem.* **2000,** *10,* 2085–2089.
[12] Clark, T., Ruiz, J. D., Fan, H., Brinker, C. J., Swanson, B. I., Parikh, A. N.,"A new application of UV-ozone treatment in the preparation of substrate-supported, mesoporous thin films" *Chem. Mater.* **2000,** *12,* 3879–3884.
[13] Hurd, A. J., Steinberg, L.,"The physics of evaporation-induced assembly of sol-gel materials" *Granular Matter.* **2001,** *3,* 19–21.
[14] Guliants, V. V., Carreon, M. A., Lin, J. Y.,"Ordered mesoporous and macroporous inorganic films and membranes" *J. Membr. Sci.* **2004,** *235,* 53–72.
[15] Brinker, C. J., Scherer, G. W., Sol–Gel Science. The Physics and Chemistry of Sol–Gel Processing, Chapter 13 Film Formation, Academic Press, San Diego, CA, 1990.
[16] Lu, Y., Fan, H., Doke, N., Loy, D. A., Assink, R. A., LaVan, D. A., Brinker, C. J.,"Evaporation-induced self-assembly of hybrid bridged silsesquioxane film and particulate mesophases with integral organic functionality" *J. Am. Chem. Soc.* **2000,** *122,* 5258–5261.
[17] Israelachvili, J. N., Mitchell, D. J., Ninham, B. W.,"Theory of self-assembly of hydrocarbon amphiphiles into micelles and bilayers" *J. Chem. Soc. Faraday Trans. 2* **1976,** *72,* 1525–1568.
[18] Rao, G. V. R., Lopez, G. P., Bravo, J., Pham, H., Datye, A. K., Xu, H., Ward, T. L., "Monodispersed mesoporous silica microspheres formed by evaporation-induced self-assembly of surfactant templates in aerosols" *Adv. Mater.* **2002,** *14,* 1301–1304.
[19] Bore, M. T., Rathod, S. B., Ward, T. L., Datye, A. K.,"Hexagonal mesostructure in powders produced by evaporation-induced self-assembly of aerosols from aqueous tetraethoxysilane solutions" *Langmuir* **2003,** *19,* 256–264.
[20] Yu, K., Smarsly, B., Brinker, J. C.,"Self-assembly and characterization of mesostructured silica films with a 3D arrangement of isolated spherical mesopores" *Adv. Funct. Mater.* **2003,** *13,* 47–52.
[21] Zhang, Z., Yan, X., Tian, B., Yu, C., Tu, B., Zhu, G., Qiu, S., Zhao, D.,"Synthesis of ordered small pore mesoporous silicates with tailorable pore structures and sizes by polyoxyethylene alkyl amine surfactant" *Micropor. Mesopor. Mater.* **2006,** *90,* 2331.
[22] Platschek, B., Petkov, N., Bein, T.,"Tuning the structure and orientation of hexagonally ordered mesoporous channels in anodic alumina membrane hosts: A 2-D small angle X-ray scattering study" *Angew. Chem. Int. Ed.* **2006,** *45,* 1134–1138.
[23] Hatton, B. D., Landskron, K., Whitnall, W., Perovic, D. D., Ozin, G. A.,"Spin-coated periodic mesoporous organosilica thin films. Towards a new generation of low-dielectric-constant materials" *Adv. Funct. Mater.* **2005,** *15,* 823–829.

[24] (a) Soler-Illia, G. J. A. A., Sanchez, C., Lebeau, B., Patarin, J.,"Chemical Strategies To Design Textured Materials: from Microporous and Mesoporous Oxides to Nanonetworks and Hierarchical Structures" *Chem. Rev.* **2002,** *102,* 4093–4138; (b) Yang, P., Zhao, D., Margolese, D. I., Chmelka, B. F., Stucky, G. D.,"Generalized syntheses of large-pore mesoporous metal oxides with semicrystalline frameworks" *Nature* **1998,** *396,* 152–155; (c) He, X., Antonelli, D., "Recent Advances in Synthesis and Applications of Transition Metal Containing Mesoporous Molecular Sieves" *Angew. Chem. Int. Ed.* **2002,** *41,* 214–229.

[25] (a) Soler-Illia, G. J., de, A. A., Louis, A., Sanchez, C.,"Synthesis and characterization of mesostructured titania based materials through evaporation-induced self-assembly" *Chem. Mater.* **2002,** *14,* 750–759; (b) Crepaldi, E. L., Soler-Illia, G. J. de, A. A., Grosso, D., Cagnol, F., Ribot, F., Sanchez, C.,"Controlled Formation of Highly Organized Mesoporous Titania Thin Films: From Mesostructured Hybrids to Mesoporous Nanoanatase TiO_2" *J. Am. Chem. Soc.* **2003,** *125,* 9770–9786.

[26] Yun, H., Zhou, H., Honma, I.,"Synthesis of self-standing mesoporous nanocrystalline titania-phosphorous oxide composite films" *Chem. Commun.* **2004,** *24,* 2836–2837.

[27] Henderson, M. J., Gibaud, A., Bardeau, J.-F., White, J. W.,"An X-ray reflectivity study of evaporation-induced self-assembled titania-based films" *J. Mater. Chem.* **2006,** *16,* 2478–2484.

[28] Beyers, E., Cool, P., Vansant, E. F.,"Stabilisation of mesoporous TiO_2 by different bases influencing the photocatalytic activity" *Micropor. Mesopor. Mater.* **2007,** *99,* 112–117.

[29] (a) Choi, S. Y., Lee, B., Carew, D. B., Mamak, M., Peiris, F. C., Speakman, S., Chopra, N., Ozin, G. A.,"3D hexagonal (R-3m) mesostructured nanocrystalline titania thin films: synthesis and characterization" *Adv. Funct. Mater.* **2006,** *16,* 1731–1738; (b) Carreon, M. A., Choi, S. Y., Mamak, M., Chopra, N., Ozin, G. A.,"Pore architecture affects photocatalytic activity of periodic mesoporous nanocrystalline anatase films" *J. Mater. Chem.* **2007,** *17,* 82–89.

[30] Brezesinski, T., Groenewolt, M., Gibaud, A., Pinna, N., Antonietti, M., Smarsly, B. M., "Evaporation-induced self-assembly (EISA) at its limit:Ultrathin, crystalline patterns by templating of micellar monolayers" *Adv. Mater.* **2006,** *18,* 2260–2263.

[31] Fox, M. A., Dylay, M. T.,Heterogeneous photocatalysis" *Chem. Rev.* **1993,** *93,* 341–357.

[32] Yamaguchi, T.,"Recent progress in solid superacid" *Appl. Catal.* **1990,** *61,* 1–25.

[33] Hagfeldt, A., Gratzel, M.,"Light-Induced Redox Reactions in Nanocrystalline Systems" *Chem. Rev.* **1995,** *95,* 49–68.

[34] Sakatani, Y., Grosso, D., Nicole, L., Boissiere, C., Soller-Illia, G. J. de A. A., Sanchez, C., "Optimised photocatalytic activity of grid-like mesoporous TiO_2 films : effect of crystallinity, pore size distribution, and pore accessibility" *J. Mater. Chem.* **2006,** *16,* 77–82.

[35] Hou, K., Tian, B., Li, F., Bian, Z., Zhao, D., Huang, C.,"Highly crystallized mesoporous TiO_2 films and their applications in dye sensitized solar cells" *J. Math. Chem.* **2005,** *15,* 2414–2420.

[36] Lancelle-Beltran, E., Prene, P., Boscher, C., Belleville, P., Buvat, P., Sanchez, C.,"All-Solid-State Dye-Sensitized Nanoporous TiO_2 Hybrid Solar Cells with High Energy-Conversion Efficiency" *Adv. Mater.* **2006,** *18,* 2579–2582.

[37] Martinez-Ferrero, E., Sakatani, Y., Boissiere, C., Grosso, D., Fuertes, A., Fraxedas, J., Sanchez, C., "Nanostructured titanium oxynitride porous thin films as efficient visible-active photocatalysts" *Adv. Funct. Mater.* **2007,** *17,* 3348–3354.

[38] Castro, Y., Julian-Lopez, B., Boissiere, C., Viana, B., Grosso, D., Sanchez, C.,"Preparation, structural and optical characterization of rare earth doped mesoporous Y_2O_3 thin films by EISA method" *Micropor. Mesopor. Mater.* **2007,** *103,* 273–279.

[39] Brezesinski, T., Antonietti, M., Groenewolt, M., Pinna, N., Smarsly, B.,"The generation of mesostructured crystalline CeO_2, ZrO_2 and CeO_2-ZrO_2 films using evaporation-induced self-assembly" *New J. Chem.* **2005,** *29,* 237–242.

[40] Kuemmel, M., Grosso, D., Boissiere, C., Smarsly, B., Brezesinski, T., Albouy, P. A., Amenitsch, H., Sanchez, C.,"Thermally stable nanocrystalline γ-Alumina layers with highly ordered 3D mesoporosity" *Angew. Chem. Int. Ed.* **2005,** *44,* 4589–4592.

[41] Brezesinski, T., Groenewolt, M., Antonietti, M., Smarsly, B.,"Crystal-to-crystal phase transition in self-assembled mesoporous iron oxide films" *Angew. Chem. Int. Ed.* **2006,** *45,* 781–784.

[42] Grosso, D., Boissiere, C., Smarsly, B., Brezesinski, T., Pinna, N., Albouy, P. A., Amenitsch, H., Antonietti, M., Sanchez, C.,"Periodically ordered nanoscale islands and mesoporous films composed of nanocrystalline multimetallic oxides" *Nature Materials* **2004,** *3,* 787–792.
[43] Yuan, L., Bhatt, S., Beaucage, G., Guliants, V. V., Mamedov, S., Soman, R. S.,"Novel mesoporous mixed Nb-M (M=V,Mo and Sb) oxides for oxidative dehydrogenation of propane" *J. Phys. Chem B* **2005,** *109,* 23250–23254.
[44] Carreon, M. A., Guliants, V. V., Yuan, L., Hughett, A. R., Dozier, A., Seisenbaeva, G. A., Kessler, V. G.,"Mesoporous Nanocrystalline Mixed Metal Oxides from Heterometallic Alkoxide Precursors: Cobalt-Nickel Oxide Spinels for Propane Oxidation" *Eur. J. Inorg. Chem.* **2006,** *24,* 4983–4988.
[45] Pan, J. H., Lee, W. I.,"Preparation of highly ordered cubic mesoporous WO_3/TiO_2 films and their photocatalytic properties" *Chem. Mater.* **2006,** *18,* 847–853.
[46] Jasinski, P., Suzuki, T., Anderson, H. U.,"Nanocrystalline undoped ceria oxygen sensor" *Sens. Actuators B* **2003,** *95,* 73–77.
[47] Centi, G., Cavani, F., Trifiro, F., Selective Oxidation by Heterogeneous Catalysis. Fundamental and Applied Catalysis, Chapter 2. New Technological and industrial opportunities: Options; and Chapter 4. Control of the surface reactivity of solid catalysts: Industrial processes for alkane oxidation, Kluwer Academic/Plenum Publishers, New York, 2001.
[48] Wakefield, G., Holland, E., Dobson, P. J., Hutchison, J. L.,"Luminescence properties of nanocrystalline Y_2O_3:Eu" *Adv. Mater.* **2001,** *13,* 1557–1560.
[49] (a) Postula, W. S., Feng, Z., Philip, C. V., Akgerman, A., Anthony, R. G.,"Conversion of Synthesis Gas to Isobutylene over Zirconium Dioxide Based Catalysts" *J. Catal.* **1994,** *145,* 126–131; (b) Feng, Z., Postula, W. S., Philip, C. V., Anthony, R. G.,"Selective Formation of Isobutane and Isobutene from Synthesis Gas over Zirconia Catalysts Prepared by a Modified Sol-Gel Method" *J. Catal.* **1994,** *148,* 84–90.
[50] (a) Kim, Y., Kim, C., Choi, I., Rengaraj, S., Yi, J.,"Arsenic removal using mesoporous alumina prepared via a templating method" *Environ. Sci. Technol.* **2004,** *38,* 924–931; (b) Schaep, J., Vandecasteele, C., Peeters, B., Luyten, J., Dotremont, C., Roels, D.,"Characteristics and retention properties of a mesoporous γ-Al_2O_3 membrane for nanofiltration" *J. Membr. Sci.* **1999,** *163,* 229–237.
[51] (a) Poizot, P., Laruelle, S., Grugeon, S., Dupont, L., Tarascon, J.-M.,"Nano-sized transition-metal oxides as negative-electrode materials for lithium-ion batteries" *Nature* **2000,** *407,* 496–499; (b) Liaw, B. J., Cheng, D. S., Yang, B. L.,"Oxidative dehydrogenation of 1-butene on iron oxyhydroxides and hydrated iron oxides" *J. Catal.* **1989,** *118,* 312–326; (c) Pankhurst, Q. A., Pollard, R. J.,"Fine-particle magnetic oxides" *J. Phys. Condens. Matter* **1993,** *5,* 8487–8508; (d) Pelino, M., Colella, C., Cantalini, C., Faccio, M., Ferri, G., D'Amico, A.,"Microstructure and electrical properties of an α-hematite ceramic humidity sensor" *Sens. Actuators B* **1992,** 7, 464–469.
[52] Chen, X., Knez, M., Berger, A., Nielsch, K., Gosele, U., Steinhart, M.,"Formation of titania/silica hybrid nanowires containing linear mesocage arrays by evaporation-induced block-copolymer self-assembly and atomic layer deposition" *Angew. Chem. Int. Ed.* **2007,** *46,* 6829–6832.
[53] (a) Nagarajan, V., Roytburd, A., Stanishevsky, A., Prasertchoung, S., Zhao, T., Chen, L., Melngailis, J., Auciello, O., Ramesh, R.,"Dynamics of ferroelastic domains in ferroelectric thin films" *Nature Mater.* **2003,** *2,* 43–47; (b) Bhattacharya, K., Ravichandran, G.,"Ferroelectric perovskites for electromechanical actuation" *Acta Mater.* **2003,** *51,* 5941–5960; (c) Chu, M. W., Szafraniak, I., Scholz, R., Harnagea, C., Hesse, D., Alexe, M., Gösele, U.,"Impact of misfit dislocations on the polarization instability of epitaxial nanostructured ferroelectric perovskites" *Nature Mater.* **2004,** *3,* 87–90.

CHAPTER 17

Zeolite Nanocrystals: Hierarchical Assembly and Applications

Yahong Zhang, Nan Ren, *and* Yi Tang

Contents

Abstract

Zeolite nanocrystals (nanozeolites) represent a hot topic in the domain of porous materials due to their controllable colloidal character, short diffusion path and large and adjustable surface. In the first part of this chapter, various hierarchical assembly strategies including template-directed (such as polymer, carbon, mesoporous silica and bio-templates) and template-free methods as well as the construction of micro/mesoporous composite have been reviewed in detail. Although there is still a long way to go for commercialization and adapting these newly born materials in industrial processes, the application of the nanozeolite and their hierarchical assemblies have shown their potentials with some of the emerging, exciting examples in catalysis and biology. The recent progress of the application of nanozeolites and their hierarchical assemblies are summarized in the second part. Finally, a prospect of the nanozeolitic materials on the hierarchical assembly and application is envisaged which is aimed to provide a brief guidance to the readers.

Keywords: Application, Hierarchical assembly, Zeolite nanocrystals

Ordered Porous Solids
DOI: 10.1016/B978-0-444-53189-6.00017-2

1. Introduction

Zeolite materials have been widely used as heterogeneous catalysts and adsorbents in the fields of oil refining and petrochemical industry. In the recent decades, zeolite nanocrystals (donated as nanozeolites) have attracted considerable attention due to the expectation of their unique surface properties, shorter diffusion pathlengths and higher tolerances for coking during the catalytic reactions. So far, approximately half of 300 papers regarding nanozeolites was focused on the synthesis and crystallization mechanism of nanozeolites (with crystal sizes of 100 nm or less), which has been well-reviewed by Valtchev and Tosheva.[1] A variety of purely siliceous and aluminosilicate nanozeolites (e.g., Fig. 17.1) with different frameworks, such as LTA, FAU, SOD, GIS, LTL, MOR, BEA, MFI and MEL, were obtained from clear solutions, gels or within confined-space.[2–23] Meanwhile, some nanosized titanisilicates, exfoliated aluminophosphates and ultrathin layered compounds [e.g., α-zirconium phosphate, aluminophosphate and MCM-22(P)] were also prepared.[24–34] With the emergence of the vast numbers of colloidal nanozeolites, the application of the colloidal zeolites began to be concerned and explored in the fields of catalysis, sensor, optical antenna, medical diagnostics and separation,[35–42]

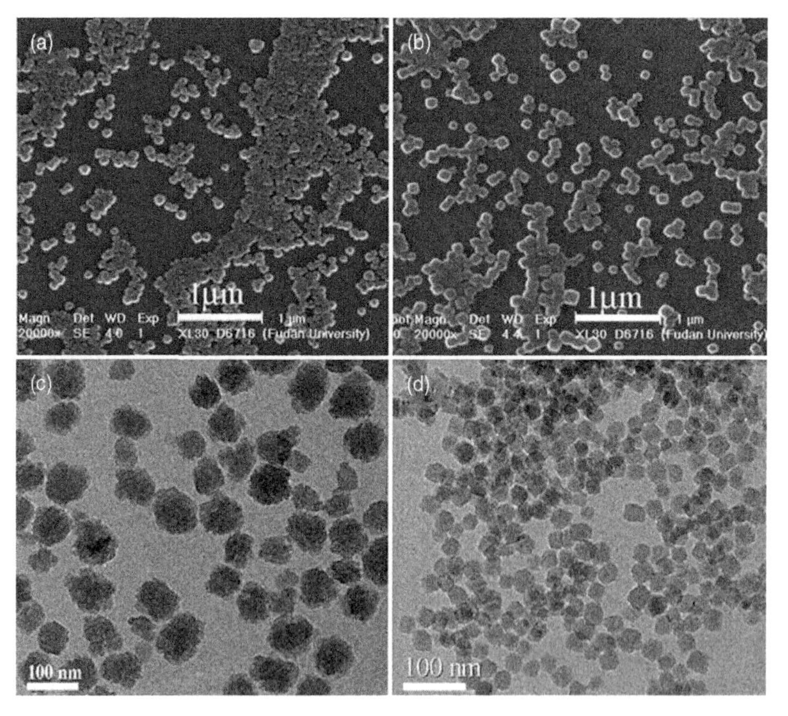

Figure 17.1 Scanning electron microscopy (SEM) (a, b) and transmission electron microscopy (TEM) (c, d) images of zeolite FAU (a), LTA (b), silicalite-1 (c) and SOD (d) nanocrystals.

owing to their characteristic surface properties, such as large external surface area, tunable surface charge and composition and regular surface silanol. At the same time, due to the superior colloidal properties of nanozeolites, they can be conveniently used as building blocks or nanoseeds to construct compact defect-free zeolite films and hierarchical structured materials.[43–47] The zeolite film prepared by pre-seeding method may cause a preferred orientation growth of zeolite crystals,[48] which is favourable for the traditional sorption and separation and has been well-covered by the previous review articles on the membranes.[48–54] One noteworthy example in this field is zeolite low-*k* film material, which was first proposed by Yan *et al.* in 2001.[55] The pure silica zeolite low-*k* material was predicted as a potential candidate for the future microprocessors. On the contrary, the hierarchical zeolite materials with specific macro-morphological features were expected to increase the manipulation of nanomaterials and even bring some new synergetic effects compared with dispersed zeolite nanocrystals.[41,56,57] In this chapter, three aspects would be highlighted. First, we would mainly review the methods for the hierarchical assembly of zeolitic materials with different structures by employing colloidal nanozeolites and/or sub-units as building blocks. Then, as the original motivation that triggers the burst of zeolite nanotechnology, the impact of nanozeolites on catalysis is discussed based on their short diffusion path and surface acidity. The third section is devoted to the emerging applications of the nanozeolitic materials in the field of biology for their unique external surface. Finally, a prospect of the nanozeolitic materials in the hierarchical assembly and applications related to catalysis and biology is predicted.

2. Hierarchical Assemblies of Zeolite Nanocrystals

Nanozeolites generally exist in the form of dispersed colloidal suspension, which often brings some serious manipulation problems, such as the tendency of spontaneous aggregation, the high pressure drop in the case of catalyst powders and the difficulty in being separated from the reaction products. Although the conventional moulding or aggregation process could solve these problems to some extent, a sharp decrease will occur on their accessible active external surface which is believed to be the most desired feature of nanomaterials. Therefore, an important issue in this field concentrates on the construction of the 'macro-structures' via rational assembly or moulding approaches. The hierarchical assembly strategy is undoubtedly the best way to realize such idea. A large and accessible surface area of nanozeolites could be mostly retained through this approach and the handling problems could be avoided.[58] More importantly, the assembled hierarchical architecture may also endow the materials with novel characteristics through integrating the multi-functionalities and porosities in one structure, which provide more opportunities for the applications ranging from catalysis to electronic devices. In this section, some of the recent progresses addressing hierarchical assembly of nanozeolites are highlighted according to different templates and methods.

2.1. Removable polymer templates

One of the general strategies for the preparation of hierarchical porous structures is on the basis of sacrificial templates or their aggregates with appropriate sizes and shapes.[1] The nanozeolites are assembled on these sacrificial templates by electrostatic or chemical driving forces, followed by removing the templates via calcination or dissolution. Polystyrene (PS) microsphere is the most widely used template for this purpose due to their characteristics, such as easy to be synthesized, monodispersed size and adjustable diameter ranging from several tens to several hundreds nanometers. Wang *et al.*[59] used PS as template to prepare a hollow sphere of nanozeolite through a layer-by-layer (LBL) technique. Typically, the PS spheres were pre-charged by sequentially depositing several layers of cationic and anionic polyelectrolytes. The nanozeolite particles and oppositely charged polyelectrolytes were then alternately deposited onto the charged PS substrates to form nanozeolite/poly(diallyldimethylammonium chloride) (PDDA) multi-layers. The PS template was finally removed by calcination. By changing the type of nanozeolites and thickness of deposited layers, a series of hollow nanozeolite microspheres were obtained.[60,61] Stein and co-workers[62] were the first to publish the work on construction of 3D-ordered macroporous zeolitic macrostructures by using PS spheres as templates. In this method, no zeolite nanocrystal was employed, and the monolithic zeolite macrostructure with bimodal pore structures was prepared by *in situ* crystallization of silicate in the presence of structure-directing agent. Rhodes *et al.*[63] took the core-shell PS microspheres with nanozeolites deposited layers to form macroscopic close-packed assemblies by centrifugation. A monolithic zeolitic material with a close macroporous structure was produced after calcination to remove the organic components. Zeolite monolithic materials with interconnective macroporous structures have also been obtained by pre-arraying PS into close-packed macrostructure via suction filtration followed by casting various nanozeolites.[64,65] However, the mechanical strength of the hollow nanozeolite spheres obtained by these methods is not high enough to be used practically. Valtchev[66–68] has proved that the secondary hydrothermal treatment in a suitable gel or clear solution could improve the mechanical stability of both the LBL-constructed hollow nanozeolite spheres and the 3D-arrayed macroporous nanozeolite monolith prepared by sedimentation. Although, the PS sphere is a widely used template in the formation of zeolitically structured materials, the obtained zeolitic materials present little feasibility for the further application in catalysis and other fields because of the difficulty to pre-capsulate active composition. Therefore, the use of PS as template for the preparation of zeolite macrostructure is not further encouraged unless a novel interest on such structure materials is found.

Macroporous anion-exchange resin beads (0.3–1.2 mm) have been used as macrotemplates for the synthesis of mesoporous zeolite spheres.[69,70] The mesoporous nanozeolite spheres were directly prepared from the clear zeolite synthesis solutions containing resin beads, which simplified the preparation process by combing the nanozeolite synthesis and assembly into one step. Notably, due to the residual ion-exchange capacity, the resin/zeolite composite beads could be further functionalized to form bifunctional composition. For example, the resin/zeolite composite beads could exchange various metal anions including $PdCl_6^{2-}$,

VO_3^- and WO_4^{2-} prior to calcination and thus the metal-containing zeolite spheres were obtained.[71,72] Due to the large size of the resin beads, the macrostructures prepared can be directly used as the catalysts, and therefore the binders for the conventional zeolite-supported catalysts are no longer required here. Very recently, a core-shell composite microsphere possessing a thick core and a thin shell were constructed by using macroporous anion exchange resin as template.[73] The macro- and mesopores in zeolite BEA core (300–500 μm) with hierarchical organization facilitate the accessibility for zeolitic micropores, whereas the thin zeolite MFI shell (<1.0 μm) provides a small diffusion limit for the guest molecules to contact with active interior. However, because the type and percentage of ion exchange resin in the synthesis solution showed a great influence on the formation of the zeolitic macrostructures; therefore, only two types of nanozeolite microspheres, that is, BEA and MFI, were reported so far. Lately, a universal and simple approach called polymerization-induced colloid aggregation (PICA) was first employed to prepare nanozeolite microspheres with various compositions and uniform diameters of 3–8 μm.[74] In brief, the different types of nanozeolites were dispersed in a mixed solution containing urea and formaldehyde, followed by the addition of hydrochloric acid to induce the polymerization of urea and formaldehyde. The polymer/nanozeolite composite microspheres were spontaneously formed owing to the driving force of surface tension at ambient temperature. The nanozeolite microspheres featured a uniform spherical morphology (Fig. 17.2) and large/adjustable secondary pore architecture (30–80 nm) after removing the polymer by combustion. The microspheres could retain the properties of the original nanozeolites to a great extent, and thereby show a protein adsorption capacity similar to that of the dispersed nanozeolites.[74] This method is expected to be applied in preparing the hierarchical microspheres with nano-scaled-composited zeolite–zeolite, zeolite–metal or zeolite–oxide functionalities.

Besides these unique spherical templates, some other polymer macrotemplate have also been used to prepare structured zeolite materials. Lee *et al.*[75] used polyurethane foams as the templates for the formation of large monolithic zeolite foams. The resulting zeolite monoliths have highly ramified networks of interconnecting macropores with tailorable sizes and shapes. Wang *et al.*[76] reported a self-supporting porous zeolite membrane with a sponge-like architecture by employing cellulose acetate filter membranes as macrotemplates. However, the mechanical stability of the membrane was low even a secondary crystallization step was adopted. A bimodal pore structured zeolite/ceramic foam with a high thermal stability and mechanical strength was developed on a non-sacrificial silicon oxycarbide microcellular foam.[77] Such template was prepared by a polymer-to-ceramic conversion occurring on pyrolysis of poly(methyl methacrylate) microbeads and silicon resin.

2.2. Digestible mesoporous silica templates

Mesoporous silica (MS) spheres were first used as a template by Dong *et al.*[78] to prepare hollow nanozeolite spheres. The preparation process was fulfilled through the vapour phase transport (VPT) treatment of the nanozeolite (seeds) coated MS spheres. In this process, MS spheres acted as not only structure templates but also

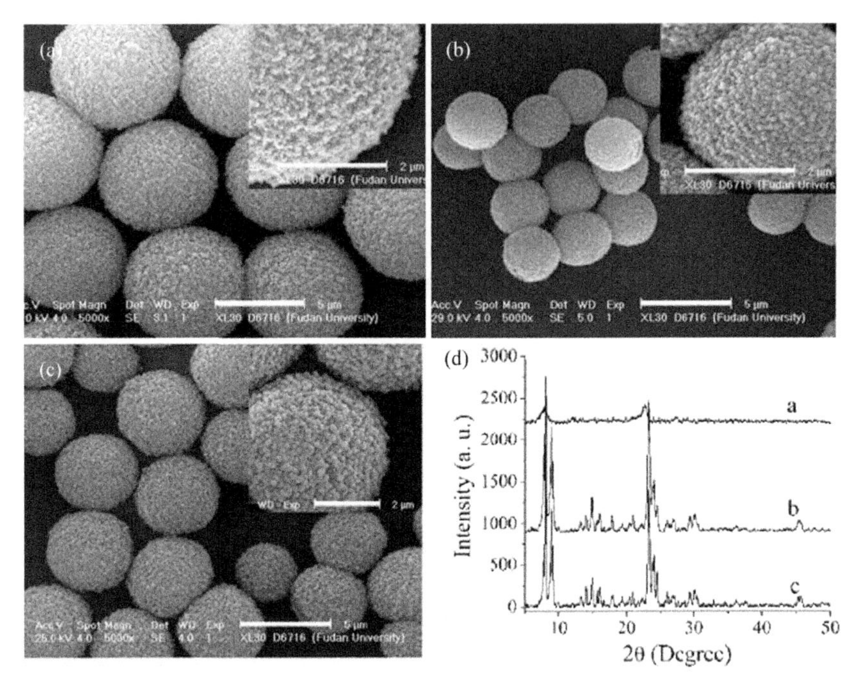

Figure 17.2 Scanning electron microscopy (SEM) images of nanozeolite microspheres derived from zeolite BEA (50) (a), silicalite-1 (b) and ZSM-5 (c) nanoparticles and their X-ray diffraction (XRD) patterns (d). The three XRD patterns in (d) correspond to the nanozeolite microspheres of zeolite BEA (50) (a), silicalite-1 (b) and ZSM-5 (c), respectively. The scale bars in the SEM images and their insets are 5 and 2 μm, respectively (Reproduced from Ref. 74 with the permission of American Chemical Society).

silica nutrient for the growth of zeolitic shell. In other words, the hollow spherical shells built of grown zeolite crystals were formed upon the complete digestion of the MS cores. Hollow zeolite microcapsules with various non-spherical shapes were also obtained through VPT treatment of pre-seeded MS with various shapes.[79] Following the above mentioned process, Song *et al.*[80] reported a hexagonal hollow ZSM-5 tube by using mesoporous silica fibre as template. The aluminium ingredient was introduced into the zeolite framework by impregnating the seeded MS fibres with $Al(NO_3)_3$ and NaCl aqueous solutions before VPT treatment. Moreover, this procedure has also been applied to prepare 3D-ordered zeolite monolith with closed macropores by the hydrothermal treatment of the array of nanozeolite pre-coated MS spheres in a clear silica-containing solution.[81] Wang *et al.*[82] modified this synthesis process to prepare 3D interconnected macroporous zeolite monolith. Different from Ref. 81, the MS spheres were pre-arrayed into 3D-ordered membrane by gravity sedimentation and calcination before they were seeded with silicalite-1 nanocrystals. The interconnected macroporous zeolitic membrane was found to be more suitable for the immobilization of enzyme. The MS templates not only provide a diversity of morphology but also permit a mechanically stable zeolitically structured material to be formed due to the synchronized processes of shell-growth and core-consumption. Moreover, it is worth noting that MS templates make the guest encapsulation much easier due to their inherent porosity. Therefore, the guest species, such as Fe_2O_3[78,79], Ag[79,83] and PdO[83] nanoparticles as well as micrometer-sized carbon and polymer[83] which had been pre-incorporated into the mesopores of the MS templates, could be successfully entrapped inside the generated capsules along with digestion of the silica in the MS cores (Fig. 17.3). The compact zeolitic shell is expected to provide a perfect protection for the active interior. Very recently, the protective performance of zeolitic microcapsule for the active inner species (Pd[56], Pt and Ag[57]) was evaluated in the reactions of Heck coupling[56] and alcohol selective oxidation.[57] As being displayed in the following Section 3.01, such novel capsular catalyst showed the outstanding reactant-selectivity, poison-resistance and reusability.

2.3. Biotemplates

Biological species is an attractive alternate template for the formation of hierarchical nanozeolite materials. The reasons are (1) they can provide thousands of possibilities for hierarchical structures due to their diverse species with intrinsically hierarchical tissue structures and morphologies and (2) they are cheap, abundant, renewable and environmentally benign. The first example of biotemplate employment was reported by Mann *et al.*[84]. A hierarchical zeolite fibre was constructed by the infiltrating nanozeolite suspension into bacterial supercellular thread. Silicalite-1 nanoparticles were aggregated specifically within the organized micro-architecture. After removing the bacterial template, an intact zeolite fibre with ordered macroporous channels lined by a wall of coalesced silicalite-1 nanoparticles (100 nm in width) was obtained. Starch gels and sponges had also been used by Mann *et al.*[85] to form silicalite monoliths and thin films with a hierarchical meso- and macroporous structure. Diatomite is another biotemplate that has been used to produce hierarchical zeolite structures. Generally, diatomite was first covered by nanozeolites, followed by a VPT process[86] of hydrothermally treated in clear synthesis solutions

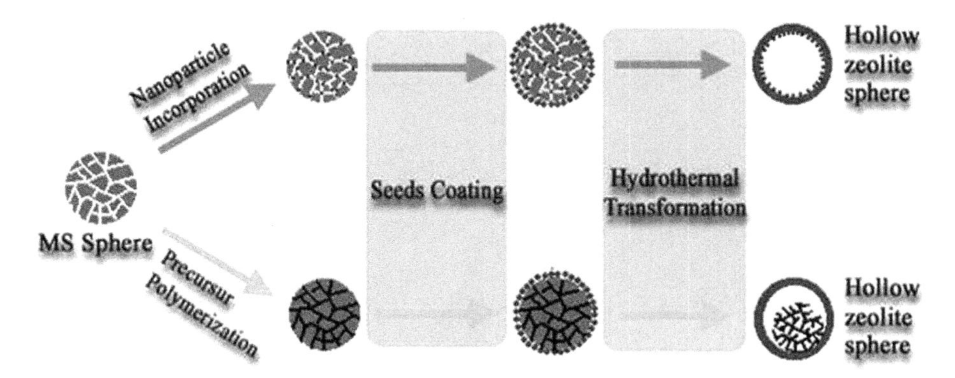

Figure 17.3 Schematic illustration of the encapsulation of nano- and micrometer sized particles into discrete hollow zeolite microspheres (Redrew from Ref. 83 with the permission of John Wiley & Sons, Inc.). (See color insert.)

containing silicon and aluminium sources.[87] The diatomaceous silica was transferred into zeolite through the above processes, while the hierarchical morphology and macroporosity of diatomite were preserved in the final products. Due to the existence of over 70,000 known diatom species, which are perfect silica and alumina sources for zeolite synthesis, it is possible to extend this method to prepare hierarchical zeolite structures with rationally tailored morphologies and porosities. In addition, the hierarchical porous materials were also obtained through direct LBL assembly of zeolite nanocrystals on diatom substrate.[88]

Dong *et al.*[89,90] used the wood cell (cedar and bamboo) as templates for the preparation of hierarchical zeolite by a seeding-growth strategy. The wood cells were deposited with nanozeolites by electrostatic adsorption, and then bore a secondary hydrothermal growth and calcination. The synthetic strategy ensured the faithful replication of the detailed cellular structure of wood in the final products (Fig. 17.4) due to the good penetrability and intrinsic crystallization habit of the colloidal nanozeolite within the wood issue. Valtchev *et al.*[91,92] reported the zeolite replica of *Equisetum arvense* by a biomineral-silica-induced mechanism. The leaves or stems of *Equisetum arvense* without any pretreatment were hydrothermally treated in a clear synthesis solution. Their zeolitization was directly induced by the biogenic silica within its structure, and so the employment of nanozeolite seeds was omitted. Moreover, such zeolitization of silica-containing plants provided a positive replica of the plant structure due to the zeolite crystallization in the organic tissue. Similar to the cases of diatomites, the tremendous hierarchical zeolite materials could be obtained by crystallizing diverse microporous aluminosilicates on silica-containing plants with large variety. Very recently, self-supporting MFI-type zeolite frameworks with hierarchical porosity and complex architecture were also prepared through an *in situ* seeding and secondary growth route in the presence of a Luffa cylindrica template.[93] The applicability of such self-supporting biomorphous zeolitic materials was attempted for the first time in a model catalytic process, that is, the cracking of *n*-hexane (see the details in Section 3.1).

Figure 17.4 Scanning electron microscopy (SEM) images of zeolitic tissues obtained through bamboo (a) and cedar (b) templates (Reproduced from Ref. 89 with the permission of John Wiley & Sons, Inc.)

2.4. Others

Carbon materials are also one of the widely used templates owing to their easy removal and tailorable porosity. Various hierarchical zeolite materials and monolith with high crystallinity were templated by carbon aerogels[94–96] and organic aerogels without a carbonization treatment.[97,98] Besides, the regular mesoporous carbon materials replicated from ordered mesoporous silica, such as CMK-3, were used as the secondary templates for the preparation of zeolitic hierarchical materials.[21,99] However, the tedious preparation process of these carbon templates seriously limited their practical applications. Very recently, the porous carbon particles generated from sucrose were used as template for the formation of mesoporous zeolite single crystals.[100,101] In addition, carbon nanoparticles,[102] carbon nanotubes[103] and carbon nanofibres [104,105] were also used as the suppliers of the mesoporosity to prepare hierarchical zeolite single crystals through a steam-induced crystallization process. These mesopore-containing zeolite crystals not only presented a better catalytic performance than their counterparts without mesoporosity[106,107] but also stabilized the metal nanoparticles against sintering by confining them in the mesopores.[108] Hollow nanozeolite fibres were also prepared by depositing nanozeolite particles on micron-sized carbon fibre templates via LBL[109] and electrophoretic assembly[110] methods. In addition, some inorganic templates with spherical and cubic morphologies, such as $CaCO_3$ and $Fe_3(SO_4)_2(OH)_5{\cdot}2H_2O$, were also used to prepare the zeolite microcapsules with active interiors by hydrothermally treating the nanozeolite pre-coated bulky templates in a zeolite precursor gel.[111] In this work, the controllable release of guest species was realized through adjusting the thickness of zeolite shells with high microporosity and thermal/chemical stablility. Hierarchical zeolite macrostructures with designed shapes were also prepared by gel-casting technique,[112] where the highly cross-linked polyacrylamide hydrogel was adopted as the template for the preparation of silicalite microtubes.

Besides employing the solid templates, hierarchical nanozeolite structures could be obtained by self-assembly of zeolite nanocrystals. Self-standing and optically transparent silicalite membranes with well-defined shapes and controlled mesoporosity have been prepared by this method followed by high-pressure compression and then secondary growth via microwave treatment.[113] The hierarchical-structured zeolite films were also assembled on the stainless steel grids either by LBL method[114] or by secondary growth after seeding in a gel solution.[115] A more systematic work was reported by Huang *et al.*,[116] in which the micropatterned films, micro- or macroporous materials, self-standing membranes and long zeolite fibres have been successfully prepared by using nanozeolites as building blocks. Very recently, a template-free method was reported to prepare hierarchical zeolite materials, where the meso- or macroporosity was generated in the zeolite matrix by the soft- or self-aggregation of the uniform zeolite precursor.[117–119] In addition, the mesoscaled cationic polymers were introduced into the hydrothermal synthesis solution which acted as co-template to form hierarchical mesoporous zeolite BEA.[120] It was claimed that the cationic polymers could effectively interact with the negatively charged inorganic silica species in alkaline media, resulting in the hierarchical mesoporosity. Compared with the conventional zeolite BEA, these hierarchical H-BEA zeolites exhibited a high reactivity and selectivity in the catalytic alkylation of benzene with propan-2-ol, and the total life of the catalyst has been greatly improved.[120] More importantly, the authors have also extended such methods to prepare mesoporous HZSM-5 zeolite which is believed to have wider range of applications in the domain of catalysis.

Micro/mesoporous zeolitic materials are gaining much attention in recent years owing to a high acidity and hydrothermal stability. In general, the preparation of these micro/mesoporous zeolitic materials could be divided into several typical routes. The earliest example involved the partial recrystallization of mesoporous walls in the presence of zeolite structure-directing agents to obtain the micro/mesoporous structure.[121–124] An alterative typical approach is the preparation of ordered mesoporous zeolitic structures by means of the self-assembly of zeolite precursors with surfactant micelles as templates for mesoporosity.[125–131] In this strategy, the zeolite precursors with the primary or secondary structure units were first prepared by hydrothermally aging the synthesis solution of conventional zeolite, and then the resultant gel mixture was assembled into mesoporous structure with the aid of surfactants. In the final mesoporous walls, however, there only contained primary or secondary zeolite units due to limitation of the wall thickness, otherwise no well-defined mesostructures could be formed. A compared approach is based on blending the pre-synthesized zeolite nanocrystallites into the mesoporous TUD-1 matrix.[35] In the final product, the size range of zeolite nanocrystallites was from 20 to 40 nm, and the diameter of the mesopores surrounding the zeolite nanoparticles was between 3 and 25 nm. In this method, the homogeneity of the zeolite nanocrystals in the mesoporous matrix was emphasized. In the third approach, the zeolite precursors or nanocrystals were directly coated onto the mesorporous materials to form hierarchical zeolite materials.[132–134] The micro/mesoporous zeolite materials could also be prepared on the basis of dual-templates, that is, micropore- and mesopore-directing agents.[135,136] However, such synthesis often resulted in phase-segment products (i.e., amorphous mesoporous material or bulk zeolite without mesoporosity, or physical mixtures of both) because of the competition of the two reactions directed by these different templates. Recently, a direct synthetic route to mesoporous zeolites with tunable mesoporous structure was reported.[137] In this method, the rationally designed amphiphilic organosilanes were used not only directing the formation of mesoporous structures but also modulating the crystallization of zeolite through its strong interaction with the growing zeolite crystals during the synthesis. The typical micro/mesoporous zeolitic structure prepared by this method could be well-identified through their electron microscopy (EM) images and the corresponding N_2 adsorption results in Fig. 17.5.

3. Emerging Applications of Zeolite Nanocrystals

The interests of scientists for preparation and assembly of nanozeolites are grounding on the facts that the reduction of crystals size could bring about higher efficiency in catalytic processes or could offer applicable opportunities in the new

Figure 17.5 Mesoporous LTA zeolite synthesized using $[(CH_3O)_3SiC_3H_6N(CH_3)_2C_{16}H_{33}]Cl$ as the structure directing agent. (a and b) SEM images with different magnifications, (c) Transmission electron microscopy (TEM) image and (d) N_2 adsorption/desorption isotherm of its Na^+ and Ca^{2+} form (inset, the pore size distribution of Na^+ form calculated from the adsorption branch) (Reproduced from Ref. 137 with the permission of Nature Publication Group).

fields. Therefore, in the recent years, with the deep understanding for the formation and assembly of nanozeolites, the scientific community gradually focused their attentions on the application of zeolite nanocrystals. Various novel applications were exploited ranging from sensor, low-*k* film and optical antenna to medical diagnostics, which have been well-summarized in Ref. 1. In this chapter, we will review the applications of nanozeolites mainly concentrating on catalysis and biology. The former involves in the application of the short diffusion path and abundant surface acidity in nanozeolites and the latter covers some new applications of their external surface property and ion-exchangeability.

3.1. Catalytic applications

Zeolites possess well-defined crystalline structures with molecular-sized pores, ion exchangeable sites and high hydrothermal stability, making them being widely used as catalysts in petrochemical processing and fine chemical industry. However, due to their low diffusion efficiency which often leads to the low availability of active sites and/or fast deactivation, the optimization of catalytic performance is still an inherent challenge for zeolites. The reduction of the zeolite crystals to nanometer is an effective way to reduce the mass transport limitations of guest molecules in the micropores by shortening the diffusion pathlength. Therefore, the exploration of the catalytic application of the nanozeolites becomes the top concern in the research on their applications. Sugimoto *et al.*[138] investigated the effect of crystal size on the catalytic properties of nanocrystalline H-ZSM-5 in the conversion of methanol to gasoline (MTG reaction), and a good correlation was observed between crystal size and their catalytic properties. In contrast to large crystallites, the aggregated H-ZSM-5 consisting of small crystallites showed long catalytic life and high selectivity for hydrocarbon and aromatics. Yamamura *et al.*[139] studied the relationship between external surface area of nanosized ZSM-5 and catalytic performance for ethylene oligomerization. It was found that the lifetime of ZSM-5 nanozeolite in the oligomerization of ethylene absolutely depended on their external surface area or crystal size. Schwarz *et al.*[140] have synthesized ZK-5 zeolites with various crystal sizes from 0.4 to 2.0 μm and found that the small-sized zeolites were more active than the large sized ones in the methylamine synthesis reactions. Loenders *et al.*[141] found that the intracrystalline diffusion limitation of zeolite BEA would be negligible in the alkylation of isobutane with 1-butene when the size of zeolite BEA crystals was smaller than 14 nm, corresponding to the external surface areas larger than ca. 280 $m^2 g^{-1}$. The alkylate yield obtained is proportional to the number of Brönsted acid sites of the zeolite catalysts in the absence of diffusion limitation. Additionally, the nanosized zeolite BEA has been used as solid acid catalyst in Friedel–Crafts acylation of aromatics and heteroaromatics as well.[142–145] It was found that the zeolite crystallite size played an important role on the improvement of the reactivity and catalyst decay, and the nanosized zeolite showed the most promising reactivity accompanying with the minimizing of the coke deposition.[144,145]

Notably, the reduction of zeolite size leads to more accessible catalytic active sites, and thereby ensures its high reactivity in the bulky molecule participated catalytic reactions, such as cracking of polyolefins with the molecular sizes larger than those of the zeolitic micropores. Serrano *et al.*[146–150] studied the catalytic properties of nanocrystalline HZSM-5 in the catalytic cracking of polyolefins into light hydrocarbon mixtures, and the high activity and selectivity towards olefinic gases were obtained over this catalyst. The amount of external surface of the catalyst was considered as one of the main factors affecting the catalyst activity. Due to the abundant acid sites on their large external surface, the nanozeolites could crack the polyolefin with high conversion. Additionally, the small molecules generated from polyolefin cracking would then diffuse into the micropores of nanozeolite and bear secondary reactions on the acid sites in the internal surface. The decrease of the crystal size also favours internal diffusion of these small molecules. In the meantime,

Mastral *et al.*[151] also employed nanocrystalline HZSM-5 in the catalytic pyrolysis of polyethylene. The use of nanocrystalline HZSM-5 allowed high yields of gas fractions and a higher selectivity to the products obtained than those achieved by thermal cracking. On the contrary, with the decrease of the crystals sizes, the shape selectivities of many reactions, such as alkylation or isomerization of aromatics, will decrease because of the increasing of external surface acidity of and the shortening of inner pore channels for the guest molecules.[152–158] Although, the decrease of crystals size will decrease their characteristic molecule-selectivity, their activity and lifetime could be probably increased because the shorter diffusion path and/or more pore-opening increased its coke tolerance ability.[159] The enhancement of the lifetime with the expense of the loss of selectivity becomes a dilemma for the application of nanoparticles of zeolites in catalysis. Accordingly, it is suggested that the small-sized zeolite crystals should be applied with blockage or removal of their external acidic sites for these diffusion-controlled reactions.[160]

Besides, the size effects of zeolite materials were also investigated in the bifunctional catalysts. Camblor *et al.*[161] supported the NiMo active species onto the zeolites with different sizes, and the best reaction results were obtained on the zeolite BEA nanocrystals (sizes in the range of 10–40 nm) supported NiMo catalyst. Since such reaction is largely depended on the surface acidity of the zeolite supports, the improvement of the reactivity here could be ascribed to the large accessible acid sites of the nanozeolites. Similar results have been obtained from Landau's research group[162] in which the Pt/H-BEA-Al_2O_3 with the size of 10–30 nm shows the high selectivity towards the desired middle distillates fraction. Arribas and Martínez [163] studied the influence of zeolite BEA crystal size on the product selectivity and sulphur resistance by the simultaneous isomerization of *n*-heptane and hydrogenation of benzene over Pt/zeolite BEA catalysts. The results showed that the catalyst prepared from a nanocrystalline zeolite displayed a high selectivity to *iso*-C7 and an improved sulphur resistance. The former could be contributed to a faster diffusion of the branched C7 isomers through the nanosized zeolite samples, thus minimizing secondary cracking reaction. The latter is due to a better dispersion of Pt on the nanocrystalline zeolite owing to its much higher surface area and mesoporosity.[163] In addition, the nanosized ZSM-5 zeolites (20–50 nm) have also been synthesized and modified with mixed rare earth oxides/Ga_2O_3. The obtained catalysts showed high activity in olefins reduction in fluid catalytic cracking gasoline with the upgraded gasoline yield of 96 wt.% and the research octane number of 90.0.[164]

Furthermore, the hierarchical assemblies of the zeolite nanocrystals are believed to be more competitive for their catalytic applications because of the integration of multi-functionalities in these assemblies. Majano *et al.*[165] prepared nanosized zeolite BEA assemblies via transformation of micron-sized grains of porous amorphous silica using a steam-assisted conversion (SAC) method.[166] Thanks to the additional acid sites in the assemblies, they showed slightly higher activity and much better C4 product yields compared to that on the zeolite BEA nanocrystal powders in the reaction of pentane hydroisomerization. Lei *et al.*[167] synthesized a hierarchically structured zeolite ZSM-5 catalyst by transforming the skeletons of a bimodal pore silica gel through the SAC method. The hierarchical macro/microporous zeolite catalyst showed a high catalytic activity for catalytic cracking of large molecules which cannot diffuse through the inner pores of ZSM-5, and the initial conversion was improved to 98% from 14.3% of microporous ZSM-5 crystals. Zampieri *et al.*[93] reported that the self-supporting biomimetic ZSM-5 macrostructure prepared by the biotemplating method (Section 28.2.3) showed a catalytic activity for *n*-hexane cracking with no need of ion-exchange and a low deactivation tendency even at 550 °C. Kustova *et al.*[168] also found that the introduction of mesoporosity into the conventional Cu-exchanged ZSM-5 and ZSM-11 by carbon-templating method[102–104] resulted in a significant improvement of their catalytic activities in direct NO decomposition. Recently, a hierarchical macro-meso-microporous structure was successfully synthesized via *in situ* assembling zeolite Y nanoclusters onto kaolin. Such a hierarchical composite catalysts exhibited an excellent catalytic performance in heavy crude oil cracking.[169] The above results indicated that the hierarchical pore structures can facilitate the transportation of reactant molecules towards the active sites, and thereby improve their catalytic performances and reduce the probability of coke formation. More interestingly, a nano-silver/zeolite film/copper grid composite catalyst (SZFC) has been prepared recently via the *in situ* electrolytic process on the zeolite film pre-assembled on the copper grids (Fig. 17.6a). The gas-phase oxidation reactions of mono- and di-alcohols were selected as the probes.[170,171] Figure 17.6b display the experimental result of 1,2-propylene glycol oxidation over SZFC catalysts. Compared with the conventional bulky electrolytic silver catalyst, the prepared SZFC catalyst showed the doubling of the selectivity of the target methyl glyoxal product (from 32.2% to 75.6%) accompanying with the slightly enhancement of the conversion (from 83.6% to 93.8%) at relative low temperature 320 °C. The outstanding catalytic performance could be attributed to the high activity of the highly dispersed silver nanoparticles (generally 2–3 nm) with abundant surface $Ag_n^{\delta+}$ species stabilized by its zeolitic film. The relative low reaction temperature also assured the low tendency of over-oxidation and cracking of polyhydric alcohols which occurred in the traditional process at high temperature.

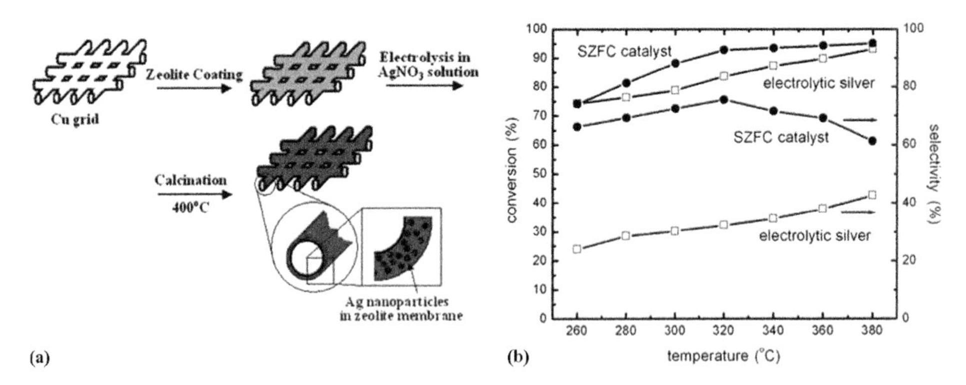

Figure 17.6 The schematic illustration for the fabrication of SZFC catalyst (a) and the catalytic performance of the SZFC catalysts in the oxidation of 1,2-propylene glycol (b) (Redrawn from Ref. 170 with the permission of The Royal Society of Chemistry). (See color insert.)

Recently, the research on zeolitically microcapsular reactor is gaining much attention as a new type of catalyst because such structure has some functionalities similar to those of 'artificial cell'. The zeolitic shell could well control the permeation of the guest molecules via its molecule-sieving character. In addition, it also protects the interiors to form a unique microenvironment. These features no doubt endow such materials with multi-functionalities for catalysis. The pioneer work of this idea was fulfilled by the assembly of a zeolitic shell onto the conventional catalyst such as Si–Al mixed oxides,[172] Pt/TiO_2 [173] and so on. He *et al.*[174,175] coated Co/SiO_2 catalyst pellets with a 10-μm-thick HZSM-5 zeolite shell via the hydrothermal treatment and have shown their functionality in Fischer–Tropsch synthesis. The zeolite-coated catalysts have been found to enhance the production of short-chain isoparaffins because the longer-chain hydrocarbons produced were further cracked and isomerized on acidic sites in the thick zeolite shell owing to their low diffusion rate in the zeolite channels. The above 'core-shell' catalyst could be regarded as the zeolitic microreactor on which the reactant selectivity and product distribution could both be controlled by the molecule-sieving effect of the zeolitic shell (Fig. 17.7).

Very recently, Ren *et al.*[56,57] prepared a series of the hollow zeolitic microcapsules catalyst (denoted as HZMC catalyst) encapsulated with different noble metal nanoparticles, for example, Pd in Fig. 17.8. The Heck coupling reaction between aryl iodides and olefins with a quasi-homogeneous mechanism was adopted as a probe reaction to evaluate the HZMC catalyst encapsulated with Pd nanoparticles (denoted as Pd@S1). During the reaction, the *in situ* generated large-sized soluble reactive intermediate could be well protected in the interior of the catalyst from leaching, resulting in the retaining of the reactivity on Pd@S1 catalyst in at least 10 reaction runs. On the contrary, the conventional Pd/C catalyst suffered a fast decay of the reactivity only in three runs due to the leaching of active intermediates (Fig. 17.9). Such a strategy of encaging the homogeneous catalytic reaction into a

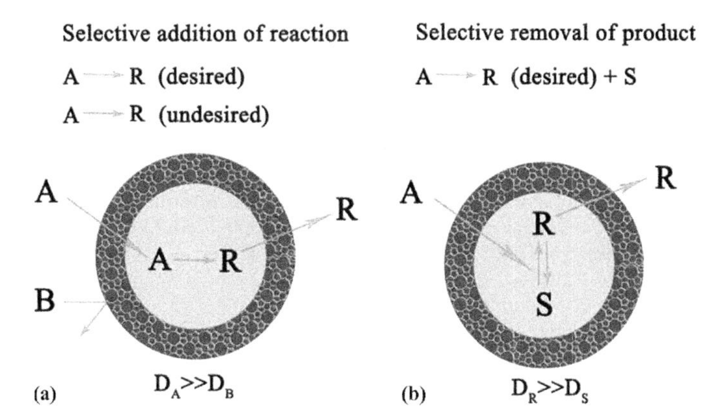

Figure 17.7 The typical model of controlling the product distribution in the zeolitic reactor by molecule-sieving effect of the zeolitic shell (Redrawn from Ref. 172 with the permission of The Royal Society of Chemistry). (See color insert.)

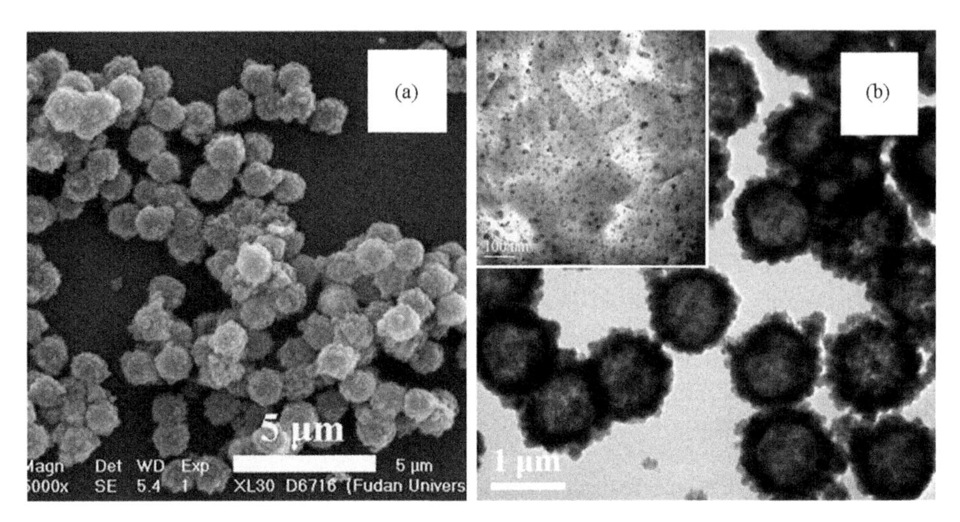

Figure 17.8 Scanning electron microscopy (SEM) (a) and Transmission electron microscopy (TEM) (b) images of the HZMC catalyst encapsulated with Pd nanoparticles (Reproduced from Ref. 56 with the permission of Elsevier B.V.).

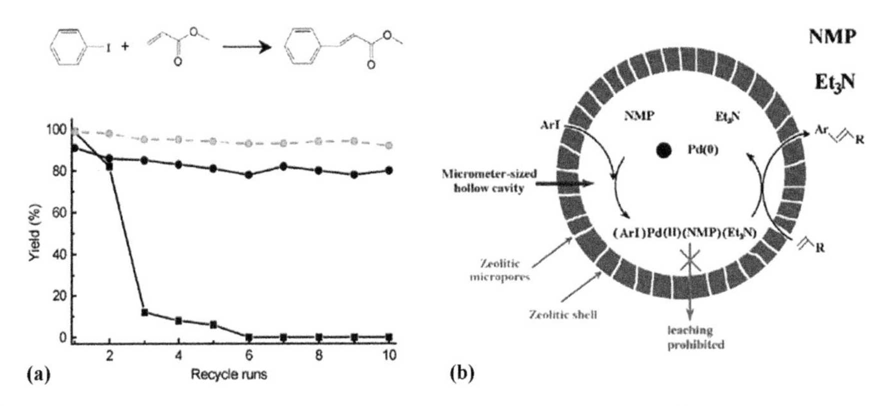

Figure 17.9 (a) Presents the Heck coupling activity and reusability of Pd@S1 (•) and Pd/C (▪) and (b) schematically illustrates the encaged quasi-homogeneous Heck coupling in Pd@S1 catalyst. Reaction conditions: 5 mmol of aryl halide, 8 mmol of olefin, 5 mmol of anhydrous triethylamine (Et_3N), 0.0125 mmol of Pd and 30 ml of N-methyl-2-pyrrolidone (NMP) under stirring at 120 °C for 3 h. The green dashed plot was obtained by doubling the amount of Pd@S1 catalyst (i.e., 0.025 mmol Pd), which shows that a higher yield of nearly 100% could be reached with higher catalyst amount (Redrawn from Ref. 56 with the permission of Elsevier B.V.). (See color insert.)

zeolitic shell may represent a new concept for the heterogenization of homogeneous catalyst. On the contrary, the Pt and Ag encapsulated HZMC catalysts (denoted as Pt@S1 and Ag@S1, respectively) were tested in the alcohol oxidation reactions both in liquid and gas phase. Besides the superior durability due to the protection of zeolitic shell, the Pt@S1 and Ag@S1 catalysts presented both the good reactant-selectivity and poison-resistivity due to the characteristic molecule-sieving effect of

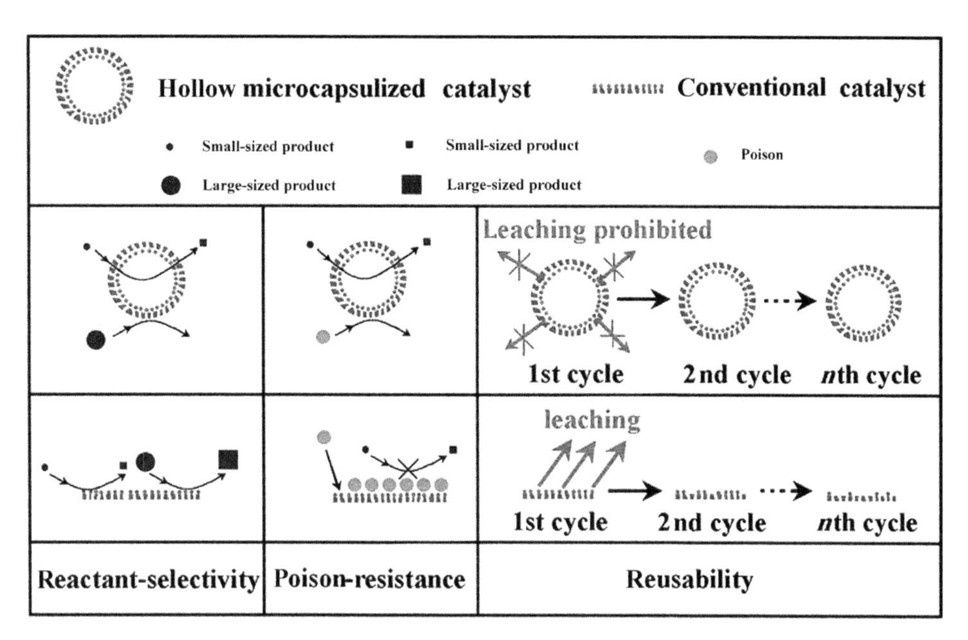

Figure 17.10 Schematic illustration of the reactant-selectivity, poison-resistance ability and reusability of hollow zeolitically microcapsulized catalyst (Reproduced from Ref. 57 with the permission of Elsevier B.V.). (See color insert.)

the zeolitic shell. These features are summarized in Fig. 17.10. Almost at the same time, the hollow iron-containing ZSM-5 tubes have also been prepared via the similar method and have been adopted in the visible light photoreduction of Cr(VI) in aqueous solution.[176] Unlike the iron-exchanged ZSM-5 hollow tubes, the iron encapsulated ones showed the good reusability, which was also considered as the result of the protecting effect of the zeolitic shell.

The meso/microporous composite hierarchical materials are another research interest in this field because such materials are expected to possess not only the mesopores for the catalytic reaction of bulky reactant molecules but also the strong acidity and the high hydrothermal stability brought by nanozeolite crystallite. The meso/microporous composite materials prepared via various methods (cf. Section 2.4) have been reported to have improved hydrothermal stability accompanying with the strong surface acidity as well as the high catalytic reactivity in the various reactions. Recently, Ryoo and co-workers[177] have studied the catalytic performance of their previous reported zeolite materials with micro/mesoporous hierarchical structures.[137] The results showed that the micro/mesoporous hierarchical MFI zeolites displayed remarkably increasing lifetime in several reactions such as isomerization of 1,2,4-trimethylbenzene, cumene cracking and esterification of benzyl alcohol with hexanoic acid, as compared with conventional MFI and mesoporous aluminosilicate MCM-41 (Fig. 17.11). Such notable retardation towards catalyst deactivation, brought by the hierarchical structures, is reported for the first time.

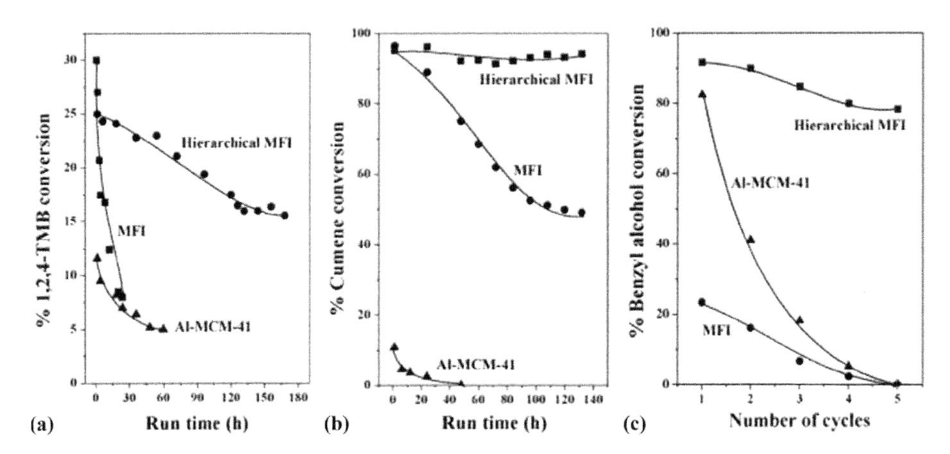

Figure 17.11 The deactivation behavior of hierarchical MFI, MFI and Al-MCM-41 in the reaction of (a) 1,2,4-trimethylbenzene isomerization, (b) cumene cracking and (c) esterification of benzyl alcohol and hexanoic acid (Reproduced from Ref. 177 with the permission of The Royal Society of Chemistry).

3.2. Emerging novel biology-related applications

With the decreasing of the zeolite crystal size to nanoscale, their external surface areas increase dramatically and thereby their surface properties exhibit more remarkably, such as adjustable surface charge, hydrophilicity/hydrophobicity and surface ion-exchangeable site. These rich surface characteristics offer attractive possibilities to explore the adsorption and reaction of bulky biomolecules on them, which may reveal the unique features of nanozeolites compared to the amorphous or non-porous nanomaterials. Moreover, these studies can also provide new opportunities for the immobilization, separation and identification of biomolecules. For example, enzyme immobilization has been one of the most important strategies for large-scaled applications due to the ease of continuous operation, catalyst recycling and product purification. The zeolite nanoparticles could provide an ideal support in the optimization of immobilized enzymes for their minimum diffusion limitation, maximum surface area and high enzyme loading.[178] Furthermore, the surface characteristics of nanozeolites could be manipulated to further improve the biocatalytic efficiency of immobilized enzyme.

The most important feature which benefits the application of nanozeolites in biology is their adjustable interaction with guest molecules on their characteristic surface. Recently, a nanozeolite-mediated enrichment procedure was proposed for proteomics research based on the adsorption ability of nanozeolites with proteins/polypeptides.[179] Three standard peptides and a typical myoglobin tryptic digest were used to evaluate the enriched performance of nanozeolites. The nanozeolites not only exhibited a 10^2–10^3 concentration factors but also displayed an excellent anti-salt ability during the enrichment. The enriched peptides could be directly identified with a high signal-to-noise ratio by MALDI (matrix-assisted laser desorption ionization)–TOF (time of flight)–MS (mass spectrometry) (Fig. 17.12). This means that

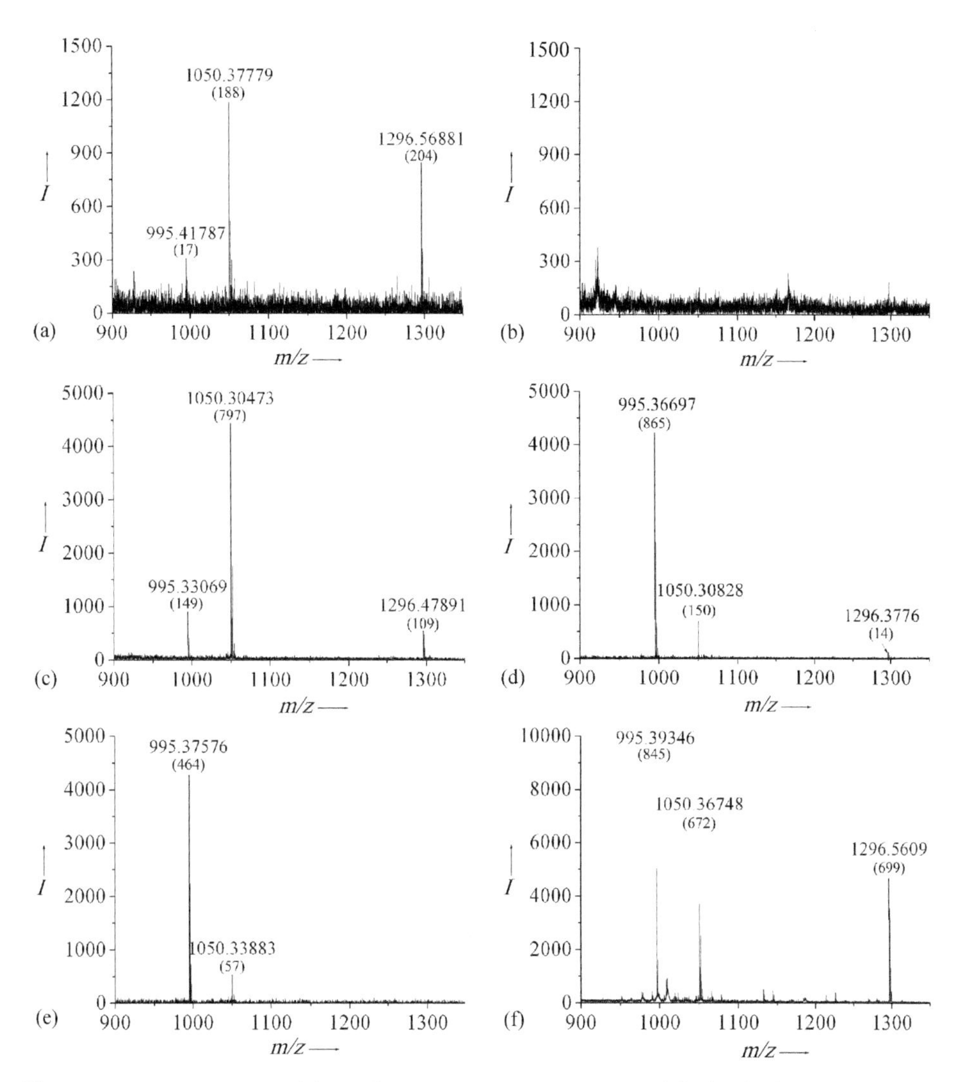

Figure 17.12 Matrix-assisted laser desorption ionization–time of flight (MALDI–TOF) mass spectra of the peptide mixtures prepared without enrichment: (a) 50 pg μl^{-1} and (b) 20 pg μl^{-1} and the peptides/nanozeolite (BEA-25) samples after enrichment in peptide mixture solutions of (c) 1 pg μl^{-1}, (d) 0.4 pg μl^{-1}, (e) 0.04 pg μl^{-1} and (f) 1 pg μl^{-1} containing 1 M $CaCl_2$. The data in parentheses are S/N of the corresponding peptides. MALDI–TOF–MS experiments were performed on a 4700 Proteomics Analyzer (Applied Biosystems, USA). Samples were desorbed with an Nd-YAG laser (355 nm) operated at a repetition rate of 200 Hz and an acceleration voltage of 20 kV. Measurements were performed in the reflector TOF detection mode (Reproduced from Ref. 179 with the permission of John Wiley & Sons, Inc.).

such a procedure allows the enrichment and believable identification of low abundant proteins. Notably, the different types of nanozeolite display different surface properties, and thereby will provide a diverse capacity of the peptide enrichment. At the same time, by utilizing their easy modifiability of surface hydroxyl groups,

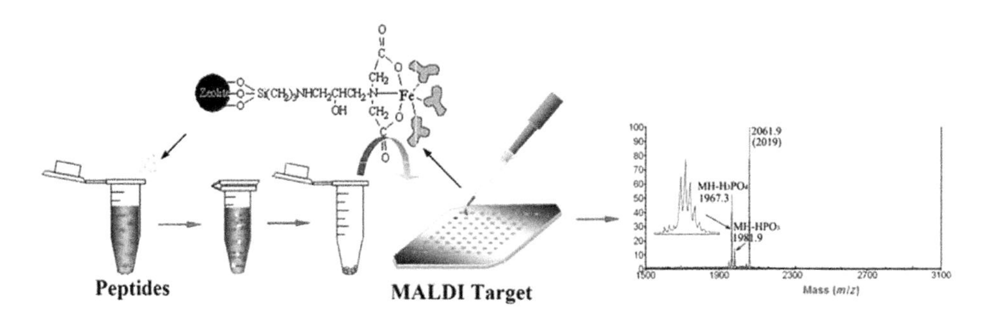

Figure 17.13 Schematic illustration of phosphopeptides enrichment procedure by a surface grafting metal chelation of zeolite beta nanoparticles with immobilized Fe^{3+} ions. (See color insert.)

zeolite BEA nanoparticles with immobilized Fe^{3+} ions were employed to enrich and isolate phosphopeptides selectively from tryptic β-casein digest by the chelation of surface grafted metal ions[180] (Fig. 17.13). Mapping protein phosphorylation is attracting great interest due to the importance of this post-translational modification in protein function. After selective enrichment by Fe^{3+}-nanozeolites, the signal of phosphopeptides (m/z 2061.9) become prominent in the MS during the identification of MALDI–TOF–MS, and the MALDI–MS/MS result of the parent ions (m/z 2061.9) affirmed its phosphorylation site at serine (Fig. 17.14).

More importantly, due to the small size of nanozeolites, they could be dispersed easily in various solutions of any volume to facilitate the enrichment/chelation of peptide/protein. Moreover, they were easily mixed with the organic matrix to improve subsequent MS identification of the species bound to their surface. Hence, the risk of sample loss from any elution steps is avoided. Furthermore, because of their small size and high adsorption amount as well as more hydrophobic interaction with biomolecules, de-aluminated Y nanozeolites were used as vehicles to carry both low molecular weight substances and macromolecules such as proteins into viable cells.[181] In addition, the delivery capability of nanozeolite for the biomolecules was utilized to study the endosomal mechanisms and pathways of a cell. Monocytes were used as target cells to verify the tracking capability of zeolite nanoparticles, and the results showed that nanozeolite was a useful and efficient tool to study cellular mechanisms by the delivery of biomolecules into viable cells.[182]

Another feature favouring the bioapplication of nanozeolites is the ion-exchangeable sites on the surface of aluminum-containing nanozeolites. Various transitional metal ions could be facilely immobilized on the surface of these nanozeolites via a simple ion exchange process, which provides a unique surface for the selective absorption and immobilization of targeted proteins or peptides with a special domain. This fact makes them promising packing sorbent in immobilized metal ion affinity chromatography (IMAC) for protein separation. Co^{2+}-exchanged nanozeolite/diatomite hierarchical materials were employed as the carriers of IMAC to separate selenoprotein-P (Se-P) proteins in mouse plasma through the chelation of their histidine residues with Co^{2+}.[183] It was concluded that the interaction between the cations on nanozeolites, and the imidazole-ring in the histidine residues of protein chains is the key factor for this separation (Fig. 17.15). Meanwhile,

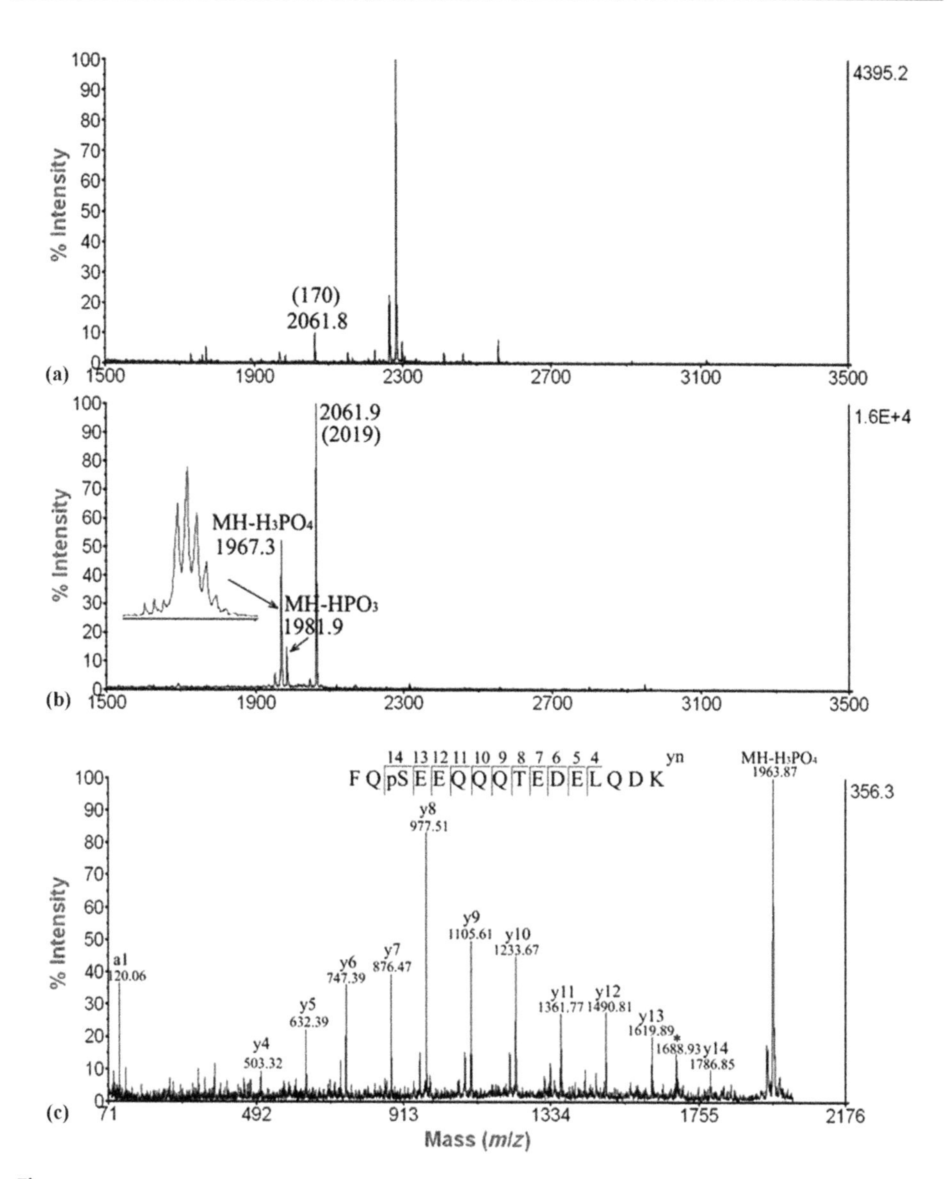

Figure 17.14 Matrix-assisted laser desorption ionization–time of flight (MALDI–TOF) mass spectra of a tryptic-casein digest (0.2 nmol ml^{-1}) without any pretreatment (a) and isolated by Fe^{3+}-nanozeolites (b). The data in parentheses are S/N of the corresponding peptides; (c) presents the MALDI–TOF mass spectra of the parent m/z 2061.9 isolated by Fe^{3+}-nanozeolites obtained on a TOF-TOF instrument. The amino acid sequence coverage is shown by yn ions. A prominent loss of 98 Da (m/z 1963.87) clearly shows its monophosphorylation. The mass difference 167 Da between fragment ions y13 (m/z 1619.9) and y14 (m/z 1786.9) corresponds to a phosphoserine residue, indicating the phosphorylation site at serine. The appearance of the peptide fragment (asterisk) by the loss of phosphoric acid (H_3PO_4, 98 Da) from y14 ion further confirms this assignment (Reproduced from Ref. 180 with the permission of The Royal Society of Chemistry).

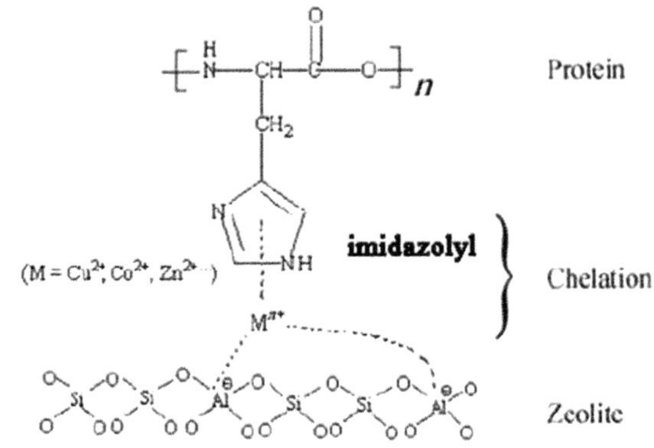

Figure 17.15 Diagrams of the interaction between the protein and zeolite coating by the chelation of imidazolyl in histidine with metal ions immobilized in zeolite (Reproduced from Ref. 184 with the permission of Elsevier B.V.).

Zn^{2+}-exchanged ZSM-5/fly ash cenosphere (FAC) hierarchical composite was used to successfully separate two model proteins (bovine serum albumin and chicken egg albumin) based on their different sequences of histidine residues.[184] Due to their high chemical and mechanical stability, hierarchical macrostructure as well as ion-exchange ability, these hierarchical composites, compared with traditional gel matrix, are expected to be of great promise as a novel and ideal IMAC sorbent for biomolecular separation, especially under high pressure.

By using nanozeolite crystal's as the building block, more useful microdevices with complex architecture could be constructed for the purpose of bioapplication. For example, zeolite nanocrystals were assembled on the electrode through LBL method to construct biocompatible interface (nanozeolite LTL immobilized cytochrome-*c* and nanozeolite BEA immobilized tyrosinase) towards sensitive biosensing.[185,186] Such enzyme-immobilized electrode provided an excellent object for the study of the interaction between enzymes and target molecules. The reason that nanozeolite particles performed as a good enzyme carrier could be attributed to their amazing enzyme adsorption capacity and adjustable interaction with enzyme. As the trypsin-immobilizing carrier, nanozeolites have been patterned in a microfluidic channel for fabricating an enzyme microreactor[187] (Fig. 17.16). The resulting trypsin

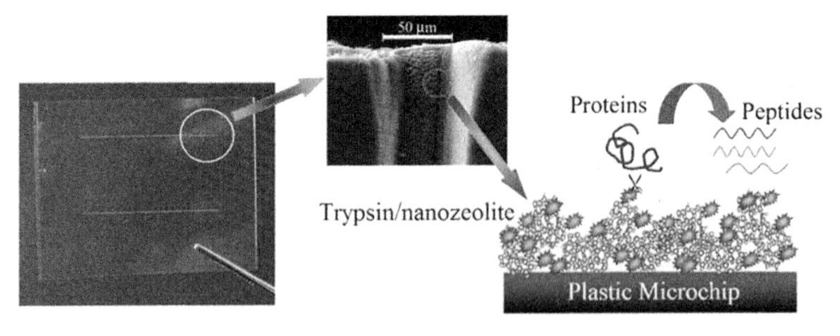

Figure 17.16 Microfluidic chip and the schematic illustration of nanozeolite-mediated microfluidic chip as enzyme bioreactor. (See color insert.)

Table 17.1 Matrix-assisted laser desorption ionization–time of flight–mass spectrometry (MALDI–TOF–MS) results obtained from digests by nanozeolote/trypsin-immobilized on-chip microreactors compared with those of in-solution digestion

	Bovine serum albumin		Cytochrome-*c*	
Protein	Microreactor	In-solution	Microreactor	In-solution
Total number of fragments	71	71	22	22
Amino acid sequence coverage (%)	27–39	29	76–85	90
Protein score (MS/MS)	257–373	302	235–306	326
Digestion time	<5 s	6 h	<5 s	6 h
Peptides matched	23	20	17	18

microreactor not only achieved the efficient digestion of proteins in a very short reaction time (<3 s) but also showed a good stability and universality for protein identification (Table 17.1). This enormous enhancement of proteolytic efficiency could be assigned to both the microscopic confinement and the high surface-to-volume ratio enhanced by the accumulation of sufficient nanozeolites in the micro-channel. Moreover, due to the miniaturization of the digestion procedures, the consumed amount (only $\sim$20 fmol per analysis) of sample is 100 times lower than that of the conventional tryptic reaction scheme. These imply a great potential in automated high-throughput analysis using a parallel-channel microchip platform. In addition, silicalite-1 nanoparticles were melted on the inner-surface of hydrophobic microfluidic channel to guarantee the formation of stable gel matrix via silica sol–gel for protein immobilization.[188] Very recently, the uniform nanozeolite microspheres with large/adjustable secondary pore architecture[74] and magnetically separable nanozeolites particles[189] have been reported for the purpose of separation or immobilization of biomolecules. These zeolite materials not only exhibited a comparable enzyme-immobilized capacity and biocatalytic performance to those of conventional zeolite nanocrystals but also could be simply manipulated or separated from reaction solution by filtration or magnet.

Besides the colloidal zeolite nanocrystals which have been proved as effective carriers for the enrichment and separation of the proteins/peptides or as excellent building blocks for the construction of nanodevices, the ultrathin layered zeolitic compound seems to be another noticeable material. On the one hand, the layered zeolite materials provide a large surface area for the attachment or entrapment of enzymes due to their nanoscale in one dimension and, on the other, they could be easily recovered in the reaction because of their micrometer scale in the other two dimensions. The delaminated laminar zeolites (ITQ-6) have been employed as supports to immobilized enzymes by electrostatic and covalent interactions.[190,191] As a result, ITQ-6, which was obtained by exfoliation of as-synthesized laminar precursor of ferrierite, was proved to be highly suitable as a reversible enzyme support. The enzyme covalently bonded to the ITQ-6 support displayed higher stability by comparing with that of the enzyme in free form and chemically bonded to amorphous silica.[191] This is actually the first example of nanozeolite material involving in bio-related application. Additionally, there were also about tens of publications about bioapplications of traditional zeolite crystals from 90s of last century. They included zeolite-containing enzyme-modified electrodes,[192–194] immobilization, purification, separation of biomolecules,[195–207] removal of surfactants or preservatives from protein solution by using zeolites,[208,209] as well as the adsorption and separation of small biomolecules such as glucose,[210] amino acids,[211] nucleosides and nucleotides[212] in the microporous channel of zeolites.

4. Conclusions and Prospects

As an important kind of nanomaterials, nanozeolites possess a series of remarkable characters, such as various crystalline frameworks, large active surface, tuneable surface charge/composition, regular pore-opening/half cages arrays, as well as surface ion-exchangeable sites, making them different both from the amorphous or non-porous materials and from the conventional micron-sized zeolites. Therefore, they are expected to have a great impact on the fields ranging from catalytic to biological applications. However, with the crystal size of zeolite being reduced to nanoscale, their poor stability and difficult manipulation would become the serious obstacles in most of their applications. Consequently, besides the great interest in synthesis and crystallization mechanism of nanozeolites, in the recent years, the assembly of the hierarchical nanozeolitic macrostructures has also attracted growing attentions. Abundance of specific macrotemplates with regular/specific morphologies, such as polymers, mesoporous silica, carbon materials and natural biotemplates were used to prepare hierarchical nanozeolite materials. Diverse hierarchical nanozeolitic macrostructures such as hollow/core-shell spheres/capsules, fibers, tubes and membranes have been built, which are opening the flourishing possibilities of the application of nanozeolite in catalysis, adsorption, biotechnology, as well as advanced material sciences.

Following the fruitful achievements of the conventional zeolites in both petrochemical and fine chemical catalysis, the application of nanozeolite in the area of catalysis became the original motivation that triggered the burst of zeolite nanotechnology. In the past decades, the effects of size and external surface brought by the decreasing of zeolite size have been deeply investigated on various catalytic reactions, particularly on those involving bulky molecules. The rich acid sites on external surface of nanozeolites make them possessing a high activity for the reaction of large molecules, while their abundant pore-opening and short diffusion path would greatly enhance their tolerance to coking. However, at the same time, such increasing the external surface acidity with the decrease of the crystals size could also decrease the shape selectivity of catalysts in those diffusion-controlled shape-selective reactions. Therefore, some optimizations among the reactive activity, lifetime and shape-selectivity still is required for a definite reaction. Furthermore, to overcome the manipulation difficulty met in most of catalytic processes, more and more scientists began to focus their attentions on the construction of catalytic nanozeolitic macrostructures. Their hierarchical macrostructure would also endow them with novel features during their applications. The capsulated zeolite microreactor with unique core-shell structure is the typical example in this field. The zeolitic microporous shell around the catalyst could not only control product distribution but also protect the encapsulated active species from being leached and poisoned. Moreover, such zeolitic microcapsules will also be expected to be useful for pharmaceutical applications for their possibility to encapsulate and controllably release active ingredients. However, it should be noted that, so far, few examples were reported on the catalytic application of the nanozeolite-built materials because of their limited mechanical strength, unmatched morphology and the less comprehension on the relation between their structure and property. Therefore, one of the most important concerns in this field would be the development of more rational assemble approach to match the requirements of the catalytic applications.

Besides the traditional catalysis, the abundant frameworks and the fascinating surface properties of the nanozeolites also inspired many scientists to explore their novel potential applications. Life science is the most stirring field in 21st century, and the achievement of nanoscience would provide opportunity to drive its development. With the deepening of the understanding on the special interaction on the surface/interface between nanozeolite and biomolecules, various nanozeolite materials including dispersed colloidal zeolites, hierarchical nanozeolite assemblies and layered zeolitic nanocompounds have been studied in their bioapplication. By adjusting the surface charge, surface grafting groups and type of the exchanged ions, the proteins and peptides could be selectively or non-selectively enriched on the surface of the nanozeolites, hence realizing their separation and identification through bio-mass spectrometry or IMAC technique. On the other hand, by using various assembly methods, some biocompatible microdevices such as the enzyme-immobilized sensors and bio-chips could be well constructed. Except for the surface properties, the unique microporosity in the zeolite crystal would be another important feature for their bioapplication. It has been reported that the micropore system of the micron-sized zeolite crystals could be used to control the microenvironment of enzymes immobilized on them due to their ability to adsorb some species.[191–196] For instance, their ability to adsorb water was particularly important when immobilized enzymes were used in organic media, such as lipases for various reactions or in the dynamic resolution of enantiomeric species. Their supplied surface acidity could also be used in bi-functional enzymatic-acid catalysts. These results assuredly provide clues for the further applications of nanozeolites as enzyme supports. Therefore, the desirable/undesirable effects of nanozeolite materials as enzyme supports on the bio-catalytic performance should be studied systemically in the future.

To sum up, zeolite nanomaterials have enjoyed tremendous success in their preparation and assembly over the past decades, and their applications are emerging in the wide areas including catalysis, life science, optics, electronics as well as bioanalysis and identification. Nevertheless, this is still the beginning of a long-lasting and fruitful period in zeolite nanotechnology. A multi-faceted and open-ended endeavor for their applications and developments will be continued far beyond the traditional catalytic and adsorption. Yet, the potential of many nano-zeolite materials including both colloidal particles and hierarchical structures on other novel applications is remaining to be discovered.

ACKNOWLEDGEMENTS

This work is supported by the NSFC (20325313, 20721063) and STCMS (06DJ14006, 075211013) and Major state basic research development program (2003CB615807).

REFERENCES

[1] Valtchev, V., Tosheva, L., "Nanozeolites: Synthesis, Crystallization Mechanism, and Applications", *Chem. Mater.* **2005,** *17,* 2494–2513.

[2] Zhang, B. Z., White, M. A., Lumsden, M., Mueller-Neuhaus, J., Robertson, K. N., Cameron, T. S., Gharghouri, M., "Control of Particle Size and Surface Properties of Crystals of NaX Zeolite", *Chem. Mater.* **2002,** *14,* 3636–3642.

[3] Mintova, S., Olson, N. H., Bein, T., "Electron Microscopy Reveals the Nucleation Mechanism of Zeolite Y from Precursor Colloids", *Angew. Chem. Int. Ed.* **1999,** *38,* 3201–3204.

[4] Zhu, G., Qiu, S. L., Yu, J., Sakamoto, Y., Xiao, F. S., Xu, R. R., Terasaki, O., "Synthesis and Characterization of High-quality Zeolite LTA and FAU Single Nanocrystals", *Chem. Mater.* **1998,** *10,* 1483–1486.

[5] Schoeman, B. J., Sterte, J., Otterstedt, J. E., "The Synthesis of Colloidal Zeolite Hydroxysodalite Sols by Homogeneous Nucleation", *Zeolites* **1994,** *14,* 208–216.

[6] Kecht, J., Mihailova, B., Karaghiosoff, K., Mintova, S., Bein, T., "Nanosized Gismondine Grown in Colloidal Precursor Solutions", *Langmuir* **2004,** *20,* 5271–5276.

[7] Hincapie, B. O., Garces, L. J., Zhang, Q., Sacco, A., Suib, S. L., "Synthesis of Mordenite Nanocrystals", *Microporous Mesoporous Mater.* **2004,** *67,* 19–26.

[8] Tsapatsis, M., Lovallo, M., Okubo, T., Davis, M. E., Sadakata, M., "Characterization of Zeolite-L Nanoclusters", *Chem. Mater.* **1995,** *7,* 1734–1741.

[9] Larlus, O., Valtchev, V., "Crystal Morphology Control of LTL-type Zeolite Crystals", *Chem. Mater.* **2004,** *16,* 3381–3389.

[10] Schoeman, B. J., Babouchkina, E., Mintova, S., Valtchev, V., Sterte, J., "The Synthesis of Discrete Colloidal Crystals of Zeolite Beta and Their application in the Preparation of This Microporous Films", *J. Porous Mater.* **2001,** *8,* 13–22.

[11] Mintova, S., Reinelt, M., Metzger, T. H., Senker, J., Bein, T., "Pure Silica BETA Colloidal Zeolite Assembled in Thin Films", *Chem. Commun.* **2003,** 326–327.

[12] Song, W., Justice, R. E., Jones, C. A., Grassian, V. H., Larsen, S. C., "Synthesis, Characterization, and Adsorption Properties of Nanocrystalline ZSM-5", *Langmuir* **2004,** *20,* 8301–8306.

[13] Li, Q., Creaser, D., Sterte, J., "The Nucleation Period for TPA-silicalite-1 Crystallization Determined by a Two-stage Varying-temperature Synthesis", *Microporous Mesoporous Mater.* **1999,** *31,* 141–150.

[14] Yang, S., Navrotsky, A., Wesolowski, D. J., Pople, J. A., "Study on Synthesis of TPA-silicalite-1 from Initially Clear Solutions of Various Base Concentrations by in Situ Calorimetry, Potentiometry, and SAXS", *Chem. Mater.* **2004,** *16,* 210–219.

[15] Valtchev, V., Faust, A. C., Lézervant, J., "Rapid Synthesis of Silicalite-1 Nanocrystals by Conventional Heating", *Microporous Mesoporous Mater.* **2004,** *68,* 91–95.

[16] Rakoczy, R. A., Traa, Y., "Nanocrystalline zeolite A: Synthesis, Ion Exchange and Dealumination", *Microporous Mesoporous Mater.* **2003,** *60,* 69–78.

[17] Valtchev, V., Bozhilov, K. N., "Transmission Electron Microscopy Study of the Formation of FAU-type Zeolite at Room Temperature", *J. Phys. Chem. B* **2004,** *108,* 15587–15598.
[18] Mintova, S., Petkov, N., Karaghiosoff, K., Bein, T., "Transformation of Amorphous Silica Colloids to Nanosized MEL Zeolite", *Microporous Mesoporous Mater.* **2001,** *50,* 121–128.
[19] Madsen, C., Jacobsen, C. J. H., "Nanosized Zeolite Crystals - convenient Control of Crystal Size Distribution by Confined Space Synthesis", *Chem. Commun.* **1999,** 673–674.
[20] Schmidt, I., Madsen, C., Jacobsen, C. J. H., "Confined Space Synthesis. A Novel Route to Nanosized Zeolites", *Inorg. Chem.* **2000,** *39,* 2279–2283.
[21] Kim, S. S., Shah, J., Pinnavaia, T. J., "Colloid-imprinted Carbons as Templates for the Nanocasting Synthesis of Mesoporous ZSM-5 zeolite", *Chem. Mater.* **2003,** *15,* 1664–1668.
[22] Naik, S. P., Chen, J. C., Chiang, A. S. T., "Synthesis of Silicalite Nanocrystals via the Steaming of Surfactant Protected Precursors", *Microporous Mesoporous Mater.* **2002,** *54,* 293–303.
[23] Wang, H. T., Holmberg, B. A., Yan, Y. S., "Synthesis of Template-free Zeolite Nanocrystals by Using in Situ Thermoreversible Polymer Hydrogels", *J. Am. Chem. Soc.* **2003,** *125,* 9928–9929.
[24] Zhang, G., Sterte, J., Schoeman, B. J., "Preparation of Colloidal Suspensions of Discrete TS-1 Crystals", *Chem. Mater.* **1997,** *9,* 210–217.
[25] Cundy, C. S., Forrest, J. O., Plaisted, R. J., "Some Observations on the Preparation and Properties of Colloidal Silicalites. Part 1: Synthesis of Colloidal Silicalite-1 and Titanosilicalite-1 (TS-1)", *Microporous Mesoporous Mater.* **2003,** *66,* 143–156.
[26] Cundy, C. S., Forrest, J. O., "Some Observations on the Preparation and Properties of Colloidal Silicalites Part II: Preparation, Characterisation and Properties of Colloidal Silicalite-1, TS-1, Silicalite-2 and TS-2", *Microporous Mesoporous Mater.* **2004,** *72,* 67–80.
[27] Mintova, S., Mo, S., Bein, T., "Nanosized $AlPO_4^{-5}$ Molecular Sieves and Ultrathin Films Prepared by Microwave Synthesis", *Chem. Mater.* **1998,** *10,* 4030–4036.
[28] Zhu, G., Qiu, S. L., Gao, F., Wu, G., Wang, R., Li, B., Fang, Q., Li, Y., Gao, B., Xu, X., Terasaki, O., "Synthesis of Aluminophosphate Molecular Sieve $AlPO_{4\text{-}11}$ Nanocrystals", *Microporous Mesoporous Mater.* **2001,** *50,* 129–135.
[29] Xu, J. S., Tang, Y., Zhang, H., Gao, Z., "Studies on the Colloidization and Stability of Layered M(IV) Phosphates in Aqueous Amine Solutions", *J. Incl. Phenom. Mol. Recognit. Chem.* **1997,** *27,* 303–317.
[30] Huang, Q., Wang, W. H., Yue, Y. H., Hua, W. M., Gao, Z., "Delamination and Intercalation of Layered Aluminophosphate with $[Al_3P_4O_{16}]^{3-}$ stoichiometry in Water/alcohol/amine Solutions", *J. Colloid. Interf. Sci.* **2003,** *257,* 268–275.
[31] Huang, Q., Wang, W. H., Yue, Y. H., Hua, W. M., Gao, Z., "Delamination and Alkylamine Intercalation of a Layered Microporous Aluminophosphate $[Al_3P_4O_{16}][CH_3(CH_2)_{(3)}NH_3]_3$", *Microporous Mesoporous Mater.* **2004,** *67,* 189–194.
[32] Wang, C., Hua, W. M., Yue, Y. H., Gao, Z., "Delamination and Aromatic Amine Intercalation of Layered Aluminophosphate with $[Al_3P_4O_{16}]^{3-}$ stoichiometry", *J. Colloid. Interf. Sci.* **2005,** *285,* 731–736.
[33] Wang, C., Hua, W. M., Yue, Y. H., Gao, Z., "Controlled Delamination and Intercalation of Layered Microporous Aluminophosphate by a Novel two-step Method", *Microporous Mesoporous Mater.* **2005,** *84,* 297–301.
[34] Corma, A., Fornes, V., Pergher, S. B., Maesen, T. L. M., Buglass, J. G., "Delaminated Zeolite Precursors as Selective Acidic Catalysts", *Nature* **1998,** *396,* 353–356.
[35] Waller, P., Shan, Z., Marchese, L., Tartaglione, G., Zhou, W., Jansen, J. C., Maschmeyer, T., "Zeolite Nanocrystals Inside Mesoporous TUD-1: A High-Performance Catalytic Composite", *Chem. Eur. J.* **2004,** *11,* 4970–4976.
[36] Mintova, S., Mo, S., Bein, T., "Humidity Sensing with Ultrathin LTA-type Molecular Sieve Films Grown on Piezoelectric Devices", *Chem. Mater.* **2001,** *13,* 901–905.
[37] Mintova, S., Bein, T., "Nanosized Zeolite Films for Vapor-sensing Applications", *Microporous Mesoporous Mater.* **2001,** *50,* 159–166.
[38] Wang, Z., Larsson, M. L., Grahn, M., Holmgren, A., Hedlund, J., "Zeolite Coated ATR Crystals for New Applications in FTIR-ATR Spectroscopy", *Chem. Commun.* **2004,** 2888–2889.

[39] Mintova, S., De Waele, V., Schmidhammer, U., Riedle, E., Bein, T., "In situ Incorporation of 2-(2-hydroxyphenyl)benzothiazole within FAU Colloidal Crystals", *Angew. Chem. Int. Ed.* **2003,** *42,* 1611–1614.
[40] Megelski, S., Calzaferri, G., "Tunning the Size and Shape of Zeolite L-based Inorganic-organic Host-guest Composites for Optical Antenna Systems", *Adv. Funct. Mater.* **2001,** *11,* 277–286.
[41] Platas-Iglesias, C., Elst, L. V., Zhou, W. Z., Muller, R. N., Geraldes, C. F. G. C., Maschmeyer, T., Peters, J. A., "Zeolite GdNaY Nanoparticles with very High Relaxivity for Application as Contrast Agents in Magnetic Resonance Imaging", *Chem. Eur. J.* **2002,** *8,* 5121–5131.
[42] Bouizi, Y., Diaz, I., Rouleau, L., Valtchev, V., "Core-shell Zeolite Microcomposites", *Adv. Funct. Mater.* **2005,** *15,* 1955–1960.
[43] Mintova, S., Bein, T., "Microporous Films Prepared by Spin-coating Stable Colloidal Suspensions of Zeolites", *Adv. Mater.* **2001,** *13,* 1880–1883.
[44] Larlus, O., Mintova, S., Valtchev, V., Jean, B., Metzger, T. H., Bein, T., "Silicalite-1/polymer Films with low-k Dielectric Constants", *Appl. Surf. Sci.* **2004,** *226,* 155–160.
[45] Stein, A., "Advances in Microporous and Mesoporous solids - Highlights of Recent Progress", *Adv. Mater.* **2003,** *15,* 763–775.
[46] Davis, M. E., "Ordered Porous Materials for Emerging Applications", *Nature* **2002,** *417,* 813–821.
[47] Cundy, C. S., Cox, P. A., "The Hydrothermal Synthesis of Zeolites: History and Development from the Earliest Days to the Present time", *Chem. Rev.* **2003,** *103,* 663–701.
[48] Snyder, M. A., Tsapatsis, M., "Hierarchical nanomanufacturing: From Shaped Zeolite Nanoparticles to High-performance Separation Membranes", *Angew. Chem. Int. Ed.* **2007,** *46,* 7560–7573.
[49] Bein, T., "Synthesis and Applications of Molecular Sieve Layers and Membranes", *Chem. Mater.* **1996,** *8,* 1636–1653.
[50] Tavolaro, A., Drioli, E., "Zeolite Membranes", *Adv. Mater.* **1999,** *11,* 975–996.
[51] Caro, J., Noack, M., Kölsch, P., Schäfer, R., "Zeolite Membranes - state of their Development and Perspective", *Microporous Mesoporous Mater.* **2000,** *38,* 3–24.
[52] Chiang, A. S. T., Chao, K. J., "Membranes and Films of Zeolite and Zeolite-like Materials", *J. Phys. Chem. Solids* **2001,** *62,* 1899–1910.
[53] Coronas, J., Santamaria, J., "State-of-the-art in Zeolite Membrane Reactors", *Top. Catal.* **2004,** *29,* 29–44.
[54] McLeary, E. E., Jansen, J. C., Kapteijn, F., "Zeolite Based Films, Membranes and Membrane Reactors: Progress and Prospects", *Microporous Mesoporous Mater.* **2006,** *90,* 198–220.
[55] Wang, Z. B., Wang, H. T., Mitra, A. P., Huang, L. M., Yan, Y. S., "Pure-silica Zeolite Low-k Dielectric Thin Films", *Adv. Mater.* **2001,** *13,* 746–749.
[56] Ren, N., Yang, Y. -H., Zhang, Y. -H., Wang, Q. -R., Tang, Y., "Heck Coupling in Zeolitic Microcapsular Reactor: A Test for Encaged Quasi-homogeneous Catalysis", *J. Catal.* **2007,** *246,* 215–222.
[57] Ren, N., Yang, Y. -H., Shen, J., Zhang, Y. -H., Xu, H. -L., Gao, Z., Tang, Y., "Novel, Efficient Hollow Zeolitically Microcapsulized Noble Metal Catalysts", *J. Catal.* **2007,** *251,* 182–188.
[58] Drews, T. O., Tsapatsis, M., "Progress in Manipulating Zeolite Morphology and Related Applications", *Curr. Opin. Colloid Interf. Sci.* **2005,** *10,* 233–238.
[59] Wang, X. D., Yang, W. L., Tang, Y., Wang, Y. J., Fu, S. K., Gao, Z., "Fabrication of Hollow Zeolite Spheres", *Chem. Commun.* **2000,** 2161–2162.
[60] Valtchev, V., Mintova, S., "Layer-by-layer Preparation of Zeolite Coatings of Nanosized Crystals", *Microporous Mesoporous Mater.* **2001,** *43,* 41–49.
[61] Yang, W. L., Wang, X. D., Tang, Y., Wang, Y. J., Ke, C., Fu, S. K., "Layer-by-layer Assembly of Nanozeolite Based on Polymeric Microsphere: Zeolite Coated Sphere and Hollow Zeolite sphere", *J. Marcomol. Sci. A* **2002,** *39,* 509–526.
[62] Holland, B. T., Abrams, L., Stein, A., "Dual Templating of Macroporous Silicates with Zeolitic Microporous Frameworks", *J. Am. Chem. Soc.* **1999,** *121,* 4308–4309.

[63] Rhodes, K. H., Davis, S. A., Caruso, F., Zhang, B., Mann, S., "Hierarchical Assembly of Zeolite Nanoparticles into Ordered Macroporous Monoliths Using Core-shell Building Blocks", *Chem. Mater.* **2000,** *12,* 2832–2834.
[64] Wang, Y. J., Tang, Y., Ni, Z., Hua, W. M., Yang, W. L., Wang, X. D., Tao, W. C., Gao, Z., "Synthesis of Macroporous Materials with Zeolitic Microporous Frameworks by Self-assembly of Colloidal Zeolites", *Chem. Lett.* **2000,** 510–511.
[65] Zhu, G. S., Qiu, S. L., Gao, F. F., Li, D. S., Li, Y. F., Wang, R. W., Gao, B., Li, B. S., Guo, Y. H., Xu, R. R., Liu, Z., Terasaki, O., "Template-assisted Self-assembly of Macro-micro Bifunctional Porous Materials", *J. Mater. Chem.* **2001,** *11,* 1687–1693.
[66] Valtchev, V., "Core-shell Polystyrene/zeolite a Microbeads", *Chem. Mater.* **2002,** *14,* 956–958.
[67] Valtchev, V., "Silicalite-1 Hollow Spheres and Bodies with a Regular System of Macrocavities", *Chem. Mater.* **2002,** *14,* 4371–4377.
[68] Valtchev, V., "Preparation of Regular Macroporous Structures Built of Intergrown Silicalite-1 Nanocrystals", *J. Mater. Chem.* **2002,** *12,* 1914–1918.
[69] Tosheva, L., Valtchev, V., Sterte, J., "Silicalite-1 Containing Microspheres Prepared Using Shape-directing Macro-templates", *Microporous Mesoporous Mater.* **2000,** *35–36,* 621–629.
[70] Tosheva, L., Mihailova, B., Valtchev, V., Sterte, J., "Zeolite Beta Spheres", *Microporous Mesoporous Mater.* **2001,** *48,* 31–37.
[71] Naydenov, V., Tosheva, L., Sterte, J., "Spherical Silica Macrostructures Containing Vanadium and Tungsten Oxides Assembled by the Resin Templating Method", *Microporous Mesoporous Mater.* **2002,** *55,* 253–263.
[72] Naydenov, V., Tosheva, L., Sterte, J., "Palladium-Containing Zeolite Beta Macrostructures Prepared by Resin Macrotemplating", *Chem. Mater.* **2002,** *14,* 4881–4885.
[73] Bouizi, Y., Majano, G., Mintova, S., Valtchev, V., "Beads Comprising a Hierarchical Porous Core and a Microporous Shell", *J. Phys. Chem. C* **2007,** *111,* 4535–4542.
[74] Kang, Y. J., Shan, W., Wu, J. Y., Zhang, Y. H., Wang, X. Y., Yang, W. L., Tang, Y., "Uniform Nanozeolite Microspheres with Large Secondary Pore Architecture", *Chem. Mater.* **2006,** *18,* 1861–1866.
[75] Lee, Y. J., Lee, J. S., Park, Y. S., Yoon, K. B., "Synthesis of large monolithic zeolite foams with variable macropore architectures", *Adv. Mater.* **2001,** *13,* 1259–1263.
[76] Wang, Y. J., Tang, Y., Dong, A. G., Wang, X. D., Ren, N., Shan, W., Gao, Z., "Self-Supporting Porous Zeolite Membranes with Sponge-Like Architecture and Zeolitic Microtubes", *Adv. Mater.* **2002,** *14,* 994–997.
[77] Zampieri, A., Colombo, P., Mababde, G. T. P., Selvam, T., Schwieger, W., Scheffler, F., "Zeolite Coatings on Microcellular Ceramic Foams: A Novel Route to Microreactor and Microseparator Devices", *Adv. Mater.* **2004,** *16,* 819–823.
[78] Dong, A. G., Wang, Y. J., Tang, Y., Yang, W. L., Ren, N., Zhang, Y. H., Gao, Z., "Hollow Zeolite Capsules: A Novel Approach for Fabrication and Guest Encapsulation", *Chem. Mater.* **2002,** *14,* 3217–3219.
[79] Dong, A. G., Wang, Y. J., Wang, D. J., Yang, W. L., Zhang, Y. H., Ren, N., Gao, Z., Tang, Y., "Fabrication of Hollow Zeolite Microcapsules with Tailored Shapes and Functionalized Interiors", *Microporous Mesoporous Mater.* **2003,** *64,* 69–81.
[80] Song, W., Kanthasamy, R., Grassian, V. H., Larsen, S. C., "Hexagonal, hollow, aluminium-containing ZSM-5 tubes prepared from mesoporous silica templates", *Chem. Commun.* **2004,** 1920–1921.
[81] Dong, A. G., Wang, Y. J., Tang, Y., Zhang, Y. H., Ren, N., Gao, Z., "Mechanically Stable Zeolite Monolith with Three-Dimensional Ordered Macropores by the Transformation of Mesoporous Silica Spheres", *Adv. Mater.* **2002,** *14,* 1506–1510.
[82] Wang, Y. J., Caruso, F., "Macroporous Zeolitic Membrane Bioreactors", *Adv. Funct. Mater.* **2004,** *14,* 1012–1018.
[83] Dong, A. G., Ren, N., Yang, W. L., Wang, Y. J., Zhang, Y. H., Wang, D. J., Hu, J. H., Gao, Z., Tang, Y., "Preparation of Hollow Zeolite Spheres and Three-Dimensionally Ordered Macroporous Zeolite Monoliths with Functionalized Interiors", *Adv. Funct. Mater.* **2003,** *13,* 943–948.

[84] Zhang, B., Davis, S. A., Mendelson, N. H., Mann, S., "Bacterial Templating of Zeolite Fibres with Hierarchical Structure", *Chem. Commun.* **2000,** 781–782.
[85] Zhang, B., Davis, S. A., Mann, S., "Starch Gel Templating of Spongelike Macroporous Silicalite Monoliths and Mesoporous Films", *Chem. Mater.* **2002,** *14,* 1369–1375.
[86] Wang, Y. J., Tang, Y., Dong, A. G., Wang, X. D., Ren, N., Gao, Z., "Zeolitization of Diatomite to Prepare Hierarchical Porous Zeolite Materials through a Vapor-Phase Transport Process", *J. Mater. Chem.* **2002,** *12,* 1812–1818.
[87] Anderson, M. W., Holmes, S. M., Hanif, N., Cundy, C. S., "Hierarchical Pore Structures through Diatom Zeolitization", *Angew. Chem. Int. Ed.* **2000,** *39,* 2707–2710.
[88] Wang, Y. J., Tang, Y., Wang, X. D., Dong, A. G., Shan, W., Gao, Z., "Fabrication of Hierarchically Structured Zeolites through Layer-by-Layer Assembly of Zeolite Nanocrystals on Diatom Templates", *Chem. Lett.* **2001,** 1118–1119.
[89] Dong, A. G., Wang, Y. J., Tang, Y., Ren, N., Zhang, Y. H., Yue, Y. H., Gao, Z., "Zeolitic Tissue through Wood Cell Templating", *Adv. Mater.* **2002,** *14,* 926–929.
[90] Daw, R., "Materials - Zeolites Branch out", *Nature* **2002,** *418,* 491–491.
[91] Valtchev, V., Smaihi, M., Faust, A. C., Vidal, L., "Biomineral-Silica-Induced Zeolitization of Equisetum Arvense", *Angew. Chem. Int. Ed.* **2003,** *42,* 2782–2785.
[92] Valtchev, V., Smaihi, M., Faust, A. C., Vidal, L., "Equisetum Arvense Templating of Zeolite Beta Macrostructures with Hierarchical Porosity", *Chem. Mater.* **2004,** *16,* 1350–1355.
[93] Zampieri, A., Mabande, G. T. P., Selvam, T., Schwieger, W., Rudolph, A., Hermann, R., Sieber, H., Greil, P., "Biotemplating of Luffa Cylindrica Sponges to Self-Supporting Hierarchical Zeolite Macrostructures for Bio-Inspired Structured Catalytic Reactors", *Mater. Sci. Eng. C* **2006,** *26,* 130–135.
[94] Tao, Y., Hanzawa, Y., Kaneko, K., "Template Synthesis and Characterization of Mesoporous Zeolites", *Colloid Surf. A* **2004,** *241,* 75–80.
[95] Tao, Y., Kanoh, H., Kaneko, K., "ZSM-5 Monolith of Uniform Mesoporous Channels", *J. Am. Chem. Soc.* **2003,** *125,* 6044–6045.
[96] Tao, Y., Kanoh, H., Kaneko, K., "Uniform Mesopore-Donated Zeolite Y Using Carbon Aerogel Templating", *J. Phys. Chem. B* **2003,** *107,* 10974–10976.
[97] Tao, Y., Kanoh, H., Kaneko, K., "Synthesis of Mesoporous Zeolite a by Resorcinol-Formaldehyde Aerogel Templating", *Langmuir* **2005,** *21,* 504–507.
[98] Li, W. C., Lu, A. H., Palkovtis, R., Schmidt, W., Spliethoff, B., Schüth, F., "Hierarchically Structured Monolithic Silicalite-1 Consisting of Crystallized Nanoparticles and Its Performance in the Beckmann Rearrangement of Cyclohexanone Oxime", *J. Am. Chem. Soc.* **2005,** *127,* 12595–12600.
[99] Yang, Z., Xia, Y., Mokaya, R., "Zeolite ZSM-5 with Unique Supermicropores Synthesized Using Mesoporous Carbon as a Template", *Adv. Mater.* **2004,** *16,* 727–732.
[100] Zhu, K., Egeblad, K., Chritensen, C. H., "Mesoporous Carbon Prepared from Carbohydrate as Hard Template for Hierarchical Zeolites", *Eur. J. Inorg. Chem.* **2007,** 3955–3960.
[101] Kustiva, M., Egeblad, K., Zhu, K., Chritensen, C. H., "Versatile Route to Zeolite Single Crystals with Controlled Mesoporosity: in situ Sugar Decomposition for Templating of Hierarchical Zeolites", *Chem. Mater.* **2007,** *19,* 2915–2917.
[102] Jacobsen, C. J. H., Madsen, C., Houzvicka, J., Schmidt, I., Carlsson, A., "Mesoporous Zeolite Single Crystals", *J. Am. Chem. Soc.* **2000,** *122,* 7116–7117.
[103] Schmidt, I., Janssen, A. H., Cusiavsson, E., Stahl, K., Pehrson, S., Dahl, S., Carlsson, A., Jacobsen, C. J. H., "Carbon Nanotube Templated Growth of Mesoporous Zeolite Single Crystals", *Chem. Mater.* **2001,** *13,* 4416–4418.
[104] Janssen, A. H., Schmidt, I., Jacobsen, C. J. H., Koster, A. J., de Jong, K. P., "Exploratory Study of Mesopore Templating with Carbon During Zeolite Synthesis", *Microporous Mesoporous Mater.* **2003,** *65,* 59–75.
[105] Kustova, M. Y., Hasselriis, P., Christensen, C. H., "Mesoporous MEL-type Zeolite Single Crystal Catalysts", *Catal. Lett.* **2004,** 205–211.
[106] Christensen, C. H., Johannsen, K., Schmidt, I., Christensen, C. H., "Catalytic Benzene Alkylation over Mesoporous Zeolite Single Crystals: Improving Activity and Selectivity with a New Family of Porous Materials", *J. Am. Chem. Soc.* **2003,** *125,* 13370–13371.

[107] Schmidt, I., Krogh, A., Wienberg, K., Carlsson, A., Brorson, M., Jacobsen, C. J. H., "Catalytic Epoxidation of Alkenes with Hydrogen Peroxide over First Mesoporous Titanium-Containing Zeolite", *Chem. Commun.* **2000,** 2157–2158.

[108] Christensen, C. H., Schmidt, I., Carlsson, A., Johannsen, K., Herbst, K., "Crystals in Crystals-Nanocrystals within Mesoporous Zeolite Single Crystals", *J. Am. Chem. Soc.* **2005,** *127,* 8098–8102.

[109] Wang, Y. J., Tang, Y., Wang, X. D., Yang, W. L., Gao, Z., "Fabrication of Hollow Zeolite Fibers through Layer-By-Layer Adsorption Method", *Chem. Lett.* **2000,** 1344–1345.

[110] Ke, C., Yang, W. L., Ni, Z., Wang, Y. J., Tang, Y., Gu, Y., Gao, Z., "Electrophoretic Assembly of Nanozeolites: Zeolite Coated Fibers and Hollow Zeolite Fibers", *Chem. Commun.* **2001,** 783–784.

[111] Wang, D. J., Zhu, G. B., Zhang, Y. H., Yang, W. L., Wu, B. Y., Tang, Y., Xie, Z. K., "Controlled Release and Conversion of Guest Species in Zeolite Microcapsules", *New J. Chem.* **2005,** *29,* 272–274.

[112] Wang, H. T., Huang, L. M., Wang, Z. B., Mitra, A. P., Yan, Y. S., "Hierarchical Zeolite Structures with Designed Shape by Gel-Casting of Colloidal Nanocrystal Suspensions", *Chem. Commun.* **2001,** 1364–1365.

[113] Huang, L. M., Wang, Z. B., Wang, H. T., Sun, J. Y., Li, Q. Z., Zhao, D. Y., Yan, Y. S., "Hierarchical Porous Structures by Using Zeolite Nanocrystals as Building Blocks", *Microporous Mesoporous Mater.* **2001,** *48,* 73–78.

[114] Wang, Y. J., Tang, Y., Wang, X. D., Shan, W., Ke, C., Gao, Z., Hu, J. H., Yang, W. L., "Fabrication of Zeolite Coatings on Stainless Steel Grids", *J. Mater. Sci. Lett.* **2001,** *20,* 2091–2094.

[115] Guo, H. L., Zhu, G. S., Li, H., Zou, X. Q., Yin, X. J., Yang, W. S., Qiu, S. L., Xu, R. R., "Hierarchical Growth of Large-scaled Ordered Zeolite Silicalite-1 Membranes with High permeability and Selectivity for Recycling CO_2", *Angew. Chem. Int. Ed.* **2006,** *45,* 7053–7056.

[116] Huang, L. M., Wang, Z. B., Sun, J. Y., Miao, L., Li, Q. Z., Yan, Y. S., Zhao, D. Y., "Fabrication of Ordered Porous Structures by Self-Assembly of Zeolite Nanocrystals", *J. Am. Chem. Soc.* **2000,** *122,* 3530–3531.

[117] Han, W., Jia, Y., Xiong, G., Yang, W., "Synthesis of Hierarchical Porous Materials with ZSM-5 Structures Via Template-Free Sol-Gel Method", *Sci. Technol. Adv. Mater.* **2007,** *8,* 101–105.

[118] Han, W., Jia, Y. X., Yao, N., Yang, W. S., He, M. Y., Xiong, G. X., "A Novel Template-Free Sol-Gel Synthesis of Silica Materials with Mesoporous Structures and Zeolitic Walls", *J. Sol–Gel Sci. Technol.* **2007,** *43,* 205–211.

[119] Holland, B. T., "Transformation of Mostly Amorphous Mesoscopic Aluminosilicate Colloids into High Surface Area Mesoporous ZSM-5", *Microporous Mesoporous Mater.* **2006,** *89,* 291–299.

[120] Xiao, F. S., Wang, L. F., Yin, C. Y., Lin, K. F., Di, Y., Li, J. X., Xu, R. R., Su, D. S., Schlögl, R., Yokoi, T., Tatsumi, T., "Catalytic Properties of Hierarchical Mesoporous Zeolites Templated with a Mixture of Small Organic Ammonium Salts and Mesoscale Cationic Polymers", *Angew. Chem. Int. Ed.* **2006,** *45,* 3090–3093.

[121] Kloetstra, K. R., van Bekkum, H., Jansen, J. C., "Mesoporous Material Containing Framework Tectosilicate by Pore-Wall Recrystallization", *Chem. Commun.* **1997,** 2281–2282.

[122] Verhoef, M. J., Kooyman, P. J., van der Waal, J. C., Rigutto, M. S., Peters, J. A., van Bekkum, H., "Partial Transformation of MCM-41 Material into Zeolites: Formation of Nanosized MFI Type Crystallite", *Chem. Mater.* **2001,** *13,* 683–687.

[123] Huang, L. M., Guo, W. P., Deng, P., Xue, Z. Y., Li, Q. Z., "Investigation of Synthesizing MCM-41/ZSM-5 Composites", *J. Phys. Chem. B* **2000,** *104,* 2817–2823.

[124] On, D. T., Kaliaguine, S., "Large-Pore Mesoporous Materials with Semi-Crystalline Zeolitic Frameworks", *Angew. Chem. Int. Ed.* **2001,** *40,* 3248–3251.

[125] Liu, Y., Zhang, W. Z., Pinnavaia, T. J., "Steam-Stable Aluminosilicate Mesostructures Assembled from Zeolite Type Y Seeds", *J. Am. Chem. Soc.* **2000,** *122,* 8791–8792.

[126] Zhang, Z. T., Han, Y., Xiao, F. -S., Qiu, S. L., Zhu, L., Wang, R. W., Yu, Y., Zhang, Z., Zou, B. S., Wang, Y. Q., Sun, H. P., Zhao, D. Y., Wei, Y., "Mesoporous Aluminosilicates with Ordered Hexagonal Structure, Strong Acidity, and Extraordinary Hydrothermal Stability at High Temperatures", *J. Am. Chem. Soc.* **2001,** *123,* 5014–5021.

[127] Liu, Y., Zhang, W. Z., Pinnavaia, T. J., "Steam-Stable MSU-S Aluminosilicate Mesostructures Assembled from Zeolite ZSM-5 and Zeolite Beta Seeds", *Angew. Chem. Int. Ed.* **2001,** *40,* 1255–1258.

[128] Liu, J., Zhang, X., Han, Y., Xiao, F. S., "Direct Observation of Nanorange Ordered Microporosity within Mesoporous Molecular Sieves", *Chem. Mater.* **2002,** *14,* 2536–2540.

[129] Kremer, S. P. B., Kirschhock, C. E. A., Aerts, A., Villani, K., Martens, J. A., Lebedev, O. I., Van Tandello, G., "Tiling Silicalite-1 Nanoslabs into 3D Mosaics", *Adv. Mater.* **2003,** *15,* 1705–1707.

[130] Carr, C. S., Kaskel, S., Shantz, D. F., "Self-Assembly of Colloidal Zeolite Precursors into Extended Hierarchically Ordered Solids", *Chem. Mater.* **2004,** *16,* 3139–3146.

[131] Han, Y., Li, N., Zhao, L., Li, D. F., Xu, X. Z., Wu, S., Di, Y., Li, C. J., Zou, Y. C., Yu, Y., Xiao, F. S., "Understanding of the High Hydrothermal Stability of the Mesoporous Materials Prepared by the Assembly of Triblock Copolymer with Preformed Zeolite Precursors in Acidic Media", *J. Phys. Chem. B* **2003,** *107,* 7551–7556.

[132] On, D. T., Kaliaguine, S., "Zeolite-coated Mesostructured Cellular Silica Foams", *J. Am. Chem. Soc.* **2003,** *125,* 618–619.

[133] On, D. T., Kaliaguine, S., "Ultrastable and Highly Acidic, Zeolite-Coated Mesoporous Aluminosilicates", *Angew. Chem. Int. Ed.* **2002,** *41,* 1036–1039.

[134] Mavrodinova, V., Popova, M., Valchev, V., Nickolov, R., Minchev, Ch., "Beta Zeolite Colloidal Nanocrystals Supported on Mesoporous MCM-41", *J. Colloid Interf. Sci.* **2005,** *286,* 268–273.

[135] Karlsson, A., Stöcker, M., Schmidt, R., "Composites of Micro- and Mesoporous Materials: Simultaneous Syntheses of MFI/MCM-41 Like Phases by a Mixed Template Approach", *Microporous Mesoporous Mater.* **1999,** *27,* 181–192.

[136] Zhang, Z. T., Han, Y., Zhu, L., Wang, R. W., Yu, Y., Qiu, S. L., Zhao, D. Y., Xiao, F. S., "Strongly Acidic and High-Temperature Hydrothermally Stable Mesoporous Aluminosilicates with Ordered Hexagonal Structure", *Angew. Chem. Int. Ed.* **2001,** *40,* 1258–1262.

[137] Choi, M., Cho, H. S., Srivastava, R., Venkatesan, C., Choi, D. -H., Ryoo, R., "Amphiphilic Organosilane-Directed Synthesis of Crystalline Zeolite with Tunable Mesoporosity", *Nat. Mater.* **2006,** *6,* 718–723.

[138] Sugimoto, M., Katsuno, H., Takatsu, K., Kawata, N., "Correlation Between the Crystal Size and Catalytic Properties of ZSM-5 Zeolites", *Zeolites* **1987,** *7,* 503–507.

[139] Yamamura, M., Chaki, K., Wakatsuki, T., Okado, H., Fujimoto, K., "Synthesis of ZSM-5 Zeolite with Small Crystal Size and Its Catalytic Performance for Ethylene Oligomerization", *Zeolites* **1994,** *14,* 643–649.

[140] Schwarz, S., Corbin, D. R., Sonnichsen, G. C., "The Effect of Crystal Size on the Methylamines Synthesis Performance of ZK-5 Zeolites", *Microporous Mesoporous Mater.* **1998,** *22,* 409–418.

[141] Loenders, R., Jacobs, P. A., Martens, J. A., "Alkylation of Isobutane with 1-butene on Zeolite Beta", *J. Catal.* **1998,** *176,* 545–551.

[142] Botella, P., Corma, A., López-Nieto, J. M., Valencia, S., Jacquot, R., "Acylation of Toluene with Acetic Anhydride over Beta Zeolites: Influence of Reaction Conditions and Physicochemical Properties of the Catalyst", *J. Catal.* **2000,** *195,* 161–168.

[143] Derouane, E. G., Schmidt, I., Lachas, H., Christensen, C. J. H., "Improved Performance of Nano-Size H-BEA Zeolite Catalysts for the Friedel-Crafts Acetylation of Anisole by Acetic Anhydride", *Catal. Lett.* **2004,** *95,* 13–17.

[144] Sartori, G., Maggi, R., "Use of Solid Catalysts in Friedel-Crafts Acylation Reactions", *Chem. Rev.* **2006,** *106,* 1077–1104.

[145] Kantam, M. L., Ranganath, K. V. S., Sateesh, M., Kumar, K. B. S., Choudary, B. M., "Friedel-Crafts Acylation of Aromatics and Heteroaromatics by Beta Zeolite", *J. Mol. Catal. A: Chem.* **2005,** *225,* 15–20.

[146] Serrano, D. P., Aguado, J., Escola, J. M., "Catalytic Cracking of a Polyolefin Mixture over Different Acid Solid Catalysts", *Ind. Eng. Chem. Res.* **2000,** *39,* 1177–1184.

[147] Serrano, D. P., Aguado, J., Escola, J. M., Rodríguez, J. M., "Nanocrystalline ZSM-5: A Highly Active Catalyst for Polyolefin Feedstock Recycling", *Stud. Surf. Sci. Catal.* **2002,** *142,* 77–84.

[148] Serrano, D. P., Aguado, J., Escola, J. M., Garagorri, E., Rodríguez, J. M., Morselli, L., Palazzi, G., Orsi, R., "Feedstock Recycling of Agriculture Plastic Film Wastes by Catalytic Cracking", *Appl. Catal. B: Environ.* **2004,** *49,* 257–265.
[149] Serrano, D. P., Aguado, J., Escola, J. M., Rodríguez, J. M., "Influence of Nanocrystalline HZSM-5 External Surface on the Catalytic Cracking of Polyolefins", *J. Anal. Appl. Pyrolysis* **2005,** *74,* 353–360.
[150] Serrano, D. P., Aguado, J., Rodríguez, J. M., Peral, A., "Catalytic Cracking of Polyethylene over Nanocrystalline HZSM-5: Catalyst Deactivation and Regeneration Study", *J. Anal. Appl. Pyrolysis* **2007,** *79,* 456–464.
[151] Mastral, J. F., Berrueco, C., Gea, M., Ceamanos, J., "Catalytic Degradation of High Density Polyethylene over Nanocrystalline HZSM-5 Zeolite", *Polym. Degrad. Stab.* **2006,** *91,* 3330–3338.
[152] Csicsery, S. M., "Shape-selective Catalysis in Zeolite", *Zeolites* **1984,** *4,* 202–213.
[153] Kaeding, W. W., "Shape-Selective Reactions with Zeolite Catalysis .5. Alkylation or Disproportionation of Ethylbenzene to Produce Para-Diethylbenzene", *J. Catal.* **1985,** *95,* 512–519.
[154] Kim, J., Namba, S., Yashima, T., "Shape Selectivity of ZSM-5 Type Zeolite for Alkylation of Ethylbenzene with Ethanol", *Bull. Chem. Soc. Jpn.* **1988,** *61,* 1051–1055.
[155] Shiralkar, V. P., Joshi, P. N., Eapen, M. J., Rao, B. S., "Synthesis of ZSM-5 with Variable Crystallite Size and Its Influence on Physicochemical Properties", *Zeolites* **1991,** *11,* 511–516.
[156] Hibino, T., Niwa, M., Murakami, Y., "Shape Selectivity over HZSM-5 Modified by Chemical Vapor-Deposition of Silicon Alkoxide", *J. Catal.* **1991,** *128,* 551–558.
[157] Beschmann, K., Riekert, L., Müller, U., "Shape Selectivity of Large and Small Crystals of Zeolite ZSM-5", *J. Catal.* **1994,** *145,* 243–245.
[158] Bhat, Y. S., Das, J., Rao, K. V., Halgeri, A. B., "Inactivation of External Surface of ZSM-5: Zeolite Morphology, Crystal Size, and Catalytic Activity", *J. Catal.* **1996,** *159,* 368–374.
[159] Zhao, G. L., Teng, J. W., Xie, Z. K., Yang, W. M., Chen, Q. L., Tang, Y., "Catalytic Cracking Reaction of C_4-Olefin over Zeolites H-ZSM-5, H-Mordenite and H-SAPO-34", *Stud. Surf. Sci. Catal.* **2007,** *170,* 1307–1312.
[160] Melson, S., Schüth, F., "The Influence of the External Acidity of H-ZSM-5 on Its Shape Selective Properties in the Disproportionation of Ethylbenzene", *J. Catal.* **1997,** *170,* 46–53.
[161] Camblor, M. A., Corma, A., Martinez, A., Martinez-Soria, V., Valencia, S., "Mild Hydrocracking of Vacuum Gasoil over Nimo-Beta Zeolite Catalysts: The Role of the Location of the Nimo Phases and the Crystallite Size of the Zeolite", *J. Catal.* **1998,** *179,* 537–547.
[162] Landau, M. V., Vradman, L., Valtchev, V., Lezervant, J., Liubich, E., Talianker, M., "Hydrocracking of Heavy Vacuum Gas Oil with a Pt/H-Beta-Al_2O_3 Catalyst: Effect of Zeolite Crystal Size in the Nanoscale Range", *Ind. Eng. Chem. Res.* **2003,** *42,* 2773–2782.
[163] Arribas, M. A., Martínez, A., "Simultaneous Isomerization of N-Heptane and Saturation of Benzene over Pt/Beta Catalysts - The Influence of Zeolite Crystal Size on Product Selectivity and Sulfur Resistance", *Catal. Today* **2001,** *65,* 117–122.
[164] Zhang, P. Q., Guo, X. W., Guo, H. C., Wang, X. S., "Study of the Performance of Modified Nano-Scale ZSM-5 Zeolite on Olefins Reduction in FCC Gasoline", *J. Mol. Catal. A: Chem.* **2007,** *261,* 139–146.
[165] Majano, G., Mintova, S., Ovsitser, O., Mihailova, B., Bein, T., "Zeolite Beta Nanosized Assemblies", *Microporous Mesoporous Mater.* **2005,** *80,* 227–235.
[166] Matsukata, M., Ogura, M., Osaki, M., Rao, T. P. R. H. P., Nomura, M., Kikuchi, E., "Conversion of Dry Gel to Microporous Crystals in Gas Phase", *Top. Catal.* **1999,** *9,* 77–92.
[167] Lei, Q., Zhao, T. B., Li, F. Y., Zhang, L. L., Wang, Y., "Catalytic Cracking of Large Molecules over Hierarchical Zeolites", *Chem. Commun.* **2006,** 1769–1771.
[168] Kustova, M. Yu., Rasmussen, S. B., Kustov, A. L., Christensen, C. H., "Direct NO Decomposition over Conventional and Mesoporous Cu-ZSM-5 and Cu-ZSM-11 Catalysts: Improved Performance with Hierarchical Zeolites", *Appl. Catal. B: Environ.* **2006,** *67,* 60–67.
[169] Tan, Q. F., Bao, X. J., Song, T. C., Fan, Y., Shi, G., Shen, B. J., Liu, C. H., Gao, X. H., "Synthesis, Characterization, and Catalytic Properties of Hydrothermally Stable Macro-Meso-Micro-Porous Composite Materials Synthesized via in Situ Assembly of Preformed Zeolite Y Nanoclusters on Kaolin", *J. Catal.* **2007,** *251,* 69–79.

[170] Shen, J., Shan, W., Zhang, Y. H., Du, J. M., Xu, H. L., Fan, K. N., Shen, W., Tang, Y., "A Novel Catalyst with High Activity for Polyhydric Alcohol Oxidation: Nanosilver/Zeolite Film", *Chem. Commun.* **2004,** 2880–2881.

[171] Shen, J., Shan, W., Zhang, Y. H., Du, J. M., Xu, H. L., Fan, K. N., Shen, W., Tang, Y., "Gas-Phase Selective Oxidation of Alcohols: In Situ Electrolytic Nano-Silver/Zeolite Film/Copper Grid Catalyst", *J. Catal.* **2006,** *237,* 94–101.

[172] Nishiyama, N., Miyamoto, M., Egashira, Y., Ueyama, K., "Zeolite Membrane on Catalyst Particles for Selective Formation of P-Xylene in the Disproportionation of Toluene", *Chem. Commun.* **2001,** 1746–1747.

[173] Nishiyama, N., Ichoika, Y., Egashira, Y., Ueyama, K., Gora, L., Zhu, W. D., Kapteijn, F., Moulijn, J. A., "Reactant-Selective Hydrogenation over Composite Silicalite-1-Coated Pt/ TiO_2 Particles", *Ind. Eng. Chem. Res.* **2004,** *43,* 1211–1215.

[174] He, J. J., Liu, Z., Yoneyama, Y., Nishiyama, N., Tsubaki, N., "Multiple-Functional Capsule Catalysts: A Tailor-Made Confined Reaction Environment for the Direct Synthesis of Middle Isoparaffins from Syngas", *Chem. Eur. J.* **2006,** *12,* 8296–8304.

[175] He, J. J., Yoneyama, Y., Xu, B. L., Nishiyama, N., Tsubaki, N., "Designing a Capsule Catalyst and its Application for Direct Synthesis of Middle Isoparaffins", *Langmuir* **2005,** *21,* 1699–1702.

[176] Kanthasamy, R., Larsen, S. C., "Visible Light Photoreduction of Cr(VI) in Aqueous Solution Using Iron-Containing Zeolite Tubes", *Microporous Mesoporous Mater.* **2007,** *100,* 340–349.

[177] Srivastava, R., Choi, M., Ryoo, R., "Mesoporous Materials with Zeolite Framework: Remarkable Effect of the Hierarchical Structure for Retardation of Catalyst Deactivation", *Chem. Commun.* **2006,** 4489–4491.

[178] Kim, J., Grate, J. W., Wang, P., "Nanostructures for Enzyme Stabilization", *Chem. Eng. Sci.* **2006,** *61,* 1017–1026.

[179] Zhang, Y. H., Wang, X. Y., Shan, W., Wu, B. Y., Fan, H. Z., Yu, X. J., Tang, Y., Yang, P. Y., "Enrichment of Low-Abundance Peptides and Proteins on Zeolite Nanocrystals for Direct MALDI-TOF MS Analysis", *Angew. Chem. Int. Ed.* **2005,** *44,* 615–617.

[180] Zhang, Y. H., Yu, X. J., Wang, X. D., Shan, W., Tang, Y., Yang, P. Y., "Zeolite Nanoparticles with Immobilized Metal Ions: Isolation and MALDI-TOF-MS/MS Identification of Phosphopeptides", *Chem. Commun.* **2004,** 2882–2883.

[181] Dahm, A., Eriksson, H. J., "Ultra-stable Zeolites - a Tool for In-cell Chemistry", *Biotechnology* **2004,** *11,* 279–290.

[182] Andersson, L. I. M., Eriksson, H., "De-aluminated Zeolite Y as a Tool to Study Endocytosis, a Delivery System Revealing Differences Between Human Peripheral Dendritic Cells", *Scand. J. Immunol.* **2007,** *66,* 52–61.

[183] Xu, F., Wang, Y. J., Wang, X. D., Zhang, Y. H., Tang, Y., Yang, P. Y., "A Novel Hierarchical Nanozeolite Composite as Sorbent for Protein Separation in Immobilized Metal-ion Affinity Chromatography", *Adv. Mater.* **2003,** *15,* 1751–1753.

[184] Wang, D. J., Zhang, Y. H., Xu, F., Shan, W., Zhu, G. B., Yang, P. Y., Tang, Y., "The Application of Zeolite/FAC Composites in Protein Separation", *Stud. Surf. Sci. Catal.* **2004,** *Part A–C 154,* 2027–2033.

[185] Yu, T., Zhang, Y. H., You, C. P., Zhuang, J. H., Wang, B., Liu, B. H., Kang, Y. J., Tang, Y., "Controlled Nanozeolite-assembled Electrode: Remarkable Enzyme-immobilization Ability and High Sensitivity as Biosensor", *Chem. Eur. J.* **2006,** *12,* 1137–1143.

[186] Zhou, X. Q., Yu, T., Zhang, Y. H., Kong, J. L., Tang, Y., Marty, J. L., Liu, B. H., "Nanozeolite-assembled Interface towards Sensitive Biosensing", *Electrochem. Commun.* **2007,** *9,* 1525–1529.

[187] Zhang, Y. H., Liu, Y., Kong, J. L., Yang, P. Y., Tang, Y., Liu, B. H., "Efficient Proteolysis System: A Nanozeolite-Derived Microreactor", *Small* **2006,** *2,* 1170–1173.

[188] Huang, Y., Shan, W., Liu, B. H., Liu, Y., Zhang, Y. H., Zhao, Y., Lu, H. J., Tang, Y., Yang, P. Y., "Zeolite Nanoparticle Modified Microchip Reactor for Efficient Protein Digestion", *Lab Chip* **2006,** *6,* 534–539.

[189] Shan, W., Yu, T., Wang, B., Hu, J. K., Zhang, Y. H., Wang, X. D., Tang, Y., "Magnetically Separable Nanozeolites: Promising Candidates for Bio-applications", *Chem. Mater.* **2006,** *18,* 3169–3172.

[190] Corma, A., Fornés, V., Jordá, J. L., Rey, F., Fernundez-Lafuente, R., Guisan, J. M., Mateo, C., "Electrostatic and Covalent Immobilisation of Enzymes on ITQ-6 Delaminated Zeolitic Materials", *Chem. Commun.* **2001,** 419–420.
[191] Corma, A., Fornés, V., Rey, F., "Delaminated Zeolites: An Efficient Support for Enzymes", *Adv. Mater.* **2002,** *14,* 71–74.
[192] Walcarius, A., "Zeolite-modified Electrodes in Electroanalytical Chemistry", *Anal. Chim. Acta* **1999,** *384,* 1–17.
[193] Marko-Varga, G., Burestedt, E., Svensson, C. J., Emneus, J., Gorton, L., Ruzgas, T., Unger, K. K., "Effect of HY-zeolites on the Performance of Tyrosinase-modified Carbon Paste Electrodes", *Electroanalysis* **1996,** *8,* 1121–1126.
[194] Liu, B. H., He, R. Q., Deng, J. Q., "Characterization of Immobilization of an Enzyme in a Modified Y Zeolite to Matrix and its Application to an Amperometric Glucose Biosensor", *Anal. Chem.* **1997,** *69,* 2343–2348.
[195] Gonqalves, A. P. V., Lopes, J. M., Lemos, F., Ramôa Ribeiro, F., Prazeres, D. M. F., Cabral, J. M. S., Aires-Barros, M. R., "Zeolites as Supports for Enzymatic Hydrolysis Reactions. Comparative Study of Several Zeolites", *J. Mol. Catal. B: Enzym.* **1996,** *1,* 53–60.
[196] Serralha, F. N., Lopes, J. M., Lemos, F., Prazeres, D. M. F., Aires-Barros, M. R., Cabral, J. M. S., Ramôa Ribeiro, F., "Zeolites as Supports for an Enzymatic Alcoholysis Reaction", *J. Mol. Catal. B: Enzym.* **1998,** *4,* 303–311.
[197] Xing, G. W., Li, X. W., Tian, G. L., Ye, Y. H., "Enzymatic Peptide Synthesis in Organic Solvent with Different Zeolites as Immobilization Matrixes", *Tetrahedron* **2000,** *56,* 3517–3522.
[198] Maugard, T., Tudella, J., Legoy, M. D., "Study of Vitamin Ester Synthesis by Lipase-catalyzed Transesterification in Organic Media", *Biotechnol. Prog.* **2000,** *16,* 358–362.
[199] Carvalho, R., Lemos, H. F., Cabral, J. M. S., Ramôa Ribeiro, F., "Influence of the Presence of NaY Zeolite on the Activity of Horseradish Peroxidase in the Oxidation of Phenol", *J. Mol. Catal. B: Enzym.* **2007,** *44,* 39–41.
[200] Ghose, S., Mattiasson, B., "Protein Adsorption to Hydrophobic Zeolite Y: Salt Effects and Application to Protein Fractionation", *Biotech. Appl. Biochem.* **1993,** *18,* 311–320.
[201] Klint, D, Arvidsson, P., Blum, Z., Eriksson, H., "Purification of Proteins by the Use of Hydrophobic Zeolite-Y", *Protein Expr. Purif.* **1994,** *5,* 569–576.
[202] Klint, D, Eriksson, H., "Conditions for the Adsorption of Proteins on Ultrastable Zeolite Y and its Use in Protein Purification", *Protein Expr. Purif.* **1997,** *10,* 247–255.
[203] Klint, D., Karlsson, G., Bovin, J. O., "Cryo-TEM Snapshots of Ferritin Adsorbed on Small Zeolite Crystals", *Angew. Chem. Int. Ed.* **1999,** *38,* 2560–2562.
[204] Huang, Y. C., Yu, Y. C., Lee, T. Y., "Purification of Antibodies by Zeolite-A", *Enzyme Microb. Tech.* **1995,** *17,* 564–569.
[205] Yu, Y. C., Huang, Y. C., Lee, T.Y, "Purification of Antibodies from Protein Mixtures and Mouse Ascites Fluid Using Zeolite X", *Biotechnol. Progr.* **1998,** *14,* 332–337.
[206] Matsui, M., Kiyozumi, Y., Yamamoto, T., Mizushina, Y., Mizukami, F., Sakaguchi, K., "Selective Adsorption of Biopolymers on Zeolites", *Chem. Eur. J.* **2001,** *7,* 1555–1560.
[207] Nygaard, S., Wendelbo, R., Brown, S., "Surface-Specific Zeolite-Binding Proteins", *Adv. Mater.* **2002,** *14,* 1853–1856.
[208] Eriksson, H., Green, P., "The Use of Zeolite Y in the Purification of Intracellular Accumulated Proteins from Genetically Engineered Cells", *Biotechnol. Tech.* **1992,** *6,* 239–244.
[209] Eriksson, H., "Removal of Toxic Preservatives in Pharmaceutical Preparations of Insulin by the Use of Ultra-Stable Zeolite Y", *Biotechnol. Tech.* **1998,** *12,* 329–334.
[210] Cheng, Y. L., Lee, T. Y., "Separation of Fructose and Glucose Mixture by Zeolite-Y", *Biotechnol. Bioeng.* **1992,** *40,* 498–504.
[211] Munsch, S., Hartmann, M., Ernst, S., "Adsorption and Separation of Amino Acids from Aqueous Solutions on Zeolites", *Chem. Commun.* **2001,** 1978–1979.
[212] Fisher, K., Huddersman, K., "Separation of Nucleosides and Nucleotides Using Cation-Exchanged Zeolites", *New J. Chem.* **2002,** *26,* 1698–1701.

CHAPTER 18

Bioinspired Porous Materials

Valentin Valtchev *and* Svetlana Mintova

Contents

Abstract

The present chapter focuses on the synthesis of bioinspired porous solids. Section 1 provides a brief historical overview of the synthesis of porous solids and the use of templating species. The main body of the chapter comprises three inter-connected parts presenting basic information related to the synthesis of main group porous solids, namely micro-, meso- and macroporous. The interest in hierarchical porous constructions is also addressed because most of the bioinspired solids possess two or three pore size regimes. The concluding section reviews different types of biological templates employed in the synthesis of porous materials and the prospects for bioinspired porous solids.

Keywords: Biotemplating, Porous, Replication, Hierarchical Structures

Glossary

Each unique zeolite framework type has a three-letter code (MFI, FAU and so on) assigned by the Structural commission of the International Zeolite Association.
ZSM-5, Zeolite Socony Mobil no. 5.

1. Introduction

Complex materials are distinguished from simple crystalline solids in that they generally possess molecular or structural length scales much greater than the simple constituents. In order to combine different characteristics like, for instance, strength with light weight or toughness and rigidity, combinations of materials and their

Ordered Porous Solids
DOI: 10.1016/B978-0-444-53189-6.00018-4

processing under specific conditions are required. Further, the multi-level organisation of a synthetic material is usually obtained by a sequential assembly of basic components. Considering the expensive, time-consuming and specific equipment needed for the processing of synthetic materials, the ability of nature to build complex biological specimens is simply amazing. The latter is based on the remarkable capability of biological organisms to recognise, sort and process different species by highly functional assemblies of proteins, nucleic acids and other macro-molecules that carry out complicated tasks that are still challenging for us to emulate. Thus, the biological concept gives inspiration for creating a new generation of synthetic materials and devices with advanced structures and functions. It should also be mentioned that the increasing interest in building hybrid materials based on precise binding of biological molecules to technologically relevant surfaces. The inter-disciplinary research projects at the frontier between materials science and biology increased tremendously during the last decade. Consequently, a number of review articles and text books devoted to this rapidly growing discipline appeared recently.[1–10]

Besides the astonishing ability of self-organisation and complex functionality, biological specimens are attractive with their shapes and patterns that are unapproachable by the synthetic materials. These complex architectures combined with multi-level organisation and narrow specialisation of different parts inspires modern materials science. The organic chemistry, especially supramolecular chemistry, has been very successful in creating large superstructures of often stunning morphology. In contrast, inorganic chemistry is still lagged behind. The intricate organic–inorganic reactions in biological specimens leading to multi-level inorganic constructions are far from being understood and simulated. Thus, the only possibility to obtain inorganic materials with multi-level organisation and complex morphology is to replicate biological samples.

The present chapter deals with the porous materials inspired or prepared by employing bio-precursors/templates. Special attention will be paid to the hierarchical porous structures that are difficult to obtain using synthetic templates. Hierarchical porous structures are expected to show a better performance in areas such as heterogeneous catalysis, chromatography and separation of biomolecules, gases and liquids. Their presumable higher efficiency is a result of the combination of high selectivity and improved kinetics with respect to corresponding bulk materials. Various combinations of hierarchical porous structures, for example micro-/mesoporous,[11–13] meso-/macroporous,[14–16] micro-/macroporous[17–21] and micro-/meso-/macroporous,[22] have been synthesised by extension of the template strategy that was first employed for the synthesis of microporous zeolite-type materials.[23] Small organic molecules used in the synthesis of zeolitic materials have been substituted by surfactant micelles to synthesise ordered mesoporous materials[24,25] and lately by arrays of uniform spheres for the preparation of macroporous solids with monomodal pore size distribution.[26,27] The utilisation of templates with specific macro-morphological features would allow the control of both the shape and the size of the solid. Although some morphological constructions with hierarchical porosity, for example beads,[12] have been obtained with synthetic templates, the morphological features of porous materials are far from being controlled. Therefore, templates that could provide specific morphological constructions with hierarchical organisation at the nanolevel scale are highly desirable. As mentioned, the ability to combine complex morphology with hierarchical architecture is typical of biological specimens. Examples of the transformation of such specimens into inorganic structures are also provided by nature, for example, the process of petrification in which a supersaturated mineral containing solution precipitates within the interstices of the organic tissues filling them with the mineral.[28] After the decay of the organic tissue, a negative replica of the biological structure is obtained. This process is able to replicate the intimate organisation including the cell structure. However, the petrification is a long and difficult to completely reproduce under laboratory conditions process. Hence, methods applicable to laboratory timescale and conditions will have to be developed in order to obtain molecular-scale replicas of biological templates.

This chapter reviews the open literature devoted to the materials with distinct morphological features and porous organisation templated or inspired by biological specimens.

2. Hierarchical Porous Structures

Most of the porous materials issued from biological templates exhibit hierarchical porous organisation. However, for the sake of simplicity in the present chapter, they are divided into three categories, that is, micro-, meso- and macroporous, depending on the dominant porosity type.

2.1. Microporous materials

Microporous zeolite-type crystalline solids are generally synthesised under mild hydrothermal conditions. The network of pores and cavities in the zeolite structure is structured by templating species that can be removed after the zeolite formation. Alkali and alkali-earth cations with their hydration spheres are the structure directing agents governing the formation of natural and low silica synthetic zeolites. In the laboratory, alkali–water complexes are often replaced by organic cations or other polar organic species that results in the formation of zeolites with a higher Si/Al ratio or new framework types. High silica (Si/Al $>$ 10) zeolite materials are formed in the presence of large variety of organic molecules. Among them, the alkyl-, arylamonium hydroxides and salts, amines, and alcohols are the most widely used to the best of our knowledge. Organic species of biological origin were not involved in the formation of microporous zeolite-type materials. The zeolites–biomatter crossing point is namely related to the employment of natural specimens as macro-templates for the preparation of morphological polycrystalline constructions. Two approaches have been developed in order to overcome the incompatibility between the organic support and the inorganic precursors yielding microporous solids: (1) pre-synthesised microporous nanoparticles have been assembled on the biological supports; and (2) the crystallisation on the biotemplate has been induced by zeolite seeds initially adsorbed on the support, which on hydrothermal treatment with a fresh synthesis solution grow into a dense zeolite layer.

The first approach has been employed for the preparation of zeolite fibres, produced by infiltration of zeolite nanocrystals into the ordered void spaces of macroscopic bacterial threads.[29] *Bacilus subtilus* and preformed silicalite-1 nanoparticles were, respectively, the organic and inorganic part involved in this preparation. The swelling procedure gave a highly compacted network of silicalite-1 crystals after air-drying. Calcination at 600 °C removed both the structure directing agent (tetrapropylammonium) used in the zeolite synthesis and the supercellular template. The all-silicalite-1 replica retained the fibre-like morphology of the bacterial template, where organised arrays of ca. 0.5-μm-wide channels parallel to the fibre axis with channel walls of about 100 nm can be observed. Thus, a micro-/macroporous material with fibre-like morphology was obtained.

The above approach, that is, to use preformed zeolite nanocrystals and a biological template in order to prepare complex materials was further extended by Mann and co-workers.[30] Potato starch and silicalite-1 nanocrystals (ca. 50 μm) were used in the preparation of films, monoliths and sponge-like structures. Depending on the desired material, the starch/silicalite-1 weight ratio was varied between 1 and 33 for different preparations. The films were prepared from initial mixtures with low (~2 wt.%) starch concentration, whereas the monoliths were obtained from much more concentrated (up to 40 wt.%) gels. Dry film and monoliths could be cut to desired sizes and shapes. The calcination at 600 °C did not result in significant shrinkage or loss of structural integrity. Both, film and monoliths, were built of closely packed crystals; however, the monolith contained macropores that varied with the starch/silicalite-1 weight ratio. The same study reported the preparation of sponge-like starch structures with 50–200 μm pores and 5–10 μm walls. The obtained 3D macroporous network was immersed into a colloidal silicalite-1 suspension. A layer of zeolite nanoparticles with thickness between 5 and 20 μm was deposited on the walls of the sponge-like framework. Removal of the starch framework by calcination produced a macroporous all-zeolite monolith. Thus, a series of materials with hierarchical porosity employing zeolite nanoblocks and sacrificial starch templates was obtained.

A sponge-like zeolite membrane was prepared by a similar approach, where a cellulose acetate filter membrane was employed as a template.[31] A commonly used technique, that is, reversal of the surface charge of the support by a cationic polymer followed by electrostatic adsorption of nanoparticles of opposite charge was employed to seed the surface of the fibre templates. Calcination of the material resulted in an all-zeolite membrane with limited strength. Obviously, the low degree of zeolite coating and the weak interaction between zeolite particles are not sufficient to obtain a robust membrane. In order to increase the membrane strength, the authors subjected the adsorbed zeolite nanocrystals to a secondary growth under hydrothermal conditions. The inter-growth of zeolite crystallites provided thicker and firmer walls and consequently much better mechanical strength. Besides zeolite micropores, the membrane contained two types of macropores, tubular pores related to the combustion of the cellulose template and interconnected macropores along the zeolite microtubes. This type membrane is expected to show high efficiency for the immobilisation of enzymes after grafting of organic functional groups (e.g., amidocyanogen, imine, aldehyde) on the zeolite microtube walls.

Nanoseeds and secondary growth were also applied in the replication of wood cells by Dong *et al.*[32] Two kinds of tissues, cedar with relatively uniform pores of about 10–20 μm and bamboo with non-uniform 2–20 μm sized pores, were employed in this study. Again a polyelectrolyte agent was used to adsorb the seeds on the template surface. The nanoseeds induced the formation of zeolite film during the subsequent hydrothermal treatment. The biotemplate was removed by calcination providing an all-zeolite replica that retained the macro-morphological features of the wood cells. Thus, zeolitic tissues comprising macropores of interest for catalytic and separation applications were obtained.

Rattan palm stem templates were used in the preparation of zeolite-containing SiC monoliths.[33,34] Rattan palm is particularly appropriate for the preparation of hierarchical pore monoliths because of the multi-modal pore distribution with straight channels, which are not interrupted by branches or season rings. A multi-step procedure was developed to transform the palm structure into an inorganic porous composite. First, the native stem was pyrolysed in an inert atmosphere to obtain a carbon preform. Second, the biocarbon preform was infiltrated with different amounts of liquid Si at 1550 °C to obtain β-SiC or SiSiC. Finally, the latter was subjected to hydrothermal treatment in a synthesis solution that transformed a part of the silicon into a microporous MFI-type material. Thus, the SiSiC composite acted not only as a carrier but also as a silicon source for the zeolite synthesis. What is noteworthy for this material is the high mechanical strength that is not typical for materials prepared from biological templates.

Further extension of the use of biological tissues in the preparation of an inorganic porous architecture was reported by Zampierri *et al.*[35] Luffa cylindrical sponge was replicated by a zeolite. Again MFI-type material was synthesised for this purpose. The difference with the previous investigations was that a two-step crystallisation procedure was implemented in order to obtain a faithful replica of the biotemplate. The first synthesis step, where a dilute precursor solution was involved, provided nanoseeds uniformly distributed on the sponge tissues. Secondary growth, performed with a viscous gel precursor, yielded several microns thick zeolite film. During the synthesis, the nutrient solution penetrated into the vascular system of the Luffa sponge and crystallised in the inner cellular network with channels ranging between 10 and 20 μm in diameter. Depending on the synthesis conditions, hollow microchannels or rods were obtained. The combined SEM/adsorption measurements revealed a complex hierarchical pore network that resembled the original spongy architecture. After combustion of the macro-template, the obtained all-zeolite monoliths were tested in *n*-hexane cracking in a fixed bed reactor. The bioinspired hierarchical porous monolith showed a low deactivation tendency even at 550 °C and high attrition resistance.

The above studies present the preparation of all-zeolite or zeolite-containing hierarchical porous structures by employing biological tissues as sacrificial templates. On the other hand, inorganic skeletons of different organisms remaining after the decay of the organic matter have also been used in the preparation of zeolite-containing composite with multi-level porous organisation. Probably the most widely used bio-supports for such preparations are the silica skeletons of diatoms. The diatoms are microscopic single-cell algae plants enclosed in intricate silica shell. When these aquatic plants die, their shells collect on the ocean or lake bottom. The material is collected as diatomaceous earth or light-weight rock called diatomite. The latter is often used as a row source of silica for the synthesis of different types of zeolites.[36,37]

Diatoms were seeded with MFI-type nanoparticles by sonication in a colloidal suspension and then placed in a synthesis mixture and hydrothermally treated to grow an extended inter-grown zeolite layer on the diatoms surface.[38,39] XRD, SEM and nitrogen adsorption measurements confirmed the formation of a MFI-type material on the diatom supports. Based on the specific surface area measurements, the zeolite content was about 5 wt.%. The zeolite crystals were extremely well bound to the zeolite surface as shown by SEM after thermal and hydrothermal tests. Finally, the diffusion of water through packed zeolite/diatomite columns was tested and compared with that of a packed bed of silicalite-1 crystals. As can be expected, the flow rate through the hierarchical porous material was much higher in respect to the packed zeolite crystals.

Wang *et al.* have also seeded the diatoms prior to film growth.[40] The diatomite surface was modified and the nanoseeds were electrostatically adsorbed prior to the growth process. The subsequent synthesis procedure differed substantially from the hydrothermal treatment employed in the previous study. Seeded diatoms were treated in a vapour-phase containing water, ethylendiamine and triethylamine. Depending on the synthesis conditions (5–10 days at 180 °C), materials with different degrees of conversion into zeolite of the diatom-building silica were obtained. For instance, after 10 days treatment, a material with 210 m^2/g was obtained that corresponds to about 50% transformation into MFI-type material. Nevertheless, the obtained material retained the morphological organisation of the diatomite and its macroporous character. Zeolite nanoparticles (silicalite-1, TS-1 and ZSM-5) were also deposited on diatomite supports through layer-by-layer assembly, applying a cationic polymer.[41] However, the electrostatic adsorption provided a material with limited mechanical strength and thus the zeolite nanocrystals could easily be detached from the surface of the support.

The fact that diatom skeletons are essentially built of silica suggests that under appropriate conditions a zeolite layer may be grown without preliminary seeding. By varying the amount of tetrapropylammonium hydroxide and the reactivity of synthesis solution, we successfully synthesised extended uniform silicalite-1 layers on a diatom skeleton.[42] The XRD pattern clearly demonstrated the presence of MFI-type material. The amount was limited since only the most intense peaks appeared in the XRD pattern of the material. The zeolite content was evaluated by N_2 adsorption, namely by the increase of the specific surface area. The S_{BET} of the initial diatoms was 0.5 m^2/g, whereas after the hydrothermal treatment and the calcination, a material with 43 m^2/g was obtained, which corresponds to about 10 wt.% zeolite in the material. The formation of a silicalite-1 layer was also revealed by SEM inspection of the initial and hydrothermally treated diatoms (Fig. 18.1). A uniform continuous layer covered the surface of the diatom skeletons, which retained the

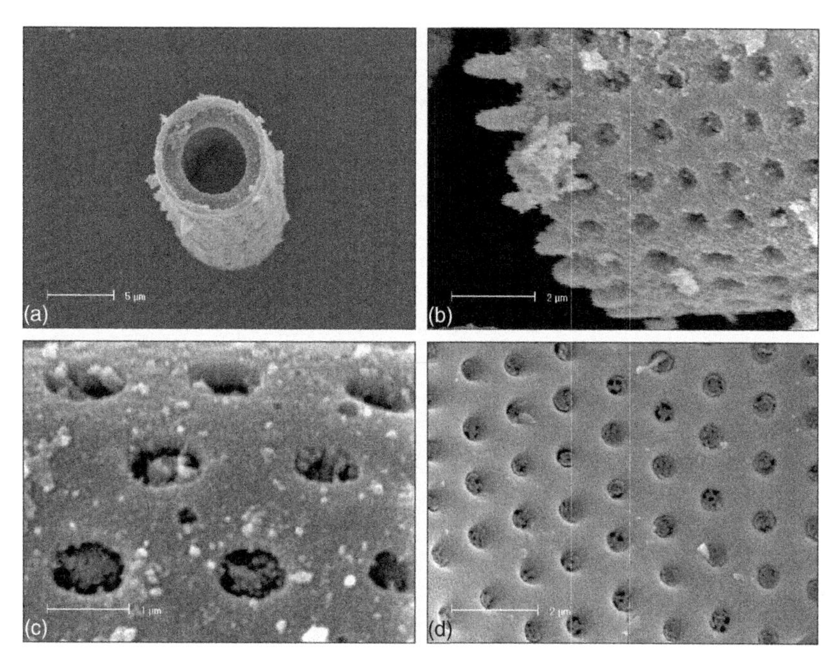

Figure 18.1 SEM images of a diatom subjected to hydrothermal treatment in a silicalite-1 yielding solution: (a) low-magnification view of a treated diatom skeleton, (b, c) high-magnification images of the external surface of the diatom with the silicalite-1 layer and (d) close view of a non-treated diatom.

macro-morphological features of the biological support. Zeolite A has also been crystallised on diatom skeletons without preliminary seeding.[43]

Replication of a template usually retains the macro-morphological feature of the specimen, but represents a negative replica of the lower level structural elements. In order to obtain a positive replica, a second replication is necessary, which, however, is difficult and rarely provides satisfactory results. Faithful zeolitisation of the micro-architecture of the plant structure would not be achieved if the mineralisation process did not take place within the plant tissue, thus resembling the petrification of biological cells in nature.[44] Such an opportunity is offered by some plants, for example, those from the Equisetaceae family, whose members are rich in amorphous silica deposited at discrete knobs and rosettes at the epidermal surface.[45,46] Recently, it was found that biogenic silica deposited in the epidermal surface of *Equisetum arvense* (horsetail) promotes zeolite crystallisation.[47] This opened up routes for faithful laboratory zeolitisation of silica-containing fresh plants yielding advanced porous materials. This is a unique case where a single-step treatment resulted in a positive replica of a biological specimen.

Dry leaves of *E. arvense* were subjected to hydrothermal treatment in a silicalite-1 precursor solution without any preliminary treatment. In a second step, the biotemplate was eliminated by high-temperature treatment at 600 °C. According to the XRD analysis, a highly crystalline MFI-type material was obtained. The type I adsorption/desorption isotherm of the calcined material confirmed the microporous character of the materials. A steep rise in the uptake at low relative pressures corresponds to the filling of micropores with N_2. In contrast to pure microporous materials, the steep uptake at low relative pressures is not followed by a flat curve. Instead, an inclination of the curve with increase of the pressure can be observed. At high relative pressure, the upward turn with a hysteresis loop is indicative of the generated inter-crystalline mesoporosity. Thus, the zeolitisation of *E. arvense* provided a material with combined micro-/mesoporosity. Figure 18.2 shows general and close views of zeolitised *E. arvense* leaves and stems after the hydrothermal treatment and calcination procedure. The morphology of the leaves is maintained

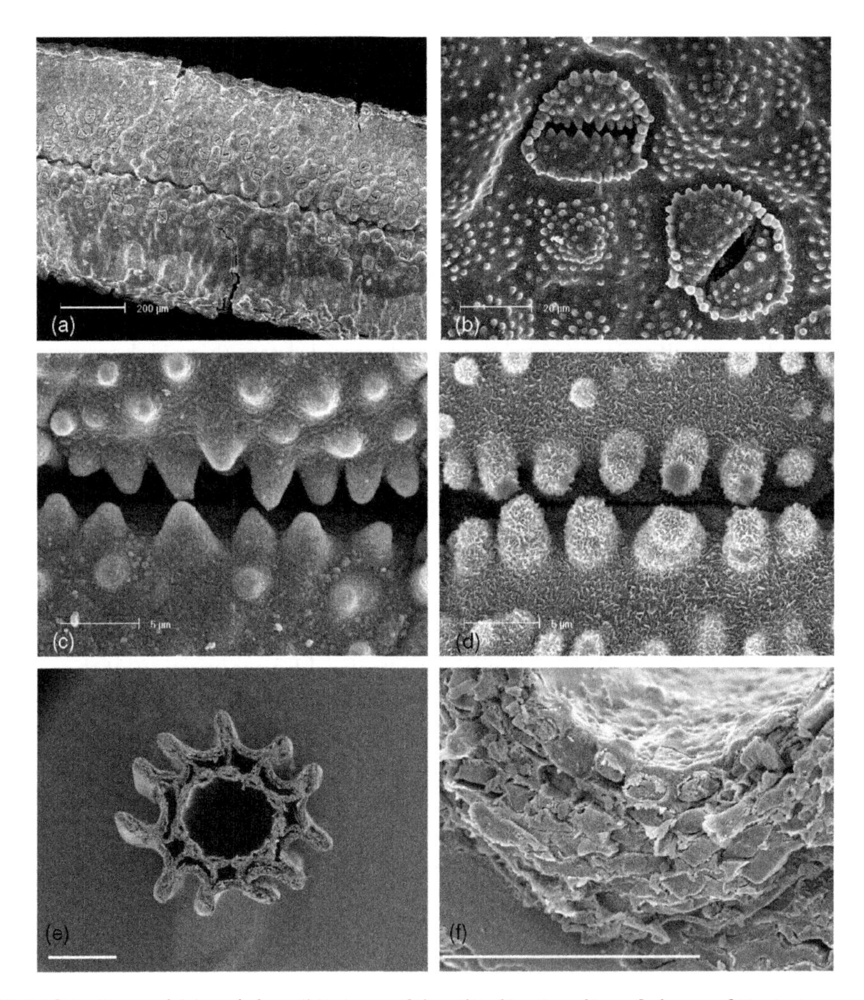

Figure 18.2 General (a) and close (b) views of the silicalite-1 replica of a leave of *Equisetum arvense* (horsetail). The silicalite-1 replica of a stomata structure (c) is compared with the initial plant structure (d), where fine filaments covering the leaf can be seen. Cross section of an initial stem of *Equisetum arvense* (e) and a higher magnification view of a zeolite β-stem replica (f) (Scale bars = 1 mm).

after the combustion of the organic tissues (Fig. 18.2a and b). As can be seen, in the silicalite-1 replica, all details of the original plant structure are completely preserved (Fig. 18.2c and d). Another important finding is the extremely small-sized silicalite-1 crystallites grown on the surface. According to the dynamic light scattering measurements, the size of the silicalite-1 crystals found in the solution after the hydrothermal treatment is 90 nm. The TEM investigation showed that the crystallites grown in the epidermal surface of *E. arvense* are much smaller, ranging from 20 to 40 nm. This study revealed also that the crystals are fairly uniform in size. This is a sound proof that the reactive bio-mineral silica promotes the zeolite nucleation in the epidermal surface. Thus, a very homogeneous fine silicalite-1 layer is formed and even the nanometre scale details of the morphology of the plant are replicated. ZSM-5 replicas of *E. arvense* were also obtained.[48]

The zeolitisation of stems of *E. arvense* stems (Fig. 18.2e) using a BEA-type yielding synthesis mixture resulted in the formation of micro-/meso-/macroporous material.[49] The stems of *E. arvense* are hollow with walls containing smaller channels, jointed with very distinct nodes. It was found that the synthesis time determined the amount of zeolite that crystallised within the template. Zeolite crystallisation (20–40 nm crystals) within the tissue of the biotemplate was followed by the formation of larger zeolite particles on the epidermal surface and filling with crystalline material of the small and larger channels surrounding the central hole (Fig. 18.2f). This consecutive filling of the stem channels with zeolite may be used to control the level of BEA-type zeolite loading and the meso-/macroporosity of the ultimate replica.

This study showed that fresh silica-containing plants can be transformed into materials with hierarchical porosity that retained the morphological features of the biotemplate. The use of different zeolites and various parts of a silica-containing plant shows that the method is extraordinarily basic and general. The various possibilities offered by the biotemplates were demonstrated by the synthesis of micro-/meso- and micro-/meso-/macroporous zeolite macro-structures by using leaves and stems of *E. arvense* as templates, respectively. Uniform abundant nucleation of zeolite within the bulk volume of the plant tissues induced by the biomorphic silica provided faithful replication of the plant structure and very high concentration of zeolite per unit mass of the material. The application of silica-rich biological templates may result in a large variety of morphological constructions built up of various types of zeolites and other crystalline silica-containing solids that crystallise under mild hydrothermal conditions. The possibilities of building biomorphic materials based on biological templates are tremendous, for example, only the *Equisetaceae* family comprises about 15 members, each of them with specific morphological features and sizes varying between several centimetres and 2 m.

2.2. Mesoporous materials

A large variety of surfactants and amphiphilic polymers have been used to structure ordered porous materials by self-assembly process.[50–52] This family of solids with pores ranging from a few to tens of nanometres filled up the gap between micro- and macroporous materials. However, the large well-defined pore dimensions and the complex architecture observed in biological materials are still out of reach for synthetic chemists. Hence, the bio-mineralisation leading to open structures received increasing attention. The interest in the natural silica productions is particularly high because it may open possibilities for the synthesis of hierarchically structured elements under mild conditions. Chemically produced silica is usually prepared at elevated temperatures, high pressure and/or strongly alkaline or acidic conditions. Thus, the mimic of biological conditions providing inorganic materials is highly desired from both environmental and economic points of view.

In the previous section, the preparation of micro-/macroporous structures using diatom frameworks, which are combined with a zeolitic material, was described. The intriguing architectures of diatom shells were not only employed as supports for zeolitisation but also inspired many scientists to obtain better control over the structure and morphology of synthetic porous materials.[3,53–56] Thus, diatoms were employed as model organisms in the efforts to understand the mechanism of bio-mineralisation and biopolymer-mediated silica formation in particular.

Kröger *et al.* demonstrated that peptides extracted from diatoms cell walls could lead to the rapid precipitation of silica.[57,58] The authors studied the isoforms of Silafin-1A protein, which are small polycationic proteins. The study revealed that only silafins whose lysine groups are modified with oligo-*N*-methylpropyleneamines are able to precipitate silica. An analogous function played tripropylenetetramine in the synthesis of silica spheres as shown by Noll *et al.*[59] The obtained spheres exhibited a surface features that resemble the exoskeleton of diatoms from the genus *Coscinodiscus*.[60]

In a series of papers, Livage and co-workers have investigated biomolecule-induced polymerisation of silica species. The effect of amino acids and peptides on the polymerisation of aqueous solutions was firstly studied.[61] The authors found out that the amino acids and peptides can interact with a silicate solution through hydrogen bonds and electrostatic interactions. Another important finding of this study was the more pronounced effect of peptides with respect to that of amino acids. In other words, the electrostatic interactions or hydrogen bonds with a single molecule are not sufficient to favour condensation reactions. Thus, the authors assumed that the distribution of charged side-groups and hydrogen-bonded chains brings reactive species close enough for condensation to occur. In particular, the positively charged NH_3^+ groups along the peptide chain could induce the condensation of negatively charged silica oligomers, which is a key point in understanding the role of silaffins in biosilification. Further, the authors studied in detail the interactions of amino-containing peptides, that is, lysine, arginine and their homopeptides, with sodium silicate and colloidal silica.[62] The results of this study clearly demonstrated that the polypeptides may serve as substrates for silica formation as the key parameter is the length of the polymer chain and therefore the number of NH_3^+ binding sites. Two possible scenarios, depending on the nature of the silicate species, were presented. In the case of monomeric species, the amino groups favour the oligomerisation process, followed by the gelation of these species. The second mechanism includes binding of silica particles to the polyelectrolyte chains to form peptide–silica assemblies, whose precipitation occurs via the bridging of aggregates by additional polymer. On the basis of the latter studies, Coradin *et al.* successfully synthesised multi-scale porous silica in the presence of arginine-based surfactants.[63] The employed bio-surfactants provided solids with a specific surface area between 510 and 720 cm^2/g and pore volume of about 1.0 cm^3/g. Three different pore scales were revealed by α-plot analysis: (1) micropores representing 1–2% of the total pore volume; (2) mesopores, which account for 25% of the total pore volume and about 50% and of the specific surface area and (3) large meso- and macropores. TEM inspection of the material showed that the obtained solids consist of about 50 nm particles. At high magnification, the typical framework structure of wormhole mesoporous silica was observed.

Pollen grains from *Taraxacum*, *Trifolium*, *Papaver* and *Brassica genera* were used to obtain SiO_2, $CaCO_3$ and $CaHPO_4$ replicas.[64] Tree pollen grains are appropriate biotemplate because of the tough outer shell (exine), which is amenable to inorganic mineralisation without consequent loss of fine structure. Besides, a high degree of specific morphological complexity for various species, pollen offers the advantage to be ubiquitous and inexpensive material. The employed preparation procedure is extremely simple, namely the freeze-dried pollen was soaked with an inorganic solution and then subjected to high-temperature treatment to eliminate the biological template. Nitrogen adsorption measurements of the silica replica showed a relatively high specific surface area (817 m^2/g) and a type IV isotherm with a characteristic hysteresis due to mesoporosity. According to the authors, the high surface area and mesoporosity are a consequence of the complex foam-like surface morphology of the native pollen. The material was further functionalised with silver and iron oxide nanoparticles, thus presenting interest in heterogenous catalysis. It is worth mentioning that only the SiO_2 replica showed a high specific area, the calcium carbonate and calcium phosphate replica did not exhibit such a characteristic.

Tobacco mosaic virus (TMV) is a very stable nanotube complex and thus is of interest as a host for chemical reactions. Knez *et al.* employed TMV as a chemically functionalised template for binding metal ions, namely different chemical groups of the coat protein were used as ligands or to electrostatically attach metal ions.[65] A range of different TMV–metal composites were prepared from virus suspensions in water or in phosphate buffer by contacting with Pd^{2+}, Pt^{2+} and Au^{3+} solutions and subsequent electroless deposition of nickel and cobalt. Specific ligand–metal ion interactions, where the ligand is a functional group of the virion structure, were proposed. The proper choice of the metal ion, pH and the duration of the treatment leads to the preference of certain groups. These groups can be concentrated on the external surface or in the channel depending on the nature of the virion. In this way, metal ions and their reduction products, clusters, can be placed on selected regions of the viral nanotube.

2.3. Macroporous materials

Different bio-organic species and technological approaches have been used in the preparation of macroporous materials. Macroporosity is usually obtained by replication of isolated or structured cells of the biotemplate. Nanoscale reproduction is not common for macroporous biotemplated materials. Thus, various ceramics and ceramic composites exhibiting macroporous features were obtained using natural moulding materials.[8] Herein, the obtained materials will be presented according to the chemical composition since the diversity of the employed biotemplates makes their classification difficult.

The most often used biologically derived materials for processing ceramic composites are wood and natural fibres (cotton, cellulose and so on). Biological specimens are usually pyrolysed in an inert atmosphere in order to decompose the building polyaromatic hydrocarbon polymers to carbon. Pyrolysis of natural tissues results in substantial shrinkage and weight losses of the original biological form. Nevertheless, the cellular structure was preserved and thus the porous organisation (Fig. 18.3). In a second step, the carbon preform is infiltrated with an inorganic

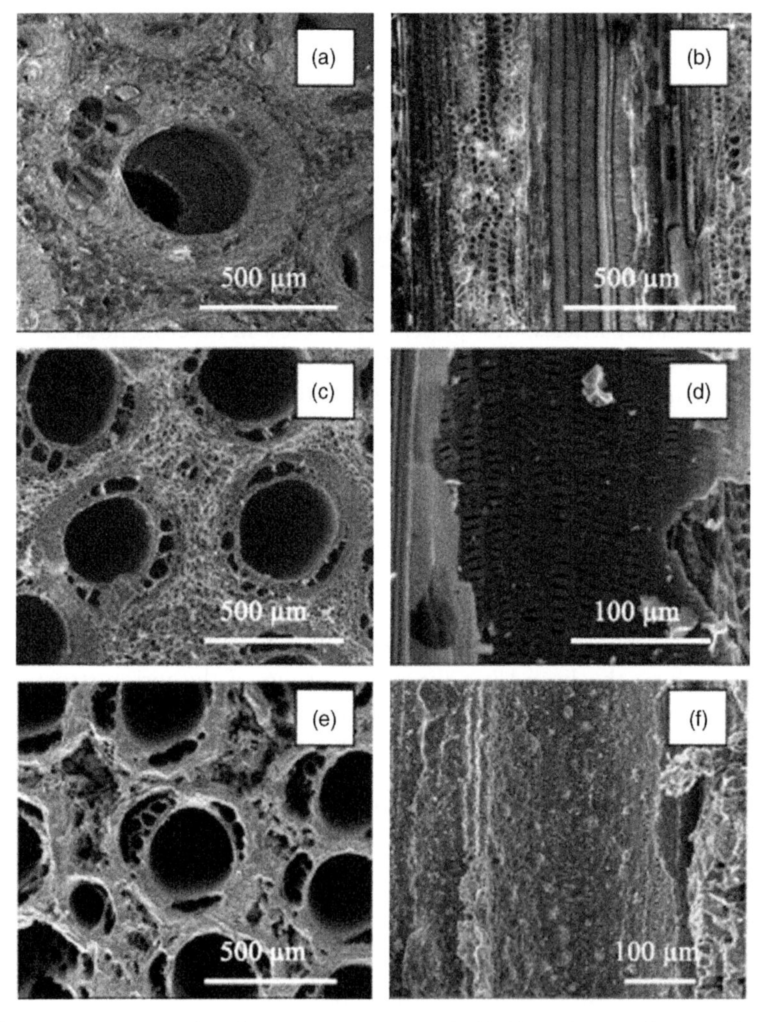

Figure 18.3 SEM micrographs of a native Rattan stem (a, b), biocarbon perform (c, d) and SiSiC replica (e, f). (Reproduced from Ref. 34 with permission from Elsevier.)

precursor and subjected to appropriate high-temperature treatment to obtain a biomorphic ceramics. Using this approach, different types of wood were transformed into SiC- and SiSiC-ceramics.[66] Liquid and gaseous Si infiltration was employed in order to incorporate the silicon. Both ways provided β-SiC grains with a mean particle size of about 5 μm. The liquid approach yielded SiSiC-ceramics, while the Si-gaseous infiltration resulted in pure SiC without residual Si. It is worth noting that the macroporous features of the material were preserved in both cases.

A different approach was employed by Shin *et al.* in the preparation of SiC ceramics using pine and poplar wood.[67] Dried wood precursors were first soaked with HCl to leach out the lignin and then treated with TEOS/HCl/EtOH/H_2O solution. Finally, the resulting mixture was heated in Ar atmosphere at 1400 °C. The morphological analysis showed that the cellular structure of the biotemplate was retained. The specific surface area of the final cellular ceramics varied between 60 and 100 m^2/g with pore sizes randomly distributed in the nanometre to micrometre range. The materials were stable up to 1400 °C in air. According to the energy dispersive spectroscopy analysis, the atomic ratio Si/C was 1.0/1.04, thus showing no excess of Si.

The effect of the starting Si content on the properties and structure of biomorphic SiC ceramics was studied by Huo *et al.*[68] Beech templates and liquid Si infiltration were employed in this investigation. Different amounts of liquid Si were infiltrated into the biocarbon template. Depending on the Si content materials with a Si/C = 1 ratio, and with excess of carbon or silicon were obtained. The analysis of the characteristics of the prepared series of materials revealed that the key factors affecting their properties were the conversion degree from biocarbon to SiC and amount of the residual silicon. With increasing starting Si contents, the porosity decreased whereas the bending strength and the toughness improved.

Biomorphic porous β-SiC was derived from natural millet.[69] Millet grains are characterised by complex hierarchical porous structures and composed of aligned cells, which allows infiltration of inorganic precursors. Again, a two-step procedure including carbonisation of the millet and siliconisation of carbon was employed. A carbonised preform was covered with silicon powder and heated at 1600 C in argon. The resulting SiC faithfully replicated the microstructure and morphology of the carbon template. The pore characteristics of the SiC ceramics and the biotemplate were fairly similar, that is, pores of 93 and 87 μm, respectively. The specific surface area of the ceramic composite was lower compared to the millet grains.

The great interest in this type of ceramics is exemplified by the number of publications reporting the transformation of different biotemplates into SiC. A list of some recent studies is presented in Table 18.1, where besides the literature references and the template used, the characteristics of the obtained ceramics are provided.

The synthesis of oxides with controlled pore characteristics is of great interest due to their technical importance in different fields such as catalysis, optics, functional coatings, electronics and medicine. Aluminium oxides are among the most widely used materials in this field and the fine control of their textural characteristics is of great technological importance. Consequently, their preparation using biotemplates has gained attention recently.

Table 18.1 Templates employed, specific characteristics and corresponding literature references for macroporous bioinspired SiC ceramics

Template	Specific characteristics	References
Basswood	50 % porosity	70
Pine wood	Macropores (14 μm) + mesopores (50 nm)	71
Bamboo	β-SiC + SiO_2	72
Wood	Microtubes	73
Paper	3D honeycomb structure	74
Pine	Hybrid pore structure; β-SiC + α-SiC	75
Paper	Methyltrichlorosiliane infiltration	76
Beech/Pine wood	SiSiC	77
Pine	Highly porous	78
Oak	β-SiC	79
Ebony, Maple, Oak, Balsa, Pine	Study of the mechanical properties Light weight structural materials	80

Yeast cells were employed in the synthesis of alumina with combined macro-/mesoporosity. The interest in the yeast cell is due to the solid cell walls whose hardness is sufficient to withstand the pretreatment procedure and fulfil the function of a template. In addition, the cells are flexible and soft which is a required feature when a material is subjected to sintering. Exploring the potential of yeast cells, Towata and Sivakumar prepared meso-/macroporous alumina using the sol–gel method.[81,82] Aluminium hydroxide sols were synthesised by hydrolysis of aluminium isopropoxide in an aqueous dispersion of yeast cells. The precursor composite of yeast cells–aluminium hydroxide was recovered by centrifugation. The yeast cells were then removed by heat treatment at temperatures between 500 and 1000 °C, which led also to the formation of γ-Al_2O_3. The obtained material was transformed into α-Al_2O_3 on treatment at 1200 °C. The resulting alumina had a bimodal pore size distribution with distinct macropores of 1.5–2 μm and mesopores ranging between 2 and 20 nm. Some control of the alumina wall thickness could be achieved by varying the yeast/aluminium isopropoxide ratio. This ratio also determined the specific surface area of the material that was between 30 and 118 m^2/g.

Biomorphic alumina fibres were synthesised using cotton as a biotemplate.[83] Dried cotton fibres were immersed in an aluminium chloride solution and desiccated, which resulted in an uniform coating of aluminium hydrous oxide. In a second step, a high-temperature treatment sintered the alumina precursor and removed the cotton template. Thus, hollow alumina fibres were obtained. Depending on the sintering temperature, materials with a specific surface area of 127 m^2/g (800 °C), 125 m^2/g (1000 °C) and 10 m^2/g (1200 °C) were obtained. The corresponding alumina polymorphs were γ-Al_2O_3 (800 and 1000 °C) or α-Al_2O_3 (1200 °C).

A simple and effective method for the preparation of hierarchically ordered porous γ-Al_2O_3 was published by Li and He.[84] Stems of *Pueraria lobata*, which exhibit a hierarchical porous structure, were employed as templates. The plant is native to Eastern Asia and belongs to the bean family. Again aluminium chloride was employed as a precursor. After immersion into the aluminium precursor, the slices of the stem were exposed to NH_3 gas and then calcined to eliminate the organic template and transform the precursor into γ-alumina. The replica retained the macroporous structure of the stem template. In addition, distinct mesopores with a maximum at 6 nm were revealed by BJH pore size analysis. The meso-/macropore organisation and the relatively high specific surface area (175 m^2/g) make the material suitable for catalyst supports. Consequently, Pt nanoparticles were deposited on such a support and annealed at different temperatures. The mean diameter and the standard deviation of the Pt nanocrystals were estimated to be 3.24 and 0.93 nm, respectively. The sharp particle size distribution suggests that the Pt nanoparticles probably occupied the small mesopores in the biomorphic replica.

Shin *et al.* employed pine and poplar wood for the preparation of hierarchical meso-/macroporous silica ceramics.[85] Surfactant-templated sol–gel process was developed to mineralise the ordered cellular structures in the wood tissues. The process allowed to adjust easily the hydrolysis rate by changing the solvent ratio and the acidity of the solution and thus to avoid bulk precipitation or gelation of the silicate species. Consequently, the silicate species penetrated the cell wall structures and hydrolysed and condensed around the cellular tissues. At the same time, the surfactant micellar structures were incorporated into the silica network which produced organised nanoporous channels during the calcination. The mesoporous channels of the biotemplate played a critical role in maintaining the large-scale structural integrity during the thermal decomposition of the wood template because they provided pathways for the removal of the decomposed organics without degradation of the structure. After the combustion of the organic template, the macroporous cellular structure of the wood was preserved. In addition, the surfactant template provided mesopore channels in the silica network. Thus, a meso-/macroporous material with a relatively high surface area (350–650 m^2/g) and mesopore volume (0.12–0.26 cm^3/g) was obtained. One order of magnitude lower values were obtained for the material prepared without surfactant.

Bio-mimetic growth of silica nanotubes in confined media was reported by Gautier *et al.*[86] Multiple silica impregnation at pH 5 of a polymer membrane was used in this preparation. The method mimics silica growth in confined space, which is typical of biological systems. At the end of the process, the polycarbonate matrix was dissolved in an excess of $CHCl_3$. The diameter of the obtained nanotubes closely fitted the membrane dimensions 0.3–0.4 and 0.1–0.2 μm, respectively.

A great concern in the replication of natural specimens is the deterioration of the delicate biological structure. For instance, wet chemistry and high-temperature methods may cause considerable damage to the fragile tissues. Consequently, methods that allow faithful replication without damaging are highly desired. Cook *et al.* reported a mild chemical vapour deposition (CVD) method which allowed the replication of delicate biological structures.[87] The method consists in the controlled vapour-phase oxidation of silanes on the surface of biological structures. Indeed, conventional CVD processes create a stream of oxide particles in gas-phase reactions that cannot uniformly coat 3D objects because of shadowing. Other techniques based on surface reactions at elevated temperatures are also not applicable to delicate biological tissues. The mild CVD method reported by Cook *et al.*[87] operates at room temperature and under a pressure between 1 and 5 kPa, where the silane reacts with an excess of vapourised hydrogen peroxide to deposit silica on almost any material by a surface-phase process. The produced silica clusters have extraordinary flow properties and are capable to penetrate into the smallest gaps within the substrate, which makes the method adequate for replication of intricately biological objects. The method was exemplified by the replication of peacock butterfly wings. CVD silica replica accurately replicated the details of the wings architectures. More important, the process allows the replication of protective surfaces designed by nature to be self-cleaning like for instance the wings of housefly and the ultrahydrophobic leaves of *Colosalia esclulenta*.

Butterfly wings were also chosen as a natural template in the preparation of biomorphic ZnO.[88] A water-free ethanol solution of $Zn(NO_3)_2$ was employed as a precursor. It was transformed into ZnO at 500 °C using very low heating and cooling rates. After this treatment, the samples remained intact. However, about 50% shrinkage was measured after removal of the biotemplate and crystallisation of ZnO. The XRD study revealed that the crystalline material was zincite. The morphology and microstructure features of the ZnO replica are presented in Figure 18.4. Detailed replication of the structure to the micrometre level and the presence of macropores can be seen at high magnification (Fig. 18.4c and d).

Cotton fibres were used in the preparation of biomorphic SnO_2 microtubules.[89] Thus, dried cotton fibres were infiltrated with tin alkoxide solution using a spraying method. After drying, the composite was treated at 600, 700 or 800 °C. As a result of

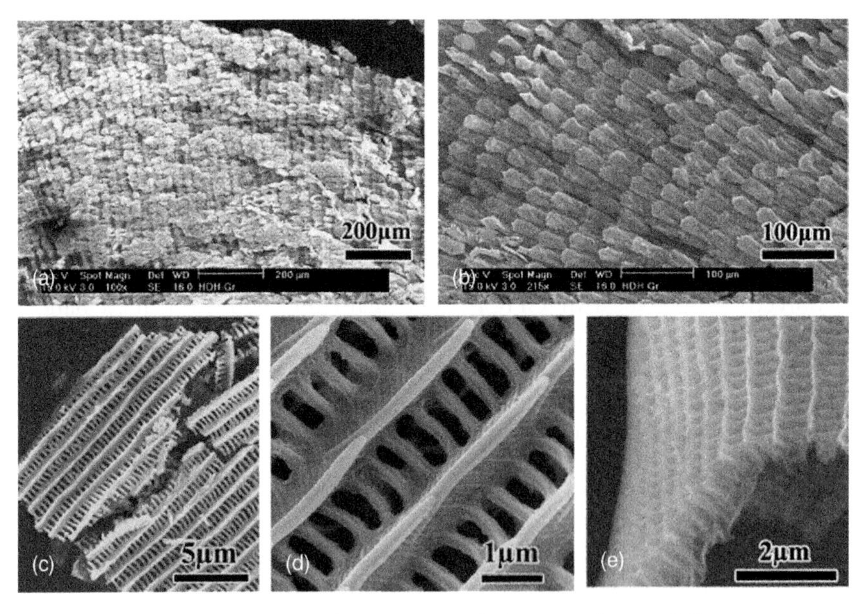

Figure 18.4 SEM micrographs (a–e) of ZnO replica of *Thaumantis diores* at different magnifications. (Reproduced from Ref. 88 with permission from Elsevier.)

this sintering process, the cotton templates were removed and biomorphic tin oxide microtubules were obtained. The diameter of the final SnO_2 microtubules ranged between 5 and 15 μm. A type III adsorption/desorption isotherm was recorded, where the capillary condensation at high relative pressure is characteristic of large meso- and macropores. The materials exhibited a relatively low specific surface area (6–24 m^2/g).

A new process for replication of biomorphic structures was developed by Li *et al.*[90] The novelty concerned the transport of inorganic material source to the biotemplate, namely a supercritical CO_2 transportation of titanium tetrabutyloxide was employed. The method was exemplified by mimicking bamboo inner shell membrane. After incorporation of the titanium precursor, the biotemplate was treated in an autoclave at 40 °C and then transferred into a muffle furnace. High-temperature treatment (500 °C for 5 h) eliminated the bamboo template and transformed the amorphous titania into anatase. The obtained materials showed much better pore characteristics in respect to a reference sample prepared by the sol–gel technique. For instance, the pore volume of the material was 0.179 cm^3/g, while the sample issued from sol–gel preparation showed 0.056 cm^3/g. A specific surface area of 108 m^2/g was measured for the material prepared under supercritical CO_2 material, which is more than two times higher in respect to the reference sample. It is worth mentioning that the macro-scale characteristics of the bamboo membrane were better preserved in the replica obtained by the new technique.

Iron oxide replicas of wood templates were also reporetd.[91] Wood templates were soaked with ethanolic solution of ferric nitrate at 60 °C for 3 days and then sintered at high temperature. The elimination of the wood tissues provided a well-defined macropore structure with pores in the range 10–15 μm. BJH pore size analysis revealed the presence of textural mesopores with sizes between 2 and 50 nm.

Wood templates were also used for the preparation of macroporous Cr_2O_3,[92] MgO,[93] NiO[94] and ytria-stabilised ZrO_2[95] biomorphic ceramics.

Macroporous echinoid (sea urchins) plates with a pore size of the order of 15 μm were used to template porous gold structures.[96] The experiments were carried out with 1-mm-thick clean skeletal plates. The plate was placed in a gold paint and after removing it was heated with hot air gun to burn the protective organic matter of the paint. The cycle was repeated 10 times and then the gold was annealed in a furnace at 400 °C. All gold replica was obtained by dissolution of the calcite support. The sample retained the morphological features of the template, but it was too fragile and difficult to handle without damaging the structure. The method is quite general and the preparation of other metal replicas can be envisaged.

A new possible application of these highly ordered biological structures was demonstrated by Losic *et al.*[97] Diatom exoskeletons were used for fabrication of micro- and nanoscale patterns that retained the morphological features of the template. Polymer replicas were fabricated from three different diatom species demonstrating that this approach is not limited to one particular species or pattern. The positive replica of diatom templates showed a similar pore structure and surface relief as the master pieces. According to the authors, many other solidified structures including gels, precursors to ceramics and carbons, luminescent phosphors, salts and colloids could also be patterned using biological templates.

Again a diatom, but in this case covalently functionalised with DNA, was used as a template for the sequence-specific assembly of DNA-functionalised nanoparticles.[98] Diatoms were first amino-silane functionalised and then coupled to fluorophore-labelled thiolate DNA. The ability of pre-functionalised treated diatoms to assemble nanoparticles was demonstrated by exposing them to 13 nm Au particles. The analysis showed that the particles cover both the interior and the exterior of the cell walls without masking the pore organisation of the template. It was also shown that DNA can program the assembly of multiple layers of nanoparticles onto the template.

After the wood and diatom templates, probably the most largely explored template is the eggshell membrane. The latter consists mainly of proteins such as collagen (types I, V and X), osteopontin and sialoprotein. Consequently, the membrane is stable in water and alcoholic solutions and thus different wet chemistry procedures could be applied in the preparation of inorganic replicas. The interest in eggshell membrane replication is mainly due to its peculiar morphology, that is, a network of fibres. Consequently, the replication of eggshell membranes provides a network of hollow tubes with diameter of several micrometres. Different types of macroporous oxide networks were obtained employing this template, for instance TiO_2,[99–101] SnO_2,[102] ZrO_2[103] and ZnO.[104] Thin films of $CaCO_3$ were grown on eggshell membranes and compared with the ones grown on a synthetic (Nylon 66) membrane.[105] Barium sulphate tubes were also prepared using an eggshell membrane template.[106]

The list of porous structures obtained via duplication or by using given parts of biological specimens could be further extended. However, the presented examples provide a fairly accomplished picture of the present state of the art.

3. Concluding Remarks

Table 18.2 summarises the main classes of biological specimens employed and porous solids obtained using natural templates. Without being excessive, this summary gives a clear idea about the extreme diversity of natural species already used in the synthesis of porous materials. As can be seen, they range from biomolecules, viruses and bacteria to parts of giant trees and from single-cell algae to skeleton of animal and insects.

Wide diversity of biological specimens employed in the synthesis of porous solids resulted in materials with various porous characteristics and morphological features. A great part of synthesised materials exhibit hierarchical porosity due to the multilevel organisation of natural specimen. Most of them possess two types of pores, namely meso- and macropores. Morphological constructions comprising up to three pore size regimes were also obtained. Pore size organisation and specific surface area of porous solids depend not only on the templating specimen, but also on the crystallised inorganic phase. For instance, silica replica of pollen grains showed much higher specific surface area in respect to the calcite counterpart. Zeolite replication of biological species resulted in the formation of materials with hierarchical porous structure possessing well-defined micropores. Actually, microporous

Table 18.2 Examples of biotemplates employed in the preparation of porous materials, characteristics of the obtained porous solids and corresponding literature references

Template				
Class	Species	Chemical composition of porous solid	Porosity type	References
DNA	*Escherichica coli*	SiO_2	Meso-/Macro-	107
Bacteria	*Bacilus subtilis*	SiO_2	Micro-/Meso-	29
Surfactant	Arginine-based	SiO_2	Meso-/Macro-	63
Virus	Tobacco mosaic virus	Metal hybrides (Pd, Pt, Au)	Meso- (nanotubes)	65
Pollen grains	*Taraxacum*, *Trifolium*, *Papaever*, *Brassica* genera	SiO_2, $CaCO_3$, $CaHPO_4$	Meso-/Macro-	64
Grain	Millet	SiC	Meso-/Macro-	69
Cell	Yeast	Al_2O_3	Meso-/Macro-	81
Fibre	Cotton	Al_2O_3	Meso-/Macro-	83
Fibre	Cotton	SnO_2	Macro	89
Plant	*Pueraria lobita*	Al_2O_3	Meso-/Macro-	84
Plant	*Equisetum arvense*	SiO_2	Micro-/Meso-	47
Plant	*Equisetum arvense*	SiO_2-, Al_2O_3	Micro-/Meso-/ Macro-	48
Egg shell membrane	Collagen, Osteropontin, Sialoprotein	TiO_2	Macro-	99
Lepidoptera	Peacock butterfly	SiO_2	Macro-	87
Lepidoptera	*Thaumantis diores*	ZnO	Macro-	88
Wood	Pine, Poplar	SiC	Macro-	67
Wood	Paulownia, Pine, Lauan, Fir	Fe_2O_3	Meso-/Macro-	91
Palm	Rattan	SiO_2	Micro-/Macro	33
Bamboo	Inner shell membrane	TiO_2	Macro-	90
Skeleton	Echinoid	Au	Macro-	96
Diatoms	*Coscinodiscus* sp.	Polymers	Macro-	97

materials obtained by natural templating are not reported until now. Hence, the incorporation of zeolites in morphological constructions issued from natural templates is the unique way to obtain a hierarchical porous structure exhibiting distinct micropores.

Chemical compositions of bioinspired porous solids also vary in a very wide range as the oxides are largely presented. Among them, silica is probably the most broadly explored in the replication of different specimens due to its technological importance. The preparation of aluminium, titanium, iron, tin, zinc and other metal oxides replicas was also reported. Further, calcium carbonate and phosphate porous structures were obtained using the templating approach. Following the silica, probably the SiC and SiSiC ceramics issued from wood templates were the most widely studied. Noble metals, for instance Au, Pt, Pd, have been employed in the preparation of meso- and macroporous constructs. Finally, carbon and polymer replicas of biotemplates were obtained. Combinations of the above substances were also employed in the replication of biotemplates.

Large diversity of morphological constructions with hierarchical porous organisation and varieties of compositions offer enormous possibilities for bioinspired materials. Besides the complex morphology and multi-level organisation, which at present cannot be reproduced by synthetic templates, the natural specimens are environmentally benign, inexpensive and renewable. Nevertheless, industrial application of biotemplated porous materials is not envisaged at present. Probably, the main obstacle to the application of this material is their fragility. There are a few examples of materials issued from biotemplates that exhibit a reasonable mechanical strength. Most of the materials disintegrate easily and even simple laboratory operations are difficult to be performed. Another drawback is the difficulty to achieve standardisation for natural materials. In other words, depending of the conditions (climate, region and so on), substantial changes in the structure of a biological specimen may take place. Consequently, porous inorganic counterparts of biological templates from different origin may differ in their characteristics.

Although, there is a long way to the use of inorganic porous material prepared via utilisation of sacrificial natural templates, the interest in such materials is extremely high. The latter is proved by the constantly increasing publication activity in the field. Thus, it is not risky to predict that the interest in the generation of structures with complexity and functionality close to biological specimens will not decline. Hence, one may expect substantial advances and new exciting materials to be available in near future.

ACKNOWLEDGEMENT

The authors thank Dr. Henri Kessler for helpful discussion.

REFERENCES

[1] Mann, S. *Biomineralization, the Inorganic-Organic Interface, and Crystal Engineering, in Biomimetics: Design and Processing of Materials*, Sarikaya, M., Aksay, I. A. (Eds.), AIP Press, New York, **1995**.
[2] Mann, S., *Biomineralization: Principle and Concepts in Bioinorganic Materials Chemistry*. Oxford University Press, Oxford, **2001**.
[3] Dujardin, E., Mann, S., *Adv. Mater.* **2002,** *14,* 775–788.
[4] Niemeyer, C. M., *Angew. Chem. Int. Ed.* **2001,** *40,* 4128–4158.
[5] Zampieri, A., Schwieger, W., Zollfrank, C., Greil, P., in: *Handbook of Biomineralization: Biomimetic and Bioinspired Chemistry,* Wiley-VCH Verleg, Weinheim, Germany, **2007**, pp. 255–288.

[6] Takai, O., *Annals of the New York Academy of Sciences,* **2000,** *1093* (Progress in Convergence), 84–97.
[7] Weber, A., Gruber-Traub, C., Herold, M., Borchers, K., Tovar, G. E. M., *NanoS* **2006,** *2,* 20–27.
[8] Sieber, H., *Mater. Sci. Eng. A* **2005,** *412,* 43–47.
[9] Fratzl, P., *J. R. Soc. Interface* **2007,** *4,* 637–642.
[10] Chen, R., Hunt, J. A., *J. Mater. Chem.* **2007,** *17,* 3974–3979.
[11] Wang, H., Wang, Z., Huang, L., Mitra, A., Holmberg, B., Yan, Y., *J. Mater. Chem.* **2001,** *11,* 2307–2311.
[12] Tosheva, L., Valtchev, V., Sterte, J., *Microporous Mesoporous Mater.* **2000,** *35–36,* 621–627.
[13] Liu, Y., Pinnavaia, T. J., *Chem. Mater.* **2002,** *14,* 3–6.
[14] Velev, O. D., Tessier, P. M., Lenhoff, A. M., Kaler, E. W., *Nature* **1999,** *401,* 548–551.
[15] Lebeau, B., Fowler, C. E., Mann, S., Farcet, C., Charleux, B., Sanchez, C., *J. Mater. Chem.* **2000,** *10,* 2105–2109.
[16] Yu, C., Tian, B., Fan, J., Stucky, G. D., Zhao, D., *Chem. Lett.* **2002,** *31,* 62–64.
[17] Lee, Y.-J., Lee, J. S., Park, Y. S., Yoon, K. B., *Adv. Mater.* **2001,** *13,* 1259–1264.
[18] Holland, B. T., Abrams, L., Stein, A., *J. Am. Chem. Soc.* **1999,** *121,* 4308–4311.
[19] Valtchev, V., *J. Mater. Chem.* **2002,** *12,* 1914–1918.
[20] Wang, Y. J., Tang, Y., Ni, Z., Hua, W. M., Yang, W. L., Wang, X. D., Tao, W. C., Gao, Z., *Chem. Lett.* **2000,** *29,* 510–512.
[21] Zhu, G., Qin, S., Gao, F., Li, D., Li, Y., Wang, R., Gao, B., Li, B., Guo, Y., Xu, R., Liu, Z., Terasaki, O., *J. Mater. Chem.* **2001,** *11,* 1687.
[22] Rhodes, K. H., Davis, S. A., Caruso, F., Zhang, B., Mann, S., *Chem. Mater.* **2000,** *12,* 2832.
[23] Szostak, R., *Molecular Sieves,* 2nd edition, Blackie Academic & Professional, London, **1998**.
[24] Vartuli, J. C., Roth, W. J., Beck, J. S., McCullen, S. B., Kresge, C. T., in: Karge, G., Weitcamp, J. (Eds.), *Molecular Sieves: Science and Technology*, Vol. I, Springer, Berlin, **1998**, pp. 97–119.
[25] Lee, J., Kim, J., Kim, J., Jia, H., Kim, M. I., Kwak, J. H., Jin, S., Dohnalkova, A., Park, H. G., Chang, H. N., Wang, P., Grate, J. W., Hyeon, T., *Small* **2005,** *1,* 744–753.
[26] Velev, O. D., Keler, E. W., *Adv. Mater.* **2000,** *12,* 531–536.
[27] Stein, A., *Microporous Mesoporous Mater.* **2001,** *44–45,* 227–234.
[28] Iler, R. K., in: *The Chemistry of Silica,* John Wiley & Sons, New York, **1979**.
[29] Zhang, B., Davis, S. A., Mendelson, N. H., Mann, S., *Chem. Commun.* **2000,** 781–782.
[30] Zhang, B., Davis, S. A., Mann, S., *Chem. Mater.* **2002,** *14,* 1369–1375.
[31] Wang, Y., Tang, Y., Dong, A., Wang, X., Ren, N., Shan, W., Gao, Z., *Adv. Mater.* **2002,** *14,* 994–997.
[32] Dong, A., Wang, Y., Tang, Y., Ren, N., Zhang, Y., Yue, Y., Gao, Z., *Adv. Mater.* **2002,** *14,* 926–929.
[33] Zampierri, A., Sieber, H., Selvam, T., Mabande, G. T. P., Schwieger, W., Greil, P., *Adv. Mater.* **2005,** *17,* 344–348.
[34] Zampierri, A., Kullmann, S., Selvam, T., Bauer, J., Schwieger, W., Sieber, H., Fey, T., Greil, P., *Micropor. Mesopor. Mater.* **2006,** *90,* 162–174.
[35] Zampierri, A., Mabande, G. T. P. S., Selvam, T., Schwieger, W., Rudolph, A., Hermann, R., Sieber, H., Fey, T., Greil, P., *Mater. Sci. Eng. C* **2006,** *26,* 130–135.
[36] Ghost, B., Agrawal, D. C., Bhatia, S., *Ind. Eng. Chem. Res.* **1994,** *33,* 2107–2110.
[37] Inui, S., Kimura, S., Nonaka, S., *Jap. Pat.* 04187515, **1992**.
[38] Anderson, M. A., Holmes, S. M., Hanif, N., Cundy, C. S., *Angew. Chem. Int. Ed.* **2000,** *39,* 2707–2710.
[39] Holmes, S. M., Plaisted, R. J., Crow, P., Foran, P., Cundy, C. S., Anderson, M. W., *Stud. Surf. Sci. Catal.* **2001,** *135,* 3340–3340.
[40] Wang, Y., Tang, Y., Dong, A., Wang, X., Ren, N., Gao, Z., *J. Mater. Chem.* **2002,** *12,* 1812–1818.
[41] Wang, Y. J., Tang, Y., Wang, X. D., Dong, A. G., Shan, W., Gao, Z., *Chem. Lett.* **2001,** 1118–1119.
[42] Valtchev, V., *unpublished data.*

[43] Sugawara, T., Saruta, S., Sugawara, K., *Zeoraito* **1994,** *11,* 73–79.
[44] Iler, R. K., in: *The Chemistry of Silica,* John Wiley & Sons, New York, **1979,** p. 88.
[45] Berterma, R., Tacke, R., *Zeitschrift fuer Naturforschung, B* **2000,** *55,* 459–466.
[46] Kaufman, P. B., Wilber, W. C., Schmid, R., Najati, N. S., *Am. J. Bot.* **1971,** *58,* 309–316.
[47] Valtchev, V., Smaihi, M., Faust, A.-C., Vidal, L., *Angew. Chem. Int. Ed.* **2003,** *42,* 2783–2785.
[48] Valtchev, V., Smaihi, M., Faust, A.-F., Vidal, L., *Stud. Surf. Sci. Catal.* **2004,** *154A,* 588–592.
[49] Valtchev, V., Smaihi, M., Faust, A.-F., Vidal, L., *Chem. Mater.* **2004,** *16,* 1350–1355.
[50] Kresge, C. T., Leonowicz, M. E., Roth, W. J., Varuli, J. C., Beck, J. S., *Nature* **1992,** *359,* 710–712.
[51] Sayari, A., Hamoudi, S., *Chem. Mater.* **2001,** *13,* 3151–3168.
[52] John, V. T., Simmons, B., McPherson, G. L., Bose, A., *Curr. Opin. Colloid Interface Sci.* **2002,** *7,* 288–295.
[53] Vrieling, E. G., Beelen, T. P. M., Van Santen, R. A., Gieskes, W. W. C., *J. Phycol.* **2000,** *35,* 1044–1053.
[54] Cha, J. N., Stucky, G. D., Morse, D. E., Deming, T. J., *Nature* **2000,** *403,* 289–292.
[55] Volkmer, D., Tuguli, S., Fricke, M., Nielsen, T., *Angew. Chem. Int. Ed.* **2003,** *42,* 56–61.
[56] Vrieling, E. G., Beelen, T. P. M., Van Santen, R. A., Gieskes, W. W. C., *J. Biotechnol.* **1999,** *70,* 41–53.
[57] Kröger, N., Deutzmann, R., Sumper, M., *Science* **1999,** *286,* 1129–1132.
[58] Kröger, N., Deutzmann, R., Sumper, M., *J. Biol. Chem.* **2001,** *276,* 26066–26070.
[59] Noll, F., Sumper, M., Hampp, N., *NanoLett* **2002,** *2,* 91–95.
[60] Crawford, S. A., Higgins, M. J., Mulvaney, P., Wetherbee, R., *J. Phycol.* **2001,** *37,* 543–554.
[61] Coradin, T., Livage, J., *Colloids Surf. B: Biointerfaces* **2001,** *21,* 329–336.
[62] Coradin, T., Durupthy, O., Livage, J., *Langmuir* **2002,** *18,* 2331–2336.
[63] Coradin, T., Roux, C., Livage, J., *J. Mater. Chem.* **2002,** *12,* 1242–1244.
[64] Hall, S. R., Bolger, H., Mann, S., *Chem. Commun.* **2003,** 2784–2785.
[65] Knez, M., Sumser, M., Bittner, A. M., Wege, C., Jeske, H., Martin, T. P., Kern, K., *Adv. Func. Mater.* **2004,** *14,* 116–124.
[66] Sieber, H., Hoffmann, C., Kaindl, A., Greil, P., *Adv. Eng. Mater.* **2000,** *2,* 105–109.
[67] Shin, Y., Wang, C., Exarhos, G. J., *Adv. Mater.* **2005,** *17,* 73–76.
[68] Huo, G., Jin, Z., Qian, J., *J. Mater. Proc. Techn.* **2007,** *182,* 34–48.
[69] Wang, Q., Jin, G.-Q., Wang, D.-H., Guo, X.-Y., *Mater. Sci. Eng. A* **2007,** *459,* 1–6.
[70] Qian, J.-M., Wang, J.-P., Jin, Z.-H., *Mater. Chem. Phys.* **2003,** *82,* 648–653.
[71] Vyshnyakova, K., Yushin, G., Pereselentseva, L., Gigotsi, Y., *Appl. Ceram. Tech.* **2006,** *3,* 485–490.
[72] Cheung, T. L. Y., Ng, D. H. L., *J. Am. Ceram. Soc.* **2007,** *90,* 559–564.
[73] Kim, J.-W., Myong, S.-W., Kim, H.-C., Lee, J.-H., Jung, Y.-G., Jo, Ch.-Y., *Mater. Sci. Eng. A* **2006,** *A434,* 171–177.
[74] Streitwieser, D. A., Popovska, N., Gerhard, H., *J. Eur. Ceram. Soc.* **2006,** *26,* 2381–2387.
[75] Qian, J.-M., Jin, Z.-H., *J. Eur. Ceram. Soc.* **2006,** *26,* 1311–1316.
[76] Streitwieser, D. A., Popovska, N., Gerhard, H., Emig, G., *J. Eur. Ceram. Soc.* **2005,** *25,* 817–828.
[77] Zollfrank, C., Sieber, H., *J. Eur. Ceram. Soc.* **2004,** *24,* 495–506.
[78] Sieber, H., Vogli, E., Muller, F., Greil, P., Popovska, N., Gerhard, H., *Key Eng. Mater.* **2002,** *206–213,* 2013–2016.
[79] Vogli, E., Mukerji, J., Hoffman, C., Kladny, R., Sieber, H., Greil, P., *J. Am. Ceram. Soc.* **2001,** *84,* 1236–1240.
[80] Greil, P., Lifka, T., Kaindl, A., *J. Eur. Ceram. Soc.* **1998,** *18,* 1975–1983.
[81] Towata, A., Sivakumar, M., Yasui, K., Tuziuti, T., Kozuka, T., Ohta, K., Iida, Y., *J. Surf. Sci. Nanotech.* **2005,** *3,* 405–411.
[82] Towata, A., Sivakumar, M., Yasui, K., Tuziuti, T., Kozuka, T., Iida, Y., *J. Ceram. Soc. Jpn.* **2005,** *113,* 4696–4699.
[83] Fan, T., Sun, B., Gu, J., Zhang, D., Lau, L. W. M., *Scripta Mater.* **2005,** *53,* 893–897.
[84] Li, C., He, J., *Langmuir* **2006,** *22,* 2827–2831.
[85] Shin, Y., Liu, J., Chang, J. H., Nie, Z., Exarhos, G. J., *Adv. Mater.* **2001,** *13,* 728–732.

[86] Gautier, C., Lopez, P. J., Hemadi, M., Livage, J., Coradin, T., *Langmuir* **2006,** *22,* 9092–9095.
[87] Cook, G., Timms, P. L., Göltner-Spickermann, C., *Angew. Chem. Int. Ed.* **2003,** *42,* 557–559.
[88] Zhang, W., Zhang, D., Fan, T., Ding, J., Guo, Q., Ogawa, H., *Micropor. Mesopor. Mater.* **2006,** *92,* 227–233.
[89] Sun, B., Fan, T., Xu, J., Zhang, D., *Mater. Lett.* **2005,** *59,* 2325–2328.
[90] Li, J., Shi, X., Wang, L., Liu, F., *J. Colloid Interface Sci.* **2007,** *315,* 230–236.
[91] Liu, Z., Fan, T., Zhang, W., Zhang, D., *Micropor. Mesopor. Mater.* **2005,** *85,* 82–88.
[92] Fan, T., Li, X., Liu, Z., Gu, J., Zhang, D., *J. Am. Ceram. Soc.* **2006,** *89,* 3511–3515.
[93] Li, X., Fan, T., Liu, Z., Ding, J., Guo, Q., Zhang, D., *J. Eur. Ceram. Soc.* **2006,** *26,* 3657–3664.
[94] Liu, Z., Fan, T., Zhang, D., *J. Am. Ceram. Soc.* **2006,** *89,* 662–665.
[95] Sieber, H., Ramboo, C., Cao, J., Vogli, E., Greil, P., *Key Eng. Mater.* **2002,** *206–213,* 2009–2012.
[96] Meldrum, F. C., Seshadri, R., *Chem. Commun.* **2000,** 29–30.
[97] Losic, D., Mitchell, J. G., Lal, R., Voelcker, N. H., *Adv. Func. Mater.* **2007,** *17,* 2439–2446.
[98] Rosi, N. L., Thaxton, C. S., Mirkin, C. A., *Angew. Chem. Int. Ed.* **2004,** *43,* 5500–5503.
[99] Yang, D., Qi, L., Ma, J., *Adv. Mater.* **2002,** *14,* 1543–1546.
[100] Dong, Q., Su, H., Zhang, D., liu, Z., Lai, Y., *Micropor. Mesopor. Mater.* **2007,** *98,* 344–351.
[101] Dong, Q., Su, H., Cao, W., Zhang, D., Guo, Q., Lai, Y., *J. Solid State Chem.* **2007,** *180,* 949–955.
[102] Dong, Q., Su, H., Zhang, D., Zhu, N., Guo, X., *Scripta Mater.* **2006,** *55,* 799–802.
[103] Yang, D., Qi, L., Ma, J., *J. Mater. Chem.* **2003,** *13,* 1119–1123.
[104] Dong, Q., Su, H., Xu, J., Zhang, D., Wang, R., *Mater. Lett.* **2007,** *61,* 2714–2717.
[105] Ajikumar, P. K., Lakshminarayanan, R., Valiyaveettil, S., *Cryst. Growth Design* **2004,** *4,* 331–335.
[106] Liu, J.-K., Wu, Q.-S., Ding, Y.-P., Wang, S.-Y., *J. Mater. Res.* **2004,** *19,* 2803–2806.
[107] Numata, M., Sugiyasu, K., Hasegawa, T., Shinkai, S., *Angew. Chem. Int. Ed.* **2004,** *43,* 3729–3783.

CHAPTER 19

Strategies Towards the Assembly of Preformed Zeolite Crystals into Supported Layers

Lubomira Tosheva *and* Aidan M. Doyle

Contents

Abstract

This chapter reviews the methods used for the assembly of preformed zeolite crystals onto various supports. Methods for the arrangement of both colloidal and micron-sized crystals of different morphologies are considered. The supports used for zeolite crystal deposition can be two- or three-dimensional and the applicability of the different methods to supports of various shapes is discussed. The major application of the seeded supports is for the preparation of dense zeolite films by secondary growth. The influence of the seeded layers on the properties of the secondary grown films is presented. Other potential applications of zeolite seed layers are also mentioned.

Keywords: Assembly, Methods, Seeded supports, Zeolite crystals

1. Introduction

The assembly of preformed zeolite crystals into supported layers is a very important research topic in modern zeolite science. The progress in this area is a result of the tremendous development of the synthesis of zeolite materials in non-conventional forms. Typically, synthetic zeolites are produced as a crystalline powder of micron-sized crystallites by hydrothermal treatment of synthesis gels or solutions. In many applications, the performance of the zeolite is limited by its powder form. For instance in catalysis, slow diffusion of reactants and products in the

Ordered Porous Solids
DOI: 10.1016/B978-0-444-53189-6.00019-6

zeolite micropores and pore blockage caused by the formation of heavy components within the micropores are serious limitations toward the efficient use of zeolites. Another disadvantage of zeolite powders is that they are difficult to handle, causing loss of material during process set-up or zeolite regeneration. One possibility to optimize the technological use of zeolites is by the formation of zeolite macro particles such as extrudates, tablets, and spheres, meeting the process requirements for activity, pressure drop, and attrition resistance. However, zeolite macroparticles are prepared by the addition of amorphous non-zeolitic binders, which affect the zeolite performance. This has initiated research into the preparation of alternative zeolite materials to fully exploit the extraordinary zeolite properties. Another catalyst for this research is the technological development and increased awareness of environmental issues generating a constant search for new materials to replace existing technological solutions. The potential of zeolites prepared in the form of tailored all-zeolite or composite structures has been realized for many advanced applications. The research can be divided into three areas: (i) deposition of thin zeolite layers onto various types of two-dimensional (2D) supports, (ii) preparation of all-zeolite or composite macroscopic structures using 3D supports, and (iii) synthesis of mesoporous zeolites or zeolite/mesoporous material composites. The third class is not relevant to the present topic and will not be discussed here. 2D supports are considered here to be dense flat or curved supports (Fig. 19.1 (a) and (b)). The zeolite crystal layers on such supports are deposited onto the support surface. Such supports are discussed in the present book in Chapter 12 Rebrov *et al.* The 3D supports contain large pores and voids, and the zeolite crystals are assembled not only on the external support surface but also within the bulk of the supports (Fig. 19.1(c)). In some cases, the seeded 3D supports can then be transferred into all-zeolite macroscopic structures.

Zeolite crystallization is a complex process depending on many parameters such as temperature, pH, composition of the reaction mixture, nature of reactants, and pre-treatment.[1] In this chapter, preformed crystals are considered and thus zeolite synthesis protocols will be discussed only when relevant to the crystal deposition methods. Generally, zeolite crystals can be micron-sized (conventional) or nano-sized (colloidal). The micron-sized crystals have well-developed crystal faces with a morphology depending on the type of zeolite structure and the conditions of synthesis. Colloidal zeolites often have spherical morphology and sizes $\leq$ 100 nm.

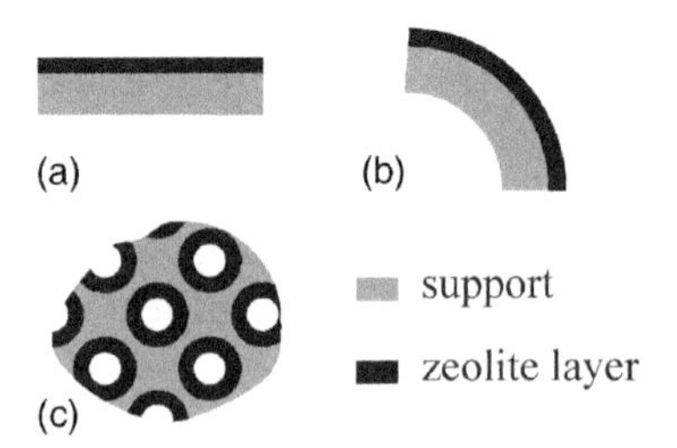

Figure 19.1 Schematic illustration of layers of preformed zeolite crystals assembled on (a) flat, (b) curved supports, and (c) within the bulk structure of porous substrates.

Zeolites in the form of stable colloidal suspensions (the dispersed crystals do not settle under gravity with time) have been developed to broaden the zeolite post-synthesis manipulation tools using the knowledge from colloid chemistry.[2] There has been an intense research effort in developing and manipulating colloidal zeolites.[3] At present, there are several established deposition methods, some of which are only applicable to colloidal zeolites. However, only a limited number of zeolite structures have been prepared in the form of stable colloidal suspensions. Thus, new developments in the synthesis and manipulation of colloidal zeolites have somewhat slowed down in recent years. At the same time, the number of reports dealing with the assembly of micron-sized crystals has increased owing to the greater possibilities in synthesis and flexibility in tuning the crystal properties. In addition to crystal size and morphology, the other important factors in the crystal assembly process are chemical composition of crystals and surface selectivity (hydrophobicity/hydrophilicity), the crystal phase purity, and degree of crystallinity. Another very important parameter is the quality of the zeolite crystals. Best results are achieved with crystals of uniform morphology and narrow crystal size distribution. Other factors specific for the particular methods used can influence the properties of the assembled layers. In the following sections, the most popular methods used for the deposition of preformed zeolite crystals will be described.

2. Methods for the Assembly of Preformed Zeolite Crystals

2.1. Single-step methods

The methods considered in this section are simple and do not involve any substrate modification (other than cleaning) or chemical modification of the zeolite crystals. Generally, the zeolite crystals deposited by such methods are randomly oriented. Also, sometimes several deposition cycles are needed to achieve a high density of seeded layers. Another drawback is the inferior adherence of the zeolite crystals to the support. The advantage of the single-step methods is the simplicity in terms of method protocols and additional chemicals involved.

The simplest method for attaching zeolite seeds is by rubbing the substrate with zeolite water slurries or dried zeolite powders. Rubbing has been used to prepare seeded porous supports such as mullite,[4,5] and α-Al_2O_3 tubular supports,[6] or disks.[7] The zeolites deposited in these examples were Silicalite-1, ZSM-5, T and Y with the size of the zeolite aggregates up to 4 μm. However, the rubbing procedure has been mentioned in the papers without detailed experimental description or comments on reproducibility.

2.1.1. Dip-coating, slip-casting, and washcoating

These three methods generally consist of immersion of the support into zeolite slurry, or sol, followed by optional blowing of the excess solvent, and finally drying.[8] More specifically, in the dip-coating process, the substrate is immersed and withdrawn from the slurry at a constant speed. In the slip-casting process, the substrate and the slurry are kept in contact for a certain period of time and the excess solvent is then removed by blowing. The three methods are not clearly differentiated in zeolite papers, especially when the seed deposition method is not the primary subject of the research. For instance, often in an article's experimental section, the method is described as dip-coating but no information about immersion or withdrawal speeds are given. The supports used can be non-porous, and flat (glass, Si wafers) or ceramic honeycomb monolith (cordierite) or porous (α-Al_2O_3 tubes or disks). In the latter case, the size of the zeolite crystals used should be similar or larger to the pores of the support to avoid plugging of the support pores.

The support coverage, zeolite load, zeolite layer thickness, and homogeneity of the zeolite crystals deposited by immersion methods have been studied for different systems. Beers *et al.* have studied the influence of the amount of zeolite in the zeolite slurry, the presence of binders and surfactants, the solvent and the substrate properties on the preparation of BEA-coated structured supports.[9] The supports used were macroporous monoliths (silica and cordierite) and wire gauze packing, and the zeolite slurry consisted of zeolite Beta (5–20 μm) in water or butyl acetate with a concentration up to 25 wt.%. The as-made coated supports were tested directly afterwards as catalysts and the results showed that the most important characteristic of the coatings was the zeolite loading. The use of the organic solvent resulted in relatively low loadings owing to high viscosity. Also, the addition of a surfactant to the slurry did not affect the loading substantially. Colloidal silica (1 wt.%) was added as a binder to the water slurry so as to improve the adherence of the crystals to the support. The role of the binder is to fill the gaps between the zeolite crystals and to increase the layer stability after calcination via bond formation. Increasing the amount of binder above 5 wt.% was found to deteriorate the catalytic activity of the zeolite coated monoliths. Different factors influencing the zeolite loading of washcoated cordierite honeycomb reactors have also been studied.[10,11] Zeolites ZSM-5, mordenite, and ferrierite of broad particle size distribution in the range up to 20 μm for the first two zeolites and up to 100 μm for the third zeolite were used.[10] The thickness and homogeneity of the coated layers were dependent on the zeolite concentration and correspondingly the relative viscosity of the slurry, the number of immersion steps, and the flow rate during blowing of the excess solvent. The use of more diluted solutions (20–30%) and increasing the number of immersions improved the zeolite layer homogeneity. Stability of the coatings was tested by ultrasonic treatment in acetone for 1 h and calculating the material weight loss. The stability was dependent on the type of zeolite and the size of zeolite aggregates: smaller aggregates yielded more stable coatings. The adherence could also be improved by the addition of 3 wt.% colloidal SiO_2 binder. The authors have also studied the influence of the solvents for the preparation of ZSM-5 washcoated cordierite.[11] Water, different aliphatic alcohols, ethyl acetate, and acetone were used as solvents and best results were obtained from water slurries.

Slip-casting has been used to prepare Silicalite-1 layers on porous α-Al_2O_3 tubes or disks.[12–15] The influence of the seed size has been studied using Silicalite-1 crystals with sizes in the range of 100 nm–7.5 μm and with spherical to coffin-shape morphology.[12] Continuous seed layers were formed at a casting time of 8 s using 100 nm, 600 nm, and 1.5 μm crystals, whereas those prepared with 3.0 and 7.5 μm crystals contained large voids. Wong *et al.* have shown that the thickness of the slip-cast layer prepared from 150 nm Silicalite-1 suspensions can be adjusted by the solid content and the slip-casting time.[13] The slip-casting procedure has also been used to prepare coatings inside tubular supports.[14] The bottom end of the tubes was tapped, the tubes were filled with the Silicalite-1 colloidal suspension (0.65 wt. %, 50–60 nm), and kept in contact for 3 h. In some cases, the dipping–drying–calcination cycle has been repeated several times to optimize the properties of the seeded layer.[15] The influence of the pH of the colloidal zeolite suspension on the density and homogeneity of dip-coated zeolite Y films has been studied as well.[16] Best results were obtained at pH 11.5 and explained by the z-potential measurements data; at that pH the z-potential was the lowest (−63 mV, maximum colloidal stability). Densely packed homogeneous layers were obtained after only one dip-coating cycle. Another parameter shown to influence the crystal density of dip-coated films is the tilt angle.[17] The amount of seed crystals (zeolite A, 10 μm) increased with tilting of the substrate to 45 °. However, only one side of the substrate could be coated on the tilted substrate requiring the repetition of the procedure to coat the other side when necessary. The zeolite A layers were oriented, which is not surprising considering the cubic morphology of the crystals. Oriented zeolite A layers by dip-coating have also been obtained by Tsapatsis and co-workers on glass substrates.[18,19] Colloidal zeolite A suspensions with an average particle size of 100–300 nm were used and the dip-coating cycle was repeated three times to achieve high coverage. Limited areas of the substrate were covered with ordered zeolite arrays at low concentrations of the dipping solution by convective flow mechanism.[19] Recently, the same group has reported the preparation of colloidal crystal layers of 150–200 nm ZSM-2 anisometric particles (hexagonal plates) by convective assembly.[20] The method consists of dip-coating with controlled evaporation. Continuous or striped films were obtained depending on the concentration of the suspensions and the substrate angle.

2.1.2. Spin-coating

Spin-coating is widely used for the preparation of thin films on flat substrates. This method was first applied using colloidal Silicalite-1 suspensions to obtain spin-coated films on Si wafers.[21,22] The Silicalite-1 crystals were dispersed in ethanol prior to spin-coating. The thickness of the films was controlled by the concentration of the zeolite suspension and the number of spin-on cycles. The spin-coated films were smooth and of multilayer character with preferred orientation when crystals of non-spherical morphology were used. Spin-coated films of other zeolites such as Beta,[23] $AlPO_4$-18,[24] and Silicalite-2[25] have also been obtained from corresponding colloidal suspensions.

The acceleration rate, spinning rate, and time can influence the film properties of spin-coated films. However, the influence of these parameters has not been studied by zeolite researchers. Spinning rates of 1500–3000 rpm and spinning times of 15–60 s have been used in the earlier examples. Another important factor is the solvent type. The influence of the dispersion medium on the properties of spin-coated Silicalite-1 films has been studied using different aliphatic alcohols.[26] Methanol and

2-propanol were found to be superior for the ordering and density of the spin-coated films compared with ethanol.

The stability of the spin-coated zeolite crystals by ultrasonic tests has not been studied. The adherence of spin-on films has been evaluated using laser spallation, and spin-coated films have shown inferior adhesion compared with zeolite films prepared by *in situ* or seeded growth.[27] Thus, spin-coating is preferred for the assembly of colloidal zeolite crystals onto flat supports owing to its simplicity and time needed for the whole process. However, the stability and density of as-made spin-coated films are not always suitable for direct application and further processing is necessary.

2.1.3. Zeolite crystals assembly driven by external forces

Holmes *et al.* have used ultrasonic treatment to assemble 100 nm Silicalite-1 crystals on stainless steel mesh supports.[28] The authors treated the supports with a 6 wt.% aqueous colloidal suspension of Silicalite-1 for 6 h, during which the temperature rose to ~70 °C. The ultrasonic treatment was found to be essential for the preparation of stable coated layers. The coatings could not be washed off by simple immersion in distilled water, which was not the case for coatings prepared under similar conditions with agitation. The same group later used sonication to seed diatoms.[29] Ultrasonic treatment during seed deposition has also been used by Lai and Gavalas to prepare seeded porous α-Al_2O_3 plates and tubes.[30] The authors, however, did not comment on the exact role of the ultrasonic treatment. The method was used for 0.4 and 2 μm colloidal Silicalite-1 suspensions and the optimum pH of the suspensions was about 8. The ultrasonic aided zeolite crystal deposition is a simple method, but can be time-consuming with poor reproducibility of the experimental conditions (temperature rise control, ultrasonic bath models, etc.). Also, because this method is used in only a limited number of works, no conclusions about the applicability of the method for different zeolite/support systems can be drawn.

A method developed for the deposition of preformed zeolite crystals on porous tubular supports is the vacuum seeding method.[31,32] In this method, the bottom end of a tube immersed in the zeolite suspension is sealed and the upper end is connected to a water pump. The zeolite crystals are assembled onto the support surface, owing to the pressure difference between the two sides of the supports upon opening the valve to the vacuum. Homogeneous seeded layers can be fabricated by the method from colloidal zeolite suspensions, which do not have a concentration gradient caused by particle sedimentation. Haung *et al.* have used three zeolite A colloidal zeolite suspensions with broad particle size distributions in the range of 300–3000 nm to seed α-Al_2O_3 tubular supports.[31] The influence of the seed particle size, suspension concentration, pressure difference, and coating times were studied and it was possible to prepare uniform, compact, and continuous seeding layers by adjusting these parameters. The size of the zeolite crystals and in particular the size relative to the size of the support pores is a very important characteristic of the vacuum seeding method. For instance, the use of 30–50 nm sodalite crystals resulted in the deposition of the seeds not only onto the support surface but also within the support pore structure.[32]

Charged colloidal zeolite particles have been assembled onto different supports by electrophoretic deposition.[33–35] Generally, the support in this method acts as one of the electrodes (most often as the anode) and the zeolite crystals are deposited on the support under an applied electrical potential. The supports used were carbon fibers,[33] carbon discs,[34] and stainless steel.[35] The coverage and thickness of the seeded layers was dependent on the applied voltage, the zeolite suspension concentration, the zeolite type, the dispersion medium, and the time of deposition. Organic solvents such as acetyl-acetone were found to be more efficient than water.[35] The thickness of the seeded layers could be increased by increasing the voltage, deposition time, and the concentration of the zeolite suspension. The method was applied for different type zeolites such as Silicalite-1, ZSM-5, Beta, sodalite, L and A. Drawbacks of the method are the limited number of supports that can be used for crystal deposition (only supports that can act as electrodes) and the small-scale coating areas.

Another method based on applied electric field has been reported for the preparation of preferentially oriented films of $AlPO_4$-5.[36] The method is based on the fact that anisotropically shaped particles such as fibres can be oriented in an electric field with the longest axis parallel to the direction of the field. A colloidal suspension of $AlPO_4$-5 (1 μm width×40–80 μm length) in fluorocarbon oil was loaded onto a glass slide and an electric field was applied. The organized fibres sedimented onto the glass and excess solvent were removed. This procedure was repeated several times to achieve high coverage.

Another method for the deposition of preformed zeolite crystals that can be mentioned in this sub-section is the laser ablation method.[37–39] The method has been applied for both flat and spherical substrates for zeolites of different composition and structure. The zeolite powders were pressed into free-standing 2.5 cm diameter targets for laser ablation experiments. This is a rather sophisticated method for zeolite crystal assembly, which limits its wide application. Also, the deposited films were amorphous according to X-ray diffraction measurements. Nevertheless, well-adhered and oriented zeolite films could be prepared after secondary growth of the coated substrates.

2.1.4. Self-assembly and Langmuir–Blodgett zeolite films

The spontaneous adsorption of nanoparticles onto a substrate can be used for the fabrication of self-assembled films. Cho *et al.* have prepared TS-1 films on Si wafers using this approach.[40] The method consisted of a simple addition of hexanoic acid to the TS-1 aqueous colloidal suspension (75 nm particles) at pH under 4. The hexanoic acid was adsorbed on the surface of the zeolite particles, as well as on the support surface, and a self-assembled film formed as a result of the particle coagulation on the substrate. The reaction time was between 6 and 48 h at 25–70 °C and the resultant films were uniform and comparatively dense. Yan and co-workers have prepared self-assembled Silicalite-1 films on Si wafers.[41,42] Purified Silicalite-1 nanocrystals (30–80 nm) were dispersed in ethanol, then a drop of the suspension was placed on the substrate and pressed with a polydimethylsilane stamp for 12 h to allow ethanol to evaporate.[41] As a result, the zeolite nanoparticles self-assemble into micro-patterned films containing close-packed Silicalite-1 crystals. When the

Silicalite-1 nanocrystals were prepared with increased amount of ethanol in the synthesis solution and used without purification, the particles self-assembled on the Si wafer surface into patterned knotted rope web or wrinkled honeycomb films.[42] The self-assembly of Silicalite-1 nanocrystals (20 and 74 nm) on glass from water or ethanol suspensions has also been reported.[43] The films were formed at low Silicalite-1 concentrations (0.67 wt.%) and contained several layers of particles. Discontinuous monolayer films were obtained by decreasing the zeolite concentration to 0.13 wt.%. Recently, the self-assembly of dye-loaded zeolite L crystals (6 μm length$\times$2 μm diameter) on the petals of different flowers has been reported.[44] The flower petals were immersed into the zeolite aqueous suspension and, following removal, were dried. The zeolite crystals aligned into patterns such as hexagons and parallel lanes depending on the flower species.

The Langmuir–Blodgett method is another method used for the assembly of preformed zeolite crystals on flat substrates.[45–48] In this method, the zeolite particles are dispersed in a suitable solvent and spread at the air–water interface in a Langmuir–Blodgett trough. The floating zeolite film, formed upon compression to a certain surface pressure, is then transferred onto the support by vertical lifting. The method allows the preparation of highly ordered monolayer films. It has been applied to both micron-sized crystals[45,46] and colloidal zeolites[46–48] of different composition (pure silica, aluminosilicate, and aluminophosphate molecular sieves) and morphology Fig 19.2.

A drawback of the method is the use of a special equipment (the Langmuir–Blodgett trough). The properties of the floating films were dependent on the zeolite crystal characteristics (hydrophobicity, composition, size, and morphology) as well as the spreading solvent. For instance, Langmuir–Blodgett films on Si wafers were prepared from colloidal Silicalite-1 dispersed in methanol, ethanol, 2-propanol, and 1-butanol.[47] Methanol was found to be the best solvent and monolayer densely packed continuous films were obtained from methanol suspensions. The Langmuir–Blodgett zeolite films were stable enough to allow further processing such as secondary growth.[48]

2.2. Multi-step methods

Multi-step methods include suitable modification of the support or both the support and the zeolite crystals to allow assembly of the zeolite crystals on the support. Owing to their high reproducibility and applicability to unlimited numbers of

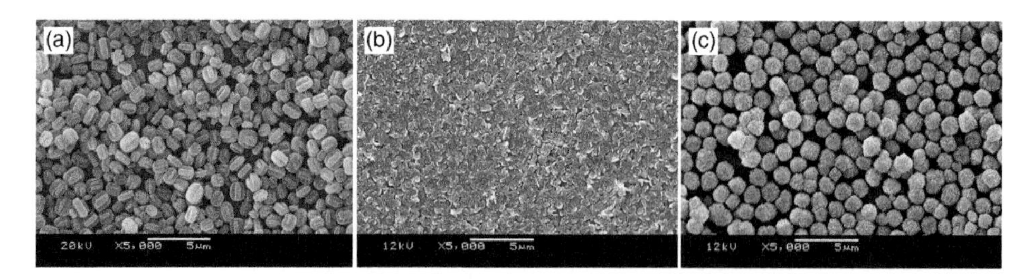

Figure 19.2 Langmuir–Blodgett films of (a) Silicalite-1, (b) $AlPO_4$-18, and (c) Beta.

supports with different chemical composition, shape, and porous structure, these methods are most often used to deposit preformed zeolite crystal onto supports. Another advantage of the multi-step methods in this section is the possibility to build zeolite multilayers, which are used in bottom-up nanofabrication of advanced materials.

2.2.1. Layer-by-layer deposition

The Layer-by-layer (LbL) technique involves an easy and inexpensive process for mono- and multilayer assembly of charged particles through electrostatic interactions. In this method, conventional polyelectrolytes such as poly(diallyldimethylammonium chloride) (PDDA, cationic) and poly(sodium 4-styrenesulfonate) (PSS, anionic) are firstly adsorbed on the support followed by the adsorption of oppositely charged particles; this cycle can be repeated many times to build the desired number of layers. The method is very simple – the polyelectrolytes are commercially available and beakers and tweezers are the only apparatus required. Also, the technique is versatile with application in many areas and for many systems. The assembling mechanism is still not fully understood and the film structure and quality depend on the experimental conditions. Different aspects of the LbL method, including factors that influence the assembly process, insights into the LbL mechanism, technical developments and applications, have been recently reviewed by Ariga *et al.*[49]

The LbL technique was first adopted for the assembly of zeolite nanocrystals by Sterte and co-workers, the so-called film seed method.[50,51] Polished silicon and alumina wafers were seeded with Silicalite-1 and zeolite A colloidal crystals, respectively. The synthesis procedure consisted of the following steps: (i) immersion of the cleaned substrate in a 0.4 wt.% aqueous solution of cationic polymer (Berocell 6100, Akzo Nobel AB, repeating unit $[CH_2CHOHCH_2NMe_2]^+{}_n$, M_w=50,000 g mol^{-1}) at pH 8.0 for 5 min, then rinsing with a 0.1 M ammonia solution and distilled water to remove excess polymer, and (ii) immersion of the modified support into 2.5 wt.% colloidal zeolite aqueous suspension at pH $>$ 10 for 5 min followed by rinsing with ammonia solution and distilled water. During the first step, the charge of the support surface was reversed to positive and in the second step the negatively charged zeolite crystals were adsorbed. The procedure could be repeated several times to improve the crystal density. The group applied this method to different colloidal zeolite systems and different substrates to explore the possibilities of the method. The primary objective in their studies was to evaluate the properties of the dense zeolite films prepared by secondary growth of the seeded supports. Thus, the seed deposition experimental procedure was not varied substantially and the influence of the parameters that could influence the seed deposition has not been studied in detail. The film seed method has been used to assemble zeolite particles on dense polished supports such as Si wafers, quartz, steel, alumina, and tantalum plates from Silicalite-1, Y and Beta colloidal suspensions.[52–56] Hedlund *et al.* used Silicalite-1 seeds with three different sizes, 60, 160, and 350 nm, to seed Si wafers.[52] They found that the crystals were homogeneously distributed throughout the support surface with varying crystal density depending on the crystal size/number of adsorption steps. The small crystals resulted in higher crystal coverage and dense layers could be obtained after 2 adsorption steps, whereas four adsorption steps were needed when larger seeds were used. In the case of 70 nm zeolite Y crystals, 30% coverage was achieved after one adsorption step, which increased to 60% after a second adsorption of seeds.[55] Porous α-Al_2O_3 substrates have been used to prepare zeolite films by the film seed method.[57] The 60 nm Silicalite-1 crystals, however, penetrated within the support porous structure. In order to avoid plugging of the support pores, a masking procedure has been developed.[58] In this procedure, a polymer film was formed on the support surface, the pores were then filled with a wax, the protective polymer film was dissolved and then the film seeded method was applied. After the formation of the zeolite film, the wax was removed from the support pores by calcination.

The LbL method has also been modified for the fabrication of zeolite films on noble metal substrates (e.g., gold).[59,60] The gold substrate was treated with a 20 mM methanol solution of γ-mercaptopropyltrimethoxysilane for 3 h followed by immersion in 0.1 M HCl for 15 h to hydrolyze the silane. Positively charged Silicalite-1 particles obtained by treatment with a strongly acidic ion exchange resin were then adsorbed on the negatively charged modified Au surface.[59] Alternatively, cationic polymer was firstly adsorbed on the modified support surface followed by the adsorption of negatively charged zeolite crystals.[60] In the latter case, Silicalite-1 crystals with a size of 60, 165, and 320 nm were used and the results for the crystal density on the support surface were similar to the results obtained on Si wafers.[52]

The film seed method has also been applied to fibrous materials.[61,62] Fibres of different composition were used such as carbon, ZrO_2, Al_2O_3, mullite, ceramic and pyrex glass fibres, and high seed coverage was achieved on all fibres independently of their chemical compositions. Other structured supports such as cordierite monoliths, porous alumina spheres, ceramic foams, and packing steel supports have been used successfully in the seed film method.[63,64] Different colloidal zeolites have been assembled via the LbL method on polystyrene beads.[65] Examples of seeded substrates by the film seed method are shown in Fig. 19.3.

Because of its versatility and flexibility, the LbL method has been widely applied to the preparation of zeolite hierarchical structures by templating methods. Zeolite hierarchical materials are macroscopic structures with a porous structure organized at different pore size levels, for instance micro/meso, micro/macro, micro/meso/macro, etc. They can be all-zeolite or zeolite in combination with other material(s).

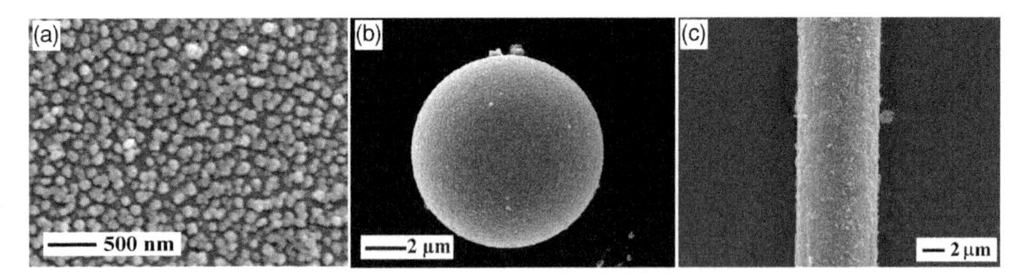

Figure 19.3 Silicalite-1 nanocrystals assembled by the LbL method on (a) flat substrate, (b) bead, and (c) fiber.

Different templates, usually organic materials that can be removed after the synthesis by calcination, can be used to control the macroscopic/porous structure of the product materials. In addition, silicate materials, which can serve as nutrients for the zeolite crystallization, can also be applied. The consumption of the silicate species during secondary growth results in the formation of pores of controlled size. Owing to their chemical nature and complex morphological features, it is rarely possible to prepare zeolite hierarchical structures by *in situ* crystallization methods using such templates. Most often, the macroscopic templates are seeded with zeolite crystals prior to secondary crystallization and the LbL method is most popularly used for the deposition of zeolite seeds. Diatomite has been seeded by the LbL method using cationic PDDA modifier and negatively charged 80 nm Silicalite-1 and 45 nm Beta colloidal zeolites.[66] The diatomite surface was covered with a complete and homogeneous seed layer after only one adsorption step but the cycle was repeated up to 3 times to increase the amount of seeds. Materials of complex morphologies such as cedar and bamboo wood cells and cellulose acetate filter membranes have also been seeded with 80 nm Silicalite-1 using the same method.[67,68] Polystyrene beads[69] and mesoporous silica spheres[70] have been coated with Silicalite-1 layers by the LbL technique. The beads were initially modified with a tri- PDDA/PSS/PDDA layer to provide uniform positively charged surfaces and subsequent layers were built using only PDDA as a cationic modifier. The seeded beads were then organized into 3D arrays to prepare all-zeolite ordered macroporous monoliths. These are only a few examples of the application of the LbL method for the preparation of hierarchical zeolitic structures but they are indicative of the great flexibility of the method.

The simplicity of the LbL method and its applicability to supports of various shape, chemical composition and structure, has made its use universal by zeolite researchers. The only restriction of the method is that colloidal zeolite suspensions are used. However, considering that some of the most important zeolites can be prepared in colloidal form, the LbL is likely to remain the most widely used method for the assembly of preformed zeolite crystals.

2.2.2. Assembly via covalent and ionic linkages

The method for the assembly of zeolite microcrystals via covalent linkages has been developed by Prof. Yoon and his group. They explored a number of ways to modify the substrate, or the substrate and the zeolite surfaces, to facilitate the bonding of the zeolite crystals to the support. It is beyond the scope of this chapter to review the tremendous amount of results obtained and only certain aspects of their work will be discussed here. Recently, the work was summarized by Yoon and readers can refer to his review for more information.[71]

The zeolite assembly via covalent linking was first reported for the preparation of monolayers of micron-sized zeolite A (~0.5 μm, cubic) and ZSM-5 (1.6–13 μm, coffin-shaped) crystals on glass, mica, alumina and other zeolite supports.[72,73] The procedure included either modification of both the substrate and the zeolite crystals[72] or only the support.[73] The modification of the zeolite crystals was performed by treatment with 3-aminopropyltriethoxysilane (AP-TES) at 110 °C for 1 h under argon. The glass was treated with a suitable silylating agent (e.g., [3-(2,3-epoxypropoxy)propyl]trimethoxysilane (EPP)-TMS and 3-chloropropyltrimethoxysilane

(CP-TMS)). The treatment consisted of immersion in silane solution in toluene and reflux under argon for 3 h. The modified support was then introduced into the zeolite dispersed in toluene and the mixture was refluxed under argon for 3 h. After the reaction, the support was sonicated in toluene for 20 s to remove loosely attached zeolite crystals. Monolayers of closely packed zeolite crystal with preferred orientation were obtained by both methods. The binding of the zeolite crystals to the support was via bridges formed between the terminal zeolite amine and the support epoxide groups or the terminal zeolite hydroxyl and the support chloride groups. The binding strength of the zeolite to the support was evaluated by monitoring the progressive loss of zeolite crystals from the support upon ultrasonic treatment in toluene. The loss of materials was determined from SEM images and weighing the samples on a microbalance. The zeolite crystals were strongly attached to the crystals and could survive sonication times > 5 min.

Different approaches were explored to further increase the binding strength of the zeolite microcrystals to the support. Fullerene was used to modify AP-TES-treated glass plates followed by reaction with AP-TES functionalized zeolite A crystals.[74] The zeolite crystals assembled onto the glass substrate via amine-fullerene amine linkages. Different polyamine (dendritic polyamine and polyethylenimine (PEI)) were inserted between the EP-tethered zeolite and glass support.[75] The zeolite monolayers showed increased binding strength and up to 75% of the zeolite crystals could survive sonication for 1 h. In addition, the method was found useful for the preparation of zeolite coated optical fibres. The binding strength was further increased through ionic linkages between the zeolite crystals and the support.[76,77] For instance, trimethylpropylammonium iodide ($TMPA^+I^-$) tethered ZSM-5 crystals were assembled onto sodium butyrate (Na^+Bu^-) tethered glass by ion exchange of the Na^+ ions with the $TMPA^+$ ions. Typical SEM images of zeolite (ZSM-5) microcrystals assembled on glass support, via the method developed by Yoon's group, are shown in Fig. 19.4. Similar close packing and high ordering of the zeolite crystals on the support were achieved using both covalent and ionic linkages.

Zeolite microcrystals were also assembled as closely packed monolayers on conductive substrates such as platinum, gold and indium-tin oxide glass.[78] The zeolite crystals were modified with AP-TES or CP-TMS and deposited on bare or polymeric amine pre-coated substrates. Micropatterning of oriented zeolite monolayers was also possible using suitable methods to obtain glass substrates with patterned modifier-functionalized areas.[77,79] The deposition of the zeolite microcrystals occurred only on the tethered glass areas.

The time needed to achieve close to 100% degree of coverage, when the reaction between the modified support and the zeolite crystals was performed under reflux and stirring, was more than 24 h. This time could be decreased to several minutes by introducing ultrasonic agitation.[80] Two modes were used: sonication without stacking (SW), and sonication with stacking (SS) in which the tethered glass was stacked between two bare glass plates. The results showed that SS was the best method for preparing monolayer assemblies of zeolite microcrystals on glass substrates. An interesting observation was that the attached microcrystals were continuously replaced by crystals from the solution and the replacement rate was higher under SW compared to SS conditions. The SS method was used to assemble zeolite

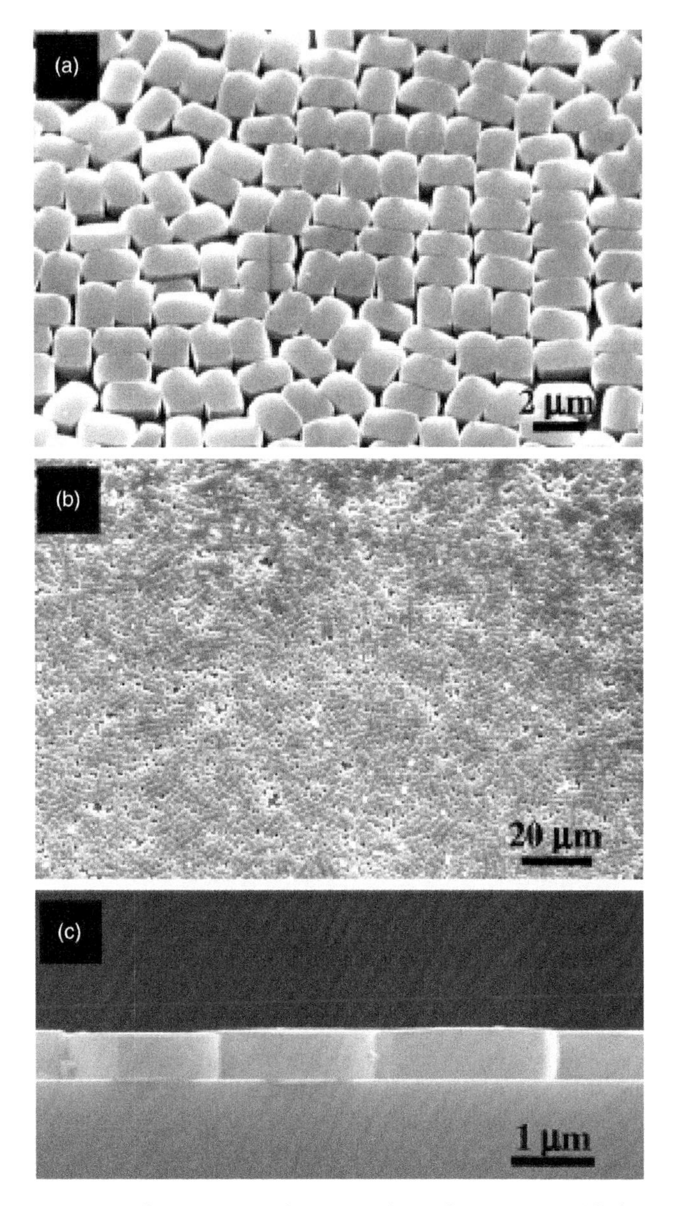

Figure 19.4 SEM images of ZSM-5 monolayers on glass substrates assembled via ionic linkages. Reprinted from Park *et al.*,[77] with permission from Elsevier.

L microcrystals with cylindrical and hexagonal columnar morphology via covalent linkers.[81] The cylindrical zeolite L crystals were deposited vertically on the support whereas the hexagonal crystals assembled horizontally after 2 min of sonication. The crystal alignment was explained by the tendency of the crystal to deposit onto its largest crystal face. Recently, the authors found that high quality monolayers of Silicalite-1 and ETS-10 crystals could be prepared by manual rubbing for 10–20 s.[82]

Different functionalization of the zeolite crystals and the support was used such as ion- and PEI-tethering. The manual assembly is very simple compared to wet SS methods and allows easy scale-up. The uniformity of the zeolite crystals was essential for the preparation of closely packed monolayers by rubbing.

The method for assembly of zeolite microcrystals via covalent and ionic linkages is quite complicated compared to the LbL method. It consists of many steps, toxic solvents are employed and special conditions are required. However, it allows the fabrication of monolayers of aligned zeolite microcrystals with good binding strength to the support. The synthesis of zeolite microcrystals is more efficient compared to colloidal zeolites with greater possibilities for manipulation of the crystal morphology. This makes the method very promising for the preparation of tailored materials and wide utilization and further developments of the method can be expected. The CP-TMS modification of the support to facilitate the deposition of zeolite microcrystals is the simplest variation of the covalent linkage method.[73] This method has been used by Tsapatsis and co-workers to seed α-alumina supports with Silicalite-1 crystals.[83,84] Cylindrical zeolite L crystals with length to diameter ratio of 1 and 0.3 were assembled using different covalent linkers according Yoon's method.[85] Closely packed and comparatively dense monolayers strongly bound to the surface were obtained with the zeolite channels oriented perpendicular to the substrate surface. However, the crystal ordering was somewhat inferior compared to the zeolite L monolayers prepared by Lee *et al.*[81], probably because the crystals used had broader particle size distribution. MCM-22 flake crystals of uniform size and diameter around 1 μm were assembled by the CP-TMS modification method under reflux or SS conditions on aluminium alloy sheets and α-Al_2O_3.[86] The MCM-22 crystals were attached with the large basal *ab*-plane parallel to the substrate. The zeolite layer was then coated with silica and calcined. The MCM-22 deposition, silica coating and calcination was repeated up to 3 times to improve the performance of the coatings. Recently, a modification of the CP-TMS method was reported for coating the confined microchannels structured supports with colloidal zeolite crystals (80 ± 30 nm zeolite A spherical particles).[87] The support was treated with an ethanol solution of CP-TMS in an autoclave at 100° for 4 h and the zeolite seeds were then attached to the modified substrate by a second treatment in an autoclave of the substrate immersed in the colloidal zeolite suspension in ethanol. The multi-channel plate was uniformly coated with zeolite seeds but the assembled crystals were not closely-packed. The work indicates that the covalent linkage method can be applied to colloidal zeolite systems as well. However, it is not clear whether there are any benefits of applying this method compared to the facile LbL colloidal seed deposition.

3. Applications of Supported Layers of Preformed Zeolite Crystals

Zeolite seeded supports are mainly applied for the preparation of zeolite films by secondary growth. *In situ* crystallization has been used to prepare zeolite films. However, the *in situ* crystallization mechanism on the support is not well understood. The method can be applied to a limited number of supports depending on their chemical composition. Reproducibility of the results is often questionable and control over the film smoothness, thickness, and crystal orientation is difficult. The film formation by secondary growth of pre-deposited zeolite crystals offers an elegant way to circumvent these problems. First, the seeds reduce the effect of the surface chemistry of the support and sometimes can minimize the corrosion of the support during secondary treatment. Second, pre-seeding can reduce or completely eliminate the induction time needed for zeolite nucleation on the support.[88] The seeds can also prevent the transformation of the zeolite into other types of zeolites in the course of the hydrothermal treatment.[89]

The drastic influence of the presence of seeds prior to secondary growth is illustrated in Fig. 19.5 for zeolite Beta films prepared in fluoride media on Si wafers. In the absence of seeds, the zeolite films were very thick (> 100 μm), rough and substrate etching can be observed in the SEM images (Fig. 19.5(a)). When the films were grown from seeded Si wafers, smooth and homogeneous films with a thickness of ca. 20 μm were obtained (Fig. 19.5(b)).

For applications such as selective membranes, where the molecular sieving properties of the zeolite are utilized, control over zeolite film orientation and presence of defects are very important for the performance of the membranes. Defects can be created during film preparation and their number can be reduced to a certain extent by using seeds. However, stress-induced cracks can be formed during calcination of the films to remove the organic structure directing templates and other methods should be used to eliminate those cracks. The most important feature of seeded syntheses is that films of controlled orientation can be obtained.[88,90] The zeolite films prepared by secondary growth of colloidal seeds assembled by the LbL method are usually only partially oriented. In addition, the preferred orientation can be lost by prolonged secondary growth. Better control of the orientation of the secondary growth films can be achieved using micron-sized crystals of tailored morphology as seeds. For example, Choi *et al.* have obtained uniformly *a*-oriented Silicalite-1 films using Silicalite-1 seeds synthesized with a trimer-tetrapropylammonium template and assembled on the substrate via covalent linkages.[84]

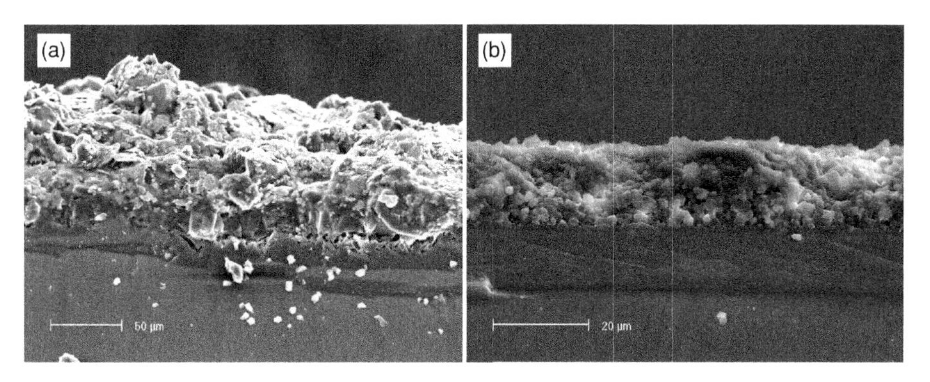

Figure 19.5 Zeolite Beta films prepared in F media by (a) *in situ* crystallization and (b) secondary growth of the pre-seeded support.

Preformed zeolite crystals deposited on different zeolite type crystals or structures can also be used for the fabrication of zoned zeolite materials. For instance, large zeolite Beta crystals were seeded with Silicalite-1 nanocrystals by the film seed method and, upon secondary crystallization to densify the outer layer, zoned Beta/Silicalite was obtained.[91] Seed crystal deposition has also been widely used for the preparation of hierarchical zeolite structures.[29,66–70]

Assembled zeolite layers have been evaluated for potential applications without any further processing. Zeolite washcoated structured supports have been tested in catalytic applications.[9,10] Spin-coated pure silica zeolite films have been suggested as promising low-*k* dielectrics.[21,25] Further information about low-*k* film application of zeolites can be found in Chapter 13 of this book. The multilayer MCM-22/silica films prepared by covalent linking of the zeolite crystals have been found to be a promising alternative for corrosion protection as well as gas separation.[86] Aligned zeolite L crystals have been used as a host for preparing supramolecularly organized systems with exciting properties.[81,85] Zeolite L and A monolayers were assembled on gold-coated glass or gold foil through molecular linkages, modified with AgCl, and the photocatalytic activity of AgCl in water oxidation was tested.[92] The zeolite supported AgCl showed enhanced oxygen production, which was higher for the zeolite A supported AgCl because of the smaller size of the deposited AgCl. Zeolite L nanocrystals were assembled on indium tin oxide (ITO) glass electrode surface by the LbL method followed by enzyme (cytochrome c) immobilization.[93] The enzyme/zeolite/ ITO electrodes showed excellent biosensor properties, long lifetime, and wide pH stability.

4. Concluding Remarks

Different methods can be used to assemble preformed zeolite crystals on substrates. Choosing an appropriate method depends on the aim of the synthesis and targeted application. The most popular deposition method is the LbL method applicable to colloidal zeolites. The technique is inexpensive, simple and reproducible and can be applied to unlimited numbers of supports independently of their shape, structure, or chemical composition. The LbL method is particularly useful for the preparation of zeolite negative replicas of structures with complex morphology such as biological systems. However, densification of the seeded layers by secondary growth is almost always needed. The introduction of the method for crystal deposition via covalent or ionic linkages is expected to have a large impact on the application of zeolite materials. First, the method allows the preparation of densely packed oriented monolayers of zeolite microcrystals with enhanced binding strength to the support. The monolayer character of the films and the zeolite crystal orientation enables the fabrication of exciting advanced materials by incorporation of guest molecules in the oriented zeolite channels. Second, the perfect zeolite ordering can be used for the synthesis of completely oriented zeolite films by secondary growth. The high-binding strength also makes the as-made materials interesting for several applications without any necessity for secondary growth of the films. Considering the progress made in manipulating the zeolite morphology,[94] improvements of the assembly method are highly desirable. Such improvements could lead to the deposition of dense layers of preformed zeolite crystals with tailored zeolite channel orientation in respect of the support and film thickness, both achieved by manipulation of the zeolite crystal morphology. Providing that the density of the crystals is high enough and the adherence of the films is acceptable, the secondary growth could be avoided, which will be a substantial breakthrough in zeolite science.

REFERENCES

[1] Barrer, R. M., *Hydrothermal Chemistry of Zeolites*, Academic Press, London, 1982.
[2] Schoeman, B. J., Sterte, J., Otterstedt, J.-E., *Zeolites* **1994,** *14,* 568–575.
[3] Tosheva, L., Valtchev, V. P., *Chem. Mater.* **2005,** *17,* 2494–2513.
[4] Lin, X., Kita, H., Okamoto, K., *Ind. Eng. Chem. Res.* **2001,** *40,* 4069–4078.
[5] Cui, Y., Kita, H., Okamoto, K., *J. Membr. Sci.* **2004,** *236,* 17–27.
[6] Kusakabe, K., Kuroda, T., Murata, A., Morooka, S., *Ind. Eng. Chem. Res.* **1997,** *36,* 649–655.
[7] Bonhomme, F., Welk, M. E., Nenoff, T. M., *Microporous Mesoporous Mater.* **2003,** *66,* 181–188.
[8] Avila, P., Montes, M., Miró, E. E., *Chem. Eng. J.* **2005,** *109,* 11–36.
[9] Beers, A. E. W., Nijhuis, T. A., Aalders, N., Kapteijn, F., Moulijn, J. A., *Appl. Catal. A* **2003,** *243,* 237–250.
[10] Zamaro, Z. M., Ulla, M. A., Miró, E. E., *Chem. Eng. J.* **2005,** *106,* 25–33.
[11] Zamaro, Z. M., Ulla, M. A., Miró, E. E., *Catal. Today* **2005,** *107–108,* 86–93.
[12] Zhang, X., Liu, H., Yeung, K. L., *Mater. Chem. Phys.* **2006,** *96,* 42–50.
[13] Wong, W. C., Au, L. T. Y., Ariso, C. T., Yeung, K. L., *J. Membr. Sci.* **2001,** *191,* 143–163.
[14] Motuzas, J., Julbe, A., Noble, R. D., van der Lee, A., Beresnevicius, Z. J., *Microporous Mesoporous Mater.* **2006,** *92,* 259–269.
[15] Kanezashi, M., O'Brien, J., Lin, Y. S., *J. Membr. Sci.* **2006,** *286,* 213–222.
[16] Kuzniatsova, T., Kim, Y., Shqau, K., Dutta, P. K., Verweij, H., *Microporous Mesoporous Mater.* **2007,** *103,* 102–107.
[17] Ban, T., Ohwaki, T., Ohya, Y., Takahashi, Y., *Angew. Chem. Int. Ed.* **1999,** *38,* 3324–3326.
[18] Boudreau, L. C., Tsapatsis, M., *Chem. Mater.* **1997,** *9,* 1705–1709.
[19] Boudreau, L. C., Kuck, J. A., Tsapatsis, M., *J. Membr. Sci.* **1999,** *152,* 41–59.
[20] Lee, J. A., Meng, L., Norris, D. J., Scriven, L. E., Tsapatsis, M., *Langmuir* **2006,** *22,* 5217–5219.
[21] Wang, Z., Mitra, A., Wang, H., Huang, L., Yan, Y., *Adv. Mater.* **2001,** *13,* 1463–1466.
[22] Mintova, S., Bein, T., *Adv. Mater.* **2001,** *13,* 1880–1883.
[23] Mintova, S., Reinelt, M., Metzger, T. H., Senker, J., Bein, T., *Chem. Commun.* **2003,** 326–327.
[24] Vilaseca, M., Mintova, S., Karaghiosoff, K., Metzger, T. H., Bein, T., *Appl. Surf. Sci.* **2004,** *226,* 1–6.
[25] Li, Z., Lew, C. M., Li, S., Medina, D. I., Yan, Y., *J. Phys. Chem. B* **2005,** *109,* 8652–8658.
[26] Wee, L. H., Tosheva, L., Vasilev, C., Doyle, A. M., *Microporous Mesoporous Mater.* **2007,** *103,* 296–301.
[27] Hu, L., Wang, J., Li, Z., Li, S., Yan, Y., *J. Mater. Res.* **2006,** *21,* 505–522.
[28] Holmes, S. M., Markert, C., Plaisted, R. J., Forrest, J. O., Agger, J. R., Anderson, M. W., Cundy, C. S., Dwyer, J., *Chem. Mater.* **1999,** *11,* 3329–3332.
[29] Anderson, M. W., Holmes, S. M., Hanif, N., Cundy, C. S., *Angew. Chem. Int. Ed.* **2000,** *39,* 2707–2710.
[30] Lai, R., Gavalas, G. R., *Ind. Eng. Chem. Res.* **1998,** *37,* 4275–4283.
[31] Huang, A., Lin, Y. S., Yang, W., *J. Membr. Sci.* **2004,** *245,* 41–51.
[32] Lee, S.-R., Son, Y.-H., Julbe, A., Choy, J.-H., *Thin Solid Films* **2006,** *495,* 92–96.
[33] Ke, C., Yang, W. L., Ni, Z., Wang, Y. J., Tang, Y., Gu, Y., Gao, Z., *Chem. Commun.* **2001,** 783–784.

[34] Berenguer-Murcia, Á., Morallón, E., Cazorla-Amorós, D., Linares-Solano, Á., *Microporous Mesoporous Mater.* **2003,** *66,* 331–340.
[35] Shan, W., Zhang, Y., Yang, W., Ke, C., Gao, Z., Ye, Y., Tang, Y., *Microporous Mesoporous Mater.* **2004,** *69,* 35–42.
[36] Lin, J.-C., Yates, M. Z., Petkoska, A. T., Jacobs, S., *Adv. Mater.* **2004,** *16,* 1944–1948.
[37] Balkus, K. J., Jr., Muñoz, T., Jr., Gimon-Kinsel, M. E., *Chem. Mater.* **1998,** *10,* 464–466.
[38] Washmon-Kriel, L., Balkus, K. J., Jr., *Mocroporous Mesoporous Mater.* **2000,** *38,* 107–121.
[39] Xiong, C., Coutinho, D., Balkus, K. J., Jr., *Microporous Mesoporous Mater.* **2005,** *86,* 14–22.
[40] Cho, G., Lee, J.-S., Glatzhofer, D. T., Fung, B. M., Yuan, W. L., O'Rear, E. A., *Adv. Mater.* **1999,** *11,* 497–499.
[41] Huang, L., Wang, Z., Sun, J., Miao, L., Li, Q., Yan, Y., Zhao, D., *J. Am. Chem. Soc.* **2000,** *122,* 3530–3531.
[42] Wang, H., Wang, Z., Huang, L., Mitra, A., Yan, Y., *Langmuir* **2001,** *17,* 2572–2574.
[43] Song, W., Grassian, V. H., Larsen, S. C., *Microporous Mesoporous Mater.* **2006,** *88,* 77–83.
[44] Bossart, O., Calzaferri, G., *Microporous Mesoporous Mater.* **2008,** *109,* 392–397.
[45] Morawetz, K., Reiche, J., Kamusewitz, H., Kosmella, H., Ries, R., Noack, M., Brehmer, L., *Colloids Surf. A* **2002,** *198–200,* 409–414.
[46] Tosheva, L., Valtchev, V. P., Mihailova, B., Doyle, A. M., *J. Phys. Chem. C* **2007,** *111,* 12052–12057.
[47] Tosheva, L., Wee, L. H., Wang, Z., Mihailova, B., Vasilev, C., Doyle, A. M., *Stud. Surf. Sci. Catal.* **2007,** *170,* 577–584.
[48] Wang, Z., Wee, L. H., Mihailova, B., Edler, K. J., Doyle, A. M., *Chem. Mater.* **2007,** *19,* 5806–5807.
[49] Ariga, K., Hill, J. P., Ji, Q., *Phys. Chem. Chem. Phys.* **2007,** *9,* 2319–2340.
[50] Hedlund, J., Schoeman, B. J., Sterte, J., *Stud. Surf. Sci. Catal.* **1997,** *105,* 2203–2210.
[51] Hedlund, J., Schoeman, B., Sterte, J., *Chem. Commun.* **1997,** 1193–1194.
[52] Hedlund, J., Mintova, S., Sterte, J., *Microporous Mesoporous Mater.* **1999,** *28,* 185–194.
[53] Mintova, S., Hedlund, J., Valtchev, V., Schoeman, B. J., Sterte, J., *J. Mater. Chem.* **1998,** *8,* 2217–2221.
[54] Wang, Z., Hedlund, J., Sterte, J., *Microporous Mesoporous Mater.* **2002,** *52,* 191–197.
[55] Lassinantti, M., Hedlund, J., Sterte, J., *Microporous Mesoporous Mater.* **2000,** *38,* 25–34.
[56] Schoeman, B. J., Babouchkina, E., Mintova, S., Valtchev, V., P Sterte, J., *J. Porous Mater.* **2001,** *8,* 13–22.
[57] Hedlund, J., Noack, M., Kölsch, P., Creaser, D., Caro, J., Sterte, J., *J. Membr. Sci.* **1999,** *159,* 263–273.
[58] Hedlund, J., Jareman, F., Bons, A.-J., Anthonis, M., *J. Membr. Sci.* **2003,** *222,* 163–179.
[59] Sterte, J., Mintova, S., Zhang, G., Schoeman, B. J., *Zeolites* **1997,** *18,* 387–390.
[60] Engström, V., Mihailova, B., Hedlund, J., Holmgren, A., Sterte, J., *Microporous Mesoporous Mater.* **2000,** *38,* 51–60.
[61] Valtchev, V., Hedlund, J., Schoeman, B. J., Sterte, J., Mintova, S., *Microporous Mater.* **1997,** *8,* 93–101.
[62] Valtchev, V., Schoeman, B. J., Hedlund, J., Mintova, S., Sterte, J., *Zeolites* **1996,** *17,* 408–415.
[63] Öhrman, O., Hedlund, J., Sterte, J., *Appl. Catal. A* **2004,** *270,* 193–199.
[64] Sterte, J., Hedlund, J., Creaser, D., Öhrman, O., Zheng, W., Lassinantti, M., Li, Q., Jareman, F., *Catal. Today* **2001,** *69,* 323–329.
[65] Valtchev, V., Mintova, S., *Microporous Mesoporous Mater.* **2001,** *43,* 41–49.
[66] Wang, Y., Tang, Y., Dong, A., Wang, X., Ren, N., Gao, Z., *J. Mater. Chem.* **2002,** *12,* 1812–1818.
[67] Dong, A., Wang, Y., Tang, Y., Ren, N., Zhang, Y., Yue, Y., Gao, Z., *Adv. Mater.* **2002,** *14,* 926–929.
[68] Wang, Y., Tang, Y., Dong, A., Wang, X., Ren, N., Shan, W., Gao, Z., *Adv. Mater.* **2002,** *14,* 994–997.
[69] Rhodes, K. H., Davis, S. A., Caruso, F., Zhang, B., Mann, S., *Chem. Mater.* **2000,** *12,* 2832–2834.

[70] Dong, A., Wang, Y., Tang, Y., Zhang, Y., Ren, N., Gao, Z., *Adv. Mater.* **2002,** *14,* 1506–1510.
[71] Yoon, K. B., *Acc. Chem. Res.* **2007,** *40,* 29–40.
[72] Kulak, A., Lee, Y.-J., Park, Y. S., Yoon, K. B., *Angew. Chem. Int. Ed.* **2000,** *39,* 950–953.
[73] Ha, K., Lee, Y.-J., Lee, H. J., Yoon, K. B., *Adv. Mater.* **2000,** *12,* 1114–1117.
[74] Choi, S. Y., Lee, Y.-J., Park, Y. S., Ha, K., Yoon, K. B., *J. Am. Chem. Soc.* **2000,** *122,* 5201–5209.
[75] Kulak, A., Park, Y. S., Lee, Y.-J., Chun, Y. S., Ha, K., Yoon, K. B., *J. Am. Chem. Soc.* **2000,** *122,* 9308–9309.
[76] Lee, G. S., Lee, Y.-J., Yoon, K. B., *J. Am. Chem. Soc.* **2001,** *123,* 9769–9779.
[77] Park, J. S., Lee, G. S., Yoon, K. B., *Microporous Mesoporous Mater.* **2006,** *96,* 1–8.
[78] Ha, K., Park, J. S., Oh, K. S., Zhou, Y.-S., Chun, Y. S., Lee, Y.-J., Yoon, K. B., *Microporous Mesoporous Mater* **2004,** *72,* 91–98.
[79] Ha, K., Lee, Y.-J., Jung, D.-Y., Lee, J. H., Yoon, K. B., *Adv. Mater.* **2000,** *12,* 1614–1616.
[80] Lee, J. S., Ha, K., Lee, Y.-J., Yoon, K. B., *Adv. Mater.* **2005,** *17,* 837–841.
[81] Lee, J. S., Lim, H., Ha, K., Cheong, H., Yoon, K. B., *Angew. Chem. Int. Ed.* **2006,** *45,* 5288–5292.
[82] Lee, J. S., Kim, J. H., Lee, Y. J., Jeong, N. C., Yoon, K. B., *Angew. Chem. Int. Ed.* **2007,** *46,* 3087–3090.
[83] Lai, Z., Bonilla, G., Diaz, I., Nery, J. G., Sujaoti, K., Amat, M. A., Kokkoli, E., Terasaki, O., Thompson, R. W., Tsapatsis, M., Vlachos, D. G., *Science* **2003,** *300,* 456–460.
[84] Choi, J., Ghosh, S., Lai, Z., Tsapatsis, M., *Angew. Chem. Int. Ed.* **2006,** *45,* 1154–1158.
[85] Ruiz, A. Z., Li, H., Calzaferri, G., *Angew. Chem. Int. Ed.* **2006,** *45,* 5282–5287.
[86] Choi, J., Lai, Z., Ghosh, S., Beving, D. E., Yan, Y., Tsapatsis, M., *Ind. Eng. Chem. Res.* **2007,** *46,* 7096–7106.
[87] Yang, G., Zhang, X., Liu, S., Yeung, K. L., Wang, J., *J. Phys. Chem. Solids* **2007,** *68,* 26–31.
[88] Gouzinis, A., Tsapatsis, M., *Chem. Mater.* **1998,** *10,* 2497–2504.
[89] Xu, X., Yang, W., Liu, J., Lin, L., *Microporous Mesoporous Mater.* **2001,** *431,* 299–311.
[90] Hedlund, J., Jareman, F., *Curr. Opin. Colloid Interface Sci.* **2005,** *10,* 226–232.
[91] Bouizi, Y., Diaz, I., Rouleau, L., Valtchev, V. P., *Adv. Funct. Mater.* **2005,** *15,* 1955–1960.
[92] Reddy, V. R., Currao, A., Calzaferri, G., *J. Mater. Chem.* **2007,** *17,* 3603–3609.
[93] Yu, T., Zhang, Y., You, C., Zhuang, J., Wang, B., Liu, B., Kang, Y., Tang, Y., *Chem. Eur. J.* **2006,** *12,* 1137–1143.
[94] Drews, T. O., Tsapatsis, M., *Curr. Opin. Colloid Interface Sci.* **2005,** *10,* 233–248.

SECTION III

APPLICATIONS AND PROSPECTS OF ORDERED POROUS MATERIALS

CHAPTER 20

Electrochemistry with Micro- and Mesoporous Silicates

Alain Walcarius

Contents

Abstract

An overview is presented on the implication of microporous zeolites and ordered mesoporous silica-based materials in electrochemistry. After a brief discussion on the interest of these materials for the electrochemical applications, the various strategies applied to confine such insulating solids at an electrode surface are described. The basic electrochemical behavior of zeolite-modified electrodes (ZMEs), including the electron transfer mechanisms, and the voltammetric response of redox probes in mesoporous (organo) silica particles and mesostructured thin films are then discussed. Afterwards, the exploitation of the attractive properties of the micro and mesoporous silicates for electrochemical applications is described via several examples: voltammetric analysis after preconcentration and permselectivity, electrocatalysis, bioelectrochemistry, indirect amperometric detection, and some others.

Keywords: Bioelectrochemistry and biosensors, Electrocatalysis, Electrochemical detection and sensing, Mass transport in porous media, Microporous aluminosilicates, Organic-inorganic hybrid materials, Ordered mesoporous silica, Silica modified electrodes, Thin films, Zeolite modified electrodes

Ordered Porous Solids
DOI: 10.1016/B978-0-444-53189-6.00020-2

1. Introduction

The wish to combine in a single device an intelligently designed chemical component in close connection to an electrode surface, enabling to couple the intrinsic properties of the modifier to a particular redox process, has generated a huge amount of researches in designing integrated chemical systems called 'chemically modified electrodes' (CMEs).[1] CMEs have been classified into three main categories: (1) monolayer modified electrodes, (2) homogeneous multimolecularly layered electrodes (mainly based on polymer films), and (3) composite electrodes based on multicomponent heterogeneous matrixes that are constituted by either (bio)organic or inorganic modifiers. Electrodes modified with micro- and mesoporous silicate-based materials form subclasses of this category. Several reviews are available, dealing with the implication of microporous zeolites[2–6] or layered aluminosilicate clays[7–11] in electrochemistry, or with the intersection between the chemistry of silica-based porous solids or sol–gel-derived materials and electrochemical science.[12–25] In this chapter, an overview is given on microporous zeolite- and mesoporous silica (MPS)-modified electrodes.

There are numerous reasons why zeolites and mesoporous silicates are attractive for electrochemical applications. Zeolites constitute a unique family of materials displaying both ion exchange capacity and size selectivity in a single component.[26,27] This would impart both charge and size selectivity properties. For example, it will be possible to discriminate between positively-charged electroactive species on the basis of their size, to distinguish between redox cations and anions, or to promote charge transfer in microporous media by self-organized electron transfer chains. Encapsulated organometallic complexes in zeolite cages by physical entrapment (e.g., by ship-in-a-bottle synthesis[28]) can also be used as supported electrocatalysts. The most commonly used zeolites in connection to electrochemistry are those characterized by a rather high ion exchange capacity, such as zeolites A, X, Y, L, mordenite, among some others. Silica-based materials, and especially organic-inorganic hybrids, exhibit chemical and physico-chemical features that might be readily exploited when used to modify electrode surfaces. They indeed combine in a single porous solid both the mechanical stability of a rigid inorganic framework with the particular reactivity of organo-functional groups. They can be used as a support for immobilized or covalently grafted organic ligands that enable solid–liquid extraction of various species (e.g., heavy metals) present at trace levels in aqueous media, they can be manufactured as composites based on interpenetrating inorganic-organic polymers exhibiting either ionic or electronic conductivity, and the sol–gel process enables the encapsulation of bioactive molecules, alone or together with charge-transfer mediators or catalysts, which is attractive for applications in electrochemical biosensors. Of particular interest are the ordered mesoporous materials prepared by the surfactant template route[29] because of their exceptionally uniform and open structure that might be easily functionalized by grafting or co-condensation.[30–33] The regular spatial arrangement of mesopore channels of monodisperse size is thus expected to promote high diffusion rates and enhanced access of target analytes to a large number of accessible binding sites, which often constitute the rate-determining steps in electrochemical sensing devices based on modified electrodes. This rather novel class of materials offers the possibility to design advanced nanoreactors at electrode/solution interfaces.

Zeolite modified electrodes (ZMEs) have now been studied for more than 25 years, including investigations on their basis electrochemical behaviour, discussions on electron transfer mechanisms, as well as description of various advanced applications.[2–6] The topic of MPS-modified electrodes is more recent (past decade) and this young field developing at the intersection between the chemistry of ordered MPS-based materials with electrochemical science has been recently reviewed.[24] A brief summary is provided here on the main advances made with ZMEs and MPS-MEs, by focussing on what these materials have brought to the fields of electrochemistry, to understand how it is possible to perform electrochemical transformations with such electronically insulating materials, to show how their attractive properties can be exploited for target applications and, sometimes, how electrochemistry can be applied to their characterization or even synthesis.

2. Confinement of Microporous Zeolites and Mesoporous Silica Materials onto Electrode Surfaces

Because zeolites and MPSs are electronic insulators, their use in connection to electrochemistry requires a close contact to an electrode feeder material. Various strategies have been applied to build suitable integrated electrochemical devices, but most of them are based on two generic approaches (with some variants): (1) the dispersion of solid particles in carbon-based conductive composite matrices and (2) their deposition as thin films on solid electrode surfaces. More details on the preparation of such modified electrodes are given elsewhere.[3,6,24]

Modified carbon paste electrodes (Fig. 20.1) can be prepared by mixing as homogeneously as possible micrometric-size particles of zeolite or mesoporous (organo)silica and carbon powder together with a mineral oil acting as a binder. This mixture is then sunk into a holder equipped with an electrical contact and the electrode surface is smoothed by mechanical polishing. The method is intrinsically simple and, as far as all components of the modified paste are uniformly dispersed in the composite, the electrode can reach a good level of reproducibility after mechanical renewal of its surface, as common for other chemically modified carbon paste electrodes.[34,35] This has been widely applied to prepare ZMEs and MPS-MEs, which were then applied for electroanalysis (see, e.g., Ref. [36–41] for zeolites and Ref. [42–48] for mesoporous (organo) silicas). In the goal of getting more convenient devices for practical applications, this carbon-based composite approach has been extended to the production of disposable electrodes by the thick-film technology (screen-printing) by using carbon inks doped with either zeolites[49,50] or MPS particles.[51,52]

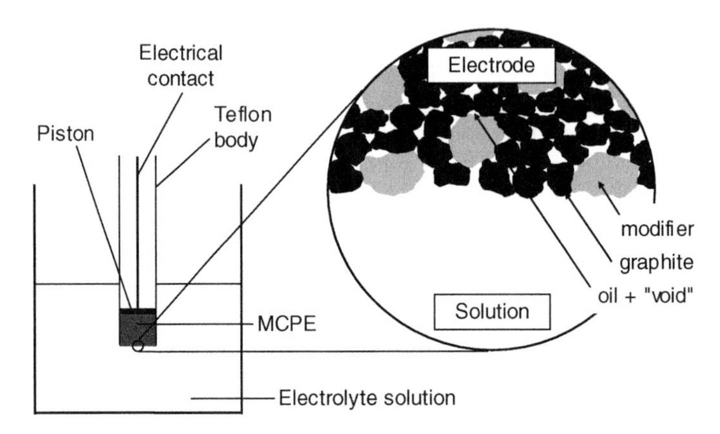

Figure 20.1 Schematic representation of a modified carbon paste electrode (MCPE).

Other composite *electrodes* have been also proposed. Zeolite-containing carbon–polystyrene–divinylbenzene composite electrodes[53] constitute a robust alternative to zeolite-modified carbon pastes. Dry graphite–zeolite mixtures, pressed either on a stainless steel grid or as pellets, was applied to prepare binder-free ZMEs that are likely to be used in organic medium.[54,55] Pressed pellets of pure zeolite inserted between two solid electrodes were also used[56]. Directly arising from the sol–gel technology, ceramic–carbon composite electrodes consisting in a continuous bulk silica framework comprising physically entrapped carbon particles can be obtained from gelation of a starting sol containing graphite particles in suspension in a water–alcohol medium in the presence of a tetraalkoxysilane precursor,[20,57] but this approach was not yet applied to the preparation of ordered MPS-modified electrodes.

Films based on zeolite and MPS particles are shown in Fig. 20.2. A straightforward way to confine silicates at an electrode/solution interface is their deposition onto the electrode surface. Zeolite and MPS particles thus have been deposited on solid electrodes by evaporation of an organic suspension containing the dispersed mineral particles. Most often, a dissolved organic polymer (e.g., polystyrene, poly (vinyl alcohol)) was added to the suspension in order to increase the adhesion of particles to the electrode surface and between them. This has led to zeolite-polymer[58–62] and MPS-polymer[63,64] composite films. Sometimes, carbon particles were added along with the mineral powder to improve the conductivity of the device. A related approach is to deposit first a layer of materials particles on the electrode surface and then to cover it with a thin porous polymer film to ensure its mechanical stability.[65–68] Dense monograin layers of close-packed monodisperse zeolite particles have also been reported[69] and zeolites can be covalently attached to some electrode surfaces (indium-tin oxide or gold) via silane and/or thiol reagents.[70,71]

Continuous zeolite- and ordered *MPS films* on electrode surfaces constitute probably the major advances made in the preparation of ZMEs and MPS-MEs in

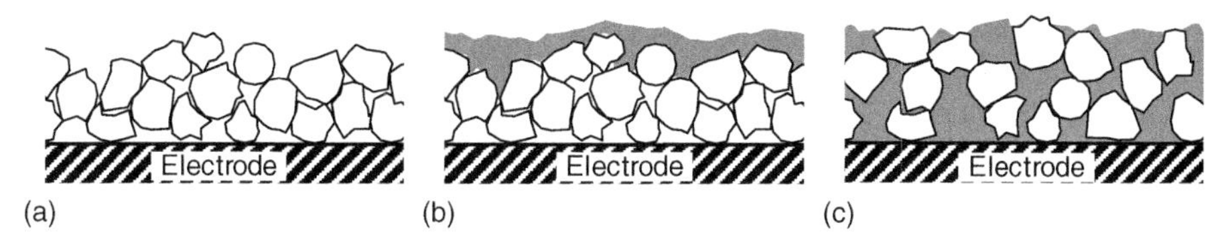

Figure 20.2 Schematic representation of the way to confine silicate particles (white) onto the surface of solid electrode: (a) particles deposited from a suspension of pure material, (b) same as (a) but covered with a thin polymer film (grey) and (c) silicate-polymer composite film obtained by evaporation of material particles suspended in a polymer solution.

the past few years. The so-called 'secondary growth' method was applied to form continuous oriented pure zeolite films on electrode surfaces on the basis of a seed and feed approach.[72] The process involves deposition of colloidal zeolite nanocrystals on the underlying electrode surface which is transferred after drying to a feedstock gel of given composition (adjusted as a function of the type of desired zeolite structure) where further growth and intergrowth of the seed crystals into a continuous film is induced by hydrothermal treatment. The effectiveness of the process is illustrated on Fig. 20.3. Zeolite films of faujasite and A types on various electrodes (Pt, C, mixed oxides) have been obtained via the secondary growth method.[72–77] An *in situ* crystallisation method was also applied to form continuous MFI type zeolite films by exploiting the preferred *b*-orientation and the monocrystal nature of the film,

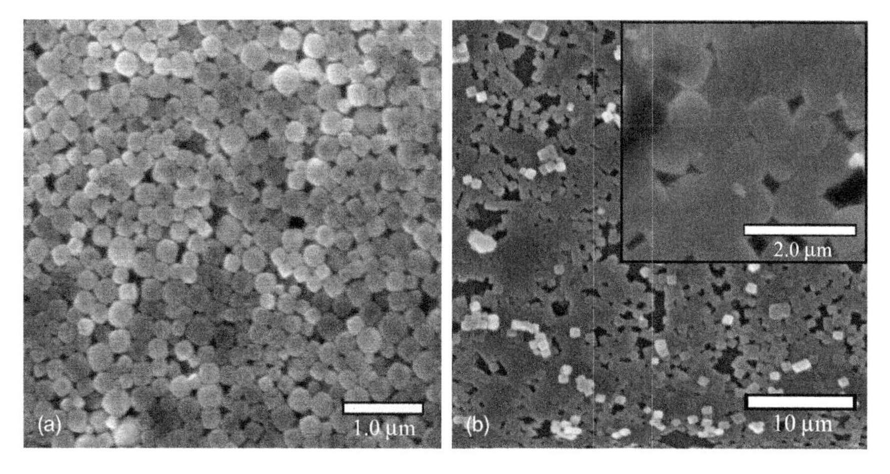

Figure 20.3 (a) Alignment and close packing of zeolite crystals deposited on an yttria-stabilized zirconia surface. (b) Film growth demonstrated after a single day of regrowth in a zeolite precursor solution. Reprinted from Rauch and Liu[73], Copyright (2003), with permission from Springer.

giving rise to homogeneous coatings without detectable defects or grain boundaries.[78] It is known for a long time that the sol-gel process provides a versatile way to prepare a wide range of silica-based materials in the form of thin films on various supports including electrodes.[12,23] When using a surfactant template, ordered MPS films can be obtained by depositing the 'precursor+surfactant' sol by spin-coating or dip-coating onto the electrode surface where the mesostructures are formed by evaporation induced self-assembly (EISA).[79–84] For details on the EISA technique, the interested reader is directed to references.[85,86] Adjusting the experimental conditions enables the elaboration of various kinds of mesostructures, for example with 2D hexagonal, 3D hexagonal or cubic geometry.[86,87] The film porosity is revealed after template removal, which can be performed by calcination in case of pure silica mesoporous films but this process is prevented in case of organically-modified silica films for which solvent extraction must be applied. Pore orientation is a very important parameter at it governs the accessibility of solution-phase analytes to the underlying electrode surface. This is the reason why a novel approach called 'Electro-Assisted Self-Assembly' (EASA) was developed recently to produce mesostructured silica films on electrodes with mesopore channels oriented normal to the support.[88] This was realized by applying a cathodic potential to an electrode immersed in a hydrolyzed sol solution containing the surfactant, in order to generate OH^- species locally at the electrode/solution interface and to induce polycondensation of the silane precursors. The electrochemically-driven cooperative self-assembly of surfactant micelles and silica formation thus resulted in well-packed mesopore channels growing perpendicularly to the electrode surface (Fig. 20.4), a configuration which has proven to be very difficult to obtain by EISA.[89] Homogeneous deposits can be obtained over wide areas (cm^2), even on non flat surfaces, with thicknesses typically in the 50–200 nm range (see parts C and D on Fig. 20.4). Perpendicular orientation with accessible pores from the surface is expected to have beneficial effects on improving sensitivity in electrochemical detections, via enhancement of mass transport rates through the film.

Membranes and solid electrolytes constitute the last categories of zeolite- and MPS-modified electrodes. Incorporation of zeolite particles into organic polymers, such as siloprene, poly(tetrafluoroethylene-co-ethylene-co-vinylacetate) or poly (vinyl chloride) has been applied to design potentiometric sensing devices.[90–92] Nafion-zeolite composites are solid electrolytes useful for direct methanol fuel cells.[93] Conducting polymers were also used for embedding the silicate particles in the form of composite membranes.[94,95]

The choice of the electrode configuration is of critical importance regarding the target application. Indeed, film-based devices will require mass and/or charge-transfer reactions to occur across the material layer from the solution to the electrode surface while bulky composites will offer both the modifier and the conductive part of the electrode in direct contact to the solution. The first configuration would thus be of interest for permselective detection whereas the second one would be more appropriate for electrocatalysis or for voltammetric detection subsequent to analyte accumulation.

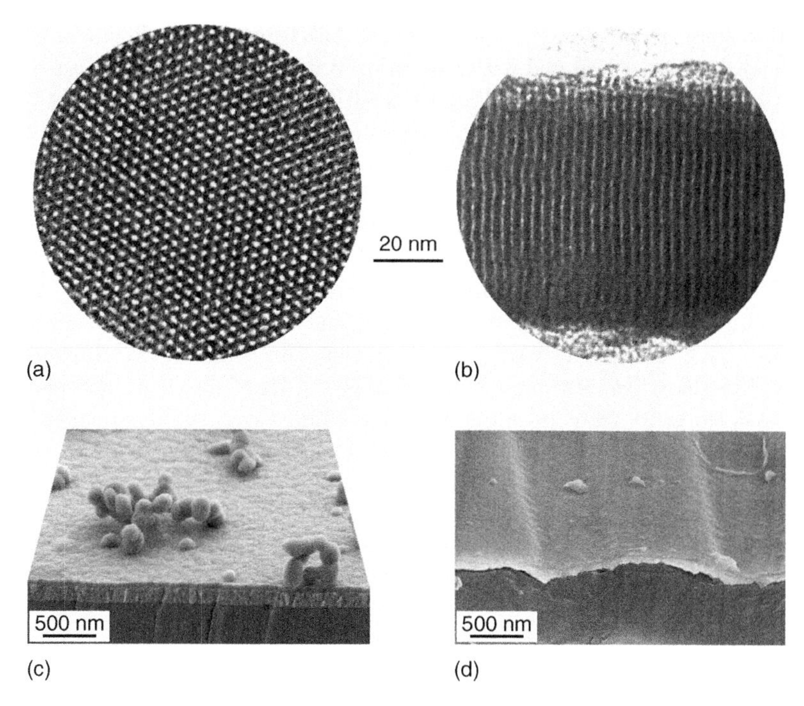

Figure 20.4 (a,b) TEM images of CTAB-templated mesoporous silica films formed by electrodeposition on glassy carbon: high-magnification top view (a) and cross-sectional view (b). (c) Cross-sectional view obtained by FE-SEM for a film prepared as in (a,b), but electrodeposited this time on ITO (after vertical cleaving of the substrate glass/ITO/mesoporous silica (MPS)). (d) Cross-sectional view obtained by FE-SEM for a MPS film electrodeposited on a gold CD-trode displaying a streaked morphology at the μm level. Figures (a–c) have been adapted from Walcarius *et al.*[88], with permission of Nature Publishing Group.

3. Basic Electrochemical Behaviour

An intriguing point when considering the use of insulating materials as zeolites or MPS in electrochemistry is related to the role(s) they are expected to play with respect to electron transfer reactions. One has to distinguish various cases depending on the location of the redox probe (initially as solution-phase species or situated within the porous material) and on the electrode configuration (bulk or film). A brief overview is given below on the effect of the presence of zeolite or MPS on the electrode response and on the main parameters influencing the behaviour of guest redox species.

3.1. Electrochemical response of zeolite-modified electrodes

Due to the cation exchange properties of zeolites, their presence at an electrode/solution interface is likely to induce accumulation of cations while rejecting anions. Numerous redox cations (Ag^{+}, Cu^{2+}, Pb^{2+}, Cd^{2+}, $Ru(NH_3)_6^{3+}$, methyl- and heptylviologen, dopamine) have been preconcentrated at ZMEs, resulting in increased voltammetric signals. The observed voltammetric peak currents can be much larger than those sampled with using unmodified electrodes. The ratio of peak current density observed at a ZME to that at a corresponding unmodified electrode[96] is called 'enhancement factor,' which can be used to quantify the effectiveness of concentration effects.[97] Enhancement factors are affected by several parameters, such as the ion exchange capacity and pore size of zeolites, the charge and size of the redox analyte, their concentration in solution and soaking time of the electrode, as well as the ZME configuration. Some illustrative data are given in Fig. 20.5, comparing the enhancement factors measured for four redox probes of various charge and size (Ag^{+}, Cu^{2+}, dopamine, methylviologen), three zeolite types (A, X and Y) and two electrode configurations (bulk ZMCPE composite and ZMGCE film-based electrodes). They enable to draw the main conclusions as follow: peak currents were in general higher when using zeolites displaying high cation exchange capacity, large pores and small particle size, for cationic analytes of small hydrated size and great positive charge, and for long soaking time especially at film ZMEs. The fact that bulk ZME (i.e., ZMCPE) gave faster enhancement behaviour compared to zeolite film-based electrodes is explained by the presence of zeolite particles in direct contact with the electroactive analyte in the former case

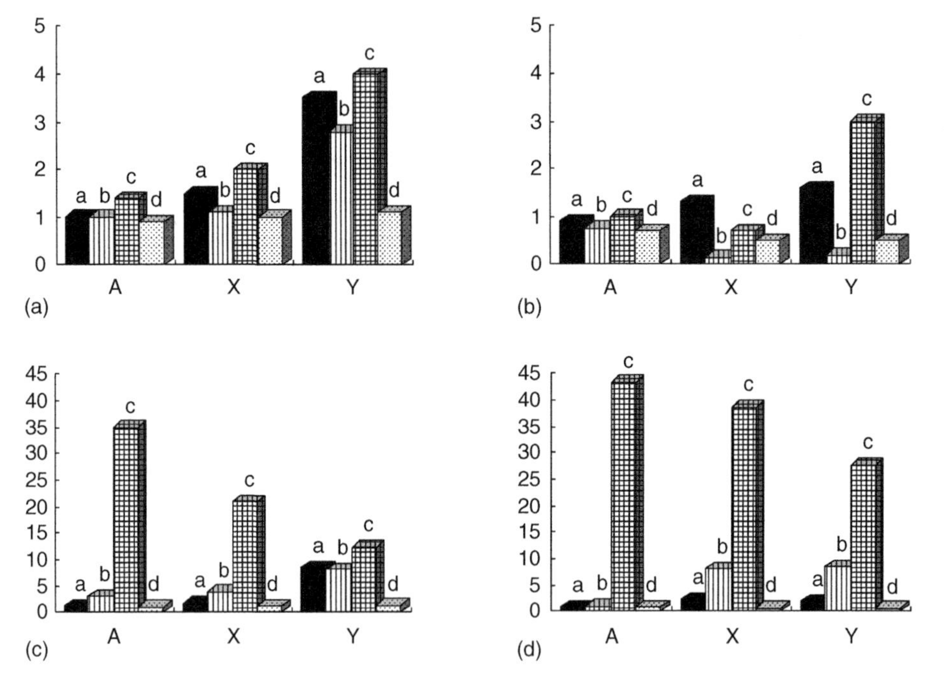

Figure 20.5 Enhancement factors calculated from cyclic voltammograms recorded in solutions containing 1 mM of analyte (a – MV^{2+}; b – Cu^{2+}; c – Ag^{+}; d – DA^{+}), by using zeolite-modified carbon paste electrodes (ZMCPE) and zeolite films deposited on glassy carbon electrodes (ZMGCE), for zeolites A, X and Y; (a,b) factors calculated for 1 min immersion of ZMCPE (a) or ZMGCE (b) into the solution; (c,d) factors for 15 min immersion of ZMCPE (c) or ZMGCE (d). Reproduced from Walcarius *et al.*[97] Copyright (1997), with permission from Elsevier.

upon immersion of the electrode into solution, whereas reaching the zeolite layer might be delayed in the latter case due to the hydrophobic character of the polymer binder (or overlayer).[97] Size-excluded electroactive cations (those of size larger than the zeolite pore aperture) and charge-excluded species (anions) did not result in any enhancement of the voltammetric peaks.[96] Preconcentration efficiency was low in the presence of high concentration of other cations in the medium (e.g., electrolyte cations) because of competing effects at the exchanging sites.

Of special interest are the recently described continuous zeolite films on electrodes as they ensure the absence of intergrain diffusion, the redox probes being forced to cross the microporous zeolite framework to reach the electrode surface. For example, ITO electrodes respectively unmodified or covered with continuous zeolite A films of different pore apertures (3 Å for KA and 4 Å for NaA) gave rise to responses to molecular oxygen (diameter equal to 3.145 Å) which were dramatically decreased when the electrode is covered with NaA due to diffusional restrictions and totally suppressed when using a KA film due to 'perfect' molecular sieving of the film.[72] Another example is illustrated in Fig. 20.6 showing that a MFI zeolite film displaying a pore size of 5.5 Å is likely to accommodate $Ru(NH_3)_6^{3+}$ species (Fig. 20.6A) of approximately the same size whereas the bigger $Co(phen)_3^{2+}$ cations (Ø~13 Å) are excluded from reaching the electrode surface (Fig. 20.6B).[78] The permselective behaviour was also observed when the redox probes were present in mixture in the solution (Fig. 20.6C). This advantage only belongs to the continuous zeolite phases deposited on electrode surface as previously reported film-type ZMEs (zeolite/polymer deposits or zeolite monograin layers) were characterized by interparticle mesopores and did not exhibit such molecular sieving effects (Fig. 20.6D).

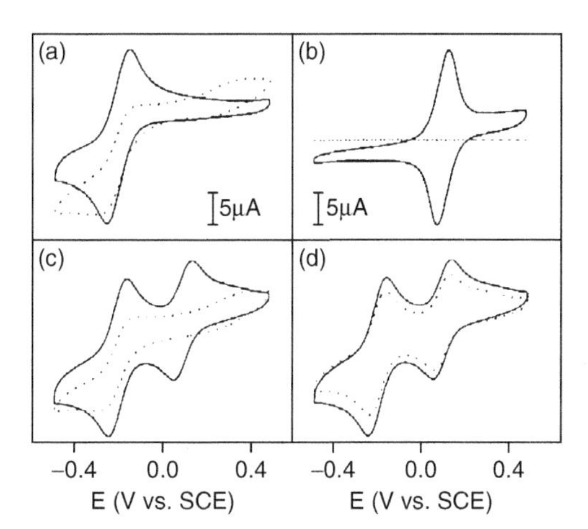

Figure 20.6 CV of $Ru(NH_3)_6^{3+}$ and/or $Co(Phen)_3^{2+}$ in aqueous solutions. Solid line, bare electrode; dotted line, film electrode. (a) $Ru(NH_3)_6^{3+}$ on MFI film; (b) $Co(Phen)_3^{2+}$ on MFI film; (c) $Ru(NH_3)_6^{3+}$ + $Co(Phen)_3^{2+}$ on MFI film; (d) $Ru(NH_3)_6^{3+}$ + $Co(Phen)_3^{2+}$ on a film made from zeolite particles (MFI film means *b*-oriented MFI). Reprinted from Li *et al.*[78] Copyright (2004) American Chemical.

Beside the above observations, the electrochemical responses of ZMEs prepared with zeolite particles previously loaded with redox species have been widely studied in order to understand why such species located in an insulating environment are likely to undergo charge-transfer reactions and how the microporous host would affect their electrochemical behaviour. Discussions on that point require to make distinction between the small redox species incorporated within the zeolite framework by ion exchange (highly mobile species) and those, bigger, which were physically entrapped as 'ship-in-a-bottle' complexes in the zeolite cages (species with restricted motion).

Immersing a ZME made of zeolite particles containing pre-exchanged small electroactive ions into a fresh electrolyte solution (free of redox-active species) usually led to the observation of well-defined voltammetric signals, but their intensity was found to progressively decrease either upon continuous cycling potentials or by increasing the soaking time of the electrode in the solution before recording the CV curve. Indeed, as long as the electrode is soaked in the electrolyte solution, redox species are exchanged for the electrolyte cations in the zeolite particles, leading to their continuous depletion at the electrode surface.[38,58,96,98–101] The leaching process is rather fast and promoted by increasing the electrolyte concentration and/or when using electrolyte cations of smaller hydrated size, which led to voltammetric currents of smaller intensities.[38,66,100,102] Even lower, nearly undetectable, response was observed when using an electrolyte solution with size-excluded cations because they are too big to enter the zeolite pores, and this resulted in negligible leaching by ion exchange.[102] The electrochemistry of ZMEs is thus critically affected by ion exchange reactions, which was also confirmed by voltammetric experiments performed at various scan rates indicating diffusion-controlled behaviour.[103] As a consequence, the voltammetric response of redox probes pre-exchanged in ZMEs resulted in peak currents as high as fast were the mass transport processes (optimal in the more open structures and for the smaller cations) and as high was the content of probe in the zeolite. However, the ZME behaviour can be seriously complicated as distinct ion exchange sites can co-exist in the same zeolite.[26,27] This is notably the case of silver(I) in zeolites A, X, and Y, for which distinct voltammetric peaks have been observed depending on the cation location, the zeolite type, the nature of supporting electrolyte, and the type of ZME.[66,98,99,102] Such distinct ion exchange sites are also likely to affect the electrode sensitivity as kinetics associated to the exchange process might be very different. This is illustrated on Fig. 20.7 for Ag^+ in faujasite-type zeolites. Part A on Fig. 20.7 compares the electrochemical response of a zeolite Y containing few Ag^+ ions distributed randomly in the zeolite structure (sample $Ag_6Na_{50}Y$) with another Y sample holding a little bit more Ag^+ species but almost exclusively in the small channels ($Ag_{16}Cs_{40}Y$). Voltammetric signals observed with the zeolite sample containing more silver are surprisingly much lower, and this is explained by the easy exchange of Ag^+ located in the large cages while those located in the small channels are less accessible to the electrolyte cation.[102] Similar behaviour is observed with zeolite X for which a slight increase in the Ag^+ content from 3.4 to 5.4 ions by unit cell results in a 50-fold enhancement of the voltammetric response recorded at ZME (Fig. 20.7B). This is due to the fact that Ag^+ ions are almost exclusively located in the small-channel

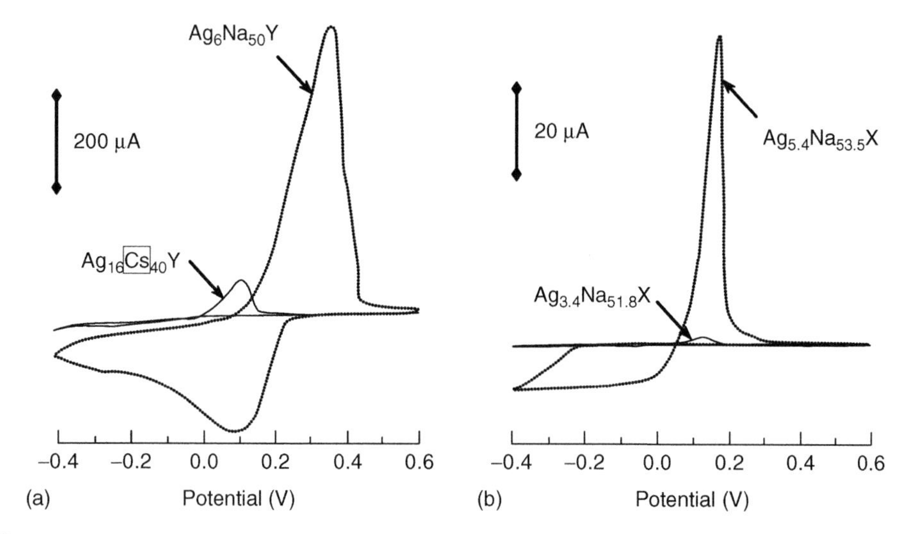

Figure 20.7 (a) Cyclic voltammetry (CV) of Ag_6Na_{50}-Y and $Ag_{16}Cs_{40}$-Y modified electrodes in water containing 0.1 M KNO_3. Reference electrodes were either SCE ($Ag_6Na_{50}Y$) or Pt quasi-reference ($Ag_{16}Cs_{40}Y$). (b) CV of $Ag_{3.4}Na_{51.8}$-X and $Ag_{5.4}Na_{53.5}$-X in 0.1 M $NaNO_3$. Scan rate 20 mV/s. Reproduced from Baker *et al.*[102] and Baker and Senaratne[104], respectively, by permission of The Royal Society of Chemistry.

network in $Ag_{3.4}Na_{51.8}$-X while some supercage sites are occupied in $Ag_{5.4}Na_{53.5}$-X.[104] The voltammetric response of silver species ion exchanged in supercages is governed by the size of hydrated electrolyte cation but the faradic currents due to silver initially located in the small cages is monitored by the ionic radii and dehydration energies of the electrolyte cations.[99] Finally, one has to mention that solvent can also affect the electrochemical response of ZMEs, often by acting on the rate of ion exchange processes.[65]

Redox species, larger than the zeolite pore entrance, can be also accommodated to the aluminosilicate microstructure by physical entrapment within the zeolite cages. The electrochemical behaviour of such encapsulated probes is fundamentally different as that of small ion-exchanged cations. In spite of their strongly restricted mobility, these encapsulated complexes gave rise to non negligible electrochemical response.[4,105] Most studied species are positively-charged organometallic complexes, such as metal *tris*-bipyridine (metal = Fe, Ru, Os, Co), metal phthalocyanine (metal = Fe, Co, Cu, Rh), and metal-schiff base complexes.[54,55,61,106–112] They can be located either in the bulk zeolite, or simply bound to the external surface of zeolite particles, or even occluded within the first layer of complete or broken cages situated at the particle boundary. Those species located at the outermost surfaces are expected to contribute to a large extent to the electron transfer reactions, as those located deeply in the bulky isolating structure can only be reached *via* electron hopping between the adjacent redox species, which implies a high density of these sites.[113] That species located at the outermost surface of zeolite grains can promote charge transfer to the complexes encapsulated deeper in the porous medium in case of favourable spatial arrangement.[70,114,115] Despite this electrocatalytic behaviour, it is

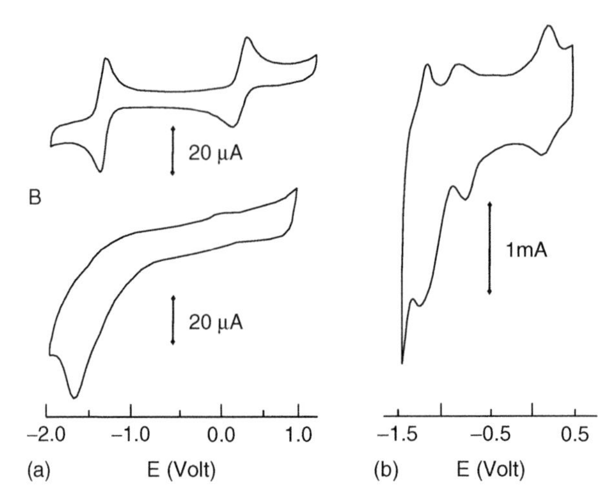

Figure 20.8 CV of Co(salen) in acetonitrile. (c) recorded at 50 mV/s in Co(salen) solution (+0.1 M $LiClO_4$), using unmodified GCE. (B) recorded at 50 mV/s in 0.1 M $LiClO_4$ using a graphite + Co(salen)-NaY film coated on GCE (+polymer). (c) recorded in 0.1 M NBu_4BF_4 at 10 mV/s, using a graphite + Co(salen)-NaY pressed powder electrode. From senaratne *et al.*[65] (a, b) and Bedioui *et al.*[106] (c), by permission of The Royal Society of Chemistry (a, b) and the American Chemical Society (c).

worth to mention that most complexes encapsulated in the bulk zeolite are electrochemically 'silent'. As illustrated in Fig. 20.8, the strong interactions between the organometallic complex and the zeolite lattice may significantly affect the voltammetric response, via shift in peak potentials (or peak splitting) with respect to the behaviour of the same redox species at bare electrodes, with changes very dependent on the preparation mode of the entrapped complex and type of ZME.[65,106,116]

Finally, one has to mention that some zeolites or zeotypes containing electroactive centres in their framework lattice (intrinsic composition) exhibit voltammetric responses that have been attributed to redox processes involving the framework species. Examples are available for titanosilicalites,[117–119] iron-silicalites[62,120] and vanadium-silicalites.[121,122] Anodically induced degradation of mordenite was also reported.[123] Finally, electrochemistry involving the molecular sieve alone (without solvent) was possible at elevated temperature (i.e., 200–500 °C) where extra-framework metal cations become electroactive.[56,124]

3.2. Electron transfer mechanisms at zeolite-modified electrodes

Probably the most important question arising when considering the electrochemistry with zeolites relies on the mechanism(s) responsible for electron transfer processes occurring at ZMEs as zeolites are intrinsically insulating materials. This question has led to intriguing observations and controversial discussions in the literature (see an overview in Ref.[125]), but one can actually consider that at least three mechanisms are operating to explain the electrochemical response of redox species incorporated in ZMEs, as represented by the following equations.[1–5]

Mechanism I:

$$E^{m+}_{(Z)} + ne^{-} + nC^{+}_{(S)} \leftrightarrows E^{(m-n)+}_{(Z)} + nC^{+}_{(Z)} \tag{1}$$

Mechanism II:

$$E^{m+}_{(Z)} + mC^{+}_{(S)} \leftrightarrows E^{m+}_{(S)} + mC^{+}_{(Z)} \tag{2}$$

$$E^{m+}_{(S)} + ne^{-} \leftrightarrows E^{(m-n)+}_{(S)} \tag{2b}$$

Mechanism III (3 subgroups):

$$E^{m+}_{(Z,surf)} + ne^{-} + nC^{+}_{(S)} \leftrightarrows E^{(m-n)+}_{(Z,surf)} + nC^{+}_{(Z,surf)} \tag{3a}$$

$$E^{(m-n)+}_{(Z,surf)} + nC^{+}_{(Z,surf)} + E^{m+}_{(Z,bulk)} \leftrightarrows E^{(m-n)+}_{(Z,bulk)} + nC^{+}_{(Z,bulk)} + E^{m+}_{(Z,surf)} \tag{3b}$$

$$M^{m+}_{(S)} + ne^{-} \leftrightarrows M^{(m-n)+}_{(S)} \tag{4a}$$

$$M^{(m-n)+}_{(S)} + nC^{+}_{(S)} + E^{m+}_{(Z,surf)} \leftrightarrows E^{(m-n)+}_{(Z,surf)} + nC^{+}_{(Z,surf)} + M^{m+}_{(S)} \tag{4b}$$

$$M^{m+}_{(Z,surf)} + ne^{-} + nC^{+}_{(S)} \leftrightarrows + M^{(m-n)+}_{(Z,surf)} + nC^{+}_{(Z,surf)} \tag{5a}$$

$$M^{(m-n)+}_{(Z,surf)} + nC^{+}_{(Z,surf)} + E^{m+}_{(Z,surf)} \leftrightarrows E^{(m-n)+}_{(Z,bulk)} + nC^{+}_{(Z,bulk)} + M^{m+}_{(Z,surf)} \tag{5b}$$

where E is the electroactive species with charge $m+$, C^{+} represents the electrolyte cation (chosen as monovalent for convenience), M is a mediator (chosen with charge $m+$ for convenience), the subscripts z and s refer to the zeolite phase and the solution, respectively, and the subscripts *surf* and *bulk* to the zeolite surface (either external surface or outermost sub-surface layer of cages) and bulk ion exchange sites.

One can thus distinguish three electron transfer mechanism at ZMEs. Mechanism I is purely intrazeolitic, where the electroactive species undergoes intracrystalline electron transfer while charge balance is maintained by solution-phase electrolyte cation entering the zeolite framework (Eq. (1)). This mechanism does not distinguish between species located deeply in the bulk zeolite and those situated in the boundary region of the zeolite grains. Mechanism II is purely extrazeolitic, and involves the ion exchange of the electroactive probes for the electrolyte cations (Eq. (2a)) prior to their electrochemical transformation in the solution phase (Eq. (2b)). The group of mechanisms III distinguishes between the electroactive probes located in the bulk zeolite and those situated at the external boundary of the particle. The first case is the direct electron transfer to electroactive species situated at the outer surface of the zeolite particles (i.e., those easily accessible to the electrons), charge compensation being ensured by the electrolyte cation (Eq. (3a)). This step can be (but is not necessarily) followed by electron hopping to the probes located in the bulk of the solid, with concomitant migration of the electrolyte cation inside the zeolite structure (Eq. (3b)). In the presence of a charge-transfer mediator either dissolved in solution or adsorbed on the zeolite surface, electrochemical transformation (Eqs. (4a) and (4b)) can lead to indirect charge transfer to either surface-confined or bulk-located electroactive probes (Eqs. (5a) and (5b)).

Evidences to support one of these mechanisms, depending of the electrode systems, have been described in the literature.[125] Briefly, one can consider that the extrazeolitic mechanism is the main pathway responsible for explaining charge transfer at ZMEs incorporating ion exchangeable electroactive species.[58,66,102,103,126–128] The case of encapsulated complexes is less unequivocal, but must involve in great majority the electroactive species located at the outermost boundary of zeolite particles, which are ascribed either as intrazeolitic (because interacting with the zeolite framework) or extrazeolitic (because located outside of the bulk zeolite), depending on the authors.[3,105,108,129] Extrazeolite electron transfers always gave rise to much higher currents than extrazeolite ones. This was further confirmed when studying lead-loaded zeolites at various sulfidation levels for which more intense response was observed for the more mobile ion-exchanged species (Pb^{2+} in this case) in comparison to that of the immobilized ones (i.e., PbS after sulfidation).[130]

3.3. Electrochemical response of redox probes in mesoporous silica

As for zeolites, the presence of mesoporous (organo)silica particles at an electrode surface is likely to affect the electrochemical response of solution-phase redox probes, mostly via enrichment (or rejection) of these species at the electrode/solution interface. Such accumulation behaviour can be monitored *in situ* via multisweep cyclic voltammetry.[131,132] Fig. 20.9 illustrates the interest of ion exchangers made of MPS particles functionalized with sulfonate or quaternary ammonium groups, used as modifiers of carbon paste electrodes, to induce dramatic increase in peak currents (respectively for Cu^{2+} and $Fe(CN)_6^{3-}$ detection), in comparison to the small responses at bare electrode. Similar behaviour was also reported for preconcentration of the ammoniacal complexes of $Cu^{(II)}$ (or $Hg^{(II)}$) within MCM-41-modified CPE via ion/ligand exchange on the silanol groups,[42] for chemisorption of $Hg^{(II)}$ species on the same modified electrode,[133] as well as for $Cu^{(II)}$ binding to aminopropyl-grafted MCM-41-modified CPE via complexation to amine moieties.[134]

Increase in peak currents can be also simply achieved by soaking the electrode in diluted solutions for some time prior to record the cyclic voltammetric curves, as reported for accumulation of metal tris-bipyridine complexes at MCM-41-modified electrodes.[79] It is interesting to notice that the use of ordered mesoporous sorbents gave rise to peak currents higher than those achieved with the corresponding amorphous solids (e.g., silica gels),[135] most probably because of less restricted mass transfer in case of ordered structures (see section 4.1.). Most of the above investigations have been performed using modified carbon paste electrodes because of faster

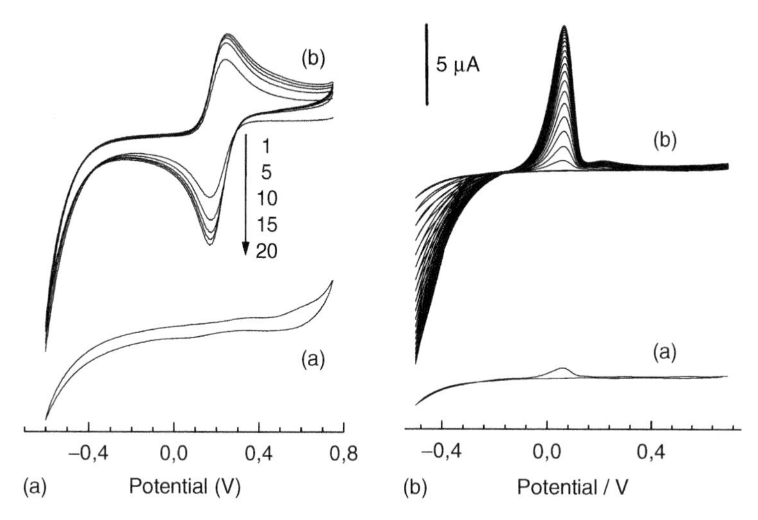

Figure 20.9 Multisweep CV recorded in solutions containing either (a) 5×10^{-5} M $K_3Fe(CN)_6$ (in 0.01 M KCl) or (B) 5×10^{-5} M $Ru(NH_3)_6Cl_3$ (in 0.01 M $NaNO_3$) using CPEs containing (curve a) unmodified MCM-41 silica particles and (curve b) a trimethylpropylammonium-functionalized MCM-41 sample (a) or a sulfonic acid functionalized MCM-41 (b). Scan rate: 100 mV/s. The curves did not vary upon multiple potential scans for electrodes based on non-functionalized MCM-41. In part (a, curve b), only the 1st, 5th, 10th, 15th and 20th curves are depicted. From Ganesan and Walcarius[131] and Walcarius and Ganesan,[132] Copyright (2004 & 2006) American Chemical Society.

response time in comparison to film-based devices (the analyte is not forced to cross the entire material to reach the electrode surface as both carbon and silica particles are present at the modified CPE/solution interface). Even if the redox probes are readily incorporated into the mesoporous material, the fact that peak potentials did not change with respect to those obtained for solution-phase species at bare electrode suggest an extra-material electron transfer mechanism in which the probe would diffuse to a conductive part of the electrode to be detected.

This is better evaluated by considering the electrochemical response of redox species that have been accommodated beforehand to the mesoporous material, for which additional signals (or shifts in peak potentials) have been reported for a series of electroactive species, including metal *tris*-bipyridines, iron phenanthroline, ruthenium hexamine, phenothiazins, bis(diphenyl-pyrylium)phenylene or open and closes spiropyrans derivatives, as well as ferrocene-based dendrimers.[67,136–140] Such complications in voltammetric signals result from redox probes affected by the mesoporous environment. It is even possible to modulate the reactivity of charge-transfer mediators via incorporation in distinct mesoporous materials, as exemplified for the electrocatalytic activity of nicotinamide adenine dinucleotide (NADH) associated to two mesoporous aluminosilicates, i.e., of MCM-41 and SBA-15 types, with respect to the oxidation of 1,4-dihydrobenzoquinone (H_2Q).[141] Both supported systems were characterized by NADH electrochemistry differing from that of solution-phase NADH at bare electrodes (electron transfer featuring a boundary-associated mechanism), but the electrocatalytic process was different for the two supports (surface-confined NADH-H_2Q adduct with SBA-15 and surface reaction/regeneration with MCM-41). Other catalysts such as 1:12-phosphomolybdic ($PMo_{12}O_{40}^{3-}$) or 12-tungstophosphoric (PW_{12}) heteropolyanions can be incorporated in ammonium-functionalized MPS via favourable electrostatic interactions and the resulted immobilized species are reported to keep their electrocatalytic properties.[142,143] Charge transfer mediators can be also immobilized in surfactant-containing MPS (i.e., without template extraction) while maintaining their electroactivity. This is notably the case of methylene blue (MB) embedded in a surfactant-silica hybrid MCM-41 mesophase for which a stable voltammetric response was obtained, with characteristic peak potentials very close to those obtained for MB simply adsorbed on MCM-41 and for solution-phase MB at the bare electrode so that it is thought that MB species are most likely located in the hydrophilic surfactant/silica interface.[140] Stronger interactions are however observed when the redox probe can be dissolved in the liquid crystalline phase filling the mesoporous material, as reported for ferrocene methanol (FcMeOH) in CTAB-based MCM-41 for which a well-defined signal was observed for FcMeOH oxidation at a potential higher by *ca.* 200 mV with respect to that recorded on bare electrode.[144]

When redox-active moieties are immobilized in the mesoporous material via covalent binding, their physical mobility is restricted or even suppressed so that their electrochemical response is expected to be dramatically modified with respect to solution-phase species. This has been investigated with hexagonally organized MPS with 4,4′-bipyridinium units (viologen derivatives) integrated via covalent bonds within the silica walls,[145] and with ferrocene moieties grafted on the mesopore walls of as-synthesized MCM-41 materials, either by post-functionalization through ester or amide bonds on pre-functionalized MCM-41 samples[143] or by direct grafting of MCM-41 by a ferrocene-containing organotrialkoxysilane.[48] These redox derivatives were electrochemically accessible and the mechanism responsible for charge transfer is thought to be electron hopping between adjacent sites. The process was as more effective as high was the functionalization level because the density of organofunctional groups must be high enough to enable effective charge propagation through the MPS. Ordered mesoporous materials are claimed to be advantageous over silica gels homologues to impart larger voltammetric response to the immobilized redox probes.[48]

3.4. Electrochemistry of mesoporous silica films

This section focuses especially on highly ordered continuous mesoporous (organo) silica thin films obtained from the one-step EISA (or EASA) procedure (see section 2) as this configuration offers the most 'ideal' system for characterizing their permeability (and/or permselective) properties without interference from interparticle mass transport (solution-phase analytes must cross the mesoporous structure to reach the electrode surface). Several parameters are expected to affect this access. First, the presence of the template in a homogeneous crack-free MPS film totally blocks the response of the electrode to solution-phase redox probes (unless they can solubilise into the liquid crystalline mesophase, as for ferrocene methanol[144])

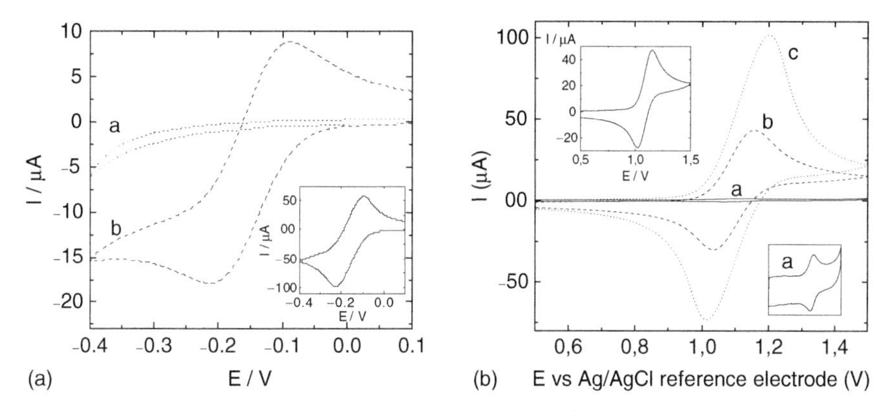

Figure 20.10 (a) CV recorded at 50 mV/s in 5 mM $Ru(NH_3)_6^{3+}$ at a GC electrode covered with a CTAB-templated silica film (a) before and (b) after surfactant extraction. Inset: electrochemical response at the bare electrode.From Etienne *et al.*[144], reproduced with permission of WILEY-VCH Verlag GmbH. (b) CV recorded in 0.5 mM $Ru(bpy)_3^{2+}$ using ITO electrodes covered by MPS thin films (curve a)) 2D hexagonal, (curve b) 3D hexagonal and (curve c) cubic structures. Bottom-right inset: enlargement of curve a; top-left inset: signal obtained at bare ITO. Reprinted from Etienne and Walcarius[87], Copyright (2007) American Chemical Society.

because of the non porous character of the templated hydrophobic film (see part A on Fig. 20.10). This can be used to check if the deposited MPS material covers well the entire electrode area without significant defect. Second, after removal of the template, the film becomes accessible to external reagents and the intensity of the voltammetric response is directly related to the rate of mass transport through the film (voltammetric signals are indeed diffusion-controlled for reversible redox probes[146]). This has been reported to be very dependent on both the type of MPS (structure, pore size and orientation) and the probe characteristics (size, charge, concentration) as these parameters are likely to affect mass transport rates for the probe in the interior of the film.[79,87,88,147,148] An illustration is given on part B of Fig. 20.10, indicating that better permeability was observed with cubic mesostructures, giving rise to higher peak currents than for a 3D hexagonal film, whereas the voltammetric response of the 2D hexagonal layer was dramatically much lower.[87] The low sensitivity of the electrode covered by a 2D hexagonal mesostructure has been attributed to unfavourable orientation of the mesopore channels parallel to the surface[79] (and possible crashing of the mesostructure[87]). The resort to highly ordered 2D hexagonal structure with channels well-oriented normal to the electrode surface, as obtained by electrodeposition (EASA), is therefore an attractive way to ensure easy access of the redox probe to the electrode surface because of fast diffusion across the film.[88] Ordered silica deposits exhibiting a periodic bimodal macro-mesostructure (with combination of meso- and macropores of variable dimension) constitute an alternative mean to enhance diffusion rates as a result of their open structure made of highly interconnected pores.[147] The nature of the probe was also found to affect its permeation through the film. For example, positively-charged redox probes such as $Ru(NH_3)_6^{3+}$ or $Ru(bpy)_3^{2+}$ were found to accumulate in the mesoporous structure due to favourable electrostatic interactions with the internal

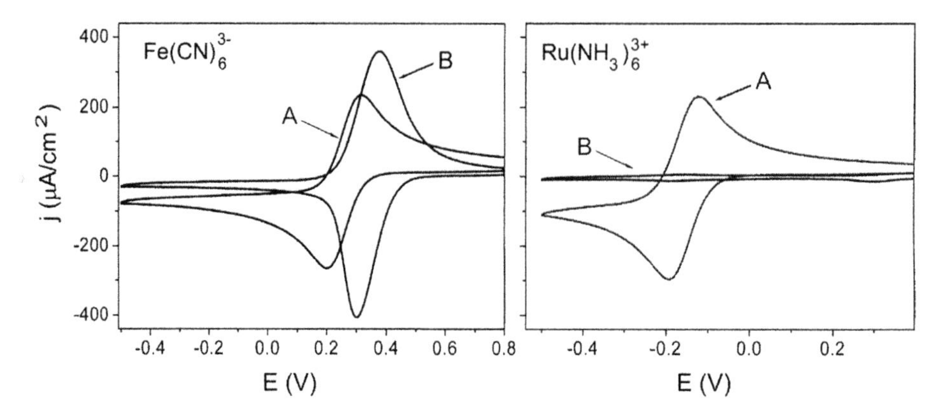

Figure 20.11 Voltammetric response of FTO electrodes: bare (A) and coated with APTES-functionalized (B) mesoporous SiO_2 films in 1 mM solutions of $Fe(CN)_6^{3-}$ (left) or $Ru(NH_3)_6^{3+}$ (right) at pH 3. Reprinted from Fattakhova-Rohlfing *et al.*[84], Copyright (2007) American Chemical Society.

surface of mesochannels (the silica surface is negatively-charged at pH above 2[149]). This led to enhanced peak currents while the response to anionic probes was much lower as a consequence of electrostatic repulsion.[79,87,88]

The presence of organo-functional groups attached to the walls of the mesopore channels is likely to change the permeation characteristics of the mesoporous film and to tune its permselective properties. For example, deposition of a MPS film on F-doped tin oxide (FTO) substrate, which was then grafted with aminopropyl groups that were protonated to form the corresponding ammonium moieties, led to significant increase in the electrode response to $Fe(CN)_6^{3-}$ (accumulation by ion exchange) while blocking that of $Ru(NH_3)_6^{3+}$ (electrostatic repulsion), as shown on Fig. 20.11.[84] This is opposite to the behaviour observed on pristine silica films, indicating the possible tuning of charge selectivity in mesoporous films by appropriate functionalization. Another nice example is based on the use of an azobenzene-containing organosilane to form a periodic organosilica framework on a transparent indium-tin oxide substrate, which was subjected to photoregulation of mass transport.[81] Indeed, the reversible photoresponsive conformation change (cis or trans) of the azobenzene groups covalently bonded to the mesopore walls enables to tune the entrance to mesopore channels by opening and closing a nanoscale valve, which resulted in corresponding variations in current response to solution-phase redox probes, with sensitivities inversely proportional to the size of the probe (as demonstrated using ferrocene dimethanol and ferrocene dimethanol diethylene glycol probes).

4. Applications

Advanced applications of electrodes modified with microporous zeolites and ordered mesoporous (organo)silica materials were mostly developed in the field of electroanalysis.[5,6,23,24] They exploit the remarkable and sometimes unique properties of the modifier, which can be advantageously coupled to redox processes for practical applications. Some examples are briefly described in the following sections.

4.1. Preconcentration electroanalysis and permselectivity

CMEs have been widely used for the voltammetric detection of target analytes subsequent to open-circuit accumulation.[1] In doing so, the recognition properties of the modifier are exploited for analyte preconcentration on the electrode surface prior to its electrochemical determination. Both electrodes modified with zeolites and functionalized MPSs have been used for that purpose. The sensitivity of such detection scheme is primarily related to the effectiveness of the accumulation process, which is expected to be as high as fast would be the transport of the analyte from the dilute solution to the active sites located in the porous material. The selectivity of the process is connected to the nature of the interaction and particular affinity between the modifier and the analyte.

Preconcentration electroanalysis at ZMEs involves the accumulation of cationic analytes by ion exchange from diluted solutions in order to increase their concentration at the electrode/solution interface and thereby to enhance the sensitivity of the subsequent voltammetric detection. Materials used for that purpose are aluminium-rich zeolites, i.e., A, X & Y types, because they are characterized by high exchange capacities. Several analytes have been determined according to this detection scheme, including Ag^{+}, Cu^{2+}, Hg^{2+}, Zn^{2+}, Cd^{2+}, Pb^{2+}, paraquat, diquat, or dopamine.[36,49,75,97,150–155] The preferred electrode configuration was most often zeolite-modified carbon paste electrode (ZMCPE) because of fast response time and ease of regeneration by mechanical polishing. The only drawback of ZMCPE is possible memory effects due to the presence of residual solution in the bulk electrode in case of prolonged exposure to the preconcentration medium.[38] Performance can be improved by the resort to screen-printed zeolite-modified electrodes prepared by the thick-film technology, which resulted in much less solution imbibition and enhanced accumulation efficiency.[49] Another approach to reduce memory effects at ZMCPE is to replace the conventionally used mineral oil binder by solid paraffin, which makes the composite more hydrophobic and thus less subject to impregnation by aqueous solutions,[156] as illustrated in Fig. 20.12 for ZMCPEs highly charged with Cu^{2+} ions.

Ordered MPS materials, and especially those containing organo-functional groups, have been largely used as preconcentration agents prior to voltammetric determinations. Actually, the well-ordered mesoporous (organo)silica materials are of particular interest to ensure great accessibility to active centres and fast mass transport due to a regular porosity and open structures as well as good mesopore connectivity in some mesostructures. The uniform pore structure of these mesoporous solids allows vastly improved access of guest species to the binding centres in comparison to their non-ordered silica gel homologues, as demonstrated for $Hg^{(II)}$ binding to thiol-functionalized samples, either obtained by post-synthesis grafting[157–159] or by the co-condensation route,[160,161] for

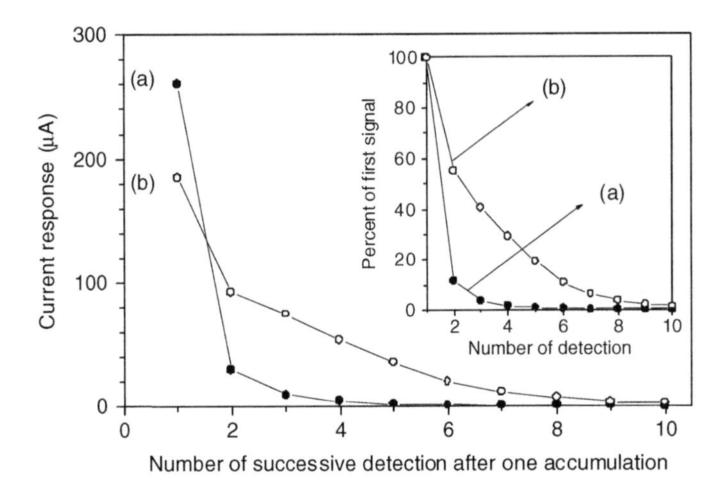

Figure 20.12 Peak current response for release of Cu^{2+} from ZMCPEs into 0.1 M KNO_3, after one open-circuit accumulation from a 0.01 M Cu^{2+} for 30 min. ZMCPEs were prepared with (a) solid paraffin and (b) mineral oil. Relative normalized responses are given in inset. From Walcarius *et al.*[156], with permission of Springer-Verlag GmbH.

which 100% accessibility (i.e., all the SH binding sites complexed with Hg^{II}) has been reported. This is better than the less-than-complete filling levels always observed for corresponding non-ordered adsorbents. This advantage is however limited to the case of neutral complexes (e.g., $\equiv SiO_2-(CH_2)_3-S-HgOH$) as restricted access was reported when charged moieties (e.g., $\equiv SiO_2-(CH_2)_3-S-Hg^+$) are formed on the walls of long mesopore channels due to local repulsive effects.[162] Highly ordered mesoporous materials are also very attractive because they contribute to enhance significantly diffusion rates of host species in a constrained environment, in comparison to more hindered mass transport in non-ordered functionalized silica gel homologues.[159,163] This is illustrated in part A of Fig. 20.13 for $Hg^{(II)}$ uptake by four different silica materials grafted with thiol groups, characterized by various pore sizes, and exhibiting a well-defined mesostructure or a disordered one (see figure legend for explanation). One can clearly notice faster mass transfer rates in the uniformly mesostructured materials in comparison to more restricted motion in the non-ordered homologues materials. Applying a suitable diffusion model,[159,163] it is possible to evaluate quantitatively the apparent diffusion coefficients for $Hg^{(II)}$ in these porous solids and the values confirm the above conclusion as diffusion rates were higher by more than one order of magnitude when passing from a thiol-functionalized silica gel of 6 nm pore size to a mesostructured material (SBA-15 type) of the same pore aperture.[159] Note that even faster diffusion rates have been observed in wormlike mesostructures, suggesting long-range versus short-range order-disorder effects.[161] The main consequence of these results is illustrated in part B of Fig. 20.13 for $Hg^{(II)}$ detection at carbon paste electrodes modified with these four materials, showing a 45 times increase in peak currents when passing from a non-ordered thiol-functionalized silica gel of 6 nm pore size to an ordered

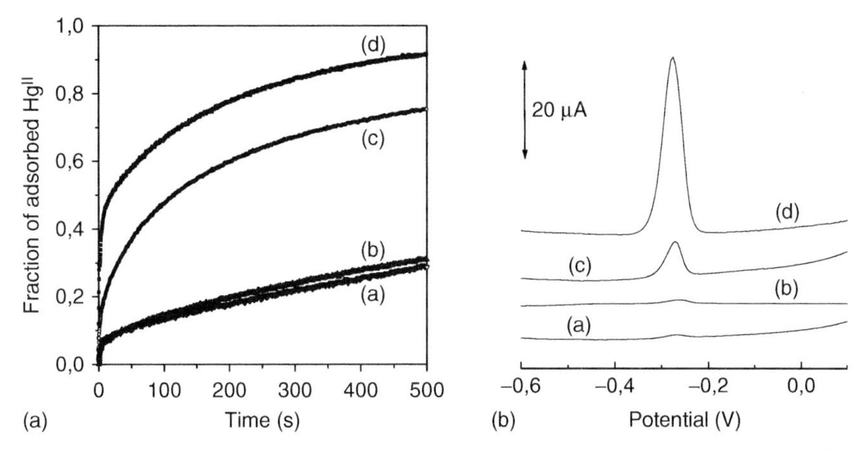

Figure 20.13 (a) Uptake of $Hg^{(II)}$, as a function of time, by (a) MPS-K100, (b) MPS-K60, (c) MPS-MCM30 and (d) MPS-MCM60 from a 0.1 mM $Hg(NO_3)_2$ solution (in 0.5 M HCl); $Hg^{(II)}$ was in excess with respect to the –SH groups. MPS: mercaptopropyl-grafted silica; K100: silica gel with 100 Å pore size (Kieselgel 100); K60: silica gel with 60 Å pore size (Kieselgel 60); MCM30: MPS of the MCM-41 type with 30 Å pore size; MCM60: MPS of the MCM-41 type with 60 Å pore size. (b) Electrochemical curves obtained with the corresponding materials incorporated in carbon paste electrodes, after 2 min accumulation in 1 μM $Hg^{(II)}$, recorded after transfer to an analyte-free electrolyte solution (5% thiourea in 0.1 M HCl) by applying anodic stripping voltammetry. Higher accumulation efficiencies are obtained for the ordered mesoporous structures and within that group for larger pore materials. Reprinted from Walcarius *et al.*[135], reproduced with permission of WILEY-VCH Verlag GmbH.

mesostructure of the same pore aperture. Even more overwhelming, the sensitivity obtained with a MCM-41 material displaying channels two-times smaller in diameter (3 nm) was higher (by a factor 8) than that achieved with the 6-nm pore size amorphous silica gel adsorbent. This demonstrates the definite advantage of mesostructured materials for increasing sensitivity in preconcentration electroanalysis.[135]

Numerous MPSs containing various organo-functional groups covalently attached to the internal surface of mesochannels (scheme 1) have been used as electrode modifiers and then applied to the accumulation of target species via complexation or ion exchange prior to their detection. They include functions like amine, thiol, quaternary ammonium, sulfonate, glycinyl-urea, salicylamide, carnosine, acetamide phosphonic acid or benzothiazolethiol. The resulting modified electrodes have been applied to the determination of mercury,[164–166] lead,[51,80,164,167] copper,[42,88] cadmium,[168] silver,[144] uranium[169] or europium,[52] alone or in mixture (see, e.g., Ref. [44,46]). The electrochemical devices were based either on carbon paste or screen-printed electrodes. Selectivity of the final detection is somewhat connected to the selectivity of the recognition event and interference may arise from competitive binding of species other than the target analyte (decrease of its electrochemical response) or from redox species that are electroactive in the potential region where the analyte is expected to be detected (undesired increase of the signal intensity). Note that electrochemistry enables the

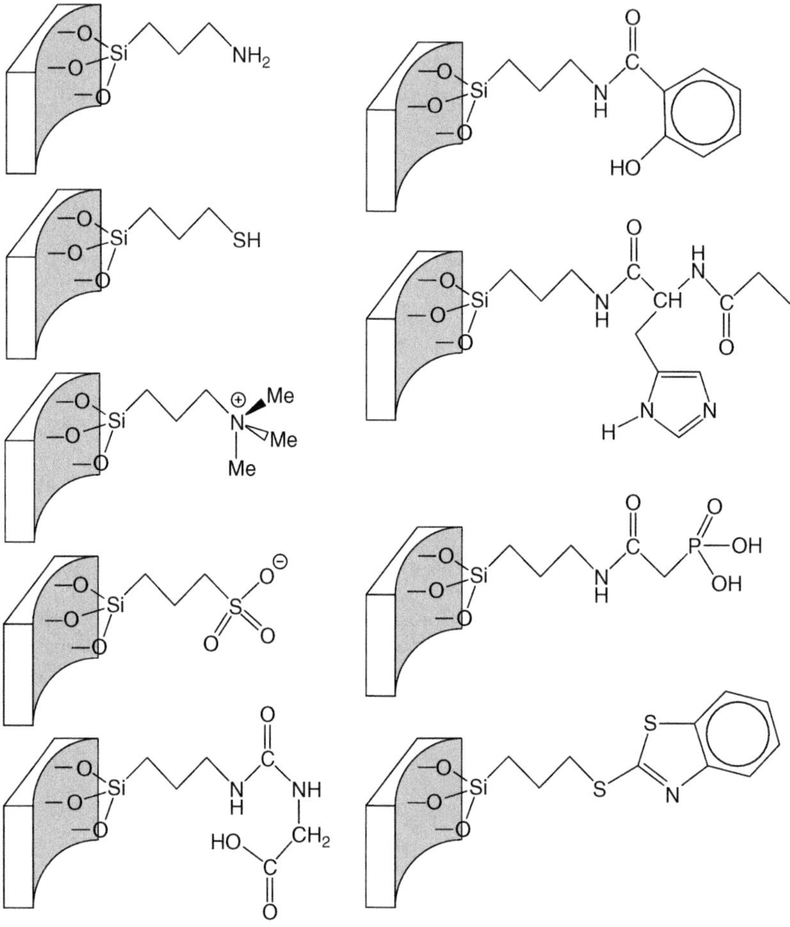

Scheme 1

analysis of metal species in mixture, by providing discrimination on the basis of distinct redox behaviour of the analytes (peak potentials located at different values).[44,46] As aforementioned, high sensitivity is ensured by the regular mesostructure, but this parameter can be also influenced by the electrode configuration, with lower detection limits usually reached with carbon paste composites than with film-based devices.[80,164]

Nevertheless, zeolite and MPS film electrodes have been also applied to electrochemical sensing by providing attractive features due to their permselective properties. For example, the ion exchange properties and size selectivity at the molecular level of continuous zeolite films on electrodes have been exploited for the selective detection of dopamine while rejecting the common interference of ascorbic acid (Fig. 20.14).[77] As shown, the voltammetric response to ascorbic acid is totally suppressed due to the presence of the continuous zeolite film onto the electrode surface (part B of the figure), but the expected preconcentration behaviour of the

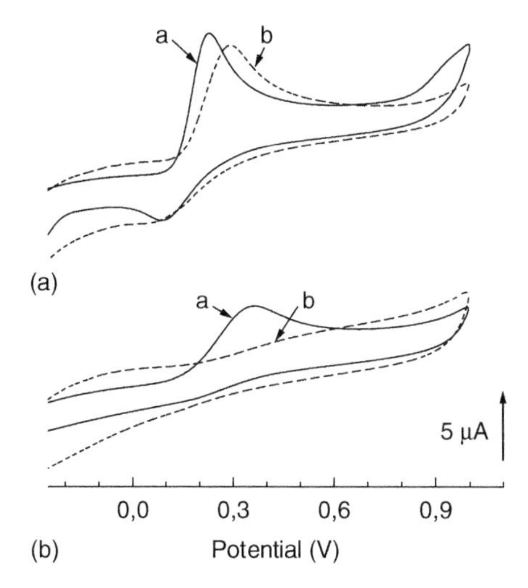

Figure 20.14 Cyclic voltammograms of (a) 1 mM dopamine and (b) 1 mM ascorbic acid, recorded in phosphate buffer (0.1 M, pH 7) respectively at (a) unmodified and (b) zeolite film modified glassy carbon electrodes. Scan rate: 50 mV/s.. From Walcarius *et al.*[77], reproduced with permission of WILEY-VCH Verlag GmbH.

film towards the positively-charged dopamine species did not result in any noticeable enhancement of voltammetric signals, which could be explained by restricted diffusion rates across the microporous film.

Much higher sensitivities were observed in ordered mesoporous deposits. Ion-permselective properties can be induced in these materials using organo-functional groups bearing a net charge, as illustrated for MPS films grafted with aminopropyl groups that were protonated/deprotonated by pH switching.[84] Anionic redox probes were favourably accumulated in acidic medium (amine groups protonated) while rejecting positively-charged species, whereas the material turned cation selective in alkaline medium because of favourable electrostatic interactions with the negatively-charged silica surface.

4.2. Electrocatalysis

Microporous zeolites and MPS materials are attractive hosts for charge-transfer mediators. The latter are necessary to lower overpotentials that may occur in some electrochemical transformations due to limitations in the rate of electron transfer. The principle of mediated electrocatalysis is described by Eqs. (6)–(8) in the case of the reduction of a substrate S_{Ox} into a product P_{Red}. Basically, if the electrochemical reduction of S_{Ox} (Eq. (6)) occurs at high overpotentials ($E_{\mathrm{S}}^{\circ\prime} = E_{\mathrm{S}}^{\circ}$), the use of a mediator couple ($M_{\mathrm{Ox}}/M_{\mathrm{Red}}$) allows to lower the potential barrier by electrogenerating M_{Red} at $E_M^{\circ\prime} = E_M^{\circ}$ (Eq. (8)), which then reacts chemically with S_{Ox} (Eq. (7)). The overall process results therefore in the electrochemical transformation of S_{Ox} at a much lower potential value than the initial substrate in the absence of mediator ($E_M^{\circ\prime} \gg E_{\mathrm{S}}^{\circ\prime}$).

$$S_{Ox} + ne^- \rightarrow P_{Red}(E_S^{\circ\prime} \ll E_S^{\circ}) \tag{6}$$

$$S_{Ox} + M_{Red} \rightarrow P_{Red} + M_{Ox}(\text{occurring if} E_S^{\circ} > E_M^{\circ}) \tag{7}$$

$$M_{Ox} + ne^- \rightarrow M_{Red}(E_M^{\circ\prime} \cong E_M^{\circ}) \tag{8}$$

The mediator can of course be added directly in solution but, for practical reasons, it is more convenient to immobilize it at an electrode surface, to get a reagent-free device. Electrodes modified with zeolites and MPS have been used to this purpose, in the fields of both synthesis and sensing.

Metal complexes encapsulated in zeolites can contribute to electro-assisted oxidation or reduction of organic substrates.[106,110,111,170,171] For example, cobalt-salen complexes entrapped in zeolite Y were found to facilitate the reduction of benzyl bromide and benzyl chloride, with a higher efficiency than that observed for the same catalyst in solution.[106,110] Iron phthalocyanine encapsulated in the same zeolite was a good electrocatalyst for either hydrazine oxidation or molecular oxygen reduction.[111] Oxygen reduction was also improved by Mn-salen[170] or methylviologen[172] supported within zeolite Y particles. Some electrocatalyzed syntheses were also performed using microheterogeneous dispersions (i.e., nanocrystalline particles suspended in solution between feeder electrodes), as exemplified with zeolite-associated complex Co(salen)-Y which was applied to electrocatalyze the reaction of benzyl chloride with CO_2.[110]

Other applications rely on the ability of the immobilized catalysts to improve the selectivity of electroanalytical determinations by lowering the overpotential of the target analyte while turning most interfering species electrochemically 'silent', or to enhance the sensor sensitivity by increasing the current response by continuous electrogeneration-consumption of the active intermediate. The quest for long-term stability of the resulting sensing device requires durable immobilization of the catalyst to avoid its leaching out of the electrode into the external solution, but the mediator must keep enough mobility to act as an effective electron shuttle between the electrode surface and the target analyte. Examples with ZMEs are available for the determination of various analytes, including ascorbic acid or uric acid using electrodes modified with iron(III) exchanged zeolite Y,[173,174] ascorbic acid at mordenite doped with MB,[175] dopamine using the zeolite Y exchanged with triphenylpyrylium cations,[176] and hydrogen peroxide with methylene green-containing zeolite X.[177] Amine-functionalized MPS samples have been protonated to host 12-Tungstophosphoric heteropolyacid and 1:12 phosphomolybdic anions via favourable electrostatic interactions, which have been respectively used as charge-transfer co-factors for the amperometric detection of NO_2^-[178] and ClO_3^-/BrO_3^-.[142] The electrocatalytic reduction of nitrite ions was also achieved by using a redox polymer based on the $[Os(bpy)_2Cl]^+$-poly(4-vinylpyridine) complex immobilized onto a MPS.[179] Titanium-doped MCM-41 particles have been used to improve the electrochemical response to NADH at low overpotential.[64] NADH encapsulated within MPS was also characterized by significant electrocatalytic activity towards the oxidation of 1,4-dihydrobenzoquinone.[141]

4.3. Bioelectrochemistry

Hosting properties of ordered MPS-based materials can be exploited to encapsulate enzymes and other proteins without preventing their biological activity.[180,181] This has generated a huge amount of work on the application of this novel family of materials in bioelectrochemical devices.

Many efforts were first directed to induce direct electrochemistry of heme proteins (i.e., haemoglobin,[182–185] myoglobin,[186,187] cytochrome c[188,189]) that were incorporated in MPS particles or encapsulated in continuous mesoporous layers (both of them deposited as thin films on electrode surfaces). These immobilized proteins retain their biological activity and give rise to well-defined voltammetric signals, indicating that direct electron transfer was indeed possible. Heme proteins immobilized in the mesoporous matrix were also sensitive to the presence of hydrogen peroxide, as a consequence of their peroxidase activity.[182,184–186,189] This resulted in nice electrocatalytic responses towards hydrogen peroxide, as illustrated on Fig. 20.15, and the sensitivity of the device was slightly dependent on the electrode configuration. The biosensor response was notably improved by addition of gold nanoparticles or CdTe quantum dots in the biomaterial layer.[183,187] Of related interest is the use of gold nanoparticles[190] or carbon nanostructures[191] to increase the conductivity in glucose oxidase-based composite film electrodes, as applied to glucose sensing. Other enzymes such as horseradish peroxidase or bi-enzymatic systems made of tyrosinase-peroxidase have been also immobilized in MPS matrices, which were then deposited as thin composite films onto glassy carbon electrodes and applied to hydrogen peroxide biosensing[192] or to the mediator-free detection of phenol, catechol and cresol derivatives,[193] respectively. Another application of MPS in bioelectrochemistry is the electrochemical immunoassay of cardiac

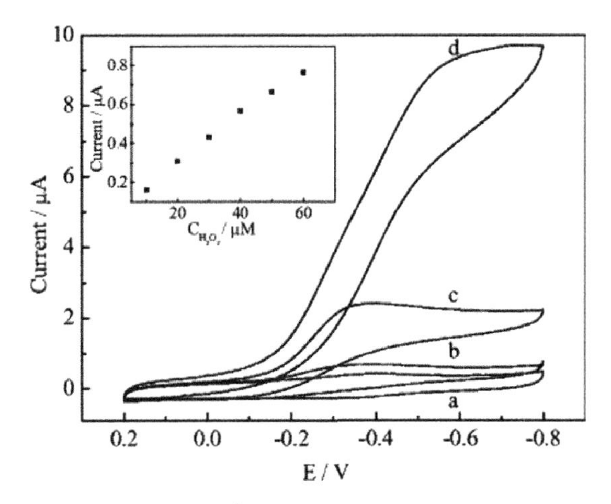

Figure 20.15 CV of SBA-15 with 5×10^{-4} M H_2O_2 (curve a) and Hb/SBA-15/GCE in pH 7.0 PBS with different H_2O_2 concentration at 100 mV/s; 5×10^{-5} M (curve b); 5×10^{-4} M (curve c); 1×10^{-3} M (curve d). Inset: The linear fitting program of the cathodic peak current of Hb for H_2O_2. Reprinted from Liu *et al.*[184], Copyright (2007), with permission from Springer.

troponin I combining the concepts of dual monoclonal antibody 'sandwich' principle, the silver enhancement on gold nanoparticles associated to the antibody immobilized in the mesoporous material, and the anodic stripping voltammetry detection mode.[194] Its sensitivity was directly related to the pore size of the host material, giving rise to better signals with the large pore SBA-15 sample than with the smaller pore MCM-41 material.

Electrochemical immunoassay of cardiac troponin I was also designed using zeolites to immobilize the antibody, but the resulting system was less sensitive than when using mesoporous hosts due to difference in porosity.[194] Cytochrome *c* was also immobilized on electrode surfaces by impregnation within the porous environment provided by zeolite films previously deposited on ITO or carbon electrodes.[195,196] This device not only exhibited direct electron transfer of the immobilized biomolecules but also displayed an excellent response to the reduction of hydrogen peroxide down to 3 nM.[196]

Notwithstanding the impossibility to physically entrap biomolecules in the micropores of zeolites due to size exclusion, other bioelectroanalytical applications of ZMEs have emerged. On one hand, the hydrophilic character of zeolite particles has been exploited to increase the analyte exposition to higher active enzyme quantities in carbon paste electrodes, improving thereby the sensitivity of the amperometric biosensor,[39] and on the other hand the ion exchange capacity of zeolites was applied to concentrate positively-charged mediators likely to enhance the bioelectrochemical event.[197,198] More recent examples involve synergetic effects arising from the incorporation of ferrocene-doped zeolite particles in both glucose oxidase- or dehydrogenases-modified CPEs,[199,200] which led to increased sensitivity of the biosensor response as well as improved stability unlike previous enzyme-based ZMCPEs.

4.4. Indirect amperometric detection

The amperometric detection of non electroactive species is intrinsically not possible by a direct way. As this detection mode is highly desirable (because it is more sensitive than potentiometry), efforts have been made to propose indirect methods enabling detection of non-redox analytes by amperometry. ZMEs have been reported very useful for that purpose, by exploiting the fact that zeolites combine in a single material an exchange capacity and size selectivity properties at the molecular level. The principle of indirect amperometric detection at ZME is directly related to the extrazeolite electron transfer mechanism (Eqs. 2(a) and 2(b)), see section 3.2.) implying that size-excluded electrolyte cations (i.e., bigger than the zeolite pore aperture) did not enable the redox probe to be exchanged, resulting in negligible electroactivity This is illustrated in part A of Fig. 20.16 for the example of K^+ detection at Cu^{2+}-doped ZMCPE. When polarizing the electrode at a potential value likely to reduce Cu^{2+} ions, no detectable current response was observed in a supporting electrolyte made of a large cation (i.e., tetrabutylammonium, TBA^+) because TBA^+ did not enter the zeolite pores, whereas a nice amperometric response was observed each time a sample of solution containing a small cation (i.e., K^+) was injected onto the electrode surface (Fig. 20.16B), as a result of Cu^{2+}

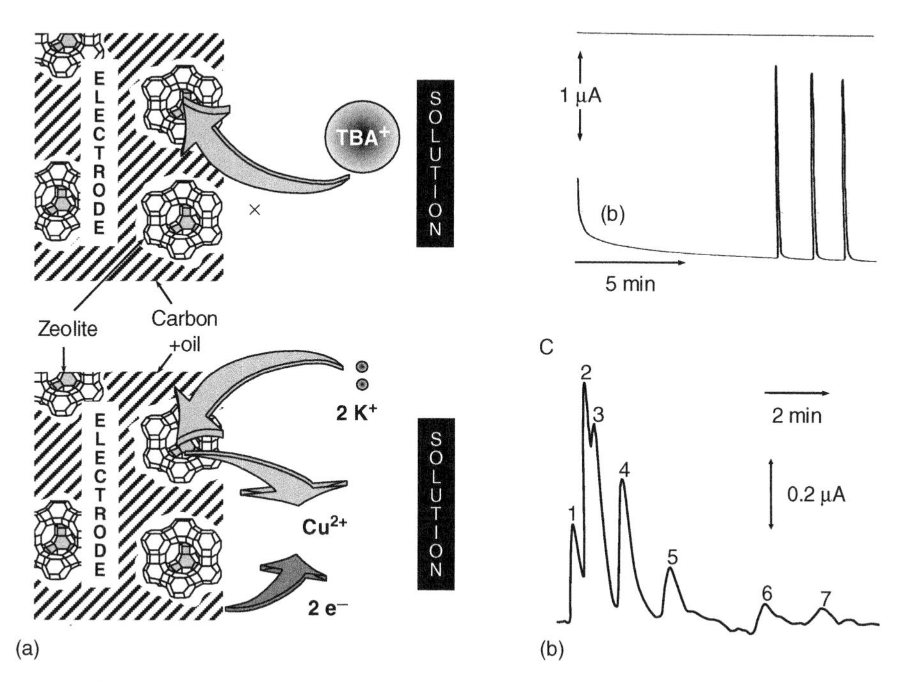

Figure 20.16 (a) Schematic representation of the detection principle for indirect amperometry of K^+ at Cu^{2+}-doped ZMCPE. (b) FIA responses obtained at (curve a) an undoped zeolite Y modified electrode and (curve b) Cu^{2+}-exchanged zeolite Y modified electrode, for three successive injections of 1 mM Na^+ (into 10 mM tetrabutylammonium bromide, TBABr); carrier, 10 mM TBABr; flow rate, 5 ml/min; applied potential, −0.4 V. (c) Chromatogram obtained using a Cu^{2+}-exchanged ZMCPE for a standard mixture of cations: 0.7 mM Li^+ (1), 0.2 mM Na^+ (2), 0.3 mM NH_4^+ (3), 0.13 mM K^+ (4), 0.04 mM Cs^+ (5), 0.2 mM Mg^{2+} (6), and 0.13 mM Ca^{2+} (7); mobile phase, 3 mM HNO_3 + 0.1 mM EDTA; flow rate, 1 ml/min; potential, −0.4 V. From Walcarius[201] (a, b) and Walcarius *et al.*[202] (c), reproduced with permission from Elsevier Science (a, b) and WILEY-VCH Verlag GmbH (c).

exchange for K^+ and subsequent Cu^{2+} reduction. A detailed study on the factors affecting such indirect amperometric detection of nonelectroactive cations is available,[201] reporting that the sensitivity of this method is mainly governed by diffusion of both the electron transfer co-factor (i.e., Cu^{2+} in Fig. 20.16A) and the cationic analyte (i.e., K^+ in Fig. 20.16A). Better sensitivity was thus obtained by using more open zeolites (such as faujasite) and small mediators (such as Ag^+) and the selectivity series for monovalent cations in aqueous medium was determined from their hydrated size, their response being as large as small was the hydrated cation ($Cs^+ > K^+ > Na^+ > Li^+$). Bivalent cations gave usually rise to signals higher than the monovalent ones because at the same concentration they are supposed to exchange twice as much mediator. Methylviologen-exchanged zeolite Y modified CPE was applied to both the separate detection of alkali- and alkaline earth metal cations by FIA[101] as well as to their detection in mixture after separation in suppressor-free ion chromatography using a mobile phase containing only size-excluded cations.[203]

Indirect amperometric detection can be performed in the absence of supporting electrolyte, the resulting current response being the sum of a capacitive component (coming from conductivity change in the mobile phase when the analyte pass on the electrode surface) and a faradic one (due to the reduction of the redox mediator from the zeolite).[202] This has allowed developing a novel detector for ion chromatography working with conventional mobile phases and suppressor devices (see an illustrative chromatogram on Fig. 20.16C). Finally, of related interest is the use of a zeolite membrane as an interface between two immiscible electrolyte solutions, which once polarized led to the observation of an amperometric response to quaternary ammonium ions.[204] Selective detection of small cations in the presence of bigger ones was achieved owing to the size selectivity of the aluminosilicate combined with the presence of non porous silica preventing intergrain diffusion of size-excluded species, by using a zeolite Y membrane prepared from pressed discs healed with tetraethoxysilane.[205]

4.5. Miscellaneous

Some other applications involving zeolites and MPS in electrochemistry have been reported, including gas sensors, potentiometry, electrochemiluminescence and photoelectrochemistry.

MPS films deposited on electrode surfaces are sensitive to gas adsorption, resulting in measurable conductivity changes that can be exploited to detect gaseous analytes. Examples are available for monitoring relative humidity changes or alcohol vapours via the fast resistive-type response of MPS films deposited on interdigitated electrodes[82,206–208] or for the determination of gaseous CO, H_2, or CH_4 using pressed pellets of SnO_2-MCM-41 particles.[209] NO and NOx sensing via surface photovoltage measurements was achieved with the aid of metal-insulator-semiconductor devices comprising a MPS film.[210,211]

Zeolite-polymer composite membranes as well as binder-free zeolite membranes have been prepared for applications in potentiometric sensing. Several alkali-metal and metal cations have been potentiometrically detected using zeolite/epoxy membranes but significant selectivity of Cs^+ ions over other ions was usually observed due to the preferential hosting properties of the microporous solid for this small (hydrated) cation.[212–214] Nernstian or nearly-Nernstian responses for analyte concentrations extended from 1 to 10^{-4} M, with a detection limit of about 10^{-5} M, using mordenite type zeolites. Improving the zeolite/epoxy adhesion led to rise to ideal Nernstian responses.[215] The great affinity of clinoptilolite for ammonium cations was exploited for the sensitive potentiometric sensing of this analyte (detection limit down to 10 nM for a zeolite membrane deposited onto an ion-sensitive field-effect transistor device[216]), which was then applied in an urea-based electrochemical biosensor.[90] Binder-free pressed discs of zeolite powder and free-standing membranes of inter-grown zeolite crystals (either formed by a one-step synthesis or applied to heal defects in the pressed structures) showed selective potentiometric response to cations likely to exchange freely in the interior of the zeolite structure whereas no detectable signal was observed for size-excluded cations, with better selectivity for coherent films because of the absence of intergrain defects.[217]

Sensitive potentiometric detection of Cd^{2+} and Pb^{2+} ions (down to 1 μM) was also achieved with free-standing zeolite membranes obtained by treating pressed discs of zeolite with tetraethoxysilane.[218]

Electrogenerated chemiluminescence (ECL) applications have been developed on the basis of MPS-modified electrodes. CPEs modified with $Ru(bpy)_3^{2+}$-doped sulfonic acid-functionalized MCM-41 particles gave rise to well-defined ECL response, which was much more sensitive when using an ionic liquid as the carbon paste binder instead of the more conventional silicone oil binder (Fig. 20.17). The ECL response was found to be very sensitive to the presence of tripropylamine and this was exploited for the flow-injection analysis of this analyte in the concentration range extended from 2.2×10^{-8} to 1×10^{-5} M, with a detection limit of 7.2 nM.[219] Covalent attachment of $Ru(bpy)_3^{2+}$ to the walls of MPS deposited onto the surface of an ITO electrode exhibited both photovoltaic behaviour and sensitive ECL signals.[220] Periodic mesoporous organosilica structures of MCM-41 or SBA-15 type containing 9,10-diarylanthracene units have been prepared and used for the construction of light-emitting diodes for which much higher ECL sensitivity was observed with smaller pore material.[68]

The last category concerns some photoelectrochemical applications based on ZMEs. On the basis of the photocatalytic behaviour of AgCl-coated electrodes in water (by the electrochemical re-oxidation of metallic silver produced during the process[221]), silver-doped zeolite A-modified electrodes immersed in a chloride

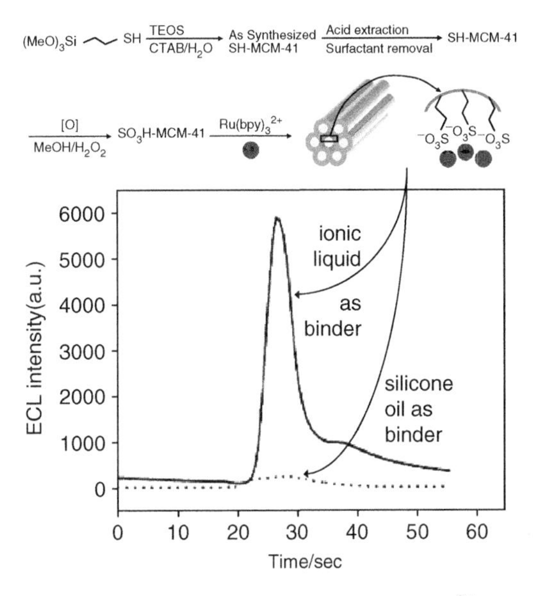

Figure 20.17 Schematic illustration of the preparation of $Ru(bpy)_3^{2+}$ ion-exchanged in sulfonic acid-functionalized MCM-41 particles (Top) and ECL intensity of this system incorporated in carbon paste electrodes using either a silicone oil (dotted line) or an ionic liquid (solid line) as binder, for a solution containing 10 μM tripropylamine in 0.1 M phosphate buffer (pH 7.5) at a scan rate of 50 mV/s. Adapted from Li *et al.*[219], Copyright (2007), with permission from the Royal Society of Chemistry.

solution have been applied to the photocatalytic oxidation of water to molecular oxygen.[222] Silver(0)-Y zeolite exchanged with methylviologen was reported to enable photoreduction of intrazeolitic methylviologen by silver clusters stabilized in/on zeolite Y.[223] Finally, zeolite particles exchanged with $Ru(bpy)_3^{2+}$ and doped with TiO_2 nanoparticles have been applied to electron transfer photocatalysis[224] for which the role of TiO_2 was to ensure electron relay to electroactive species entrapped in the microporous zeolite.

5. Conclusions

The attractive properties of microporous zeolites and ordered MPS-based materials can be advantageously exploited for electrochemical applications. In spite of their insulating characteristics, these solids have been found to be suitable electrode modifiers that bring interesting features at an electrode/solution interface. This has opened the door to novel advanced applications in various fields of electrochemistry, especially in electroanalysis. Most electron transfer reactions involved associated mass transport so that the sensitivity of the devices was often limited by diffusion processes. One can state that ZMEs have attained their maturity whereas the implication of ordered mesoporous materials in electrochemical science has just started and would lead to significant further developments in view if the particularly rich chemistry likely to provide a wide variety of materials with tailor-made properties.

REFERENCES

[1] Murray, R. W., *Molecular design of electrodes surfaces*, in: *Techniques of Chemistry*, Vol. 22, Wiley, New York, 1992.
[2] Rolison, D. R., *Chem. Rev.* **1990,** *90,* 867.
[3] Rolison, D. R., *Stud. Surf. Sci. Catal.* **1994,** *85,* 543.
[4] Bedioui, F., *Coord. Chem. Rev.* **1995,** *144,* 39.
[5] Walcarius, A., *Electroanalysis* **1996,** *8,* 971.
[6] Walcarius, A., *Anal. Chim. Acta* **1999,** *384,* 1.
[7] Fitch, A., *Clays Clay Minerals* **1990,** *38,* 391.
[8] Macha, S. M., Fitch, A., *Mikrochim. Acta* **1998,** *128,* 1.
[9] Navratilova, Z., Kula, P., *Electroanalysis* **2003,** *15,* 837.
[10] Zen, J. M., Kumar, A. S., *Anal. Chem.* **2004,** *76,* 205A.
[11] Mousty, C., *Appl. Clay Sci.* **2004,** *27,* 159.
[12] Lev, O., Wu, Z., Bharathi, S., Glezer, V., Modestov, A., Gun, J., Rabinovich, L., Sampath, S., *Chem. Mater.* **1997,** *9,* 2354.
[13] Alber, K. S., Cox, J. A., *Mikrochim. Acta* **1997,** *127,* 131.
[14] Lin, J., Brown, C. W., *Trends Anal. Chem.* **1997,** *16,* 200.
[15] Collinson, M. M., *Mikrochim. Acta* **1998,** *129,* 149.
[16] Walcarius, A., *Electroanalysis* **1998,** *10,* 1217.
[17] Wang, J., *Anal. Chim. Acta* **1999,** *399,* 21.
[18] Collinson, M. M., *Crit. Rev. Anal. Chem.* **1999,** *29,* 289.
[19] Walcarius, A., *Chem. Mater.* **2001,** *13,* 3351.
[20] Rabinovich, L., Lev, O., *Electroanalysis* **2001,** *13,* 265.

[21] Walcarius, A., *Electroanalysis* **2001,** *13,* 701.
[22] Collinson, M. M., *Trends Anal. Chem.* **2002,** *21,* 30.
[23] Walcarius, A., Mandler, D., Cox, J., Collinson, M. M., Lev, O., *J. Mater. Chem.* **2005,** *15,* 3716.
[24] Walcarius, A., *C. R. Chim.* **2005,** *8,* 693.
[25] Collinson, M. M., *Acc. Chem. Res.* **2007,** *40,* 777.
[26] Breck, D. W., *Zeolite Molecular Sieves. Structure, Chemistry and Uses*, Wiley, New York, 1974.
[27] Dyer, A., *An Introduction to Zeolite Molecular Sieves*, Wiley, Chichester, 1988.
[28] Balkus, K. J., Jr., Gabrielov, A. G., *J. Inclusion Phenom. Mol. Recognit. Chem.* **1995,** *21,* 159.
[29] Kresge, C. T., Leonowicz, M. E., Roth, W. J., Vartuli, J. C., Beck, J. S., *Nature* **1992,** *359,* 710.
[30] Moller, K., Bein, T., *Chem. Mater.* **1998,** *10,* 2950.
[31] Burkett, S. L., Sims, S. D., Mann, S., *Chem. Commun.* **1996,** 1367–1368.
[32] Macquarrie, D. J., *Chem. Commun.* **1996,** 1961–1962.
[33] Hoffmann, F., Cornelius, M., Morell, J., Fröba, M., *Angew. Chem. Int. Ed.* **2006,** *45,* 3216.
[34] Kalcher, K., Kauffmann, J.-M., Wang, J., Svancara, I., Vytras, K., Neuhold, C., Yang, Z., *Electroanalysis* **1995,** *7,* 5.
[35] Kalcher, K., Svancara, I., Metelka, R., Vytras, K., Walcarius, A., in: Grimes, C. A., Dickey, E. C., Pishko, M. V. (Eds.), *Encyclopedia of Sensors,* Vol. 4, American Scientific Publishers, Stevenson Ranch, California, 2006, pp. 283–430.
[36] Wang, J., Martinez, T., *Anal. Chim. Acta* **1988,** *207,* 95.
[37] El Murr, N., Kerkeni, M., Sellami, A., Bentaarit, Y., *J. Electroanal. Chem.* **1988,** *246,* 461.
[38] Walcarius, A., Lamberts, L., Derouane, E. G., *Electrochim. Acta* **1993,** *38,* 2257.
[39] Wang, J., Walcarius, A., *J. Electroanal. Chem.* **1996,** *404,* 237.
[40] Marko-Varga, G., Burestedt, E., Svensson, C. J., Emnéus, J., Gorton, L., Ruzgas, T., Lutz, M., Unger, K. K., *Electroanalysis* **1996,** *8,* 1121.
[41] Walcarius, A., Vromman, V., Bessière, J., *Sens. Actuators B* **1999,** *56,* 136.
[42] Walcarius, A., Despas, C., Trens, P., Hudson, M. J., Bessière, J., *J. Electroanal. Chem.* **1998,** *453,* 249.
[43] Sayen, S., Etienne, M., Bessiere, J., Walcarius, A., *Electroanalysis* **2002,** *14,* 1521.
[44] Yantasee, W., Lin, Y., Fryxell, G. E., Busche, B. J., *Anal. Chim. Acta* **2004,** *502,* 207.
[45] Walcarius, A., Etienne, M., Delacôte, C., *Anal. Chim. Acta* **2004,** *508,* 87.
[46] Yantasee, W., Fryxell, G. E., Conner, M. M., Lin, Y., *J. Nanosci. Nanotechnol.* **2005,** *5,* 1537.
[47] Sayen, S., Walcariu, A., *J. Electroanal. Chem.* **2005,** *581,* 70.
[48] Delacôte, C., Bouillon, J.-P., Walcarius, A., *Electrochim. Acta* **2006,** *51,* 6373.
[49] Walcarius, A., Rozanska, S., Bessière, J., Wang, J., *Analyst* **1999,** *124,* 1185.
[50] Li, J.-P., Peng, T.-Z., Fang, C., *Anal. Chim. Acta* **2002,** *455,* 53.
[51] Yantasee, W., Deibler, L. A., Fryxell, G. E., Timchalk, C., Lin, Y., *Electrochem. Commun.* **2005,** *7,* 1170.
[52] Yantasee, W., Fryxell, G. E., Lin, Y., *Analyst* **2006,** *131,* 1342.
[53] Shaw, B. R., Creasy, K. E., *Anal. Chem.* **1988,** *60,* 1241.
[54] Bedioui, F., de Boysson, E., Devynck, J., Balkus, K. J., Jr., *J. Electroanal. Chem.* **1991,** *315,* 313.
[55] Briot, E., Bedioui, F., *Curr. Top. Electrochem.* **1997,** *4,* 87.
[56] Creasy, K. E., Shaw, B. R., *J. Electrochem. Soc.* **1990,** *137,* 2353.
[57] Gun, J., Tsionsky, M., Rabinovich, L., Golan, Y., Rubinstein, I., Lev, O., *J. Electroanal. Chem.* **1995,** *395,* 57.
[58] Gemborys, H. A., Shaw, B. R., *J. Electroanal. Chem.* **1986,** *208,* 95.
[59] Baker, M. D., Senaratne, C., *Anal. Chem.* **1992,** *64,* 697.
[60] Li, J., Calzaferri, G., *J. Electroanal. Chem.* **1994,** *377,* 163.
[61] Ganesan, V., Ramaraj, R., *Langmuir* **1998,** *14,* 2497.
[62] Domenech, A., Perez-Ramirez, J., Ribera, A., Mul, G., Kapteijn, F., Arends, I. W. C. E., *J. Electroanal. Chem.* **2002,** *519,* 72.
[63] Villemure, G., Pinnavaia, T. J., *Chem. Mater.* **1999,** *11,* 789.
[64] Dai, Z., Lu, G., Bao, J., Huang, X., Ju, H., *Electroanalysis* **2007,** *19,* 604.
[65] Senaratne, C., Zhang, J., Baker, M. D., Bessel, C. A., Rolison, D. R., *J. Phys. Chem.* **1996,** *100,* 5849.

[66] Brouwer, D. H., Baker, M. D., *J. Phys. Chem. B* **1997,** *101,* 10390.
[67] Domenech, A., Garcia, H., Carbonell, E., *J. Electroanal. Chem.* **2005,** *577,* 249.
[68] Alvaro, M., Benitez, M., Cabeza, J. F., Garcia, H., Leyva, A., *J. Phys. Chem. C* **2007,** *111,* 7532.
[69] Lainé, P., Seifert, R., Giovanoli, R., Calzaferri, G., *New J. Chem.* **1997,** *21,* 453.
[70] Li, Z., Lai, C., Mallouk, T. E., *Inorg. Chem.* **1989,** *28,* 178.
[71] Jiang, Y.-X., Si, D., Chen, S.-P., Sun, S.-G., *Electroanalysis* **2006,** *18,* 1173.
[72] Kornic, S., Baker, M., *Chem. Commun.* **2002,** 1700.
[73] Rauch, W. L., Liu, M., *J. Mater. Sci.* **2003,** *38,* 4307.
[74] Zhang, Y., Chen, F., Zhuang, J., Tang, Y., Wang, D., Wang, Y., Dong, A., Ren, N., *Chem. Commun.* **2002,** 2814.
[75] Zhang, Y., Chen, F., Shan, W., Zhuang, J., Dong, A., Cai, W., Tang, Y., *Microporous Mesoporous Mater.* **2003,** *65,* 277.
[76] Berenguer-Murcia, A., Morallon, E., Cazorla-Amoros, D., Linares-Solano, A., *Microporous Mesoporous Mater.* **2003,** *66,* 331.
[77] Walcarius, A., Ganesan, V., Larlus, O., Valtchev, V., *Electroanalysis* **2004,** *16,* 1550.
[78] Li, S., Wang, X., Beving, D., Chen, Z., Yan, Y., *J. Am. Chem. Soc.* **2004,** *126,* 4122.
[79] Song, C., Villemure, G., *Microporous Mesoporous Mater.* **2001,** *44/45,* 679.
[80] Yantasee, W., Lin, Y., Li, X., Fryxell, G. E., Zemanian, T. S., Viswanathan, V. V., *Analyst* **2003,** *128,* 899.
[81] Liu, N., Dunphy, D. R., Atanassov, P., Bunge, S. D., Chen, Z., Lopez, G. P., Boyle, T. J., Brinker, C. J., *Nano Lett.* **2004,** *4,* 551.
[82] Bertolo, J. M., Bearzotti, A., Generosi, A., Palummo, L., Albertini, V. R., *Sens. Actuators B* **2005,** *111–112,* 145.
[83] Etienne, M., Walcarius, A., *Electrochem. Commun.* **2005,** *7,* 1449.
[84] Fattakhova-Rohlfing, D., Wark, M., Rathousky, J., *Chem. Mater.* **2007,** *19,* 1640.
[85] Brinker, C. J., Lu, Y., Sellinger, A., Fan, H., *Adv. Mater.* **1999,** *11,* 579.
[86] Grosso, D., Cagnol, F., Soler-Illia, G. J. A. A., Crepaldi, E. L., Amenitsch, H., Brunet-Bruneau, A., Bourgeois, A., Sanchez, C., *Adv. Funct. Mater.* **2004,** *14,* 309.
[87] Etienne, M., Quach, A., Grosso, D., Nicole, L., Sanchez, C., Walcarius, A., *Chem. Mater.* **2007,** *19,* 844.
[88] Walcarius, A., Sibottier, E., Etienne, M., Ghanbaja, J., *Nat. Mater.* **2007,** *6,* 602.
[89] Brinker, C. J., Dunphy, D. R., *Curr. Opin. Coll. Interface Sci.* **2006,** *11,* 126.
[90] Hamlaoui, M. L., Reybier, K., Marrakchi, M., Jaffrezic-Renault, N., Martelet, C., Kherrat, R., Walcarius, A., *Anal. Chim. Acta* **2002,** *466,* 39.
[91] Arvand-Barmchi, M., Mousavi, M. F., Zanjanchi, M. A., Shamsipur, M., *Sens. Actuators B* **2003,** *96,* 560.
[92] Sohrabnejad, Sh., Zanjanchi, M. A., Arvand, M., Mousavi, M. F., *Electroanalysis* **2004,** *16,* 1033.
[93] Baglio, V., Di Blasi, A., Arico, A. S., Antonucci, V., Antonucci, P. L., Nannetti, F., Tricoli, V., *Electrochim. Acta* **2005,** *50,* 5181.
[94] Takei, T., Yoshimura, K., Yonesaki, Y., Kumada, N., Kinomura, N., *J. Porous Mater.* **2005,** *12,* 337.
[95] Malkaj, P., Dalas, E., Vitoratos, E., Sakkopoulos, S., *J. Appl. Polym. Sci.* **2006,** *101,* 1853.
[96] Shaw, B. R., Creasy, K. E., Lanczycki, C. J., Sargeant, J. A., Tirhado, M., *J. Electrochem. Soc.* **1988,** *135,* 869.
[97] Walcarius, A., Barbaise, T., Bessière, J., *Anal. Chim. Acta* **1997,** *340,* 61.
[98] Baker, M. D., Zhang, J., *J. Phys. Chem.* **1990,** *94,* 8703.
[99] Senaratne, C., Baker, M. D., *J. Phys. Chem.* **1994,** *98,* 13687.
[100] Li, J., Calzaferri, G., *J. Electroanal. Chem.* **1994,** *377,* 163.
[101] Walcarius, A., Lamberts, L., Derouane, E. G., *Electroanalysis* **1995,** *7,* 120.
[102] Baker, M. D., Senaratne, C., Zhang, J., *J. Chem. Soc. Faraday Trans.* **1992,** *88,* 3187.
[103] Walcarius, A., Lamberts, L., Derouane, E. G., *Electrochim. Acta* **1993,** *38,* 2267.
[104] Baker, M. D., Senaratne, C., *Phys. Chem. Chem. Phys.* **1999,** *1,* 1673.
[105] Roué, L., Briot, E., Bedioui, F., *Can. J. Chem.* **1998,** *76,* 1886.
[106] Bedioui, F., de Boysson, E., Devynck, J., Balkus, K. J., Jr., *J. Chem. Soc. Faraday Trans.* **1991,** *87,* 3831.

[107] Gaillon, L., Bedioui, F., Devynck, J., *J. Mater. Chem.* **1994,** *4,* 1215.
[108] Bedioui, F., Devynck, J., Balkus, K. J., Jr., *J. Phys. Chem.* **1996,** *100,* 8607.
[109] Briot, E., Bedioui, F., Balkus, K. J., Jr., *J. Electroanal. Chem.* **1998,** *454,* 83.
[110] Bessel, C. A., Rolison, D. R., *J. Am. Chem. Soc.* **1997,** *119,* 12673.
[111] Vinod, M. P., Das, T. K., Chandwadkar, A. J., Vijayamohanan, K., Chandwadkar, J. G., *Mater. Chem. Phys.* **1999,** *58,* 37.
[112] Doménech, A., Formentin, P., Garcia, H., Sabater, M. J., *Eur. J. Inorg. Chem.* **2000,** 1339.
[113] Vitale, M., Castagnola, N. B., Ortins, N. J., Brooke, J. A., Vaidyalingam, A., Dutta, P. K., *J. Phys. Chem. B* **1999,** *103,* 2408.
[114] Li, Z., Mallouk, T. E., *J. Phys. Chem.* **1987,** *91,* 643.
[115] Li, Z., Wang, C. M., Persaud, L., Mallouk, T. E., *J. Phys. Chem.* **1988,** *92,* 2592.
[116] Rolison, D. R., Stemple, J. Z., *J. Chem. Soc. Chem. Commun.* **1993,** 25.
[117] de Castro-Martins, S., Khouzami, S., Tuel, A., Bentaarit, Y., El Murr, N., Sellami, A., *J. Electroanal. Chem.* **1993,** *350,* 15.
[118] de Castro-Martins, S., Tuel, A., Bentaarit, Y., *Zeolites* **1994,** *14,* 130.
[119] Bodoardo, S., Geobaldo, F., Penazzi, N., Arrabito, M., Rivetti, F., Spano, G., Lamberti, C., Zecchina, A., *Electrochem. Commun.* **2000,** *2,* 349.
[120] Prandi, L., Bodoardo, S., Penazzi, N., Fubini, B., *J. Mater. Chem.* **2001,** *11,* 1495.
[121] Venkatathri, N., Vinod, M. P., Vijayamohanan, K., Sivasanker, S., *J. Chem. Soc. Faraday Trans.* **1996,** *92,* 473.
[122] Bedioui, F., Briot, E., Devynck, J., Balkus, K. J., Jr., *Inorg. Chim. Acta* **1997,** *254,* 151.
[123] Shi, G., Xue, G., Hou, W., Dong, J., Wang, G., *J. Electroanal. Chem.* **1993,** *344,* 363.
[124] Petranovic, N., Susic, M. V., *Zeolites* **1983,** *3,* 271.
[125] Walcarius, A. in: Aurbach, S. M., Carrado, K. A., Dutta, P. K. (Eds.), *Handbook of Zeolite Catalysts and Microporous Materials,* Marcel Dekker, 2003, Chap. 14, pp. 721–783.
[126] Baker, M. D., Senaratne, C., Zhang, J., *J. Phys. Chem.* **1994,** *98,* 1668.
[127] Baker, M. D., Zhang, J., McBrien, M., *J. Phys. Chem.* **1995,** *99,* 6635.
[128] Xiong, W., Baker, M. D., *J. Phys. Chem. B* **2005,** *109,* 13590.
[129] Rolison, D. R., Bessel, C. A., Baker, M. D., Senaratne, C., Zhang, J., *J. Phys. Chem.* **1996,** *100,* 8610.
[130] Walcarius, A., *J. Solid State Electrochem.* **2006,** *10,* 469.
[131] Ganesan, V., Walcarius, A., *Langmuir* **2004,** *20,* 3632.
[132] Walcarius, A., Ganesan, V., *Langmuir* **2006,** *22,* 469.
[133] Walcarius, A., Bessière, J., *Chem. Mater.* **1999,** *11,* 3009.
[134] Walcarius, A., Lüthi, N., Blin, J.-L., Su, B.-L., Lamberts, L., *Electrochim. Acta* **1999,** *44,* 4601.
[135] Walcarius, A., Etienne, M., Sayen, S., Lebeau, B., *Electroanalysis* **2003,** *15,* 414.
[136] Jiang, Y. X., Song, W. B., Liu, Y., Wei, B., Cao, X. C., Xu, H. D., *Mater. Chem. Phys.* **2000,** *62,* 109.
[137] Ganesan, R., Viswanathan, B., *J. Mol. Catal. A* **2002,** *181,* 99.
[138] Ganesan, R., Viswanathan, B., *J. Mol. Catal. A* **2004,** *223,* 21.
[139] Jiang, Y. X., Ding, N., Sun, S. G., *J. Electroanal. Chem.* **2004,** *563,* 15.
[140] Bodoardo, S., Borello, L., Fiorilli, S., Garrone, E., Onida, B., Otero Arean, C., Penazzi, N., Turnes Palomino, G., *Microporous Mesoporous Mater.* **2005,** *79,* 275.
[141] Domenech, A., Garcia, H., Marquet, J., Bourdelande, J. L., Herance, J. R., *Electrochim. Acta* **2006,** *51,* 4897.
[142] Li, L., Li, W., Sun, C., Li, L., *Electroanalysis* **2002,** *14,* 368.
[143] Fattakhova-Rohlfing, D., Rathousky, J., Rohlfing, Y., Bartels, O., Wark, M., *Langmuir* **2005,** *21,* 11320.
[144] Etienne, M., Cortot, J., Walcarius, A., *Electroanalysis* **2007,** *19,* 129.
[145] Domenech, A., Alvaro, M., Ferrer, B., Garcia, H., *J. Phys. Chem. B* **2003,** *107,* 12781.
[146] Bard, A. J., Faulkner, L. R., *Electrochemical Methods, Fundamentals and Applications* (2nd edition), Wiley, New York, 2001.
[147] Sel, O., Sallard, S., Brezesinski, T., Rathousky, J., Dunphy, D. R., Collord, A., Smarsly, B. M., *Adv. Funct. Mater.* **2007,** *17,* 3241.
[148] Wei, T.-C., Hillhouse, H. W., *Langmuir* **2007,** *23,* 5689.

[149] Iler, R. K., *The Chemistry of Silica*, Wiley, New York, 1979.
[150] Hernandez, P., Alda, E., Hernandez, L., *Fresenius J. Anal. Chem.* **1987,** *327,* 676.
[151] Wang, J., Walcarius, A., *J. Electroanal. Chem.* **1996,** *407,* 183.
[152] Bing, C., Kryger, L., *Talanta* **1996,** *43,* 153.
[153] Chen, B., Goh, N.-K., Chia, L.-S., *Electrochim. Acta* **1997,** *42,* 595.
[154] Mogensen, L., Kryger, L., *Electroanalysis* **1998,** *10,* 1285.
[155] Kilinc Alpat, S., Yuksel, U., Akcay, H., *Electrochem. Commun.* **2005,** *7,* 130.
[156] Walcarius, A., Mariaulle, P., Lamberts, L., *J. Solid State Electrochem.* **2003,** *7,* 671.
[157] Feng, X., Fryxell, G. E., Wang, L. Q., Kim, A. Y., Liu, J., Kemner, K. M., *Science* **1997,** *276,* 923.
[158] Mercier, L., Pinnavaia, T. J., *Adv. Mater.* **1997,** *9,* 500.
[159] Walcarius, A., Etienne, M., Lebeau, B., *Chem. Mater.* **2003,** *15,* 2161.
[160] Brown, J., Richer, R., Mercier, L., *Microporous Mesoporous Mater.* **2000,** *37,* 41.
[161] Walcarius, A., Delacôte, C., *Chem. Mater.* **2003,** *15,* 4181.
[162] Walcarius, A., Delacôte, C., *Anal. Chim. Acta* **2005,** *547,* 3.
[163] Walcarius, A., Etienne, M., Bessiere, J., *Chem. Mater.* **2002,** *14,* 2757.
[164] Yantasee, W., Lin, Y., Zemanian, T. S., Fryxell, G. E., *Analyst* **2003,** *128,* 467.
[165] Etienne, M., Delacôte, C., Walcarius, A., in: Nunez, M. (Ed.), *Progress in Electrochemistry Research,* Nova Science Publishers, Hauppauge, New York, 2005, pp. 145–184.
[166] Jieumboué Tchinda, A., Ngameni, E., Walcarius, A., *Sens. Actuators B* **2007,** *121,* 113.
[167] Yantasee, W., Timchalk, C., Fryxell, G. E., Dockendorff, B. P., Lin, Y., *Talanta* **2005,** *68,* 256.
[168] Cesarino, I., Marino, G., do Rosario Matos, J., Cavalheiro, E. T. G., *J. Braz. Chem. Soc.* **2007,** *18,* 810.
[169] Yantasee, W., Lin, Y., Fryxell, G. E., Wang, Z., *Electroanalysis* **2004,** *16,* 870.
[170] Gaillon, L., Sajot, N., Bedioui, F., Devynck, J., Balkus, K. J., Jr., *J. Electroanal. Chem.* **1993,** *345,* 157.
[171] Balkus, K. J., Jr., Khanmamedova, A. K., Dixon, K. M., Bedioui, F., *Appl. Catal. A* **1996,** *143,* 159.
[172] Creasy, K. E., Shaw, B. R., *Electrochim. Acta* **1988,** *33,* 551.
[173] Jiang, Y., Zou, M., Yuan, K., Xu, H., *Electroanalysis* **1999,** *11,* 254.
[174] Mazloum Ardakani, M., Akrami, Z., Kazemian, H., Zare, H. R., *J. Electroanal. Chem.* **2006,** *586,* 31.
[175] Arvand, M., Sohrabnezhad, Sh., Mousavi, M. F., Shamsipur, M., Zanjanchi, M. A., *Anal. Chim. Acta* **2003,** *491,* 193.
[176] Doménech, A., Doménech-Carbo, M. T., Garcia, H., Galletero, M. S., *Chem. Commun.* **1999,** 2173.
[177] Gligor, D., Muresan, L. M., Dumitru, A., Popescu, I. C., *J. Appl. Electrochem.* **2007,** *37,* 261.
[178] Li, W., Li, L., Wang, Z., Cui, A., Sun, C., Zhao, J., *Mater. Lett.* **2001,** *49,* 228.
[179] Xie, F., Li, W., He, J., Yu, S., Fu, T., Yang, H., *Mater. Chem. Phys.* **2004,** *86,* 425.
[180] Hartmann, M., *Chem. Mater.* **2005,** *17,* 4577.
[181] Yiu, H. H. P., Wright, P. A., *J. Mater. Chem.* **2005,** *15,* 3690.
[182] Dai, Z., Liu, S., Ju, H., Chen, H., *Biosens. Bioelectron.* **2004,** *19,* 861.
[183] Xian, Y., Xian, Y., Zhou, L., Wu, F., Ling, Y., Jin, L., *Electrochem. Commun.* **2007,** *9,* 142.
[184] Liu, Y., Xu, Q., Feng, X., Zhu, J.-J., Hou, W., *Anal. Bioanal. Chem.* **2007,** *387,* 1553.
[185] Zhang, L., Zhang, Q., Li, J., *Electrochem. Commun.* **2007,** *9,* 1530.
[186] Dai, Z., Xu, X., Ju, H., *Anal. Biochem.* **2004,** *332,* 23.
[187] Zhang, Q., Zhang, L., Liu, B., Lu, X., Li, J., *Biosens. Bioelectron.* **2007,** *23,* 695.
[188] Washmon-Kriel, L., Jimenez, V. L., Balkus, K. J., Jr., *J. Mol. Catal. B* **2000,** *10,* 453.
[189] Zhang, X., Wang, J., Wu, W., Qian, S., Man, Y., *Electrochem. Commun.* **2007,** *9,* 2098.
[190] Bai, Y., Yang, H., Yang, W., Li, Y., Sun, C., *Sens. Actuators B* **2007,** *124,* 179.
[191] Wu, S., Ju, H., Liu, Y., *Adv. Funct. Mater.* **2007,** *17,* 585.
[192] Dai, Z., Ju, H., Chen, H., *Electroanalyis* **2005,** *17,* 862.
[193] Dai, Z., Xu, X., Wu, L., Ju, H., *Electroanalyis* **2005,** *17,* 1571.
[194] Guo, H., He, N., Ge, S., Yang, D., Zhang, J., *Microporous Mesoporous Mater.* **2005,** *85,* 89.
[195] Dai, Z., Liu, S., Ju, H., *Electrochim. Acta* **2004,** *49,* 2139.

[196] Yu, T., Zhang, Y., You, C., Zhuang, J., Wang, B., Liu, B., Kang, Y., Tang, Y., *Chem. Eur. J.* **2006,** *12,* 1137.
[197] Liu, B., Yan, F., Kong, J., Deng, J., *Anal. Chim. Acta* **1999,** *386,* 31.
[198] Chen, C. F., Wang, C. M., *J. Electroanal. Chem.* **1999,** *466,* 82.
[199] Serban, S., El Murr, N., *Anal. Lett.* **2003,** *36,* 1739.
[200] Serban, S., El Murr, N., *Biosens. Bioelectron.* **2004,** *20,* 161.
[201] Walcarius, A., *Anal. Chim. Acta* **1999,** *388,* 79.
[202] Walcarius, A., Mariaulle, P., Louis, C., Lamberts, L., *Electroanalysis* **1999,** *11,* 393.
[203] Walcarius, A., Lamberts, L., *Anal. Lett.* **1998,** *31,* 585.
[204] Dryfe, R. A. W., Holmes, S. M., *J. Electroanal. Chem.* **2000,** *483,* 144.
[205] Senthilkumar, S., Dryfe, R. A. W., Saraswathi, R., *Langmuir* **2007,** *23,* 3455.
[206] Innocenzi, P., Martucci, A., Guglielmi, M., Bearzotti, A., Traversa, E., *Sens. Actuators B* **2001,** *76,* 299.
[207] Bertolo, J. M., Bearzotti, A., Falcaro, P., Traversa, E., Innocenzi, P., *Sensor Lett.* **2003,** *1,* 64.
[208] Bearzotti, A., Bertolo, J. M., Innocenzi, P., Falcaro, P., Traversa, E., *J. Eur. Ceram. Soc.* **2004,** *24,* 1969.
[209] Li, G., Kawi, S., *Sens. Actuators B* **1997,** *59,* 1.
[210] Zhou, H. S., Yamada, T., Asai, K., Honma, I., Uchida, H., Katsube, T., *Jpn. J. Appl. Phys.* **2001,** *40,* 7098.
[211] Yamada, T., Zhou, H. S., Uchida, H., Tomita, M., Ueno, Y., Honma, I., Asai, K., Katsube, T., *Microporous Mesoporous Mater.* **2002,** *54,* 269.
[212] Johansson, G., Risinger, L., Falth, L., *Anal. Chim. Acta* **1980,** *119,* 25.
[213] Evmiridis, N. P., Demertzis, M. A., Vlessidis, A. G., *Fresenius J. Anal. Chem.* **1991,** *340,* 145.
[214] Arvand-Barmchi, M., Mousavi, M. F., Zanjanchi, M. A., Shamsipur, M., *Sens. Actuators B* **2003,** *96,* 560.
[215] Demertzis, M., Evmiridis, N. P., *J. Chem. Soc. Faraday Trans. 1* **1986,** *82,* 3647.
[216] Hamlaoui, M. L., Kherrat, R., Marrakchi, M., Jaffrezic-Renault, N., Walcarius, A., *Mater. Sci. Eng. C* **2002,** *21,* 25.
[217] King, A. J., Lillie, G. C., Cheung, V. W. Y., Holmes, S. M., Dryfe, R. A. W., *Analyst* **2004,** *129,* 157.
[218] Senthilkumar, S., King, A. J., Holmes, S. M., Dryfe, R. A. W., Saraswathi, R., *Electroanalysis* **2006,** *18,* 2297.
[219] Li, J., Huang, M., Liu, X., Wei, H., Xu, Y., Xu, G., Wang, E., *Analyst* **2007,** *132,* 687.
[220] Font, J., de March, P., Busque, F., Casas, E., Benitez, M., Teruel, L., Garcia, H., *J. Mater. Chem.* **2007,** *17,* 2336.
[221] Lanz, M., Calzaferri, G., *J. Photochem. Photobiol. A Chem.* **1997,** *109,* 87.
[222] Calzaferri, G., Gfeller, N., Pfanner, K., *J. Photochem. Photobiol. A Chem.* **1995,** *87,* 81.
[223] Szulbinski, W. S., *Inorg. Chim. Acta* **1998,** *269,* 253.
[224] Bossmann, S. H., Turro, C., Schnabel, C., Pokhrel, M. R., Payawan, L. M., Jr., Baumeister, B., Wörner, M., *J. Phys. Chem. B* **2001,** *105,* 5374.

CHAPTER 21

Nanoparticle Doped Photopolymers for Holographic Applications

Izabela Naydenova *and* Vincent Toal

Contents

Abstract

This chapter presents a review of holographic recording materials and their main areas of application. The principles of holographic recording in photopolymers are discussed and the advantages of adding nanoparticles to the photopolymer are demonstrated with supporting data from experiments with these nanocomposite materials. The applications of holography, which particularly benefit from the addition of nanoparticles, are highlighted, including optical data storage and gratings for spectroscopic instruments and sensors.

Keywords: Data storage, Diffraction gratings, Holography, Holographic recording materials, Photopolymers, Nanocomposites, Nanoparticles, Photopolymerizable, Sensors, Zeolites

Ordered Porous Solids
DOI: 10.1016/B978-0-444-53189-6.00021-4

1. Introduction

Holography as a method for full, 3D recording of images has been known since it is discovered by Gabor in 1947.[1] There are two main driving factors determining the progress of optical holography—the development of powerful coherent light sources with a wide range of available wavelengths and the development of holographic recording materials. As optical holography requires coherent light sources its progress towards more practical applications started with the appearance of the laser in 1960.[2,3] Demonstrations of the first holographically recorded images using laser light were reported by Emmett Leith, Juris Upatnieks,[4,5] and Yuri Denisyuk.[6,7] Since then lasers have undergone tremendous developments and nowadays relatively cheap lasers operating at large variety of wavelengths from about 670 to 325 nm, characterised by a long coherence length, stability, and sufficient output intensity, are available. This makes holography affordable for many scientific laboratories in the world and contributes to rapid growth of several holographic applications. Some of these applications are holographic optical memories,[8–14] holographic sensors,[15–17] holographic optical elements,[18,19] holographic electro-optical switchable devices,[20–24] security holograms,[25–28] holographic systems for non-destructive testing of mechanical and biological systems,[29–32] and art holography involving full colour holography.[33,34] The second factor playing a crucial role in the progress of different holographic applications is the development of novel holographic recording materials with greatly improved characteristics. The list of requirements for properties of holographic recording material is highly demanding and includes the following:

- high sensitivity ($\sim 10^3$ cm/J)
- low scatter (bidirectional scattering distribution function, BSDF $< 10^{-5}$)
- spatial resolution ranging from hundreds up to several thousands of lines per mm for reflection holography
- large dynamic range related to large photoinduced changes in the material's refractive index ($\Delta n > 10^{-2}$)
- low shrinkage (<0.1 %)
- temporal stability
- long shelf life
- convenient format allowing easy handling—dry layers with controlled thickness

Some of the requirements depend more on the actual holographic application. Irreversible recording is required for most of the applications while reversibility is sought for write-many, read-many holographic memories and some light controlled holographic optical elements. For example holographic optical memory requires highly sensitive, low scattering, thick layers with negligible shrinkage and large dynamic range while for holographic sensors large thickness is not necessary and shrinkage may not be an issue.

A large variety of the available holographic recording materials[35–37] include silver-halide emulsions,[38–40] dichromated gelatine,[41–43] photoresists,[44–46] photo-thermoplastics,[47–49] photochromics,[50–53] photorefractives,[54–56] photopolymers,[57–66] and azodye-containing polymers.[67–70] Each group of materials has different characteristics and is more suitable for a specific application. For instance silver-halide emulsions and dichromated gelatin are suitable for art holography[71,72] as they posses extreme sensitivity and very high spatial resolution of up to 10,000 l/mm[73,74] but are not suitable for holographic memory applications as they require wet processing after the recording.

Photoresists are suitable for mass production of holograms by hot stamping of metal foil using a nickel coated master hologram recorded in the photoresist.[75] Photothermoplastics are mainly used in holographic interferometry applications as they can be processed *in situ*.[76] Post processing is also required for the photoresists and the photothermoplastics and their spatial resolution is somewhat limited—3000 l/mm[77] and 1200 l/mm, respectively.[78,79] Photochromic materials require no post-processing but owing to the nature of the recording processes involved, their diffraction efficiency is very low.[80,81] They are suitable for designing sensors and photomodulated devices.[82] Photorefractive materials are characterised by high spatial resolution but relatively low sensitivity and the inorganic versions require post-exposure treatment when permanent holographic recording is sought.[83] Photorefractives are broadly used for real-time holographic applications such as optical signal processing.[84–86] Because they can be produced as thick crystals, photorefractive materials have been extensively studied as media for holographic optical memories.[87] Azodye-containing polymers, photochromics, and the photorefractives can be distinguished from other materials because they are sensitive not only to light intensity but also to the polarisation of the recording light.[88–92] This makes them suitable for polarisation holographic applications and for the design of holographic optical elements to control the polarisation of light.[93,94] Photopolymers are holographic recording materials that perhaps have the potential to cover almost all of the long list of requirements.[95] They are characterised by high sensitivity, high spatial frequency response, and can be processed easily in layers with thickness from a few hundreds of nanometres to a few millimetres and their dynamic range can be modified by addition of different dopants.[96] The self-processing versions do not require any additional treatment after the recording step. In recent years there is an increasing interest in the development of new photopolymerisable materials based on photopolymers doped with nanosize dopants.[97] This approach offers an extremely flexible tool for the improvement of the properties of the resulting nanocomposites by fine independent tuning of the properties of the host photopolymers and the properties of the, usually inorganic, nanosize component. This chapter summarises recent developments in nanoparticle doped photopolymer design and its potential for different holographic applications. It will review different nanodopants including porous and solid nanoparticles and their influence on the properties of photopolymerisable nanocomposites for holographic recording. It will also present some results from studies for holographic memory applications, holographic optical sensors, and the design of holographic optical elements.

2. Holographic Recording

During holographic recording the interference pattern formed in the intersection area of two coherent waves (object and reference waves) is recorded in a photosensitive medium (Fig. 21.1(a)).[98,99] As the illuminated material is photosensitive, its optical properties change, and the variations of the intensity of the recording light field are copied by a variation of the absorption coefficient, the refractive index, or of the thickness of the recording medium. Alternatively variations in the polarisation of the recording light field are copied by the recording medium's birefringence and/or dichroism.[100] An amplitude (when the absorption coefficient of the medium is altered) or a phase hologram (when the refractive index or the thickness of the medium changes) is formed (Fig. 21.1(b)). The hologram can be considered as a complex, spatially varying diffraction grating, which, when again illuminated with the reference wave, will diffract the light and produce a second wave, reconstructing the original object wave (Fig. 21.1(c)).

Owing to its nature, a hologram preserves information not only about the amplitude variation of the object wave but also about its phase.[98–100] This causes the realistic 3D appearance of the holographic image of an object.

In the simplest case when both recording waves are plane waves the result is a simple diffraction grating with a spatial period depending on the angle between them (Fig. 21.2). The relation between the spatial period, Λ, the angle between the two recording waves, θ, and the wavelength of recording, λ, is given by the

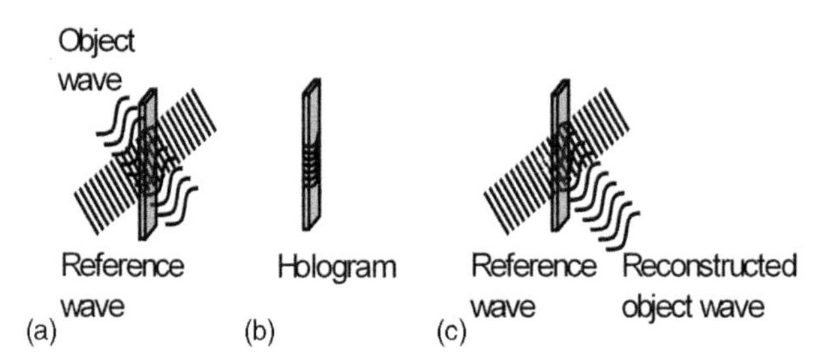

Figure 21.1 Recording and reconstruction of a hologram.

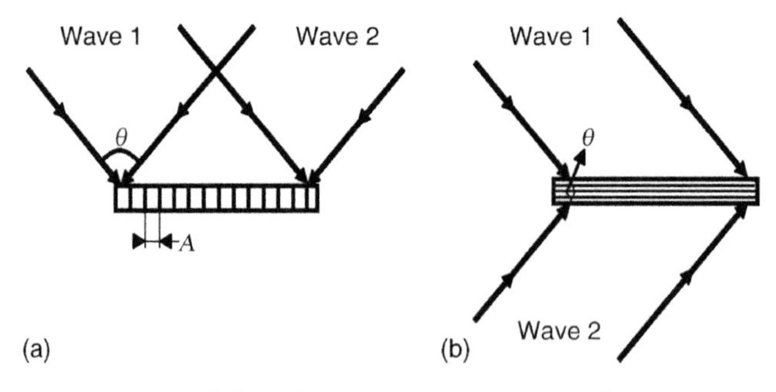

Figure 21.2 (a) Transmission and (b) reflection geometry of recording.

Bragg equation.[98,99] For transmission recording geometry (Fig. 21.2(a)), when object and reference waves are incident on the recording layer from the same side, the relation is

$$2\Lambda \sin(\theta/2) = \lambda, \tag{21.1}$$

and for the reflection mode of recording (Fig. 21.2(b)), when the object and the reference waves are incident on the recording layer from the opposite sides, the relation is

$$2n\Lambda \sin(\theta/2) = \lambda, \tag{21.2}$$

where n is the refractive index of the recording medium, θ is the angle between the two recording beams inside the medium, and λ is the wavelength.

During the reconstruction process, when the transmission or reflection hologram is illuminated by wave 1, the diffracted wave propagates in the same direction as wave 2 (Fig. 21.3) and bears all its characteristics.

The effectiveness of the recorded hologram is characterised by its diffraction efficiency η:

$$\eta = \frac{I_{\text{diffracted}}}{I_{\text{probe}}}, \tag{21.3}$$

where $I_{\text{diffracted}}$ is the intensity of the diffracted wave (wave 2 in Fig. 21.3) and I_{probe} is the intensity of reconstructing wave (wave 1 in Fig. 21.3). The effectiveness of the hologram is determined by the values of the changes introduced in the optical properties of the recording material. By measuring the diffraction efficiency, the values of the light induced refractive index modulation and/or the values of the absorption coefficient modulation can be determined. For thick phase holograms that are usually of interest for most holographic applications, the diffraction efficiency is given with a good approximation by the coupled wave theory.[101]

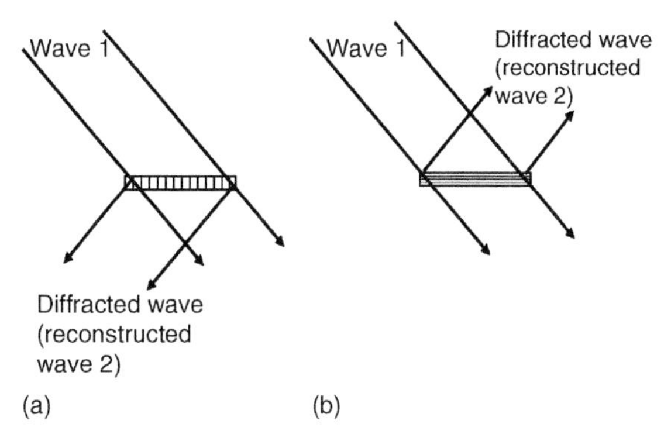

Figure 21.3 Reconstruction process in transmission (a) and reflection (b) geometry of recording.

The refractive index modulation amplitude for a transmission phase hologram is given by

$$\Delta n = \frac{\lambda \cos\theta \cdot \text{Arc}\sin\sqrt{\eta}}{\pi d} \tag{21.4}$$

and for a reflection phase hologram, is given by

$$\Delta n = \frac{\lambda \cos\theta \cdot \text{Arc}\tanh\sqrt{\eta}}{\pi d}., \tag{21.5}$$

where d is the grating's thickness; λ, reading beam wavelength; θ, reading beam incidence angle, and η, the diffraction efficiency.

An important difference between the two recording geometries—transmission and reflection—is related to the spatial frequency of the pattern that is recorded. The spatial frequency is determined by $1/\Lambda$ and is commonly measured in l/mm. When the recording is carried out in transmission mode the spatial frequency usually does not exceed 2000 l/mm, whereas the spatial frequencies in reflection mode of recording can easily exceed 5000 l/mm. This automatically imposes a requirement for a high spatial frequency response of the recording material when reflection holograms are to be recorded.

2.1. Holographic recording in photopolymers

As this chapter focuses on photopolymerisable nanocomposites, a brief introduction to the properties and the mechanism of holographic recording in photopolymers will be useful to understand the concept and the benefits of introducing nanodopants in photopolymers and the development of photosensitive nanocomposites.

Photopolymers generally consist of monomer/monomers, a free radical generator, a photosensitizing dye and, optionally, a polymer binder,[102,103] which facilitates the preparation of dry layers. The steps involved in holographic recording in photopolymers are as follows. Upon illumination of the photopolymer with a non-uniform light field of appropriate wavelength the sensitizing dye absorbs a photon and reacts with the free radical generator to produce free radicals. These initiate free radical chain polymerisation where the light was absorbed. The polymer chains continue growing (polymer chain propagation) until this process is terminated by encounters with other free radicals. The changes in the density and the molecular polarizability, which accompany the polymerisation, lead to a change in the local photopolymer refractive index and a phase hologram is recorded. According to the theoretical models describing holographic recording in photopolymer systems, another important process in hologram formation is monomer diffusion from the dark to bright fringe areas.[104–111] As the monomer consumption rate is different in the bright and the dark fringes, a monomer concentration gradient is created during the recording process, leading to diffusion of the monomer molecules. The monomer diffusion rates vary significantly from one photopolymer system to another. For instance in some photopolymer systems the monomer diffusion constants are relatively low (6.51×10^{-11} cm^2/s for Omnidex DuPont photopolymers[105] and 3.57×10^{-14} cm^2/s for the system developed by Bell Labs[106]), whereas in other systems diffusion constants as high as 1.6×10^{-7} cm^2/s have been measured.[112] In the same study, it was observed that in systems where rapid monomer diffusion is measured, a second diffusion process in the opposite direction—from bright to dark fringe areas—occurs on a much slower time scale characterised by a diffusion constant of 6.35×10^{-10} cm^2/s.[112] The second diffusion process is ascribed to diffusion of terminated or unterminated short polymer chains and, although it occurs mainly at the beginning of the holographic recording, cannot be ignored when high spatial frequency response is sought. The significance of the short polymer chain diffusion is also emphasised in Ref. [111]. A simplified picture of the processes involved in the holographic recording when two plain waves interfere is shown in Fig. 21.4. Here, the interference pattern is characterised by a sinusoidal spatial variation of light intensity and initially the monomer in the bright fringe areas is converted to polymer, followed by monomer diffusion from dark to bright areas and counter diffusion of short polymer chains from bright to dark areas.

When a more detailed picture of the holographic recording process is sought, the dependence of both diffusion processes and the polymerisation rate on time and, consequently, on the extent of polymerisation should be taken into account.[105,106] As the monomer and the short polymer chain diffusion processes make opposite contributions to the final refractive index modulation, careful design of the permeability of the binder matrix is required to achieve maximum refractive index modulation. For instance at low spatial frequencies of recording, where the fringe spacing is in order of a few micrometers, fast diffusion rates are required to achieve high refractive index modulation. At high spatial frequencies of recording, where

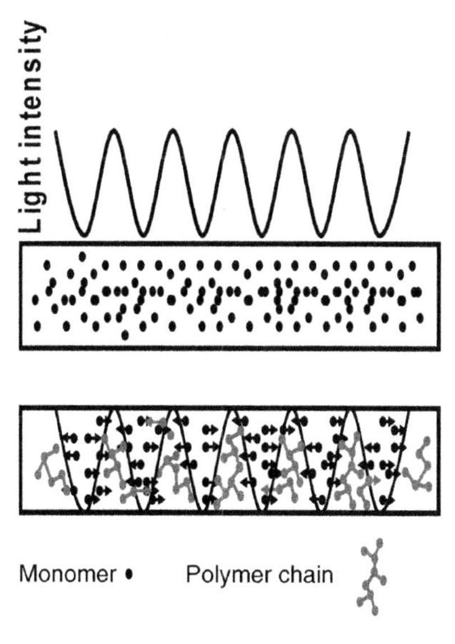

Figure 21.4 Light induced redistribution of photopolymer components.

the fringe spacing can be less than 200 nm, high permeability of the matrix and consequently fast diffusion rates are not desirable as they decrease the refractive index modulation owing to smearing of the recorded pattern.[112–114]

Indeed we have demonstrated that when a less permeable binder[115] is used in the photopolymer composition, excellent results in reflection mode of recording can be obtained.[116] An alternative explanation of the generally poor high spatial frequency response of the photopolymerisable materials is given in Ref. [117]. The authors ascribe this phenomenon to non-local polymerisation—a growth of the polymer chain away from the location where the photon is initially absorbed and the free radical is generated. Following this model a decrease of the polymer chain size might be expected to improve the high spatial frequency response. However no experimental evidence for this has been published.

2.2. Holographic recording in photopolymerisable nanocomposites

The existence of light induced mass transport in photopolymerisable materials inspired many authors to utilise these processes in new approaches for improving material properties.[118–123] The basic idea behind the concept of the photopolymerisable nanocomposites is to introduce nanodopants with significantly different refractive index from that of the host material and to achieve redistribution of the nanodopants during the holographic recording. Such redistribution causes significant increase in the ultimate refractive index modulation and, consequently, improvement of the dynamic range of the material. The variety of nanoparticles used to improve the properties of holographic recording materials is large. Solid nanoparticles such as TiO_2, SiO_2, and more recently ZrO_2 have been used by several groups.[96,97,119–121] The use of porous nanoparticles is somewhat restricted at the moment and reported by only one group.[123] A picture of the nanoparticles redistribution during the holographic recording is shown in Fig. 21.5.

A simplified version of holographic recording using two plane waves is used in this figure. As in the case of recording in photopolymer material shown in Fig. 21.4, owing to the photoinduced concentration gradient, the monomer molecules diffuse from dark to bright areas and the short polymer chains diffuse in the opposite direction. The nanoparticles are a non-reactive component and do not participate in the process of photopolymerisation but they are excluded from the bright fringe areas.

Strong experimental evidence that such redistribution indeed occurs is shown in Refs. [124,125]. The authors used electron-probe microanalyser analysis (EPMA) to study the diffusion processes during holographic exposure of SiO_2 nanoparticle doped photopolymer film.[124] The average diameter of the nanoparticles in this nanocomposite is 13 nm and the bulk refractive index is 1.46. By comparing the spatial distribution of Si and O atoms, with the spatial distribution of S atoms, which are characteristic for their monomer, and by showing that the two are out of phase with each other, the authors have provided decisive evidence for the exclusion of nanoparticles from the bright fringe areas as shown schematically in Fig. 21.4. Another study by the same group utilised real-time optical measurement of the phase shift between the light interference pattern and a recorded hologram in a

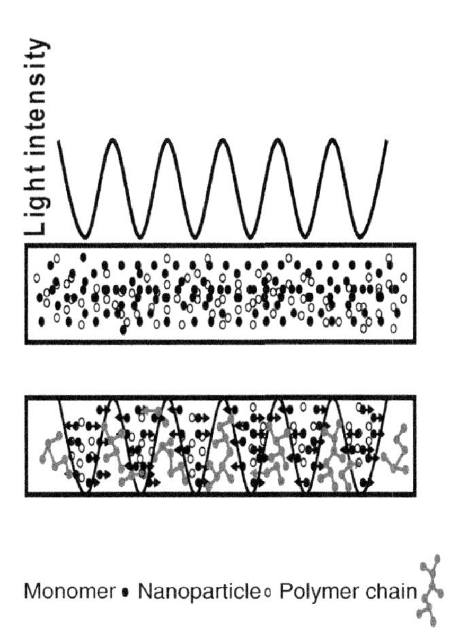

Figure 21.5 Light induced redistribution of nanoparticles.

photopolymer dispersed nanoparticles.[126] Two nanocomposites were studied, one containing TiO_2 nanoparticles with average diameter of 15 nm and a bulk refractive index of 2.55, higher than the refractive index of the photopolymer host system (1.55 in liquid and 1.59 in solid phase). The second nanocomposite contained SiO_2 nanoparticles with average diameter of 13 nm and bulk refractive index of 1.46, lower than the refractive index of the photopolymer system. In this comparative study it is demonstrated that in the nanocomposites containing TiO_2 nanoparticles a phase shift of 180° between the intensity profile of the light interference pattern and the refractive index profile of the recorded hologram is observed showing that the refractive index in the dark fringes is higher than that in the bright fringes. In contrast, in the nanocomposites containing SiO_2 nanoparticles the measured phase shift is 0°, corresponding to lower refractive index in the dark areas than in the bright ones. These observations strongly support the model based on the exclusion of the nanodopants from the bright fringes.

More direct evidence for nanodopant redistribution is given in.[118] In their study the authors demonstrated that the holographic recording in photopolymerisable materials is a flexible tool for spatial arrangement of nanoparticles. They studied different types of nanoparticles with average diameters ranging from 5 to 260 nm. TEM micrographs of microtomed sections of a transmission holographic grating, recorded in a nanocomposite containing gold nanoparticles, show lines of single nanoparticles separated by approximately 800 nm wide polymer stripes. The exact location of the nanoparticles with respect to the bright fringe areas is not determined. The authors speculate that the gold nanoparticles are excluded from the areas where the polymer is formed.

Indirect evidence for nanodopant redistribution can be found in the reports of significant increase in the dynamic range of holographic recording in photopolymerisable nanocomposites. If it is assumed that, as a result of the holographic recording of a grating, the simplest hologram used to characterise the material's properties, a square-wave particle distribution is introduced then the refractive index modulation amplitude can be expressed by[118]

$$\Delta n = \frac{2f_{\text{nanodopants}}}{\pi}(n_{\text{nanodopant}} - n_{\text{host}})\sin(\alpha\pi), \tag{21.6}$$

where $f_{\text{nanodopants}}$ is the volume fraction of nanoparticles in the nanoparticle-rich region, $n_{\text{nanodopant}}$ is the refractive index of the nanodopants, n_{host} is the refractive index of the host organic matrix, and α is the fraction of the period spacing with a rich content of nanoparticles. By measuring the diffraction efficiency and determining the refractive index modulation amplitude one can estimate the volume fraction of nanoparticles, $f_{\text{nanodopants}}$. Values from 0.038 to 0.05 have been reported for this parameter.[118,120,121] For acrylamide-based photopolymer systems doped with Si-MFI nanoparticles reported in,[123] if α is assumed to be 0.5 the value of $f_{\text{nanodopants}}$ would be 0.055, quite comparable with the values reported for the other photopolymerisable nanocomposite systems.

3. Mechanism for the Holographic Redistribution of Nanoparticles

Different mechanisms have been suggested to explain the redistribution of the nanodopants during holographic recording in photopolymerisable nanocomposites.[118–122] The redistribution is explained by nanoparticle segregation due to anisotropic polymerisation.[118] Among proposed driving forces for nanoparticle segregation are a thermodynamically favoured phase separation and aggregation of colloids as the molecular weight of the matrix is increased during the polymerisation, spatially periodic shrinkage of the volume upon monomer polymerisation, and concentration gradient driven monomer and oligomer diffusion. In their model the authors consider the importance of the balance between the polymerisation rate and the mobility of the nanoparticles. The mobility of the nanoparticles must be fast enough compared with both the polymerisation rate and the propagation rate of the polymerisation front from the bright towards the dark fringe area, in order to avoid trapping of the nanoparticles in the polymerised regions. Otherwise no segregation will be observed. Mutual concentration driven diffusion of monomers and nanoparticles is considered as the main reason for nanoparticle redistribution.[96,97,120,126] The mechanism of nanoparticle redistribution in a photopolymer system containing two monomers, one a multifunctional, highly reactive monomer, the other a monofunctional, less reactive monomer is described.[121] Highly reactive monomer is found to facilitate the segregation of the nanoparticles. After the holographic recording begins the highly reactive monomer is consumed in the bright regions leading to the creation of a concentration gradient and diffusion from dark to bright fringe areas. The monofunctional monomers together with the nanoparticles counter-diffuse to the dark regions. The higher compatibility of the less reactive monomer with the nanoparticles perhaps enhances nanoparticles diffusion. Similar concentration driven mutual diffusion is suggested as a mechanism for nanoparticle redistribution.[123] The addition in this model is the consideration of the diffusion of oligomers and short polymer chains from bright to dark areas and its role in nanoparticle redistribution.

3.1. Improvement of the dynamic range

The dynamic range of the holographic recording material is determined by the extent to which the refractive index of the material can be changed. For standard photopolymerisable materials the maximum achievable refractive index modulation is in the order of 10^{-3}.[127] In Table 21.1 different photopolymerisable nanocomposites containing solid and porous inorganic nanoparticles are compared. The addition of nanodopants with much different refractive index than that of the host polymer matrix is found to increase the refractive index modulation. The largest increase, R, of the refractive index modulation in doped layers compared with non-doped layers reported in the literature is in order of 4 for the nanocomposites reported by Sanchez *et al.*[121] and by Suzuki *et al.*[128] The lowest value of R (1.25) is reported by Kim and the authors propose an alternative to the nanoparticle redistribution mechanism for the improvement in their material.[122] They present some evidence that the improvement is mostly due to interfacial hydrogen bonding-induced polarizability change due to the hydrophilic nature of the silica nanoparticles used as nanodopants in their material. If this is indeed the explanation for the improvement it would also explain the low value of the improvement factor R.

Increase in refractive index modulation was observed for dry and liquid nanoparticle doped systems. The permeability of the organic matrix must be in such a way to allow mass transport of the nanoparticles in order to achieve significant improvement of the dynamic range.

Another important characteristic of the holographic recording material is its scattering properties. The initial reports of the scattering properties of nanocomposite systems reveal a significant scattering in the order of 20% for 40–50 μm layers.[96,97] Different approaches have been adopted to solve this problem. One approach is to decrease the size of the nanopants.[119–121] Another approach is to functionalise their surfaces in order to avoid aggregation after adding the nanodopants to the photopolymer system.[120] In addition to the two approaches mentioned earlier the use of nanoparticles with smaller refractive index such as ZrO_2 nanoparticles ($n = 2.1$ at 589 nm) than TiO_2 ($n = 2.55$ at 589 nm) proves beneficial for the decrease of the scattering from nanocomposite layers.[120,128]

When compared with the solid nanoparticle version of the photosensitive nanocomposites, the zeolite doped photopolymers show similar improvement of the refractive index modulation and their scattering properties are much more favourable for optical applications as the scattering in a 40 μm thick Si-MFI doped layer (8 wt.% 60 nm average diameter of the nanoparticles) does not exceed 6%. The

Table 21.1 Comparison of different photopolymerizable nanocomposites

Reference	Host photopolymer	Nanoparticles composition and size, nm	Phase of the recording material	Maximum refractive index modulation Δn and improvement factor R*	Scattering properties
Solid nanodopants					
Vaia *et al.*[118]	Acrylate based photoreactive syrup	Gold, 5 nm	Liquid syrup	15.4×10^{-3}**, at 1250 l/mm	n/a
Suzuki *et al.*[96]	Methacrylate, photopolymer	TiO_2, 15 nm	Dry layer	5.1×10^{-3}, at 1000 l/mm	20% at 633 nm, in 40 μm layer doped with, 15 vol.% nanoparticles
Suzuki *et al.*[97]	Methacrylate, photopolymer	SiO_2, 36 nm	Dry layer	14×10^{-3}, $R = 3{,}5$, at 1000 l/mm	20% at 633 nm, in 50 μm layer doped with, 6 vol.% nanoparticles
Sanchez *et al.*[121]	UV sensitive acrylates	TiO_2, 4 nm	Liquid syrup	15.5×10^{-3}, $R = 4$, at 2000 l/mm	12% at 633 nm, in 15 μm layer doped with, 30 wt % nanoparticles
Suzuki *et al.*[128]	Acrylate photopolymer	ZrO_2, 3 nm	Dry layer	4×10^{-3}, $R = 4$, at 1000 l/mm	<1% at 633 nm, in 40 μm layer, doped with, 15 vol.% nanoparticles
Kim *et al.*[122]	Acrylate, photopolymer	SiO_2, 10–12 nm	Dry layer	$<10^{-3}$, $R = 1{,}25$, at 1130 l/mm	n/a
Sakhno *et. al.*[120]	Acrylate photopolymer	TiO_2 and ZrO_2, 6–8 nm	Liquid composition	$16{,}6 \times 10^{-3}$ (TiO_2), 12.5×10^{-3} (ZrO_2), at 1050 l/mm	At 633 nm in 20 μm layer containing 30 wt.% of dopant, 10% (TiO_2), 8% (ZrO_2)
Porous nanodopants					
Naydenova *et al.*[123]	Acrylamide-based photopolymer	Si-MFI, 60 nm	Dry layer	3.5×10^{-3}, $R = 2.3$, at 2000 l/mm	At 633 nm in 40 μm layer, <6%

* R indicates the ratio of the maximum refractive index modulation achieved with nanodopants to the maximum refractive index modulation achieved in undoped layers.
** Calculated using the data published in the paper.

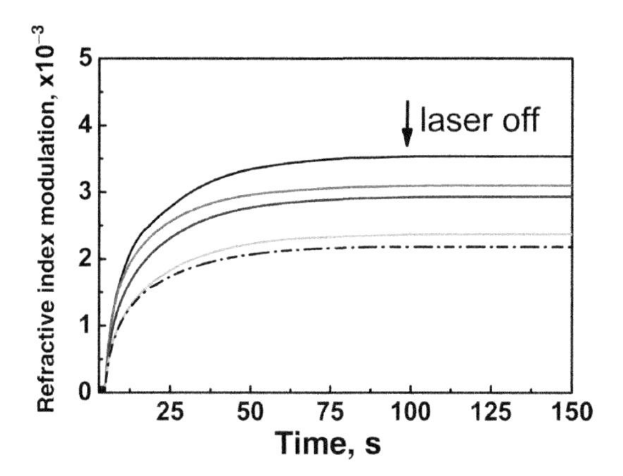

Figure 21.6 Dynamics of recording in 40 μm thick layers at spatial frequency of 1000 l/mm. Recording intensity is 5 mW/cm^2. The concentrations of the Si-MFI nanoparticles are as follows: 0 wt.% (dash-dotted black), 1.5 wt.% (solid light grey), 5 wt.% (solid dark grey), 4.5 wt.% (solid grey), 6 wt.% (solid black).

synthesis of different types of colloidal zeolites is described the review by Tosheva and Valtchev.[129]

In the following section more detailed data is presented on Si-MFI doped acrylamide-based photopolymer. The improvement of the dynamic range by addition of Si-MFI porous nanoparticles is reported.[123] Figure 21.6 shows the real-time growth of the refractive index modulation induced during the recording of a grating with a spatial frequency of 1000 l/mm in 40 μm thick photopolymer layer. when the concentration of the nanoparticles increases from 0 to 6 wt.% the refractive index modulation increases from 1.8×10^{-3} to 3.5×10^{-3}. A similar increase was observed at 2000 l/mm but no improvement was observed at a higher spatial frequency of 4600 l/mm (Fig. 21.7).

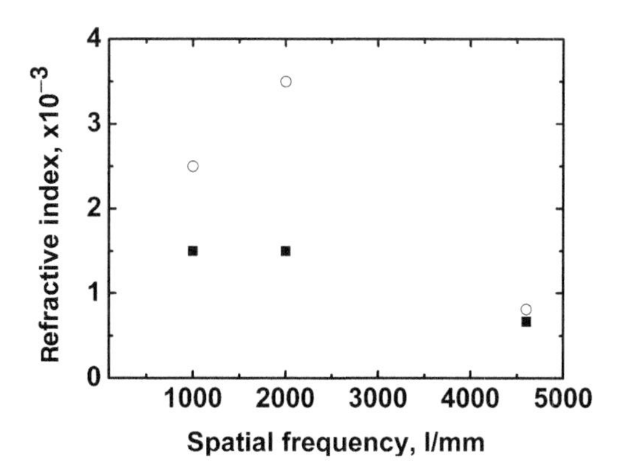

Figure 21.7 Maximum refractive index modulation achieved at different spatial frequencies of recording in photopolymer doped with 60 nm sized Si-MFI nanoparticles (circles) and undoped layers (squares). The concentration of the nanoparticles in the dry layer is 6 wt.%.

The study of the holographic properties of acrylamide-based photopolymer doped with Si-MFI nanoparticles shows a significant increase of the material's dynamic range at spatial frequencies of recording between 1000 and 2000 l/mm. This effect could possibly be due to redistribution of the porous silica nanoparticles during the holographic recording and/or restricted diffusion of short polymer chains caused by the nanoparticle structures. As the refractive index of the Si-MFI nanoparticles is lower than that of the photopolymer such an increase of refractive index modulation would be expected if the nanoparticles are excluded from the bright fringes and concentrated into the dark fringes. However, no improvement of the light induced refractive index modulation at low spatial frequency of 200 l/mm was observed. This observation implies that at such large fringe spacing of 5 μm no significant redistribution of the nanoparticles occurs.

The study of the influence of the size and morphology of the porous nanoparticles shows no significant difference in the refractive index modulation for 40 and 60 nm size nanoparticles, Fig. 21.8.[130]

3.2. Improvement of the shrinkage properties

A study of the photopolymer shrinkage was carried out by recording a single slanted holographic grating in a photopolymer layer. A slant angle of 6° between the photopolymer layer normal and the bisector of the angle between the recording beams was used. The spatial frequency of the grating was 1300 lines/mm. Exposure energy of 8 mJ/cm^2 was delivered from an Argon ion laser operating at 514 nm. Once the grating was recorded, it was probed by the reading beam (RB in Fig. 21.9) at the same wavelength of 514 nm but with reduced intensity of 80 μJ/cm^2. The dependence of the intensity of the diffracted light (OB) on the incident angle was measured and the level of shrinkage was determined from the angular shift in the direction of the maximum intensity.

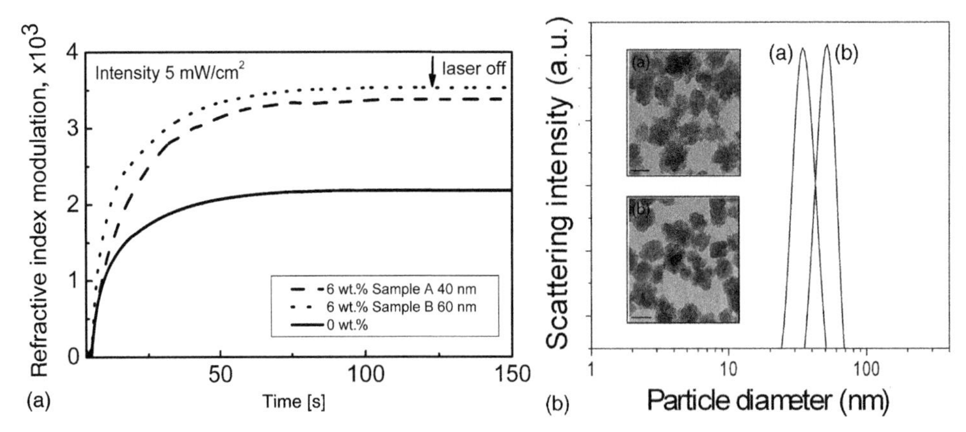

Figure 21.8 (a) Refractive index modulation for photopolymer nanocomposite doped with 40 and 60 nm large nanoparticles; (b) TEM picture and results from DLS characterisation of the two nanodopants.

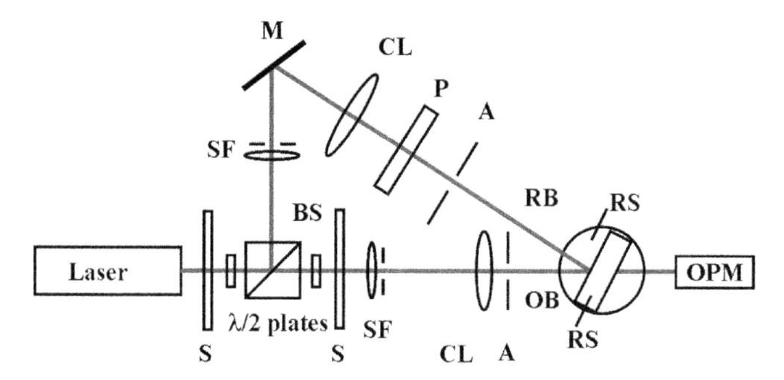

Figure 21.9 Experimental set-up for shrinkage measurements; S – shutter, SF – spatial filter, CL – collimating lens, BS – beam splitter, P – polarizer, RS – rotational stage, PS – photopolymer holder, OPM – optical power meter, M – mirror, RB – reading beam, OB – object beam and A – aperture.

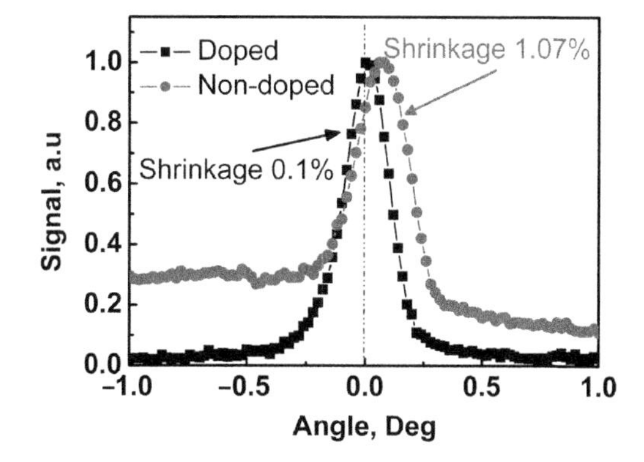

Figure 21.10 Angular selectivity curves for Si-MFI-containing (4.5 wt.% in the dry layer) photopolymer thick sample (~ 650 μm) (black) and non-doped sample (grey).

The study shows that the incorporation of Si-MFI nanoparticles leads to a significant decrease of the shift of the angular selectivity curve (see Fig. 21.10).

The fractional change Δd in a material with thinckness d_0 can be obtained by knowing the initial slant angle of the grating (ϕ_0) and the final slant angle (ϕ_1), i.e.:

$$\Delta d = d_0 \left[\frac{\tan \phi_1}{\tan \phi_0} - 1 \right]. \tag{21.7}$$

The shrinkage was found to be dependent on the thickness of the layer; the percentage shrinkage increases as the sample thickness decreases (Fig. 21.11).[131] In all cases the percentage shrinkage of the layers containing nanoparticles was significantly lower than that of the pure photopolymer samples.

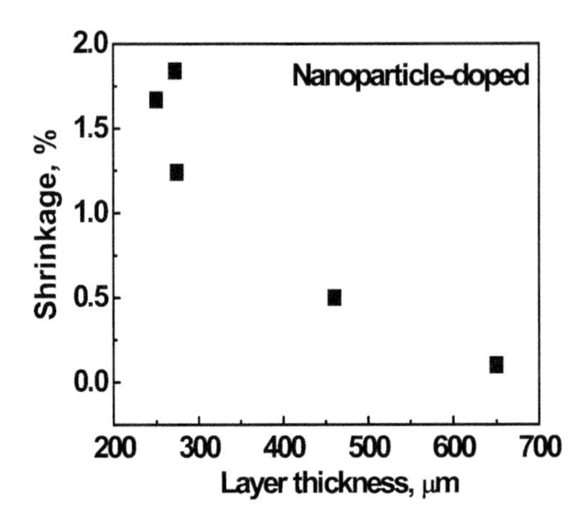

Figure 21.11 Dependence of the shrinkage on the film thickness. The shrinkage was estimated after recording of gratings with low diffraction efficiency ($< 5\%$), as required for holographic data storage applications.

4. Holographic Applications

The variety of holographic applications constantly grows with the advances in optics, optoelectronics, and holographic recording materials design. In the following sections we will focus on three applications—holographic data storage, holographic optical elements, and holographic sensors and how they can benefit by the development of photopolymerisable nanocomposites.

4.1. Holographic data storage

Conventional memory technologies such as optical disks, magnetic hard drives, and semiconductor memories, are based on two-dimensional surface storage of the information. The main approach to increasing the capacity and readout rates of the data stored in these devices is to reduce the physical dimensions of the information bits. For reason the development of new data storage systems is approaching some fundamental limits, such as the thermal stability of the recorded bits and the optical resolution of the laser systems used for recording. Holographic data storage (HDS) is an optical storage technique, which offers densities as high as tens of TB/cm^2 with data transfer rates of many GB/s by utilizing not only the surface but also the volume of the recording medium.

Furthermore the data can be stored in form of pages instead of bits (Fig. 21.12).[8–14] The pages consist of pixels, each pixel corresponding to a bit. Millions of bits can be written or retrieved at a single step and the data access rate can reach 1 Gbit/s compared with 20 Mbits/s for optical digital videodisks—DVD. HDS has unique advantages, such as the ability to perform parallel data searches (associative recall) and implement novel encryption techniques.

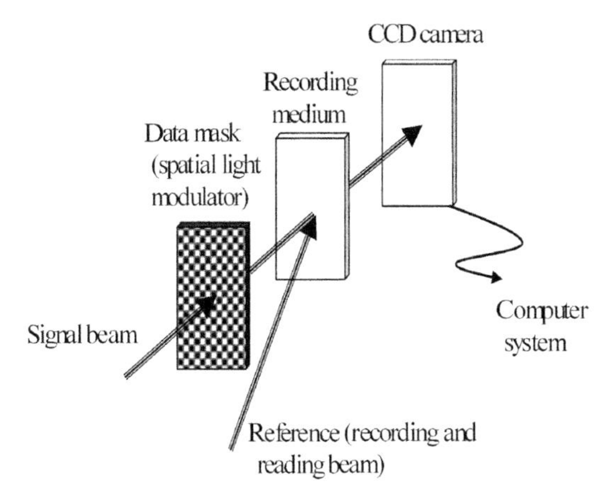

Figure 21.12 Page format holographic data storage.

The potential of holographic storage technology arises from an intrinsic property of the recorded hologram—it is selective with respect to the wavelength and the direction of propagation of the reading beam.[98,99] This makes it possible to record multiple holograms in the same volume of the medium and to reconstruct them selectively. The selectivity of the reconstruction process increases approximately linearly with the thickness of the recording material. Consequently the number of the multiplexed holograms and the density of stored information are dependent on the thickness of the medium. Producing a thick photosensitive layer characterised by low scatter appears to be a major difficulty.

Photopolymers are one of the best candidates for optical data storage, in particular for write-once read-many times memories. They are characterized by the large dynamic range of the photoinduced changes. Additionally they are inexpensive and samples with different shapes and thicknesses (up to a few hundred micrometers) are relatively easy to produce. The main disadvantage of the most currently investigated photopolymers is that they suffer from post recording shrinkage.[132–135]

It has been demonstrated that the nanoparticle dopants in photopolymers can lead to significant decrease of the material shrinkage.[96,123] An almost threefold decrease of the fractional change of the film thickness was observed when TiO_2 nanoparticles were used as dopants.[96] The shrinkage in the doped layers was 2.9% while the undoped layer was characterised by 8% shrinkage. Significant decrease of the level of shrinkage was observed when zeolite nanoparticles were used as nanodopants[123,130] Zeolite A, Zeolite Y, and Silicalite 1 were the nanoparticles studied in this experiment. It was observed that incorporation of Zeolite Y type nanoparticles leads to lowest shrinkage in photopolymer layers of thickness of 90 μm. The shrinkage in these layers was 1.2% compared to 2.2% in the undoped layers. In much thicker layers (650 μm), even better improvement of the shrinkage properties of the doped layers was observed. Shrinkage of 1.1% for the non-doped layers and 0.1% for the doped layers was measured. The last observation places the zeolite doped nanocomposites among the materials characterised by the lowest shrinkage and makes them very promising candidates as materials for holographic data storage. In addition to the shrinkage properties another important parameter of the materials for holographic data storage is their dynamic range. The dynamic range of a material for holographic data storage refers to the total response of the recording medium when divided up among many holograms multiplexed in the material volume. In terms of holographic data storage it is usually parameterised as M/#.[136] In general, the larger the M/#, the greater the number of holograms having a specified diffraction efficiency that can be recorded in a given volume. One way of viewing the M/# is in terms of how the dynamic range scales when different numbers N of equalised holograms are recorded. The square root of the average diffraction efficiency for any one set of equalised holograms is graphed against the reciprocal of the number of holograms recorded in the set. This is repeated for sets containing different numbers of holograms. The slope of a linear fit applied to the data, is equal to the M/#.

$$M/\# = N\sqrt{\bar{\eta}}. \quad (21.8)$$

$M/\#$ is directly related to the quantity of data that may be stored in a specific volume of medium, i.e., bit density. Using peristrophic multiplexing, the $M/\#$ in the layers characterised by the lowest shrinkage, i.e. layers doped with zeolite Y nanoparticles, was characterised and compared with the $M/\#$ of the undoped layers. Zeolite Y doped layers was found to shrank by 1.25% ± 0.06% and had an $M/\#$ of 1.74 for 176 μm thickness layer. This result still compares favourably with the standard composition which has a shrinkage of 2.19% ± 0.1% and a $M/\#$ of 1.89 for layer of thickness 223 μm as, if we assume a simple proportionality of the $M/\#$ to the film thickness, an $M/\#$ of 2.2 would be expected for the 223 μm thick zeolite Y doped layers.

4.2. Holographic optical elements

Holographic optical elements (HOEs) are basically holograms that can play a particular role in an optical system. Some examples of such elements are holographic lenses, holographic diffraction gratings, holographic filters, holographic beam splitters, and fan-out devices. The main advantages of the HOEs are that they are compact and light even for large aperture elements, they provide the possibility to correct optical system aberrations, and their function is essentially independent of the substrate.[98] In this section the benefits of utilizing photopolymerisable nanocomposites in the design of HOEs based on transmission holographic diffraction gratings will be discussed. These gratings are used as dispersive elements in spectrometers. For spectroscopic applications an important characteristic of the diffraction grating is the spectral range of operation. In other words it is important to know the dependence of diffraction efficiency of the grating on the wavelength when the angle of incidence is fixed and satisfies the Bragg condition (1). To characterise this dependence, a parameter ($\Delta\lambda_{FWHM}$) called the spectral Bragg envelope is introduced. This is the wavelength range in which the diffraction efficiency is equal to or above half of the maximum diffraction efficiency

$$\frac{\Delta\lambda_{FWHM}}{\lambda} \propto \frac{\Lambda}{d}\cot(\alpha), \quad (21.9)$$

where Λ is the fringe spacing of the diffraction grating, d is the diffraction grating thickness and α is the incidence angle.

It is seen from Eq. (21.9) that the spectral range of operation can be controlled by the spatial frequency and the thickness of the grating. In Figs. 21.13 and 21.14 the experimentally measured spectral responses of holographic gratings recorded at two spatial frequencies, 1200 and 600 l/mm, in photopolymerisable layers of different thicknesses, are presented.

It is clearly seen that at spatial frequencies of 1200 l/mm (Fig. 21.13) and 600 l/mm (Fig. 21.14), when the thickness is decreased, a much larger spectral range can be covered.

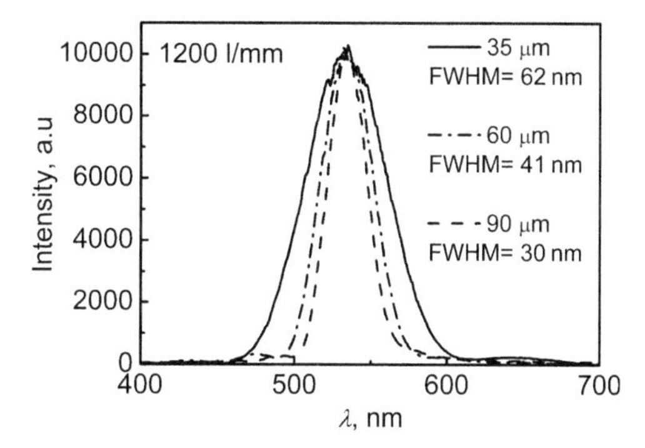

Figure 21.13 Dependence of the spectral response on the thickness of the grating. Spatial frequency of recording is 1200 l/mm.

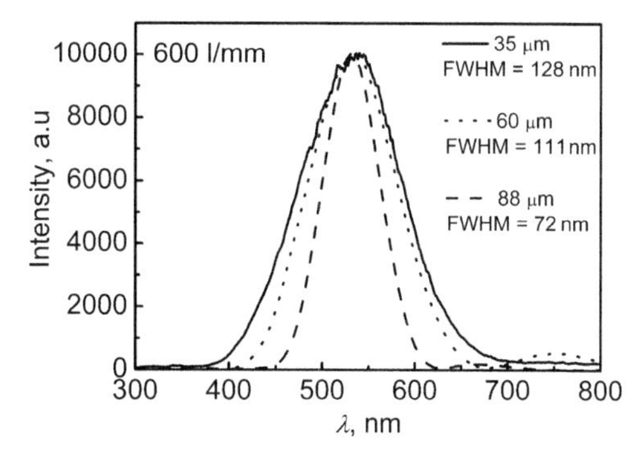

Figure 21.14 Dependence of the spectral response on the thickness of the grating. Spatial frequency of recording is 600 l/mm.

The measurements presented in Figs. 21.13 and 21.14 were carried out at a central wavelength of 532 nm. The gratings under study were unslanted, i.e., the bisector of the angle between the two recording beams coincided with the sample normal. Here the angle of diffraction is the same as the angle of incidence and it is fixed for given spatial frequency. From Eq. (21.9) one would expect a linear dependence of the spectral Bragg envelope on the thickness of the grating. The observed dependencies for the two spatial frequencies are presented in Fig. 21.15.

The spatial frequency determines the linear dispersion of the grating and consequently the resolution of the spectrometer and it is usually the first parameter to be determined. The actual spectral range of operation can be controlled by the thickness of the grating. A thinner grating will result in a broader spectral range of operation. At the same time the decrease of the grating's thickness is restricted by the requirement for high diffraction efficiency. As seen from Eq. (21.10)

$$\eta = \sin^2\left(\frac{\pi n_1 d}{\lambda \cos\theta}\right), \tag{21.10}$$

where n_1 is the refractive index modulation, d is the grating's thickness, λ is the probe beam wavelength, and θ is the Bragg angle for the probe beam. The diffraction efficiency depends on the optical path difference ($n_1 d$).

To avoid decrease of the diffraction efficiency when increased spectral range is sought by reducing the thickness of the grating, an accompanying increase in the refractive index modulation must be provided. The photopolymerisable nanocomposites with their increased dynamic range can prove an excellent solution to this problem. If one takes the maximum reported refractive index modulation in photopolymerisable nanocomposites of 16×10^{-3}, we can estimate that the thickness of the diffraction grating providing 100% diffraction efficiency at 532 nm is just 17 μm.[120] Of course such a low thickness of the hologram would mean that only at very high spatial frequencies will the grating behave as a thick one and this should be taken into account if the optical system design requires a thick phase hologram.

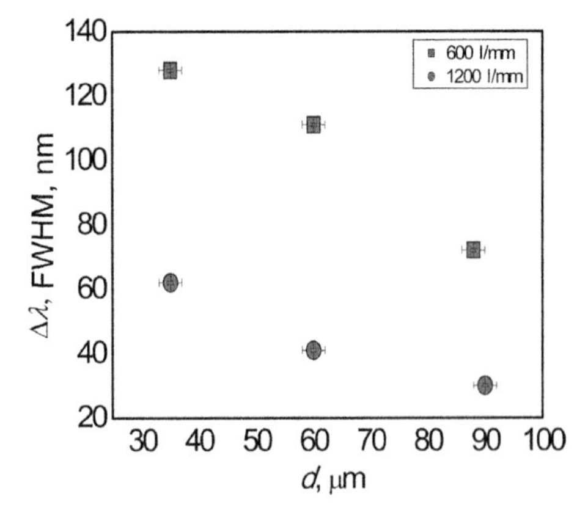

Figure 21.15 Dependence of the Bragg envelope spectral width on the grating thickness.

A simple extrapolation of the results in Fig. 21.15 shows that the spectral Bragg envelope would be 150 nm for a grating of 600 l/mm spatial frequency and 80 nm for a grating of 1200 l/mm. In addition to the large operational spectral range such thin holographic gratings would be characterised by a very low angular selectivity, in other words there would be a large tolerance towards deviations of the reconstructing beam incidence angle from the Bragg angle. A grating with such characteristics is of great interest for design of different optical systems. One example is an electronic speckle pattern shearing interferometry (ESPSI) system for direct measurements of displacement derivatives. Such systems can be used for evaluation of objects under mechanical or thermal stress, or vibrating objects.[18,19,31,137,138] The use of holographic diffraction gratings in ESPSI and the challenges related to it are discussed in Ref [19].

4.3. Holographic sensors

Another exciting application of holography, gaining significant attention in recent years, is the development of novel holographic sensors. Basically, a holographic sensor consists of a holographic optical element disposed throughout a support medium. The holographic optical element is fabricated by exposing a suitable photosensitive material to the optical interference pattern produced when two coherent light beams, usually from a laser, intersect.[98] The material records the variation in light intensity as a variation in refractive index, absorption or thickness, and a corresponding holographic optical element is produced.

In this section the principles of operation of two different types of holographic sensors recorded in photopolymerisable nanocomposites are described. In both cases nanodopants that are sensitive to specific analytes are used. The main difference between the two sensors is that the operating principle of the first one depends on the spatial redistribution of the analyte-sensitive nanodopants while the second one requires the presence of analyte-sensitive nanodopants but does not require their spatial redistribution. The first type of sensor can be designed to operate in transmission or reflection mode, while the second type of sensor requires operation in reflection mode.

The first example (Fig. 21.16) is a transmission holographic grating recorded in a nanozeolite doped photopolymerisable material. In Fig. 21.16(a) the holographic recording of the sensor is shown. A given amount of photopolymerisable material is deposited on a glass or plastic substrate and left to dry. The solid layer is exposed to two interfering beams of light. The spatial frequency and the thickness of the layer are chosen so that a thick transmission hologram is recorded. This provides the existence of only one order of diffraction, whose intensity can be monitored. As the holographic grating is a thick volume grating, the formula (21.4) relating the refractive index modulation and the diffraction efficiency of the grating is applicable. Special attention must be given to the properties of the photopolymerisable nanocomposite in order to ensure that a spatial redistribution of the nanodopants occurs as a result of the recording process (Fig. 21.16(a)). When illuminated at the appropriate angle by one beam the hologram produces a second diffracted beam, whose intensity depends on the refractive index modulation introduced during the recording.

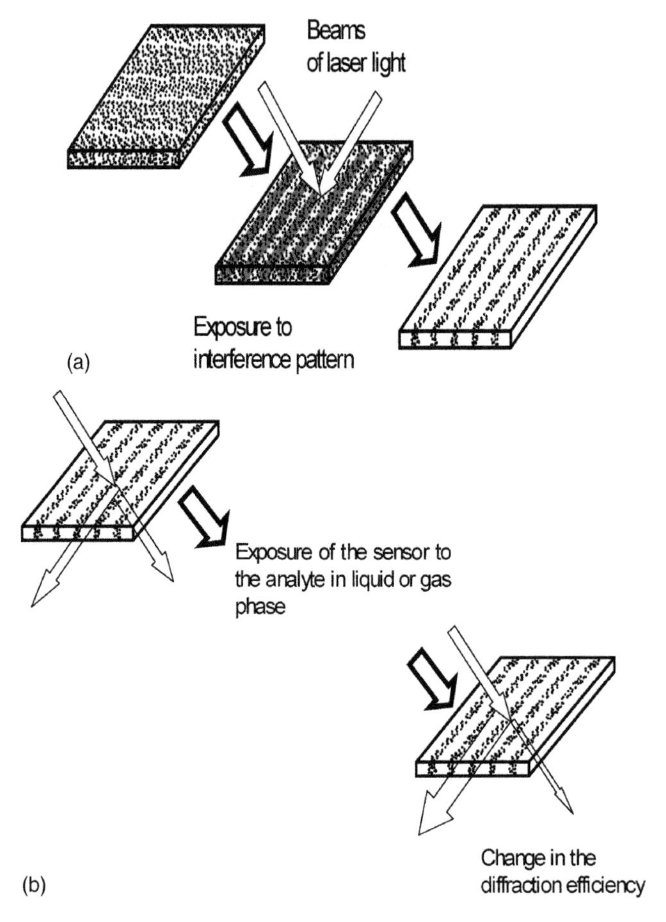

Figure 21.16 (a) Recording of the holographic sensor; (b) Reading of the holographic sensor.

If we assume the nanoparticle redistribution profile to be square, the refractive index modulation is given by Eq. (21.6). If it is a sinusoidal spatial redistribution the formula will be simplified to become[96]:

$$n_1 = f_{\text{nanodopants}}(n_{\text{nanodopants}} - n_{\text{host}}), \tag{21.11}$$

where $f_{\text{nanodopants}}$ is the volume fraction of nanoparticles in the nanoparticle-rich region, $n_{\text{nanodopant}}$ is the refractive index of the nanodopants, and n_{host} is the refractive index of the host organic matrix.

Zeolites are well known for their ability to adsorb different chemicals accompanied by a change in refractive index.[139,140] For every detector, a careful consideration of the zeolite/analyte pair is required so that when the sensor is exposed to an analyte-containing environment, the analyte is adsorbed by the zeolite nanoparticles. In order for this to happen one more conditions must be fulfilled, namely that the host matrix is characterised by sufficient permeability assuring access by the analyte not only to the nanoparticles on the surface but also to the nanoparticles dispersed in the volume of the layer containing the holographic grating. The adsorption of the analyte leads to a change of the refractive index of the zeolite nanodopants. If we assume that the refractive index of the host matrix remains constant, the change in the refractive index modulation n_1 is proportional to the change of the refractive index of the zeolite as follows:

$$\Delta n_1 = f_{\text{nanodopants}} \cdot \Delta n_{\text{nanodopants}}. \quad (21.12)$$

By differentiating Eq. (21.10):

$$\Delta \eta = \sin\left(\frac{2n_1 d\pi}{\lambda}\right)\pi\cos\theta\left[\left(\frac{d}{\lambda}\right)\Delta n_1 + \left(\frac{n_1}{\lambda}\right)\Delta d - \left(\frac{n_1}{\lambda^2}\right)\Delta\lambda\right] \quad (21.13)$$

and assuming that the dimensions of the hologram and the probe wavelength remain constant ($\Delta d = 0$ and $\Delta\lambda = 0$), one can estimate the maximum change in the diffraction efficiency $\Delta\eta$ that can be expected for a particular change in the refractive index Δn_1. If, for example, $f_{\text{nanodopants}} = 0.05$ (the upper limit of the experimentally reported data for this parameter), $\lambda = 532$ nm, the sensor is 50 μm in thickness and $\Delta n_{\text{nanodopants}}$ is 10^{-3} the maximum change in the diffraction efficiency that can be observed is ~1.5%. If $\Delta n_{\text{nanodopants}}$ increases by one order of magnitude to 10^{-2} the observed diffraction efficiency change would be 15%. In both cases such changes in the diffraction efficiency can be easily detected by a photodetector.

An example of a holographic sensor not requiring redistribution of the zeolite nanodopants is shown in Fig. 21.17.

Basically this type of sensor is a reflection hologram with properties that change owing to a change in the overall refractive index of the zeolite-containing nanocomposite as a result of adsorption of the analyte by the zeolite nanoparticles. The step of recording the sensor is straightforward recording of a reflection hologram, but in contrast to the first type of sensor, nanoparticle redistribution does not occur. As mentioned in Section 3.1, the recording conditions governing the polymerisation rate and the actual chemical composition of the material can be adjusted in order to achieve this. A reflection hologram of this kind, when illuminated with a white light, will reflect a particular wavelength in a particular direction and, if the hologram is thick enough, only one particular colour is observable at a time. By differentiating formula (21.2) one can analyse the reasons that can cause the change in colour in the particular direction of observation.

$$\Delta\lambda = 2\sin\theta(\Lambda\Delta n + n\Delta\Lambda), \quad (21.14)$$

where θ is the Bragg angle and the angle of illumination for obtaining maximum diffraction efficiency, Λ is the fringe spacing, n is the average refractive index, and $\Delta\Lambda$ and Δn are the variations of these two parameters that can cause a change of the reflected wavelength and consequently the colour of the hologram. If one assumes

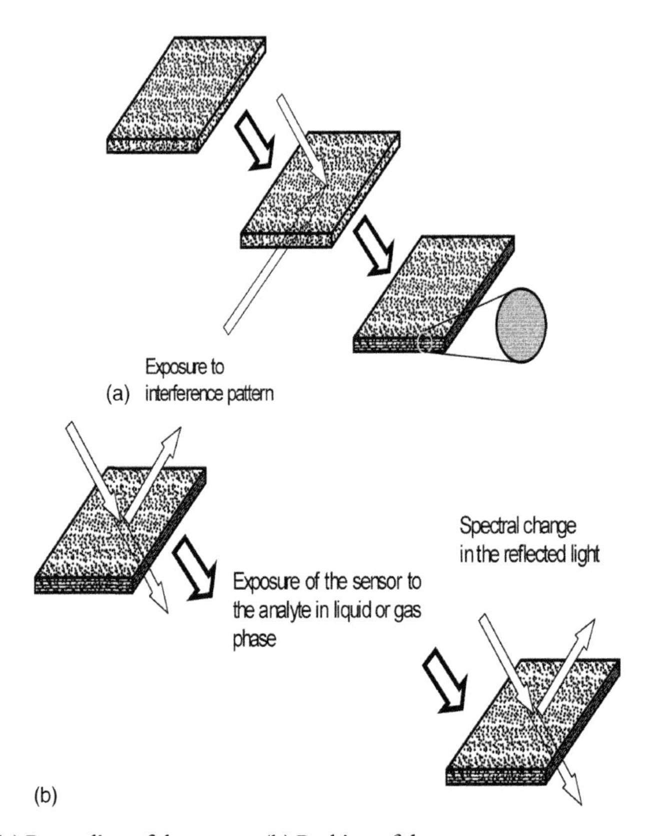

Figure 21.17 (a) Recording of the sensor; (b) Probing of the sensor.

that the dimensions of the hologram remain constant and no change in the fringe spacing occurs because of interaction of the hologram with the analyte, the formula (21.14) simplifies to

$$\Delta\lambda = 2\sin\theta(\Lambda\Delta n). \tag{21.15}$$

Now, one can estimate what change in the wavelength of the reconstructed light can be expected as a result of exposing the sensor to an environment containing the targeted analyte. The fringe spacing of a reflection hologram can be varied significantly depending on the recording geometry but if we assume that it is 250 nm (corresponding to 4000 l/mm) and the change in the average refractive index due to adsorption of the analyte by the zeolite nanoparticles is 10^{-2}, the maximum expected change in the wavelength in this case will be 5 nm. Such a change can be readily detected by a spectral analyser. Changes in the average refractive index modulation of the nanocomposite of the order of 5×10^{-2} would be sufficient to obtain an easily observable change of the colour of the light diffracted by the hologram.

Both sensors described in this section operate because of a change of the properties of zeolite nanoparticles on analyte adsorption. The nanoparticles in these two cases are located in the bulk of the host photopolymer. Note that the variety of holographic sensors is not restricted to these two cases. It is also possible to design sensors in which the zeolite nanoparticles are deposited on the photopolymer surface after it has been holographically patterned. It has been observed that optical resonant properties of periodical surface relief structures can be changed by surface deposition of layers of nanoparticles with different effective refractive index.[141] We believe that this principle could be used to design sensors based on refractive index change in surface deposited zeolite nanoparticles by analyte adsorption.

5. Summary

In summary, we have presented a review of the recent advances in the development of photopolymerisable nanocomposite materials. The basic principle for improving the recording properties of photopolymerisable materials by adding nanodopants is discussed. The main challenges in designing these new holographic recording materials are identified. Different systems containing solid and porous nanodopants developed for holographic recording are compared. Special attention is paid to the improvement in dynamic range and shrinkage properties. Holographic applications, which benefit by the improved properties of the nanoparticle doped photopolymers, such as holographic data storage and design of holographic optical elements and sensors are discussed. The principle of operation of two novel holographic sensors based on zeolite doped photpolymerisable nanocomposites is described and an estimate of their sensitivity is given.

This review demonstrates that photopolymerisable nanocomposite design is a flexible and promising approach for improving the properties of holographic recording materials.

Acknowledgements

The studies on zeolite nanocomposites reported in the review were supported by Science Foundation Ireland (Grant No. 065/RFP/PHY085).

References

[1] Gabor, D., "A new microscopic principle", *Nature* **1948,** *161,* 777.
[2] Maiman, T. H., "Stimulated optical radiation in ruby", *Nature* **1960,** *187,* 493.
[3] Hecht, J. B., *The Race to Make the Laser,* Oxford Univ. Press, New york, **2005** ISBN 0-19-54210-1.
[4] Leith, E., Upatnieks, J., "Reconstructed wavefronts and communication theory", *JOSA* **1962,** *52,* 1123.
[5] Leith, E., Upatnieks, J., "Wavefront reconstruction with continuous-tone effects", *JOSA* **1962,** *53,* 1377.

[6] Denisyuk, Y., "Photographic reconstruction of the optical properties of an object in its own scattered radiation field", *Sov. Phys. – Doklady* **1962,** *7,* 543.
[7] Denisyuk, Y., "On the reproduction of the optical properties of an object by the wave field of its scattered radiation", *Opt. Spectr.* **1963,** *15,* 279.
[8] Psaltis, D., Mok, F., "Holographic memories", *Sci. Am.* **1995,** *273,* 70.
[9] Shelby, R. M., Hoffnagle, J. A., Burr, G. W., Jefferson, C. M., Bernal, M.-P., Coufal, H., Grygier, R. K., Gunther, H., Macfarlane, R. M., Sincerbox, G. T., "Pixel-matched holographic data storage with megabit pages", *Opt. Lett.* **1997,** *22,* 1509.
[10] Hesselink, L., Orlov, S., Bashaw, M., "Holographic data storage systems", *Proc. IEEE* **2004,** *92,* 1231.
[11] Ashley, J., Bernal, M.-P., Burr, G. W., Coufal, H., Guenther, H., Hoffnagle, J. A., Jefferson, C. M., Marcus, B., Macfarlane, R. M., Shelby, R. M., Sincerbox, G. T., "Holographic data storage", *IBM J. Res. Dev.* **2000,** *44,* 341.
[12] Denz, C., Muller, K.-O., Visinka, F., Tschudi, T., *A Demonstration platform for phase-coded multiplexing,* in: Coufal, H., Psaltis, D., Sincerbox, G. (Eds.), *Holographic Data Storage,* Springer, New york, **2000**.
[13] Eichler, J. J., Kuemmel, P., Orlic, S., Wappelt, A., "High-density disk storage disk storage by multiplexed microholograms", *IEEE J. Select. Top. Quant. Electr.* **1998,** *4,* 840.
[14] Guattari, F., Maire, G., Contreras, K., Arnaud, C., Pauliat, G., Roosen, G., Jradi, S., Carre, C., "Balanced homodyne detection of Bragg microholograms in photopolymer for data storage", *Opt. Express* **2007,** *15,* 2234.
[15] Kabilan, S., Marshall, A. J., Sartain, F. K., Lee, M. C., Hussain, A., Yang, X. P., Blyth, J., Karangu, N., James, K., Zeng, J., Smith, D., Domschke, A., Lowe, C. R., "Holographic glucose sensors", *Biosens. Bioelectron.* **2005,** *20,* 1602.
[16] Marshall, A. J., Blyth, J., Davidson, C. A. B., Lowe, C. R., "pH-sensitive holographic sensors", *Anal. Chem.* **2003,** *75,* 4423.
[17] Naydenova, I., Jallapuram, R., Toal, V., Martin, S., in: Varadan, V. K. (Ed.), "Hologram-based humidity indicator for domestic and packaging applications", *Nanosensors, Microsensors, and Biosensors and Systems, Proc. SPIE,* **2007,** *6528,* 652811.
[18] Guntaka, S., Toal, V., Martin, S., "Holographically recorded photopolymer diffractive optical element for holographic and electronic speckle-pattern interferometry", *Appl. Opt.* **2002,** *41,* 7475.
[19] Mihaylova, E., Naydenova, I., Duignan, B., Martin, S., Toal, V., "Photopolymer Diffractive Optical Elements in Electronic Speckle Pattern Shearing Interferometry", *Opt. Lasers Eng.* **2006,** *44,* 965.
[20] Pavani, K., Naydenova, I., Martin, S., Raghavendra, J., Howard, R., Toal, V., "Electro-optical switching of liquid crystal diffraction gratings by using surface relief effect in the photopolymer", *Opt. Comm.* **2007,** *273,* 267.
[21] Lucchetta, D. E., Criante, L., Simoni, F., "Optical characterization of polymer dispersed liquid crystals for holographic recording", *J. Appl. Phys.* **2003,** *93,* 9669.
[22] Sutherland, R. L., "Polarization and switching properties of holographic polymer-dispersed liquid-crystal gratings. I. Theoretical model", *JOSA B* **2002,** *19,* 2995.
[23] Sutherland, R. L., Natarajan, L. V., Tondiglia, V. P., Chandra, S., Shepherd, C. K., Randelik, D. M. B., Siwecki, S. A., Bunning, T. J., "Polarization and switching properties of holographic polymer-dispersed liquid-crystal gratings. II. Experimental investigations", *JOSA B* **2002,** *19,* 3004.
[24] Ramsey, R. A., Sharma, S. C., "Switchable holographic gratings formed in polymer-dispersed liquid-crystal cells by use of a He-Ne laser", *Opt. Lett.* **2005,** *30,* 592.
[25] Blyth, J., "Security display hologram to foil counterfeiting", *Proc. Soc. Photo-Opt. Instr. Eng.* **1985,** *523,* 18.
[26] Lai, S. C., "Security holograms using an encoded reference wave", *Opt. Eng.* **1996,** *35,* 2470.
[27] Kumar, R., Mohan, D., Aggarwal, A. K., "Interferometric key readable security holograms with secrete-codes", *Pramana J. Phys.* **2007,** *68,* 443.
[28] Deng, S. G., Liu, L. R., Lang, H. T., Zhao, D., Liu, X. M., "Watermarks encrypted in the cascaded Fresnel digital hologram", *OPTIK* **2007,** *118,* 302.

[29] Dirksen, D., Droste, H., Kemper, B., Delere, H., Deiwick, M., Scheld, H., von Bally, G., "Lensless Fourier holography for digital holographic interferometry on biological samples", *Opt. Laser Eng.* **2001,** *36,* 241.
[30] Sirohi, R. S., "Optical methods in non-destructive testing", *Insight* **2001,** *43,* 230.
[31] Mihaylova, E., Naydenova, I., Martin, S., Toal, V., "Electronic spackle pattern shearing interferometer with a photopolymer holographic grating", *App. Opt.* **2004,** *43,* 2439.
[32] Herriau, J. P., Delboulbe, A., Huignard, J. P., "Non destructive testing using real-time holographic interferometry in BSO crystals", *Proc. Soc. Photo-Opt. Instr. Eng.* **1983,** *398,* 123.
[33] Hammond, A. L., "Holography – beginnings of a new art form or at least of an advertising bonanza", *SCIENCE* **1973,** *180,* 484.
[34] Bjelkhagen, H. I., Jeong, T. H., Vukicevic, D., "Color reflection holograms recorded in a panchromatic ultrahigh-resolution single-layer silver halide emulsion", *J. Im. Sci. Techn.* **1996,** *40,* 134.
[35] Benton, S. A., Stevenson, S. H.,Trout, T. J. (Eds.), *"Practical Holography XVI and Holographic Materials VIII"*, Proc. SPIE, **2002.** *4659,* ISBN: 9780819443991.
[36] Meerholz, K. (Ed.), *"Organic Holographic Materials and Applications III"*, Proc. SPIE, **2005,** *5939,* ISBN: 9780819459442.
[37] Orlic, S., Meerholz, K. (Eds.), *"Organic Holographic Materials and Applications IV"*, Proc. SPIE, **2006,** *6335.*
[38] Thiry, H., "Preparation and properties of ultra-fine grain AgBr emulsions", *J. Photogr. Sci.* **1987,** *35,* 150.
[39] Iwasaki, M., Kubota, T., Tanaka, T., "Preparation of ultra-fine grain emulsion for holography", *Selected Papers on Holographic Recording Materials,* in: Bjelkhagen, H. (Ed.), *SPIE Milestone series,* **1996,** *130,* 123.
[40] Bjelkhagen, H. "Silver halide emulsions for Lippmann photography and holography", *Springer Series in Optical Science, 66* , Springer-Verlag, Heidelberg, New York, **1993**.
[41] Curran, R. K., Shankoff, T. A., "The mechanism of hologram formation in dichromated gelatine", *Appl. Opt.* **1970,** *9,* 1651.
[42] Samoilovich, D. M., Zeichner, A., Friesem, A., "The mechanism of volume hologram formation in dichromated gelatine", *Selected Papers on Holographic Recording Materials,* in: Bjelkhagen, H. (Ed.), *SPIE Milstone series,* **1996,** *130,* 272.
[43] Kubota, T., "Recording of high quality colour holograms", *Appl. Opt.* **1986,** *25,* 4141.
[44] Beesly, M. J., Ccastedine, J. G., "The use of photoresist as a holographic recording medium", *Appl. Opt.* **1970,** *9,* 2720.
[45] Bartolini, R. A., "Characterisation of relief phase holograms recorded in photoresists", *Appl. Opt.* **1974,** *13,* 129.
[46] Cvetkovich, T., in: Jeong, T. H. (Ed.), "Holography in photoresist materials: 1991 update", *Proc. SPIE Int. Symp. on Display Holography,* **1992,** *1600,* 60.
[47] Credelle, T. L., Spong, F. W., "Thermoplastic media for holographic recording", *RCA Rev.* **1972,** *33,* 206.
[48] Lee, T. C., "Holographic recording in thermoplastic films", *Appl. Opt.* **1974,** *13,* 888.
[49] Colburn, W. S., Tompkins, E. N., "Improved thermoplastic-photoconductor devices for holographic recording", *Appl. Opt.* **1974,** *13,* 2934.
[50] Judeinstein, P., Oliveira, P. W., Krug, H., "Photochromic organic-inorganic nanocomposites as holographic storage media", *Adv. Mat. Opt. Electr.* **1997,** *7,* 123.
[51] Miniewicz, A., Kochalska, A., Mysliwiec, J., "Deoxyribonucleic acid-based photochromic material for fast dynamic holography", *Appl. Phys. Lett.* **2007,** *91,* Art. No. 041118.
[52] Fu, S. C., Liu, Y. C., Lu, Z. F., "Photo-induced birefringence and polarization holography in polymer films containing spirooxazine compounds pre-irradiated by UV light", *Opt. Comm.* **2004,** *242,* 115.
[53] Hampp, N., Thoma, R., Oesterhelt, D., "Biological photochrome bacteriorhodopsine and its genetic variant ASP96 – ASN as media for optical-pattern recognition", *Appl. Opt.* **1992,** *31,* 1834.
[54] Kamshilin, A. A., Frejlich, J., Garcia, P. M., "Electro-photochromic gratings in photorefractive $Bi_{12}TiO_{20}$ crystals", *Appl. Opt.* **1992,** *31,* 1787.

[55] Uhrich, C., Hesselink, L., "Submicrometer defect enhancement in periodic structures by using photorefractive holography", *Opt. Lett.* **1992,** *17,* 1087.
[56] Ford, J. E., Ma, J., Fainman, Y., "Multiplex holography in strontium barium niobate with applied field", *JOSA A* **1992,** *9,* 1183.
[57] Colburn, W. S., Hains, K. A., "Volume hologram in photopolymer material", *Appl. Opt.* **1971,** *10,* 1636.
[58] van Renesse, R. L., "Photopolymers in holography", *Opt. Laser Tech.* **1972,** *4,* 24.
[59] Wopschall, R., Pampalone, T., "Dry photopolymer film for recording holograms", *Appl. Opt.* **1972,** *11,* 2096.
[60] Booth, B. L., "Photopolymer material for holography", *Appl. Opt.* **1975,** *14,* 593.
[61] Calixto, S., "Dry polymer for holographic recording", *Appl. Opt.* **1987,** *26,* 3904.
[62] Fernandez, E., Garcia, C., Pascual, I., Ortuno, M., Gallego, S., Belendez, A., "Optimization of a thick polyvinyl alcohol-acrylamide photopolymer for data storage using a combination of angular and peristrophic holographic multiplexing", *Appl. Opt.* **2006,** *45,* 7661.
[63] Hsu, K. Y., Lin, S. H., Hsiao, Y. N., Whang, W. T., "Experimental characterization of phenanthrenequinone-doped poly(methyl methacrylate) photopolymer for volume holographic storage", *Opt. Eng.* **2003,** *42,* 1390.
[64] Carre, C., Lougnot, D. J., "Photopolymers for holographic recording: from standard to seld-processing materials", *J. Phys. (Paris) III* **1993,** *3,* 1445.
[65] Martin, S., Leclere, P., Renotte, Y., Toal, V., Lion, Y., "Characterisation of an acrylamide-based photopolymer holographic recording material", *Opt. Eng.* **1992,** *33,* 3942.
[66] Martin, S., Feely, C., Toal, V., "Holographic Characteristics of an Acrylamide Based Recording Material", *Appl. Opt.* **1997,** *36,* 5757.
[67] Kozlovsky, M. V., "Chiral polymers with photoaffected phase behaviour for photo-optical and optoelectronic applications, *Synt. Met.* **2002,** *127,* 67.
[68] Naydenova, I., Nikolova, L., Ramanujam, P. S., Hvilsted, S., "Light-induced circular birefringence in cyanoazobenzene side-chain liquid-crystalline polyester films", *J. Opt. A: Pure Appl. Opt.* **1999,** *1,* 438.
[69] Guo, H. Y., Si, J. H., Qian, G. D., "Photoinduced birefringence in bulk azodye-doped hybrid inorganic-organic materials by a feratosecond laser", *Chem. Phys. Lett.* **2003,** *378,* 553.
[70] Nunzi, J. M., Barille, R., Kandjani, S. A., Kucharski, S., "Self-Organization of Materials with Walking and Talking Molecules", *Nonlinear Opt. Quant. Opt.* **2006,** *35,* 209.
[71] Bjelkhagen, H. I., Vu kicevic, D., in: Jeong, T. H. (Ed.), "Lippman colour holography in a single-layer silver-halide emulsion", *Proc. SPIE,* **1994,** *2333,* 34.
[72] Kubota, T., "Recording of high quality colour holograms", *Appl. Opt.* **1986,** *25,* 4141.
[73] Bjelkhagen, H., Vukicevic, D., "Investigation of silver-halide emulsions for holography", in: Jeong, T. (Ed.), *Proc. SPIE,* **1994,** *2043,* 20.
[74] Crespo, J., Pardo, M., Satorre, M. A., Quintana, J. A., "Ultraviolet spectrally responsive holograms in dichromated gelatine", *App. Opt.* **1993,** *32,* 3068.
[75] Werlich, H., Sincerbox, G., Yung, B., "Fabrication of high-efficiency surface-relief holograms", *J. Techn.* **1984,** *10,* 105.
[76] Friesem, A. A., Katzir, Y., Ravnoy, Z., Sharon, B., "Photoconductor-t thermoplastic Devices for Holographic Non Destructive Testing", *Opt. Eng.* **1980,** *19,* 659.
[77] Bartolini, R., "Characteristics of relief phase holograms recorded in photoresists", *Appl. Opt.* **1974,** *13,* 129.
[78] Lo, D. S., Johnson, L. H., Honebrink, R. W., "Spatial frequency response of thermoplastic films", *Appl. Opt.* **1975,** *14,* 820.
[79] Banyasz, I., "The effect of the finite spatial resolution of thermoplastic recording materials on holographic image", *Appl. Opt.* **1998,** *37,* 2081.
[80] Shibaev, V., Bobrovsky, A., Boiko, N. N., "Photoactive liquid crystalline polymer systems with light-controllable structure and optical properties", *Progr. Polym. Sci.* **2003,** *28,* 729.
[81] Chen, Y., Wanga, C., Fana, M., Yaob, B., Menkeb, N., "Photochromic fulgide for holographic recording", *Opt. Mat.* **2004,** *26,* 75.
[82] Pieroni, O., Fissii, A., Popova, G., "Photochromic polypeptides", *Progr. Polym. Sci.* **1998,** *23,* 81.

[83] Volk, T., Wohlecke, M., "Thermal fixation of the photorefractive holograms recorded in lithium niobate and related crystals", *Crit. Rev. Solid State Mat. Sci.* **2005,** *30,* 125.
[84] Hong, J. H., Norman, J. B., Chang, T. Y., "Photorefractive integrator characterisation", *Appl. Opt.* **1995,** *34,* 6775.
[85] Volodin, B. L., Meerholz, K. S., Kippelen, B., Kukhtarev, N., Peyghambarian, N., "Highly efficient photorefractive polymers for dynamic holography", *Opt. Eng.* **1995,** *34,* 2213.
[86] Matoba, O., Javidi, B., "Secure ultrafast communication with spatial-temporal converters", *Appl. Opt.* **2000,** *39,* 2975.
[87] Gu, C., Xu, Y., Liu, Y. S., Pan, J. J., Zhou, F. Q., He, H., "Applications of photorefractive materials in information storage, processing and communication", *Opt. Mat.* **2003,** *23,* 219.
[88] Nikolova, L., Todorov, T., Ivanov, M., Ramanujam, P. S., Hvilsted, S., "Polarization holographic gratings in side-chain azobenzene polyesters with linear and circular photoanisotropy", *Appl. Opt.* **1996,** *35,* 3835.
[89] Naydenova, I., Nikolova, L., Todorov, T., Holme, N. C. R., Ramanujam, P. S., Hvilsted, S., "Diffraction from polarization holographic gratings with surface relief in side-chain azobenzene polyesters", *JOSA B* **1998,** *15,* 1257.
[90] Holme, N. C. R., Nikolova, L., Ramanujam, P. S., Hvilsted, S., "An analysis of the anisotropic and topographic gratings in a side-chain liquid crystalline azobenzene polyester", *Appl. Phys. Lett.* **1997,** *70,* 1518.
[91] Kumar, G. S., Neckers, D. C., "Photochemistry of azobenzene-containing polymers", *Chem. Rev.* **1989,** *89,* 1915.
[92] Natansohn, A., Rochon, P., Gosselin, J., Xie, S., "Azo polymers for reversible optical storage. 1 poly [4′-[[2-(acryloyloxy)ethyl]ethylamino]-4-nitroazobenzene]", *Macromolecules* **1992,** *25,* 2268.
[93] Nikolova, L., Todorov, T., "Diffraction efficiency and selectivity of polarization holographic recording", *Opt. Acta* **1984,** *31,* 579.
[94] Todorov, T., Nikolova, L., Stoyanova, K., "Polarization holography. 3 Some applications of polarization holographic recording", *Appl. Opt.* **1985,** *24,* 785.
[95] Dhar, L., Hale, A., Katz, H. E., Schilling, M. L., Schnoes, M. G., Schilling, F. C., "Recording media that exhibit high dynamic range for digital holographic data storage", *Opt. Lett.* **1999,** *24,* 487.
[96] Suzuki, N., Tomita, Y., Kojima, T., "Holographic recording in TiO2 nanoparticle-dispersed methacrylate photopolymer films", *Appl. Phys. Lett.* **2002,** *81,* 4121.
[97] Suzuki, N., Tomita, Y., "Silica-nanoparticle-dispersed methacrylate photopolymers with net diffraction efficiency near 100%", *Appl. Opt.* **2004,** *43,* 2125.
[98] Hariharan, P., *Optical Holography, Principles, Techniques and Applications* 2nd edition, Cambridge University Press, **1996**.
[99] Collier, R., Burckhardt, C., Lin, L., *Optical Holography*, Academic Press, New York, **1971**.
[100] Kakichashvili, S. D., "Method of recording phase polarization holograms", *Kvantovaya Elecron.* **1974,** *1,* 1435.
[101] Kogelnik, H., "Coupled wave theory for thick hologram gratings", *Bell Syst. Tech. J.* **1969,** *48,* 2909.
[102] Colburn, W., Haines, K., "Volume hologram formation in photopolymer materials", *Appl. Opt.* **1971,** *10,* 1636.
[103] Trout, T., Schmieg, J., Gambogi, W., Weber, A., "Optical photopolymers: Design and Applictions", *Adv. Mater.* **1998,** *10,* 1219.
[104] Zhao, G., Mourolis, P., "Diffusion model of hologram formation in dry photopolymer materials", *J. Mod. Opt.* **1994,** *41,* 1929.
[105] Moreau, V., Renotte, Y., Lion, Y., "Characterisation of DuPont photopolymer: determination of kinetic parameters in a diffusion model", *Appl. Opt.* **2002,** *41,* 3427.
[106] Colvin, V., Larson, R., Harris, A., Schilling, M., "Quantitative model of volume hologram formation in photopolymers", *J. Appl. Phys.* **1997,** *81,* 5913.
[107] Kwon, J. H., Hwang, H. C., Woo, K. C., "Analysis of temporal behaviour of beams diffracted by volume gratings formed in photopolymers", *JOSA B* **1999,** *16,* 1651.
[108] Piazzolla, S., Jenkins, B., "First harmonic diffusion model for holographic grating formation in photopolymers", *JOSA B* **2000,** *17,* 1147.

[109] Lawrence, J., O'Neill, F., Sheridan, J., "Adjusted intensity nonlocal diffusion model of photopolymer grating formation", *JOSA B* **2002,** *19,* 621.
[110] Aubrecht, I., Miler, M., Koudela, I., "Recording of holographic diffraction gratings in photopolymers: theoretical modelling and real-time monitoring of grating growth", *J. Mod. Opt.* **1998,** *45,* 1465.
[111] Caputo, R., Sukhov, A. V., Tabirian, N. V., Umeton, C., Ushakov, R. F., "Mass transfer process induced by inhomogeneous photo - polymerisation in a multi-component medium", *Chem. Phys.* **2001,** *271,* 323.
[112] Naydenova, I., Martin, S., Jallapuram, R., Howard, R., Toal, V., "Investigations of the diffusion processes in self-processing acrylamide-based photopolymer system", *Appl. Opt.* **2004,** *43,* 2900.
[113] Martin, S., "A new photopolymer recording material for holographic applications: Photochemical and holographic studies towards an optimized system", *Ph.D. Thesis,* University of Dublin, **1995**.
[114] Martin, S., Naydenova, I., Toal, V., Jallapuram, R., Howard, R. G., "Two way diffusion model for the recording mechanism in a self developing dry acrylamide photopolymer", *Proc. SPIE* **2006,** *6252,* 37.
[115] Jallapuram, R., Naydenova, I., Toal, V., Martin, S., Howard, R., "Spatial frequency response of acrylamide holographic photopolymer" in: Shakher, C., Mehta, D. S. (Eds.), *Proc. of International Conference on Laser Applications and Optical Metrology,* Anamaya Publishers, New Delhi, **2003,** 275.
[116] Naydenova, I., Sherif, H., Martin, S., Jallapuram, R., Toal, V., "A holographic sensor", WO2007060648, **2007**.
[117] Lawrence, J. R., O'Neill, F. T., Sheridan, J. T., "Photopolymer holographic recording material", *OPTIK* **2001,** *112,* 449.
[118] Vaia, R., Dennis, C., Natarajan, L., Tondiglia, Tomlin, V. D., Bunning, T., "One step, Micrometer-scale Organization of Nano- and Mesoparticles Using Holographic Photopolymerisation: A generic technique", *Adv. Mat.* **2001,** *13,* 1570.
[119] Tomita, Y., Nishibiraki, H., "Improvement of holographic recording sensitivities in the green in SiO_2 nanoparticles-dispersed methacrylate photopolymers doped with pyrromethene dyes", *Appl. Phys. Lett.* **2003,** *83,* 410.
[120] Sakhno, O., Goldenberg, L., Stumpe, J., Smirnova, T., "Surface modified ZrO_2 and TiO_2 nanoparticles embedded in organic photopolymers for highly effective and UV-stable volume holograms", *Nanotechnology* **2007,** *18,* 105704.
[121] Sanchez, C., Escuti, M., Heesh, C., Bastiaansen, C., Broer, D., Loos, J., Nussbaumer, R., "TiO_2 Nanoparticle – Photopolymer Composites for Volume Holographic recording", *Adv. Funct. Mater.* **2005,** *15,* 1623.
[122] Kim, W., Jeong, Y., Park, J., "Nanoparticle- induced refractive index modulation of organic-inorganic hybrid photopolymer", *Opt. Expr.* **2006,** *14,* 8976.
[123] Naydenova, I., Sherif, H., Mintova, S., Martin, S., Toal, V., "Holographic recording in nanoparticle-doped photopolymer", *Proc. SPIE* **2006,** *6252,* 45.
[124] Tomita, Y., Chikama, K., Nohara, Y., Suzuki, N., Furushima, K., Endoh, Y., "Two dimensional imaging of atomic distribution morphology created by holographically induced mass transfer of monomer molecules and nanoparticles in a silica-nanoparticle-dispersed photopolymer film", *Opt. Lett.* **2006,** *31,* 1402.
[125] Tomita, Y., Suzuki, N., Chikama, K., "Holographic manipulation of nanoparticle distribution morphology in nanoparticle-dispersed photopolymers", *Opt. Lett.* **2005,** *30,* 839.
[126] Suzuki, N., Tomita, Y., "Real-time phase-shift measurement during formation of a volume holographic grating in nanoparticles-dispersed photopolymers", *Appl. Phys. Lett.* **2006,** *88,* 011105.
[127] Colburn, W. S., "Review of materials for holographic optics", *J. Im. Sci. Tech.* **1997,** *41,* 443.
[128] Suzuki, N., Tomita, Y., Ohmori, K., Hidaka, M., Chikama, K., "Highly transparent ZrO_2 nanoparticles-dispersed acrylate photopolymers for volume holographic recording", *Opt. Expr.* **2006,** *14,* 12712.
[129] Tosheva, L., Valtchev, V. P., "Nanozeolites: synthesis, crystallization mechanism and applications", *Chem. Mater.* **2005,** *17,* 2494.

[130] Naydenova, I., Dalton, C., Sherif, H., Larlus, O., Martin, S., Toal, V., Mintova, S., to be submitted in *Appl. Phys. Lett.*
[131] Sherif, H., "Characterisation of an Acrylamide-Based Photopolymer for Holographic Data Storage", *MPhil Thesis,* Dublin Institute of Technology, Dublin, Ireland, **2005**.
[132] Rhee, U. S., Caulfield, H. J., Shamir, J., "Characteristics of the Du-pont photopolymer for angularly multiplexed page – oriented holographic memories", *Opt. Eng.* **1993,** *32,* 1839.
[133] Karrer, P., Corbel, S., Andre, J. C., "Shrinkage effects in photopolymerisable resins containing filling agents – applications to stereophotolithography", *J. Polym. Sci. A – Polym. Chem.* **1992,** *30,* 2715.
[134] Wu, S. D., Glytsis, E. N., "Characteristics of DuPont photopolymers for slanted holographic grating formations", *JOSA B – Opt. Phys.* **2004,** *21,* 1722.
[135] Rochtchanovitch, P., Kostrov, N., Goulanian, E., "Method of characterization of effective **shrinkage** in reflection holograms", *Opt. Eng.* **2004,** *43,* 1160.
[136] Mok, F. H., Burr, G. W., Psaltis, D., "System Metric For Holographic Memory Systems", *Opt. Lett.* **1996,** *21,* 12.
[137] Mihaylova, E., Naydenova, I., Martin, S., Toal, V., "Comparison of three electronic speckle pattern shearing interferometers using photopolymer holographic optical elements", *Proc. SPIE* **2006,** *6252,* 466.
[138] Mihaylova, E., Naydenova, I., Martin, S., Toal, V., "Electronic speckle pattern shearing interferometry using photopolymer diffractive optical elements for vibration measurement", Proceedings of 6th International conference on Vibration Measurements by Laser Techniques: Advance Applications, Ancona, Italy, **2004**, 73.
[139] Klingstedt, F., Kalantar, Neyestanaki, A., Lindfors, L.-E., Salmi, T., Heikkilä, T., Laine, E., "An investigation of the activity and stability of Pd and Pd-Zr modified Y-zeolite catalysts for the removal of PAH, CO, CH_4 and NOx emissions", *Appl. Catal. A: Gen.* **2003,** *239,* 229.
[140] Wang, Y., Zhou, S., Xia, J.-R., Xue, J., Xu, J.-H., Zhu, J.-H., "Trapping and egradation of volatile nitrosamines on cyclodextrin and zeolites", *Microporous Mesoporous Mater.* **2004,** *75,* 247.
[141] Nazarova, D., Mednikarov, B., Sharlandjiev, P., "Resonant optical transmission from a one-dimensional relief metalized subwavelength grating", *Appl. Opt.* **2007,** *46,* 8250.

CHAPTER 22

Inorganic Sulphur Pigments Based on Nanoporous Materials

Stanisław Kowalak *and* Aldona Jankowska

Contents

Abstract

Zeolites and other molecular sieves can be used as matrices for pigments. Ultramarine is the most spectacular example of a pigment based on a sodalite structure containing sulphur radicals (chromophores) encapsulated inside β-cages. The conventional synthesis of Ultramarine from kaolin can be altered by using zeolites as starting materials. The sulphur species can be introduced into zeolites either during their crystallization in the presence of sulphur compounds or by post-synthesis treatment with sulphur radical precursors. The chromophores (mainly anion radicals $S_3^{\bullet -}$, $S_2^{\bullet -}$) are formed by heating (300–900 °C) the sulphur-containing zeolites so that they are entrapped inside the cages.

Ordered Porous Solids
DOI: 10.1016/B978-0-444-53189-6.00022-6

The resulting pigments maintain the parent zeolite structure or the zeolite matrix can recrystallize (mostly to SOD) upon thermal treatment. Non-aluminosilicate (e.g., $AlPO_4$-20) matrices can be used too. The application of zeolites allows to attain a broader range of product colours than during conventional synthesis from kaolin and substantially lowers the emission of polluting gases.

Keywords: Encapsulation, Molecular sieves, Pigments, Radicals, Sulphur, Zeolites

1. Introduction

A well ordered nano-scale pore system of the molecular sieves offers a tremendous variety in their application. In addition to large scale processes involving ion-exchange, adsorption, catalysis, agriculture and building industry, growing attention of fundamental research and industry is being paid to the more advanced applications of these materials, such as the preparing of sensors, optical and microelectronic devices, luminophores, membranes and pigments.

Zeolites and other micro- and mesoporous molecular sieves can host the molecules and ions inside their inner voids owing to chemical bonding to the framework, or occlusion and encapsulation of chosen species. Very finely dispersed and well insulated guest moieties can reveal novel properties and often much higher stability. The encapsulated species can play the role of catalytic centers (*zeozymes*), where they combine the benefits of the high activity of introduced species with higher stability and the shape selectivity effect resulting from the ordered geometry of intracrystalline channels or cavities of molecular sieves. Metallophthalocyanines, porfirines, Schiff base ligand complexes of transition metal cations encapsulated inside the molecular sieves appear to be very interesting catalysts.[1] The molecular sieve matrices are also used for sensors of different kinds. Zeolites and other molecular sieves can accommodate the polymers of specific properties (e.g., molecular wires). The light sensitive species (e.g., rare earth cations) can achieve more pronounced luminescent properties after accommodation and separation inside the molecular sieve matrices. Encapsulated dye molecules can be applied to microelectronic switchers by a hole burning effect. Separation in zeolite cavities allows effective semiconductors to be prepared (e.g., CdS). These applications are widely investigated and numerous reviews concerning this topic are available.[2–5]

The inner voids of molecular sieves are also highly suitable for the accommodation of molecules playing a role of chromophores. Owing to their perfect dispersion and insulation within the cages or cavities the resulting products attain extraordinary stability and efficient colour tinting strength.

Synthetic zeolites as well as mesoporous materials are mostly colourless. Some colouration of natural zeolites (yellow, beige) results from the presence of iron or other transition metal cations. Some zeolitic materials (e.g., lazurite,[6] bystrite[7]) show a distinct colouration owing to the encapsulation of sulphur salts. Some transition metals introduced into tetrahedral framework positions result in distinct colouration of the resulting materials (e.g., CoAPO[8]). An ion modification of zeolites with the transition metal cations (Cu, Co, Cr, Fe,) often causes a noticeable colouration of the obtained derivatives, although their colour intensity is usually not very high and they are not considered potential pigments. The colour intensity of such cation modification can be increased markedly by introducing appropriate ligands surrounding the cations.

Very intense colouration can be attained by treating the dehydrated zeolites (SOD, LTA, FAU) with alkali metals (mainly with their vapours). The metal ionic clusters (Me_4^{3+}) combined with an unpaired electron (*electride*) are relatively stable inside the β-cages and they decompose only after extended extraction with water. The colouration and colour intensity depends on the number of clusters introduced and on the kind of alkaline metal introduced into the zeolite.[9–11]

2. Molecules Encapsulated in Zeolites

2.1. Metallophtalocyanines

The main aim of encapsulation (*ship in a bottle*)[12] of metallophthalocyanines into zeolites was to obtain efficient heterogeneous catalysts for mild oxidation. On the other hand metallophtallocyanine (particularly Cu-Pc) is commercially produced as a valuable pigment. The incorporation of Me-Pc into zeolites can increase the resistibility of the pigment and the zeolite host can extend its application. Besides forming the Pc in zeolite from adsorbed dicyanobenzene the Me-Pc can be encapsulated into molecular sieves (including zeolites A) by crystallization around the guest molecules.[13]

2.2. Styrene

Styrene adsorbed on zeolite H-ZSM-5 results in a pink-violet colouration of a sample already at room temperature.[14,15] Styrene undergoes oligomerization (mostly dimerization) and the resulting carbocations are compensated by a negative charge of the zeolite framework.[16–19] Cation-radicals can be identified by ESR spectroscopy.[17] Small oligomers are formed only in zeolites with medium sized channels (H-ZSM-5, H-mordenite, or H-ferrierite), but not in H-form of faujasites, since large pores and cavities do not hinder the formation of bulkier oligomers. The polyvalent cation modified zeolites due to their acid sites also form coloured products with styrene.[17] These coloured products are very stable and they decompose only after heating at ~400 °C. The treatment with alkalis results in decomposition of coloured species. The colouration of styrene oligomer cations can be substantially changed by using substituted styrene derivatives. Particularly interesting is adsorbed trans-stilbene forming blue products in H-ZSM-5 or H-mordenite channels. Stilbene cation radical is suggested to form first and then the so-called electron-hole pair is generated.[19]

2.3. Salts encapsulated in zeolites

Encapsulated salts can be introduced into cavities or cages of zeolites and zeotypes (e.g., gallosilicate, aluminogermanate) during their hydrothermal crystallization in the presence of chosen compounds.[20–23] The salts of intense colouration such as permanganates, manganates, or chromates can be introduced into sodalite (violet MnO_4^-), zeolite A (violet MnO_4^-), nosean (yellow, CrO_4^{2-}), or cancrinite (green MnO_4^{2-}). The resulting samples do not fade on washing.

2.4. Thiocyanates

Using the hydrothermal synthesis of sodalite Weller[24] introduced SCN^- anions into the sodalite cages. The resulting samples were modified with cations such as Li, Ag and K. Thiocyanate underwent oxidation upon heating and sulphur anion radicals ($S_3^{\bullet-}$, $S_2^{\bullet-}$) were generated, which was reflected in the remarkable colours of the products (green, blue).

Regarding the successful encapsulation of various coloured salts in zeolites it was tempting to entrap iron thiocyanate inside the zeolite matrices. The latter is known as a very intense purple coloured compound, applied in the classical qualitative test for indicating the presence of iron. Unfortunately, despite many attempts (such as incorporation of SCN^- into sodalite during its crystallization followed by treatment with iron salts or treating of Fe-modified zeolites with thiocyanate solutions) the coloured complexes never remained in the zeolite phase. Usually washing of the samples resulted in total discolouration of the zeolite phase owing to removal of iron thiocyanate.[17]

3. Sulphur Chromophores in Zeolite Matrices

3.1. Lazurite and natural ultramarine (historical background)

The historical background of lazurite and ultramarine pigments has been presented in the excellent reviews by Seel[25] and Mertens.[26] The natural lazurite (*lapis lazuli*) has been known and prized since ancient times (5000 BC). It was used as a semi-precious gem for artistic, decorative and jewellery purposes in the developed cultures of that time (Mesopotamia, Egypt) Fig. 22.1.

The intense blue and very stable colouration of this mineral made it an attractive raw material for preparing a pigment for artistic painting. The name *ultramarine* appeared in the middle ages following the trade journeys of Marco Polo who brought it from Afghanistan and its name (in Italian) referred to the overseas origin of this mineral. The preparation of pigment from raw mineral was very tedious and comprised of many steps (milling, extraction, separation) and eventually the finest product never made more than 10% of parent mineral. Prof. Seel quoted in his review[25] opinion and suggestion of Italian painter Cennino Cennini (1370–1440 AD) expressed in the chapter of his book *Trattato della Pittura* devoted to preparation of ultramarine pigment: *Remember that it needs a special skill to prepare it*

Figure 22.1 (a) Natural lazurite and (b) Golden mask of Tutankhamun (1361–1352 BC). (See color insert.)

well. And you must know that the beautiful young women know better how to make it than the men, because they stay at home constantly and because they are more patient and have more gentle hands. But beware of the older ones. Perhaps, this opinion should be taken into an account even now when the new students are chosen to join the research on pigments.

The properties of ultramarine pigment were greatly appreciated by the masters of painting, although its very high price limited its application only to the most valuable works. Besides Afghanistan other large deposits of natural ultramarine were found later in Chile and Siberia (near Lake Baikal).

Ultramarine blue remained a luxury product until the beginning of the nineteenth century. The great civilizing progress resulting from the Industrial Revolution created an enormous demand for new materials or for larger amounts of already known products. Dyes and pigments were among them. The high cost and limited supply of natural lazurite as well as laborious preparation of the pigment made this procedure unable to fulfill the demand of the market and searching for synthetic substitute of natural ultramarine became an important challenge. There is not certain information on the first successful synthesis of artificial ultramarine, but it was noticed that occasionally a blue admixture was formed during manufacturing of soda by the Leblanc process. Similar product was sometimes obtained in lime kilns in Sicily from the raw material rich in sulphur and it was applied as substitute of ultramarine. Very crucial for the modern production of ultramarine was in 1824 an initiative of the Societe d'Encouragement pour l'Industriale Nationale in Paris who offered a prize for invention of an inexpensive method to prepare a pigment analogous to the ultramarine obtained from natural lapis lazuli, and which would cost no more than 300 francs per pound. The search for synthetic analogues of natural ultramarine became more realistic and promising after the discovery of the chemical composition of lazurite by Clement and Desormes[27] in 1806, who indicated its main components (SiO_2 – 35.8%, Al_2O_3 – 34.8%, Na_2O – 23.2%, S – 3.1%, $CaCO_3$ – 3.1%). The prize was received by Jean Baptiste Guimet in 1828, although the estimated cost of his product was more than twice as much as stipulated in 1824. The invented procedure for the preparation of artificial ultramarine consisted in long high temperature treatment of a mixture containing mainly kaolin clay, sulphur, sodium carbonate and charcoal. Almost simultaneously and independently a synthetic procedure was developed by the German chemist Christian Gottlob Gmelin. The first industrial production of ultramarine according to Gmelin's method started in Germany in 1834. Soon afterwards many other ultramarine factories opened in Europe and their number approached 100 by the end of the nineteenth century. Artificial ultramarine became a very common and inexpensive product applied mostly for wall painting, as an optical brightener for laundry detergents, widely used in the paper industry, and later for the colouration of plastics. Contrary to many other inorganic pigments (particularly those based on heavy metal compounds) the ultramarine blue is non toxic and can be applied also for cosmetic products or even as an admixture to food (e.g., sugar). One of the first producers of ultramarine was Isaac Reckitt, who used it as an additive to starch (*dolly tubes*) as an optical brightener and afterwards he opened (in 1884) a factory in Hull (Reckitt's Colours) that has become the world's largest producer of ultramarine. The company still exists (since 1994 as

Holliday Pigments, a division of Yule Catto) and two factories in Hull (England) and Comines (France) produce almost 20,000 tons of ultramarine per year. The second world producer is Nubiola (Spain) supplying ~15,000 of ultramarine yearly. Considerable amounts of ultramarine are produced in India and China. The total world annual production approaches 50,000 tons.

As illustrated in Fig. 22.2 the synthesis of ultramarine consists of several steps. The main substrate kaolin clay is calcined first at 450 °C and after its dehydration metakaolin is obtained. Then metakaolin is mixed and milled with elemental sulphur, sodium carbonate and a reductive agent (charcoal, pitch or heavy oil). Usually almost equal amounts of metakaolin, soda and sulphur are applied for the initial mixture. Sometimes additional quartz or feldspar are added. Attempts to use fly ash instead of kaolin for ultramarine synthesis have been reported.[29,30] Another sources of sulphur such as sodium sulfates (with higher loading of reducing agents) or sulfides[31] can be also used for ultramarine synthesis. The use of the latter results in lower emission of polluting gases.

The starting mixture is formed in bricks or put into ceramic crucibles and transferred into kilns. In some cases the initial mixture is heated without either vessels or bricks. A continuous method of production was also tried.[28] The loose powder initial mixture was continuously admitted into the tubular rotary electrically heated furnace (similar to that applied in the cement industry) and the final product was received after passing the furnace tube. The thermal method commences with the raising temperature (first step) up to 800 °C. Usually oil or gas are used as heating fuel (sometimes coal is still applied). The consumption of oxygen from the kiln

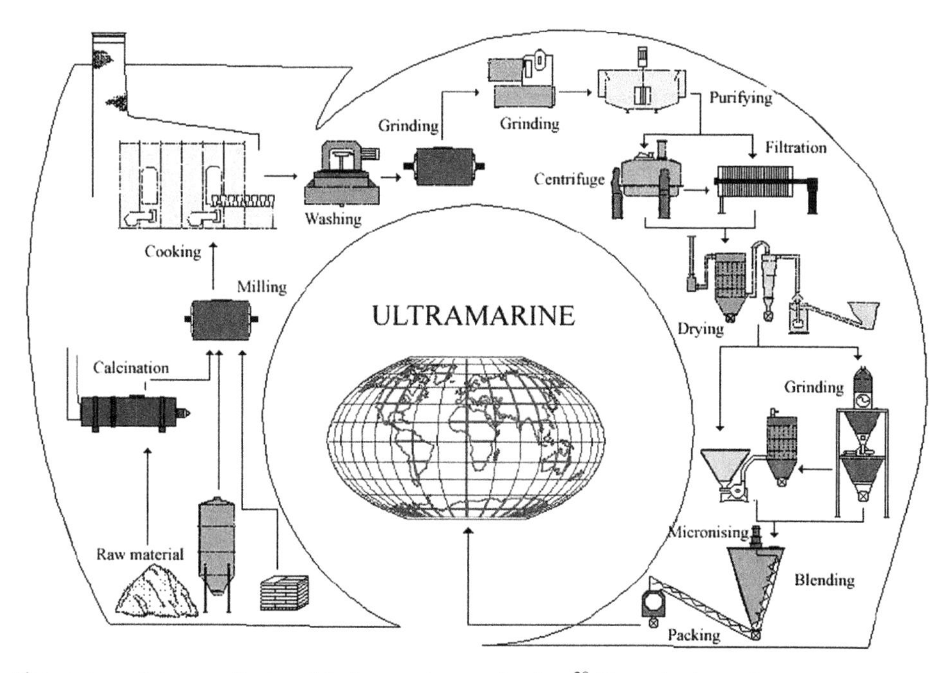

Figure 22.2 Scheme of industrial ultramarine production.[28] (See color insert.)

chamber upon fuel burning is beneficial in providing a reductive atmosphere and subsequently to protect sulphur from total combustion, although some SO_2 is formed. The mixture is maintained at high temperature for several days (second step) and then it is cooled down (third step). Upon cooling, when the temperature drops down to 600 °C the gases (air or SO_2) are admitted to the kiln in order to oxidize the primary product (usually green) and form the final stable intense blue form of ultramarine. The whole thermal process takes about 20 days. After cooling, the product is crushed, washed with water in order to remove the soluble salts, filtered, ground, cleaned, separated, blended and packed. The final product is a fine powder with the particle size of ~1 μm. It always contains more than 10% of sulphur. The following general chemical formula of ultramarine can be presented: $Na_{6+n}(Al_6Si_6O_{24})S_m$. In the case of the typical blue pigment it can be expressed as: $Na_7(Al_6Si_6O_{24})S_3$.

Generally the elemental sulphur applied for synthesis undergoes reduction from the valence state 0 to the formal level −1/3 in the resulting radical $S_3^{\bullet-}$. The heating of the mixtures is conducted first in a reductive atmosphere, and raw green ultramarine contains substantial amounts of radicals $S_2^{\bullet-}$ with a formal sulphur valence −1/2. Subsequent oxidation results in forming anion radicals $S_3^{\bullet-}$:

$$S_8(0) \xrightarrow{\text{reduction}} {}^{\bullet}S_2^-(-1/2) \xrightarrow{\text{oxidation}} {}^{\bullet}S_3^-(-1/3)$$

It is believed that reaction between sulphur and alkalis in the first step of thermal process results in forming alkaline oligosulfides:

$$Na_2CO_3 + S + C(\text{reductor}) \rightarrow Na_2S + CO_2 + CO,$$

or as result of a disproportionation of sulphur in the alkaline medium:

$$2Na_2CO_3 + 3S \rightarrow 2Na_2S + SO_2 + 2CO_2$$

The resulting Na_2S reacts with excess of sulphur to form oligosulfides. SO_2 can react with alkalis forming Na_2SO_3 and subsequently with sulphur to form $Na_2S_2O_3$.

The second step at a temperature above 700 °C involves the reaction of metakaolin with alkalis and alkaline sulfides. Aluminosilicate with an SOD structure is formed. The role of oligosulfides is very crucial at this stage, since calcinations of kaolin with plain alkalis (NaOH or Na_2CO_3) never leads to the formation of a sodalite at these temperatures. It is likely that sulphur species play a role of structure directing agents. Moreover, they remain encapsulated inside formed β-cages and eventually they are transferred into radicals there ($S_3^{\bullet-}$ and considerable amount of $S_2^{\bullet-}$).

The last step of the thermal process consists in the mild oxidation of raw green ultramarine where radicals $S_2^{\bullet-}$ are oxidized to $S_3^{\bullet-}$.

The industrial procedure of ultramarine production has not been changed substantially since the nineteenth century and its main drawback is the evolution of considerable amounts of polluting volatile sulphur compounds (SO_2 and H_2S)

in the waste gases. The law regulation concerning the environmental protection has become much more restrictive in the twentieth century and the factories were forced to neutralize the polluting gases, which markedly affected the cost of production. Many small old factories were unable to fulfill the established law requirements preventing air pollution and subsequently they have had to be closed.

Ultramarine pigments are produced mainly as ultramarine blue with a green or red shade. The violet and pink modifications are prepared by means of post-synthesis treatment of ultramarine blue with chlorine, dry HCl or with ammonium nitrate at elevated temperatures.[32] Colour intensities of the latter modifications of ultramarine are always markedly lower than that of primary ultramarine blue. The colouration of ultramarine can also be modified by an ion-exchange procedure (particularly in solid state) in which the original sodium cations are replaced by other cations.[33–35] The introduction of potassium cations either by cation exchange of the product (or more markedly) by admitting some K_2CO_3 into the initial mixture results in a more distinct red shade of the resulting material.[36]

Other chalcogen chromophores such as selenium (Se_2^- and Se_2) and tellurium (Te_2^- and Te_2) can be encapsulated in sodalite cages and result in forming intense red (Se) or green (Te) products.[37–40] The attempts to prepare mixed S-Se systems were not successful.[37–40] Although the first syntheses of selenium ultramarine analogues were described[41] as early as 1877 their use in commercial processes is unviable because of the strong toxicity of selenium (and tellurium) compounds.

Despite the development of industrial ultramarine production in the nineteenth century, the structure and particularly the nature of chromophores remained unknown for a long time. Although the chemical composition of natural lazurite has been established for two hundred years,[27] the crystalline structure of natural and synthetic ultramarine was established only in 1929 by F.M. Jaeger.[42] It has been proven by XRD analysis that the crystalline structure of natural and artificial ultramarine show the same SOD structure as sodalite, nosean, hauyn. The further more precise SS MAS NMR measurement indicated that the ordering of the sodalite framework atoms (Si, Al) in synthetic ultramarine is less perfect than that in natural lazurite.[43]

The nature of the colour carrier in ultramarine was a matter of speculation for a long time. Although it was clear that sulphur compounds are responsible for intense colouration, their chemical nature remained unexplained. Many different compounds were considered the potential chromophores (Fig. 22.3).[44] It was noticed that ultramarines of different shades show different S/Na ratios in their compositions. Chemical analysis of ultramarine always indicated more sodium than aluminum atoms which suggested that part of sodium cations had to be compensated by sulphur bearing anions. The treatment of ultramarine with acids always liberated H_2S which could suggest the presence of sulfides in ultramarine. Since some cages contained several sulphur atoms it was very likely that oligosulfides could contribute in structure of chromophores. Sulfides or oligosulfides are, however, not blue.

Significant progress in understanding the nature of the colour carrier has been attained by studies on alkaline oligosulfides dissolved in aprotic solvents such as acetone, DMF, DMSO and hexamethylphosphoric acid triamide (HMPT).[30,42,43,45–49]

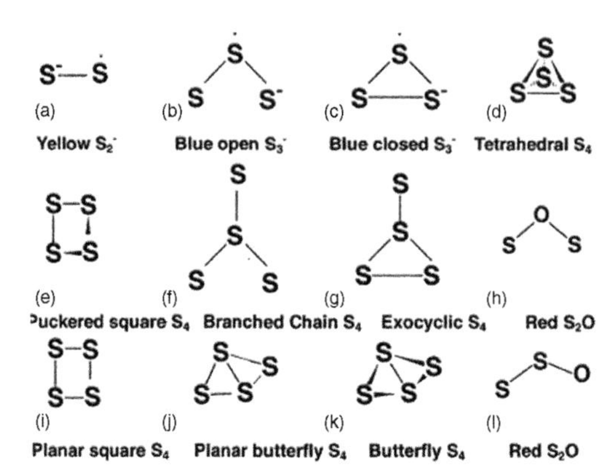

Figure 22.3 Potential sulphur chromophores.[44]

The solid divalent oligosulfides are always yellow, not blue, neither red, while their aprotic solvent solutions indicate intense colouration. The solutions of short sulphur chain oligosulfides are green, but they turn intense blue on increasing in sulphur content and finally they become red and dark red upon further increase in sulphur loading. The solution prepared in oxygen-free conditions showed, that the yellow-green colour of Na_2S solutions turned intense blue after increasing the content of sulphur up to a level corresponding with the formula Na_2S_6. This led to the conclusion that oligosulfide chains undergo the homolytic scission of the S-S bond, which results in the generation of oligosulfide anion radicals. Thus, the anion radical $S_3^{\cdot -}$ could be considered as the blue chromophore. It is believed that anion radicals are encapsulated inside β-cages and they are combined with sodium cations (Fig. 22.4).

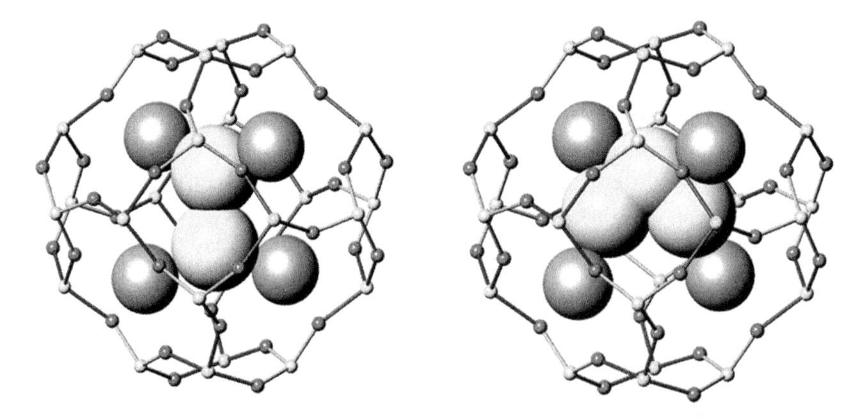

Figure 22.4 Models of sodalite cages with entrapped anion radicals $S_2^{\bullet -}$ (left) or $S_3^{\bullet -}$ (right).[50] (See color insert.)

The decisive arguments were provided by ESR spectroscopy[34,51–54] that indicated a signal with a g-tensor value of 2.029 for natural and synthetic ultramarine. The same ESR spectra (although with a much broader signal) have been recorded for a blue solution of oligosulfide in aprotic solvent (Fig. 22.5).

The opened isomer $S_3^{\bullet -}$ (Fig. 22.6) is believed to be the most stable and is responsible for blue colouration, while the closed one represents the transition state.[29,30,55,56] The continues wave EPR mode at X-band frequencies is mostly applied for determining the components of g–tensor. For more detailed measurements the electronuclear double resonance (ENDOR) and electron spin-echo envelope modulation (ESEEM) techniques are applied.[55,57]

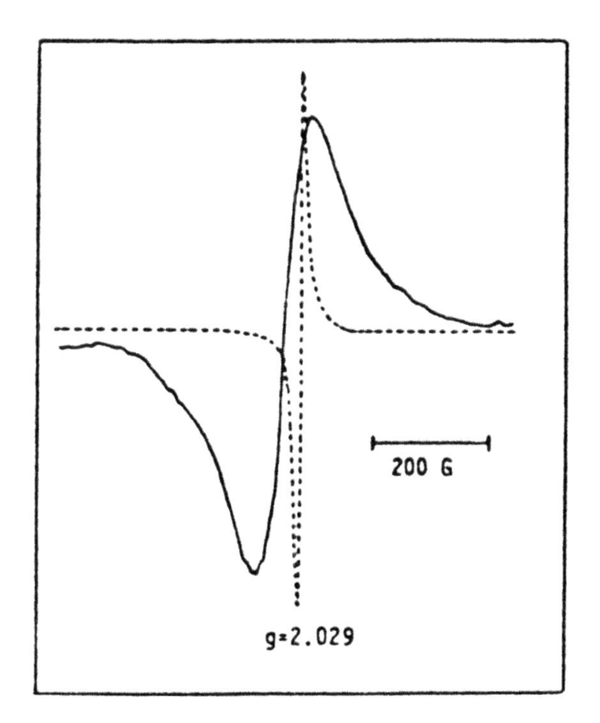

Figure 22.5 EPR spectra of ultramarine (narrow signal) and of anion radical $S_3^{\bullet -}$ in HMPT solution.[25]

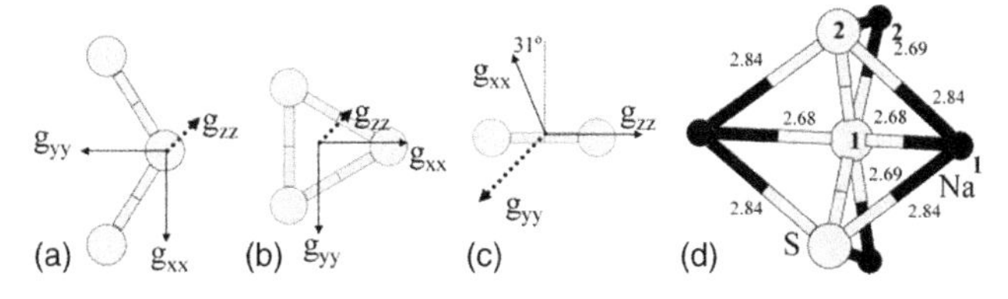

Figure 22.6 The radical models applied for g–tensor DFTcalculations: (a) $S_3^{\bullet -}$ open form (C_{2v}), (b) $S_3^{\bullet -}$ closed form (D_{3h}), (c) $S_2^{\bullet -}$ (C_1), d–$[Na_4S_3]^{3+}$ (C_{2v}).[55]

The anion radical $S_2^{\bullet-}$ is considered a yellow chromophore.[25,48,49,53] Their ESR spectra are, however, not univocally interpreted in literature. The various values of g-tensor in the range 2.0–2.4 were attributed to this radical in earlier works,[25,53] but in the more recent papers it is rather ascribed to impurities[58] and some authors doubt[55,58] whether any $S_2^{\bullet-}$ radicals could be detected by means of ESR spectroscopy because of relaxation effects.

The nature of red chromophore is not certain either. Number of various compounds were considered as potential red colour carrier (S_3, SOS, SOS^-, S_2O, S_2O^-, S_3Cl, S_3Cl^-, S_2Cl, S_4, S_4^-).[29,44] The bulkier sulphur moieties (S_4, S_4^-) were believed the most possible chromophores.[48,49] The ESR signal with $g = 2.0342$ was attributed to anion radical $S_4^{\bullet-}$.[25] The more recent publications[29,30] provide the strong arguments that rather neutral S_4 molecule (particularly the chain *cis* isomer) is responsible for red colouration. According to the calculation this chromophore can be accommodated inside the β-cages.[29]

The electronic spectra of various types of ultramarine (Fig. 22.7) indicate the typical peaks: at ~600 nm for ultramarine blue, that reflect the anion radical $S_3^{\bullet-}$ (U-Bl in Fig. 22.7a), at ~380 nm ascribed to $S_2^{\bullet-}$ for ultramarine yellow (U-Ye in Fig. 22.7a) and both peaks for ultramarine green (U-Gr in Fig. 22.7a). The ultramarine red (Fig. 22.7b U-Re) shows a predominant peak at ~500 nm and the violet ultramarine indicates the main peaks at ~500 and 600 nm.

The IR spectroscopy is not very useful for the identification of sulphur chromophores. The band at 550–580 cm^{-1} is assigned to the $S_3^{\bullet-}$ anion radical,[25] but due to an frequent overlapping with structural bands of zeolite structure it is not very distinct.

Much more suitable and important tool for identification of sulphur species is Raman spectroscopy, which allows the particular radicals and elemental sulphur to be distinguished. The band at 550 cm^{-1} is assigned to radical $S_3^{\bullet-}$, the radical $S_2^{\bullet-}$ is reflected by the band at ~600 cm^{-1} (Fig. 22.8)[25,59] and the peaks at 153, 218, 473 cm^{-1} are attributed to elemental sulphur S_8.[60]

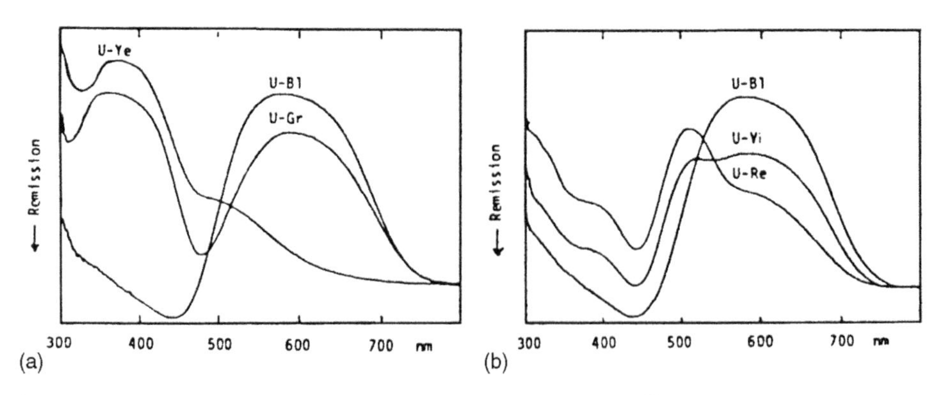

Figure 22.7 UV–vis spectra of various types of ultramarine[25] (U-Bl–blue, U-Gr–green, U-Ye–yellow, U-Vi–violet, U-Re–red).

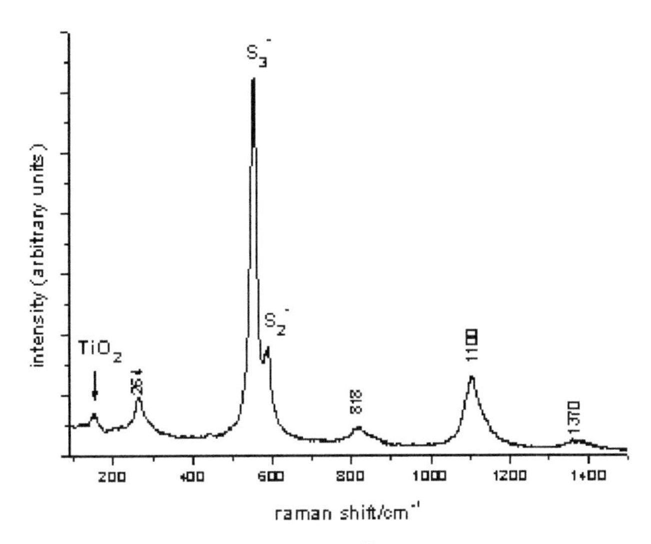

Figure 22.8 Typical Raman spectra of ultramarine.[59]

3.2. Use of synthetic zeolites for the preparation of ultramarine analogues

As mentioned already, ultramarine is an aluminosilicate sodalite containing sulphur anion radicals combined with sodium cations and embedded inside the β-cages. Sodalite, regardless of its very narrow (~0.3 nm) pore openings and low adsorption capacity, is usually included in the list of zeolites. The sodalite units are constituents of the structures of several types of zeolites (e.g., SOD, LTA, FAU, EMT, LTN).

The SOD type of structure is formed during the high temperature process of conventional ultramarine synthesis and the sulphur species are surrounded by the aluminosilicate upon forming the sodalite structure. It was tempting to apply the already existing sodalite cages of zeolites for entrapping sulphur radicals inside their voids or to introduce the radicals or rather their precursors during hydrothermal crystallization of zeolites.

The first but not very successful attempts to apply zeolites for the production of ultramarine took place in the thirties of the twentieth century.[61] Synthetic zeolites were used as starting material by Kummins[62,63] at the end of nineteen forties. The authors used sodium oligosulfide mixed with zeolite (type of zeolite was not indicated) and heated the mixture first in nitrogen and then in air. The product showed a pale colouration because of its low sulphur content. Much better results were obtained when elemental sulphur (56%) and sodium acetate (9.3%) were mixed with zeolite (34%) and soap (1%) and thermal treatment was conducted at 800 °C in an oxygen-free atmosphere and then mild oxidation was carried out at 500 °C in air. The method of preparing circular particles of ultramarine[64] by thermal treatment of zeolite P (GIS) with sulphur was developed. The authors stressed the short heating period and high colour intensity of the product and the maintenance of the morphology (circular particles) of the parent zeolite.

Ishida et al.[65,66,67] applied zeolites 4A and 5A for preparing ultramarine analogues. The authors impregnated zeolites with Na_2S and then heated them with sulphur vapours at 820 °C and finally they oxidized the samples at 500 °C. They did not foresee any potential industrial application of the method owing its relatively low colour intensity and the high cost of zeolite A at this time.

The further study of ultramarine preparation could be summarized in several groups: attempts to introduce the sulphur radicals from aprotic solutions, introduction of radical precursors into zeolite cages during hydrothermal synthesis of zeolites, thermal treatment of chosen zeolites with sulphur radical precursors, generation of radicals by irradiation of zeolites containing sulphur compounds application of various alkaline cations, introduction of sulphur radicals into non-sodalite cages and in non-aluminosilicate matrices.

3.3. Attempts to introduce sulphur radicals into zeolites from their aprotic solutions

It has been mentioned that sulphur anion radicals can be obtained by dissolving oligosulfides in aprotic solvents (acetone, DMSO, DMF, HMPA, liquid ammonia, amines). It seemed conceivable that treatment of activated zeolites (free of water and other adsorbed species) with sulphur radical solution could be viable for the accommodation of radicals inside the cages (e.g., sodalite cages). Treatment of zeolites (A, X) with coloured solutions (in DMSO) of oligosilfides resulted in formation of the coloured suspension, but subsequent evacuation of the solvent always resulted in discolouration of the zeolite.[68] The removal of the aprotic solvent which was responsible for a stabilization and protection of the radicals in solution caused the immediate recombination of the latter before reaching the cages and finally the ordinary oligosulfides were formed in the zeolite cavities. The experiments conducted in autoclaves at elevated temperatures did not lead to encapsulation of radicals either. The above results supported the believe that generation of radicals in ultramarine has already taken place in β-cages. The radical precursors have to be introduced into the cages before forming the radicals. The precursors can enter the cages during zeolite (i.e. sodalite) crystallization (sulphur species are surrounded by cages in the stage of their forming) or sulphur compounds can be encapsulated by entering the cages of small molecules (sulfides) and small sulphur particles (e.g., S_2) and the subsequent forming of oligosulfides inside the voids.[59]

3.4. Introduction of radical precursors into zeolite cages during hydrothermal synthesis of zeolites

The conventional synthesis of ultramarine comprises a high temperature, solvent-free crystallization of SOD structure. It is well known that various molecules can be encapsulated into the β-cages[10] during the hydrothermal crystallization of sodalite or other types of zeolites. Cancrinite is very prone to occlude the salt molecules in its channels during its formation by crystallization from gel or recrystallization from other zeolites (e.g., from zeolite A).

Weller[24] demonstrated a hydrothermal synthesis of sodalite in the presence of thiocyanate, which was encapsulated inside the formed cages. The samples were modified with various cations and then thermally treated at different temperatures (400–900 °C) over various periods time (10–2850 min). The products obtained at 500 °C and 600 °C showed blue-green and blue colouration, respectively. The authors believe that sulphur species entrapped in sodalite cages during crystallization are transformed into radicals upon heating in the presence of SO_2.

Vaughan et al.[55] prepared sodalite by means of hydrothermal crystallization of initial gel formed by merging solutions of sodium silicate, tetramethylammonium hydroxide (template) and then by adding solutions of sodium aluminate and aluminum sulfate. The latter was a source of both Al and of sulphur. The resulting samples were heated within a range of 893–1273 K in air for 5 h. The sulfate anions encapsulated inside β-cages were reduced by means of TMA cations (also encapsulated in sodalite) upon heating. The resulting samples showed blue colouration and their electronic spectra indicated the presence of $S_2^{\bullet-}$ and $S_3^{\bullet-}$ radicals. The ESR and ENDOR spectra distinguished three types of $S_3^{\bullet-}$ radicals. Two of them showed the same rhombic g-tensor and it is believed that they are located in β-cages surrounded by cages lacking the sulphur radicals. The third type indicated an isotropic g-tensor and it was inferred that in the latter case the *cluster* of $S_3^{\bullet-}$ radicals occupied the adjacent sodalite cages.

Crystallization of sodalite[31] or cancrinite[69] in the presence of sodium sulfide or oligosulfide led to the formation of zeolite structures with a noticeable amount of introduced sulphur species. The products always showed pale yellow-green colouration and their ESR spectra indicated some signals reflecting the presence of radicals. The intensity of colouration increased considerably after heating the product at 800 °C.

Another method of incorporation of sulphur moieties into zeolite cages consisted in the hydrothermal recrystallization of zeolites A with aqueous solutions of sodium oligosulfides or with alkaline solutions and elemental sulphur. The process was carried out at temperatures above 150 °C and usually led to the formation of sodalite or cancrinite (under higher alkalinity) types of structure. The products showed yellow-green colouration similar to the products obtained by the hydrothermal synthesis of zeolites from initial aluminosilicate gels in the presence of sulphur compounds. Further heating at higher temperature results in intense blue products.[31]

3.5. Irradiation of occluded sulphur species

Zeolites A and X impregnated with sodium oligosulfides were irradiated with γ-rays or accelerated electrons. For zeolites A, a noticeable colouration of the samples was observed, particularly when water was removed, although their ESR spectra did not resemble that of ultramarine.[70]

3.6. Thermal treatment of various zeolites with sulphur radical precursors

More systematic investigation of common synthetic zeolites (A, X, Y, sodalite) indicated a possibility to obtain ultramarine analogues by thermal treatment of the mixture of zeolites (Fig. 22.9) with sulphur radical precursors such as alkaline oligosulfides or elemental sulphur and alkalis (NaOH, Na_2CO_3). Zeolites A appeared to be the most promising starting material for the syntheses. The relatively low price of the latter resulting from large scale production is an additional factor encouraging its potential industrial application.

3.6.1. Syntheses based on zeolite A

Ishida et al.[66] impregnated zeolites 4A or 5A with Na_2S and after drying treated the mixture with sulphur vapours at 500 °C in an N_2 stream. Then the temperature gradually increased up to 820 °C and afterwards the samples were cooled to 500 °C. The heating at 500 °C was conducted in an air atmosphere. The products obtained from zeolites 4A and 5A were green or sky-blue, respectively.

Besides the impregnation of zeolites A with sulfides a simple mixing and grinding of the components was used too.[35,71] Then the mixtures were heated over a broad range of temperatures (300–900 °C). The experiments were always conducted in laboratory scale (several grams of zeolite) and a short time of thermal treatment (several hours) was sufficient for obtaining coloured products. The heating was always conducted in covered ceramic crucibles. In some experiments the sealed evacuated ampoules were used.[72,73]

Similar results were obtained when instead of oligosulfides, the elemental sulphur and alkalis were used as radical precursors. The weight loss upon heating in the case of mixtures of zeolites with sodium oligosulfides resulted mostly from the release of water (from $Na_2S{\cdot}10H_2O$ used for forming oligosulfides), while mixtures with sulphur and Na_2CO_3 were caused mostly by the evolution of CO_2 upon carbonate decomposition.[71] It has been found that the length of the oligosulfide chain, which reflects various Na_2/S ratio, significantly affects the colouration of the resulting products and also their structure. The features of the products are also influenced substantially by the temperature of heating and by the time of the thermal treatment.

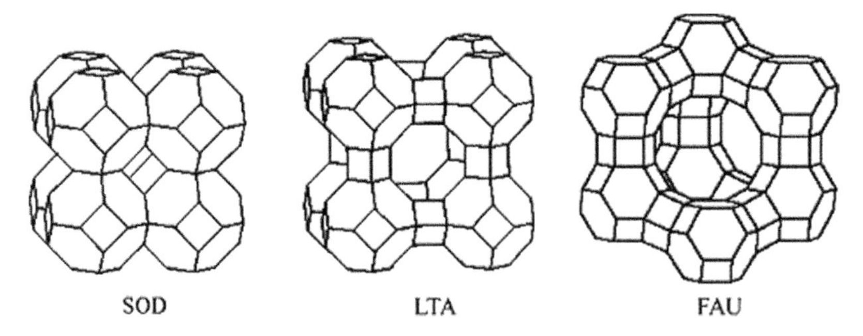

Figure 22.9 Models of indicated zeolite structures.

Generally, the thermal treatment of mixtures with a low Na_2/S ratio does not affect the original structure of parent zeolite, whereas the higher alkalinity of the mixture results in structure transformation towards sodalite (Figs. 22.10 and 22.11). Mixtures with medium alkalinity indicate a distorted LTA structure and a considerable decrease in crystallinity. The XRD patterns of the above samples are reminiscent of those of nepheline hydrate II (Fig. 22.10b), described by Barrer,[76] where most of the typical LTA reflections remained, but low angle reflections are missed.

The structure transformation depends very much on treatment temperature.[74] The range of mixture alkalinity (Na_2/S ratio) leading to SOD increases with temperature very markedly (Fig. 22.12).

The colouration of the products depends very much on the alkalinity of the mixture and is also affected by treatment temperature (Fig. 22.12). It is interesting that at low temperatures (500 °C) colouration of samples changes with increasing alkalinity from blue (for the lowest Na_2/S ratio) to green and yellow green (for the highest Na_2/S ratio), while the opposite correlation is noticed for samples heated at 800 °C. The correlation shown in Fig. 22.12 is based on intensities of two main bands in UV–vis spectra, i.e., at ~600 nm from the blue chromophore (radical $S_3^{\bullet -}$) and at 410 nm from the yellow chromophore (radical $S_2^{\bullet -}$). The latter band is sometimes split and the shoulder at 380 nm reflects a contribution of elemental sulphur.[60,77] The same mixture heated at different temperatures leads to products of different colouration. As seen in Fig. 22.13 the contribution of blue colour component increases with growing treatment temperature at the expense of yellow chromophore.

The different courses of thermal reactions at different temperatures probably results from different chemistry. Two main pathways of the radical generation can be taken into an account.

The formal valence of sulphur in anion radical $S_3^{\bullet -}$ is −1/3, which means that elemental sulphur applied for synthesis should be reduced to attain the required valence level. If the oligosulfides are used as the radical precursors the homolytic scission of S-S bond (*cracking*) can be considered as a decisive step in forming the anion radicals:

$$S_4^{2-} \rightarrow 2S_2^{\bullet -}, \quad S_6^{2-} \rightarrow 2S_3^{\bullet -}, \quad S_8^{2-} \rightarrow 2S_4^{\bullet -}$$

The homolytic scission can result in breaking the chain in various positions and additionally the subsequent reaction of the step products can take place:

$$S_6^{2-} \rightarrow S_4^{\bullet -} + S_2^{\bullet -}, \quad S_8^{2-} \rightarrow 2S_3^{\bullet -} + 2S_{el}$$

The reaction with elemental sulphur as a radical precursor involves its disproportionation in the presence of alkalis:

$$3S + 6NaOH \rightarrow 2Na_2S + Na_2SO_3 + 3H_2O$$

Resulting sodium sulfide reacts with an excess of sulphur to form oligosulfides.

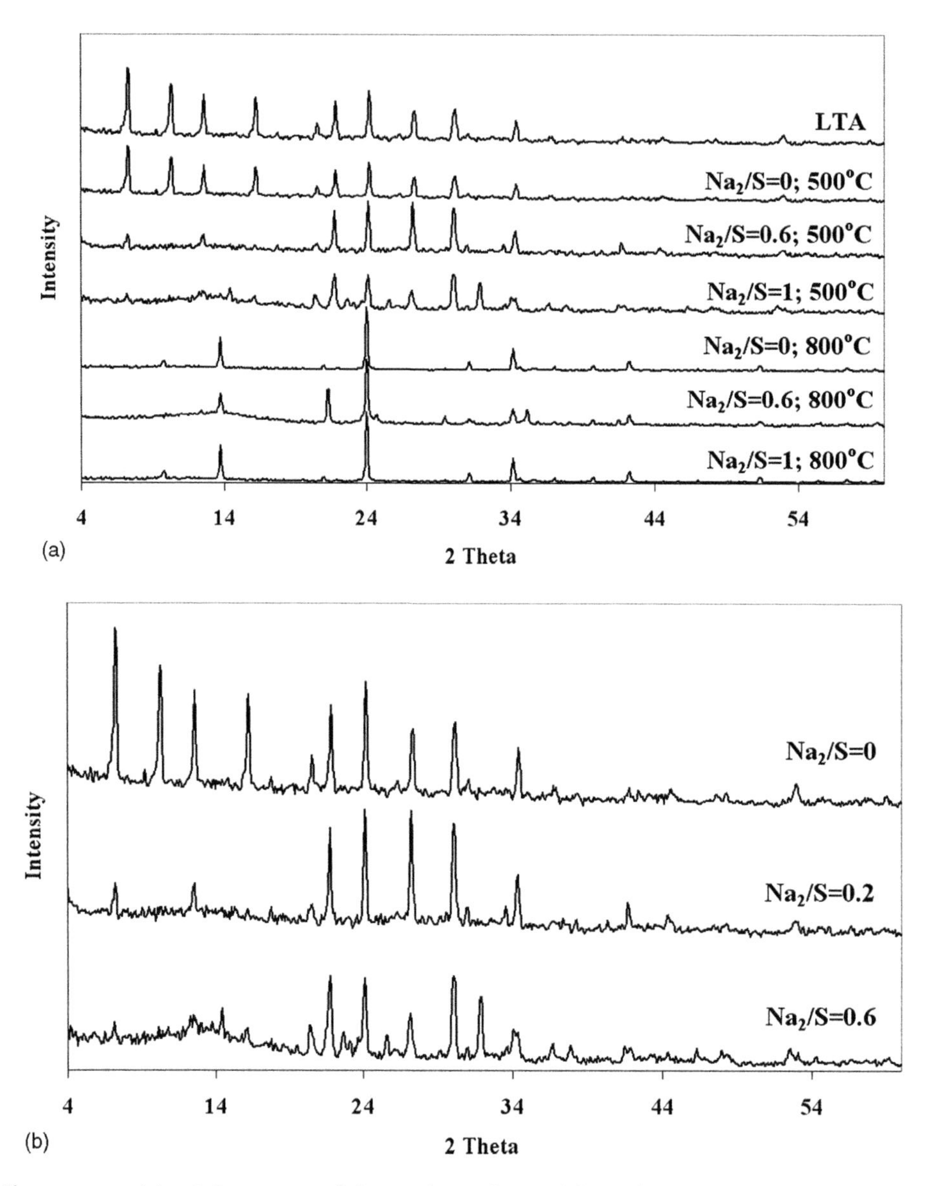

Figure 22.10 (a) XRD patterns of the products obtained from the mixtures of zeolite A with the same loading of sulphur and various content of Na_2CO_3 (expressed as Na_2/S) after heating at 500 or 800 °C. (b) Structure changes of zeolite A in mixtures of different alkalinity after treatment at 500 °C.[74]

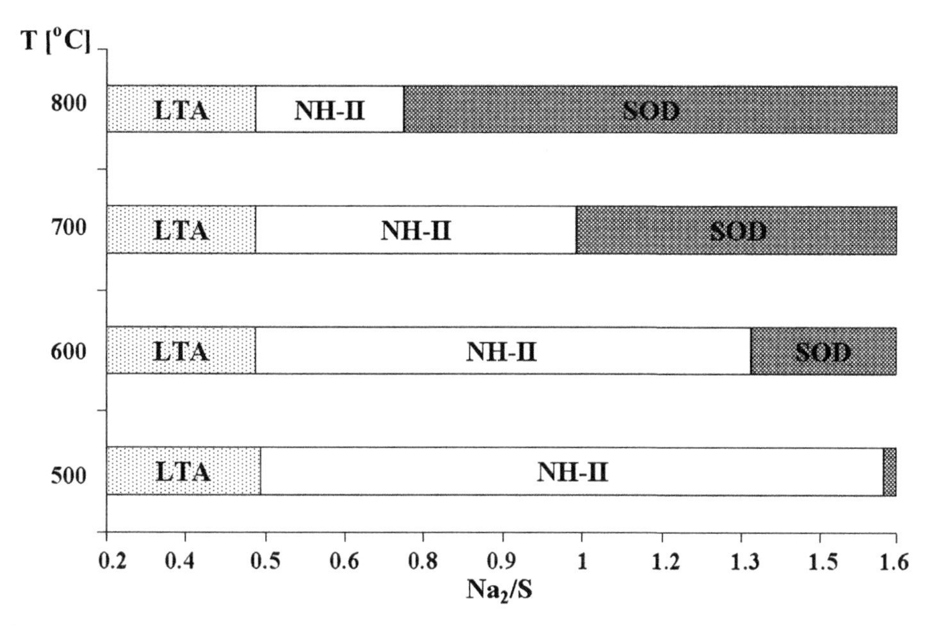

Figure 22.11 Structure changes of zeolite A mixed with sulphur and various amount of Na_2CO_3 (Na_2/S) at indicated temperatures (NH II stands for nepheline hydrate).[74,75]

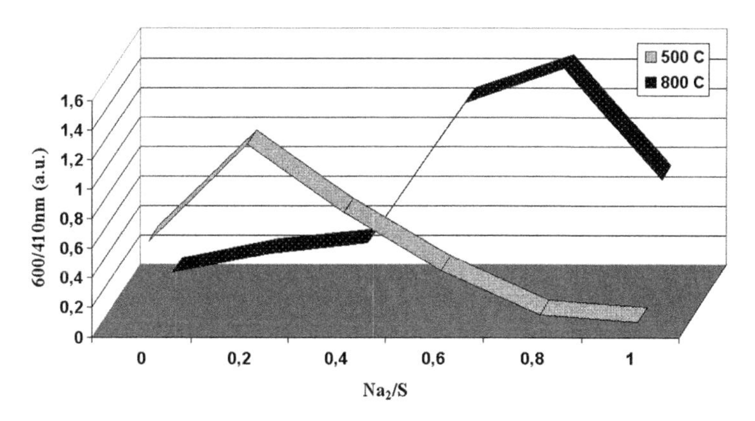

Figure 22.12 Correlation between alkalinity of the mixtures containing zeolite A, sulphur and Na_2CO_3 and colouration of the products obtained at low (500 °C) and high (800 °C) temperatures (expressed as intensity quotient of UV–vis bands at 600 and 410 nm).[75]

Another pathway of radical generation can result from the partial oxidation of sulfides or oligosulfides by means of oxygen, elemental sulphur or sulphur oxides:

$$2S_3^{2-} + 1/2O_2 \rightarrow 2S_3^{\bullet -} + O^{2-}$$

$$2S_3^{2-} - 1/8S_2 \rightarrow 2S_3^{\bullet -} + S^{2-}$$

$$13S^{2-} + 5S^{4+} \rightarrow 6S_3^{\bullet -}$$

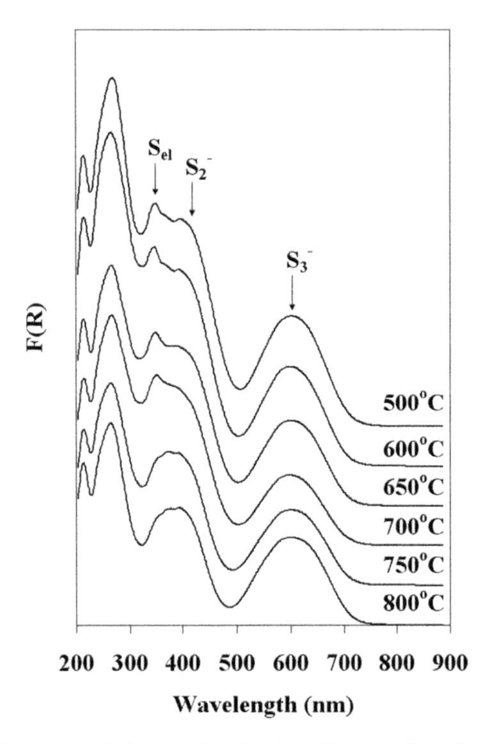

Figure 22.13 Electronic spectra of the samples obtained from zeolite Na-A mixed with Na_2S (sulphur content—20 wt.% compared to zeolite weight) and pitch after heating at indicated temperatures.[72]

The penetration of sulphur moieties within the intracrystalline voids of applied zeolites during the thermal process is not quite clear. It is possible that small sulfide anions and/or small sulphur molecules (dissociation of S_8 molecules can take place at high temperature) enter the cages and homolytic cleavage of resulting oligosulfides occurs in the cages to generate the radicals.

The problem of sulphur species reaction becomes even more complex if the structure transformations of the parent zeolites are concerned. The original structure of zeolites can be maintained only under mild synthesis conditions (i.e., low temperature, low alkalinity of mixtures), whereas the more severe conditions result in transformation of the zeolite structure mostly (particularly when sodium is used as an alkali source) towards sodalite. The sulphur radicals already encapsulated inside the β-cages of the parent zeolite A at low temperatures are probably involved in the consequent structure changes upon further temperature rise. It might be taken into an account that sodalite segments with entrapped radicals undergo rearrangement to form SOD structure. The XRD patterns of the samples prepared at various temperatures illustrate (Fig. 22.10) that intensity of reflection declines with heating temperature and some low angle reflections vanish (nepheline hydrate II). However, no total amorphization of the zeolite structure was ever noticed at any temperature.

Instead, the XRD reflections of samples prepared at higher temperatures (above 700 °C) became more intense and indicated SOD structure.

3.6.2. FAU (X,Y)

Zeolite with FAU structure can be also applied for the preparation of ultramarine analogues. A similar thermal procedure to that of zeolites A was applied. Low heating temperature (300–500 °C) and low alkalinity of mixture allowed the original structure to be maintained, but usually products of pale colouration or sometimes colourless samples were obtained.[31,68,75] Higher temperatures and higher mixtures alkalinity resulted in structure transformation to sodalite and in much more intense colouration (blue, turquoise). The products obtained from zeolite X showed always more intense colouration than those prepared from zeolite Y under the same conditions.

3.6.3. Sodalite

Using synthetic sodalite (mixed with sulphur radical precursors) for the thermal synthesis of ultramarine leads to the formation of coloured products (blue, turquoise), although the colour intensity is not higher than that of samples prepared under the same conditions from zeolite A.[31] The parent sodalite does not undergo any considerable structure transformation, but some impurities of carnegeite are noticed in the products. Lower colour intensity can be caused by limited diffusion of sulphur species within the sodalite structure during the thermal process.

3.6.4. Cancrinite

As mentioned in Section 3.4 the sulphur species were introduced into cancrinite during its synthesis and they were thermally transformed into radicals.[31,78,79] The structure of cancrinite consists of 11-hedral ($4^6 6^5$) ε-cages surrounding the 12-ring channels (Fig. 22.14). The cancrinite cages can be considered as matrices hosting the sulphur radicals similarly as sodalite cages, although their smaller size can affect the kind of generated radicals. Synthetic cancrinite was also applied to the procedures involving thermal incorporation of sulphur moieties into its structure. Sulphur and various amounts of sodium carbonate were mixed with zeolite and heated at 500 °C or 800 °C. The treatment at lower temperature resulted in the formation of yellow or green products with a preserved CAN structure.[80] A substantial contribution of the yellow chromophore (radical $S_2^{\bullet-}$) was noticeable also for the samples prepared at 800 °C, particularly those obtained from low alkaline mixtures. The samples prepared from the highest alkaline mixtures underwent structure transformation to sodalite, which was reflected in a more distinct blue colouration. The cancrinite applied for syntheses can be crystallized from zeolite A in the presence of various anions (S_2^{2-}, $S_2O_3^{2-}$, NO_3^-, CO_3^{2-}) which are always occluded in the channels after synthesis.[31] The introduced anions affected the further thermal reaction with sulphur and alkalis. Sulfides and thiosulfites caused an increase in product colour owing to enhanced sulphur content, whereas carbonates disturbed the formation of coloured products at 500 °C and some colouration was seen only for samples obtained at 800 °C from the most alkaline mixtures that show SOD structure.

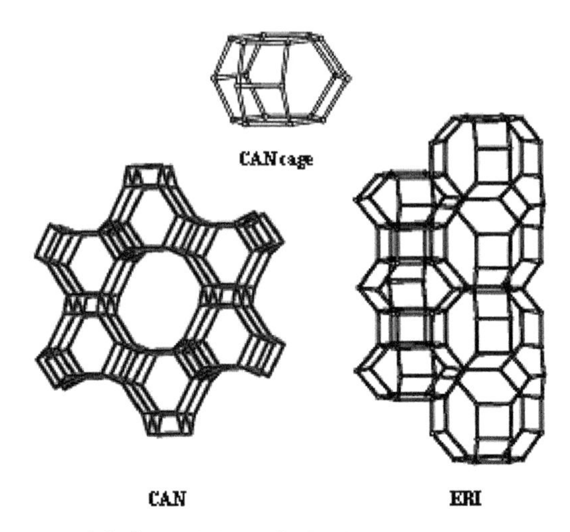

Figure 22.14 Structure model of cancrinite and erionite.

The results of syntheses based on cancrinite confirm the supposition that the size of the zeolite cage influences the contribution of the particular type of sulphur radical. It seems that smaller ε-cages in cancrinite favor a generation of smaller radicals $S_2^{\bullet-}$ at the expense of bigger $S_3^{\bullet-}$.

3.6.5. Erionite

The structure of erionite[81,82] also consists of ε-cages fused with double 6-rings and with erionite cavities (Fig. 22.14). The erionite cavities are circumscribed by erionite cages and D6R prisms.

The synthesis procedure comprised mixing with sulphur and alkalis and then heating at 500 °C or 800 °C. The products obtained at 500 °C maintained the original structure, while heating at 800 °C with Na_2CO_3 resulted in forming sodalite. Replacement of Na_2CO_3 by K_2CO_3 led to an unknown structure upon thermal treatment. Both chromophores ($S_2^{\bullet-}$ and $S_3^{\bullet-}$) were seen in the products, but the predominance of $S_2^{\bullet-}$ was less pronounced than in the case of cancrinite.

3.6.6. Losod

The unit cell of the losod structure contains two cancrinite cages combined with a 17-hedral losod cage ($4^6 6^{11}$), which is much larger (Fig. 22.15) than the sodalite cage.

Syntheses with erionite and (particularly) cancrinite suggested an influence of cage size on the distribution of sulphur radicals (S_2, S_3). It was conceivable that cages larger than those of conventional sodalite could facilitate the generation and stabilization of bulkier radicals $S_4^{\bullet-}$ or molecules S_4 (potential red chromophores). However, the thermal treatment with the radical precursors (sodium oligosulfides or elemental sulphur with alkalis) never resulted in red products. The samples prepared under mild conditions maintained the original structure, but the colouration of the resulting samples (blue or green) was not very intense.[83,84] It seems that even if the large losod cages could first accommodate several radicals, they

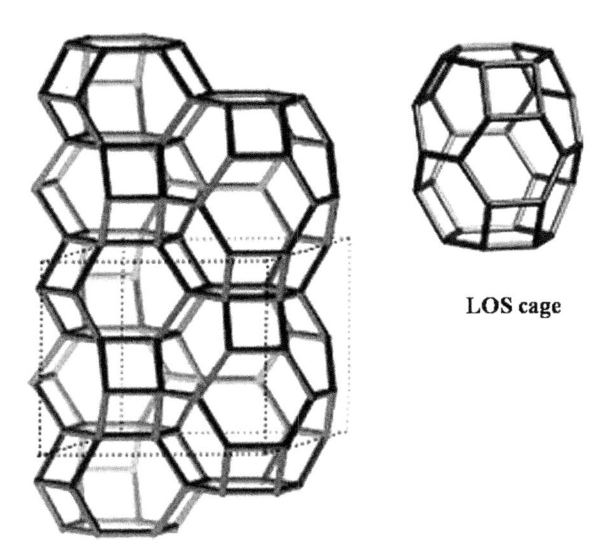

Figure 22.15 Structure model of losod.

subsequently easily undergo recombination to form ordinary oligosulfides. Probably, the observed colouration of the samples resulted from radicals located exclusively in cancrinite cages. Attempts to reduce slightly the size of losod cages by means of cation modification with K^+ or Cs^+ cations always resulted in degradation of the losod structure during thermal procedure.

3.6.7. ESR spectra of sulphur radicals embedded in non-sodalite cages

The ESR spectra recorded at room temperature for the coloured products obtained from cancrintite and also to some extent erionite[82,85] and losod[84] with maintained original structure of parent zeolites show the anisotropic signal (Fig. 22.16a). It is in contrast to the spectra of conventional ultramarine and those of the coloured samples prepared from zeolites with sodalite units in their structures (i.e. SOD, LTA), where always isotropic signals are recorded at room temperature. The anisotropic signals appear in the EPR spectra of the latter pigments only at low temperatures which retard the motion of radicals (Fig. 22.16b). It was tempting to attribute the additional signals recorded at room temperature for the samples with CAN, ERI or LOS structure to $S_2^{\bullet-}$ radicals, because of substantial contribution of yellow chromophore in these products. It could be assumed that radicals $S_2^{\bullet-}$ localized inside the small ε-cages could be recorded due to retarded mobility, but it is more likely that the anisotropy results from radicals $S_3^{\bullet-}$ confined in small ε-cages.

3.7. Natural zeolites

Some natural zeolites such as chabazite, stilbite, clinoptilolite and mordenite were used for thermal synthesis of pigment with sodium trisulfide as a sulphur source and pitch tar as a reducing agent.[72] Only samples with chabazite and stilbite turned blue after heating and the parent zeolites recrystallized to sodalite structure. Zeolites with

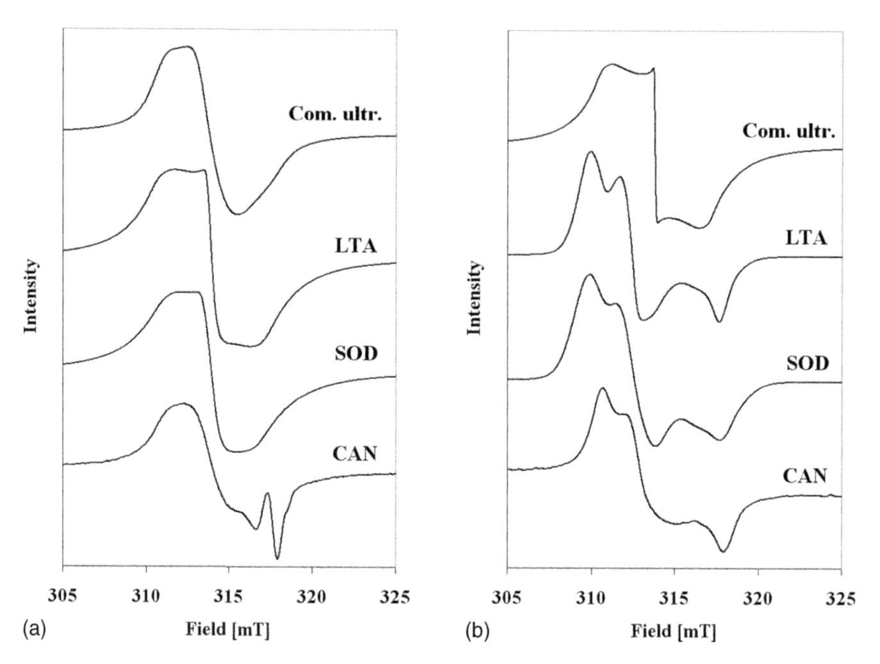

Figure 22.16 EPR spectra of commercial ultramarine and of coloured products obtained from zeolites. The crystalline structures of the products is indicated. (a)–spectra measured at room temperature, (b)–spectra of the same samples recorded in liquid nitrogen.[85]

higher Si/Al ratios (mordenite, clinoptilolite) also underwent structure transformation upon heating with sulfides which led to unknown structures but the products did not show any remarkable colouration.

3.8. Application of various alkaline cations

Sodium cations are always used as components of the initial mixture (mostly as Na_2CO_3) in the conventional synthesis of ultramarine from kaolin and usually the sodium forms of zeolites were applied for preparing the ultramarine analogues. It has been known[33–35] that modification of synthetic ultramarine with salts of various cations affects its colouration, but limited ion-exchange ability of sodalite does not facilitate this procedure. The solid state cation modification seems more efficient.[35] The coloured products obtained from zeolites with maintained original structure (e.g., LTA) are more prone to such a modification.

The influence of various cations contributing in thermal ultramarine synthesis from zeolites on the properties and structure of the resulting products has been reported in several papers.[35,36] Initial zeolites (always sodium forms) were modified with various cations (Li^+, K^+, Cs^+, NH_4^+, Ca^{2+}, Zn^{2+}, Cu^{2+}) and then mixed with sulphur and variable amounts of alkalis (e.g., sodium carbonate) and eventually heated at high temperature.[35] When Na_2CO_3 is used as an alkali source the influence of cations introduced into parent zeolites on colouration and structure of products was especially conspicuous for samples with low mixture alkalinity. The presence of cations other than sodium affected both colouration and structure of the products. Zeolites modified with NH_4^+, Zn^{2+}, Cu^{2+} underwent an amorphization upon thermal treatment with sulphur (without any additional alkalis) at 800 °C. Zeolite Li-A in mixture with sulphur and Na_2CO_3 did not recrystalize to sodalite structure at 500 °C even at the highest alkali loading, whereas at 800 °C it transformed into sodalite already under very low alkalinity of the mixture. The cation modified zeolites A mixed only with sulphur remain colourless after heating which is in contrast to zeolite Na-A. The only exception was Cu-A, who always formed brown products regardless of alkali loading. The colour of the latter resulted, however, from CuS, while the sulphur radicals never were detected. Under high Na_2CO_3/S ratios the impact of the zeolite cations was less noticeable in view of a considerable sodium excess. The influence of various cations was much more pronounced when sodium was almost totally eliminated from the starting mixtures by replacing Na_2CO_3 by carbonates of respective cations or by CaO. The size of the alkali cation plays an important role in the colouration of the products. As illustrated in Fig. 22.17 the big potassium cations seem to favor a generation of smaller radicals $S_2^{\bullet -}$ (yellow chromophore) within the β-cages, whereas the small Li cations facilitate forming larger $S_2^{\bullet -}$ radicals.

In the case of Cs-A,[86] where the size of pore opening of zeolite is considerably reduced (below 0.3 nm) by bulky cations the access to the large cavities is limited similarly as in small cages (e.g., β-cage). Thus, the large cavity (also reduced in its volume after modification with Cs) could be suitable for radical accommodation and protection. Its volume is still larger than typical β-cage and it should be able to contain bigger sulphur moieties (e.g., S_4). Indeed, the product obtained from mixture of Cs-A, sulphur and high loading of Cs_2CO_3 showed pink colouration (Fig. 22.17). Its ESR spectrum was much different than that of typical ultramarine and the main signal could reflect the presence of sulphur radical (e.g., $S_4^{-\cdot}$) responsible for red colour inside the cavities. Use of CaO as an alkali source did not result in any colouration of the resulting products and no sulphur radicals were detected. The zeolite structure thermal transformations depend very much on the nature of the cations. Mixtures lacking sodium cations never form the SOD structure.[35]

Thermal synthesis of pigments from zeolites A with ammonium oligosulfides at 800 °C resulted in forming a product with a preserved LTA structure that exhibited a reversible thermochromic effect.[87] The samples were almost colourless at room temperature, but they turned blue upon heating at 530 °C. Probably thermal dissociation of S_6^{2-} resulted in forming two radicals and their recombination to hexasulfide took place on cooling.

3.9. Ultramarine with non-aluminosilicate matrices

Natural and synthetic ultramarine is an aluminosilicate and the cages (mostly sodalite cages) host the sulphur anion radicals. It is known that these radicals can be entrapped in the crystalline structure of some halides.[53] Some coloured products obtained from zeolites (particularly with contribution of cations other than sodium) showed a non-zeolite structure of kaliophilite or carnegeite. The sulphur radicals

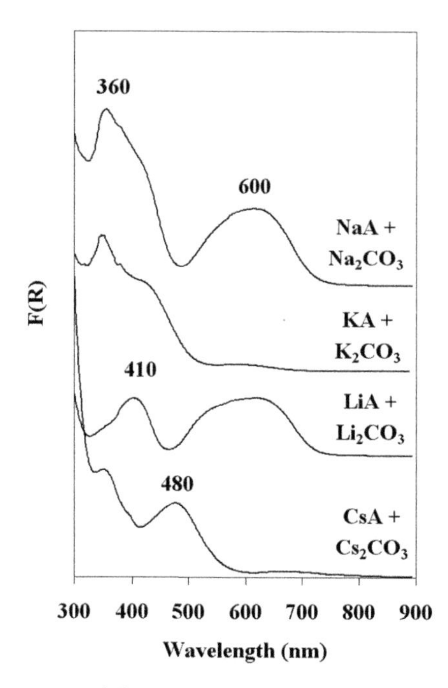

Figure 22.17 Electronic spectra of the samples prepared from zeolite modified with indicated cations and mixed with sulphur and carbonates of respective alkaline cations.[35,86]

could be accommodated in layered materials, although the chromophores were not perfectly stabilized there and colours of the samples faded very fast.[88]

It was interesting to check whether the radicals can be stabilized in the cages of non-aluminosilicate zeotypes such as $AlPO_4$. The important difference between zeolites and $AlPO_4$ materials consists in a lack of framework charge in the latter and subsequently, a lack of cations. Regardless of this material $AlPO_4$–20 with SOD structure was successfully applied for synthesis of the ultramarine analogue by means of thermal treatment (800 °C) of prepared aluminum phosphate with sodium trisulfide and pitch. The resulting product showed a distinct blue colouration and good stability.[31,75] The incorporation of sulphur radical precursors during the crystallization of $AlPO_4$–20 was less successful because the sulphur compounds present in the initial mixture affected the synthesis and the product showed low crystallinity, nevertheless it turned blue after heating.[75]

Sulphur and selenium were encapsulated[60] into $AlPO_4$–5 and SAPO-44. Encapsulated sulphur (S_8) showed the shifted bands in the Raman spectra, whereas the Raman spectra of entrapped selenium indicated formation of various Se_n chains and rings.

Weller et al.[21] synthesized gallosilicate and aluminogermanate sodalite and cancrinite with introduced anions (e.g., SCN^-, $SeCN^-$, MnO_4^-). Thermal treatment of samples containing thiocyanates and selenocyanates resulted in coloured products. The attempts of Vaughan[89] to prepare sodalites with various combinations of T atoms (Ga/Si, Ge/Si, Al/Ge, Al/Ga) and with entrapped SO_4^{2-} or SeO_4^{2-} did not result in the formation of coloured products after heating, as had been noticed earlier for aluminosilicate sodalite.[55]

The entrapment of sulphur moieties in SOD structure of zincophosphate[90] was not fully successful either. The low thermal stability of zincophosphate made high temperature treatment with sulphur radical precursors impossible. The attempt to introduce sulfides or oligosulfides during the spontaneous tribochemical crystallization failed too because of high affinity of zinc cations and sulfide anions which resulted in formation of ZnS precipitate. On the other hand, the spontaneous crystallization of zincophosphate sodalite in the presence of elemental sulphur or CdS powder led to SOD structure and substantial part of the mixture components (S, CdS) have been proven to be encapsulated into the cages.[91]

4. Conclusions

The use of zeolites and other molecular sieves for preparing sulphur pigments analogous to ultramarine seems very promising. The conventional procedure with kaolin as a starting material can be supplemented by syntheses based on zeolites. The syntheses can be conducted in a broad range of temperatures including relatively low ones (even ~300 °C). Much lower (compared to kaolin) content of sulphur (or sulphur compounds) in mixture with zeolites suffices to obtain the products of intense colouration. The amount of volatile sulphur compounds developing upon thermal procedure is markedly lower than that in conventional synthesis. The modification of synthesis conditions (composition of initial mixture, temperature, duration of thermal treatment, admixture of reductive agents) allows to attain broad variety of colours in resulting products.

The sulphur radical precursors can be introduced into zeolites either during their hydrothermal crystallization in the presence of sulphur compounds or during thermal treatment of zeolites mixed with sulphur (together with alkalis) or alkali sulfides. Zeolite A seems most suitable for the syntheses, but coloured products can be also obtained with other zeolites, particularly those containing sodalite cages in their structure. It is also possible to obtain ultramarine analogues from zeolites furnished with other cages (cancrinite, losod) or even with those which do not contain any cages (chabazite, stilbite). Non-aluminosilicate matrices ($AlPO_4$-20, gallosilicalites, aluminogermanates) can also host the sulphur and selenium chromophores and the resulting products show the colouration and properties similar to ultramarine.

The thermal treatment of zeolites with radical precursors results in the formation of coloured products with maintained zeolite structures if low temperatures and low alkalinity of mixtures are applied. More drastic conditions always cause the structure transformation towards sodalite, provided that sodium is the only, or a very prevalent alkali cation in the mixture. Lack or deficit of Na^+ cations results in recrystallization towards other structures (e.g., kaliophilite, carnegeite). The radicals $S_2^{\bullet -}$ and $S_3^{\bullet -}$ are always recorded in the products, their contribution depends on the alkalinity of the mixtures as well as on heating temperature and duration of treatment.

The size and geometry of the cages in applied zeolites affects the colouration of the products. Samples with preserved ε-cages in their structures (e.g., CAN and also ERI) show a higher contribution of yellow chromophore ($S_2\bullet^-$), which can result from the better fitting of smaller radicals in smaller cages.

The cations introduced into zeolites or applied as alkali sources influence the structure and colouration of the products. The sulphur pigments obtained from zeolites (particularly that with maintained original structure) are more susceptible than the conventional ultramarine for a post-synthesis treatment with cations which results in a colour modification.

ACKNOWLEDGEMENTS

The critical reading of the manuscript by Prof. A.B. Wieckowski and stimulating discussions with him are very much appreciated.

REFERENCES

[1] (a) Jacobs, P. A., *Stud. Surf. Sci. Catal.* **2005,** *157,* 289–310 and literature quoted therein.; (b) Dioos, B. M. L., Sels, F., Jacobs, P. A., *Stud. Surf. Sci. Catal.* **2007,** *168,* 915–946 and literature quoted therein.
[2] Bein, T., *Stud. Surf. Sci. Catal.* **2007,** *168,* 611–657 and literature quoted therein.
[3] Behrens, P., Stucky, G. D., Novel Materials Based on Zeolites in Comprehensive Supramolecular Chemistry, vol. 7 Solid-State Supramolecular Chemistry: Two- and Three-Dimensional Inorganic Networks, (Vol. Editors: Alberti, G., Bein, T.), Pergamon, Elsevier, Amsterdam 1996, 721–772 and literature quoted therein.
[4] Bein, T., *Stud. Surf. Sci. Catal.* **2007,** *168,* 579–619.
[5] Ramamurthy, V., Garcia-Garibay, M. A., *Stud. Surf. Sci. Catal.* **2007,** *168,* 693–719.
[6] Depmeir, W., *Rev. Mineral. Geochem.* **2005,** *57,* 203–240.
[7] Pobedimskaya, E. A., Terenteva, L. F., Sapozhnikov, A. N., Kashaev, A. A., Dorokhova, G. I., *Sov. Phys. Dokl.* **1991,** *36,* 553–555.
[8] Girnus, I., Janke, K., Vetter, R., Richter-Mendau, J., Caro, J., *Zeolites* **1995,** *12,* 33–39.
[9] Barrer, R. M., Cole, J. F., *J. Phys. Chem.* **1968,** *29,* 1755–1758.
[10] Stucky, G. D. Srdanov, V. I. Harrison, W. T. A. Gier, T. E. Keder, N. I. and Metiu, H. I., Three – Dimensional Periodic Packing, Sodalite, a Model Sysyem In Supramolecular Architecture; Synthetic Control in Thin Films and Solids (Ed. T. Bein) ACS Symposium Series, Washington, DC **1992**, *499*, 294–313.
[11] Windiks, R., Sauer, J., *Phys. Chem. Chem. Phys.* **1999,** *1,* 4505–4513.
[12] Tolman, C. H., Herron, N., *Catal. Today* **1988,** *3,* 235–243.
[13] Balkus, K. J., Kowalak, S., Methods for the preparation of molecular sieves, including zeolites, using metal chelate complexes. U.S. Patent 5,167,942, June 1992.
[14] Pollack, S. S., Sprecher, F. R., Formel, E. A., *J. Mol. Catal.* **1991,** *66,* 195–203.
[15] Kowalak, S., Pawłowska, M., Więckowski, A. B., Goslar, J., *Stud. Surf. Sci. Catal.* **1994,** *83,* 179–185.
[16] Cano, M. L., Corma, A., Fornes, V., Garcia, H., *J. Phys. Chem.* **1995,** *99,* 4241–4246.
[17] Kowalak, S., Jankowska, A., Pietrzak, N., Stróżyk, M., *Stud. Surf. Sci. Catal.* **2001,** *135,* 363.
[18] Kox, M. H. F., Stavitski, E., Weckhuysen, B. M., *Angew. Chem. Int. Ed.* **2007,** *46,* 3652–3655.
[19] Vezin, H., Moissette, A., Hureau, M., Bremard, C., *Chem. Phys. Chem.* **2006,** *7,* 2474–2477.

[20] Hughes, E. M., Kurten, D. M., Weller, M. T., in: Treacy, M. M. J., Marcus, B. K., Bisher, M. E., Higgins, J. B. (Eds.), Proceedings of the 12th International Zeolite Conference, Baltimore, Maryland, USA, July 5–10, 1998, pp. 2087–2094, *Mater. Res. Soc.* **1998**.
[21] Johnson, G. M., Mead, P. J., Weller, M. T., *Micropor. Mesopor. Mater.* **2000,** *38,* 445–460.
[22] Hund, F., *Z. Anorg. Allg. Chem.* **1984,** *511,* 225–230.
[23] Hund, F., Kohl, P., Kemper, J., Reiner, D., *Z. Anorg. Allg. Chem.* **2002,** *628,* 1457–1458.
[24] (a) Weller, M. T., Wong, G., Adamson, C. L., Dodd, S. M., Roe, J. J., *J. Chem. Soc. Dalton Trans.* **1990,** 593–597; (b) Buhl, J. -Ch., Gesing, Th. M., Gurris, Ch., *Micropor. Mesopor. Mater.* **2001,** 50, 25–32.
[25] Seel, F., *Stud. Inorg. Chem.* **1984,** *5,* 67–89.
[26] Mertens, J., *AMBIX* **2004,** *L1,* 219–244.
[27] Clement, F., Desormes, J. B., *Annales de Chimie* **1806,** *57,* 317–326.
[28] *Ultramarine Blue, Technical Handbook, Nubiola* 2nd edition, Barcelona, **2001**.
[29] Landman, A. A., Ph.D. Thesis, **2003,** University of Pretoria.
[30] Landman, A. A., de Waal, D., *S. Afr. J. Chem.* **2005,** *58,* 46–52.
[31] Kowalak, S., Stróżyk, M., Pawłowska, M., Miluśka, M., Przystajko, W., Kania, J., *Colloids Surf. A* **1995,** *101,* 179–185.
[32] Zeltner, J., Process to manufacture a red ultramarine dye. German Patent No. 1, July 2, **1877**.
[33] Barrer, R. M. and Raitt, J. S., *J. Chem. Soc.* **1954,** 4641–4651.
[34] Matsunaga, Y., *Can. J. Chem.* **1959,** *37,* 994–995.
[35] Kowalak, S., Jankowska, A., Łączkowska, S., *Stud. Surf. Sci. Catal.* **2005,** *158A,* 215–222.
[36] Booth, D. G., Dann, S. E., Weller, M. T., *Dyes Pigments* **2003,** *58,* 73–82.
[37] Reinen, D., Lindner, G. G., *Chem. Soc. Rev.* **1999,** *28,* 75–84.
[38] Clark, R. J. H., Dines, T. J., Kurmoo, M., *Inorg. Chem.* **1983,** *22,* 2766–2772.
[39] Schlaich, H., Lindner, G. G., Feldmann, J., Göbel, E. O., Reinen, D., *Inorg. Chem.* **2000,** *39,* 2740–2747.
[40] Lindner, G. G., Witke, K., Schlaich, D., Reinen, D., *Inorg. Chim. Acta* **1966,** *252,* 39–45.
[41] Plique, J., *Bl. Soc. Chim. Paris N.S.* **1877,** *28,* 522.
[42] Jeager, F. M., *Trans. Faraday Soc.* **1929,** *25,* 320–345.
[43] Klinowski, J., Carr, S. W., Tarling, S. E., Barnes, B., *Nature* **1987,** *330,* 56–58.
[44] Landman, A. A., de Waal, D., *Cryst. Eng.* **2001,** *4,* 159–169.
[45] Giggenbach, W., *J. Inorg. Nucl. Chem.* **1968,** *30,* 3189–3201.
[46] Giggenbach, W., *Inorg. Chem.* **1971,** *10,* 1306–1308.
[47] Giggenbach, W., *J. Chem. Soc. Dalton Trans.* **1973,** *7,* 729–731.
[48] Seel, F., Güttler, H. J., *Angew. Chem.* **1973,** *85,* 416–417.
[49] Seel, F., Güttler, H. J., *Angew. Chem. Int. Ed. Engl.* **1973,** *12,* 420–421.
[50] Weller, M. T., *J. Chem. Soc. Dalton Trans.* **2000,** *23,* 4227–4240.
[51] Gardner, D. M., Frankel, G. K., *J. Am. Chem. Soc.* **1955,** *77,* 6399–6400.
[52] Więckowski, A. B., Mechandzijew, D., *Prace Kom. Mat. Przyr.* **1968,** *9,* 3.
[53] Morton, J. R., in: *At. Mol. Etud. Radio Elec. Proc. 15th Colloque Ampere,* Sept. 1968, Amsterdam, North Holland, **1968**.
[54] Więckowski, A. B., Wojtowicz, W., Śliwa-Nieściór, J. *J. Magn. Reson. Chem.* **1999,** *37,* 150–153.
[55] Arieli, A., Vaughan, D. E. W., Goldfarb, D., *J. Am. Chem. Soc.* **2004,** *126,* 5776–5788.
[56] Koch, W., Natterer, J., Henemann, C., *J. Chem. Phys.* **1995,** *102,* 6159–6167.
[57] Schweiger, A., Jaschke, G., *Principles of Pulse Electron Paramagnetic Resonance,* Oxford University Press, Oxford, **2001**.
[58] Gobeltz, N., Demortier, A., Lelieur, J. P., Lorriaux, A., Duhayon, C., *New J. Chem.* **1996,** *20,* 19–22.
[59] Gobeltz, N., Demortier, A., Lelieur, J. P., Duhayon, C., *J. Chem. Soc. Faraday Trans.* **1998,** *94,* 2257–2260.
[60] (a) Li, G., Chen, J., Xu, R., in : Treacy, M. M. J., Marcus, B. K., Bisher, M. E., Higgins, J. B. (Eds.), Proceedings of the 12th International Zeolite Conference, Baltimore, Maryland, USA, July 5–10, 1998, pp. 2147–2154.; (b) Poborchi, V. V., Kolobov, A. V., Oyanagi, H., Caro, J., Zhuravlev, V. V.,

Tanaka, K., Treacy, M. M. J., Marcus, B. K., Bisher, M. E., Higgins, J. B., (Eds.), Proceedings of the 12th International Zeolite Conference, Baltimore, Maryland, USA, July 5–10, 1998, pp. 2217–2224.
[61] Singer, J., *Z. Anorg. Chem.* **1932,** *204,* 232–237.
[62] Kumins, C. A., Gessler, A. E., *Ind. Eng. Chem.* **1953,** *45,* 567–572.
[63] Kumins, C. A., Process for making ultramarine pigment. U.S. Patent 2,544,695, March 13, **1951**.
[64] Galastianin, W. D., Nadiarin, A. K., Karakhanian, S. S., Oganiesjan, E. B., Szakhazarian, F. S., Grigoroan, C., Soviet Patent 16,381,447, **1991**.
[65] Ishida, I., Fujimura, Y., Fujiyoshi, K., Wakamatsu, M., *Ceram. Soc. Jpn.* **1983,** *91,* 53–62.
[66] Ishida, I., Fujimura, Y., Fujiyoshi, K., Wakamatsu, M., *Ceram. Soc. Jpn.* **1982,** *90,* 326–327.
[67] Ishida, I., Fujimura, Y., Fujiyoshi, K., Satoh, Y., Wakamatsu, M., *Ceram. Soc. Jpn.* **1984,** *92,* 579–585.
[68] Kowalak, S., Stróżyk, M., Pawłowska, M., Miluśka, M., Kania, J., *Stud. Surf. Sci. Catal.* **1997,** *105A,* 237–244.
[69] Jankowska, A., Zeidler, S., Kowalak, S., in: Gil, B., Kukulska-Zając, E., Kozyra, P., Makowski, W. (Eds.),, *Annual Meeting of the Polish Zeolite Association,* Proceedings of XII Zeolite Forum, Polańczyk, Poland, Sept. 10–15, 2006, Polish Zeolite Association, Kraków, Poland.
[70] Więckowski, A. B., Stuglik, Z., Stróżyk, M., Kowalak, S., *Mol. Phys. Rep.* **2000,** *28,* 25–29.
[71] Kowalak, S., Jankowska, A., Łączkowska, S., *Catal. Today* **2004,** *90,* 167–172.
[72] Kowalak, S., Jankowska, A., *Micropor. Mesopor. Mater.* **2003,** *61,* 213–222.
[73] Loera, S., Ibarra, I. A., Laguna, H., Lima, E., Bosch, P., Lara, V., Haro-Poniatowski, E., *Ind. Eng. Chem. Res.* **2006,** *45,* 9195–9200.
[74] Kowalak, S., Jankowska, A., *Eur. J. Mineral.* **2005,** *17,* 861–867.
[75] Jankowska, A., PhD thesis, **2004**, Faculty of Chemistry, Adam Mickiewicz University.
[76] Barrer, R. M., White, E. A. D., *J. Chem. Soc.* **1952,** *2,* 1561–1571.
[77] Seff, K., *J. Phys. Chem.* **1972,** *76,* 2601–2605.
[78] Hund, F., *Z. Anorg. Allg. Chem.* **1984,** *509,* 153–160.
[79] Lindner, G. G., Massa, W., Reinen, D., *J. Solid State Chem.* **1995,** *117,* 386–391.
[80] Kowalak, S., Jankowska, A., Zeidler, S., *Micropor. Mesopor. Mater.* **2006,** *93,* 111–118.
[81] Kowalak, S., Jankowska, A., *Micropor. Mesopor. Mat.* **2008,** *110,* 570–578.
[82] Kowalak, S., Jankowska, A., Zeidler, S., Więckowski, A. B., *J. Solid State Chem.* **2007,** *180,* 1119–1124.
[83] Jankowska, A., Mikołajska, E., Kowalak, S., in: Derewiński, M., Burkat-Dulak, A., Pashkova, V. (Eds.), *Annual Meeting of the Polish Zeolite Association,* Proceedings of XIV Zeolite Forum, Kocierz, Poland, Sept. 16–21, 2007, Polish Zeolite Association, Kraków, Poland, **2007**.
[84] Kowalak, S., Jankowska, A., Mikołajska, E. in: Aleksandrov, H., Petrova, G., Vayssilov, G. (Eds.), *Advanced Micro- and Mesoporous Materials,* Proceedings of Second International Symposium, Varna, Bulgaria, Sept. 6–9, 2007, Bulgaria, **2007**.
[85] Kowalak, S., Jankowska, A., *Polish J. Chem.* **2008,** *81,* 131–140.
[86] Jankowska, A., Kowalak, S., *Stud. Surf. Sci. Catal.* **2008,** 174, in press.
[87] Kowalak, S., Stróżyk, M., *J. Chem. Soc., Faraday Trans* **1996,** *92,* 1639–1642.
[88] Ogawa, M., Saito, F., *Chem. Lett.* **2004,** *33,* 1030–1031.
[89] Vaughan, D. E. W., *Stud. Surf. Sci. Catal.* **2007,** *170A,* 193–198.
[90] Kowalak, S., Jankowska, A., Baran, E., *Chem. Commun.* **2001,** 575–576.
[91] Kowalak, S., Jankowska, A., Janiszewska, E., Frydrych, E., *Eur. J. Mineral.* **2005,** *17,* 853–860.

CHAPTER 23

Advances in the Use of Carbon Nanomaterials in Catalysis

Benoît Louis, Dominique Bégin, Marc-Jacques Ledoux, *and* Cuong Pham-Huu

Contents

Abstract

Nanoscience is concerned with materials that have dimensions on the nanoscale and possess size-dependent properties differing from those of the bulk. Catalysis is one of the primary branches of nanoscience because many nanomaterials have found widespread applications in chemical reactions: nanoparticles of metal, metal oxide, zeolites and organised carbon nanomaterials. This chapter covers the literature from the early 1990s until

Ordered Porous Solids
DOI: 10.1016/B978-0-444-53189-6.00023-8

beginning 2007 and presents a brief overview of carbon nanotubes (CNT) and carbon nanofibres (CNF) chemistry, with a deeper focus on their applications in catalysis, both as a support and as a catalyst. This chapter is composed of four sections: the first one briefly introduces the general features of CNT and CNF. In a second section, an attempt is made to present an overview of the advantages gained from the structure and properties of carbon nanomaterials for their subsequent use in catalysis, while also taking into account their limits. The use of the carbon nanomaterial itself as a catalyst is also discussed, together with the so-called confinement effect. The third part presents the main fields of application to date in catalysis of metal or active phase supported on CNT/CNF. The last section consists in the immobilisation or the assembly of these nanomaterials, thus preparing structured catalysts that help to overcome the problems encountered with bulky CNT or CNF.

Keywords: Carbon Nanotubes, Carbon Nanofibers, CVD, Heterogeneous Catalysis, Macroassembly, Structured Catalyst

Glossary

Carbon nanofibre	any carbon structure at the nanometric scale build by graphene layers arranged as stacked cones, cups or plates.
Carbon nanotube	carbon nanostructure having a cylindrical or tubular configuration consisting of rolled graphite basal planes.
Catalyst	a substance, usually used in small amounts relative to the reactants, that modifies and increases the rate of a reaction without being consumed in the process.
CVD or chemical vapour deposition	chemical process used to produce high-purity, high-performance solid materials. The process is often used in the semi-conductor industry to produce thin films. In a typical CVD process, the substrate is exposed to one or more volatile precursors that react and/or decompose on the substrate surface to produce the desired deposit.
Heterogeneous catalysis	chemistry term that describes a catalytic process where the catalyst is in a different phase (solid, liquid or gas) than the reactants. Heterogeneous catalysts provide a surface for the chemical reaction to take place on.
Macroscopic assembly	shaping in a desired morphology of any material (nanotube or nanofibre for instance) at the reactor scale.

1. Carbon Nanotubes and Carbon Nanofibres Properties: Towards Smart and Versatile Materials

1.1. Generalities

Carbon nanotubes (CNT) and carbon nanofibres (CNF) are cylindrical or tubular carbon formation with radii on the nanometre scale and lengths of up to several microns.[1,2] Carbon nanostructures (1D carbon), nanotubes and nanofibres, represent the smallest organised form of carbon, and have been the target of an increasing scientific interest,[2–6] since the discovery of multi-walled CNT in the arc discharge materials in 1991.[2] CNT, single-wall (SWNT) and multi-walled (MWNT), consist of rolled graphite basal planes with low surface energy and an open channel in the middle. CNF are similar in shape and length to CNT but consist of graphite planes oriented via an angle with respect to the fibre axis; hence, no tubule is present in the middle, that is, a pile of 'Chinese hats', which by projection shows a fishbone arrangement. Figure 23.1 presents the difference between prismatic and basal planes in graphite. The orientation angle of the graphene planes with respect to the fibre axis could also be finely tuned by modifying the synthesis conditions.[7–9] The exposed external surface of the CNF is constituted of prismatic planes with high reactivity.

Typically, the length of a nanotube varies from one to several microns, with diameters ranging from 0.5 nm for SWNT to more than 100 nm for MWNT (depending on the number of walls). The inner diameter of the CNT can vary from a few to several tenths of nanometres depending on the synthesis conditions. For the MWNT, the distance between two neighbouring layers of graphene is slightly above 0.34 nm, that is, the inter-planar distance of d_{0002} plane of graphite. Indeed, the curvature of the graphene planes induces the presence of constraints.[10] The electronic density of the CNT is also modified compared with that of planar graphite. It is expected that the rolling up of a graphene sheet to form the tube creates re-hybridisation of a carbon orbital with a simultaneous modification of the electronic density.

These nanomaterials are also characterised by their extremely high aspect ratio. The aspect ratio is defined as the length of the major axis divided by the width or diameter of the minor axis. According to this definition, spheres have an aspect ratio of 1, while CNT or CNF exhibit an aspect ratio ranging from a few tenths to several thousands. The high aspect ratio of these materials confers them a significant external surface area compared to classical material.

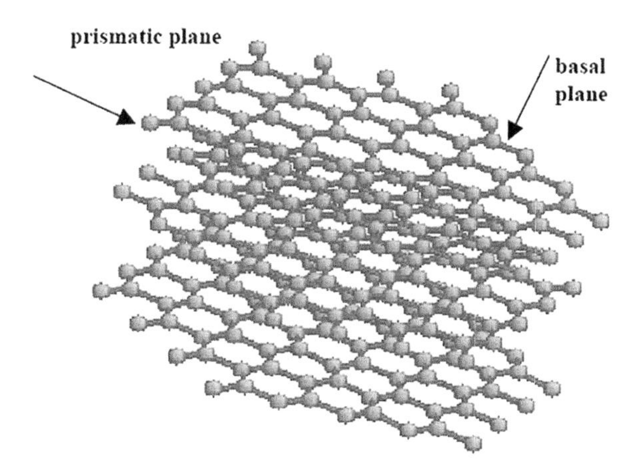

Figure 23.1 Prismatic and basal planes in graphite.

1.2. Synthesis of CNT and CNF

The first CNT structure was discovered in 1991 by Iijima in the carbon soot produced by an arc discharge machine.[2] The first attempt to produce CNT with high yield was initiated by Ebbesen and Ayajan by modifying the deposition parameters of the arc discharge technique.[11] Generally to produce CNT, an arc is maintained between two cylindrical electrodes under an He atmosphere at extremely high temperature, that is, 3700 °C. The tubes are formed as a soft sooty fibrous core on the cathode. Arc discharge methods only allow operating in a discontinuous mode, thus rendering the scaling-up difficult. Another drawback remains the low selectivity towards CNT and the large amount of carbon nanoparticles or soot that are formed inducing a need for post-synthesis purification, which is time and yield consuming.[12,13] It is therefore of interest to find new synthesis methods that allow to overcome the drawbacks of physical synthetic routes (arc discharge and laser ablation). The most promising method to produce CNT with high selectivity and high yield consists in the catalytic assisted synthesis.[14,15] The catalytic process consists of contacting a gaseous carbon source, that is hydrocarbons, CO or organic complexes, with a catalytic active phase (Ni, Co, Fe and alloys) at mild temperatures ranging from 600 to 1200 °C. The ability of these metals to generate carbon nanostructures from gaseous carbon sources is related to their catalytic activity in the decomposition of a carbon source, combined with the ability to transform it into metastable carbide forms. Hence, these metal catalysts present a high affinity to form a solid solution with carbon, allowing its diffusion throughout the metal particle, thus producing ordered CNT (or CNF).[16–19] The catalytic route allows the formation of large amounts of CNT with respect to the starting weight of catalyst along with a high selectivity. Pure CNT can be recovered after a simple acidic or basic treatment of the final product to remove the catalytic metal and the support.[20] To illustrate the large production of CNT (several grams per hour), Figure 23.2 shows the experimental set-up before synthesis (Fe/Al_2O_3 catalyst) and

Figure 23.2 (A) Reactor filled with an Fe/Al_2O_3 catalyst (1 g) and (B) after CNTgrowth under a mixture of C_2H_6/H_2 at 680 °C for 2 h. The carbon yield was about 6000 wt.% with respect to the iron catalyst weight. (See color insert.)

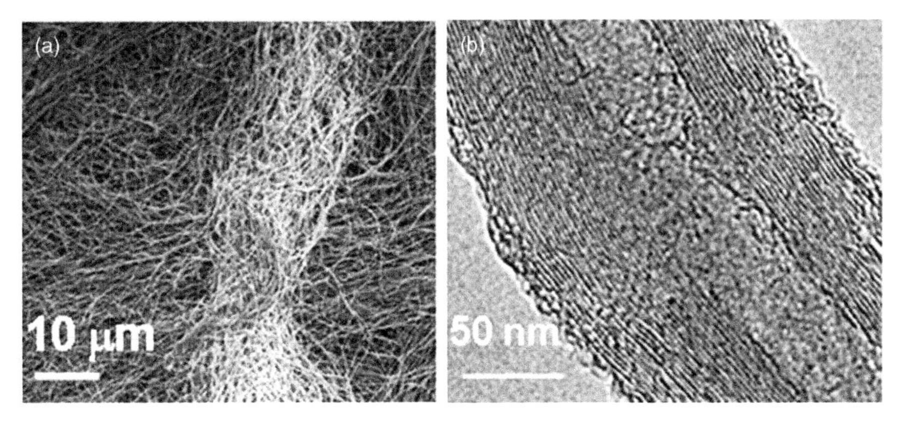

Figure 23.3 (A) SEM micrograph of the CNT synthesised via CVD of a mixture of C_2H_6/H_2 at 680 °C. (B) TEM micrograph showing the relatively homogeneous diameter of the CNT along with topological defects in the graphene planes.

after flowing an ethane/hydrogen mixture during 2 h through the reactor. Several grams of a black fluffy powder were produced per gram of catalyst per hour.[21]

The selectivity towards CNT was investigated by statistical SEM observations of the as-synthesised material. Electron micrographs are presented in Fig. 23.3A and B. The MWNTs formed were extremely pure as no trace of other carbon nanoparticles were observed in the sample. On the other hand, the MWNTs were extremely homogeneous in diameter, 40 nm. The average aspect ratio value experimentally determined from statistical SEM analysis was higher than 1000. Synthesis carried out at higher temperature, 800 °C instead of 680–750 °C, keeping other reaction conditions similar was less selective and led to the formation of carbon nanoparticles in the sample.

Ermakova *et al.*[22] have reported that the formation of these carbon impurities was strongly dependent on the interaction between the active phase and the support. It is thought that the interaction between the active phase and the support was relatively low in our sample. TEM observation reveals that the tube walls have a relatively low crystallinity and contains a high number of defects within the graphene plane. The mechanism(s) of CNT and CNF formation via CVD, or more exactly CCVD (catalytic chemical vapour deposition), process is (are) still under debate.[16,19,22–37]

The porosity of the multi-walled CNT is mainly due to a mesoporous inner hollow channel with an average pore size ranging from 5 to 60 nm and aggregated pores ranging from 20 to 100 nm. No micropores were found in these carbon nanomaterials whereas micropores are generally present in activated charcoal.[3,4,38–40]

1.3. Structural and electronic features

Before presenting the use of CNT and CNF in catalysis, it is important to highlight their specific structure and main features while keeping in mind the catalytic process requirements. CNT can be divided in two categories: SWNT and

MWNT. The former are made of a perfect graphene sheet, that is a mono-atomic layer made of a hexagonal display of sp^2 hybridised polyaromatic carbon atoms, rolled up into a cylinder and closed by two semi-fullerene caps. The inner diameter of these SWNT structures varies between 0.4 and 2.5 nm having several microns in length. MWNT can be regarded as concentric SWNT with increasing diameter and coaxially disposed. The external diameter can reach 100 nm or more depending on the number of walls (from two in double-walled to several tenths). The concentric walls are regularly spaced by approximately the inter-graphene distance (0.34 nm).

The main difference between CNT and CNF comes from the lack of a hollow channel in the latter as observed in Fig. 23.4. CNF represents a general term referring to graphitic structures with other orientations of the graphene plane: ribbon-like (parallel to the growth axis), platelet (perpendicular) and herring-bone (obliquely).[7,34,41] The fibrous structure gives rise to textural properties, mesoporosity (0.5–2 cm^3/g) and specific surface areas ranging from 10 to 200 m^2/g.[10]

SWNT are semi-conductors that behave like pure quantum wires (1D-system) where the electrons are confined along the tube axis.[42] Hence, two factors mainly govern the electronic properties: the tube diameter and its helicity; the latter is defined by the way the graphene layer is rolled up (armchair, zigzag or chiral). The curvatures of the graphene sheets strongly modify the π-electron cloud.[43]

MWNT and CNF possess similar electronic properties; at high temperature, their electrical conductivity is comparable to graphite, whereas at low temperature, they reveal 2D-quantum features. CNF have been added to polymers, which enable enhanced conductivity at very low carbon loading.

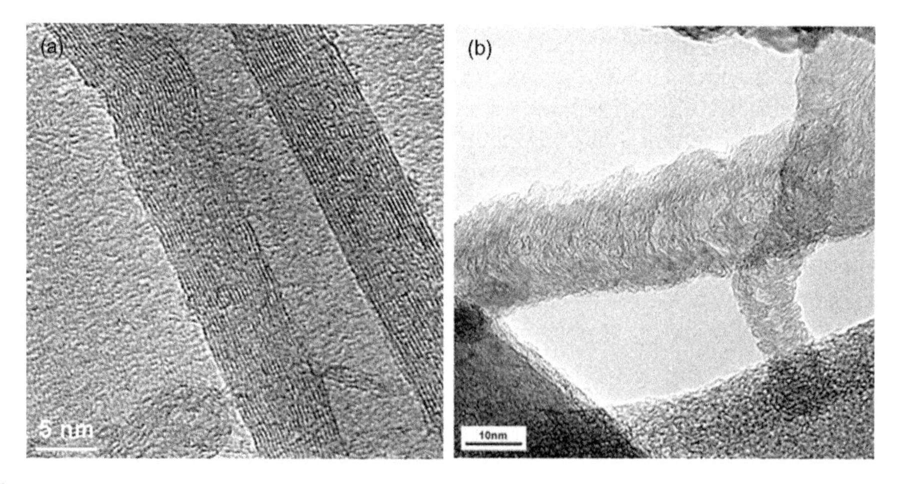

Figure 23.4 High resolution TEM images of (A) a multi-walled carbon nanotube and (B) carbon nanofibres.

Finally, special metal–carbon interactions are created after coating a metal on these conductive supports, thus creating major differences when compared with graphite,[40] and being of primary importance for catalysis.

1.4. Mechanical and thermal properties

CNT in theory are the most resistant fibres, possessing a Young's modulus on the tera-Pascal order[44] and a resistance to traction close to 250 GPa.[45] For comparison, the resistance to traction displayed by steel is usually 2–3 GPa while weighing more than CNT. Furthermore, CNT are flexible and can be bent several times at 90° without undergoing structural changes.[46] CNF also display unique mechanical properties that have triggered the search for applications in composite materials, especially in the reinforcement of materials. As an illustration, Fig. 23.5 shows a composite material where CNF were grown on malleable graphite felt (microfibrous network). The synthesis and applications of these composites will be detailed in Section 4.

Another important feature of CNT and CNF is the thermal stability under reaction conditions (under oxidative atmosphere). It is noteworthy that organised carbon nanomaterials are more stable towards oxidation (about 650 °C) than activated carbon but more reactive than graphite.[40] However, the presence of residual metal particles (growth catalyst) and surface defects can catalyse carbon gasification. Among the different classes of tubes made of organic or inorganic materials, carbon nanostructures exhibit attractive electronic, mechanical and structural properties. Thanks to their extraordinary properties, CNT and CNF are used in different nanotechnological applications, in light weight and high strength composites, or as adduct to polymers. The combination of their high external surface area due to the high aspect ratio (length-to-diameter ratio) and their high thermal conductivity confer them several advantages compared with the traditional supports such as alumina, silica or activated charcoal. Finally, the presence of an empty

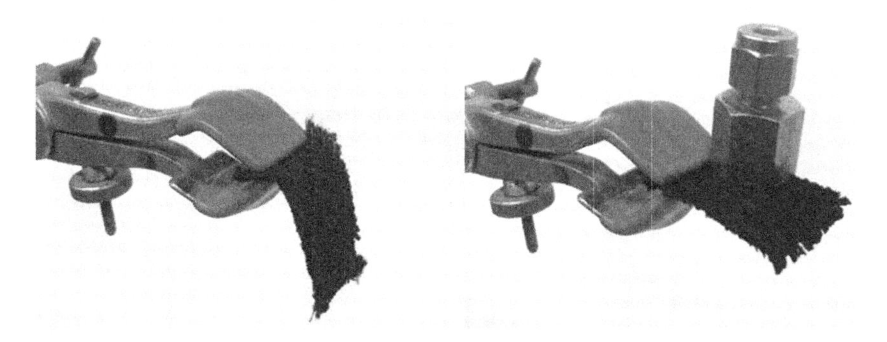

Figure 23.5 Photos showing the mechanical improvement of the graphite cloth after deposition of 100 wt.% of carbon nanofibres inside its matrix. (See color insert.)

channel, in the case of CNT, rises new physical properties due to fluids modification during their diffusion, inducing a so-called confinement effect.

2. Carbon Nanomaterials in Catalysis: A Laughing Choice?

2.1. What does a catalyst require?

The catalyst role is to allow the transformation of reactants into desired products with a high selectivity, reducing at the same time the cost linked to the operating conditions and separation steps. Usually, industrial catalysts consist of an active phase coated on high surface area supports (alumina, silica or activated carbon) in order to increase its dispersion and hence the number of active sites. Catalytic nanomaterials are of primary importance because of their reactivity. To find practical applications, nanomaterials must be dispersed in some medium or on some support, such as a matrix or a solid surface.

The application and development of an industrial catalyst starts with research and control of materials at the molecular level following by laboratory evaluation, then a pilot plant, which pave the way to the real industrial catalyst. The understanding of the macroscopic phenomena involved in catalysis seems to be relatively well controlled for a large part of the catalytic processes, whereas the understanding at a nanoscopic scale has to be improved, to design new generations of catalysts with higher activity and selectivity. Decreasing the size of the active phase significantly increases the effective contact surface of the reactants with it, thus contribute to lower the cost of the catalyst per unit weight of active phase. Hence, this leads to include catalysis into novel and over-increasing nanotechnology science. However, during the catalytic reaction, the starting size of the catalytic site is modified by aggregation that made the size control a real technological and scientific challenge.

One of the most common way to avoid sintering is to coat an active phase within nanocages or nanochannels, that is zeolites or mesoporous silica. However, the diffusion and the accessibility of the reactants to these constraint catalytic sites can be affected by the tortuous structure of the support and consequently have a significant effect on the selectivity of the reaction, rendering such system unable to operate under harsh conditions.[47] In addition, it is thought that the tailoring of nanostructured catalysts could lead to new electronic and catalytic properties for an improvement of catalytic activity and selectivity in view of the coming legislations on waste reduction. Indeed, nanoparticles are made by the assembly of few atoms, with few nanometres in size. Nanoparticles hold a large number of metastable atoms with respect to the bulk structure. It is thought that these metastable atoms could play an important role during the catalyst action while their stabilisation, that is through electronic or chemical interactions with the support underlying surface, allows one to explore new catalytic systems with peculiar activity and selectivity.

Recently, new development in the field of 1D and conductive materials, that is carbon and silicon carbide, shed a new light on the use of such materials in catalysis in place of traditional macroscopic supports.[4,6,10,38,39,48,49] Metals supported on these carbon nanostructures exhibit unusual catalytic activity and selectivity patterns when compared to those encountered with traditional catalyst supports. The extremely high external surface area displayed by these nanoscale materials significantly reduces the mass transfer limitations, and the peculiar interaction between the deposited active phase and the surface of the support were advanced to explain their catalytic behaviours. Many of their fundamental and astonishing physical properties are known and present efforts are largely devoted to produce CNT with high selectivity and in a large scale in order to decrease their price and allow their exploitation in a wide range of applications.

2.2. Peculiarities of carbon nanomaterials

The interaction of CNT with their environment and particular with gases or active species adsorbed either on their internal or external surfaces attracts increasing attention due to the possible influence of the adsorption on some of the tube properties, and to the possibility of using CNT hollow channel as nanoreactor or as exo-template material. The exposed surface of the CNF is constituted by prismatic planes with high surface reactivity compared with the basal planes of graphite that constitutes the external surface of CNT. Recent reviews have shown that CNF are attractive in several applications such as electronic, medical, light weight reinforcement materials and catalysis.[3,10,35,40,49–51] The results obtained have significantly extended our knowledge of nanoscale science and created numerous applications for future nanotechnologies. In catalysis, the possibility to maintain the active phase in a nanoscopic size without excessive aggregation allows one to get more insight in the understanding of such active phase. Among their different applications, catalysis seems to be promising according to the latest results reported to date in the open literature.[7,10,14,17,24,25,28–30,34,40,52–54] The peculiar interactions between a coated active phase and the 1D carbon surface were advanced to explain their high catalytic performance. Catalysis plays an important role in the processing of several vital compounds; therefore, a large research is ongoing to enhance catalytic activity and selectivity with the aim of reducing waste and by-products. In parallel with research conducted on the development of new active phase formulations, numerous studies have also been focused on the development of new catalyst supports that can improve the catalytic performance of existing processes.

The possibility for using nanostructured carbon in catalysis seems to be nearer, owing to the possibility of producing them in large quantities, which significantly decreases the investment cost. The most important drawback concerning the use of these carbon nanomaterials is linked to their nanoscopic size that renders their daily use problematic.

2.3. Confinement effect

The fluid dynamics within the inner channel of CNT completely differs from macroscopic systems. Indeed, the fluid mobility is governed not only by its intrinsic properties but also by its strong interaction with the CNT walls, which create a nanoreactor.[48] Preliminary studies from Gogotsi and co-workers dealt with the unusual behaviour of entrapped water, which rapidly reached its supercritical state under heating because of so-called confinement effect.[67,68] It is expected that the tubular morphology and the high aspect ratio of CNT induce a confinement effect on the gas or liquids trapped inside the central channel, thus leading to completely different physical behaviour when compared to conventional bulk materials. CNT have been filled with various materials: fullerenes,[69] metal halides,[70,71] oxides,[72] alloys[14,73] and zeolites.[48,74] These encapsulated materials can be subsequently used in several potential domains including magnetism, catalysis, linear spin chains, electronics and polymers. To conclude, the confinement effect should therefore be taken into account to explain the amazing properties of these unidimensional carbon nanomaterials.[72,75]

2.4. Metal–carbon interactions: Importance of surface functional groups

Interactions between a metal particle and its C-support can be of physical nature, where the size of the particle is determined by its carrier.[55] In contrast, chemical interactions involve charge transfer between the two protagonists. The latter can take place via oxidation/reduction or acid/base reactions. These donor/acceptor interactions are linked to the corresponding Lewis acid and base characters of the materials involved,[56] thus modifying the electronic structure and hence the dispersion of the metal. For instance, CNT/CNF are oxidised along their side walls or open ends[57] to produce oxygen-containing surface groups: hydroxyls, carbonyls, carboxylic, quinoic and lactones on the external surface (Fig. 23.6).

Although, such functionalisation will cause certain changes in the delocalised π-orbital system of CNT/CNF. These O-species are of primary importance to enhance metal–carbon interactions.[15,58] Indeed, these O-containing functional groups can guide the manner but also the size of coated metal nanoparticles. Because the catalytic performance is often directly connected to the size of the metal nanoparticles, like in the Fischer–Tropsch synthesis,[59,60] it is important to quantify and to tailor these O-sites. Furthermore, these species can act themselves as a co-catalyst or even as a catalyst for the synthesis of styrene.[61,62] On the contrary, the rate

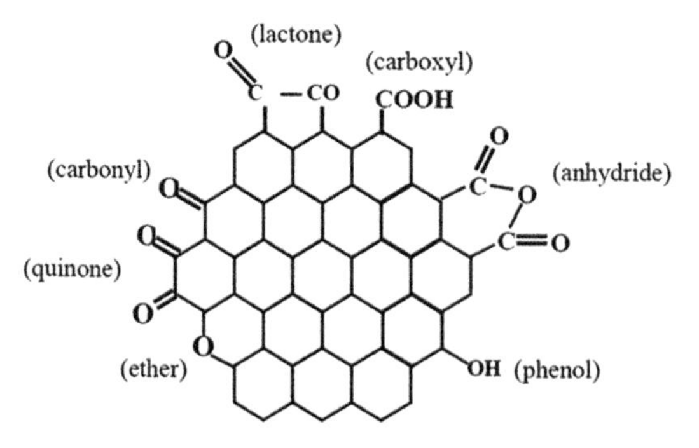

Figure 23.6 Different oxygen-containing surface groups.

of hydrogenation of cinnamaldehyde can be enhanced about 20 times when these O-groups are completely removed.[15,58] The direct catalytic effect of these O-active sites is based on the adsorption of organic reactants.[63,64] While decreasing the number of O-surface groups, the carbon surface gradually changes from polar to non-polar, thus controlling the adsorption mode of reactants, thereby influencing both activity and selectivity. A refined model developed by Koningsberger that implies an indirect influence of these O-groups on catalysis via the electronic state of the metal explains the interaction between metal nanoparticles <2 nm and different supports.[65,66] With a decrease in the electron richness of the support O-atoms, the density of states and the Fermi level of metal particles shift towards higher binding energy, thus suggesting that the presence of these O-species would give rise to a change in the electronic structure of the metal. Several groups tend to quantify both the amount and the nature of these hydrophilic groups present on the surface of carbon nanomaterials.

2.5. Summary

Recently, CNT were extensively studied as catalyst support in several fields of catalysis from gas-phase to liquid-phase processes. CNT hold several advantages when they are used as catalyst support: (1) their small dimension that significantly increases the external surface area of the catalyst especially for liquid-phase reactions where diffusion becomes rate limiting; (2) their high thermal conductivity provides a homogeneous and rapid heat transfer to the catalyst support that avoid the formation of surface hot spots, which could modify the overall selectivity of the reaction; (3) their structure that exhibits a high specific surface area, generally >100 m^2/g, along with a large pore volume without micropores. The existence of an electronic interaction (π-interaction) between the carbon nanomaterial surface and the coated species could also affect in a significant manner the final properties of supported species.

3. Unidimensional Carbon Nanomaterials in Catalysis?

3.1. Preparation of nanocarbon decorated with metal nanoparticles

In the case of highly exothermic reactions, the safety of a whole process can be put in jeopardy due to hot spots in the catalyst bed, leading sometimes to temperature runaway in the reactor. One strategy to overcome such problem is to prepare a catalyst where the active sites are coated on a support with high chemical inertness and high thermal conductivity. The high chemical inertness allows an easier active phase surface modification during the catalytic reaction, while the thermal conductivity plays a role of heat disperser from a single site through the entire matrix of the support. In the case of endothermic reactions, the heat conductivity of the support renders easier the supply of heat from the reactor walls to the active sites.

It is expected that 1D carbon nanomaterials will be extensively employed in numerous fields in the future, ranging from energy, healthcare, polymers, environment, to catalysis. Some examples of catalytic application of these 1D materials will be addressed in the following section. Hence, their performance is compared with that obtained on traditional catalysts.

Several methods such as ion exchange, organometallic grafting, incipient wetness impregnation, electron beam evaporation and deposition/precipitation have been used to prepare metal nanoparticles on CNT and CNF. The carbon supports are usually treated chemically (generally with acids, oxidants) or thermally to increase the number of surface functional groups, thus influencing the metal dispersion.[76,77] Besides the classical acid treatments, other aggressive reagents were used to functionalise CNT. Rao *et al.* have used OsO_4, $KMnO_4$ and HF-BF_3 systems to open nanotubes.[78] Recently, while treating MWNT with superacids, we were even able to modify the morphology of the tubes (Fig. 23.7), thus leading to special interactions with deposited metal.[70]

3.2. Carbon nanomaterials in catalysis

3.2.1. Selective hydrogenations: C=C bonds, nitrobenzene

Hydrogenation catalysts usually consist of a homogeneous dispersion of palladium nanoparticles inside CNT. The palladium was introduced into the CNT support via an incipient wetness impregnation using an aqueous solution of palladium nitrate. After impregnation, the wet solid was allowed to dry and further reduced in flowing

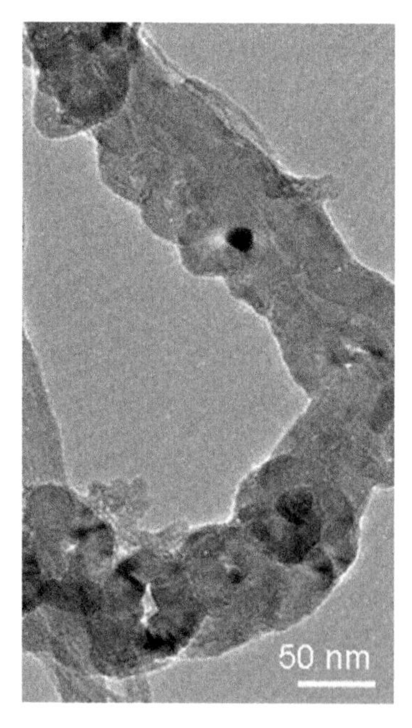

Figure 23.7 TEM image of MWNT treated several hours in BBr_3 at room temperature.

hydrogen at 400 °C for 2 h. The preferential location of the metal particles inside the tubes could be explained by the fact that during the impregnation process, the liquid completely filled the inner part of the tube by capillarity.[79] Indeed, CNT could be easily filled with low surface tension liquids, that is <190 mN/m.[75] Water, with a surface tension of 72 mN/m, would thus wet and rapidly fill the inner part of the tube. The filling is also strongly favoured by the presence of surface oxygenated groups that render the tube surface more hydrophilic. The ability of the low surface tension liquid to fill the CNT also strongly depends on the tube diameter according to work reported by Ugarte *et al.*[80]: the smaller the inner channel, the lower the filling efficiency. It is important to note that the precise location of the metal nanoparticles can only be evaluated by a TEM 3D technique.[81,82]

This Pd/CNT catalyst was tested in the liquid-phase selective hydrogenation of the C=C bond in the α,β-unsaturated cinnamaldehyde to investigate the benefits of a CNT support use versus activated charcoal-based catalysts.[83] The catalyst was suspended in dioxane under continuous stirring during hydrogen supply via bubbling into the liquid at 80 °C. The hydrogenation performance (expressed in terms of conversion) and the selectivity of the Pd/CNT catalyst are presented in Table 23.1 and compared with those obtained on a commercial Pd/AC (activated charcoal).

CNT-based catalysts exhibit a higher hydrogenation activity compared with that observed on activated charcoal-based catalysts under similar conditions.[84] According to TEM analysis that did not show any significant difference in the metal particles size distribution, the higher hydrogenation rate observed on the CNT-based catalyst was explained by the high surface-to-volume ratio compared with the activated charcoal-based catalyst due to mass transfer limitations. Because the inner diameter of the CNT channel, 40–60 nm, mostly belongs to the mesoporous range, one should expect the absence of any diffusion problems that are generally linked with activated charcoal micropores, that is <5 nm. As discussed above, CNT can be efficiently filled by capillarity force when the solvent has a low surface tension, that is <200 mN/m. The high capillarity forces existing inside the mesopores and open ends CNT also greatly favour the penetration of fluids inside the channel. A large difference was observed on the product selectivity, that is almost exclusively C=C hydrogenation on the CNT-based catalyst whereas both saturated alcohol (corresponding to the hydrogenation of C=O and C=C bonds) and aldehyde were produced in similar yields with the activated charcoal-based catalyst. The absence of microporosity and the high mass transfer rate in the CNT-based catalyst

Table 23.1 Cinnamaldehyde conversion and the C=C bond hydrogenated product yield as a function of time on stream on the Pd/CNT and Pd/AC catalysts

Catalyst	Pd/CNT			Pd/AC		
Time on stream (h)	1	18	32	1	12	39
Conversion (%)	4	44	99	2	30	95
Yield in C = C bond hydrogenation (%)	4	34	88	1	18	50

were probably responsible for the significant improvement of the C=C bond hydrogenation selectivity. In the activated charcoal-based catalyst, the large amount of micropores could lead to diffusion problems and successive hydrogenation giving the saturated alcohol. Similar results have also been reported elsewhere during the hydroformylation of 1-hexene over Rh-based catalyst.[40] Again, the lower mass transfer process due to the microporous character of the activated charcoal was advanced to explain the change in selectivity.

The selective hydrogenation of nitrobenzene to aniline was also studied for comparison. In this reaction, the catalyst consisted of Pd 5 wt.%/MWNT with 60% of palladium particles being located inside the CNT channel. The catalytic activity obtained in a stirring tank reactor at room temperature is presented in Table 23.2.

The MWNT-based catalyst exhibits a higher conversion than the commercial Pd/AC catalyst especially when the initial concentration of nitrobenzene inside the reactor was increased. The high hydrogenation activity observed on the MWNT-based catalyst could be attributed to the nanoscopic size of the catalyst and the complete absence of micropores, which provided an efficient contact surface between the reactants and the active phase particularly in liquid-phase medium (where the mass transfer phenomenon becomes significant). In the case of the activated charcoal, the large part of the active phase located inside the micropores was probably not accessible resulting to a lower activity despite the large difference observed on the specific surface area between the two catalysts, that is 900 m^2/g for the activated charcoal against 10 m^2/g for the CNT catalyst.

3.2.2. CNT as a catalyst itself: The case of oxidative dehydrogenation

The oxidative dehydrogenation (ODH) reaction of ethylbenzene to styrene in the presence of oxygen is one of the top ten industrial processes.[85] The catalytic performances of nanocarbons were higher than those using carbon black and graphite as catalysts.[86] The combustion stability of graphite, CNT and onion-like carbons gave stable catalytic performances on stream. Such amazing catalysis was explained by the presence of delocalised π-electrons in graphitic carbons, together with a high density of reactive edge plans, which can immobilise reactant species on the graphite surface.[87] A reaction pathway for ODH on graphitic carbon catalysts involving O-species present on the surface as active sites was proposed by Pereira *et al.*[88] They claimed that pre-oxidised MWNT are more active for ODH during the initial stages of the reaction, highlighting the importance of oxygenated surface

Table 23.2 Liquid-phase hydrogenation of nitrobenzene over Pd/CNT and Pd/AC catalysts at room temperature and under 20 bar total pressure of hydrogen

Catalyst	Pd/MWNT				Pd/AC			
Time (min)	100	270	350	450	100	270	350	450
Conversion (%)	32	72	85	93	22	61	70	75

The initial nitrobenzene concentration in dioxane was 2 mol/litre.

groups.[88] The following mechanism was advanced to explain the ODH sequence on these carbon-based catalysts: the hydrocarbon was first adsorbed on an available surface site next to basic oxygen groups and underwent dehydrogenation with a concomitant formation of surface OH groups.[61] The basic quinoidic groups were generated by reacting the surface OH groups with the dissolved oxygen with a release of water into the reaction medium.

ODH of di-hydroanthracene in liquid phase was carried out over different CNT-based materials. CNT acted directly as a support and active phase for the liquid-phase ODH of di-hydroanthracene to anthracene at low-temperature. The test was also performed on an exfoliated graphite material for comparison. The results clearly showed that the reaction was chemically selective for whatever catalyst was used (only traces of di-anthraquinone formed). The catalytic dehydrogenation rate was relatively low on as-synthesised CNT (90% anthracene yield after 120h). The same catalyst after heat treatment at high temperature exhibits a huge dehydrogenation activity increase, that is 99% instead of 30% for the same period of test (24 h). Moreover, only 50% of anthracene was produced after 170h with exfoliated carbon. Hence, this indicates a positive effect of the graphitisation on the dehydrogenation activity. The higher catalytic activity observed on heat-treated CNT versus as-synthesised materials could be due to an enhancement of their adsorption ability due to a lower amount of adsorbed oxygen on their surface. It was shown that both O_2 and aromatic compounds (like anthracene) can be easily adsorbed on the CNT surface with an adsorption energy of 18.5 kJ/mol for O_2[89] and 9.4 kJ/mol for C_6H_6.[90] The adsorption of aromatic molecules is favoured by the delocalisation and the hybridisation of π electrons between the nanotube and adsorbed molecules. These results are in line with previous studies on ODH of ethylbenzene into styrene over various kinds of carbon-based materials which have shown that during the reaction, part of the catalyst surface, that is the defects surface, play the role of oxygen adsorption site and thus generate strong basic surface oxygen groups that participate in the ODH reaction.[91]

The catalytic activity of untreated or treated MWNT could also be due to a curvature effect, even if it is weak because of the small tube diameters (30–50 nm). This is confirmed by the lower catalytic activity observed on exfoliated carbon, which does not have the same morphologic properties (absence of curvature). Nakamichi *et al.*[92] observed a similar conversion, reaching 93% over activated charcoal under the same conditions. It is also possible that the adsorption strength of organic molecules became weaker on the graphitised surface because of the reduction of the number of surface defects with high adsorption energy.

3.2.3. Grafting of organometallic complexes

The immobilisation of organometallic complexes on CNT or CNF requires a preliminary surface modification step before complexation reaction.[93,94] The coordination mode of metal complexes, Wilkinson-type for instance [$RhCl(PPh_3)_3$], led often to hexacoordinated species on pre-oxidised CNT surface.[95] All types of reactions from classical organic and inorganic chemistry should be feasible on the reactive surface groups of CNT/CNF: COOH, OH, C=O, $-NH_2$. A multi-step synthesis has to be often used[96]: to transform carboxylic groups into more reactive acid chloride groups, to sulphonate, to create nucleophilic groups or COO^- basic groups.[97] A sophisticated method was reported by Terrones *et al.* for the deposition of gold nanoparticles on *in situ* nitrogen-doped CNT (CN_x).[98] It appears well established that classical organic chemistry opens efficient routes for the chemist to prepare supported metal complexes on CNT/CNF, following the methodology described in the review of Ma and Zaera with different surface types.[99] Whereas it seems clear that oxidative treatments improve the further functionalisation step, one has to really produce an active site on the CNT/CNF surface rather than creating detached amorphous fragments after severe treatments.[100]

3.2.4. Problems linked to the use of nanoscopic carbon nanomaterials particles

Carbon nanomaterials have an average diameter ranging from a few to several tenths of nanometres with lengths up to several hundreds of nanometres. These nanoscopic dimensions significantly increase the reactants accessibility due to the high external surface area that contributes to an improvement of the catalytic activity, while the high diffusion rate of the products reduces many secondary reactions. In addition, attrition problems are also low for these nanoscopic materials. However, severe drawbacks are also linked to the use of these nanoscopic powders in conventional reactors: in liquid-phase medium, the small dimension renders the recovery almost impossible by conventional means, which can lead to the blocking of filtres and valves by the solid, whereas in a fixed-bed reactor, the nanoscopic size induces huge pressure drops. The macroassembly of these carbon nanomaterials for catalytic use in conventional reactors is of considerable interest. There have been some developments on the use of direct assembly of CNF by allowing them to grow on alumina or silica beads containing the growth catalyst[101] or on metallic foam.[102] A clever reactor system was developed by Ruta *et al.* to immobilise ionic liquids on CNF,[103] thus opening new routes towards macrostructured carbon nanomaterials. However, the attrition may be significantly increased due to a weak anchorage of the CNF onto alumina or silica surfaces. The weight of alumina, silica or metallic support for the CNF growth also represents a handicap for these composites. Moreover, these supports could also react with the reactants or products leading to a loss of selectivity or could even be destroyed by the reaction medium. Recent works have described the use of ceramic foam as a macroscopic shape to anchor carbon nanomaterials for catalytic uses.[104] It is thought that a direct macroscopic shaping of carbon nanomaterials should open a large field of applications for these nanoscopic materials in catalysis. Herein we report two direct methods to assemble carbon nanomaterials into controlled macroscopic size and shape depending on downstream applications. The macroscopic support should not alter the physical properties of the carbon nanostructures coated on it, that is its high mechanical strength, in order to avoid plugging of the catalyst bed. It should also preserve the high specific volume in order to afford a high space velocity of gaseous reactants, high thermal conductivity, which is a necessary condition for a catalyst which operated in highly exothermic or endothermic reactions, and finally, insure a high chemical resistance to be used in aggressive environments.

4. Carbon Nanomaterials with Controlled Macroscopic Shapes: Novel Versatile Structured Catalysts

Carbon nanostructures, *nanotubes* and *nanofibres*, with a macroscopic shaping can be successfully employed as catalyst supports either in gas-phase or in liquid-phase operation modes. Such assembly allows an ease of handling without altering their intrinsic physical and chemical properties. Moreover, the preparation of such structured catalysts renders them able to be used in a reactor without diffusion and pressure drop problems. Examples of catalytic reactions involving these materials are described in this section: decomposition of hydrazine for satellite path corrections and selective oxidation of H_2S into elemental sulphur.

4.1. Challenges

The first need for a practical use of these nanoscale materials is their proper fabrication on a macroscopic level. Carbon nanomaterials also follow the current scientific wave towards structured catalysts and reactors.[105,106] So the challenge consists of preparing carbon nanomaterials via coating on a pre-shaped macroscopic body at the reactor scale (carbon felt, cloth, monolith or foam) or by designing the synthesis itself using a confinement synthesis method. It is expected that the combination of the exceptional thermal and mechanical stability of these carbon nanomaterials, their high specific surface area and their simple and effective synthesis may allow them to fully compete with traditional materials in several domains of applications: separation, filtration and catalyst support. The absence of micropores inside these materials also render them attractive compared to the traditional activated charcoal where diffusional phenomena usually lowers the diffusion rate of reactants/products towards the active sites.

4.2. CNF with macroscopic assembly

The synthesis was carried out using a graphite felt as a macroscopic host structure. The graphite felt consists of an entangled network of microfilaments (10 μm in diameter) with a smooth surface and low specific surface area, 10 m^2/g. The graphite felt was cut into an appropriate macroscopic shape, depending on the required purpose. A nickel salt was deposited on the graphite disk by incipient wetness impregnation method with a solution of water and ethanol. The metal loading was chosen to be high enough for the production of CNF, but sufficiently low to avoid a further purification step to remove any residual metal.[107] It is worth noting that after the CNF synthesis, a major part of the metal catalyst was completely encapsulated by the carbon layers, thus being no longer accessible in the downstream catalytic use of the composite. The CNF synthesis was performed by loading the macroscopic piece containing the nickel growth catalyst inside a tubular quartz reactor (100 mm inner diameter and 1200 mm in length) and placed in the middle part of an electrical furnace controlled by two thermocouples (Fig. 23.2). The nickel oxide was then reduced *in situ* in flowing hydrogen (50 STP ml/min) at atmospheric pressure and 400 °C for 1 h. After reduction, the H_2 flow was replaced by a mixture of C_2H_6/H_2 (40/60 vol.%) with a total flow rate of 100 ml/min and the reactor temperature was increased from 400 to 670 °C (heating rate of 20 °C/min) and held at this temperature for 2 h. The material was cooled to room temperature under the reactant mixture and then discharged in air for further characterisation. The CNF growth rate in the present work was about 50 g/g of Ni/h. It is significant to note that the CNF yields obtained in our laboratory were among the highest reported in the literature taking into account the extremely low amount of metal catalyst loading, that is 1 wt.%. Up to now, the highest yield of CNF by CVD techniques reported in the open literature was about *50 grams of carbon per gram of metal catalyst.*[17,22,108] The corresponding SEM images of the CNF composite are presented in Fig. 23.8 and clearly evidence the change in the morphology of the starting macroscopic host structure. The low-magnification SEM image shows the complete coverage (>95%) of the starting microfilaments by a dense and entangled network of CNF (Fig. 23.2B). The thickness of the CNF network was about a few microns. According to SEM images, the CNF possess very homogeneous diameters of 30 nm with length generally exceeding several hundreds of nanometres. The synthesis was extremely selective towards CNF formation as no carbon nanoparticles were observed on the sample after the synthesis. Such nanoscopic materials with meso- and macroporous structure represent an interesting catalyst support as they combine an entangled network of nanoscopic solid with high external surface area (>100 m^2/g) and a large pore volume without micropores and ink-like bottled pores as usually observed with traditional supports. The high yield and coverage efficiency of the

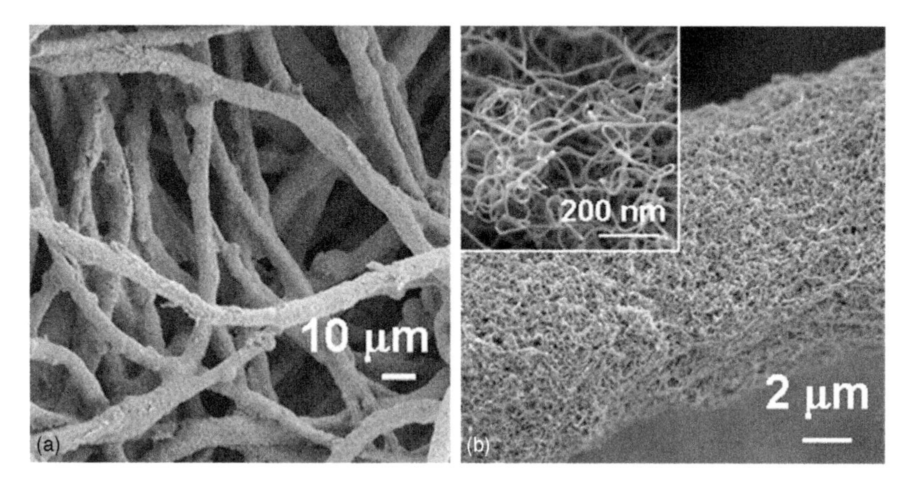

Figure 23.8 SEM micrographs of the CNF composite synthesised by CVD over the graphite felt macroscopic host structure. (A) Smooth surface of the starting microfilaments became rough after CNF growth for 2h. (B) Graphite filaments that constitute the starting graphite felt were completely covered by a dense network of CNF. SEM image (inset of Fig. 23.2B) shows the average diameter of CNF, 30 nm.

formed CNF were explained by an octopus-like growth mechanism of these nanomaterials.[17,109] High-resolution TEM image also confirmed the microstructure of the CNF, that is fishbone or Chinese hat structure. So, the CNF produced from the present synthesis were completely filled and no empty channel was observed along the longitudinal axis. The inter-planar distance between the two adjacent graphene planes was about 0.335–0.34 nm, in agreement with the value reported for CNF in the literature.

Some of the different macroscopic shapes of the CNF composite that can be synthesised according to the present method depending on downstream applications are presented in Fig. 23.9. The CNF formation allows significant improvement of the mechanical strength of the composite as shown in Fig. 23.5. This improvement was directly attributed to the formation of numerous micro- and nanobridges throughout the composite. The CNF are strongly anchored on the macroscopic host structure as no weight loss was observed on the composite after 30 min of sonication in water. The high mechanical anchorage of the CNF can be explained by an octopus-like growth process that allows part of the formed CNF to penetrate the microfilamentous graphite structure leading to a strong anchorage of the nanoscopic structure on the surface of the macroscopic piece.

4.3. Self-supported MWNT

Processing CNT with a controlled macroscopic shape to obtain materials with a practical use is a major challenge in the field of advanced materials. The synthesis and the uses of free nanotubes or fibres could be hazardous for health. The fluffy powder can be shaped while using a binder but this does not solve the hazard during synthesis of the pristine tubes. In addition, the binder can block the accessibility to the channel or the inter-tubes spacing. The assembly process allows the direct use of these 1D carbon materials in conventional catalytic reactors that could be further optimised as an integrated process. The direct use of nanoscopic materials even with exceptional performance is indeed not simple and generally barely possible due to the problems linked with handling and pressure drop. CNT with controlled macroscopic shapes were efficiently synthesised in the laboratory by modifying the shape of the reactor (constraint synthesis) or by modifying the nature of the carbon source and the catalyst ($FeCp_2$ pyrolysis) or by patterning in a periodic structure the growth catalyst

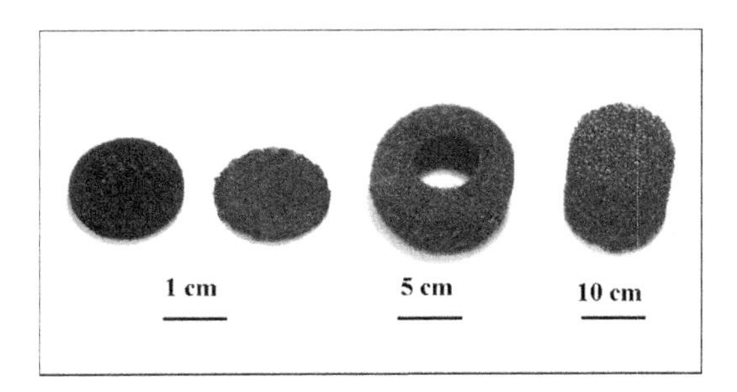

Figure 23.9 Carbon nanofibre composites with different sizes and shapes.

before allowing the contact with a gaseous carbon source. It is expected that the macroscopic assembly of such nanostructured materials will present a real opportunity for their use as a catalyst support in competition with traditional catalyst carriers.

4.3.1. Self-supported CNT via constraint synthesis

In the case of a synthesis under constraint, the catalyst consisted in Fe 20 wt.%/ Al_2O_3, coated inside a tubular reactor, where the two ends are closed with a permeable carbon felt allowing the gaseous reactants to diffuse through it and parallely confining the growth space of the CNT.[110–112] The MWNT synthesised under these conditions displayed a significantly higher density compared to those obtained without constraint as shown in Fig. 23.10. The synthesis under constraint allowed the apparent density to significantly increase up to a value typically close to 200 kg/m^3 (against 20 kg/m^3 in absence of confinement). SEM and TEM analysis confirm the similarity in the morphology and microstructure of these CNT than those obtained in an open reactor. In the former case, the CNT were denser due to the constraint generated in the reactor, in line with the high density observed. The shape can be varied at will by changing the size and shape of the reactor.

The TEM analysis of the MWNT showed the complete absence of any amorphous carbons in the sample, confirming the high selectivity of the synthesis. The diameter of MWNT typically ranged between 30 and 50 nm, with length reaching several micrometres. A relatively high number of structural defects were observed along the tube walls. These defects were attributed to the low synthesis temperature of the CVD method that could lead to the insertion of pentagons (positive curvature) or heptagons (negative curvature) inside the hexagonal structure of the graphene layers.

The great advantage of the present synthesis compared to those previously reported in the literature[113,114] is based on the simplicity of the process that allows

Figure 23.10 (Left) Self-supported MWNTsynthesised according to the confinement synthesis. (Right) Self supported MWNTcompared to the same sample synthesised without constraint.

easy scaling-up with minimum cost investment and avoids any addition of foreign agents during the assembly.

The density of as-synthesised self-supported MWNT can be modified by wetting them with water or an organic solvent, followed by a slow evaporation of the solvent. Surface tension effects during the liquid evaporation shrink the cylinder into a dense cylinder with smaller dimensions. Furthermore, the piece of self-supported MWNT is strong enough to be easily manipulated after synthesis and cut to suit the desired shape for downstream applications. In the present work, the synthesis of pure MWNT self-supported in a macroscopic shape (cylinder, disk, cube and so on) with centimetre sizes using a CVD synthesis under macroscopic constraints have been performed and one example of a confined reactor can be seen in Fig. 23.11.

4.3.2. Aligned CNT patterns

Zhu *et al.*[115] have reported the synthesis of a 20-cm-long thread of CNT using a pyrolysis process of a mixture of hexane, ferrocene and thiophene. The group of Windle has reported the direct spinning of fibres made of CNT from hot zone chemical vapour deposition.[116] Fibres made of SWNT with microscopic diameters were successfully obtained by Smalley and co-workers[117] using SWNT dispersed in a superacid, followed by an extrusion and coagulation. From this technique, a continuous spool of microscopic SWNT fibres with well-aligned orientation was obtained with a relatively simple process.

Hata *et al.*[118] have reported the synthesis of aligned SWNT of millimetre sized-bundles on a silicon wafer, using the catalytic decomposition of ethylene in the presence of a controlled amount of steam. However, as far as the assembly of CNT is of concern, almost no process dealing with the one-step synthesis of CNT with a 3D macroscopic shape of a centimetre size or more has so far been described. Recently,

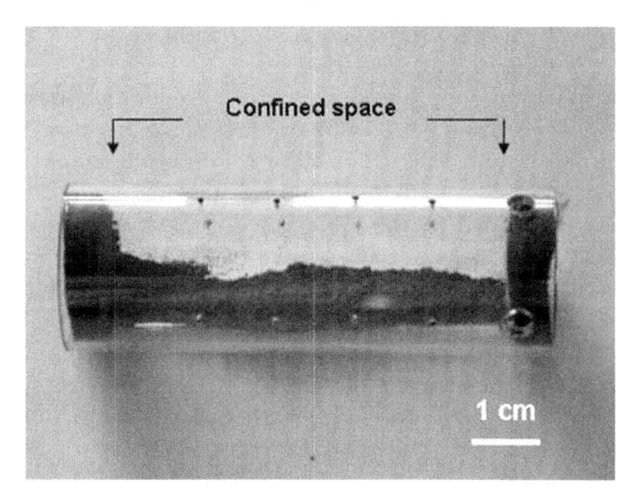

Figure 23.11 Example of a confined reactor used in the synthesis of self-supporting MWNT. The macroscopic shapes can be tailored at will.

the group of Ajayan has pointed out the possibility of making micro- and macroscopic arrangement of CNT, that is 3D brushes consisting of growing CNT on a Si rod or aligned patterns of CNT embedded in a flexible polymer matrix.[3]

4.4. Applications

4.4.1. Catalytic decomposition of hydrazine

Hydrazine decomposition catalysts are commonly employed in microthrusters used for satellite altitude control and path corrections. The usual catalyst is iridium metal supported on stabilised γ-alumina (shell catalyst, 30–36 wt.% Ir/Al_2O_3). The hydrazine decomposition pathway can be described as follows: first, it decomposes into ammonia and nitrogen, followed by ammonia conversion into hydrogen and nitrogen. The former reaction is a structure insensitive reaction, independent of the particle size, while the latter is a structure sensitive reaction that depends on the average size or diameter of dispersed metal. Therefore, the characteristics of a hydrazine decomposition catalyst depend mainly on its activity in ammonia decomposition.

The iridium catalyst was loaded on the support via incipient wetness impregnation using a solution of water and ethanol (80/20 volume) containing a nitrate salt of iridium. The iridium loading, expressed in terms of metal, was set to be 30 wt.%.

The decomposition of hydrazine was carried out in a 2 N microthruster (inner diameter of 6.7 mm, length of 20 mm) designed for the testing of Ir/Al_2O_3 catalysts. The microthruster was installed in a vacuum chamber kept under 0.1 mbar pressure and attached to a thrust balance in order to reproduce as closely as possible the actual operating conditions. The hydrazine injection pressures during the tests were kept at 22 and 5.5 bar, corresponding to the initial and final operating pressure values during the satellite life, respectively. For the bench-scale test, that is 2 N microthruster (Fig. 23.12), the weight of the commercial catalyst was 1.144 g (332 mg of Ir) whereas the weight of the CNF-based catalyst was only 0.145 g (44 mg of Ir).

Altitude control with monopropellant hydrazine engines requires operation over a wide range of duty cycles and pulse widths. Figure 23.13 compares the multi-pulse average thrust of two tested catalysts for 100 runs under different monopropellant injection pressures and pulse-on/pulse-off electrovalve modes. These results show a higher performance obtained on the CNF-based catalyst than on the commercial catalyst. The thrust pulses delivered during the hydrazine decomposition over the Ir/CNF composite were invariably higher than those observed on the commercial Ir/Al_2O_3 catalyst, regardless of the difference between the iridium loadings in the propulsion chamber. The performance of the Ir/CNF catalyst was attributed to the high accessibility to the catalytic sites by the reactant due to the high external surface area of the support and the absence of any closed porosity inside the catalyst. On the alumina-based catalyst, due to the presence of a micro- and mesoporous network, only one fraction of hydrazine reached the active sites, thus leading to flow maldistribution. In addition, the high thermal conductivity of the CNF also allows a rapid homogenisation of the temperature in the entire catalyst bed in contrast to the

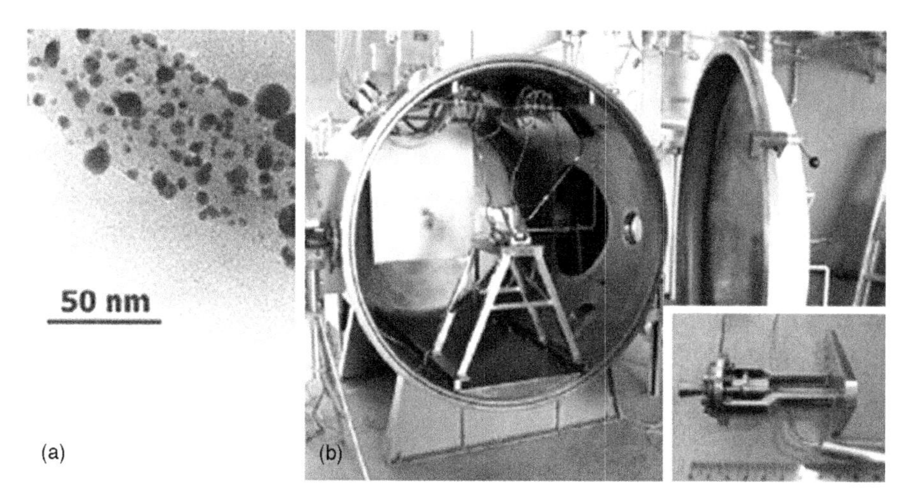

Figure 23.12 (A) TEM micrograph of the Ir/CNF catalyst after aging at 500 °C in flowing hydrogen saturated with steam for 24 h. The average particle size of the iridium measured by TEM was around 7 nm. (B) Bench-scale micropropulsion setup experiment allowing the catalytic experiments under vacuum that reproduces the real working conditions (InpE, Brazil). Inset: Close view of the 2 N microthruster.

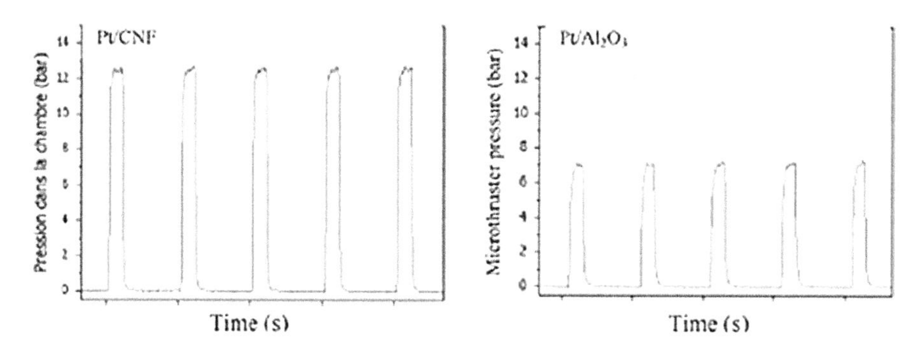

Figure 23.13 Pressure generated by the catalytic decomposition of the liquid hydrazine in the vacuum chamber of the propulsion setup experiment over the Ir/CNF composite and the Ir/Al_2O_3 catalysts.

alumina-based catalyst. The pressure response time is defined as the time measured from the propellant electrovalve activation signal to up to 90% of the steady state reached in the combustion chamber. The time was 46 ms for the Ir/CNF catalyst and 62 ms for the commercial catalyst at 5.5 bar, and 41 ms for the Ir/CNF catalyst and 43 ms for the commercial catalyst at 22 bar. The CNF-based catalyst rises above the commercial catalyst for its spontaneous restarting ability. The time response of the commercial catalyst increases when the propellant injection pressure decreases because the pressure is not sufficient enough to push the propellant inside the alumina porous network.

4.4.2. Selective oxidation of H_2S into elemental sulphur

Because of its high toxicity, H_2S must be removed as much as possible before releasing off-gas from natural gas plants and petroleum refineries into the atmosphere. The catalytic selective oxidation of traces of H_2S in air has been extensively studied and details concerning this process were summarised in the work of Pham-Huu.[119] CNF composites with defined macroscopic assembly were used as supports for the catalytic selective oxidation of H_2S into elemental S at low temperature and relatively high space velocity. The catalyst was NiS_2 with a metal loading of 5 wt.%. Details of the catalyst preparation and characterisation can be found in previous studies.[120,121] The catalytic performance of the NiS_2/CNF catalyst was also compared with NiS_2/SiC catalyst in grain form. The catalytic desulphurisation reaction conditions were as follows: H_2S, 2500 ppm; O_2, 5000 ppm; H_2O, 30 vol.% and balance helium, which are typically the industrial working concentrations in a discontinuous mode. At the reaction temperature of 60 °C, the steam condensed on the reactor walls and the reaction was carried out in a trickle-bed mode. The results obtained with the different catalysts as a function of time on stream are presented in Fig. 23.14. On the NiS_2/CNF composite, the H_2S conversion was complete during the first 20 h on stream and then a slight deactivation was observed that could be attributed to active site encapsulation by solid sulphur. It should be noted that H_2S conversion still remained close to 50% even after 80 h of reaction and more than 150 wt.% of solid sulphur deposition on the catalyst. On the other catalyst, H_2S conversion never reached 100%, while deactivation occurred rapidly after about 20 h on stream and after 50 h the H_2S conversion almost vanished. The slow increase of the desulphurisation activity was explained by the progressive steam

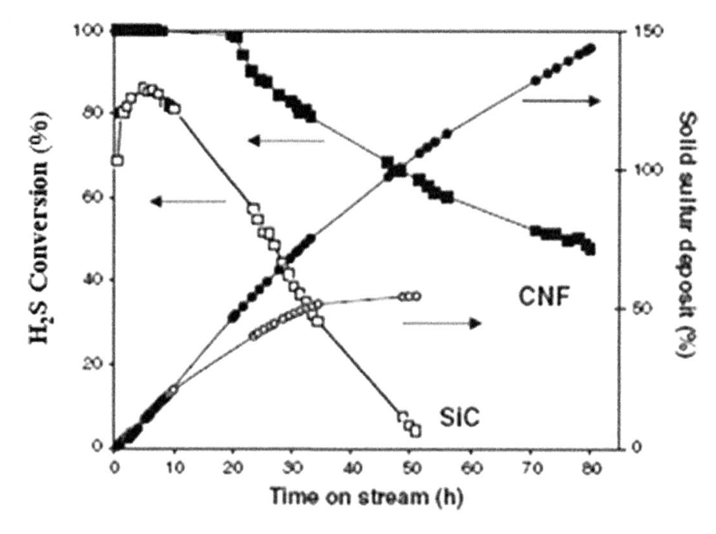

Figure 23.14 Desulphurisation activity and solid sulphur deposition on the NiS_2/CNF (CNF) composite and NiS_2/SiC grains (SiC) at 60 °C. *Filled square:* H_2S conversion on NiS_2/CNF catalyst, *Filled circle:* sulphur deposition (wt.%) on the NiS_2/CNF catalyst, *Open square:* H_2S conversion on NiS_2/SiC catalyst, *Open circle:* sulphur deposition (wt.%) on the NiS_2/SiC catalyst.

condensation inside the porosity of the support that slowly evacuates coated solid sulphur and favours access of the reactants to the active sites. The high desulphurisation activity observed on the NiS_2/CNF catalyst was attributed to the high external surface of the support and ink-bottled pores that strongly favour the reactants accessibility to the active centres and high rate of products escaping.

According to SEM investigations, the morphology of the catalyst after the desulphurisation test was investigated and solid sulphur was formed in a discrete manner on the surface of the catalyst, while a large area of the catalyst surface was still accessible for the reactants, thus explaining the high activity of the catalyst despite the high amount of sulphur coated on it.

5. Conclusions

In conclusion, we have shown that carbon nanomaterials (nanotubes and nanofibres) with different macroscopic assembly can be prepared in large quantities by two very simple and reproducible methods: supported on a macroscopic host for CNF and direct self-support for CNT. The catalytic CVD processes allow the formation of CNF and CNT with high selectivity and few impurities and are more suitable to large-scale processing of these 1D materials.

The carbon surface can be decorated with different types of standard catalysts such as metal, oxide or even homogeneous catalysts. The high external surface area and the absence of ink-bottled pores inside these materials significantly improve the reactants and products diffusion that contribute to an increase in the catalytic activity and selectivity. The existence of a strong interaction between the coated active phase and the CNF surface could also lead to a peculiar electronic alteration of the supported active phase and, finally, to unexpected catalytic behaviour.

The different catalytic applications presented above confirm the potential of the carbon nanomaterials to be used as catalyst supports either in gas-phase or in liquid-phase media. We believe that our understanding concerning the role of CNF surface on the catalytic performance represents the most simple but reasonable starting point, although future work is undoubtedly required to quantify and allows a better understanding of the advantages that these carbon nanomaterials can bring to the catalysis field. Indeed, after carefully examination of numerous concluding remarks from the literature, it is apparent that there are still problems that should be solved before introducing these 1D materials into real-life applications. For instance, the confinement effect that results from the empty nanoscopic channel of CNT on the catalytic activity still needs to be elucidated. It is also worth noting that apart from the direct use of these nanomaterials as catalyst supports other useful applications of these materials, especially self-supported CNT, and may also be expected in the synthesis of several 1D materials (metal, oxide or ceramic) which could find some new applications in catalysis but also in several areas of the emerging nanotechnology field. In view of the large body of scientific and industrial reports in the literature on this exciting field, the references cited in this chapter are not meant to be exhaustive but are merely representative of the subject up to date.

ACKNOWLEDGEMENTS

The work presented here is issued from fruitful collaborations that involved a large number of collaborators both inside and outside the laboratory. The authors address their special acknowledgements to Dr. Ricardo Vieira (InpQ, Brazil). The authors thank the IPCMS (UMR 7504 du CNRS) for SEM and TEM facilities.

REFERENCES

[1] Nolan, P. E., Lynch, D. C., Cutler, A. H., *J. Phys. Chem. B* **1998,** *102,* 4165–4175.
[2] Iijima, S., *Nature* **1991,** *354,* 56.
[3] Ajayan, P. M., *Chem. Rev.* **1999,** *99,* 1797–1800.
[4] Dresselhaus, M. S., Dresselhaus, G., Eklund, P. C., *Science of Fullerenes and Carbon Nanotubes*, Academic Press, London, 1996.
[5] Ebbesen, T. W., *Carbon Nanotubes:Preparation and Properties*, CRC Press, Boca Raton, 1997.
[6] Harris, P. J. F., *Carbon Nanotubes and Related Structures, New Materials for the 21st Century*, Cambridge University Press, Cambridge, 2000.
[7] Ochoa-Fernandez, E., Chen, D., Yu, Z., Totdal, B., Ronning, M., Holmen, A., *Catal. Today* **2005,** *102–103,* 45–49.
[8] Yu, Z., Chen, D., Totdal, B., Holmen, A., *J. Phys. Chem. B* **2005,** *109,* 6096–6102.
[9] Yoon, S. H., Lim, S., Hong, S. H., Qiao, W., Duayne-Whitehurst, D., Mochida, I., An, B., Yokoyama, K., *Carbon* **2005,** *43,* 1828–1838.
[10] Jong, K. P. D., Geus, J. W., *Catal. Rev. Sci. Eng.* **2000,** *42,* 481–510.
[11] Ebbesen, T. W., Ajayan, P. M., *Nature* **1992,** *358,* 220–222.
[12] Cao, A., Zhang, X. F., Xu, C. L., Liang, J., Wu, D. H., Wei, B. Q., *J. Mater. Res.* **2001,** *16,* 3107.
[13] Ebbesen, T. W., Ajayan, P. M., Hiura, H., Tanigaki, K., *Nature* **1994,** *367,* 519–520.
[14] Pham-Huu, C., Keller, N., Roddatis, V. V., Mestl, G., Schlögl, R., Ledoux, M. J., *Phys. Chem. Chem. Phys.* **2002,** *4,* 514–521.
[15] Toebes, M. L., Prinsloo, F. F., Bitter, J. H., Dillen, A. J. V., Jong, K. P. D., *J. Catal.* **2003,** *214,* 78–87.
[16] Emmenegger, C., Bonard, J. M., Mauron, P., Sudan, P., Lepora, A., Grobety, B., Zuttler, A., Schapbach, L., *Carbon* **2003,** *41,* 539–547.
[17] Pham-Huu, C., Vieira, R., Louis, B., Carvalho, A., Amadou, J., Dintzer, T., Ledoux, M. J., *J. Catal.* **2006,** *240,* 194–202.
[18] Snoeck, J. W., Froment, G. F., Fowles, M., *J. Catal.* **1997,** *169,* 240–249.
[19] Zaikovskii, V. I., Chesnokov, V. V., Buganov, R. A., *Kinet. Catal.* **1999,** *40,* 612.
[20] Rinzler, A. G., Liu, J., Nikolaev, P., Huffman, C. B., Rodriguez-Macias, F. J., Boul, P. J., Lu, A. H., Heymann, D., Colbert, D. T., Lee, R. S., Fischer, J. E., Rao, A. M., Eklund, P. C., Smalley, R. E., *Appl. Phys. A* **1998,** *67,* 29–37.
[21] Louis, B., Gulino, G., Vieira, R., Amadou, J., Dintzer, T., Galvagno, S., Centi, G., Ledoux, M. J., Pham-Huu, C., *Catal. Today* **2005,** *102–103,* 23–28.
[22] Ermakova, M. A., Ermakov, D. Y., Chuvilin, A. L., Kuvshinov, G. G., *J. Catal.* **2001,** *201,* 183–197.
[23] Alstrup, I., *J. Catal.* **1988,** *109,* 241–251.
[24] Baker, R. T. K., Harris, P. S., Thomas, R. B., Waite, R. J., *J. Catal.* **1973,** *30,* 86–95.
[25] Planeix, J. M., Coustel, N., Coq, B., Brotons, V., Kumbhar, P. S., Dutartre, R., Geneste, P., Bernier, P., Ajayan, P. M., *J. Am. Chem. Soc.* **1994,** *116,* 7935–7936.
[26] Rodriguez, N. M., Kim, M. S., Baker, R. T. K., *J. Phys. Chem.* **1994,** *98,* 108.
[27] Mojet, B. L., Hoogenraad, M. S., Dillen, A. J. V., Geus, J. W., Koninsberger, D. C., *J. Chem. Soc. Faraday Trans.* **1997,** *93,* 4371.

[28] Park, C., Baker, R. T. K., *J. Phys. Chem. B* **1998,** *102,* 5168–5177.
[29] Park, C., Baker, R. T. K., *J. Catal.* **1998,** *179,* 361–374.
[30] Park, C., Keane, M. A., *J. Catal.* **2003,** *221,* 386–399.
[31] Wal, R. L. V. D., Hall, L. J., *Carbon* **2003,** *41,* 659–672.
[32] Rainer, D. R., Goodman, D. W., *J. Mol. Catal. A* **1998,** *131,* 259–283.
[33] Ting, J. M., Liu, R. M., *Carbon* **2003,** *41,* 601–603.
[34] Chen, D., Christensen, K. O., Ochoa-Fernandez, E., Yu, Z., Totdal, B., Latorre, N., Monzon, A., Holmen, A., *J. Catal.* **2005,** *229,* 82–96.
[35] Dai, H., *Acc. Chem. Res.* **2002,** *35,* 1035–1044.
[36] Laurent, C., Flahaut, E., Peigney, A., Rousset, A., *New J. Chem.* **1998,** *22,* 1229–1238.
[37] Perez-Cabero, M., Romeo, E., Royo, C., Monzon, A., Guerrero-Ruiz, A., Rodriguez-Ramos, I., *J. Catal.* **2004,** *224,* 197–205.
[38] Andrews, R., Jacques, D., Qian, D., Rantell, T., *Acc. Chem. Res.* **2002,** *35,* 1008–1017.
[39] Sun, Y. P., Fu, K., Lin, Y., Huang, W., *Acc. Chem. Res.* **2002,** *35,* 1096–1104.
[40] Serp, P., Corrias, M., Kalck, P., *Appl. Catal. A* **2003,** *253,* 337–358.
[41] Vieira, R., Bernhardt, P., Ledoux, M. J., Pham-Huu, C., *Catal. Lett.* **2005,** *99,* 177–180.
[42] Bethune, D. S., Klang, C. H., Vries, M. S. D., Gorman, J., *Nature* **1993,** *363,* 605–607.
[43] Ouyang, M., Huang, J. L., Lieber, C. M., *Acc. Chem. Res.* **2002,** *35,* 1018–1025.
[44] Lukic, B., Seo, J. W., Bacsa, R. R., Delpeux, S., Beguin, F., Bister, G., Fonseca, A., Nagy, J. B., Kis, A., Jeney, S., Kulik, A. J., Forro, L., *Nano Lett.* **2005,** *5,* 2074–2077.
[45] Treacy, M. M. J., Ebbesen, T. W., Gibson, J. M., *Nature* **1996,** *381,* 678–681.
[46] Forro, L., *Nature* **2006,** *441,* 414–415.
[47] Farrauto, R. J., Bartholomew, C. H., *Fundamentals of Industrial Catalytic Processes,* Blackie Academic & Professional, London, 1997.
[48] Nhut, J. M., Pesant, L., Tessonnier, J. P., Wine, G., Guille, J., Pham-Huu, C., Ledoux, M. J., *Appl. Catal. A* **2003,** *254,* 345–363.
[49] Pham-Huu, C., Ledoux, M. J., *Topic. Catal.* **2006,** *40,* 49–63.
[50] Lau, K. T., Hui, D., *Composites* **2002,** *B33,* 263.
[51] Zhou, O., Shimoda, H., Gao, B., Oh, S., Fleming, L., Yue, G., *Acc. Chem. Res.* **2002,** *35,* 1045–1053.
[52] Coq, B., Planeix, J. M., Brotons, V., *Appl. Catal. A* **1998,** *173,* 175–183.
[53] Ledoux, M. J., Pham-Huu, C., *Catal. Today* **2005,** *102–103,* 2–14.
[54] Park, C., Rodriguez, N. M., Kim, M. S., Baker, R. T. K., *J. Catal.* **1997,** *169,* 212–227.
[55] Wal, R. L. V. D., Ticich, T. M., Curtis, V. E., *Carbon* **2001,** *39,* 2277–2289.
[56] Dupuis, A. C., *Prog. Mater. Sci.* **2005,** *50,* 929.
[57] Tsang, S. C., Chen, Y. K., Harris, P. J. F., Green, M. L. H., *Nature* **1994,** *372,* 159–162.
[58] Toebes, M. L., Zhang, Y., Hajek, J., Nijhuis, T. A., Bitter, J. H., Dillen, A. J. V., Murzin, D. Y., Koningsberger, D. C., Jong, K. P. D., *J. Catal.* **2004,** *226,* 215–225.
[59] Bezemer, G. L., Bitter, J. H., Kuipers, H. P. C. E., Oosterbeek, H., Holewijn, J. E., Xu, X., Kapteijn, F., Dillen, A. J. V., Jong, K. P. D., *J. Am. Chem. Soc.* **2006,** *128,* 3956–3964.
[60] Bezemer, G. L., Radstake, P. B., Koot, V., Dillen, A. J. V., Geus, J. W., Jong, K. P. D., *J. Catal.* **2006,** *237,* 291–302.
[61] Mestl, G., Maksimova, N. I., Keller, N., Roddatis, V. V., Schlögl, R., *Angew. Chem. Int. Ed.* **2001,** *113,* 2122.
[62] Maciá-Agulló, J. A., Cazorla-Amorós, D., Linares-Solano, A., Wild, U., Su, D. S., Schlögl, R., *Catal. Today* **2005,** *102–103,* 248–253.
[63] Bitter, J. H., Seshan, K., Lercher, J. A., *J. Catal.* **1997,** *171,* 279–286.
[64] Haruta, M., Tsubota, S., Kobayashi, T., Genet, M. J., Delmon, B., *J. Catal.* **1993,** *144,* 175–192.
[65] Koningsberger, D. C., Oudenhuijzen, M. K., Graaf, J. D., Bokhoven, J. A. V., Ramaker, D. E., *J. Catal.* **2003,** *216,* 178–191.
[66] Ramaker, D. E., Graaf, J. D., Veen, J. A. R. V., Koningsberger, D. C., *J. Catal.* **2001,** *203,* 7–17.
[67] Gogotsi, Y., Naguib, N., Libera, J. A., *Chem. Phys. Lett.* **2002,** *365,* 354.
[68] Libera, J., Gogotsi, Y., *Carbon* **2001,** *39,* 1307–1318.

[69] Kuzmani, H., Winther, J., Burger, B., *Synth. Met.* **1997,** *85,* 1173.
[70] Louis, B., Tobias, G., Ward, M., Green, M. L. H., in preparation. 2007.
[71] Xu, C., Sloan, J., Brown, G., Bailey, S., Williams, V. C., Friedricks, S., Coleman, K. S., Flahaut, E., Hutchinson, J. L., Dunin-Borkowski, R. E., Green, M. L. H., *Chem. Commun.* **2000,** 2427.
[72] Ajayan, P. M., Stephan, O., Redlich, P., Colliex, C., *Nature* **1995,** *375,* 564–567.
[73] Tessonnier, J. P., Estournes, G. W. C., Leuvrey, C., Ledoux, M. J., Pham-Huu, C., *Catal. Today* **2005,** *102–103,* 29–33.
[74] Lacroix, M., Louis, B., Pham-Huu, C., Ledoux, M. J., *Stud. Surf. Sci. Catal.* **2005,** *158B,* 169.
[75] Ebbesen, T. W., *Acc. Chem. Res.* **1998,** *31,* 558–566.
[76] Kuznetsova, A., Popova, I., Yates, J. T., Bronikowski, M. J., Huffman, C. D., Liu, J., Smalley, R. E., Hwu, H. H., Chen, J. G., *J. Am. Chem. Soc.* **2001,** *123,* 10699–10704.
[77] Ros, T. G., Dillen, A. G. V., Geus, J. W., Koningsberger, D. C., *Chem. Eur. J.* **2002,** *5,* 1151.
[78] Satishkumar, B. C., Govindaraj, A., Mofokeng, G., Subbanna, G. N., Rao, C. N. R., *J. Phys. B* **1996,** *29,* 4925.
[79] Ye, H., Naguib, N., Gogotsi, Y., Yazicioglu, A. G., Megaridis, C. M., *Nanotechnology* **2004,** *15,* 232.
[80] Ugarte, D., Chatelain, A., de Herr, W. A., *Science* **1996,** *274,* 1897.
[81] Ersen, O., Werckmann, J., Houlle, M., Ledoux, M. J., Pham-Huu, C., *Nano Lett.* **2007,** *7,* 1898–1907.
[82] Janssen, A. H., Yang, C. M., Wang, Y., Schuth, F., Koster, A. J., de Jong, K. P., *J. Phys. Chem. B* **2003,** *107,* 10552–10556.
[83] Pham-Huu, C., Keller, N., Charbonniere, L. J., Ziessel, R., Ledoux, M. J., *Chem. Commun.* **2000,** *19,* 1871.
[84] Semagina, N., Renken, A., Kiwi-Minsker, L., *Chem. Eng. Sci.* **2007,** *62,* 5344.
[85] Baerns, M., Hofmann, H., Renken, A., *Chemische Reaktionstechnik,* Georg Thieme Verlag Stuttgart, New York, 1999.
[86] Su, D. S., Maksimova, N., Delgado, J. J., Keller, N., Mestl, G., Ledoux, M. J., Schlögl, R., *Catal. Today* **2005,** *102–103,* 110–114.
[87] Su, D. S., Müller, J.-O., Jentoft, R. E., Rothe, D., Jacob, E., Schlogl, R., *Top. Catal.* **2004,** *30*–31, 241–245.
[88] Pereira, M. F. R., Figueiredo, J. L., Órfão, J. J. M., Kalck, P., Kihn, Y., *Carbon* **2004,** *42,* 2807–2813.
[89] Ulbricht, H., Moos, G., Hertel, T., *Phys. Rev. B* **2002,** *66,* 075404.
[90] Zhao, J., Lu, J. P., *Appl. Phys. Lett.* **2003,** *82,* 3746–3748.
[91] Ago, H., Kuglet, T., Cacialli, F., Salaneck, W. R., Schaffer, M. S. P., Windle, A. H., Friend, R. H., *J. Phys. Chem. B* **1999,** *103,* 8116–8121.
[92] Nakamichi, N., Hirotoshi, K., Masahiko, H., *J. Org. Chem.* **2003,** *68,* 8272–8273.
[93] Liang, C., Xia, W., Soltani-Ahmadi, H., Schlüter, O., Fischer, R. A., Muhler, M., *Chem. Commun.* **2005,** 282–283.
[94] Zhang, Y., Zhang, H. B., Lin, G. D., Chen, P., Yuan, Y. Z., Tsai, K. R., *Appl. Catal. A* **1999,** *187,* 213–224.
[95] Banerjee, S., Wong, S. S., *J. Am. Chem. Soc.* **2002,** *124,* 8940–8948.
[96] Baleizao, C., Gigante, B., Garcia, H., Corma, A., *Tetrahedron* **2004,** *60,* 10461–10468.
[97] Giordano, R., Serp, P., Kalk, P., Kihn, Y., Schreiber, J., Marhic, C., Duvail, J. L., *Eur. J. Inorg. Chem.* **2003,** *4,* 610–617.
[98] Jiang, K., Eitan, A., Schadler, L. S., Ajayan, P. M., Siegel, R. W., Grobert, N., Mayne, M., Reyes-Reyes, M., Terrones, H., Terrones, M., *Nano Lett.* **2003,** *3,* 275–277.
[99] Ma, Z., Zaera, F., *Surf. Sci. Rep.* **2006,** *61,* 229–281.
[100] Salzmann, C. G., Llewellyn, S. A., Tobias, G., Ward, M. A. H., Huh, Y., Green, M. L. H., *Adv. Mater.* **2007,** *19,* 883–887.
[101] Huang, W., Zhang, X. B., Tu, J., Kong, F., Ning, Y., Xu, J., Tendeloo, G. V., *Phys. Chem. Chem. Phys.* **2002,** *4,* 5325–5329.
[102] Tribolet, P., Kiwi-Minsker, L., *Catal. Today* **2005,** *105,* 337–343.

[103] Ruta, M., Yuranov, I., Dyson, P. J., Laurenczy, G., Kiwi-Minsker, L., *J. Catal.* **2007,** *247,* 269–276.
[104] Cordier, A., Flahaut, E., Viaizzi, C., Laurent, C., Peigney, A., *J. Mater. Chem.* **2005,** *15,* 4041–4050.
[105] Cybulski, A., Moulijn, J. A., *Structured Catalysts and Reactors,* Marcel Dekker, New York, 1998, p. 670.
[106] Lousis, B., Kiwi-Minsker, L., Reuse, P., Renken, A., *Ind. Eng. Chem. Res.* **2001,** *40,* 1454–1459.
[107] Vieira, R., Netto, D. B., Ledoux, M. J., Pham-Huu, C., *Appl. Catal. A* **2005,** *279,* 35–40.
[108] Lee, M. K. V. D., Dillen, A. J. V., Geus, J. W., Jong, K. P. D., Bitter, J. H., *Carbon* **2006,** *44,* 629–637.
[109] Lousis, B., Vieira, R., Carvalho, A., Amadou, J., Ledoux, M. J., Pham-Huu, C., *Top. Catal.* **2007,** *45,* 75.
[110] Amadou, J., Begin, D., Nguyen, P., Tessonnier, J. P., Dintzer, T., Vanhaecke, E., Ledoux, M. J., Pham-Huu, C., *Carbon* **2006,** *44,* 2587–2589.
[111] Cao, A., Veedu, V. P., Li, X., Yao, Z., Ghasemi-Nejhad, M. N., Ajayan, P. M., *Nat. Mater.* **2005,** *4*(7), 540–545.
[112] Jung, Y. J., Kar, S., Talapatra, S., Soldano, C., Viswanathan, G., Li, X., Yao, Z., Ou, F. S., Avadhanula, A., Vajtai, R., Curran, S., Nalamasu, O., Ajayan, P. M., *Nano Lett.* **2006,** *6,* 413–418.
[113] Zhang, M., Fang, S., Zakhidov, A. A., Lee, S. B., Aliev, A. E., Williams, C. D., Atkinson, K. R., Baughman, R. H., *Science* **2005,** *309,* 1215–1219.
[114] Kappenstein, C., Balcon, S., Rossignol, S., Gengembre, E., *Appl. Catal. A* **1999,** *182,* 317–325.
[115] Zhu, H., Jiang, B., Xu, C., Wu, D., *Chem. Commun.* **2002,** 1858–1859.
[116] Li, Y. L., Kinloch, I. A., Windle, A. H., *Science* **2004,** *304,* 276–278.
[117] Ericson, L. M., Fan, H., Peng, H., Davis, V. A., Zhou, W., Sulpizio, J., Wang, Y., Booker, R., Vavro, J., Guthy, C., Parra-Vasquez, A. N. G., Kim, M. J., Ramesh, S., Saini, R. K., Kittrell, C., Lavin, G., Schmidt, H., Adams, W. W., Bilups, W. E., Pasquali, M., Hwang, W. F., Hauge, R. H., Fischer, J. E., Smalley, R. E., *Science* **2004,** *305,* 1447–1450.
[118] Hata, K., Futaba, D. N., Mizuno, K., Namai, T., Yumura, M., Iijima, S., *Science* **2004,** *306,* 1362–1364.
[119] Nhut, J. M., Nguyen, P., Pham-Huu, C., Keller, N., Ledoux, M. J., *Catal. Today* **2004,** *91–92,* 91–97.
[120] Keller, N., Pham-Huu, C., Crouzet, C., Ledoux, M. J., Savin-Poncet, S., Nougayrède, J. B., Bousquet, J. L., *Catal. Today* **1999,** *53,* 535–542.
[121] Ledoux, M. J., Pham-Huu, C., Keller, N., Nougayrède, J. B., Savin-Poncet, S., Bousquet, J. L., *Catal. Today* **2000,** *61,* 157–163.

CHAPTER 24

STRONG BRØNSTED ACIDITY IN ALUMINA-SILICATES: INFLUENCE OF PORE DIMENSION, STEAMING AND ACID SITE DENSITY ON CRACKING OF ALKANES

Jeroen A. van Bokhoven

Contents

Abstract

The intrinsic activity of alkane cracking over zeolites was determined using monomolecular cracking of alkane. Because the rate-limiting step is the protonation of the alkane, the intrinsic activation energy reflects the energy that is required to transfer the proton from zeolite to reactant, which can be regarded as a measure of acid strength. This barrier was independent of alkane chain length and varied only slightly for different zeolites, although the rates differed considerably. A true compensation relation was observed that originated from the compensation between the heat and entropy of adsorption, explaining variations in rate because of variation in synthesis method and post-synthesis treatments such as steaming and poisoning.

Keywords: Acid strength, Alkane cracking, Amorphous silica alumina, Activation energy, Intrinsic activity, Monomolecular cracking, Zeolite

Ordered Porous Solids
DOI: 10.1016/B978-0-444-53189-6.00024-X

1. Introduction

Brønsted acid sites in alumina-silicates are the catalytically active centres in many acid catalysed reactions. These materials are widely applied in the chemical industry with over 120,000 tonnes of zeolites used per year in catalysis. The most important application by far is zeolite Y in catalytic cracking, which accounts for over 90% of the zeolites used. In catalytic cracking, large hydrocarbon molecules are cracked into smaller ones, thus upgrading the bottom of the oil barrel. Acid catalysts are required that have sufficiently large pore openings, so the reactants can interact with the active sites. Solid acid catalysts are the catalyst of choice for many reactions in the oil refinery,[1] such as the above-mentioned catalytic cracking, and isomerisation, alkylation and oligomerisation reactions. Applications in the production of fine chemicals are also routinely reported.[2] The catalytically active sites show high activity and the crystalline porous zeolite structures are responsible for unique selectivity, which can be divided into three types: reactant, product and transition-state selectivity. Zeolites are crystalline structures. Amorphous silica aluminas (ASAs) also show catalytic activity, although they lack any long-range order.

2. Structure of the Active Site

The structure of the catalytically active sites in zeolites is well established and results from protons that balance the negative framework charge, which originates from aluminium atoms replacing silicon atoms. The aluminium–hydroxyl bond length is significantly longer than those of the three silicon–oxygen ones.[3–5] Zeolite structure types vary in pore topology and often Si/Al ratio. These properties affect the catalytic performance, and unique activity and selectivity are observed. The stretching vibration of the bridging hydroxyl groups are directly observed between 3500 and 3630 cm^{-1} in infrared spectroscopy. The exact wave number is a function of zeolite framework type, pore dimension and Si/Al ratio of the framework.[6,7] Although ASAs are active in Brønsted acid catalysed reactions, the stretching vibrations of the acidic hydroxyls are only seldom observed in infrared spectroscopy[8,9] and the nature of the catalytically active sites is debated.[10]

3. Reacting Species

Positively charged carbocations are transition states in acid-catalysed reactions by zeolites. The possible species are shown in Scheme 24.1. They can be carbenium ions, which are tri-coordinated carbon atoms, and carbonium ions, which are five-coordinated. The carbenium ions are sp^2 hybridised. These species are generally transition-state species, although various stable carbenium ions that are stabilised by the zeolite walls have been reported.[11,12] Carbenium ions generally form surface alkoxide species on one of the oxygen atoms surrounding the aluminium.

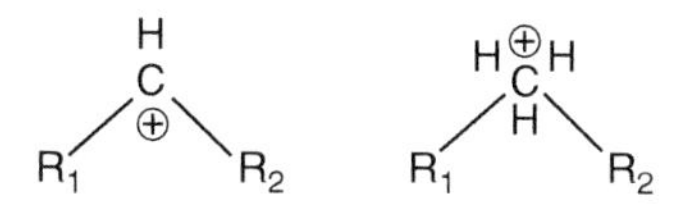

Scheme 24.1 Positively charged carbocations are intermediates in acid-catalysed reactions: carbenium ion (left) and carbonium ion (right).

Carbonium ions are formed by direct protonation of the hydrocarbon, which may then undergo cracking or dehydrogenation.[13] Alkoxides are formed by protonation of an olefin and formation of a bond between a carbon and an oxygen atom from the zeolite framework. Alternatively, it can be formed by hydride abstraction from a hydrocarbon by an alkoxide species. The olefin can be formed by decomposition of the carbonium ion or be present in the feed. Carbenium ion-like species are the transition states of reactions of alkoxide species in reactions such as isomerisation and β-scission.

4. Determining Reaction Constants

Differences in catalytic activity between different zeolite framework types are often assumed to originate from different acidity of the acid sites.[14,15] However, no spectroscopic or temperature-programmed method has produced a universal scale of acidity to which zeolites and ASA can be held. In many zeolites, there may also be high heterogeneity of the acid strength of the sites. In this chapter, a ranking of the zeolite acid strength based on the intrinsic reactivity of a demanding reaction could be determined.

When reaction rates per weight of catalyst of a reaction over various zeolites are compared, one has to carefully observe what factors contribute to the rate. The number of sites that participate in a reaction must be taken into account. Moreover, reactions that have an order in one of the reactants and/or products have additional terms in the kinetic rate expression. For a simple first order reaction A → B, one can write in Eq. (24.1):

$$r = k'[\mathrm{A}]_{\mathrm{gas}} = k\theta_{\mathrm{zeolite}} = k\frac{K[\mathrm{A}]_{\mathrm{gas}}}{1 + K[\mathrm{A}]_{\mathrm{gas}}} \tag{24.1}$$

where $'k$ is the apparent rate constant, k the intrinsic rate constant, $\theta_{\mathrm{zeolite}}$ the coverage of A in the zeolite pores, $[\mathrm{A}]_{\mathrm{gas}}$ the concentration of reactant A in the gas phase and K the enrichment factor or Henry constant of A in the pores of the zeolite. In the first order regime, where the coverage of A on the catalytically active sites is very low, the equation simplifies into Eq. (24.2):

$$r = kK[\mathrm{A}]_{\mathrm{gas}} = k'[\mathrm{A}]_{\mathrm{gas.}} \tag{24.2}$$

The observed rate thus contains the intrinsic reaction constant, which reflects the true activation barrier that must be overcome and an entropic term, and the enrichment factor, K, which presents the sorption of reactant A in the pores of the zeolite. Variation in K affects the observed rates without changing the intrinsic reaction constant. A concentration effect on the rates of reaction in zeolites was shown by Dessau in the Diels-Alder formation of 4-vinylcyclohexene from buta-1,3-diene.[16] This reaction is thermally activated and of second order in buta-1,3-diene. Although neither acidic nor basic catalysts were known to activate this reaction, large pore zeolites in their sodium form displayed enhanced activity. Because zeolites enhanced the local concentration of the reactants, thus increasing K, an enhancement in rate was observed at temperatures as high as 775 K.[17] An additional effect was the proper orientation of the molecules. The role of energy and entropy on the adsorption in zeolites was early on recognised.[18]

Equation (24.3) indicates that the reaction constant consists of a temperature-dependent term, comprising the intrinsic activation energy, E_{act}^{int}, and a temperature-independent term, the pre-exponential term A:

$$k = A_{int} e^{-E_{act}^{int}/RT}. \tag{24.3}$$

The sorption term K also has a temperature-dependent and -independent term, which is shown in Eq. (24.4).

$$\Delta G_{ads} = \Delta H_{ads} - T\Delta S_{ads} = -RT \ln K \tag{24.4}$$

with ΔG_{ads}, ΔH_{ads} and ΔS_{ads} the free Gibbs energy, enthalpy and entropy of adsorption, respectively.

Rearrangement of Eqs. (24.3) and (24.4) gives the true activation barrier from the apparent activation barrier and the heat of adsorption. Equation (24.5) is called Tempkin equation[19]:

$$E_{act}^{int} = E_{act}^{app} - n\sum \Delta H_{ads} \tag{24.5}$$

where n is the order in reactant. Upon adsorption, heat is released and the heat of adsorption is generally negative and the apparent activation barrier is lower than the true activation barrier. For a zero-order reaction, the heat of adsorption terms disappear because the surface is completely covered and the reaction becomes independent of the gas phase reactant concentration. The thermodynamic data of adsorption of alkanes into the pores of zeolites have been established experimentally[20] and theoretically.[21] Although individual reports generally show internally consistent data, the absolute values of the sorption data vary significantly. Because of the importance of adsorption, a paragraph will be spent on the sorption of alkanes in the pores of alumina-silicates.

The influence of sorption via the K term on reaction constant k' and the reaction rate r can be substantial and cannot be neglected. Thus, when comparing different zeolites, the intrinsic reaction rates must be compared. In this chapter, the intrinsic activation barriers and reaction constants for a reaction for which the rate-limiting step is the protonation of the reactant will be described.

Catalytic cracking of saturated hydrocarbons over zeolites and ASA is dominated by two reaction mechanisms. The first is the monomolecular cracking and the second is the bimolecular cracking mechanism. In the first, an alkane is protonated forming a carbonium ion that cracks into an alkane and an olefin or dehydrogenates into an olefin and hydrogen. The rate-limiting step of the cracking reaction is the protonation of the alkane. In the second mechanism, an alkane is activated via hydride transfer of an adsorbed alkoxide species that is followed by secondary reactions such as isomerisation and β-scission. Because the rate-limiting step in the monomolecular cracking of alkanes is the protonation of the alkane, the intrinsic reaction constant of this reaction can be considered as a good candidate for determining an acidity scale of zeolites. The intrinsic activation barrier presents the energy that is required to transfer the proton from the zeolite to the alkane. This latter term will be compared for zeolites of different framework type.

5. Cracking of Alkanes Over ZSM5

All framework aluminium atoms in ZSM5 contribute to acid sites that are catalytically active,[22,23] which make it a good choice to analyse the reactivity of the catalytically active sites. Table 24.1 gives the turnover frequency of monomolecular cracking of alkanes that increased with increasing chain length.[24,25] Figure 24.1a shows the Arrhenius plots of the monomolecular cracking of these alkanes. The slope of the line that represents propane is the steepest, and those of the longer alkanes are progressively less steep. The apparent activation energies were determined from the slopes of these lines and are also given in Table 24.1. Propane showed the highest activation barrier, hexane the lowest. The data in Fig. 24.1a were extrapolated to higher temperature and all lines intersected in a single point. This point is called the isokinetic point. Isokinetic points are evidence of true compensation relations and are often observed in kinetic data of similar reactions over a single catalyst and of several similar catalysts in a single reaction.[26] Compensation relations are also characterised by a linear relation between the apparent activation energy and logarithm of the pre-exponential term. Such a plot is called a constable plot. Figure 24.1b shows that there is a linear relation between the apparent activation energy and the natural logarithm of the pre-exponential term of cracking of alkanes. Haag has also observed such linear relation for the

Table 24.1 Relative turnover frequency, apparent activation energy, heat of adsorption and intrinsic activation energy for the monomolecular cracking of short alkanes over zeolite ZSM5

Reactant[a]	Relative TOF	E_{act}^{app} (kJ/mol)	$-\Delta H_{ads}$ (kJ/mol)	E_{act}^{int} (kJ/mol)
Propane	1	155	43	198
n-Butane	6.5	135	62	197
n-Pentane	34	120	74	194
n-Hexane	114	105	92	197

[a] Data taken from reference [34].

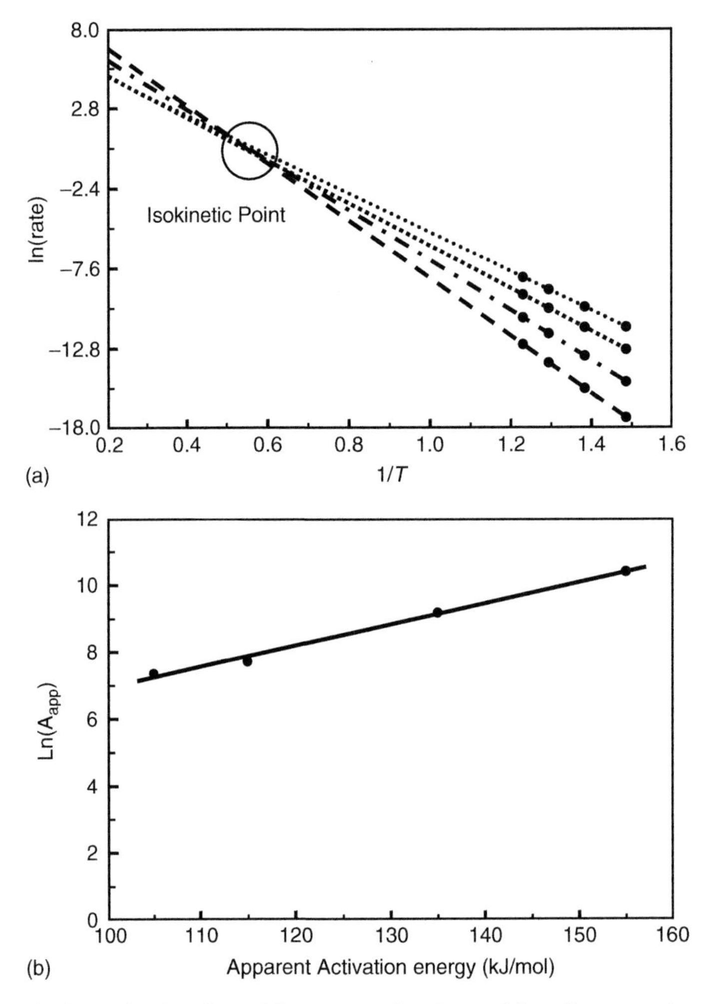

Figure 24.1 (a) The Arrhenius plots of the monomolecular cracking of propane, butane, pentane and hexane over zeolite ZSM5 intersect in a single point, the isokinetic point and (b) linear constable plot of the apparent kinetic parameters of the monomolecular cracking over zeolite ZSM5. The plot has been constructed. Reprinted from reference [41] with permission from Elsevier.

monomolecular cracking of butane till decane over ZSM5.[24] This linear relation and the isokinetic point in Figure 24.1a suggests that the differences in reaction constant have a common origin.

Table 24.1 also gives the true activation barriers determined from the apparent activation energy and the heat of adsorption using Tempkin equation [Eq. (24.5)]. Consistent with the lower slopes in the Arrhenius plot (Figure 24.1a), the apparent activation energy decreased with increasing chain size. Because the heat of adsorption increased in the reverse order, the true activation barriers were similar for all chain lengths.[24,25] The activation barrier to transfer the proton from the Brønsted acid site to the alkane to form the carbocationic transition state is rather independent of the chain length; the differences in the apparent activation energy are virtually completely dominated by the heats of adsorption.

Figure 24.2a shows the Arrhenius plots of experimentally determined monomolecular cracking of alkanes of longer chain length.[21,27,28] A constant trend of decreasing slope with increasing chain length was observed and for the longest chains, even a positive slope was detected, which means that there is a negative activation barrier for this system. The adsorption terms, *K*, for all these alkanes were theoretically calculated as function of temperature.[21] Figure 24.2b shows that linear plots were obtained when plotting *K* against the inverse temperature. Increasing the chain length increased the slope of the lines. Extraction of the intrinsic reaction constant from the lines in Fig. 24.2a and b produced the intrinsic reaction constants as function of inverse temperature, which is given in Fig. 24.2c. All the lines are overlapping, except for that of octane, which slightly deviated. The intrinsic reaction constants were virtually independent of the chain length, although they showed large differences in reaction rates. The large differences that were observed in the Arrhenius plots in Fig. 24.2a have virtually disappeared which shows that the differences were dominated by the Henry constants *K*. The monomolecular cracking of alkanes is determined by the different degree of adsorption in the pores. The determination of the true activation energy using Eq. (24.5) is the most easily accessible parameter to evaluate the influence of sorption on the rate of reaction.

6. Adsorption of Alkanes in Zeolite Pores

The microporous structure and very large surface areas of zeolites makes them excellent adsorbents for gases and liquids because the size of the pores is in the range of small- to medium-sized molecules. As shown above, the sorption of reactants in the zeolite pores affects the experimental rates in non-zero order reactions. Hydrocarbon adsorption into the zeolite pores has been extensively described experimentally and theoretically. The adsorption of normal alkanes into the pores of a zeolite depends strongly on the chain length and the pore dimension. The interaction is dominated by dispersive forces, the van der Waals interaction, in addition to a smaller contribution of a dipole–dipole-induced interaction with the Brønsted acid sites in case of proton-exchanged zeolites. Other cations in the zeolite pores also interact with the adsorbate. Because the van der Waals interaction is dominating, longer alkanes adsorb with a higher heat of adsorption compared to shorter ones. A linearly increasing heat of adsorption with increasing chain length has been repeatedly reported.[29,30] The better the fit of the alkane in the pore, the higher the heat of adsorption is Fig. 24.3 schematically shows an alkane adsorbed in two pores of different size. The confinement of the alkane in the pore of middle size results in the highest heat of adsorption. When the size of the zeolite pore decreases to close to or below the kinetic diameter of the alkane, repulsive forces start dominating, which makes the heat of adsorption less favourable or prevent the alkane from adsorption, which is shown by the smallest zeolite pore in Fig. 24.3. The pore shape will also affect the interaction between alkane and pore wall. Table 24.2 gives the kinetic diameters of various hydrocarbon

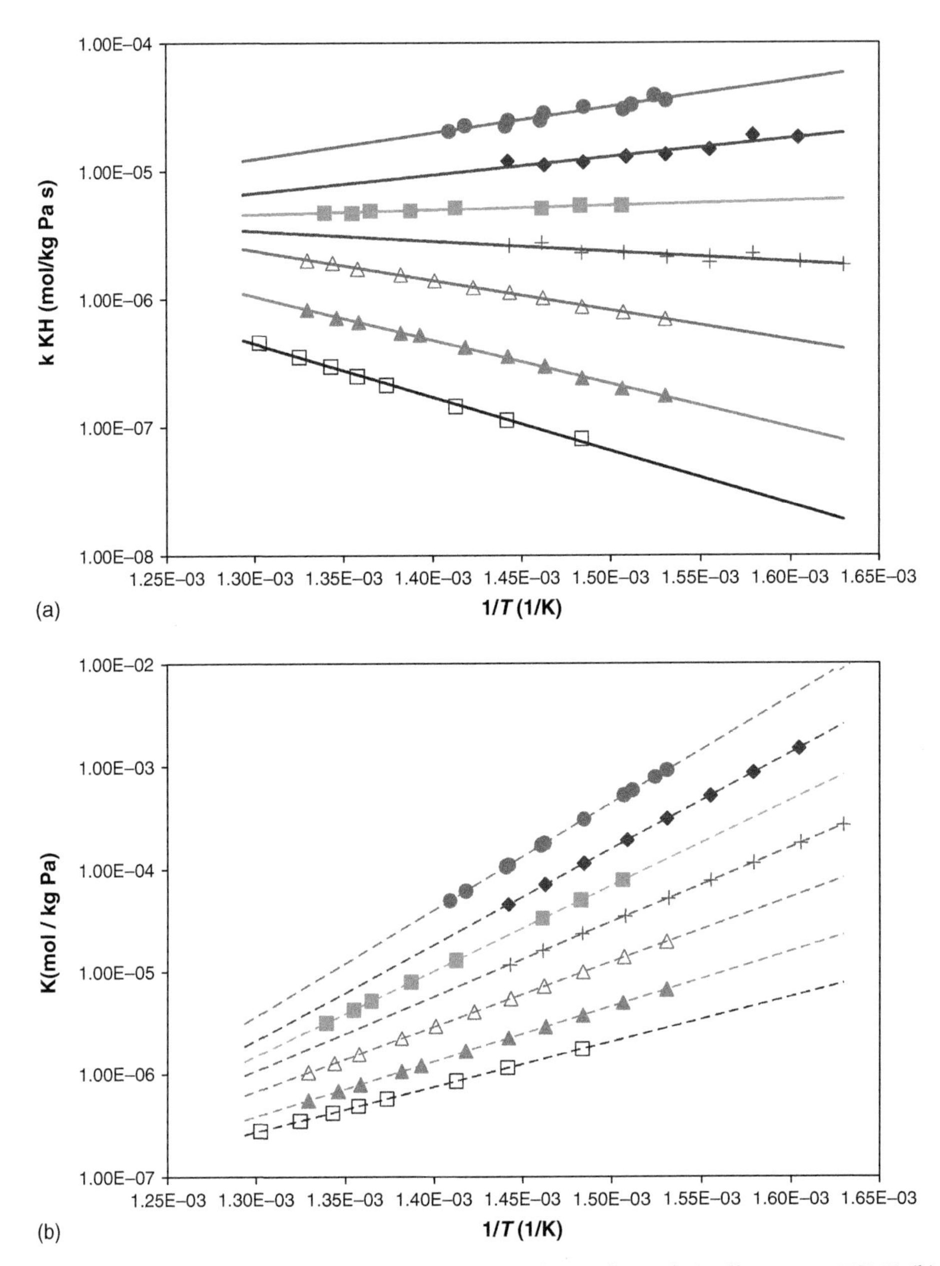

Figure 24.2 (a) Arrhenius plots of monomolecular cracking of long chain alkanes over ZSM5; (b) theoretical adsorption constants, K_{ads} of long chain alkanes in zeolite ZSM5 as function of inverse temperature and

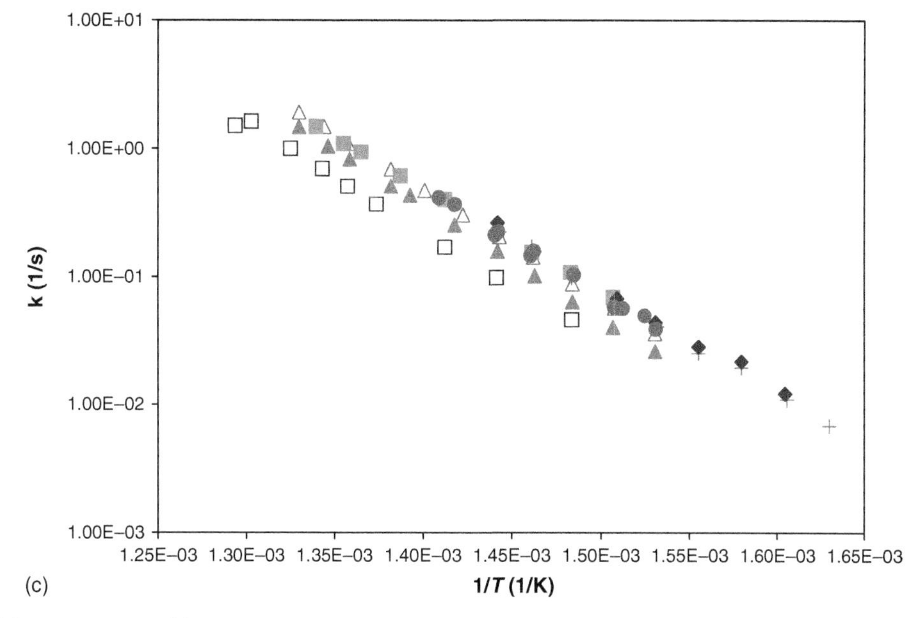

Figure 24.2 cont'd (c) intrinsic rate constants of cracking of alkanes over zeolite ZSM5 from combined plots a and b. •, C_{20}; ◆, C_{18}; ▪, C_{16}; +, C_{14}; △, C_{12}; ▲, C_{10} and □, C_8. Reprinted from reference [21] with permission from Elsevier.

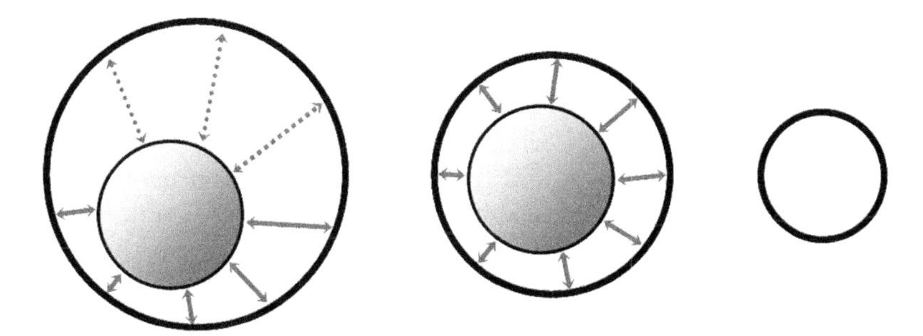

Figure 24.3 Adsorption of an alkane in zeolite pores of different size. The smallest diameter of the alkane is represented by the filled sphere; the black circles represent the openings of a zeolite pore. The heat of adsorption is affected by the fit of the alkane in the pore.

molecules and the diameter of the pore size in often-used zeolites.[31] The kinetic diameters of various molecules are given in Table 24.3. As an example, the heat of adsorption of *n*-butane is also provided. A maximum in the heat of adsorption as function of pore size is observed. The confinement of alkanes in larger pores is worse than that in smaller ones.[32,33] In zeolites with the smallest pores, the repulsive forces start playing an important role. In zeolites with a cage structure, the size of the cages may determine the adsorption strength, more so than the opening of the window. Zeolites

Table 24.2 Zeolite pore size and heat of adsorption of *n*-butane

Zeolite	number of T atoms in a ring	Ring size	$-\Delta H_{ads}$ *n*-butane (kJ/mol)
SOD	6	–	
LTA	8	4.1 × 4.1	
FER	8	3.5 × 4.8	59
	10	4.2 × 5.4	
TON	10	4.6 × 5.7	60
MFI	10	5.1 × 5.5	58
	10	5.3 × 5.6	
MOR	8	2.6 × 5.7	50
	12	6.5 × 7.0	
FAU	12	7.4 × 7.4	39

Table 24.3 Kinetic diameter of some molecules

Molecules	Kinetic diameter (Å)
n-Parraffin	4.3
i-Parraffin	5.3
Benzene/toluene/*p*-xylene	5.7
o-,*m*-xylene	6.3

with narrow windows and a cage structure may show so-called cage effects. The cage effect is responsible for the preferential adsorption of smaller alkanes in the cages because longer alkanes do not fit in the cages and they pay a thermodynamic penalty.[21]

The gain in energy upon adsorption of an alkane into the zeolite pores is accompanied by a loss of entropy and there is a linear relation between the heat and entropy of adsorption of a series of alkanes of increasing length,[29] when the pores of the zeolite are sufficiently large so repulsive forces play no role. This compensation relation is responsible for the linear lines of the Henry coefficients in Fig. 24.2b that virtually intersect through a single point. The heats of adsorption affect the slopes of the lines according to Eq. (24.4). The relation between enthalpy and entropy also affected the apparent kinetic reaction rates in Fig. 24.2a; correction for these terms produced almost constant intrinsic reaction constants.

7. Cracking of Alkanes Over Different Zeolites

Having established the role of sorption on the kinetic parameters of alkane cracking and keeping in mind that the size of the pores affects this sorption, the true kinetic parameters over zeolites of varying structure type must be determined to assess the intrinsic reactivity of zeolites of different framework type. Because the true

Table 24.4 Rate of monomolecular cracking of propane over zeolites with various Si/Al ratios

Zeolite	Rate of cracking $\times 10^{-6}$ (g s bar)[a]	E_{act}^{app} (kJ/mol)
39ZSM5[b]	9.9	146
4.9MOR	32	145
9.9MOR	20	145
16.7MOR	6.9	149
10.5Beta	6.2	157
27Beta	3.2	159
110Beta	0.5	157
2.6Y	0.29	165
3.3Y	1.3	166
3.6Y	5.4	165

[a] Rate at 823 K.
[b] The number represents the Si/Al ratio of the zeolite framework.

activation barrier reflects the energy that is required to transfer the proton to the alkane and the heats of adsorption can be consistently determined experimentally and theoretically, these barriers are determined.

The rates and activation barriers of the monomolecular cracking of propane were determined for zeolites beta, mordenite, Y and ZSM5.[34] Table 24.4 summarises the rates per gram, which shows the dependence of the rates on structure and Si/Al ratio. In zeolites beta and Y, the frameworks were synthesised containing different aluminium content. All aluminium atoms were shown to be in the framework and display a sharp tetrahedral signal in ^{27}Al MAS NMR. Zeolites mordenite and beta showed decreasing rates with increasing Si/Al ratio. The rate of zeolite Y with an Si/Al ratio of 2.6 was very low, and in contrast to the other zeolites, the activity increased 5-fold at an Si/Al ratio of 3.3 and 13-fold at an Si/Al ratio of 3.6. The general trend was the larger the zeolite pore, the less the rate of reaction. The apparent activation energies were inversely related to the rates of reaction when comparing different framework types; ZSM5 and mordenite that showed the highest rates of reaction had the lower apparent activation barrier, and zeolite Y, which was the least active, showed the highest activation barrier. A very important observation was that the activation energy was structure type dependent and varied only slightly with the Si/Al ratio of the zeolite. Mordenite with Si/Al ratios between 4.9 and 16.7 displayed activation barriers between 145 and 149 kJ/mol, beta with Si/Al ratios between 10.5 and 110 had values between 157 and 159 kJ/mol and zeolite Y with Si/Al ratios between 2.6 and 3.6 varied from 165 to 166 kJ/mol.

This independence of the activation barrier on Si/Al ratio illustrated that the energy to transfer the proton from the active site of the zeolite to the alkane is independent of the Si/Al ratio of the framework. Because there is no reason to assume that the heat of adsorption and the true activation energy change equal amounts, but in opposite direction, the acid sites that participate in the reaction show identical behaviour. When assuming that the heat of adsorption depends on structure and not on the Si/Al ratio, the true activation barriers were determined and

Table 24.5 Intrinsic activity of propane monomolecular cracking over zeolite samples

Zeolite	Pore dimension	$-\Delta H_{ads}$ (kJ/mol)	E_{act}^{app} (kJ/mol)	E_{act}^{int} (kJ/mol)
MFI	10 MR 5.1 × 5.5 Å 5.3 × 5.6 Å	−46	146	192
MOR	8 MR 2.6 × 5.7 Å 12 MR 6.5 × 7.0 Å	−41	145–149	186–190
BEA	12 MR 6.6 × 6.7 Å 5.6 × 5.6 Å	−42	157–159	199–201
FAU	12 MR 7.4 × 7.4 Å	−31	165–166	196–197

are given in Table 24.5. The variation in the true activation barriers was much smaller than in the apparent ones, which shows that the different heats of adsorption are responsible for the largest changes in the apparent activation barrier and that sorption plays a very important role in determining the catalytic rates of reaction. The heats of adsorption are responsible for a compensation of the activation barriers.

In ZSM5, all aluminium atoms form an acid site, which equally contributes to catalytic activity. In any of the other zeolites in Table 24.4, such relation is less straightforward.[35] In mordenite, some of the acid sites are positioned in the eight-membered ring side pockets. These sites may contribute differently to activity than those in the 12-membered channels. Moreover, a limited number of Brønsted acid sites in zeolite beta contribute to activity.[35] Nonetheless, there is a clear relationship between aluminium content and reactivity in zeolite mordenite and beta. The situation is very different for zeolite Y. Decreasing the framework aluminium content increased the rate per gram. This observation differs from earlier reports that had suggested that aluminium that was partially dislodged from the framework is essential to enhance the catalytic rate of zeolite Y.[36] The data presented here suggested that zeolite Y that only contains framework aluminium can show enhanced activity in the monomolecular cracking of an alkane. The role of extra framework aluminium will be addressed in a separate paragraph.

Beaumont and Barthomeuf have predicted that the maximum number of sites that show effective Brønsted acidity increases with Si/Al until a ratio of about five, depending on the framework type (six for zeolite Y).[37] They developed a model that suggested that the strength of the Brønsted acid sites depends on the number of next nearest aluminium neighbours. Only aluminium atoms with no next nearest aluminium neighbours exhibit strong acidity. A large amount of theoretical modelling and experimental measurements have since shown that catalytic activity shows a maximum at about 30 framework aluminium atoms per unit cell.[38–40] An Si/Al ratio of 6 corresponds to 27 aluminium atoms. The zeolite Y samples that had higher Si/Al ratios from synthesis followed this trend (Table 24.4). It was proposed that these samples contain higher rates because of a higher number of isolated acid sites in the framework, which was corroborated by ^{29}Si MAS NMR that showed a higher content of silicon atoms with fewer aluminium neighbours. Although the total aluminium content was lower, the number of sites that participated in the reaction was higher. This agreed to the identical activation energy that was observed over all zeolite Y samples (Table 24.4), which shows that the energy that was required to protonate the alkane was independent of the Si/Al ratio. This implies that a large fraction of the acid sites in zeolite Y do not contribute to activity in the monomolecular cracking of alkanes. Poisoning of Brønsted acid sites in zeolite Y has been shown to decrease the catalytic activity much faster than the number of sites, which showed that not all sites in zeolite Y show catalytic activity.

The compensation that was observed between the apparent activation energy and natural logarithm of the pre-exponential term when reacting alkanes of increasing length over zeolite ZSM5 (Figs. 24.1 and 24.2) and the relation between heat of adsorption and apparent activation energy over zeolites of different framework type (Table 24.5) suggest that there is a general compensation relation in the kinetic parameters of alkane cracking over zeolites. Such a correlation has been observed for a series of well-characterised zeolite samples of different structure type, varying Si/Al ratio, different extends of Na poisoning and steaming for the cracking of hexane.[41] The line in the constable plot in Fig. 24.4 strongly supports such a compensation, which finds its origin in the compensation between the heat and entropy of adsorption.[42] Differences in the apparent activation energy between different zeolite samples or alkanes of different length are caused by variation in the heat of adsorption. Likewise, variations in the pre-exponential terms are caused by changes in the entropy of adsorption. Unlike the case for cracking of alkanes of increasing length, which was quantitatively interpreted as shown in Fig. 24.2, no such complete analysis has been put forward for an alkane reacting over zeolites of different structure type, beyond the quantitative interpretation of the activation energy. The compensation relation is expected to break down when the pore sizes and shapes are such that cage effects play a role. Such cage effects are particularly found when small port zeolites, like RHO and LTA, contain cages in which the longer alkanes pay a thermodynamic penalty for adsorbing into the cages, because of their

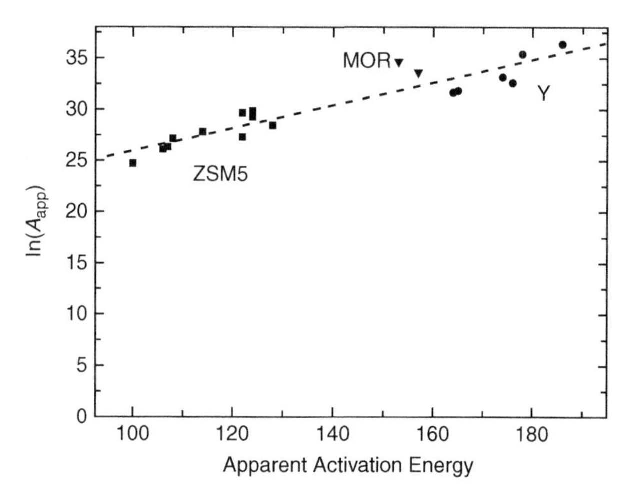

Figure 24.4 Linear constable plot for the monomolecular cracking of hexane over zeolites ZSM5, mordenite and Y. Reprinted from reference [41] with permission from Elsevier.

size relative to the cage size and the narrow windows. Because smaller alkanes easily fit into such cages, there is preferential adsorption and thus conversion of these smaller alkanes.

All of these observations show that the role of sorption must not be overlooked when interpreting zeolite reactivity of reactions that are non-zero order, such as given by Eqs. (24.1)–(24.5). The intrinsic kinetic parameters of cracking over zeolites are much less dependent on alkane length and pore structure and dimension than the adsorption thermodynamic terms. The intrinsic activation barrier varies only mildly with changing pore size.

8. Cracking of Alkanes Over Steamed Zeolites and Amorphous Silica-Alumina

Steaming increases the rate of reaction of zeolites,[43,44] the origin of which is strongly debated. An often-heard explanation is the enhanced acid strength of Brønsted acid sites that is formed by the interaction of extra-framework aluminium with the framework Brønsted acid sites generating unique acid sites.[44,45] Such sites have been observed in the stretching hydroxyl region in infrared spectra: A new vibration band at about 3600 cm^{-1} has often been assigned to such species and extensive experimental data suggested that extra framework aluminium is dispersed over the zeolite crystals that could be responsible for the enhanced acid strength.[46] However, various authors have strongly opposed the existence of sites of enhanced acid strength.[36,41] No correlation between the heats of adsorption of strong bases on the Brønsted acid sites and catalytic activity was found.[47]

While in steamed zeolites, a higher activity is often ascribed to an enhancement of the acid strength, the low activity of ASA compared with zeolites is generally attributed to the presence of acid sites of lower acid strength. Unlike zeolites, the hydroxyl stretching region in infrared spectra of ASA does not show a clear band of a bridging hydroxyl acid site.[48,49] The general interpretation is that the concentration of such sites is very low or that a bridging hydroxyl group is unstable in the framework in the amorphous structure. Acidity has been ascribed to silanol groups that are enhanced in strength by interaction with a nearby aluminium atom.[50] Recently, the adsorption of a water molecule on a strong Lewis acid site has been suggested to show high acid strength, which could be the site that is responsible for high catalytic activity. Negative bands of the symmetric and asymmetric stretching modes of water at 3697 and 3611 cm^{-1} were observed after the adsorption of carbon monoxide. Shifts of the bands similar to that of a zeolitic hydroxyl group were observed, which suggested that the corresponding acid sites show strong acidity.[51]

Table 24.6 summarises the rates of cracking and the apparent activation energies for zeolite ultrastable Y and various ASAs, which were synthesised by co-precipitation of $AlCl_3 \cdot 6H_2O$ and water glass according to a patent of Chevron (US 4988659 [1991]). This resulted in a tetrahedral aluminium coordination only. The rate per gram was about 50 times higher for HUSY; per total aluminium content, this increased to more than 100.

Table 24.6 Rate and apparent activation energy of monomolecular cracking of propane over HUSY and ASAs with various Si/Al ratios

Sample	Rate of cracking $\times 10^{-6}$ (g s bar)	E_{act}^{app} (kJ/mol)
15ASA[a]	0.040	182
7ASA	0.062	186
3ASA	0.047	183
HUSY	2.0	166

[a] The number represents the Si/Al ratio of the zeolite framework.

The rate of HUSY was about seven times more active than zeolite Y with an Si/Al ratio of 2.6. The apparent activation energy of HUSY was 166 kJ/mol, which is virtually identical to that of the zeolite Y samples that did not contain extra framework aluminium. This indicates that the nature of the catalytically active site that is responsible for the protonation of the alkane is independent of the presence of extra framework aluminium. Steaming dealuminates the framework, which enlarges the number of isolated acid sites that are responsible for the cracking activity. This effect of steaming is very similar to that in zeolite Y that was synthesised with a higher Si/Al framework ratio (Table 24.4).[52]

The activation energies of the ASAs varied between 182 and 186 kJ/mol, which is higher than those of any of the zeolites. To determine the true activation barrier, the heat of adsorption of propane was determined.[10] The adsorption of propane on the surface of ASA was much lower than on the surface of zeolites, which already indicated that the sorption strength of ASA is much lower than that of zeolites. Measuring the adsorption at two temperatures enabled the determination of the heat of adsorption, which was about 11 kJ/mol, significantly lower than over zeolite Y and ZSM5, that show 31 and 46 kJ/mol, respectively. Assuming that the heat of adsorption is independent of the Si/Al ratio of ASA, the true activation barrier over ASA varied between 193 and 197 kJ/mol, which is very close to the values found for zeolites. The very low activity of ASA originates from the very low alkane coverage and from a very low number of sites that are active in this demanding reaction. It remains unclear what the relation is between the species that show bands in the stretching hydroxyl region in infrared spectra and the catalytically active structures.[10,50]

9. Concluding Remarks

Figure 24.5 shows the energy scheme of monomolecular cracking over alumina-silicates, which is a graphic representation of Eq. (24.5) and the results presented above. The reactant adsorbs on the surface of the alumina-silicate and the active site protonates the alkane forming the transition state. This transition state then decomposes into methane and ethylene. Upon adsorption, a heat of adsorption equal to ΔH_{ads} will be released. The differences between alkanes of different length and reaction over zeolites of different framework type are dominantly found in the

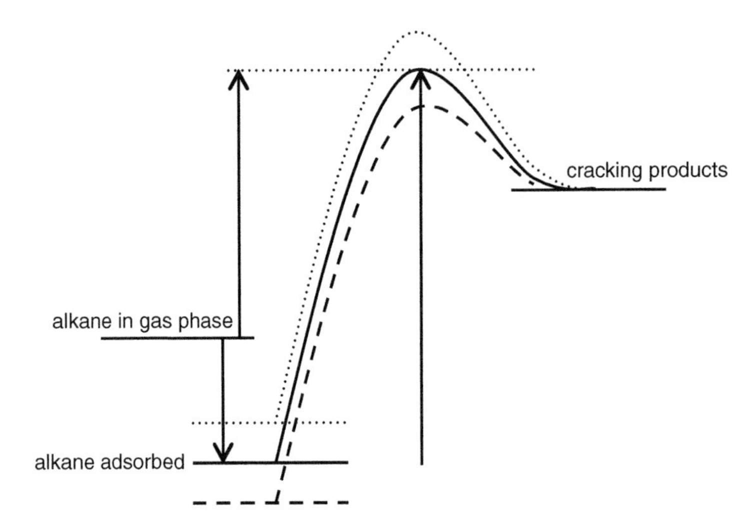

Figure 24.5 Energy scheme of monomolecular cracking of alkanes over zeolites. The intrinsic activation barrier is independent of the alkane length and shows only minor variation with framework structure.

heat of adsorption. The intrinsic activation barrier E_{act}^{int} shows significantly much less variation. The energy to protonate an alkane at high temperature does not depend on the length of the alkane and varies slightly for zeolites with different framework structures. The pore size affects the heat of adsorption and via the adsorption term K the rate of reaction.

REFERENCES

[1] Sie, T., in: van Bekkum, H., Jacobs, P. A., Flanigen, E. M., Jansen, J. C. (Eds.), Hydrocarbon processing with zeolites, *Introduction to Zeolite Science and Practice,* Elsevier, **2001**. Amsterdam, 2001*Stud. Surf. Sci. Catal. 137,* 747–819.

[2] Holderich, W. F., van Bekkum, H., in: van Bekkum, H., Jacobs, P. A., Flanigen, E. M., Jansen, J. C. (Eds.), Zeolites and related materials in organic syntheses. Brønsted and Lewis catalysis, *Introduction to Zeolite Science and Practice,* Elsevier, **2001**, Amsterdam, 2001*Stud. Surf. Sci. Catal. 137*, 821–910.

[3] Eichler, U., Brändle, M., Sauer, J., Predicting absolute and site specific acidities for zeolite catalysts by a combined quantum mechanics interatomic potential function approach, *J. Phys. Chem. B* **1997,** *101,* 10035–10050.

[4] Eheresmann, J. O., Wang, W., Herreros, B., Luigi, D.-P., Venkatraman, T. N., Song, W., Nicholas, J. B., Haw, J. F., Theoretical and experimental investigation of the effect of proton transfer on the AI-27 MAS NMR line shapes of zeolite-adsorbate complexes: An independent measure of solid acid strength, *J. Am. Chem. Soc.* **2002,** *124,* 10868–10874.

[5] Kentgens, A. P. M., Iagu, J., Kalwei, M., Koller, H., Direct observation of Brønsted acidic sites in dehydrated zeolite H-ZSM5 using DFS-enhanced AI-27 MQMAS NMR spectroscopy, *J. Am. Chem. Soc.* **2001,** *123,* 2925–2926.

[6] Beaumont, R., Barthomeuf, D., X, Y, aluminum-deficient and ultrastable faujasite-type zeolites I. Acidic and structural properties, *J. Catal.* **1972,** *26,* 218–225.

[7] Sohn, J. R., DeCanio, S. J., Fritz, P. O., Lunsford, J. H., Acid catalysis by dealuminated zeolite-Y. 2. The roles of aluminum, *J. Phys. Chem.* **1986,** *90,* 4847–4851.
[8] Boehm, H. P., Knözinger, H., Nature and estimation of functional groups on solid surfaces. In: Anderson, J.R., Boudart, M., Editors. *Catal. Sci. Technol.* **1983,** *4,* Springer-Verlag, 39–207.
[9] Bevilacqua, M., Montanari, T., Finocchio, E., Busca, G., Are the active sites of protonic zeolites generated by the cavities? *Catal. Today* **2006,** *116,* 132–142.
[10] Xu, B., Sievers, C., Lercher, J. A., van Veen, R. A. L., Prins, R., van Bokhoven, J. A., Strong Brønsted acidity in amorphous silica-aluminas, *J. Phys. Chem. C* **2007,** *111,* 12075–12079.
[11] Song, W. G., Nicholas, J. B., Haw, J. F., A persistent carbenium ion on the methanol-to-olefin catalyst HSAPO-34: Acetone shows the way, *J. Phys. Chem B* **2001,** *105,* 4317–4323.
[12] Clark, L. A., Sierka, M., Sauer, J., Stable mechanistically-relevant aromatic-based carbenium ions in zeolite catalysts, *J. Am. Chem. Soc.* **2003,** *125,* 2136–2141.
[13] Olah, A. G., Halpern, Y., Shen, J., Mo, K. Y., Electrophilic reactions at single bonds. 12. Hydrogen-deuterium exchange, protolysis (deuterolysis), and oligocondensation of alkanes with superacids, *J. Am. Chem. Soc.* **1973,** *95,* 4960–4970.
[14] Jacobs, P. A., Acid zeolites – an attempt to develop unifying concepts, *Catal. Rev. Sci. Eng.* **1982,** *24,* 415–440.
[15] Rabo, J. A., Gajda, G. J., Acid function in zeolites – recent progress, *Catal. Rev. Sci. Eng.* **1989,** *31,* 385–430.
[16] Dessau, R. M., Catalysis of Diels-Alder reactions by zeolites, *J. Chem. Soc. Chem. Commun.* **1986,** *15,* 1167–1168.
[17] Venuto, P. B., Organic molecules and zeolite crystal - interface, *Chem. Tech.* **1971,** 215–224.
[18] Barrer, R. M., in: *Zeolites and Clay Minerals as Sorbents and Molecular Sieves,* Academic Press, New York, **1978**. 1–497.
[19] Temkin, M., Kinetics of some industrial heterogeneous catalytic reactions, in: Advances in Catalysis, *28,* Academic Press, New York, **1979,** 173–291.
[20] Eder, F., Lercher, J. A., On the role of the pore size and tortuosity for sorption of alkanes in molecular sieves, *J. Phys. Chem. B* **1997,** *101,* 1273–1278.
[21] Maesen, T. L. M., Beerdsen, E., Calero, S., Dubbeldam, D., Smit, B., Understanding cage effects in the n-alkane conversion on zeolites, *J. Catal.* **2006,** *237,* 278–290.
[22] Olson, D. H., Haag, W. O., Lago, R. M., Chemical and physical-properties of the ZSM-5 substitutional series, *J. Catal.* **1980,** *61,* 390–396.
[23] Haag, W. O., Lago, R. M., Weisz, P. B., The active site of acidic aluminosilicate catalysts, *Nature* **1984,** *309,* 589–591.
[24] Haag, W. O., Catalysis by zeolite – Science and technology, *Stud. Surf. Sci. Catal.* **1994,** *84,* 1375–1394.
[25] Narbeshuber, T. F., Vinek, H., Lercher, J. A., Monomolecular conversion of light alkanes over H-ZSM-5, *J. Catal.* **1995,** *157,* 388–395.
[26] Bond, G. C., Keane, M. A., Lercher, J. A., Compensation phenomena in heterogeneous catalysis: General principles and a possible explanation, *Catal. Rev.-Sci. Eng.* **2000,** *42,* 323–383.
[27] Wei, J., Non-linear phenomena in zeolite diffusion and reaction, *Ind. Eng. Chem. Res.* **1994,** *33,* 2467–2472.
[28] Wei, J., Adsorption and cracking of N-alkanes over ZSM-5: Negative activation energy of reaction, *Chem. Eng. Sci.* **1996,** *51,* 2995–2999.
[29] Eder, F., Stockenhuber, M., Lercher, J. A., Brønsted acid site and pore controlled siting of alkane sorption in acidic molecular sieves, *J. Phys. Chem. B* **1997,** *101,* 5414–5419.
[30] Garcıa-Perez, E., Dubbeldam, D., Maesen, T. L. M., Calero, S., Influence of cation Na/Ca ratio on adsorption in LTA 5A: A systematic molecular simulation study of alkane chain length, *J. Phys. Chem. B* **2006,** *110,* 23968–23976.
[31] http://www.iza-structure.org/databases
[32] Derouane, E. G., Andre, J.-M., Lucas, A. A., Surface curvature effects in physisorption and catalysis my microporous solids and molecular-sieves, *J. Catal.* **1988,** *110,* 58–73.
[33] Mortier, W. J., Electronegativity equalization and solid state chemistry of zeolites, *Stud. Surf. Sci. Catal.* **1988,** *37,* 253-268.

[34] Xu, B., Sievers, C., Hong, S. B., Prins, R., van Bokhoven, J. A., Catalytic activity of Brønsted acid sites in zeolites: Intrinsic activity, rate-limiting step, and influence of the local structure of the acid sites, *J. Catal.* **2006,** *244,* 163–168.
[35] Kotrel, S., Rosynek, M. P., Lunsford, J. H., Intrinsic catalytic cracking activity of hexane over H-ZSM-5, H-beta and H-Y zeolites, *J. Phys. Chem. B* **1999,** *103,* 818–824.
[36] Beyerlein, R. A., ChoiFeng, C., Hall, J. B., Huggins, B. J., Ray, G. J., Effect of steaming on the defect structure and acid catalysis of protonated zeolites, *Top. Catal.* **1997,** *4,* 27–42.
[37] Beaumont, R., Barthomeuf, D., X,Y, aluminum-deficient and ultrastable faujasite-type zeolites. 1. acidic and structural properties, *J. Catal.* **1972,** *26,* 218–225.
[38] Fritz, P. O., Lunsford, J. H., The effect of sodium poisoning on dealuminated Y-type zeolites, *J. Catal.* **1989,** *118,* 85–98.
[39] Lónyi, F., Lunsford, J. H., The development of strong acidity in hexafluorosilicate-modified Y-type zeolites, *J. Catal.* **1992,** *136,* 566–577.
[40] Beyerlein, R. A., McVicker, G. B., Yacullo, L. N., Ziemiak, J. J., Influence of framework and nonframework aluminum on the acidity of high-silica, proton-exchanged FAU-framework zeolites, *J. Phys. Chem.* **1988,** *92,* 1967–1970.
[41] van Bokhoven, J. A., Williams, B. A., Miller, J. T., Koningsberger, D. C., Observation of a compensation relation for monomolecular alkane cracking by zeolites: the dominant role of reactant sorption, *J. Catal.* **2004,** *224,* 50–59.
[42] Ramachandran, C. E., Williams, B. A., van Bokhoven, J. A., Miller, J. T., Observation of a compensation relation for n-hexane adsorption in zeolites with different structures: implications for catalytic activity, *J. Catal.* **2005,** *233,* 100–108.
[43] Miradatos, C., Barthomeuf, D., Mirodatos Superacid sites in zeolites, *J. Chem. Soc. Chem. Commun.* **1981,** *39,* 39–40.
[44] Lombardo, E. A., Sill, G. A., Hall, K. W., The assay of acid sites on zeolites as measured by ammonia poisoning, *J. Catal.* **1989,** *119,* 426–440.
[45] Suzuki, K., Noda, T., Katada, N., Niwa, M., IRMS-TPD of ammonia: Direct and individual measurement of Brønsted acidity in zeolites and its relationship with the catalytic cracking activity, *J. Catal.* **2007,** *250,* 151–160.
[46] Sun, Y., Chu, P. J., Lunsford, J. H., The origin of strong acidity in H-ZSM-20 zeolites, *Langmuir* **1991,** *7,* 3027–3033.
[47] Lee, C., Parrillo, D. J., Gorte, R. J., Farneth, W. E., Relationship between differential heats of adsorption and Brønsted acid strengths of acidic zeolites: H-ZSM-5 and H-Mordenite, *J. Am. Chem. Soc.* **1996,** *118,* 3262–3268.
[48] Boehm, H. P., Knözinger, H. Nature and estimation of functional groups on solid surfaces. In: Anderson, J.R., Boudart, M., Editors. *Catal. Sci. Technol.* **1983,** *4,* Springer-Verlag, 39–207.
[49] Trombetta, M., Busca, G., Lenarda, M., Storaro, L., Pavan, M., An investigation of the surface acidity of mesoporous Al-containing MCM-41 and of the external surface of ferrierite through pivalonitrile adsorption, *Appl. Catal. A: Gen.* **1999,** *182,* 225–235.
[50] Bevilacqua, M., Montanari, T., Finocchio, E., Busca, G., Are the active sites of protonic zeolites generated by the cavities? *Catal. Today* **2006,** *116,* 132–142.
[51] Garrone, E., Onida, B., Bonelli, B., Busco, C., Ugliengo, P., Molecular water on exposed Al3+ cations is a source of acidity in silicoaluminas, *J. Phys. Chem. B.* **2006,** *110,* 19087–19092.
[52] Xu, B., Bordiga, S., Prins, R., van Bokhoven, J. A., Reversibility of structural collapse in zeolite Y: Alkane cracking and characterization, *Appl. Catal. A Gen.* **2007,** doi: 10.1016/j.apcata.2007.09.018. 66–73.

CHAPTER 25

Catalysis by Mesoporous Molecular Sieves

Jiří Čejka *and* Ajayan Vinu

Contents

Abstract

Mesoporous molecular sieves of various chemical composition, structural types, and dimensionality of their channels are interesting ordered materials with a high potential in adsorption but particularly in catalysis. This review covers various aspects of the catalytic behavior of mesoporous molecular sieves, in which these mesoporous molecular sieves operate both as catalysts or as catalyst supports for highly active oxidic or organometallic species. In the first part, examples of catalytic mesoporous molecular sieves are focused on single heteroatoms located in siliceous matrix as active sites for acid-catalyzed reactions, oxidation with hydrogen peroxide by Baeyer–Villiger reaction, Meerwein–Ponndorf–Verley reduction, esterification, and etherification. The second part describes catalytic systems based on mesoporous silica or alumina supports modified with transition metal oxides for hydrodesulfurization (MoO_3) and metathesis of olefins and unsaturated esters (MoO_3 or Re_2O_7). Immobilization of organometallic complexes onto the surface of mesoporous materials and their catalytic activities are discussed in the final chapter.

Keywords: Alkylations, Acylations, Catalysis, Immobilization, Mesoporous, Metathesis, Molecular sieves, Nanomaterials, Redox reactions

Ordered Porous Solids
DOI: 10.1016/B978-0-444-53189-6.00025-1

1. Introduction to Mesoporous Molecular Sieves

Successful synthesis of mesoporous molecular sieves has opened a new fascinating research area in inorganic chemistry strongly connected with organometallic chemistry and solid state chemistry and has potential applications in adsorption, drug delivery and catalysis.[1,2] At present, more than 20 mesoporous materials are available differing in the size of the channels, their connectivity, chemical composition, and structure directing agents applied for their synthesis.[3,4] A number of reviews covering synthesis, characterization, and catalytic properties of mesoporous molecular sieves can be found, and some of them are cited here.[4–6] Although not all original expectations of these materials were fullfiled, particularly due to the lack of strong Broensted acid sites and lower (hydro)thermal stability when compared with zeolites, ordered mesoporous materials formed an important group of porous materials and attracts a growing interest of researchers in many different areas.

Some emerging areas of material and application chemistry like micro/mesoporous composite materials[7,8] or metal-organic frameworks[9] are closely related with mesoporous molecular sieves; however, they are out of the scope of this chapter.

The term mesoporous defines materials with pores in the range 2–50 nm[10] but its most common usage implies that such materials but with uniform pores are often ordered to produce a distinct X-ray diffraction pattern. These patterns usually consist of only several diffraction lines located at low angles owing to the absence of crystalline phases but long-range ordering. At present, such mesoporous materials constitute an integral and significant portion of what was originally zeolite science,[11,12] which enormously expanded and diversified into the area of microporous and mesoporous molecular sieves.[13] This has occurred within the last 15 years, prior to that ordered mesoporous materials with uniform pores were entirely unknown and unrecognized. On the basis of a large number of citations of relevant publications[14] that the great initial interest and subsequent explosive expansion of the mesoporous materials were doubtless caused by the corresponding patents and publications coming from Mobil in the early nineties.[1,2] The discovery and recognition of the zeolite-like (but with significant differences, *vide supra*) large pore materials occurred while pursuing pore expansion beyond available zeolite pore sizes. Analogous to the zeolite framework formation through incorporation of organic molecules as mono-molecular structure directing agents, the use of self-assembling surfactant molecules resulted in supramolecular templating by organic aggregates producing ordered inorganic frameworks with large uniform pores. The exploitation of surfactants for generating these novel materials produced a wealth of new and diverse products with a potential for being used as catalysts and sorption media.[15–17]

The original work from Mobil referred to the new materials as M41S family and identified three basic silicate-based structures akin to surfactant lyotropic liquid crystal phases, namely hexagonal, cubic, and lamellar with the corresponding designation MCM-41, MCM-48, and MCM-50.[1] MCM-41 consisted of one-dimensional (1-D) parallel channels with pores up to 10 nm and large internal pore area exceeding 1000 m^2/g.[1,2] The ease of preparation made MCM-41 the most widely studied mesoporous material up till now. MCM-48 contains 3-D channel system, which made it more attractive than MCM-41. However, it is more difficult to synthesize and only after some convenient preparation routes were identified the catalytic applications were more systematically pursued. The potential advantage of more open nature of MCM-48 in catalysis compared with that of 1-D MCM-41 has been addressed in some articles but additional work on this critical issue may be warranted.[18]

The M41S materials were obtained under conditions similar to zeolite synthesis, however, with cationic surfactants. For zeolite and zeotype synthesis, individual organic molecules are used or the synthesis is carried out even with purely inorganic system, while for the synthesis of ordered mesoporous materials, formation of micelles in water solution is needed. Subsequent effort extended synthesis regime to diverse reagents, compositions, such as many nonsilica-alumina elements, and especially various types of surfactants including neutral and anionic.[19–21] Using different surfactants afforded many new structures expanding the dimensions and nature of the available pore systems. With regard to catalytic application the one-dimensional hexagonal structures and the cubic one have been the primary focus. They remain dominant with addition of SBA-15 obtained with block co-polymers, a hexagonal thick-wall material with microprosity connecting the channels.[19] Another frequently used class of the solids was prepared under mild conditions with neutral amine templates, e.g., MSU-type.[20]

Mesoporous materials have a periodic structure like zeolites but their framework is made of amorphous solids, which makes their activity much lower than zeolites. Aluminosilicate versions of the mesoporous materials were studied and developed first showing significant activity in catalytic conversions of hydrocarbons.[22] The most important question concerned their performance compared with both zeolites and amorphous solids. The first and conventional use and preparation of mesoporous materials as catalysts were in a manner similar to zeolites: binding, removal of template, exchange, adding metal function, etc. The large pores and surface areas combined with amorphous walls containing many hydroxyl groups allowed for additional functionalization of vastly expanding potential for generating catalytic materials. Not only the active metal centers were introduced but other techniques like grafting were also applied. The grafted moieties could be active themselves, like mercaptopropyl siloxanes, which can selectively adsorb heavy elements like mercury. Additionally, mercaptopropylsiloxanes on MCM-41 can be easily oxidized to sulfonic groups and serve as highly active acid catalysts for transesterification.[23–25] Alternatively, they can serve as ligands for tethering active complexes as a method of heterogenizing homogeneous catalysts.

A direct approach to mesoporous materials with organic functional groups involves so-called Periodic Mesoporous Organosilicas.[26] They are prepared from substituted orthosilicates, optionally in the presence of tetraethyl orthosilicate (TEOS) or analogous hydrolyzable moieties. Hydrolysis of such compounds in the presence of surfactants results in an assembly of large pore materials with silica and organic functional groups in the walls. Template removal makes these groups available for reactants and adsorbates.

As a general rule the synthesis of mesoporous materials involves the use of surfactant molecules as structure directing agents, typically in relatively large amounts.

As shown above the mesoporous materials developed in the past 15 years are extremely diverse in terms of structures and compositions. The objective of this chapter is to discuss recent trends and advances of mesoporous molecular sieves in catalysis and to highlight the potential of these catalysts particularly in the field of organic reactions, which properly modified mesoporous catalysts are capable to catalyze. In the following discussion of catalytic application we will initially concentrate on silicate-based single site materials for acid-base and redox reactions. Further, we will show catalytic behavior of mesoporous silicates and aluminas modified with active transition metal oxides. Finally, utilization of mesoporous silicates bearing metallic nanoparticles or immobilized organometallic complexes will be related to their activities in C–C bond forming reactions, metathesis reactions, etc.

2. Catalysis by Single Atoms

2.1. Acid-catalyzed reactions

The exploration of catalytic applications of mesoporous materials has been very extensive and diverse expanding beyond traditional porous materials domain.[27] There have been many general and specialized reviews concerning the subject. A few comprehensive and detailed reviews have appeared recently[28–31] presenting up to date overview of the subject. The intrinsic acid activity of mesoporous materials is similar to amorphous solids and quite weaker than the highly active zeolites. This is mainly due to the absence of strong bridging Si-OH-Al groups, typical for zeolites. On the basis of these reasons various approaches to enhance the acidity, especially of the Broensted type, were investigated during the last decade.[8]

This low acidity was quite evident in various hydrocarbon conversion processes where mesoporous materials were measured against zeolites for activity and selectivity.[22] The studies included early on the catalytic cracking, a process with highest potential impact where USY zeolite is the primary active catalyst component. MCM-41 was as found to be less active and less gasoline selective than USY in a formulated and mildly steamed catalyst.[22] The conversion of heavier fractions appeared to be higher with MCM-41, which could be explained as the consequence of substantially larger pores. A study of the fresh MCM-41 suggested higher amount of gasoline and lower coke but the product steamed under FCC conditions apparently collapsed.[32] Compared with amorphous silica-alumina, mildly steamed MCM-41 was superior in producing more gasoline at equivalent coke yield and being more selective for heavy oil conversion. In summary, under certain conditions the uniform pores of MCM-41 did seem to offer benefits, especially when compared with amorphous silica-alumina. However, a low activity and above all low hydrothermal stability prevented consideration of use in this important commercial process.

Despite the initial failure of mesoporous materials to show interesting performance under realistic conditions in FCC the potential benefits were too rewarding to abandon further attempts. A significant effort went into addressing these apparent instability and activity deficiencies of mesoporous material, MCM-41 in particular. Numerous approaches, many novel and quite ingenious, were proposed and tried as described later.

A systematic work on MCM-41 synthesis showed that increasing severity of the conditions, i.e. temperature and time, affords more robust structures, e.g., as-ascertained by contraction of the unit cell upon calcination.[33] The discovery of SBA-15, possessing several nanometer thick walls, represented another way to prepare mesoporous materials possessing more stable frameworks.[19] All apparently hydrothermally stable structures based on silica were affected adversely by the presence of aluminum. An opposite trend was postulated when aluminum was introduced post-synthesis and the apparent hydrothermal stability was increasing with rising Al content with maximum at around 20/1 = Si/Al.[34] However, as the stability was tested by boiling in water rather than by steaming the structure preservation could be explained by reduced solubility of silica in the presence of Al.[35] By analogy to microporous materials it was postulated that a crystalline wall in mesoporous material could alleviate the problem of both activity and hydrothermal stability. Since most zeolites are not hydrothermally stable under FCC conditions the selection of preferred zeolite for the wall was also of some importance. The most often explored zeolites in this category were faujasite, ZSM-5, and zeolite beta.[8] No method to produce fully crystalline mesoporous structure has been discovered yet. Various strategies were designed to produce materials with both crystalline and amorphous domains and as a class are named micro-mesoporous composites or hierarchic systems. Some of the examples are presented below while a detailed discussion has been carried out by Trong-On[29] and Liu and Pinnavaia[7]:

1. Two structure directing agents, one for a zeolite and the other (a surfactant) for mesoporous phase, are used together in the synthesis mixture. The product is a mixture of the two phases in various degrees of intimacy.[36]
2. Zeolite synthesis mixture is crystallized for a certain period of time to generate zeolite fragments, preferably before they are detectable by X-ray powder diffraction. Surfactant is added to assemble the mesoporous structure with the zeolite seeds, protozeolite clusters, etc, into mesoporous frameworks.[37]
3. Simultaneous crystallization of zeolite and mesoporous materials precursor phases.[38]
4. Using a thick wall mesoporous material such as SBA-15 a secondary crystallization of microporous domain is generated by addition of an appropriate template.[39]
5. Mesopore material is treated with a dilute solution containing small zeolite fragments such as ZSM-5 nanoslabs with a size of ca. 2.8 nm. The subsequent steam stability assessment showed that these zeolite coated mesoporous molecular sieves were capable of significant structure retention in contrast to the original mesoporous matrix.[40]

These composites have been extensively characterized demonstrating both increased activity and hydrothermal stability by physicochemical methods and model reactions.[7] Further possibility is to use carbon black particles to synthesize mesoporous zeolites, which will combine the acidity of zeolites with mesoporous features.

There are many examples of mesoporous materials being comparable or even superior to commercial hydrotreating catalysts. Potential for improvement due to the presence of mesoporous materials has been demonstrated in hydrotreatment processes including HDS, HDN, and hydrocracking. An MCM-41 support was tested against USY and amorphous silica-alumina, all loaded with NiMo, in hydrotreatement of vacuum gas oil at 350–450 °C.[41] It showed a better activity in HDS and HDN, and better selectivity towards middle distillate in mild hydrocracking. MCM-41 showed potential for increasing hydrotreating activity, shown with dithiobenzene through increase of CoMo metal loading. Doubling of the metal loading from 13.5% MoO_3 and 2.9% CoO resulted in almost 3-fold increase in the conversion.[42] In contrast, γ-alumina supported catalyst showed no effect due to increased loading.

In addition, NiW-USY/MCM-41/Alumina was found superior in activity with comparable selectivity to several commercial distillate selective hydrocracking catalysts.[22] The beneficial effect of high alumina in MCM-41 (6.7 Si/Al) addition was shown directly by incorporation of increasing amounts in NiMo HDS catalyst supported on γ-alumina.[43] As the content of mesoporous component in the support was increased the HDS thiophene activity also increased. The effect was attributed to the interaction of the NiMo phase with highly dispersed oxyaluminum species inside MCM-41. A CoMo catalyst showed similar performance enhancement with increasing amount of MCM-41 towards HDS of dithiobenzene.

A third mode, in which MCM-41 like material can benefit in hydrotreating catalytic performance, was observed in a demetallation process.[44] Ni or Mo MCM-41 was found active for removing trace elements, with particular selectivity for Fe, V, Ni, and As. The activity increased with increasing pore size of the mesoporous support and the best results were obtained with the largest pore catalyst (8 nm). In a similar way but a different process, i.e., FCC, MCM-41 showed potential for trapping metals like V and Ni, which made them less harmful to the catalyst.[22]

It is evident from many examples provided earlier that new mesoporous materials typically exhibit much lower conversions of various hydrocarbon fractions compared with zeolites and, therefore, are less competitive in established applications. On the other hand, they offer advantages in areas where lower acidity is suitable and especially when large molecules are involved. One such area is catalysis of fine chemical synthesis. Taguchi and Schüth list a number of examples of organic transformations that were carried out using various forms, i.e., mainly with different heteroelements, of mesoporous catalysts.[28] These processes include Friedel–Crafts alkylations, acylations, acetalizations, Diels–Alder condensations, Beckmann rearrangement, aldol condensations, and others. The representative example of the large pore benefit is Friedel–Crafts alkylation of 2,4-di-*tert*-butylphenol with cinnamyl alcohol to give dihydrobenzopyran using Al-MCM-41 with a pore diameter of 3.0 nm. It provides a higher yield than HY zeolite ($<$1%), USY zeolite (mesopore diameter of 2.0–6.0 nm resulting from the ultrastabilization treatment), the conventional catalysts, such as H_2SO_4 (12%) and amorphous silica–alumina (6%) catalysts.

Alkylation of aromatic hydrocarbons is an industrially important reaction because many alkylated phenols are important intermediates for the manufacture of antioxidants, ultraviolet absorbers, phenolic resins, polymerization inhibitors, lube additives, and heat stabilizers for polymeric materials.[45] The alkylation of phenol with *tert*-butylalcohol (TBA) has been intensively studied using mesoporous acidic catalysts[46–49] and the reaction scheme is given in Scheme 25.1. *tert*-Butyl derivatives of phenol are precursors for a number of commercially important antioxidants and intermediates in the synthesis of various agrochemicals, fragrants, thermoresistant polymers, and protecting agents for plastics.[45] Selvam et al. have reported the tert-butylation of phenol over AlMCM-41 and FeMCM-41 and proposed that the moderate acidity of these catalysts is favorable for the formation of the p-isomer (4-TBP).[46–48]

Vinu *et al.* also reported the tert-butylation of phenol over FeAlMCM-41 and reported higher conversion and selectivity to 4-TBP as compared with other mesoporous catalysts.[49] Recently, the same reaction was also studied over AlSBA-15, which possesses two dimensional highly ordered pore structure with a high acidity. The catalyst AlSBA-15 (Si/Al = 45) showed excellent performance in the acid-catalyzed tertiary-butylation of phenol employing *tert*-butanol as the alkylation agent.[50] A high phenol conversion of 89.5 is observed for AlSBA-15 (Si/Al = 45), which is very high as compared with other mesoporous materials like FeMCM-41, AlMCM-41, and FeAlMCM-41,[46–48] even at a low reaction temperature of 150 °C.

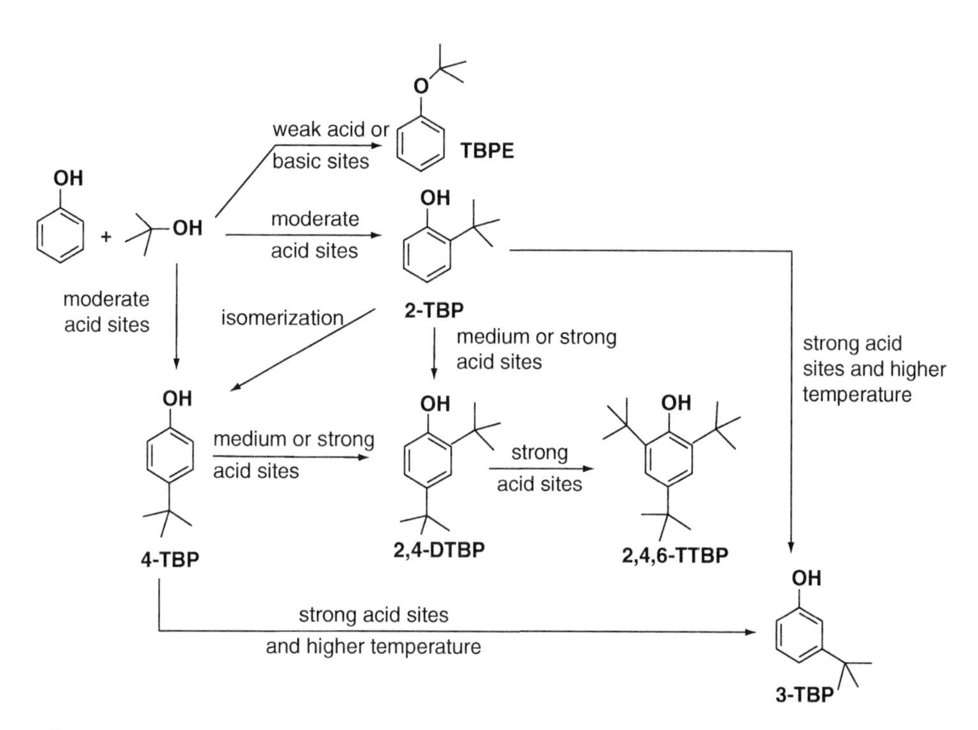

Scheme 25.1 Reaction pathways of phenol alkylation with tert-butyl alcohol.

The AlSBA-15 catalysts showed higher selectivity to dialkylated product rather than monoalkylated product. On the other hand, three-dimensional cage type mesoporous ferrosilicate catalysts (FeSBA-1) with different iron contents was found to be highly active with higher selectivity towards the monoalkylated product.[51,52] This demonstrated that the catalysts with three-dimensional pore structure show better performance in this reaction. This performance is based on the fact that this catalyst provides a high number of active sites and limited pore blocking, allowing for a faster diffusion of the reactant and product molecules.[51,52] Heteropolyacids supported over mesoporous materials have received significant attention in the recent years because of their potential applications in acid-catalyzed reactions.[53,54] Recently, the catalytic activity of tungstophosphoric acid over (TPA)/ZrO_2 supported over mesoporous silica MCM-41 and SBA-15 was demonstrated in Friedel–Crafts alkylation of phenol with benzyl alcohol under liquid phase conditions.[54]

Mesoporous hybrid catalysts are one of the most interesting materials and have been used for acid catalysis. Sulfonic acid functionalized MCM-41 and HMS materials prepared by coating for silylation were used as acid catalysts for the esterification of glycerol with lauric acid.[24] It was demonstrated that the sulfonated-HMS materials showed 52% yield after 10 h of reaction while the HMS materials gave 53% yield after 24 h of reaction time. The higher yield of HMS materials could be due to the presence of more open structure. Diaz et al. also displayed a higher yield using MCM-41-type materials functionalized with methyl and the sulfonic acid groups.[24] On the other hand, a higher yield could be obtained in the presence of lauric acid instead of oleic acid. The sulfonic acid functionalized MCM-41 or HMS were also employed for other acid-catalyzed reactions such as alcohol dehydration and hydroxylation–condensation reactions and these catalysts were found to be highly active.[55–57]

Esterification of alcohols by carboxylic acids is one of the most important reactions and was studied over heteropoly acid supported mesoporous acid catalysts. Recently, Vinu and his coworkers demonstrated the esterification activity of silicotungstenic acid (STA)/ZrO_2 supported over mesoporous silica MCM-41 and SBA-15 in the esterification of isoamyl alcohol with acetic acid and the results were compared with the zeolite catalysts and neat heteropoly acids.[58] The mesoporous material has an advantage in the formation of nanosized and catalytically active STA/ZrO_2 by stabilizing zirconia in tetragonal phase at higher temperatures, which provided better catalytic activity than the neat STA/ZrO_2 in an esterification reaction. Among the catalysts, 15 wt.% STA/ZrO_2/SBA-15 was found to be the most active and acidic catalyst, showing higher conversion and selectivity than neat 15 wt.% STA/ZrO_2 in esterification of isoamyl alcohol with acetic acid. The reaction was found to be heterogeneously catalyzed and no contribution from homogeneous (leached) STA into the medium under the reaction conditions was observed. Moreover, STA/ZrO_2/MS were employed for esterification of benzyl alcohol with acetic acid (see Scheme 25.2). It was found that the catalyst with 15wt. % loading of STA is highly active and exhibited high benzyl alcohol conversion with selectivity for benzyl acetate as high as 96% within 2 h of reaction time.[59] In conclusion, this is seen as an illustration of the potentials offered by mesoporous materials in fine chemical synthesis with bulky substrates.

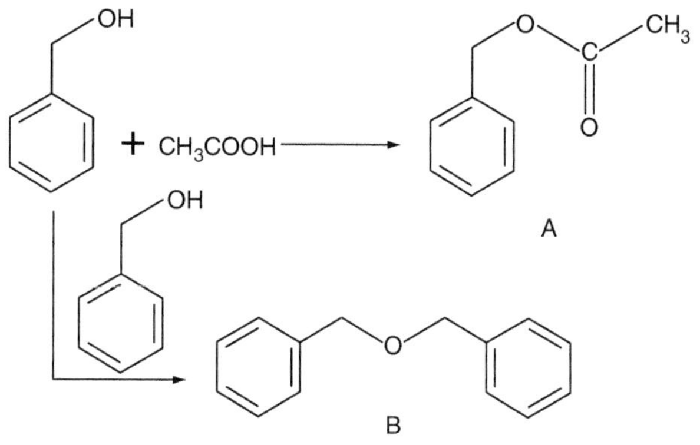

where, A : Benzyl acetate and
B : Dibenzyl ether

Scheme 25.2 Esterification (A) and etherification (B) of benzyl alcohol.

Recent reports showed the benefits of periodic pores structural surfaces wherein the ethane linker was involved in the enhanced selectivity and activity in esterification reactions.[60] The authors also investigated the catalytic application of sulfonic acid functionalized mesoporous benzene-silica with crystalline pore wall structures, where the catalytic sites are designed on the mesoporous hybrid solid surface along with the hydrophobic benzene sites providing a better catalytic environment. These catalysts were found to be highly reactive in esterification of acetic acid with ethanol and the catalytic results showed higher conversion compared with commercially available H-Nafion.[61] The higher catalytic activity of mesoporous benzene-silica materials functionalized via grafting was explained in terms of the fact that sulfonic acid sites on the surface or near the pore mouth are easily accessible for the reactants in the catalytic reaction. In addition, Nakajima et al. showed the esterification and pinacol–pinalcolone rearrangement reactions using their recently developed hybrid mesoporous solid acid catalyst, wherein the surface ethenylene sites were modified to phenylene sulfonic acid via Diels–Alder reaction.[62]

The ability to tailor the pore size of mesoporous materials, at the border with microporous systems, offers unique advantages as shown recently by Iwamoto et al. in the acetalization of cyclohexanone on MCM-41.[63] The maximum catalytic activity was obtained for MCM-41 with a pore diameter of 1.9 nm, and the reaction rate constant decreased for materials with both smaller and larger pores. The optimal activity of the 1.9 nm pore size catalyst was explained as stemming form a co-operative action of surface silanols pointing towards the center. Smaller pores were believed to exhibit diffusion limitations while for larger ones the break down of co-operative action was postulated.

An advantage arising from lower acidity of mesoporous catalysts than zeolites was in acetalization of heptanal, 2-phenylpropanal or diphenylacetaldehyde with trimethyl orthoformate. Here both siliceous and Al-substituted MCM-41 were used and compared to different zeolite catalysts.[64] The latter, represented by zeolites Beta and Y were initially more active than mesoporous materials, but deactivated more rapidly. In contrast, both siliceous and Al-containing mesoporous catalysts performed similarly suggesting adequate acidity stemming from silanol groups for this process. The pore size appeared the most important as indicated by mesoporous catalysts superiority with increasing size of the reactant.

Aromatic ketones are important substrates in pharmaceutical, agrochemical, and fragrance industry. They are mainly produced by Friedel–Crafts acylation of respective ketones with carboxylic acids, their anhydrides or chlorides both in liquid and gaseous phase.[65]

Anisole was acylated over mesoporous supported Nafion and heteropolyacid catalyst in a continuous slurry reactor at 70 °C.[66] Turn-over numbers of 400 were achieved before deactivation and complete activity of regenerated catalyst indicates this type of catalysts as rather promising.

The catalytic activity of mesoporous (Al)MCM-41 and (Al,Zn)MCM-41 was evaluated in vapor phase acylation of *m*-cresol with ethyl acetate.[67] The reaction conditions were optimized to achieve high yield of 4,7-dimethyl coumarin, which is formed by 'one-pot'ring acylation followed by esterification and subsequent intramolecular condensation.

Halogen free acylation of toluene with acetic anhydride was investigated over FeSBA-1.[68] Also mesoporous sulfated zirconia catalysts prepared by precipitation of $Zr(OH)_4$ were tested in acylation of anisole with benzoic anhydride[69] and veratrole.[70] These catalysts are shown to be highly active even at low reaction temperatures (30–50°C). Recently, Vinu and coworkers also demonstrated the catalytic activity of TPA/ZrO_2 supported over mesoporous silica MCM-41 and SBA-15 in acetylation of veratrole with acetic anhydride (Scheme 25.3).[53] The 15 wt.% TPA/ZrO_2/MCM-41 catalyst gave highest catalytic activity with a maximum conversion of acetic anhydride (43.9%) and 100% selectivity for acetoveratrone (3,4-dimethoxyacetophenone).

Very recently, Kapoor et al. have shown that sulfonic acid functionalized derivatives of three-dimensional (*Pm*-3*n*) cubic phenylene-bridged hybrid mesoporous silica material derived from 1,4-bis(triallyl)phenylene precursor are effective in Friedel-Craft acylation reactions.[71,72]

2.2. Base catalyzed reactions

Trong-On[29] lists basically three base-catalyzed reactions studied in the literature:

(i) Knoevenagel condensation (carbon–carbon bond formation through addition of an active hydrogen compound to a carbonyl group followed by dehydration; the product is often an alpha, beta conjugated enone),
(ii) monoglyceride synthesis and Michael additions (the most useful method for the mild formation of C–C bonds,
(iii) addition of carbanion to an alpha, beta unsaturated carbonyl compound). The products of these reactions are specialty chemicals used in perfumes, pharmaceuticals, polymers, etc.

These reactions are catalyzed by various amine compounds usually via homogeneous catalysis. The mesoporous materials provide support for attachment of the basic functions group enabling processes under heterogeneous conditions. Another

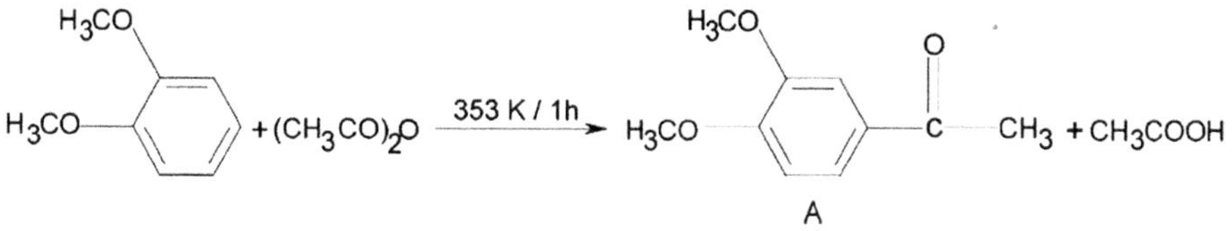

$$H_3CO\text{-}C_6H_4(OCH_3) + (CH_3CO)_2O \xrightarrow{353\ K\ /\ 1h} (H_3CO)_2C_6H_3\text{-}C(=O)\text{-}CH_3 + CH_3COOH$$

A

Scheme 25.3 Synthesis of acetyl veratrole by acylation of veratrole with acetic anhydride.

type of mesoporous basic catalysts is ale silica/aluminosilicate matrices exchanged with alkali metals. In this case, Knoevenagel condensation was studied by the group of Ernst showing high activity for Cs-exchanged (Al)MCM-41.[71]

Base catalysts prepared via amine functionalization of the mesoporous silicas via post-synthesis grafting or direct silylation were also investigated in esterification[72] and Knoevenagel condensation reaction. Jaenicke et al.[73] displayed a yield of about 68% in the esterification of glycidol in the presence of 3-chloropropyltriethoxysilane functionalized MCM-41 materials treated with 4,4,0-triazabicyclo-5-decene. Knoevenagel condensation of ethylcyano-acetate (malononitile) with benzylaldehyde was also performed with high yield.[74] In addition, the earlier-mentioned catalysts were employed in a trans-esterification reaction. Kapoor et al. reported polyamidoamine dendritic mesoporous silicas and polyamidoamine dendritic mesoporous benzene-silicas as very effective catalysts in the Knoevenagel condensation due to enhanced hydrophobic interactions in the pore framework system.[75]

2.3. Ti, Zr, Nb, and Sn containing mesoporous materials

Zeolites substituted with Ti and other early transition metals proved to be effective catalysts for selective oxidation of organic compounds, including some very important reactions such as phenol hydroxylation.[75] As with previously discussed processes the mesoporous analogues of zeolite oxidation catalysts were expected to be useful and show advantage in catalysis involving larger molecules, both substrates and oxidants. Ti-substituted mesoporous materials appear to be the earliest and the most studied after silicates and aluminosilicates.[27,28,29,31] The preparation was usually carried out by direct synthesis with Ti included in the synthesis mixture or by post-synthesis grafting.[76] Purity of the reaction mixture as well as the absence of alkali metal cations are very important to achieve a high degree of Ti incorporation into the silicate matrix. However, as was pointed out by Schüth[16] that 'in many cases it is very difficult to judge the performance ... because the reaction conditions are often not comparable with each other and no proper benchmarks are analyzed.'

The performance of Ti-MCM-41 is best judged compared with zeolite Ti-beta, which is large pore zeolite allowing the use of bulky oxidants and reactants. It is illustrated by the oxidation of α-terpineol using tetrabutyl hydroperoxide (TBHP) as an oxidant, in which Ti-MCM-41 was found to give higher alkene conversion (62.2 mol%) and epoxide selectivity (30.2%) than Ti-β zeolite (32.7 and 12.8 mol%, respectively).[77] The larger organic oxidants offer hydrophobicity, which may enable better interaction with the reactants and also with the mesoporous support.

The latter may be modified towards more hydrophobic Ti-MCM-41 becoming a more efficient catalyst, e.g., in epoxidation and oxidative cyclization using TBHP.[28] More hydrophobic materials exhibited high activity in catalytic epoxidation of unsaturated alcohols followed by cyclization to cyclic ethers. The oxidation activity increased extremely when TBHP was used instead of H_2O_2.[78] This can be attributed to the preferred uptake of the reagents in the hydrophobic pore system.

Silanols groups are ubiquitous in mesoporous materials, which contributes to reduced hydrophobicity of the support and may affect catalysis with organic oxidants. The product with decreased surface silanols was superior to conventional Ti-MCM-41.[79] Similarly, when a Ti-mesoporous material was modified by attachment of trimethylsilyl groups the activity in liquid phase oxidation was greatly increased.[80] The capability to use large organic oxidants with Ti-MCM-41 may also be of benefit in preventing titanium leaching, which can cause problems in liquid phase oxidation with hydrogen peroxide. It was demonstrated upon replacement of hydrogen peroxide with *tert*-butyl hydroperoxide in oxidation of crotylalcohol using Ti-β and Ti-MCM-41.[81] This substitution resulted in reduced Ti-leaching and selective epoxide formation was observed.

Owing to the existence of the Ti-ZSM-5 zeolite, TS-1, the idea of mesoporous materials containing zeolitic 'seeds' in the framework could be exploited for the oxidation catalysts as well. The corresponding titanosilicate material has been named MTS-9. It was obtained from TS-1 seeds and with triblock-copolymer, Pluronic P123, as the structure directing agent in acidic medium and therefore can be considered related to SBA-15. It was found very active towards phenol hydroxylation with hydrogen peroxide showing 26% conversion similar to TS-1. It was also tested for hydroxylation of bulkier molecule such as 2,3,6-trimethylphenol, which is not suitable for oxidation with both TS-1 or Ti-MCM-41. MTS-9 showed a high conversion of ca. 19% proving potential for oxidizing larger molecules.[29,82]

Recent studies of Ti-containing mesoporous materials include examples that go beyond traditional exploration of redox properties. Vinu et al examined performance of Ti-SBA-15 with very high Ti loading.[83] Increased Ti incorporation was possible by decreasing acid concentration in the synthesis mixture. It enabled preparation of highly ordered product with Si/Ti down to 1.9. The catalytic performance was appraised based on oxidation of styrene with hydrogen. A related work demonstrated preparation of Ti-MCM-41 in the range $100 > Si/Ti > 2$ from TEOS, Ti-alkoxides, cationic surfactant and ammonia.[84] The syntheses were accomplished at room temperature and well structured and catalytically active materials were obtained. The activity was tested using cyclohexane oxidation with *tert*-butylhydroperoxide. The highest conversion and excellent selectivity were observed for $100 > Si/Ti > 30$; it is believed that they correspond to the concentration of isolated and tetrahedral Ti sites.

The presence of Al, which imparts acidity, is often detrimental to the performance of oxidation catalysts. Some recent work demonstrated promising activity upon incorporation of both Ti and Al in mesoporous materials.[85] One example is provided by Ti-Si-HMS materials with Ti/Si up to 4 mol% and Al/Si up to 1 mol%. In propylene epoxidation with molecular oxygen at 250°C the catalysts showed up to 48% propylene conversion and ~30% propylene oxide (PO) selectivity.

The presence of both Ti and Al in the framework was important for PO yield. Extra-framework Ti appeared indifferent towards PO yield, while extra-framework Al had a detrimental influence on PO selectivity. The catalyst deactivated apparently due to coking but could be regenerated by calcination.

Lapisardi et al. reported preparation of bifunctional Ti-AlSBA-15 by Ti introduction into Al-SBA-15 obtained directly or through post-synthesis Al activation.[86] The resulting catalyst was used for 'one-pot' synthesis of adipic acid from cyclohexene and *tert*-butylhydroperoxide. The yield up to 80% was reached after 24 h under mild conditions, i.e. temperature 80 °C.

Zr containing mesoporous materials represent a natural extension of the Ti-containing oxidation catalyst. The number of reported studies is significantly smaller. In general, the Zr mesoporous materials apparently exhibit activity and selectivity comparable to Ti materials for oxidation of organic molecules with hydrogen peroxide and TBHP.[30] However, they appear less selective for epoxidation reactions, which were attributed to strong Lewis acidity of $Zr(^{+4})$.

Recently, Corma's group showed that Sn incorporated into the structure of siliceous zeolites or mesoporous molecular sieves is active and highly selective in Baeyer–Villiger oxidation of cyclic ketones to the respective lactones, being so important for flavor industry.[87–89] History of this reaction together with application of traditional and less traditional catalysts was described in a nice review.[90] Number of different cyclic ketones including cyclohexanone, adamantanone, and cyclopentanone were easily transformed to their lactones at low reaction temperatures. It is expected that Sn operates as Lewis acid site being capable to activate carbonyl group of the cyclic ketone and to initiate ring enlargement by introducing oxygen atom from the hydrogen peroxide being used as oxidating agent. Schemes 25.4 and 25.5 show two typical examples of Baeyer–Villiger reaction, transformation of cyclohexanone and adamantanone with hydrogen peroxide, respectively.[91] Later on, it was shown that Sn-mesoporous catalysts can be prepared not only via direct but

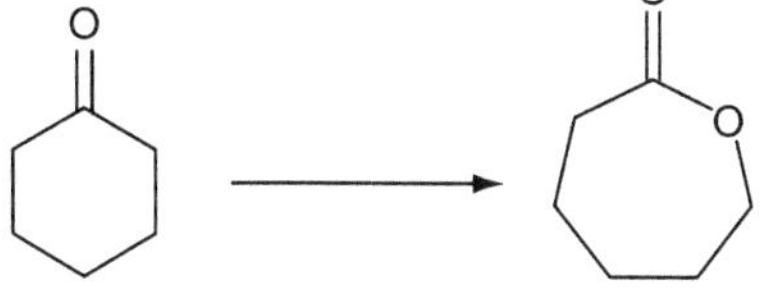

Scheme 25.4 Baeyer–Villiger oxidation of cyclohexanone.

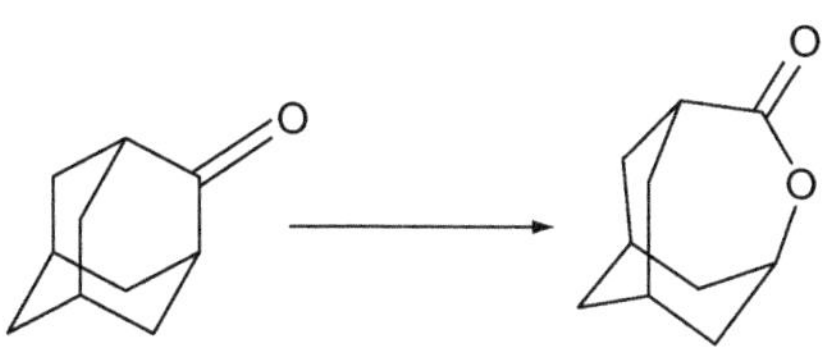

Scheme 25.5 Baeyer–Villiger oxidation of adamantanone.

also post-synthesis methods and the concentration of Sn can be enhanced by microwave heating.[92,93]

Further extension of Baeyer–Villiger reaction describes application of (Fe)MCM-41 catalysts in the presence benzaldehyde with molecular oxygen as oxidation agent.[94] This catalyst was found highly reusable without any particular loss of activity and selectivity. Also modification of hexagonal mesoporous silica with SbFe provided highly active catalyst for Baeyer–Villiger oxidation of cyclohexanone to caprolactone.[95] Clearly, these mesoporous catalysts are very perspective materials for the Baeyer–Villiger reaction when the respective cyclic ketones are available. The traditional routes to synthesize these lactones usually consist of several tedious steps with low yields, thus, Baeyer-Villiger reaction is surely preferable solution.

The introduction of niobium into the structure of mesoporous molecular sieves has a particular advantage of producing efficient and long-lived catalysts for a 'one-pot' synthesis of the precursor of nylon-6,6-cyclohexene epoxide.[96,97] Combination of Nb with Sn in MCM-41 or MCM-48 clearly differentiates catalytic behavior of Nb and Sn. While Nb oxidizes double bond of unsaturated hydrocarbons to the respective epoxides, Sn is not active at all for this reaction. Just opposite situation is in oxidation of cyclic ketones to respective lactones, it means in Baeyer–Villiger reaction.[98]

Corma and coworkers investigated the oxidation behavior of methyl-bearing Ti-MCM-41 materials in the cyclohexene epoxidation.[99] The catalyst showed 94% selectivity to epoxides with 98% conversion of cyclohexene. Organometallic complexes encapsulated aminopropyl- or chloropropyl-modified MCM-41-type mesoporous materials were also employed as the efficient oxidative catalysis of several substrates, including alkenes, ketones, fatty esters, and hydrocarbons, using hydrogen peroxide or *tert*-butyl hydroperoxide as oxidation agents.[79,100–105] Jia and Theil et al. reported the epoxidation activity of functionalized hybrids for the catalytic epoxidation of cyclooctene.[101–103] The grafted molybdenum ligand 3-trimethoxysilylpropyl[3-(2-pyridyl)-1-pyrazolyl]acetamide, in the presence of TBHP, possessed enhanced activity compared to the analogous solution phase system. The increased activity was proposed to derive from additional and cooperative interactions between incorporated catalysts and the pore wall or the pore wall and the substrate TBHP. Inagaki et al. also explored the catalytic properties of organic inorganic mesoporous hybrid materials, and reported the advantage of their use in several oxidation reactions.[104]

3. Transition Metal Oxides Supported on Mesoporous Molecular Sieves

Although introduction of single metal ions into the structure of mesoporous molecular sieves can be done via direct as well as post-synthesis procedures, their modification with metal oxides is limited only on post-synthesis methods. These methods include, e.g., impregnation or thermal spreading. With regard to that both mesoporous silicas as well as aluminas were employed in these studies.

3.1. Hydrodesulfurization with MoO_3

Sulphur removal from various feedstocks by hydrodesulphurization is one of the most important industrial reactions at the moment. Sulphur obtained by this reaction forms the main raw material for the production of sulphuric acid. Both silica-based and alumina-based mesoporous molecular sieves were used as supports for the preparation of hydrodesulfurization catalysts.[104]

Čejka et al. described catalytic properties of molybdenum oxide supported on organized mesoporous aluminas of different pore diameters in hydrodesulphurization of thiophene.[105,106] Thermal spreading method was used to prepare the catalysts and X-ray diffraction did not provide any diffraction lines evidencing well-dispersed molybdenum oxide species. Important conclusions can be drawn based on these results: both catalysts supported on mesoporous alumina (with up to 30 wt.% of molybdenum oxide) exhibited a higher thiophene conversion compared with commercial catalyst (15 wt.%) and the size of mesoporous alumina pores seems to influence the transport of reactants. Further investigations of catalytic activity of scaffolding and lathlike mesoporous γ-aluminas in hydrodesulfurization of dibenzothiophene were performed by Pinnavaia and his group.[107] Incipient wetness impregnation was used to modify the mesostructured alumina with Mo and Co. Resulting conversions depended on the loading and morphology of different mesostructured aluminas but were comparable with those obtained for commercial HDS catalysts.

3.2. Metathesis using MoO_3 and Re_2O_7

New type of heterogeneous catalysts for metathesis of linear olefins and their functional derivatives has been developed using mesoporous aluminas and silicas supported of rhenium(VII) oxide and molybdenum(VI) oxide, respectively. These catalysts were tested in metathesis of olefins, and unsaturated esters and ethers.[108–110] For organized mesoporous aluminas modified with rhenium oxide it was clearly shown that these catalysts are superior when compared with conventional alumina in 1-decene metathesis, see Scheme 25.6.

The results of this investigation evidenced that organized mesoporous alumina-based catalysts are more active than catalysts with conventional aluminas as supports, and the catalyst activity increases with increasing pore size diameter. The selectivity to 9-octadecene was in all cases at least 95%, which was recently confirmed by other groups.[111–113]

Only a small portion of Re-oxide species of probably lower than original valency is active in metathesis reactions.[114] In relation to the surface and chemical properties of organized mesoporous aluminas used as support, higher activities of mesoporous

$$2\,CH_3(CH_2)_7CH{=}CH_2 \rightleftarrows CH_2{=}CH_2 + CH_3(CH_2)_7CH{=}CH(CH_2)_7CH_3$$

Scheme 25.6 Metathesis of 1-decene.

based catalysts could be related to a higher presence of pentacoordinated Al atoms in mesoporous aluminas probably coordinating Re oxide species, a higher level of Lewis acid site, and a lower fraction of basic OH groups in this support.[115] Despite the substantial effort to understand the reasons for higher activity of Re_2O_7 catalysts supported on organized mesoporous aluminas compared to conventional ones the origin of this higher activity is still a matter of debate.

Further extension of metathesis over mesoporous alumina-based catalysts was focused on the activity of these catalysts in transformations of functional derivatives of olefins.[110] Re(VII) oxide on alumina is known to catalyze also the metathesis of oxygen-containing substrates, however, only in combination with co-catalyst like Me_4Sn.[114] The catalysts were particularly investigated in metathesis of *p*-allyl anisole (Scheme 25.7) and di-allyl-di-ethyl malonate (Scheme 25.8).

Further substrates include, e.g., methyl undecenoate and 6-hexenyl acetate. It was shown that a high selectivity to desired products in metathesis of unsaturated esters or ethers was achieved with all catalysts used. Generally, the specific activity (TON) is lower in comparison with unsubstituted substrates, which evidences some competition between metathesis and catalyst deactivation (strong adsorption) by polar substrates. For catalysts with the same loading of Re oxide the activities were very similar as soon as the pore size was larger than 4 nm. It can be inferred that organized mesoporous as well as crystalline mesostructured aluminas exhibit similar surface chemical properties, which result in the similar interaction with Re oxide.

In a similar way, metathesis of 1-olefins was investigated on mesoporous silicas modified with MoO_3. The results showed that (i) with MoO_3/MCM-41 the reaction proceeds about 8 times rapidly than with MoO_3/SiO_2 achieving higher final conversion and high selectivity, (ii) changing MoO_3/MCM-41 for MoO_3/Al-MCM-41, the reaction rate increases even more, but the selectivity falls considerably (probably due to the presence of acidic sites causing isomerization of starting 1-octene followed by cross-metathesis).[116] It was clearly evidenced that the activity of MoO_3/mesoporous catalysts depends strongly on the dispersion of Mo species. Raman spectra of different MoO_3 catalysts showed the presence of at least three types of molybdenum species on the surface (i) isolated molybdenum oxide species (signal at 981 cm^{-1}), (ii) surface polymolybdate (959 cm^{-1}) and bulk MoO_3 (995 and 819 cm^{-1}). The population of these individual species depends strongly on the

2 CH_3O–C₆H₄–CH₂CH=CH₂ $\xrightarrow{-C_2H_4}$ CH_3O–C₆H₄–CH₂CH=CHCH₂–C₆H₄–OCH_3

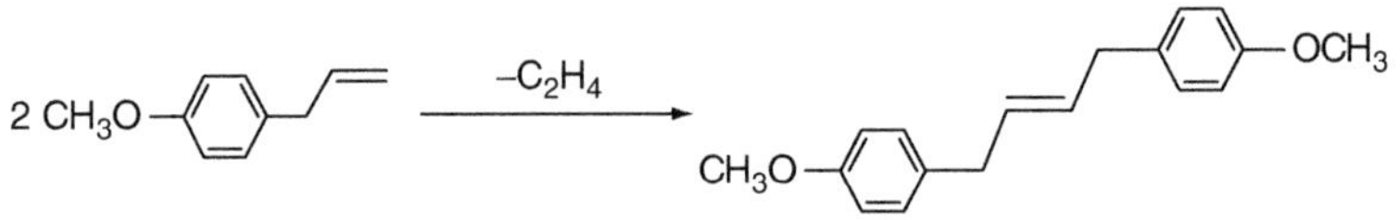

Scheme 25.7 Metathesis of *p*-allyl anisole.

(CH₂=CHCH₂)₂C(COOEt)₂ ⟶ cyclopentene-C(COOEt)₂ + $H_2C=CH_2$

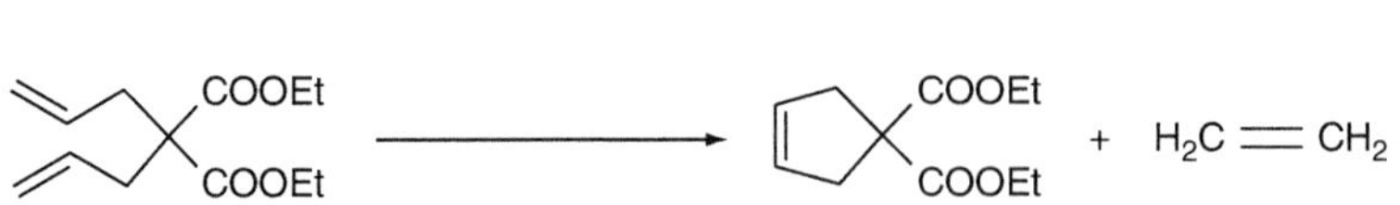

Scheme 25.8 Metathesis of di-allyl-di-ethyl malonate.

catalyst loading and on the support type. It can be inferred that particularly isolated MoO_3 species could serve as precursors to the active sites for 1-olefin metathesis.[116] In a similar to metathesis catalysts based on Re_2O_7 supported on organized mesoporous aluminas also a high activity and complete description of MoO_3 active sites on mesoporous silicas remain unclear.

4. Immobilized Organometallic Complexes

4.1. Grubbs and Schrock catalysts for metathesis

Immobilization of well-defined metathesis catalysts on suitable supports is intended to prepare new metathesis catalysts easily separable from products, reusable and applicable in flow systems. Some attempts to anchor Schrock and Grubbs catalysts on silica support were already reported.[117–119] However, mesoporous molecular sieves were rarely used for this purpose. First generation Grubbs catalyst was anchored on phosphinated MCM-41 and used in ring-opening metathesis polymerization of norbornene and ring-closing metathesis of diethyl diallyl malonate.[120] Similarly, Ru Schiff base complexes were immobilized on MCM-41 and investigated in the same reactions after activation with trimethylsilyl diazomethane.[121]

When molybdenum Schrock type catalysts were anchored on siliceous MCM-41, MCM-48 and SBA-15 via a ligand exchange reaction with surface OH groups (Scheme 25.9),[122] a high activity and selectivity in metathesis of 1-heptene was achieved. No substantial changes in the activity of individual catalysts were observed owing to different textural properties and pore sizes of mesoporous substrates under study. The selectivity to 6-dodecene was in all cases higher than 98%.

It should be noted that the activity of immobilized Schrock catalyst approached the activity of parent homogeneous complex[122] and this catalyst was resistant against leaching of the active phase. Further testing of immobilized Schrock catalyst proved its high activity in metathesis of 5-hexenyl acetate in benzene at room temperature

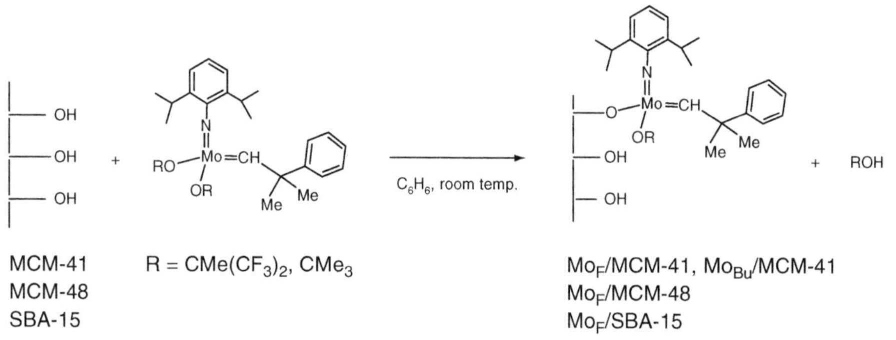

MCM-41
MCM-48
SBA-15

R = $CMe(CF_3)_2$, CMe_3

C_6H_6, room temp.

Mo_F/MCM-41, Mo_{Bu}/MCM-41
Mo_F/MCM-48
Mo_F/SBA-15

Scheme 25.9 Immobilization of Schrock complex over mesoporous silica.

providing 76% conversion and 100% selectivity. Under the same conditions, ring-closing metathesis of diethyl diallyl malonate gave 60% conversion and 100% selectivity.

There is a still open question what is the best way of immobilization of this type of catalyst on the siliceous mesoporous support. In Scheme 25.9 the exchange with one OR group could result in a very short attachment of the complex to the surface. Thus, further studies should be directed to the covalent bonding of the respective complexes to the surface via proper linkers (e.g., aminopropyl group). On the other, this would require the modification of the organometallic complex.

4.2. Cu-complexes for hydroxyalkylation

Corma's group recently reported the application of chiral Cu(II) bisoxazoline covalently bonded to mesoporous MCM-41 for enantioselective Friedel–Crafts hydroxyalkylation of 1,3-methoxybenzene with 3,3,3-trifluoropyruvate.[123] Relatively long linker enabled proper anchoring of the Cu-complex on the surface of MCM-41 (Scheme 25.10). As a result, a highly active catalyst was prepared providing *ee* values higher than 95% and even higher than for unsupported complex. No leaching of the complex under reaction conditions appeared.

4.3. C–C bond formation over Pd nanoparticles

Homogeneous palladium catalysts gained enormous importance in various coupling reactions such as Heck, Suzuki, Sonogashira, etc. The properties of Pd nanoparticles can be adjusted to provide high reaction rates, high turnover numbers, selectivities and yields. Recent development in ligand-free Pd catalysts attracted enormous interest as homogenous catalysts have number of drawbacks, in particular problems with catalyst recovery and recycling and necessity to separate catalysts from the reaction mixtures.[124] Thus, immobilization of Pd nanoparticles on proper inorganic supports appears as a promising way to overcome these shortages. Palladium particles

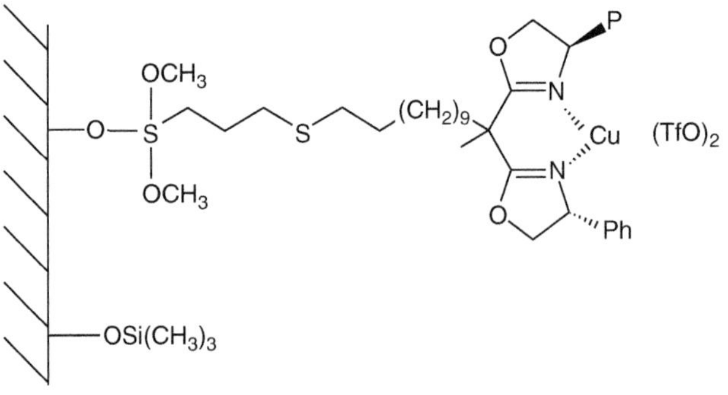

Scheme 25.10 Anchoring of the Cu-complex to the surface of MCM-41.

can be immobilized on the surface of, e.g., activated carbon, various oxides, zeolites, as well as mesoporous molecular sieves.[125–127] Zeolites and porous oxides represent one of the most interesting supports due to the well-defined properties. Pd(II)/zeolite Y was investigated in Suzuki reaction of aryl bromides with phenylboronic acid.[128–130]

Mesoporous molecular sieves offer a nice possibility to modify the accessible surface with, e.g., aminoalkyl or mercaptoalkyl groups. This approach does not lead to a substantial decrease in the pore size; in addition efficient reusable catalysts for Heck reaction can be prepared. Examples show high activities and selectivities in reaction of 4-bromoacetophenone with ethyl acrylate[131] or non-activated aryl bromides with styrene.[132] These examples of immobilized Pd particles clearly show very broad application potential of these novel catalysts in the field of C–C bond forming reactions. Mercaptomodified SBA-15 was found as a material capable of adsorbing Pd from organic and aqueous solutions.[133] Resulting catalyst was highly active in Suzuki–Miyaura and Mizoroki–Heck coupling reactions. Hot filtration experiments proved that reaction is occurring predominantly on surface-bonded Pd.

FSM-16 was used as proper mesoporous support for mercaptopropylsiloxane Pd(II) complex as a stable and recyclable heterogeneous catalyst for the Heck reaction of 4-bromoacetophenone with ethyl acrylate and for the Suzuki reaction of 4-bromoanisole with phenylboronic acid.[134] Similar catalysts based on Ru and Pt were successfully tested in Heck vinylation of various aryl iodides with methyl acrylate using Ru(III) and Pt(IV) complexes immobilized on functionalized FSM-16.[135]

Palladium–Schiff base complexes immobilized on mesoporous and dealuminated silicas were shown and efficient and recyclable catalysts for the Heck reaction and Suzuki coupling under phosphine-free conditions. The advantage of this type of catalyst is high stability under oxygen and moisture conditions, which makes them rather attractive.[136] Also rather mild reaction conditions represent an important advantage of these catalysts.

We have prepared a series of Cs^+- and K^+-exchanged mesoporous aluminosilicate molecular sieves (Al)MCM-41 differing by the way how the Al-centres were incorporated into the parent MCM-41 scaffold. The obtained basic supports were converted to supported palladium catalysts by grafting of palladium nanoparticles and the obtained *bifunctional catalysts* tested in the Heck reaction between butyl acrylate and bromobenzene to give butyl (*E*)-cinnamate under both hydrothermal[137,138] and microwave heating.[139] Microwave heating was shown to enhance the rate of the Heck reaction, however, a faster agglomeration of Pd particles was also observed. The catalysts containing both catalytically active palladium particles and basic sites were found to be active in that reaction even without added base or a co-catalyst, which simplifies the subsequent product separation. Notably, the activity of the newly devised catalytic systems exceeded that reported previously.

In addition, it was reported that ball-shaped Pd particles located in MCM-48 nanocatalyst exhibit a high hydrogenolysis and hydrogenation activity.[140]

5. Conclusions

Particular interest has been paid to mesoporous molecular sieves since their first synthesis by Mobil researchers because they showed peculiar properties, mainly well-ordered structure possessing narrow pore size distribution and large surface areas although with amorphous walls (MCM-41 and MCM-48). Further on, the pore sizes and structural diversity of ordered mesoporous materials were substantially enlarged by applying other synthesis methods like that used for SBA-15 type materials. All these issues are important for application of these materials as adsorbents, catalysts, and catalysts supports. In recent years, number of mesoporous molecular sieves increased not only for different structural types but also for their chemical composition. In this review we focused only on silicate mesoporous materials with some exceptions of organized mesoporous alumina.

Mesoporous molecular sieves doubtless accelerated the development of new interesting catalysts in acid, base, and redox catalysis. In addition, large surface areas together with pore sizes in mesopore range enable the immobilization of highly active and selective homogeneous catalysts. This results in the preparation of heterogenized systems, when properly synthesized, benefiting from their robustness and operating in a liquid phase without leaching. Such catalysts expand the application potential of mesoporous materials.

In this review we tried to show some nice examples of mesoporous molecular sieves-involving acid-catalyzed reactions (acylations, alkylations, esterifications, base-catalyzed reactions (Knoevenagel condensation), hydrodesulfurization, metathesis, hydroxyalkylations, and C–C bond forming reactions (Heck reaction or Suzuki reaction). Activity of synthesized catalysts strongly depends on the type of mesoporous support under investigation, way of its modification, interaction of active species with the surface, and reaction conditions. Detailed description of all mesoporous catalysts and recent publications on reactions carried out with them is out of the scope of this review.

Finally, based on the annual number of published articles and corresponding citations, it is clear that mesoporous molecular sieves are very interesting and important group of inorganic porous materials with a high application potential.

Acknowledgements

J.Č. thanks the Grant Agency of the Academy of Sciences of the Czech Republic (project A4040411, IAA400400805) and the Grant Agency of the Czech Republic (104/05/0192, 203/05/0197 and 104/07/0383) for financial support of his research.

References

[1] Kresge, C. T., Leonowicz, M. E., Roth, W. J., Vartuli, J. C., Beck, J. S., *Nature* **1992,** *359,* 710–712.

[2] Beck, J. S., Vartuli, J. C., Roth, W. J., Leonowicz, M. E., Kresge, C. T., Schmidt, K. D., Chu, C. T. W., Olson, D. H., Sheppard, E. W., McCullen, S. B., Higgins, J. B., Schlenker, J. L., *J. Am. Chem. Soc.* **1992,** *114,* 10834–10843.
[3] Kruk, M., Jaroniec, M., Sayari, A., *Langmuir* **1997,** *13,* 6267–6273.
[4] Sayari, A., *Chem. Mater.* **1996,** *8,* 1840–1852.
[5] Oye, G., Glomm, W. R., Vralstad, T., Volden, S., Magnusson, H., Stöcker, M., Sjöbolm, J., *Adv. Colloid Interface Sci.* **2006,** *123–126,* 17–32.
[6] Vartuli, J. C., Degnan, T. F., *Stud. Surf. Sci. Catal.* **2007,** *168,* 837–854.
[7] Liu, Y., Pinnavaia, T. J., *J. Mater. Chem.* **2002,** *12,* 3179–3190.
[8] Čejka, J., Mintova, S., *Catal. Rev.* **2007,** *49,* 457–509.
[9] Ferey, G., *Stud. Surf. Sci. Catal.* **2007,** *168,* 327–374.
[10] Sing, K. S. W., Everett, D. H., Haul, R. A. W., Moscou, L., Pierotti, R. A., Rouquerol, J., Siemieniewska, T., *Pure Appl. Chem.* **1985,** *57,* 603–619.
[11] van Bekkum, H., Flanigen, E. M., Jansen, J. C. (Eds.), *Stud. Surf. Sci. Catal.,* **1991,** *58.*
[12] Čejka, J., van Bekkum, H., Corma, A., Schüth, F. (Eds.), *Stud. Surf. Sci. Catal.* **2007,** *168.*
[13] Galarneau, A., Di Renzo, F., Fajula, F., Vedrine, J. (Eds.), *Stud. Surf. Sci. Catal.* **2001,** *135.*
[14] Kresge, C. T., Vartuli, J. C., Roth, W. J., Leonowicz, M. E., *Stud. Surf. Sci. Catal.* **2004,** *148,* 53–72.
[15] Terasaki, O. (Ed.), *Stud. Surf. Sci. Catal.* **2004,** *148.*
[16] Schüth, F., *Stud. Surf. Sci. Catal.* **2001,** *135,* 1–10.
[17] Roth, W. J., Vartuli, J. C., *Stud. Surf. Sci. Catal.* **2005,** *157,* 91–110.
[18] De Bruyn, M., Limbourg, M., Denayer, J., Baron, G. V., Parvulescu, V., Grobet, P. J., de Vos, D. E., Jacobs, P. A., *Appl. Catal. A* **2003,** *254,* 189–201.
[19] Zhao, D., Huo, Q., Feng, J., Chmelka, B. F., Stucky, G. D., *J. Am. Chem. Soc.* **1998,** *120,* 6024–6036.
[20] Tanev, P. T., Chibwe, M., Pinnavaia, T. J., *Nature* **1994,** *368,* 321–324.
[21] Che, S., Garcia-Bennett, A. E., Yokoi, T., Sakamoto, K., Kunieda, H., Terasaki, O., Tatsumi, T., *Nat. Mater.* **2003,** *2,* 801–805.
[22] Vartuli, J. C., Roth, W. J., Degnan, T. F., Jr., in: Schwartz, J. A., Contescu, C., Putyera, K. (Eds.), *Dekker Encyclopedia of Nanoscience and Nanotechnology,* Marcel Dekker, 2004, P. 1791.
[23] Diaz, I., Marquez-Alvarez, C., Mohino, F., Perez-Pariente, J., Sastre, E., *J. Catal.* **2000,** *193,* 283–294.
[24] Diaz, I., Marquez-Alvarez, C., Mohino, F., Perez-Pariente, J., Sastre, E., *J. Catal.* **2000,** *193,* 295–302.
[25] Perez-Pariente, J., Diaz, I., Mohino, F., Sastre, E., *Appl. Catal. A* **2003,** *254,* 173–188.
[26] Inagaki, S., *Stud. Surf. Sci. Catal.* **2004,** *148,* 109–116.
[27] Corma, A., *Chem. Rev.* **1997,** *97,* 2373–2419.
[28] Taguchi, A., Schüth, F., *Micropor. Mesopor. Mater.* **2005,** *77,* 1–45.
[29] Trong-On, D., in: Pandalai, S. G. (Ed.), *Recent Research Developments in Catalysis,* Research Signpost, Kerala (India), 2003, p. 171.
[30] Trong-On, D., Desplantier-Giscard, D., Danumah, C., Kaliaguine, S., *Appl. Catal. A* **2001,** *222,* 299–357.
[31] Thomas, J. M., Raja, R., *Stud. Surf. Sci. Catal.* **2004,** *148,* 163–170.
[32] Corma, A., Grande, M. S., Gonzalez-Alfaro, V., Orchilles, A. V., *J. Catal.* **1996,** *159,* 375–382.
[33] Roth, W. J., Vartuli, J. C., *Stud. Surf. Sci. Catal.* **2004,** *135,* 134–139.
[34] Mokaya, R., *Chem. Commun.* **2001,** 633–634.
[35] Mokaya, R., *Chem. Commun.* **2001,** 933–934.
[36] Karlsson, A., Stocker, M., Schmidt, R., *Micropor. Mesopor. Mater.* **1999,** *27,* 181–192.
[37] (a) Liu, Y., Zhang, Y., Pinnavaia, T. J., *J. Am. Chem. Soc.* **2000,** *122,* 8791–8797. (b) Zhang, Z., Han, Y., Xiao, F.-S., Qiu, S., Zhu, L., Wang, R., Yu, Y., Zhang, Z., Zou, B., Wang, Y., Sun, H., Zhao, D., Wei, Y., *J. Am. Chem. Soc.* **2001,** *123,* 5021–5024.
[38] Prokešová, P., Mintova, S., Čejka, J., Bein, T., *Micropor. Mesopor. Mater.* **2003,** *64,* 165–174.
[39] Verhoef, M. J., Kooyman, P. J., van der Waal, J. C., Rigutto, M. S., Peters, J. A., van Bekkum, H., *Chem. Mater.* **2001,** *13,* 683–687.
[40] Trong-On, D., Kaliaguine, S., *J. Am. Chem. Soc.* **2003,** *125,* 618–619.

[41] Corma, A., Martinez, A., Martinez-Soria, V., Monton, J. B., *J. Catal.* **1995,** *153,* 25–31.
[42] Song, C., Reddy, K. M., *Appl. Catal. A* **1999,** *176,* 1–10.
[43] Klimova, T., Ramirez, J., Calderon, M., Dominguez, J. M., *Stud. Surf. Sci. Catal.* **1998,** *117,* 493–500.
[44] Vartuli, J. C., Shigh, S., Kresge, C. T., Beck, J. S., *Stud. Surf. Sci. Catal.* **1998,** *117,* 13–24.
[45] Knop, A., Pilato, L. A., Phenolic Resin Chemistry, Springer, Berlin, 1985.
[46] Sakthivel, A., Badamali, S. K., Selvam, P., *Micropor. Mesopor. Mater.* **2000,** *39,* 457–463.
[47] Badamali, S. K., Sakthivel, A., Selvam, P., *Catal. Lett.* **2000,** *65,* 153–157.
[48] Sakthivel, A., Saritha, N., Selvam, P., *Catal. Lett.* **2001,** *72,* 225–228.
[49] Vinu, A., Usha Nandhini, K., Murugesan, V., Böhlmann, W., Umamaheswari, V., Pöppl, A., Hartmann, M., *Appl. Catal. A* **2004,** *265,* 1–10.
[50] Vinu, A., Devassy, B. M., Halligudi, S. B., Hartmann, M., *Appl. Catal. A* **2005,** *281,* 207–213.
[51] Vinu, A., Krithiga, T., Murugesan, V., Hartmann, M., *Adv. Mater.* **2004,** *16,* 1817–1820.
[52] Vinu, A., Krithiga, T., Balasubramanian, V. V., Asthana, A., Srinivasu, P., Mori, T., Ariga, K., Ramanath, G., Ganesan, P. G., *J. Phys. Chem. B* **2006,** *110,* 11924–11931.
[53] Sawant, D. P., Vinu, A., Jacob, N. E., Lefebvre, F., Halligudi, S. B., *J. Catal.* **2005,** *235,* 341–352.
[54] Dhanashri, P., Sawant, D. P., Vinu, A., Lefebvre, F., Halligudi, S. B., *J. Mol. Catal. A* **2007,** *262,* 98–108.
[55] Van Rhijn, W. M., De Vos, D. E., Sels, B. F., Bossaert, W. D., Jacobs, P. A., *Chem. Commun.* **1998,** 317–318.
[56] Lim, M. H., Blanford, C. F., Stein, A., *Chem. Mater.* **1998,** *10,* 467–472.
[57] Harmer, M. A., Farneth, W. E., Sun, Q., *Adv. Mater.* **1998,** *10,* 1255–1260.
[58] Sawant, D. P., Vinu, A., Mirajkar, S. P., Lefebvre, F., Ariga, K., Anandan, S., Mori, T., Nishimura, C., Halligudi, S. B., *J. Mol. Catal. A* **2007,** *271,* 46–56.
[59] Sawant, D. P., Vinu, A., Justus, J., Srinivasu, P., Halligudi, S. B., *J. Mol. Catal. A* **2007,** *276,* 149–157.
[60] Yang, Q., Kapoor, M. P., Shirokura, N., Ohashi, M., Inagaki, S., Kondo, J. N., Domen, K., *J. Mater. Chem.* **2005,** *15,* 666–673.
[61] Yang, Q., Kapoor, M. P., Inagaki, S., Shirokura, N., Kondo, J. N., Domen, K., *J. Mol. Catal. A* **2005,** *230,* 85–89.
[62] Nakajima, K., Tomita, I., Hara, M., Hayashi, S., Domen, K., Kondo, J. N., *Adv. Mater.* **2005,** *17,* 1839–1845.
[63] Iwamoto, M., Tanaka, Y., Sawamura, N., Namba, S., *J. Am. Chem. Soc.* **2003,** *125,* 13032–13033.
[64] Climent, M. J., Corma, A., Iborra, S., Navarro, M. C., Primo, J., *J. Catal.* **1996,** *161,* 783–789.
[65] Sartori, G., Maggi, R., *Chem. Rev.* **2006,** *106,* 1077–1104.
[66] Sarsani, V. S. R., Lyon, C. J., Hutchenson, K. W., Harmer, M. A., Subramaniam, B., *J. Catal.* **2007,** *245,* 184–190.
[67] Shanmugapriya, K., Palanichamy, M., Balasubramanian, V. V., Murugesan, V., *Micropor. Mesopor. Mater.* **2006,** *95,* 272–278.
[68] Vinu, A., Krithiga, T., Gikulakrishnan, N., Srinivasu, P., Anandan, S., Ariga, K., Murugesan, V., Balasubramanian, V. V., Mori, T., *Micropor. Mesopor. Mater.* **2007,** *100,* 87–94.
[69] Zane, F., Melada, S., Signoretto, M., Pinna, F., *Appl. Catal. A* **2006,** *299,* 137–144.
[70] Breda, A., Signoretto, M., Ghedini, E., Pinna, F., Cruciani, G., *Appl. Catal. A* **2006,** *308,* 216–222.
[71] Ernst, S., Bongers, T., Casel, C., Munsch, S., *Stud, Surf. Sci. Catal.* **1999,** *125,* 367–374.
[72] Cauvel, A., Renard, G., Brunel, D., *J. Org. Chem.* **1997,** *62,* 749–751.
[73] Jaenicke, S., Chuah, G. K., Lin, X. H., Hu, X. C., *Micropor. Mesopor. Mater.* **2000,** *35,* 143–153.
[74] Lin, X. H., Chuah, G. K., Jaenicke, S., *J. Mol. Catal. A* **1999,** *150,* 287–294.
[75] Perego, G., Millini, R., Bellussi, G. in: Karge, H. G., Weitkamp, J. (Eds.), Molecular Sieve Science and Technology, Springer, New york, **1998**, p. 187.
[76] Corma, A., Navarro, M. T., Pérez-Pariente, J., *J. Chem. Soc. Chem. Commun.* **1994,** 147–148.
[77] Blasco, T., Corma, A., Navarro, M. T., Pérez-Pariente, J., *J. Catal.* **1995,** *156,* 65–74.
[78] Bhaumik, A., Tatsumi, T., *J. Catal.* **2000,** *189,* 31–39.

[79] Maschmeyer, T., Rey, F., Sankar, G., Thomas, J. M., *Nature* **1995,** *378,* 159–162.
[80] Amore, B., Schwarz, M. B., Chem, S., *Chem. Commun.* **1999,** 121–122.
[81] Davies, L. J., McMorn, P., Bethell, D., Page, P. C. B., King, F., Hancock, F. E., Hutchings, G. J., *J. Mol. Catal. A* **2001,** *165,* 243–247.
[82] Xiao, F.-S., Han, Y., Yu, Y., Meng, X., Yang, M., Wu, S., *J. Am. Chem. Soc.* **2002,** *124,* 888–889.
[83] Vinu, A., Srinivasu, P., Miyahara, M., Ariaga, K., *J. Phys. Chem. B* **2006,** *110,* 801–806.
[84] Galacho, C., Ribeiro, R., Carrott, M., Carrott, P. J., *Micropor. Mesopor. Mater.* **2007,** *100,* 312–321.
[85] Liu, Y., Murata, K., Inaba, M., Mimura, N., *Appl. Catal. A* **2006,** *309,* 91–105.
[86] Lipisardi, G., Chiker, F., Launay, F., Nogier, J.-P., Bonardet, J.-L., *Catal. Commun.* **2004,** *5,* 277–281.
[87] Corma, A., Nemeth, L. T., Renz, M., Valencia, S., *Nature* **2001,** *412,* 423–425.
[88] Renz, M., Blasco, T., Corma, A., Fornes, V., Jensen, R., Nemeth, L., *Chem. Eur. J.* **2002,** *8,* 4708–4717.
[89] Corma, A., Navarro, M. T., Renz, M., *J. Catal.* **2003,** *219,* 242–246.
[90] Ten Brink, G. J., Arends, I. W. C. E., Sheldon, R. A., *Chem. Rev.* **2004,** *104,* 4105–4123.
[91] Corma, A., Navarro, M. T., Nemeth, L. T., Renz, M., *Chem. Commun.* **2001,** 2190–2191.
[92] Nekoksová, I., Žilková, N., Čejka, J., *Stud. Surf. Sci. Catal.* **2005,** *156,* 779–786.
[93] Nekoksová, I., Žilková, N., Čejka, J., *Stud. Surf. Sci. Catal.* **2005,** *158,* 1589–1596.
[94] Kawabata, T., Ohishi, Y., Itsuki, S., Fujisaki, N., Shishido, T., Takaki, K., Zhang, Q., Qang, Y., Takehira, K., *J. Mol. Catal. A* **2005,** *234,* 99–106.
[95] ambert, A., MacQuarrie, D. J., Carr, G., Clark, J. H., *New J. Chem.* **2000,** *24,* 485–488.
[96] owak, I., Kilos, B., Ziolek, M., Lewandowska, A., *Catal. Today* **2003,** *78,* 487–498.
[97] Nowak, I., Ziolek, M., *Micropor. Mesopor. Mater.* **2005,** *78,* 281–288.
[98] Nowak, I., Feliczak, A., Nekoksová, I., Čejka, J., *Appl. Catal. A* **2007,** *321,* 40–48.
[99] Corma, A., Jorda, J. L., Navarro, M. T., Rey, F., *Chem. Commun.* **1998,** 1899–1900.
[100] Bhaumik, A., Tatsumi, T., *Catal. Lett.* **2000,** *66,* 181–188.
[101] Jia, M., Theil, W. R., *Chem. Commun.* **2002,** 2392–2393.
[102] Jia, M., Seifert, A., Theil, W. R., *Chem. Mater.* **2003,** *15,* 2174–2179.
[103] Theil, W. R., Priermeier, T., *Angew. Chem. Int. Ed.* **1995,** *34,* 1737–1738.
[104] Marquez-Alvarez, C., Žilková, N., Perez-Pariente, J., Čejka, J., *Catal. Rev.* **2008,** in press.
[105] Čejka, J., Žilková, N., Kaluža, L., Zdražil, M., *Stud. Surf. Sci. Catal.* **2008,** *50,* 222–286.
[106] Kaluža, L., Zdražil, M., Žilková, N., Čejka, J., *Catal. Commun.* **2002,** *3,* 151–157.
[107] Hicks, R. W., Castagnola, N. B., Zhang, Z. R., Pinnavaia, T. J., Marshall, C. L., *Appl. Catal. A* **2003,** *254,* 311–318.
[108] Balcar, H., Hamtil, R., Žilková, N., Čejka, J., *Catal. Lett.* **2004,** *97,* 25–29.
[109] Hamtil, R., Žilková, N., Balcar, H., Čejka, J., *Appl. Catal. A* **2006,** *302,* 193–200.
[110] Balcar, H., Hamtil, R., Žilková, N., Zhang, Z., Pinnavaia, T. J., Čejka, J., *Appl. Catal. A* **2007,** *320,* 56–63.
[111] Onaka, M., Oikawa, T., *Chem. Lett.* **2002,** 850–851.
[112] Oikawa, T., Ookoshi, T., Tanaka, T., Yamamoto, T., Onaka, M., *Micropor. Mesopor. Mater.* **2004,** *74,* 93–103.
[113] Aguado, J., Escola, J. M., Castro, M. C., Paredes, B., *Appl. Catal. A* **2005,** *284,* 47–57.
[114] Mol, J. C., *Catal. Today* **1999,** *51,* 289–299.
[115] Balcar, H., Žilková, N., Bastl, Z., Dědeček, J., Hamtil, R., Brabec, L., Zukal, A., Čejka, J., *Stud. Surf. Sci. Catal.* **2007,** *170,* 1145–1152.
[116] Topka, P., Balcar, H., Rathouský, J., Žilková, N., Verpoort, F., Čejka, J., *Micropor. Mesopor. Mater.* **2006,** *96,* 44–54.
[117] Buchmeiser, M. R., *New J. Chem.* **2004,** *28,* 549–557.
[118] Mayr, M., Buchmeiser, M. R., Wurst, K., *Adv. Synth. Catal.* **2002,** *344,* 712–719.
[119] Rhers, B., Salameh, A., Baudouin, A., Quadrelli, E. A., Taoufik, M., Copéret, Ch., Lefebvre, F., Basset, J. M., Solans-Monfort, X., Eisenstein, O., Lukens, W. W., Lopez, L. P. H., Sinha, A., Schrock, R. R., *Organometallics* **2006,** *25,* 3554–3557.
[120] Melis, K., Vos, D., De Jacobs, P., Verpoort, F., *J. Mol. Catal. A* **2001,** *169,* 47–56.

[121] De Clercq, B., Lefebvre, F., Verpoort, F., *New J. Chem.* **2002,** *26,* 1201–1208.
[122] Balcar, H., Žilková, N., Sedláček, J., Zedník, J., *J. Mol. Catal A* **2005,** *232,* 53–58.
[123] Corma, A., Garcia, H., Moussaif, A., Sabater, M. J., Zniber, R., Redouane, A., *Chem. Commun.* **2002,** 1058–1059.
[124] Yin, L., Liebscher, J., *Chem. Rev.* **2007,** *107,* 133–173.
[125] Bhanage, B. M., Arai, M., *Catal. Rev.* **2001,** *43,* 315–344.
[126] Mehnert, C. P., Weaver, D. W., Ying, J. Y., *J. Am. Chem. Soc.* **1998,** *120,* 12289–12296.
[127] Hagiwara, H., Shimizu, Y., Hoshi, T., Suzuki, T., Ando, M., Ohkubo, K., Yokoyama, C., *Tetraherdron Lett.* **2001,** *42,* 4349–4351.
[128] Corma, A., Garcia, H., Leyva, A., *Appl. Catal. A* **2002,** *236,* 179–185.
[129] Bulut, H., Artok, L., Yilmaz, S., *Tetrahedron Lett.* **2003,** *44,* 289–292.
[130] Artok, L., Bulut, H., *Tetrahedron Lett.* **2004,** *45,* 3881–3884.
[131] Uozumi, Y., *Top. Curr. Chem.* **2004,** *242,* 77–112.
[132] Mandal, S., Roy, D., Chaudhari, R. V., Sastry, M., *Chem. Mater.* **2004,** *16,* 3714–3718.
[133] Crudden, C. M., Sateesh, M., Lewis, R., *J. Am. Chem. Soc.* **2005,** *127,* 10045–10050.
[134] Shimizu, K., Koizumi, S., Hatamachi, T., Yoshida, H., Komai, S., Kodama, T., Kitayama, Y., *J. Catal.* **2004,** *228,* 141–151.
[135] Horniakova, J., Nakamura, H., Kawase, R., Komura, K., Kubota, Y., Sugi, Y., *J. Mol. Catal. A* **2005,** *233,* 49–54.
[136] Gonzalez-Arellano, C., Corma, A., Iglesias, M., Sanchez, F., *Adv. Synth. Catal.* **2006,** *346,* 1758–1765.
[137] Demel, J., Čejka, J., Štěpnička, P., *J. Mol. Catal. A* **2007,** *263,* 259–265.
[138] Demel, J., Čejka, J., Štěpnička, P., *J. Mol. Catal. A* **2007,** *274,* 127–132.
[139] Demel, J., Park, S.-E., Čejka, J., Štěpnička, P., *Catal. Today* **2008,** in press **2008,** *132,* 63–67.
[140] Lee, H.-Y., Ryu, S., Kang, H., Jun, Y.-W., Cheon, J., *Chem. Commun.* **2006,** 1325–1327.

CHAPTER 26

Catalytic Phases Embedded in Mesostructured Matrices and their Nanocasts: Effects of Spatial Dimension and Assembling Mode on Activity

Miron V. Landau *and* Leonid Vradman

Contents

Abstract

Insertion of catalytic phases inside the pores of mesostructured host matrices allows simultaneous control of the catalytic phase size and assembling modes: nanoparticles ensemble, nanowires or coating layers. The surface area of the catalytic phase-mesostructured host matrix composite can exceed that of pure catalytic phase with $D = 10$ nm only with a nanoparticle ensemble. Removal of the mesostructured host matrix from preformed composites allows exceeding the surface area of pure 10-nm catalytic phase with all the assembling modes. The catalytic activity follows the catalytic phase surface area in cases when the specific activity of the catalytic phase is constant at that size range. This behaviour alters in cases when the diminishing of the catalytic phase size and/or changing the shape significantly affect its specific activity in a selected

Ordered Porous Solids
DOI: 10.1016/B978-0-444-53189-6.00026-3

reaction. The available information about the directing of a catalytic phase to proper assembling modes created in mesostructured host matrices and in corresponding nanocasts and their effects on catalytic performance of catalytic phases is analyzed.

Keywords: Catalytic phases, Host–guest composites, Mesostructured matrices, Performance in catalytic reactions

Abbreviations

CP	catalytic phase,
MHM	mesostructured host matrix,
SA	surface area,
NSA	normalized surface area,
PSD	pore size distribution,
PSC	porous single crystal,
NP	nanoparticles,
CL	coating layer,
NT	nanotubes,
NW	nanowires.

1. Introduction

Catalytic phases (CPs), solids with well-defined structures, represent an important class of catalytic materials.[1,2] These phases include transition metals, their oxides, mixed hydroxides and sulfides, zeolites, pillared clays, heteropolyacids and other compounds. Atoms or groups of atoms in specific coordination at the surface of a CP (e.g. Co–Mo–S atoms at the edge planes of MoS_2 crystals with a layered structure,[3] SO_4–ZrO_x assembles at the surface of ZrO_2 with a tetragonal structure[4] or HO–W groups in a specific array of polyanions with Keggin structure[5])—the 'active sites'—display unique chemical functionality. This functionality that builds efficient catalytic cycles is predetermined by the solids' structure and can be a strong function of the size and shape of CPs' particulates.[6]

The maximization of the catalytic reaction rate (catalysts activity) requires uppermost exposure of these catalytically active atoms or atomic assembles to the reactants. Therefore, the general requirement to the preparation of CP-based catalytic materials is maximal CP surface area. The bulk nanoparticulates of inorganic compounds of different shape can be produced by precipitation (co-precipitation), sol–gel processing, microemulsion and other methods.[7–9] But thermolysis/stabilization needed for the formation of required crystalline structure of these CPs decreases the materials' surface area. Therefore, most of the transition metal oxide phases with well-defined structure can be synthesized without stabilizers with the size parameter of $\geq$10 nm.[7] The bulk nanoparticulates of transition metals can be obtained at a lower size of 1–7 nm by colloidization, nanocasting and other methods.[8,9] But they suffer from sintering at elevated temperatures required for catalytic reactions.

Conducting the same synthetic processes in the confined space of mesopores inside the mesostructured host matrices (MHMs) provides efficient stabilization of required CP structures at high dispersion level.[10,11] Insertion of CPs into the pores of an MHM—silicas of MCM-41, MCM-48, SBA-15, SBA-16 types obtained by surfactant-assisted self-assembling or silica-MHM templated nanocasted carbons like CMK-3 and CMK-5—opens wide opportunities for tailoring the size/shape characteristics of CPs as shown in Fig. 26.1.

The surface area increases with decreasing the dimension parameter of the CP in nanometric range: particle size, nanowires diameter or thickness of the coating layer at the surface of the supporting matrix. The further removal of the matrix provides an opportunity to increase the total CP loading in the material, producing corresponding nanocasts with higher available specific surface area of active phases.[12] It allows to take full advantage of the maximized CP's surface area in materials loaded to catalytic reactors at standard weight/volume amounts in cases where they do not sinter at catalytic reaction conditions.

The rate (W) of a given heterogeneous reaction catalyzed by a solid material is determined by several parameters:

$$W = \mathrm{SA} \times [\mathrm{S}] \times k_0 \exp\left(\frac{-E_\mathrm{a}}{RT}\right) \times F([\mathrm{R_i}], [\mathrm{P_i}]) \qquad (1)$$

where SA—surface area of CP ($m^2\ g^{-1}$), [S]—surface concentration of 'active sites' in CP ($mol\ m^{-2}$), k_o—pre-exponent constant, E_a—activation energy of the limiting step in the catalytic cycle and F—a function defining the effects of reagents' (R)/ products' (P) concentrations beyond the catalyst's surface. Only the surface area is directly determined by the dimension/shape characteristics of the CP, while the other parameters are determined by the structure of its particulates. The composition, electronic states and geometric arrangement of 1–10 surface atom assemblies involved in making and breaking the chemical bonds in adsorbed reagents molecules determine the 'quality' of active sites. It governs the reaction's limiting step (k_o, E_a) and mechanism (F). The quality and concentration of active sites [S] are established

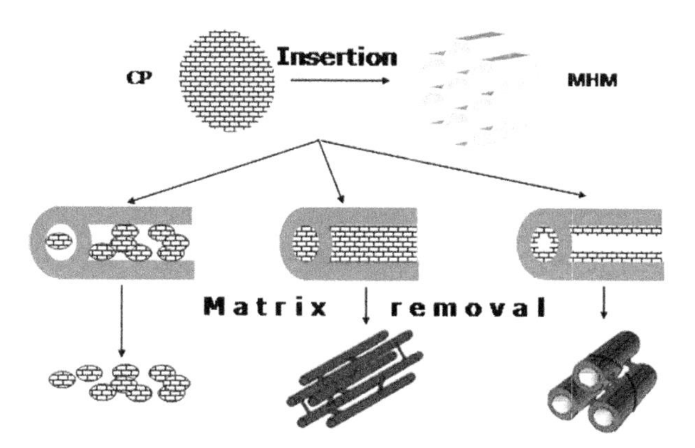

Figure 26.1 Tailoring the shape of CP particulates in the mesostructured host matrix followed by its removal.

by the bulk structure of CP particulates. They may vary in a very complicated way with decreasing size/shape parameters of different CPs depending on many factors.

The structure of atoms in nanoparticulates does not always follow that in the infinite bulk structure taking the configuration that minimizes their volume, crystal facets and edge contributions to the total energy at a distinct size.[13] Decreasing the size of faceted nanoparticles causes an increase in the surface concentration of coordinatively unsaturated atoms located at the crystal edges and corners on expense of the terrace sites. This contribution changes significantly by passing to small spherical nanoparticles, nanorods (nanowires) or nanometric layers (nanotubes) of CPs. Decreasing the dimension of CPs' particulates can alter their wetting ability relative to the surface of the MHM, further affecting their shape and concentration of active sites. Decreasing the metal's crystal size in the 1–10 nm range causes a shift from the bulk metal properties to the insulator electronic properties, creating a gap between the highest occupied and the lowest unoccupied states known as a quantum size effect.[6,8,9] For semiconductors, decreasing the dimension parameter of CP changes the value of this energy band gap, affecting the reactivity of active sites.[6,9] Additional reasons for alteration of catalytic activity at the nanoscale range of CPs' spatial dimensions could be increase in concentration of anion vacancies and defect sites on the crystallite surface, higher electrical conductivity and ion mobility of nanocrystalline surfaces, increase in the crystallographic strain as crystallites become smaller and change the shape and alteration of the exposure of different crystallographic planes of greatly different activities.[15] Therefore, moving forward to the nanostructured catalytic materials based on the CP with the highest surface area, it is necessary to clarify the main routes at this road.

The critical questions for selecting the right preparation strategies of CP/MHM host/guest composites are as follows:

- What shape of CP, nanoparticles, nanowires or nanometric layers converted to corresponding nanotubes after matrix elimination, will give the maximal surface area at equal dimension?
- What is the relative thermostability of these three CP shapes that can determine their selection?
- How to direct the CP in the composite to the desired shape with minimal blocking of MHM pores?
- What is the significance of the dimension/shape effects for the specific activity, and at what size range they are predominant for the practically important combinations 'CP—catalytic reaction'?

To our knowledge, the first three questions were not considered systematically by other authors. The last question was widely debated in the literature in connection with the structure sensitivity effects—correlations between the specific activity and CP crystal size.[14] The new aspects here are the effect of CP surface curvature clearly observed experimentally for spherical nanoparticles, nanowires and nanotubes and opportunities to control the structure–functional relations of a CP by its confined crystallization inside the pores of the MHM.

In the present chapter, we will relate to the first question based on the geometrical modelling of CP surface area in CP/MHM composites and corresponding nanocasts. The other questions will be considered based on the available results of experimental measurements for different CPs aiming to derive some general rules that have proper theoretical basis. The consideration includes only those catalytic systems where the CPs were clearly identified as particulates with distinct atomic structure. The catalysts containing active species in forms of small molecular clusters with less than 40 atoms, isolated metal complexes inserted by surface grafting and anchoring or metal ions inserted by isomorphous substitution of silicon atoms are beyond the scope of the study.

2. Approximation of the CP Surface Area in CP/MHM Composites and Corresponding Nanocasts with a Geometric Model

2.1. Methodology and computational details

The geometrical modelling of the specific surface area of the CP particulates embedded in the MHM was conducted based on the following assumptions that allow fixing the system and clearly comparing the CP shape effects at similar dimension:

- The dimension parameter of the CP particulate (i.e. diameter of the nanoparticles or nanowires; the thickness of the coating layer or the nanotubes walls) is in the range of 1–10 nm.
- The density of MHM-embedded CP particulate is equal to the theoretical density of the CP with infinite bulk structure.
- The CP particulates are non-porous, i.e. there are neither closed nor open pores inside the nanocrystals, nanowires or coating layer.
- The degree of CP particulates agglomeration is negligible; all CP particulates contribute equally to the surface area, excluding the surfaces of nanowires and coating layers attached to the inner surface of MHM pore walls.
- Uniform size distribution of the CP particulates, i.e. all CP particulates are similar in dimension parameter.
- MHM consists of the nanotubular channels of cylindrical form with the uniform diameter corresponding to mesostructured silicas of MCM-41 or SBA-15 with hexagonal array of tubular mesopores.
- In order to get the numerical results for further comparison in case of coating layer (nanotube) CP assembling mode, the MHM is assumed to possess a surface area of 1000 m^2 $g^{-1}$and a pore diameter of 6 nm.
- The surface areas calculated for CPs embedded in the MHM were normalized per gram of catalytic material—CP/MHM, reflecting the catalytic performance of composites. For reference, pure CP and CP nanocasts not containing MHMs, the surface area, were normalized per gram of CP.

The surface area of the spherical or cubic particulates (crystals) of a CP can be calculated according to the following equation:

$$SA_{\text{particle}} = \frac{6000}{\rho \times D} \tag{2}$$

where SA—specific surface area of the CP ($m^2\ g^{-1}$), ρ—CP's theoretical density ($g\ cm^{-3}$), D—particle (crystal) size (nm). For the CP in the form of nanowires with diameter D (nm) or nanotubes with wall thickness t (nm), the surface area was calculated according to Eqs. (3) and (4), respectively:

$$SA_{\text{wire}} = \frac{4000}{\rho \times D} \tag{3}$$

$$SA_{\text{tube}} = \frac{4000}{\rho \times 2 \times t} \tag{4}$$

Notice that Eq. (4) takes into account both inner and outer surface areas of nanotubes. The resulting surface area of the nanotubes is a function of the CP density and wall thickness being independent of the nanotubes' diameter.

For spherical or cubic particulates (crystals) of a CP stabilized inside the MHM, the surface area of the CP per gram of a composite catalytic material CP/MHM was calculated according to the following equation:

$$SA_{\text{particles in MHM}} = \frac{6000}{\rho \times D} \cdot x \tag{5}$$

where x—CP's loading (weight fraction). For a particular CP with a given ρ, the surface area is a function of the x/D ratio. For nanoparticles assembling mode, the CP surface area in the CP/MHM composite does not depend on the MHM parameters and is determined exclusively by particles' dimension, density and loading.

For a coating layer of a CP that could be formed at the internal surface of MHM mesopores, the geometrical model has to take into account the surface area (SA_{MHM}, $m^2\ g^{-1}$) and the pore diameter of the parent MHM (d, nm), the thickness of the covering layer (t, nm) and the fraction of the MHM surface covered with CP (coverage θ, where $0 < \theta \leq 1$). The layer thickness and the coverage are interrelated since at a certain CP loading, a thin layer will cover larger part of the MHM surface compared with the thick layer. As the MHM pores are assumed to be cylindrical with a uniform diameter d and the outer surface of MHM particles is negligible, the surface area of an MHM is equivalent to the inner surface of a single cylinder with length L:

$$SA_{\text{MHM}} = \pi \times d \times L \tag{6}$$

where L represents the sum of lengths of all pores in 1 g of an MHM. Similarly, the CP covering layer surface area could be calculated according to following expression:

$$\mathrm{SA}_{\mathrm{layer}} = \pi \times (d - 2 \times t)L \times (1 - x)\theta \tag{7}$$

Combining Eqs. (6) and (7) to exclude L results in

$$\mathrm{SA}_{\mathrm{layer}} = \mathrm{SA}_{\mathrm{MHM}}(1 - x) \times \frac{(d - 2 \times t)}{d} \times \theta \tag{8}$$

Similar to the layer's surface area, the volume of the layer can be expressed from the geometry of the cylinder or from the CP's density and loading:

$$\theta \times \left(\frac{\pi \times d^2}{4} - \frac{\pi \times (d - 2 \times t)^2}{4} \right) \frac{\mathrm{SA}_{\mathrm{MHM}} \times (1 - x)}{\pi \times d} = \frac{x}{\rho \times 100^3} \tag{9}$$

From this equation, the coverage θ could be calculated for any layer thickness $t < d/2$. As was already mentioned, the coverage and the layer thickness are interrelated. In addition, the minimal CP layer thickness is 1 nm. Thus, to calculate the surface area of the layer according to Eq. (8), it was assumed that $t = 1$ nm for $\theta \leq 1$, where θ is calculated from Eq. (9). At the loadings x higher than required for the full coverage ($\theta = 1$, $t = 1$ nm), the layer thickness increases beyond 1 nm and can be calculated from simple transformation of Eq. (9):

$$t = \frac{d - \sqrt{d^2 - \frac{4 \times d \times x}{(1-x) \times \rho \times 100^3 \times \mathrm{SA}_{\mathrm{MHM}}}}}{2} \tag{10}$$

2.2. CP surface area as a function of its assembling mode

Comparison of the surface area of the bulk CP in the form of nanoparticles (nanocrystals) assemble with $D = 10$ nm [Eq. (2)] with the surface area of CP particles of different dimension D inside an MHM [Eq. (5)] yields the condition for preparation of a CP/MHM composite with the CP's specific surface area higher than that in the bulk reference CP:

$$\frac{6000}{\rho \times D} x > \frac{6000}{\rho \times 10} \quad \text{or} \quad x/D > 0.1 \tag{11}$$

This condition is illustrated in Fig. 26.2: increased CP surface area of the CP/MHM relative to the reference pure CP with a particle size of 10 nm (represented by the solid line $D = 10$ nm at $x/D = 0.1$, i.e. $x = 1$) may be achieved by combination of smaller particle size ($D < 10$ nm) with CP loadings beyond $x = 0.1$ (shadowed area). Increasing the CP dispersion at low loadings (white area under the line $D = 10$ nm),

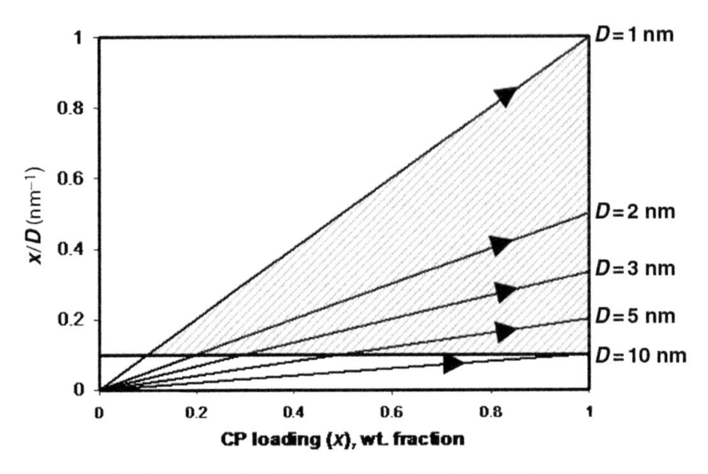

Figure 26.2 Relationship between the CP loading (x) and x/D ratio at different levels of CP dispersion particle size D) for nanoparticles (nanocrystals) assembling mode of CP in the CP/MHM composites.

being a common practice in preparation of supported catalysts, does not allow one to exceed the surface area of the reference bulk material. The CP loading x increases when the matrix is removed from the CP/MHM composite, as a part of the nanocasting strategy. Assuming that the MHM removal does not change the CP's particle size D (moving according to the lines in Fig. 26.2 in directions indicated by swords) and neglecting the effect of particles' aggregation on the CP's surface area, the x/D ratio will increase linearly with a decrease in the matrix content (increasing x), thus increasing the active surface area in the material. At the limiting loading of $x = 1$, the x/D ratio approaches its maximal values as a function of CP's particle size D. Clearly, partial or complete removal of an MHM is an attractive method to obtain high performance catalytic materials with high CP surface area.

Figure 26.3 presents the surface area of a CP in the form of a coating layer as a function of CP loading x for CPs with different densities. The surface area of the

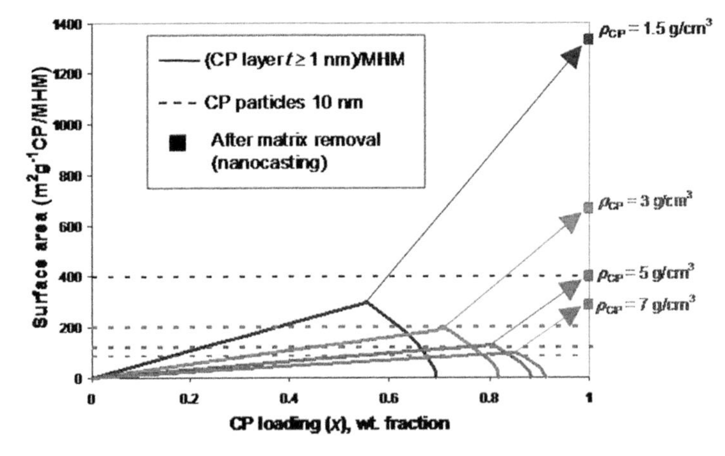

Figure 26.3 Relationship between the CP loading (x) and surface area of CP for the coating layer assembling mode. (See color insert.)

layer was calculated according to Eq. (8) assuming $t = 1$ nm for $\theta \leq 1$, where θ is calculated from Eq. (9). The CP's surface area in CP/MHM composite increases with rise in the CP's loading (x) until full coverage of the MHM surface ($\theta = 1$) with the film of $t = 1$ nm. In addition, it increases with a decrease in the CP's density. After reaching the full MHM surface coverage, any further increase in CP's loading can be achieved only by increasing the layer thickness beyond 1 nm. The corresponding CP's surface areas were calculated by Eq. (10). As a result, the surface area of a CP layer decreases and finally turns to zero when the layer thickness becomes equal to $d/2$ corresponding to the total filling of MHM mesopores with the CP. Figure 26.3 clearly shows that the surface area of the reference bulk CP (indicated by dotted lines for different ρ_{CP}) can hardly be exceeded by forming 1-nm-thick coating layer of the same CP at the walls of MHM mesopores. The surface area of the coating layer can be slightly higher than that of the bulk reference material only for the CP with density higher than 5 g/cm^3 and at loadings higher than 80 wt.%. Gradual MHM removal after full coverage of the MHM surface will move the CP surface area along the straight lines in directions indicated by swords in Fig. 26.3. After matrix removal, as estimated for the full coverage of the MHM surface by a CP layer with 1 nm thickness, the obtained nanotubes with a wall thickness of 1 nm will display the surface area substantially higher compared with that of a CP in form of 10-nm nanoparticles. Increasing the layer (nanotube walls) thickness (t) will proportionally decrease the surface area of nanocasted tubes according to Eq. (4).

In nanorods (nanowires) assembling mode, the CP will completely fill the MHM pores, and so the CP's surface will be accessible for reacting molecules only after matrix removal. In this case, the diameter of the nanowires is equal to the MHM pore diameter. Thus, the nanowires' surface area is a function of the CP's density and matrix (Fig. 26.4).

The surface area of a nanocasted CP in the form of nanorods or nanowires will be higher than that of a CP in the form of 10-nm nanoparticles, only if the matrix diameter is lower than a certain critical value. Comparison of surface area calculated from Eqs. (1) and (2) yields this critical matrix diameter: $4000/(6000/10) = 6.667$ nm, independent on the CP's density. Nanotubes with a wall thickness of 1 nm prepared in the MHM with a pore diameter of 2 nm have no inner surface, being actually nanowires with a

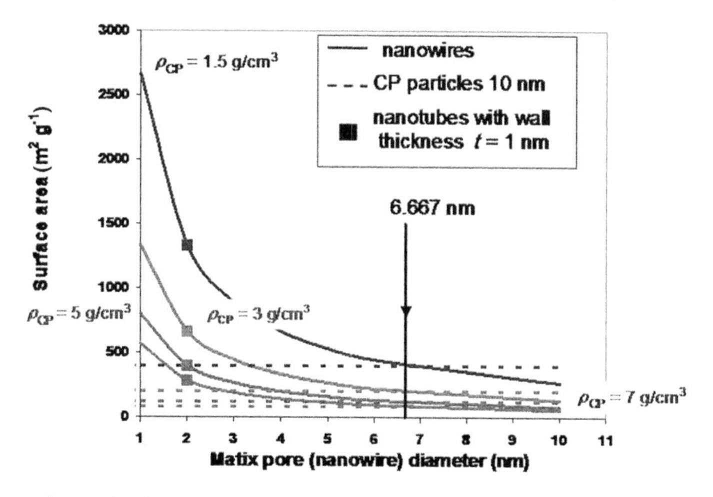

Figure 26.4 Relationship between the MHM pore diameter and surface area of nanocasted wires completely filling the MHM pores before matrix removal. (See color insert.)

2-nm diameter. Their surface areas calculated according to Eq. (4), which in this case yields similar results with Eq. (3), are shown also in Fig. 26.4 as filled squares. It is important to remember, however, that the surface area of nanotubes does not depend on the matrix diameter and is a function only of the layer (wall) thickness and CP's density [Eq. (3)]. The surface area of casted (MHM templated) nanotubes is much higher than that of nanowires prepared by the complete filling of the MHM pores with a CP in a wide range of MHM pore diameters with $d > 2$ nm. But in case of non-complete filling of MHM pores, so that $D_{wires} < d$, the surface area of nanocasted CP tubes became lower than that of corresponding nanowires when $t > D_{wires}/2$.

2.3. Conditions for surface area maximization

The calculations presented above provide a firm basis for formulating the conditions for maximization of a CP's surface area as a function of its dimension parameters and assembling mode inside the MHM and after its elimination. The comparison of dimension parameters and expressions for CP's surface area for different CP assembling modes shown in Table 26.1 allows ranking of the CP assembling modes at equal CP densities and values of dimension parameters.

For CP/MHM composites, the surface area of a CP per gram of composite will decrease in the sequence:

Nanoparticles > Nanowires > Coating layer

For example, the numerical values obtained in this sequence assuming $x = 0.5$ and $D_{part} = D_{wire} = t_L = 1$ nm for a CP with a theoretical density of 5 g cm^{-3} are 600 (nano particles), 400 (nanowires) and 80 (coating layer) m^2 g^{-1}. The reference pure CP in the form of nanoparticles (nanocrystals) with $D_{part} = 10$ nm has a surface area 120 m^2 g^{-1}. For pure CPs fabricated by nanocasting after MHM elimination, the surface area decreases in the sequence:

Nanoparticles > Nanowires > Nanotubes

The numerical values under the same assumptions and $x = 1$ are 1200 (nanoparticles), 800 (nanowires) and 400 (nanotubes) m^2 g^{-1}. At $D_{part} = D_{wire} = t_L = 3$ nm,

Table 26.1 Dimension parameters and expressions for CP surface area for different CP assembling modes

CP assembling mode	Dimension parameter	CP Surface Area (SA): CP/MHM	CP Surface Area (SA): CP nanocast
Nanoparticles (NP)	D_{part}.	$6000{\star}x/\rho{\star}D_{part}$	$6000/\rho{\star}D_{part}$
Coating layer★) (CL)	t_L★★★★)	$SA_{MHM}{\star}(1-x){\star}(d\text{-}2t_L){\star}\theta/d$	–
Nanotubes (NT)	t_T★★★★)	–	$4000/2{\star}\rho{\star}t_T$
Nanowires (NW)	D_{wire}	$4000{\star}x/\rho{\star}D_{wire}$★★★)	$4000/\rho{\star}D_{wire}$★★)

★ d—MHM pore diameter
★★ $D \leq d$
★★★ $D < d$
★★★★ $t_L, t_T < d/2$

a maximal dimension that could be equal for all CP assembling modes in an MHM with $d = 6$ nm, the corresponding values of surface area are 400 (nanoparticles), 267 (nanowires) and 133 (nanotubes) $m^2 g^{-1}$. So, the surface area of a nanocasted CP always exceeds that of the reference CP in the form of nanoparticles with $D_{part} = 10$ nm.

From the previous consideration it follows that direction of the CP assembling to the nanoparticles mode, besides decreasing its dimension parameter, is the condition for maximization of the CP's surface area in both CP/MHM host/guest composites and corresponding nanocasts. But in practice, nanoparticles can yield lower equilibrated CP surfaces due to higher sintering resistance of nanowires and nanotubes assemblies compared with nanoparticles. Therefore, the selection of the optimal CP assembling mode should be done specifically for a given CP/ MHM pair, guest loading and reaction conditions.

The geometrical modelling and conclusions about CP surface area maximization, described above for an MHM with hexagonal array of mesopores, are valid also for other MHMs such as 2D nanocasted carbons built by ordered arrays of nanowires (CMK-3) or nanotubes (CMK-5)[16], or SBA-16 with 3D cubic array of interconnected cavities[17]. In all cases, formation of CP nanoparticles or nanowires yields CPs whose surface area is determined by the same Eqs. (2) and (3), respectively [or Eq. (5) in case of nanoparticles inside any MHM]. On the contrary, creation of a coating layer at the surface of SBA-16 and CMK carbons cannot be handled by Eqs. (7–10) because the surface area of these MHMs cannot be approximated by single cylinder assumption [Eq. (6)]. However, the surface of a coating layer actually duplicates the surface of the parent MHM to lessen the decrease in pore size (due to the thickness of the layer) and to lessen the dilution of the MHM by a CP. Thus, the surface area of a coating layer is a function mostly of MHM surface area and much less depends on the MHM pores' shape and arrangement. Therefore, the general conclusions made above about the surface area of the layer coating SBA-15 type MHMs are valid also for MHMs with more complex pore structure like CMKs and SBA-16.

The provided geometrical consideration does not take into account the microporosity of the MHM mesopore walls in silica MHM SBA-15 and SBA-16 that also can accumulate CPs. In these cases, two corrections should be done in calculations conducted above in order to obtain the right values of CPs' surface area: to consider the surface of only MHM mesopores [SA_{MHM} in Eqs. (7–10)] and to correct the CP loading x for the amount of catalytic material plugged in the walls as proposed in Baca *et al*[18]. These corrections do not change the general conclusions made above about the effect of dimension/shape parameters of CPs assembled in the MHM mesopores.

3. Control of CP Assembling Mode in a Mesostructured Matrix

The CP/MHM composites are prepared by insertion of corresponding CPs or their precursors in the mobile form of solution, vapour or melt to the pore system of an MHM at its synthesis stage or by postsynthetic treatments followed by special activation procedures. In general case, it yields the nanoparticulate shape of CPs where nanoparticles of various sizes (0.1–20 nm) are randomly distributed in mesopores or/and micropores and at the external surface of MHM crystals. The common practice is to direct the nanoparticles' location to the MHM intraporous space where the pore walls limit their growth. A large variety of preparation strategies were successfully developed for this purpose. For insertion of CP precursors, they use highly enhanced host–guest interactions including *chelation forces*, *capillary forces* and *hydrophilic/hydrophobic affinity* of the MHM pore walls enriched with silanols or grafted with hydrophilic/hydrophobic moieties. After the insertion step, the nanoparticles are fixed inside the pores as a result of precursor's solidification by gelation, crystallization or decomposition. Nearly all the known CPs were successfully inserted as nanoparticles in different MHMs, and their characteristics are considered in many comprehensive reviews.[10–12,19–23] Depending on the physicochemical properties of the selected CP, its nanoparticles can be inserted into the MHM by addition of corresponding precursors to the sol–gel mixture for MHM synthesis, impregnation of the MHM with the related solution followed by thermolysis, reduction or other treatment, chemical vapour deposition or chemical solution decomposition, interaction with the functional groups at the MHM pore walls, ion exchange with metal cations or even incorporation of prefabricated nanoparticles.[11,19] There are developed strategies for the insertion of nanoparticles exclusively to the micropores located in the walls of silica MHMs[24,25] and selective deposition at the external surface of MHM crystals.[26]

The proper control of the CP shape inside the MHM, that is, directing from the nanoparticles to the nanorods (nanowires) or coating layer, is the issue that was extensively developed in the last years. It was not considered in details in previous reviews, and will be analysed below as an attempt to classify the preparation strategies according to CP assembling modes.

3.1. Nanowires (Nanorods)

The elongated generally cylindrical nanoparticles of CPs that differ by the aspect ratios: 3–5 for nanorods and >10 for nanowires, can be formed inside the mesoporous channels of MHMs. After MHM removal, they appear as isolated disordered or interconnected particles with mesoscopic order. The latter are often called porous single crystals (PSC). The two most successful strategies for formation of nanowires of inorganic phases (CP) inside the mesoporous channels of MHMs are (i) *bulk filling synthesis* and (ii) *scrolling of internal coating layers*.

The *bulk filling* approach was pioneered by Martin,[27] Moskowitz[28] and Searson[29] for preparation of isolated nanorods/nanowires of conductive polymers, metals and semiconductors. It consisted in complete or partial filling with CPs or their precursors the bulk of the pores (0.02–0.5 μm) in anodic alumina membranes, sheets of single-crystal mica or polycarbonate membraness treated by nuclear track etching. This *bulk filling* was conducted by electrodeposition technique using a galvanostatic electroplating circuit. The plated templates were placed at the working electrode with the pores' axes perpendicular to the electrode surface and inserted in the excess of corresponding electrolyte or monomers solution. The nanowires' diameter was always equal to that of templates' pores, and the length can be controlled by the electrodeposition duration. Later, this technique was successfully applied for formation of nanowires of CdS[30] and Pd[31] in thin films of SBA-15 MHM. The careful control of the electrodeposition process allowed stopping the pores filling process when the length of the nanowires achieved the thickness of the template film excluding deposition of the excess CP layer at the external surface of the MHM. The nanowires appeared as oriented cylindrical monocrystals of corresponding phases with 6.5 nm diameter and completely filled the MHM pores at high CP loadings of 40 wt.% (CdS) or higher for Pd. They followed the swirling mesostructure of the MHM film after liberation from the silica matrix. This method is highly reproducible, yielding uniform and complete filling of all MHM mesopores with CP. However, its application is limited only to the MHM thin films. Therefore, other techniques were developed for *bulk filling* of the mesopores that can be applied to MHM crystal particulates.

The main problem that should be solved here is how to achieve the dense radial and axial filling of MHM mesoporous channels at least at the part of their length with the CP or its precursor, rather than uniform (but not dense enough) CP dispersion. The simplest way for *bulk filling* of mesopores is inserting of large amounts of guest materials inside the MHM working at high loadings. But even in this case the measures should be taken against moving the corresponding mobile hosts (solutions, melts) from mesopores to the external MHM surface caused by capillary forces at the drying-calcination step, and their splitting inside the mesopores into nanodroplets being precursors of isolated nanoparticles. These two targets can be achieved by combination of several techniques with a complementary action. They include hydrophilization/hydrophobization of the pore walls' surface via removing the surfactant by microwave-induced digestion in HNO_3-H_2O_2-H_2O solution that increases the surface silanols' concentration[32] or grafting the moieties containing Si–CH_3 groups.[33] These surface modifications increase the affinity of MHM pores to the solutions of CP's precursors in water or organic solvents, respectively. Multiple MHM impregnations with solutions of CP's precursors followed by solvents' evaporation is a common practice for reaching the high (50–70 wt.%) loadings of CP nanowires.[34–37,54] In order to minimize the deposition of CP nanoparticles at the external surface of MHM crystals, it was proposed to dissolve the excess of precursor solidified after solution evaporation in a small amount of CH_2Cl_2, so that the formed solution was inserted back into the mesopores by the capillary forces.[35,38] This permitted one to obtain Pt nanowires that fill only part of mesoporous channels in MHM crystals at low CP loadings. Another original technique providing better control of high mesopores filling with CPs nanowires is the so-called 'two solvents method'.[39–41] Pioneered by Davidson,[39] it consists of insertion of dehydrated MHMs in the organic solvent (n-C_6) poorly miscible with water followed by contacting the obtained suspension with the volume of aqueous CP precursor solution set equal to the MHM pore volume. This brings about insertion of all the aqueous solution to the MHM pores due to their hydrophilic affinity. After separation of organics, slow drying at room temperature and transformation of corresponding precursor to the CP, the *bulk filling* of MHM pores with MnO_2,[39] Cr_2O_3[40] or $ZnFe_2O_4$[41] nanowires can reach 97%.

Direct impregnation of the MHM with the same aqueous solution yielded CP nanoparticles (MnO_2) inside the pores and at the external surface of the MHM.[39]

The infiltration of the melted CP precursors to the MHM[42–45] is a viable alternative to the *bulk filling* of its mesoporous channels with precursors solutions. The viscous hydrated melts have lower mobility inside the MHM mesopores at elevated temperatures required for their decomposition and CP solidification. Because of low surface tension they tend to the *bulk filling* of the MHM mesopore channels yielding nanowires after decomposition–solidification without separation into droplets as do aqueous solutions. The main requirement here is that the melting point of the CP precursor should be lower than its decomposition temperature. So, the successful formation of Cr-, Ni, Co- and Ce-oxide nanowires in SBA-15 and KIT-6 MHM was observed using corresponding nitrate melts, while the Pb-nitrate and $(NH_4)_2Cr_2O_7$ salts decomposing before melting yielded oxide nanoparticles at the external MHM surface.[42] It looks like the intermediate arrangement of the precursors melt inside the pores during evaporation of the solvent after MHM impregnation with corresponding solutions favoured the formation of CP nanowires in many cases without being manifested by the authors.[32,36,40,41,46–48] In this respect, the effect of the solvent used for preparation of CP precursors impregnating solutions should be mentioned.[49,50] Its low surface tension, C_2H_5OH (22 mN m^{-1}) versus H_2O (72 mN m^{-1}), was critical for preparation of Ag-nanowires in SBA-15 MHM using $AgNO_3$ solution.[49] The *bulk filling* of the mesopores with C_2H_5OH solution and then with the melt after solvents evaporation yielded nanowires after salts decomposition at 300 °C, while separated droplets of impregnated aqueous solution gave isolated spherical Ag nanoparticles. Most likely, for the same reason, the addition of 2-propanol to an aqueous solution of H_2PtCl_6 precursor was needed to get Pt nanowires inside FSM-15 and FSM-16 MHMs after photoreduction at room temperature.[50]

The additional measures directing the assembling of guests CPs in the MHM to the nanorods/nanowires mode are intraporous seeding and induced low-temperature solidification. After *bulk filling* of the tubular mesopore with the mobile CP precursor or its solution, the solidification of the CP should be conducted in a way yielding continuous nanowires. In the pioneer work in which the nanowire of a CP (Pt) was obtained for the first time by Ryoo and Ko,[51] they first formed small Pt-clusters at 0.1 wt.% loading in MCM-41 by cationic exchange-H_2 reduction and then inserted significant amount of Pt (5 wt.%) by impregnation–evaporation. The small Pt clusters formed first served as seeds for the formation of Pt wires at the subsequent H_2-reduction step. After initial isotropic growth, the wire diameter reaches the MHM pore size and so they can only proceed to grow along the pore axis. The efficiency of this strategy was further confirmed in publications dealing with CP loadings up to 60–70 wt.%.[34,36,52] The formation of Pt nanowires inside the MHM pores by the photoreduction of H_2PtCl_6 was attributed to reduction of $PtCl_6{}^{2-}$ ions catalyzed by the small Pt nanoparticles formed first.[50] This approach was modified by inserting palladium into SBA-15 structure at its crystallization step.[53] Pd nanoclusters formed after H_2-reduction served as catalysts in the electroless deposition of Au nanowires inside the pores after immersing the MHM in excess of $HAuCl_4$–NH_2OH aqueous solution.[53] A similar method was used for the fabrication of Ni and Cu nanowires in SBA-15 channels inserting first Pd catalyst and then conducting the electroless deposition from Cu- and Ni-nitrates solution by reaction with formaldehyde.[64]

Since the formation of nanowires is favoured by the low-temperature solidification inside the pores, minimizing the mobility of CP precursors and solvents evaporation, many techniques were developed for doing it. Besides photoreduction of platinum[50] and catalytic reduction of Au[53] mentioned above, the CP nanowires were produced inside the MHM pores by treatment of $CeCl_3$ solution inside SBA-15 with gaseous NH_3 at 25 °C to deposit solid Ce-hydroxide converted then to CeO_2 at 700 °C.[54] MnO_2 nanowires were deposited by sonication of aqueous $KMnO_4$ solution inside the pores of SBA-15[55] and oxidation of Mn^{2+} ions to Mn^{4+} by reaction with surface silanols of SBA-15 at 80–120 °C.[39,56] Ag nanowires were obtained inside SBA-15 channels at room temperature by reaction $AgNO_3 \rightarrow Ag + NO + O_2$ in the chamber of electronic microscope under high-energy electronic beam.[43]

The *scrolling of internal coating layers* strategy for production of nanowires inside the MHM pores requires chemical modification of the matrix pores surface with $-NH_2$,[55,57–60] $-SH$[61] or vinyl[37] groups. Acid–base interaction between surface $-NH_2$ groups and acidic molecules ($H_3PW_{12}O_{40}$,[57,95] $H_2Cr_2O_7$,[59] $H_4PVMo_{11}O_{40}$, $H_5PV_2PMo_{10}O_{40}$[60]) or chelation of transition metal ions (MnO_4^-,[55] Fe^{3+}[58] with $-NH_2$ groups, $AuCl_4^-$[61] with $-SH$ groups and Co^{2+}[37] with vinyl groups) from impregnation solution form a very thin layer of corresponding molecules or ions on the mesoporous surface. This enhances the diffusion of the additional ions or molecules into the MHM channel system[61] and favours their condensation reactions. So, the formed layer acts as both nanoreactor and reactant.[62] As a result, the nanowires are growing inside the MHM mesopores due to the folding-scrolling of formed compact CP layer and its filling with other reactants that are transported from the solution into scrolls by diffusion and heat movement.[62,63] In such a way, nanowires of WO_3,[57,95] Cr_2O_3,[59] MoO_3,[60] MnO_2,[55] Fe_2O_3,[58] Co_3O_4[37] and Au^o were synthesized.[61]

Many catalytically active metals, metal oxides, sulfides and multimetallic oxide phases were prepared in the form of nanorods/nanowires hosts combining several discussed synthetic protocols relevant to the physico-chemical properties of MHMs, CP precursors and CP. Regardless of the preparation strategy used, CP composition and loading, the nanowires' diameter always corresponds to the pore diameter of the MHM with the accuracy of one atomic layer. So, the spatial dimension of nanowires varies in the range of 2.5–8.0 nm. The aspect ratio of these particles ranges from 15 to 200–300, which corresponds to nanowires and not to nanorods. In most cases, the nanowires have a monocrystalline structure with no specific relation between the crystal orientation and the orientation of mesopore structure (Fig. 26.5a). After MHM removal, the mesoscopic order is preserved and PSC materials are formed when MHMs with interconnected mesopores are used as templates (Fig. 26.5 a,b). Disordered nanowires are obtained if synthesized in isolated mesopores like in MCM-41(Fig. 26.5c). The disordered nanowires can be obtained using MHMs with interconnected pores like SBA-15 when the microporosity responsible for nanowires connection is masked during the silane modification.[61] When

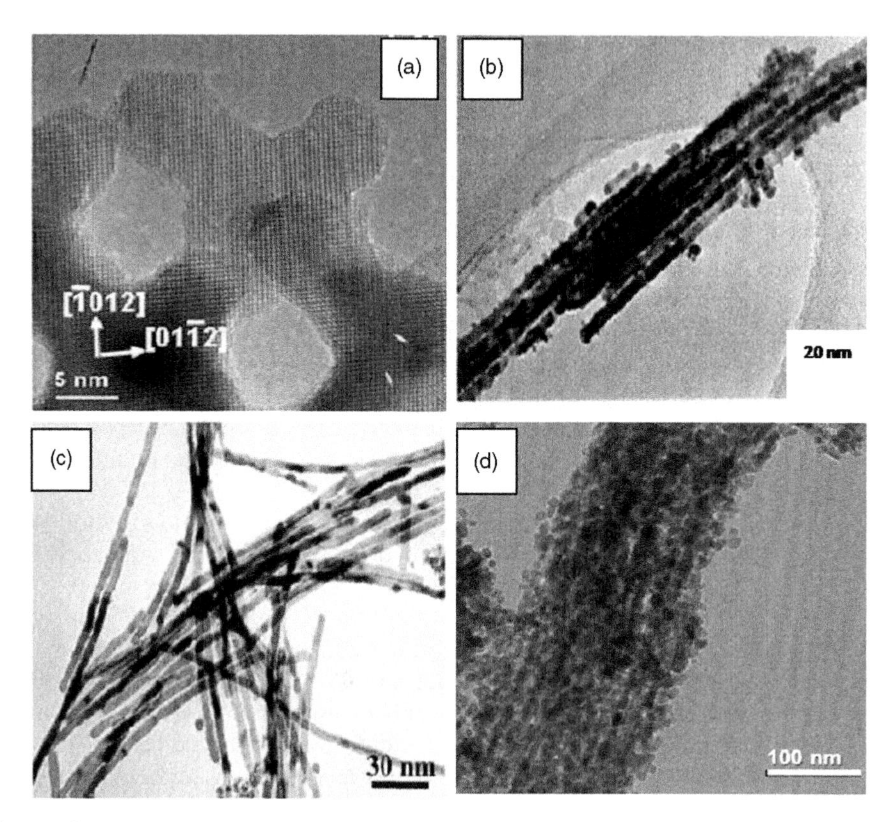

Figure 26.5 Mesoscopic arrays of nanowires mediated by formation of CP/MHM composites: (a) Cr_2O_3 PSC (KIT-6);[40] (b) Pt PSC (SBA-15);[38] (c) disordered Pt nanowires (MCM-41);[34] (d) jointed nanoparticles arrays $ZnFe_2O_4$ (SBA-15).[41]

crystallization of a CP requires calcination at high temperatures beyond the stability limit of the MHM structure, the nanowires represent ordered arrays of jointed globular nanocrystals ($ZnFe_2O_4$, 900 °C[41]; Fig. 26.5d) or dense arrays of short nanorods (900 °C, GaN[33]). The ordered arrays of jointed Pt and Au nanocrystals were also obtained inside the MHM film by the impregnation-H_2 reduction approach.[65] It was explained by the involvement of a 3D hexagonal structure in the major 2D hexagonal network of the mesoporous thin film.

There is no information about the thermostability of metal oxide, sulphide or mixed oxide nanowires or corresponding PSC isolated from the parent MHMs. But the metallic, that is, Pd, nanowires isolated from MCM-41 MHM demonstrated relatively low thermal stability.[44] Their shapes began to deteriorate at 200 °C, and their complete melting was observed at 300 °C, which is more than 1100 °C lower compared to the bulk melting point of Pd. The Pt nanowires began to agglomerate upon heating in He at 300 °C, significantly lower than the bulk melting point, because of the nanoscale dimension.[34]

The nanowire arrays of CPs liberated from parent MHMs by treatment with aqueous NaOH or HF solutions demonstrated relatively low surface areas: 55 m^2g^{-1}

($ZnFe_2O_4$[41]), 142 m^2 g^{-1} (MnO_2[56]), 91 m^2 g^{-1} (MnO_2[55]), 47 m^2 g^{-1} (NiO[47]), 56 m^2 g^{-1} (NiO[32]), 92 m^2 g^{-1} (Co_3O_4[46]), 122 m^2 g^{-1} (Co_3O_4[37]), 82 m^2 g^{-1} (Co_3O_4[32]), 74 m^2 g^{-1} (Cr_2O_3[46]), 58 m^2 g^{-1} (Cr_2O_3[59]), 86 m^2 g^{-1} (Cr_2O_3[40]), 65 m^2 g^{-1} (Cr_2O_3[32]), 137 m^2 g^{-1} (Fe_2O_3[32]), and 70 m^2 g^{-1} (In_2O_3[32]). The reported surface areas are in good agreement with the predictions obtained from Eq. (3) using theoretical density of corresponding CPs and existing nanowires' diameters. In these calculations, it can be assumed that part of the nanowires' surface at their interconnections in PSCs is not accessible for nitrogen adsorption.

In parallel with these achievements, the nanowires of many industrially important CPs like oxides of tin, zirconium, titanium and aluminium were not synthesized in mesoporous MHM templates. The scientific basis for formation of CP nanowires inside MHMs should be further developed. One of the questions that should be clarified is the integrity of nanowires' assembly inside the microcrystals of the MHM hard template. Since the density of the CP precursors (i.e. the corresponding salts melts or solutions) is significantly lower then that of the final CP, part of the MHM pore volume remains CP-free after precursors conversion to the final state. So, the PSC of many CPs, detected by high resolution transmitting electron microscopy (HRTEM) after MHM removal, are fragments formerly disintegrated in the volume of parent MHM particles. The high integrity of PSC is important for application in microelectronic devices, sensors, etc., challenging this question for further investigations.

3.2. Coating layer

A layer of CP with a thickness of several atoms can be obtained at the surface of the mesoporous channels inside the MHM by three general techniques: (i) *walls decoration with CP,* (ii) *grafting of CP precursor* and (iii) *spreading CP on MHM surface*. The first strategy consists in directing the labile precursor molecules to the surface of mesoporous channels via chemical interaction with the pore walls, resulting in *immediate* formation of the desired CP layer. A layer of metallic Pd with thickness of 0.6 nm was synthesized at the surface of SBA-15 mesoporous channels functionalized with grafted thrimethoxysilane species after the removal of surfactants.[66–68] The mobile Pd^{2+}cations from the impregnation solution of $PdAc_2$ reacted with the surface Si–H groups according to reaction 2Si–H + $Pd(OAc)_2 \rightarrow 2\ Si(OAc) + Pd^{o} + H_2$ yielding a coating layer of metallic palladium. The external surface of MHM crystals was passivated by Si-CH_3 groups before surfactant removal. Reaction of Sn vapour with the silanols at the surface of the SBA-15 MHM channels according to reaction Sn + Si–OH $\rightarrow$ SiO_2–SnO_2 + H_2 at 700 °C coated the pores' surface with a thin layer of SnO_2.[69] Tin existed in this layer in forms of tetrahedral, octahedral and twofold coordinated oxygen-deficient ions bonded to silica atoms through oxygen bridges, and hexacoordinated ions of amorphous tin oxide formed as a result of condensation of tin ions with oxygen. The $Mo(CO)_6$ in decalin solution containing dissolved oxygen reacted with the surface of the pore walls in Al-MCM-41 MHM under ultasonication, forming a layer of hydrated molybdenum oxide species according to the scheme presented in Fig. 26.6.[73] Based on MAS NMR data, it was

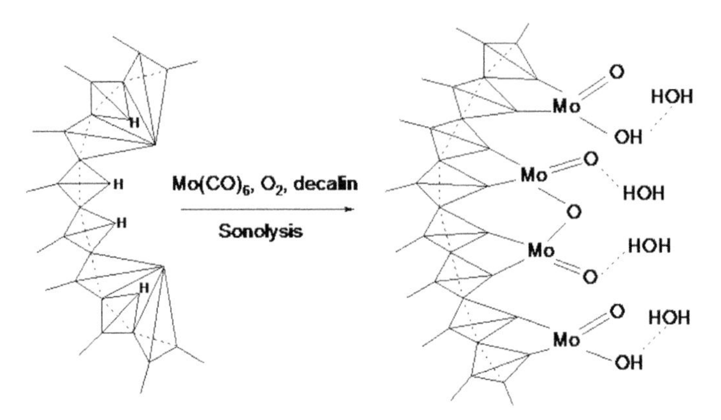

Figure 26.6 Scheme of sonolysis of $Mo(CO)_6$ at the Al-MCM-41 surface.[73]

concluded that sonication promotes the splitting of siloxane bridges in strained Q^4 silica units, which increases the MHM capacity for anchoring the MoOx species beyond that expected from the surface concentration of silanols after surfactant removal. The strong interaction between $W(CO)_6$ and pore walls during chemical solution decomposition of this CP precursor inside the SBA-15 MHM yielded a WO_3 coating layer.[80] But decomposition of Pt-carbonyl inside the FSM-15 and FSM-16 MHMs pores yielded discrete Pt^o nanocrystals due to poor interaction of this metallic CP with the pore walls.[50]

Another walls *decoration* approach consists in insertion of CP precursors to the starting gel for crystallization of MHMs. Specific chemical interaction with surfactant molecules directs the dissolved mobile precursor species to the surfactant–silica interface. This facilitates the formation of the CP coating layer at the channels surface during the subsequent surfactants removal and interaction of the precursor with silica walls. In such a way, the SBA-15 MHM was coated with SnO_2[70] and MgO[71,72] layers. It was found that the acidity (HCl concentration) of the crystallization solution is critical for the shape of the formed SnO_2 layer.[70] This is a result of changing the protonation extent of hydrated PEO surfactant components and the charge on silica species that control the chemical interaction of both of them with tin chloride (precursor of tin oxide). At highly acidic conditions the coating layer consisted of agglomerated tin oxide nanoparticles, while at higher pH the layer was uniform, being formed in the hydrophilic corona area of the surfactant micelles. The arrangement of the MgO layer was mediated by the formation of helical crown-ether-like metal POE complexes binding the mobile Mg^{2+} ions from $MgAc_2$ solution to the hydrophilic corona area of surfactants micells.[71] The concentration of surface basic sites with the basic strength corresponding to that in pure MgO linearly increased with an increase in the MgO loading in the MgO/SBA-15 composite from 10 to 30 wt.%. This can reflect the growing of MHM surface coverage in this loading range. It was also shown that formation of the MgO layer efficiently protects the SBA-15 MHM structure from degradation by treating with KNO_3.[72] This is evident for uniformity of the coating layer.

The *grafting* technique consists of reacting the precursor's molecules with silanols in the MHM, which stabilizes them at the mesopores surface by covalent bonds. In this case, in contrast to decoration, an additional treatment is needed in order to convert the grafted precursors to the final CP coating layer. Several examples of coating of the MHM surface with TiO_2, WO_3, MoO_3, VO_x and Al_2O_3 CP's monolayers are reported in literature based on the following *grafting* chemistry: nSi–OH + $Ti(iPrO)_4$ → $(Si–O)_n$ – $Ti(i\text{-}PrO)_{4-n}$ + nPrOH,[74,77,81,82] nSi–OH + $TiCl_4$ → $(Si–O)_n$ – $TiCl_{4-n}$ + nHCl,[75] nSi–OH + WCl_6 → $(Si–O)_n$ – WCl_{6-n} + nHCl,[76,77] nSi–OH + $V(EtO)_3$ → $(Si–O)_n$ – $V(EtO)_{3-n}$ + nEtOH,[77] nSi–OH + $MoOCl_4$ → $(Si–O)_n$ – $MoOCl_{4-n}$ + nHCl,[77] nSi–OH + $Al(sec\text{-}BuO)_3$ → $(Si–O)_n$ – $Al(sec\text{-}BuO)_{3-n}$ + $n$sec–BuOH,[78] nSi–OH + $Al(i\text{-}PrO)_3$ → $(Si–O)_n$ – Al(i-PrO)$_{3-n}$ + nPrOH.[18,79] At the second step of the synthetic protocol, the grafted species were converted to corresponding oxide CPs by hydrolysis, leading to formation of surface hydroxides followed by their thermal dehydration,[74,75,78] or directly converted to metal oxides by calcination in air.[18,76,77,79,81,82] In order to increase the MHM capacity for CP in form of the coating layer, it was proposed to treat the surfactant-containing MHM at the first *grafting* stage.[79] This increased the surface concentration of silanols in the MHM. The *grafting* approach allows a reliable control of the thickness of the coating layer by repetition of the *grafting*–hydrolysis–dehydration steps, so that at later *grafting* stages the precursor is chemically bonded already to the CP spread on the MHM surface.[75,78] This yielded a gradual decrease in the pore diameter in the SBA-15 MHM from 6.7 to 3.2 nm corresponding to the final thickness of the TiO_2 coating layer of 1.75 nm.[78] The chemical interaction of precursors with the MHM surface is critical for creation of a coating layer assembly: hydrolysis or thermal decomposition of titanium alkoxide dissolved in the solvent that filled the MHM pores yielded TiO_2 in the form of discrete nanoparticles.[83,84] Therefore, the unreacted precursors should be carefully removed from the MHM after the first grafting step conducted with an excess of solution.[18,74–79] As an alternative, the precursor can be inserted by incipient-wetness impregnation keeping the balance between its amount and the amount of silanols in the MHM.[81,82]

Formation of a solid CP film at the surface of the MHM can be considered in terms of wetting the MHM surface with CP and *spreading* of the latter along the MHM surface. The theoretical basis for this phenomenon related to the zero contact angle between catalytic and MHM (support) phases is considered in the comprehensive review by Knözinger and Taglauer.[85] The smaller CP particles spread on the supports surface more easily. For every CP/MHM pair there exists a critical radius of CP particle for spreading below which the CP can spread completely across the MHM as a monolayer patch. It becomes clear that the *decoration and grafting* strategies described above yield the coating layer assembling because they force CPs to appear at a very high dispersion level. But there are several examples when the coating layer is obtained by the synthetic protocols that are not purposely designed to produce very small clusters of CPs at the MHM surface.

The multistep impregnation of SBA-15 MHM with aqueous yttrium nitrate solution followed by calcination at 600 °C yielded a coating layer of yttrium oxide with the thickness of 0.4 nm at loadings up to 30 wt.%.[86] Implementation of this strategy yields nanowires and/or nanoparticles assembly for many other CPs. The authors explained the observed effect by the strong interaction between SBA-15 MHM and YO_x CP. Yttriia belongs to the family of rare earth oxides that strongly interact with silica forming stable mixed oxides. Increasing the CP/MHM interaction energy should enlarge the critical CP radius for spreading at elevated calcination temperature,[85] decreasing the demand for the initial dispersion level. The mixed oxide layer of about 0.5 nm thickness was obtained after deposition of mixed Ce–Mn–oxide phase inside SBA-15 MHM channels by internal co-gelation of corresponding hydroxides followed by calcination at 350–600 °C.[87] The CP/MHM interaction in this case was so strong that the crystallization of corresponding Mn_2O_3 and CeO_2 oxide phases occurred only after heating to 700 °C. Similarly, the deposition of Zr-hydroxide inside the pores of SBA-15 MHM by internal hydrolysis of $Zr(n\text{-}PrO)_4$[83] or interaction of $ZrOCl_2$ with urea in aqueous solution[89] yielded a thick amorphous zirconia layer stable against crystallization at high temperature. Chemical solution decomposition of $Zr(n\text{-}PrO)_4$ dissolved in the solvent that filled the MHM pores yielded crystalline ZrO_2 in the form of discrete nanoparticles with well-defined tetragonal structure.[83] This can be explained by strong interaction of the Zr-hydroxyde phase with SBA-15 silica favouring its spreading on the MHM surface in contrast to the ZrO_2 phase. In such a way can be rationalized the observation of coating layers of WO_3 and W_2C CP in SBA-15 after impregnation of the MHM with the aqueous ammonium paratungstate solution followed by calcination at 550 °C and carburization at 700 °C.[88] The strong interaction of WO_3 and W_2C with silica favours the formation of coating layer by spreading of these CPs on the MHM surface without special measures for their initial dispergation. Noticeably, the isolation of WO_3 CP from the silica surface at the preparation stage by MHM sililation yields nanowires due to scrolling of the internal oxide layer.[57]

The thickness of the CP layer in the MHM estimated directly from HRTM images or from comparison of pore size distribution in the MHM before and after coating varied in the range of 0.4–2.0 nm at the loading of CPs 5–40 wt.%. The normalized surface area calculated as proposed in Vradman *et al.*[90] is in the range of 0.9–1.1, which is evident for minimal MHM pores blocking. There are not yet available data about the successful liberation from the MHM of CPs arranged as coating layers to form nanotubes assemblies. The surface areas of coated CP/MHM composites are relatively high depending on the MHM and CP type, CP loading and coating methods: 379 $m^2\ g^{-1}$ (Pd[66]), 532–733 $m^2\ g^{-1}$ (Al_2O_3[78]), 290–490 $m^2\ g^{-1}$ (Al_2O_3[18]), 452–916 $m^2\ g^{-1}$ (Al_2O_3[79]), 691–835 $m^2\ g^{-1}$ (TiO_2[81]), 900 $m^2\ g^{-1}$ (TiO_2[82]), 619 $m^2\ g^{-1}$ (TiO_2[76]), 665–700 $m^2\ g^{-1}$ (TiO_2[74]), 150–300 $m^2\ g^{-1}$ (TiO_2[75]), 395 $m^2\ g^{-1}$ (ZrO_2[83]), 483 $m^2\ g^{-1}$ (MoO_3[73]), 622 $m^2\ g^{-1}$ (VO_x[76]), 586 $m^2\ g^{-1}$ (WO_3[76]), 713–1277 $m^2\ g^{-1}$ (SnO_2[70]), 150–400 $m^2\ g^{-1}$ (YO_x[86]), 336–629 $m^2\ g^{-1}$ (MgO[71]). These figures could be hardly correlated with the estimations made using Eqs.(7–10). Being measured by non-selective N_2-adsorption, these values reflect the total surface area of CP/MHM composites including the unknown silica surface areas not covered with CP and the micropore surface in MHMs. The latter can reach 28–40% of the total surface area of CP/MHM composite.[18,70]

4. Evaluation of CP Assembling Mode in CP/MHM Composites and Their Nanocasts

The assembling mode of a CP is usually evaluated by combining several characterization techniques including N_2 adsorption–desorption (N_2-AD), small and wide angle X-ray diffraction (XRD), HRTEM and scanning electron microscopy (SEM) both united with the energy dispersive X-ray spectrometry (EDS). XRD provides structural parameters such as CP particle size and MHM pore wall thickness as well as both CP and MHM crystallographic arrangement. However, it does not provide direct information about the CP location and assembling mode. From N_2-AD measurements, textural parameters such as pore size distribution, micro- and mesopores volume and specific surface area are derived. Proper interpretation of the changes in these parameters as well as the changes in the shape of N_2–AD isotherms after CP insertion provides vital information about the location and assembling mode of a CP in an MHM. Electron microscopy offers direct spatially resolved information on the actual location and distribution of the CP phase within the MHM support. Although TEM is a very powerful technique, it reflects only a very small fraction of material at relevant magnifications. In addition, the resulting images are a 2D projection of a 3D structure. Complex 3D structures can be imaged using electron tomography (ET), though there are only few examples in the literature that apply ET for characterization of CP/MHM composites.[91] In addition to those methods, various spectroscopic techniques (X-ray photoelectron spectroscopy, XPS, solid state MAS NMR, UV-vis, Raman, FTIR) are occasionally employed mainly to explore the chemical state of a CP and the nature of its interactions with the MHM support. The use of various techniques for the evaluation of CP assembling mode is illustrated below.

In characterization of the nanocrystal ensembles inside the MHM, besides the size and the structure of the CP, two quite complicated issues have to be resolved: whether the CP crystals are located inside or outside the MHM pores and what is the extent of MHM pores blocking by the CP. TEM and XRD provide information about the spatial dimension, the structure and, in some cases, location of the nanocrystals. TEM is especially useful when the distance between the atomic layers in the nanocrystals is large enough to allow taking clear images of nanocrystals inside the pores. For example, Fig. 26.7a clearly shows the WS_2 nanoparticles occluded within the SBA-15 nanotubes.[92] Parallel fringes running across the nanoparticle images have a periodicity of 6.2 Å, which corresponds to the relatively large distance of 6.13 Å between the atomic layers packed along the *c*-axis in the hexagonal WS_2 structure. In addition, the local EDS yielded W contents similar to the average value obtained from SEM–EDS. No particles of pure WS_2 phase were detected outside the SBA-15 pores in WS_2/SBA-15 composite taking the micrographs from several different area of the sample. This confirms that all WS_2 particles are located only inside the SBA-15 pores.

In contrast to that, the crystal domain size (XRD) of the metallic Ni^o phase in the Ni/SBA-15 composite was much higher than the pore diameter, indicating that at least part of the Ni^o-phase was located outside the MHM pores.[90] Clear images of

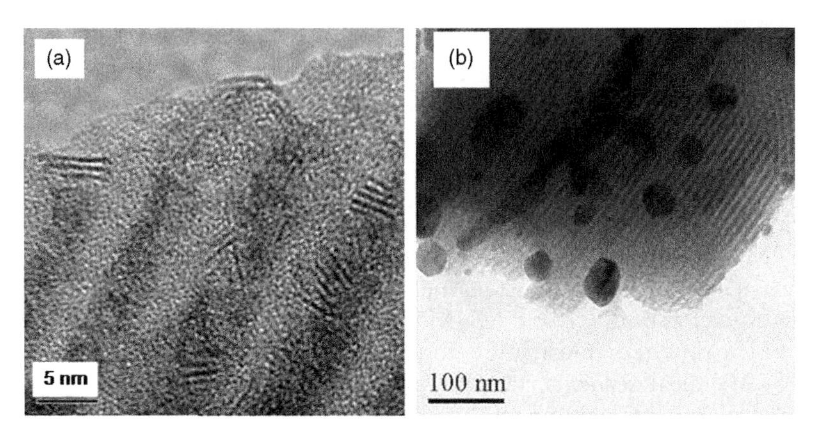

Figure 26.7 TEM micrographs of the WS_2/SBA-15[92](a) and Ni^o/SBA-15[90] (b).

globular particles of a size much larger than the pore diameter of SBA-15 were obtained by HRTEM (Fig. 26.7b). The local EDS analysis taken from the areas of these particles showed a very high concentration of Ni, while the analysis taken from empty SBA-15 channels showed very low Ni concentration. This completed the XRD and TEM data, showing that at least part of Ni^o-phase is located outside the SBA-15 pores. These examples show that a proper combination of characterization techniques such as TEM, EDS and XRD can distinguish the inside and outside location of a CP.

N_2-AD data can be used to uncover the effect of MHM pores blocking by a CP. The specific surface of the CP/MHM composite is always lower than that of the parent MHM as a result of dilution of the MHM with the CP. Thus, in order to estimate the pore blocking effect, the surface area of the composite should be normalized with the content of the MHM. The normalized surface area (NSA)[73,90] analysis also helps to identify the CP assembling mode:

- NSA < 1 indicates significant blocking of the pores (Fig. 26.8—samples MoO_3/MCM–EV, WO_3/SBA–EV32, ZrO_2/SBA–EV50). CP in these composites is assembled in the form of large particles that block MHM pores.
- NSA $\sim$1 indicates the absence of pores blocking and that the CP exists in the form of a layer that covers the MHM surface (Fig. 26.8—samples MoO_3/MCM-US and WO_3/SBA-CSD). It should be noticed that the mechanical mixtures of the CP with MHM powders also represent material with NSA $\sim$1.[90] Indeed, the contribution of the bulk CP with its usually large particles to the total composite surface area is so low that the surface area of the remaining parent MHM is actually measured. Thus, NSA data can characterize the coating layer assembling mode only together with corresponding XRD, TEM and spectroscopic data (see below).
- NSA > 1 indicates not only the absence of the blocking of the MHM pores but also significant contribution of the CP nanocrystals ensembles to the total surface area of the CP/MHM material (Fig. 26.8—samples TiO_2/SBA-IH, TiO_2/SBA-CSD, NiO/SBA-IG, ZrO_2/SBA-CSD).

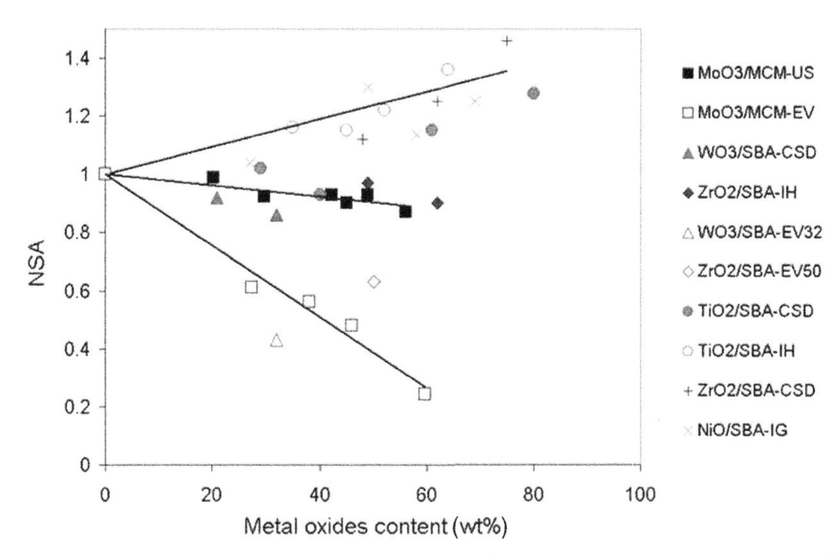

Figure 26.8 Normalized surface area (NSA) measured for different CP/MHM composites.[90]

The coating layer assembling mode is the most complicated for characterization. The main reason for this is that the layer is usually very thin and exhibits XRD-amorphous nature. Thus, XRD and TEM mainly give proof of the high dispersion of the CP and do not provide data about its location and assembling mode, which makes it difficult to distinguish several possible cases: monolayer covering the MHM surface, multilayer or even amorphous particles. The narrowing of mesopore channels and increasing the thickness of the pore walls that can be observed at the HRTEM micrographs can serve as an evidence for formation of the CP coating layer.[73,80,83,86,87,89] The formation of the Pd metallic layer coating SBA-15 surface increased the intensity of small angle XRD peaks, since metal coating increases the scattering contrast between pores and walls.[66–68] Spectroscopic techniques may be used to identify CP/MHM interactions that serve as indirect evidence of the formation of the coating layer that interacts with the MHM surface. For example, XPS and NMR applied to the amorphous Mo-oxide phase deposited on the wide pore Al-MCM-41 detected chemical interactions between the Mo-phase and surface silica atomic layer forming surface silicates (Fig. 26.6).[73] In addition to the detected chemical interactions, N_2–AD pointed on the decrease of pore diameter from 8.3 nm in the parent MHM to 7.7 nm in the Mo-oxide/MHM composite. This corresponded to the calculated thickness of the closed-packed Mo-oxide monolayer. The NSA decreased only as a result of the narrowing of the MHM mesopores (Fig. 26.8—samples denoted as MoO_3/MCM-US) and was close to unity at all loadings. The pore size distributions (PSDs) of the MoO_3/Al-MCM-41 samples were very similar to that measured for the parent Al-MCM-41 with a small shift to lower pore diameters. This shift is a result of the formation of Mo-oxide phase monolayer on the surface of the MHM mesopores. Thus, all the characterization methods suggested that the amorphous Mo-oxide phase was located inside the MHM mesopores in the form of a closed-packed monolayer. Interestingly, the results of characterization of amorphous Zr-oxide phase deposited on the SBA-15 by internal hydrolysis were very similar to those of MoO_3/Al-MCM-41 samples: amorphous nature (XRD, TEM) and NSA close to unity (Fig. 26.8—sample ZrO_2/SBA-IH). Only careful analysis of N_2–AD data distinguished the assembling mode of Mo-oxide and Zr-oxide amorphous phases. Insertion of ZrO_2 into SBA-15 led to a marked change in the shape of the hysteresis loop. The hysteresis loop of the loaded material was characteristic of a percolation effect caused by nanoparticles settling within the mesopores, effectively forming ink-bottle–type pores.[89] PSD shifted significantly to the lower pore diameters, reflecting the filling of the SBA-15 mesopores with nanoparticles (or thick non-uniform layer) and not the formation of a monolayer. This example indicates the vital role of N_2–AD measurements in characterization of CP/MHM composites.

The approach for characterization of nanowires is in many aspects similar to that of nanocrystals inside the MHM except two main differences: the evaluation of continuous nature of CP nanowires and characterization of nanowires after matrix removal. TEM is a common tool to evaluate a continuous nature of CP nanowires. They appear on the TEM images as dark continuous rod-like objects inside the MHM following the direction of the channels. The nanowires have uniform spatial dimensions consistent with the pore diameters of the MHM. TEM also clearly shows the free-standing nanowires after MHM removal: nanowires diameter and length, arrangement and linkage between them (Fig. 26.5). Electron[42,46,95] or XRD[34,58,59] diffraction patterns of MHM-free nanowires arrays evaluate their structure and arrangement, which normally corresponds to the order in the parent MHM. The N_2–AD isotherms are helpful in evaluation of the texture characteristics of nanowires assemblies liberated from MHMs. Their position and shape also indicate the formation of nanowires inside the MHM channels. After CP insertion they are shifted down to the lower amounts of adsorbed nitrogen without changing their shape and position parallel to P/P_o axis. Such behaviour without shifting of PSD is interpreted as a result of complete filling of part of mesoporous channels leaving other channels CP-free.[40]

5. Effect of CPs' Spatial Dimension and Assembling Mode on Their Catalytic Performance

When CPs are embedded in the MHM as NP with the spatial dimension of >3 nm that excludes the quantum size effects, their catalytic activity in many cases is proportional to the specific surface area of the corresponding phase.[10,11,22,23] The effects of nanoscopic assembly of CPs inside the MHMs on their performance in catalytic reactions were never studied systematically. The investigations that correctly compare the catalytic activity–selectivity–stability of any CP in any catalytic reaction varying the assembling mode as NP-NW-CL at fixed CP spatial dimension are not published yet. Today is built the basis for such measurements since several CPs are already synthesized in the same MHM in all the three assembling modes like metallic Pd (CL;[66–68] NW;[31,44,45] NP[24,93,94]) or WO_3 (CL;[76,80] NW; [57,95] NP[80]). Significant efforts should be made in order to make possible such comparison for other practically important CPs. Now this information is limited by experiments aiming to evaluate the performance of NW or NL of different CPs relative to their NP assembling mode without fixing the spatial dimension parameters.

In spite of lower CP surface area, the NW assembling can display proper advantages relative to NP shape. Three reasons were established for it: (i) modification of the chemical state of CP surface atoms relative to that in NP due to the larger contacting interface and stronger interaction with the MHM pore walls; (ii) selective exposure of specific crystallographic planes at the NW surface due to the oriented growth of NW along the tubular MHM pores and (iii) better diffusion/transport of reagents and products in the regular mesopores of PSC relative to disordered pore structure in NP nanocasts. Additional performance improvement by directing the CP from the NP to the NW shape was obtained for the composite catalytic materials prepared with a binder. In this case, implementation of PSC reduces the contacting of the binder with the active CP surface.

Comparison of the catalytic activity of Fe_2O_3 NW with that of Fe_2O_3 NP embedded in SBA-15 MHM in hydroxylation of phenol by H_2O_2 showed the advantage of NP assembly.[58] In order to compensate the lower surface area of NW and get similar phenol conversions at equal products selectivities, the CP loading was increased twice in the case of NW. But the NW assembly displayed substantially higher leaching stability compared with NP. It was attributed to the stronger interaction of the Fe_2O_3 phase with MHM walls. Significantly higher catalytic activity of Pt NW embedded in FSM-16 MHM compared with Pt NP at similar loading and spatial dimension of 2.5 nm was measured in two reactions: hydrogenolysis of *n*-butane [96,97] and water–gas shift reaction.[50] In spite of 2.4–2.9 times lower Pt surface area measured for NWs by H_2 and CO chemisorption, they displayed 12.4 or 36× higher TOF in butane hydrogenolysis at 606 K calculated per total or surface Pt atoms, respectively. A similar effect was observed in $CO + H_2O \leftrightarrow CO_2 + H_2$ reaction at 373 K. The TOF calculated per surface Pt atoms was 10× higher and the CO_2 production rate about 3× higher for NW compared with NP.[50] The observed effects were explained by changing the chemical state of surface Pt atoms in NW, making them electron deficient due to strong interaction with MHM pore walls. This was confirmed by XPS, XANES and FTIR characterization. Another possible explanation is predominant exposure of Pt [111] plane at the NW surface detected by HRTEM. The NWs of both CPs–Fe_2O_3 and Pt–were not liberated from MHM before testing of their catalytic performance. Since the NW diameter was comparable with the diameter of mesoporous channels, it remained unclear how the reacting molecules reached the CP surface. The walls between the channels in SBA-15 MHM contain micropores that can hardly serve for molecules transport. The FSM-16 MHM synthesized from kanematite is an analogue of MCM-41 and consist of non-microporous walls. May be formation of NW inside the curved MHM channels causes their local degradation, which increases the accessibility of the active surface? This question should be further investigated.

Significantly higher reactivity of MnO_2 NW liberated as PSC from SBA-15 MHM relative to the NP assembly templated by disordered silica-gel was measured in electrochemical redox cycle $MnO_2 + Na^+ + e^- \leftrightarrow MnOONa$.[56] This effect observed at similar surface areas of MnO_2 materials was accredited to the two advantages of NW assembly: the selective exposure of [111] crystallographic planes (d = 0.469 nm) at the surface of MnO_2 NW compared with [100] plane (d = 0.241 nm) in NP assembly, and about 3× higher pore volume of MnO_2 PSC. This facilitates the reaction at the crystals surface possessing larger net space for Na^+ insertion between the atomic layers and diffusion of Na^+ ions into and/or out of the MnO_2 matrix. A similar effect was measured comparing the performance of NiO NW and NP assemblies in the electrochemical redox cycle $NiO + OH^- \leftrightarrow NiOOH + e^-$.[47] The advantage of Pt-Ru NW assembly as a catalyst in the membrane electrode of the direct methanol fuel cell versus NP of Pt-Ru black was observed in the electrooxidation of methanol.[35] The NWs were liberated as PSC from SBA-15 MHM and had 1.5× lower surface area compared with corresponding NPs. In spite of it, the higher catalytic activity was measured with Pt-Ru NW yielding about 3× higher current density at potentials >0.7 V compared with NP. It was explained by better reagents/products mass transport in mesopores PSC and smaller contact of catalysts surface with Nafion binder possessing low electrical conductivity.

The main difference of CL from NP and NW shapes of CPs is significantly stronger interaction of the former with the MHM pore walls. It powerfully affects the CP structure in the CL, and this effect is extended beyond the monolayer thickness. Therefore, in most cases the catalytic performance of CPs CL is significantly different from that displayed by CPs of the same composition in 'bulk' NP and NW shapes. It also determines the strong dependence of the CPs' structure and performance on the CP loading. The nature of CP moieties changes along with the surface coating and layer thickness. The CL assembling provides substantial advantages in catalytic performance due to minimal blocking of the MHM pore system with CPs when their favourable structure can be properly controlled.

Testing the SBA-15 MHM covered with the WO_x layer by grafting method in dehydration of methanol revealed a volcano shape of the curve, representing the reaction rate dependence on CP loading.[76] The isolated monotungstate species grafted at the surface of MHM pore walls displayed low activity. The reaction rate reached a maximum at surface coverage slightly beyond monolayer due to condensation of surface species into polytungstate layer, and then strongly dropped at 200% monolayer coverage after formation of 3D WO_x NPs. The WO_x-SBA-15 composite prepared by impregnation displayed 2.5 lower activity at the same CP loading due to sintering of CL in presence of water vapour produced in catalytic reaction that changed the CP assembling mode to NPs. Similar volcano plots were observed by increasing the CP loading by coating the MCM-41 MHM with Al_2O_3[18,78] and alumina-grafted SBA-15 with $H_3PW_{12}O_{40}$.[98] The nature of observed maxima is determined by the requirements of the testing reaction to the chemical functionality of active sites. For cumene cracking and isopropanol dehydration with Al_2O_3-MHM composites, it was a result of formation of Brønsted acid sites due to insertion of tetrahedraly coordinated Al atoms in silica network at low CP loadings and subsequent blocking of these sites with non-active (cumene) or less active (i-PrOH) Al_2O_3 species.[18,78] In contrast to this, a gradual increase in catalytic activity with an increase in the Al_2O_3 loading up to 38 wt.% was observed in alkylation of phenol with methanol.[78] In this case, the catalytic activity of the alumina layer with ~2 nm thickness was 2.3× higher compared with the assembly of γ-Al_2O_3 NPs of the same spatial dimension and similar total surface area. This is a result of strong increase in the medium strength acid sites concentration at the surface of alumina CL with amorphous ill-defined structure. The higher activity of alumina CP in the form of CL is a result of substantially different local arrangement of Al atoms with unusually high concentration of pentacoordinated species confirmed by NMR.[99] The amorphous structure of Mn-oxide in a CL of mixed MnO_x–CeO_2 oxides in SBA-15 MHM caused low activity in catalytic wet oxidation of 2,4,6-trichlorophenol.[87] Conversion of this CL to agglomerates of well-crystallized Mn_2O_3–CeO_2 NPs of the same spatial dimension and their further liberation from the MHM increased the catalysts' activity by a factor of 3. This is an example when stabilization of unfavourable CP structure in a coating layer deteriorates the catalysts' performance.

Implementation of CP coating layer with the same structure as in CP NPs is favourable for catalytic activity in cases when it is impossible to stabilize the NPs with the same spatial dimension inside the MHM. The Pd CP in the form of CL with 0.5 nm thickness at the surface of SBA-15 MHM displayed significantly higher activity in Heck carbon–carbon coupling reactions of aryl halides compared with Pd deposited on Nb-MCM-41 by CVD.[66,68] The CVD yielded a material with 22.3 wt.% Pd loading and 32% dispersion (140 $m^2\ g^{-1}$ Pd) that corresponds to the particle size of 3.6 nm [Eq. (2)].[100] The Pd coating layer displayed equal reagent conversions and product yields at CP loading of only 5 wt.%.[68] This corresponds to more than 4× higher activity per gram of loaded Pd. Formation of MoO_x monolayer in MCM-41 MHM as precursor for further modification with Co and conversion to a mixed sulphide CP increased its available surface area by a factor of 4 compared with that obtained after insertion the CP as NPs by impregnation method.[73] This caused a twice higher catalytic activity in hydrodesulfurization of dibenzothiophene at equal CP loadings.

6. Conclusions

The geometric modelling showed that surface areas of catalytic phases in MHMs and corresponding nanocasts are determined by their assembling mode. At the same value of CP spatial dimension it decreases in the sequence nanoparticles > nanowires > coating layer (nanotubes). The consideration of methods for regulation of the assembling modes of CPs and their spatial dimensions in MHMs established some general rules that allow one to direct the catalytic phase to the shape of nanoparticles, nanowires or coating layer. In special cases, defined based on the available experimental data, the nanowires and coating layer assembling modes can present advantages in catalysts activity–selectivity–stability in spite of lower surface area of CPs. The control of catalytic phase assembling mode in catalytic phase–mesostructured host matrix composites and their nanocasts is a viable tool for improvement of catalytic performance. Its proper implementation requires systematic study of the shape effects on the properties of nanostructured catalytic phases templated by MHMs. This study is now at its initial step. Many questions like availability of nanowires surface for reacting molecules inside the mesostructured host matrix, exact structures of CPs in largely amorphous ill-defined coating layers, structure–functional relations governed by catalytic phases assembling modes and formation mechanisms of different CPs shapes in MHMs of diverse origin should be carefully investigated.

ACKNOWLEDGEMENTS

This study was supported by the Israeli Science Foundation, grant 739/06. The authors thank Mrs. Diana Olvovsky for help in selection and classification of the literature data.

REFERENCES

[1] Hagen, J., *Industrial Catalysis. A Practical Approach*, Wiley-VCH, Weinheim, 1999.
[2] Chorkendorff, I., Niemantsverdriet, J. W., *Concepts of Modern Catalysis and Kinetics*, Wiley-VCH GmbH& Co. KGaA, 2003.
[3] TØpsoe, H., Clausen, B. S., Massoth, P. E., in: Anderson, J. R., Boudart, M. (Eds.), *Hydrotreating Catalysis*, Springer Series CATALYSIS – Science and Technology, 1996,Vol. 112, pp.155–230.
[4] Song, X., Sayari, A., Sulfated Zirconia-Based Strong Solid-Acid Catalyst: Recent Progress. *Catal. Rev. Sci. Eng.* **1996,** *38,* 329–412.
[5] Misono, M., Unique acid catalysis of heteropolycompounds (heteropolyoxometalates) in the solid state. *Chem. Commun.* **2001,** 1141–1152.
[6] van Santen, R. A., Neurock, M., *Molecular Heterogeneous Catalysis*, Wiley-VCH GmbH& Co. KGaA, 2006.
[7] Landau, M. V., Transition metal oxides, in: Schüth, F., Sing, K. S. W., Weitkamp, J. (Eds.), *Handbook of Porous Solids*, Vol. 3, Wiley-VCH, Weinheim, 2002, pp. 1677–1765.
[8] Cushing, B. L., Kolesnichenko, V. L., O'Connor, C. J., Recent Advances in the Liquid-Phase Syntheses of Inorganic Nanoparticles. *Chem. Rev.* **2004,** *104,* 3893–3946.
[9] Thomas, P. J., O'Brien, P. O., Recent Developments in the Synthesis, Properties and Assemblies of Nanocrystals. in: Rao, C. N. R., Miller, A., Anthony, K. (Eds.), *Nanomaterials Chemistry,* Wiley-VCH, Weinheim, 2007, pp. 1–44.
[10] Taguchi, A., Schüth, F., Ordered mesoporous materials in catalysis. *Micropo. Mesopor. Mater.* **2004,** *77,* 1–45.
[11] Landau, M. V., Vradman, L., Wolfson, A., Rao, P. M., Herskowitz, M., Dispersions of transition metal-based phases in mesostructured silica matrix: preparation of high-performance catalytic materials. *Cumpt. Rend. Chimie* **2005,** *8,* 679–691.
[12] Valdes-Solis, T., Fuertes, A. B., High-surface area inorganic compounds prepared by nanocasting techniques. *Mater. Res. Bull.* **2006,** *41,* 2187–2197; Yang, H., Zhao, D., Synthesis of replica mesostructures by nanocasting strategy. *J. Mater. Chem.* **2005,** *15,* 1217–1231.
[13] Uppenbrink, J., Wales, D. J., Structure and energetics of model metal clusters. *J. Chem. Phys.* **1992,** *96,* 8520–8534.
[14] Freund, H.-J., Model studies on heterogeneous catalysts at the atomic level. *Catal. Today* **2005,** *100,* 3–9; Somorjai, G. A., Molecular concepts of heterogeneous catalysis. *THEOCHEM* **1998,** *424,* 101–117; Haber, J, Concept of structure-sensitivity in catalysis by oxides. *Stud. Surf. Sci. Catal.* **1989,** *48,* 447–467.
[15] Moser, W. R., Find, J., Emerson, S. C., Krausz, I. M., Engineered synthesis of nanostructured materials and catalysts. *Adv. Chem. Eng.* **2001,** *27,* 1–48.
[16] Kruk, M., Jaroniec, M., Kim, T. W., Ryoo, R., Synthesis and Characterization of Hexagonally Ordered Carbon Nanopipes. *Chem. Mater.* **2003,** *15,* 2815–2823.

[17] Sakamoto, Y., Kanjedo, M., Terasaki, O., Zhao, D. Y., Ki, J. M., Stucky, G. A., Shin, H. J., Ryoo, R., Direct imaging of the pores and cages of three-dimensional mesoporous materials. *Nature* **2000,** *408,* 449–451.
[18] Baca, M., de la Rochefoucauld, E., Ambroise, E., Krafft, J.-M., Hajjar, R., Man, P. P., Carrier, X., Blanchard, J., Characterization of mesoporous alumina prepared by surface alumination of SBA-15. *Micropo. Mesopor. Mater.* **2008,** *110,* 232–241.
[19] Bronstein, L. M., Nnoparticles made in mesoporous solids. *Top. Curr. Chem.* **2003,** *226,* 55–89.
[20] Chen, C. L., Mou, C. Y., Mesoporous Materials as Catalysts Supports. in: Zhou, B., Hermans, S., Somorjai, G. A. (Eds.), *Nanotechnology in Catalysis,* Kluwer Acad. Plenum Publishers, 2004, pp. 313–327.
[21] Shi, J., Hua, Z., Zhang, L., Nanocomposites from ordered mesoporous materials. *J. Mater. Chem.* **2004,** *14,* 795–806.
[22] On, D. T., Desplantier, D., Danumah, C., Kaliaguine, S., Perspectives in catalytic applications of mesostructured materials. *Appl. Catal. A* **2003,** *253,* 545–602.
[23] Schüth, F., Wingen, A., Sauer, J., Oxide loaded ordered mesoporous oxides for catalytic applications. *Micropor. Mesopor. Mater.* **2001,** *44–45,* 465–476.
[24] Yuranov, I., Kiwi-Minsker, L., Buffat, P., Renken, A., Selective Synthesis of Pd Nanoparticles in Complementary Micropores of SBA-15. *Chem. Mater.* **2004,** *16,* 760–761.
[25] Wang, W., Song, M. L., Synthesis and characterization of SBA-15 with titania clusters entrapped in micropores. *J. Nonryst. Solids* **2006,** *352,* 3153–3157.
[26] Rao, P. M., Landau, M. V., Wolfson, A., Shapira-Tchelet, A. M., Herskowitz, M., Cesium salt of a heteropolyacid in nanotubular channels and on the external surface of SBA-15 crystals: preparation and performance as acidic catalysts. *Micropor. Mesopor. Mater.* **2005,** *80,* 43–55.
[27] Foss, C. A., Tierney, M. J., Martin, C. R., Template Synthesis of Infrared-Transparent Metal Microcylinders: Comparison of Optical Properties with the Predictions of Effective Medium Theory. *J. Phys. Chem.* **1992,** *96,* 9001–9007.
[28] Preston, C. K., Moskowits, M., Optical Characterization of Anodic Aluminum Oxide Films Containing Electrochemically Deposited Metal Particles. 1. Gold in Phosphoric Acid Anodic Aluminum Oxide Films. *J. Phys. Chem.* **1993,** *97,* 8495–8503.
[29] Whitney, T. M., Searson, P. C., Jiang, J. S., Chien, C. L., Fabrication and magnetic properties of arrays and metallic nanowires. *Science* **1993,** *261,* 1316–1319.
[30] Gu, J., Shi, J., Chen, H., Xiong, L., Shen, W., Ruan, M., Periodic Pulse Electrodeposition to Synthesize Ultra-high *Density* CdS Nanowire Arrays Templated by SBA-15. *Chem. Lett.* **2004,** *33,* 828–829.
[31] Wang, D., Zhou, W. L., McCaughy, B. F., Hampsey, J. E., Ji, L., Jiang, Y.-B., Xu, H., Tang, J., Schmehl, R. H., O'Connor, C., Brinker, C. J., Lu, Y., Electrodeposition of Metallic Nanowire Thin Films Using Mesoporous Silica Tmplates. *Adv. Mater.* **2003,** *15,* 130–132.
[32] (a)Tian, B., Liu, X., Yang, H., Xie, S., Yu, C., Tu, B., Zhao, D., General Synthesis of Ordered Crystallized Metal Oxide Nanoarrays Replicated by Microwave-Digested Mesoporous Silica. *Adv. Mater.* **2003,** *15,* 1370–1374; (b) Tian, B., Liu, X., Solovyov, L. A., Liu, Z., Yang, H., Zhang, Z., Xie, S., Zhang, F., Tu, B., Yu, C., Terasaki, O., Zhao, D., Facile Synthesis and Characterization of Novel Mesoporous and Mesorelief Oxides with Gyroidal Structures. *J. Am. Chem. Soc.* **2004,** *126,* 865–875.
[33] Yang, C.-T., Huang, M. H., Formation of Arrays of Gallium Nitride Nanorods within Mesoporous Silica SBA-15. *J. Phys. Chem. B* **2005,** *109,* 17842–17847.
[34] Shin, H. J., Ko, C. H., Ryoo, R., Synthesis of platinum networks with nanoscopic periodicity using mesoporous silica as template. *J. Mater. Chem.* **2001,** *11,* 260–261.
[35] Choi, W. C., Woo, S. I., Bimetallic Pt-Ru nanowire network for anode material in a direct-methanol fuel cell. *J. Power Sources* **2003,** *124,* 420–425.
[36] Liu, Z., Terasaki, O., Ohsuna, T., Hiraga, K., Shin, H. J., Ryoo, R., An HRTEM Study of Channel Structures in Mesoporous Silica SBA-15 and Platinum Wires Produced in the Channels. *Chem. Phys. Chem.* **2001,** *2,* 229–231.
[37] Wang, Y., Yang, C.-M., Schmidt, W., Spliethoff, B., Bill, E., Schüth, F., Weakly Ferromagnetic Ordered Mesoporous Co_3O_4 Synthesized by Nanocasting from Vinyl-Functionalized Cubic Ia3d Mesoporous Silica. *Adv. Mater.* **2005,** *17,* 53–56.

[38] Han, Y.-J., Kim, J. M., Stucky, G. D., Preparation of Noble Metal Nanowires Using Hexagonal Mesoporous Silica SBA-15. *Chem. Mater.* **2000,** *12,* 2068–2069.
[39] Imperor-Clerc, M., Bazin, D., Appay, M.-D., Beaunier, P., Davidson, A., Crystallization of β-MnO_2 Nanowires in the Pores of SBA-15 Silicas: In Situ Investigation Using Synchrotron Radiation. *Chem. Mater.* **2004,** *16,* 1813–1821.
[40] Jiao, K., Zhang, B., Yue, B., Ren, Y., Liu, S., Yan, S., Dickinson, C., Zhou, W., He, H., Growth of porous single-crystal Cr_2O_3 in a 3D mesopore system. *Chem. Commun.* **2005,** 5618–5620.
[41] Liu, S., Yue, B., Jiao, K., Zhou, Y., He, H., Template synthesis of one-dimensional nanostructured spinel zinc ferrite. *Mater. Lett.* **2006,** *60,* 154–158.
[42] Yue, W., Zhou, W., Synthesis of Porous Single Crystals of Metal Oxides via a Solid-Liquid Route. *Chem. Mater.* **2007,** *19,* 2359–2363.
[43] Ding, X., Briggs, G., Zhou, W., Chen, Q., Peng, L.-M., In Situ growth and characterization of Ag and Cu Nanowires. *Nanotechnology* **2006,** *17,* S376–S380.
[44] Lee, K.-B., Lee, S.-M., Cheon, J., Size-Controlled Synthesis of Pd Nanowires Using a Mesoporous Silica Template via Chemical Vapor Infiltration. *Adv. Mater.* **2001,** *13,* 517–520.
[45] Kang, H., Jun, Y.-W., Park, J.-I., Lee, H.-B., Cheon, J., Synthesis of Porous Palladium Superlattice, Nanoballs and Nanowires. *Chem. Mater.* **2000,** *12,* 3530–3532.
[46] Dickinson, C., Zhou, W., Hodkins, R. P., Shi, Y., Zhao, D., He, H., Formation Mechanism of Porous Single-Crystal Cr_2O_3 and Co_3O_4 Templated by Mesoporous Silica. *Chem. Mater.* **2006,** *18,* 3088–3095.
[47] Wang, Y.-g., Xia, Y.-y., Electrochemical capacitance characterization of NiO with ordered mesoporous structure synthesized by template SBA-15. *Electrochim. Acta* **2006,** *51,* 3223–3227.
[48] Arbiol, J., Cabot, A., Morante, J. R., Distributions of noble metal Pd and Pt in mesoporous silica. *Appl. Phys. Lett.* **2002,** *81,* 3449–3451.
[49] Huang, M. H., Choudray, A., Yang, P., Ag nanowire formation within mesoporous silica. *Chem. Commun.* **2000,** 1063–1064.
[50] Fukuoka, A., Higashimoto, N., Sakamoto, Y., Sasaki, M., Sugimoto, N., Inagaki, S., Fukushima, Y., Ichikawa, M., Ship-in-bottle synthesis and catalytic performances of platinum carbonyl clusters, nanowires and nanoparticles in micro- and mesoporous materials. *Catal. Today* **2001,** *66,* 23–31.
[51] Ko, C. H., Ryoo, R., Imaging the channels in mesoporous molecular sieves with platinum. *Chem. Commun.* **1996,** 2467–2468.
[52] Liu, Z., Sakamoto, Y., Ohsuna, T., Hiraga, K., Terasaky, O., Ko, C. H., Shin, H. J., Ryoo, R., TEM Studies of Platinum Nanowires Fabricated in Mesoporous Silica MCM-41. *Angew. Chem. Int. Ed.* **2000,** *39,* 3107–3110.
[53] Gu, J., Shi, J., Xiong, L., Chen, H., Li, L., Ruan, M., A new strategy to incorporate high density gold nanowires into channels of mesoporous silica thin films by electroless deposition. *Solid State Sci.* **2004,** *6,* 747–752.
[54] Laha, S. C., Ryoo, R., Synthesis of thermally stable mesoporous cerium oxide with nanocrystalline frameworks using mesoporous silica templates. *Chem. Commun.* **2003,** 2138–2139.
[55] Zhu, S., Zhou, Z., Zhang, D., Wang, H., Synthesis of mesoporous amorphous MnO_2 from SBA-15 via surface modification and ultrasonic waves. *Microporous Mesoporous Mater.* **2006,** *95,* 257–264.
[56] Chen, H., Dong, X., Shi, J., Zhao, J., Hua, Z., Gao, J., Ruan, M., Yan, D., Templated synthesis of hierarchically porous manganese oxide with a crystalline nanorod framework and its high electrochemical performance. *J. Mater. Chem.* **2007,** *17,* 855–860.
[57] Zhu, K., He, H., Xie, S., Zhang, X., Zhou, W., Jin, S., Yue, B., Crystalline WO_3 nanowires synthesized by templating method. *Chem. Phys. Lett.* **2003,** *377,* 317–321.
[58] Jiao, F., Yue, B., Zhu, K., Zhao, D., He, H., α-Fe_2O_3 Nanowires. Confined Synthesis and Catalytic Hydroxylation of Phenol. *Chem. Lett.* **2003,** *32,* 770–771.
[59] Zhu, K., Yue, B., Zhou, W., He, H., Preparation of three-dimensional chromium oxide porous single crystals templated by SBA-15. *Chem. Commun.* **2003,** 98–99.
[60] Yue, B., Tan, D.-J., Yan, S.-R., Zhou, Y., Zhu, K.-K., Pan, J.-F., Zhuang, J.-H., He, H.-Y., Preparation of MoO_3-V_2O_5 Nanowires with Controllable Mo/V Ratios inside SBA-15 Channels Using a Chemical Approach with Heteropolyacid. *Chin. J. Chem.* **2005,** *23,* 32–36.
[61] Petkov, N., Stock, N., Bein, T., Gold Electroless Reduction in Nanosized Channels of Thiol-Modified SBA-15 Material. *J. Phys. Chem. B.* **2005,** *109,* 10737–10743.

[62] Xiong, Y., Xie, Y., Li, Z., Li, X., Gao, S., Aqueous-Solution Growth of GaP and InP Nanowires: A General Route to Phosphide, Oxide, Sulfide and Tungstate Nanowires. *Chem. Eur. J.* **2004,** *10,* 654–660.
[63] Yada, M., Hiyoshi, H., Ohe, K., Machida, M., Kijima, T., Synthesis of Aluminum-Based Surfactant Mesophases Morphologically Controlled through a Layer to Haxagonal transition. *Inorg. Chem.* **1997,** *36,* 5565–5569.
[64] Zhang, Z., Dai, S., Blom, D. A., Shen, J., Synthesis of Ordered Metallic Nanowires inside Ordered Mesoporous Materials through Electroless Deposition. *Chem. Mater.* **2002,** *14,* 965–968.
[65] Fukuoka, A., Araki, H., Sakamoto, Y., Sugimoto, N., Tsukada, H., Kumai, Y., Akimoto, Y., Ichikawa, M., Template Synthesis of Nanoparticles Arrays of Gold and Platinum in Mesoporous Silica Films. *Nano Lett.* **2002,** *2,* 793–795.
[66] Li, L., Shi, J.-L., Yan, J.-N., A highly efficient heterogeneous catalytic system for Heck reactions with a palladium colloid layer reduced in situ in the channel of mesoporous silica materials. *Chem. Commun.* **2004,** 1990–1991.
[67] Li, L., Shi, J.-L., Zhang, L.-X., Xiong, L.-M., Yan, J.-N., A Novel and Simple In-Situ Reduction Route for the Synthesis of an Ultra-Thin Metal Nanocoating in the Channels of Mesoporous Silica Materials. *Adv. Mater.* **2004,** *16,* 1079–1082.
[68] Li, L., Zhang, L.-X., Shi, J.-L., Yan, J.-N., Liang, J., New and efficient heterogeneous catalytic system for Heck reaction: palladium colloid layer in situ reduced in the channel of mesoporous silica materials. *Appl. Catal. A* **2005,** *283,* 85–89.
[69] Liu, Z. C., Chen, H. R., Huang, W. M., Gu, J. L., Bu, W. B., Hua, Z. L., Shi, J. L., Synthesis of a new SnO_2/mesoporous silica composite with room-temperature photoluminescence. *Micropor. Mesopor. Mater.* **2006,** *89,* 270–275.
[70] Shah, P., Ramaswamy, A. V., Lazar, K., Ramaswamy, V., Direct hydrothermal synthesis of mesoporous Sn-SBA-15 materials under weak acidic conditions. *Micropor. Mesopor. Mater.* **2007,** *100,* 210–226.
[71] Wei, Y. L., Wang, Y. M., Zhu, J. H., Wu, Z. Y., In-Situ Coating of SBA-15 with MgO: Direct Synthesis of Mesoporous Solid Bases from Strong Acidic Systems. *Adv. Mater.* **2003,** *15,* 1943–1945.
[72] Wu, Z. Y., Jiang, Q., Wang, Y. M., Wang, H. J., Sun, L. B., Shi, L. Y., Xu, J. H., Wang, Y., Chun, Y., Zhu, J. H., Generating Superbasic Sites on Mesoporous Silica SBA-15. *Chem. Mater.* **2006,** *18,* 4600–4608.
[73] Landau, M. V., Vradman, L., Herskowitz, M., Koltypin, Y., Gedanken, A., Ultrasonically controlled deposition -precipitation. Co-Mo-HDS catalyst deposited on wide-pore MCM material. *J. Catal.* **2001,** *201,* 22–36.
[74] Sun, D., Liu, Z., He, J., Han, B., Zhang, J., Huang, Y., Surface sol-gel modification of mesoporous silica molecular sieve SBA-15 with TiO_2 in supercritical CO_2. *Mesopor. Micropor. Mater.* **2005,** *80,* 165–171.
[75] Mahurin, S., Bao, L., Yan, W., Liang, C., Dai, S., Atomic layer deposition of TiO_2 on mesoporous silica. *J. Non-Cryst. Solids* **2006,** *352,* 3280–3284.
[76] Herrera, J. E., Kwak, J. H., Hu, J. Z., Wang, Y., Peden, C. H. F., Macht, J., Iglesia, E., Synthesis, characterization and catalytic function of novel highly dispersed tungsten oxide catalyst on mesoporous silica. *J. Catal.* **2006,** *239,* 200–211.
[77] Herrera, J. E., Kwak, J. H., Hu, J. Z., Wang, Y., Peden, C. H. F., Synthesis of nanodispersed oxides of vanadium, titanium, molybdenum, and tungsten on mesoporous silica using atomic layer deposition. *Top. Catal.* **2006,** *39,* 245–255.
[78] Landau, M. V., Dafa, E., Kaliya, M. L., Sen, T., Herskowitz, M., Mesoporous catalytic material prepared by grafting of wide-pore Al-MCM-41with Alumina Multilayer, M. *Micropor. Mesopor. Mater.* **2001,** *49,* 65–81.
[79] Pan, Y.-S., Lin, H.-P., Cheng, H. H., Cheng, C.-F., Tang, C.-Y., Lin, C.-Y., Chemical Coating of Aluminum Oxide onto As-synthesized Mesoporous Silicas. *Chem. Lett.* **2006,** *35,* 608–609.
[80] Vradman, L., Peerr, Y., Mann-Kiperman, A., Landau, M. V., Thermal decomposition-precipitation inside the nanoreactors. High loading of W-oxide nanoparticles into the nanotubes of SBA-15. *Studies in Surface Science and Catalysis* **2003,** *146,* 121–124.
[81] Luan, Z., Maes, E. M., van der Heide, P. A. W., Zhao, D., Czernuszewicz, R. S., Kevan, L., Incorporation of Titanium into Mesoporous Silica Molecular Sieve SBA-15. *Chem. Mater.* **1999,** *11,* 3680–3686.

[82] Luan, Z., Kevan, L., Characterization of titanium-containing mesoporous silica molecular sieve SBA-15 and génération of paramagnetic hole and electrone centers. *Micropor. Mesopor. Mater.* **2001,** *44–45,* 337–344.
[83] Landau, M. V., Vradman, L., Wang, X., Titelman, L., High loading TiO_2 and ZrO_2 nanocrystals ensembles inside the mesopores of SBA-15: preparation, texture and stability. *Micropor. Mesopor. Mater.* **2005,** *78,* 117–129.
[84] van Grieken, R., Aguado, J., Lopez-Munoz, M. J., Marugan, J., Synthesis of size-controlled silica-supported TiO_2 photocatalyst. *J. Photochem. Photobiol. A Chem.* **2002,** *148,* 315–322.
[85] Knözinger, H., Taglauer, E. Spreading and Wetting. in: Ertl, G., Knözinger, H., Weitkamp, J. (Eds.), *Preparation of Solid Catalysts,* Wiley-VCH, Weinheim, 1999, pp. 501–526.
[86] Sauer, J., Marlow, B., Spliethoff, B., Schüth, F., Rare Earth Oxide Coating of the Walls of SBA-15. *Chem. Mater.* **2002,** *14,* 217–224.
[87] Abecassis-Wolfowich, M., Landau, M. V., Brenner, A., Herskowitz, M., Low-temperature Combustion of 2,4,6-trichlorophenol in Catalytic Wet Oxidation with Nanocasted Mn-Ce-oxide Catalyst. *J. Catal.* **2007,** *247,* 201–213.
[88] Hu, L., Ji, S., Xiao, T., Guo, C., Wu, P., Nie, P., Preparation and Characterization of tungsten Carbide Confined in the Channels of SBA-15 Mesoporous silica. *J. Phys. Chem. B* **2007,** *111,* 3599–3608.
[89] Schüth, F., Wingen, A., Sauer, J., Oxide loaded ordered mesoporous oxides for catalytic applications. *Micropor. Mesopor. Mater.* **2001,** *44-45,* 465–476.
[90] Vradman, L., Landau, M. V., Kantorovich, D. Y., Koltypin, Y., Gedanken, A., Evaluation of metal oxide phase assembling mode inside the nanotubular pores of mesostructured silica. *Micropor. Mesopor. Mater.* **2005,** *79,* 307–318.
[91] Friedrich, H., Sietsma, J. R. A., de Jongh, P. E., Verkleij, A. J., de Jong, K. P., Measuring Location, Size, Distribution and Loading of NiO Crystallites in Individual SBA-15 Pores by Electron Tomography. *J. Am. Chem. Soc.* **2007,** *129,* 10249–10254.
[92] Vradman, L., Landau, M. V., Herskowitz, M., Ezersky, V., Talianker, M., Nikitenko, S., Koltypin, Y., Gedanken, A., High loading of short WS_2 Slabs inside the SBA-15. Promotion with nickel and performance in hydrodesulfurization and hydrogenation. *J. Catal.* **2003,** *213,* 163–175.
[93] Yuranov, I., Moeckli, P., Suvorova, E., Buffat, P., Kiwi-Minsker, L., Renken, A., Pd/SiO_2 catalysts: synthesis of Pd nanoparticles with the controlled size in mesoporous silica. *J. Mol. Catal. A Chem.* **2003,** *192,* 239–251.
[94] Li, J.-J., Xu, X.-Y., Jiang, Z., Hao, Z.-P., Hu, C., Nanoporous Silica-Supported Nanometric Palladium: Synthesis, Characteriztion and Catalytic Deep Oxidation of Benzene. *Environ. Sci. Technol.* **2005,** *39,* 1319–1323.
[95] Yue, B., Tang, H., Kong, Z., Zhu, K., Dickinson, C., Zhou, W., He, H., Preparation and characterization of three-dimensional mesoporous crystals of tungsten oxide. *Chem. Phys. Lett.* **2005,** *407,* 83–86.
[96] Fukuoka, A., Higashimoto, N., Sakamoto, Y., Inagaki, S., Fukushima, Y., Ichikawa, M., Preparation and catalysis of Pt and Rh nanowires and particles in FSM-16. *Micropor. Mesopor. Mater.* **2001,** *48,* 171–179.
[97] Sasaki, M., Osada, M., Higashimoto, N., Yamamoto, T., Fukuoka, A., Ishikawa, M., Templating fabrication of platinum nanoparticles and nanowires using mesoporous channels of FSM-16 – their structural characterization and catalytic performances in water gas shift reaction. *J. Mol. Catal. A Chem.* **1999,** *141,* 223–240.
[98] Rao, P. M., Wolfson, A., Kababya, S., Vega, S., Landau, M. V., Immobilization of molecular $H_3PW_{12}O_{40}$ heteropolyacid catalyst in alumina-grafted silica- gel and mesostructured SBA-15 silica matrices. *J. Catal.* **2005,** *232,* 210–225.
[99] Goldbourt, A., Landau, M. V., Vega, S., Aluminum Species in Alumina Multilayer grafted inside mesoporous MCM-41 and characterized with ^{27}Al FAM(II)-MQMAS NMR. *J. Phys. Chem. B.* **2003,** *107,* 724–731.
[100] Mehnert, C. P., Weaver, D. W., Ying, J. Y., Heterogeneous Heck Catalysis with Palladium-Grafted Molecular Sieves. *J. Am. Chem. Soc.* **1998,** *120,* 12289–12296.

CHAPTER 27

Nanoporous Materials—Catalysts for Green Chemistry

Duncan J. Macquarrie

Contents

Abstract

The ability of nanoporous materials to reduce the environmental impact of chemical processes is discussed. The chapter begins with a discussion of methods such as Life Cycle Analysis and green chemistry metrics for measuring the impact of a process, and continues to discuss how nanoporous catalytic materials can be utilized to improve processes. Considerations for minimizing the impact of nanoporous catalyst synthesis follow. Diffusion plays an important role in catalytic reactions involving nanoporous catalysts, and pertinent results on diffusion behaviour are briefly reviewed. This is followed by a section discussing the key advances in catalyst design and utilization; for example the control of surface properties by incorporation of secondary groups, and the development of bifunctional systems which allow two consecutive steps of a reaction to proceed together are covered. The final section of the chapter discusses reactor design as it affects nanoporous catalysis. Continuous reactors such as trickle bed reactors, spinning disc reactors and membrane reactors are discussed as options which can best demonstrate the benefits of nanoporous catalysis for green chemistry.

Keywords: Green chemistry, catalysis, bifunctional, life cycle, heterogeneous catalysis, metrics

Ordered Porous Solids
DOI: 10.1016/B978-0-444-53189-6.00027-5

Glossary

Life cycle analysis	method for assessing all the inputs and outputs of a process with a view to quantifying the overall environmental burden associated with the process.
Green chemistry	an approach to chemistry which embraces the concept of minimising waste streams and process hazard through designing processes to be as effective as possible avoiding unnecessary waste streams.
Green chemistry metrics	methods of quantification which provide measures of the environmental efficiency of a process. Examples include Atom economy/atom efficiency, which measure the amount of atoms of reagents which end up in the product; ideally 100% are incorporated, but typically fewer. E-factor is a measure of the total waste produced per unit product.
Environmental Footprint	a measure of the environmental impact of a process.
EPR	electron paramagnetic resonance.
Starbon	partially carbonised expanded starch.
Trickle Bed reactor	a continuous reactor, typically tubular, where feed is passed downward using gravity, over a bed of catalyst particles.

1. Introduction

Over the last 15 years, green chemistry has come to prominence and is now shaping much of how we do chemistry. This field combines a wide range of techniques and methodologies into a philosophy of how we design and implement chemical processes. Moreover, it drives us to design not only processes, but products to minimize environmental impact and to attain a sustainable and safer chemical production .[1,2] This encompasses not only driving improvements in the processing of existing products, but also changing raw materials from crude oil-derived feedstocks to renewable, plant-based platform molecules (often currently seen as waste products).[3–6] In a few cases (e.g., chitin and chitosan from crabshells, squid pens, etc.) the waste may be from animals.[7,8] After choosing the raw materials and the desired product, the synthetic route must then be chosen to be the most atom efficient that is available (ideally, addition reactions are preferred, since they combine all the atoms of the reactants into the product leaving none behind to become waste streams, but in practice that is often not possible). Subsequently, the reaction (or more accurately, the process) must be carried out in the most chemically- and energy-efficient manner possible, using catalysis wherever possible to drive temperatures (and therefore energy) down and the yields and selectivity up. Product isolation must be kept as simple as possible, and catalyst recovery and reuse is important. These requirements for highly efficient catalysis and the importance of a simple recovery of catalyst and simple isolation of products helps us to broadly define the ideal catalyst. Having the catalyst in a different phase to the reactants and product(s) will aid in separation of both the catalyst (with a view to reuse) and the product, reducing the waste generated during isolation (often the largest source of waste products in a chemical process is the separation and purification of product—if a catalyst has to be recovered and recycled, then this is an additional processing step with its own waste). Therefore, solid catalysts are generally preferable to soluble catalysts, given the similar activity and selectivity. For a solid catalyst to function well, it should have a regular structure to ensure that the catalytic sites on the surface are as similar to each other as possible, helping to keep selectivity high; it should also have as high a surface area as possible, allowing the maximum number of catalytic sites per unit material. This requirement leads us naturally to nanoporous materials with well-defined channels of molecular dimensions. The zeolites have been used for decades for this purpose, but their relatively narrow pore diameters restrict their application to quite small molecules. The larger pore Micelle Templated Silicas, with pore sizes up to 10 nm, are thus ideal candidates for clean catalytic systems. In particular, the regular pores of these materials have low tortuosity and thus are likely to possess good diffusion properties—an important factor for catalytic activity in a solid–liquid or solid–gas system.

When designing a green process, considering all the parameters in the process is necessary, rather than just changing one part. The most successful examples all bear the hallmark of a careful and holistic redesign. Having said this, we could assume that such a redesign relies on the availability of well demonstrated alternatives for each of the component parts of the process. To achieve this, a wide variety of additional techniques and technologies are available—reactions in supercritical fluids, in microwave reactors, in intensive processing rigs and continuous reactors all can play a part; alternative solvents such as supercritical fluids, ionic liquids, water or solvent free systems have all been researched; biotechnological process such as fermentation can produce raw materials from renewable resources and provide us with new routes to the basic platform molecules that we require. Intelligent engineering of the appropriate reactors and processing facilities can enhance the best reactions and allow us to derive the maximum benefit from the chemistry. Combining the earlier-mentioned technologies with nanoporous materials allows us to derive the maximum benefit from these materials and to develop the best process possible.

2. Metrics—Measurement and Comparison of Processes

Assessing the benefits of the chosen route is essential to decide how much of an improvement can actually be achieved. This is far from being an easy task—in principle a full Life Cycle Assessment (LCA) should be carried out to quantify all the resources used, and this should then be compared with alternatives to decide which is 'greenest'.[9]

In practice this is very difficult, especially at a early stage of development, where many key parameters are not well known. So a series of metrics have been developed, with which we can guide our choices and assess strengths and weaknesses of an approach.[10,11] Of these, E factor is probably the most powerful and among the simplest to use, being simply a measure of the mass of waste produced per kg of product. Other key metrics include atom economy (100% atom efficiency means that all the atoms of the starting material are incorporated in the final product).

One of the central concepts to developing cleaner production processes for a range of chemistries is the use of heterogeneous catalysis. This has been a mainstay of the petrochemical industries for decades, largely due to the unique and remarkable properties of the zeolites. The extension to liquid phase reactions and fine chemicals manufacture has been gaining momentum over the last 20 years, initially using amorphous or partly ordered supports such as silica, carbon, clay and alumina, and more recently using the highly ordered mesoporous materials, which are the focus of this book. Although zeolites have also played a role in liquid phase transformations, their microporosity has often limited their application to the larger molecules that are often involved. The use of heterogeneous catalysts can improve the environmental performance of a reaction by reducing energy requirements (lower activation energy), improving selectivity, and helping with separation and recovery/reuse of the catalyst (often very difficult with homogeneous catalysts). The last point also simplifies the isolation of the product. Lowering activation parameters/speeding up reactions, improving selectivity and isolation minimize E, C, and W in Fig. 27.1, stage 4. Improving the activity and lifetime of the catalyst reduce all the parameters. Heterogeneous catalyst can also be incorporated into continuous reactors, allowing reaction and separation to occur simultaneously.

It is generally the case, however, that the preparation of a heterogeneous catalyst generates wastes that would not otherwise have been generated, and these must be factored into the overall environmental footprint of the synthesis of the product. Therefore, a clean method for production of the material must be developed, taking into account all the influences above, and the lifetime and final fate of the catalyst must also be considered. Clearly, the longer the lifetime of the catalyst, and the more often it can be recovered and reused, the lower will be the impact of the waste generated during its synthesis. In Life Cycle terms, the fate of the spent catalyst (i.e., its destruction or disposal) also has to be factored in as a contribution to the waste associated with the process per unit product.

The particular benefits of templated mesoporous structures with very regular pore systems relate to their very high surface area and the ability to attach a range of catalytic groups as well as tailor surface chemistry. Diffusion through the pores is aided by the regularity of the pores, and selectivity may be enhanced compared to amorphous materials, where each attached catalytic unit might find itself in a different environment both in terms of physical constraints and surface chemistry.

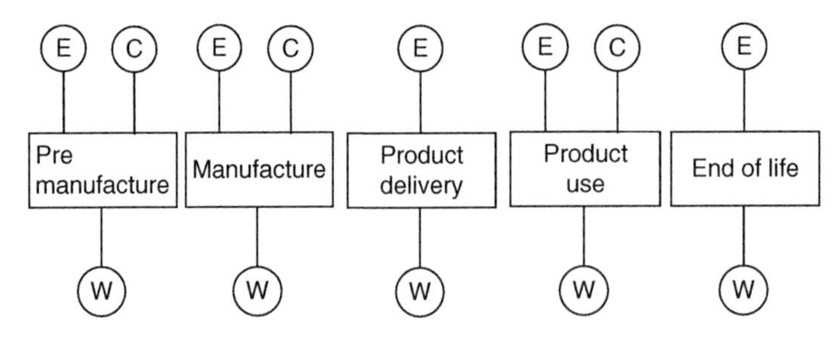

Figure 27.1 Simple schematic life cycle of a product. E – energy input; C – consumable input; W–Waste output.

3. Green Aspects in the Synthesis of Nanoporous Catalysts

One of the often overlooked aspects of developing green catalytic systems is the synthesis of the catalyst itself. Clearly, the impact of the synthesis on the overall performance of the resultant process will decrease with turnover number, but energy inputs and, of particular relevance for nanoporous materials, recovery of template are important factors which should be considered. Nonetheless, the synthesis should always be as efficient and clean as it can be. A simple schematic is shown below, which illustrates the main material flows in a generic catalyst lifecycle (Fig. 27.2). Obviously, a full LCA would include the inputs and wastes associated with the provision of the template, solvents and acids/bases etc required for the synthesis.

In this context, the ability to wash out and recover the template has a significant impact on the system as a whole, and has been demonstrated for the neutral amine systems developed by Pinnavaia,[12] where the interfacial interactions with the material are H-bonding in nature and relatively easily disturbed. Ethanol solutions of the template are recovered and can be used directly, without removal of ethanol, to produce a subsequent batch of material. Polyethers are also relatively readily recovered, but can in some cases leave some residual template behind. Quaternary templates can be removed but, as they are partly counterions to surface charges, the use of acidic ethanol is necessary—this will eventually produce salt waste.

The first stage of the LCA is also important, and the inputs required to prepare the raw materials for catalyst synthesis should not be overlooked. Depending on, e.g., the nature of the Si source, these inputs can be very different. Water glass, $(RO)_4Si$ and waste silica from other processes (paper production and the burning of plant waste both generate significant amounts of silica-rich ash) will all have very different environmental footprints.

The simplicity of the process to prepare the catalysts is also important, and a direct one-stage process is likely to be ideal from the point of view of consumables, energy and waste, and low temperature, relatively rapid preparations are likely to be beneficial, all other things being equal.

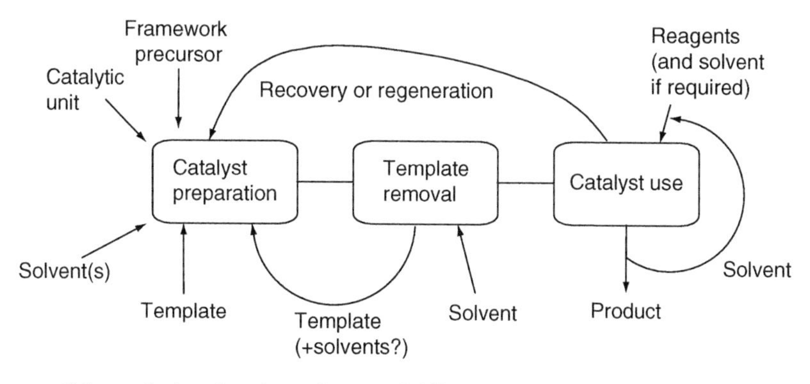

Figure 27.2 Schematic showing the main material flows in synthesis and use of a catalyst.

Although there is a wide range of materials prepared via templating routes, the vast majority of catalytic applications have come from the templated silicas and related materials such as the Periodic Mesoporous Organo Silicas (PMOS), which incorporate organic functionality into the pore walls.[13,14] The majority of these materials have little functionality in the wall groups—most of the precursors are simple organic groups such as $-CH_2CH_2-$ or $-C_6H_4-$, but more recently, more functional systems have been utilized. For example, ferrocenyl and ethers/sulphides are known, where the heteroatom could be used as a ligand for a metal.[15,16] Additionally, the incorporation of functional monosilanes (of which there is a very wide variety) has led to some potentially very interesting catalytic materials.[17,18] Clearly, the synthesis of the necessary bis-silanes can often be difficult and wasteful, but nonetheless, the performance of the catalysts does appear to be very good, and thus the activity and lifetime may outweigh the synthetic difficulties.

4. Catalysis Using Mesoporous Materials

4.1. Diffusion and sorption kinetics

As outlined in Section 1, the potential for nanoporous catalysts to play a role in the development of green processes is enormous. The combination of controllable and regular structures with the possibility of controlling the density of surface sites is powerful, as is the ability to tailor pore size for the application (allowing larger, complex catalytic sites to be anchored, and larger molecules to be processed). The potential to fix these catalysts in reactors, allowing for continuous processing and simultaneous separation of product and catalyst promises to enhance the value of such processes even more. One of the most important criteria (compared to homogeneous counterparts) of a solid catalyst, especially one with catalytic sites in pores, is the diffusion in and out of the pore and the adsorption of the reactants at the active site. Desorption of product from the active site after reaction to liberate the site for subsequent reaction cycles is also clearly very important. Simplistically, one might assume that this was related in a straightforward manner to the relationship between pore size and molecular size, pore connectivity and pore tortuosity (i.e., how twisted the pore is). In reality, the situation appears to be more complex, and solvated reactants and products have to be considered, as do surface energy, flow, polarity, hydrophilicity and wettability also play roles. Several studies indicate that the motion of molecules through nanosized channels does not conform to the simple fluid dynamic models developed for larger spaces.

For example, Hansen *et al.*[19] have demonstrated, by pulsed field gradient NMR techniques, that the diffusion of *n*-hexane is significantly more rapid in smaller channels. Their conclusions are that changes to the surface chemistry are more important than the reduction in space within the channel. Stallmach *et al.*[20] have also noted increased diffusion rates of benzene in MCM-41 pores (greater even than the corresponding bulk diffusion). They have attributed this unusual behaviour to gas-phase diffusion taking place within the pores of the material, even under conditions where liquid-state diffusion would be expected. For less volatile molecules such as long chain hydrocarbons, this effect was absent.

Okazaki and Toriyama have also demonstrated that mixed liquids can have additional diffusional factors to consider.[21] By using a combination of NMR and EPR measurements, they showed that a mixture of the hydrophobic cyclohexane and the hydrophilic, more polar 2-propanol are not homogeneously distributed throughout a mesoporous pore system. The cyclohexane molecules prefer to reside more centrally in the nanochannels than the 2-propanol molecules, which tend to occupy sites predominantly on or close to the surface of the channel. They also demonstrated that molecules can be displaced from the surface by other, more polar, molecules and that the mechanism of movement of molecules in nanochannels should be considered to be due to temperature-induced pressure changes—a process they call 'collective diffusion'. These studies have clear implications for the choice of solvents in catalytic reactions involving nanoporous materials, and indeed for reactions involving multiple species (including solvent) as partitioning of the relevant species at the active site may be strongly influenced by the nature of the material. Indeed, earlier studies on non-structured organosilicas have shown that substantial solvent effects can be expected which may be related to the phenomena described by Okazaki and Toriyama. For example, the catalysis of the Knoevenagel reaction by aminopropyl silica and aminopropyl-functionalized Micelle Templated silica was shown to be highly solvent dependent, with subtly different solvent dependencies, but broadly in line with the phenomena described by Okazaki and Toriyama.[22,23]

In systems consisting of organically modified mesoporous silicas and aqueous solutions, Walcarius found that ionic adsorbates were adsorbed in a manner which was actually relatively predictable from simple adsorbate size/pore size considerations.[24]

The protonation of aminopropyl containing materials and the adsorption of Cu(II) on the same materials, as well as the adsorption of Hg(II) on mercaptopropyl materials all were slower as the size of the pore dropped and as the counterion size increased. Diffusion dropped as a function of extent of adsorption, to be expected as the pore fills up with adsorbed material. It was shown that the highly structured mesoporous materials had significantly enhanced adsorption kinetics compared to amorphous materials of similar average pore size, especially at low levels of adsorption (i.e., before pore blocking became relevant).[25]

4.2. Examples of catalytic processes occurring in nanoporous catalysts

There are literally hundreds of papers dealing with catalysis using organically functionalized silicas in a wide range of organic transformations. These have been reviewed thoroughly,[26–33] and it is not the intention to cover the whole range of catalytic transformations here. Rather, some of the more interesting developments which have come to light recently, and which have particular relevance to green chemistry will be covered.

The development of a series of active catalysts for a range of reaction types has stimulated further research to improve their activity, lifetime and selectivity. One of the concepts employed here is to focus not on the catalytic site itself, but to work to ensure the environment surrounding the active site is as conducive as possible to the catalytic process. In this respect, the surface chemistry can be modified to provide a less polar surface, which may have benefits in terms of adsorption of reactants and desorption of products. As described earlier, such changes to the nature of the surface may also play an important role in the nature and rate of diffusion through the channels, and may play a more profound role in activity and selectivity than is often realized. Poisoning, which is often due to the build up of polar side products, may also be reduced, leading to longer lifetimes. The heterogenization of catalysts designed to work in organic solvents of relatively low polarity is likely to be favoured by a surface which is of relatively low polarity, and this may also be a factor in designing a catalyst. Factors affecting diffusion and sorption kinetics, as outlined in Section 4.1. may also be strongly influenced by such modifications.

The examples give below demonstrate the utility of the procedure using different methods of surface modification. It is possible to add so-called 'spectator groups', typically short chain alkyl or phenyl groups, which are themselves catalytically inactive, but which improve the catalytic activity, presumably by modification of the surface polarity and diffusion properties of the system. However, changes to the inherent nature of the functional groups (e.g., acid strength) or to the distribution of sites on incorporation of a second silane have not been measured, and we cannot exclude the possibility that some of the behavioural changes are related to such fundamental differences in, for example, acidity of basicity, or to effects of site isolation. A more far-reaching approach is to replace the silica source with a bifunctional organosilane (e.g., $(EtO)_3Si$-C_6H_4-$Si(OEt)_3$) so that the organic group becomes part of the walls of the material. A final approach is the post-functionalization of the material by, e.g., Me_3Si groups attached after synthesis. This is often effective, but adds an extra step to the synthesis, and can partially restrict access to the catalytic sites. The extraction of template using alcohols such as EtOH also leads to the formation of SiOEt surface groups which potentially would make the surface less polar by replacing some of the silanols. These shorter alcohols would have limited hydrolytic stability; however, longer alcohols form a more hydrophobic and more stable surface.

The concept of surface modification to improve catalytic function has been successfully demonstrated for acid, base and metal-catalyzed reactions, and the incorporation of spectator groups has been achieved by post-modification, *in situ* co-condensation and by the use of organosilicas.

The use of solid acid catalysts in esterification (and transesterification) reactions is important since it avoids the need for an aqueous work-up which can hydrolyze products back to the reagents. Perez-Pariente *et al.* [34,35] have used this approach in the synthesis of glycerol monoesters, which have interest as surface active agents derived from renewable resources. They showed that the activity of their sulphonic acid catalyst could be trebled by incorporating methyl groups (via a methyl silane) into its structure during co-condensation synthesis. Selectivity was also improved in the monoesterification of glycerol with fatty acids. The acylation of anisole, and the alkylation of furan with acetone were both found to be significantly accelerated using perfluorosulphonic acid/propyl silicas in comparison to systems containing only the catalytically active perfluorosulphonic acid group.[36] (Fig. 27.3)

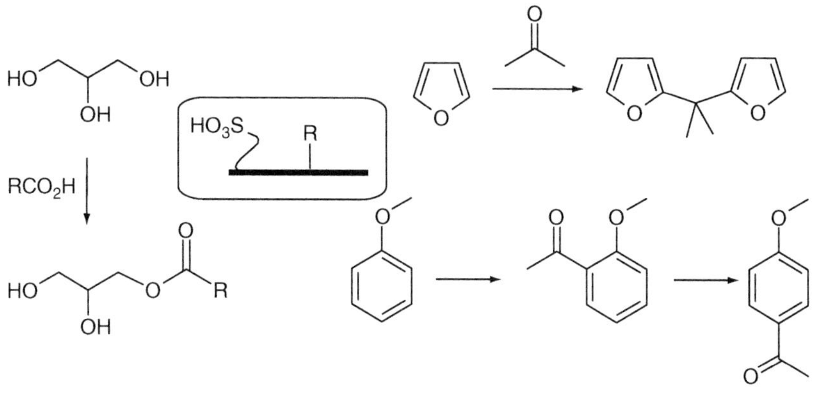

Figure 27.3 Reactions catalyzed by sulphonic acid–spectator systems.

Similar effects have been demonstrated in base catalysis. Sullivan[37] demonstrated the value of bridged PMOS made from (bis(silyl)arenes and -alkanes and aminopropylsilanes) in the base catalyzed Knoevenagel reactions, her catalysts proving to be exceptionally active, and better than those based on aminopropyl-mesoporous silicas containing phenyl spectator groups[38] which were themselves better than those lacking the phenyl functionality (Fig. 27.4).

Interestingly, a 4–5 fold improvement in turnover number was achieved by preparing simple aminopropyl silicas using the direct co-condensation method instead of the alternative grafting process.[39] This was ascribed to the lack of nucleophilicity in the amine groups prepared by the co-condensation route,[40] which meant that the usual poisoning mechanism—amide formation between the cyanoacetate ester and the amine groups of the catalyst—was eliminated. Poisoning in this case was found to be due to the slow build up of a polar by-product on the surface. Indeed, the attachment of non-polar spectator groups to the surface in this case not only increased the reaction rate, but increased the lifetime of the catalysts by around an order of magnitude. Significant differences in the behaviour of functional groups as a consequence of preparation route have also been noted for enzyme-based systems (see below).

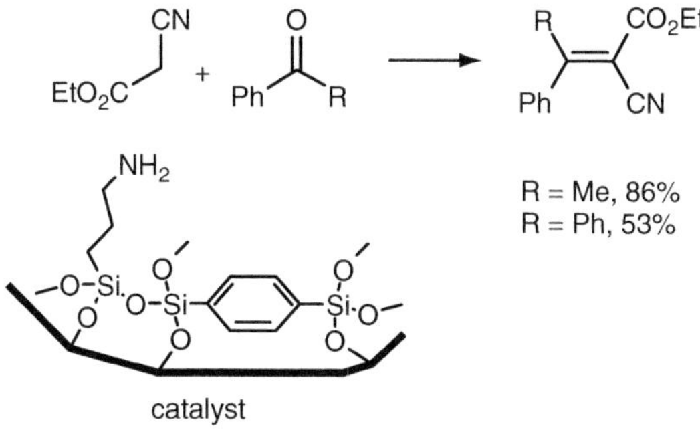

Figure 27.4 PMOS base catalysis of the Knoevenagel reaction.

Inagaki has also demonstrated that a sulphonic acid functionalized PMOS is a very active acid catalyst, being more active than Naf ion in the esterification of ethanoic acid with ethanol.[41] Interestingly, the catalyst prepared in a one step co-condensation was much more active than the post-modified system. This was ascribed as being partly due to the lower number of acid sites in the latter, but possibly also to the less well-defined porosity after grafting of the silane in the post-modified system (Fig. 27.5).

Li *et al.* have published details of an MCM-41-based material which combines Pd clusters and phenyl groups on the walls of the material.[42] This material is an efficient catalyst for the Ullmann coupling of haloarenes to give biaryls. They demonstrated that the incorporation of phenyl groups into the material caused a modest increase in activity, but a dramatic increase in selectivity, leading to a doubling of product yield. While the modification of surface chemistry is important here, as in the other examples, the possibility that there is a direct interaction between the Ph groups and the Pd nanoclusters cannot be ruled out. Interestingly, the phenyl materials were stable in, and functioned effectively in water. With the best catalyst and the most appropriate conditions, reduction of iodobenzene to benzene was suppressed (Fig. 27.6).

Given that the increased level of organic incorporation in these mesoporous silicas led to general increases in rate of several reaction types, it seems logical to move away from silicas, and investigate mesoporous carbons as supports for catalysts. Such carbons are available via the mesoporous silica-templated route first described

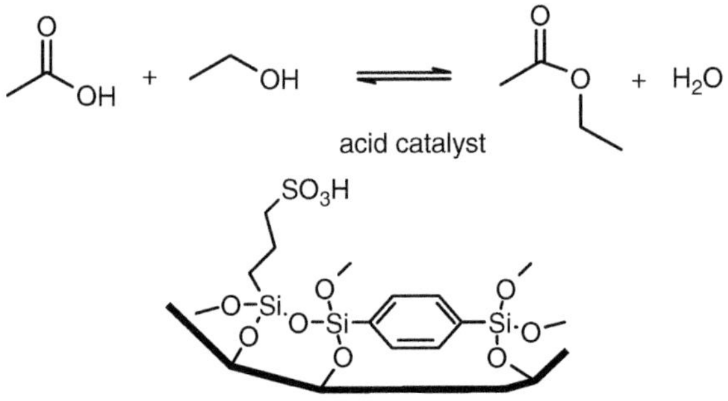

Figure 27.5 PMOS-sulphonic acid catalysis of esterification reactions.

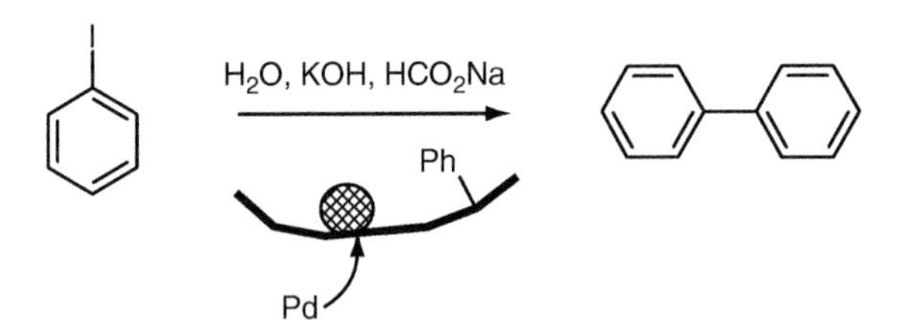

Figure 27.6 Ullmann coupling using Ph modified Pd-silica.

by Ryoo *et al.*[43] This involves the filling of (typically) SBA-15 pores with a carbonizable precursor such as sucrose, furfuryl alcohol or polyacrylonitrile, carbonizing the precursor at high temperature in an inert atmosphere, and finally removing the silica with either HF or NaOH. Careful working ensures a very good replication of the silica pore space as a mesoporous carbon. While there may be few functional groups to attach catalytic groups to, some functionality may be produced by post-treatment. For example, Lázaro *et al.*[44] incorporated O functionality by oxidation with nitric acid under conditions where structural integrity was maintained. Interestingly, H_2O_2 treatment led to collapse of the porous material.[45]

Feng *et al.*[46] have recently demonstrated that CMK-5 functionalized with aryl sulphonic acids is a very active catalyst for both esterification and acid catalyzed alkylation of phenol with acetone to give bis-phenol A. In the latter case, conversions were relatively low, but significantly better than either ethylene bridged periodic mesoporous organosilica or a functionalized SBA-15. Functionalization of the carbon was carried out using diazonium chemistry.

Schüth *et al.* have demonstrated the incorporation of sub-nanometer Pd clusters in a mesoporous carbon prepared, in the presence of Pd, using polyacrylonitrile as carbon source.[47] It is likely that the N atoms in the precursor and in the product play an important role in the distribution and stabilization of the Pd within the structure, helping to minimize clustering. The catalysts are very active and exceptionally selective in the oxidation of alcohols to aldehydes in supercritical CO_2 using O_2 as oxidant, in a particularly clean and effective oxidation. (Fig. 27.7)

Particularly remarkable is the lack of overoxidation of the aldehydes to acids, normally a serious problem in the selective oxidation of primary alcohols.

A related series of nanoporous materials, the 'starbons',[48,49] have been prepared from starch through an expansion and carbonization process, and have demonstrated similar activity in aqueous phase catalytic reactions. While these materials are less ordered than the CMK-5 system, they have a reasonably regular pore structure, are very easily prepared—requiring no template—and are likely to be considerably cheaper to prepare, being derived from a very cheap renewable source. Additionally, their preparation is relatively clean. They also have tunable surface chemistry (achieved via variation in carbonization temperature leading to controllable generation and removal of, e.g., alcohol/carbonyl functionalities) and thus don't require oxygenation as a pretreatment. After starch expansion, acid catalyzed carbonization

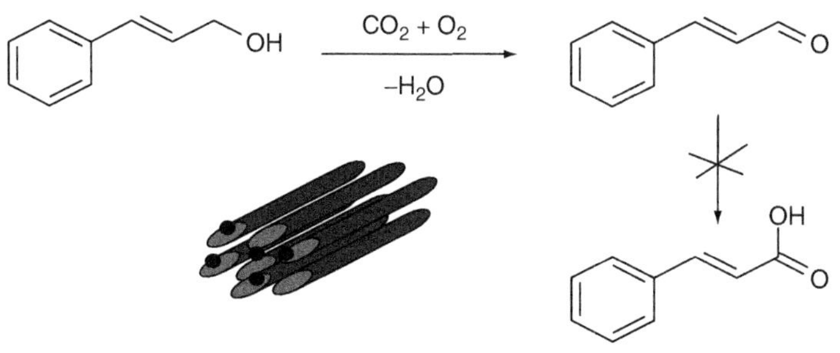

Figure 27.7 Selective oxidation of cinnamyl alcohol over Pd-mesoporous carbon.

is carried out at relatively low temperatures, preserving and generating significant surface functionality. Higher carbonization temperatures lead to reduction in the oxygen content of the material, and lead to a more carbon-like material, with a more hydrophobic surface. It is possible to exploit this surface chemistry to attach additional surface groups such as sulphonic acids. The sulphonic acid functionalized materials have been used for esterification reactions, and have the remarkable property of being able to esterify acids in aqueous environments. This is thought to be due to their hydrophobic nature reducing the ingress of water to the active sites, allowing esterification to take place in relatively water-free conditions; the product, having been displaced from the catalyst surface is relatively stable in solution, where there are no acid groups to catalyze the hydrolysis of the product. There is a clear optimum temperature for activity; this is considered to be due to a combination of sufficient functional group density on the surface and the requirement for a relatively hydrophobic material, although this has not been studied in depth.

Aqueous phase chemistry has clear benefits from a green standpoint, and the provision of key platform molecules (many of which are acids[50]) from fermentation systems means that manipulations directly in the aqueous environment of a fermentation broth will be particularly relevant in the future. The main drawback to fermentative processes is the difficult isolation of products from the complex mixture of aqueous salts/amino acids, unconverted feedstock and cells. Direct conversions in such media are clearly challenging, but may lead to more easily separable products and, overall, simpler processing. For example, succinic acid, which can be prepared by fermentation, can be converted to the (water insoluble) diethyl succinate directly using starbons in aqueous ethanol. The isolation of this valuable platform molecule is therefore considerably simplified.

5. Towards One-Pot Multistage Syntheses

One strategy which is important in multistep syntheses is the combination of steps. A significant amount of process waste is generated every time an intermediate product is formed in a multistep synthesis. While this has the (beneficial) effect of purifying the intermediate before the next reaction, it is often the case that this purification is actually unnecessary, and the reaction mixture can be directly used in the next step. One elegant approach which has been demonstrated recently is the development of bifunctional catalysts based on mesoporous silicas. In one case, the catalysts have been used to carry out two consecutive stages of a synthesis without isolation of the intermediate, whereas the other has been used to activate both components of the reaction mixture simultaneously, leading to unusually efficient catalysis.

Sanchez *et al.*[51] developed a combined base and hydrogenation catalyst for the one-pot Knoevenagel reaction followed by direct hydrogenation of the alkene thus formed (Fig. 27.8). Their catalyst very efficiently carried out both transformations under mild conditions, and avoided the isolation of the intermediate unsaturated cyanoester, saving time, energy and isolation waste. Additionally, the rapid

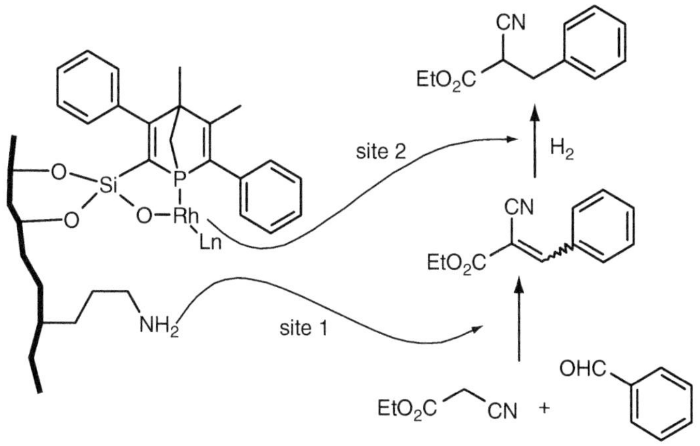

Figure 27.8 Combined base catalysis and hydrogenation on a single catalyst.

reduction of the intermediate prevents undesirable side reactions, such as the Michael addition of a nucleophile β to the ester group in the product, a reaction such intermediates are prone to undergo.[52]

Clearly, such an approach could be generally very useful to react further a relatively unstable or reactive intermediate.

Lin *et al* have published details of a bifunctional catalyst system, where basic groups activate a nucleophile, while urea sites provide mild acid catalysis via H-bonding activation of the electrophile, typically an aldehyde.[53] Turnover numbers were 2–3 times higher where both functional groups were present, compared to just the base catalytic groups. This behaviour was observed for the Henry reaction of nitromethane with benzaldehydes, the aldol reaction (via enamine formation) and the cyanosilylation reaction (Fig. 27.9).

Sharma and Asefa have published details of a similarly bifunctional system, which relies on the spatial separation of aminopropyl groups on a surface, interspersed with silanols.[54] The weak acidity of the silanols activates substrates towards the Henry

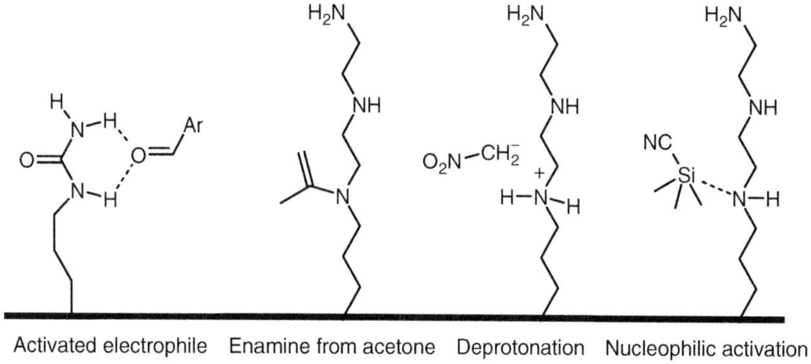

Figure 27.9 Activation modes for bifunctional acid/base catalyst.

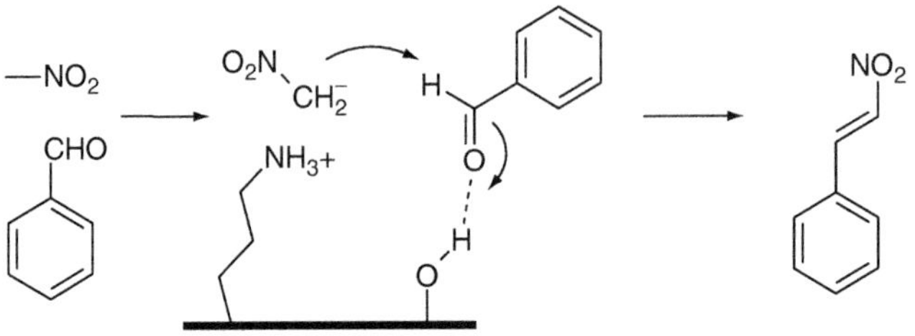

Figure 27.10 Henry reaction using both mild base and acid sites in proximity.

reaction of nitroalkanes with aldehydes (Fig. 27.10). Again, the key feature is the H-bonding activation of the aldehyde carbonyl with simultaneous base activation of the nitroalkane. A balance between the relative numbers of the two sites was shown to be important for optimal activity, with dramatic rate and yield enhancements for the optimal system. The best system provided a 99% conversion in 15 min compared to 53% and 3% for less active catalysts, both of which had higher numbers of amine sites (by a factor of ca. 3). The blank silica led to no reaction, indicating that both sites are relevant to the catalysis.

The ability to provide mild acid/base pairs on a surface is therefore relatively straightforward. The same can be achieved in solution—indeed many Knoevenagel reactions (see Fig. 27.8) are known to be best catalyzed by mixtures of amines and carboxylic acids in solution, relying on the moderate amounts of free acid and base present at equilibrium in such systems.

Clearly, in solution, the use of a strong acid and a base (or strong base and mild acid) would lead to complete protonation of the base to form salt, and no acid-base catalysis would be expected. However, on a solid surface these two functionalities can co-exist if they are spatially separated.

This has been achieved via the use of bis-silanes and monosilanes in a clever synthetic strategy.[55] The authors utilized $(MeO)_3Si(CH_2)_3SS(CH_2)_3Si(OMe)_3$ and a protected aminopropyl silane $(MeO)_3Si(CH_2)_3NHBoc$ to prepare a material with disulfide links in the material walls and pendant amino chains. The disulfides could then be oxidized to sulphonic acids, producing a material with both strong acid sites and basic sites in the same structure, but separated. They demonstrated that, even in the presence of the strong acid, the amine groups were not protonated, by proving that they would react as an efficient nucleophile with acrylamide. (Fig. 27.11)

Davis' group has also presented very interesting work along similar lines, again showing the possibility of having strong acids and bases present simultaneously, and also showing that they can behave cooperatively under the correct conditions. They prepared a series of amine-functionalized SBA-15 materials which contained sulphonic,[56] phosphonic and carboxylic groups.[57] With the sulphonic acid systems, they found that excellent activity in the aldol reaction could be achieved, beyond that achieved with similar mono-functional catalysts. By addition of *p*-toluenesulphonic acid, they could protonate the amine and achieve purely acid catalysis; conversely they could 'switch off' the acidity of the sulphonic acids by addition of propylamine to deprotonate the bound sulphonic acids to give base catalysts. Of particular note was their finding that the activity of their bifunctional catalysts

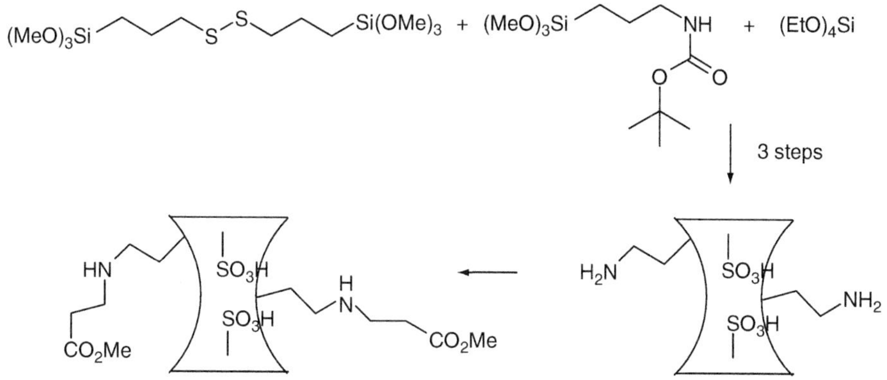

Figure 27.11 Bifunctional acid/base material and its nucleophilic reactivity.

depended on the strength of the acid group. For the aldol reaction, the relatively weak carboxylic acid (pka: ca. 5)/amine systems were better than the moderately acidic phosphonic acid (pKa: ca. 3)/amine systems, which were in turn better than the sulphonic acid (pKa: ca. −2)/amine systems. This indicates that the correct balance must be struck for a given application. This balance clearly is likely to vary depending on the demands of the reaction and the specific reaction partners. This level of fine-tuning brings many challenges in optimization, but does promise exceptional control over catalyst design and process implementation. (Fig. 27.12)

A final example of a bifunctional catalyst providing exceptional activity is provided by Bonardet *et al.*[58] They developed a bifunctional SBA-15 material containing acid sites (Al) and oxidation sites (Ti). They applied this material to the synthesis of adipic acid from cyclohexene. There have been many attempts to develop green processes to adipic acid from a range of starting materials, as the current production methods generate considerable amounts of waste, requiring oxidants such as nitric acid, and releasing considerable amounts of NO_x as by-products.

They prepared their catalyst by taking an Al-containing SBA-15 and reacting it with $TiCl_4$ to provide the Ti sites. Their process takes cyclohexene directly, and converts it, via four steps into adipic acid, with an impressive overall yield of up to

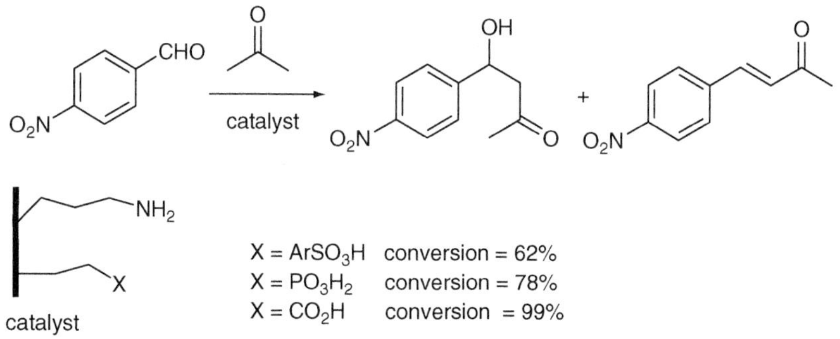

X = $ArSO_3H$ conversion = 62%
X = PO_3H_2 conversion = 78%
X = CO_2H conversion = 99%

Figure 27.12 Bifunctional acid/base catalysis of the aldol reaction.

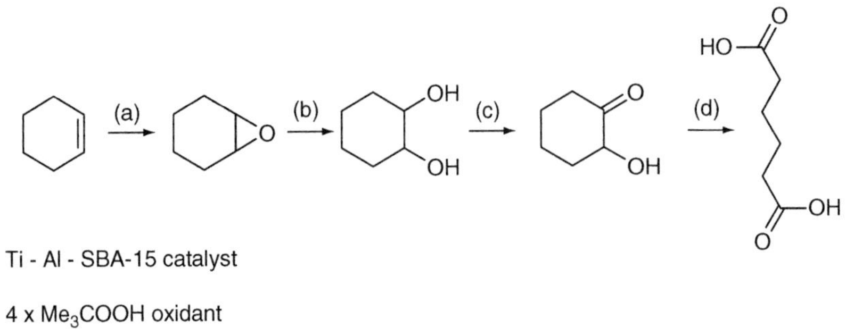

(a) (b) (c) (d)

Ti - Al - SBA-15 catalyst

4 x Me_3COOH oxidant

Figure 27.13 One pot oxidation of cyclohexene to adipic acid: (a) oxidation catalysis; (b) acid catalyis; (c) oxidation and acid catalysis; (d) oxidation catalysis.

88%, and very good selectivity. Their oxidant is an organic hydroperoxide such as *t*-BuOOH, and the process runs in acetonitrile, which will solubilize all the components of the reaction mixture. Overall four equivalents of the peroxide are used, generating t-BuOH as co-product. Ideally, hydrogen peroxide would be used, but the process is a significant step in this direction already (Fig. 27.13).

6. Enzyme Immobilization in Mesoporous Hosts

There are many advantages to using enzymes as catalysts—they operate in aqueous environments at moderate temperatures, and are highly selective. Improving their stability by crosslinking or immobilization is beneficial, and helps in their separation from reaction products. The ability to produce mesopores with pore diameters of several nanometers allows enzymes to be immobilized within the pores of mesoporous silicas such as SBA-15, and the resultant supported enzymes can be used as catalysts. Simply supporting enzymes on/in mesoporous silica results in materials of limited stability with only H-bonding interactions to hold the enzyme in place. Effective adsorption of enzymes leading to stable binding normally requires the correct pore size and the presence of functional groups on the material to provide either physical stabilization or chemical binding. Care must be taken to ensure that the functional groups used are not damaging to the activity of the enzyme; indeed it has been demonstrated that the correct functional groups can actually enhance activity over that of the natural enzyme.

An example of the use of physical constraints on the enzyme has been provided by Balkus *et al.*[59] They immobilized the redox-active cytochrome c in ether SBA-15 or MCM-48 (the latter giving higher loadings) with functionalization of the pore entrances with either aminopropyl groups or cyanopropyl groups. In both cases, the enzyme was effectively trapped in the pore system, but the basic amine groups caused a reduction in the electrochemical activity of the enzyme. The cyanopropyl system, on the other hand, had no negative impact on the enzyme, and its electrochemical activity was retained.

Wright *et al.*[60,61] have shown that –SH groups have a positive effect on the stability of protein attachment to SBA-15. The leaching observed on unfunctionalized silica disappeared when the –SH groups were present. This is likely to be due to covalent links with thiol groups in the protein. They investigated various different functional groups, attached by post-functionalization and found that other groups helped to stabilize the enzyme (trypsin) within the structure, whereas –Ph had essentially no effect. Dramatic differences were seen on changing the mode of preparation of the SBA-15—*in situ* co-condensation gave better performance that post-modified in terms of stability, in some cases, 5–10 times better. The behaviour of the enzyme in hydrolysis reactions showed a broadly similar trend, with *in situ* systems generally being better than unfunctionalized SBA-15, which was in turn significantly better than post-modified materials. (Fig. 27.14 and Table 27.1.)

Using a different enzyme (*Penicillin Acylase*), Chong and Zhao found significant changes in activity with aminopropyl-functionalized material giving the highest activity per gram *material*—essentially double that of the natural enzyme; expressed per unit *enzyme*, the best material was one functionalized with vinyl groups, which

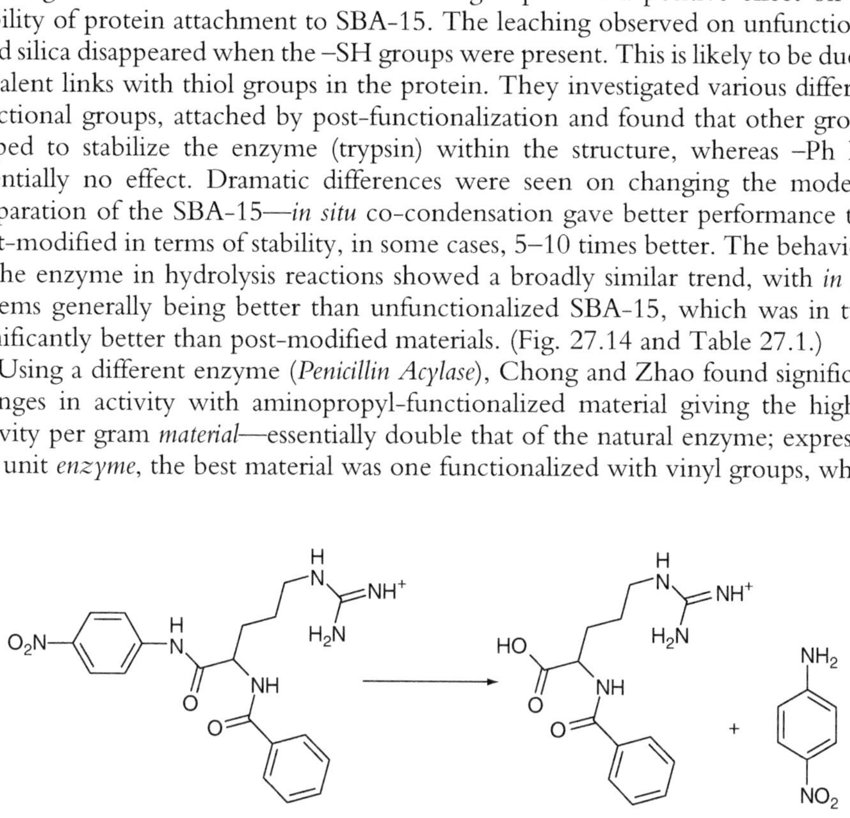

Figure 27.14 Hydrolysis of an amide by immobilized trypsin.

Table 27.1 Relative leaching and activity of trypsin immobilized on functionalized SBA-15

Functional group		% Leaching	Relative activity
None		47	0.33
–SH	*In situ*	1	0.82
	Post-modified	15	0.04
–Cl	*In situ*	1	0.27
	Post-modified	0	0.21
$-NH_2$	*In situ*	23	0.03
	Post-modified	19	0.04
–	*In situ*	2	0.58
CO_2H	Post-modified	10	0.12
–Ph	*In situ*	18	0.41
	Post-modified	54	0.16

Reaction as outlined in Fig. 27.14.

was around three times more active than the native enzyme in the hydrolysis of Penicillin G. [62]

While it is clear that supporting enzymes in an appropriately functionalized material will produce a stable and very active material with clear advantages over the enzyme in the natural state, it is far from clear what the nature of the interactions and the source of the positive effects are. Much work remains to be done in a very fascinating field which has clear potential for green chemical transformations.

7. Incorporation of Nanoporous Solids in Flow Reactors

The vast majority of examples of liquid/solid two phase reactions has involved the use of stirred flasks as reactors. While this is a simple and intuitive model for the majority of organic transformations, it does not necessarily provide the optimum performance of a process. This is particularly true, as diffusion processes and mass transport are rarely very efficient in such systems. Reactors offering better mixing and mass transport may well provide significant improvements in rate, catalyst lifetime and selectivity, as well as providing a safer option for processing larger volumes, as the quantities of reaction mixture present in the reaction zone at a given time are much lower.

Jaenicke *et al.*[63] have demonstrated that $AlCl_3$ supported on MCM-41 performs well in the formation of 2-(2,4-difluorophenyl)propane (an intermediate in the synthesis of fluconazone) from 1,3 difluorobenzene and 2-chloropropane in a continuous packed bed reactor. They found that passing the reactant mixture over a short plug of catalyst produced a good yield of product is a short time. Compared to experiments carried out in the batch reactor, a much greater catalyst lifetime was achieved. This relates to the very moisture sensitive nature of the catalyst, which makes isolation from batch reactions difficult, as exposure to moisture rapidly deactivates the catalyst. In a continuous reactor, the catalyst remains active and does not need to be exposed to the atmosphere at any stage (including after reaction). Unfortunately, the greener options of using propene or propanol failed owing to polymerization of the former, and deactivation caused by the production of water in the latter.

Claus *et al.*[64] have investigated the hydrogenation of aqueous solutions of glucose to sorbitol over nanoporous Ni and Ru catalysts in a continuous trickle bed reactor. They investigated the behaviour of a series of catalysts and found good activity and selectivity for Ru catalysts supported on mesoporous silicas. In agreement with the diffusion studies reported by Walcarius (see earlier text), for aqueous systems rates of reaction dropped with reduction in pore size. Conversion and selectivity were enhanced in the continuous regime.

The combination of immobilized catalysts in a flow reactor using supercritical CO_2 as solvent is a particularly attractive route to catalytic transformations. In particular, the excellent diffusion properties of this medium should be ideally suited for use with nanoporous catalysts. At this stage, most of the work carried out has been with mesoporous amorphous systems, rather than templated nanoporous materials. Nonetheless, the work described in the literature indicate that this approach has real potential to get the best from nanoporous catalytic processes. Poliakoff's group has been active recently in developing such technology, and has worked with the fine chemicals manufacturer Thomas Swann to commercialize a process for continuous hydrogenation using a solid catalyst.[65] This work has been extended to asymmetric hydrogenation using bisphosphine ligands with some success.[66]

Baiker *et al.* have also investigated the hydrogenation of citral in supercritical CO_2.[67] They compared the performance of their reaction in both batch and continuous set ups, and found that selectivity was much greater in continuous mode. This was attributed to the much greater catalyst: substrate ratio found in a continuous reactor.

A Pd based catalyst has also been used to carry out Suzuki coupling reactions in a continuous manner using supercritical CO_2.[68] Initial results were encouraging with supercritical reactions giving slightly higher conversion.

A different option is to use the Spinning Disc Reactor, consisting of a rapidly rotating plate which is heated (or cooled) to the desired temperature. Reactants are fed onto the disc at the centre, and flow with high shear mixing towards the edge, where they fall off and are collected. The mixing efficiency and the efficiency of heat transfer are both much higher than can be achieved in batch processes. Vicevic *et al.* [69,70] have demonstrated the attachment of nanoporous catalysts to the surface of the plate using epoxy resin binders, and have also shown that such systems are capable of excellent performance in the very challenging Lewis acid catalyzed isomerization of α-pinene, giving the commercially very important campholenic aldehyde. (Fig. 27.15)

This isomerization, which can give as many as 100 products[71] can be carried out in batch mode with selectivities of the order of only 60% using zinc triflate supported on Micelle Templated Silica. Moving to the catalytic spinning disc reactor, the selectivity can be increased to 75% with very short residence times (<0.3 s) helping to reduce the damaging reactions of the product. In both spinning disc reactor and batch conditions, the Micelle Templated Silica catalyst outperformed other versions based on amorphous silicas.

A second example of the use of a Spinning Disc Reactor is the cationic polymerization of styrene using supported BF_3 catalysts.[72] Here, the use of the spinning disc allows polymerization of styrene to occur at much higher concentrations and temperatures than in a batch reactor. This is due to the far better heat transfer controlling the exotherm—in the batch system the exotherm is difficult to control

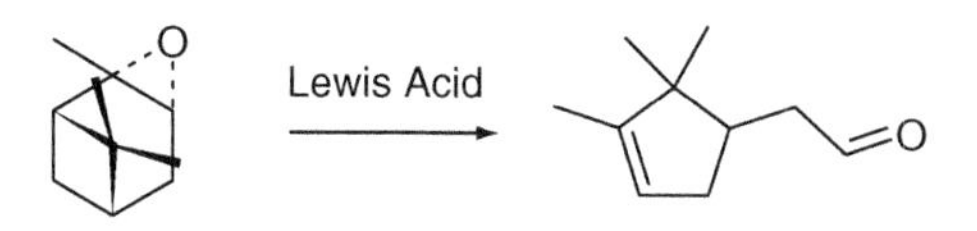

Figure 27.15 Isomerization of α-pinene to campholenic aldehyde.

unless the reaction is run very dilute, which slows down reaction rates considerably, as well as introducing much more solvent into the system. A further benefit of operating under spinning disc conditions is the excellent molecular weight distributions obtained. This was considered to be due to the much better heat transfer and temperature control. In this example, the use of a catalyst immobilized on the walls of the reactor has the advantages of simultaneous reaction and separation, lower solvent volumes required (with lower energy demands for heating solvent) better reaction control and better product quality, as well as a much safer operating environment, another key green chemical requirement.

Membrane reactors are yet another example of a potentially beneficial continuous reaction/separation concept which could be very well suited to the incorporation of nanoporous catalysts and to the simplification of processing. While such membranes can be fabricated directly from nanoporous materials, or could be designed such that nanoporous particulates could be held within a membrane, such that the membrane itself was the active part of the system, relatively little appears to have been done on catalytic applications of such systems. Prouzet and Bossière recently published a review on the synthesis of MSU-X silicas and their successful formation as part of a membrane for filtration.[73] Other groups have also worked on the successful production and characterization of nanoporous silica membranes using a variety of synthetic methods. DeSisto *et al.* have formed membranes of mesoporous silica using Pluronic P123 as template. Porosity of the magnitude of a few nanometers was readily obtained—this was somewhat smaller than that obtained from the corresponding system in particulate form.[74]

Yi *et al.* have described the preparation of a membrane consisting of mesoporous silica functionalized with aminopropyl groups.[75] They have used this membrane for the adsorption of Cu(II) from aqueous solution (see Section 4.1). Their system allowed for very good adsorption of Cu(II) from solution as the aqueous system passed through the membrane. Given the well established catalytic activity of aminopropyl silicas, it seems an obvious extension to utilize these membranes as reactors.

8. Conclusions

The use of heterogeneous catalysts is an important feature of green chemistry, and the provision of well-structured materials containing a wide range of functionality can provide significant opportunities for the development of a green process. The ability to tailor such materials has reached unprecedented levels over the last decade or so, and many research groups have embraced this technology to provide clean and efficient catalytic reactions and processes using them. The main part of this work has been on heterogenized versions of existing catalysts, but the real potential of the materials is beginning to be exploited, with subtle and clever systems such as bifunctional catalysis being developed, where surface interactions can be modified and optimized alongside the catalytic site itself.

As reflected in this overview, the vast majority of work has been carried out on silicas, or closely related materials. Several other materials appear to be well suited for catalytic applications, and we may expect that these will feature more in the coming years. Likewise, the ability to design and prepare materials with ever more finely controlled features is an important driver in the design of cleaner processes.

Telescoping synthetic routes by providing a material which will carry out two consecutive steps directly without any requirement for isolation has been demonstrated—there must be many more examples where such an approach can be powerful in simplifying and streamlining processes. More than that, the rapid conversion of ‘difficult’ (i.e., unstable or toxic) intermediates *in situ* can have a significant impact on many processes. However, having established that nanoporous materials have great potential in these areas, researchers working in this area should apply metrics as rigorously as possible to measure just how green their processes are.

Finally, the importance of the correct presentation of the catalyst is beginning to be seen, with correct reactor design, significant improvements to both the reaction and the process are possible. All of these factors mean that nanoporous materials have a major role to play in green chemistry, and in the development of sustainable new processes.

REFERENCES

[1] Anastas, P. A., Warner, J. C. (Eds.), *Green Chemistry; Theory and Practice,* Oxford University Press, Oxford, 2000.
[2] Clark, J. H., Macquarrie, D. J. (Eds.), *Handbook of Green Chemistry and Technology,* Blackwell, Oxford, 2002.
[3] Kamm, B., Production of platform chemicals and synthesis gas from biomass, *Angew. Chem. Int. Ed.* **2007,** *46,* 5056–5058.
[4] Clark, J. H., Budarin, V. L., Deswartes, F. E. I., Hardy, J. J. E., Kerton, F. M., Hunt, A. J., Luque, R., Macquarrie, D. J., Milkowski, K., Rodriguez, A., Samuel, O., Tavener, S. J., White, R. J., Wilson, A. J., Green chemistry and the biorefinery: A partnership for a sustainable future, *Green Chem.* **2006,** *8,* 853–860.
[5] Gallezot, P., Catalytic routes from renewables to fine chemicals, *Catal. Today* **2007,** *121,* 76–91.
[6] Gallezot, P., Process options for converting renewable feedstocks to bioproducts, *Green Chem.* **2007,** *9,* 295–302.
[7] Macquarrie, D. J., Hardy, J. J. E., Applications of functionalized chitosan in catalysis, *Ind. Eng. Res. Chem.* **2005,** *44,* 8499–8520.
[8] Guibal, E., Heterogeneous catalysis on chitosan-based materials: a review, *Prog. Polym. Sci.* **2005,** *30,* 71–109.
[9] Jiminez-Gonzalez, C., Curzons, A. D., Constable, D. J. C., Cunningham, V. L., Cradle-to-gate life cycle inventory and assessment of pharmaceutical compounds, *Int. J. Life Cycle Assess.* **2004,** *9,* 114–121.
[10] Constable, D. J. C., Curzons, A. D., Cunningham, V. L., Metrics to ‘green’ chemistry – which are the best? *Green Chem.,* **2002,** *4,* 521–527.
[11] Andraos, J., Unification of reaction metrics for green chemistry: Applications to reaction analysis, *Org. Process Res. Dev.* **2005,** *9,* 149–163.
[12] Tanev, P. T., Pinnavaia, T. J., A neutral templating route to mesoporous molecular-sieves, *Science* **1995,** *267,* 865–867.
[13] Hatton, B., Landskron, K., Whitnall, W., Perovic, D., Ozin, G. A., Past, present, and future of periodic mesoporous organosilicas – The PMOs, *Acc. Chem. Res.* **2005,** *38,* 305–312.

[14] Hunks, W. J., Ozin, G. A., Challenges and advances in the chemistry of periodic mesoporous organosilicas (PMOs), *J. Mater. Chem.* **2005,** *15,* 3716–3724.
[15] Hunks, W. J., Ozin, G. A., Periodic mesoporous phenylenesilicas with ether or sulfide hinge groups – a new class of PMOs with ligand channels, *Chem. Comm.* **2004,** 2426–2427.
[16] Zhu, H., Jones, D. J., Zajac, J., Dutartre, R., Rhomari, M., Rozière, J., Synthesis of periodic large mesoporous organosilicas and functionalization by incorporation of ligands into the framework wall, *Chem. Mater.* **2002,** *14,* 4886–4894.
[17] Burleigh, M. C., Markowitz, M. A., Spector, M. S., Gabor, B. P., Direct synthesis of periodic mesoporous organosilicas: Functional incorporation by co-condensation with organosilanes, *J. Phys. Chem. B* **2001,** *105,* 9935–9942.
[18] Wahab, M. A., Imae, I., Kawakami, Y., Ha, C.-S., Periodic mesoporous organosilica materials incorporating various organic functional groups: Synthesis, structural characterization, and morphology, *Chem. Mater.* **2005,** *17,* 2165–2174.
[19] Hansen, E. W., Courivaud, F., Karlsson, A., Kolboe, S., Stöcker, M., Effect of pore dimension and pore surface hydrophobicity on the diffusion of n-hexane confined in mesoporous MCM-41 probed by NMR - a preliminary investigation, *Microporous Mesoporous Mater.* **1998,** *22,* 309–320.
[20] Stallmach, F., Gräser, A., Kärger, J., Krause, C., Jeschke, M., Oberhagemann, U., Spange, S., Pulsed field gradient NMR studies of diffusion in MCM-41 mesoporous solids, *Microporous Mesoporous Mater.* **2001,** *44–45,* 745–753.
[21] Okazaki, M., Toriyama, K., Inhomogeneous distribution and collective diffusion of solution molecules in the nanochannel of mesoporous silica, *J. Phys. Chem. B* **2003,** *107,* 7654–7658.
[22] Macquarrie, D. J., Clark, J. H., Lambert, A., Priest, A., Mdoe, J. E. G., Mdoe Catalysis of the Knoevenagel reaction by gamma-aminopropylsilica, *React. Funct. Polym.* **1997,** *35,* 153–158.
[23] Macquarrie, D. J., Jackson, D. B., Aminopropylated MCMs as base catalysts: a comparison with aminopropylated silica, *Chem. Comm.* **1997,** 1781–1782.
[24] Walcarius, A., Etienne, M., Bessière, J., Rate of access to the binding sites in organically modified silicates. 1. Amorphous silica gels grafted with amine or thiol groups, *Chem. Mater.* **2002,** *14,* 2757–2766.
[25] Walcarius, A., Etienne, M., Lebeau, B., Rate of access to the binding sites in organically modified silicates. 2. Ordered mesoporous silicas grafted with amine or thiol groups, *Chem. Mater.* **2003,** *15,* 2161–2173.
[26] Zhao, D. M., Zhao, J. Q., Zhao, S. S., Application of functionalized MCM-41 in organic synthesis, *Prog. Chem.* **2007,** *19,* 510–519.
[27] Thomas, J. M., Raja, R., The advantages and future potential of single-site heterogeneous catalysts, *Top. Catal.* **2006,** *40,* 3–17.
[28] Oye, G., Glomm, W. R., Vralstad, T., Volden, S., Magnusson, H., Stocker, M., Sjoblom, J., Synthesis, functionalisation and characterisation of mesoporous materials and sol-gel glasses for applications in catalysis, adsorption and photonics, *Adv. Coll. Interface Sci.* **2006,** *123,* 17–32.
[29] Clark, J. H., Macquarrie, D. J., Tavener, S. J., The application of modified mesoporous silicas in liquid phase catalysis, *Dalton Trans.* **2006,** 4297–4309.
[30] Taguchi, A., Schüth, F., Ordered mesoporous materials in catalysis, *Microporous Mesoporous Mater.* **2005,** *77,* 1–45.
[31] Cejka, J., Organized mesoporous alumina: synthesis, structure and potential in catalysis, *App. Cat. A. Gen.* **2003,** *254,* 327–338.
[32] Stein, A., Melde, B. J., Schroden, R. C., Hybrid inorganic-organic mesoporous silicates - Nanoscopic reactors coming of age, *Adv. Mater.* **2000,** *12,* 1403–1419.
[33] Sayari, A., Catalysis by crystalline mesoporous molecular sieves, *Chem. Mater.* **1996,** *8,* 1840–1852.
[34] Diaz, I., Marques-Alvarez, C., Mohino, F., Perez-Pariente, J., Sastre, E., Combined alkyl and sulfonic acid functionalization of MCM-41-type silica - Part 1. Synthesis and characterization, *J. Catal.* **2000,** *193,* 283–294.
[35] Diaz, I., Marques-Alvarez, C., Mohino, F., Perez-Pariente, J., Sastre, E., Combined alkyl and sulfonic acid functionalization of MCM-41-Type silica - Part 2. Esterification of glycerol with fatty acids, *J. Catal.* **2000,** *193,* 295–302.

[36] Macquarrie, D. J., Tavener, S. J., Harmer, M. A., Novel mesoporous silica-perfluorosulfonic acid hybrids as strong heterogeneous Bronsted catalysts, *Chem. Commun.* **2005,** 2363–2365.
[37] Al-Haq, N., Ramnauth, R., Kleinebiekel, S., Ou, D. L., Sullivan, A. C., Wilson, J., Aminoalkyl modified polysilsesquioxanes; synthesis, characterisation and catalytic activity in comparison to related aminopropyl modified silicas, *Green Chem.* **2002,** *4,* 239–244.
[38] Macquarrie, D. J., Organically modified hexagonal mesoporous silicas - Clean synthesis of catalysts and the effect of high loading and non-catalytic second groups on catalytic activity of amine-derivatised materials, *Green Chem.* **1999,** *1,* 195–198.
[39] Macquarrie, D. J., Jackson, D. B., Aminopropylated MCMs as base catalysts: a comparison with aminopropylated silica, *Chem. Commun.* **1997,** 1783–1784.
[40] Macquarrie, D., Rocchia, M., Onida, B., Garrone, E., Brunel, D., Blanc, A. C., Fajula, F., Lentz, P., Nagy, J. B., Comparison of 3-aminopropylsilane linked to MCM-41 and HMS type silicas synthesized under biphasic and monophasic conditions *Stud. Surf. Sci. Catal.* **2001,** *135,* 319–319.
[41] Yang, Q., Kapoor, M. P., Inagaki, S., Shirokura, N., Kondo, J. N., Domen, K., Catalytic application of sulfonic acid functionalized mesoporous benzene-silica with crystal-like pore wall structure in esterification, *J. Mol. Cat. A* **2005,** *230,* 85–89.
[42] Wan, Y., Chen, J., Zhang, D., Li, H., Ullmann coupling reaction in aqueous conditions over the Ph-MCM-41 supported Pd catalyst, *J. Mol. Catal. A* **2006,** *258,* 89–94.
[43] Ryoo, R., Joo, S. H., Jun, S., Synthesis of highly ordered carbon molecular sieves via template-mediated structural transformation, *J. Phys. Chem. B* **1999,** *103,* 7743–7746.
[44] Lázaro, M. J., Calvillo, L., Bordejé, E. G., Moliner, R., Juan, R., Ruiz, C. R., Functionalization of ordered mesoporous carbons synthesized with SBA-15 silica as template, *Microporous Mesoporous Mater.* **2007,** *103,* 158–165.
[45] Lu, A.-H., Li, W.-C., Muratova, N., Spleithoff, B., Schüth, F., Evidence for C-C bond cleavage by H_2O_2 in a mesoporous CMK-5 type carbon at room temperature , *Chem. Commun.* **2005,** 5184–5186.
[46] Wang, X., Liu, R., Waje, M. M., Chen, Z., Yan, Y., Bozhilov, K. N., Feng, P., Sulfonated ordered mesoporous carbon as a stable and highly active protonic acid catalyst, *Chem. Mater.* **2007,** *19,* 2395–2397.
[47] Lu, A. H., Li, W. C., Hou, Z., Schüth, F., Molecular level dispersed Pd clusters in the carbon walls of ordered mesoporous carbon as a highly selective alcohol oxidation catalyst, *Chem. Commun.* **2007,** 1038–1040.
[48] Budarin, V. L., Clark, J. H., Hardy, J. J. E., Luque, R., Milkowski, K., Tavener, S. J., Wilson, A. J., Starbons: New starch-derived mesoporous carbonaceous materials with tunable properties, *Angew. Chem. Int. Ed.* **2006,** *45,* 3782–3786.
[49] Budarin, V. L., Clark, J. H., Luque, R., Macquarrie, D. J., Versatile mesoporous carbonaceous materials for acid catalysis, *Chem. Commun.* **2007,** 634–636.
[50] Werpy, T., Peterson, G., Top value added chemicals from biomass feedstock – Volume I: Results of screening for potential candidates from sugars and synthesis gas, US Department of Energy, August 2004.
[51] Goettmann, F., Grosso, D., Mercier, F., Mathey, F., Sanchez, C., New P - O ligand grafted on periodically organised mesoporous silicas for one-pot bifunctionnal catalysis: Coupling of base catalysed Knoevenagel condensation with in situ Rh catalysed hydrogenation, *Chem. Commun.* **2004,** 1240–1241.
[52] Sartori, G., Bigi, F., Maggi, R., Sartorio, R., Macquarrie, D. J., Lenarda, M., Storaro, L., Coluccia, S., Martra, G., Catalytic activity of aminopropyl xerogels in the selective synthesis of (E)-nitrostyrenes from nitroalkanes and aromatic aldehydes, *J. Catal.* **2004,** *222,* 410–418.
[53] Huh, S., Chen, H.-T., Wiench, J. W., Pruski, M., Lin, V. S.-Y., Cooperative catalysis by general acid and base bifunctionalized mesoporous silica nanospheres, *Angew. Chem. Int. Ed.* **2005,** *44,* 1826–1830.
[54] Sharma, K. K., Asefa, T., Efficient bifunctional nanocatalysts by simple postgrafting of spatially isolated catalytic groups on mesoporous materials, *Angew. Chem. Int. Ed.* **2007,** *46,* 2879–2882.
[55] Alauzan, J., Mehdi, A., Reyé, C., Corriu, R. J. P., Mesoporous materials with an acidic framework and basic pores. A successful cohabitation, *J. Am. Chem. Soc.* **2006,** *128,* 8718–8719.

[56] Zeidan, R. K., Hwang, S.-J., Davis, M. E., Multifunctional heterogeneous catalysts: SBA-15-containing primary amines and sulfonic acids, *Angew. Chem. Int. Ed.* **2006,** *45,* 6332–6335.
[57] Zeidan, R. K., Davis, M. E., The effect of acid-base pairing on catalysis: An efficient acid-base functionalized catalyst for aldol condensation, *J. Catal.* **2007,** *247,* 379–382.
[58] Lapisardi, G., Chiker, F., Launey, F., Nogier, J. P., Bonardet, J. L., Preparation, characterisation and catalytic activity of new bifunctional Ti-AlSBA15 materials. Application to a "one-pot" green synthesis of adipic acid from cyclohexene and organic hydroperoxides, *Microporous Mesoporous Mater.* **2005,** *78,* 289–295.
[59] Washmon-Kriel, L., Jiminez, V. L., Balkus, K. J., Jr., Cytochrome c immobilization into mesoporous molecular sieves, *J. Mol. Catal. B Enzym.* **2000,** *10,* 453–469.
[60] Yiu, H. H. P., Botting, C. H., Botting, N. P., Wright, P. A., Size selective protein adsorption on thiol-functionalised SBA-15 mesoporous molecular sieve, *Phys. Chem. Chem. Phys.* **2001,** *3,* 2983–2985.
[61] Yiu, H. H. P., Wright, P. A., Botting, N. P., Enzyme immobilisation using SBA-15 mesoporous molecular sieves with functionalised surface, *J. Mol. Catal. B Enzym.* **2001,** *15,* 81–92.
[62] Chong, A. S. M., Zhao, X. S., Functionalized nanoporous silicas for the immobilization of penicillin acylase, *Appl. Surf. Sci.* **2004,** *237,* 398–404.
[63] Hu, X., Chuah, G. K., Jaenicke, S., Solid acid catalysts for the efficient synthesis of 2-(2,4-difluorophenyl) propane, *Appl. Catal. A* **2001,** *209,* 117–123.
[64] Kusserow, B., Schimpf, S., Claus, P., Hydrogenation of glucose to sorbitol over nickel and ruthenium catalysts, *Adv. Synth. Catal.* **2003,** *345,* 289–299.
[65] Licence, P., Ke, J., Sokolova, M., Ross, S. K., Poliakoff, M., Chemical reactions in supercritical carbon dioxide: from laboratory to commercial plant, *Green Chem.* **2003,** *5,* 99–104.
[66] Stevenson, P., Kondor, B., Licence, P., Scovell, K., Ross, S. K., Poliakoff, M., Continuous asymmetric hydrogenation in supercritical carbon dioxide using an immobilised homogeneous catalyst, *Adv. Synth. Catal.* **2006,** *348,* 1605–1610.
[67] Burgener, M., Furrer, R., Mallat, T., Baiker, A., Hydrogenation of citral over Pd/alumina: comparison of "supercritical" CO_2 and conventional solvents in continuous and batch reactors, *App. Catal. A Gen.* **2004,** *268,* 1–8.
[68] Leeke, G. A., Santos, R. C. D., Al-Duri, B., Seville, J. P. K., Smith, C. J., Lee, C. K. Y., Holmes, A. B., McConvey, I. F., Continuous-flow Suzuki-Miyaura reaction in supercritical carbon dioxide, *Org. Proc R and D* **2007,** *11,* 144–148.
[69] Vicevic, M., Boodhoo, K. V. K., Scott, K., Catalytic isomerisation of alpha-pinene oxide to campholenic aldehyde using silica-supported zinc triflate catalysts I. Kinetic and thermodynamic studies, *Chem. Eng. J.* **2007,** *133,* 31–41.
[70] Vicevic, M., Boodhoo, K. V. K., Scott, K., Catalytic isomerisation of alpha-pinene oxide to campholenic aldehyde using silica-supported zinc triflate catalysts II. Performance of immobilised catalysts in a continuous spinning disc reactor, *Chem. Eng. J.* **2007,** *133,* 43–57.
[71] Hölderich, W. F., Roseler, J., Heitmann, G., Liebens, A. T., The use of zeolites in the synthesis of fine and intermediate chemicals, *Catal. Today* **1997,** *37,* 353–363.
[72] Boodhoo, K. V. K., Dunk, W. A., Vicevic, M., Jachuck, R. J., Sage, V., Macquarrie, D. J., Clark, J. H., Classical cationic polymerization of styrene in a spinning disc reactor using silica-supported BF3 catalyst, *J. Appl. Polym. Sci.* **2006,** *101,* 8–19.
[73] Prouzet, E., Bossière, C., A review on the synthesis, structure and applications in separation processes of mesoporous MSU-X silica obtained with the two-step process C. R. Chimie, **2005,** *8,* 579–596.
[74] Higgins, S., Kennard, N., Hill, R., DiCarlo, J., DeSisto, W. J., Preparation and characterization of non-Ionic block co-polymer templated mesoporous silica membranes, *J. Memb. Sci.* **2006,** *279,* 669–674.
[75] Oh, S., Kang, T., Kim, H., Moon, J., Hong, S., Yi, J., Preparation of novel ceramic membranes modified by mesoporous silica with 3-aminopropyltriethoxysilane (APTES) and its application to Cu2+ separation in the aqueous phase, *J. Memb. Sci.* **2007,** *301,* 118–125.

CHAPTER 28

The Fascinating Chemistry of Iron- and Copper-Containing Zeolites

Gerhard D. Pirngruber

Contents

Abstract

Iron and copper zeolites have remarkable catalytic properties in the decomposition of N_2O and NO, in the selective oxidation of benzene to phenol and in the transformation of methane to methanol. Especially, their ability to oxidize methane to methanol has provoked comparisons with transition metal-based enzyme catalysts (methane mono-oxygenase, MMO). This chapter collects the available information about the nature of surface oxygen species in iron and copper zeolites and explains the fascinating chemistry of these materials. It focuses entirely on mechanistic aspects and touches application-related issues only very briefly.

Keywords: Fe-ZSM-5, Cu-ZSM-5, α-oxygen, N_2O, NO, Selective oxidation, Phenol, Methanol

Ordered Porous Solids
DOI: 10.1016/B978-0-444-53189-6.00028-7

1. Introduction

It has been observed a long time ago that transition-metal cations, which are embedded in a zeolite host matrix, do not have the same catalytic properties as the corresponding transition-metal oxides. Iron- and copper-containing zeolites are among the most remarkable examples. They catalyse a number of different reactions, the most fascinating being the selective oxidation of very inert hydrocarbon substrates. Fe-ZSM-5 and some other iron zeolites oxidize benzene to phenol with very high selectivity when using N_2O as oxidant.[1] Iron oxide based catalysts are totally unselective for the same reaction.[2] Moreover, Fe-ZSM-5[3] and Cu-ZSM-5[4] are capable of converting methane to methanol, although the reaction is stoichiometric and not catalytic. These features have provoked comparisons of the zeolite catalysts with the metalloprotein methane mono-oxygenase (MMO),[5] which catalyses the oxidation of alkanes to alcohols.[6,7,8] The active core of MMO contains a dimer of iron or of copper. Although Fe/Cu-ZSM-5 and MMO contain the same metal and catalyse the same reaction, they function very differently. In enzymes, amino acids are the ligands of the active core, and the nature of the amino acid ligands strongly influences the catalytic properties. For example, two other metalloproteins have a very similar diiron core as MMO,[9] but differ strongly in their catalytic behaviour: Ribonucleotide reductase (RNR-R2) generates tyrosyl radicals for the reduction of ribonucleotides in DNA biosynthesis and fatty acid desaturase (Δ-9 ACP) introduces a double bond into saturated fatty acids. This demonstrates that not only the structure of the iron core but also the more distant environment of the active centre governs the catalytic activity. The amino acids surrounding the active site create a binding site for the substrate, called hydrophobic pocket in the case of MMO. It is also believed that the flexible binding of carboxylate ligands to the iron core is crucial for the activation of the oxidizing agent, which is O_2. In a zeolite, the analogues of the amino acid ligands are the lattice oxygen atoms, which bind the transition-metal cation. The 'binding site' of the substrate is determined by the adsorption in the pores. The zeolite host is rigid and the electron donor/acceptor properties of the coordinating lattice oxygen atoms can only be varied in a very limited range (by changing the Si/Al ratio and/or the nature of extra-framework co-cations). A fine-tuning of the catalytic and the adsorption properties as a function of the environment of the active as in the enzyme is therefore not possible.

Nevertheless, the observation that transition-metal cations in zeolites can activate very inert substrates, like methane or benzene, and oxidize them with very high selectivity is remarkable. This chapter will attempt to identify the factors that give some iron and copper zeolites these particular, 'enzyme-like' properties. We will focus on iron zeolites and their capability to oxidize benzene to phenol. Because this reaction is performed using N_2O as the oxidant, we will also analyse the chemistry of N_2O decomposition on iron zeolites. Finally, analogies and differences between iron and copper zeolites will be discussed.

2. The Oxidation of Benzene to Phenol over Iron Zeolites

The direct oxidation of benzene to phenol is a difficult reaction. Because of high stability of the benzene ring rather severe reaction conditions are required, which lead to total instead of selective oxidation. That is why the production of phenol is done via an indirect route, that is, the alkylation of benzene to cumene, followed by oxidation of cumene to phenol and acetone. The first success in obtaining phenol directly from benzene was achieved by Iwamoto *et al.*,[10] who tested several transition-metal oxide catalysts, using N_2O instead of O_2 as oxidant. V_2O_5 supported on silica achieved 10% conversion at 70% selectivity to phenol at 823 K. Most other transition-metal oxides (with the exception of MoO_3 and WO_3) were unselective and led mainly to total oxidation. Some time later, the surprising discovery was made that H-ZSM-5 is a highly selective catalyst for the oxidation of benzene to phenol,[2,11,12] its selectivity being close to 100% at 15–30% conversion. A pilot plant with a ZSM-5 catalyst was constructed by Solutia Inc., but industrial breakthrough seems to be hampered by problems of catalyst deactivation.

In the early days, it was not very clear whether the activity of ZSM-5 is due to acid sites (either Brønsted[11,13] or Lewis[14,15,16]) or due to traces of iron, which are always present in commercial ZSM-5 catalysts. Meanwhile, the crucial role of iron in the reaction is not questioned any more.[17] Very small amounts of iron (at the level of several hundred ppm) suffice to achieve a high activity. These iron levels are always present in zeolites unless special precautions are taken to use extremely pure starting materials for the zeolite synthesis.[18] Too high concentrations of Fe are even detrimental for the selectivity and accelerate deactivation.[19] Moreover, it was observed that a pretreatment of the samples at high temperatures, especially in the presence of steam, increases the catalytic activity. These two phenomena, that is, the very low concentration of active sites and the beneficial effect of harsh pretreatment conditions, make the Fe-ZSM-5 system very special. We will make an attempt to explain this particular behaviour of Fe-ZSM-5 (and of other iron zeolites) based on a detailed analysis of the available literature. We discuss the effect of the composition of the zeolite matrix, the role of acidity and the effect of pretreatments at high temperatures. Most catalysts for the oxidation of benzene deactivate with time on stream. It is therefore necessary to consider the initial, intrinsic activity of the catalyst and its rate of deactivation when comparing different data.

2.1. The effect of the zeolite matrix

Not all iron zeolites are active in benzene oxidation. The best results were achieved with ZSM-5, ZSM-11[20] and zeolite beta.[21–23] Only a few studies show data on other (iron) zeolites. FER[24] and Fe-MCM-22[25] are moderately active, but a significant phenol formation has neither been observed with zeolites Y, X and A nor with erionite, clinoptilolite, mordenite and EU-1.[1,13] Note that all low-silica zeolites, which were tested (A, X, Y, erionite, clinoptilolite), were inactive and/or unselective. Yet, the Si/Al ratio of the zeolite is not the only criterion. EU-1 and mordenite have similar Si/Al ratios as some of the best ZSM-5 catalysts, but do not work well in benzene oxidation. If a zeolite is inactive for the selective oxidation of benzene, two explanations are possible: either the active sites necessary for the selective oxidation are not created or the catalyst deactivates very quickly. In the case of EU-1 and mordenite, both of which have a one-dimensional pore system, rapid deactivation may be the main reason for the low phenol productivity.

When comparing MFI-type aluminosilicates (Al-Si), gallosilicates (Ga-Si), titanosilicates (Ti-Si), and borosilicates (B-Si) of the same iron content, the aluminum or gallium-containing zeolites have the highest activity.[26] In titano-, boro- and ferrosilicates,[27] much higher iron contents are required to achieve the same phenol productivity. Aluminum and gallium as framework elements lead to strong Brønsted acidity, whereas titanium and boron do not. This suggests that the acidity of the zeolite matrix plays an important role. In the case of the aluminosilicate ZSM-5, it was observed that the activity increases with decreasing Si/Al ratio,[28] that is, with the concentration ofBrønsted acid sites (see also section 2.4). On the contrary, acidity also accelerates deactivation of the catalyst. Purely siliceous samples or ZSM-5 catalysts with a very high Si/Al ratio are usually more stable with time on stream.

2.2. The effect of pretreatments at high temperature

In most zeolite catalysts used for benzene oxidation, iron is incorporated into the zeolite, intentionally or not, during the hydrothermal synthesis. If the synthesis is carried out carefully, iron occupies a T-atom position in the zeolite framework at the end of the hydrothermal synthesis. Already upon calcination at 773 K, some of the bonds of Fe to framework oxygen atoms are broken, and the coordination of Fe changes from a regular tetrahedron to an irregular geometry with higher Fe–O coordination number. We can imagine the process to be similar to the formation of framework-associated octahedral Al (*vide infra*). Approximately 20% of the iron atoms leave the regular tetrahedral positions in Fe-silicalite-1,[18,29] but ~40% in Fe-ZSM-5.[18] This gives us a first hint why Al has a positive effect on the catalytic activity.

What happens when more severe pretreatments are applied, for example, treatments in air, vacuum, inert gas, H_2 or steam at 873 K, or even higher temperatures? All of these treatments were reported to increase the catalytic activity in benzene oxidation.[15,26,30–32] Often the increase in activity is directly related to the severity of the treatment, that is, steaming is more effective than inert gas, vacuum or air.[30] Yet 'over-steaming' of a sample may reduce its catalytic activity again,[33] that is, there is an optimum in the severity of the pretreatment. We will come back to this point later.

The above-mentioned treatments at high temperature not only change the nature of the iron species but also affect the zeolite host structure, in particular, concerning the formation of extra-framework aluminum (or gallium, boron, etc.) species. In the next section, we therefore discuss the fate of the framework aluminum species during treatments at high temperature, according to the state of the art in zeolite science. This will take us away from the chemistry of iron for a moment, but it is necessary for the understanding of the structural changes in the zeolite host.

2.2.1. Dehydroxylation and dealumination of high-silica zeolites

During treatment of a zeolite at very high temperatures, two processes take place in parallel, dehydroxylation of the zeolite lattice and dealumination of the zeolite lattice. Figure 28.1 given below represents dehydroxylation.

Dehydroxylation creates an oxygen vacancy in the zeolite lattice. It is difficult to measure the absolute concentration of vacancies, but electrical conductivity measurements show that the number of lattice defects in ZSM-5 increases linearly with temperature in the range of 650–950 K.[34] The framework Al atom next to the oxygen vacancy becomes a three-coordinated Al Lewis acid site. Three-coordinated Al atoms have indeed been directly observed by Al-EXAFS. Their concentration was estimated to be lower than 10% of the total Al content (for zeolite beta, after treatment at 975 K).[35]

The Al–O framework bonds may also be broken by hydrolysis. In that case, one speaks of dealumination instead of dehydroxylation. Dealumination takes place in the presence of water, for example, during the steaming of the sample or during the calcination of the template. Figures 28.2 and 28.3 show two plausible mechanisms of dealumination that have been proposed in the literature.

The extra-framework aluminum remains as a cationic species in the zeolite pores (Fig. 28.4). These species can be washed out by treatment with HCl. Prolonged steaming leads to an agglomeration of the extra-framework aluminum species to an oxidic phase.[38]

Partial hydrolysis of the Al-framework bonds leads to aluminum cations, which are still associated to the framework, but have octahedral coordination geometry. These partially hydrolysed Al cations can return to a tetrahedrally coordinated framework position by dehydration or by adsorption of a strong base, for example, ammonia.

Campbell *et al.* reported a very detailed study of the dealumination/dehydroxylation of ZSM-5 as a function of different thermal treatments and as function of the

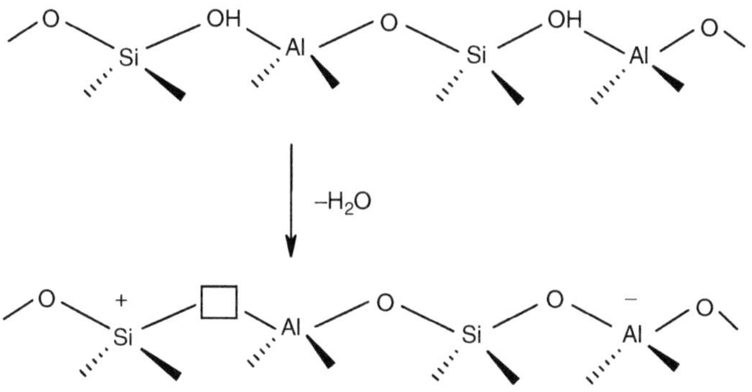

O Si OH Al O Si OH Al O

$-H_2O$

O Si $^{+}$ □ Al O Si O $^{-}$ Al O

Figure 28.1 Dehydroxylation of the zeolite lattice during high temperature treatment.

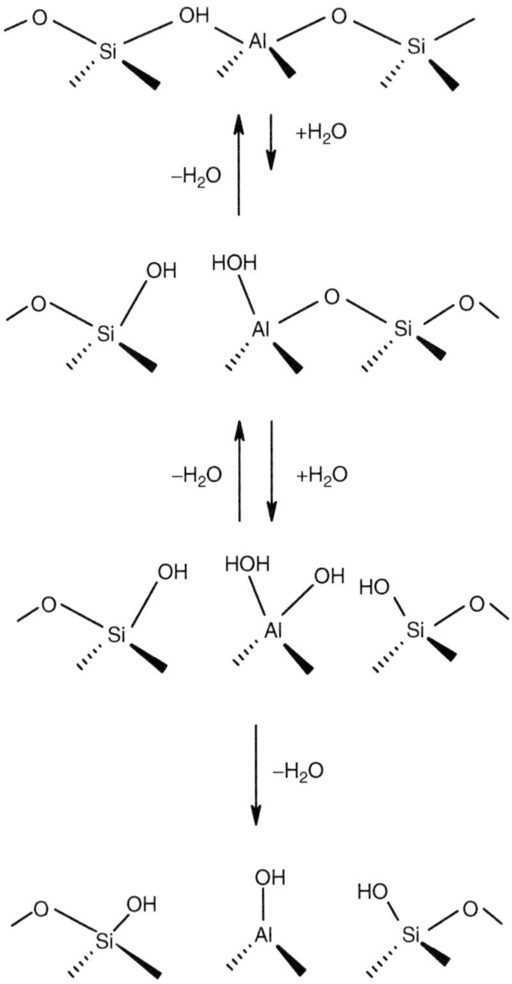

Figure 28.2 Plausible mechanism of dealumination by treatment in steam. Adapted from Ref. [36].

Si/Al ratio.[40] They compared treatment in inert gas at 998 K with steaming at 873 and 1023 K. Steaming at 1023 K dislocated more aluminum atoms from their ideal tetrahedral framework position than treatment in inert gas, that is, it is more severe. The most important result of their work is, however, the observation that ZSM-5 samples with high Si/Al ratio are significantly more resistant to dealumination or dehydroxylation. Already in earlier work, the view was expressed that dealumination and dehydroxylation are favoured by the presence of vicinal Al–O–Si–O–Al pairs. For example, in Fig. 28.3 the cationic extra-framework Al species is stabilized by the vicinal Al atom, which provides an ion exchange site. Obviously, the probability to find vicinal Al pairs increases with decreasing Si/Al ratio.[36]

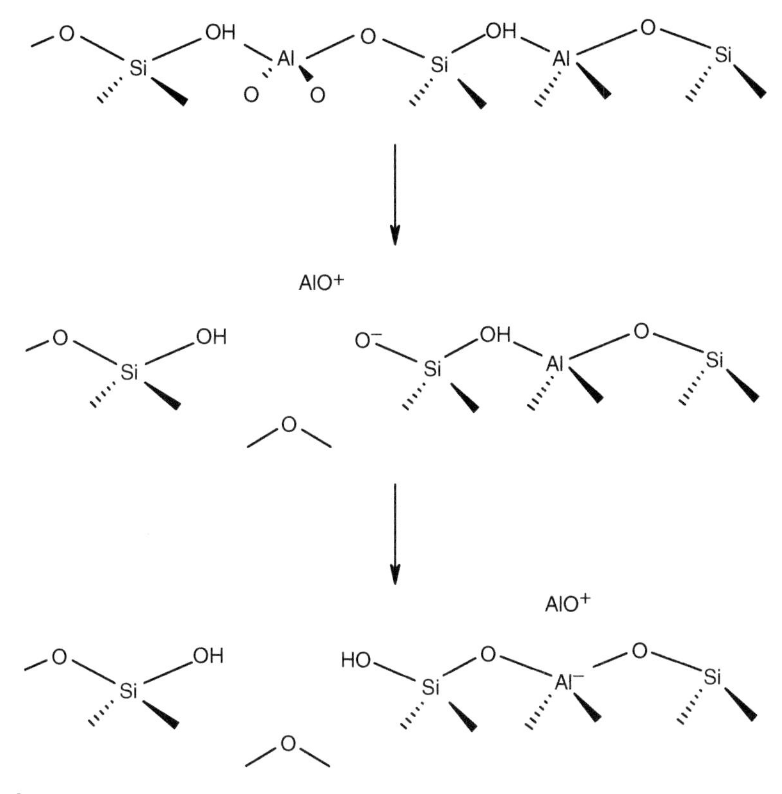

Figure 28.3 Plausible mechanism of dealumination. Adapted from Ref. [37].

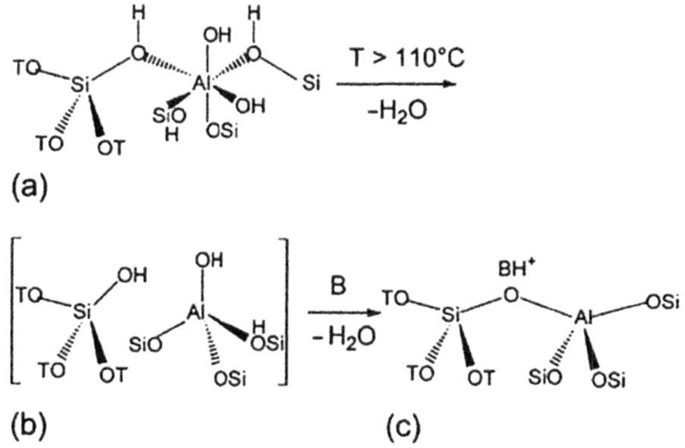

Figure 28.4 Reversible transformation of framework-associated octahedral Al (a) into tetrahedral Al by dehydration (b) and/or by adsorption of a base B (c), the base B being, for example, NH_3. Reproduced with permission from Ref. [39].

2.2.2. The transformations of the iron species at high temperatures

The transformations of the framework iron species at high temperatures are, in principal, analogous to the transformations of the framework aluminum species. As mentioned above, Fe is partially removed from its tetrahedrally coordinated framework position already during calcination at 773 K. The ultraviolet-visible (UV-Vis) spectra indicate that the Fe–O coordination number increases and this change has been attributed to the formation of extra-framework Fe species. Adsorption of ammonia, however, restores the initial tetrahedral coordination.[28] This behaviour is similar to that of framework-associated aluminum, and by analogy, we can assume that the Fe–O framework bonds are only partially hydrolysed upon calcination and that the Fe cations remain associated to the framework. More severe treatments will continue the rupture of the bonds of Fe to the framework and generate first isolated extra-framework cations and then clustered iron species.[18,28,41] The charged extra-framework iron species are most probably stabilized at the ion exchange positions in the zeolite, that is, they are stabilized close to the framework aluminum atoms.[42]

Extra-framework iron prefers an octahedral coordination. This is achieved by coordination of water molecules to the iron cation. Above 673 K, in dry atmosphere, the water molecules desorb and coordinatively unsaturated iron centres are generated. Moreover, the Fe^{3+} cations are spontaneously reduced to Fe^{2+}. We will come back to the dehydroxylation and autoreduction of the iron centres in Section 3.2.1. Both play an important role in determining the catalytic activity.

2.3. The properties of the active sites for benzene oxidation—α-sites

In the previous section, we have described the transformation of iron and aluminum species during pretreatments at high temperatures in general. It is now time to discuss which of these species are active in the oxidation of benzene to phenol. This question is not easy to answer and a spectroscopic identification of the active sites has not yet been given. We therefore start by describing the characteristic properties of the active sites and try to deduce structural information from these data.

The active sites for benzene oxidation are Fe^{2+} centres.[30] They dissociate N_2O to N_2 and a surface oxygen species, and are thereby re-oxidized to Fe^{3+}. The reaction can be formally described as:

$$Fe^{2+} + N_2O \rightarrow Fe^{3+} - O^{-}\bullet + N_2 \tag{1}$$

The surface oxygen atom then reacts with benzene to yield phenol. The dissociation of N_2O by a redox reaction with transition-metal cations is not specific for iron. Most transition-metal cations catalyse the dissociation of N_2O. Specific for iron zeolites is, however, the observation that the active sites for benzene oxidation are oxidized by reaction with N_2O, but not by reaction with O_2. This feature distinguishes the active sites for benzene oxidation from other Fe^{2+} catalysts, which are readily oxidized by O_2. Panov and co-workers,[30] who first described the above-mentioned properties of the active sites for benzene oxidation, named the active sites α-sites and the deposited oxygen α-oxygen. For reasons of simplicity, we will follow that nomenclature. α-Oxygen is stable on the catalyst surface up to 573 K, then it desorbs in the form of O_2. The concentration of α-sites/α-oxygen atoms is therefore measured by reacting N_2O with the catalyst at ~523 K and quantifying the amount of N_2 evolved. Unfortunately, only very few research groups have quantified the concentration of α-sites using this method, but those who did all found a (usually linear) correlation between the concentration of α-sites and the activity in benzene oxidation.[32,43] We can therefore be confident that α-sites are indeed the active sites for benzene oxidation, and we can relate catalyst activity to the concentration of these sites (even when their concentration was not actually determined).

The highest α-site concentrations that have been reported are in the order of 10^{19} sites/g, but often the concentration is much lower. A concentration of 10^{19} sites/g corresponds to 17 μmol/g or ~1000 ppm Fe. The concentration of α-sites is always much lower than the iron loading of the sample (Figs. 28.5 and 28.6). Only at very low iron loading (<500 ppm) and after a very severe pretreatment (inert gas at 1323 K or steam at 923 K) the α-site concentration comes close to the total iron content of the sample.[42] At high iron loadings, only a small fraction of the iron atoms is active.[19]

What is known about the structure of the α-sites? From the observation that the concentration of α-sites increases with the severity of the pretreatment, we can deduce that the α-sites are related to iron species, which were dislocated from their original position in the zeolite framework, that is, they are most probably extra-framework cations. Moreover, the inertness of α-sites towards O_2 and the observation that the α-oxygen atoms are stable on the surface up to 573 K indicates that the α-sites are isolated from each other.[5] Panov and group proposed that α-sites are binuclear iron cluster and provided evidence by Mössbauer spectroscopy, but other

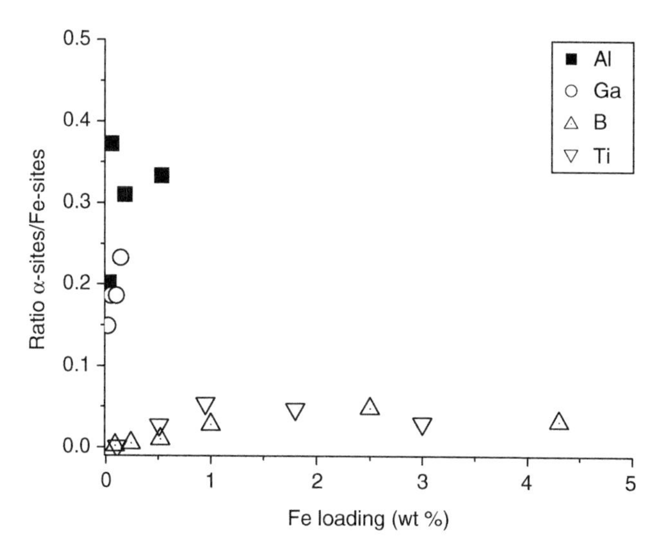

Figure 28.5 Ratio of α-sites to the total iron content for a series of alumino-, gallo-, boro- or titanosilicates with the MFI topology. All samples were steamed at 923 K in 50% H_2O. The figure was constructed using data of Ref. [26].

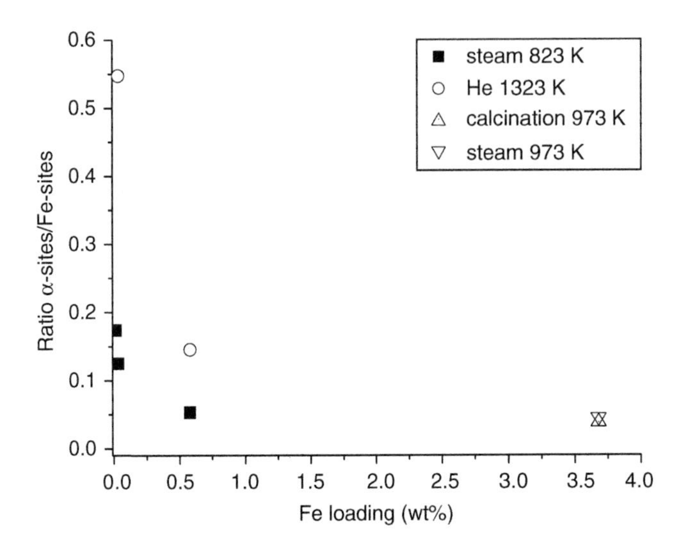

Figure 28.6 Ratio of α-sites to the total iron content for a series of Fe-ZSM-5 samples with varying iron loading and pretreated under different conditions. The figure was constructed using data of Refs. [19] and [43].

groups have questioned the interpretation of the spectra.[44] The idea of a binuclear α-site is certainly tempting because of the analogy with MMO, but some experimental observations make an isolated iron cation seem more likely. In particular, it is hard to understand why a binuclear Fe^{2+} site would not react with O_2 (we will see later that binuclear Cu sites do indeed react with O_2). Moreover, we should keep in mind that α-sites can be formed on catalysts with extremely low iron loadings (100 ppm of Fe correspond to 0.01 Fe per unit cell), which makes the formation of dimers highly unlikely.

How can we explain that only a small fraction of the iron atoms is transformed into an α-site? Part of the answer is that the transformation of Fe to extra-framework species is often not complete. We saw that it is more difficult to generate α-sites in purely siliceous samples where the extraction of Fe from tetrahedrally coordinated framework positions is less pronounced. On the contrary, α-sites can also be generated by post-synthesis introduction of iron and in that case all the iron cations are in extra-framework position. Even in these samples only a small fraction of the iron cations becomes an α-site. The amount of extra-framework Fe is, therefore, not the crucial parameter. It was proposed that a special arrangement of the extra-framework iron cations in the vicinity of extra-framework aluminum (or gallium?) cations is required to generate an active site.[45] This could explain why alumino-silicates and gallo-silicates are more active than silicates. We will come back to this point in the next sections, which will provide additional information concerning the structure of α-sites.

We can understand at this point why too harsh pretreatments (over-steaming) may lead to a decrease of the catalyst activity. Very severe steaming over prolonged time leads to a clustering of the extra-framework iron species as well as of the extra-framework aluminum (gallium, etc.) species. From the above discussion, it is clear that oxide clusters will not have the catalytic properties of α-sites. Over-steaming, therefore, reduces the concentration of α-sites. The optimum pretreatment conditions must therefore be chosen so as to extract the totality of the iron from the framework, to reduce it to Fe^{2+}(only Fe^{2+}can react with N_2O) and to generate extra-framework aluminum in the vicinity of iron, while avoiding clustering of the extra-framework Fe and Al species. The optimum pretreatment conditions differ from sample to sample because the stability of the framework depends on the concentration of Fe and Al (and of other non-silicon framework elements).

2.4. The role of acidity

While it is clear that acid sites are not the active sites for the oxidation of benzene for phenol, the acidity of the catalyst has an influence on the catalytic activity. As the severity of the pretreatment increases the concentration of α-sites increases, but the concentration of Brønsted acid sites decreases. This indicates that Brønsted sites are, a priori, not necessary for the reaction. Once, the optimum pretreatment severity is reached, the α-site concentration and the concentration of Brønsted acid sites decrease in parallel (Fig. 28.7).[32] This led Notté to propose that Brønsted acid sites play an important role in the generation of α-sites. However, the decrease of the α-sites in the left part of Fig. 28.7 could be attributed to a clustering of the iron sites, due to the increasing severity of the pretreatment, rather than to the loss of Brønsted acidity. We therefore believe that Brønsted OH-groups are indeed not part of the active site. On the contrary, it is difficult to create α-sites in zeolite matrices that are not or only weakly Brønsted acidic (silicalite, borosilicate, titanosilicate) sites.[26]

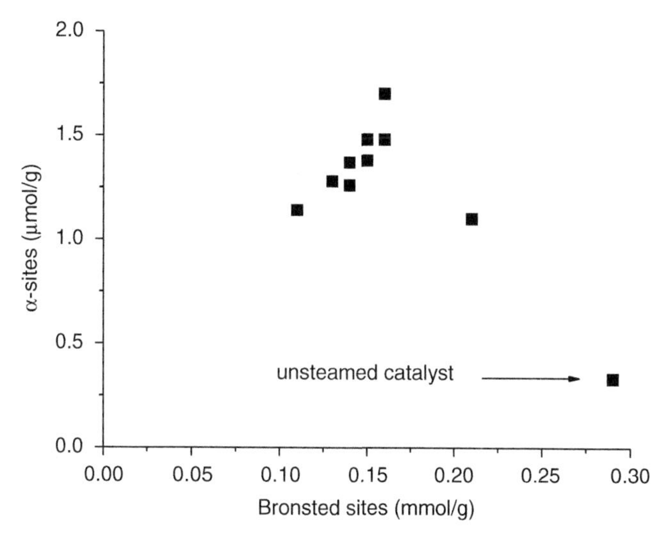

Figure 28.7 Relation between α-sites and Brønsted acid sites. Adapted from Ref. [33]. The catalyst underwent different pretreatments, unfortunately not specified in the article.

Hence, an important role of the Brønsted acid sites may be to provide ion exchange sites for the stabilization of cationic extra-framework iron species.

Concerning Lewis acidity, most studies agree that it is positively correlated with the activity in benzene oxidation.[16] Lewis acid sites created by steaming and by dehydroxylation do not behave in the same way.[31] Lewis acid sites created by dealumination/dehydroxylation certainly do not act alone in benzene oxidation, but they may act in combination with an extra-framework iron cation.[32,44] If a special arrangement of iron cations in the vicinity of a Lewis acid site is required to create an α-site, we can understand why it is difficult to transform 100% of the iron atoms into α-sites.

2.5. Catalysts prepared by post-synthesis introduction of iron

The discussion in the preceding sections was entirely focused on catalysts prepared by direct synthesis, that is, iron was introduced at the stage of the hydrothermal synthesis. Very few studies describe the behaviour of catalysts prepared by post-synthetic introduction of iron, that is, by ion exchange or impregnation of an iron salt. It seems that the behaviour of the ion-exchanged catalysts is not fundamentally different from those prepared by hydrothermal synthesis.[46] In principle, ion-exchanged catalysts should allow a better fine-tuning of the catalytic activity in benzene oxidation because the iron sites are already present as extra-framework cations. Unfortunately, this route has not been explored in detail. Fe-ZSM-5 catalysts prepared by chemical vapour deposition of $FeCl_3$ have very high iron loadings. The iron is mainly present in the form of small iron oxide clusters, each containing a few iron atoms. Only a tiny fraction of the iron atoms could be identified as α-sites. Because of high concentration of oxide clusters, a lot of total oxidation takes place in parallel to production of phenol and the catalysts deactivate quickly.[19]

3. The Decomposition of N_2O by Iron Zeolites

We already mentioned in the introduction that iron zeolites are not only used as catalysts for the selective oxidation of benzene to phenol with N_2O as the oxidant but also for the decomposition of N_2O. N_2O has a strong greenhouse effect (310 times more than CO_2). It is emitted in the tail gas of nitric acid plants. While the emissions are very small in volume (compared to CO_2), the high greenhouse factor makes the contribution of N_2O emissions to global warming non-negligible. *Uhde GmbH* has therefore recently commercialized a process for the combined removal of N_2O and NO_x using a Fe-ZSM-5 catalyst (EnviNO_x process). Two plants are already in operation (in Austria and Egypt) and several others are under construction. It should be noted that iron zeolites are a priori not the most active catalysts for N_2O decomposition. Noble metal zeolites have much higher intrinsic activities than Fe,[47] but they are strongly inhibited by NO_x, which is always present in the tail gas together with N_2O. Iron zeolites, on the contrary, are promoted by the presence of

NO_x.[48] A detailed description of the promotion of iron zeolites by NO_x is out of the scope of this chapter. We will concentrate on the decomposition of pure N_2O, with the objective to make analogies and links to the previous section about the oxidation of benzene.

N_2O decomposition and benzene oxidation by N_2O are related because they share the initial reaction step, the dissociation of N_2O into N_2 and a surface oxygen atom (reaction 1). The fate of the deposited surface oxygen atom is, however, very different. In benzene oxidation, it reacts with benzene to yield phenol, that is, the objective is to stabilize the surface oxygen in a highly reactive state on the catalyst surface until it meets a benzene molecule. In N_2O decomposition, on the contrary, the objective is to destabilize the deposited oxygen atoms and to make them desorb rapidly as O_2. The best catalyst in benzene oxidation will therefore not be the best catalyst in N_2O decomposition and vice versa. In contrast to benzene oxidation, the catalysts for N_2O decomposition are usually prepared by ion exchange (although direct hydrothermal synthesis also works) and have rather high iron loadings.

3.1. The mechanism of N_2O decomposition over iron zeolites

As in the oxidation of benzene, the first reaction step of N_2O decomposition is the dissociation of an N_2O molecule on a Fe^{2+}site.

$$N_2O \rightarrow N_2 + O_s \tag{2}$$

This step is relatively fast and not rate limiting (at least at low temperatures).[49] The rate-determining step of N_2O decomposition is the formation of O_2. Two options are possible.

(1) The deposited surface oxygen reacts directly with N_2O in an Eley–Rideal type reaction:

$$O_s + N_2O \rightarrow N_2 + O_2 \tag{3}$$

(2) Two deposited surface oxygen atoms react with each other and desorb as O_2:

$$2O_s \rightarrow O_2 \tag{4}$$

Transient response experiments have shown that O_2 formation is always strongly delayed with respect to N_2 formation[47] and also takes place in the absence of N_2O in the gas phase.[50] Also isotope-labelling experiments are not compatible with an Eley–Rideal-type reaction of surface oxygen atoms with N_2O. They prove that the O_2 is formed by recombination of two surface oxygen atoms.[51]

There is little or no inhibition of N_2O decomposition by O_2 (remember that the inertness towards O_2 is a characteristic property of α-sites). Because desorption of O_2 from the catalyst is very slow and rate limiting, one could expect O_2 to be very strongly adsorbed on the active sites and to inhibit the catalyst activity. Recent kinetic data indicate that desorption of O_2 itself is not the rate-determining step, but the recombination of two surface oxygen atoms to an O_2 precursor on the surface (reaction 5).[52] The O_2 precursor is not stable on the surface and desorbs rather rapidly.

$$O_s + O_s \rightarrow O_{2,s} \quad \text{slow} \tag{5}$$

$$O_{2,s} \rightarrow O_2 \quad \text{fairly rapid} \tag{6}$$

Isotope labelling experiments show that the oxygen atoms in the produced O_2 do not only originate from N_2O but also from the catalyst surface.[49,53] There is an intense mixing of the oxygen atoms from N_2O with the oxygen atoms of the catalyst before O_2 is desorbed. The mobility of the extra-lattice oxygen atoms is high above 673 K. They can migrate away from the iron sites and occupy positions in the zeolite framework instead. The mechanistic picture that arises for N_2O decomposition is therefore the following: Two oxygen atoms are deposited on iron sites. The deposited oxygen atoms can exchange with lattice oxygen atoms surrounding the iron site, and they can also migrate over the catalyst surface to a neighbouring site where they recombine to an O_2 precursor and desorb into the gas phase. The recombination is favoured if the distance between two iron centres is not too long, that is, if the concentration of the Fe in the sample is high. The formation of small iron oxide clusters further facilitates the recombination because two oxygen atoms can be deposited on the same cluster. This is a marked difference to benzene oxidation where the formation of metal oxide clusters was unfavourable because it provoked over-oxidation and deactivation.

Are the sites where the oxygen atoms from N_2O are deposited identical to the α-sites discussed in the previous section and are the deposited oxygen atoms identical to α-oxygen? There are certainly analogies but also differences between benzene oxidation and N_2O decomposition. The active sites for N_2O decomposition are Fe^{2+} sites, as in benzene oxidation. Yet α-sites react with N_2O in a 1:1 stoichiometry, that is, they formally generate an O^- radical anion (reaction 1). On N_2O decomposition catalysts with relatively high iron loading, we did not observe a 1:1 stoichiometry between Fe^{2+} and N_2O, but the classical 2:1 ratio.[54] The ratio 2:1 is expected for the oxidation of Fe^{2+} to Fe^{3+} by N_2O, which formally generates an O^{2-} anion.

$$2Fe^{2+} + N_2O \rightarrow 2Fe^{3+} + O^{2-} + N_2 \tag{7}$$

Yet a whole body of evidence shows that the deposited oxygen atoms are a lot more reactive than conventional O^{2-} anions in the lattice of the catalyst. Possibly, the O^{2-} anion is deposited as an extra-lattice anion bridging two iron cations, that is,

$$2Fe^{2+} + N_2O \rightarrow [Fe - O - Fe]^{4+} + N_2 \tag{8}$$

and thereby obtains a high reactivity. In that case, however, the 2:1 stoichiometry of reaction 7 would be strictly related to the concentration of iron dimers, which is not in agreement with the experimental data. The exact chemistry of the deposition of N_2O on the catalyst is, therefore, still not understood. The data suggest that Fe sites, which are not too distant, communicate with each other and react like a dimer (even if they are not directly bonded to each other). We have to regard the solid as entity instead of looking at isolated active sites.

3.2. The effect of pretreatments at high temperature

As in the oxidation of benzene, pretreatments of the iron zeolite at high temperatures increase the activity in N_2O decomposition.[55,56] In the case of benzene oxidation catalysts, the effect of the pretreatments at high temperatures was attributed to three factors: (1) the extraction of iron from the framework to an extra-framework position, (2) the creation of extra-framework Al Lewis sites in the vicinity of iron and (3) the reduction to Fe^{2+}. Most catalysts used for N_2O decomposition are prepared by post-synthesis ion exchange, that is, the extraction of iron from the framework is not a relevant issue. We will therefore focus our discussion on the reduction to Fe^{2+} and the role of extra-framework Al sites and of defects.

3.2.1. The autoreduction of Fe^{3+} to Fe^{2+}

When iron zeolites are heated in inert gas to temperatures above 773 K, they release O_2 into the gas phase.[57] At the same time, Fe^{3+} is reduced to Fe^{2+}.

$$4Fe^{3+} + 2O^{2-} \rightarrow 4Fe^{2+} + O_2 \qquad (9)$$

The desorption of O_2 leaves four electrons behind on the catalyst, that is, four Fe^{3+} must be reduced to Fe^{2+} for each molecule of O_2. This suggests that autoreduction can only take place on large iron oxide clusters, but not on isolated iron cations. Paradoxically, that is not observed experimentally. Fe-ZSM-5 catalysts containing iron oxide clusters autoreduce less than samples containing well-dispersed cations.[52] Fe^{3+} is more stabilized in a large iron oxide cluster (by the Madelung energy) than as an isolated cation. Therefore, autoreduction is thermodynamically favoured for isolated sites. The high kinetic barrier for autoreduction of isolated sites is overcome by the high mobility of the lattice oxygen atoms at the temperatures where the reduction occurs.

Autoreduction increases with increasing pretreatment temperature and its extent is higher in vacuum than in inert gas than in air. The catalytic activity of the samples follows the same order and can be perfectly correlated to the extent of autoreduction. Surprisingly, treatment in steam leads to more autoreduction than treatment in inert gas.[58] Also in that case, the catalytic activity is exclusively correlated to the extent of autoreduction. We will come back to a possible relationship between autoreduction and dealumination in the next section.

The extent of autoreduction is, in general, relatively small. In the case of catalysts containing isolated sites, treatment in inert gas at 873 K reduces less than 20% of the iron sites. Steaming at the same temperature reduces up to 50% of the iron atoms.

For clustered iron species the extent of autoreduction is much less. Very harsh conditions, that is, temperatures above 1150 K, are required to reduce the totality of the iron sites.

3.2.2. The mechanism of autoreduction and the creation of lattice defects

It is not easy to understand how isolated Fe^{3+} cations at extra-framework positions can release O_2 into the gas phase. There are hardly any experimental data on the mechanism of autoreduction, and we can only make speculations. The first question concerns the origin of the oxygen atoms that are released into the gas phase. At low temperatures (up to at least 673 K), the Fe^{3+} cations are hydrated, that is, they are present in the form of $[Fe(OH)_2]^+$. The hydroxyl groups bound to the iron cation are the most probable source of O_2. We can therefore write a hypothetical equation of autoreduction as:

$$4[Fe(OH)_2]^+ \rightarrow 4[Fe(OH)]^+ + 2H_2O + O_2 \qquad (10)$$

According to this equation the autoreduction is related to the dehydroxylation of the iron cations. However, we have mentioned already in the preceding section that autoreduction by treatment in inert gas at 873 K (which could be associated to reaction 10) is only very limited. Much harsher conditions, that is, steaming or much higher temperatures in inert gas are required to increase the extent of autoreduction further. This indicates that further autoreduction is related to the dealumination/dehydroxylation of the zeolite host structure. Once the iron sites are entirely dehydroxylated, the source of oxygen atoms for further autoreduction must be lattice oxygen atoms, and it is therefore not surprising that 'deep' autoreduction is linked to creation of defects in the zeolite lattice. This also brings us back to the idea of α-sites being Fe^{2+} cations in the vicinity of Al Lewis acid sites. Indeed, IR spectroscopy shows that steaming or treatment in inert gas at very high temperatures (>1100 K) generates Fe^{2+} sites with reduced electron density,[23] suggesting that they are situated in the vicinity of an Al Lewis site. The Al Lewis site is probably generated by the removal of lattice oxygen atoms for the autoreduction of Fe^{3+} to Fe^{2+}. We can speculate that the active sites for benzene oxidation (the α-sites) have a similar structure.

4. Active Oxygen Atoms in Copper Zeolites

Copper zeolites have been used for the same reactions as iron zeolites, for example, the decomposition of N_2O, the oxidation of benzene and the transformation of methane to methanol. For both transition metals, that is, iron and copper, binuclear metal sites (dimers) have been invoked to explain the remarkable catalytic properties. While in the case of iron, the evidence for the role of dimers is rather circumstantial (*vide supra*), the following sections will show that there is solid proof that binuclear copper sites are involved in the catalytic reactions. Unfortunately, copper zeolites are very sensitive (a lot more than iron zeolites) to the presence of

H_2O or SO_2 in the feed, which is a serious drawback for their practical application. We will, however, not discuss such application-related issues, but concentrate on mechanistic aspects and on the chemistry of the oxygen species. We first discuss the decomposition of NO, the catalytic cycle of which is strongly related to the cycle of N_2O decomposition.

4.1. Copper zeolites for the direct decomposition of NO and N_2O

The catalytic destruction of NO in exhaust gas streams usually relies on the use of reductants (ammonia, urea or hydrocarbons) to transform NO into N_2. This process is called selective catalytic reduction (SCR). In principle, the use of a reductant should not be necessary because the direct decomposition of NO into the elements is thermodynamically feasible. Cu-ZSM-5 is one of the few materials that efficiently catalyse this reaction. The activity of copper zeolites as a function of the exchange level of copper follows an S-shaped curve,[59] and it was clearly demonstrated that the strong increase in activity above a certain copper concentration is related to the formation of copper dimers.[60,61]

Figure 28.8 shows the reaction mechanism that has been proposed for direct NO decomposition.[62] The active site is a Cu^+ dimer, which was created by autoreduction during the pretreatment of the catalyst. Copper zeolites reduce much more readily than iron zeolites. Simple heating to 773 K in inert gas reduces a majority of the Cu^{2+} ions to Cu^+ ions, whereas the same treatment reduces only very little Fe^{3+} to Fe^{2+}. The Cu^+ dimer adsorbs two molecules of NO. This intermediate then transforms into an oxygen-bridged $[CuOCu]^{2+}$ species and N_2O. Subsequently, N_2O reacts with the oxygen-bridged copper dimer to yield a bis(μ-oxo)dicopper core, which finally releases O_2 and closes the catalytic cycle. Spectroscopic studies have provided very strong evidence for the existence of the bis(μ-oxo)dicopper core and also for its implication in the catalytic cycle of NO decomposition. N_2O

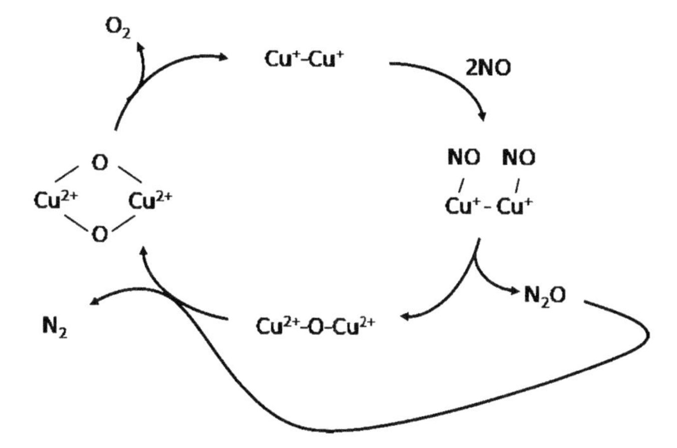

Figure 28.8 Reaction mechanism of direct NO decomposition. Adapted from Ref. [62].

decomposes via a similar catalytic cycle: the $[CuOCu]^{2+}$ species is formed via direct reaction of the Cu^+ dimer with N_2O, the rest of the cycle is analogous to the decomposition of NO.

An alternative reaction mechanism has been proposed[63] (Figure 28.9) in which the $[CuOCu]^{2+}$ intermediate does not react with N_2O, but with NO to form NO_2. The NO_2 transports an oxygen atom to a neighbouring site (Fig. 28.9), and the resulting nitrate species decomposes to a bis(μ-oxo)dicopper core, which releases O_2. Also in this mechanism, which may act in parallel with the one shown in Fig. 28.8, the copper dimer is essential for the recombination of two oxygen atoms to O_2.

4.2. The oxidation of methane to methanol

The active bis(μ-oxo)dicopper species in Cu-ZSM-5 is similar to the active core found in some copper-enzymes which catalyse oxidation reactions. Indeed, Groothaert *et al.*[4] found that Cu-ZSM-5 is also capable of oxidizing methane to methanol, as does the enzyme MMO. They pretreated Cu-ZSM-5 in O_2 at 723 K to generate bis(μ-oxo)dicopper species and then reacted them with CH_4 at medium temperatures (between 398 and 498 K). The bis(μ-oxo)dicopper species disappeared, but no products were detected in the reactor effluent. Extraction of the catalyst by a water/acetonitrile mixture showed, however, that methanol was formed inside the pores. The amount of methanol correlated with the initial concentration of bis(μ-oxo)dicopper species. Because of the strong adsorption of

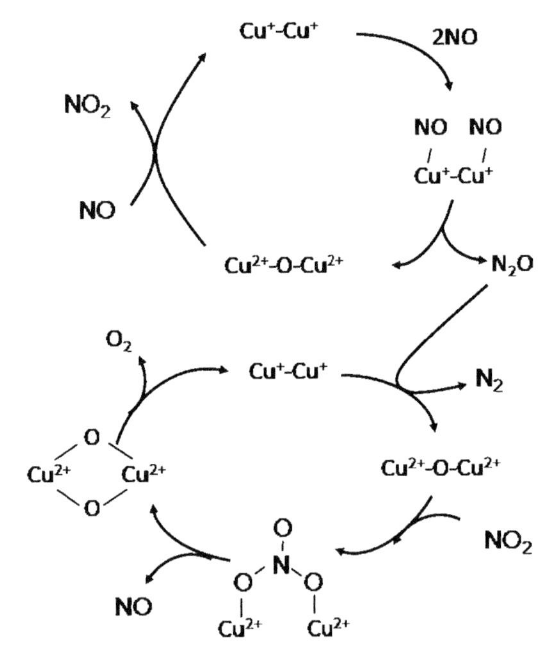

Figure 28.9 Alternative reaction mechanism of direct NO decomposition. Adapted from Ref. [63].

methanol in the catalyst, however, the reaction was not catalytic, that is, only a single turn-over to methanol was achieved, and then the active sites were blocked by methanol. Attempts to desorb methanol at higher temperatures led to its decomposition to CO_2.

An analogous behaviour was observed for iron zeolites. Also these catalysts can oxidize methane to methanol at low temperatures, but methanol remains adsorbed in the pores and has to be extracted with a solvent.[3] In the case of iron zeolites the active sites are the α-sites, which were discussed in Section 2.3. Because α-sites are inert towards O_2, the active oxygen species for the oxidation of CH_4 has to be created by reaction with N_2O.

4.3. The oxidation of benzene to phenol

Like Fe-ZSM-5, Cu-ZSM-5 is capable of oxidizing benzene directly to phenol, but O_2 is used as an oxidant instead of N_2O. In contrast to Fe-ZSM-5, the best results were achieved with a Cu-ZSM-5 catalyst with high copper loading (exchange level 100%).[64] The yield and selectivity of phenol are, however, significantly smaller than in the case of Fe-ZSM-5/N_2O.[65] For a tentative explanation, we remind the reader of a very general concept of selective oxidation, that is, the relation between selectivity and metal-oxygen bond strength. Because copper zeolites autoreduce a lot easier than iron zeolites, their metal-oxygen bond strength seems to be weaker, and this may explain why they are less selective in the oxidation of benzene.

5. Conclusions—Iron Versus Copper Zeolites

In the preceding sections, we discussed the catalytic properties of iron and copper zeolites and tried to relate them to structural information. We recall the most important features of the two catalysts.

Iron zeolites work best as selective oxidation catalysts at low iron concentrations. Moreover, the active sites for selective oxidation react specifically with N_2O, but not with O_2. This indicates that site isolation is an important premise for obtaining selective iron zeolite catalysts. The activity of iron zeolites increases when they are pretreated under very harsh conditions. These harsh conditions are required to enforce the reduction of Fe^{3+} to Fe^{2+}, and there are indications that this autoreduction goes in parallel with the creation of lattice defects. The Fe^{2+} sites in the vicinity of lattice defects are most probably the sites where the surface oxygen atoms, which were generated by dissociation of N_2O, are stabilized in a highly reactive form.

All iron zeolites that were identified as selective catalysts for benzene oxidation are high-silica zeolites. In low-silica zeolites, the harsh conditions required for the autoreduction of iron lead to too much dealumination. However, the role of the zeolite host structure certainly merits further investigation. It has the dual function of generating and stabilizing the active sites and of adsorbing the substrate in the pores.

Copper zeolites catalyse similar reactions as their iron analogues, but there are some distinct differences: Cu zeolites work best at high copper loadings and their catalytic activity is strongly related to the formation of Cu dimers. The active site is a bis(μ-oxo)dicopper species, which is generated by reaction of a Cu^{+} dimer with N_2O, NO and O_2. The ability to react with O_2 strongly distinguishes the active sites in copper zeolites from the selective oxidation sites in iron zeolites.

In contrast to metal oxide catalysts, the zeolite host structure stabilizes the transition-metal cations in monomeric or dimeric form, allows them to switch easily between two oxidation states and to generate highly reactive extra-lattice oxygen species. These features are responsible for the remarkable catalytic properties of the transition-metal zeolites. Compared to transition-metal catalysts in enzymes, however, the inorganic materials lack the ability to fine-tune the activation of oxidant and the binding of the substrate and will certainly never achieve the level of sophistication found in nature's catalysts.

REFERENCES

[1] Panov, G. I., Kharitonov, A. S., Sobolev, V. I., "Oxidative Hydroxylation Using Dinitrogen Monoxide - a Possible Route For Organic-Synthesis Over Zeolites." *Appl. Catal. A-Gen.* **1993,** *98,* 1–20.

[2] Panov, G. I., Sheveleva, G. A., Kharitonov, A. S., Romannikov, V. N., Vostrikova, L. A., "Oxidation of Benzene to Phenol By Nitrous-Oxide Over Fe-ZSM-5 Zeolites." *Appl. Catal. A-Gen.* **1992,** *82,* 31–36.

[3] Dubkov, K. A., Sobolev, V. I., Panov, G. I., "Low-temperature oxidation of methane to methanol on FeZSM-5 zeolite." *I. Kinet. Catal.* **1998,** *39,* 72–79.

[4] Groothaert, M. H., Smeets, P. J., Sels, B. F., Jacobs, P. A., Schoonheydt, R. A., "Selective oxidation of methane by the bis(μ-oxo)dicopper core stabilized on ZSM-5 and mordenite zeolites." *J. Am. Chem. Soc.* **2005,** *127,* 1394–1395.

[5] Panov, G. I., Uriarte, A. K., Rodkin, M. A., Sobolev, V. I., "Generation of active oxygen species on solid surfaces. Opportunity for novel oxidation technologies over zeolites." *Catal. Today* **1998,** *41,* 365–385.

[6] Dalton, H., "Biological methane activation - lessons for the chemists." *Catal. Today* **1992,** *13,* 455.

[7] Baik, M.-H., Newcomb, M., Friesner, R. A., Lippard, S. J., "Mechanistic Studies on the Hydroxylation of methane by methane monooxygenase." *J. Am. Chem. Soc.* **2003,** *103,* 2385.

[8] Costas, M., Mehn, M. P., Jensen, M. P., "Dioxygen activation at mononuclear nonheme iron active sites: Enzymes, models and intermediates." *Chem. Rev.* **2004,** *104,* 939–986.

[9] Whittington, D. A., Lippard, S. J., "Crystal structures of the soluble methane monooxygenase hydroxylase from Methylococcus capsulatus (Bath) demonstrating geometrical variability at the dinuclear iron active site." *J. Am. Chem. Soc.* **2001,** *123,* 827–838.

[10] Iwamoto, M., Matsukami, K., Kagawa, S., "Catalytic Oxidation by Oxide Radical Ions. 1. One-Step Hydroxylation of Benzene to Phenol over Group 5 and 6 Oxides Supported on Silica Gel." *J. Phys. Chem.* **1983,** *87,* 903–905.

[11] Suzuki, E., Nakashiro, K., Ono, Y., "Hydroxylation of Benzene with Dinitrogen Monoxide over H-ZSM-5 Zeolite." *Chem. Lett.* **1988,** 953–956.

[12] Gubelmann, M., Tirel, Ph. FR Patent 2 630 735, **1988.**

[13] Burch, R., Howitt, C., "Investigation of zeolite catalysts for the direct partial oxidation of benzene to phenol." *Appl. Catal. A-Gen.* **1993,** *103,* 135–162.

[14] Zholobenko, V. L., Senchenya, I., Kustov, L. M., Kazansky, V. B., "Complexation and decomposition of N_2O on Bronsted acid sites in high-silicon zeolites: spectral and quantum chemical study." *Kinet. Catal.* **1991,** *32,* 132–137.

[15] Kustov, L. M., Tarasov, A. L., Bogdan, V. I., Tyrlov, A. A., Fulmer, J. W., "Selective oxidation of aromatic compounds on zeolites using N_2O as a mild oxidant. A new approach to design active sites." *Catal. Today* **2000,** *61,* 123–128.
[16] Motz, J. L., Heinichen, H., Hölderich, W. F., "Direct hydroxylation of aromatics to their corresponding phenols catalysed by H-Al-ZSM-5 zeolite." *J. Mol. Catal. A* **1998,** *136,* 175–184.
[17] Sobalik, Z., Kubanek, P., Bortnovsky, O., Vondrova, A., Tvaruzkova, Z., Sponer, J. E., Wichterlova, B., "On the necessity of a basic revision of the redox properties of H-zeolites." *Stud. Surf. Sci. Catal.* **2002,** *142,* 533–540.
[18] Hensen, E. J. M., Zhu, Q., Janssen, R. A. J., Magusin, P. C. M. M., Kooyman, P. J., van Santen, R. A., "Selective oxidation of benzene to phenol with nitrous oxide over MFI zeolites 1. On the role of iron and aluminum." *J. Catal.* **2005,** *233,* 123–135.
[19] Zhu, Q., van Teeffelen, R. M., van Santen, R. A., Hensen, E. J. M., "Effect of high-temperature treatment on Fe/ZSM-5 prepared by chemical vapor deposition of $FeCl_3$ (II). Nitrous oxide decomposition, selective oxidation of benzene to phenol, and selective reduction of nitric oxide by isobutane." *J. Catal.* **2004,** *221,* 575–583.
[20] Pirutko, L. V., Dubkov, K. A., Solovyeva, L. P., Panov, G. I., "Effect of ZSM-11 crystallinity on its catalytic performance in benzene to phenol oxidation with nitrous oxide." *React. Kinet. Catal. Lett.* **1996,** *58,* 105–110.
[21] Starokon, E. V., Dubkov, K. A., Pirutko, L. V., Panov, G. I., "Mechanisms of iron activation on Fe-containing zeolites and the charge of alpha-oxygen." *Top. Catal.* **2003,** *23,* 137–143.
[22] Centi, G., Genovese, C., Giordano, G., Katovic, A., Perathoner, S., "Performance of Fe-BEA catalysts for the selective hydroxylation of benzene with N_2O." *Catal. Today* **2004,** *91–92,* 17–26.
[23] Yuranov, I., Bulushev, D. A., Renken, A., Kiwi-Minsker, L., "Benzene to phenol hydroxylation with N_2O over Fe-Beta and Fe-ZSM-5: Comparison of activity per Fe-site." *Applied Catalysis A: General* **2007,** *319,* 128–136.
[24] Shevade, S. S., Rao, B. S., "Gas-phase hydroxylation of benzene over H-Ga-FER zeolite." *Catal. Lett.* **2000,** *66,* 99–103.
[25] Meloni, D., Monaci, R., Rombi, E., Guimon, C., Martinez, H., Fechete, I., Dumitriu, E., "Synthesis and characterization of MCM-22 zeolites for N_2O oxidation of benzene to phenol." *Stud. Surf. Sci. Catal.* **2002,** *142,* 167–173.
[26] Pirutko, L. V., Chernyavsky, V. S., Uriarte, A. K., Panov, G. I., "Oxidation of benzene to phenol by nitrous oxide - Activity of iron in zeolite matrices of various composition." *Appl. Catal. A-Gen.* **2002,** *227,* 143–157.
[27] Meloni, D., Monaci, R., Solinas, V., Berlier, G., Bordiga, S., Rossetti, I., Oliva, C., Forni, L., "Activity and deactivation of Fe-MFI catalysts for benzene hydroxylation to phenol by N_2O." *J. Catal.* **2003,** *214,* 169–178.
[28] Hensen, E. J. M., Zhu, Q., van Santen, R. A., "Selective oxidation of benzene to phenol with nitrous oxide over MFI zeolites 2. On the effect of the iron and aluminum content and the preparation route." *J. Catal.* **2005,** *233,* 136–146.
[29] Bordiga, S., Buzzoni, R., Geobaldo, F., Lamberti, C., Giamello, E., Zecchina, A., Leofanti, G., Petrini, G., Tozzola, G., Vlaic, G., "Structure and reactivity of framework and extraframework iron in Fe-silicalite as investigated by spectroscopic and physicochemical methods." *J. Catal.* **1996,** *158,* 486–501.
[30] Sobolev, V. I., Dubkov, K. A., Paukshtis, E. A., Pirutko, L. V., Rodkin, M. A., Kharitonov, A. S., Panov, G. I., "On the role of Bronsted acidity in the oxidation of benzene to phenol by nitrous oxide." *Appl. Catal. A-Gen.* **1996,** *141,* 185–192.
[31] Dubkov, K. A., Ovanesyan, N. S., Shteinman, A. A., Starokon, E. V., Panov, G. I., "Evolution of iron states and formation of alpha-sites upon activation of FeZSM-5 zeolites." *J. Catal.* **2002,** *207,* 341–352.
[32] Kubanek, P., Wichterlova, B., Sobalik, Z., "Nature of active sites in the oxidation of benzene to phenol with N_2O over H-ZSM-5 with low Fe concentrations." *J. Catal.* **2002,** *211,* 109–118.
[33] Notté, P. P., "The AlphOx process or the one-step hydroxylation of benzene into phenol by nitrous oxide. Understanding and tuning the ZSM-5 catalyst activities." *Top. Catal.* **2000,** *13,* 387–394.

[34] Balint, I., Springuel-Huet, M. A., Aika, K.-I., Fraissard, J., "Evidence for oxygen vacancy formation in HZSM-5 at high temperature." *Phys. Chem. Chem. Phys.* **1999,** *1,* 3845–3851.
[35] van Bokhoven, J. A., van der Eerden, A. M. J., Koningsberger, D. C., "Three-coordinate aluminum in zeolites observed with in situ X- ray absorption near-edge spectroscopy at the AlK-edge: Flexibility of aluminum coordinations in zeolites." *J. Am. Chem. Soc.* **2003,** *125,* 7435–7442.
[36] Sano, T., Ikeya, H., Kasuno, T., Wang, Z. B., Kawakami, Y., Soga, K., "Influence of crystallinity of HZSM-5 zeolite on its dealumination rate." *Zeolites* **1997,** *19,* 80–86.
[37] Sonnemans, M. H. W., Heijer, C. D., Crocker, M., "Studies of the acidity of mordenite and ZSM-5. 2. Loss of Bronsted acidity by dehydroxylation and dealumination." *J. Phys. Chem.* **1993,** *97,* 440–445.
[38] Shannon, R. D., Gardner, K. H., Stadey, R. H., Bergeret, G., Gallezot, P., Auroux, A., "The nature of nonframework aluminum species formed during dehydroxylation of H-Y" *J. Phys. Chem.* **1985,** *89,* 4778.
[39] Omegna, A., van Bokhoven, J. A., Prins, R., "Flexible aluminum coordination in alumino-silicates. Structure of zeolite H-USY and amorphous silica-alumina." *J. Phys. Chem. B* **2003,** *107,* 8854.
[40] Campbell, S. M., Bibby, D. M., Coddington, J. M., Howe, R. F., Meinhold, R. H., "Dealumination of HZSM-5 zeolites - 1. Calcination and Hydrothermal treatment." *J. Catal.* **1996,** *161,* 338–349.
[41] Perez-Ramirez, J., Mul, G., Kapteijn, F., Moulijn, J. A., Overweg, A. R., Domenech, A., Ribera, A., Arends, I., "Physicochemical characterization of isomorphously substituted FeZSM-5 during activation." *J. Catal.* **2002,** *207,* 113–126.
[42] Berlier, G., Zecchina, A., Spoto, G., Ricchiardi, G., Bordiga, S., Lamberti, C., "The role of Al in the structure and reactivity of iron centers in Fe-ZSM-5-based catalysts: a statistically based infrared study." *J. Catal.* **2003,** *215,* 264–270.
[43] Yuranov, I., Bulushev, D. A., Renken, A., Kiwi-Minsker, L., "Benzene hydroxylation over FeZSM-5 catalysts: which Fe sites are active?" *J. Catal.* **2004,** *227,* 138–147.
[44] Taboada, J. B., Overweg, A. R., Craje, M. W. J., Arends, I., Mul, G., van der Kraan, A. V., "Systematic variation of Fe-57 and Al content in isomorphously substituted ^{57}FeZSM-5 zeolites: preparation and characterization." *Microporous Mesoporous Mater.* **2004,** *75,* 237–246.
[45] Hensen, E. J. M., Zhu, Q., van Santen, R. A., "Extraframework Fe-Al-O species occluded in MFI zeolite as the active species in the oxidation of benzene to phenol with nitrous oxide." *J. Catal.* **2003,** *220,* 260–264.
[46] Jia, J., Pillai, K. S., Sachtler, W. M. H., "One-step oxidation of benzene to phenol with nitrous oxide over Fe/MFI catalysts." *J. Catal.* **2004,** *221,* 119–126.
[47] Li, Y., Armor, J. N., "Catalytic decomposition of nitrous oxide on metal exchanged zeolites." *Appl. Catal. B-Environ.* **1992,** *1,* L21.
[48] Perez-Ramirez, J., Kapteijn, F., Mul, G., Moulijn, J. A., "NO-assisted N_2O decomposition over Fe-based catalysts: Effects of gas-phase composition and catalyst constitution." *J. Catal.* **2002,** *208,* 211–223.
[49] Wood, B. R., Reimer, J. A., Bell, A. T., Janicke, M. T., Ott, K. C., "Nitrous oxide decomposition and surface oxygen formation on Fe-ZSM-5." *J. Catal.* **2004,** *224,* 148–155.
[50] Pirngruber, G. D., Roy, P. K., "A look into the surface chemistry of N_2O decomposition on iron zeolites by transient response experiments." *Catal. Today* **2005,** *110,* 199–210.
[51] Pirngruber, G. D., Roy, P. K., "The mechanism of N_2O decomposition on Fe-ZSM-5: an isotope labeling study." *Catal. Lett.* **2004,** *93,* 75.
[52] Kondratenko, E. V., Perez-Ramirez, J., "Mechanism and kinetics of direct N_2O decomposition over Fe-MFI zeolites with different iron speciation from temporal analysis of products." *J. Phys. Chem. B* **2006,** *110,* 22586–22595.
[53] Valyon, J., Millman, W. S., Hall, W. K., "The desorption of O_2 during NO and N_2O decomposition on Cu- and Fe-zeolites." *Catal. Lett.* **1994,** *23,* 215.
[54] Pirngruber, G. D., Roy, P. K., Prins, R., "The role of autoreduction and oxygen mobility in N_2O decomposition over Fe-ZSM-5." *J. Catal.* **2007,** *246,* 147–157.

[55] Zhu, Q., Mojet, B. L., Janssen, R. A. J., Hensen, E. J. M., van Grondelle, J., Magusin, P., van Santen, R. A., "N_2O decomposition over Fe/ZSM-5: effect of high-temperature calcination and steaming." *Catal. Lett.* **2002,** *81,* 205–212.
[56] Perez-Ramirez, J., Kapteijn, F., Mul, G., Moulijn, J. A., "Superior performance of ex-framework FeZSM-5 in direct N_2O decomposition in tail-gases from nitric acid plants." *Chem. Commun.* **2001,** 693–694.
[57] Voskoboinikov, T. V., Chen, H. Y., Sachtler, W. M. H., "On the nature of active sites in Fe/ZSM-5 catalysts for NOx abatement." *Appl. Catal. B-Environ.* **1998,** *19,* 279–287.
[58] Roy, P. K., Prins, R., Pirngruber, G. D., "The effect of pretreatment on the reactivity of Fe-ZSM-5 catalysts for N_2O decomposition: Dehydroxylation vs. steaming." *Appl. Catal. B Environmental* **2008,** *80,* 226–236.
[59] Yahiro, H., Iwamoto, M., "Copper ion-exchanged zeolite catalysts in deNOx reaction." *Appl. Catal. A-Gen.* **2001,** *222,* 163–181.
[60] Groothaert, M. H., van Bokhoven, J. A., Battiston, A. A., Weckhuysen, B. M., Schoonheydt, R. A., "Bis(μ-oxo)dicopper in Cu-ZSM-5 and its role in the decomposition of NO: A combined in situ XAFS, UV-Vis-Near-IR, and kinetic study." *J. Am. Chem. Soc.* **2003,** *125,* 7629–7640.
[61] Da Costa, P., Moden, B., Meitzner, G. D., Lee, D. K., Iglesia, E., "Spectroscopic and chemical characterization of active and inactive Cu species in NO decomposition catalysts based on Cu-ZSM5." *Phys. Chem. Chem. Phys.* **2002,** *4,* 4590–4601.
[62] Groothaert, M. H., Lievens, K., Leeman, H., Weckhuysen, B. M., Schoonheydt, R. A., "An operando optical fiber UV-vis spectroscopic study of the catalytic decomposition of NO and N_2O over Cu-ZSM-5." *J. Catal.* **2003,** *220,* 500–512.
[63] Moden, B., Da Costa, P., Fonfe, B., Lee, D. K., Iglesia, E., "Kinetics and mechanism of steady-state catalytic NO decomposition reactions on Cu-ZSM5." *J. Catal.* **2002,** *209,* 75–86.
[64] Kubacka, A., Wang, Z., Sulikowski, B., Corberan, V. C., "Hydroxylation/oxidation of benzene over Cu-ZSM-5 systems: optimization of the one-step route to phenol." *J. Catal.* **2007,** *250,* 184–189.
[65] Shibata, Y., Hamada, R., Ueda, T., Ichihashi, Y., Nishiyama, S., Tsuruya, S., "Gas-phase catalytic oxidation of benzene to phenol over Cu-impregnated HZSM-5 catalysts." *Industrial & Engineering Chemistry Research* **2005,** *44,* 8765–8772.

INDEX

C

D

I

Q

R

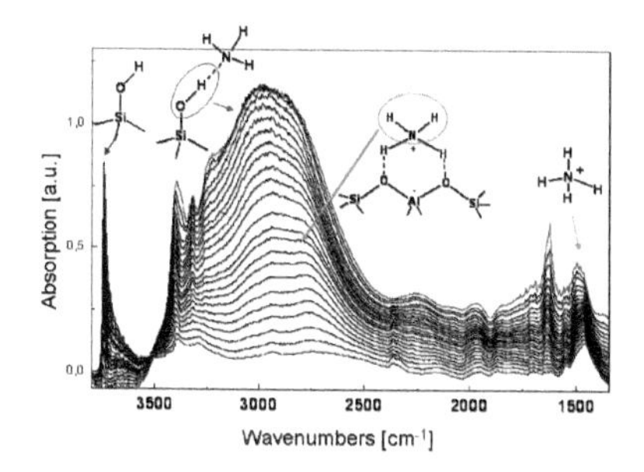

Color Plate 1.23 NH_3 adsorption spectra of zeolite Beta at room temperature (varying NH_3 dosage), showing typical adsorption sites/species.

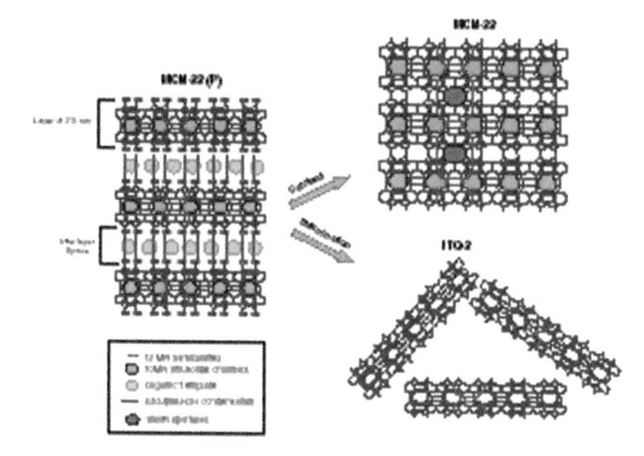

Color Plate 1.25 Structural schematic representation of MWW-type zeolites and ITQ-2 delaminated zeolite. Reproduced from Daz *et al.*[58] by permission from Elsevier.

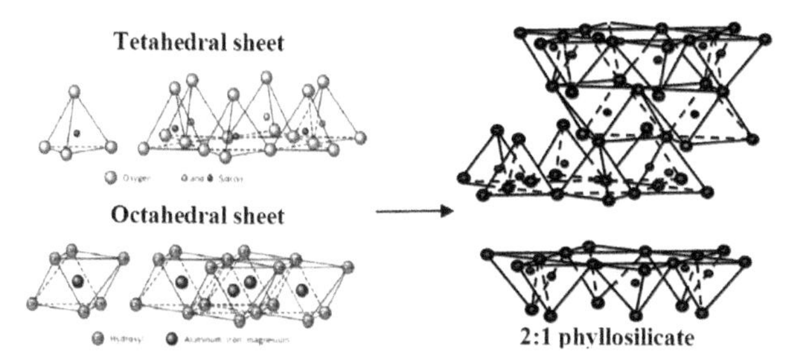

Color Plate 2.1 Schematic presentation of tetrahedral and octahedral sheets in layered materials (left) and a 2:1 phyllosilicate structure (right).

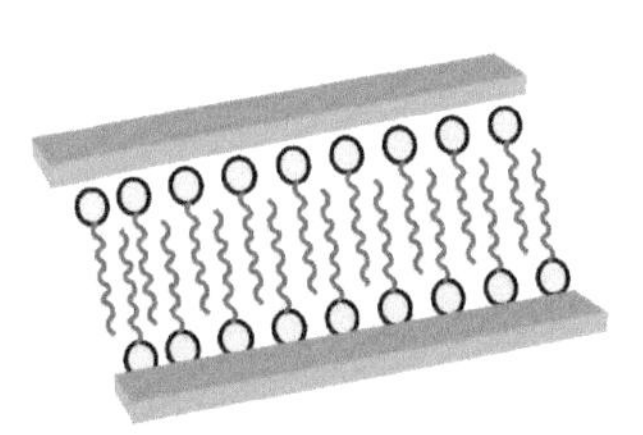

Color Plate 2.2 Schematic presentation of an alkylammonium compound intercalated between two clay layers.

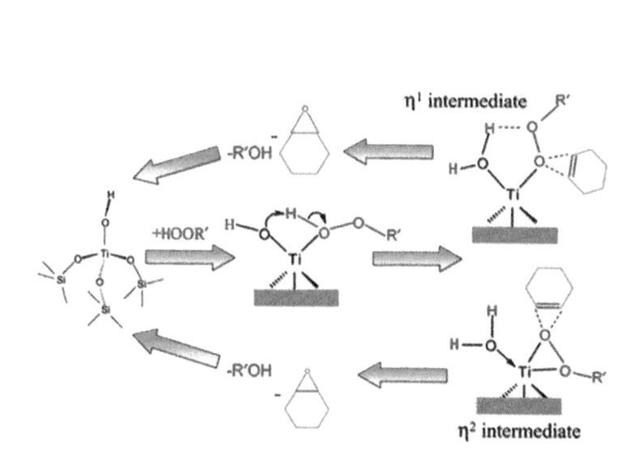

Color Plate 3.3 Epoxidation reactions via Ti (η^1-OOH) and/or Ti(η^2-OOH) species. (Reproduced from Ref. 20 by permission of The Royal Society of Chemistry.)

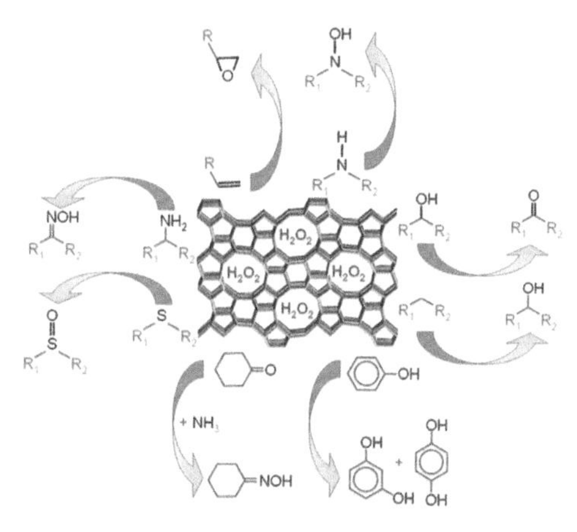

Color Plate 3.4 Oxidation reactions catalysed by TS-1 in the presence of 30% aqueous H_2O_2. (From Ref. 25 by permission of Springer-Verlag.)

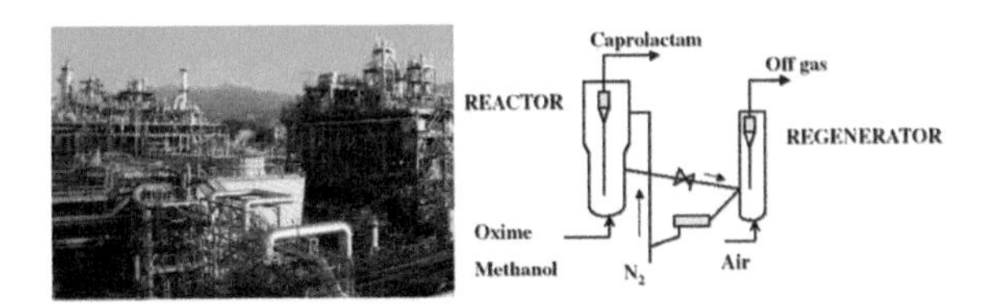

Color Plate 3.5 Production plant for ϵ-caprolactam and scheme of the fluidized bed reactor for the caprolactam production. (Reproduced from Ref. 31 by permission of The Chemical Society of Japan.)

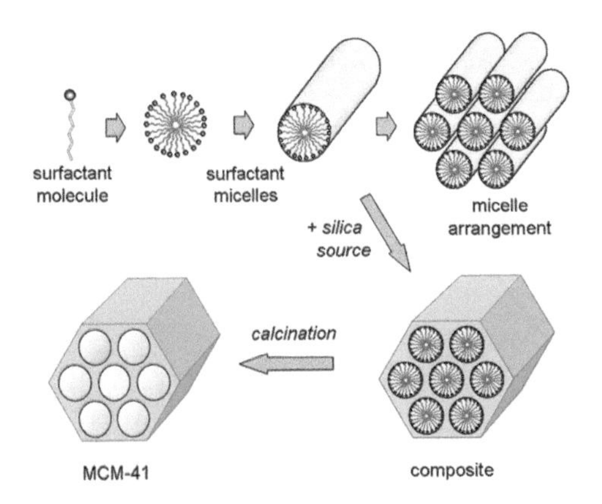

Color Plate 3.7 Formation of MCM-41 type materials.

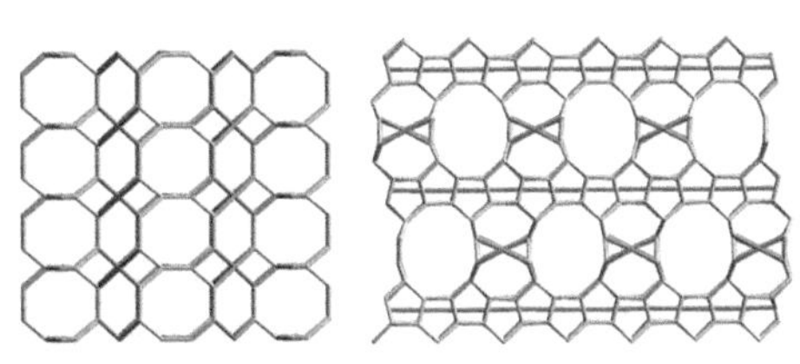

Color Plate 3.10 Structure frameworks of ETS-4 (left) and ETS-10 (right); oxygen atoms are omitted for clarity, that is, the lines represent directly connections between silicon tetrahedra with titanium atoms in octahedra and connections between neighbouring titanium atoms in the titania chains.

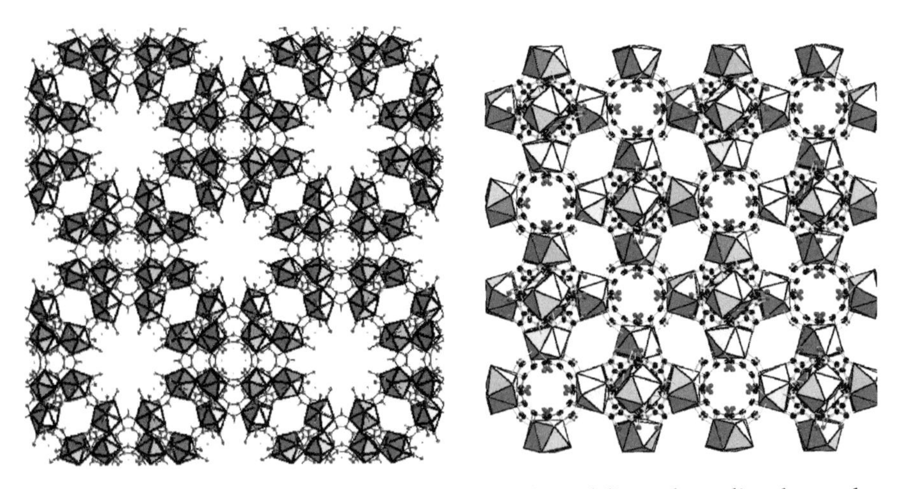

Color Plate 4.3 View of the structures of the ZMOFs with a sodalite and a zeolite-rho topology.

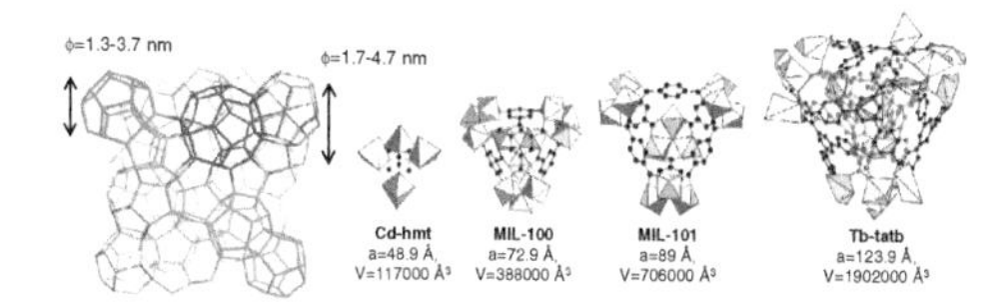

Color Plate 4.4 MOFs with a zeotype MTN architecture (left); hybrid supertetrahedra (right).

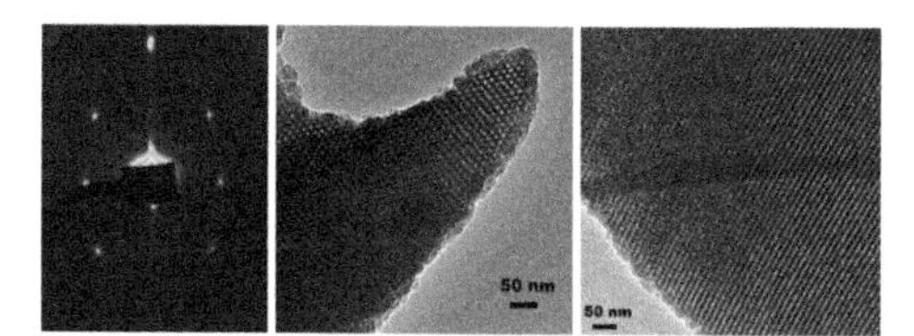

Color Plate 5.6 Grazing Incidence Small Angle X-ray Scattering (GISAXS) pattern of the F127-templated aluminophosphate thin film treated at 400 °C (left) and its transmission electron microscopy (TEM) images along [111] pore direction (middle) and [110] pore direction.

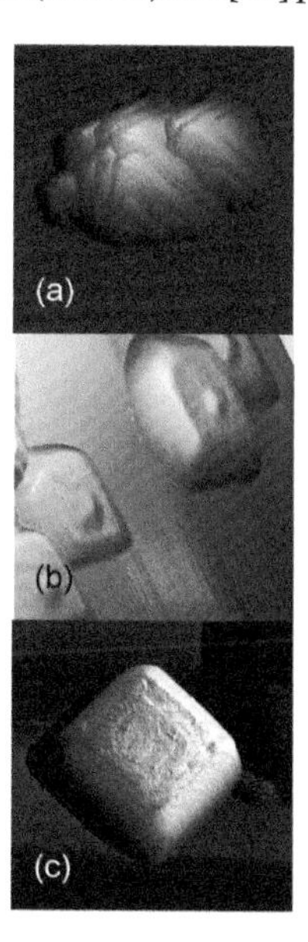

Color Plate 6.23 AFM images of (A) the 'pyramidal-shaped' particles obtained in the solid phase drawn off the reaction mixture (hydrogel) after its hydrothermal treatment at 80 °C for 30 min, (B) near-cubic-shaped crystals contained in the solid phase drawn off the reaction mixture after its hydrothermal treatment at 80 °C for 90 min and (C) typical cubic crystals of zeolite A contained in the solid phase drawn off the reaction mixture after its hydrothermal treatment at 80 °C for 240. (Reproduced from Ref. 179 with permission.)

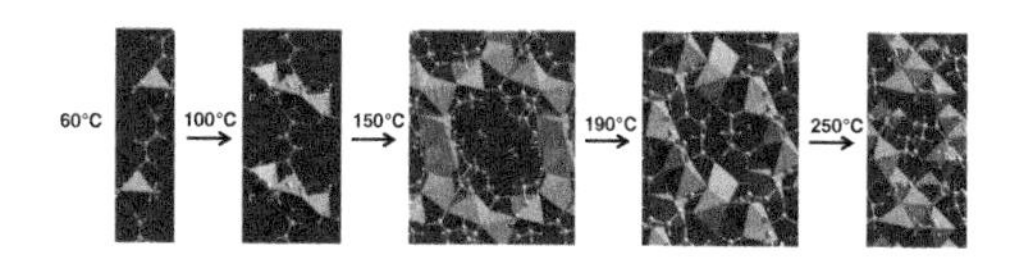

Color Plate 4.7 Progression of the five phases of cobalt succinate from low to high temperature.

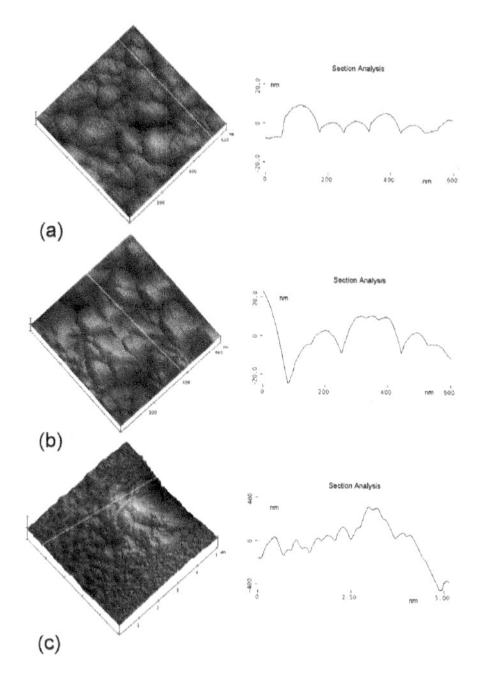

Color Plate 6.22 AFM images of the X-ray amorphous aluminosilicate (Na_2O:Al_2O_3:2.576-SiO_2:2.28H_2O) separated from freshly prepared aluminosilicate hydrogel (3.51Na_2O:Al_2O_3:2.1SiO_2:85.2H_2O) showing (A) aggregate of disc-shaped particles (having the mean diameter of about 80 nm and mean height of about 15 nm, (B) 'transition', probably partially crystalline, features (particles of 'quasi-crystalline' phase and (C) aggregates of 'pyramidal-shaped' features which look like fully crystalline material (see Fig. 37C). (Reproduced from Ref. 179 with permission.)

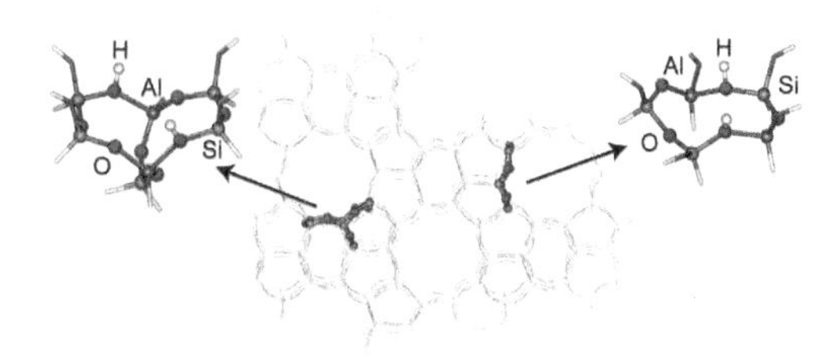

Color Plate 8.1 Example for selection of isolated cluster model as a fragment from the zeolite framework.

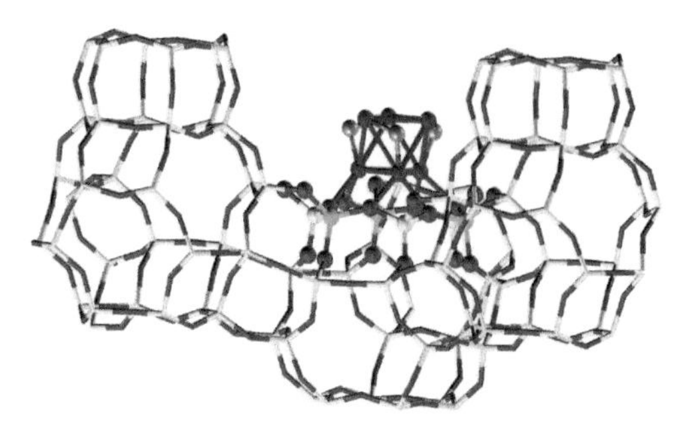

Color Plate 8.2 Schematic representation of QM/MM hybrid model: atoms included in the inner part (I) are shown as balls, while the outer part (O) is shown only as a framework.

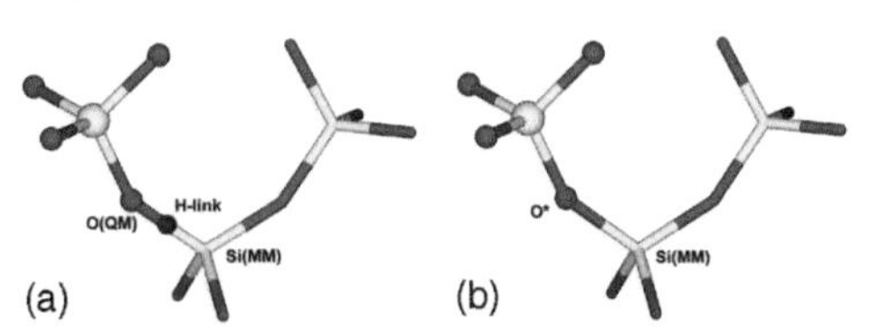

Color Plate 8.3 Schematic representation of QM/MM border: termination by H as link atoms (a); termination by special border centres as used in covEPE scheme (b).

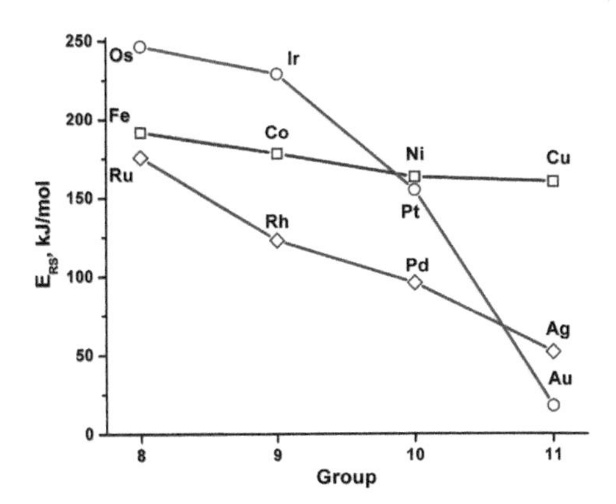

Color Plate 8.4 Energy E_{RS} (per OH group) gained due to reverse H spillover on M_6/zeo(3H) according to the values reported in Vayssilov and Rosch[91]: $E_{RS} = \{E[Me_6/Zeo(3H)]-E[Me_6H_3/Zeo]\}/3$.

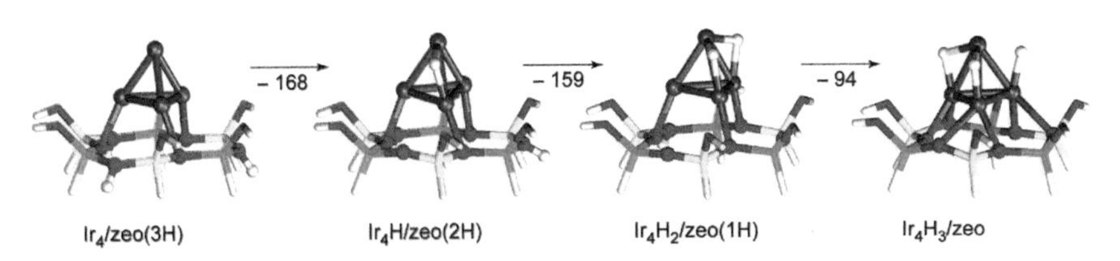

Color Plate 8.5 Stepwise proton transfer from the zeolite fragment, zeo(3H), to the supported Ir_4 cluster. Reaction energies in kJ/mol, according to the values reported in Petrov et al.[93]

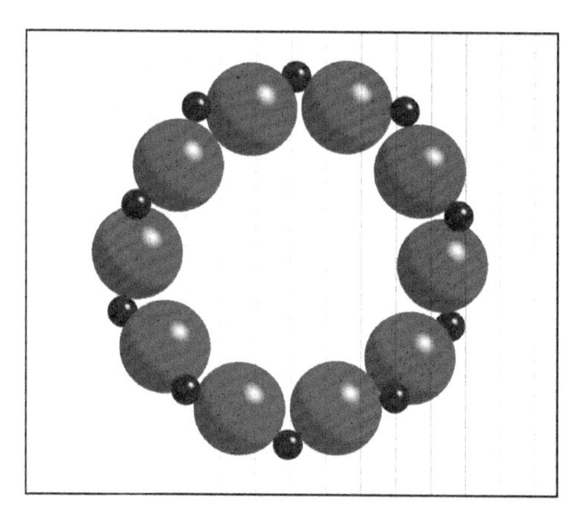

Color Plate 9.1 Illustration of the pore size of a 10-ring zeolite (in this case zeolite ZSM-5). The effective pore opening is 5.5Å and the oxygen-to-oxygen distance is 8.1 Å.

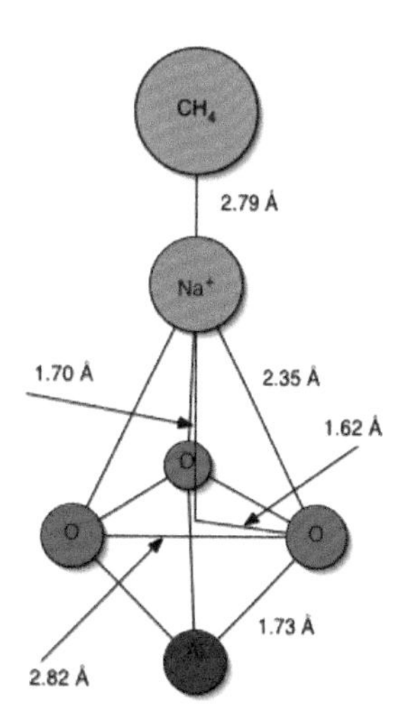

Color Plate 9.5 Schematic of the coordination of sodium cations to a model $AlO_{4/2}^{-}$ tetrahedron. Distances are not to scale. Since this site has a threefold axis of symmetry, the Al–O bond distance is 1.73 Å and the angle between the O–Al–O atoms is 109.47° (the tetrahedral angle), all distances can be calculated using elementary geometry.

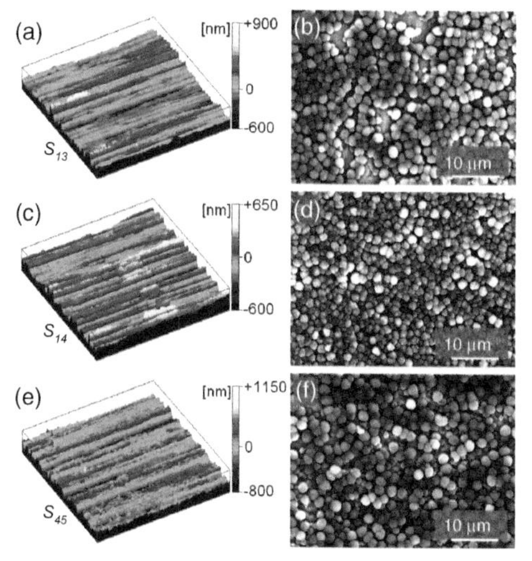

Color Plate 12.2 Laser scanning confocal microscopy scans (scan area is 320 × 320 μm^2) of the substrates: (a) S_0, (c) S_4, (e) S_{55} and the corresponding scanning electron micrographs of the as synthesized zeolite Beta coatings (b, d and f, respectively). The subscripts denote nanometric surface roughness corresponding to a weight loss of 0, 4, and 55 g m^{-2} after the etching treatment. The black arrows in the laser scanning confocal microscopy scans indicate the direction of the lay of the substrate parallel to which the 2-D height profiles were digitized for the determination of the R_a value.

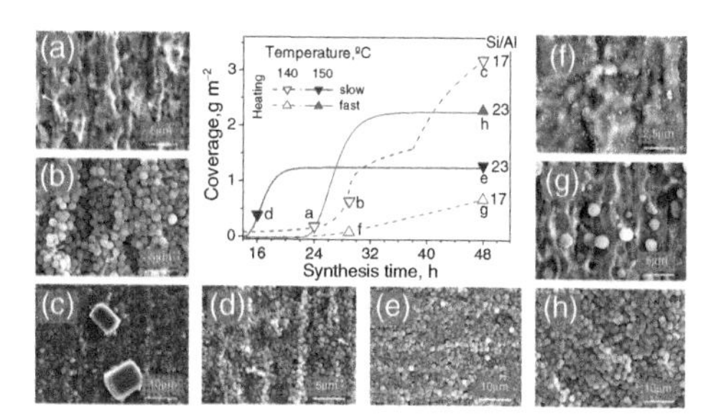

Color Plate 12.5 Crystallization curves and corresponding SEM images of the coatings grown from the synthesis gels with Si/Al ratios of 17 and 23 at 140 and 150 °C after different synthesis times.

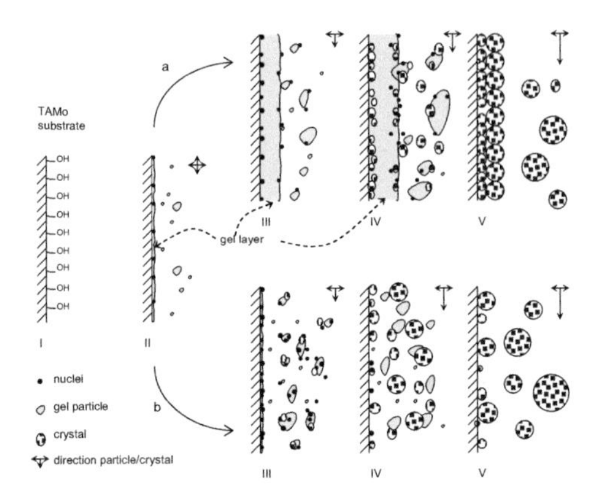

Color Plate 12.6 Two growth mechanisms of zeolite Beta coatings: (a) from a gel layer and (b) from a diluted precursor mixture. A good wetting of the super-hydrophilic surface of a metal substrate with the precursor mixture is reached after the UV treatment (Step I). A thin homogeneous amorphous aluminosilicate gel layer is deposited on the surface of the substrate in the initial stage of the synthesis process (Step II). Then, heterogeneous nucleation occurs at the substrate/gel interface. Depending on the synthesis conditions, either (a) the gel particles agglomerate in the bulk liquid into large lumps which in turn are responsible for the growth of the gel layer at the surface (a-III and a-IV) or (b) the nucleation and crystallization are enhanced in the bulk liquid (b-III, b-IV), which prevents the formation of a stable gel layer at the substrate and results in a discontinuous coating with a large crystal size distribution (b-V). In the first case, however, nuclei are formed not only in the bulk liquid but also at the both sides of the gel layer (a-III and a-IV). This is followed by fast crystallization at the gel/liquid interface, and a relatively slow crystallization at the substrate/gel interface (a-V). As a result, the crystals show a narrow crystal size distribution.

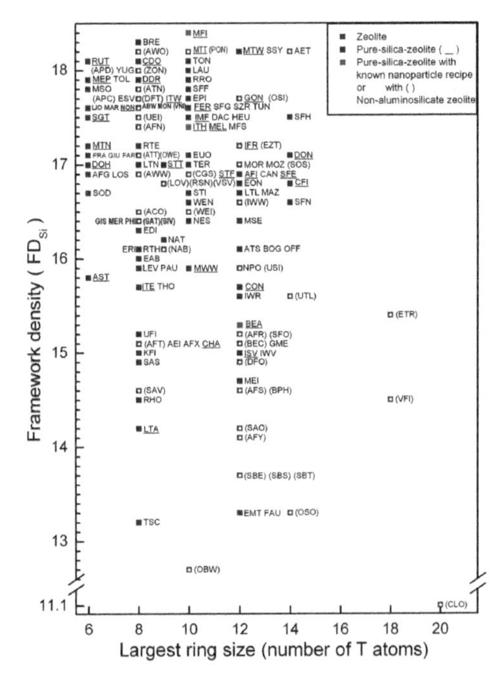

Color Plate 13.4 Road map for the development of zeolite low-*k* materials. Pure-silica-zeolites (PSZs) with a known nanoparticle recipe are marked with red boxes, other PSZs are marked with blue boxes with—, non-aluminosilicate zeolites are marked with hollow boxes with ( ), and all others are marked as dark boxes.[48] (Adapted with permission from *The Journal of Physical Chemistry B*, 2005, 109 (18):8653. Copyright 2005 American Chemical Society.)

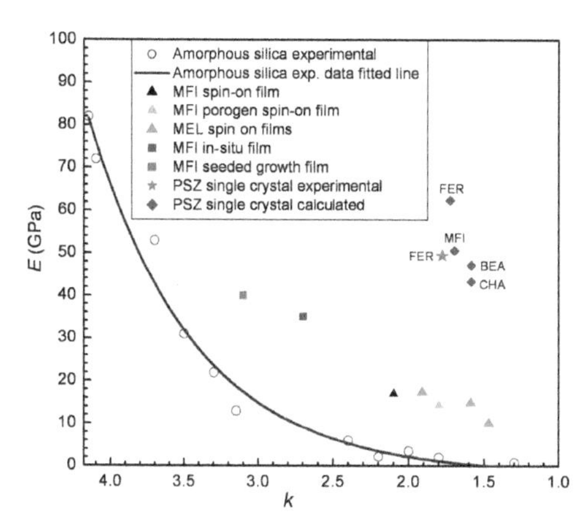

Color Plate 13.7 *E* vs. *k* for amorphous silicas and pure-silica-zeolites (PSZs). The open circles and solid data fitted line correspond to experimental data taken from Xu et al.[22] Squares correspond to experimental data for polycrystalline PSZ MFI films; triangles correspond to experimental data for spin-on PSZ MFI and MEL films; diamonds correspond to calculations of PSZ single crystals; and the star corresponds to experimental data for FER single crystals.[56] (Z.J. Li, M.C. Johnson, M.W. Sun, E.T. Ryan, D.J. Earl, W. Maichen, J.I. Martin, S. Li, C.M. Lew, J. Wang, M.W. Deem, M.E. Davis and Y.S. Yan: "Mechanical and Dielectric Properties of Pure Silica Zeolite Low-*k* Materials". *Angewandte Chemie International Edition.* 2006. Volume 45. Pages 6329–6332. Copyright Wiley-VCH Verlag GmbH & Co. KGaA. Reproduced with permission.)

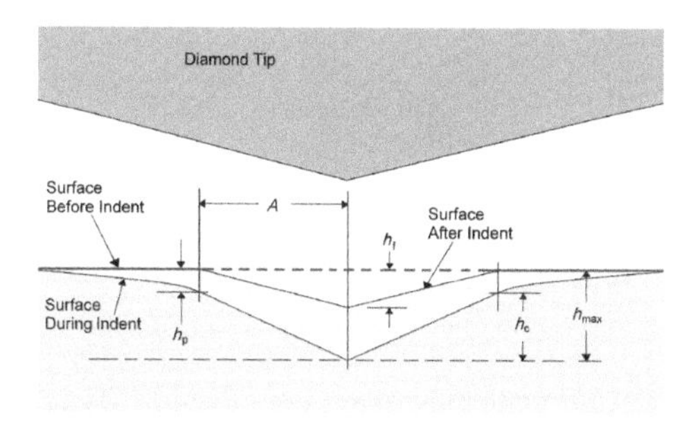

Color Plate 13.10 Schematic of the film surface before, during and after an indentation.[104]

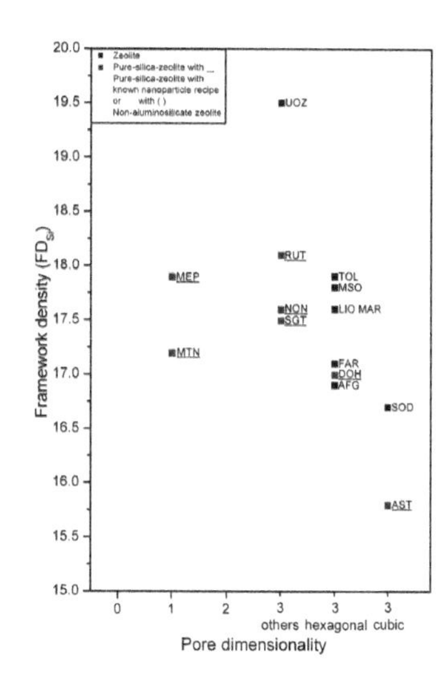

Color Plate 13.12 Pore dimensionality vs. framework density for zeolite low-*k* materials with the largest ring size of 6. Pure- silica-zeolites (PSZs) with a known nanoparticle recipe are marked with red boxes, other PSZs are marked with blue boxes with _, non-aluminosilicate zeolites are marked with hollow boxes with ( ) and all others are marked as dark boxes.

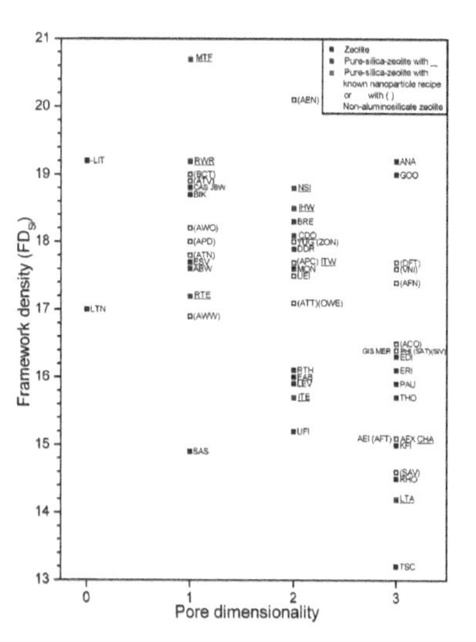

Color Plate 13.13 Pore dimensionality vs. framework density for zeolite low-*k* materials with the largest ring size of 8. Pure-silica-zeolites (PSZs) with a known nanoparticle recipe are marked with red boxes, other PSZs are marked with blue boxes with _, non-aluminosilicate zeolites are marked with hollow boxes with ( ) and all others are marked as dark boxes.

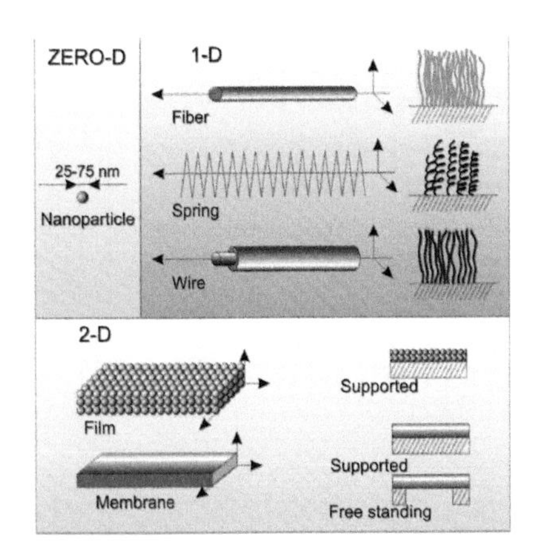

Color Plate 15.1 Nanoporous structures usually prepared by *bottom-up* procedures: from zero-dimensional to 2-D assemblies.

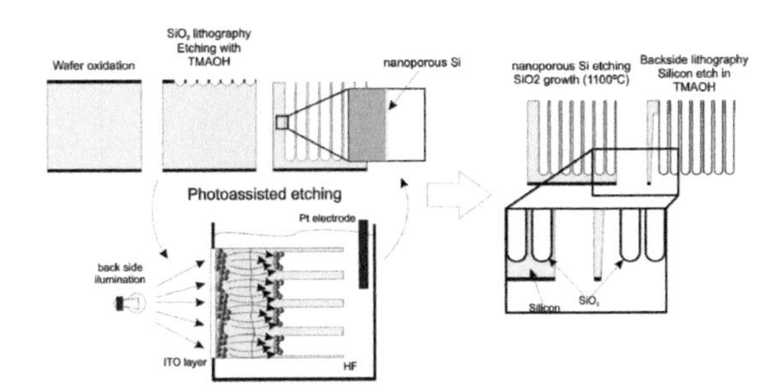

Color Plate 15.2 An example of a *top-down* approach for the fabrication of nanoporous structures: from an Si wafer to an array of silicon nanoporous pillars (adapted from Rodriguez *et al.*[23]).

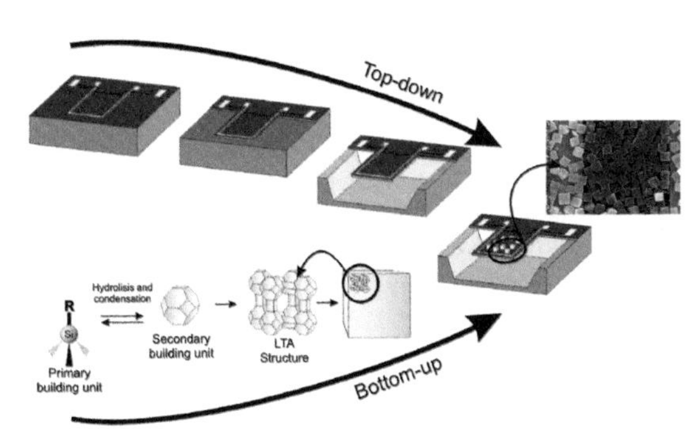

Color Plate 15.3 A hybrid approach to prepare humidity sensors: Si cantilevers are fabricated with standard *top-down* procedures. Then a selective coating is deposited on the cantilever using NaA zeolite crystals prepared by a *bottom-up* synthesis.

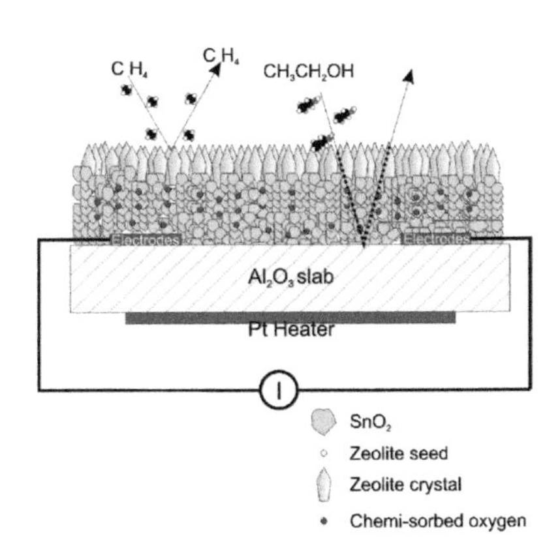

Color Plate 15.5 Schematic cross-section of a zeolite-modified Pd/SnO_2 sensor in operation. The zeolite layer eliminates the sensor sensitivity to methane, even in high concentrations.[73]

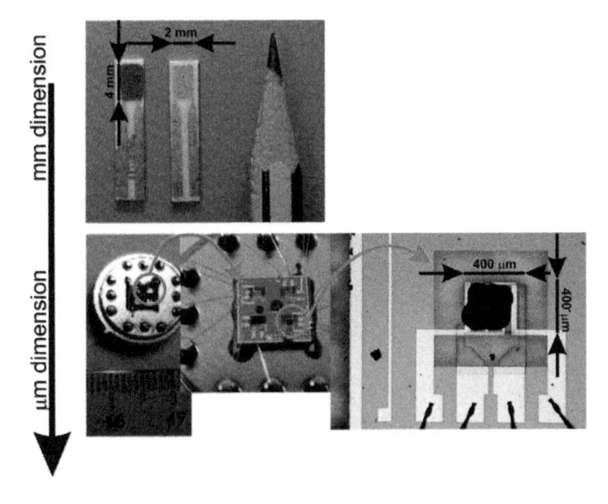

Color Plate 15.6 Zeolite-modified Pd/SnO_2 sensors of different dimensions. Top: a 2 mm × 4 mm sensor screen-printed on alumina. On top of the Pd/SnO_2 surface, a zeolite layer was grown hydrothermally.[73] Bottom: Detail of a micromachined sensor containing four microsensors, each 400 microns of lateral size. In this case, zeolite crystals were deposited by a microdropping technique.[74]

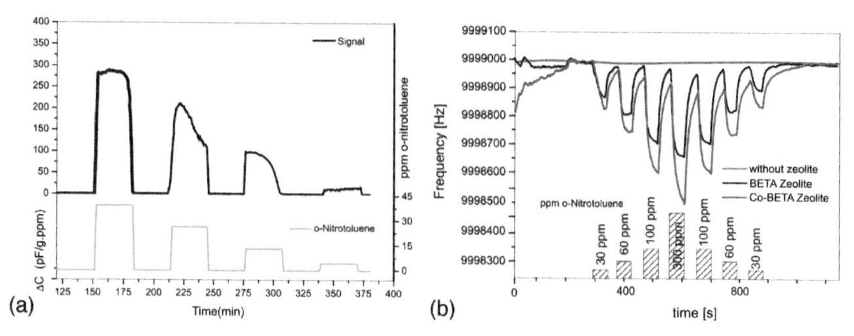

Color Plate 15.7 Room temperature capacitance of a BEA zeolite-modified IDC-type capacitor as a function of the concentration of *o*-nitrotoluene vapours.

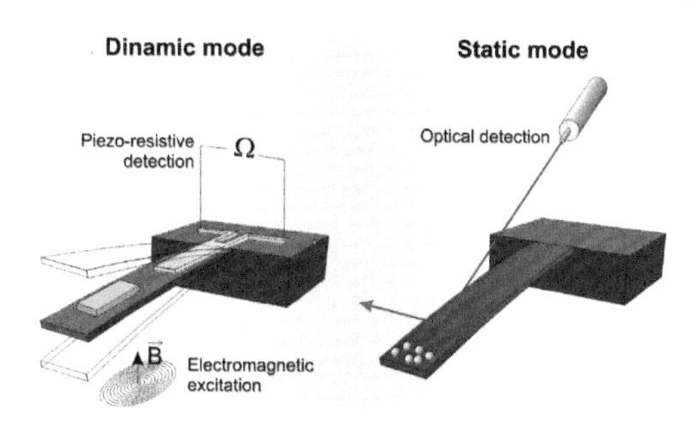

Color Plate 15.8 Scheme showing cantilever mass sensors operating in the resonant-dynamic (left) or deflection (right) modes.

Color Plate 15.9 Left: Optical micrograph of a spade-shaped (the spade form increases the sensible area) cantilever sensor. Right: SEM micrographs detailing the zeolite-coated surface of the cantilever.

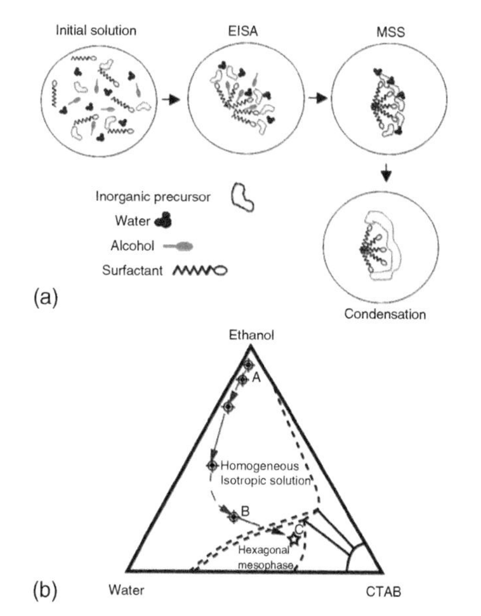

Color Plate 16.2 (a) Schematic representation of evaporation-induced-self-assembly for the preparation of long-range ordered mesostructures. (b) Typical EISA trajectory for the formation of mesostructured films by dip-coating.[7b]

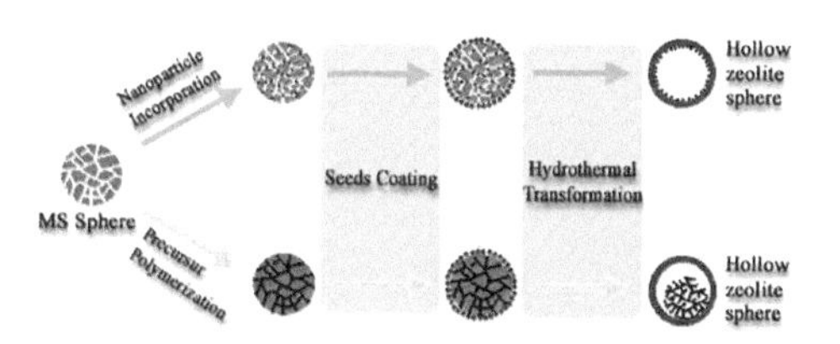

Color Plate 17.3 Schematic illustration of the encapsulation of nano- and micrometer sized particles into discrete hollow zeolite microspheres (Redrew from Ref. 83 with the permission of John Wiley & Sons, Inc.).

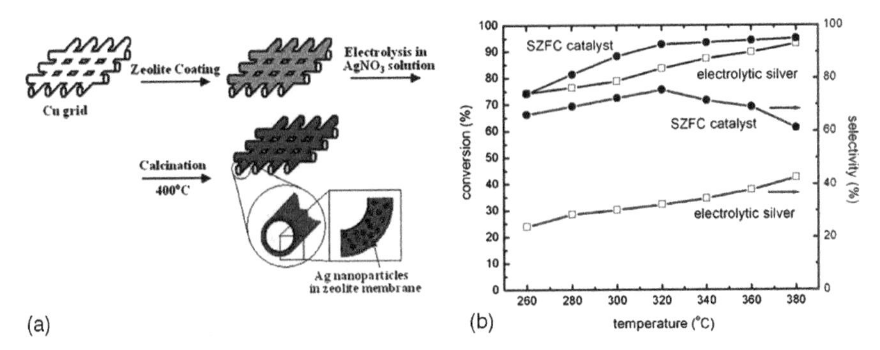

Color Plate 17.6 The schematic illustration for the fabrication of SZFC catalyst (A) and the catalytic performance of the SZFC catalysts in the oxidation of 1,2-propylene glycol (B) (Redrawn from Ref. 170 with the permission of The Royal Society of Chemistry).

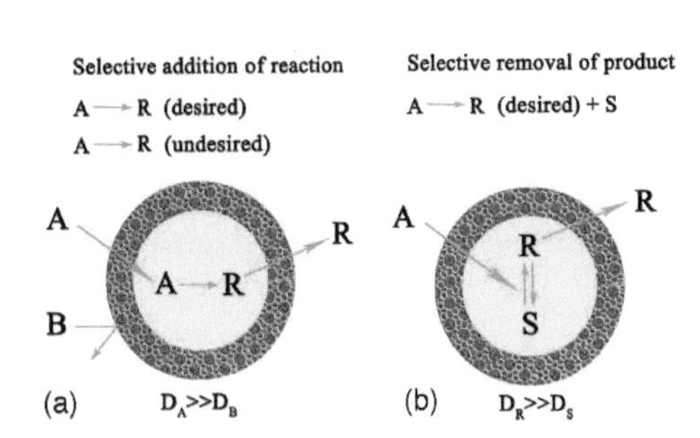

Color Plate 17.7 The typical model of controlling the product distribution in the zeolitic reactor by molecule-sieving effect of the zeolitic shell (Redrawn from Ref. 172 with the permission of The Royal Society of Chemistry).

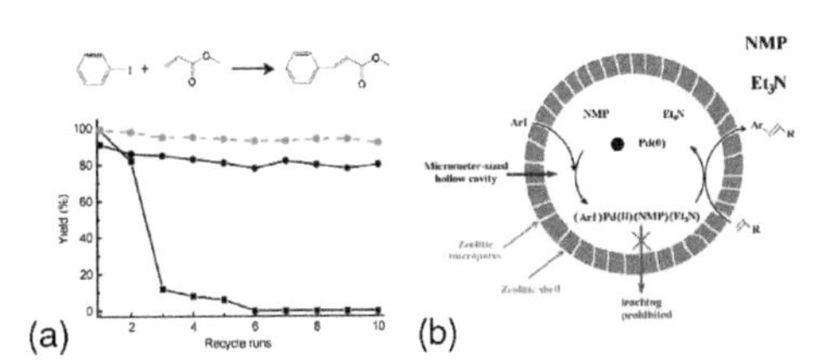

Color Plate 17.9 (A) Presents the Heck coupling activity and reusability of Pd@S1(•) and Pd/C (▪) and (B) schematically illustrates the encaged quasi-homogeneous Heck coupling in Pd@S1 catalyst. Reaction conditions: 5 mmol of aryl halide, 8 mmol of olefin, 5 mmol of anhydrous triethylamine (Et_3N), 0.0125 mmol of Pd and 30 ml of NMP under stirring at 120 °C for 3 h. The green dashed plot was obtained by doubling the amount of Pd@S1 catalyst (i.e., 0.025 mmol Pd), which shows that a higher yield of nearly 100% could be reached with higher catalyst amount (Redrawn from Ref. 56 with the permission of Elsevier B.V.).

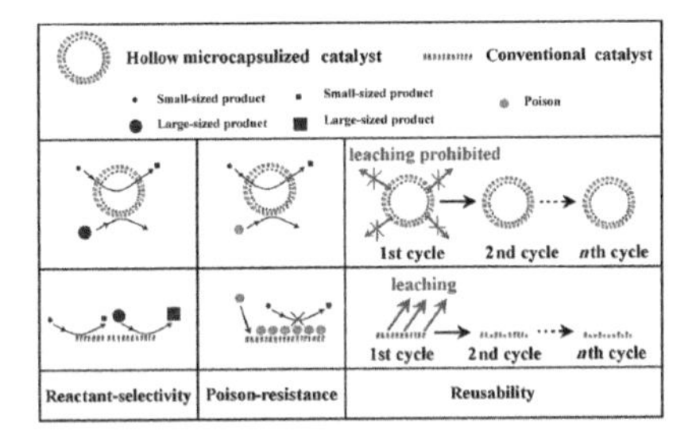

Color Plate 17.10 Schematic illustration of the reactant-selectivity, poison-resistance ability and reusability of hollow zeolitically microcapsulized catalyst (Reproduced from Ref. 57 with the permission of Elsevier B.V.).

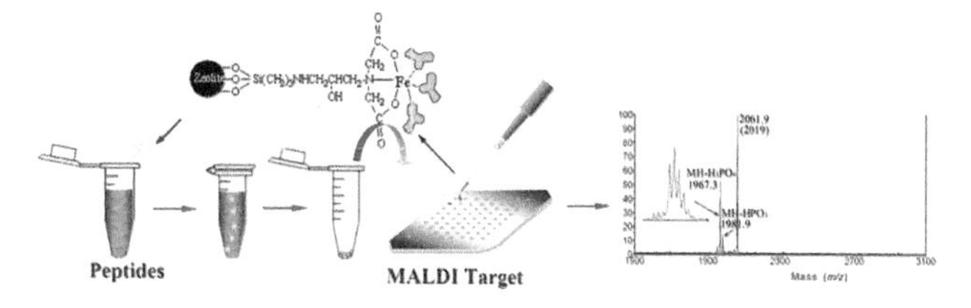

Color Plate 17.13 Schematic illustration of phosphopeptides enrichment procedure by a surface grafting metal chelation of zeolite beta nanoparticles with immobilized Fe^{3+} ions.

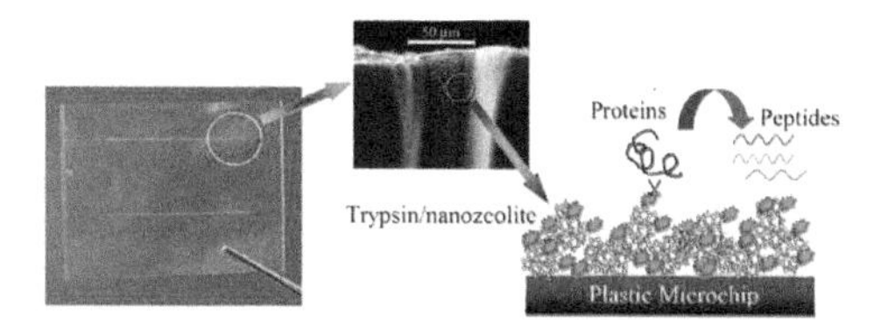

Color Plate 17.16 Microfluidic chip and the schematic illustration of nanozeolite-mediated microfluidic chip as enzyme bioreactor.

Color Plate 22.1 (a) Natural lazurite and (b) Golden mask of Tutankhamun (1361–1352 BC).

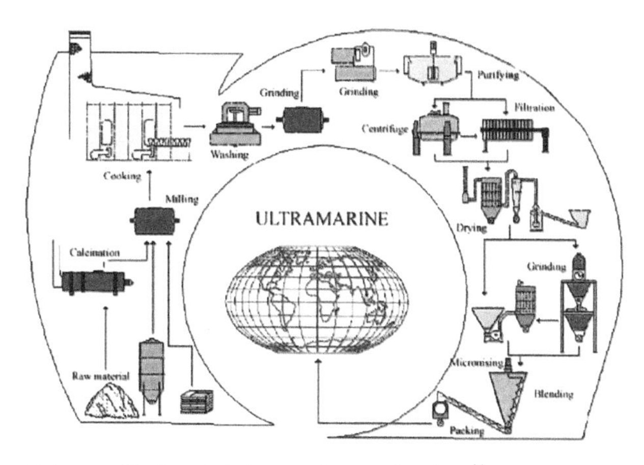

Color Plate 22.2 Scheme of industrial ultramarine production.[28]

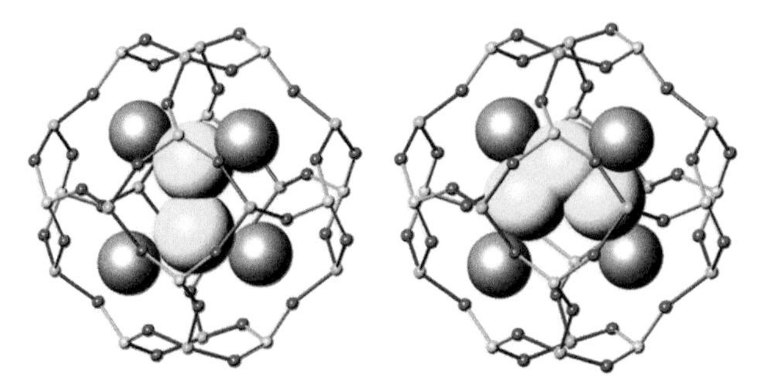

Color Plate 22.4 Models of sodalite cages with entrapped anion radicals $S_2^{\bullet -}$ (left) or $S_3^{\bullet -}$ (right).[50]

Color Plate 23.2 (A) Reactor filled with an Fe/Al_2O_3 catalyst (1 g) and (B) after CNT growth under a mixture of C_2H_6/H_2 at 680 °C for 2 h. The carbon yield was about 6000 wt.% with respect to the iron catalyst weight.

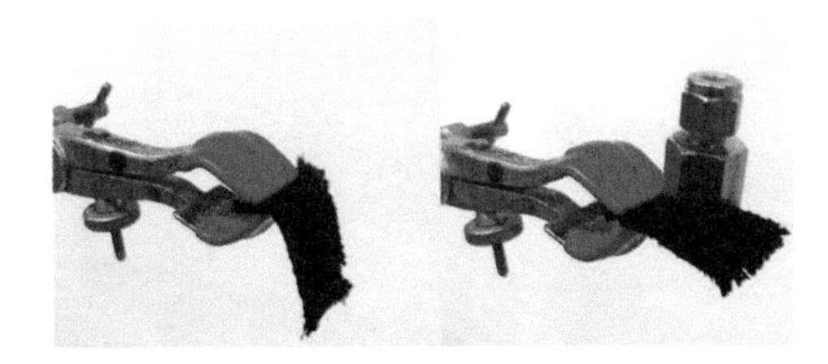

Color Plate 23.5 Photos showing the mechanical improvement of the graphite cloth after deposition of 100 wt.% of carbon nanofibres inside its matrix.

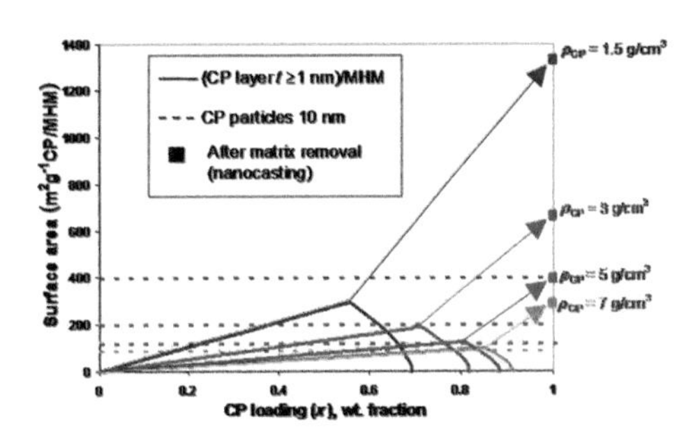

Color Plate 26.3 Relationship between the CP loading (x) and surface area of CP for the coating layer assembling mode.

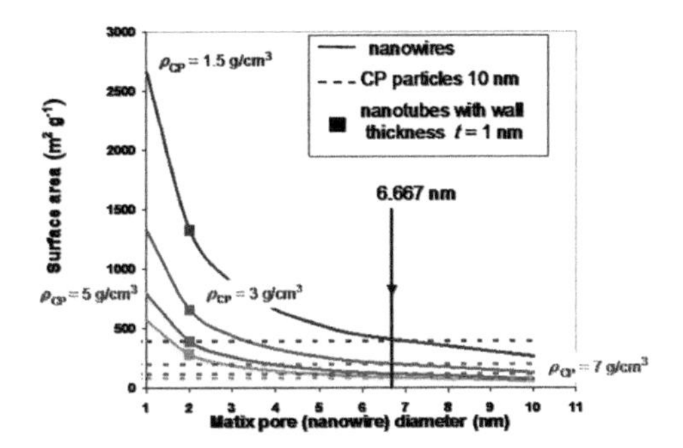

Color Plate 26.4 Relationship between the MHM pore diameter and surface area of nanocasted wires completely filling the MHM pores before matrix removal.

Printed and bound by CPI Group (UK) Ltd, Croydon, CR0 4YY
08/05/2025
01864814-0002